A Guide to Using This Handbook

The design density and area of application equals 0.26 gpm/ft^2 over 3900 ft^2 (min over 363 m^2).

7-3.2.1.1 Protection for Class I through Class IV commodities in the following conf shall be provided in accordance with this chapter:

(1) Nonencapsulated commodities that are solid pile, palletized, or bin box stor 30 ft (9.1 m) in height

(2) Nonencapsulated commodities on shelf storage up to 15 ft (4.6 m) in height

(3)* Encapsulated commodities that are solid pile, palletized, bin box, or shelf stor 15 ft (4.6 m) in height

A-7-3.2.1.1(3) Full-scale tests show no appreciable difference in the number of sprin open for either nonencapsulated or encapsulated products up to 15 ft (4.6 m) high. T not available for encapsulated products stored higher than 15 ft (4.6 m). Howeve storage tests involving encapsulated storage 20 ft (6.1 m) high, increased protection w over that for nonencapsulated storage.

The protection specified in Chapter 6 contemplates a maximum of 10-ft (3-m) c from top of storage to sprinkler deflectors for storage heights of 15 ft (4.6 m) and h

1 This handbook contains the complete text of the 1999 editions of three standards: NFPA 13, *Installation of Sprinkler Systems*, NFPA 13D, *Sprinkler Systems in One- and Two-Family Dwellings and Manufactured Homes*, and NFPA 13R, *Installation of Sprinkler Systems in Residential Occupancies up to and Including Four Stories in Height*. The standard and non-mandatory appendix material are printed in black. For convenience, explanatory material from Appendix A immediately follows the standard provision to which it relates. This non-mandatory material from the standards is printed in black and identified by the block letter A.

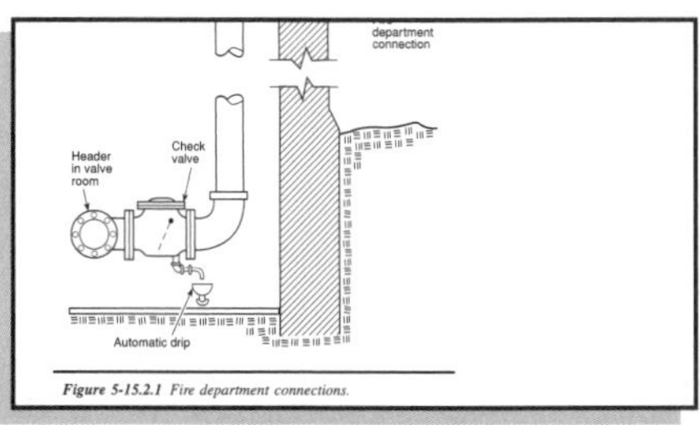

Figure 5-15.2.1 Fire department connections.

2 Figures from the standards and appendices are printed in black. Figure numbering corresponds to the relevant paragraph in the standards. The wide margins are designed for the reader's notes. Illustrations in the commentary are called "exhibits" and are numbered sequentially within each chapter. Like commentary text, captions for commentary exhibits appear in green ink.

to other buildings as indicated in the Formal Interpretation to Section 5-2. This F Interpretation implies that each building is to have its own sprinkler system.

Formal Interpretation

Reference: 5-2

Question: Would the 52,000 sq ft maximum area on one floor apply to several buildi being supplied by one riser assembly? Can the 52,000 sq ft be exceeded if the individual b are not in excess of 52,000 sq ft and are separated by a clear space of 20 ft or more?

Answer: No. The 52,000 sq ft area limitation and the 20 ft separating distance are not to the problem. Each building should have its individual system riser.

Issue Edition: 1976

Reference: 3-3.1

Date: August 1977

3 Formal Interpretations are NFPA's official response to questions about standard requirements. Printed in a shaded black box and marked by a gavel icon, Formal Interpretations are easy to identify.

protection of exhaust ducts, hood exhaust duct collars, and hood exhaust plenum chambers.

Exception: Sprinklers or automatic spray nozzles in ducts, duct collars, and plenum chambers shall not be required where all cooking equipment is served by listed grease extractors.

A-4-9.2 See Figure A-4-9.2.

Sprinklers are an effective way of controlling fires in cooking equipment, filters, and ducts if they are located to cover the cooking surfaces and all areas of an exhaust system to which fires could spread. If all cooking equipment is served by listed grease extractors, the sprinkler protection can be limited to the cooking surfaces. This arrangement is subject to acceptance by the authority having jurisdiction. Some manufacturers of exhaust systems that incorporate listed grease extractors provide listed built-in water spray fire protection for cooking surfaces in a preengineered package ready for connection to the sprinkler system.

4-9.3 Exhaust ducts shall have one sprinkler or automatic spray nozzle located at the top of each vertical riser and at the midpoint of each offset. The first sprinkler or automatic spray nozzle in a horizontal duct shall be installed at the duct entrance. Horizontal exhaust ducts shall

4 Expert commentary explains the reasoning behind requirements and helps users understand and apply them. Commentary text is printed in green ink to distinguish it from text of the standards. Note that the commentary is not part of the standards and is not enforceable.

pended or Floor-Mounted Vertical Obstructions. The distance from sprinklers tains, free standing partitions, room dividers, and similar obstructions in light ncies shall be in accordance with Table 5-6.5.2.3 and Figure 5-6.5.2.3.

e distances given in Table 5-6.5.2.3 were determined through tests in which s with either a solid fabric or close mesh [¼ in. (6.4 mm)] top panel were roader-mesh top panels–for example, ½ in. (13 mm) or greater measured on he obstruction of the sprinkler spray is not likely to be severe and the authority tion might not need to apply the requirements in 5-6.5.2.3.

nces specified in Table 5-6.5.2.3 were derived from NBSIR 80-2097, *Full-Tests with Automatic Sprinklers in a Patient Room*. The distance requirements

nkler Systems Handbook 1999

5 Both standards and the commentary feature icons to draw the reader's attention to certain types of material. An icon of an open book alerts the reader to references to other published material. Calculations are highlighted by an icon of a calculator.

6 For the first time, the handbook's back endsheets show commonly used tables and graphs.

Automatic Sprinkler Systems Handbook

EIGHTH EDITION

Automatic Sprinkler Systems Handbook

EIGHTH EDITION

EDITED BY

Milosh T. Puchovsky, P.E.

With the complete text of the 1999 edition of NFPA 13, *Standard for the Installation of Sprinkler Systems;* NFPA 13D, *Standard for the Installation of Sprinkler Systems in One- and Two-Family Dwellings and Manufactured Homes;* and NFPA 13R, *Standard for the Installation of Sprinkler Systems in Residential Occupancies up to and Including Four Stories in Height*

NFPA®

National Fire Protection Association,
Quincy, Massachusetts

Product Manager: Pam Powell
Project Editor: Kimberly Cervantes
Copy Editor: Cara Grady
Text Processing: Marilyn Lupo
Composition: Modern Graphics
Art Coordinator: Nancy Maria
Illustrations: Todd K. Bowman and George Nichols
Cover Design: Twist Creative Group
Manufacturing Buyer: Ellen Glisker
Printer: R. R. Donnelley

Notice Concerning Liability: Publication of this handbook is for the purpose of circulating information and opinion among those concerned for fire and electrical safety and related subjects. While every effort has been made to achieve a work of high quality, neither the NFPA nor the contributors to this handbook guarantee the accuracy or completeness of or assume any liability in connection with the information and opinions contained in this handbook. The NFPA and the contributors shall in no event be liable for any personal injury, property, or other damages of any nature whatsoever, whether special, indirect, consequential, or compensatory, directly or indirectly resulting from the publication, use of, or reliance upon this handbook. This handbook is published with the understanding that the NFPA and the contributors to this handbook are supplying information and opinion but are not attempting to render engineering or other professional services. If such services are required, the assistance of an appropriate professional should be sought.

Notice Concerning Code Interpretations: This eighth edition of the *Automatic Sprinkler Systems Handbook* is based on the 1999 editions of NFPA 13, *Standard for the Installation of Sprinkler Systems;* NFPA 13D, *Standard for the Installation of Sprinkler Systems in One- and Two-Family Dwellings and Manufactured Homes*; and NFPA 13R, *Standard for the Installation of Sprinkler Systems in Residential Occupancies up to and Including Four Stories in Height.* All NFPA codes, standards, recommended practices, and guides are developed in accordance with the published procedures of the NFPA technical committees comprised of volunteers drawn from a broad array of relevant interests. The handbook contains the complete text of NFPA 13, NFPA 13D, and NFPA 13R, and any applicable Formal Interpretations issued by the Association. These documents are accompanied by explanatory commentary and other supplementary materials.

The commentary and supplementary materials in this handbook are not a part of the Standards and do not constitute Formal Interpretations of the NFPA (which can be obtained only through requests processed through the responsible technical committee in accordance with the published procedures of the NFPA). The commentary and supplementary materials, therefore, solely reflect the personal opinions of the editor or other contributors and do not necessarily represent the official position of the NFPA or its technical committees.

NFPA No.: F9-13HB99
ISBN: 0-87765-443-3
Library of Congress Card Catalog No.: 99 075956

Printed in the United States of America
04 03 02 01 5 4 3

This handbook is dedicated to all those volunteers
who gave their time, resources, and energy
to completing the 1999 edition of NFPA 13.

The development of the 1999 edition of NFPA 13 was a very ambitious endeavor that required a massive effort. Collectively, committee members and other individuals from around the world have spent many thousands of hours working through NFPA's open consensus process in preparing the most comprehensive standard on sprinkler systems.

Contents

Preface

The first automatic fire extinguishing system on record was patented in England in 1723 and consisted of a cask of water, a chamber of gunpowder, and a system of fuses. In about 1852 the perforated pipe system represented the first form of a sprinkler system used in the United States. In 1874, Henry S. Parmelee of New Haven, Connecticut, patented the first practical automatic sprinkler.

C.J.H. Woodbury of the Boston Manufacturers Mutual Fire Insurance Company and F.E. Cabot of the Boston Board of Fire Underwriters completed a study on the performance of sprinklers for the Factory Mutual Fire Insurance Company in 1884. This study was the basis for the first set of rules for the installation of automatic sprinkler systems that were developed by John Wormald of the Mutual Fire Insurance Corporation of Manchester, England, in 1885. In 1887, similar rules were prepared in the United States by the Factory Improvement Committee of the New England Insurance Exchange.

By 1895, the commercial growth and development of sprinkler systems were so rapid that a number of different installation rules had been adopted by various insurance organizations. Within a few hundred miles of Boston, Massachusetts, nine radically different standards for the size of piping and sprinkler spacing were being used. This problem led to the creation of NFPA 13 and the formation of the National Fire Protection Association in 1896.

The 1999 edition places NFPA 13 on the threshold of a new millennium — one that is sure to include an accelerated rate of technological discoveries and access to greater amounts of information. We will be required to make even more choices in an age of information overload. The design and installation of sprinkler systems will not be exempted from these challenges.

In many respects, the issues that led to the development of the first edition of NFPA 13 are relevant today. The unprecedented development of sprinkler system products, design techniques, and installation practices over the past several years is offering numerous options for effective system design. While this increased flexibility provides numerous advantages, it also requires more diligence by those designing, installing, and approving sprinkler systems as the rules for various system components become less uniform.

As has been the case for over 100 years, the intent of NFPA 13 is to provide a means for analyzing sprinkler system information and presenting it in a form that will lead to effective system designs and installations. This task continues to become increasingly demanding as scientific and other discoveries generate information at an increasingly accelerated rate. In response to these challenges, NFPA has expanded the scope of NFPA 13 so that it is now the most comprehensive document addressing sprinkler systems. Additionally, NFPA has established new volunteer technical committees to deal with rapidly developing technologies and other discoveries.

NFPA 13 now addresses sprinkler system installations for all types of facilities regardless of the type of fire hazards present. This reorganization includes the centralization of sprinkler system information from over 40 NFPA codes and standards. A major part of this effort included the incorporation of system information for storage operations and the installation of buried pipe. System information for specialized hazards from other documents has been copied or specifically referenced. Additionally, numerous technical changes have been incorporated.

As the scope of NFPA 13 has expanded, so has that of the *Automatic Sprinkler Systems Handbook*. This edition includes new commentary on underground piping installation and rack and on-floor storage applications of various commodities including plastics, rubber tires, baled cotton, wooden and plastic pallets, and roll paper. In addition, the handbook includes updated commentary on those portions of NFPA 13 that have been revised. Those portions dealing with hanging and bracing of system piping are especially noteworthy. A new supplement addressing the history of sprinklers, sprinkler systems, and NFPA standards is also included.

Acknowledgments

Over the past several years, I have had the good fortune to work with some of the most knowledgeable people in the sprinkler industry. The help of these individuals has been invaluable in completing this handbook. I would especially like to acknowledge and thank the following people for their efforts with this project:

Ed Budnick
Russ Fleming
Joe Hankins
Steve Hoover
Roland Huggins
Morgan Hurley
Ken Isman
Rolf Jensen
Larry Keeping
Ken Linder
Chris Lummus
Dan Madrzykowski
Maurice Pilette
Todd Schumann
Richard Scott
Peter Smith
Bill Thomas
Jack Walsh

I would also like to acknowledge those who contributed their time, effort, and knowledge in developing previous editions of the handbook. Many are still involved with the sprinkler project. Their work has created the solid foundation on which to build this and future editions.

I would be remiss if I did not acknowledge others at NFPA with whom I worked very closely on this project. Both Kim Cervantes and Pam Powell provided key input throughout various stages of the book's development.

I am particularly grateful to my wife, Debi, my daughter, Angelica, and my new son, Noah, as they made sure there were plenty of smiles and laughs to go around, and helped me set my priorities. As always, Shelby Marie continues to amuse.

In closing, sprinkler systems have established an enviable record for achieving life safety and property protection. This handbook is intended to provide a better understanding of the requirements of NFPA 13, 13D, and 13R and how to implement them. I hope that this edition helps people make better decisions about fire safety, and that it advances the legacy of automatic sprinkler systems worldwide.

Milosh T. Puchovsky, P.E.

PART ONE

NFPA 13, *Standard for the Installation of Sprinkler Systems,* with Commentary

Part One of this handbook includes the complete text and illustrations of the 1999 edition of NFPA 13, *Standard for the Installation of Sprinkler Systems.* The text and illustrations from the standard are printed in black and are the official requirements of NFPA 13. Line drawings and photographs from the standard are labeled "Figures."

Paragraphs that begin with the letter "A" are extracted from Appendix A of the standard. Although printed in black ink, this nonmandatory material is purely explanatory in nature. For ease of use, this handbook places Appendix A material immediately after the standard paragraph to which it refers.

Part One also includes Formal Interpretations on NFPA 13; these apply to previous and subsequent editions for which the requirements remain substantially unchanged. Formal Interpretations are not part of the standard and are printed in shaded gray boxes.

In addition to standard text, appendixes, and Formal Interpretations, Part One includes commentary that provides the history and other background information for specific paragraphs in the standard. This insightful commentary takes the reader behind the scenes, into the reasons underlying the requirements.

To make commentary material readily identifiable, commentary text, captions, and tables are all printed in green. So that the reader can easily distinguish between line drawings and photographs from the standard and from the commentary, line drawings and photographs in the commentary are labeled "Exhibits." The distinction between figures in the standard and exhibits in the commentary is new for this edition of the *Automatic Sprinkler Systems Handbook.*

This edition of the handbook also includes one Supplement. Part Four is devoted to a Supplement that documents the history of the NFPA sprinkler standards and links those standards to advances in the industry.

CHAPTER 1

General Information

1-1 Scope

This standard provides the minimum requirements for the design and installation of automatic fire sprinkler systems and exposure protection sprinkler systems, including the character and adequacy of water supplies and the selection of sprinklers, fittings, piping, valves, and all materials and accessories, including the installation of private fire service mains..This standard encompasses "combined service mains" used to carry water for both fire service and other uses as well as mains for fire service use only.

NFPA 13 addresses the design and installation of sprinkler systems employing automatic or open sprinklers that discharge water to suppress or control a fire. NFPA 13 provides only the minimum requirements for satisfactory sprinkler system performance during a fire. The overall level of fire protection provided by sprinkler systems can be enhanced through the use of additional provisions, such as supplemental supervisory signaling systems.

As an installation standard, NFPA 13 does not specify the buildings or structures that require sprinkler systems. The purpose of NFPA 13 is to provide for the proper design and installation of a sprinkler system and to specify acceptable system components when a sprinkler system is required. In other words, NFPA 13 specifies how to properly design and install a sprinkler system using the proper components. The standard does not identify when a system is required. The local building code, NFPA *101®*, *Life Safety Code®*, or insurance regulations typically specify the buildings and structures that require sprinkler systems.

Effective with the 1999 edition, all requirements pertaining to the design and installation of automatic sprinkler systems were, to the extent possible, centralized in NFPA 13. The scope of NFPA 13 was expanded to include sprinkler system design and installation requirements for storage facilities, formerly located in the NFPA 231 series of documents. The design and installation requirements for underground piping and private fire service mains, from NFPA 24, were also included.

The 1999 edition also contains sprinkler system information extracted from other NFPA standards to assist NFPA 13 users with the design and installation of automatic sprinkler systems for hazards and facilities not previously addressed by NFPA 13. Sprinkler system information from over 40 NFPA codes, standards, and recommended practices is cited in NFPA 13. For example, 7-10.5 contains sprinkler system discharge criteria from NFPA 40, *Standard for the Storage and Handling of Cellulose Nitrate Motion Picture Film.* In a few instances where extracted text could be confusing or misleading, NFPA 13 only references the appropriate NFPA standard. For example, 7-10.1 references rather than extracts sprinkler system discharge criteria from NFPA 30, *Flammable and Combustible Liquids Code.*

NFPA 13 also references other documents that address equipment that is essential to the proper and effective operation of the sprinkler system. For example, Chapter 9 establishes the types of acceptable water supply sources. However, Chapter 9 does not contain installation or performance requirements for those water supply sources. Other NFPA standards such as NFPA 20, *Standard for the Installation of Stationary Pumps for Fire Protection*, and NFPA 22, *Standard for Water Tanks for Private Fire Protection*, provide specific installation requirements for fire pumps and water storage tanks.

Since NFPA 13 only pertains to sprinkler systems, other NFPA standards should be referenced for design and installation requirements of other types of water-based fire protection systems. The following identifies some of these standards:

NFPA 11, *Standard for Low-Expansion Foam*

NFPA 11A, *Standard for Medium- and High-Expansion Foam Systems*

NFPA 13D, *Standard for the Installation of Sprinkler Systems in One- and Two-Family Dwellings and Manufactured Homes*

NFPA 13R, *Standard for the Installation of Sprinkler Systems in Residential Occupancies up to and Including Four Stories in Height*

NFPA 14, *Standard for the Installation of Standpipe and Hose Systems*

NFPA 15, *Standard for Water Spray Fixed Systems for Fire Protection*

NFPA 16, *Standard for the Installation of Foam-Water Sprinkler and Foam-Water Spray Systems*

NFPA 750, *Standard on Water Mist Fire Protection Systems*

1-2* Purpose

The purpose of this standard is to provide a reasonable degree of protection for life and property from fire through standardization of design, installation, and testing requirements for sprinkler systems, including private fire service mains, based on sound engineering principles, test data, and field experience. This standard endeavors to continue the excellent records that have been established by sprinkler systems while meeting the needs of changing technology. Nothing in

this standard is intended to restrict new technologies or alternate arrangements, provided the level of safety prescribed by this standard is not lowered. Materials or devices not specifically designated by this standard shall be utilized in complete accord with all conditions, requirements, and limitations of their listings.

A-1-2 Since its inception, this document has been developed on the basis of standardized materials, devices, and design practices. However, Section 1-2 and other subsections such as 3-3.5 and 5-4.9 allow the use of materials and devices not specifically designated by this standard, provided such use is within parameters established by a listing organization. In using such materials or devices, it is important that all conditions, requirements, and limitations of the listing be fully understood and accepted and that the installation be in complete accord with such listing requirements.

NFPA 13 encourages innovative and economically feasible measures that provide life safety and property protection. Section 1-2 allows for the use of increasingly available specially listed materials and products and promotes the continued development of new sprinkler-related technologies. Products and system arrangements not specifically covered by the standard can be used, provided it can be demonstrated that these products or arrangements do not lower the level of safety provided by the standard or alter the standard's intent.

Section 1-2 also alerts the NFPA 13 user that specialized products often have specific requirements or limitations, which are not addressed by NFPA 13. The listing information and the relevant manufacturer's literature must be referenced. For example, nonmetallic pipe is an acceptable alternative to traditional materials, such as steel or copper, for certain applications. Initially tested and listed in 1984 for use only in systems designed in accordance with NFPA 13D, nonmetallic pipe is now listed for use in any light hazard occupancy.

As with most life safety systems, it is difficult to precisely quantify the overall level of protection to life and property provided by sprinkler systems. For example, accurate mathematical predictions that everyone exposed to a rapidly spreading flammable liquids fire in a fully sprinklered processing plant would escape without harm or that property damage could be limited to a specific dollar value or to a percentage of the overall building area cannot be made. However, life safety and property protection in buildings are both known to be greatly enhanced by the presence of an automatic sprinkler system complying with NFPA 13.

Scientific understanding of fire, sprinklers, and their interaction continues to increase. This knowledge allows for better system performance and the development of more effective new products. For example, residential sprinklers were developed to protect people in the room of fire origin within a dwelling unit who are not intimate with ignition, provided the fire load is typical of that found in a residential-type occupancy. More information on residential sprinklers is provided in NFPA 13D. Another example includes the development of early suppression fast response (ESFR) sprinklers with various K-factors. The application of this relatively new technology allows for fire suppression of various commodities stored up to 40 ft (12.2 m) in height with relatively low discharge pressures. See 7-9.5 for more information.

Fire data collected and analyzed by NFPA's Fire Analysis and Research Division has lead to the following conclusions:

"When sprinklers are present, the chances of dying in a fire and the average property loss per fire are both cut by one-half to two-thirds, compared to fires where sprinklers are not present. What's more, this simple comparison understates the potential value of sprinklers because it lumps together all sprinklers, regardless of type, coverage, or operational status, and is limited to fires reported to fire departments. If unreported fires could be included and if complete, well maintained, and properly installed and designed systems could be isolated, sprinkler effectiveness would be seen as even more impressive."

Furthermore, "NFPA has no record of a fire killing more than two people in a completely sprinklered building where the system was properly operating, except in an explosion or flash fire or where industrial fire brigade members or employees were killed during fire suppression operations."[1] In some cases, victims of fatal fires in sprinklered properties were involved in the ignition of the fire and received their injuries prior to the operation of the sprinklers or were unable to escape due to a physical or mental impairment.

With regard to unsatisfactory sprinkler system performance, the cited NFPA study indicates that this occurrence is rare, and that the reason almost always involves some type of human error. Examples of human error include failure to properly maintain the system or failure to upgrade an existing system when the hazard or occupancy in which it is located changes. Overall, the NFPA study states that the major reason why a sprinkler system would not operate satisfactorily is that the water supply was shut off.

1-2.1 A sprinkler system and private fire service mains are specialized fire protection systems and require knowledgeable and experienced design and installation.

The requirements of NFPA 13 were developed through the application of engineering principles, fire test data, and field experience. During its 100-plus-year history, the technical committees on automatic sprinkler systems have reviewed, analyzed, and evaluated sprinkler system–related information and presented it in a useful form. While those who developed and updated NFPA 13 possess numerous credentials regarding their ability to design, install, review, or evaluate sprinkler systems, NFPA 13 does not specify particular qualifications for its users. However, as with any specialized subject, a good understanding of the basic principles as well as a continued effort to keep current with developing technologies are essential. NFPA 13 is a design and installation standard. It is not a how-to manual or a textbook.

Although NFPA 13 does not specify qualifications for those who design and install sprinkler systems, state and local government agencies typically do. For example, some states require that registered fire protection engineers or those who design sprinkler systems be National Institute for Certification of Engineering Technicians (NICET) certified and that those who install sprinkler systems be licensed. These requirements

[1]Rohr, K. D., 1998. "U.S. Experience with Sprinklers: Who Has Them? How Well Do They Work?," Quincy, MA: National Fire Protection Association Fire Analysis and Research Division.

are similar to the required qualifications for those who design and install electrical systems that affect public safety.

1-3 Retroactivity Clause

The provisions of this document are considered necessary to provide a reasonable level of protection from loss of life and property from fire. They reflect situations and the state of the art at the time the standard was issued. Unless otherwise noted, it is not intended that the provisions of this document be applied to facilities, equipment, structures, or installations that were existing or approved for construction or installation prior to the effective date of this document.

Exception: In those cases where it is determined by the authority having jurisdiction that the existing situation involves a distinct hazard to life or property, this standard shall apply.

The retroactivity clause appears in many NFPA codes and standards. Its main purpose is to reinforce the premise that any sprinkler system installed in accordance with the applicable edition of NFPA 13 is considered to comply with the standard for the system's lifetime as long as no system modifications are made and the fire hazard remains unchanged. In other words, an existing system is not required to be reviewed for compliance with every new edition of the standard. For example, the 1999 edition requires that means be provided downstream of pressure-reducing valves so that flow tests at system demand can be conducted. This requirement is not intended to retroactively apply to a system that was installed in accordance with previous editions of the standard that did not contain the requirement.

Omission of this clause would require building owners, code enforcers, insurance companies, and installers to undertake the never-ending task of updating and revising their sprinkler systems every time a new edition of NFPA 13 was published. While newer editions often include updated requirements that enhance the level of protection, older editions should not be interpreted as unsafe. In those instances where a severe deficiency is discovered, the exception to Section 1-3 provides latitude for the authority having jurisdiction (AHJ). For example, it could be discovered that an existing system would not be able to perform satisfactorily for the fire hazard presented.

Where the hazard within a sprinklered building or space has changed, an evaluation of the existing sprinkler system needs to be conducted to verify its adequacy. For example, it is highly unlikely that a sprinkler system designed and installed to protect 8-ft (2.4-m) high on-floor storage would be capable of protecting 18-ft (5.4-m) high rack storage of the same commodity. Sprinkler protection for the high rack storage arrangement needs to be upgraded in compliance with NFPA 13 provision for such storage.

Routine inspections and testing of a sprinkler system in accordance NFPA 25, *Standard for the Inspection, Testing, and Maintenance of Water-Based Fire Protection Systems*, uncover many of the impairments that lead to unsuccessful sprinkler system performance. Proper knowledge of system design and installation requirements as well

as a vigilant maintenance program are also critical in ensuring that sprinkler systems remain one of the most reliable and effective defenses against fire.

1-4 Definitions

Many of the terms used throughout NFPA 13 are unique to sprinkler systems and are defined in this section. Due to the large number and types of terms used throughout this standard, the terms are organized into one of eight categories to facilitate their search. These categories include terms specific to NFPA, terms associated with general fire protection, terms describing various types of sprinkler systems, terms addressing individual sprinkler system components, terms associated with automatic sprinklers, terms that characterize building construction features, terms pertaining to private water supply piping, and terms related to storage operations.

1-4.1 NFPA Definitions.

Approved.* Acceptable to the authority having jurisdiction.

A-1-4.1 Approved. The National Fire Protection Association does not approve, inspect, or certify any installations, procedures, equipment, or materials; nor does it approve or evaluate testing laboratories. In determining the acceptability of installations, procedures, equipment, or materials, the authority having jurisdiction may base acceptance on compliance with NFPA or other appropriate standards. In the absence of such standards, said authority may require evidence of proper installation, procedure, or use. The authority having jurisdiction may also refer to the listings or labeling practices of an organization that is concerned with product evaluations and is thus in a position to determine compliance with appropriate standards for the current production of listed items.

In the context of NFPA 13, the term *approved* has a different meaning than the term *listed*. A component that is approved is not necessarily listed. Components critical to the proper operation of a sprinkler system, such as alarm valves, dry pipe valves, sprinklers, and hangers are required to be both listed and approved. Noncritical components, such as drain valves, are not required to be listed but are required to be approved.

Components required to be approved are necessary to maintain an acceptable level of system reliability. However, their impairment would not render the sprinkler system out of service. Components such as pressure gauges, drain valves, and water meters fit into this category.

Authority Having Jurisdiction.* The organization, office, or individual responsible for approving equipment, materials, an installation, or a procedure.

A-1-4.1 Authority Having Jurisdiction. The phrase "authority having jurisdiction" is used in NFPA documents in a broad manner, since jurisdictions and approval agencies vary, as do their responsibilities. Where public safety is primary, the authority having jurisdiction may be a federal, state, local, or other regional department or individual such as a fire chief; fire marshal;

chief of a fire prevention bureau, labor department, or health department; building official; electrical inspector; or others having statutory authority. For insurance purposes, an insurance inspection department, rating bureau, or other insurance company representative may be the authority having jurisdiction. In many circumstances, the property owner or his or her designated agent assumes the role of the authority having jurisdiction; at government installations, the commanding officer or departmental official may be the authority having jurisdiction.

Listed.* Equipment, materials, or services included in a list published by an organization that is acceptable to the authority having jurisdiction and concerned with evaluation of products or services, that maintains periodic inspection of production of listed equipment or materials or periodic evaluation of services, and whose listing states that either the equipment, material, or service meets appropriate designated standards or has been tested and found suitable for a specified purpose.

A-1-4.1 Listed. Evaluation of the product or service should address reliable operation and performance for the intended function. The means for identifying listed equipment may vary for each organization concerned with product evaluation; some organizations do not recognize equipment as listed unless it is also labeled. The authority having jurisdiction should utilize the system employed by the listing organization to identify a listed product or device.

One listing agency uses the designation *classified* to indicate that a specific product meets its testing and evaluation requirements. Materials with this designation meet the intent of *listed.* As indicated in the appendix material, evaluation of the product should address reliable operation for the intended function. Most components that are critical to system performance are required to be listed. Such components include, but are not limited to, pipe, alarm valves, hangers, and sprinklers. However, this rule has some exceptions. Steel pipe, for example, that meets specific industry standards is not required to be listed because it has a long-established track record of acceptable performance.

Shall. Indicates a mandatory requirement.

This term indicates a requirement of this standard and mandates that a specific provision of NFPA 13 be followed. When the term *shall* is attached to a specific provision of the standard, compliance with that provision is not optional. For example, 3-2.5.1 states that sprinklers shall have their frame arms color coded a certain way. Color coding of the deflector instead of the frame arms is not permitted. The term *should*, which is the next term discussed, refers to a recommendation of this standard rather than a requirement.

Any exceptions to a requirement of this standard are specifically stated. With regard to color coding of sprinkler frame arms, NFPA 13 permits the deflector to be color coded under specific conditions as described in Exception No. 1 of 3-2.5.1. The use of the term *shall* in the exception indicates that it is only under the conditions of the exception that sprinkler deflectors are permitted to be painted. Mandatory requirements of NFPA 13 are found in the main body of the standard—that is, Chapters 1 through 13.

Should. Indicates a recommendation or that which is advised but not required.

This term indicates a recommendation of this standard. A provision of this standard associated with the term *should* is intended to be advisory, and its use is limited to the appendices. The term identifies a good idea or a better practice. If the recommendation is not followed, the sprinkler system is still expected to perform satisfactorily.

For example, A-5-15.4.2 recommends that the inspector's test connection be installed at the end of the hydraulically most remote branch line for a wet pipe sprinkler system. Placing the inspector's test connection at some other location will not negatively impact the performance of the system. An arrangement with the inspector's test not at the end of the hydraulically most remote branch line is still in compliance with the standard.

Any section number preceded by a letter is an appendix item. Terms such as *should* and *recommend* are prevalent in these sections.

Standard. A document, the main text of which contains only mandatory provisions using the word "shall" to indicate requirements and which is in a form generally suitable for mandatory reference by another standard or code or for adoption into law. Nonmandatory provisions shall be located in an appendix, footnote, or fine-print note and are not to be considered a part of the requirements of a standard.

1-4.2 General Definitions.

Compartment. A space completely enclosed by walls and a ceiling. The compartment enclosure is permitted to have openings to an adjoining space if the openings have a minimum lintel depth of 8 in. (203 mm) from the ceiling.

This term is primarily associated with the use of residential sprinklers. For a system utilizing residential sprinklers, the hydraulic demand of the system depends on the number of sprinklers required to operate simultaneously in a compartment.

Openings in a compartment cannot extend to the ceiling. A minimum 8-in. (203-mm) lintel ensures that heat from a fire collects at the ceiling in the room where the fire operates and results in faster operation of the sprinklers nearest to the fire.

Drop-Out Ceiling. A suspended ceiling system, which is installed below the sprinklers, with listed translucent or opaque panels that are heat sensitive and fall from their setting when exposed to heat.

A drop-out ceiling allows the installation of a false ceiling beneath an existing sprinkler system. Drop-out ceilings provide an alternative to relocating sprinklers beneath a new ceiling.

Drop-out ceilings are evaluated to ensure that they do not contribute to fire growth and that they do not significantly delay the operation of the sprinkler system. Once installed, building owners must ensure that any ceiling panels replaced during the life of the system are of the same type originally installed.

Dwelling Unit. One or more rooms arranged for the use of one or more individuals living together, as in a single housekeeping unit normally having cooking, living, sanitary, and sleeping facilities. For purposes of this standard, dwelling unit includes hotel rooms, dormitory rooms, apartments, condominiums, sleeping rooms in nursing homes, and similar living units.

Dwelling units are a series of connected rooms within a residential occupancy. Residential occupancies are identified in A-2-1.1 as light hazard. The residential classification encompasses all dwelling units included in the definition. The residential classification allows for the use of listed residential sprinklers. In addition, some spaces within a dwelling unit are identified by certain exceptions as not requiring sprinkler coverage. NFPA 13 precisely defines those spaces in those specific types of residential occupancies, such as hotels, in which sprinkler coverage can be omitted. See 5-13.9 for more information on dwelling units.

Fire Control. Limiting the size of a fire by distribution of water so as to decrease the heat release rate and pre-wet adjacent combustibles, while controlling ceiling gas temperatures to avoid structural damage.

Since the inception of the first sprinkler system in the 1870s, the focus of most sprinkler systems has been to control a fire rather than to extinguish it. While sprinkler systems have extinguished numerous fires, sprinkler systems generally are designed to limit the size of a developing fire and prevent it from growing and spreading beyond its general area of origin. In some cases fires are shielded from sprinkler system discharge, which makes complete extinguishment difficult.

A sprinkler system's ability to extinguish or suppress a fire has not been widely discussed until fairly recently because the interaction between sprinkler discharge and a fire had not been sufficiently understood. Phenomena associated with fire growth rates, mass loss of the burning fuel package, and the rate of heat release produced by a fire need to be considered. Only within the past two decades have research efforts provided a greater understanding of these phenomena. The result was the development of nontraditional sprinkler devices such as the large drop sprinkler and the ESFR sprinkler.

Fire Suppression. Sharply reducing the heat release rate of a fire and preventing its regrowth by means of direct and sufficient application of water through the fire plume to the burning fuel surface.

The concept of suppression was fully realized in sprinkler system technology in the spring of 1988 when the first ESFR sprinkler was introduced. The definition of *fire suppression* is qualitative in nature because it does not specify to what extent—for example, 10 percent or 50 percent—the rate of heat release must be reduced in order to achieve suppression.

A sprinkler can achieve a 100 percent reduction of the heat release rate—that is, complete extinguishment—of a free-burning cellulose material, such as wood, located in the center of a room, if the fire, especially the area near the ignition, is not shielded from sprinkler discharge. In contrast, if a plasticized product stored 10 ft (3 m) high is burning with the ignition source near the floor, and the fire is partially shielded from the sprinkler because the material is stored on wood pallets, then a less than 100 percent reduction in the heat release rate is more likely. The degree to which a fire is shielded from the sprinkler and its discharge is a key factor affecting a sprinkler system's ability to suppress a fire.

Exhibit 1.1 shows a simplified heat release rate curve and how it would be impacted by a sprinkler system designed for fire control versus one designed for fire suppression. Currently, the only device that can achieve fire suppression is the ESFR sprinkler.

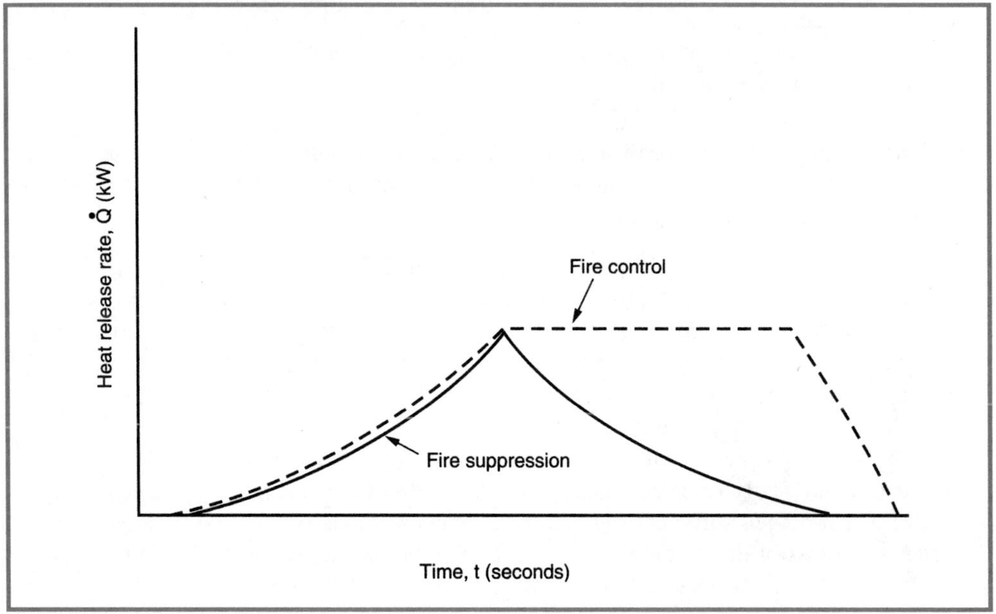

Exhibit 1.1 *Simplified fire control/fire suppression analogy.*

High-Challenge Fire Hazard. A fire hazard typical of that produced by fires in combustible high-piled storage.

The occupancy hazard descriptions given in 2-1.1, 2-1.2, and 2-1.3 are a means of categorizing the types of fires that can be protected by sprinkler systems installed in accordance with the occupancy hazard requirements of NFPA 13. To a large degree occupancy hazards are dependent on the function of the space. Occupancy hazards have been the basis for the sprinkler design criteria found in NFPA 13 since 1947. Section 2-1 provides additional information on occupancy hazard classifications.

Since about 1947, the quantity and arrangement of materials stored in buildings have changed quite dramatically. For example, warehouses that accommodate 70-ft (21-m) high rack storage now exist as do warehouse-type spaces used for bulk retail operations. These buildings can contain a wide range and a large quantity of commodities such as plastics, flammable liquids, aerosols, and automobile tires. These kinds of facilities often create fire hazards that exceed the hazards characterized by the occupancy hazard descriptions.

Prior to the 1999 edition of NFPA 13, these facilities fell outside the scope of NFPA 13. The sprinkler protection of high-piled commodities such as plastics and automobile tires was covered by the NFPA 231 series of documents. However, due to the centralization of sprinkler system requirements, the sprinkler protection of storage facilities is now addressed by NFPA 13.

Rather than categorizing the associated fire hazards in storage facilities according to the occupancy hazard approach, the hazards are categorized according to the types of commodities stored. See Chapter 2 and Sections 7-3 through 7-8 for additional information. With regard to hazards presented by other commodities or processes, the appropriate sprinkler system information is now either cited or referenced in NFPA 13. For example, where aerosols are stored, 7-10.2 references NFPA 30B, *Code for the Manufacture and Storage of Aerosol Products.*

Unless specifically stated to the contrary, regulations governing specific installation requirements and information pertaining to hydraulic calculations are the same for systems protecting high-challenge fire hazards as they are for light, ordinary, and extra hazard occupancies. For example, rules for positioning sprinklers, techniques for hydraulic calculations, selection of appropriate pipe materials, and requirements for determining the coverage area for a sprinkler are the same.

The term *high challenge* implies that provisions in excess of those required for light, ordinary, and extra hazard occupancies are necessary to achieve fire control or fire suppression. An additional provision could include the need for higher sprinkler discharge densities, a larger design area, or the installation of additional devices or components.

High-Piled Storage. Solid-piled, palletized, rack storage, bin box, and shelf storage in excess of 12 ft (3.7 m) in height.

Storage heights exceeding 12 ft (3.7 m) are considered high piled by NFPA 13. The scope of the 1999 edition was revised to include high-piled storage. See Sections 2-2 and 7-3 through 7-8 for the relevant sprinkler system criteria. High-piled storage was previously addressed in the NFPA 231 series of documents. Exhibit 1.2 illustrates an example of high-piled storage consisting of on-floor, palletized, and future rack storage.

Hydraulically Designed System. A calculated sprinkler system in which pipe sizes are selected on a pressure loss basis to provide a prescribed water density, in gallons per minute per square foot (mm/min), or a prescribed minimum discharge pressure or flow per sprinkler, distributed with a reasonable degree of uniformity over a specified area.

A sprinkler system designed using hydraulic analysis is preferable over those systems designed using a pipe schedule approach. Hydraulic designs provide for a more accurate analysis of the piping system and allow for the selection of the most suitable pipe sizes. A hydraulic analysis also demonstrates if the water supply is adequate for the sprinkler system demand. The availability of an increasing number of product-specific

Exhibit 1.2 Example of high-piled storage.

sprinkler system components all but necessitates the hydraulic design of sprinkler systems.

A hydraulic analysis determines the sprinkler system's demand and the necessary water supply. An acceptable system is one in which the system's demand is less than the available water supply. The results of hydraulic analysis are often presented in graphical form. Graphical representation of the hydraulic analysis provides easier review and evaluation. The graph usually contains two curves—one that represents the system demand and the other that represents the available water supply. Figure A-8-3.2(d) illustrates this graphical representation. Where fire pumps or other equipment supplement the available water supply, their effect on the water supply must also be presented graphically, resulting in additional curves on the graph.

To hydraulically design a sprinkler system, the following information must be obtained:

(a) The design density and the anticipated area of sprinkler system operation. For light, ordinary (Groups 1 and 2), and extra hazard (Groups 1 and 2) occupancies, this information can be obtained from Figure 7-2.3.1.2. For most other types of hazards, the design density and anticipated area of sprinkler system operation are found in Chapter 7 under the hazard heading. In a few instances, such as for flammable and combustible liquids (NFPA 30) and aerosol products (NFPA 30B), references are made to specific NFPA standards for this information. Some insurance companies set their own requirements for sprinkler discharge densities and area of sprinkler system operation for the properties they insure. This topic is discussed in more detail in Chapters 7 and 8.

(b) The characteristics of the available water supply. The types of acceptable water supplies (see also Chapter 9) are summarized as follows:

(1) Connection to a waterworks system. Information on the characteristics of the water supply can be obtained from the water department, fire department, or insurance companies or by conducting a waterflow test as described in A-9-2.1.
(2) Gravity tanks. The available pressure is determined by the tank's height. The friction loss through the water distribution system is based on the required flow and is calculated through hydraulic analysis. The determination of water supply characteristics for gravity tanks is not considered to be a complicated procedure. More information on gravity tanks can be found in NFPA 22.
(3) Fire pumps. A fire pump increases the pressure and flow from an available water supply such as a municipal water supply system. The effect of the fire pump on the available water supply curve can be presented graphically. It is important to note that fire pumps do not create additional water capacity but rather boost the pressure and flow of an existing water supply. For example, if a fire pump is connected to a municipal water supply, the fire pump cannot create a larger quantity of water than what already exists in the municipal supply. Instead, the fire pump will boost the pressure from the municipal supply so that additional flow is available. The function of the fire pump is to raise the water supply curve in terms of pressure and flow so that it exceeds the sprinkler system demand.

In the case of a suction tank located on the ground, the fire pump provides all the necessary pressure and resultant waterflow. The water tank provides the necessary quantity of water. When pumps take suction from tanks, the elevation of the bottom of the tank with respect to the elevation of the fire pump needs to be considered since the effects of gravity impact overall performance. Where pump suction is from water mains, the fire pump curve is added to the water supply curve for the water main. This addition can be done mathematically or graphically. NFPA 20 provides more information about fire pumps.

(4) Pressure tanks. In this type of water supply the designer defines the characteristics required, chooses a tank of proper volume, and establishes the air pressure from the procedure given in A-9-2.3.3. For more information see NFPA 22.

NFPA 13 does not require a safety factor to be applied to either the water supply or the system demand. While NFPA 13 does not specifically require the application of safety factors in any part of a sprinkler system design or installation, safety factors are implicitly imbedded in many of the document's requirements. Some designers, however, leave a certain margin between the water supply and the system demand to account for future use or deterioration of the water supply. For instance, the construction of additional buildings drawing water from a common municipal supply can negatively impact the available water supply. This situation could result in a sprinkler system deficiency because the water supply can no longer meet the system demand. Designers who do not account for future potential development by leaving a margin between the system demand and the water supply run an increased risk of having an inadequate sprinkler system in future years. The use of a safety factor is considered good practice but is not required.

In addition, some areas of the United States develop critical water shortages during certain times of the year, causing some water purveyors to initiate water supply reduction programs. In other cases, available water distribution system pressures and flows are intentionally reduced by trimming pumps and valves resulting in lower static and residual pressures. A safety factor increases the likelihood that these systems will perform adequately if gradual reductions in the water supply occur during the life of the sprinkler system. An appropriate size for the safety factor takes into account the existing water supply and any anticipated future development of the site.

Limited-Combustible Material. As applied to a building construction material, a material not complying with the definition of noncombustible material that, in the form in which it is used, has a potential heat value not exceeding 3500 Btu per lb (8141 kJ/kg) and complies with one of the following, (a) or (b). Materials subject to increase in combustibility or flame spread rating beyond the limits herein established through the effects of age, moisture, or other atmospheric condition shall be considered combustible. (a) Materials having a structural base of noncombustible material, with a surfacing not exceeding a thickness of ⅛ in. (3.2 mm) that has a flame spread rating not greater than 50. (b) Materials, in the form and thickness used, other than as described in (a), having neither a flame spread rating greater than 25 nor evidence of continued progressive combustion and of such composition that surfaces that would be exposed by cutting through the material on any plane would have neither a flame spread rating greater than 25 nor evidence of continued progressive combustion.

This definition is from NFPA 220, *Standard on Types of Building Construction*, and is used in portions of Chapter 5, specifically Section 5-13, where describing those areas in which sprinklers are not required. The presence of wall materials that are noncombustible or limited-combustible is necessary to meet many of the exceptions that allow for the omission of sprinklers in certain areas.

Miscellaneous Storage.* Storage that does not exceed 12 ft (3.66 m) in height and is incidental to another occupancy use group. Such storage shall not constitute more than 10 percent of the building area or 4000 ft^2 (372 m^2) of the sprinklered area, whichever is greater. Such storage shall not exceed 1000 ft^2 (93 m^2) in one pile or area, and each such pile or area shall be separated from other storage areas by at least 25 ft (7.62 m).

A-1-4.2 Miscellaneous Storage. The sprinkler system design criteria for miscellaneous storage at heights below 12 ft (3.7 m) is covered by this standard in Chapters 5 and 7. Paragraph 7-2.3.2.2 describes design criteria and Section 5-2 describes installation requirements (area limits). These requirements apply to all storage of 12 ft (3.7 m) or less in height.

The miscellaneous storage concept applies to a building where storage constitutes only a part of the building's use, such as the back room of a mercantile facility. Limitations are imposed on the amount and arrangement of storage that is permitted.

Where one of the limitations cannot be met or the building is used exclusively for storage, sprinkler discharge criteria for storage such as that found in Section 7-3 need to be used.

The concept of miscellaneous storage was developed to address those situations where only a relatively small portion of a building is used for storage and where such storage does not exceed 12 ft (3.7 m) in height. One example is a manufacturing operation that uses a portion of its facility to store small amounts of finished product and raw materials. As with most facilities, the fire hazard associated with the storage is likely to differ from that of the manufacturing operation. Prior editions of NFPA 13 only addressed storage if it was considered miscellaneous. However, this edition of NFPA 13 treats storage under 12 ft (3.7 m) in height and miscellaneous storage virtually the same.

The 10 percent limit is taken from certain accessory use definitions as found in the model building codes. The 4000-ft^2 (372-m^2) limit is derived by taking 10 percent of the maximum allowable area of coverage for a sprinkler system riser protecting high-piled storage. (See Section 5-2 for system protection area limitations.)

Noncombustible Material. A material that, in the form in which it is used and under the conditions anticipated, will not ignite, burn, support combustion, or release flammable vapors when subjected to fire or heat. Materials that are reported as passing ASTM E 136, *Standard Test Method for Behavior of Materials in a Vertical Tube Furnace at 750°C*, shall be considered noncombustible materials.

This definition is from NFPA 220. The use of noncombustible or limited-combustible materials is necessary to meet many of the exceptions that allow for the omission of sprinklers in certain areas. Construction materials that fall under this category include certain types of acoustical ceiling tiles and insulation materials.

Pipe Schedule System. A sprinkler system in which the pipe sizing is selected from a schedule that is determined by the occupancy classification and in which a given number of sprinklers are allowed to be supplied from specific sizes of pipe.

A pipe schedule system describes one of the design approaches permitted by Chapter 7 and is the oldest design approach. A version of it appeared in the first edition of NFPA 13 in 1896. While still permitted by the standard under limited conditions, pipe schedule systems are much less common than they used to be.

Hydraulically designed systems, which were introduced in 1969, have become more popular because they allow for a more thorough and accurate analysis of the system, which typically leads to more effective and cost-efficient designs. Computer technology and the availability of numerous sprinkler hydraulic programs have also contributed to the decline of pipe schedule systems.

A pipe schedule design is often referred to as a cookbook design approach because the pipe size is predetermined based on the number of sprinklers being supplied.

Shop-Welded. As used in this standard, *shop* in the term *shop-welded* means either (1) a sprinkler contractor's or fabricator's premise or (2) an area specifically designed or authorized for welding, such as a detached outside location, maintenance shop, or other area (either temporary or permanent) of noncombustible or fire-resistive construction free of combustible and flammable contents and suitably segregated from adjacent areas.

Paragraph 3-6.2.2 requires sprinkler system piping to be shop-welded. However, under carefully controlled circumstances, the pipe can be welded in place. This definition establishes the necessary criteria for the shop-welding of pipe.

Small Rooms. A room of light hazard occupancy classification having unobstructed construction and floor areas not exceeding 800 ft^2 (74.3 m^2) that are enclosed by walls and a ceiling. Openings to the adjoining space are permitted if the minimum lintel depth is 8 in. (203 mm) from the ceiling.

Fires in small rooms of a light hazard occupancy present a lesser challenge to the sprinkler system. As a result, sprinklers installed in rooms that meet the criteria of small rooms are given an exception to the spacing rules of Chapter 5.

Small room openings do not need to be protected with doors if the openings are equipped with lintels that have a minimum depth of 8 in. (203 mm). A lintel with a depth of at least 8 in. (203 mm) ensures adequate collection of the heat from a fire at the ceiling of the room of fire origin and promotes faster operation of the sprinklers nearest to the fire.

Sprinkler System.* For fire protection purposes, an integrated system of underground and overhead piping designed in accordance with fire protection engineering standards. The installation includes one or more automatic water supplies. The portion of the sprinkler system aboveground is a network of specially sized or hydraulically designed piping installed in a building, structure, or area, generally overhead, and to which sprinklers are attached in a systematic pattern. The valve controlling each system riser is located in the system riser or its supply piping. Each sprinkler system riser includes a device for actuating an alarm when the system is in operation. The system is usually activated by heat from a fire and discharges water over the fire area.

A-1-4.2 Sprinkler System. A sprinkler system is considered to have a single system riser control valve. The design and installation of water supply facilities such as gravity tanks, fire pumps, reservoirs, or pressure tanks are covered by NFPA 20, *Standard for the Installation of Centrifugal Fire Pumps,* and NFPA 22, *Standard for Water Tanks for Private Fire Protection.*

The sprinkler system is denoted by the presence of an alarm check valve or a waterflow device and a control valve. It is important to note that the sprinkler system definition includes water supplies and underground piping. While design and installation requirements for tanks and pumps are covered by other standards, these components are used in connection with sprinkler piping and are considered an integral part of the sprinkler system. They are critical to successful sprinkler system performance and must be given serious consideration.

Sprinkler systems can be designed according to hydraulic calculation techniques or according to predetermined pipe sizes conforming to pipe schedules (see Chapter 8). However, NFPA 13 severely restricts the use of pipe schedule systems.

System Working Pressure. The maximum anticipated static (nonflowing) or flowing pressure applied to sprinkler system components exclusive of surge pressures.

This term was added to the 1999 edition and is intended to apply to the pressure in the sprinkler system when installed. The term is not intended to be used as criteria for evaluating individual system components.

Thermal Barrier. A material that will limit the average temperature rise of the unexposed surface to not more than 250 F (121 C) after 15 minutes of fire exposure, which complies with the standard time–temperature curve of NFPA 251, *Standard Methods of Tests of Fire Endurance of Building Construction and Materials.*

This term is used in 5-13.9.1 to establish a condition necessary to omit sprinklers from certain types of dwelling unit bathrooms.

1-4.3 Sprinkler System Type Definitions.

Antifreeze Sprinkler System. A wet pipe sprinkler system employing automatic sprinklers that are attached to a piping system that contains an antifreeze solution and that are connected to a water supply. The antifreeze solution is discharged, followed by water, immediately upon operation of sprinklers opened by heat from a fire.

This system is often considered auxiliary to a wet pipe system and is installed on the system side of the control valve of a wet pipe system. The antifreeze sprinkler system is considered part of a wet pipe system and the two systems as a whole must satisfy all appropriate rules, such as those pertaining to protection area limitations and hydraulic calculations, for the wet pipe system. In addition, all requirements for the antifreeze system must also be satisfied. Each auxiliary antifreeze system will require its own separate control valve, drains, and so forth. Alarms on the wet pipe system might not be adequate for the auxiliary system, necessitating additional alarms. The use of antifreeze systems will usually be regulated by the municipal water department. See Section 4-5 for specific requirements pertaining to antifreeze systems.

Circulating Closed-Loop Sprinkler System. A wet pipe sprinkler system having non–fire protection connections to automatic sprinkler systems in a closed-loop piping arrangement for the purpose of utilizing sprinkler piping to conduct water for heating or cooling, where water is not removed or used from the system but only circulated through the piping system.

The possibility of utilizing sprinkler system piping for other building functions has been contemplated for quite some time. This definition allows sprinkler piping to be used for functions other than fire protection, such as the heating and air conditioning of a building. A basic criterion used in the development of this definition was that the auxiliary functions be added in such a manner and with sufficient controls to ensure that there would be no reduction in overall fire protection. It was determined that this criterion could be controlled more adequately with totally closed systems. The inherent characteristics of a closed system provide a means of ensuring the necessary system performance. See 4-6.1 for specific requirements pertaining to circulating closed-loop systems.

Closed-loop systems possess a self-supervisory feature. System problems often become obvious to building occupants when, for example, building temperatures become too hot or too cold, and building occupants become uncomfortable and complain. This situation would typically result in the dispatch of maintenance personnel to make repairs while ensuring that the sprinkler system remained in service.

Closed-loop systems are often heat pump systems in which water is circulated through heating and air-conditioning equipment, utilizing sprinkler pipe as the primary means of circulating water. Heat is either added to or removed from the circulating water as dictated by requirements of the system. Water on the other hand is neither added to nor removed from the piping system.

Combined Dry Pipe-Preaction Sprinkler System. A sprinkler system employing automatic sprinklers attached to a piping system containing air under pressure with a supplemental detection system installed in the same areas as the sprinklers. Operation of the detection system actuates tripping devices that open dry pipe valves simultaneously and without loss of air pressure in the system. Operation of the detection system also opens listed air exhaust valves at the end of the feed main, which usually precedes the opening of sprinklers. The detection system also serves as an automatic fire alarm system.

Combined dry pipe-preaction systems are employed primarily where more than one dry pipe system is required due to the size of the space, and where it is not possible to install supply piping to each dry pipe valve through a heated or protected space. Primary examples where combined dry pipe-preaction systems are used include piers, wharves, and very large cold storage warehouses. These systems are special application systems and should be used as such. Many cold storage rooms can be adequately protected with a conventional dry pipe system or a double interlock preaction system.

Deluge Sprinkler System. A sprinkler system employing open sprinklers that are attached to a piping system that is connected to a water supply through a valve that is opened by the operation of a detection system installed in the same areas as the sprinklers. When this valve opens, water flows into the piping system and discharges from all sprinklers attached thereto.

A deluge system's mode of activation is similar to that for a preaction system. With the exception of the non-interlock preaction system, the activation of these two types of systems is dependent on the operation of a supplemental detection system. The difference between the two types of systems is that preaction systems employ automatic sprinklers that respond to heat, and deluge systems use open sprinklers that have no fusible element. Operation of the detection system in a preaction system fills the system piping with water once certain interlocks are triggered. Water is not discharged from the system until a sprinkler operates. For a deluge system, operation of the detection system results in flow from all system sprinklers. Deluge systems are normally used for high-hazard areas requiring an immediate application of water over the entire hazard. An aircraft hangar is an example of a facility that is likely to use deluge systems. Exhibit 1.3 illustrates an example of a deluge system using a supplemental electronic heat or smoke detection system.

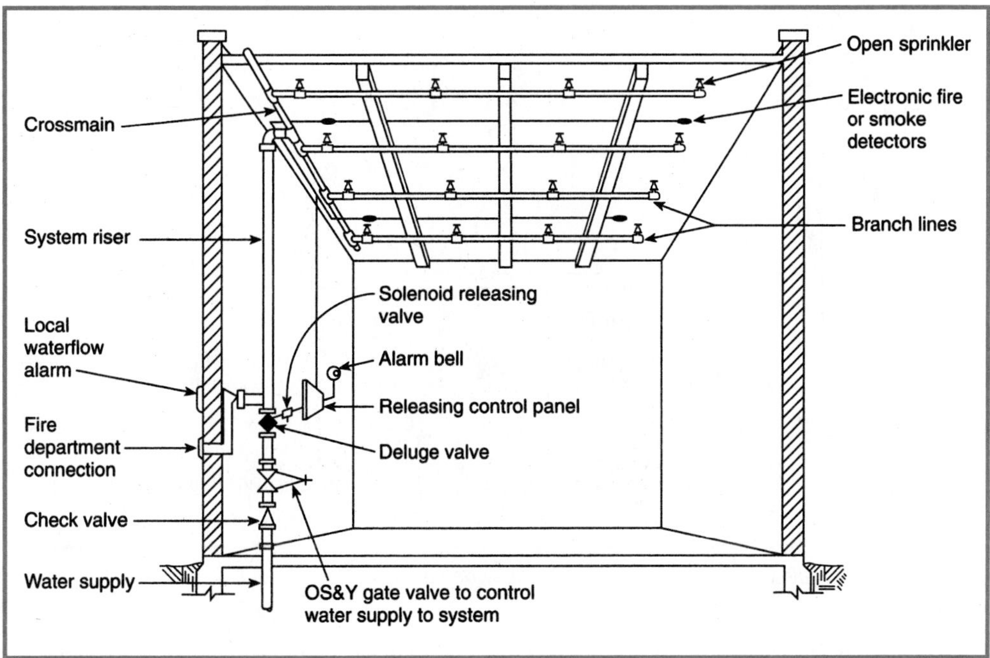

Exhibit 1.3 *Example of a deluge system.*

Dry Pipe Sprinkler System. **A sprinkler system employing automatic sprinklers that are attached to a piping system containing air or nitrogen under pressure, the release of which (as from the opening of a sprinkler) permits the water pressure to open a valve known as a dry pipe valve, and the water then flows into the piping system and out the opened sprinklers.**

Dry pipe systems should only be installed where pipe is subject to freezing. Paragraph 5-14.3.1.1 requires dry pipe systems to be installed where the building environment cannot be maintained at or above 40°F (4°C). Dry pipe systems should not be used to reduce water damage from pipe breakage or leakage, since they operate too quickly to be of value for this purpose. An interlock preaction system should be used for this purpose. A dry pipe system is typically filled with compressed air or nitrogen and is activated when a drop in air pressure occurs. A drop in air pressure can be caused by the activation of a single sprinkler or by damage to the sprinkler system piping. A drop in air pressure causes the dry pipe valve to open and allows water to flow through the system. Exhibit 1.4 illustrates an example of a dry pipe system.

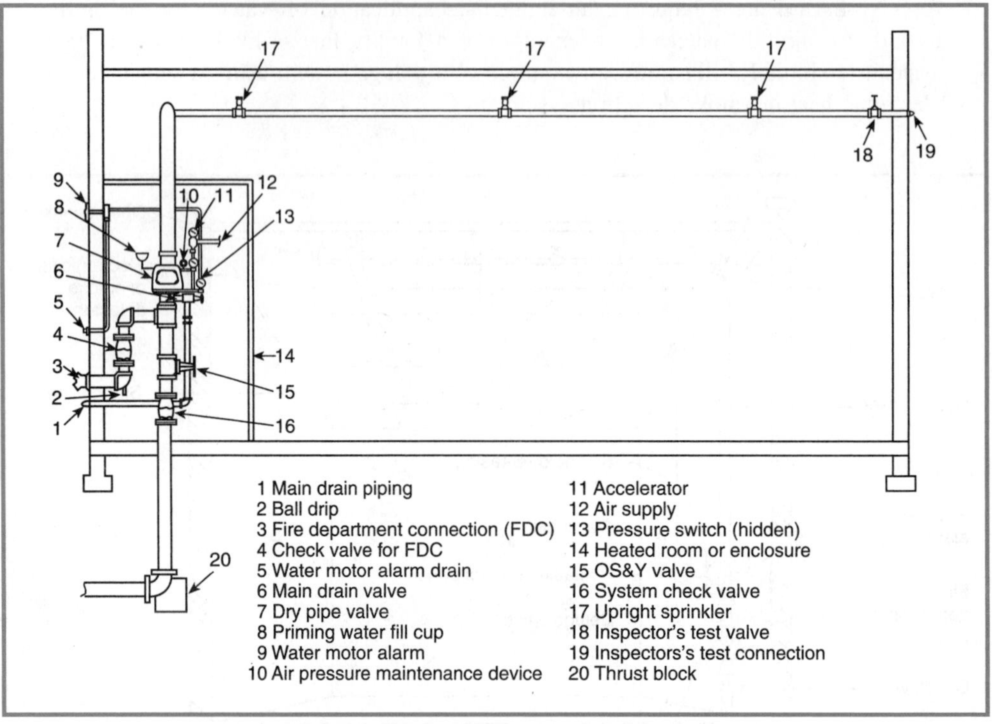

Exhibit 1.4 *Example of a dry pipe system. (Courtesy of the Viking Corporation.)*

Gridded Sprinkler System.* A sprinkler system in which parallel cross mains are connected by multiple branch lines. An operating sprinkler will receive water from both ends of its branch line while other branch lines help transfer water between cross mains.

A-1-4.3 Gridded Sprinkler System. See Figure A-1-4.3(a).

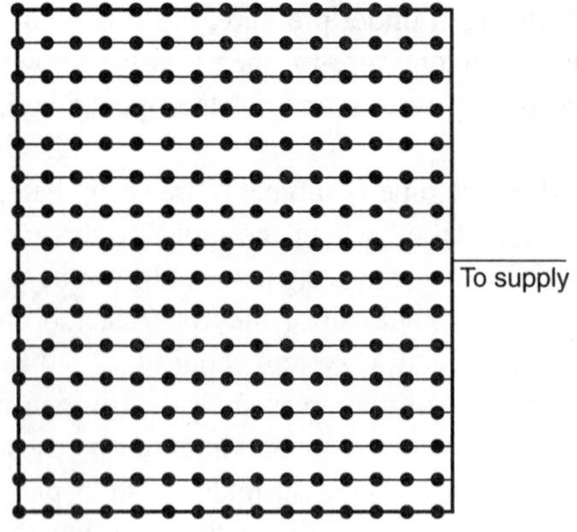

Figure A-1-4.3(a) *Gridded system.*

Grid-type systems are preferred by some designers because of their attractive hydraulic characteristics. Since a number of flow paths to the sprinklers are available, pressure drops through the system piping are lower in a gridded system compared to other system configurations. Due to the very complex nature of the hydraulic calculations involved, the use of computer hydraulic programs is almost always necessary in evaluating the piping grid and determining the pressure requirements of the system.

While the gridded system possesses highly advantageous hydraulic characteristics, certain limitations and design conditions associated with its use are as follows:

(1) Gridded systems are not permitted for dry pipe systems and certain preaction systems because excessive amounts of air can remain trapped in the system piping, which significantly delays water from reaching the operating sprinklers (see 4-2.3.2 and 4-3.2.5).
(2) Gridded systems must be equipped with a relief valve to remove trapped air from the piping. (See 4-1.2 and 5-14.1.2.1.)
(3) Gridded systems must be "peaked" to verify that the hydraulically most demanding combination of sprinklers was selected (see 8-4.4.2).

Looped Sprinkler System.* A sprinkler system in which multiple cross mains are tied together so as to provide more than one path for water to flow to an operating sprinkler and branch lines are not tied together.

A-1-4.3 Looped Sprinkler System. See Figure A-1-4.3(b).

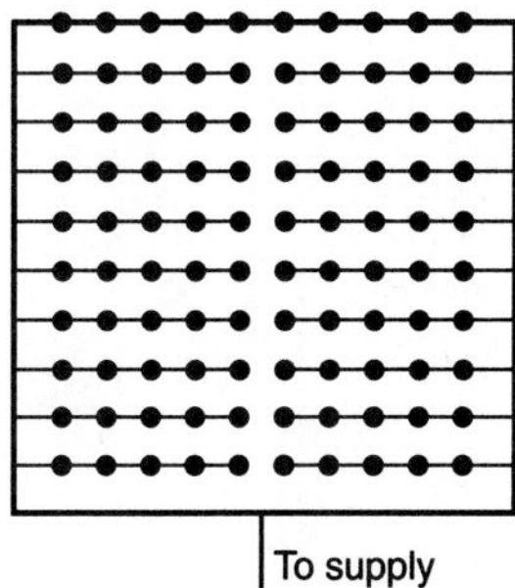

Figure A-1-4.3(b) *Looped system.*

Looped systems possess better hydraulic characteristics than do tree systems, but their hydraulic characteristics are not as good as those of a gridded system. The hydraulic calculations associated with a looped system are not as tedious and complicated as those for a gridded system. The calculations for a looped system can be conducted without the use of a computer program although the hand calculation process can be quite laborious. Additionally, looped systems do not have the same limitations and design considerations as do gridded systems.

Preaction Sprinkler System.* A sprinkler system employing automatic sprinklers that are attached to a piping system that contains air that might or might not be under pressure, with a supplemental detection system installed in the same areas as the sprinklers.

A-1-4.3 Preaction Sprinkler System The actuating means of the valve are described in 4-3.2.1. Actuation of the detection system and sprinklers in the case of double-interlocked systems opens a valve that permits water to flow into the sprinkler piping system and to be discharged from any sprinklers that are open.

The three types of preaction systems are indicated in 4-3.2.1. The user should be aware that the double interlock version is subject to the same time delay problems as a dry pipe system and, likewise, is subject to comparable dry pipe system limitations.

Preaction systems are normally used to protect properties where accidental water discharge is a significant concern. Even though accidental or premature discharge of sprinkler systems is extremely rare, some property owners still prefer these systems

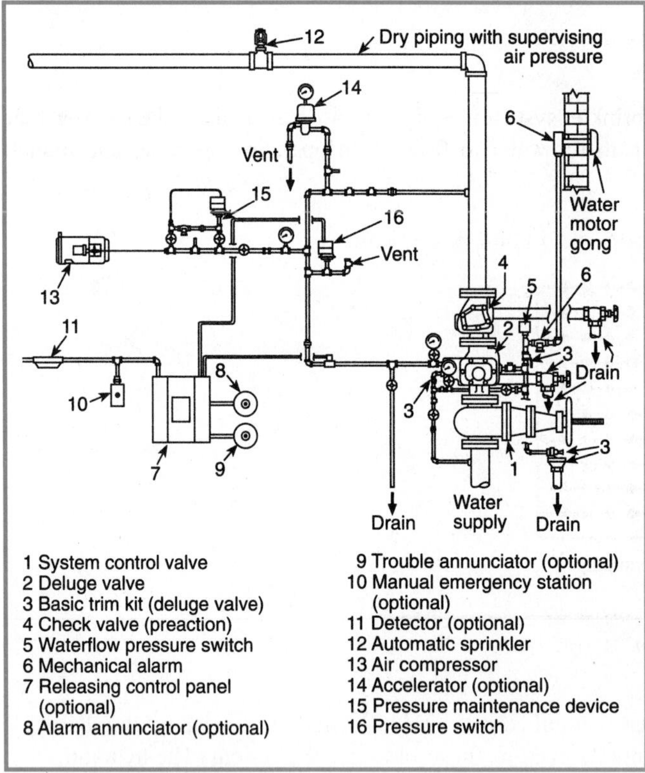

Exhibit 1.5 *Example of a preaction system. (Courtesy of The Reliable Automatic Sprinkler Co., Inc.)*

in electronic equipment areas. Exhibit 1.5 shows an example of a double interlock preaction system with electric activation. Preaction systems can also be activated through pneumatic means—that is, a dry pilot sprinkler line. Section 4-3 addresses the requirements for preaction systems. Formal Interpretation 89-1 provides additional information on the supplemental detection system to be used with a preaction system.

Formal Interpretation 89-1

Reference: 1-4.3, 1-4.2

Question 1: Does " . . . a supplemental fire detection system . . ." as described in 1-4.3 mean any type of detection system (heat, smoke, flame or gas) installed in accordance with NFPA 72?

Answer: Yes.

Question 2: Would a smoke detection system installed in accordance with NFPA 72 comply with 1-4.3?

Answer: Yes.

Question 3: Would a Preaction System or Combined Dry-Pipe and Preaction System activated by a smoke detection system, installed in accordance with NFPA 72, meet the definition of a sprinkler system as described in Section 1-4.2?

Answer: Yes.

Issue Edition: 1989

Reference: 1-3, 5-4.1.1

Issue Date: July 27, 1989

Effective Date: August 16, 1989

Wet Pipe Sprinkler System. A sprinkler system employing automatic sprinklers attached to a piping system containing water and connected to a water supply so that water discharges immediately from sprinklers opened by heat from a fire.

Wet pipe systems are the simplest and most reliable of all sprinkler systems. Operation of one sprinkler activates a wet pipe system. It is important to note that only those sprinklers activated by the heat from a fire discharge water. Exhibit 1.6 illustrates an example of a wet pipe system utilizing an alarm check valve. Section 4-1 addresses requirements for wet pipe systems.

All other types of systems addressed by NFPA 13 require the use of additional equipment such as dry pipe valves, preaction valves, or supplemental detectors for successful operation. As the number of events needed for system activation increases—that is, the activation of both sprinklers and heat detectors or the use of more elaborate devices such as dry pipe valves—the greater the need for a comprehensive inspection,

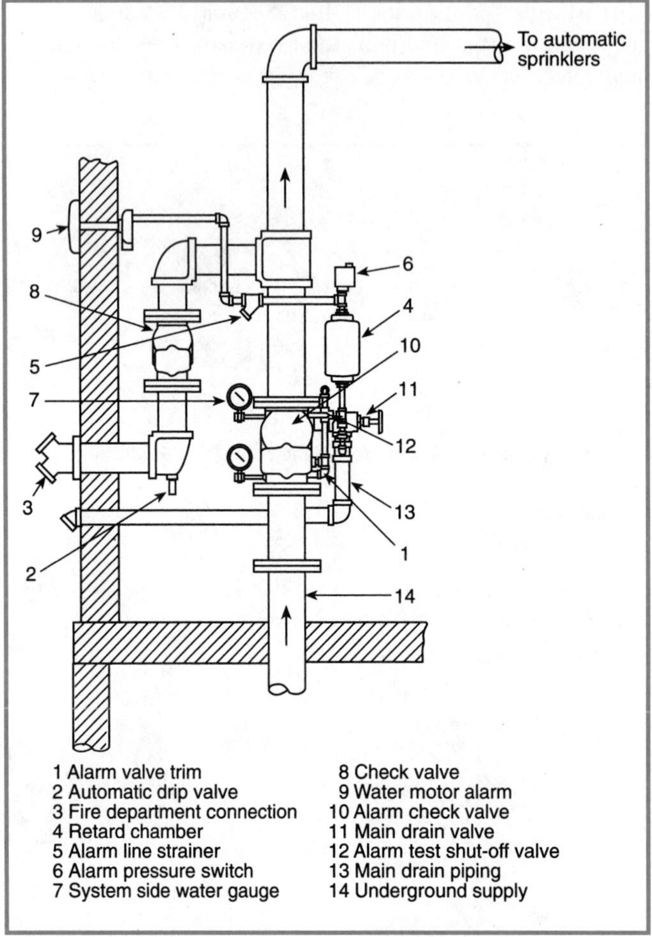

Exhibit 1.6 *Wet pipe sprinkler riser with alarm check valve.*

testing, and maintenance program. In other words, the more complicated a system, the greater the likelihood that something can go wrong and the greater the effort needed to keep the system in proper working order.

Wet pipe systems should always be the first type of system considered. Only where a wet pipe system could not properly protect a space should another type of system be contemplated. For example, a wet pipe system is not a good candidate for a cold storage room because the cold temperature is likely to cause the water in the sprinkler piping to freeze, resulting in damage to the piping and an impairment of the sprinkler system. A dry pipe or an antifreeze system is a better choice for the protection of cold storage rooms.

1-4.4* System Component Definitions.

A-1-4.4 See Figure A-1-4.4.

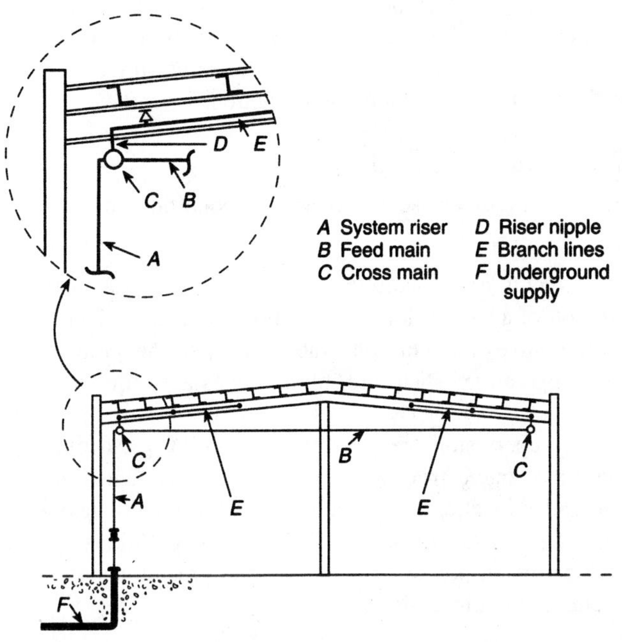

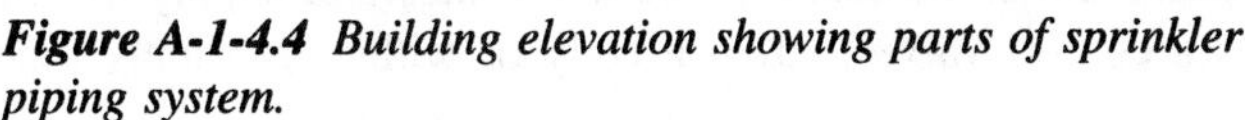

Figure A-1-4.4 Building elevation showing parts of sprinkler piping system.

Branch Lines. The pipes in which the sprinklers are placed, either directly or through risers.

Branch lines are usually the smallest diameter pipes installed on a system and normally have sprinklers attached to them.

Cross Mains. The pipes supplying the branch lines, either directly or through risers.

Feed Mains. The pipes supplying cross mains, either directly or through risers.

Flexible Listed Pipe Coupling. A listed coupling or fitting that allows axial displacement, rotation, and at least 1 degree of angular movement of the pipe without inducing harm on the pipe. For pipe diameters of 8 in. (203.2 mm) and larger, the angular movement shall be permitted to be less than 1 degree but not less than 0.5 degree.

This type of coupling is specified for use by 6-4.2 to maintain system integrity during seismic events. However, its use for other applications is also possible. Listed flexible couplings will permit differential movement of the pipe, thereby reducing the potential for pipe breakage.

Risers. The vertical supply pipes in a sprinkler system.

The section of aboveground pipe directly connected to the water supply is referred to as the system riser and is equipped with an alarm-actuating device such as an alarm check valve or waterflow switch (see 1-4.4 for definition of *System Riser*). Other vertical pipes within the system are also generically referred to as risers. In most cases, however, these smaller vertical sections of system piping are denoted as sprigs, riser nipples, or drop nipples, depending on their orientation and function.

Sprig-up. A line that rises vertically and supplies a single sprinkler.

Supervisory Device. A device arranged to supervise the operative condition of automatic sprinkler systems.

NFPA 13 requires that certain elements and conditions of the sprinkler system be supervised to decrease the likelihood of a system impairment. For example, a planned or inadvertent closure of one of the valves that controls water supply to the sprinkler system would render the system inoperative. Therefore, these valves are required to be supervised.

While several means of valve supervision are permitted by NFPA 13, a more sophisticated means includes the use of an electronically monitored tamper switch that sounds an alarm at a constantly attended location. The installation of this type of device is addressed by NFPA 72, *National Fire Alarm Code®*. Other system conditions that are required to be supervised include the air pressure in a dry pipe or preaction system and the water temperature in a circulating closed-loop system.

System Riser. The aboveground horizontal or vertical pipe between the water supply and the mains (cross or feed) that contains a control valve (either directly or within its supply pipe) and a waterflow alarm device.

A system riser is more than just a subset of the term *riser,* which is broadly defined as any vertical piping within the sprinkler system. As indicated by the definition, a system riser can be any aboveground pipe in a vertical or horizontal orientation installed between the water supply and the system mains that contains specific devices.

1-4.5 Sprinkler Definitions.

This subsection places sprinkler definitions into one of the following four categories:

(1) Characteristics that define a sprinkler's ability to control or suppress a fire
(2) Sprinkler design features
(3) Orientation
(4) Special application uses

As little as 25 years ago, most sprinklers were of the upright, pendent, or sidewall type and were classified as either having frangible bulb or solder link operating mechanisms. Since then, many advances have been made with regard to our scientific understanding of fire, sprinklers, and their interaction.

These technological advances are producing an increasingly large variety of sprinkler types and styles. Sprinklers are now manufactured and designed to allow systems to perform specific functions. This subsection describes the features that differentiate sprinklers from one another.

1-4.5.1* The following are characteristics of a sprinkler that define its ability to control or extinguish a fire.

(a) Thermal sensitivity.A measure of the rapidity with which the thermal element operates as installed in a specific sprinkler or sprinkler assembly. One measure of thermal sensitivity is the response time index (RTI) as measured under standardized test conditions.

(1) Sprinklers defined as fast response have a thermal element with an RTI of 50 (meters-seconds)$^{1/2}$ or less.
(2) Sprinklers defined as standard response have a thermal element with an RTI of 80 (meters-seconds)$^{1/2}$ or more.

(b) Temperature rating.

(c) Orifice size *(see Chapter 2).*

(d) Installation orientation *(see 1-4.5.3).*

(e) Water distribution characteristics (i.e., application rate, wall wetting).

(f) Special service conditions *(see 1-4.5.4).*

A-1-4.5.1 The response time index (RTI) is a measure of the sensitivity of the sprinkler's thermal element as installed in a specific sprinkler. It is usually determined by plunging a sprinkler into a heated laminar airflow within a test oven. The plunge test is not currently applicable to certain sprinklers.

The RTI is calculated using the following:

(1) The operating time of the sprinkler
(2) The operating temperature of the sprinkler's heat-responsive element (as determined in a bath test)
(3) The air temperature of the test oven
(4) The air velocity of the test oven
(5) The sprinkler's conductivity *(c)* factor, which is the measure of conductance between the sprinkler's heat-responsive element and the sprinkler oven mount

Other factors affecting response include the temperature rating, sprinkler position, fire exposure, and radiation.

ISO standard 6182-1 currently recognizes the RTI range of greater than 50 (meters-seconds)$^{1/2}$ and less than 80 (meters-seconds)$^{1/2}$ as special response. Such sprinklers can be recognized as special sprinklers under 5-4.9.1.

It should be recognized that the term *fast response* (like the term *quick response* used to define a particular type of sprinkler) refers to the thermal sensitivity within the operating element of a sprinkler, not the time of operation in a particular installation. There are many other factors, such as ceiling height, spacing, ambient room temperature, and distance below ceiling, that

affect the time of response of sprinklers. In most fire scenarios, sprinkler activation times will be shortest where the thermal elements are located 1 in. (25.4 mm) to 3 in. (76.2 mm) below the ceiling. A fast response sprinkler is expected to operate quicker than a standard response sprinkler in the same installation orientation. For modeling purposes, concealed sprinklers can be considered equivalent to pendent sprinklers having a similar thermal response sensitivity installed 12 in. (305 mm) below smooth unobstructed ceilings, and recessed sprinklers can be considered equivalent to pendent sprinklers having a similar thermal response sensitivity installed 8 in. (203 mm) below smooth unobstructed ceilings.

The key characteristics that affect a sprinkler's ability to control or suppress a fire are specifically addressed by this standard. These characteristics include temperature rating, orifice size, installation orientation, water distribution patterns, hydraulic criteria, wall-wetting characteristics, and thermal sensitivity. Since the 1996 edition, a greater emphasis has been placed on a sprinkler's thermal sensitivity.

Because sprinkler manufacturers and researchers investigated the effect of modifying certain sprinkler characteristics, sprinklers with faster operating times, broader spray patterns, and deeper water penetration capabilities are becoming more widely available.

With regard to item (a) of 1-4.5.1, thermal sensitivity is considered a measure of the rapidity with which the sprinkler's thermal element operates when mounted within the sprinkler frame. A common measure of thermal sensitivity is the response time index (RTI). This value is usually determined by plunging a sprinkler into a heated laminar (nonturbulent) airflow within a test oven. However, this test is not readily applicable to certain sprinklers such as recessed or concealed sprinklers.

As indicated in A-1-4.5.1, RTI is not necessarily a measure of the responsiveness of a given sprinkler when installed in the field. The RTI of a sprinkler is a function of the convective heat transfer coefficient, specific heat of the operating element, mass of the operating element, and the surface area of the operating element. Other factors that can affect a sprinkler's thermal sensitivity in the field include temperature rating, position from the ceiling, attachment to piping, and anticipated fire exposure. However, a fast-response sprinkler can be expected to operate faster than a standard response sprinkler given the same conditions. RTI is used to comparatively describe the sensitivity of the sprinkler link for any given sprinkler.

NFPA 13 addresses the concept of thermal sensitivity by classifying sprinklers as either fast response or standard response. Fast-response sprinklers include those sprinklers that have a thermal element with an RTI of 50 (meters-seconds)$^{1/2}$ [90 (ft-sec)$^{1/2}$] or less. Standard response sprinklers include those sprinklers that have a thermal element with an RTI of 80 (meters-seconds)$^{1/2}$ [145 (ft-sec)$^{1/2}$] or more.

Sprinklers having operating elements with an RTI of greater than 50 (meters-seconds)$^{1/2}$ [90 (ft-sec)$^{1/2}$] but less than 80 (meters-seconds)$^{1/2}$ [145 (ft-sec)$^{1/2}$] are considered special sprinklers. This RTI classification is consistent with that of the International Standards Organization (ISO), which considers these sprinklers as "special response." The gap between these two RTI values is intentional in order to provide a clear distinction between fast-response and standard response sprinklers.

The definitions of several sprinkler types such as quick response, residential, and early suppression fast response clearly reflect the emphasis on a sprinkler's thermal sensitivity.

Exhibit 1.7 identifies the various types of fast-response sprinklers.

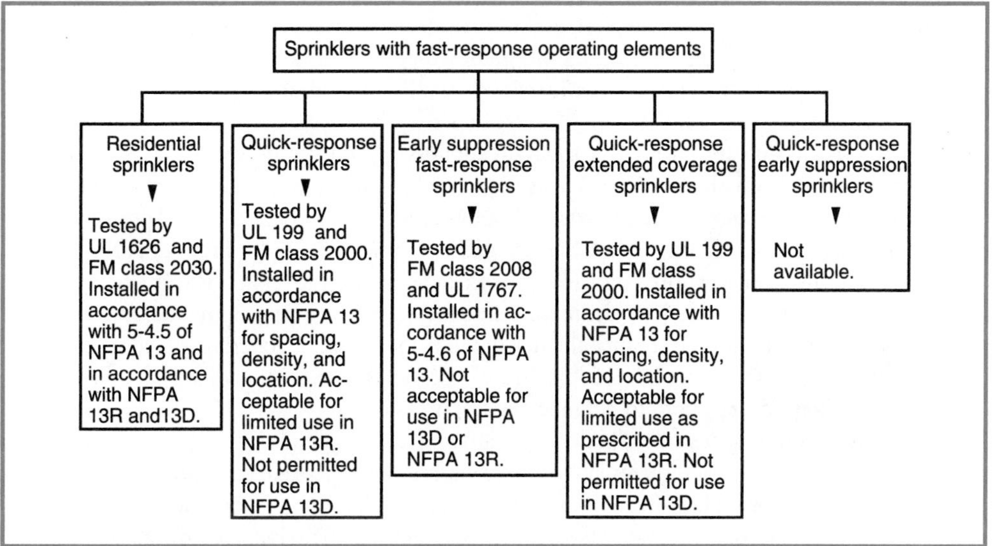

Exhibit 1.7 *Comparison of sprinklers with fast-response operating elements.*

With regard to item (c) of 1-4.5.1, the size of a sprinkler's orifice influences certain performance features such as the amount of water discharging from the sprinkler and the size of the water droplets. The size and shape of the orifice is indicated through the sprinkler's K-factor. A provision in Chapter 3 provides a means of control with respect to the manufacture and listing of sprinklers with K-factors different than the ones currently identified in the standard. See 3-2.3.1 and Table 3-2.3.1 for sprinkler discharge characteristics.

In addition, NFPA 13 requires that large-drop and ESFR sprinklers have a K-factor of at least 11.2. This limitation establishes a level of consistency with regard to these sprinkler's operational features and functional objectives.

1-4.5.2 The following sprinklers are defined according to design and performance characteristics.

Early Suppression Fast-Response (ESFR) Sprinkler.* A type of fast-response sprinkler that meets the criteria of 1-4.5.1(a)(1) and is listed for its capability to provide fire suppression of specific high-challenge fire hazards.

A-1-4.5.2 ESFR Sprinkler. It is important to realize that the effectiveness of these highly tested and engineered sprinklers depends on the combination of fast response and the quality

and uniformity of the sprinkler discharge. It should also be realized that ESFR sprinklers cannot be relied upon to provide fire control, let alone suppression, if they are used outside the guidelines specified in 7-9.5.

ESFR sprinklers evolved by examining how the effects of sprinkler sensitivity and water distribution characteristics could be combined to achieve early fire suppression. The concept is that if a sufficient amount of water can be discharged in the early phases of a fire and penetrate the developing fire plume, then fire suppression can be achieved. Once the fire plume reaches a significant size, the likelihood of suppression is greatly reduced.

A fast and hot fire plume can frustrate suppression efforts in two ways. A strong updraft is characteristic of a severe fire plume, which decreases the amount of water that can reach the burning area. Water droplets traveling through this fire plume are either vaporized before they reach the fire or blown away from the fire. A severe fire plume can also cause sprinklers distant from the fire area to open. These excessive sprinkler actuators result in a decreased amount of water being delivered from the sprinklers nearest to the fire. Since the discharge from sprinklers operating at a distance from the fire source cannot reach the fire or prewet the area immediately around it, these sprinklers do not help to achieve fire suppression.

Because other types of sprinklers operate more slowly and do not have the same water distribution characteristics as ESFR sprinklers, sprinklers such as spray sprinklers operate on the concept of fire control. With spray sprinklers, a more significant fire plume develops prior to the operation of the first sprinkler. The intensity of the fire can also remain at an elevated state until fire control is achieved. The concept of fire control is to provide sufficient sprinkler discharge so that the fire does not spread beyond the sprinkler system design area. Spray sprinklers are expected to prevent the fire from spreading by slowly reducing its intensity and by sufficiently prewetting the surrounding combustibles so they do not ignite. Concurrently, the sprinkler discharge is expected to absorb heat and cool the surrounding space, including the area containing structural members thus preventing building collapse.

ESFR sprinklers, on the other hand, operate earlier in the fire than standard sprinklers and provide adequate discharge to suppress the fire before a severe fire plume develops. In principle, early suppression is determined by the following three factors:

(1) Thermal sensitivity
(2) Required delivered density (RDD)
(3) Actual delivered density (ADD)

An ESFR sprinkler must possess specific properties related to these three factors, and systems that have not been designed to address all three of these factors cannot be relied on to achieve early fire suppression. All three of these factors can be independently measured.

Thermal sensitivity is a measurable expression of the sensitivity or responsiveness of a sprinkler's operating element when exposed to fire conditions. The most common measure of thermal sensitivity is the RTI as indicated in the commentary to A-1-4.5.1. The more responsive the element is to the effects of a fire, the lower the RTI value. The RTI of a sprinkler generally varies little with the sprinkler's temperature rating. (See 1-4.5.1.)

A sprinkler's response time for a given fire situation is a function of the thermal sensitivity of its operating element, its temperature rating, and its distance relative to the fire. Because of the thermal lag associated with the mass of the operating element, however, the gas temperature near the sprinkler can reach a value that is higher than the temperature rating of the sprinkler prior to the activation of the sprinkler. This can negatively affect the ability of a sprinkler with a high RTI to achieve early fire suppression, because the fire has had a chance to develop a significant fire plume before the activation of the first sprinkler.

Smaller fires are generally easier to suppress than larger ones. The sooner the sprinkler operates, the less the amount of water needed to suppress the fire. For sprinklers with the same temperature rating, those with a lower RTI value will operate sooner in a rapidly growing fire than sprinklers with a higher RTI value.

Required delivered density (RDD) is the measure of the amount of water needed to suppress a fire. Its value depends on the size of the fire at the time of sprinkler operation.

Actual delivered density (ADD) is a measurement of the amount of water discharged from the sprinklers that actually reaches the fire. ADD is determined during a fire test by measuring the amount of water that collects in a pan on the top horizontal surface of a burning combustible array. ADD is a means of characterizing a sprinkler's distribution pattern and droplet penetration capability during a fire event. The longer it takes a sprinkler to actuate, the faster the fire grows and the smaller the value of ADD. ADD is a function of fire plume velocity, water drop momentum and size, and the distance the water drops must travel from the sprinkler.

These three measurements—RTI, RDD, and ADD—are the controlling factors that define the time-dependent nature of early fire suppression. In theory, the earlier the water is applied to a growing fire, the lower the RDD will be and the higher the ADD will be. In other words, the faster the sprinkler response (lower RTI), the lower the RDD and the higher the ADD. Conversely, the later the water is applied (higher RTI), the higher the RDD and the lower the ADD. When the ADD is less than the RDD, the sprinkler discharge is no longer effective enough to achieve early fire suppression. Thus, it is clear that early fire suppression depends on the correct installation of ESFR sprinklers that have been shown to meet the performance criteria related to the RTI, ADD, and RDD factors for the hazard being protected. Exhibit 1.8 shows the generalized concept that early suppression is achieved when the ADD exceeds the RDD.

Implicit in the preceding theory is the expectation that water discharging from the first operating ESFR sprinkler reaches the burning material. It is critical with suppression mode sprinklers that the requirements of 5-11.5 be followed to minimize obstructions.

ESFR sprinklers are predominately used to protect storage occupancies. Caution must be exercised to avoid confusing ESFR sprinklers with other types of sprinklers

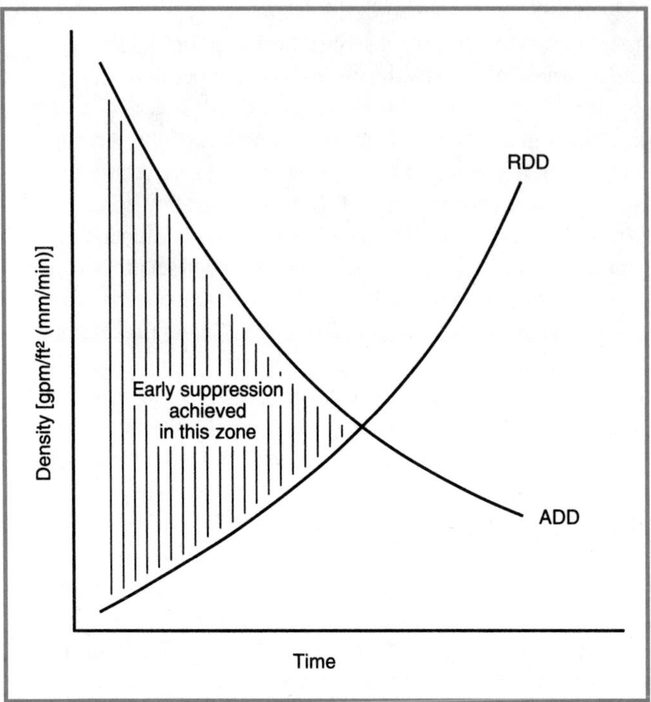

Exhibit 1.8 Generalized RDD-ADD relationship.

that are equipped with fast-response operating elements. A non-ESFR sprinkler with a fast-response element is not specifically designed to achieve fire suppression. The relationship among thermal sensitivity, ADD, and RDD needs to be considered.

Several types of ESFR sprinklers available are illustrated in Exhibits 1.9 and 1.10. Exhibit 1.9 shows three pendent-type ESFR sprinklers. The sprinklers on the right have a nominal K-factor of 14, while the sprinkler on the left has a nominal K-factor of 25.2. Exhibit 1.10 illustrates an upright ESFR sprinkler with a nominal K-factor of 11.2. It is important to note that the installation and discharge criteria for these different ESFR sprinklers are not uniform. See Chapters 5 and 7 for additional information.

Extended Coverage Sprinkler. A type of spray sprinkler with maximum coverage areas as specified in Sections 5-8 and 5-9 of this standard.

Up until the 1996 edition of NFPA 13, extended coverage (EC) sprinklers were considered special sprinklers and their installation rules were not addressed by NFPA 13. EC sprinklers are available in the pendent, upright, and sidewall configurations. The advantage of EC sprinklers is that their areas of coverage are greater than those established for standard spray and sidewall sprinklers. Exhibits 1.11 and 1.12 illustrate pendent and sidewall types of EC sprinklers. EC sprinklers can often look similar to

Exhibit 1.9 *Pendent-type ESFR sprinklers.*

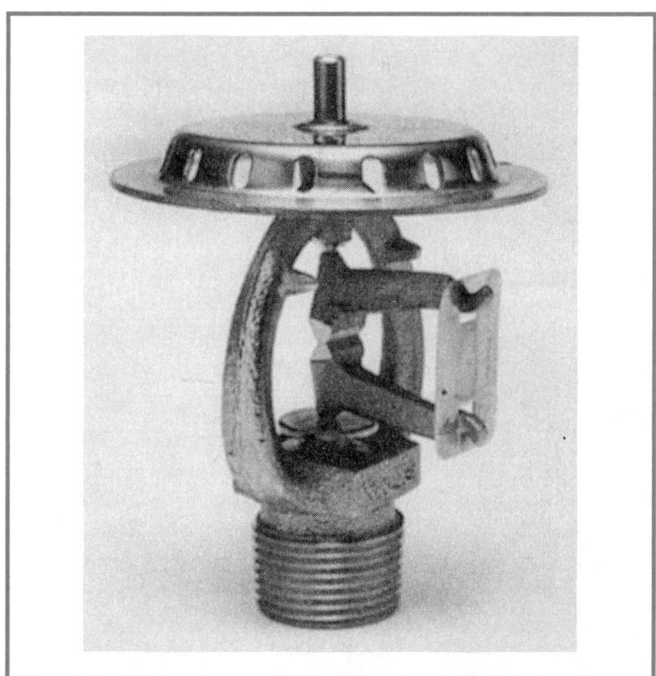

Exhibit 1.10 *Upright-type ESFR sprinkler. (Courtesy of the Viking Corporation.)*

Exhibit 1.11 *Grinnell F895 pendent-type extended coverage sprinkler.*

Exhibit 1.12 *Sidewall-type extended coverage sprinkler.*

standard coverage sprinklers, so the listing information should be consulted to confirm that the sprinklers used are in fact extended coverage sprinklers.

Large Drop Sprinkler. A type of sprinkler that is capable of producing characteristic large water droplets and that is listed for its capability to provide fire control of specific high-challenge fire hazards.

A large drop sprinkler is a control mode sprinkler primarily used for the protection of storage facilities. Exhibit 1.13 illustrates a large drop sprinkler.

Exhibit 1.13 *Viking Model A large drop sprinkler.*

Nozzles. A device for use in applications requiring special water discharge patterns, directional spray, or other unusual discharge characteristics.

Nozzles are special application devices that have unique characteristics designed to meet specific needs. These devices are equipped with heat-responsive elements or are open to the atmosphere. Nozzles are designed to spray water in a specific direction. NFPA 15 provides more information on these devices. See Exhibit 1.14.

Old-Style/Conventional Sprinkler. A sprinkler that directs from 40 percent to 60 percent of the total water initially in a downward direction and that is designed to be installed with the deflector either upright or pendent.

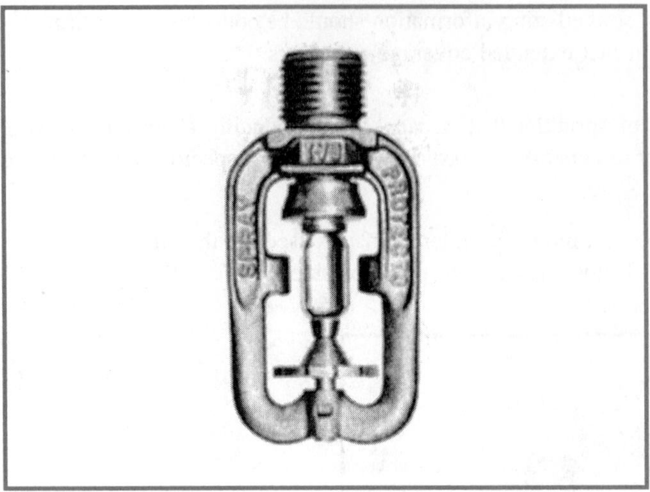

Exhibit 1.14 *Grinnell automatic Protectospray™ nozzle.*

Old-style/conventional sprinklers were predominantly used in North America prior to the development of spray sprinklers. The major difference between old-style/conventional and standard spray sprinklers is the design of their deflectors. Old-style/conventional sprinklers, whether upright or pendent, direct approximately 40 percent of their water discharge up against the ceiling with the remaining discharge directed downward as indicated in Exhibit 1.15. Spray sprinklers direct 100 percent of their water discharge downward as indicated in Exhibit 1.16.

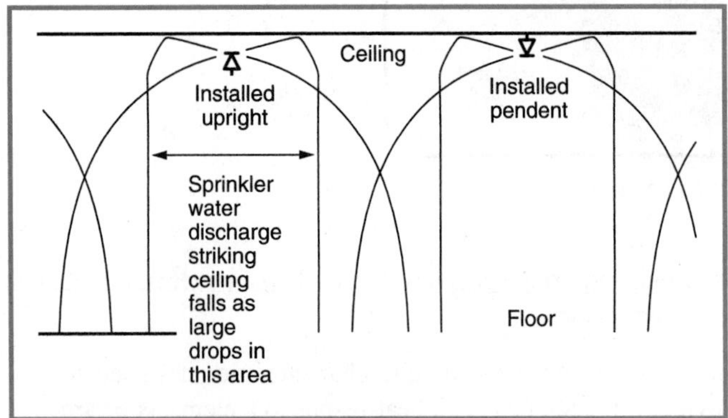

Exhibit 1.15 *Principal distribution pattern of water from old-style/conventional sprinklers (in use before 1953).*

Old-style/conventional sprinklers are still available and are used in Europe. These sprinklers are also used in North America for special applications in which an upward

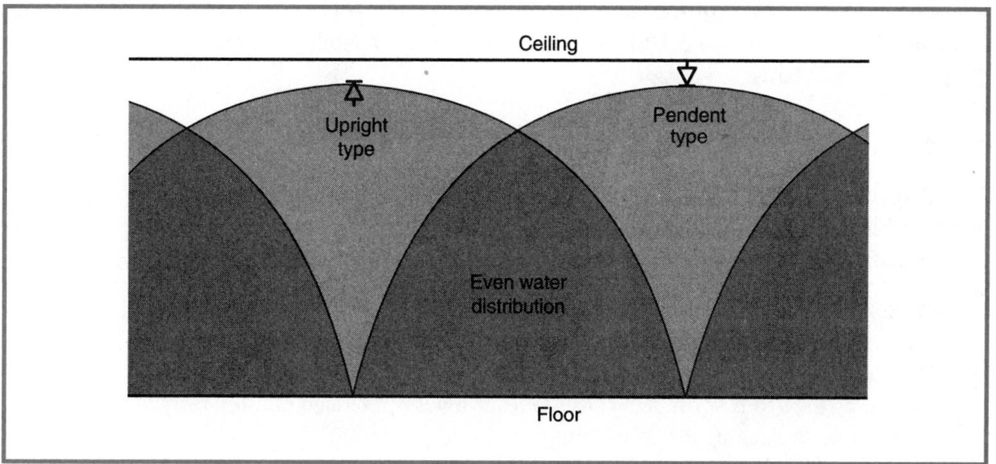

Exhibit 1.16 *Principal distribution pattern of water from standard sprinklers (introduced in 1953).*

sprinkler discharge is desired, such as for fur storage vaults and for pier and wharf protection.

Open Sprinkler. A sprinkler that does not have actuators or heat-responsive elements.

Open sprinklers are used predominately on deluge or water spray systems. See Exhibit 1.17.

Exhibit 1.17 *Grinnell open sprinkler.*

Quick-Response Early Suppression (QRES) Sprinkler.* A type of quick-response sprinkler that meets the criteria of 1-4.5.1(a)(1) and is listed for its capability to provide fire suppression of specific fire hazards.

A-1-4.5.2 QRES Sprinkler.

Research into the development of QRES sprinklers is continuing under the auspices of the National Fire Protection Research Foundation. It is expected that the proposed design criteria will be added to the standard when a thorough analysis of the test data is completed.

The concept of this sprinkler was proposed some time ago. The definition and appendix material appear in NFPA 13 to introduce the concept, since the sprinkler does not exist. The concept of the QRES sprinkler is based largely on information discovered in developing the ESFR sprinkler. The QRES sprinkler is intended to bring fire suppression to occupancies such as office buildings and manufacturing facilities.

A four-sprinkler design concept was targeted. The relevant combinations of fuel load, ceiling height, operating pressure, and sprinkler spacing need to be verified. Once any data becomes available and the results can be reviewed by the sprinkler committees, an amendment to NFPA 13 will likely be considered.

Quick-Response Extended Coverage Sprinkler. A type of quick-response sprinkler that meets the criteria of 1-4.5.1(a)(1) and complies with the extended protection areas defined in Chapter 5.

This type of sprinkler is specifically listed as having a fast-response operating element. Some standard response EC sprinklers utilize fast-response operating elements in order to achieve a desired performance. However, those EC sprinklers must be considered standard response unless specifically listed as quick-response extended coverage sprinklers. Exhibit 1.18 illustrates a type of listed quick-response extended coverage sprinkler.

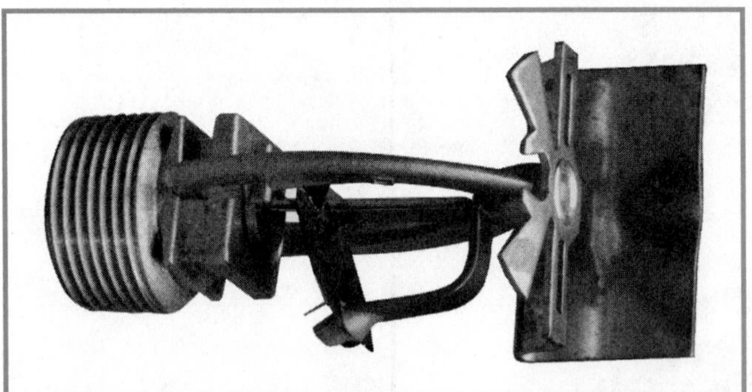

Exhibit 1.18 Quick-response extended coverage sprinkler.

Quick-Response (QR) Sprinkler. A type of spray sprinkler that meets the criteria of 1-4.5.1(a)(1) and is listed as a quick-response sprinkler for its intended use.

Quick-response (QR) sprinkler technology was developed from residential sprinkler technology. These sprinklers are similar to standard spray sprinklers except that they possess a fast-response operating element. QR sprinklers are tested against the same criteria as standard-response sprinklers. The commentary to A-1-4.5.1 provides additional discussion on thermal sensitivity and fast-response technology. Exhibit 1.19 shows a QR sprinkler adjacent to a standard-response sprinkler. The difference in the size of the operating elements should be noted. Where glass bulbs are used for standard spray sprinklers, the bulb of a QR sprinkler is typically thinner than that of a standard-response sprinkler. Where a metallic alloy is used, the operating element is usually larger in a QR sprinkler.

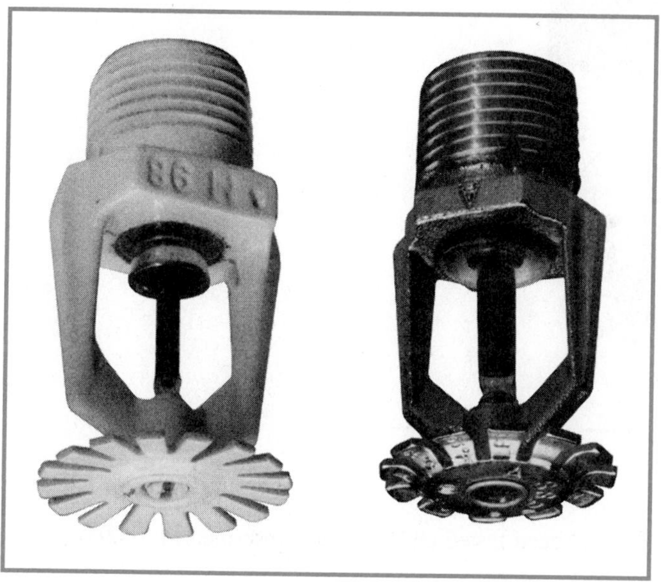

Exhibit 1.19 Grinnell Model A standard-response spray sprinkler (left) and Viking Model A quick-response standard spray sprinkler (right).

Residential Sprinkler. A type of fast-response sprinkler that meets the criteria of 1-4.5.1(a)(1) that has been specifically investigated for its ability to enhance survivability in the room of fire origin and is listed for use in the protection of dwelling units.

The residential sprinkler was introduced in the early 1980s with the primary objective of achieving life safety in dwelling units. Residential sprinklers are installed to provide a tenable space for a period of time necessary for an occupant to evacuate. Additionally, these sprinklers are intended to protect persons in the room of fire origin who are not intimate with ignition. This sprinkler possesses a fast-response operating element and produces a spray pattern that discharges water higher than the standard spray sprinkler.

Residential sprinklers are tested for their ability to meet a specified life safety criteria. This criteria includes temperature, oxygen, and carbon monoxide levels mea-

sured at a height 5 ft (1.5 m) above the floor. The commentary to NFPA 13D further describes the testing and developmental programs that led to the introduction of the residential sprinkler. Exhibit 1.20 shows a listed residential sprinkler.

Exhibit 1.20 *Viking listed residential sprinkler.*

Special Sprinkler. A sprinkler that has been tested and listed as prescribed in 5-4.9.

NFPA 13 provides detailed information for this category of sprinklers. This type of sprinkler does not fall into the other categories as it is usually intended to protect specific hazards or construction features. Special sprinklers are required to be evaluated and listed for their intended purpose. Fire test performance, spray pattern distribution with respect to both the wetting of walls and floors, operation with obstructions, thermal sensitivity, and performance under horizontal or sloped ceilings all need to be considered. While a good deal of latitude is provided for the development of special sprinklers, their K-factor and temperature rating must comply with the provisions identified by this standard. Also, their maximum area of coverage cannot exceed 400 ft^2 (36 m^2) for light and ordinary hazard occupancies and 196 ft^2 (18.2 m^2) for extra hazard occupancies and high-piled storage applications.

The special sprinkler category often serves as a starting point for the introduction of a new kind of sprinkler. Once a sufficient comfort level and overall acceptance of a special sprinkler is achieved, the need to establish a new sprinkler category is evident. Establishing a new sprinkler category is also desirable because special sprinklers do not always receive broad-based acceptance as mainstream devices. This was the case with the EC sprinkler, which is no longer considered a special sprinkler according to NFPA 13. Another example is the upright ESFR sprinkler, which in the past was classified as a special sprinkler. Examples of special sprinklers include specific application and cycling sprinklers. Exhibit 1.21 shows various types of special sprinklers.

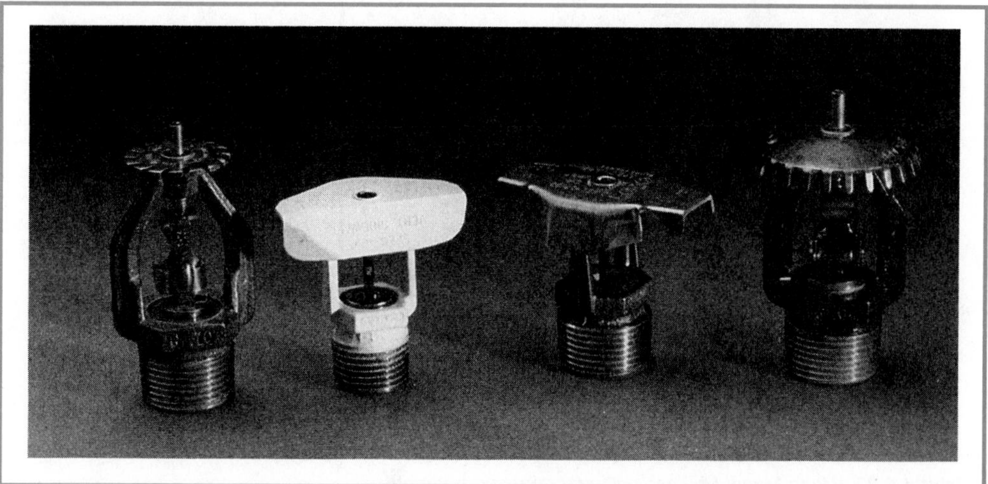

Exhibit 1.21 Listed special sprinklers, including a Reliable Model G VELO sprinkler, a Central Model WS Window sprinkler, a Central Attic sprinkler, and a Reliable model G XLO sprinkler.

Spray Sprinkler. A type of sprinkler listed for its capability to provide fire control for a wide range of fire hazards.

The spray sprinkler was introduced in the early 1950s and was the successor to the old-style/conventional sprinkler. The old-style sprinkler continues to serve the sprinkler industry. Standard spray, pendent, standard spray upright, sidewall, EC, and EC sidewall sprinklers are categories of spray sprinklers.

Standard Spray Sprinkler. A spray sprinkler with maximum coverage areas as specified in Sections 5-6 and 5-7 of this standard.

This control mode spray sprinkler is installed in accordance with the coverage area limitations in Sections 5-6 and 5-7. It is available in pendent, upright, and sidewall configurations. Because this sprinkler is proven to be effective for a broad range of hazards and applications, it is very popular and, to a certain degree, serves as the benchmark for sprinkler measurement and performance. (See Exhibit 1.22.)

1-4.5.3 The following sprinklers are defined according to orientation.

The devices described in 1-4.5.3 are based on the geometry of how the sprinkler is installed. Representative discharge patterns for these devices allow them to be categorized as concealed, flush, pendent, and recessed sprinklers, all of which have similar

Exhibit 1.22 Viking Model M glass bulb standard spray upright sprinkler.

discharge patterns and are mounted to the bottom of a branch line or pipe drop. Sidewall sprinklers are typically installed along a wall or lintel and discharge water away from the wall into the room or space. Sidewall sprinklers can be mounted on the side, bottom, or top of a branch line, as specified in their listings. Upright sprinklers have a spray pattern that appears similar to that of a pendent sprinkler. The difference is that upright sprinklers are mounted to the top of branch lines or sprigs.

Concealed Sprinkler. A recessed sprinkler with cover plates.

Exhibit 1.23 illustrates a concealed sprinkler. The cover plate drops away when exposed to a certain amount of heat. The fusible elements holding the cover plate are designed to operate prior to the activation of the sprinkler's thermal element. The cover plate is included as part of the listing.

Flush Sprinkler. A sprinkler in which all or part of the body, including the shank thread, is mounted above the lower plane of the ceiling.

Exhibits 1.24 and 1.25 illustrate flush sprinklers.

Pendent Sprinkler. A sprinkler designed to be installed in such a way that the water stream is directed downward against the deflector.

Exhibit 1.26 illustrates a standard spray pendent sprinkler.

Recessed Sprinkler. A sprinkler in which all or part of the body, other than the shank thread, is mounted within a recessed housing.

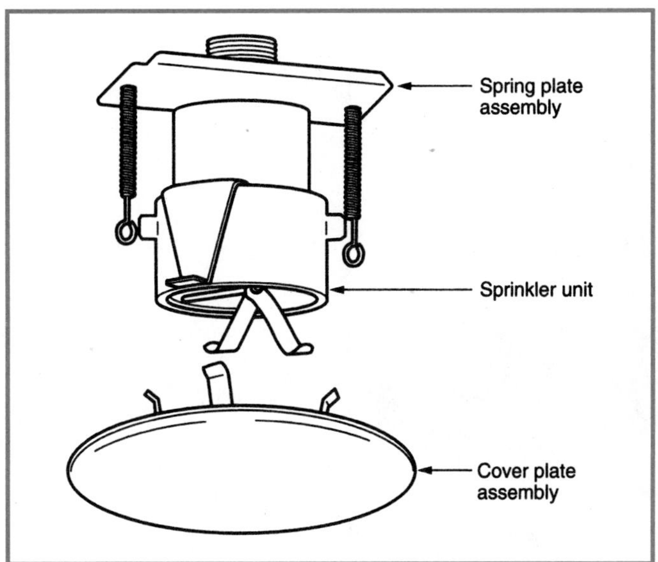

Exhibit 1.23 *Standard Model G concealed ceiling sprinkler.*

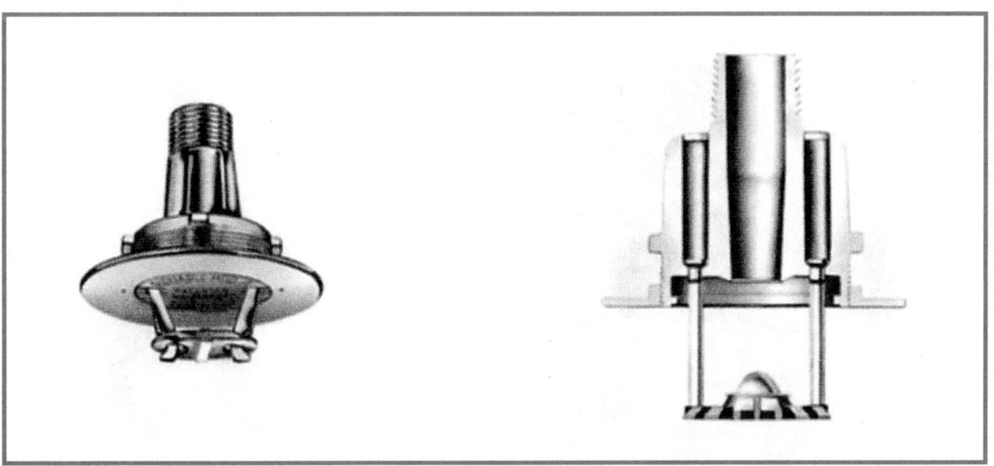

Exhibit 1.24 *Reliable Model A flush-type automatic sprinkler before operation (left) and after operation (right).*

Sidewall Sprinkler. A sprinkler having special deflectors that are designed to discharge most of the water away from the nearby wall in a pattern resembling one-quarter of a sphere, with a small portion of the discharge directed at the wall behind the sprinkler.

Exhibit 1.27 illustrates a horizontal sidewall sprinkler.

Exhibit 1.25 *Viking Model H standard spray pendent flush-mount sprinkler.*

Exhibit 1.26 *Grinnell standard spray pendent sprinkler.*

Exhibit 1.27 Viking Model M quick-response, horizontal sidewall sprinkler.

Upright Sprinkler. A sprinkler designed to be installed in such a way that the water spray is directed upwards against the deflector.

Exhibit 1.28 illustrates an upright sprinkler.

1-4.5.4 The following sprinklers are defined according to special application or environment.

The devices described in 1-4.5.4 are intended for use in specific environments. These devices are either covered with a special coating or designed for a specific function. Coatings can only be applied by the sprinkler manufacturer.

Exhibit 1.28 Central K17 special application upright spray sprinkler..

Corrosion-Resistant Sprinkler. A sprinkler fabricated with corrosion-resistant material, or with special coatings or platings, to be used in an atmosphere that would normally corrode sprinklers.

Corrosion-resistant sprinklers are usually standard sprinklers with a corrosion-resistant coating, such as wax or lead, applied to them. Any coating of this type can only be applied by the manufacturer. Exhibit 1.29 illustrates a wax-coated corrosion-resistant sprinkler.

Exhibit 1.29 Central Model A upright wax-coated corrosion-resistant sprinkler.

Dry Sprinkler.* A sprinkler secured in an extension nipple that has a seal at the inlet end to prevent water from entering the nipple until the sprinkler operates.

A-1-4.5.4 Dry Sprinkler. Under certain ambient conditions, wet pipe systems having dry-pendent (or upright) sprinklers can freeze due to heat loss by conduction. Therefore, due consideration should be given to the amount of heat maintained in the heated space, the length of the nipple in the heated space, and other relevant factors.

Dry sprinklers are intended to extend into an unheated area from a wet pipe system or to be used on a dry pipe system.

The dry upright sprinkler is similar to the dry pendent, except that the upright uses an upright deflector. Exhibit 1.30 shows a dry-pendent sprinkler. When the glass bulb is heated it breaks apart. The inner tube, which also serves as an orifice, drops to a

Exhibit 1.30 Viking Model E dry-pendent sprinkler.

predetermined position, allowing the sealing elements to pass through the tube and away from the sprinkler. Water flows through the tube and strikes the deflector, which distributes it in a spray pattern comparable to a standard spray sprinkler.

Intermediate Level Sprinkler/Rack Storage Sprinkler. A sprinkler equipped with integral shields to protect its operating elements from the discharge of sprinklers installed at higher elevations.

Ornamental/Decorative Sprinkler. A sprinkler that has been painted or plated by the manufacturer.

Some ornamental sprinklers can be nothing more than a corrosion-resistant sprinkler with a special coating similar in color to that of the interior building finish. The same holds true for the covers of concealed sprinklers. As noted in 3-2.6, any coating for decorative or protective purposes must be applied by the sprinkler manufacturer.

1-4.6 Construction Definitions.

The numerous construction terms in NFPA 13 were reduced to two basic categories for the 1991 edition. Ceiling construction impacts sprinkler installation as it is a factor in determining the position of the sprinkler deflector in relation to the ceiling and in determining the allowable area of coverage for a given sprinkler.

Sprinklers must be positioned so that they are located in the hot gas layer that develops near the ceiling during a fire. This sprinkler placement allows for timely operation. Depending on the ceiling construction type, sprinklers are permitted to be located various distances from the ceiling so that they activate shortly after the fire begins.

Sprinklers must also be positioned so that their discharge pattern is not adversely impacted by construction elements at the ceiling. If water cannot reach the combustible surfaces, the effectiveness of the sprinkler system is compromised. To a large degree, potential obstructions to sprinkler discharge are addressed by the obstruction rules of 5-5.5. Ceiling construction also impacts sprinkler discharge. Therefore, the maximum area of coverage for a specific type of sprinkler is dependent on the type of ceiling construction under which the sprinkler is installed.

The 10 construction types identified in previous editions of NFPA 13 were retained in A-1-4.6 as an aid in determining whether a given ceiling construction should be classified as either obstructed or unobstructed. The appendix material describes particular ceiling features that influence the placement of sprinklers.

Obstructed Construction.* Panel construction and other construction where beams, trusses, or other members impede heat flow or water distribution in a manner that materially affects the ability of sprinklers to control or suppress a fire.

A-1-4.6 Obstructed Construction. The following are examples of obstructed construction. The definitions are provided to assist the user in determining the type of construction feature.

(a) *Beam and Girder Construction.* The term *beam and girder construction* as used in this standard includes noncombustible and combustible roof or floor decks supported by wood beams of 4 in. (102 mm) or greater nominal thickness or concrete or steel beams spaced 3 ft to 7½ ft (0.9 m to 2.3 m) on center and either supported on or framed into girders. [Where supporting a wood plank deck, this includes semi-mill and panel construction, and where supporting (with steel framing) gypsum plank, steel deck, concrete, tile, or similar material, this includes much of the so-called noncombustible construction.]

Beam and girder construction consists of roof or floor decks supported by beams spaced 3 ft to 7½ ft (0.9 m to 2.3 m) on center that are either framed onto or into girders. This construction type results in the formation of ceiling bays or panels. Exhibit 1.31 illustrates an example of beam and girder construction in which the beams are framed into the girders. Exhibit 1.32 shows the bays formed by the beams.

The close spacing of the beams creates a greater likelihood of sprinkler spray pattern interference. Therefore, the positioning and maximum area of coverage rules for obstructed construction are different than the rules for unobstructed construction.

Beam and girder construction causes heat from a fire to accumulate within the bays or panels formed by the supporting members. As a fire continues to generate heat, the heat flows down the length of the members, activating only sprinklers within that bay or panel. Once the capacity of the bays or panels is reached, fire gases spill over into adjacent bays or panels and allow sprinklers in other bays and panels to activate. Because of this phenomenon, the deflector positioning rules in Chapter 5

Exhibit 1.31 Beam and girder construction.

provide a degree of flexibility in positioning the deflector in relation to the ceiling and the supporting members.

(b) *Composite Wood Joist Construction*. The term *composite wood joist construction* refers to wood beams of "I" cross section constructed of wood flanges and solid wood web, supporting a floor or roof deck. Composite wood joists can vary in depth up to 48 in. (1.2 m), can be spaced up to 48 in. (1.2 m) on centers, and can span up to 60 ft (18 m) between supports. Joist channels should be firestopped to the full depth of the joists with material equivalent to the web construction so that individual channel areas do not exceed 300 ft^2 (27.9 m^2). *[See Figure A-1-4.6(a) for an example of composite wood joist construction.]*

NFPA 13 provides guidance on the placement of sprinklers under composite wood joists only where the joists do not exceed a total depth of 22 in. (559 mm).

The firestopping between the composite wood joists that form the 300-ft^2 (27.9-m^2) pocket is needed to trap the heat and accelerate sprinkler operation. [See Figure A-1-4.6(a).]

Composite wood joists are typically referred to as wood "I" beams. The predominant use of these members was initially found in western North America, but now they are quite common in other areas as well.

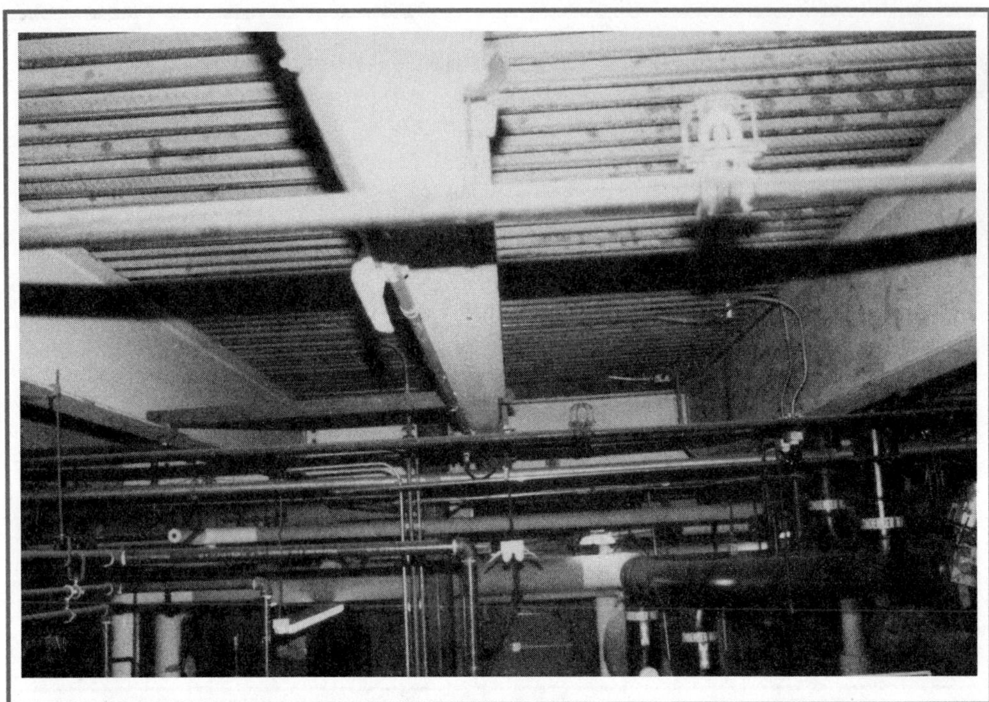

Exhibit 1.32 *Bays formed by beam and girder construction.*

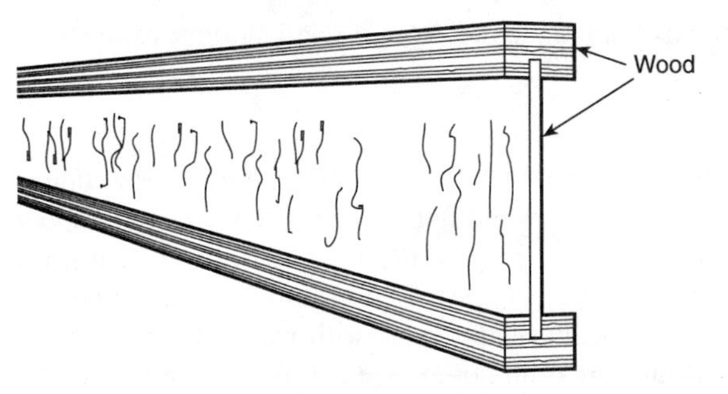

Figure A-1-4.6(a) *Typical composite wood joist construction.*

(c) *Panel Construction.* The term *panel construction* as used in this standard includes ceiling panels formed by members capable of trapping heat to aid the operation of sprinklers and limited to a maximum of 300 ft^2 (27.9 m^2) in area. Beams spaced more than 7½ ft (2.3 m) apart and framed into girders qualify as panel construction, provided the 300-ft^2 (27.9-m^2) area limitation is met.

Panel construction can be considered a version of beam and girder construction in which beams are framed into girders and form a ceiling panel with a maximum area of 300 ft^2 (27.9 m^2). The 1999 edition specifically indicates that panel construction with members spaced more than 7½ ft (2.3 m) apart can be considered obstructed construction, provided the ceiling panels do not exceed an area of 300 ft^2 (27.9 m^2). Where ceiling panels consist of an area in excess of 300 ft^2 (27.9 m^2), they cannot be considered obstructed construction. See Exhibits 1.31 and 1.32.

(d) *Semi-Mill Construction.* The term *semi-mill construction* as used in this standard refers to a modified standard mill construction, where greater column spacing is used and beams rest on girders.

(e) *Wood Joist Construction.* The term *wood joist construction* refers to solid wood members of rectangular cross section, which can vary from 2 in. to 4 in. (51 mm to 102 mm) nominal width and can be up to 14 in. (356 mm) nominal depth, spaced up to 3 ft (0.9 m) on centers, and can span up to 40 ft (12 m) between supports, supporting a floor or roof deck. Solid wood members less than 4 in. (102 mm) nominal width and up to 14 in. (356 mm) nominal depth, spaced more than 3 ft (0.9 m) on centers, are also considered as wood joist construction.

This type of light combustible construction utilizes nominal 2-in. to 4-in. (51-mm to 102-mm) wide solid wood members on edge to support a ceiling or roof deck. Sheathing such as drywall cannot be fastened to the underside of the joists. Sheathed joists are considered unobstructed construction. Most construction assemblies utilizing solid wood members less than 4 in. (102 mm) thick are classified as wood joists. Composite wood joists and wood trusses should not be considered solid wood members.

Unobstructed Construction.* Construction where beams, trusses, or other members do not impede heat flow or water distribution in a manner that materially affects the ability of sprinklers to control or suppress a fire. Unobstructed construction has horizontal structural members that are not solid, where the openings are at least 70 percent of the cross-section area and the depth of the member does not exceed the least dimension of the openings, or all construction types where the spacing of structural members exceeds 7½ ft (2.3 m) on center.

A-1-4.6 Unobstructed Construction. The following are examples of unobstructed construction. The definitions are provided to assist the user in determining the type of construction feature.

This construction type does not contain features that are expected to impede the flow of heat from a fire. Additionally, once the sprinklers have activated, the likelihood of their spray pattern being interrupted by construction features at or near the ceiling is greatly reduced. Examples of traditional unobstructed construction are identified in this appendix section.

(a) *Bar Joist Construction.* The term *bar joist construction* refers to construction employing joists consisting of steel truss-shaped members. Wood truss-shaped members, which consist of wood top and bottom chord members not exceeding 4 in. (102 mm) in depth with steel tube or bar webs, are also defined as bar joists. Bar joists include noncombustible or combustible roof or floor decks on bar joist construction. *[See Figures A-1-4.6(b) and A-1-4.6(c) for examples of bar joist construction.]*

Also known as open-web steel joist construction, this construction type utilizes a top and bottom wood or steel chord. Unlike solid wood or composite wood joists, this construction type allows heat and fire gases to pass though it and spread out across the ceiling. Obstruction to water distribution is minimal as inferred by the spacing and positioning rules of Chapter 5. The maximum protection area per sprinkler for standard spray upright and pendent under unobstructed construction can be as high as 200 ft^2 (18.6 m^2) for pipe schedule systems and 225 ft^2 (20.9 m^2) for hydraulically calculated systems [see Table 5-6.2.2(a)]. Figure A-1-4.6(b) and Exhibit 1.33 show examples of bar joist construction.

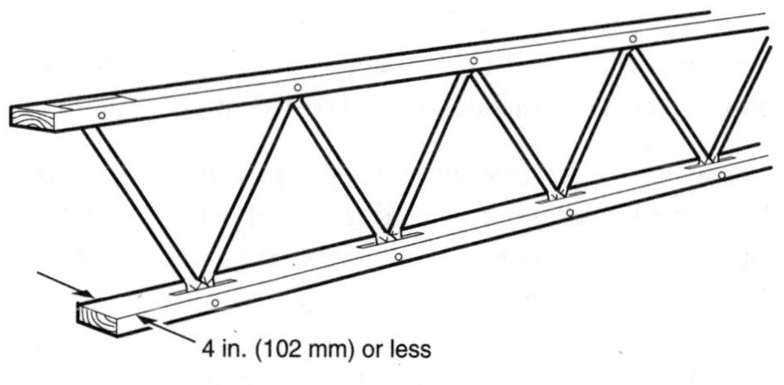

Figure A-1-4.6(b) *Wood bar joist construction.*

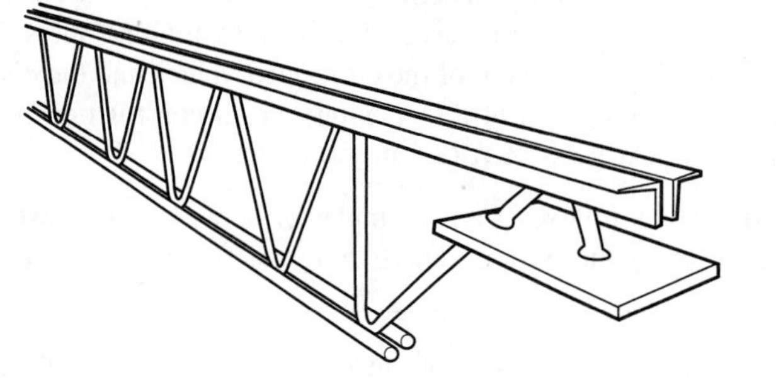

Figure A-1-4.6(c) *Open-web bar joist construction.*

(b) *Open-Grid Ceilings.* The term *open-grid ceilings* as used in this standard are ceilings in which the openings are ¼ in. (6.4 mm) or larger in the least dimension, the thickness of the ceiling material does not exceed the least dimension of the openings, and the openings constitute at least 70 percent of the ceiling area.

Exhibit 1.33 Bar joist construction.

(c) *Smooth Ceiling Construction.* The term *smooth ceiling construction* as used in this standard includes the following:

(1) Flat slab, pan-type reinforced concrete
(2) Continuous smooth bays formed by wood, concrete, or steel beams spaced more than 7½ ft (2.3 m) on centers—beams supported by columns, girders, or trusses
(3) Smooth roof or floor decks supported directly on girders or trusses spaced more than 7½ ft (2.3 m) on center
(4) Smooth monolithic ceilings of at least ¾ in. (19 mm) of plaster on metal lath or a combination of materials of equivalent fire-resistive rating attached to the underside of wood joists, wood trusses, and bar joists
(5) Open-web-type steel beams, regardless of spacing
(6) Smooth shell-type roofs, such as folded plates, hyperbolic paraboloids, saddles, domes, and long barrel shells
(7) Suspended ceilings of combustible or noncombustible construction
(8) Smooth monolithic ceilings with fire resistance less than that specified under item (4) attached to the underside of wood joists, wood trusses, and bar joists

In general, smooth ceiling construction does not incorporate supporting members that are less than 7½ ft (2.3 m) on center or have members that would interfere with the distribution of water from sprinklers. The 7½-ft (2.3-m) length serves as a threshold value. This length is the maximum dimension a standard spray sprinkler can be placed from a wall for light hazard and ordinary hazard occupancies. (In small rooms standard spray sprinklers can be placed up to 9 ft (2.7 m) from one wall. See the exception to 5-6.3.2.1.) Exhibit 1.34 illustrates smooth ceiling construction consisting of a suspended ceiling.

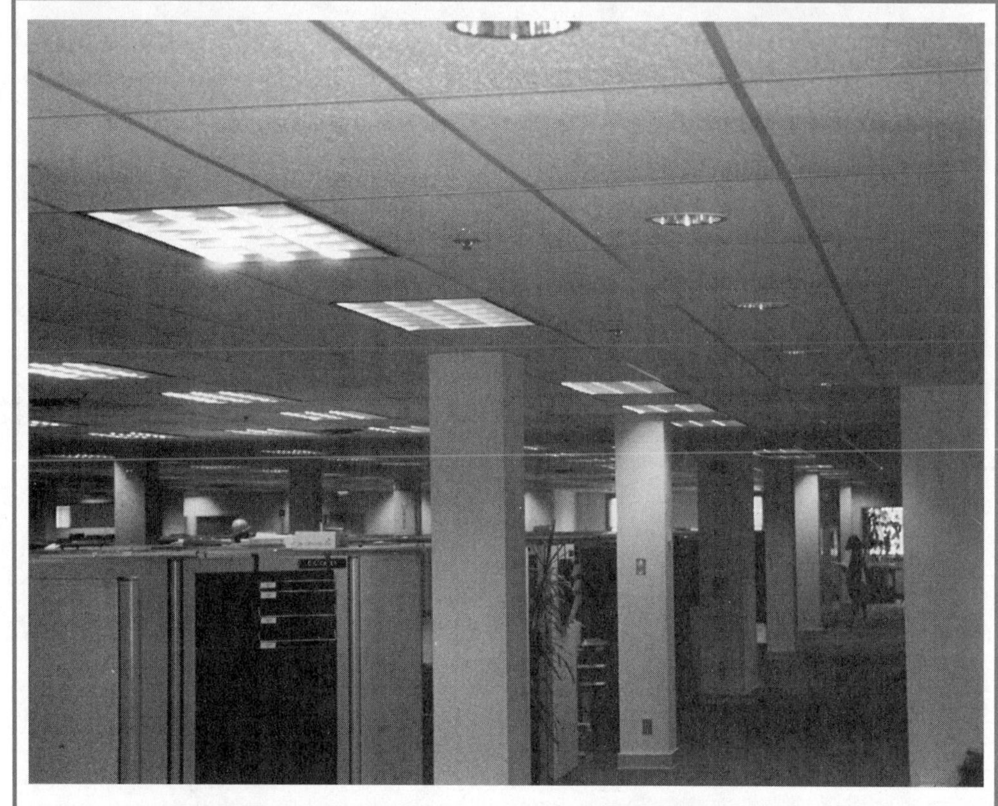

Exhibit 1.34 *Smooth ceiling construction consisting of a suspended ceiling.*

Exhibit 1.35 illustrates smooth ceiling construction consisting of flat slab, pan-type reinforced concrete. While this type of construction is typically considered smooth ceiling, it is not considered a smooth, flat ceiling as required for sidewall or extended coverage sprinklers. See 5-4.2 and 5-4.3 for specific requirements.

(d) Combustible or noncombustible floor decks are permitted in the construction specified in A-1-4.6(c)(2) through (6). Item (2) would include standard mill construction.

See Exhibits 1.36 and 1.37 on pages 58 and 59.

Exhibit 1.35 *Flat slab, pan-type reinforced concrete ceiling.*

(e) *Standard Mill Construction.* The term *standard mill construction* as used in this standard refers to heavy timber construction as defined in NFPA 220, *Standard on Types of Building Construction.*

(f) *Wood Truss Construction.* The term *wood truss construction* refers to parallel or pitched wood chord members connected by open wood members (webbing) supporting a roof or floor deck. Trusses with steel webbing, similar to bar joist construction, having top and bottom wood chords exceeding 4 in. (102 mm) in depth, should also be considered wood truss construction. *[See Figure A-1-4.6(d).]*

This construction type is similar to bar joist construction except that the top and bottom chords are typically of a heavier wood construction and the chords are typically connected by wood or steel web members rather than steel bars.

Past editions of NFPA 13 included wood truss construction in the same category as bar joist construction. This classification caused some confusion, and separate definitions were introduced in the 1989 edition. However, with only two broad categories of construction considered by NFPA 13, the differences between wood truss and bar joist construction in terms of sprinkler system installation became less important.

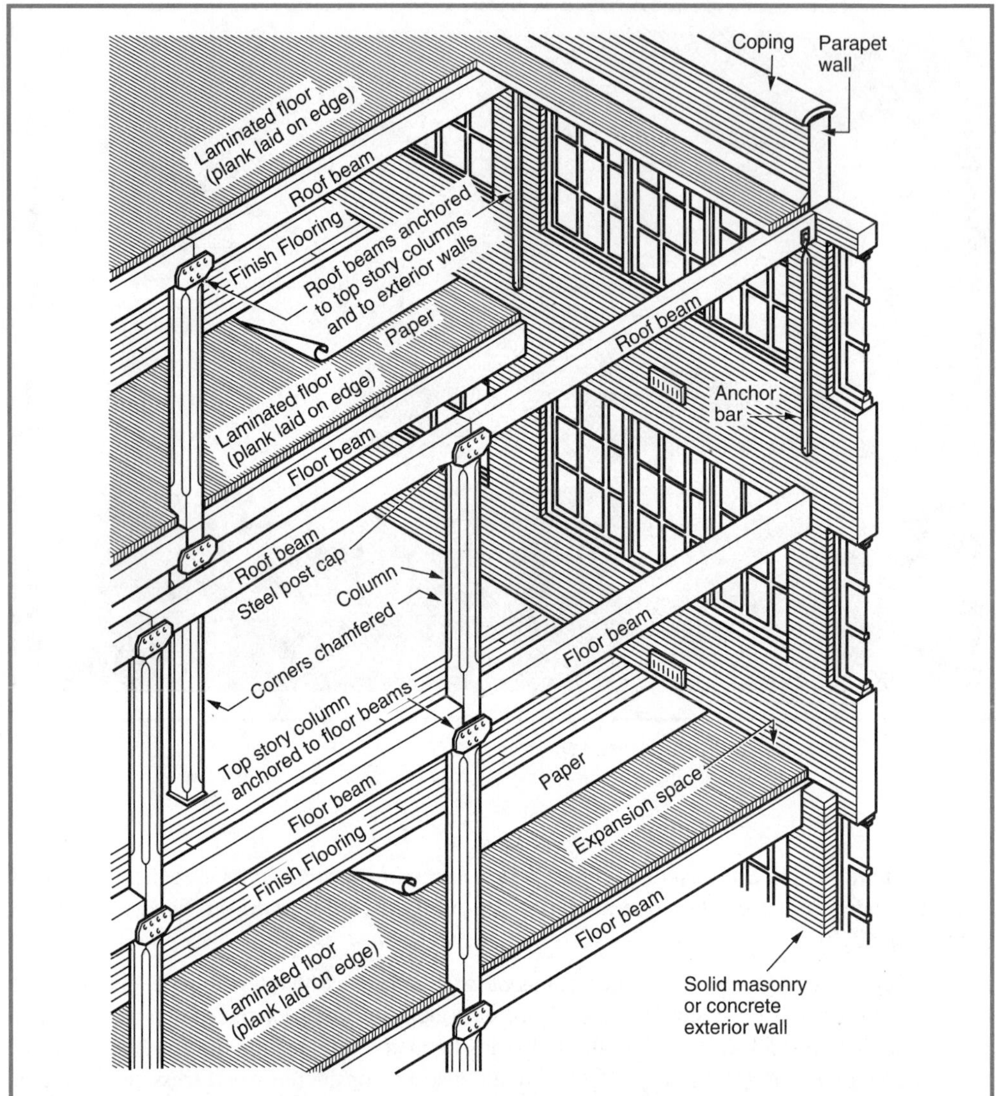

Exhibit 1.36 *Heavy timber construction of the laminated floor and beam type. (Courtesy of National Forest Products Association.)*

1-4.7 Private Water Supply Piping Definitions.

Private Fire Service Main.* Private fire service main, as used in this standard, is that pipe and its appurtenances on private property (1) between a source of water and the base of the riser for water-based fire protection systems, (2) between a source of water and inlets to foam-making systems, (3) between a source of water and the base elbow of private hydrants or

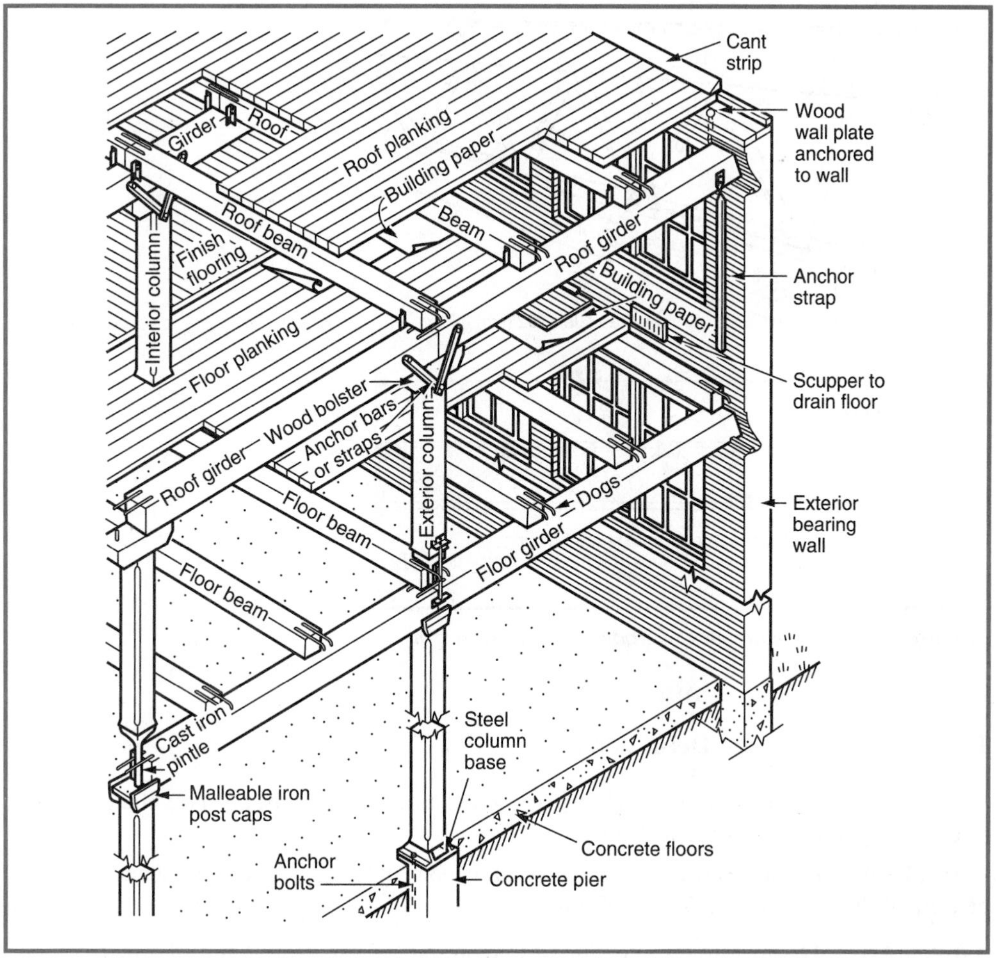

Exhibit 1.37 *Components of a heavy timber building that shows floor framing with components of a type known as semi-mill identified.*

monitor nozzles, and (4) used as fire pump suction and discharge piping, (5) beginning at the inlet side of the check valve on a gravity or pressure tank.

A-1-4.7 Private Fire Service Main. See Figure A-1-4.7.

In the context of sprinkler systems, a private fire service main is the piping that is either buried or above ground, connecting the water supply to the sprinkler system. A private fire serve main is also the piping that supplies water to private hydrants and other devices used for manual fire fighting. See Exhibit 1.38 on page 62. The installation criteria for private fire service mains from NFPA 24 was incorporated into the 1999 edition.

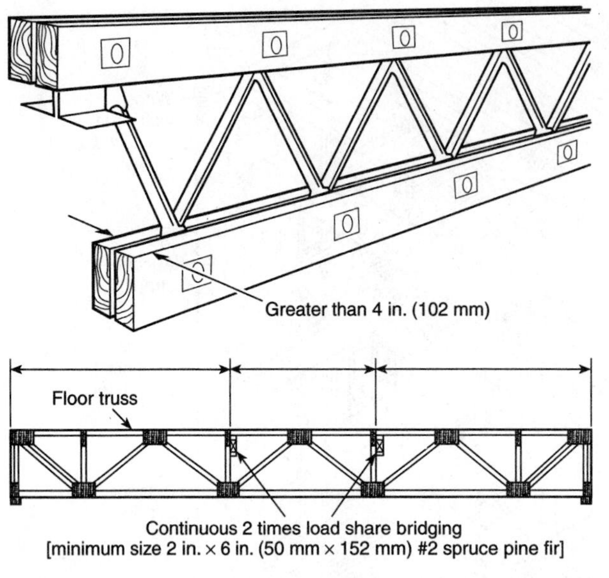

Figure A-1-4.6(d) Examples of wood truss construction.

1-4.8 General Storage Definitions.

Array, Closed. A storage arrangement where air movement through the pile is restricted because of 6-in. (152-mm) or less vertical flues.

Array, Open.* A storage arrangement where air movement through the pile is enhanced because of vertical flues larger than 6 in. (152 mm).

A-1-4.8 Array, Open. Fire tests conducted to represent a closed array utilized 6-in. (152-mm) longitudinal flues and no transverse flues. Fire tests conducted to represent an open array utilized 12-in. (305-mm) longitudinal flues.

Available Height for Storage.* The maximum height at which commodities can be stored above the floor and still maintain adequate clearance from structural members and the required clearance below sprinklers.

A-1-4.8 Available Height for Storage. For new sprinkler installations, the maximum height of storage is the height at which commodities can be stored above the floor where the minimum required unobstructed space below sprinklers is maintained. For the evaluation of existing situations, the maximum height of storage is the maximum existing height, if space between the sprinklers and storage is equal to or greater than required.

Bin Box Storage. Storage in five-sided wood, metal, or cardboard boxes with open face on the aisles. Boxes are self-supporting or supported by a structure so designed that little or no horizontal or vertical space exists around boxes.

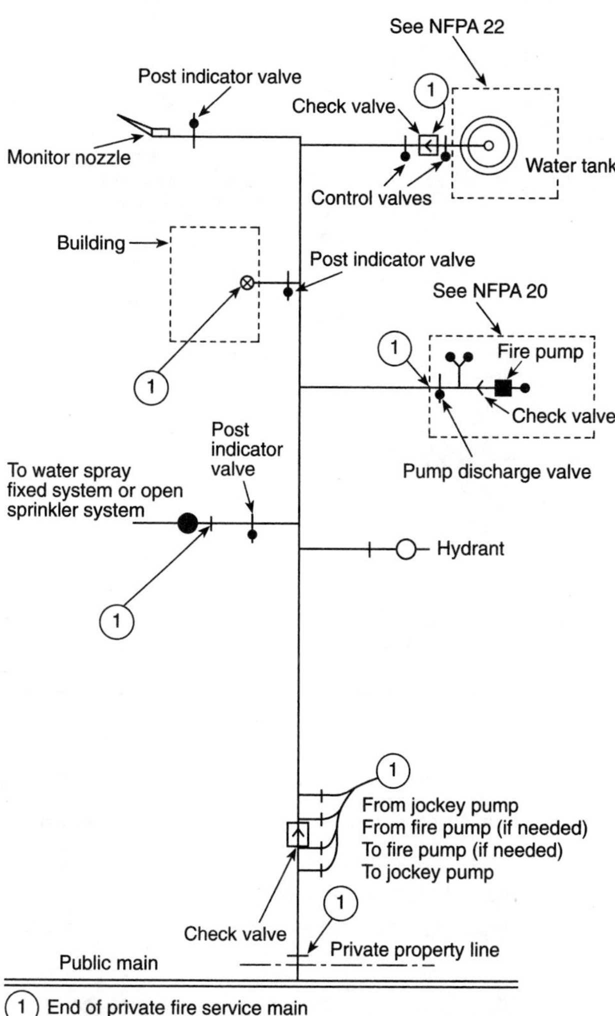
See NFPA 22
Post indicator valve
Check valve
1
Monitor nozzle
Water tank
Control valves
Building
Post indicator valve
See NFPA 20
1
Fire pump
1
Check valve
Post indicator valve
To water spray fixed system or open sprinkler system
Pump discharge valve
Hydrant
1
1
From jockey pump
From fire pump (if needed)
To fire pump (if needed)
To jockey pump
1
Check valve
Private property line
Public main
1 End of private fire service main
Note: The piping (aboveground or buried) shown is specific as to the end of the private fire service main and schematic only for illustrative purposes beyond. Details of valves and their location requirements are covered in the specific standard involved.

Figure A-1-4.7 *Typical private fire service main.*

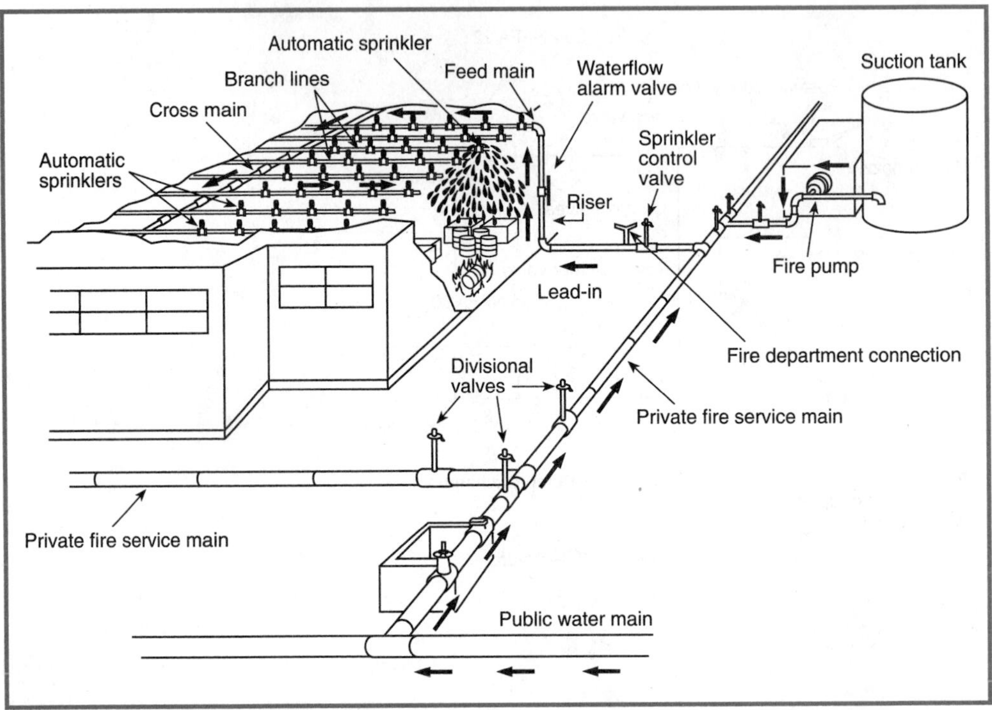

Exhibit 1.38 Private fire service main. (Adapted from Factory Mutual Systems.)

Ceiling Height. The distance between the floor and the underside of the ceiling above (or roof deck) within the storage area.

Clearance. The distance from the top of storage to the ceiling sprinkler deflectors.

Commodity. Combinations of products, packing material, and container upon which the commodity classification is based.

Compartmented.* The rigid separation of the products in a container by dividers that form a stable unit under fire conditions.

A-1-4.8 Compartmented. Cartons used in most of the Factory Mutual–sponsored plastic tests involved an ordinary 200-lb (90.7-kg) test of outside corrugated cartons with five layers of vertical pieces of corrugated carton used as dividers on the inside. There were also single horizontal pieces of corrugated carton between each layer.

Other tests sponsored by the Society of Plastics Industry, Industrial Risk Insurers, Factory Mutual, and Kemper used two vertical pieces of carton (not corrugated) to form an "X" in the carton for separation of product. This arrangement was not considered compartmented, as the pieces of carton used for separations were flexible (not rigid), and only two pieces were used in each carton.

Container (Shipping, Master, or Outer Container).* A receptacle strong enough, by reason of material, design, and construction, to be shipped safely without further packaging.

A-1-4.8 Container. The term *container* includes items such as cartons and wrappings. Fire-retardant containers or tote boxes do not by themselves create a need for automatic sprinklers unless coated with oil or grease. Containers can lose their fire-retardant properties if washed. For obvious reasons, they should not be exposed to rainfall.

Encapsulation. A method of packaging consisting of a plastic sheet completely enclosing the sides and top of a pallet load containing a combustible commodity or a combustible package or a group of combustible commodities or combustible packages. Combustible commodities individually wrapped in plastic sheeting and stored exposed in a pallet load also are to be considered encapsulated. Totally noncombustible commodities on wood pallets enclosed only by a plastic sheet as described are not covered under this definition. Banding (i.e., stretch-wrapping around the sides only of a pallet load) is not considered to be encapsulation. Where there are holes or voids in the plastic or waterproof cover on the top of the carton that exceed more than half of the area of the cover, the term *encapsulated* does not apply. The term *encapsulated* does not apply to plastic-enclosed products or packages inside a large, nonplastic, enclosed container.

The chairs shown in Exhibit 1.39 are considered encapsulated since the chairs are completely enclosed by the plastic sheet. The wrapped commodities in Exhibit 1.40 are not considered encapsulated because the plastic does not cover the top of the pallet loads.

Exhibit 1.39 *Encapsulated chairs.*

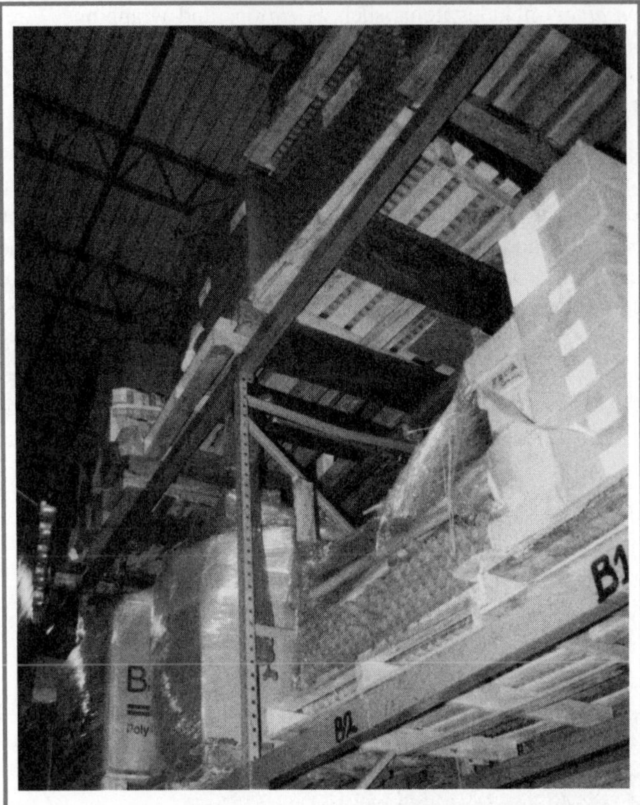

Exhibit 1.40 Commodity wrapped in a plastic sheet.

Expanded (Foamed or Cellular) Plastics. Those plastics, the density of which is reduced by the presence of numerous small cavities (cells), interconnecting or not, dispersed throughout their mass.

Exposed Group A Plastic Commodities. Those plastics not in packaging or coverings that absorb water or otherwise appreciably retard the burning hazard of the commodity. (Paper wrapped or encapsulated, or both, should be considered exposed.)

Free-Flowing Plastic Materials. Those plastics that fall out of their containers during a fire, fill flue spaces, and create a smothering effect on the fire. Examples include powder, pellets, flakes, or random-packed small objects [e.g., razor blade dispensers, 1-oz to 2-oz (28-g to 57-g) bottles].

Packaging. A commodity wrapping, cushioning, or container.

Palletized Storage. Storage of commodities on pallets or other storage aids that form horizontal spaces between tiers of storage.

Pile Stability, Stable Piles.* Those arrays where collapse, spillage of content, or leaning of stacks across flue spaces is not likely to occur soon after initial fire development.

A-1-4.8 Pile Stability, Stable Piles. Pile stability performance has been shown to be a difficult factor to judge prior to a pile being subjected to an actual fire. In the test work completed, compartmented cartons *(see A-1-4.8, Compartmented)* have been shown to be stable under fire conditions. Tests also indicated cartons that were not compartmented tended to be unstable under fire conditions.

Storage on pallets, compartmented storage, and plastic components that are held in place by materials that do not deform readily under fire conditions are examples of stable storage.

Pile Stability, Unstable Piles.* Those arrays where collapse, spillage of contents, or leaning of stacks across flue spaces occurs soon after initial fire development.

A-1-4.8 Pile Stability, Unstable Piles. Leaning stacks, crushed bottom cartons, and reliance on combustible bands for stability are examples of potential pile instability under a fire condition. An increase in pile height tends to increase instability.

Pile collapse needs to occur before or within 1 to 2 minutes after the first sprinkler operates to be a positive factor in fire control.

Roof Height. The distance between the floor and the underside of the roof deck within the storage area.

Shelf Storage. Storage on structures less than 30 in. (76.2 cm) deep with shelves usually 2 ft (0.6 m) apart vertically and separated by approximately 30-in. (76.2-cm) aisles.

The 30-in. (76.2-cm) maximum width is measured from aisle to aisle as indicated in Exhibit 1.41. Two back-to-back shelves with an overall width greater than 30 in. (76.2 cm) should be considered solid shelf racks.

Solid Unit Load of a Nonexpanded Plastic (Either Cartoned or Exposed). A load that does not have voids (air) within the load and that burns only on the exterior of the load; water from sprinklers might reach most surfaces available to burn.

Storage Aids. Commodity storage devices, such as pallets, dunnage, separators, and skids.

Unit Load. A pallet load or module held together in some manner and normally transported by material-handling equipment.

1-4.9 Rack Storage Definitions.

Aisle Width.* The horizontal dimension between the face of the loads in racks under consideration. *[See Figure A-1-4.9(a).]*

A-1-4.9 Aisle Width. See Figure A-1-4.9(a).

Bulkhead. A vertical barrier across the rack.

Cartoned. A method of storage consisting of corrugated cardboard or paperboard containers fully enclosing the commodity.

Conventional Pallets.* A material-handling aid designed to support a unit load with openings to provide access for material-handling devices. *[See Figure A-1-4.9(b).]*

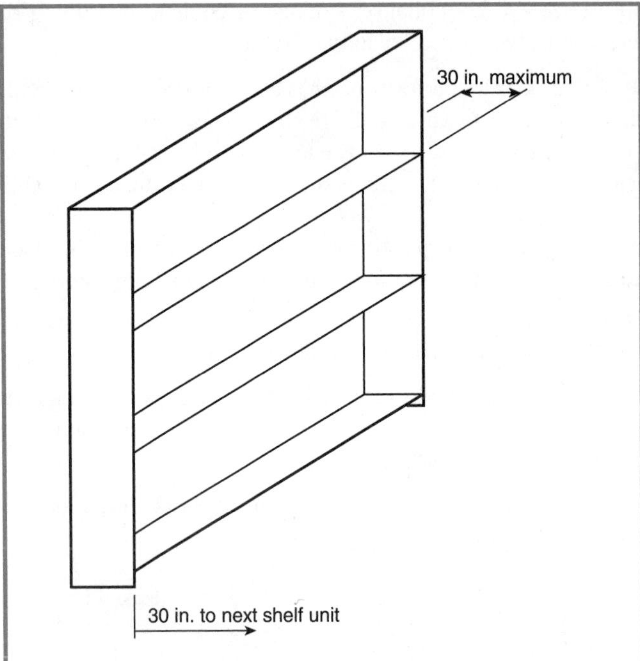

Exhibit 1.41 *Shelf storage.*

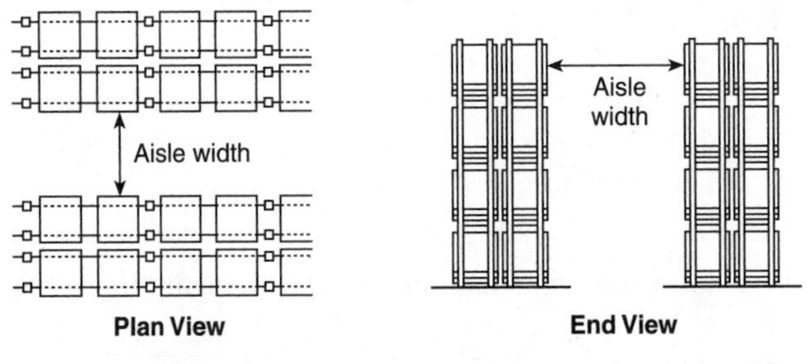

Figure A-1-4.9(a) *Illustration of aisle width.*

A-1-4.9 Conventional Pallets. See Figure A-1-4.9(b).

Face Sprinklers. Standard sprinklers that are located in transverse flue spaces along the aisle or in the rack, are within 18 in. (0.46 m) of the aisle face of storage, and are used to oppose vertical development of fire on the external face of storage.

Face sprinklers must be located at transverse flues within racks. These sprinklers are not effective if located outside the flue space, even if they are within 18 in. (0.46 m)

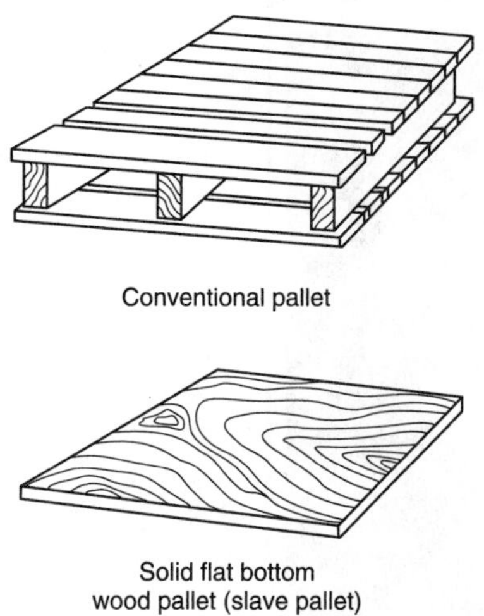

Figure A-1-4.9(b) Typical pallets.

of the rack face. Exhibit 1.42 illustrates a face sprinkler that is not located directly in the transverse flue. Note the wood slat and pallet load just above the sprinkler.

Horizontal Barrier. A solid barrier in the horizontal position covering the entire rack, including all flue spaces at certain height increments, to prevent vertical fire spread.

Longitudinal Flue Space.* The space between rows of storage perpendicular to the direction of loading. *[See Figure A-1-4.9(c).]*

A-1-4.9 Longitudinal Flue Space. See Figure A-1-4.9(c).

Rack.* Any combination of vertical, horizontal, and diagonal members that supports stored materials. Some rack structures use solid shelves. Racks can be fixed, portable, or movable *[see Figures A-1-4.9 (a) through (m)]*. Loading can be either manual—using lift trucks, stacker cranes, or hand placement—or automatic—using machine-controlled storage and retrieval systems.

A-1-4.9 Rack. Rack storage as referred to in this standard contemplates commodities in a rack structure, usually steel. Many variations of dimensions are found. Racks can be single-row, double-row, or multiple-row, with or without solid shelves. The standard commodity used in most of the tests was 42 in. (1.07 m) on a side. The types of racks covered in this standard are as follows.

(a) *Double-Row Racks.* Pallets rest on two beams parallel to the aisle. Any number of pallets can be supported by one pair of beams. *[See Figures A-1-4.9(d) through (g).]*

(b) *Automatic Storage-Type Rack.* The pallet is supported by two rails running perpendicular to the aisle. *[See Figure A-1-4.9(h).]*

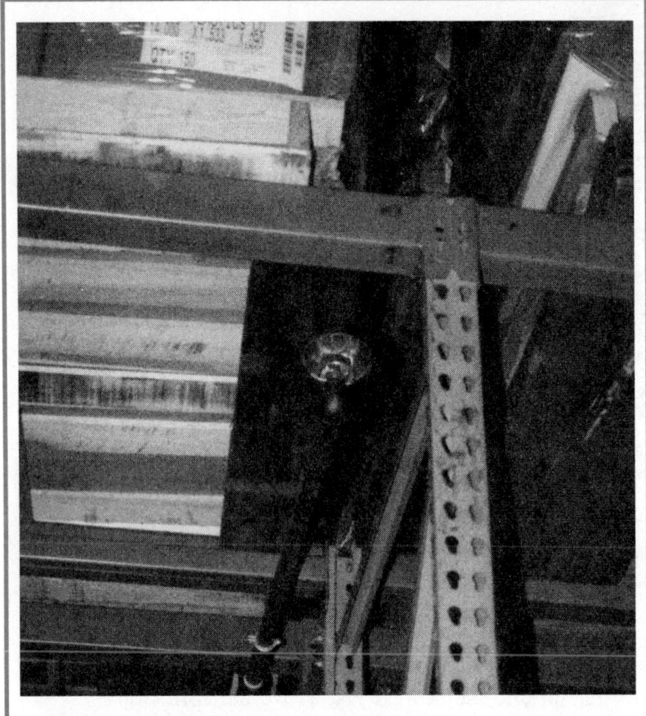

Exhibit 1.42 *Face sprinkler not properly located within the transverse flue.*

(c) *Multiple-Row Racks More than Two Pallets Deep, Measured Aisle to Aisle.* These racks include drive-in racks, drive-through racks, flow-through racks, portable racks arranged in the same manner, and conventional or automatic racks with aisles less than 42 in. (1.07 m) wide. *[See Figures A-1-4.9(i) through (n).]*

(d) *Movable Racks.* Movable racks are racks on fixed rails or guides. They can be moved back and forth only in a horizontal, two-dimensional plane. A moving aisle is created as abutting racks are either loaded or unloaded, then moved across the aisle to abut other racks. *[See Figure A-1-4.9(n).]*

(e) *Solid Shelving.* Conventional pallet racks with plywood shelves on the shelf beams *[see Figures A-1-4.9(f) and (g)]*. These racks are used in special cases. *(See Chapter 5.)*

(f) *Cantilever Rack.* The load is supported on arms that extend horizontally from columns. The load can rest on the arms or on shelves supported by the arms. *[See Figure A-1-4.9(m).]*

(g) Load depth in conventional or automatic racks should be considered a nominal 4 ft (1.22 m). *[See Figure A-1-4.9(e).]*

Racks, Double-Row. Two single-row racks placed back-to-back having a combined width up to 12 ft (3.7 m), with aisles at least 3.5 ft (1.1 m) on each side.

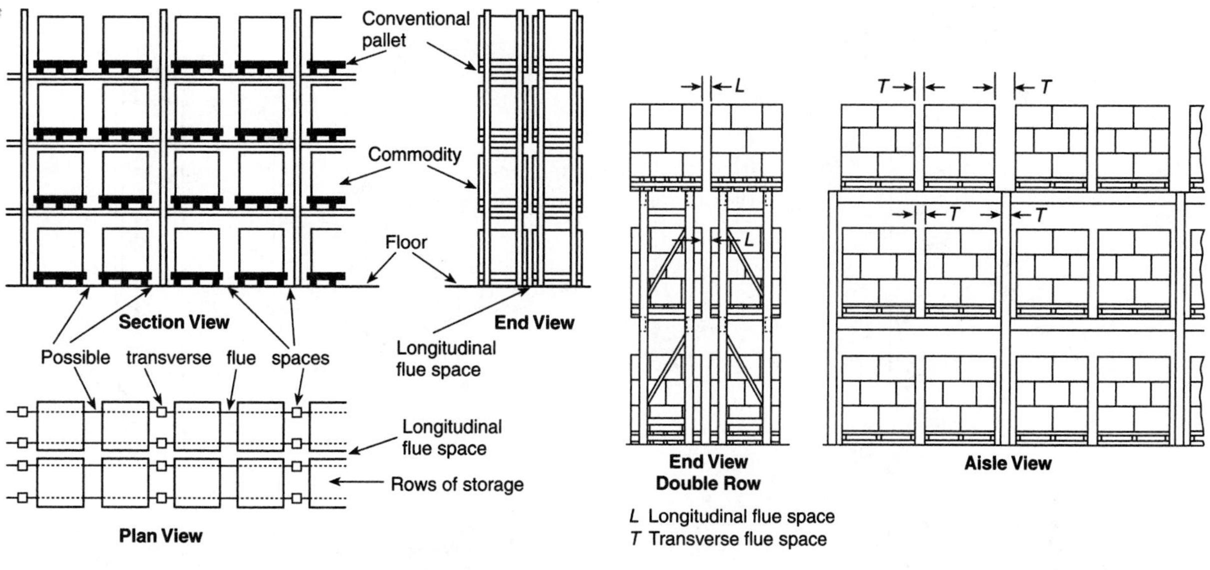

Figure A-1-4.9(c) *Typical double-row (back-to-back) rack arrangement.*

Figure A-1-4.9(d) *Conventional pallet rack.*

Racks, Movable. Racks on fixed rails or guides. They can be moved back and forth only in a horizontal, two-dimensional plane. A moving aisle is created as abutting racks are either loaded or unloaded, then moved across the aisle to abut other racks.

Racks, Multiple-Row. Racks greater than 12 ft (3.7 m) wide or single- or double-row racks separated by aisles less than 3.5 ft (1.1 m) wide having an overall width greater than 12 ft (3.7 m).

Racks, Portable. Racks that are not fixed in place. They can be arranged in any number of configurations.

Racks, Single-Row. Racks that have no longitudinal flue space and that have a width up to 6 ft (1.8 m) with aisles at least 3.5 ft (1.1 m) from other storage.

Slave Pallet. A special pallet captive to a material-handling system. *[See Figure A-1-4.9(b).]*

Solid Shelving. Solid, slatted, and other types of shelving located within racks that obstruct sprinkler water penetration down through the racks.

Exhibit 1.43 illustrates a storage rack with solid shelving.

Transverse Flue Space. The space between rows of storage parallel to the direction of loading. *[See Figure A-1-4.9(c).]*

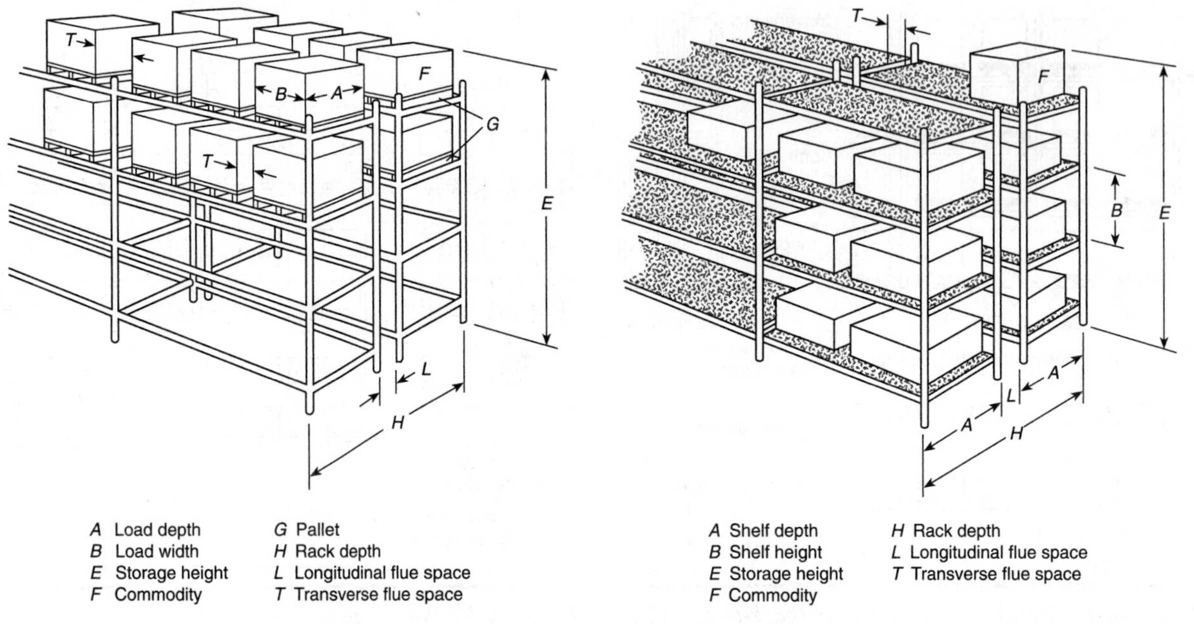

Figure A-1-4.9(e) Double-row racks without solid or slatted shelves.

Figure A-1-4.9(f) Double-row racks with solid shelves.

1-4.10 Rubber Tire Storage Definitions.

Banded Tires. A storage method in which a number of tires are strapped together.

Horizontal Channel. Any uninterrupted space in excess of 5 ft (1.5 m) in length between horizontal layers of stored tires. Such channels can be formed by pallets, shelving, racks, or other storage arrangements.

Laced Tire Storage. Tires stored where the sides of the tires overlap, creating a woven or laced appearance. *[See Figure A-1-4.10.1(g).]*

Miscellaneous Tire Storage.* The storage of rubber tires that is incidental to the main use of the building. Storage areas shall not exceed 2000 ft^2 (186 m^2). On-tread storage piles, regardless of storage method, shall not exceed 25 ft (7.6 m) in the direction of the wheel holes. Acceptable storage arrangements include (a) on-floor, on-side storage up to 12 ft (3.7 m) high; (b) on-floor, on-tread storage up to 5 ft (1.5 m) high; (c) double-row or multirow fixed or portable rack storage on-side or on-tread up to 5 ft (1.5 m) high; (d) single-row fixed or portable rack storage on-side or on-tread up to 12 ft (3.7 m) high; and (e) laced tires in racks up to 5 ft (1.5 m) in height.

A-1-4.10 Miscellaneous Tire Storage. The limitations on the type and size of storage are intended to identify those situations where tire storage is present in limited quantities and incidental to the main use of the building. Occupancies such as aircraft hangars, automobile

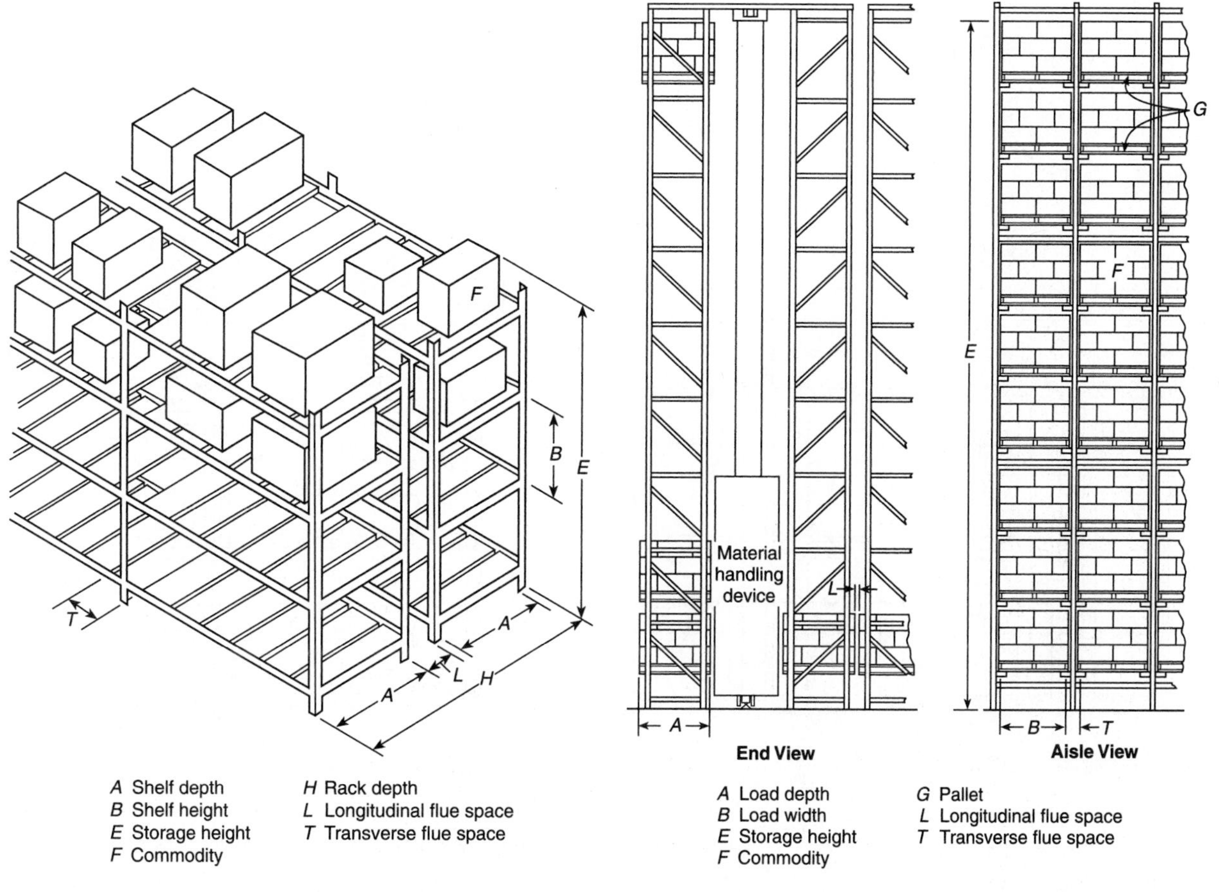

Figure A-1-4.9(g) *Double-row racks with slatted shelves.*

Figure A-1-4.9(h) *Automatic storage-type rack.*

dealers, repair garages, retail storage facilities, automotive and truck assembly plants, and mobile home assembly plants are types of facilities where miscellaneous storage could be present.

On-Side Tire Storage. Tires stored horizontally or flat.

On-Tread Tire Storage. Tires stored vertically or on their treads.

Palletized Tire Storage. Storage on portable racks of various types utilizing a conventional pallet as a base.

Pyramid Tire Storage. On-floor storage in which tires are formed into a pyramid to provide pile stability.

Rubber Tires. Pneumatic tires for passenger automobiles, aircraft, light and heavy trucks, trailers, farm equipment, construction equipment (off-the-road), and buses.

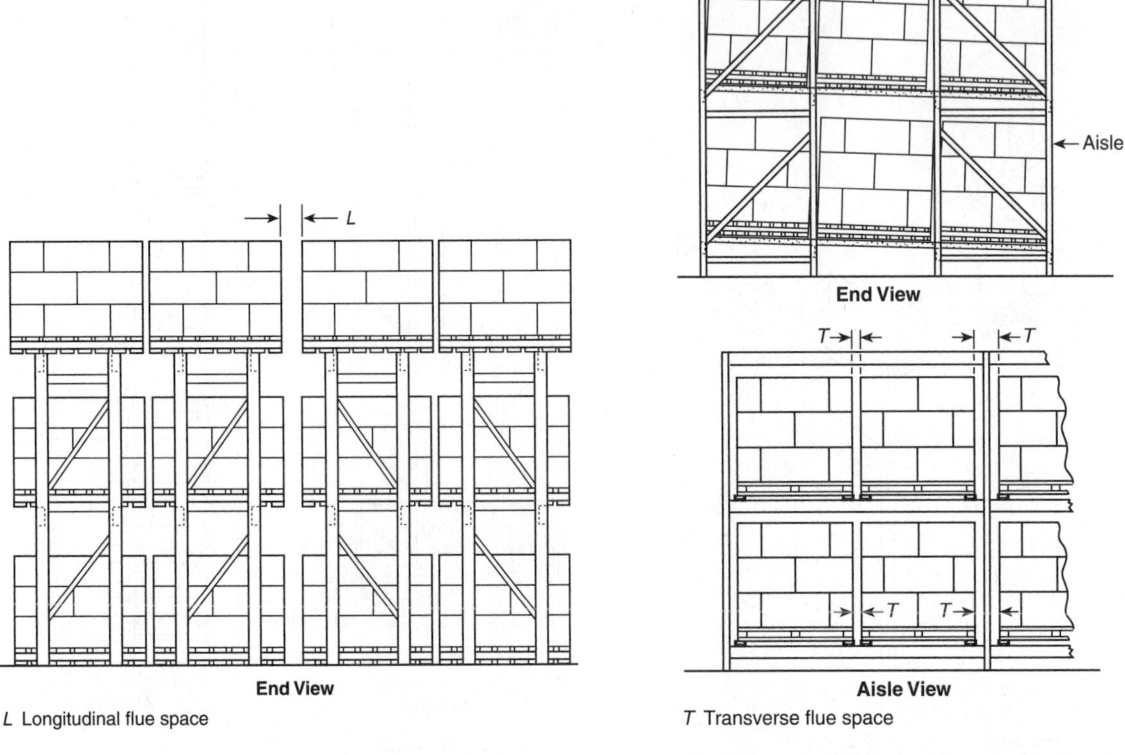

Figure A-1-4.9(i) Multiple-row rack to be served by a reach truck.

Figure A-1-4.9(j) Flow-through pallet rack.

1-4.10.1* Rubber Tire Rack Illustrations. See Figures A-1-4.10.1(a) through (g).

A-1-4.10.1 Figures A-1-4.10.1(a) through (g) do not necessarily cover all possible rubber tire storage configurations.

1-4.11 Baled Cotton Definitions.

Baled Cotton.* A natural seed fiber wrapped and secured in industry-accepted materials, usually consisting of burlap, woven polypropylene or sheet polyethylene, and secured with steel, synthetic or wire bands, or wire; can also include linters (lint removed from the cottonseed) and motes (residual materials from the ginning process). *(See Table A-1-4.11.)*

A-1-4.11 Baled Cotton. See Table A-1-4.11.

Block Cotton Storage. The number of bales closely stacked in cubical form and enclosed by aisles or building sides, or both.

Cold Cotton. Baled cotton five or more days old after the ginning process.

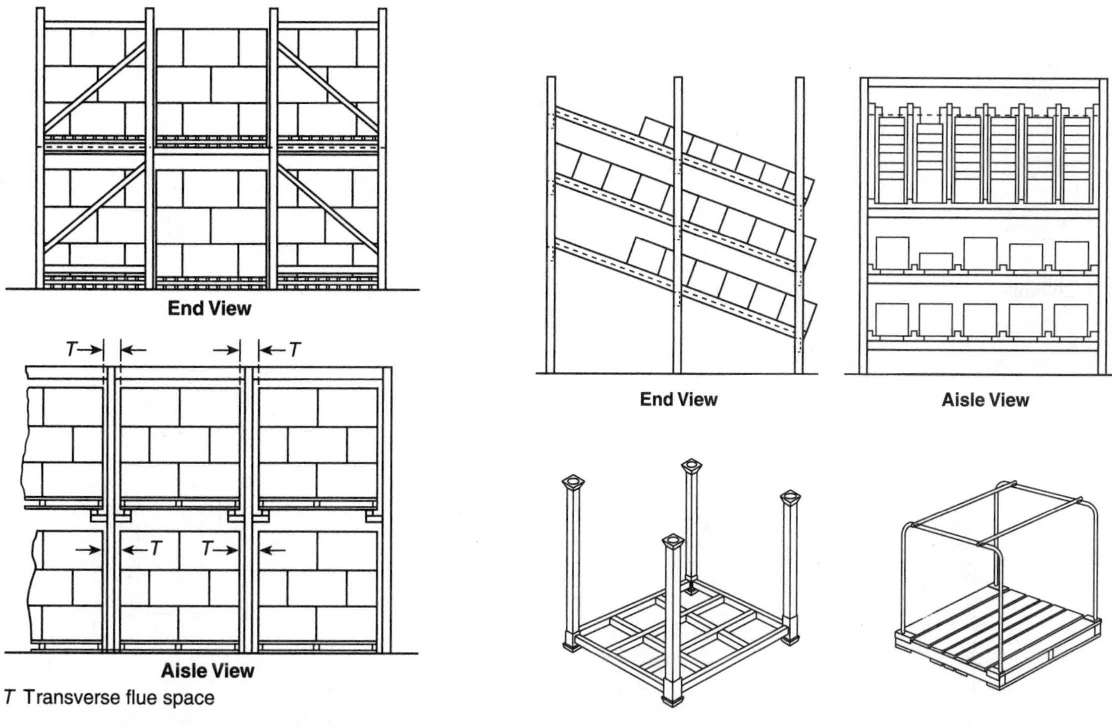

Figure A-1-4.9(k) Drive-in rack—two or more pallets deep (fork truck drives into the rack to deposit and withdraw loads in the depth of the rack).

Figure A-1-4.9(l) Flow-through racks (top) and portable racks (bottom).

Fire-Packed. A bale within which a fire has been packed as a result of a process, with ginning being the most frequent cause.

Naked Cotton Bale. A bale secured with wire or steel straps without wrapping.

1-4.12 Roll Paper Definitions.

Array, Closed (Paper). A vertical storage arrangement in which the distances between columns in both directions are short [not more than 2 in. (50 mm) in one direction and 1 in. (25 mm) in the other].

Array, Open (Paper). A vertical storage arrangement in which the distance between columns in both directions is lengthy (all vertical arrays other than closed or standard).

Array, Standard (Paper).* A vertical storage arrangement in which the distance between columns in one direction is short [1 in. (25 mm) or less] and is in excess of 2 in. (50 mm) in the other direction.

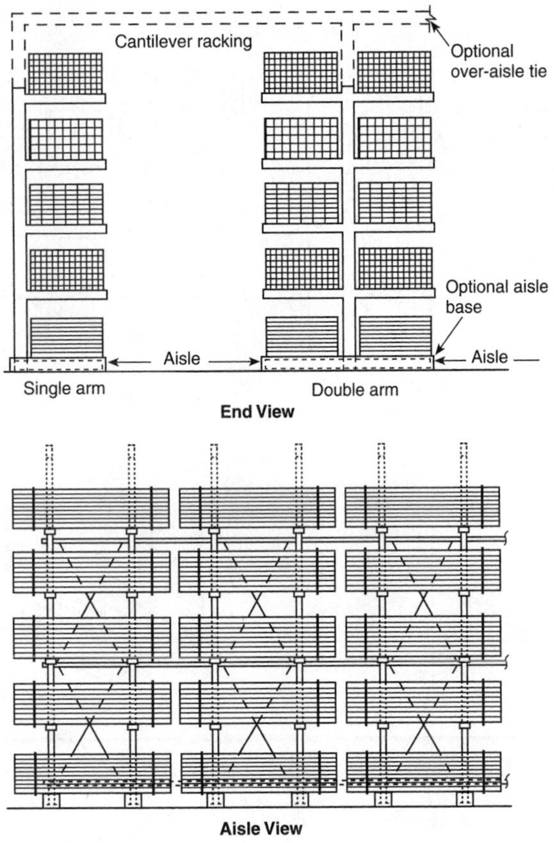

Figure A-1-4.9(m) Cantilever rack.

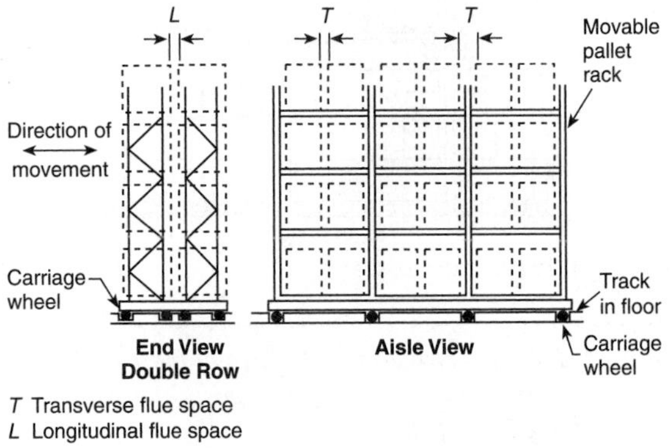

Figure A-1-4.9(n) Movable rack.

A-1-4.12 Array, Standard (Paper). The occasional presence of partially used rolls on top of columns of otherwise uniform diameter rolls does not appreciably affect the burning characteristics.

Banded Roll Paper Storage. Rolls provided with a circumferential steel strap [3/8 in. (9.5 mm) or wider] at each end of the roll.

Column. A single vertical stack of rolls.

Core. The central tube around which paper is wound to form a roll.

Paper (General Term). The term for all kinds of felted sheets made from natural fibrous materials, usually vegetable but sometimes mineral or animal, and formed on a fine wire screen from water suspension.

Roll Paper Storage, Horizontal. Rolls stored with the cores in the horizontal plane (on-side storage).

Exhibit 1.43 *Storage rack with solid shelving.*

Roll Paper Storage, Vertical. Rolls stored with the cores in the vertical plane (on-end storage).

Roll Paper Storage, Wrapped.* Rolls provided with a complete heavy kraft covering around both sides and ends.

A-1-4.12 Roll Paper Storage, Wrapped. Rolls that are completely protected with a heavy-weight kraft wrapper on both sides and ends are subject to a reduced degree of fire hazard. Standard methods for wrapping and capping rolls are outlined in Figure A-1-4.12.

In some cases, rolls are protected with laminated wrappers, using two sheets of heavy kraft with a high-temperature wax laminate between the sheets. Where using this method, the overall weight of wax-laminated wrappers should be based on the basis weight per 1000 ft^2 (92.9 m^2) of the outer sheet only, rather than on the combined basis weight of the outer and inner laminated wrapper sheets. A properly applied wrapper can have the effect of changing the class of a given paper to essentially that of the wrapper material. The effect of applying a wrapper to tissue has not been determined by test.

Roll Paper Storage Height.* The maximum vertical distance above the floor at which roll paper is normally stored.

A-1-4.12 Roll Paper Storage Height.* The size of rolls and limitations of mechanical handling equipment should be considered in determining maximum storage height.

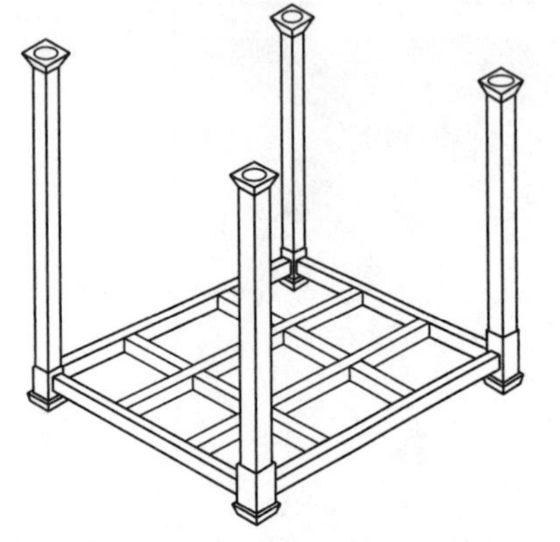

Figure A-1-4.10.1(a) Typical open portable tire rack unit.

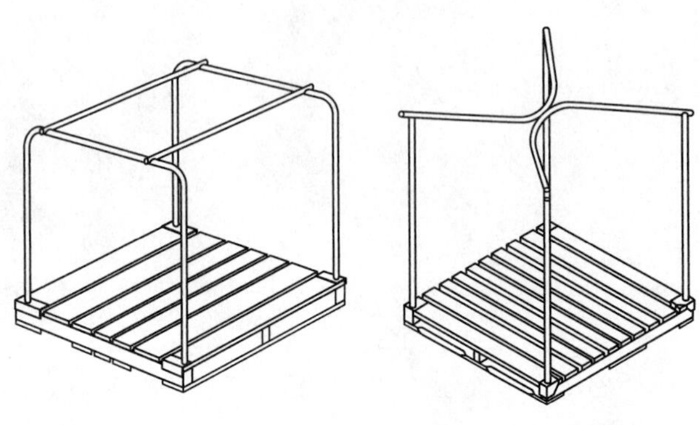

Figure A-1-4.10.1(b) Typical palletized portable tire rack units.

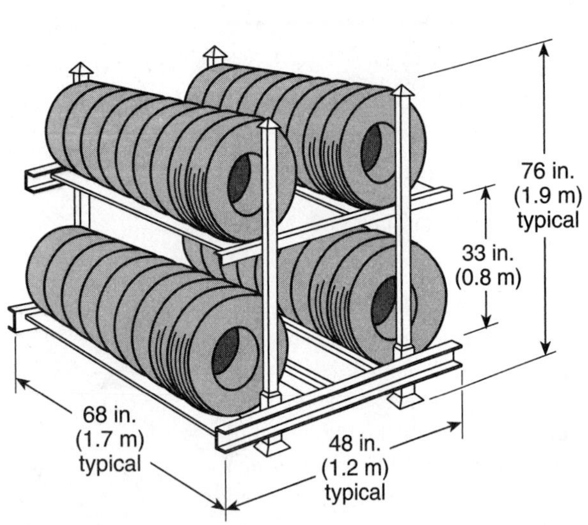

Figure A-1-4.10.1(c) Open portable tire rack.

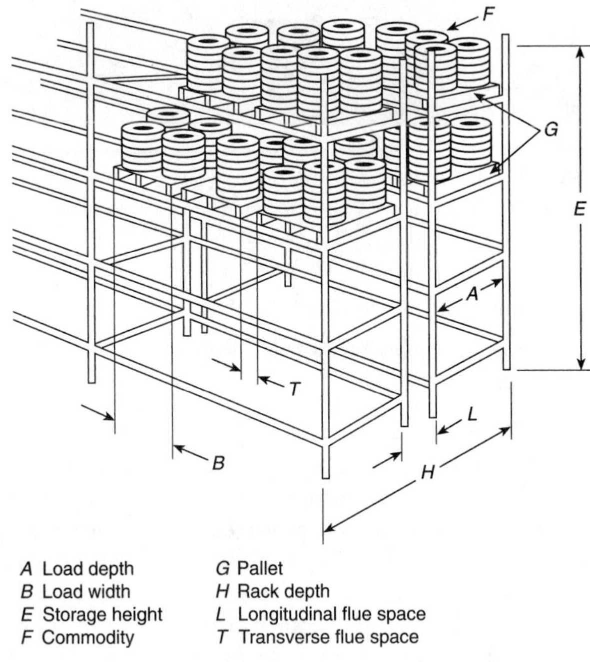

Figure A-1-4.10.1(d) Double-row fixed tire rack storage.

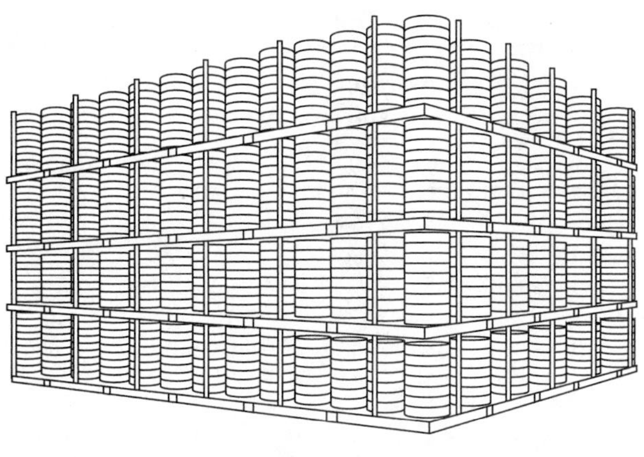

Figure A-1-4.10.1(e) *Palletized portable tire rack, on-side storage arrangement (banded or unbanded).*

Figure A-1-4.10.1(f) *On-floor storage; on-tread, normally banded.*

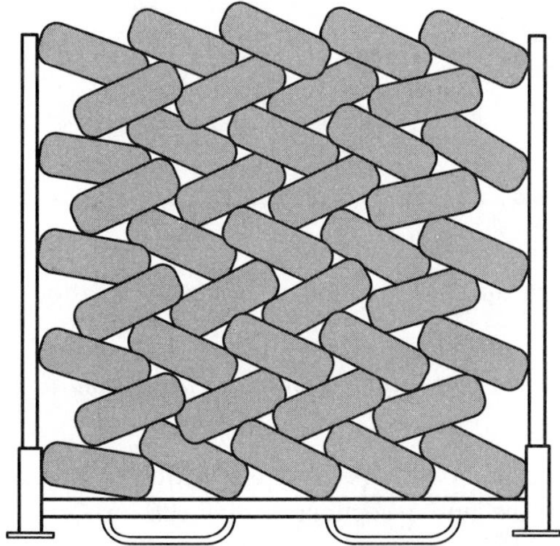

Figure A-1-4.10.1(g) *Typical laced tire storage.*

Table A-1-4.11 Typical Cotton Bale Types and Approximate Sizes

Bale Type	Dimensions		Average Weight		Volume		Density	
	in.	mm	lb	kg	ft^3	m^3	lb/ft^3	kg/m^3
Gin, flat	55 × 45 × 28	1397 × 1143 × 711	500	226.8	40.1	1.13	12.5	201
Modified gin, flat	55 × 45 × 24	1397 × 1143 × 610	500	226.8	34.4	0.97	14.5	234
Compressed, standard	57 × 29 × 23	1448 × 736 × 584	500	226.8	22.0	0.62	22.7	366
Gin, standard	55 × 31 × 21	1397 × 787 × 533	500	226.9	20.7	0.58	24.2	391
Compressed, universal	58 × 25 × 21	1475 × 635 × 533	500	226.8	17.6	0.50	28.4	454
Gin, universal	55 × 26 × 21	1397 × 660 × 533	500	226.8	17.4	0.49	28.7	463
Compressed, high density	58 × 22 × 21	1473 × 559 × 533	500	226.8	15.5	0.44	32.2	515

1-5 Abbreviations

The standard abbreviations in Table 1-5 shall be used on the hydraulic calculation form discussed in Chapter 8.

1-6 Level of Protection

1-6.1 A building, where protected by an automatic sprinkler system installation, shall be provided with sprinklers in all areas.

Exception: This requirement shall not apply where specific sections of this standard permit the omission of sprinklers.

The oldest and most important design rule of NFPA 13 is that sprinklers should be installed in all areas of a building. The few exceptions to this rule that exist are associated with specific conditions and are relatively new revisions to the standard. Some standard users are under the impression that certain areas such as electrical equipment rooms, closets, or walk-in coolers do not require sprinklers. However, unless a specific paragraph or exception in Chapter 5 gives explicit permission to omit sprinklers, sprinklers are required in all spaces of a building.

The success of a sprinkler system is largely dependent on the size of the fire when the first few sprinklers activate. If the fire is permitted to grow to a relatively large size before sprinkler operation occurs, the likelihood of successful sprinkler system performance in terms of fire control or fire suppression is significantly reduced. Sprinkler systems designed in accordance with NFPA 13 are not intended to prevent a fire in an unsprinklered area from spreading into a sprinklered area. The need to provide sprinklers throughout a building should be obvious.

Wrapper Exterior wrapper Body wrapper	General term for protective wrapping of sides and ends on roll.
Body wrap Sleeve wrap Wrap — do not cap	Wrapper placed around circumference of roll. No heads or caps needed.
Heads Headers	Protection applied to the ends of the rolls (*A* and *B*). Heads do not lap over the end of the roll.
Inside heads	Protection applied to the ends of the rolls next to the roll itself (*B*). The wrapper of the rolls is crimped down over these heads.
Outside heads	Protection applied to the ends of the rolls on the outside (*A*). This head is applied after the wrapper is crimped.
Edge protectors Edge bands	Refers to extra padding to prevent damage to roll edges (*C*).
Overwrap	The distance the body wrap or wrapper overlaps itself (*D*).
Roll cap	A protective cover placed over the end of a roll. Edges of cap lap over the end of the roll and are secured to the sides of the roll.

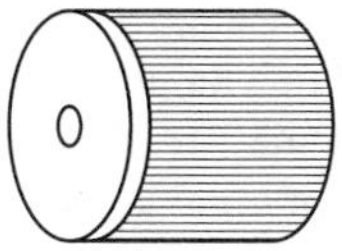

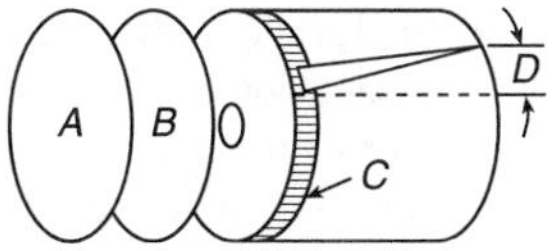

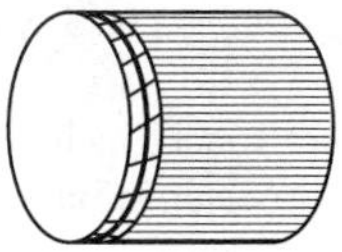

Figure A-1-4.12 *Wrapping and capping terms and methods.*

1-6.2 Limited Area Systems.

When partial sprinkler systems are installed, the requirements of this standard shall be used insofar as they are applicable. The authority having jurisdiction shall be consulted in each case.

This paragraph acknowledges that certain codes or local ordinances might permit the installation of a partial or limited area sprinkler system. For example, NFPA *101*

Table 1-5 Hydraulic Symbols

Symbol or Abbreviation	Item
p	Pressure in psi
gpm	U.S. gallons per minute
q	Flow increment in gpm to be added at a specific location
Q	Summation of flow in gpm at a specific location
P_t	Total pressure in psi at a point in a pipe
P_f	Pressure loss due to friction between points indicated in location column
P_e	Pressure due to elevation difference between indicated points. This can be a plus value or a minus value. If minus, the (−) shall be used; if plus, no sign need be indicated.
P_v	Velocity pressure in psi at a point in a pipe
P_n	Normal pressure in psi at a point in a pipe
E	90° ell
EE	45° ell
Lt.E	Long-turn elbow
Cr	Cross
T	Tee-flow turned 90°
GV	Gate valve
BV	Butterfly (wafer) check valve
Del V	Deluge valve
ALV	Alarm valve
DPV	Dry pipe valve
CV	Swing check valve
WCV	Butterfly (wafer) check valve
St	Strainer
psi	Pounds per square inch
v	Velocity of water in pipe in feet per second

recognizes a limited area sprinkler system in certain residential occupancies. This type of system is installed in the corridors along with one sprinkler just inside the door of each dwelling unit. This system is intended to provide additional time for the occupants to escape. Such a system cannot be expected to control a fire so as to minimize property damage and, therefore, property damage would most likely be quite severe.

1-7 Units

Metric units of measurement in this standard are in accordance with the modernized metric system known as the International System of Units (SI). Two units (liter and bar), outside of but recognized by SI, are commonly used in international fire protection. These units are listed in Table 1-7 with conversion factors.

Table 1-7 SI Units and Conversion Factors

Name of Unit	Unit Symbol	Conversion Factor
liter	L	1 gal = 3.785 L
millimeter per minute	mm/min	1 gpm/ft^2 = 40.746 mm/min = 40.746 (L/min)/m^2
cubic decimeter	dm^3	1 gal = 3.785 dm^3
pascal	Pa	1 psi = 6894.757 Pa
bar	bar	1 psi = 0.0689 bar
bar	bar	1 bar = 105 Pa

For additional conversions and information, see ASTM SI 10, *Standard for Use of the International System of Units (SI): The Modern Metric System.*

The English units for sprinkler design densities are gallons per minute per square foot (gpm/ft^2). The direct metric equivalent to these English units is liters per minute per square meter [(L/min)/m^2]. However, the more common units for measuring sprinkler discharge densities in those countries that employ the metric system are millimeters per minute (mm/min), and this designation is used throughout the standard.

1-7.1 If a value for measurement as given in this standard is followed by an equivalent value in other units, the first stated is to be regarded as the requirement. A given equivalent value might be approximate.

1-7.2 The conversion procedure for the SI units has been to multiply the quantity by the conversion factor and then round the result to the appropriate number of significant digits.

REFERENCES CITED IN COMMENTARY

National Fire Protection Association, 1 Batterymarch Park, P.O. Box 9101, Quincy, MA 02269-9101.

NFPA 11, *Standard for Low-Expansion Foam*, 1998 edition.
NFPA 11A, *Standard for Medium- and High-Expansion Foam Systems*, 1999 edition.

NFPA 13D, *Standard for the Installation of Sprinkler Systems in One- and Two-Family Dwellings and Manufactured Homes*, 1999 edition.
NFPA 13R, *Standard for the Installation of Sprinkler Systems in Residential Occupancies up to and Including Four Stories in Height*, 1999 edition.
NFPA 14, *Standard for the Installation of Standpipe and Hose Systems*, 1996 edition.
NFPA 15, *Standard for Water Spray Fixed Systems for Fire Protection*, 1996 edition.
NFPA 16, *Standard for the Installation of Foam-Water Sprinkler and Foam-Water Spray Systems*, 1999 edition.
NFPA 20, *Standard for the Installation of Stationary Pumps for Fire Protection*, 1999 edition.
NFPA 22, *Standard for Water Tanks for Private Fire Protection*, 1998 edition.
NFPA 25, *Standard for the Inspection, Testing, and Maintenance of Water-Based Fire Protection Systems*, 1998 edition.
NFPA 30, *Flammable and Combustible Liquids Code*, 1996 edition.
NFPA 30B, *Code for the Manufacture and Storage of Aerosol Products*, 1998 edition.
NFPA 40, *Standard for the Storage and Handling of Cellulose Nitrate Motion Picture Film*, 1997 edition.
NFPA 72, *National Fire Alarm Code®*, 1999 edition.
NFPA *101®*, *Life Safety Code®*, 1997 edition.
NFPA 220, *Standard on Types of Building Construction*, 1999 edition.
NFPA 750, *Standard on Water Mist Fire Protection Systems*, 1996 edition.

Underwriters Laboratories Inc., 333 Pfingsten Road, Northbrook, IL 60062.

UL 199, *Standard for Safety for Automatic Sprinklers for Fire-Protection Service*, 10th edition, 1997.
UL 1626, *Residential Sprinklers for Fire-Protection Service*, 2nd edition, 1994.
UL 1767, *Early-Suppression Fast-Response Sprinklers,* 2nd edition, 1995.

Factory Mutual Research Corporation, 1151 Boston-Providence Turnpike, Norwood, MA 02061.

FM 2008, *Approval Standard for Early Suppression Fast Response (ESFR) Automatic Sprinklers*, 1996 edition.

CHAPTER 2

Classification of Occupancies and Commodities

Chapter 2 combines information on the classification of occupancy hazards previously found in Chapter 1 and the description of storage commodities formerly found in NFPA 231, NFPA 231C, and NFPA 231F.

2-1* Classification of Occupancies

Occupancy classifications for this standard shall relate to sprinkler design, installation, and water supply requirements only. They shall not be intended to be a general classification of occupancy hazards.

A-2-1 Occupancy examples in the listings as shown in the various hazard classifications are intended to represent the norm for those occupancy types. Unusual or abnormal fuel loadings or combustible characteristics and susceptibility for changes in these characteristics, for a particular occupancy, are considerations that should be weighed in the selection and classification.

The light hazard classification is intended to encompass residential occupancies; however, this is not intended to preclude the use of listed residential sprinklers in residential occupancies or residential portions of other occupancies.

To a large degree, the occupancy hazard classifications form the basis of the design and installation criteria of NFPA 13. The occupancy hazards provide a convenient means of categorizing the fuel loads and fire severity associated with certain building operations. The classifications also present a relationship between the burning characteristics of these fuels and the ability of a sprinkler system in controlling the associated types of fires. The likelihood of ignition is not considered in the occupancy classifications.

The occupancy hazards presented in NFPA 13 should not be interpreted as a generic description or quantification of fire hazards and are not intended to parallel those identified in other fire safety regulations such as NFPA *101®*, *Life Safety Code®*, or building codes.

A specific operation might present more or less of a hazard, depending on the combustibility, quantity, and arrangement of the building contents. For instance, paperback books in boxes stacked to a height of 6 ft (1.83 m) in the back room of a bookstore present a lower fire hazard than the same books stacked to a height of 12 ft (3.7 m) throughout a warehouse. Accordingly, NFPA 13 categorizes the 6-ft (1.83-m) high stockpile as an ordinary hazard (Group 1) occupancy. The 12-ft (3.7-m) high warehouse storage is considered a Class III commodity and needs to be protected in accordance with the criteria for ordinary hazard (Group II) occupancies. (See Section 7-3 for more information.)

Proper classification of a fire hazard is critical to the overall success of the sprinkler system. The determination of the type of occupancy hazard influences system design and installation considerations such as sprinkler discharge criteria, sprinkler spacing, and water supply requirements. The operations of a given facility can vary significantly over time and change the overall fire hazard. These potential fluctuations in building operations must be properly accounted for when designing a new sprinkler system or when evaluating the adequacy of an existing sprinkler system.

The proper occupancy hazard classification for a given building operation should be determined by carefully reviewing the descriptions of each occupancy hazard and by evaluating the quantity, combustibility, and heat release rate of the associated contents. To aid in the classification of a building operation, examples for each type of occupancy hazard are provided in Appendix A (see A-2-1.1, A-2-1.2.1, A-2-1.2.2, A-2-1.3.1, and A-2-1.3.2.) Experience and good judgment play a critical role in properly determining the hazard classification, just as they are in evaluating the adequacy of the entire sprinkler system.

The occupancy hazard classifications are presented as qualitative descriptions rather than quantifiable measurements. Ideally, quantification of key factors such as the fire safety goals of the standard, the likely hazards contained within a specific space, the effect of building geometry and ventilation, and the interaction of sprinkler discharge with the fire would form the basis for NFPA 13. Under a performance-based approach, designers and enforcers would have more flexibility in complying with NFPA 13. While much of the technology needed for a generic widespread performance-based approach for sprinkler system design is not yet available, testing and research continue to fill the gaps. For instance, performance-based-type approaches have been applied in the development of new types of sprinklers such as the early suppression fast-response (ESFR) sprinkler and residential sprinklers.

2-1.1* Light Hazard Occupancies.

Light hazard occupancies shall be occupancies or portions of other occupancies where the quantity and/or combustibility of contents is low and fires with relatively low rates of heat release are expected.

A-2-1.1 Light hazard occupancies include occupancies having uses and conditions similar to the following:

A

Churches

Clubs

Eaves and overhangs, if of combustible construction with no combustibles beneath

Educational

Hospitals

Institutional

Libraries, except large stack rooms

Museums

Nursing or convalescent homes

Offices, including data processing

Residential

Restaurant seating areas

Theaters and auditoriums, excluding stages and prosceniums

Unused attics

Of the five occupancy hazard classifications, light hazard occupancies represent the least severe fire hazard since the fuel loads associated with these occupancies are low, and relatively small rates of heat release are expected. Generally, no processing, manufacturing, or storage operations are included, and fixtures and furniture remain in fairly permanent arrangements. Light hazard occupancies typically consist of institutional, educational, religious, residential, and office facilities. Sprinkler systems designed to protect against light hazard occupancies, therefore, have less demanding water supply requirements. Additionally, more design flexibility is possible.

2-1.2 Ordinary Hazard Occupancies.

Prior to the 1991 edition, NFPA 13 included three categories of ordinary hazard occupancies. The technical committee believed that the small differences between the "old" ordinary 2 and 3 curves had little impact on fire control. See the area/density curve comparison in Exhibit 2.1. The wide range of occupancies addressed by the overall ordinary hazard classification and the associated water supply requirements are better served by two categories of ordinary hazard rather than three.

Within the ordinary hazard classification, Group 1 occupancies present the least severe fire. Group 1 occupancies are mostly light manufacturing and service industries where the use of flammable and combustible liquids or gases is either nonexistent or very limited. Stockpiles of combustible commodities typically found within an ordinary hazard (Group 1) occupancy cannot exceed a height of 8 ft (2.4 m). Additionally, the quantity and arrangement of the stockpiles cannot exceed the limitations of miscellaneous storage as defined in Chapter 1. If the miscellaneous storage limitations are exceeded, then the Chapter 7 provisions related to storage apply.

The ordinary hazard (Group 2) classification addresses those ordinary hazard occupancies that do not meet the criteria of the Group 1 classification. Group 2 occupancies present a more severe fire hazard and demand more from the sprinkler system to achieve fire control than do Group 1 and light hazard occupancies. Ordinary hazard (Group 2) occupancies include those types of manufacturing and processing operations in which the amount and combustibility of contents is greater than that for Group 1. The Group 2 category also addresses the miscellaneous storage of contents stored up to and including 12 ft (3.7 m) in height.

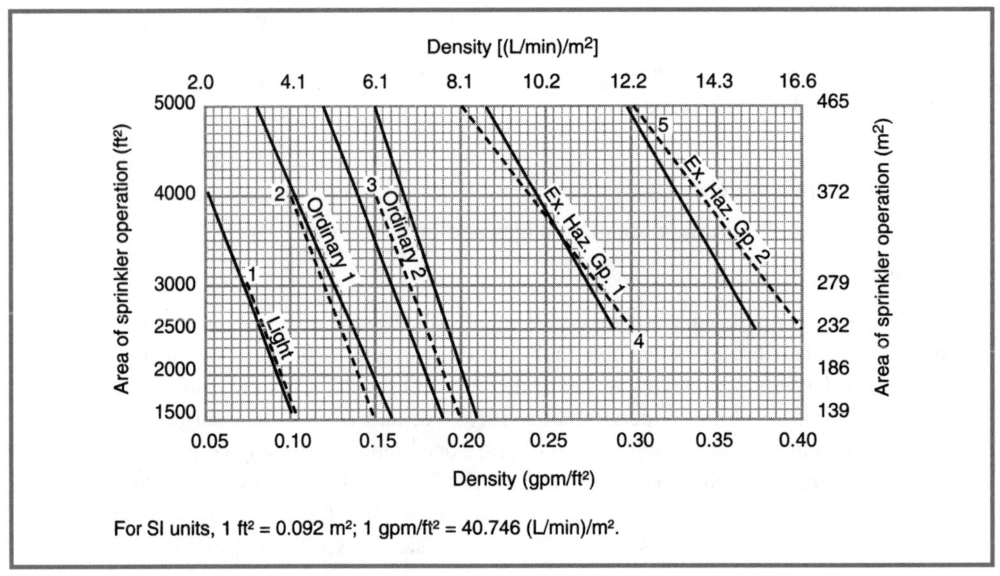

Exhibit 2.1 *Area/density curve comparison.*

2-1.2.1* Ordinary Hazard (Group 1). Ordinary hazard (Group 1) occupancies shall be occupancies or portions of other occupancies where combustibility is low, quantity of combustibles is moderate, stockpiles of combustibles do not exceed 8 ft (2.4 m), and fires with moderate rates of heat release are expected.

A-2-1.2.1 Ordinary hazard occupancies (Group 1) include occupancies having uses and conditions similar to the following:

Automobile parking and showrooms

Bakeries

Beverage manufacturing

Canneries

Dairy products manufacturing and processing

Electronic plants

Glass and glass products manufacturing

Laundries

Restaurant service areas

2-1.2.2* Ordinary Hazard (Group 2). Ordinary hazard (Group 2) occupancies shall be occupancies or portions of other occupancies where the quantity and combustibility of contents is moderate to high, stockpiles do not exceed 12 ft (3.7 m), and fires with moderate to high rates of heat release are expected.

A-2-1.2.2 Ordinary hazard occupancies (Group 2) include occupancies having uses and conditions similar to the following:

Cereal mills

Chemical plants—ordinary

Confectionery products

Distilleries

Dry cleaners

Feed mills

Horse stables

Leather goods manufacturing

Libraries—large stack room areas

The segregation between the ordinary hazard (Group 2) category for library stack rooms and the light hazard occupancy category for library reading areas is not an appropriate separation in most situations. The light hazard separation is based on the library having an area composed of tables, chairs, and perhaps some magazine racks and individual bookshelves. When this type of seating area exists, the rules for light hazard apply. Those areas of the library used primarily for book storage on shelves are considered stack areas, and the rules for ordinary hazard are applicable.

Machine shops

Metal working

Mercantile

Paper and pulp mills

Paper process plants

Piers and wharves

Post offices

Printing and publishing

Repair garages

Resin application area

Stages

Textile manufacturing

Tire manufacturing

Tobacco products manufacturing

Wood machining

Wood product assembly

2-1.3 Extra Hazard Occupancies.

Extra hazard occupancies represent the most severe fire conditions addressed under the occupancy hazard classifications in NFPA 13 and present the most severe challenge to sprinkler protection. The extra hazard (Group 1) occupancies include those with hydraulic machinery or systems with flammable or combustible hydraulic fluids under pressure. Ruptures and leaks in piping or fittings have resulted in fine spray discharge of such liquids, causing intense fires. Those occupancies with process machinery that use flammable or combustible fluids in closed systems are extra hazard (Group 1). Also in this group are occupancies that have dust and lint in suspension or that contain moderate amounts of combustible cellular foam materials. Buildings used for textile manufacturing are an example of an extra hazard occupancy.

Extra hazard (Group 2) occupancies contain more than small amounts of flammable or combustible liquids, usually in open systems where rapid evaporation can occur when these liquids are subjected to high temperatures.

The extra hazard (Group 2) occupancy classification also applies where ceiling sprinklers are severely obstructed by occupancy, not structural, conditions and where water discharged by sprinklers may not reach the burning material because of the shielding.

The extra hazard occupancy examples are classified on the basis of actual field experience with sprinkler system operations in occupancies having conditions similar to those identified.

2-1.3.1* Extra Hazard (Group 1). Extra hazard (Group 1) occupancies shall be occupancies or portions of other occupancies where the quantity and combustibility of contents is very high and dust, lint, or other materials are present, introducing the probability of rapidly developing fires with high rates of heat release but with little or no combustible or flammable liquids.

A-2-1.3.1 Extra hazard occupancies (Group 1) include occupancies having uses and conditions similar to the following:

Aircraft hangars (except as governed by NFPA 409, *Standard on Aircraft Hangars*)

Combustible hydraulic fluid use areas

Die casting

Metal extruding

Plywood and particle board manufacturing

Printing [using inks having flash points below 100°F (38°C)]

Rubber reclaiming, compounding, drying, milling, vulcanizing

Saw mills

Textile picking, opening, blending, garnetting, or carding, combining of cotton, synthetics, wool shoddy, or burlap

Upholstering with plastic foams

2-1.3.2* Extra Hazard (Group 2). Extra hazard (Group 2) occupancies shall include occupancies with moderate to substantial amounts of flammable or combustible liquids or occupancies where shielding of combustibles is extensive.

A-2-1.3.2 Extra hazard occupancies (Group 2) include occupancies having uses and conditions similar to the following:

Asphalt saturating

Flammable liquids spraying

Flow coating

Manufactured home or modular building assemblies (where finished enclosure is present and has combustible interiors)

Open oil quenching

Plastics processing

Solvent cleaning

Varnish and paint dipping

2-1.4* Special Occupancy Hazards.

A-2-1.4 Other NFPA standards contain design criteria for fire control or fire suppression *(see 2-1.4 and Chapter 13).* While these can form the basis of design criteria, this standard describes the methods of design, installation, fabrication, calculation, and evaluation of water supplies that should be used for the specific design of the system.

Other NFPA standards contain sprinkler system design criteria for fire control or suppression of specific hazards. This information has been either referenced or copied into Chapters 5 and 7 using NFPA's extract policy.

NFPA 13 was reorganized to provide the basis for methods of sprinkler system design, installation, fabrication, calculation, and evaluation of water supplies for specific designs. Under this reorganization, a detailed review of the fire sprinkler design requirements in other NFPA standards was performed by the sprinkler technical committees. As a result of these reviews, information from these other standards is cited in Chapters 5 and 7 or referenced in order to facilitate the user's access to relevant requirements and related information.

Many buildings now contain several different types of occupancies. A simple example of this is a restaurant, where for purposes of sprinkler protection, food preparation areas are treated as ordinary hazard (Group 1), while the public seating areas are handled as light hazard. On a larger scale, a grocery store contains a public shopping area, which is usually treated as an ordinary hazard (Group 2) occupancy. In many cases, however, the back room areas not within the mercantile operation will likely contain some quantity of bulk storage. While it is reasonable to expect that a higher degree of fire protection would be needed in the back room, older editions of NFPA 13 did not provide guidance.

The storage of bulk items in a back room is similar to a warehouse operation, although on a smaller scale, and requires special consideration. Depending on the size of the back room area and the type, quantity, and arrangement of items stored, the space needs to be protected in accordance with the provisions for miscellaneous storage in 7-2.3.2.2 or Section 7-3 on storage. The concept of miscellaneous storage is included to allow for correlation between occupancy and commodity classifications.

The protection schemes shown in Table 7-2.3.2.2 were developed by completing a comparative analysis of design curves for 12-ft (3.7-m) storage of the appropriate commodity classes as defined in Sections 2-2 and 7-3. In all cases, the design criteria for commodity storage could be approximated by a design point in Figure 7-2.3.1.2.

For example, wood window frames are stored in cardboard boxes on wood pallets to a height of 10 ft (3 m). This storage is incidental to another occupancy (a hardware store).

According to Table A-2-2.3.3, wood window frames are considered a Class III commodity. Assuming the use of 165°F (74°C) sprinklers, the design criteria from Section 7-3 results in an area/density protection scheme of 0.16 gpm/ft^2 [6.5 (L/min)/m^2] over a 2000-ft^2 (186-m^2) area of operation. It is important to note that an adjustment was made to the density for a 12-ft (3.7-m) storage height, since the criteria in Section 7-3 terminates at a 12-ft (3.7-m) minimum storage height.

Table 7-2.3.2.2 refers the designer to the criteria for ordinary hazard (Group 2). Figure 7-2.3.1.2, which shows the ordinary hazard (Group 2) curve, requires a density of 0.19 gpm/ft^2 [7.74 (L/min)/m^2] for a 2000-ft^2 (186-m^2) area of operation.

While the methodology used to establish the information contained in Table 7-2.3.2.2 is not exact, it does provide a good correlation between the commodity storage requirements and the general occupancy hazard requirements for most commodity classes. Some of the protection rows were intentionally left blank because no test data supports the use of any precise design criteria for certain combinations of storage and clearance.

2-2*† Commodity Classification

A-2-2 Specification of the type, amount, and arrangement of combustibles for any commodity classification is essentially an attempt to define the potential fire severity, based on its burning characteristics, so the fire can be successfully controlled by the prescribed sprinkler protection for the commodity class. In actual storage situations, however, many storage arrays do not fit precisely into one of the fundamental classifications; therefore, the user needs to make judgments after comparing each classification to the existing storage conditions. Storage arrays consist of thousands of products, which make it impossible to specify all the acceptable variations for any class. As an alternative, a variety of common products are classified in this appendix based on judgment, loss experience, and fire test results.

Table A-2-2 provides examples of commodities not addressed by the classifications in Section 2-2.

Table A-2-2.3 is an alphabetized list of commodities with corresponding classifications.

Tables A-2-2.3.1 through A-2-2.3.4 and A-2-2.4.1 provide examples of commodities within a specific class.

Table A-2-2 Examples of Commodities Not Addressed by the Classifications in Section 2-2

Boxes, Crates
- Empty, wood slatted*
Lighters (butane)
- Loose in large containers (Level 3 aerosol)

*Should be treated as idle pallets.

2-2.1 General.

Protection standards for stored commodities are based on testing that primarily involved the standard Class II and standard plastic commodities. The standard Class II commodity is a double tri-wall corrugated cardboard carton with a metal liner on a wood pallet. The standard plastic commodity is polystyrene cups in a compartmented corrugated cardboard carton on a wood pallet. Exhibit 2.2 illustrates these test commodities.

Protection for Class I, III, and IV commodities is based on limited testing and extrapolation from testing on Class II commodities. (No Class I tests, two Class III tests, and two Class IV tests were done.)

Classification of actual commodities is primarily based on comparing the commodity to the definitions for the various commodity classes. Commodity classification can also be done by testing, but it is important that such testing involve sprinklers or some other method to simulate water application by sprinklers and be large enough in scale to produce meaningful results. Bench-scale testing is not useful for making precise commodity classifications.

2-2.1.1* Classification of Commodities. Commodity classification and the corresponding protection requirements shall be determined based on the makeup of individual storage units (i.e., unit load, pallet load).

A-2-2.1.1 Commodity classification is governed by the types and amounts of materials (e.g., metal, paper, wood, plastics) that are a part of a product and its primary packaging. However, in a storage or warehousing situation, classification is also affected by such factors as the primary storage or shipping container material, the amount of air space, and the location of the more hazardous materials within the container. For example, a Group A plastic product enclosed in a five- or six-sided metal container can be considered Class II, while a ceramic product heavily wrapped in tissue paper and placed in a corrugated carton could be Class III.

Exhibit 2.2 Typical test commodities. Standard plastic commodity on left and standard Class II commodity on right. (Courtesy of Factory Mutual Research Corporation.)

2-2.1.2 Mixed Commodities. Protection requirements shall not be based on the overall commodity mix in a fire area. Mixed commodity storage shall be protected by the requirements for the highest classified commodity and storage arrangement.

Exception No. 1: Up to 10 pallet loads of a higher hazard commodity, as described in 2-2.3 and 2-2.4, shall be permitted to be present in an area not exceeding 40,000 ft^2 (3716 m^2). The higher hazard commodity shall be randomly dispersed with no adjacent loads in any direction (including diagonally). If the ceiling protection is based on Class I or Class II commodities, then the allowable number of pallet loads for Class IV or Group A plastics shall be reduced to five.

Exception No. 2: The higher hazard material shall be permitted to be confined to a designated area and properly protected for that area.

The fire area anticipated during sprinkler design is often mistakenly assumed to be the same as the sprinkler operating area. In fact, the fire area for successfully controlled storage fires seldom exceeded 200 ft^2 (18.6 m^2) in testing, and often was less than 100 ft^2 (9.3 m^2). Regardless of the overall mix of commodities in a storage area, the possibility that 100 ft^2 to 200 ft^2 (9.3 m^2 to 18.6 m^2) of the highest hazard commodity will accumulate always exists.

Factory Mutual did testing to evaluate the effect of mixing commodities in rack storage. In one test, 20 ft (6.1 m) of storage, encompassing four tiers, consisted of the standard plastic commodity in the bottom tier with the standard Class II commodity in the top three tiers. The second test was similar to the first, except that the plastic commodity was located in the top tier rather than the bottom. In the third test, plastic and Class II commodities were homogeneously mixed in the rack. The testing showed that protection for plastic commodities was needed for all three cases.

2-2.2 Pallet Types.

When loads are palletized, the use of wooden or metal pallets shall be assumed in the classification of commodities. When plastic pallets are used, the classification of the commodity unit shall be increased one class (i.e., Class III will become Class IV and Class IV will become Group A plastics). No increase shall be required for Group A plastic commodity.

Exception: When specific test data is available, the data shall take precedence in determining classification of commodities.

This requirement is based on fire products collector testing at Factory Mutual. There are listed plastic pallets that can be treated as standard wood pallets for determining the commodity type. There are also sprinklers listed for storage for which the classification does not have to be increased for plastic pallets.

2-2.3* Commodity Classes.

A-2-2.3 See Table A-2-2.3.

A

2-2.3.1* Class I. A Class I commodity shall be defined as a noncombustible product that meets one of the following criteria:

(1) Placed directly on wooden pallets
(2) Placed in single layer corrugated cartons, with or without single-thickness cardboard dividers, with or without pallets
(3) Shrink-wrapped or paper-wrapped as a unit load with or without pallets

A-2-2.3.1 See Table A-2-2.3.1.

2-2.3.2* Class II. A Class II commodity shall be defined as a noncombustible product that is in slatted wooden crates, solid wood boxes, multiple-layered corrugated cartons, or equivalent combustible packaging material, with or without pallets.

A-2-2.3.2 See Table A-2-2.3.2.

2-2.3.3* Class III. A Class III commodity shall be defined as a product fashioned from wood, paper, natural fibers, or Group C plastics with or without cartons, boxes, or crates and with or without pallets. Such a product shall be permitted to contain a limited amount (5 percent by weight or volume) of Group A or Group B plastics.

A-2-2.3.3 See Table A-2-2.3.3.

Table A-2-2.3 Alphabetized Listing of Commodity Classes

Commodity	Commodity Class
Aerosols	
Cartoned or uncartoned	
- Level 1	Class III
Alcoholic Beverages	
Cartoned or uncartoned	
- Up to 20 percent alcohol in metal, glass, or ceramic containers	Class I
- Up to 20 percent alcohol in wood containers	Class II
Ammunition	
Small arms, shotgun	
- Packaged, cartoned	Class IV
Appliances, Major (i.e., stoves, refrigerators)	
- Not packaged, no appreciable plastic exterior trim	Class I
- Corrugated, cartoned, (no appreciable plastic trim)	Class II
Baked Goods	
Cookies, cakes, pies	
- Frozen, packaged in cartons[1]	Class II
- Packaged, in cartons	Class III
Batteries	
Dry cells (nonlithium or similar exotic metals)	
- Packaged in cartons	Class I
- Blister-packed in cartons	Class II
Automobile	
- Filled[2]	Class I
Truck or larger	
- Empty or filled[2]	Group A plastics
Beans	
Dried	
- Packaged, cartoned	Class III
Bottles, Jars	
Empty, cartoned	
- Glass	Class I
- Plastic PET (polyethylene terephthalate)	Class IV
Filled noncombustible powders	
- Plastic PET	Class II
- Glass, cartoned	Class I
- Plastic, cartoned [less than 1 gal (3.8 L)]	Class IV
- Plastic, uncartoned (other than PET), any size	Group A plastics
- Plastic, cartoned or exposed [greater than 1 gal (3.8 L)]	Group A plastics
- Plastic, solid plastic crates	Group A plastics
- Plastic, open plastic crates	Group A plastics

Table A-2-2.3 Continued.

Commodity	Commodity Class
Filled noncombustible liquids	
- Glass, cartoned	Class I
- Plastic, cartoned [less than 5 gal (18.9 L)]	Class I
- Plastic, open or solid plastic crates3	Group A plastics
- Plastic, PET	Class I
Boxes, Crates	
- Empty, wood, solid walls	Class II
- Empty, wood, slatted[4]	Outside of scope
Bread	
Wrapped cartoned	Class III
Butter	
Whipped spread	Class III
Candles	
Packaged, cartoned	
- Treat as expanded plastic	Group A plastics
Candy	
Packaged, cartoned	Class III
Canned Foods	
In ordinary cartons	Class I
Cans	
Metal	
- Empty	Class I
Carpet Tiles	
Cartoned	Group A plastics
Cartons	
Corrugated	
- Unassembled (neat piles)	Class III
- Partially assembled	Class IV
Wax coated, single walled	Group A plastics
Cement	
Bagged	Class I
Cereals	
Packaged, cartoned	Class III
Charcoal	
Bagged	
- Standard	Class III
Cheese	
- Packaged, cartoned	Class III
- Wheels, cartoned	Class III
Chewing Gum	
Packaged, cartoned	Class III

(continues)

Table A-2-2.3 Continued.

Commodity	Commodity Class
Chocolate	
Packaged, cartoned	Class III
Cloth	
Cartoned and not cartoned	
- Natural fiber, viscose	Class III
- Synthetic[5]	Class IV
Cocoa Products	
Packaged, cartoned	Class III
Coffee	
- Canned, cartoned	Class I
- Packaged, cartoned	Class III
Coffee Beans	
Bagged	Class III
Cotton	
Packaged, cartoned	Class III
Diapers	
- Cotton, linen	Class III
- Disposable with plastics and nonwoven fabric (in cartons)	Class IV
- Disposable with plastics and nonwoven fabric (uncartoned), plastic wrapped	Group A plastics
Dried Foods	
Packaged, cartoned	Class III
Fertilizers	
Bagged	
- Phosphates	Class I
- Nitrates	Class II
Fiberglass Insulation	
- Paper-backed rolls, bagged or unbagged	Class IV
File Cabinets	
Metal	
- Cardboard box or shroud	Class I
Fish or Fish Products	
Frozen	
- Nonwaxed, nonplastic packaging	Class I
- Waxed-paper containers, cartoned	Class II
- Boxed or barreled	Class II
- Plastic trays, cartoned	Class III
Canned	
- Cartoned	Class I
Frozen Foods	
Nonwaxed, nonplastic packaging	Class I
- Waxed-paper containers, cartoned	Class II
- Plastic trays	Class III

Table A-2-2.3 Continued.

Commodity	Commodity Class
Fruit	
Fresh	
- Nonplastic trays or containers	Class I
- With wood spacers	Class I
Furniture	
Wood	
- No plastic coverings or foam plastic cushioning	Class III
- With plastic coverings	Class IV
- With foam plastic cushioning	Group A plastics
Grains—Packaged in Cartons	
- Barley	Class III
- Rice	Class III
- Oats	Class III
Ice Cream	Class I
Leather Goods	Class III
Leather Hides	
Baled	Class II
Light Fixtures	
Nonplastic	
- Cartoned	Class II
Lighters	
Butane	
- Blister-packed, cartoned	Group A plastics
- Loose in large containers (Level 3 aerosol)	Outside of scope
Liquor	
100 proof or less, 1 gal (3.8 L) or less, cartoned	
- Glass (palletized)[6]	Class IV
- Plastic bottles	Class IV
Marble	
Artificial sinks, countertops	
- Cartoned, crated	Class II
Margarine	
- Up to 50 percent oil (in paper or plastic containers)	Class III
- Between 50 percent and 80 percent oil (in any packaging)	Group A plastics
Matches	
Packaged, cartoned	
- Paper	Class IV
- Wood	Group A plastics
Mattresses	
- Standard (box spring)	Class III
- Foam (in finished form)	Group A plastics

(continues)

Table A-2-2.3 Continued.

Commodity	Commodity Class
Meat, Meat Products	
- Bulk	Class I
- Canned, cartoned	Class I
- Frozen, nonwaxed, nonplastic containers	Class I
- Frozen, waxed-paper containers	Class II
- Frozen, expanded plastic trays	Class II
Metal Desks	
- With plastic tops and trim	Class I
Milk	
- Nonwaxed-paper containers	Class I
- Waxed-paper containers	Class I
- Plastic containers	Class I
- Containers in plastic crates	Group A plastics
Motors	
- Electric	Class I
Nail Polish	
- 1-oz to 2-oz (29.6-ml to 59.1-ml) glass, cartoned	Class IV
- 1-oz to 2-oz (29.6-ml to 59.1-ml) plastic bottles, cartoned	Group A plastics
Nuts	
- Canned, cartoned	Class I
- Packaged, cartoned	Class III
- Bagged	Class III
Paints	
Friction-top cans, cartoned	
- Water-based (latex)	Class I
- Oil-based	Class IV
Paper Products	
- Books, magazines, stationery, plastic-coated paper food containers, newspapers, cardboard games, or cartoned tissue products	Class III
- Tissue products, uncartoned and plastic wrapped	Group A plastics
Paper, Rolled	
In racks or on side	
- Medium- or heavyweight	Class III
In racks	
- Lightweight	Class IV
Paper, Waxed	
Packaged in cartons	Class IV
Pharmaceuticals	
Pills, powders	
- Glass bottles, cartoned	Class II
- Plastic bottles, cartoned	Class IV
Nonflammable liquids	
- Glass bottles, cartoned	Class II

Table A-2-2.3 Continued.

Commodity	Commodity Class
Photographic Film	
- Motion picture or bulk rolls of film in polycarbonate, polyethylene, or metal cans; polyethylene bagged in cardboard boxes	Class II
- 35-mm in metal film cartridges in polyethylene cans in cardboard boxes	Class III
- Paper, in sheets, bagged in polyethylene, in cardboard boxes	Class III
- Rolls in polycarbonate plastic cassettes, bulk wrapped in cardboard boxes	Class IV
Plastic Containers (except PET)	
- Noncombustible liquids or semiliquids in plastic containers less than 5 gal (18.9 L) capacity	Class I
- Noncombustible liquids or semiliquids (such as ketchup) in plastic containers with nominal wall thickness of 1/4 in. (6.4 mm) or less and larger than 5 gal (18.9) capacity	Class II
- Noncombustible liquids or semiliquids (such as ketchup) in plastic containers with nominal wall thickness greater than 1/4 in. (6.4 mm) and larger than 5 gal (18.9 L) capacity	Group A plastics
Polyurethane	
- Cartoned or uncartoned expanded	Group A plastics
Poultry Products	
- Canned, cartoned	Class I
- Frozen, nonwaxed, nonplastic containers	Class I
- Frozen (on paper or expanded plastic trays)	Class II
Powders	
Ordinary combustibles—free flowing	
- In paper bags (i.e., flour, sugar)	Class II
PVA (polyvinyl alcohol) Resins	
PVC (polyvinyl chloride)	
- Flexible (e.g., cable jackets, plasticized sheets)	Class III
- Rigid (e.g., pipe, pipe fittings)	Class III
- Bagged resins	Class III
Rags	
Baled	
- Natural fibers	Class III
- Synthetic fibers	Class IV
Rubber	
- Natural, blocks in cartons	Class IV
- Synthetic	Group A plastics
Salt	
- Bagged	Class I
- Packaged, cartoned	Class II
Shingles	
- Asphalt-coated fiberglass	Class III
- Asphalt-impregnated felt	Class IV

(continues)

Table A-2-2.3 Continued.

Commodity	Commodity Class
Shock Absorbers	
- Metal dust cover	Class II
- Plastic dust cover	Class III
Signatures	
Books, magazines	
- Solid array on pallet	Class II
Skis	
- Wood	Class III
- Foam core	Class IV
Stuffed Toys	
Foam or synthetic	Group A plastics
Syrup	
- Drummed (metal containers)	Class I
- Barreled, wood	Class II
Textiles	
Natural fiber clothing or textile products	Class III
Synthetics (except rayon and nylon)—50/50 blend or less	
- Thread, yarn on wood or paper spools	Class III
- Fabrics	Class III
- Thread, yarn on plastic spools	Class IV
- Baled fiber	Group A plastics
Synthetics (except rayon and nylon)—greater than 50/50 blend	
- Thread, yarn on wood or paper spools	Class IV
- Fabrics	Class IV
- Baled fiber	Group A plastics
- Thread, yarn on plastic spools	Group A plastics
Rayon and nylon	
- Baled fiber	Class IV
- Thread, yarn on wood or paper spools	Class IV
- Fabrics	Class IV
- Thread, yarn on plastic spools	Group A plastics
Tobacco Products	
In paperboard cartons	Class III
Transformers	
Dry and oil filled	Class I
Vinyl-Coated Fabric	
Cartoned	Group A plastics
Vinyl Floor Coverings	
- Tiles in cartons	Class IV
- Rolled	Group A plastics
Wax-Coated Paper	
Cups, plates	
- Boxed or packaged inside cartons (emphasis on packaging)	Class IV
- Loose inside large cartons	Group A plastics

Table A-2-2.3 Continued.

Commodity	Commodity Class
Wax	
Paraffin, blocks, cartoned	Group A plastics
Wire	
- Bare wire on metal spools on wood skids	Class I
- Bare wire on wood or cardboard spools on wood skids	Class II
- Bare wire on metal, wood, or cardboard spools in cardboard boxes on wood skids	Class II
- Single- or multiple-layer PVC-covered wire on metal spools on wood skids	Class II
- Insulated (PVC) cable on large wood or metal spools on wood skids	Class II
- Bare wire on plastic spools in cardboard boxes on wood skids	Class IV
- Single- or multiple-layer PVC-covered wire on plastic spools in cardboard boxes on wood skids	Class IV
- Single, multiple, or power cables (PVC) on large plastic spools	Class IV
- Bulk storage of empty plastic spools	Group A plastics
Wood Products	
- Solid piles—lumber, plywood, particleboard, pressboard (smooth ends and edges)	Class II
- Spools (empty)	Class III
- Toothpicks, clothespins, hangers in cartons	Class III
- Doors, windows, wood cabinets, and furniture	Class III
- Patterns	Class IV

[1]The product is presumed to be in a plastic-coated package in a cor rugated carton. If packaged in a metal foil, it can be considered Class I.

[2]Most batteries have a polypropylene case and, if stored empty, should be treated as a Group A plastic. Truck batteries, even where filled, should be considered a Group A plastic because of their thicker walls.

[3]As the openings in plastic crates become larger, the product be haves more like a Class III commodity. Conversely, as the openings become smaller, the product behaves more like a plastic.

[4]These items should be treated as idle pallets.

[5]Tests clearly indicate that a synthetic or synthetic blend is consid ered greater than Class III.

[6]When liquor is stored in glass containers in racks, it should be con sidered a Class III commodity; where it is palletized, it should be con sidered a Class IV commodity.

Table A-2-2.3.1 Examples of Class I Commodities

Alcoholic Beverages
  Cartoned or uncartoned
  - Up to 20 percent alcohol in metal, glass, or ceramic containers
Appliances, Major (i.e., stoves, refrigerators)
  - Not packaged, no appreciable plastic exterior trim
Batteries
  Dry cells (nonlithium or similar exotic metals)
  - Packaged in cartons
  Automobile
  - Filled*
Bottles, Jars
  Empty, cartoned
  - Glass
  Filled noncombustible liquids
  - Glass, cartoned
  - Plastic, cartoned [less than 5 gal (18.9 L)]
  - Plastic, PET
  Filled noncombustible powders
  - Glass, cartoned
Canned Foods
  In ordinary cartons
Cans
  Metal
  - Empty
Cement
  Bagged
Coffee
  Canned, cartoned
Fertilizers
  Bagged
  - Phosphates
File Cabinets
  Metal
  - Cardboard box or shroud
Fish or Fish Products
  Frozen
  - Nonwaxed, nonplastic packaging
  Canned
  - Cartoned
Frozen Foods
  Nonwaxed, nonplastic packaging
Fruit
  Fresh
  - Nonplastic trays or containers
  - With wood spacers
Ice Cream
Meat, Meat Products
  - Bulk
  - Canned, cartoned
  - Frozen, nonwaxed, nonplastic containers
Metal Desks
  - With plastic tops and trim
Milk
  - Nonwaxed-paper containers
  - Waxed-paper containers
  - Plastic containers
Motors
  - Electric
Nuts
  - Canned, cartoned
Paints
  Friction-top cans, cartoned
  - Water-based (latex)
Plastic Containers
  - Noncombustible liquids or semiliquids in plastic containers less than 5 gal (18.9 L) capacity
Poultry Products
  - Canned, cartoned
  - Frozen, nonwaxed, nonplastic containers
Salt
  Bagged
Syrup
  Drummed (metal containers)
Transformers
  Dry and oil filled
Wire
  Bare wire on metal spools on wood skids

*Most batteries have a polypropylene case and, if stored empty, should be treated as a Group A plastic. Truck batteries, even where filled, should be considered a Group A plastic because of their thicker walls.

Table A-2-2.3.2 Examples of Class II Commodities

Alcoholic Beverages
  Up to 20 percent alcohol in wood containers
Appliances, major (e.g., stoves)
  Corrugated, cartoned (no appreciable plastic trim)
Baked Goods
  Cookies, cakes, pies
  - Frozen, packaged in cartons*
Batteries
  Dry cells (nonlithium or similar exotic metals) in blister pack in cartons
Bottles, Jars
  Filled noncombustible powders
  - Plastic PET
Boxes, Crates
  Empty, wood, solid walls
Fertilizers
  Bagged
  - Nitrates
Fish or Fish Products
  Frozen
  - Waxed-paper containers, cartoned
  - Boxed or barreled
Frozen Foods
  Waxed-paper containers, cartoned
Leather Hides
  Baled
Light Fixtures
  Nonplastic
  - Cartoned
Marble
  Artificial sinks, countertops
  - Cartoned, crated
Meat, Meat Products
  - Frozen, waxed-paper containers
  - Frozen, expanded plastic trays
Pharmaceuticals
  Pills, powders
  - Glass bottles, cartoned
  Nonflammable liquids
  - Glass bottles, cartoned
Photographic Film
  - Motion picture or bulk rolls of film in polycarbonate, polyethylene, or metal cans; polyethylene bagged in cardboard boxes
Plastic Containers
  Noncombustible liquids or semiliquids (such as ketchup) in plastic containers with nominal wall thickness of 1/4 in. (6.4 mm) or less and larger than 5 gal (18.9 L) capacity
Poultry Products
  Frozen (on paper or expanded plastic trays)
Powders (ordinary combustibles—free flowing)
  In paper bags (i.e., flour, sugar)
Salt
  Packaged, cartoned
Shock Absorbers
  Metal dust cover
Signatures
  Book, magazines
  - Solid array on pallet
Syrup
  Barreled, wood
Wire
  - Bare wire on wood or cardboard spools on wood skids
  - Bare wire on metal, wood, or cardboard spools in cardboard boxes on wood skids
  - Single- or multiple-layer PVC-covered wire on metal spools on wood skids
  - Insulated (PVC) cable on large wood or metal spools on wood skids
Wood Products
  Solid piles
  - Lumber, plywood, particle board, pressboard (smooth ends and edges)

*The product is in a plastic-coated package in a corrugated carton. If packaged in a metal foil, it can be considered Class I.

Table A-2-2.3.3 Examples of Class III Commodities

- Aerosols
  - Cartoned or uncartoned
    - Level 1
- Baked Goods
  - Cookies, cakes, pies
    - Packaged, in cartons
- Beans
  - Dried
    - Packaged, cartoned
- Bread
  - Wrapped, cartoned
- Butter
  - Whipped spread
- Candy
  - Packaged, cartoned
- Cartons
  - Corrugated
    - Unassembled (neat piles)
- Cereals
  - Packaged, cartoned
- Charcoal
  - Bagged
    - Standard
- Cheese
  - Packaged, cartoned
  - Wheels, cartoned
- Chewing Gum
  - Packaged, cartoned
- Chocolate
  - Packaged, cartoned
- Cloth
  - Cartoned and not cartoned
    - Natural fiber, viscose
- Cocoa Products
  - Packaged, cartoned
- Coffee
  - Packaged, cartoned
- Coffee Beans
  - Bagged
- Cotton
  - Packaged, cartoned
- Diapers
  - Cotton, linen
- Dried Foods
  - Packaged, cartoned
- Fish or Fish Products
  - Frozen
    - Plastic trays, cartoned
- Frozen Foods
  - Plastic trays
- Furniture
  - Wood
    - No plastic coverings or foam plastic cushioning
- Grains—Packaged in cartons
  - Barley
  - Rice
  - Oats
- Margarine
  - Up to 50 percent oil (in paper or plastic containers)
- Mattresses
  - Standard (box spring)
- Nuts
  - Packaged, cartoned
  - Bagged
- Paper Products
  - Books, magazines, stationery, plastic-coated paper food containers, newspapers, cardboard games, cartoned tissue products
- Paper, Rolled
  - In racks or on side
    - Medium or heavyweight
- Photographic Film
  - 35-mm in metal film cartridges in polyethylene cans in cardboard boxes
  - Paper, in sheets, bagged in polyethylene, in cardboard boxes
- PVC (polyvinyl chloride)
  - Flexible (e.g., cable jackets, plasticized sheets)
  - Rigid (e.g., pipe, pipe fittings)
  - Bagged resins
- Rags
  - Baled
    - Natural fibers
- Shingles
  - Asphalt-coated fiberglass
- Shock Absorbers
  - Plastic dust cover
- Skis
  - Wood
- Textiles
  - Natural fiber clothing or textile products
- Synthetics (except rayon and nylon)—
  - 50/50 blend or less
    - Thread, yarn on wood or paper spools
    - Fabrics
- Tobacco Products
  - In paperboard cartons
- Wood Products
  - Spools (empty)
  - Toothpicks, clothespins, hangers in cartons
  - Doors, windows, wood cabinets, and furniture

2-2.3.4* Class IV. A Class IV commodity shall be defined as a product, with or without pallets, that meets one of the following criteria:

(1) Constructed partially or totally of Group B plastics.
(2) Consists of free-flowing Group A plastic materials.
(3) Contains within itself or its packaging an appreciable amount (5 percent to 15 percent by weight or 5 percent to 25 percent by volume) of Group A plastics. The remaining materials shall be permitted to be metal, wood, paper, natural or synthetic fibers, or Group B or Group C plastics.

A-2-2.3.4 See Table A-2-2.3.4.

The designer needs to consider the location of the plastic commodity within a carton. A cartoned commodity that is 15 percent plastic by weight where all of the nonplastic content is surrounded by plastic behaves as a plastic commodity in a fire. Likewise, a heavy metal item encased in expanded plastic packaging that occupies 25 percent of the carton behaves as an expanded plastic commodity in a fire.

2-2.4* Classification of Plastics, Elastomers, and Rubber.

Plastics, elastomers, and rubber shall be classified as Group A, Group B, or Group C.

A-2-2.4 The categories listed in 2-2.4.1, 2-2.4.2, and 2-2.4.3 are based on unmodified plastic materials. The use of fire- or flame-retarding modifiers or the physical form of the material could change the classification.

2-2.4.1* Group A. The following materials shall be classified as Group A:

ABS (acrylonitrile-butadiene-styrene copolymer)
Acetal (polyformaldehyde)
Acrylic (polymethyl methacrylate)
Butyl rubber
EPDM (ethylene-propylene rubber)
FRP (fiberglass-reinforced polyester)
Natural rubber (if expanded)
Nitrile rubber (acrylonitrile-butadiene rubber)
PET (thermoplastic polyester)
Polybutadiene
Polycarbonate
Polyester elastomer
Polyethylene
Polypropylene
Polystyrene
Polyurethane
PVC (polyvinyl chloride—highly plasticized, with plasticizer content greater than 20 percent) (rarely found)
SAN (styrene acrylonitrile)
SBR (styrene-butadiene rubber)

Table A-2-2.3.4 Examples of Class IV Commodities

Ammunition
  Small arms, shotgun
  - Packaged, cartoned
Bottles, Jars
  Empty, cartoned
  - Plastic PET (polyethylene terephthalate)
  Filled noncombustible powders
  - Plastic, cartoned [less than 1 gal (3.8 L)]
Cartons
  Corrugated
  - Partially assembled
Cloth
  Cartoned and not cartoned
  - Synthetic[1]
Diapers
  - Disposable with plastics and nonwoven fabric (in cartons)
Fiberglass Insulation
  - Paper-backed rolls, bagged or unbagged
Furniture
  Wood
  - With plastic coverings
Liquor
  100 proof or less, 1 gal (3.8 L) or less, cartoned
  - Glass (palletized)[2]
  - Plastic bottles
Matches
  Packaged, cartoned
  - Paper
Nail Polish
  1-oz to 2-oz (29.6-ml to 59.1-ml) glass, cartoned
Paints
  Friction-top cans, cartoned
  - Oil based
Paper, Rolled
  In racks
  - Lightweight
Paper, Waxed
  Packaged in cartons
Pharmaceuticals
  Pills, powders
  - Plastic bottles, cartoned
Photographic Film
  - Rolls in polycarbonate plastic cassettes, bulk wrapped in cardboard boxes
PVA (polyvinyl alcohol) Resins
  Bagged
Rags
  Baled
  - Synthetic fibers
Rubber
  Natural, blocks in cartons
Shingles
  Asphalt-impregnated felt
Skis
  Foam core
Textiles
  Synthetics (except rayon and nylon)—50/50 blend or less
  - Thread, yarn on plastic spools
  Synthetics (except rayon and nylon)—greater than 50/50 blend
  - Thread, yarn on wood or paper spools
  - Fabrics
Rayon and nylon
  - Baled fiber
  - Thread, yarn on wood or paper spools
  - Fabrics
Vinyl Floor Coverings
  Tiles in cartons
Wax-Coated Paper
  Cups, plates
  - Boxed or packaged inside cartons (emphasis is on packaging)
Wire
  - Bare wire on plastic spools in cardboard boxes on wood skids
  - Single- or multiple-layer PVC-covered wire on plastic spools in cardboard boxes on wood skids
  - Single, multiple, or power cables (PVC) on large plastic spools
Wood Products
  Patterns

[1]Tests clearly indicate that a synthetic or synthetic blend is considered greater than Class III.

[2]Where liquor is stored in glass containers in racks, it should be considered a Class III commodity; where it is palletized, it should be considered a Class IV commodity.

A-2-2.4.1 See Table A-2-2.4.1.

2-2.4.2 Group B. The following materials shall be classified as Group B:

Cellulosics (cellulose acetate, cellulose acetate butyrate, ethyl cellulose)

Chloroprene rubber

Fluoroplastics (ECTFE—ethylene-chlorotrifluoro-ethylene copolymer; ETFE—ethylene-tetrafluoroethylene-copolymer; FEP—fluorinated ethylene-propylene copolymer)

Natural rubber (not expanded)

Nylon (nylon 6, nylon 6/6)

Silicone rubber

2-2.4.3 Group C. The following materials shall be classified as Group C:

Fluoroplastics (PCTFE—polychlorotrifluoroethylene; PTFE—polytetrafluoroethylene)

Melamine (melamine formaldehyde)

Phenolic

PVC (polyvinyl chloride—flexible—PVCs with plasticizer content up to 20 percent)

PVDC (polyvinylidene chloride)

PVDF (polyvinylidene fluoride)

PVF (polyvinyl fluoride)

Urea (urea formaldehyde)

2-2.5* Classification of Rolled Paper Storage.

For the purposes of this standard, the following classifications of paper shall apply. These classifications shall be used to determine the sprinkler system design criteria.

A-2-2.5 Paper Classification. These classifications were derived from a series of large-scale and laboratory-type small-scale fire tests. It is recognized that not all paper in a class burns with exactly the same characteristics.

Paper can be soft or hard, thick or thin, or heavy or light and can also be coated with various materials. The broad range of papers can be classified according to various properties. One important property is basis weight, which is defined as the weight of a sheet of paper of a specified area. Two broad categories are recognized by industry—paper and paperboard. Paperboard normally has a basis weight of 20 lb (9.1 kg) or greater measured on a 1000-ft^2 (92.9-m^2) sheet. Stock with a basis weight less than 20 lb/1000 ft^2 (9.1 kg/92.9 m^2) is normally categorized as paper. The basis weight of paper is usually measured on a 3000-ft^2 (278.7-m^2) sheet. The basis weight of paper can also be measured on the total area of a ream of paper, which is normally the case for the following types of printing and writing papers:

(1) *Bond paper*—500 sheets, 17 in. × 22 in. (432 mm × 559 mm) = 1300 ft^2 (120.8 m^2) per ream
(2) *Book paper*—500 sheets, 25 in. × 38 in. (635 mm × 965 mm) = 3300 ft^2 (306.6 m^2) per ream
(3) *Index paper*—500 sheets, 25$^1/_2$ in. × 30$^1/_2$ in. (648 mm × 775 mm) = 2700 ft^2 (250.8 m^2) per ream
(4) *Bristol paper*—500 sheets, 22$^1/_2$ in. × 35 in. (572 mm × 889 mm) = 2734 ft^2 (254 m^2) per ream
(5) *Tag paper*—500 sheets, 24 in. × 36 in. (610 mm × 914 mm) = 3000 ft^2 (278.7 m^2) per ream

For the purposes of this standard, all basis weights are expressed in lb/1000 ft^2 (kg/92.9 m^2) of paper. To determine the basis weight per 1000 ft^2 (92.9 m^2) for papers measured on a sheet of different area, the following formula should be applied:

$$\frac{\text{Basis weight}}{1000\text{ft}^2} = \text{basis weight} \times 1000 \text{ measured area}$$

Example: To determine the basis weight per 1000 ft^2 (92.9 m^2) of 16-lb (7.3-kg) bond paper:

$$\left(\frac{16\text{lb}}{1300\ \text{ft}^2}\right) 1000 = \frac{12.3\ \text{lb}}{1000\ \text{ft}^2}$$

Large- and small-scale fire tests indicate that the burning rate of paper varies with the basis weight. Heavyweight paper burns more slowly than lightweight paper. Full-scale roll paper fire tests were conducted with the following types of paper:

(1) *Linerboard*—42 lb/1000 ft^2 (19.1 kg/92.9 m^2) nominal basis weight

(2) *Newsprint*—10 lb/1000 ft^2 (4.5 kg/92.9 m^2) nominal basis weight

(3) *Tissue*—5 lb/1000 ft^2 (2.3 kg/ 92.9 m^2) nominal basis weight

The rate of firespread over the surface of the tissue rolls was extremely rapid in the full-scale fire tests. The rate of firespread over the surface of the linerboard rolls was slower. Based on the overall results of these full-scale tests, along with additional data from small-scale testing of various paper grades, the broad range of papers has been classified into three major categories as follows:

(1) *Heavyweight*—Basis weight of 20 lb/1000 ft^2 (9.1 kg/ 92.9 m^2) or greater
(2) *Mediumweight*—Basis weight of 10 lb to 20 lb/1000 ft^2 (4.5 kg to 9.1 kg/ 92.9 m^2)
(3) *Lightweight*—Basis weight of less than 10 lb/1000 ft^2 (4.5 kg/92.9 m^2) and tissues regardless of basis weight

Table A-2-2.4.1 Examples of Group A Plastic Commodities

Batteries
  Truck or larger
  - Empty or filled[1]
Bottles, Jars
  Empty, cartoned
  - Plastic (other than PET), any size
  Filled noncombustible liquids
  - Plastic, open or solid plastic crates[2]
  Filled noncombustible powders
  - Plastic, cartoned or uncartoned [greater than 1 gal (3.8 L)]
  - Plastic, solid plastic crates
  - Plastic, open plastic crates
Candles
  Packaged, cartoned
  - Treat as expanded plastic
Carpet Tiles
  Cartoned
Cartons
  Wax coated, single walled
Diapers
  Disposable with plastics and nonwoven fabric (uncartoned), plastic wrapped
Furniture
  Wood
  - With foam plastic cushioning
Lighters
  Butane
  - Blister-packed, cartoned
Margarine
  Between 50 percent and 80 percent oil (in any packaging)
Matches
  Packaged, cartoned
  - Wood
Mattresses
  Foam (in finished form)
Milk
  Containers in plastic crates
Nail Polish
  1-oz to 2-oz (29.6-ml to 59.1-ml) plastic bottles, cartoned
Paper Products
  Tissue products, uncartoned and plastic wrapped
Plastic Containers
  - Combustible or noncombustible solids in plastic containers and empty plastic containers
  - Noncombustible liquids or semiliquids (such as ketchup) in plastic containers with nominal wall thickness greater than ¼ in. (6.4 mm) and larger than 5 gal (18.9 L) capacity
Polyurethane
  Cartoned or uncartoned expanded
Rubber
  Synthetic
Stuffed Toys
  Foam or synthetic
Textiles
  Synthetics (except rayon and nylon)—50/50 blend or less
  - Baled fiber
  Synthetics (except rayon are nylon) greater than 50/50 blend
  - Baled fiber
  - Thread, yarn on plastic spools
  Rayon and nylon
  - Thread, yarn on plastic spools
Vinyl-Coated Fabric
  Cartoned
Vinyl Floor Coverings
  Rolled
Wax-Coated Paper
  Cups, plates
  - Loose inside large cartons
Wax
  Paraffin, blocks, cartoned
Wire
  Bulk storage of empty plastic spools

[1]Most batteries have a polypropylene case and, if stored empty, should be treated as a Group A plastic. Truck batteries, even where filled, should be considered a Group A plastic because of their thicker walls.

[2]As the openings in plastic crates become larger, the product be haves more like Class III. Conversely, as the openings become smaller, the product makeup behaves more like a plastic.

The following SI units were used for conversion of English units:

1 lb = 0.454 kg

1 in. = 25.4 mm

1 ft = 0.3048 m

1 ft^2 = 0.0929 m^2

The various types of papers normally found in each of the four major categories are provided in Table A-2.2.5.

Table A-2-2.5 Paper Classification

Heavyweight	Mediumweight	Lightweight	Tissue
Linerboards	Bond and reproduction	Carbonizing tissue	Toilet tissue
Medium	Vellum	Cigarette	Towel tissue
Kraft roll wrappers	Offset	Fruit wrap	
Milk carton board	Tablet	Onion skin	
Folding carton board	Computer		
Bristol board	Envelope		
Tag	Book		
Vellum bristol board	Label		
Index	Magazine		
Cupstock	Butcher		
Pulp board	Bag		
	Newsprint (unwrapped)		

Paragraph A-2-2.5 applies to the classification of large rolls of paper stored on end or on side in piles greater than 10 ft (3.1 m) high. This type of paper storage is typically found in the finished goods storage warehouses at paper mills, in the raw stock storage warehouses of manufacturing plants that convert paper into finished products, or in other warehouses of similar occupancy. Fire experience and full-scale fire testing of this type of paper storage, as indicated in Exhibits 2.3 and 2.4, demonstrate unique burning characteristics that vary with the type of paper and the method of wrapping. The classifications for this type of paper storage are based on a series of full-scale and small-scale fire tests as described in A-2-2.5.

2-2.5.1 Heavyweight Class. Heavyweight class shall include paperboard and paper stock having a basis weight [weight per 1000 ft^2 (92.9 m^2)] of 20 lb (9.1 kg).

2-2.5.2 Mediumweight Class. Mediumweight class shall include all the broad range of papers having a basis weight [weight per 1000 ft^2 (92.9 m^2)] of 10 lb to 20 lb (4.5 kg to 9.1 kg).

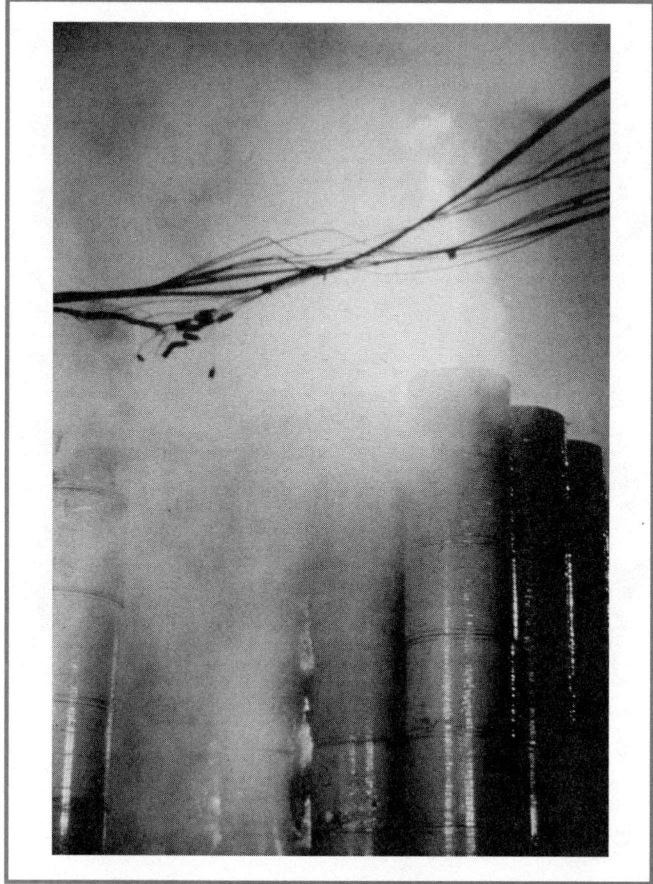

Exhibit 2.3 *Fire test of roll paper using large drop sprinklers.*

2-2.5.3 Lightweight Class. Lightweight class shall include all papers having a basis weight [weight per 1000 ft^2 (92.9 m^2)] of 10 lb (4.5 kg).

2-2.5.4 Tissue. Tissue shall include the broad range of papers of characteristic gauzy texture, which, in some cases, are fairly transparent. For the purposes of this standard, tissue shall be defined as the soft, absorbent type, regardless of basis weight—specifically, crepe wadding and the sanitary class including facial tissue, paper napkins, bathroom tissue, and toweling.

Exhibit 2.4 Fire test of roll paper using standard spray sprinklers.

REFERENCE CITED IN COMMENTARY

National Fire Protection Association, 1 Batterymarch Park, P.O. Box 9101, Quincy, MA 02269-9101.

NFPA *101®*, *Life Safety Code®*, 1997 edition.

CHAPTER 3

System Components and Hardware

3-1 General

This chapter provides requirements for correct use of sprinkler system components.

Chapter 3 identifies the types of materials and components that are acceptable for use in a sprinkler system and discusses their associated features and limitations. Information concerning proper installation of system materials and components is contained in Chapter 5.

3-1.1* All materials and devices essential to successful system operation shall be listed.

Exception No. 1: Equipment as permitted in Table 3-3.1, Table 3-5.1, and the Exceptions to 6-1.1 and 6-1.1.1 shall not be required to be listed.

Exception No. 2: Components that do not affect system performance such as drain piping, drain valves, and signs shall not be required to be listed. The use of reconditioned valves and devices other than sprinklers as replacement equipment in existing systems shall be permitted.

A-3-1.1 Included among items requiring listing are sprinklers, some pipe and some fittings, hangers, alarm devices, valves controlling flow of water to sprinklers, valve tamper switches, and gauges.

The ability of a sprinkler system to achieve fire control or suppression depends on a number of factors including the use of effective and reliable system components. To ensure a sufficient degree of sprinkler system reliability and performance, NFPA 13 requires that those components critical to system performance during a fire be listed (see 1-4.1 for definition of *Listed*). Components that would not adversely affect system performance if they were to malfunction during a fire are not required to be listed, but they must be approved.

The requirement of listed products and materials is intended to increase the likelihood that the system will perform satisfactorily when needed. For example, over 50 different types of tests are performed on a specific make of sprinkler during the listing process. These tests evaluate a number of performance features such as spray pattern distribution, temperature, pressure, and orifice and operating element.

Exception No. 1 does not require certain materials, such as Schedule 10, Schedule 30, and Schedule 40 steel pipe, copper tubing, and some of the fittings used with these pipes, to be listed for fire protection service. Pipe manufactured to particular ASTM standards and fittings made to particular ASME standards are considered sufficiently reliable. Additionally, many years of positive experience with these materials in sprinkler systems and in other types of mechanical systems supports their acceptable performance level.

Experience also demonstrates that mild steel rods of the sizes specified in NFPA 13 do not need to be listed. Exception No. 1 also recognizes that listed hangers cannot be used in certain applications and permits hanger assemblies to be designed by a registered professional engineer for these special situations.

Exception No. 2 recognizes that the listing of all sprinkler system components is unnecessary to increase overall system reliability during a fire. For example, the inspector's test valve is not critical to system performance during a fire and does not require listing.

The second sentence of Exception No. 2 allows certain components to be reconditioned and reused, provided the components retain their listing. Sprinklers are intentionally omitted from this exception, and only new sprinklers are to be used. However, when sprinklers have been installed in a building on a temporary basis—for example, prior to finish ceiling work—the same sprinklers can be utilized on that job. The same holds true for an existing system that is being lowered to accommodate a new ceiling provided the sprinklers are of the proper orientation and the associated hazard has not changed. Unfavorable past experience with the use of reconditioned sprinklers provides the basis for this requirement.

3-1.2 System components shall be rated for the maximum system working pressure to which they are exposed but shall not be rated at less than 175 psi (12.1 bar).

Since 175 psi (12.1 bar) presents a limiting criteria, many valves and other system components are listed to this pressure. Where system pressures are expected to exceed 175 psi (12.1 bar), system components and materials manufactured and listed for higher pressures must be used. Systems that do not incorporate a fire pump or are not part of a combined standpipe system do not typically experience pressures in excess of 175 psi (12.1 bar). However, each system must be evaluated on an individual basis because the presence of a fire department connection introduces the possibility of high pressures being applied by fire department apparatus.

3-2 Sprinklers

3-2.1 Only new sprinklers shall be installed.

Prompt sprinkler actuation during the early stages of a fire is critical to the successful performance of the entire sprinkler system. The sprinkler should not leak or rupture when installed, nor should it operate for any reason other than in response to a fire. A new listed sprinkler has been thoroughly evaluated to ensure that it will operate only when needed and in an appropriate manner.

3-2.2* Sprinkler Identification.

All sprinklers shall be permanently marked with a one- or two-character manufacturer symbol, followed by up to four numbers, so as to identify a unique model of sprinkler for every change in orifice size or shape, deflector characteristic, and thermal sensitivity. This rule shall become effective on January 1, 2001.

A-3-2.2 The four- to six-character identification number, with no intervening spaces, is intended to identify the sprinkler operating characteristics. The number, marked on the deflector of most sprinklers and elsewhere on decorative ceiling sprinklers, consists of one or two characters identifying the manufacturer, followed by three or four digits identifying the model.

Sprinkler manufacturers have identified their manufacturer designations for the listing organizations. Each change in orifice size, response characteristics, or deflector (distribution) characteristics results in a new model number. The model numbers do not identify specific characteristics of sprinklers but can be referenced in the database information compiled by the listing organizations. At the plan review stage, the model number should be checked against such a database or the manufacturer's literature to ensure that sprinklers are being used properly and within the limitations of their listings. Field inspections can include spot checks to ensure that the model numbers on the plans are those actually installed.

The 1999 edition introduced a requirement for sprinkler identification markings to minimize the confusion resulting from the growing number and varieties of available sprinklers. Implementation was delayed until January 1, 2000, to allow time for the manufacturers to meet compliance requirements of the testing laboratories and to retool their equipment.

3-2.3 Sprinkler Discharge Characteristics.

3-2.3.1* The K-factor, relative discharge, and marking identification for sprinklers having different orifice sizes shall be in accordance with Table 3-2.3.1.

Exception No. 1: Listed sprinklers having pipe threads different from those shown in Table 3-2.3.1 shall be permitted.

Exception No. 2: Sprinklers listed with nominal K-factors greater than 28 shall increase the flow by 100 percent increments when compared with a nominal K-5.6 sprinkler.

Exception No. 3: Residential sprinklers shall be permitted with K-factors other than those specified in Table 3-2.3.1.

Table 3-2.3.1 Sprinkler Discharge Characteristics Identification

Nominal K-factor gpm/(psi)$^{1/2}$	K-factor Range gpm/(psi)$^{1/2}$	K-factor Range dm^3/min/(kPa)$^{1/2}$	Percent of Nominal K-5.6 Discharge	Thread Type
1.4	1.3–1.5	1.9–2.2	25	½ in. NPT
1.9	1.8–2.0	2.6–2.9	33.3	½ in. NPT
2.8	2.6–2.9	3.8–4.2	50	½ in. NPT
4.2	4.0–4.4	5.9–6.4	75	½ in. NPT
5.6	5.3–5.8	7.6–8.4	100	½ in. NPT
8.0	7.4–8.2	10.7–11.8	140	¾ in. NPT or ½ in. NPT
11.2	11.0–11.5	15.9–16.6	200	½ in. NPT or ¾ in. NPT
14.0	13.5–14.5	19.5–20.9	250	¾ in. NPT
16.8	16.0–17.6	23.1–25.4	300	¾ in. NPT
19.6	18.6–20.6	27.2–30.1	350	1 in. NPT
22.4	21.3–23.5	31.1–34.3	400	1 in. NPT
25.2	23.9–26.5	34.9–38.7	450	1 in. NPT
28.0	26.6–29.4	38.9–43.0	500	1 in. NPT

A-3-2.3.1 See Table A-3-2.3.1.

Beginning with the 1999 edition, sprinkler orifice sizes will be identified only according to the sprinkler's nominal K-factor. The previous nominal orifice sizes specified in NFPA 13 are retained in Table A-3-2.3.1 to assist the user in making the transition from nominal orifice sizes to nominal K-factors.

Any given sprinkler will have a specific K-factor that will differ from the nominal K-factor sizes. However, the K-factor must lie within the ranges provided in Table 3-2.3.1. The nominal K-factor indicated in the table must be used to determine the sprinkler's flow rate at a particular pressure. This use of the nominal K-factor is of major importance for hydraulically designed systems. (See Chapter 8.)

Exhibits 3.1 and 3.2 illustrate the differences in orifice sizes among sprinklers with K-factors of 2.8, 5.6, and 25.2.

Prior to the 1999 edition, sprinklers with a nominal K-factor of other than 5.6 and a ½-in. (12.7-mm) National Pipe Thread (NPT) or sprinklers with a nominal K-factor of other than 8.0 and a ¾-in. pipe thread were identified with pintles. A pintle is a small metal protrusion extending from the sprinkler's deflector. It serves no function

Table A-3-2.3.1 Nominal Sprinkler Orifice Sizes

Nominal K-factor	Nominal Orifice Size in.	Nominal Orifice Size mm
1.4	1/4	6.4
1.9	5/16	8.0
2.8	3/8	9.5
4.2	7/16	11.0
5.6	1/2	12.7
8.0	17/32	13.5
11.2	5/8	15.9
14.0	3/4	19.0
16.8	—	—
19.6	—	—
22.4	—	—
25.2	—	—
28.0	—	—

Exhibit 3.1 *Sprinklers with K-factors of 2.8, 5.6, and 25.2.*

other than to provide some indication of the sprinkler's orifice size. Because so many sprinkler sizes are produced, the pintle is virtually meaningless and is no longer used.

According to Exception No. 2, all newly developed sprinklers need to be manufactured in increments that represent a 100 percent increase in flow when compared to a

Exhibit 3.2 *Orifices of sprinklers with K-factors of 5.6, 25.2, and 2.8.*

sprinkler with a nominal K-factor of 5.6. This exception is an effort by the technical committee to encourage new technology while, at the same time, it addresses the need for interchangeability between product lines. Specifying the K-factors that can be used by sprinklers in the future removes the possibility of the manufacture of odd sprinkler sizes that would present replacement problems.

Because residential sprinklers are intended for specific life safety objectives in certain building occupancies, these sprinklers are not subject to the K-factor range limitation specified in Table 3-2.3.1 according to Exception No. 3.

3-2.3.2 Large drop and ESFR sprinklers shall have a minimum nominal K-factor of 11.2. The ESFR sprinkler orifice shall be selected as appropriate for the hazard. *(See Chapter 7.)*

This section specifies a minimum nominal K-factor for large drop and ESFR sprinklers. This requirement establishes a degree of consistency with regard to these sprinkler's performance objectives. In addition, 3-2.3.2 alludes to the fact that there are different varieties of ESFR sprinklers and that they are not all intended to protect against the same types of hazards. Obviously, the application of an ESFR sprinkler beyond the limitations of its listing is very likely to result in an unfavorable outcome. Exhibits 1.9 and 1.10 illustrate ESFR sprinklers, and Exhibit 1.13 illustrates a large drop sprinkler.

3-2.4 Limitations.

3-2.4.1 Sprinklers shall not be listed for protection of a portion of an occupancy classification.

Exception No. 1: Residential sprinklers.

Exception No. 2: Special sprinklers shall be permitted to be listed for protection of a specific construction feature in a portion of an occupancy classification. (See 5-4.9.)

NFPA 13 encourages the development of new technologies and allows for a sufficient degree of flexibility for the pursuit of new products. However, a balance must be achieved between flexibility and practicality. The requirement of 3-2.4.1 and its exceptions attempt to define this balance. The requirement's intention is to minimize unnecessary complexities by limiting the development of devices that are narrowly limited in their application. It requires that a sprinkler be listed and usable for all hazards addressed by an occupancy classification. For instance, a sprinkler cannot be listed for the protection of only hospitals or offices but must instead be listed for all types of light hazard occupancies.

Exception No. 1 acknowledges that residential sprinklers are developed specifically for life safety purposes in dwelling units and their adjoining corridors. Residential sprinklers cannot be used in other light hazard occupancies, and this limitation has always been associated with their listing.

The purpose of Exception No. 2 is to allow for the development and use of sprinklers that address specific hazards or construction features and that are not specifically addressed by Exhibit 1.21 identifies some of these special sprinklers.

3-2.4.2 For light hazard occupancies not requiring as much water as is discharged by a sprinkler with a nominal K-factor of 5.6 operating at 7 psi (0.5 bar), sprinklers having a smaller orifice shall be permitted subject to the following restrictions:

(1) The system shall be hydraulically calculated *(see Chapter 7).*
(2) Sprinklers with K-factors of less than 5.6 shall be installed in wet systems only.

Exception: Sprinklers with K-factors of less than 5.6 installed in conformance with Section 4-7 for protection against exposure fires shall be permitted.

(3) A listed strainer shall be provided on the supply side of sprinklers with nominal K-factors of less than 2.8.

Sprinklers with nominal K-factors less than 5.6 are commonly referred to as small-orifice sprinklers. The 7-psi (0.5-bar) minimum pressure indicates that small-orifice sprinklers should only be used where the required sprinkler discharge is less than that provided by a sprinkler with a nominal K-factor of 5.6 operating at 7 psi (0.5 bar). The discharge pattern of small-orifice sprinklers becomes distorted at high pressures.

Small-orifice sprinklers are restricted to light hazard occupancies because they are not as effective in controlling a fire as are sprinklers with a nominal K-factor of at least 5.6. Small-orifice sprinklers are restricted to wet systems because of the likelihood that the sprinkler orifice will be obstructed by internal pipe scale and foreign material prevalent in dry pipe and preaction systems. Exhibits 3.1 and 3.2 show a small-orifice sprinkler adjacent to a K-5.6 sprinkler.

The strainer requirement serves to limit the introduction of foreign materials into the sprinkler system through the water supply. However, the strainer has no effect on scale buildup within the pipe and the possibility of scale being carried to the sprinkler.

As is indicated in the commentary on Table 3-2.3.1, actual orifice sizes may be slightly smaller than the sprinkler's size classification. The inclusion of the word *nominal* in this paragraph clarifies that a strainer is not required for a nominal K-2.8 sprinkler even if the actual orifice size is as small as 2.6.

3-2.4.3 Sprinklers having a K-factor exceeding 5.6 and having ½-in. (12.7-mm) National Pipe Thread (NPT) shall not be installed in new sprinkler systems.

Sprinklers that have nominal K-factors greater than 5.6 and that are equipped with a ½-in. (12.7-mm) National Pipe Thread (NPT) are only intended for use on existing systems. Such a condition could exist when a system is being upgraded because the hazard in the building has changed. This approach allows an existing ½-in. (12.7-mm) line tee to be utilized to accommodate the retrofit of sprinklers with K-factors of either 8.0 or 11.2.

3-2.5* Temperature Characteristics.

A-3-2.5 Information regarding the highest temperature that can be encountered in any location in a particular installation can be obtained by use of a thermometer that will register the highest temperature encountered; it should be hung for several days in the location in question, with the plant in operation.

The temperature rating of a sprinkler plays a critical role in achieving fire control or suppression. The selection criteria for the temperature rating of a given sprinkler is a function of the occupancy classification and ambient ceiling temperatures expected in the vicinity of the sprinkler. Determination of the highest ambient ceiling temperatures influences the selection of the proper sprinkler temperature rating and minimizes the chance of a non-fire sprinkler operation, which is typically an extremely rare event.

If sprinklers with different temperature ratings are randomly installed throughout a compartment, sprinklers remote from the fire may operate prior to sprinklers in the immediate vicinity of the fire. This phenomenon is referred to as *skipping*, which can negatively impact system performance. For this reason, sprinklers with the same temperature ratings should be used in a given area, unless otherwise specified by 5-3.1.4. A sprinkler's thermal sensitivity is another factor to consider. Note the warning in A-5-4 against mixing standard response and quick-response sprinklers in the same compartment unless a specific design objective is being addressed. As a general rule, quick-response sprinklers operate significantly faster than standard response sprinklers under the same conditions. One must be aware of the possibility of skipping if quick-response and standard response sprinklers of the same temperature rating are mixed in a compartment.

3-2.5.1 The standard temperature ratings of automatic sprinklers are shown in Table 3-2.5.1. Automatic sprinklers shall have their frame arms colored in accordance with the color code designated in Table 3-2.5.1.

Exception No. 1: A dot on the top of the deflector, the color of the coating material, or colored frame arms shall be permitted for color identification of corrosion-resistant sprinklers.

Exception No. 2: Color identification shall not be required for ornamental sprinklers such as factory-plated or factory-painted sprinklers or for recessed, flush, or concealed sprinklers.

Exception No. 3: The frame arms of bulb-type sprinklers shall not be required to be color coded.

Sprinklers are color coded in accordance with Table 3-2.5.1 to provide a ready means of identifying the temperature classifications of their operating elements. Table 3-2.5.1 indicates the temperature ranges for sprinklers in each classification and the maximum ceiling temperatures for which each classification may be installed. Exception No. 2 recognizes that traditional color codes are not applicable to specially coated sprinklers, such as decorative or ornamental sprinklers. In some cases, in order to receive a particular color finish, these devices are listed as corrosion-resistant sprinklers.

The decorative or corrosion-resistant finish applied to some sprinklers as listed by the testing laboratories requires NFPA 13 to accommodate some special rules with respect to the color-coding identification. Such accommodation is provided by Exception Nos. 1 and 2.

Table 3-2.5.1 Temperature Ratings, Classifications, and Color Codings

Maximum Ceiling Temperature		Temperature Rating		Temperature Classification	Color Code	Glass Bulb Colors
°F	°C	°F	°C			
100	38	135–170	57–77	Ordinary	Uncolored or black	Orange or red
150	66	175–225	79–107	Intermediate	White	Yellow or green
225	107	250–300	121–149	High	Blue	Blue
300	149	325–375	163–191	Extra high	Red	Purple
375	191	400–475	204–246	Very extra high	Green	Black
475	246	500–575	260–302	Ultra high	Orange	Black
625	329	650	343	Ultra high	Orange	Black

3-2.5.2 The liquid in bulb-type sprinklers shall be color coded in accordance with Table 3-2.5.1.

A large number of glass bulb sprinklers have coated frame arms for aesthetic reasons. This coating makes frame arm color coding rather confusing. Therefore, the temperature rating of glass bulb sprinklers is determined by the color of the encased liquid, which is usually a low-boiling-point alcohol. The temperature rating of this type of element is controlled by the size of a small air bubble that is trapped in the glass tube of the device. The relative size of this bubble establishes the temperature rating of the sprinkler.

A glass bulb sprinkler with a white frame arm finish but a red bulb, for example, is rated as ordinary temperature, using Table 3-2.5.1.

Paragraph 5-3.1.4 provides details on the use of sprinklers with differing temperature ratings, depending on the environment.

3-2.6 Special Coatings.

Sprinklers with special coatings or platings are specifically listed for use in atmospheres that would corrode an uncoated sprinkler. Attempts by anyone other than the manufacturer to provide sprinklers with any type of protective coating is very likely going to result in ineffective protection by the sprinkler or serious impairment of its operation. The listing of sprinklers with protective or special coatings requires that the coatings be applied at the manufacturer's facility. Field application of these platings and coatings is prohibited. Exhibit 1.29 illustrates a wax-coated corrosion-resistant sprinkler.

3-2.6.1* Listed corrosion-resistant sprinklers shall be installed in locations where chemicals, moisture, or other corrosive vapors sufficient to cause corrosion of such devices exist.

A-3-2.6.1 Examples of such locations include the following:

(1) Paper mills
(2) Packing houses
(3) Tanneries
(4) Alkali plants
(5) Organic fertilizer plants
(6) Foundries
(7) Forge shops
(8) Fumigation, pickle, and vinegar works
(9) Stables
(10) Storage battery rooms
(11) Electroplating rooms
(12) Galvanizing rooms
(13) Steam rooms of all descriptions, including moist vapor dry kilns
(14) Salt storage rooms
(15) Locomotive sheds or houses
(16) Driveways
(17) Areas exposed to outside weather, such as piers and wharves exposed to salt air
(18) Areas under sidewalks
(19) Areas around bleaching equipment in flour mills
(20) All portions of cold storage buildings where a direct ammonia expansion system is used
(21) Portions of any plant where corrosive vapors prevail

3-2.6.2* Corrosion-resistant coatings shall be applied only by the manufacturer of the sprinkler.

Exception: Any damage to the protective coating occurring at the time of installation shall be repaired at once using only the coating of the manufacturer of the sprinkler in the approved manner so that no part of the sprinkler will be exposed after installation has been completed.

A-3-2.6.2 Care should be taken in the handling and installation of wax-coated or similar sprinklers to avoid damaging the coating.

3-2.6.3* Unless applied by the manufacturer, sprinklers shall not be painted, and any sprinklers that have been painted shall be replaced with new listed sprinklers of the same characteristics, including orifice size, thermal response, and water distribution.

Exception: Factory-applied paint or coating to sprinkler frames in accordance with 3-2.5.1 shall be permitted.

A-3-2.6.3 Painting of sprinklers can retard the thermal response of the heat-responsive element, can interfere with the free movement of parts, and can render the sprinkler inoperative. Moreover, painting can invite the application of subsequent coatings, thus increasing the possibility of a malfunction of the sprinkler.

Painting is the primary, but not the only, example of a problem known as *loading*, in which a buildup of foreign material on the sprinkler acts to delay or prevent proper sprinkler activation. For example, sprinklers in textile mills are subject to loading by lint-type materials, which are by-products of the cloth-manufacturing process. Another example includes sprinklers at the bottom of elevator shafts and underneath escalators, as they are subject to loading from hydraulic oils and lubricants. These two examples are indicative of events that occur after the system is in service. Any sprinkler subject to loading of materials that cannot be readily dusted or blown away must be replaced.

3-2.6.4 Ornamental finishes shall not be applied to sprinklers by anyone other than the sprinkler manufacturer, and only sprinklers listed with such finishes shall be used.

Decorative coatings for ornamentation and other markings can only be applied by the manufacturer in accordance with the sprinkler's listing. This restriction also applies to the coverplate assembly of a listed concealed sprinkler.

3-2.6.5 Sprinklers protecting spray areas and mixing rooms in resin application areas shall be protected against overspray residue so that they will operate quickly in the event of fire. If covered, cellophane bags having a thickness of 0.003 in. (0.076 mm) or less or thin paper bags shall be used. Coverings shall be replaced frequently so that heavy deposits of residue do not accumulate. Sprinklers that have been painted or coated, except by the sprinkler manufacturer, shall be replaced with new listed sprinklers having the same characteristics.

3-2.7 Escutcheon Plates.

3-2.7.1 Nonmetallic escutcheon plates shall be listed.

3-2.7.2* Escutcheon plates used with a recessed or flush-type sprinkler shall be part of a listed sprinkler assembly.

A-3-2.7.2 The use of the wrong type of escutcheon with recessed or flush-type sprinklers can result in severe disruption of the spray pattern, which can destroy the effectiveness of the sprinkler.

Components for escutcheon plates, including cover plates, if used, are integral to proper sprinkler performance. The listing of these components ensures that the plates do not delay the operating time of the sprinkler and that any components that are designed to fall away or slide do so as intended.

Exhibit 3.3 illustrates a concealed sprinkler with the escutcheon plate and the cover plate missing. While the missing cover plate will not impair sprinkler performance, the lack of the escutcheon plate (sprinkler cup) can because the concealed sprinkler was listed for use with an escutcheon plate.

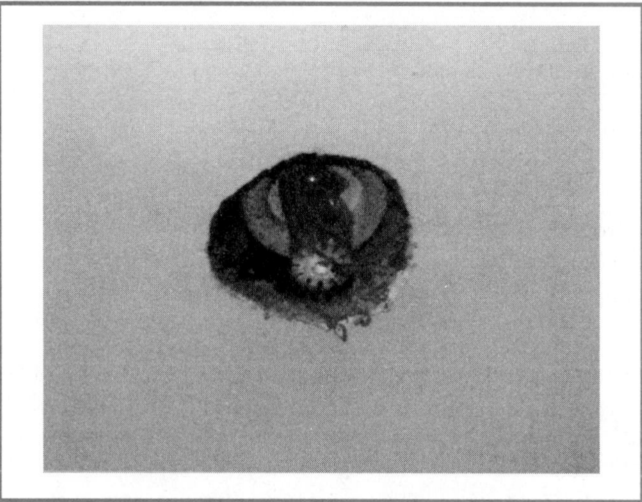

Exhibit 3.3 *Improperly installed concealed sprinkler.*

Instances have occurred where the dimensions of an unlisted recessed cup and a sprinkler were such that the deflector was not clear of the cup, which seriously impaired the spray pattern. Failure to follow all manufacturer's instructions and listing criteria can result in less than acceptable performance of the sprinkler.

3-2.8* Guards and Shields.

Sprinklers subject to mechanical injury shall be protected with listed guards.

A-3-2.8 Sprinklers under open gratings should be provided with shields. Shields over automatic sprinklers should not be less, in least dimension, than four times the distance between the shield and fusible element, except special sprinklers incorporating a built-in shield need not comply with this recommendation if listed for the particular application.

A sprinkler guard can minimize the impact from physical damage by a variety of objects, including pallet loads used in the rack storage facilities. These guards are also useful in protecting people from injury by sprinklers in areas where there is low head clearance. Exhibit 3.4 illustrates a sprinkler with a listed guard.

Any sprinkler suffering physical damage that could impact its effectiveness, such as a bent deflector, must be replaced.

Exhibit 3.4 Sprinkler with a listed guard.

Sprinklers in racks and under open gratings are examples of intermediate-level sprinklers, which must be provided with a shield to ensure their operation is not delayed due to a sprinkler that is operating above it.

3-2.9 Stock of Spare Sprinklers.

3-2.9.1 A supply of spare sprinklers (never fewer than six) shall be maintained on the premises so that any sprinklers that have operated or been damaged in any way can be promptly replaced. These sprinklers shall correspond to the types and temperature ratings of the sprinklers in the property. The sprinklers shall be kept in a cabinet located where the temperature to which they are subjected will at no time exceed 100°F (38°C).

3-2.9.2 A special sprinkler wrench shall also be provided and kept in the cabinet to be used in the removal and installation of sprinklers.

3-2.9.3 The stock of spare sprinklers shall include all types and ratings installed and shall be as follows:

(1) For systems having less than 300 sprinklers, not fewer than six sprinklers
(2) For systems with 300 to 1000 sprinklers, not fewer than 12 sprinklers
(3) For systems with over 1000 sprinklers, not fewer than 24 sprinklers

The stock of spare sprinklers required is a minimum. Spare sprinklers of all the types and ratings installed must be available. For a building with a variety of sprinklers, the number of spare sprinklers should be increased above the minimum. Having adequate stock is particularly critical where specially listed sprinklers such as those discussed in 5-4.9 are installed.

Good judgment must be used when applying the rules for the number of spare sprinklers to maintain on the premises. For instance, if a building contained 350 sprinklers, only two of which were standard spray, upright sprinklers with a temperature rating of 286°F (141°C), it is not intended that 12 such sprinklers be provided in the spare sprinkler cabinet. Two spare sprinklers are practical and acceptable to maintain, since there is no need for more than that number. Subsection 2-4.1 of NFPA 25, *Standard for the Inspection, Testing, and Maintenance of Water-Based Fire Protection Systems*, provides more information in this regard. Exhibit 3.5 illustrates a spare sprinkler cabinet equipped with a stock of spare sprinklers and a special pipe wrench.

Exhibit 3.5 *Spare sprinkler cabinet.*

3-3 Aboveground Pipe and Tube

The use of new pipe materials and the development of new methods for installing and joining pipe continue to grow. The following subsections provide details on the applications of various pipe materials used in aboveground applications. NFPA 13 currently allows for various types of steel, copper, and nonmetallic pipe to be used.

3-3.1 Pipe or tube shall meet or exceed one of the standards in Table 3-3.1 or be in accordance with 3-3.5. In addition, steel pipe shall be in accordance with 3-3.2 and 3-3.3, copper tube shall be in accordance with 3-3.4, and chlorinated polyvinyl chloride (CPVC) and polybutylene pipe shall be in accordance with 3-3.5 and with the portions of the ASTM standards specified in Table 3-3.5 that apply to fire protection service.

Table 3-3.1 Pipe or Tube Materials and Dimensions

Materials and Dimensions	Standard
Ferrous Piping (Welded and Seamless)	
Specification for Black and Hot-Dipped Zinc-Coated (Galvanized) Welded and Seamless Steel Pipe for Fire Protection Use	ASTM A 795
Specification for Welded and Seamless Steel Pipe	ANSI/ASTM A 53
Wrought Steel Pipe	ANSI B36.10M
Specification for Electric-Resistance-Welded Steel Pipe	ASTM A 135
Copper Tube (Drawn, Seamless)	
Specification for Seamless Copper Tube	ASTM B 75
Specification for Seamless Copper Water Tube	ASTM B 88
Specification for General Requirements for Wrought Seamless Copper and Copper-Alloy Tube	ASTM B 251
Fluxes for Soldering Applications of Copper and Copper-Alloy Tube	ASTM B 813
Brazing Filler Metal (Classification BCuP-3 or BCuP-4)	AWS A5.8
Solder Metal, 95-5 (Tin-Antimony-Grade 95TA)	ASTM B 32
Alloy Materials	ASTM B 446

This subsection permits the use of steel pipe and copper tube materials made to standards other than those identified in Table 3-3.1, provided the requirements of any other piping standard meet or exceed the requirements of those standards specified in Table 3-3.1. Additionally, NFPA 13 permits the use of other types of piping materials not specifically identified, provided they are listed for fire protection service.

Table 3-3.5 identifies the standards used in the manufacturing process of nonmetallic pipe. Since nonmetallic pipe is not identified in Table 3-3.1, it is required to be listed for fire protection service.

Table 3-3.1 specifies the solders, brazing alloys, and fluxes to be used when joining copper tube. The techniques and materials identified have low-corrosive characteristics and are unlikely to cause leaks through the seats of sprinklers. Leakage was a problem with some other types of piping materials and joining techniques.

The selection of a particular pipe material is often subject to the environmental conditions of the space in which the pipe is installed. For example, some industrial environments could require a corrosion-resistant pipe material. The desire to achieve a certain aesthetic appearance could also influence material selection. For instance, when piping is installed exposed, copper tubing could be used to blend with the architectural features.

3-3.2* When steel pipe listed in Table 3-3.1 is used and joined by welding as referenced in 3-6.2 or by roll-grooved pipe and fittings as referenced in 3-6.3, the minimum nominal wall thickness for pressures up to 300 psi (20.7 bar) shall be in accordance with Schedule 10 for pipe sizes up to 5 in. (127 mm), 0.134 in. (3.40 mm) for 6-in. (152-mm) pipe, and 0.188 in. (4.78 mm) for 8- and 10-in. (203- and 254-mm) pipe.

Exception: Pressure limitations and wall thickness for steel pipe listed in accordance with 3-3.5 shall be in accordance with the listing requirements.

A-3-3.2 See Table A-3-3.2.

Table A-3-3.2 Steel Pipe Dimensions

Nominal Pipe Size	Outside Diameter		Schedule 10[a] Inside Diameter		Schedule 10[a] Wall Thickness		Schedule 30 Inside Diameter		Schedule 30 Wall Thickness		Schedule 40 Inside Diamter		Schedule 40 Wall Thickness	
(in.)	in.	mm	in.	mm	in.	mm	in.	mm	in.	mm	in.	mm	in.	mm
½[b]	0.840	21.3	0.674	17.0	0.083	2.1	—	—	—	—	0.622	15.8	0.109	2.8
¾[b]	1.050	26.7	0.884	22.4	0.083	2.1	—	—	—	—	0.824	21.0	0.113	2.9
1	1.315	33.4	1.097	27.9	0.109	2.8	—	—	—	—	1.049	26.6	0.133	3.4
1¼	1.660	42.2	1.442	36.6	0.109	2.8	—	—	—	—	1.380	35.1	0.140	3.6
1½	1.900	48.3	1.682	42.7	0.109	2.8	—	—	—	—	1.610	40.9	0.145	3.7
2	2.375	60.3	2.157	54.8	0.109	2.8	—	—	—	—	2.067	52.5	0.154	3.9
2½	2.875	73.0	2.635	66.9	0.120	3.0	—	—	—	—	2.469	62.7	0.203	5.2
3	3.500	88.9	3.260	82.8	0.120	3.0	—	—	—	—	3.068	77.9	0.216	5.5
3½	4.000	101.6	3.760	95.5	0.120	3.0	—	—	—	—	3.548	90.1	0.226	5.7
4	4.500	114.3	4.260	108.2	0.120	3.0	—	—	—	—	4.026	102.3	0.237	6.0
5	5.563	141.3	5.295	134.5	0.134	3.4	—	—	—	—	5.047	128.2	0.258	6.6
6	6.625	168.3	6.357	161.5	0.134[c]	3.4	—	—	—	—	6.065	154.1	0.280	7.1
8	8.625	219.1	8.249	209.5	0.188[c]	4.8	8.071	205.0	0.277	7.0	7.981	—	0.322	—
10	10.750	273.1	10.370	263.4	0.188[c]	4.8	10.140	257.6	0.307	7.8	10.020	—	0.365	—
12	12.750	—	12.090	—	0.330	—	—	—	—	—	11.938	—	0.406	—

[a]Schedule 10 defined to 5-in. (127-mm) nominal pipe size by ASTM A 135, *Standard Specification for Electric-Resistance-Welded Steel Pipe.*

[b]These values applicable when used in conjunction with 5-13.20.2 and 5-13.20.3.

[c]Wall thickness specified in 3-3.2 and 3-3.3.

When thin-wall (Schedule 10) pipe is joined by welding or roll-grooved pipe and fittings, the groove forming or welding process cannot result in a thinner wall thickness, other than that due to the normal tolerances associated with the groove forming or welding. The term *minimum nominal wall thickness* recognizes that pipe standards permit minor tolerance variations greater or less than the stated wall thickness. Exhibit 3.6 illustrates a piece of pipe with one end roll-grooved.

The nominal ½-in. and ¾-in. pipe information in Table A-3-3.2 is provided only for those applications specified in 5-13.20. In a retrofit situation, 5-13.20 permits the use of short ½-in. or ¾-in. pipe nipples to supply sprinklers below a false ceiling from existing ½-in. or ¾-in. sprinkler outlets. This exception applies to existing systems in nonseismic areas only.

Exhibit 3.6 Section of steel pipe with one end roll-grooved.

3-3.3 When steel pipe listed in Table 3-3.1 is joined by threaded fittings referenced in 3-6.1 or by fittings used with pipe having cut grooves, the minimum wall thickness shall be in accordance with Schedule 30 pipe [in sizes 8 in. (203 mm) and larger] or Schedule 40 pipe [in sizes less than 8 in. (203 mm)] for pressures up to 300 psi (20.7 bar).

Exception: Pressure limitations and wall thickness for steel pipe specially listed in accordance with 3-3.5 shall be in accordance with the listing requirements.

When pipe is threaded or grooves are cut into it, the use of thin-wall (Schedule 10) pipe can result in an insufficient pipe wall thickness between the inside pipe diameter and the root diameter of the thread or groove. An insufficient pipe wall thickness reduces the beam strength of the pipe at those threaded or cut-grooved areas and can result in failure of the pipe and fittings. Exhibit 3.7 illustrates pipe joined with threaded fittings.

A threaded or cut-grooved assembly for pipe sizes and schedules other than those specified in NFPA 13 can be used, provided the assembly is investigated for suitability in automatic sprinkler installations and is listed for this service. See the exception to 3-6.1.2 and A-3-6.1.2.

3-3.4* Copper tube as specified in the standards listed in Table 3-3.1 shall have a wall thickness of Type K, Type L, or Type M where used in sprinkler systems.

A-3-3.4 See Table A-3-3.4.

Type M tube is the thinnest walled tubing so it has the largest inside diameter. This larger internal diameter offers a hydraulic advantage and is less expensive than other

Exhibit 3.7 *Pipe joined using threaded fittings.*

Table A-3-3.4 Copper Tube Dimensions

Nominal Tube Size (in.)	Outside Diameter		Type K Inside Diameter		Type K Wall Thickness		Type L Inside Diameter		Type L Wall Thickness		Type M Inside Diamter		Type M Wall Thickness	
	in.	mm	in.	mm	in.	mm	in.	mm	in.	mm	in.	mm	in.	mm
¾	0.875	22.2	0.745	18.9	0.065	1.7	0.785	19.9	0.045	1.1	0.811	20.6	0.032	0.8
1	1.125	28.6	0.995	25.3	0.065	1.7	1.025	26.0	0.050	1.3	1.055	26.8	0.035	0.9
1¼	1.375	34.9	1.245	31.6	0.065	1.7	1.265	32.1	0.055	1.4	1.291	32.8	0.042	1.1
1½	1.625	41.3	1.481	37.6	0.072	1.8	1.505	38.2	0.060	1.5	1.527	38.8	0.049	1.2
2	2.125	54.0	1.959	49.8	0.083	2.1	1.985	50.4	0.070	1.8	2.009	51.0	0.058	1.5
2½	2.625	66.7	2.435	61.8	0.095	2.4	2.465	62.6	0.080	2.0	2.495	63.4	0.065	1.7
3	3.125	79.4	2.907	73.8	0.109	2.8	2.945	74.8	0.090	2.3	2.981	75.7	0.072	1.8
3½	3.625	92.1	3.385	86.0	0.120	3.0	3.425	87.0	0.100	2.5	3.459	87.9	0.083	2.1
4	4.125	104.8	3.857	98.0	0.134	3.4	3.905	99.2	0.110	2.8	3.935	99.9	0.095	2.4
5	5.125	130.2	4.805	122.0	0.160	4.1	4.875	123.8	0.125	3.2	4.907	124.6	0.109	2.8
6	6.125	155.6	5.741	145.8	0.192	4.9	5.845	148.5	0.140	3.6	5.881	149.4	0.122	3.1
8	8.125	206.4	7.583	192.6	0.271	6.9	7.725	196.2	0.200	5.1	7.785	197.7	0.170	4.3
10	10.130	257.3	9.449	240.0	0.338	8.6	9.625	244.5	0.250	6.4	9.701	246.4	0.212	5.4

types of copper tube. These advantages make Type M tube an attractive copper material when pipe bending is not required.

3-3.5* Other types of pipe or tube investigated for suitability in automatic sprinkler installations and listed for this service, including but not limited to polybutylene, CPVC, and steel, differing from that provided in Table 3-3.5 shall be permitted where installed in accordance with their listing limitations, including installation instructions. Pipe or tube shall not be listed for portions of an occupancy classification. Bending of pipe conforming to 3-3.5 shall be permitted as allowed by the listing.

Exception: Pipe or tube listed for light hazard occupancies shall be permitted to be installed in ordinary hazard rooms of otherwise light hazard occupancies where the room does not exceed 400 ft^2 (13 m^2).

Table 3-3.5 Specially Listed Pipe or Tube Materials and Dimensions

Materials and Dimensions	Standard
Nonmetallic Piping Specification for Special Listed Chlorinated Polyvinyl Chloride (CPVC) Pipe	ASTM F 442
Specification for Special Listed Polybutylene (PB) Pipe	ASTM D 3309

A-3-3.5 Other types of pipe and tube that have been investigated and listed for sprinkler applications include lightweight steel pipe and thermoplastic pipe and fittings. While these products can offer advantages, such as ease of handling and installation, cost-effectiveness, reduction of friction losses, and improved corrosion resistance, it is important to recognize that they also have limitations that are to be considered by those contemplating their use or acceptance.

Corrosion studies have shown that, in comparison to Schedule 40 pipe, the effective life of lightweight steel pipe can be reduced, the level of reduction being related to its wall thickness. Further information with respect to corrosion resistance is contained in the individual listings for such pipe.

With respect to thermoplastic pipe and fittings, exposure of such piping to elevated temperatures in excess of that for which it has been listed can result in distortion or failure. Accordingly, care must be exercised when locating such systems to ensure that the ambient temperature, including seasonal variations, does not exceed the rated value.

The upper service temperature limit of currently listed CPVC sprinkler pipe is 150 F (65.5 C) at 175 psi (12.1 bar). The upper service temperature limit of currently listed polybutylene sprinkler pipe is 120 F (49 C) at 175 psi (12.1 bar).

Not all pipe or tube made to ASTM F 442, *Standard Specification for Chlorinated Poly (Vinyl Chloride) (CPVC) Plastic Pipe (SDR-PR),* and ASTM D 3309, *Standard Specification for Polybutylene (PB) Plastic Hot- and Cold-Water Distribution Systems,* as described in 3-3.5, is listed for fire sprinkler service. Listed pipe is identified by the logo of the listing agency.

Not all fittings made to ASTM F 437, *Standard Specification for Threaded Chlorinated Poly (Vinyl Chloride) (CPVC) Plastic Pipe Fittings, Schedule 80;* ASTM F 438, *Standard Specification for Socket-Type Chlorinated Poly (Vinyl Chloride) (CPVC) Plastic Pipe Fittings, Schedule 40;* and ASTM F 439, *Standard Specification for Socket-Type Chlorinated Poly (Vinyl Chloride) (CPVC) Plastic Pipe Fittings,* Schedule 80, as described in 3-5.2, are listed for fire sprinkler service. Listed fittings are identified by the logo of the listing agency.

Consideration must also be given to the possibility of exposure of the piping to elevated temperatures during a fire. The survival of thermoplastic piping under fire conditions is primarily due to the cooling effect of the discharge from the sprinklers it serves. As this discharge might not occur simultaneously with the rise in ambient temperature and, under some circumstances, can be delayed for periods beyond the tolerance of the piping, protection in the form of a fire-resistant membrane is generally required. (Some listings do provide for the use of exposed piping in conjunction with residential or quick-response sprinklers, but only under specific, limited installation criteria.)

Where protection is required, it is described in the listing information for each individual product, and the requirements given must be followed. It is equally important that such protection must be maintained. Removal of, for example, one or more panels in a lay-in ceiling can expose piping in the concealed space to the possibility of failure in the event of a fire. Similarly, the relocation of openings through protective ceilings that expose the pipe to heat, inconsistent with the listing, would place the system in jeopardy. The potential for loss of the protective membrane under earthquake conditions should also be considered.

While the listings of thermoplastic piping do not prohibit its installation in combustible concealed spaces where the provision of sprinkler protection is not required, and while the statistical record of fire originating in such spaces is low, it should be recognized that the occurrence of a fire in such a space could result in failure of the piping system.

The investigation of pipe and tube other than described in Table 3-3.1 should involve consideration of many factors, including the following:

(1) Pressure rating
(2) Beam strength (hangers)
(3) Unsupported vertical stability
(4) Movement during sprinkler operation (affecting water distribution)
(5) Corrosion (internal and external), chemical and electrolytic
(6) Resistance to failure when exposed to elevated temperatures
(7) Methods of joining (strength, permanence, fire hazard)
(8) Physical characteristics related to integrity during earthquakes

NFPA does not limit aboveground sprinkler system piping materials to the steel and copper materials identified in Table 3-3.1. However, any other materials must be listed for use in sprinkler systems. This subsection encourages development of other materials that offer cost or performance advantages. Historically, the development of new products and methods, such as threaded light wall steel pipe, copper tubing, and nonmetallic pipe, has occurred through the application of this subsection.

As indicated in 3-3.5, specially listed pipe is required to be installed in complete compliance with all conditions, requirements, and limitations of its listing.

Not all pipe made to a particular manufacturing standard is listed. Listed piping is identified by the logo of the listing agency. One example of this identification is shown in Exhibit 3.8. In this case, the Underwriters Laboratories (UL) logo is clearly visible on the pipe.

Exhibit 3.8 illustrates a piece of chlorinated polyvinyl chloride (CPVC) fire sprinkler system pipe. Unlisted piping manufactured with less exacting quality control must not be used in sprinkler systems.

Two synthetic piping materials have been listed for sprinkler system applications. These materials include CPVC pipe and polybutylene (PB) pipe, which have been

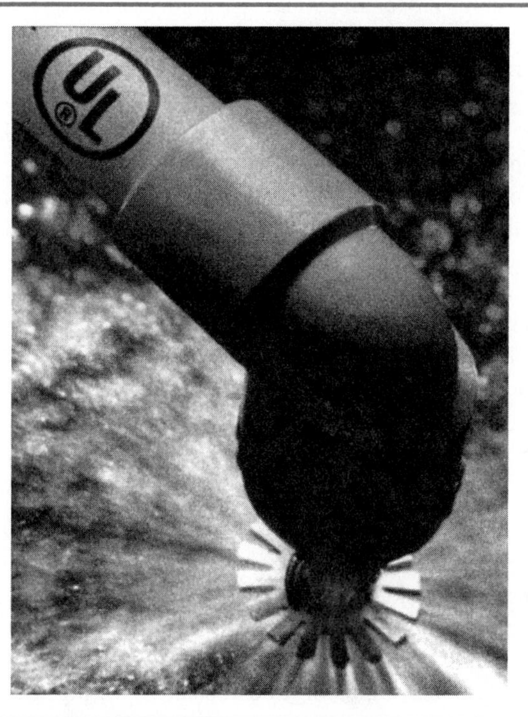

Exhibit 3.8 Listed plastic pipe.

listed for residential and light hazard occupancies. However, since April 1996, the resin used to manufacture polybutylene pipe is no longer available.

The listings of these materials include numerous restrictions. These restrictions include maximum and minimum temperature limits, use for wet pipe systems only, and use of fittings that are exclusively developed for use with the pipe.

Additionally, CPVC or PB piping is not listed for use in ordinary or extra hazard areas. While A-3-3.5 provides some cautions pertaining to the use of nonmetallic pipe, the user must refer to the listing information, including installation instructions, to accomplish correct installation. (Also see commentary following 3-5.2.) When the product requires a protective membrane, readily available materials such as gypsum wallboard, ½-in. (12.7-mm) plywood, or acoustical panels would constitute acceptable protection. Throughout the life of the building, care should be taken to ensure that the integrity of the protective membrane materials remains intact—that is, no poke-through penetrations and no removal of materials.

CPVC pipe can be installed exposed under certain circumstances. The system must be installed using listed quick-response or listed residential sprinklers. If quick-response sprinklers are used, the deflectors must be installed within 8 in. (203 mm) of the ceiling. If residential sprinklers are used, the sprinklers must be installed in accordance with their listing. To obtain these special listings, both CPVC and PB piping materials are subject to rather rigorous fire-testing procedures. Successful completion of these tests demonstrates that the material will not fail under certain fire conditions or contribute to the growth of the fire, thereby creating an acceptable confidence level with regard to these materials.

Although not shown in Table 3-3.5, some light wall steel pipe is listed for fire protection service, and its use is governed by this subsection. Light wall steel pipe is tested for its ability to provide performance characteristics similar to those of Schedule 40 steel pipe. These piping materials might be advantageous from a hydraulics standpoint since they have a larger interior diameter than other steel pipe materials. In addition, their weight is considerably less than that of Schedule 40 pipe, and their joining methods require less overall effort.

With regard to the exception, it is not the intent of NFPA 13 to prohibit the use of plastic pipe in small rooms, such as closets and mechanical rooms, in otherwise light hazard occupancies.

3-3.6 Pipe Bending.

Bending of Schedule 10 steel pipe, or any steel pipe of wall thickness equal to or greater than Schedule 10 and Types K and L copper tube, shall be permitted when bends are made with no kinks, ripples, distortions, or reductions in diameter or any noticeable deviations from round. For Schedule 40 and copper tubing, the minimum radius of a bend shall be 6 pipe diameters for pipe sizes 2 in. (51 mm) and smaller and 5 pipe diameters for pipe sizes 2½ in. (64 mm) and larger. For all other steel pipe, the minimum radius of a bend shall be 12 pipe diameters for all sizes.

NFPA 13 permits bending of all types of Schedule 10 (thin wall) and thicker-walled steel pipe as indicated in Exhibit 3.9. The minimum bending radius for Schedule 40 (standard wall) pipe is less than it is for any of the thinner-walled products. Special listed pipe having a wall thickness less than Schedule 10 cannot be bent.

Type M copper tube cannot be used for bending because the bending results in a wall that is too thin. However, the thicker Types K and L copper tubing can be used for bending because adequate material remains after bending. The listing of PB pipe permits it to be bent to a minimum radius of 10 actual pipe diameters when bent in accordance with listing limitations and manufacturer's installation instructions.

Without control over the bending radius of the pipe, excessive friction loss values for a given elbow, or similar segment of pipe that changes the direction of flow, can be expected.

3-3.7 Pipe Identification.

All pipe, including specially listed pipe allowed by 3-3.5, shall be marked continuously along its length by the manufacturer in such a way as to properly identify the type of pipe. This identification shall include the manufacturer's name, model designation, or schedule.

The variety of piping materials with varying wall thicknesses available to the sprinkler industry offers many benefits. Without the labeling requirements of 3-3.7, identification of pipe in the field would, in many instances, be extremely difficult.

Exhibit 3.9 Bent pipe.

All pipe used for sprinkler systems is required to be continuously marked, not just pipe that is listed for use in sprinkler systems. In the case of listed pipe, a testing organization determines if the pipe meets the requirements of 3-3.7 in deciding whether to list the pipe. In the case of non-listed pipe, the authority having jurisdiction makes such a determination.

The identification of some pipe materials is more obvious than others. For example, listed CPVC sprinkler pipe has a characteristic orange color, listed PB pipe has a characteristic gray color, and copper tube is easy to identify because of its color and appearance. The color schemes, however, need to be supplemented by the marking requirements established by 3-3.7.

3-4* Underground Pipe

A-3-4 Loop systems for yard piping are recommended for increased reliability and improved hydraulics. Loop systems should be sectionalized by placing valves at branches and at strategic locations to minimize the extent of impairments.

3-4.1* Piping shall be listed for fire protection service and comply with the AWWA standards in Table 3-4.1, where applicable. Steel piping shall not be used unless specifically listed for underground service in private fire service main applications.

Exception: Where externally coated and wrapped and internally galvanized, steel pipe shall be permitted to be used between the check valve and the outside hose coupling for the fire department connection.

Table 3-4.1 Manufacturing Standards for Underground Pipe

Materials	Standard
Cement Mortar Lining for Ductile Iron Pipe and Fittings for Water	AWWA C104
Polyethylene Encasement for Ductile Iron Pipe Systems	AWWA C105
Ductile Iron and Gray Iron Fittings, 3-in. Through 48-in., for Water and Other Liquids	AWWA C110
Rubber-Gasket Joints for Ductile Iron Pressure Pipe and Fittings	AWWA C111
Flanged Ductile Iron Pipe with Ductile Iron or Gray Iron Threaded Flanges	AWWA C115
Thickness Design of Ductile Iron Pipe	AWWA C150
Ductile Iron Pipe, Centrifugally Cast for Water	AWWA C151
Steel Water Pipe 6 in. and Larger	AWWA C200
Coal-Tar Protective Coatings and Linings for Steel Water Pipelines Enamel and Tape—Hot Applied	AWWA C203
Cement-Mortar Protective Lining and Coating for Steel Water Pipe 4 in. and Larger—Shop Applied	AWWA C205
Field Welding of Steel Water Pipe	AWWA C206
Steel Pipe Flanges for Waterworks Service—Sizes 4 in. Through 144 in.	AWWA C207
Dimensions for Fabricated Steel Water Pipe Fittings	AWWA C208
Reinforced Concrete Pressure Pipe, Steel-Cylinder Type for Water and Other Liquids	AWWA C300
Prestressed Concrete Pressure Pipe, Steel-Cylinder Type, for Water and Other Liquids	AWWA C301
Reinforced Concrete Pressure Pipe, Non-Cylinder Type, for Water and Other Liquids	AWWA C302
Reinforced Concrete Pressure Pipe, Steel-Cylinder Type, Pretensioned, for Water and Other Liquids	AWWA C303
Asbestos-Cement Distribution Pipe, 4 in. Through 16 in., for Water and Other Liquids	AWWA C400
Selection of Asbestos-Cement Water Pipe	AWWA C401
Installation of Ductile Iron Water Mains and Their Appurtenances	AWWA C600
Cement-Mortar Lining of Water Pipe Lines 4 in. and Larger—in Place	AWWA C602
Installation of Asbestos-Cement Water Pipe	AWWA C603
Polyvinyl Chloride (PVC) Pressure Pipe, 4 in. Through 12 in., for Water Distribution	AWWA C900
Steel Pipe—A Guide for Design and Installation	AWWA M11

A-3-4.1 Copper tubing (Type K) with brazed joints conforming to Table 3-3.1 and 3-6.4 is acceptable for underground service. Listing and labeling information, along with applicable publications for reference, follows.

(a) *Listing and Labeling.* Testing laboratories list or label cast-iron and ductile iron pipe (cement-lined and unlined, coated and uncoated), asbestos-cement pipe and couplings, steel pipe, copper pipe, fiberglass filament–wound epoxy pipe and couplings, polyethylene pipe, and polyvinyl chloride (PVC) pipe and couplings. Underwriters Laboratories Inc. lists, under re-examination service, reinforced concrete pipe (cylinder pipe, nonprestressed and prestressed).

(b) *Pipe Standards.* The various types of pipe are usually manufactured to one of the following standards:

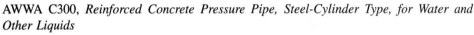

ASTM C296, *Standard Specification for Asbestos-Cement Pressure Pipe*

AWWA C151, *Ductile Iron Pipe, Centrifugally Cast for Water*

AWWA C300, *Reinforced Concrete Pressure Pipe, Steel-Cylinder Type, for Water and Other Liquids*

AWWA C301, *Prestressed Concrete Pressure Pipe, Steel-Cylinder Type, for Water and Other Liquids*

AWWA C302, *Reinforced Concrete Pressure Pipe, Non-Cylinder Type, for Water and Other Liquids*

AWWA C303, *Reinforced Concrete Pressure Pipe, Steel-Cylinder Type, Pretensioned, for Water and Other Liquids*

AWWA C400, *Standard for Asbestos-Cement Distribution Pipe, 4 in. Through 16 in., for Water and Other Liquids*

AWWA C900, *Polyvinyl Chloride (PVC) Pressure Pipe, 4 in. Through 12 in., for Water and Other Liquids*

Past experience and loss data indicate that a high potential for failure of buried steel pipe exists even where the pipe is externally coated and wrapped and internally galvanized. The integrity of the coating, wrapping, and galvanizing can be easily compromised in many underground applications. Accordingly, steel pipe cannot be installed underground to connect the primary water supply to the overhead sprinkler piping. As the exception indicates, steel pipe can be utilized for piping to a fire department connection because the fire department connection is an auxiliary rather than a primary water supply for the sprinkler system.

3-4.2* The type and class of pipe for a particular installation shall be determined through consideration of its fire resistance, the maximum system working pressure, the laying conditions under which the pipe is to be installed, soil conditions, corrosion, and susceptibility of pipe to other external loads, including earth loads, installation beneath buildings, and traffic or vehicle loads.

A-3-4.2 The following pipe design manuals can be used as guides:

AWWA C150, *Thickness Design of Ductile Iron Pipe*

AWWA C401, *Standard Practice for the Selection of Asbestos-Cement Water Pipe*

AWWA M14, *Ductile Iron Pipe and Fittings*

Concrete Pipe Handbook, American Concrete Pipe Association

3-4.3 Pipe used in private fire service shall be designed to withstand a system working pressure of not less than 150 psi (10.3 bar).

3-4.4* Lining of Buried Pipe.

All ferrous metal pipe shall be lined.

A-3-4.4 The following standards apply to the application of coating and linings:

AWWA C104, *Cement Mortar Lining For Ductile Iron Pipe and Fittings for Water*

AWWA C105, *Polyethylene Encasement for Ductile Iron Pipe Systems*

AWWA C203, *Coal-Tar Protective Coatings and Linings for Steel Water Pipelines Enamel and Tape—Hot Applied*

AWWA C205, *Cement-Mortar Protective Lining and Coating for Steel Water Pipe 4 in. and Larger—Shop Applied*

AWWA C602, *Cement-Mortar Lining of Water Pipe Lines 4 in. and Larger—in Place*

3-5 Fittings

3-5.1 Fittings used in sprinkler systems shall meet or exceed the standards in Table 3-5.1 or be in accordance with 3-5.2. In addition to the standards in Table 3-5.1, CPVC fittings shall also be in accordance with 3-5.2 and with the portions of the ASTM standards specified in Table 3-5.2 that apply to fire protection service.

Table 3-5.1 Fittings Materials and Dimensions

Materials and Dimensions	Standard
Cast Iron	
Cast Iron Threaded Fittings, Class 125 and 250	ASME B16.4
Cast Iron Pipe Flanges and Flanged Fittings	ASME B16.1
Malleable Iron	
Malleable Iron Threaded Fittings, Class 150 and 300	ASME B16.3
Steel	
Factory-Made Wrought Steel Buttweld Fittings	ASME B16.9
Buttwelding Ends for Pipe, Valves, Flanges, and Fittings	ASME B16.25
Specification for Piping Fittings of Wrought Carbon Steel and Alloy Steel for Moderate and Elevated Temperatures	ASTM A 234
Steel Pipe Flanges and Flanged Fittings	ASME B16.5
Forged Steel Fittings, Socket Welded and Threaded	ASME B16.11
Copper	
Wrought Copper and Bronze Solder Joint Pressure Fittings	ASME B16.22
Cast Bronze Solder Joint Pressure Fittings	ASME B16.18

Subsection 3-5.1 specifies the types of fittings permitted to be used in sprinkler systems. The fittings must be of the types and materials indicated in Table 3-5.1 and must be manufactured to the standards identified in the table or to other standards that meet or exceed these standards' requirements. Additionally, NFPA 13 permits the use of specially listed fittings.

The fittings specified in Table 3-5.1 are compatible with and intended for use with the piping materials listed in Table 3-3.1. Where specially listed materials are used, then specially listed fittings complying with 3-5.2 should be used.

3-5.2* Other types of fittings investigated for suitability in automatic sprinkler installations and listed for this service including, but not limited to, polybutylene, CPVC, and steel differing from that provided in Table 3-5.2, shall be permitted when installed in accordance with their listing limitations, including installation instructions.

Table 3-5.2 Specially Listed Fittings Materials and Dimensions

Materials and Dimensions	Standard
Chlorinated Polyvinyl Chloride (CPVC) Specification for Schedule 80 CPVC Threaded Fittings	ASTM F 437
Specification for Schedule 40 CPVC Socket-Type Fittings	ASTM F 438
Specification for Schedule 80 CPVC Socket-Type Fittings	ASTM F 439

A-3-5.2 Rubber-gasketed pipe fittings and couplings should not be installed where ambient temperatures can be expected to exceed 150 F (66 C) unless listed for this service. If the manufacturer further limits a given gasket compound, those recommendations should be followed.

This subsection is similar to 3-3.5, which permits alternative types of listed pipe and tube. Innovations in the technology of new fittings can be linked to new piping materials. According to 3-5.2, any new fitting listed for use in a sprinkler system is permitted by NFPA 13. However, this requirement is not intended to mean that any listed fitting can be used with any listed pipe. The user must check the testing laboratory's fire protection equipment directory to ensure that the pipe and fitting are listed as compatible. One listed fitting is the specially designed "Press Fit," which can be used on certain types of specially listed thin-wall steel pipe. The Press Fit is not to be used with nonmetallic piping materials. Exhibit 3.10 illustrates a specially listed press fit tool. Exhibit 3.11 illustrates the pipe and fitting joined using this tool.

3-5.3* Fittings shall be extra-heavy pattern where pressures exceed 175 psi (12.1 bar).

Exception No. 1: Standard weight pattern cast-iron fittings 2 in. (51 mm) in size and smaller shall be permitted where pressures do not exceed 300 psi (20.7 bar).

Exception No. 2: Standard weight pattern malleable iron fittings 6 in. (152 mm) in size and smaller shall be permitted where pressures do not exceed 300 psi (20.7 bar).

Exception No. 3: Fittings shall be permitted for system pressures up to the limits specified in their listings.

A-3-5.3 The rupture strength of cast-iron fittings 2 in. (50.8 mm) in size and smaller and malleable iron fittings 6 in. (152.4 mm) in size and smaller is sufficient to provide an adequate factor of safety.

Exception Nos. 1 and 2 permit cast-iron fittings 2 in. (51 mm) in size and smaller and malleable iron fittings 6 in. (152 mm) in size and smaller to be used in systems where

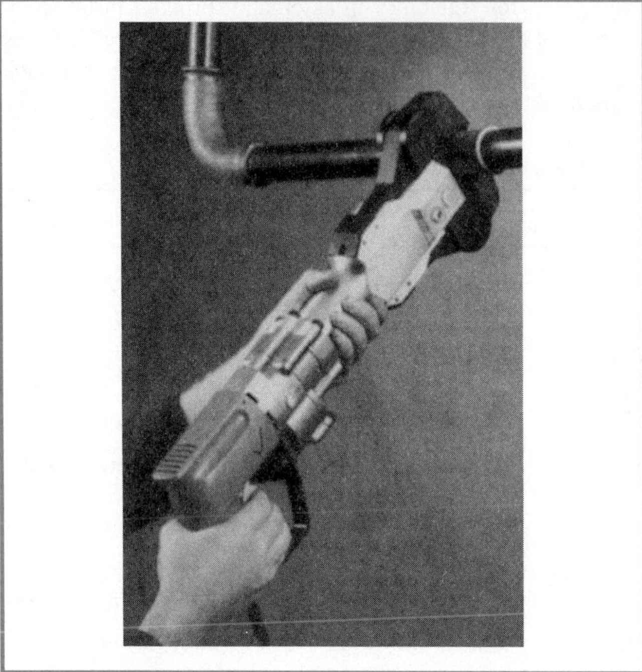

Exhibit 3.10 *Press fit tool.*

Exhibit 3.11 *Pipe and fitting joined using the press fit tool.*

pressures do not exceed 300 psi (20.7 bar). The rupture strength associated with these fittings provides an adequate level of performance. According to Exception No. 3, fittings can be used in systems up to the pressure limitation of their listing. Flanged connections must comply with 3-5.3.

3-5.4* Couplings and Unions.

Screwed unions shall not be used on pipe larger than 2 in. (51 mm). Couplings and unions of other than screwed-type shall be of types listed specifically for use in sprinkler systems.

A-3-5.4 Listed flexible connections are permissible and encouraged for sprinkler installations in racks to reduce the possibility of physical damage. Where flexible tubing is used, it should be located so that it will be protected against mechanical injury.

Screwed unions larger than 2 in. (51 mm) in size present a maintenance problem because of their tendency to develop leaks. A variety of acceptable fittings are available for use in connecting various sizes of pipe. NFPA 13 does not require that only one joining method or fitting be used throughout the system.

3-5.5 Reducers and Bushings.

A one-piece reducing fitting shall be used wherever a change is made in the size of the pipe.

Exception No. 1: Hexagonal or face bushings shall be permitted in reducing the size of openings of fittings when standard fittings of the required size are not available.

Exception No. 2: Hexagonal bushings as permitted in 5-13.20.1 are acceptable.

The intent of 3-5.5 prohibits the use of bushings where a one-piece reducing fitting is commercially available at the time the system is installed.

For other than the circumstance cited in the exception to 5-13.20.1, in which bushings are temporarily installed to facilitate an unrestricted flow to sprinklers installed below a future drop ceiling, bushings can be used only where reducing fittings are unavailable. Bushings have poorer flow characteristics than reducing fittings and have a greater tendency to leak.

The use of bushings under all but the most pressing conditions is discouraged. For example, a particular fitting may actually be manufactured but may be back ordered for several months. If the proper reducing fitting cannot be obtained after an exhaustive effort, then the use of bushings can be considered. Exhibit 3.12 illustrates the use of a bushing in place of a one-piece reducing fitting.

3-5.6* Buried Fittings.

Fittings shall be of an approved type with joints and pressure class ratings compatible with the pipe used.

A-3-5.6 Fittings generally used are cast iron with joints made to the specifications of the manufacturer of the particular type of pipe *(see the standards listed in A-3-4.1.)* Steel fittings also have some applications. The following standards apply to fittings:

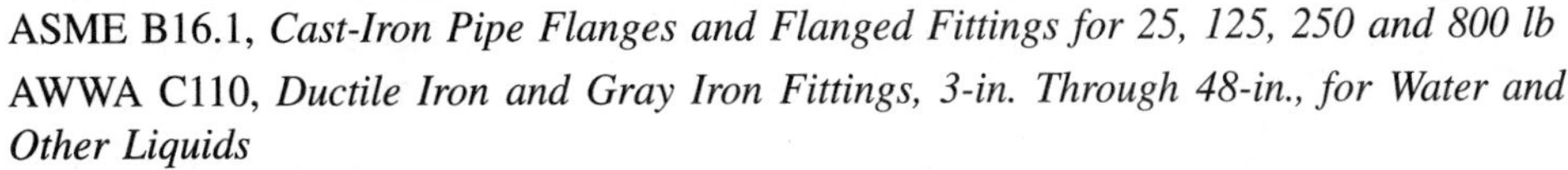

ASME B16.1, *Cast-Iron Pipe Flanges and Flanged Fittings for 25, 125, 250 and 800 lb*

AWWA C110, *Ductile Iron and Gray Iron Fittings, 3-in. Through 48-in., for Water and Other Liquids*

AWWA C153, *Ductile Iron Compact Fittings, 3 in. through 24 in. and 54 in. through 64 in. for Water Service*

AWWA C208, *Dimensions for Fabricated Steel Water Pipe Fittings*

Exhibit 3.12 A hexagonal bushing as illustrated here is discouraged and should only be used where fittings are commercially unavailable.

3-6 Joining of Pipe and Fittings

3-6.1 Threaded Pipe and Fittings.

3-6.1.1 All threaded pipe and fittings shall have threads cut to ASME B1.20.1, *Pipe Threads, General Purpose (Inch).*

Poor workmanship can result in threads that allow pipe protrusions to partially obstruct fitting openings. If such joints are permitted, they can seriously restrict system flow and greatly increase the pressure lost to friction, thereby impairing system operation.

3-6.1.2* Steel pipe with wall thicknesses less than Schedule 30 [in sizes 8 in. (203 mm) and larger] or Schedule 40 [in sizes less than 8 in. (203 mm)] shall not be joined by threaded fittings.

Exception: A threaded assembly investigated for suitability in automatic sprinkler installations and listed for this service shall be permitted.

A-3-6.1.2 Some steel piping material having lesser wall thickness than specified in 3-6.1.2 has been listed for use in sprinkler systems where joined with threaded connections. The service life of such products can be significantly less than that of Schedule 40 steel pipe, and it should be determined if this service life will be sufficient for the application intended.

All such threads should be checked by the installer using working ring gauges conforming to the "Basic Dimensions of Ring Gauges for USA (American) Standard Taper Pipe Threads, NPT," as per Table 8 of ASME B1.20.1, *Pipe Threads, General Purpose, (Inch).*

To alleviate concerns about the service life of listed thin-wall pipe, part of its listing evaluation includes an examination of its corrosion resistance ratio (CRR) with respect to Schedule 40 steel pipe. For example, a CRR of 0.21 indicates an anticipated service life of 21 percent, approximately one-fifth that of Schedule 40 pipe. When installed

AWS B2.1, *Standard for Welding Procedure and Performance Qualification*, reflects the minimum welding requirements needed for the pressures found in sprinkler systems. AWS B2.1 is the American Welding Society's replacement of discontinued specification AWS D10.9. As Formal Interpretation 78-11 indicates, methods meeting higher standards are also acceptable.

Formal Interpretation 78-11

Reference: 3-6.2.1, 3-6.2.8.1

Question 1: Do welding methods and weld inspection conforming to the requirements of ANSI B31.1-1990, *Code for Power Piping*, meet the intent of 3-6.2.1?

Question 2: Does qualification of welding procedures, welders, and welding operators to ANSI B31.1-1990, *Code for Power Piping*, and the requirements of the *ASME Boiler and Pressure Vessel Code* referenced therein meet the intent of 3-6.2.8.1?

Answer: Yes. The standard describes the minimum acceptable welding methods of procedure. Other standards requiring a higher level of weld quality, test procedures, and welder qualification meet the intent of NFPA 13.

Issue Edition: 1978

Reference: 3-12.2.1, 3-12.2.11

Date: August 1980

3-6.2.2* Sprinkler piping shall be shop-welded.

Exception No. 1: Welding of tabs for longitudinal earthquake bracing to in-place piping shall be permitted where the welding process is performed in accordance with NFPA 51B, Standard for Fire Prevention During Welding, Cutting, and Other Hot Work.

Exception No. 2: Where the design specifications call for all or part of the piping to be welded in place, welding of sprinkler piping in place shall be permitted where the welding process is performed in accordance with NFPA 51B and the mechanical fittings required by 5-13.17 and 5-13.22 are provided.

A-3-6.2.2 Cutting and welding operations account for 4 percent of fires each year in nonresidential properties and 8 percent in industrial and manufacturing properties. In-place welding of sprinkler piping introduces a significant hazard that can normally be avoided by shop-welding the piping and installing the welded sections with mechanical fittings. As a result, the standard requires that all piping be shop-welded. When such situations cannot be avoided, the exceptions outline procedures and practices that minimize the increase in hazard.

Because the welding process introduces a potential ignition source, NFPA 13 requires shop-welding of all system piping unless certain conditions are met. If the provisions of NFPA 51B, *Standard for Fire Prevention During Welding, Cutting, and Other Hot Work*, are followed, the welding process can be expected to be safely accomplished in new installations. Since other mechanical system piping can be welded in place, it is logical to expect that sprinkler pipe can also be welded in place. Prior restrictions concerning welding of pipe in place were based on the occurrence of fires as a result of the welding process. NFPA 51B contains a number of precautions that, when followed, reduce or eliminate the ignition hazard.

3-6.2.3 Fittings used to join pipe shall be listed fabricated fittings or manufactured in accordance with Table 3-5.1. Such fittings joined in conformance with a qualified welding procedure as set forth in this section shall be an acceptable product under this standard, provided that materials and wall thickness are compatible with other sections of this standard.

Exception: Fittings shall not be required where pipe ends are buttwelded.

The use of unlisted fabricated fittings is not permitted. A listed fabricated fitting has an identified equivalent length that identifies its hydraulic characteristics. The use of listed fabricated fittings also serves to increase overall system reliability.

The exception also stipulates the only condition in which a manufactured fitting or a listed fabricated fitting need not be used—that is, when two identical size pipe diameters are to be connected in the same plane with no change in direction. The pieces can then be buttwelded.

3-6.2.4 No welding shall be performed if there is impingement of rain, snow, sleet, or high wind on the weld area of the pipe product.

Welding must not be performed under conditions that introduce a hazard to the welder. Furthermore, welding performed in such adverse conditions can also result in a structurally weak product.

3-6.2.5 When welding is performed, the following procedures shall be completed.

(1) *Holes in piping for outlets shall be cut to the full inside diameter of fittings prior to welding in place of the fittings.
(2) Discs shall be retrieved.
(3) Openings cut into piping shall be smooth bore, and all internal slag and welding residue shall be removed.
(4) Fittings shall not penetrate the internal diameter of the piping.
(5) Steel plates shall not be welded to the ends of piping or fittings.
(6) Fittings shall not be modified.
(7) Nuts, clips, eye rods, angle brackets, or other fasteners shall not be welded to pipe or fittings.

Exception: Only tabs welded to pipe for longitudinal earthquake braces shall be permitted. (See 6-4.5.8.)

A-3-6.2.5(1) Listed, shaped, contoured nipples meet the definition of fabricated fittings.

Past experience indicates that quality control can be a problem with welded sprinkler piping. Conditions that are not permitted are identified in 3-6.2.5.

When welded outlets are formed, some of the concerns addressed by these requirements are as follows:

(1) The area of flow must not be restricted.
(2) Holes should be cut in a manner that provides for disc retrieval.
(3) Rough edges cause turbulence, which increases friction loss, and welding residue could obstruct waterflow through activated sprinklers.
(4) Penetration beyond the internal diameter of the pipe restricts flow and causes turbulence.

Items (5), (6), and (7) represent poor practices within the industry and are not permitted by NFPA 13. Disc retrieval mentioned in 3-6.2.5(2) is best assured by attaching the disc to the piping at the point at which it was cut as indicated in Exhibit 3.13.

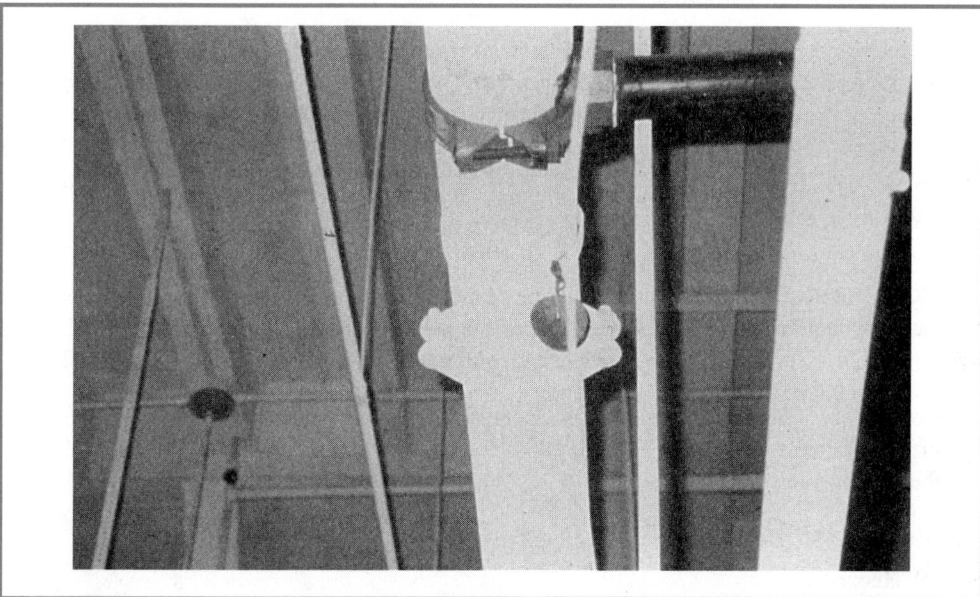

Exhibit 3.13 *Discs attached to piping from which they were cut.*

3-6.2.6 When the pipe size in a run of piping is reduced, a reducing fitting designed for that purpose shall be used.

The use of unlisted fabricated fittings is not permitted. Such fittings do not have performance characteristics comparable to fittings specifically designed for this purpose. See 3-5.5 for more information.

3-6.2.7 Torch cutting and welding shall not be permitted as a means of modifying or repairing sprinkler systems.

Torch cutting and welding are restricted to new installations, and only under those conditions specified by Exception Nos. 1 and 2 to 3-6.2.2. This restriction guards against the possible introduction of ignition sources when the sprinkler system is impaired.

3-6.2.8 Qualifications.

3-6.2.8.1 A welding procedure shall be prepared and qualified by the contractor or fabricator before any welding is done. Qualification of the welding procedure to be used and the performance of all welders and welding operators is required and shall meet or exceed the requirements of AWS B2.1, *Specification for Qualification of Welding Procedures and Welders for Piping and Tubing.*

Exception: Welding procedures qualified under standards recognized by previous editions of this standard shall be permitted to be continued in use.

As indicated in 3-6.2.1, AWS B2.1 reflects the minimum welding requirements needed for the pressures found in sprinkler systems. The welding procedures and the performance of the welders must be in conformance with AWS B2.1. This standard is the American Welding Society's replacement of discontinued specification AWS D10.9. As indicated by the exception, procedures qualified under AWS D10.9 are still applicable to work on sprinkler systems.

3-6.2.8.2 Contractors or fabricators shall be responsible for all welding they produce. Each contractor or fabricator shall have available to the authority having jurisdiction an established written quality assurance procedure ensuring compliance with the requirements of 3-6.2.5.

Contractors or fabricators are directly responsible for qualifying procedures and welders in accordance with NFPA 13, as well as for the quality of the welds performed by their employees.

3-6.2.9 Records.

3-6.2.9.1 Welders or welding machine operators shall, upon completion of each weld, stamp an imprint of their identification into the side of the pipe adjacent to the weld.

The record of a weld completed by a specific welder can be traced through the use of this procedure. Quality control problems or defective materials can be easily traced should a problem with the weld occur.

3-6.2.9.2 Contractors or fabricators shall maintain certified records, which shall be available to the authority having jurisdiction, of the procedures used and the welders or welding machine operators employed by them, along with their welding identification imprints. Records shall show the date and the results of procedure and performance qualifications.

3-6.3 Groove Joining Methods.

3-6.3.1 Pipe joined with grooved fittings shall be joined by a listed combination of fittings, gaskets, and grooves. Grooves cut or rolled on pipe shall be dimensionally compatible with the fittings.

Unlike threads, grooves are not fabricated to a national standard. Therefore, the listing of a grooved fitting includes information on the groove, the fitting, and the gasket. (See Exhibit 3.14.)

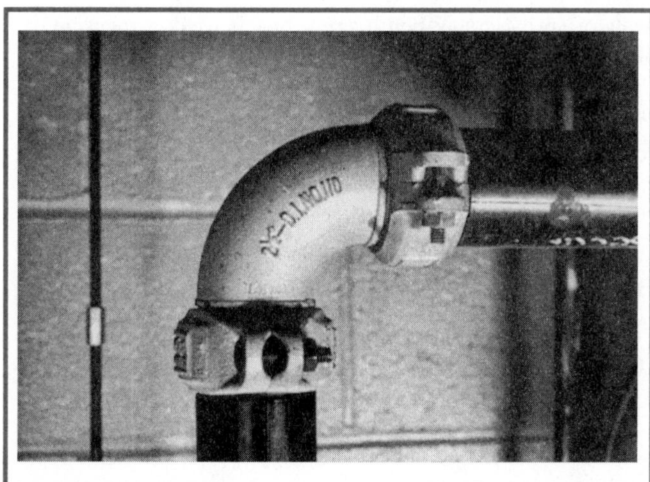

Exhibit 3.14 *Example of a grooved fitting.*

3-6.3.2 Grooved fittings including gaskets used on dry pipe systems shall be listed for dry pipe service.

Gaskets on grooved couplings have a limited degree of fire endurance where used on dry pipe systems. Therefore, the listing requirements of some grooved couplings include an evaluation of the coupling's performance in dry pipe sprinkler applications.

3-6.4* Brazed and Soldered Joints.

Joints for the connection of copper tube shall be brazed.

Exception No. 1: Solder joints shall be permitted for exposed wet pipe systems in light hazard occupancies where the temperature classification of the installed sprinklers is ordinary or intermediate.

Exception No. 2: Solder joints shall be permitted for wet pipe systems in light hazard and ordinary hazard (Group 1) occupancies where the piping is concealed, irrespective of sprinkler temperature ratings.

A-3-6.4 The fire hazard of the brazing and soldering processes should be suitably safeguarded.

Subsection 3-6.4 restricts the use of soldered joints to conditions under which the system piping is filled with water and in which heat from a fire will not reach a magnitude that can compromise the integrity of the joint.

3-6.4.1* Soldering fluxes shall be in accordance with Table 3-3.1. Brazing fluxes, if used, shall not be of a highly corrosive type.

A-3-6.4.1 Soldering fluxes manufactured to the specifications required by Table 3-3.1 are unlikely to cause damage to the seats of sprinklers. When brazing flux is used, it must be of a type not likely to damage the seats of sprinklers.

Highly corrosive fluxes have been known to react with the metal in the seat of the sprinkler, resulting in leakage of the device. Solders meeting ASTM B 813, *Standard Specification for Liquid and Paste Fluxes for Soldering Applications of Copper and Copper-Alloy Tube*, minimize that risk.

3-6.5 Other Joining Methods.

Other joining methods investigated for suitability in automatic sprinkler installations and listed for this service shall be permitted where installed in accordance with their listing limitations, including installation instructions.

This subsection allows the use of joining methods not specifically specified by NFPA 13, provided these methods are listed for use in sprinkler systems. Additionally, this subsection encourages the development and application of new technology. The requirement is similar to 3-3.5 for pipe materials and 3-5.2 for fittings.

3-6.6 End Treatment.

After cutting, pipe ends shall have burrs and fins removed.

Pipe must be reamed and deburred to minimize the reduction of the inside diameter of the pipe and to remove any rough edges from the end of the pipe. The removal of irregular edges is particularly important when the fitting utilized has internal gaskets.

3-6.6.1 Pipe used with listed fittings and its end treatment shall be in accordance with the fitting manufacturer's installation instructions and the fitting's listing.

In recent years, a large variety of new fittings were introduced, including a number that join plain end pipe. The end treatment required varies, depending on the fitting that is utilized. For example, some mechanical-type fittings require that the varnish be removed from the exterior wall of the pipe entering the joint. Failure to follow the manufacturer's procedures and listing limitations can result in an unsatisfactory installation that tends to develop leaks.

3-6.7* Buried Joints.

Joints shall be of an approved type.

A-3-6.7 The following standards apply to joints used with the various types of pipe:

ASME B16.1, *Cast-Iron Pipe Flanges and Flanged Fittings for 25, 125, 250 and 800 lb*

AWWA C111, *Rubber Gasket Joints for Ductile Iron Pressure Pipe and Fittings*

AWWA C115, *Flanged Ductile Iron Pipe with Ductile Iron or Gray Iron Threaded Flanges*

AWWA C206, *Field Welding of Steel Water Pipe*

AWWA C606, *Grooved and Shouldered Joints*

3-7* Hangers

A-3-7 See Section 6-1 for information pertaining to the type of hangers and hanger components acceptable for use on a sprinkler system.

The 1999 edition includes a new Chapter 6 that addresses the structural aspects of sprinkler systems. Information on hangers as well as earthquake bracing was moved to this chapter.

3-8 Valves

A closed valve is the primary reason why sprinkler systems do not perform adequately. The requirements in this section, as well as those in 5-14.1, are intended to minimize the impact of closed valves.

3-8.1 Types of Valves to Be Used.

3-8.1.1 All valves controlling connections to water supplies and to supply pipes to sprinklers shall be listed indicating valves. Such valves shall not close in less than 5 seconds when operated at maximum possible speed from the fully open position.

Exception No. 1: A listed underground gate valve equipped with a listed indicator post shall be permitted.

Exception No. 2: A listed water control valve assembly with a reliable position indication connected to a remote supervisory station shall be permitted.

Exception No. 3: A nonindicating valve, such as an underground gate valve with approved roadway box, complete with T-wrench, and accepted by the authority having jurisdiction, shall be permitted.

All valves controlling the flow of water to sprinklers must be listed and must incorporate a method of readily determining that the valve is open, unless the valves meet the conditions of one of the exceptions. The means of indicating the valve's position can be part of the valve itself, such as a rising stem as indicated in Exhibit 3.15, or a feature of the assembly as illustrated in Exhibit 3.16.

Exhibit 3.15 *Outside stem and yoke (OS&Y) valve with its position supervised by a lock and chain and a tamper switch.*

Exhibit 3.16 *Control valve with a position indicator.*

3-8.1.2 When water pressures exceed 175 psi (12.1 bar), valves shall be used in accordance with their pressure ratings.

The minimum working pressure for system components is established in 3-1.2 as 175 psi (12.1 bar). When working pressures are expected to exceed this value, special components such as extra-heavy valves must be used. Listed valves that can accommodate pressures as high as 300 psi (21 bar) are currently available.

3-8.1.3 Wafer-type valves with components that extend beyond the valve body shall be installed in a manner that does not interfere with the operation of any system components.

The discs of some listed wafer-type (butterfly) valves extend beyond the end of the valve body. Documented instances of such discs contacting other system components and partially obstructing the waterway exist.

3-8.2 Drain Valves and Test Valves.

Drain valves and test valves shall be approved.

3-8.3* Identification of Valves.

All control, drain, and test connection valves shall be provided with permanently marked weatherproof metal or rigid plastic identification signs. The sign shall be secured with corrosion-resistant wire, chain, or other approved means.

A-3-8.3 The intent of 3-8.3 is to provide assistance in determining the area of a building served by a particular control valve.

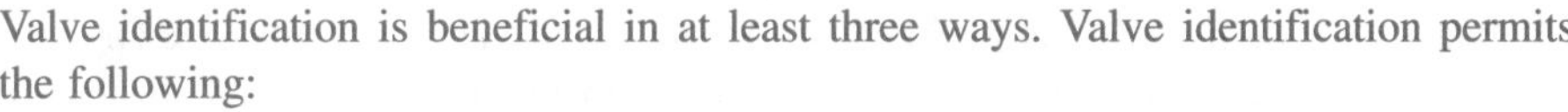

Valve identification is beneficial in at least three ways. Valve identification permits the following:

(1) Fire department personnel to locate and operate the valve
(2) Contractor to readily shut off the system to make modifications
(3) System valves to be easily identified during periodic inspections

3-9 Fire Department Connections

A fire department connection to any sprinkler system provides a desirable, auxiliary water supply to that system. Supplementing the automatic supply required by 9-1.1 increases overall system reliability. Fire department pump operators can also determine if any sprinklers have actually operated by using the fire department connection. If no water flows through the fire department connection into the system, it is likely that no sprinklers have operated. The other possibility is a closed sectional valve or piping obstruction.

3-9.1 The fire department connection(s) shall use an NH internal threaded swivel fitting(s) with an NH standard thread(s). At least one of the connections shall be the 2.5–7.5 NH standard thread, as specified in NFPA 1963, *Standard for Fire Hose Connections.*

Exception No. 1: Where local fire department connections do not conform to NFPA 1963, the authority having jurisdiction shall designate the connection to be used.

Exception No. 2: The use of threadless couplings shall be permitted where required by the authority having jurisdiction and where listed for such use.

All hose coupling threads in sprinkler systems and threads for hydrants on yard mains supplying sprinkler systems should match those of the first responding fire department. If fire department hose couplings are of the unthreaded type, such as a quick connect, the sprinkler system connections should be compatible. Other types of compatible connections are permitted because not all fire departments have switched to the NH standard thread for couplings.

3-9.2 Fire department connections shall be equipped with listed plugs or caps, properly secured and arranged for easy removal by fire departments.

The reliability of the fire department connection is dependent on its relative ease of use. This ease of use includes an evaluation of elements, such as the ability to efficiently remove the caps without special tools and to readily attach the hose. Missing caps must be replaced so the fire department connection does not become a receptacle for trash. Exhibit 5.41 illustrates a fire department connection.

3-9.3 Fire department connections shall be of an approved type.

3-10 Waterflow Alarms

3-10.1 Waterflow alarm apparatus shall be listed for the service and so constructed and installed that any flow of water from a sprinkler system equal to or greater than that from a single automatic sprinkler of the smallest orifice size installed on the system will result in an audible alarm on the premises within 5 minutes after such flow begins and until such flow stops.

NFPA 13 does not mandate the activation of a building fire alarm system when the sprinkler system is activated. If this feature is desired or specifically required by a building code, details, such as electrical circuit arrangement, time for alarm annunciation, and acceptable components, on this type of alarm system are contained in NFPA 72, *National Fire Alarm Code®*.

Where it is desired to indicate waterflow on each floor of a multistory building, a floor control assembly can be used as indicated in Figure A-5-15.4.2(b). In addition to providing a means of waterflow indication, a floor control assembly also typically contains a control valve to isolate the system piping on the floor and a means to conduct flow testing and drainage of system piping on the floor. See Section 5-15 for additional information.

3-10.2 Waterflow Detecting Devices.

3-10.2.1 Wet Pipe Systems. The alarm apparatus for a wet pipe system shall consist of a listed alarm check valve or other listed waterflow-detecting alarm device with the necessary attachments required to give an alarm.

On a wet pipe system, an alarm check valve as indicated in Exhibits 1.6 and 5.28 might be preferred to other listed waterflow-detecting alarm devices, particularly where fluctuating pressure water supplies are present. See Formal Interpretation 83-15. Fluctuating pressure can cause unwanted or nuisance alarms when other types of waterflow-detecting devices are used. On larger systems, an alarm check valve makes it possible to introduce an excess pressure higher than the supply on the system side of the alarm check valve. Waterflow is then detected by a drop in pressure on the system side of the alarm check. In other instances, the alarm check valve serves to prevent waterflow

Formal Interpretation 83-15

Reference: 3-10.2.1, 5-14.1.1.1, 5-14.1.1.2

Question 1: Would the use of a listed water flow switch and bell as the alarm apparatus meet the intent of 3-10.2.1?

Answer: Yes.

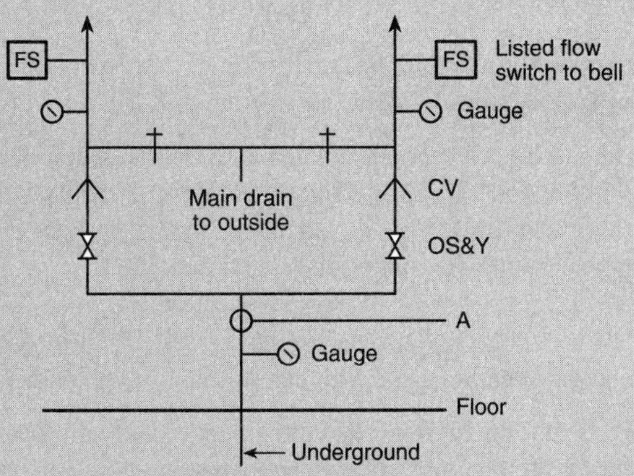

Figure 2 *Sectional Detail of Automatic Sprinkler Risers.*

Question 2: According to 5-14.1.1.1 and 5-14.1.1.2, would listed indicating valve be required at "A" in Figure 2?

Answer: A valve is not required at point "A" for the sketch provided. A more comprehensive answer would require consideration of the configuration of the outside underground piping and the control valves on it in addition to the location of the fire department connection. The pressure gauge indicated below point "A" is not required.

Issue Edition: 1983

Reference: 3-17.3.1, 3-14.2.1, 3-14.2.2

Date: March 1984

alarms on risers not involved in an actual waterflow condition when firepumps start or similar large fluctuations occur. In some instances, particularly when a system has both a large volume of dead-end piping and a backflow preventer, the alarm is not constant.

3-10.2.2 Dry Pipe Systems. The alarm apparatus for a dry pipe system shall consist of listed alarm attachments to the dry pipe valve. When a dry pipe valve is located on the system side of an alarm valve, connection of the actuating device of the alarms for the dry pipe valve to the alarms on the wet pipe system shall be permitted.

In some instances, small dry pipe systems are needed in areas of otherwise heated buildings, such as in areas under outside canopies, loading docks, unheated penthouses, and attics. In these cases, an auxiliary dry pipe system supplied from the building's wet pipe system can be more economical to install. Under these circumstances, the waterflow alarm from the dry pipe system can be connected into the wet pipe system alarm sounding or annunciating device.

3-10.2.3 Preaction and Deluge Systems. The alarm apparatus for deluge and preaction systems shall consist of alarms actuated independently by the detection system and the flow of water.

Because some preaction valves and deluge valves are not actuated by the operation of an automatic sprinkler, a supplemental detection system is required to actuate the valves. This detection system indicates fire prior to waterflow in the system. The detection system can also be utilized to initiate other fire-related activities.

3-10.2.4* Paddle-type waterflow alarm indicators shall be installed in wet systems only.

A-3-10.2.4 The surge of water that occurs when the valve trips can seriously damage the device.

In addition to damaging the device itself, the high-velocity flow could totally disengage the paddle and carry it downstream until it lodged in the piping causing an obstruction. This restriction is also applicable to a situation in which a wet pipe system acts as the supply source for a preaction, deluge, or dry pipe system. Exhibit 3.17 illustrates a paddle-type waterflow switch, and Exhibit 3.18 illustrates a paddle-type waterflow switch installed on a wet pipe riser.

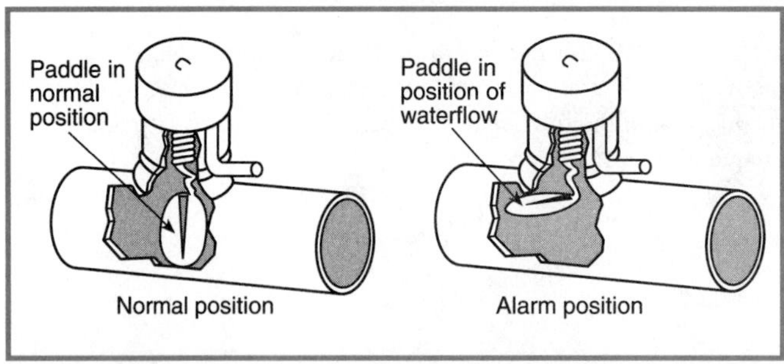

Exhibit 3.17 *Paddle-type waterflow switch.*

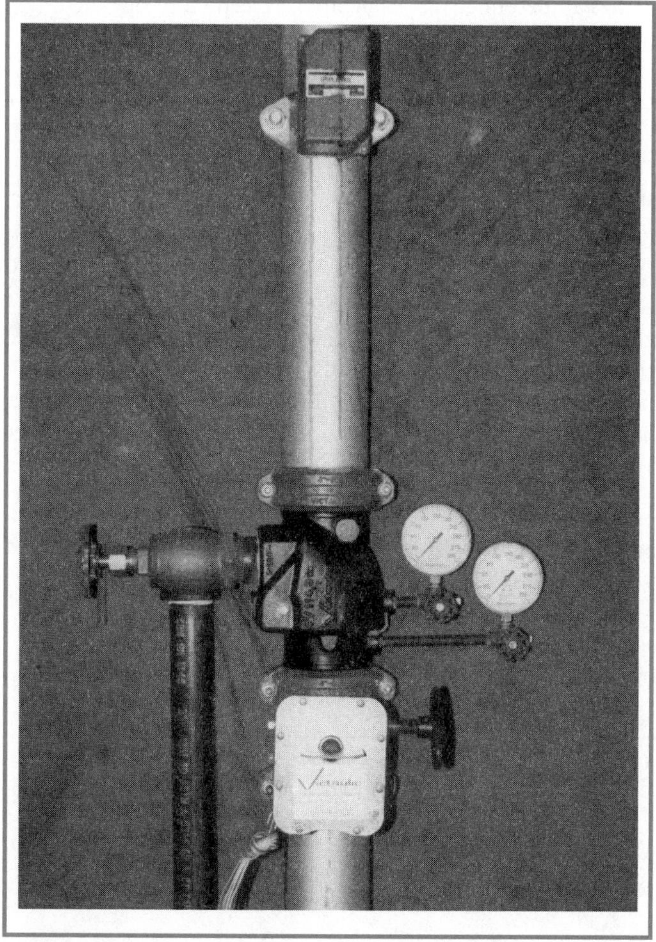

Exhibit 3.18 *Paddle-type waterflow switch installed on a wet pipe riser. (Switch located near top of riser.)*

3-10.3 Attachments—General.

3-10.3.1* An alarm unit shall include a listed mechanical alarm, horn, or siren or a listed electric gong, bell, speaker, horn, or siren.

A-3-10.3.1 Audible alarms are normally located on the outside of the building. Listed electric gongs, bells, horns, or sirens inside the building, or a combination of such used inside and outside, are sometimes advisable.

Outside alarms can be omitted where the sprinkler system is used as part of a central station, auxiliary, remote station, or proprietary signaling fire alarm system utilizing listed audible inside alarm devices.

The required audible local alarm can be either mechanically or electrically operated. NFPA 13 does not stipulate whether the alarm's location be indoors or outdoors, since the location is dependent on site-specific conditions. The location and number of audible alarms are normally dictated by the purpose of the alarms, such as for waterflow or evacuation. The location and number are also determined by normal operations within the protected premises and the locations at which the alarms would be expected to be heard. Exhibit 1.6 illustrates a water motor alarm.

3-10.3.2* Outdoor water motor–operated or electrically operated bells shall be weatherproofed and guarded.

A-3-10.3.2 All alarm apparatus should be so located and installed that all parts are accessible for inspection, removal, and repair, and such apparatus should be substantially supported.

The water motor gong bell mechanism should be protected from weather-related elements such as rain, snow, or ice. To the extent practicable, it should also be protected from other influencing factors such as birds or other small animals that might attempt to nest in such a device.

Internally moving parts of a water motor gong are expected to be adequately protected from a variety of problems such as snow, ice, birds' nests, and mechanical damage. The identified concerns are intended to give the user an idea of the types of problems to avoid. Some of these concerns can be difficult to guard against in all cases. The Formal Interpretation provides some additional insight regarding the intent of 3-10.3.2.

Formal Interpretation

Reference: 3-10.3.2

Question: Is it the intent of 3-10.3.2 to require guards against mechanical injury on all outdoor water motor gongs?

Answer: No. It is the intent of the Committee that the word "guarded" relate to the protection against birds, vermin, and debris.

Issue Edition: 1975

Reference: 3-16.4.2

Date: December 1976

3-10.4 All piping to water motor–operated devices shall be galvanized or brass or other corrosion-resistant material acceptable under this standard and of a size not less than ¾ in. (19 mm).

Most listed alarm valves, listed dry pipe valves, and listed deluge valves in NFPA 13 contain supplemental trim and pre-piped outlets for alarm lines. The trim, outlets, and associated alarm piping must be verified to be corrosion resistant. See Exhibits 1.4, 1.5 and 1.6.

3-10.5* Attachments—Electrically Operated.

A-3-10.5 Switches that will silence electric alarm-sounding devices by interruption of electric current are not desirable; however, if such means are provided, then the electric alarm-sounding device circuit should be arranged so that, when the sounding device is electrically silenced, that fact should be indicated by means of a conspicuous light located in the vicinity of the riser or alarm control panel. This light should remain in operation during the entire period of the electric circuit interruption.

3-10.5.1 Electrically operated alarm attachments forming part of an auxiliary, central station, local protective, proprietary, or remote station signaling system shall be installed in accordance with NFPA 72, *National Fire Alarm Code.*

Exception: Sprinkler waterflow alarm systems that are not part of a required protective signaling system shall not be permitted to be supervised and shall be installed in accordance with NFPA 70, National Electrical Code®, Article 760.

3-10.5.2 Outdoor electric alarm devices shall be listed for outdoor use.

3-10.6 Drains from alarm devices shall be so arranged that there will be no overflowing at the alarm apparatus, at domestic connections, or elsewhere with the sprinkler drains wide open and under system pressure. *(See 5-14.2.6.1.)*

REFERENCES CITED IN COMMENTARY

National Fire Protection Association, 1 Batterymarch Park, P.O. Box 9101, Quincy, MA 02269-9101.

NFPA 25, *Standard for the Inspection, Testing, and Maintenance of Water-Based Fire Protection Systems*, 1998 edition.

NFPA 51B, *Standard for Fire Prevention During Welding, Cutting, and Other Hot Work*, 1999 edition.

NFPA 72, *National Fire Alarm Code®*, 1999 edition.

American National Standards Institute, Inc., 11 West 42nd Street, 13th floor, New York, NY 10036.

ANSI B31.1, *Code for Power Piping*, 1990 edition.

American Society of Mechanical Engineers, Three Park Avenue, New York, NY 10016-5990.

ASME Boiler and Pressure Vessel Code, 1998 edition.

American Society for Testing and Materials, 100 Barr Harbor Drive, West Conshohocken, PA 19428-2959, 610-832-9500.

ASTM B 813, *Standard Specification for Liquid and Paste Fluxes for Soldering Applications of Copper and Copper-Alloy Tube*, 1993 edition.

American Welding Society, 550 N.W. LeJeune Road, Miami, FL 33126, 800-443-9353.

AWS B2.1, *Standard for Welding Procedure and Performance Qualification*, 1984 edition.

CHAPTER 4

System Requirements

This chapter provides the installation rules and characteristics that are unique to a particular type of system. The types of systems addressed by NFPA 13 include wet pipe, dry pipe, preaction, deluge, combined dry pipe preaction, antifreeze, circulating closed looped, outside systems for exposure protection, systems used in refrigerated spaces, and systems for commercial-type cooking equipment and ventilation. Chapter 4 covers information specific to each type of system. The information ranges from the relatively simple requirement for pressure gauges to limitations imposed on dry pipe system sizes to the design parameters for a circulating closed-loop system.

4-1 Wet Pipe Systems

Wet pipe sprinkler systems are the simplest and the most common type of sprinkler system in use. In wet pipe systems, the piping contains water at all times and is connected to a water supply so that water discharges immediately from a sprinkler when the sprinkler activates. Because wet pipe systems have relatively few components, they have an inherently higher degree of reliability than other system types. Exhibits 1.6, 3.16, 5.26, and 5.28 illustrate some of the valves and other components found on a wet pipe system.

Dry sprinklers, as defined in 1-4.5.4, can also be used on wet pipe systems where individual sprinklers are extended into spaces that are subject to freezing. The extension of sprinklers into a freezer unit is one example of this application. A minimum length for the dry sprinkler's exposed portion outside the freezer is specified to avoid freezing water at the connection point to the branch line. These minimum lengths are specified by the sprinkler manufacturer. Dry sprinklers have a maximum length of up to 4 ft (1.2 m). Special applications utilizing dry sprinklers in upright and other positions should be in accordance with their listings. For example, see Figure A-5-13.8 for the application of a dry pendent sprinkler under an exterior canopy.

4-1.1 Pressure Gauges.

A listed pressure gauge conforming to 5-15.3.2 shall be installed in each system riser. Pressure gauges shall be installed above and below each alarm check valve where such devices are present.

At least one pressure gauge is required in each system riser. When an alarm check valve is used, two gauges are required—one on the supply side of the alarm check valve and one on the system side. It is not unusual for the gauge on the system side of the valve to read a higher pressure. Pressure surges are typically trapped in the system by the check valve. Readings from these gauges can be used to establish a database on the available pressure in the water supply. The gauge on the supply side is also used during the 2-in. (51-mm) drain test required by NFPA 25, *Standard for the Inspection, Testing, and Maintenance of Water-Based Fire Protection Systems*, to establish a database of residual pressures in the water supply.

4-1.2 Relief Valves.

A gridded wet pipe system shall be provided with a relief valve not less than ¼ in. (6.4 mm) in size set to operate at pressures not greater than 175 psi (12.1 bar).

Exception No. 1: When the maximum system pressure exceeds 165 psi (11.4 bar), the relief valve shall operate at 10 psi (0.7 bar) in excess of the maximum system pressure.

Exception No. 2: Where auxiliary air reservoirs are installed to absorb pressure increases, a relief valve shall not be required.

Some situations have occurred in which the pressure buildup in gridded wet pipe systems exceeded the working pressure of the system components, which is typically 175 psi (12.1 bar). Solar heating can increase temperature differentials in a building and, because wet pipe systems are closed systems, this solar heating can cause static pressures in the system to exceed 175 psi (12.1 bar). These pressures act on system components and can cause system failure if the pressures exceed the components' pressure ratings. The location of the relief valve will affect the pressure setting needed to protect the system, since it is necessary to account for elevation pressure experienced at the base of the riser.

Since a limited number of air pockets typically exist in gridded systems, the auxiliary air reserves permitted by Exception No. 2 need to be large enough to account for any pressure buildup in the system. One method is to use an expansion chamber as discussed in 4-5.3.2.

4-1.3 Auxiliary Systems.

A wet pipe system shall be permitted to supply an auxiliary dry pipe, preaction, or deluge system, provided the water supply is adequate.

This subsection permits a combined system riser to be used when the available water supply is adequate to support such an arrangement. Exhibit 4.1 illustrates a combined system riser in which the water supply for the dry pipe system is taken from the wet pipe system. System protection area limitations specified in Section 5-2 apply when considering an auxiliary system. A control valve on the auxiliary system is desirable to facilitate maintenance and resetting of the system valve on the auxiliary system. The user needs to be aware that the combined system riser is not the same as a combined sprinkler/standpipe riser.

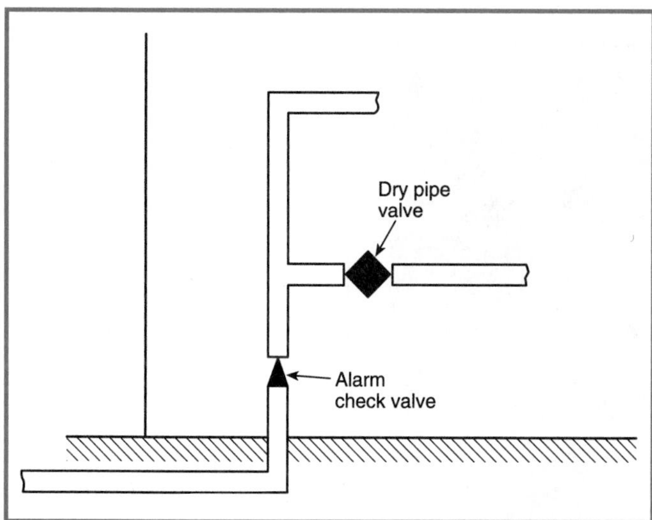

Exhibit 4.1 *Example of a combined system riser.*

4-2* Dry Pipe Systems

A-4-2 A dry pipe system should be installed only where heat is not adequate to prevent freezing of water in all parts of, or in sections of, the system. Dry pipe systems should be converted to wet pipe systems when they become unnecessary because adequate heat is provided. Sprinklers should not be shut off in cold weather.

Where two or more dry pipe valves are used, systems preferably should be divided horizontally to prevent simultaneous operation of more than one system and the resultant increased time delay in filling systems and discharging water and to prevent receipt of more than one waterflow alarm signal.

Where adequate heat is present in sections of the dry pipe system, consideration should be given to dividing the system into a separate wet pipe system and dry pipe system. Minimized use of dry pipe systems is desirable where speed of operation is of particular concern.

Dry pipe systems are usually used in buildings or in areas that are subject to cold or freezing temperatures. Dry pipe systems are typically filled with pressurized air or nitrogen rather than water. As with wet pipe systems, automatic sprinklers are used. The operation of a sprinkler causes air pressure in the system to drop. This pressure drop activates a dry pipe valve and allows water to flow through the system. Dry pipe systems are more complex than wet pipe systems and require greater attention regarding their design, installation, and maintenance. Exhibit 1.4 provides an example of a dry pipe sprinkler system, and Exhibit 4.2 illustrates some of the components found on the system riser of a dry pipe system.

Exhibit 4.2 *Dry pipe valve and associated components.*

As indicated in the Formal Interpretation for A-4-2, it is not the intent of NFPA 13 to require a building or portions of it to be heated to accommodate the sprinkler system.

Formal Interpretation

Reference: A-4-2

Question: Is it the intent of A-4-2 to recommend the use of a wet-pipe sprinkler system in the building in a cold climate where the heating system is turned off during non-working hours?

Answer: No. This section merely points out that wet-pipe systems are impractical in unheated spaces. It is not the intent of this section to recommend providing heat in any space to accommodate the sprinkler system nor to replace dry pipe systems with wet pipe systems.

Issue Edition: 1975

Reference: A-5-2.1

Date: January 1977

Reprinted for correction: January 1993

4-2.1 Pressure Gauges.

Listed pressure gauges conforming with 5-15.3.2 shall be connected as follows:

(1) On the water side and air side of the dry pipe valve
(2) At the air pump supplying the air receiver where one is provided
(3) At the air receiver where one is provided
(4) In each independent pipe from air supply to dry pipe system
(5) At exhausters and accelerators

In a dry pipe system, pressure gauges are used to monitor water pressures in the system and in the water supply and air pressures in the system.

4-2.2 Upright Sprinklers.

Only upright sprinklers shall be installed on dry pipe systems.

Exception No. 1: Listed dry sprinklers shall be permitted.*

Exception No. 2: Pendent sprinklers installed on return bends shall be permitted where both the sprinklers and the return bends are located in a heated area.

Exception No. 3: Horizontal sidewall sprinklers, installed so that water is not trapped, shall be permitted.

A-4-2.2 Exception No. 1. Installation limitations of listed dry pendent sprinklers can vary with different products. Limitations should be included in product installation instructions to warn the user of the potential accumulation of water, scale, and sediment from collecting at the sprinkler.

Since dry pipe systems are installed in areas that are subject to freezing, any water that remains in the piping after system operation should be quickly drained. The use of pendent sprinklers would allow a small amount of water to remain on the seat of the operating mechanism. This water would eventually freeze and impair the sprinkler. Therefore, pendent sprinklers are not permitted. However, if a sprinkler in the pendent position is needed, then Exception No. 1 allows the use of listed dry sprinklers as an option. Dry sprinklers prevent water from accumulating on the sprinkler's operating mechanism and in the drop nipple. This arrangement minimizes the time it takes the system to be returned to its operating condition. A limit is placed on the size of the fitting used with a dry pendent sprinkler in order to avoid trapping water on the operating mechanism. To further aid in the drainage of dry pipe systems, 5-14.2.3 requires piping to be installed at a pitch.

Another consideration to be taken into account where using dry pipe systems is the need to minimize the obstruction potential from internal pipe scale that can break loose while the piping is being charged with water. The water velocities experienced when the dry pipe valve opens are likely to dislodge pipe scale and carry it to the open sprinklers. The use of pendent sprinklers on return bends is permitted by Exception No. 2 as an alternative to dry sprinklers when the return bend, branch line piping, and sprinkler are in a heated area. The return bend serves to minimize obstructions to waterflow caused by pipe scale.

The conditions necessary to implement Exception No. 2 are rather uncommon. One example is in a cold storage warehouse that is protected with a dry pipe system. If the warehouse contained an office that presumably is heated, the sprinkler pipe in that area is also in a heated area and the return bend option could be used.

Exception No. 3 allows listed horizontal sidewall sprinklers in dry pipe, preaction, and combined dry pipe preaction systems, provided they are installed so that water is not trapped behind them. If these sprinklers are installed in such a manner as to avoid the accumulation of water, scale, and sediment, then no compelling reason to restrict their use in this application is evident.

4-2.3* Size of Systems.

A-4-2.3 The capacities of the various sizes of pipe given in Table A-4-2.3 are for convenience in calculating the capacity of a system.

4-2.3.1* Volume Limitations. Not more than 750 gal (2839 L) system capacity shall be controlled by one dry pipe valve.

Exception: Piping volume shall be permitted to exceed 750 gal (2839 L) for nongridded systems if the system design is such that water is delivered to the system test connection in not more than 60 seconds, starting at the normal air pressure on the system and at the time of fully opened inspection test connection.

A-4-2.3.1 The 60-second limit does not apply to dry systems with capacities of 500 gal (1893 L) or less, nor to dry systems with capacities of 750 gal (2839 L) or less if equipped with a quick-opening device.

Table A-4-2.3 Capacity of One Foot of Pipe (Based on Actual Internal Pipe Diameter)

Nominal Pipe Diameter (in.)	Pipe Schedule 40 (gal)	Pipe Schedule 10 (gal)	Nominal Pipe Diameter (in.)	Pipe Schedule 40 (gal)	Pipe Schedule 10 (gal)
¾	0.028		3	0.383	0.433
1	0.045	0.049	3½	0.513	0.576
1¼	0.078	0.085	4	0.660	0.740
1½	0.106	0.115	5	1.040	1.144
2	0.174	0.190	6	1.501	1.649[b]
2½	0.248	0.283	8	2.66[a]	2.776[c]

For SI units, 1 in. = 25.4 mm; 1 ft = 0.3048 m; 1 gal = 3.785 L.

[a]Schedule 30.

[b]0.134 wall pipe.

[c]0.188 wall pipe.

The requirement permitting the use of check valves to subdivide a system into volumes less than 750 gal (2839 L) was removed from NFPA 13 in 1987. Concern about air leaking by older check valves and delaying the trip time prompted this removal. Even though this practice is still identified for combined dry pipe and preaction systems (see 4-4.4.1), it is not considered applicable for dry pipe systems. However, NFPA 13 does not prohibit the use of check valves to subdivide the system so that the 60-second water delivery requirement of the exception to A-4-2.3.1 can be met. Per NFPA 25, waterflow tests must be conducted periodically. Where system volumes exceed 750 gal (2839 L), compliance with the 60-second water delivery time must be verified on a periodic basis.

The water delay associated with dry pipe sprinkler systems is a significant disadvantage of their use. This delay consists of two parts—the time it takes for the dry pipe valve to operate and the time it takes for water to reach the open sprinkler once the valve operates. To compensate for these delays, NFPA 13 requires certain provisions such as limitations on system volume, the installation of quick-opening devices (see 4-2.4), the use of more conservative C factors for hydraulically calculated systems (see 8-4.4.5), and a larger design area (see 7-2.3.2.6).

Dry pipe systems are limited to a maximum volume of 750 gal (2839 L). This volume limitation can be exceeded if water delivery to the inspector's test connection can be achieved in 60 seconds or less. The 60-second water delivery is not required where the 750-gal (2839-L) volume limitation is not exceeded. Some dry pipe systems with capacities less than 750 gal (2839 L) can take up to 3 minutes to deliver water to the inspector's test connection, which is considered acceptable.

For those designers wishing to exceed the 750-gal (2839-L) limit, *The SFPE Handbook of Fire Protection Engineering* provides a formula for estimating the length of time required to trip the dry pipe valve (see Section 4-3). Modeling the transit time of the water from the dry pipe valve to the operating sprinkler is a more complex problem. No established method to determine this delay is currently available, although research efforts are ongoing.

Designers must account for the fact that a precise method for predicting water delivery time is not readily available. When system volume exceeds 750 gal (2839 L), the only means to verify a delivery time of 60 seconds or less is by conducting a flow test through the inspector's test connection. Failure of this test can result in rejection of the system by the authority having jurisdiction.

4-2.3.2 Gridded dry pipe systems shall not be installed. *(See 5-14.2.5.3.3.)*

Gridded dry pipe systems are subject to excessive time delays for water to reach the operating sprinkler. Past experience indicates that in many gridded dry pipe systems, the times for delivering water to the inspector's test connection were excessive—in some cases as long as 10 minutes. The multiple number of flow paths the water can take to fill the grid makes this piping configuration undesirable for dry pipe systems because of the associated time delays. Additionally, excessive amounts of air can become trapped in the system.

The restriction stipulated in 4-2.3.2 is not applicable to a preaction system unless it is of the double interlock type. These types of preaction systems maintain air pressures typical to those of dry pipe systems.

The use of a looped piping arrangement is not prohibited for a dry pipe system because the associated water delivery time delays are not as severe as for a gridded system.

4-2.4 Quick-Opening Devices.

Quick-opening devices consist primarily of accelerators and exhausters. Accelerators cause the dry pipe valve to operate more quickly by redirecting system air into the valve's intermediate chamber, thus allowing water pressure to expel air from the activated sprinklers at a faster rate. Exhausters accelerate the rate at which air is discharged to atmosphere, thus reducing the time it takes for the dry pipe valve to operate and the time it takes for water to reach the open sprinklers. Exhibit 4.2 illustrates an accelerator attached to the dry pipe valve.

4-2.4.1 Dry pipe valves shall be provided with a listed quick-opening device where system capacity exceeds 500 gal (1893 L).

Exception: A quick-opening device shall not be required if the requirements of the Exception to 4-2.3.1 can be met without such a device.

Even though quick-opening devices are not required where systems are of smaller capacity [less than 500 gal (1893 L)], they can be used whenever faster operating times are desired. The exception allows omission of quick-opening devices when water delivery occurs at the inspector's test connection within 60 seconds or less. Omitting this device and failing to stay within established time constraints on larger systems creates the likelihood of a noncompliant system.

4-2.4.2 The quick-opening device shall be located as close as practical to the dry pipe valve. To protect the restriction orifice and other operating parts of the quick-opening device against submergence, the connection to the riser shall be above the point at which water (priming water and back drainage) is expected when the dry pipe valve and quick-opening device are set, except where design features of the particular quick-opening device make these requirements unnecessary.

Special circumstances can require a quick-opening device, such as an exhauster, to be located at a point remote from the dry pipe valve. In these instances, consideration should be given to protection of the device or its components against freezing. Remote devices should be accessible for servicing.

Antiflooding devices are activated when the valve trips. The devices do not stop the submergence of operating parts from the priming water.

4-2.4.3 A soft disc globe or angle valve shall be installed in the connection between the dry pipe sprinkler riser and the quick-opening device.

The shutoff valve between the dry pipe sprinkler riser and the quick-opening device is necessary, since these devices require separate maintenance activities. It is very undesirable to have the entire system out of service because of a problem with the quick-opening device. However, because quick-opening devices can be shut off separately, they are frequently left out of service. Since system design is based on proper operation of the quick-opening device, it should be maintained in proper operating condition at all times, and any necessary repairs should be promptly completed.

4-2.4.4 A check valve shall be installed between the quick-opening device and the intermediate chamber of the dry pipe valve. If the quick-opening device requires pressure feedback from the intermediate chamber, a valve type that will clearly indicate whether it is opened or closed shall be permitted in place of that check valve. This valve shall be constructed so that it can be locked or sealed in the open position.

The shutoff valve for this type of quick-opening device is much the same as the shutoff valve at the connection to the sprinkler riser. The valve should always be maintained in the open position. The check valve allows flow only from the quick-opening device to the intermediate chamber.

4-2.4.5 A listed antiflooding device shall be installed in the connection between the dry pipe sprinkler riser and the quick-opening device.

Exception: A listed antiflooding device shall not be required where the quick-opening device has built-in antiflooding design features.

Quick-opening devices and their accessory equipment, such as antiflooding devices, are often neglected when service and maintenance activities on the system are to be conducted. The manufacturer's instructions and NFPA 25 need to be carefully followed when resetting each dry pipe valve and quick-opening device. Impairments to the quick-opening device have been caused by not properly draining the equipment while resetting the dry pipe valve after it has operated. NFPA 25 contains a series of inspection, testing, and maintenance provisions that, when implemented, greatly improves the reliability of these devices.

4-2.5* Location and Protection of Dry Pipe Valve.

A-4-2.5 The dry pipe valve should be located in an accessible place near the sprinkler system it controls. Where exposed to cold, the dry pipe valve should be located in a valve room or enclosure of adequate size to properly service equipment.

These recommendations serve to reduce the amount of piping in a dry pipe system and to help minimize system capacity. Long bulk mains from dry pipe valves to system piping can use a large portion of the available system capacity, which results in a rather small area of actual sprinkler coverage. To minimize the impact of the bulk main on the system's permitted capacity or water delivery time requirement, the installation of the bulk main as underground piping below floors and the relocation of the dry pipe valve in the center of the building are sometimes considered.

However, 5-14.4.3.1 places limitations on the installation of fire mains beneath buildings. Any leaks in the buried piping underneath the building could go undetected for long periods, and any corrective action would require penetration of the floor and relocation of any adjacent equipment. The location of system valves in the center of a building also places the system controls in the center of the building, which may not be desirable during a fire event.

4-2.5.1 The dry pipe valve and supply pipe shall be protected against freezing and mechanical injury.

4-2.5.2 Valve rooms shall be lighted and heated. The source of heat shall be of a permanently installed type. Heat tape shall not be used in lieu of heated valve enclosures to protect the dry pipe valve and supply pipe against freezing.

Heat tape is not considered a permanently installed heat source for this application. A fixed heat source, such as a baseboard or unit heater, satisfies the requirements.

4-2.5.3 The supply for the sprinkler in the dry pipe valve enclosure shall be from the dry side of the system.

4-2.5.4 Protection against accumulation of water above the clapper shall be provided for a low differential dry pipe valve. An automatic high water level signaling device or an automatic drain device shall be permitted.

This requirement is critical for two reasons. If an accumulation of water extends beyond the heated enclosure, the potential for the system to freeze increases. In addition to this concern, a water column with a measurable head of pressure can develop, preventing operation of the dry pipe valve. The pressure of the water column can be greater than the pressure that would cause the valve to open.

4-2.6 Air Pressure and Supply.

Proper air pressure should be maintained in the system at all times. It is important to follow manufacturer's instructions regarding the range of air pressures to be maintained. Low air pressures could result in accidental operation of the dry pipe valve. High air pressures result in slower operation because additional air must be exhausted from the system before water can be delivered to open sprinklers.

Traditional dry pipe valves normally have a differential in water pressure to air pressure at a trip point of approximately 5.5:1. A low differential dry pipe valve usually is an alarm check valve that has been converted to a dry pipe valve. The differential in water pressure and air pressure at the trip point of these valves is approximately 1.1:1.

4-2.6.1 Maintenance of Air Pressure. Air or nitrogen pressure shall be maintained on dry pipe systems throughout the year.

The use of either air or nitrogen in dry pipe systems is allowed. The use of nitrogen can be preferable in systems where severe accumulation of condensation from air in the system is likely.

Air pressure does not have to be automatically maintained or electrically supervised. The presence of such supervision, though, does impact the frequency of maintenance inspections.

4-2.6.2* Air Supply. The compressed air supply shall be from a source available at all times and have a capacity capable of restoring normal air pressure in the system within 30 minutes.

Exception: In refrigerated spaces maintained below 5°F (−15°C), normal system air pressure shall be restored within 60 minutes.

A-4-2.6.2 The compressor should draw its air supply from a place where the air is dry and not too warm. Moisture from condensation can cause trouble in the system. **A**

A reliable and available air supply is necessary to allow the dry pipe system to stay in service and to maintain the necessary pressure differential between the water and air sides of the dry pipe valve. A compressor, a shop air supply, or a manifolded arrangement of nitrogen cylinders all constitute acceptable sources. The air maintenance device, if present, can be bypassed to meet the 30-minute filling criteria.

4-2.6.3 Air Filling Connection. The connection pipe from the air compressor shall not be less than ½ in. (13 mm) in diameter and shall enter the system above the priming water level of the dry pipe valve. A check valve shall be installed in this air line, and a shutoff valve of the renewable disc type shall be installed on the supply side of this check valve and shall remain closed unless filling the system.

4-2.6.4 Relief Valve. A listed relief valve shall be provided between the compressor and controlling valve and shall be set to relieve at a pressure 5 psi (0.3 bar) in excess of maximum air pressure carried in the system.

A relief valve on manually operated systems is needed to reduce any excess air pressure in the system. Omission of the relief valve could result in higher pressures, thus resulting in slower operating times. Additionally, high pressures have the potential to damage some of the gasket materials used on the dry pipe valve. Air maintenance devices, if installed, also provide control of the pressure level (see 4-2.6.5).

4-2.6.5 Shop Air Supply. Where the air supply is taken from a shop system having a normal pressure greater than that required for dry pipe systems and an automatic air maintenance device is not used, the relief valve shall be installed between two control valves in the air line, and a small air cock, which is normally left open, shall be installed in the fitting below the relief valve. *(See Figure 4-2.6.5.)*

The relief valve on a manual system that uses shop air supply should be set to prevent overpressurizing of the sprinkler system. As described in the commentary to 4-2.6.4, high pressures are undesirable because they result in slower operation of the system and can negatively affect system components.

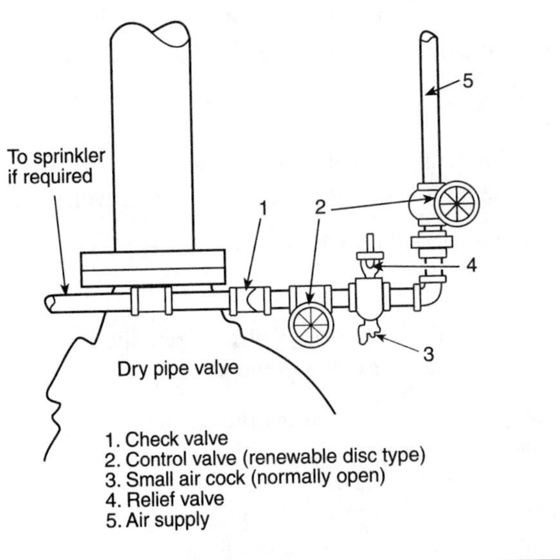

***Figure 4-2.6.5** Air supply from shop system.*

4-2.6.6 Automatic Air Compressor. Where a dry pipe system is supplied by an automatic air compressor or plant air system, any device or apparatus used for automatic maintenance of air pressure shall be of a type specifically listed for such service and capable of maintaining the required air pressure on the dry pipe system. Automatic air supply to more than one dry pipe system shall be connected to enable individual maintenance of air pressure in each system. A check valve or other positive backflow prevention device shall be installed in the air supply to each system to prevent airflow or waterflow from one system to another.

Pumping directly from an automatic air compressor through a fully opened supply pipe into the sprinkler system is not permitted. A restriction prevents the air supply system from adding air too quickly, thus preventing or slowing operation of the dry pipe valve. NFPA 13 does not require the use of a device to automatically maintain air pressure in the system. However, other means of maintaining the required air pressure can become overly burdensome. Any device used for automatic air maintenance needs to be listed for such service. Listed air maintenance devices provide both restriction of the airflow and regulation of the air pressure. A compressor used in conjunction with the air maintenance device only needs to comply with the requirements of 4-2.6.2 and does not have to be listed. Air compressors are available for use with and without air holding tanks. Compressors without air holding tanks are subject to more instances of short-cycling of the compressor. Exhibit 4.3 illustrates an air compressor without an air holding tank, and an automatic air maintenance device.

Exhibit 4.3 *Air compressor and air maintenance device for a dry pipe system.*

4-2.6.7 System Air Pressure. The system air pressure shall be maintained in accordance with the instruction sheet furnished with the dry pipe valve, or shall be 20 psi (1.4 bar) in excess of the calculated trip pressure of the dry pipe valve, based on the highest normal water pressure of the system supply. The permitted rate of air leakage shall be as specified in 10-2.3.

4-2.6.8 Nitrogen. Where used, nitrogen shall be introduced through a pressure regulator set to maintain system pressure in accordance with 4-2.6.7.

4-3 Preaction Systems and Deluge Systems

Preaction systems are more complex than wet pipe and dry pipe systems because they contain more components and equipment. The systems require specialized knowledge and experience with their design and installation, and the inspection, testing, and maintenance activities needed to ensure their reliability and functionality are more involved. Manufacturer's specifications and listing restrictions must be strictly followed. Costs associated with the initial installation, use of supplemental equipment to operate the system, and maintenance should be thoroughly evaluated when considering a preaction system.

Since the first edition of this handbook was published in 1983, different types of valves have become available that are classified for use in preaction systems. The operating characteristics of these valves cause certain types of preaction systems to have qualities similar to that of a dry pipe system, such as a double interlock preaction system. Therefore, the same rules and restrictions that apply to dry pipe systems apply to double interlock preaction systems.

4-3.1* General.

A-4-3.1 Conditions of occupancy or special hazards might require quick application of large quantities of water, and, in such cases, deluge systems might be needed.

Fire detection devices should be selected to assure operation yet guard against premature operation of sprinklers based on normal room temperatures and draft conditions.

In locations where ambient temperature at the ceiling is high from heat sources other than fire conditions, heat-responsive devices that operate at higher than ordinary temperature and that are capable of withstanding the normal high temperature for long periods of time should be selected.

Where corrosive conditions exist, materials or protective coatings that resist corrosion should be used.

To help avoid ice formation in piping due to accidental tripping of dry pipe valves in cold storage rooms, a deluge automatic water control valve can be used on the supply side of the dry pipe valve. Where this method is employed, the following also apply:

(1) Dry systems can be manifolded to a deluge valve, with the protected area not exceeding 40,000 ft^2 (3716 m^2).
(2) Where a dry system is manifolded to a deluge valve, the distance between valves should be as short as possible to minimize water hammer.
(3) The dry pipe valves should be pressurized to 50 psi (3.4 bar) to reduce the possibility of dry pipe valve operation from water hammer.

The combination dry pipe and deluge valve was one of the older approaches for a double interlock preaction valve. Current designs use a single valve.

4-3.1.1 All components of pneumatic, hydraulic, or electrical systems shall be compatible.

This requirement for compatibility is necessary to ensure that all system components function as an integrated unit. The correct coordination of detection devices, releasing equipment, and the control panel is imperative for prompt and reliable operation of the system. Specific requirements on the design of hydraulic, pneumatic, and electrical detection systems are contained in NFPA 72, *National Fire Alarm Code*®. The releasing panel or module is an essential component for system operation and is required to be listed (see 3-1.1).

4-3.1.2 The automatic water control valve shall be provided with hydraulic, pneumatic, or mechanical manual means for operation that is independent of detection devices and of the sprinklers.

In the event that the system does not automatically operate, a means for manually operating the system activation valve must be provided. This feature is usually standard on valves listed for use in preaction or deluge systems. The manual release device must be a stand-alone arrangement whose operation is ensured regardless of the potential failure of the associated detection system.

4-3.1.3 Pressure Gauges. Listed pressure gauges conforming with 5-15.3.2 shall be installed as follows:

(1) Above and below preaction valve and below deluge valve
(2) On air supply to preaction and deluge valves

4-3.1.4 A supply of spare fusible elements for heat-responsive devices, not less than two of each temperature rating, shall be maintained on the premises for replacement purposes.

4-3.1.5 Hydraulic release systems shall be designed and installed in accordance with manufacturer's requirements and listing for height limitations above deluge valves or deluge valve actuators to prevent water column.

4-3.1.6 Location and Spacing of Detection Devices. Spacing of detection devices, including automatic sprinklers used as detectors, shall be in accordance with their listing and manufacturer's specifications.

Various methods are available for activating a deluge or preaction system. Among these methods are pneumatic and hydraulic detection systems, smoke detectors, heat detectors, and detectors that respond to ultraviolet and infrared radiation. An automatic sprinkler is also recognized as a type of detector that can be used to activate an adjacent preaction or deluge system.

Although the definition of *sprinkler system* (see 1-4.2) states that these systems are "usually activated by heat," this definition should not be interpreted as prohibiting the use of smoke detection devices or other non-heat-sensing devices. Smoke, as well as ultraviolet and infrared radiation, are by-products of a fire, and their detection can be used to activate a preaction or deluge system.

As stated in 4-3.1.6, detection systems are to be installed in accordance with their listing or as directed by the detection device manufacturer. The listing criteria for the detection devices are governed by NFPA 72.

In addition to permitting a range of options concerning the types of detection devices, various combinations of operation are permitted. See the commentary following 4-3.2.1.

4-3.1.7 Devices for Test Purposes and Testing Apparatus.

4-3.1.7.1 Where detection devices installed in circuits are located where not readily accessible, an additional detection device shall be provided on each circuit for test purposes at an accessible location and shall be connected to the circuit at a point that will assure a proper test of the circuit.

4-3.1.7.2 Testing apparatus capable of producing the heat or impulse necessary to operate any normal detection device shall be furnished to the owner of the property with each installation. Where explosive vapors or materials are present, hot water, steam, or other methods of testing not involving an ignition source shall be used.

4-3.1.8 Location and Protection of System Water Control Valves.

4-3.1.8.1 System water control valves and supply pipes shall be protected against freezing and mechanical injury.

4-3.1.8.2 Valve rooms shall be lighted and heated. The source of heat shall be of a permanently installed type. Heat tape shall not be used in lieu of heated valve enclosure rooms to protect preaction and deluge valves and supply pipe against freezing.

Heat tape is not considered a permanently installed heat source. A fixed heat source, such as a baseboard or unit heater, satisfies the requirements.

4-3.2 Preaction Systems.

This subsection identifies the three basic types of preaction systems by identifying their means of operation. Specific terms that identify the mode of operation, such as *single interlock*, *non-interlock*, and *double interlock*, were added to the 1996 edition. These terms help differentiate between types of systems and allow for better reference. Besides the three basic types of systems, additional variations, such as the use of a cross-zoned smoke detection system, are possible. Other combinations can involve the use of additional devices such as an infrared detector. The detection systems for the single interlock and non-interlock systems typically have a lower temperature rating than the system sprinklers. When so configured, the single interlock and non-interlock systems are expected to operate more quickly than the double interlock or dry pipe systems. Exhibit 1.5 illustrates an example of a double interlock preaction system with electric activation.

4-3.2.1 Preaction systems shall be one of the following types.

(a) *Single Interlock System.* A single interlock system admits water to sprinkler piping upon operation of detection devices.

(b) *Non-Interlock System.* A non-interlock system admits water to sprinkler piping upon operation of detection devices or automatic sprinklers.

(c) *Double Interlock System.* A double interlock system admits water to sprinkler piping upon operation of both detection devices and automatic sprinklers.

Both the single interlock and non-interlock systems require only one event to occur before water is admitted to the system. The single interlock system is activated by the release of the detection system. Sprinkler activation does not affect this function. The non-interlock system is activated if either the detection system or a sprinkler operates.

The double interlock preaction system requires two events to occur before water is admitted to the system. One event consists of the activation of a device installed on the supplemental detection system. The other event includes the operation of a sprinkler that causes the maintained air pressure in the system to fall to a predetermined level, which is similar to that of a dry pipe system. When one of these events occurs, the system activation valve goes into a preset position. When the second event occurs, the valve opens, and water enters the system. Water does not enter the system until both events occur. These two events can occur in any order and result in the same outcome.

4-3.2.2 Size of Systems. Not more than 1000 automatic sprinklers shall be controlled by any one preaction valve.

Exception: For preaction system types described in 4-3.2.1(c), not more than 750 gal (2839 L) shall be controlled by one preaction valve unless the system is designed to deliver water to the system test connection in not more than 60 seconds, starting at the normal air pressure on the system, with the detection system operated and at the time of fully opened inspection test connection. Air pressure and supply shall comply with 4-2.6.

Because of the need to wait for the drop in air pressure, a double interlock system has characteristics that are similar to that of a dry pipe system. Double interlock preaction systems maintain air pressures similar to those found on dry pipe systems and, therefore, require the same restrictions as dry pipe systems. The restrictions include the 30 percent increase in design area (see 7-2.3.2.6), the 750-gal (2839-L) limit (see 4-2.3.1), and the prohibition of the gridded piping arrangement (see 4-2.3.2). Double interlock preaction systems are usually used for cold storage warehouses or flash-freeze facilities.

The volume of single interlock and non-interlock systems is not restricted. These systems are unrestricted because they are expected to operate more quickly than double interlock systems and do not rely on the discharge of system air in order to activate.

4-3.2.3* Supervision. Sprinkler piping and fire detection devices shall be automatically supervised where there are more than 20 sprinklers on the system. All preaction system types described in 4-3.2.1(b) and 4-3.2.1(c) shall maintain a minimum supervising air pressure of 7 psi (0.5 bar).

A-4-3.2.3 Supervision, either electrical or mechanical, as used in 4-3.2.3 refers to constant monitoring of piping and detection equipment to ensure the integrity of the system.

The term *supervised* in 4-3.2.3 refers to a method of ensuring the overall integrity of the piping system. Pressurized air usually is introduced to the preaction system pipe network. A loss in air pressure indicates a sprinkler activation or a major leak in the system. Normal supervision status registers that the system piping and sprinklers are intact. This reading, however, is not meant to verify that the system is completely free of minor leaks. The air compressor and the air maintenance device could be able to maintain the necessary air pressure even where minor leaks exist.

A single interlock preaction system, as described in 4-3.2.1(a), with more than 20 sprinklers is still required to be supervised but not at a minimum pressure. A 7-psi (0.5-bar) minimum is required on the non-interlock and double interlock systems, since they can operate similarly to a dry pipe system in that water delivery time to the open sprinklers is delayed. The minimum supervisory pressure is necessary for proper operation of the sprinkler. Proper operation entails both the lifting of the cap so the air escapes and the clearing of the mechanisms so the water discharge pattern is not affected. A minimum operating pressure at the sprinkler is a critical part of the system's activation process.

Supervision of the associated electronic detection devices is very important, since loss or failure of the detection system to an open circuit or ground fault can render the entire preaction system inoperative.

Specifics on the supervision of fire detection system circuits, now referred to as "monitoring the integrity of installation conductors," can be found in NFPA 72.

4-3.2.4 Upright Sprinklers. Only upright sprinklers shall be installed on preaction systems.

Exception No. 1: Listed dry sprinklers shall be permitted.*

Exception No. 2: Pendent sprinklers installed on return bends shall be permitted where both the sprinklers and the return bends are located in a heated area.

Exception No. 3: Horizontal sidewall sprinklers, installed so that water is not trapped, shall be permitted.

A-4-3.2.4 Exception No. 1. See A-4-2.2 Exception No. 1.

The concern regarding pipe scale, as discussed in the commentary to 4-2.2, also applies to pendent sprinklers used in preaction systems. If pendent sprinklers are required, then listed dry sprinklers must be used, or the pendent sprinklers must be installed on return bends. The return bend option applies only where the sprinkler, return bend, and branch line piping are in an area maintained at or above 40°F (4°C). Horizontal sidewall sprinklers are permitted, provided they are installed in a manner that avoids the accumulation of water, scale, and sediment at the sprinkler orifice. When installed at an angle beneath a sloped ceiling, horizontal sidewall sprinklers typically trap water.

4-3.2.5 System Configuration. Preaction systems of the type described in 4-3.2.1(c) shall not be gridded.

This restriction prohibiting a gridded pipe arrangement for double interlock preaction systems reduces the likelihood of substantial delay in the delivery of water to the open sprinklers. The 1996 edition removed the 10-psi (0.7-bar) limit for double interlock systems that existed in previous editions. Since this type of preaction system can have the same time delays associated with a dry pipe system, even at pressures below 10 psi (0.7 bar), gridded arrangements are not permitted.

4-3.3* Deluge Systems.

A-4-3.3 Where 8-in. (203-mm) piping is employed to reduce friction losses in a system operated by fire detection devices, a 6-in. (152-mm) preaction or deluge valve and a 6-in. (152-mm) gate valve between tapered reducers should be permitted.

Deluge systems are normally used in very high hazard areas. These systems have properties similar to those of a preaction system in that they rely on a supplemental detection system to operate them. However, these systems use open sprinklers and, therefore, are an open system. NFPA 409, *Standard on Aircraft Hangars*, requires the installation of foam-water deluge systems in certain aircraft hangars. NFPA does not limit the type of detection system to be used to activate a deluge system. Deluge systems may be operated by smoke, heat, ultraviolet (UV), or infrared (IR) detection systems. Exhibit 1.3 shows a schematic of a deluge system operated by electron detectors. Exhibit 4.4 illustrates a deluge valve activated by a pneumatic dry pilot sprinkler line.

4-3.3.1 The detection devices or systems shall be automatically supervised.

4-3.3.2 Deluge systems shall be hydraulically calculated.

Since all sprinklers are open, every sprinkler on the system discharges water simultaneously when the deluge valve operates. The determination of the system's area of operation is straightforward—it is the entire area protected by the deluge system.

4-4 Combined Dry Pipe and Preaction Systems

The system described in Section 4-4 is not as common as it was several decades ago. The system is intended to be applied to unusual structures, such as piers or wharves, which require unusually long runs of pipe.

4-4.1* General.

A-4-4.1 Systems described by Section 4-4 are special types of non-interlocking preaction systems intended for use in, but not limited to, structures where a number of dry pipe valves would be required if a dry pipe system were installed. These systems are primarily used in piers and wharves.

4-4.1.1* Combined automatic dry pipe and preaction systems shall be so constructed that failure of the detection system shall not prevent the system from functioning as a conventional automatic dry pipe system.

Exhibit 4.4 Example of a deluge valve.

A-4-4.1.1 See Figure A-4-4.1.1.

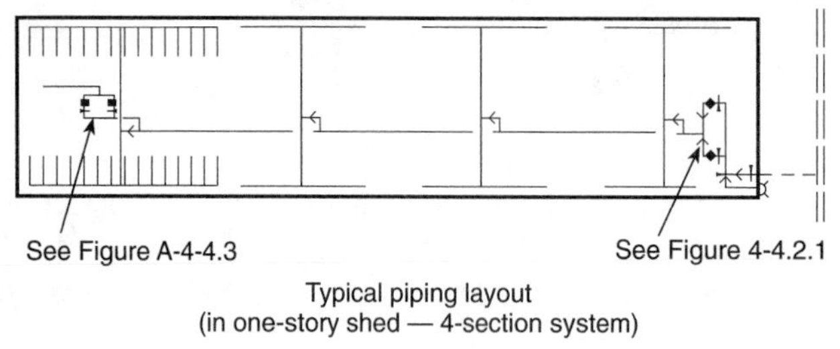

Figure A-4-4.1.1 *Typical piping layout for combined dry pipe and preaction sprinkler system.*

Conventional automatic operation of the dry pipe valve provides a highly reliable backup to the fire detection system. Since the primary purpose of the fire detection system is to cause piping to start filling sooner, loss of the detection system only results in a much slower operating mode. Although somewhat impaired, the system can still operate.

4-4.1.2 Combined automatic dry pipe and preaction systems shall be so constructed that failure of the dry pipe system of automatic sprinklers shall not prevent the detection system from properly functioning as an automatic fire alarm system.

Quite the opposite effect from that covered in 4-4.1.1 results when the detection system operates, but the dry pipe valve fails to open. With this type of failure, no water flows to the piping system and sprinklers, and only the alarm activates.

4-4.1.3 Provisions shall be made for the manual operation of the detection system at locations requiring not more than 200 ft (61 m) of travel.

Manual operation of the fire detection system provides a prompt means of filling the system with water from remote locations throughout the protected area. This method can be used in special circumstances where danger of an incident is possible and where the system can be converted to a wet pipe system prior to a fire.

4-4.1.4 Upright Sprinklers. Only upright sprinklers shall be installed on combined dry pipe and preaction systems.

Exception No. 1: Listed dry sprinklers shall be permitted.*

Exception No. 2: Pendent sprinklers installed on return bends shall be permitted where both sprinklers and return bends are located in a heated area.

Exception No. 3: Horizontal sidewall sprinklers, installed so that water is not trapped, shall be permitted.

A-4-4.1.4 Exception No. 1. See A-4-2.2 Exception No. 1.

Refer to commentary to 4-2.2 and 4-3.2.4.

4-4.2 Dry Pipe Valves in Combined Systems.

4-4.2.1 Where the system consists of more than 600 sprinklers or has more than 275 sprinklers in any fire area, the entire system shall be controlled through two 6-in. (152-mm) dry pipe valves connected in parallel and shall feed into a common feed main. These valves shall be checked against each other. *(See Figure 4-4.2.1.)*

4-4.2.2 Each dry pipe valve shall be provided with a listed tripping device actuated by the detection system. Dry pipe valves shall be cross-connected through a 1-in. (25.4-mm) pipe connection to permit simultaneous tripping of both dry pipe valves. This 1-in. (25.4-mm) pipe

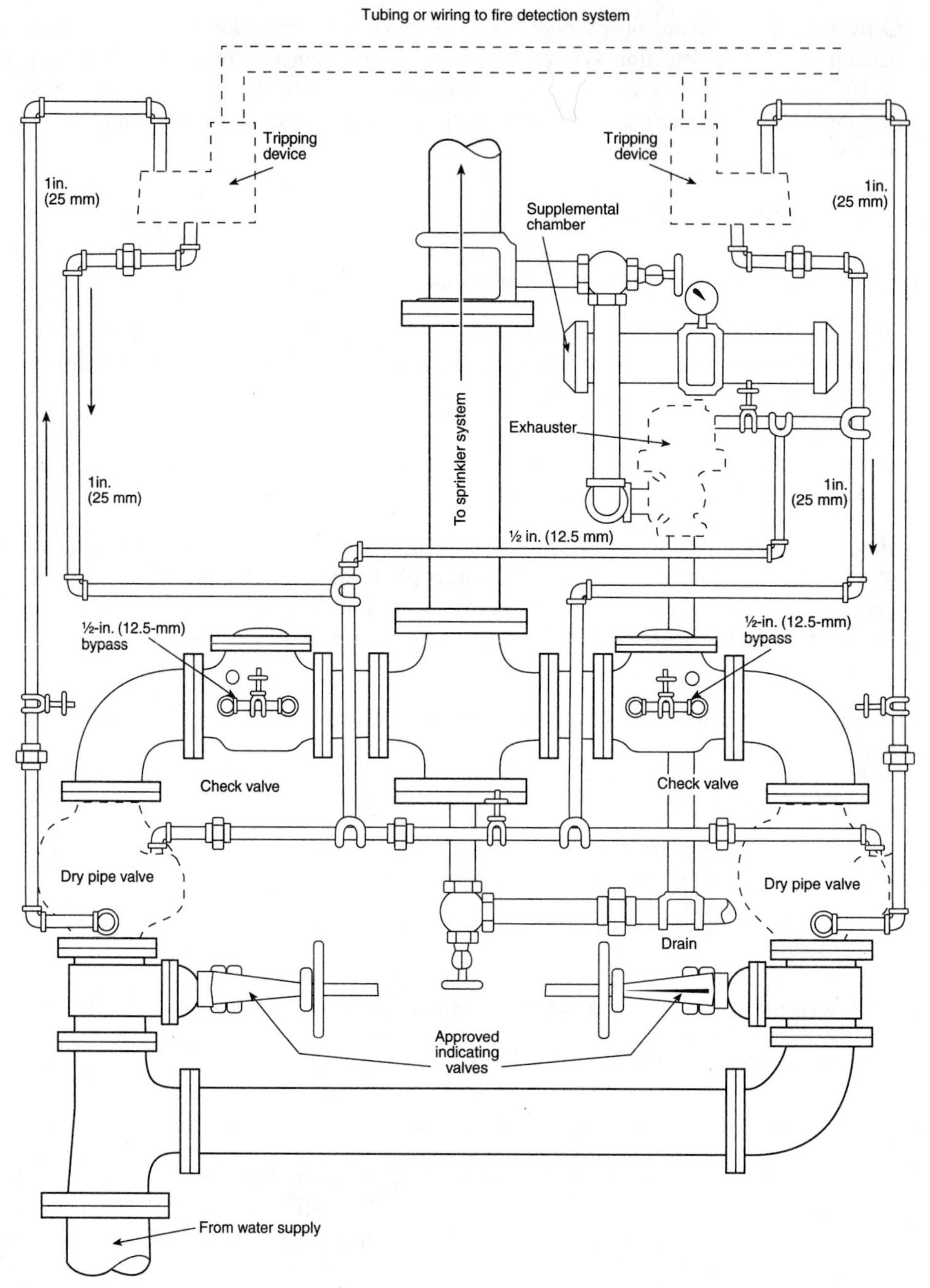

Figure 4-4.2.1 Header for dry pipe valves installed in parallel for combined systems; standard trimmings not shown. Arrows indicate direction of fluid flow.

connection shall be equipped with an indicating valve so that either dry pipe valve can be shut off and worked on while the other remains in service.

Although each dry pipe valve serves as a backup valve, simultaneous operation of both valves is desirable to increase the system's flow characteristics.

4-4.2.3 The check valves between the dry pipe valves and the common feed main shall be equipped with ½-in. (13-mm) bypasses so that a loss of air from leakage in the trimmings of a dry pipe valve will not cause the valve to trip until the pressure in the feed main is reduced to the tripping point. An indicating valve shall be installed in each of these bypasses so that either dry pipe valve can be completely isolated from the main riser or feed main and from the other dry pipe valve.

4-4.2.4 Each combined dry pipe and preaction system shall be provided with listed quick-opening devices at the dry pipe valves.

4-4.3* Air Exhaust Valves.

One or more listed air exhaust valves of 2-in. (51-mm) or larger size controlled by operation of a fire detection system shall be installed at the end of the common feed main. These air exhaust valves shall have soft-seated globe or angle valves in their intakes. Also, approved strainers shall be installed between these globe valves and the air exhaust valves.

A-4-4.3 See Figure A-4-4.3.

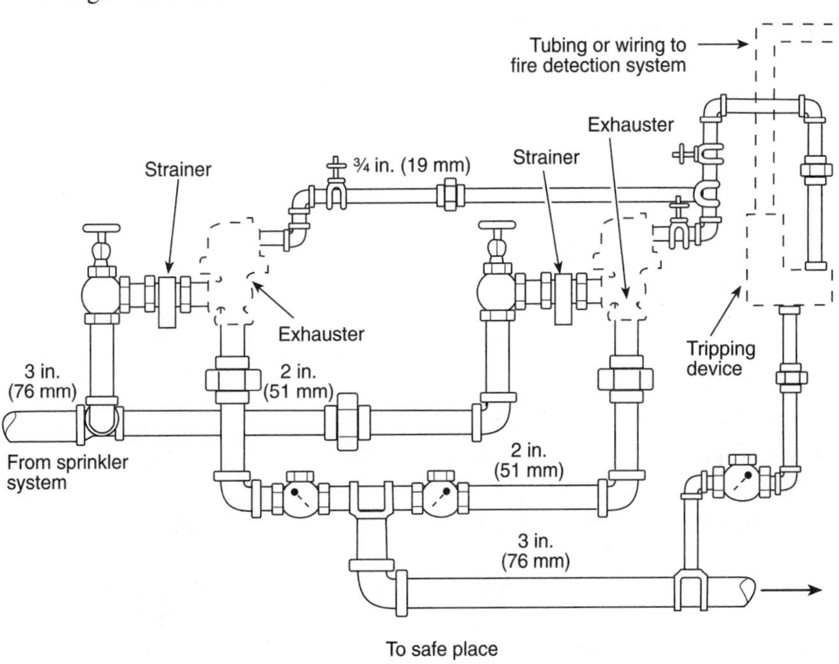

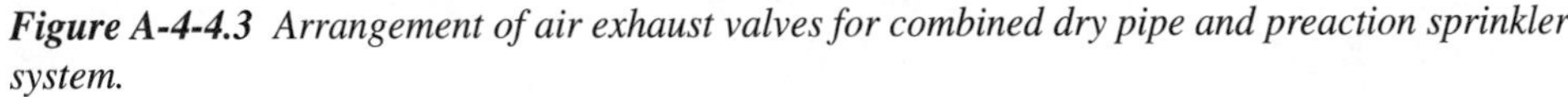

Figure A-4-4.3 Arrangement of air exhaust valves for combined dry pipe and preaction sprinkler system.

Combined dry pipe and preaction systems have a very large air capacity. The exhaust valves serve to remove air at remote ends of the piping and allow water to fill the system at a faster rate. The air exhaust valves should be located where they can be readily tested and maintained.

The exhaust valve is a quick-opening device similar to the exhauster described in the commentary to 4-2.4. A listed exhauster, which is mechanically activated by a drop in air pressure, must be used. The tripping device activated by the detection system causes the drop in air pressure experienced by the exhauster. At the time of publication, no such listed tripping devices were available.

4-4.4 Subdivision of System Using Check Valves.

4-4.4.1 Where more than 275 sprinklers are required in a single fire area, the system shall be divided into sections of 275 sprinklers or less by means of check valves. If the system is installed in more than one fire area or story, not more than 600 sprinklers shall be supplied through any one check valve. Each section shall have a 1¼-in. (33-mm) drain on the system side of each check valve supplemented by a dry pipe system auxiliary drain.

The requirements of 4-4.4.1 provide a cost-effective means to offset the large volume of air in these systems. The use of check valves is acceptable since air exhaust valves are still required to be used, and the system is a non-interlock preaction system, which is expected to be tripped by the detection system.

4-4.4.2 Section drain lines and dry pipe system auxiliary drains shall be located in heated areas or inside heated cabinets to enclose drain valves and auxiliary drains for each section.

4-4.4.3 Air exhaust valves at the end of a feed main and associated check valves shall be protected against freezing.

4-4.5 Time Limitation.

The sprinkler system shall be so constructed and the number of sprinklers controlled shall be so limited that water shall reach the farthest sprinkler within a period of time not exceeding 1 minute for each 400 ft (122 m) of common feed main from the time the heat-responsive system operates. The maximum time permitted shall not exceed 3 minutes.

These systems are a combination of the features associated with dry pipe and preaction systems and are intended for specific applications as indicated in the commentary to Section 4-4. The water delivery times associated with combined dry pipe and preaction systems, therefore, are different than those required for either dry pipe or preaction systems. Excessive delivery time for water should be reviewed with the authority having jurisdiction.

4-4.6 System Test Connection.

The end section shall have a system test connection as required for dry pipe systems.

4-5 Antifreeze Systems

Antifreeze systems are typically used as subsystems of a wet pipe system. Antifreeze systems are intended to protect small areas that could be exposed to freezing temperatures, such as outside loading docks. Antifreeze systems are also used for residential areas that are not protected against freezing temperatures, since residential sprinklers are currently listed only for wet pipe systems.

Antifreeze systems were once more attractive economically for protecting small areas than dry pipe systems. The reduction in size and cost of certain dry pipe valves combined with the added expense of backflow preventers and expansion chambers have made antifreeze systems somewhat less attractive. Even where a backflow preventer is not required, a larger area protected with a dry pipe system could still provide an economically more attractive option due to the cost of antifreeze for larger systems.

The potential effectiveness of antifreeze and water mixtures controlling a fire has come into question in the past. Antifreeze solutions are evaluated for their ability to allow for efficient fire control and to make sure that the solution does not serve as fuel for the fire. The concentrates identified in Tables 4-5.2.1 and 4-5.2.2 were tested for control mode spray sprinklers to ensure that they cause no such negative reaction.

4-5.1* Where Used.

The use of antifreeze solutions shall be in conformity with state and local health regulations.

A-4-5.1 Antifreeze solutions can be used for maintaining automatic sprinkler protection in small, unheated areas. Antifreeze solutions are recommended only for systems not exceeding 40 gal (151 L).

Because of the cost of refilling the system or replenishing to compensate for small leaks, it is advisable to use small dry valves where more than 40 gal (151 L) are to be supplied.

The current reaction by a number of water system purveyors to the use of antifreeze solutions in sprinkler systems has impacted the antifreeze system option, even though NFPA 13 only allows the use of nontoxic antifreeze solutions when the system is connected to the public water supply. Many local regulations require antifreeze systems to be equipped with a reduced-pressure zone backflow prevention device (see Exhibit 4.5) to guard against potential contamination of the public water supply. These local regulations have impacted the economic advantages offered by antifreeze systems to some degree.

A distinction is made between the use of additives for potable and nonpotable water. The solutions noted in Table 4-5.2.1 could be described as food-grade chemicals.

The size limitation of antifreeze systems is only a recommendation because of cost, not because of system performance. NFPA 13 does not place any limitations on the size of antifreeze systems.

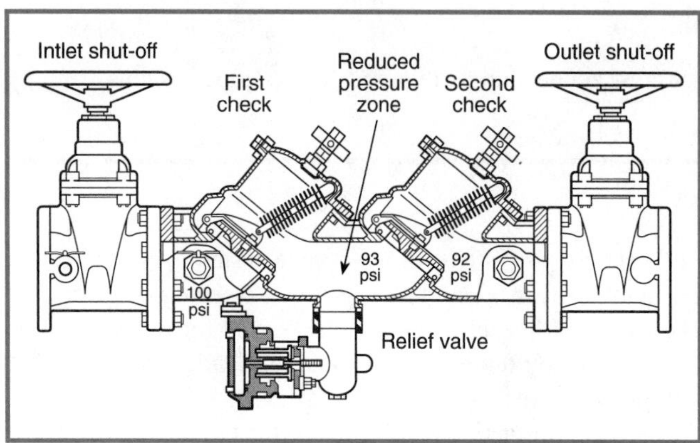

Exhibit 4.5 Backflow prevention device installed on an antifreeze system.

4-5.2* Antifreeze Solutions.

A-4-5.2 Listed CPVC sprinkler pipe and fittings should be protected from freezing with glycerine only. The use of diethylene, ethylene, or propylene glycols are specifically prohibited. Laboratory testing shows that glycol-based antifreeze solutions present a chemical environment detrimental to CPVC.

4-5.2.1 Where sprinkler systems are supplied by potable water connections, the use of antifreeze solutions other than water solutions of pure glycerine (C.P. or U.S.P. 96.5 percent grade) or propylene glycol shall not be permitted. Suitable glycerine-water and propylene glycol-water mixtures are shown in Table 4-5.2.1.

Table 4-5.2.1 Antifreeze Solutions to Be Used if Potable Water Is Connected to Sprinklers

Materials	Solution (by volume)	Specific Gravity at 60°F (15.6°C)	Freezing Point °F	Freezing Point °C
Glycerine	50% water	1.133	−15	−26.1
C.P. or U.S.P. grade*	40% water	1.151	−22	−30.0
	30% water	1.165	−40	−40.0
Hydrometer scale 1.000 to 1.200				
Propylene glycol	70% water	1.027	+9	−12.8
	60% water	1.034	−6	−21.1
	50% water	1.041	−26	−32.2
	40% water	1.045	−60	−51.1
Hydrometer scale 1.000 to 1.200 (subdivisions 0.002)				

*C.P.—chemically pure; U.S.P.—United States pharmacopoeia 96.5%.

4-5.2.2 If potable water is not connected to sprinklers, the commercially available materials indicated in Table 4-5.2.2 shall be permitted for use in antifreeze solutions.

Table 4-5.2.2 Antifreeze Solution to Be Used If Nonpotable Water Is Connected to Sprinklers

Materials	Solution (by volume)	Specific Gravity at 60°F (15.6°C)	Freezing Point °F	Freezing Point °C
Glycerine	See Table 4-5.2.1.			
Diethylene glycol	50% water	1.078	−13	−25.0
	45% water	1.081	−27	−32.8
	40% water	1.086	−42	−41.1
Hydrometer scale 1.000 to 1.120 (subdivisions 0.002)				
Ethylene glycol	61% water	1.056	−10	−23.3
	56% water	1.063	−20	−28.9
	51% water	1.069	−30	−34.4
	47% water	1.073	−40	−40.0
Hydrometer scale 1.000 to 1.120 (subdivisions 0.002)				
Propylene glycol	See Table 4-5.2.1.			

4-5.2.3* An antifreeze solution shall be prepared with a freezing point below the expected minimum temperature for the locality. The specific gravity of the prepared solution shall be checked by a hydrometer with suitable scale or a refractometer having a scale calibrated for the antifreeze solution involved. *[See Figures 4-5.2.3(a) and (b).]*

A-4-5.2.3 Beyond certain limits, an increased proportion of antifreeze does not lower the freezing point of solution (*see Figure A-4-5.2.3*).

Glycerine, diethylene glycol, ethylene glycol, and propylene glycol should never be used without mixing with water in proper proportions, because these materials tend to thicken near 32°F (0°C).

It is critical to maintain the proper concentration of antifreeze solution at all times. Pressure surges and temperature fluctuations can alter the concentration because more water either has flowed into or out of the antifreeze system. NFPA 25 requires that the antifreeze concentrations be checked periodically.

Refractometers with scales that are calibrated for a particular antifreeze solution are available for testing the concentration of antifreeze solutions. Traditional types of refractometers require that a sample of the solution be placed onto the device so that the appropriate antifreeze concentration as indicated in Figure 4-5.2.3(a) can be verified. This type of refractometer is calibrated for a particular type of antifreeze solution, so a separate device needs to be used for each type of antifreeze. See Exhibit 4.6.

Digital refractometers contain a probe that is inserted into the antifreeze solution. These refractometers provide a reading of the appropriate antifreeze concentration as identified in Figure 4-5.2.3(b). Digital refractometers do not require that their scales

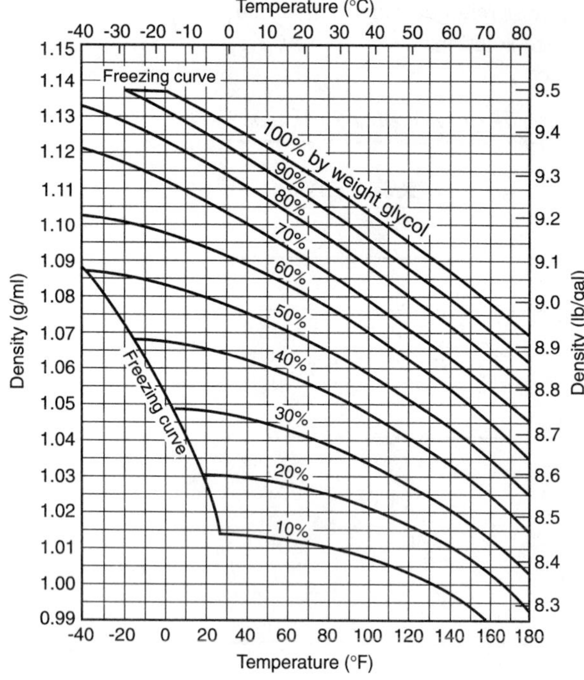

Figure 4-5.2.3(a) *Densities of aqueous ethylene glycol solutions (percent by weight).*

Temperature (°C)
-40 -30 -20 -10 0 10 20 30 40 50 60 70 80
80%
100%
90% by weight glycol
70%
60%
50%
40%
30%
20%
10%
Freezing curve
Density (g/ml)
1.09 1.08 1.07 1.06 1.05 1.04 1.03 1.02 1.01 1.00 0.99 0.98
Density (lb/gal)
9.84 9.50 9.18 9.1 9.04 9.0 8.88 8.80 8.70 8.63 8.54 8.50
-40 -20 0 20 40 60 80 100 120 140 160 180
Temperature (°F)

Figure 4-5.2.3(b) *Densities of aqueous propylene glycol solutions (percent by weight).*

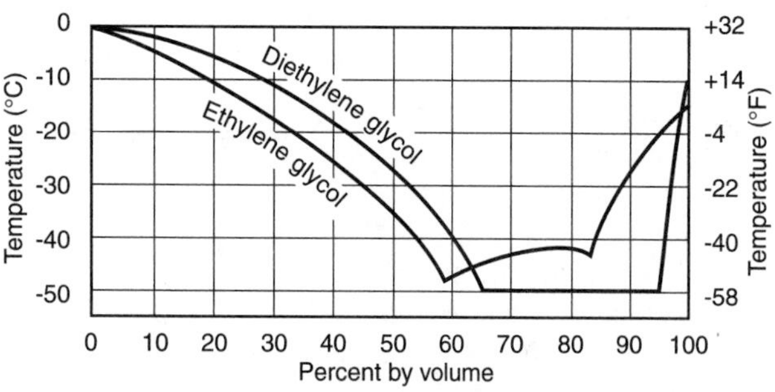

Figure A-4-5.2.3 *Freezing points of water solutions of ethylene glycol and diethylene glycol.*

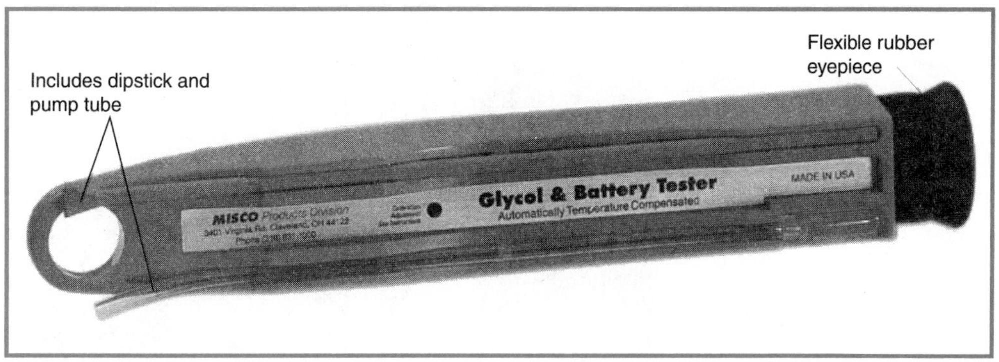

Exhibit 4.6 Refractometer.

be calibrated for each type of antifreeze, so a single device can be used to test all types of antifreeze. See Exhibit 4.7.

4-5.3 Arrangement of Supply Piping and Valves.

4-5.3.1* Where the connection between the antifreeze system and the wet pipe system does not incorporate a backflow prevention device, piping and valves shall be installed as illustrated in Figure 4-5.3.1.

A-4-5.3.1 All permitted antifreeze solutions are heavier than water. At the point of contact (interface), the heavier liquid will be below the lighter liquid, preventing diffusion of water into the unheated areas.

Since all permitted antifreeze solutions are heavier than water, an interface at which the water in the wet system will stay above the heavier antifreeze solution is created. If possible, the entire antifreeze system should be below the level of this interface, thus preventing the diffusion of water into low temperature areas. When the antifreeze system is above the interface, alternative piping arrangements and additional system components as illustrated in Figures 4-5.3.1 and 4-5.3.2 are necessary.

Figure 4-5.3.1 shows an arrangement of supply piping and valves for an antifreeze system without a backflow device and with the sprinklers above the water/antifreeze interface. In this arrangement, a check valve and a 5-ft (1.5-m) drop or U-loop is needed so that the antifreeze does not flow back into the wet pipe system. Additionally, because antifreeze solutions can expand with a rise in temperature and result in high pressures that can damage system components, a 1⁄32-in. (0.8-mm) hole drilled into the clapper of the check valve is needed. If the antifreeze system is entirely below the interface with the wet pipe system (shown at valve A), or a backflow device is used as indicated in 4-5.3.2, then the check valve and the 5-ft (1.5-m) drop are not needed.

Actual field conditions and practices could require some modification to the arrangement shown in Figure 4-5.3.1. For example, because the piping on the antifreeze system is at a higher elevation than the water supply, it is not practical to use a gravity-type filling cup at the illustrated supply location. If antifreeze is gravity-fed into the

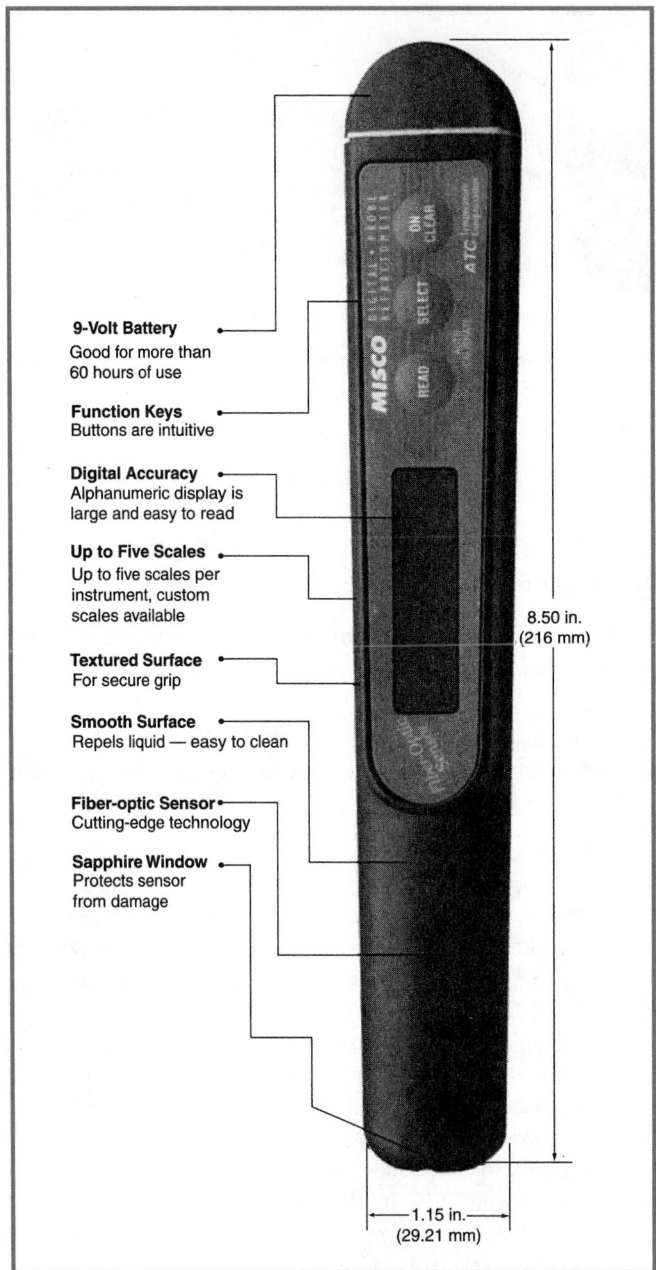
9-Volt Battery
Good for more than 60 hours of use
Function Keys
Buttons are intuitive
Digital Accuracy
Alphanumeric display is large and easy to read
Up to Five Scales
Up to five scales per instrument, custom scales available
Textured Surface
For secure grip
Smooth Surface
Repels liquid — easy to clean
Fiber-optic Sensor
Cutting-edge technology
Sapphire Window
Protects sensor from damage
8.50 in. (216 mm)
1.15 in. (29.21 mm)
MISCO
ON CLEAR
SELECT
READ
ATC
Fiber-Optic Sensor

Exhibit 4.7 *Digital refractometer.*

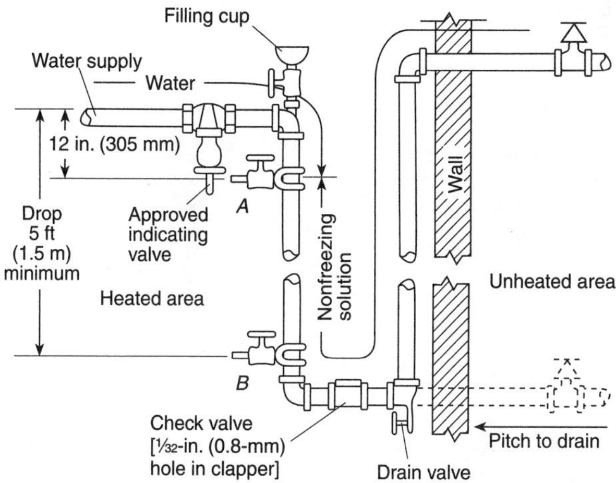

Notes:
1. Check valve shall be permitted to be omitted where sprinklers are below the level of valve *A*.
2. The 1/32-in. (0.8-mm) hole in the check valve clapper is needed to allow for expansion of the solution during a temperature rise, thus preventing damage to sprinklers.

Figure 4-5.3.1 *Arrangement of supply piping and valves.*

system, the filling cup should be relocated to the high point of the antifreeze system piping. As an alternative, antifreeze can be pumped into the system through the filling cup as presently located in Figure 4-5.3.1.

4-5.3.2* Where the connection between the antifreeze system and the wet pipe system incorporates a backflow prevention device, piping and valves shall be installed as illustrated in Figure 4-5.3.2. A listed expansion chamber of appropriate size and pre-charged air pressure shall be provided to compensate for thermal expansion of the antifreeze solution as illustrated in Figure 4-5.3.2.

A-4-5.3.2 One formula for sizing the chamber is as follows. Other methods also exist.

$$\Delta L = S_V\left(\frac{D_L}{D_H} - 1\right)$$

where:

ΔL = change in antifreeze solution volume (gal) due to thermal expansion

S_V = volume (gal) of antifreeze system, not including the expansion chamber

D_L = density (gm/ml) of antifreeze solution at lowest expected temperature

D_H = density (gm/ml) of antifreeze solution at highest expected temperature

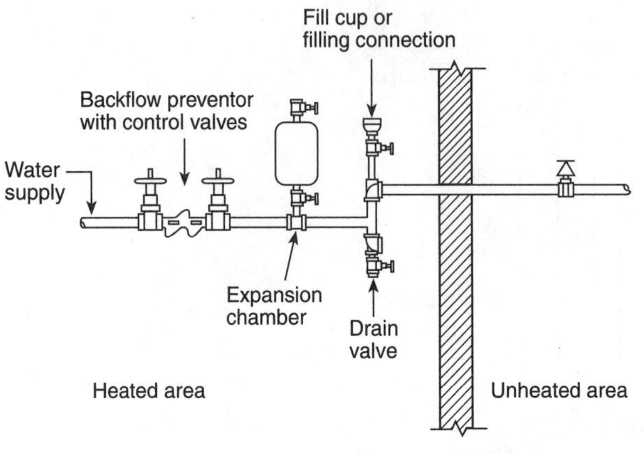

Figure 4-5.3.2 Arrangement of supply piping with backflow device.

This method is based on the following information:

$$\frac{P_0 \cdot V_0}{T_0} = \frac{P_1 \cdot V_1}{T_1} = \frac{P_2 + V_2}{T_2}$$

where:

V_{EC} = minimum required volume (gal) of expansion chamber

V_0 = air volume (gal) in expansion chamber at pre-charge (before installation)

V_1 = air volume (gal) in expansion chamber at normal static pressure

V_2 = air volume (gal) in expansion chamber at post-expansion pressure (antifreeze at high temperature)

P_0 = absolute pre-charge pressure (psia) on expansion chamber before installation

P_1 = absolute static pressure (psi) on water (supply) side of backflow preventer

P_2 = absolute maximum allowable working pressure (psi) for antifreeze system

T_0 = temperature (°R) of air in expansion chamber at pre-charge

T_1 = temperature (°R) of air in expansion chamber when antifreeze system piping is at lowest expected temperature

T_2 = temperature (°R) of air in expansion chamber when antifreeze system piping is at highest expected temperature

This equation is one formulation of the ideal gas law from basic chemistry. The amount of air in the expansion chamber will not change over time. The pressure, temperature, and volume of the air at different times will be related in accordance with this formula.

$$V_2 = V_1 - \Delta L$$

The antifreeze in the system is essentially incompressible, so the air volume in the expansion chamber will decrease by an amount equal to the expansion of the antifreeze.

It is assumed that there is no trapped air in the system piping, so the only air in the system is in the expansion chamber. This is a conservative assumption, since more air is better. In reality, there will be at least some trapped air. However, only the air in the expansion chamber can be relied upon to be available when needed.

$$V_{EC} = V_0$$

At pre-charge, the chamber will be completely full of air.

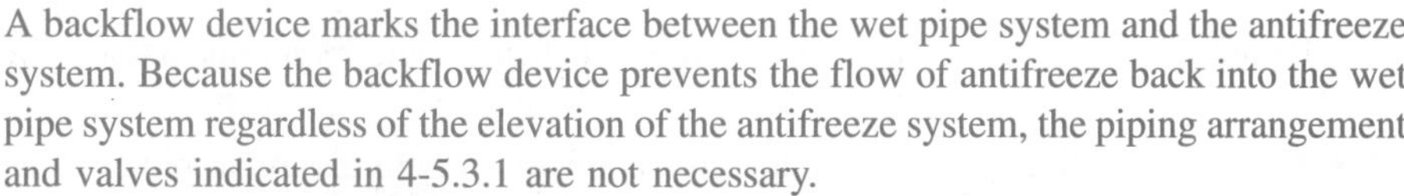

$$V_{EC} = \frac{P_1 \cdot T_0 \cdot P_2 \cdot \Delta L \cdot T_1}{P_0 \cdot T_1(P_2 \cdot T_1 - P_1 \cdot T_2)}$$

A backflow device marks the interface between the wet pipe system and the antifreeze system. Because the backflow device prevents the flow of antifreeze back into the wet pipe system regardless of the elevation of the antifreeze system, the piping arrangement and valves indicated in 4-5.3.1 are not necessary.

A reduced-pressure backflow preventer contains a relief valve that discharges water to atmosphere under some circumstances. This discharge can be a significant amount of water, and appropriate drainage needs to be provided.

To account for pressure changes that can result because of temperature differences in the space protected by the antifreeze system, an expansion chamber with pre-charged air pressure is required. The expansion chamber is required rather than some other type of pressure-relieving device, such as a relief valve, because antifreeze solution is discharged from the system when the relief valve operates. The volume of solution lost is replaced by water when the temperature cools, thus diluting the concentration of the antifreeze solution below what is required.

As indicated by the equation, calculating the appropriate size of the expansion chamber requires some understanding of the overall process involved. The antifreeze system can be exposed to a seasonal temperature fluctuation—for example, winter to summer—that causes the antifreeze solution to expand and contract. The amount of expansion depends on the total temperature change and the density of the specific solution used. The type of antifreeze and the concentration used affects the overall density of the solution in the antifreeze system.

As the solution expands, some volume flows into the expansion chamber, compressing the air within the chamber. As the air compresses, a corresponding increase of the pressure on the system occurs. The air volume must be large enough to keep the system pressure below the pressure limitations of the system components. Factors other than the amount of thermal expansion that influence the required air volume (chamber size) include seasonal temperature changes of the air within the chamber, which are often different than the temperature change of the antifreeze solution, precharged air pressure level, and static pressure of the water supply. Fluctuations of air temperature within the expansion chamber can be different than the fluctuations of the temperature of the antifreeze system, since the chamber is often located within a temperature-controlled portion of the building.

Situations, such as with a high static pressure, occur in which the system pressure cannot be kept below 175 psi (12.1 bar) with an expansion chamber.

4-6 Automatic Sprinkler Systems with Non-fire Protection Connections

4-6.1 Circulating Closed-Loop Systems.

4-6.1.1 System Components.

4-6.1.1.1 A circulating closed-loop system is primarily a sprinkler system and shall comply with all provisions of this standard such as those for control valves, area limitations of a system, alarms, fire department connections, sprinkler spacing, and so forth.

Exception: This requirement shall not apply to items otherwise specified within 4-6.1.

4-6.1.1.2 Piping, fittings, valves, and pipe hangers shall meet the requirements specified in Chapter 3.

4-6.1.1.3 A dielectric fitting shall be installed in the junction where dissimilar piping materials are joined (e.g., copper to steel).

Exception: Dielectric fittings shall not be required in the junction where sprinklers are connected to piping.

Only this paragraph in NFPA 13 requires the use of a dielectric fitting. The concern with dissimilar metals typically only applies where a system has a continuously moving fluid through it. The ionization of the area surrounding the two different materials is enhanced by the transition of the fluid across the surface of the metals. Since no other sprinkler system except this one has this condition, no further requirements concerning dielectric fittings are imposed on those systems. Exhibit 4.8 illustrates a dielectric fitting.

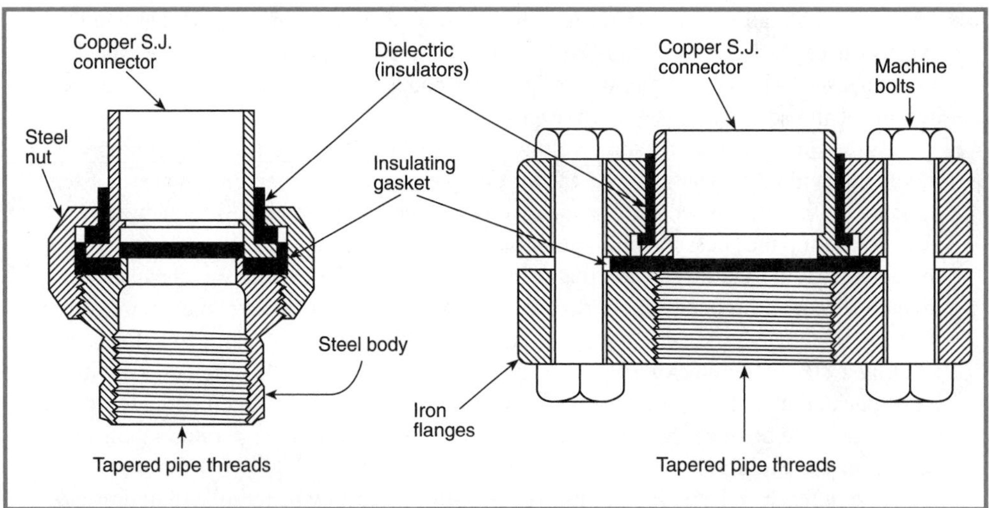

Exhibit 4.8 Dielectric fitting.

4-6.1.1.4 It shall not be required that other auxiliary devices be listed for sprinkler service; however, these devices, such as pumps, circulating pumps, heat exchangers, radiators, and luminaries, shall be pressure rated at 175 psi or 300 psi (12.1 bar or 20.7 bar) (rupture pressure of five times rated water system working pressure) to match the required rating of sprinkler system components.

Although auxiliary equipment or components are not required to be listed, they must be of materials acceptable in Chapter 3 and of sufficient quality in durability, suitability, and strength to be equal to all other sprinkler system components.

4-6.1.1.5 Auxiliary devices shall incorporate materials of construction and be so constructed that they will maintain their physical integrity under fire conditions to avoid impairment to the fire protection system.

4-6.1.1.6 Auxiliary devices, where hung from the building structure, shall be supported independently from the sprinkler portion of the system, following recognized engineering practices.

4-6.1.2* Hydraulic Characteristics. Piping systems for attached heating and cooling equipment shall have auxiliary pumps or an arrangement made to return water to the piping system in order to assure the following.

(a) Water for sprinklers shall not be required to pass through heating or cooling equipment. At least one direct path shall exist for waterflow from the sprinkler water supply to every sprinkler. Pipe sizing in the direct path shall be in accordance with the design requirements of this standard.

(b) No portions of the sprinkler piping shall have less than the sprinkler system design pressure, regardless of the mode of operation of the attached heating or cooling equipment.

(c) There shall be no loss or outflow of water from the system due to or resulting from the operation of heating or cooling equipment.

(d) Shutoff valves and a means of drainage shall be provided on piping to heating or cooling equipment at all points of connection to sprinkler piping and shall be installed in such a manner as to make possible repair or removal of any auxiliary component without impairing the serviceability and response to the sprinkler system. All auxiliary components, including the strainer, shall be installed on the auxiliary equipment side of the shutoff valves.

A-4-6.1.2 Outlets should be provided at critical points on sprinkler system piping to accommodate attachment of pressure gauges for test purposes. A

The flow to sprinklers cannot experience a reduction regardless of operation or presence of the auxiliary equipment. An enhancement of flow to the sprinkler system is allowed, but a reduction is not.

4-6.1.3 Water Temperature.

4-6.1.3.1 Maximum. In no case shall maximum water temperature flowing through the sprinkler portion of the system exceed 120°F (49°C). Protective control devices listed for this purpose shall be installed to shut down heating or cooling systems when the temperature of water flowing through the sprinkler portion of the system exceeds 120°F (49°C). When the water temperature exceeds 100°F (37.8°C), intermediate or higher temperature rated sprinklers shall be used.

4-6.1.3.2 Minimum. Precautions shall be taken to ensure that temperatures below 40°F (4°C) are not permitted.

4-6.1.4 Obstruction to Discharge. Automatic sprinklers shall not be obstructed by auxiliary devices, piping, insulation, and so forth, from detecting fire or from proper distribution of water.

4-6.1.5 Signs. Caution signs shall be attached to all valves controlling sprinklers. The caution sign shall be worded as follows:

> This valve controls fire protection equipment. Do not close until after fire has been extinguished. Use auxiliary valves when necessary to shut off supply to auxiliary equipment.

CAUTION

Automatic alarm will be sounded if this valve is closed.

4-6.1.6 Water Additives. Materials added to water shall not adversely affect the fire-fighting properties of the water and shall be in conformity with any state or local health regulations. Due care and caution shall be given to the use of additives that can remove or suspend scale from older piping systems. Where additives are necessary for proper system operation, due care shall be taken to ensure that additives are replenished after alarm testing or whenever water is removed from the system.

These additives are usually necessary to extend the life of heating units, chillers, or companion equipment associated with the heating, ventilating, and air-conditioning (HVAC) systems. It is not the intent of 10-2.2.2 to prohibit the types of materials discussed in Chapter 4.

As a courtesy to fire suppression personnel, local fire departments with primary jurisdiction over the property should be made aware of the presence of and type of water additives included in the system.

4-6.1.7 Waterflow Detection.

4-6.1.7.1 The supply of water from sprinkler piping through auxiliary devices, circulatory piping, and pumps shall not under any condition or operation, transient or static, cause false sprinkler waterflow signals.

4-6.1.7.2 A sprinkler waterflow signal shall not be impaired when water is discharged through an opened sprinkler or through the system test connection while auxiliary equipment is in any mode of operation (on, off, transient, stable).

4-7 Outside Sprinklers for Protection Against Exposure Fires

These types of systems are usually installed on the exterior surface of a building to protect against exposure fires from an adjacent building or operation. An outside system typically has two performance objectives. One objective is to significantly limit the radiation or connective heat generated by an exposure fire from entering the building through windows, doors, and other openings in the exposed walls. This protection minimizes the likelihood of the ignition of building contents and other combustibles. The other objective is to minimize the likelihood of ignition of or heat damage to exterior combustible wall surfaces, including sheathing, eaves, and cornices.

Prompt sprinkler system discharge should occur by automatic actuation of the outside sprinklers. Delayed application of water to glass surfaces once they become sufficiently heated can result in breakage from thermal shock. Delayed application of water can also result in ignition of exterior combustible surfaces or damage to heat-sensitive materials.

Outside sprinkler systems might be required to compensate for lack of physical separation between adjacent structures. NFPA 80A, *Recommended Practice for Protection of Buildings from Exterior Fire Exposures*, allows the reduction of separation distances between buildings when proper exposure protection from sprinklers is provided.

4-7.1 Applications

Exposure protection systems shall be permitted on buildings regardless of whether the building's interior is protected by a sprinkler system.

4-7.2 Water Supply and Control.

4-7.2.1* Sprinklers installed for protection against exposure fires shall be supplied from a standard water supply as outlined in Chapter 9.

Exception: Where approved, other supplies, such as manual valves or pumps or fire department connections, shall be acceptable.

A-4-7.2.1 The water supply should be capable of furnishing the total demand for all exposure sprinklers operating simultaneously for protection against the exposure fire under consideration for a duration of not less than 60 minutes.

Water should be continuously delivered at the required application rate to the outside sprinkler system for the duration of exposure fire expected from an adjacent building, structure, or other exposure source such as a flammable liquid storage tank or scrap tire storage. The required duration can exceed 60 minutes if hose streams cannot be applied in a sufficient amount of time or if the exposing fire cannot be contained during its developmental stages. Section 8-7 provides further guidance on the design parameters of an outside sprinkler system for exposure protection.

The exception permits a water supply other than the automatic water supply required in Chapter 9. The authority having jurisdiction must determine if this method of water supply, which includes fire department apparatus pumping into the fire department connection, is acceptable for the circumstances. This situation is the only one noted in NFPA 13 that does not require an automatic supply.

4-7.2.2 Where fire department connections are used for water supply, they shall be so located that they will not be affected by the exposing fire.

4-7.3 Control.

4-7.3.1 Each system of outside sprinklers shall have an independent control valve.

The division of sprinkler piping and sprinklers into separate systems should take into account the extent of a likely exposure fire from adjacent buildings or operations. Further division of sprinkler piping can be necessary if the surface to be protected is quite large.

The water supply can limit the area that can be protected by a single system. The system can be sized based on the assumption that any exposing fire will be confined to a certain size if the exposing building is compartmented by construction features such as fire walls, partitions, or enclosed stair shafts. In other words, it might not be necessary to assume that the entire exposed building wall will be engulfed in fire.

Flames from an adjacent building or structure will likely rise upward in a rectangular plume area. Each outside sprinkler system should be arranged to simultaneously discharge water over the corresponding rectangular plume area, assuming that the exposure will occur from ground level upward.

4-7.3.2 Manually controlled open sprinklers shall be used only where constant supervision is present.

Where automatic actuation of outside sprinkler systems cannot be provided, prompt system operation should be ensured through manual means by the opening of strategically located and identified valves. Personnel assigned the responsibility for and instructed in the operation and importance of these systems need to constantly monitor the exposing operations.

4-7.3.3 Sprinklers shall be of the open or automatic type. Automatic sprinklers in areas subject to freezing shall be on dry pipe systems conforming to Section 4-2 or antifreeze systems conforming to Section 4-5.

4-7.3.4 Automatic systems of open sprinklers shall be controlled by the operation of fire detection devices designed for the specific application.

Fire detection devices with weatherproof fixtures should be located on the exterior of the exposed building in such a manner as to rapidly detect heat radiation or convection from an exposing fire. Spacing and location of detectors should be closer than that specified by the manufacturer if the line of sight of the exposing building or structure is obstructed.

4-7.4 System Components.

4-7.4.1 Drain Valves. Each system of outside sprinklers shall have a separate drain valve installed on the system side of each control valve.

Exception: A separate drain valve shall not be required on open sprinkler-top fed systems arranged to facilitate drainage.

4-7.4.2 Check Valves. Where sprinklers are installed on two adjacent sides of a building, protecting against two separate and distinct exposures, with separate control valves for each side, the end lines shall be connected with check valves located so that one sprinkler around the corner will operate *[see Figures 4-7.4.2(a) and (b)]*. The intermediate pipe between the two check valves shall be arranged to drain. As an alternate solution, an additional sprinkler shall be installed on each system located around the corner from the system involved.

The "trapped" section of pipe and sprinklers located between two check valves must be drained automatically to prevent water from freezing in the piping after system operation. Drainage can be accomplished by providing a downturned tee outlet and a K-2.8 sprinkler located in the piping at a low point. If the K-2.8 sprinkler has not fused during sprinkler system operation, it must be removed to drain the piping and then replaced.

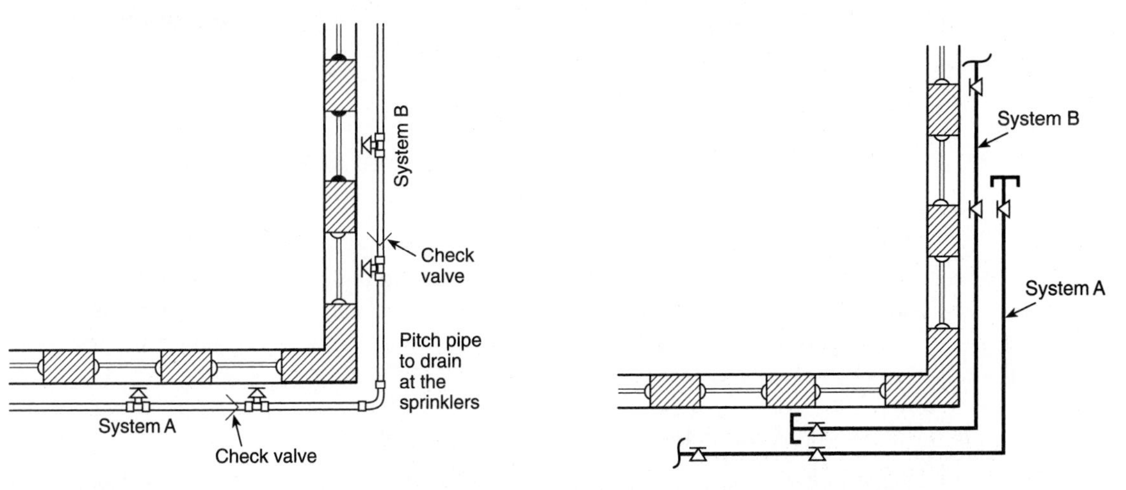

Figure 4-7.4.2(a) *Typical arrangement of check valves.*

Figure 4-7.4.2(b) *Alternate arrangement of check valves.*

4-7.4.3 System Arrangement. Where one exposure affects two sides of the protected structure, the system shall not be subdivided between the two sides but rather shall be arranged to operate as a single system.

Heat transfer from an exposing fire is typically accomplished by convection or radiation. Radiation is direct along a line of sight, while convection is via fire gases, which can be flaming and deflected by drafts or wind.

The convective gases are somewhat easier to cool due to the effect of the water spray from the sprinklers. Radiated heat is more difficult to block, since some of it can pass through the water spray and reach the exposed building. To account for this, systems should be designed so that a certain amount of water flows down the exposed building walls.

Outside sprinkler systems can be required to extend either beyond the wall area directly opposite the exposing building or structure or to surfaces other than those parallel to the exposing building wall or structure surface or both.

4-7.5 Pipe and Fittings.

Pipe and fittings installed on the exterior of the building shall be corrosion resistant.

Since outside sprinkler systems are exposed to environmental conditions, exterior surfaces of outside system components are subject to an accelerated rate of corrosion. The interior surfaces of the system are also subject to an accelerated rate of corrosion if the sprinklers are open, as in a manually or automatically controlled deluge system. Pipe scale, which forms in the interior of the system, could obstruct sprinkler orifices when the system activates. Therefore, galvanized pipe and fittings are normally used for this application. Corrosion of pipe is an important consideration, since sprinklers with nominal K-factors less than 4.2 can be used in these special systems. See the exception to 3-2.4.2(2).

4-7.6 Strainers.

A listed strainer shall be provided in the riser or feed main that supplies sprinklers having nominal K-factors smaller than 2.8 (4.0).

Scale, mud, or debris can become dislodged from piping by fast-moving water traveling through dry pipe or deluge sprinkler systems. Sprinklers with nominal K-factors less than 2.8 are more susceptible to clogging. Strainers should be listed, located adjacent to the control valve, and capable of being flushed without stopping waterflow while the outside sprinkler system is operating.

4-7.7 Gauge Connections.

A listed pressure gauge conforming with 5-15.3.2 shall be installed immediately below the control valve of each system.

4-7.8 Sprinklers.

Only sprinklers of such type as are listed for window, cornice, sidewall, or ridge pole service shall be installed for such use, except where adequate coverage by use of other types of listed sprinklers and/or nozzles has been demonstrated. Small-orifice or large-orifice sprinklers shall be permitted.

Many sprinklers approved for use in outside sprinkler systems are automatic in operation, which means that they are equipped with heat-sensitive fusible elements. The use of automatic sprinklers allows for the installation of wet pipe systems in nonfreezing climates and the installation of dry pipe or antifreeze systems in areas where outside freezing conditions can occur.

Some sprinklers approved for use in outside sprinkler systems are not equipped with heat-sensitive fusible elements. Outside sprinkler systems that use open or non-automatic sprinklers must be of the deluge type.

Outside sprinkler systems provide fire protection by removing or mitigating the effects of heat that are transmitted from an exposing fire. Water sprays from sprinklers are shown to reduce radiated heat passing through them by 50 to 70 percent, depending on waterflow rates. In addition, convected fire gases are cooled while passing through the sprinkler spray. A portion of the convected and radiated heat passes through the water spray and reaches the protected surface. However, a certain amount of sprinkler discharge should be directed so that it runs down the side of the building and cools the exposed surface. Provisions must be made so that the water remains in contact with the wall or window surface while running down the exposed surface. Additionally, consideration should be given to the effects of wind so that the surface can be properly wetted.

4-8* Refrigerated Spaces

A-4-8 Careful installation and maintenance, and some special arrangements of piping and devices as outlined in this section, are needed to avoid the formation of ice and frost inside piping in cold storage rooms that will be maintained at or below 32 F (0 C). Conditions are particularly favorable to condensation where pipes enter cold rooms from rooms having temperatures above freezing.

Whenever the opportunity offers, fittings such as those specified in 4-8.1, as well as flushing connections, should be provided in existing systems.

Where possible, risers should be located in stair towers or other locations outside of refrigerated areas, which would reduce the probabilities of ice or frost formation within the riser (supply) pipe.

Cross mains should be connected to risers or feed mains with flanges. In general, flanged fittings should be installed at points that would allow easy dismantling of the system. Split ring or other easily removable types of hangers will facilitate the dismantling.

Because it is not practical to allow water to flow into sprinkler piping in spaces that might be constantly subject to freezing, or where temperatures must be maintained at or below 40 F (4.4 C), it is important that means be provided at the time of system installation to conduct trip tests on dry pipe valves that service such systems. NFPA 25, *Standard for the Inspection, Testing, and Maintenance of Water-Based Fire Protection Systems,* contains requirements in this matter.

Dry pipe systems installed in refrigerated spaces require special care and consideration, since temperatures are below freezing and frequently in subzero ranges. If a system trips accidentally and allows the piping to fill with water, the nature of the occupancy can preclude heating this space to allow melting of ice and repair of the system. Such systems may need to have the pipe dismantled and moved to a warm area where it would be thawed, emptied, and then reinstalled. Such operations are costly and result in long periods in which systems are out of service. Where conditions are particularly severe, double interlock preaction or other types of special systems should be considered.

In refrigerated spaces, the air supply for the preaction or dry pipe system needs to be sufficiently dry to prevent the accumulation of moisture and the subsequent development of ice plugs in the system piping. Occurrences where sprinkler systems protecting freezers maintained at 5°F (15°C) or less become plugged by ice or frost deposits have increased. The plugs found can create total blockage of the feed main and severely impair the system. The 1996 edition of NFPA 13 first included provisions for decreasing the chances of ice plug or ice block formation in the air supply line or in the sprinkler piping. These provisions consist of arranging pipe and associated appurtenances in a manner that minimizes the moisture content in the air supply. Means of reducing the likelihood of ice plug formation include taking the air supply from the coldest freezer area, installing air dryers, or using a moisture-free gas such as nitrogen.

4-8.1 Spaces Maintained at Temperatures Above 32°F (0°C).

Where temperatures are maintained above 32°F (0°C) in refrigerated spaces, the requirements in this section shall not apply.

4-8.2* Spaces Maintained at Temperatures Below 32°F (0°C).

A-4-8.2 The requirements in 4-8.2 are intended to minimize the chances of ice plug formation inside sprinkler system piping protecting freezers.

4-8.2.1 Where sprinkler pipe passes through a wall or floor into the refrigerated space, a section of pipe arranged for removal shall be provided immediately inside the space. The removable length of pipe shall be a minimum of 30 in. (762 mm).

The removable section of pipe and other fittings as illustrated in Figures 4-8.2.7 and A-4-8.2.4 facilitate examination of piping to determine if ice formation is occurring. If inspections are conducted on a routine basis, the formation of ice plugs and their detrimental effect of system performance can be reduced.

4-8.2.2 A low air pressure alarm to a constantly attended location shall be installed.

Exception: Systems equipped with local low pressure alarms and an automatic air maintenance device shall not be required to alarm to a constantly attended location.

This alarm is intended to notify plant personnel that system air pressure is below acceptable limits. Plant personnel need to react to low pressure alarms by taking appropriate action to remedy the situation in order to avoid the system piping filling accidentally with water.

4-8.2.3 Piping in refrigerated spaces shall be installed with pitch as outlined in 5-14.2.3.

Any water introduced to a system protecting a very cold environment can freeze in a matter of minutes. Piping is required to be pitched to aid in the prompt removal of water from the system.

4-8.2.4* Air supply for systems shall be taken from the room of lowest temperature to reduce the moisture content of the air.

Exception No. 1: This requirement shall not apply where compressed nitrogen gas from cylinders is used in lieu of compressed air.

Exception No. 2: This requirement shall not apply where a compressor/dryer package is listed for the application using an ambient air supply.

A-4-8.2.4 A higher degree of preventing the formation of ice blocks can be achieved by lowering the moisture of the air supply entering the refrigerated space to a pressure dew point no greater than 20°F (−6.6°C) below the lowest nominal temperature of the refrigerated space. The pressure dew point of the air supply can cause moisture to condense and freeze in sprinkler pipe even when the air supply is from the freezer. One method of reducing the moisture content of the air by use of air drying systems is illustrated in Figure A-4-8.2.4.

When compressors and dryers are used for an air supply, consideration should be given to pressure requirements of the regenerative dryers, compressor size, air pressure regulator capacity, and air fill rate. Application of these factors could necessitate the use of increased air pressures and a larger air compressor.

The compressed air supply should be properly prepared prior to entering a regenerative-type air dryer, such as minimum air pressure, maximum inlet air temperature, and proper filtration of compressed air.

Taking air supply from the coldest room and then compressing it does not guarantee moisture-free air. However, this arrangement usually provides a reasonable degree of protection against the formation of ice plugs. Appendix paragraph A-4-8.2.4 includes additional information regarding the value in which the pressure dew point should be reduced. Additionally, the exceptions allow other means for reducing the moisture content of air more reliably.

4-8.2.5* An indicating-type control valve for operational testing of the system shall be provided on each sprinkler riser outside of the refrigerated space.

A-4-8.2.5 A major factor contributing to the introduction of moisture into the system piping is excessive air compressor operation caused by system leakage. Where excessive compressor operation is noted or ice accumulates in the air supply piping, the system should be checked for leakage and appropriate corrective action should be taken.

4-8.2.6* A check valve with a $\frac{3}{32}$-in. (24-mm) diameter hole in the clapper shall be installed in the system riser below the test valve required in 4-8.2.5.

Exception: When system dry pipe or preaction valves are used, if designed to completely drain all water above the seat and that are listed for installation without priming water remaining and where priming water is not used in the system riser.

A-4-8.2.6 The purpose of the check valve is to prevent evaporation of prime water into the system piping.

The check valve is intended to prevent evaporated priming water from migrating into the piping located in the freezer and causing an ice plug to form. However, dry pipe

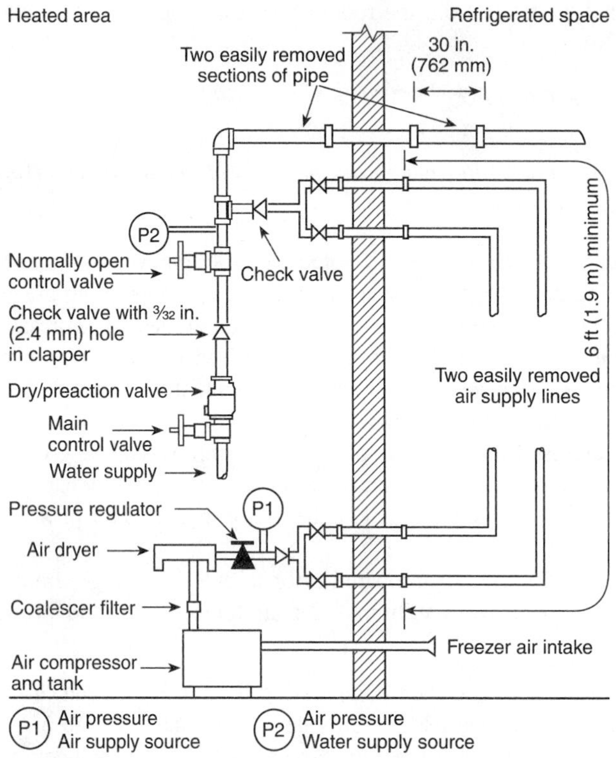

Figure A-4-8.2.4 *Refrigerator area sprinkler systems used to minimize the chances of developing ice plugs.*

and preaction valves are available that do not require priming water. If priming water is not necessary, then the check valve can be omitted as indicated by the exception. Additionally, any residual water left in the system riser after a trip test can be drained out of the drain connection at the base of the freezer system water control valve.

4-8.2.7* The air supply piping entering the freezer area shall be equipped with two easily removable supply lines at least 6 ft (1.9 m) long and at least 1 in. (25.4 mm) in diameter as shown in Figure 4-8.2.7. Each supply line shall be equipped with control valves located in the warm area. Only one air supply line shall be open to supply the system air at any one time.

Exception: Two supply lines shall not be required where compressed nitrogen gas from cylinders is used in lieu of compressed air.

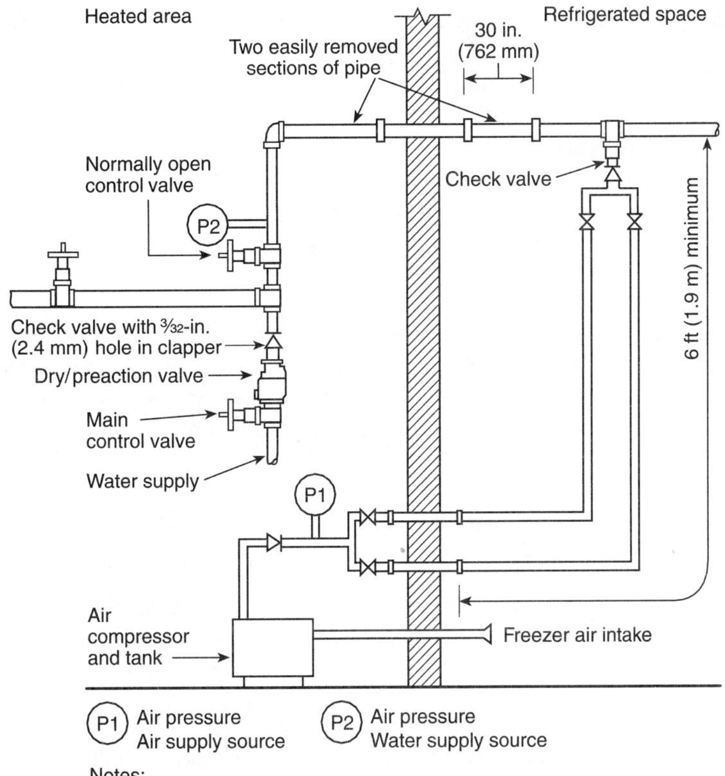

Figure 4-8.2.7 *Refrigerator area sprinkler system used to minimize the chances of developing ice plugs.*

A-4-8.2.7 The dual lines feeding the system air entering the cold area are intended to facilitate continued service of the system when one line is removed for inspection. It should be noted that, when using a system as described in Figure A-4-8.2.4, differences in the pressures at gauge P1 and gauge P2 indicate blockage in the air supply line or other malfunctions.

4-9 Commercial-Type Cooking Equipment and Ventilation

The original requirements with respect to protection of cooking equipment and ventilation systems were developed as a result of test work done by a task group of the National Fire Sprinkler Association. The requirements were introduced in 1968. Automatic sprinklers are effective for extinguishing grease and cooking oil fires because the fine droplets of the water spray lower the temperatures to below the point at which the fire can sustain itself. These current requirements were revised based on research and input from both the cooking equipment ventilation industry and the sprinkler manufacturing industry.

Large droplets sprayed onto a fire involving cooking oils in a deep fat fryer can result in sudden and rapid firespread due to the spattering effect caused by the large water droplets. In order to avoid this phenomenon, only specifically listed sprinklers can be used for the protection of deep fat fryers. (See 4-9.8.2.)

4-9.1 In cooking areas protected by automatic sprinklers, additional sprinklers or automatic spray nozzles shall be provided to protect commercial-type cooking equipment and ventilation systems that are designed to carry away grease-laden vapors unless otherwise protected. *(See NFPA 96, Standard for Ventilation Control and Fire Protection of Commercial Cooking Operations.)*

4-9.2* Standard sprinklers or automatic spray nozzles shall be so located as to provide for the protection of exhaust ducts, hood exhaust duct collars, and hood exhaust plenum chambers.

Exception: Sprinklers or automatic spray nozzles in ducts, duct collars, and plenum chambers shall not be required where all cooking equipment is served by listed grease extractors.

A-4-9.2 See Figure A-4-9.2.

Sprinklers are an effective way of controlling fires in cooking equipment, filters, and ducts if they are located to cover the cooking surfaces and all areas of an exhaust system to which fires could spread. If all cooking equipment is served by listed grease extractors, the sprinkler protection can be limited to the cooking surfaces. This arrangement is subject to acceptance by the authority having jurisdiction. Some manufacturers of exhaust systems that incorporate listed grease extractors provide listed built-in water spray fire protection for cooking surfaces in a preengineered package ready for connection to the sprinkler system.

4-9.3 Exhaust ducts shall have one sprinkler or automatic spray nozzle located at the top of each vertical riser and at the midpoint of each offset. The first sprinkler or automatic spray nozzle in a horizontal duct shall be installed at the duct entrance. Horizontal exhaust ducts shall have such devices located on 10-ft (3-m) centers beginning no more than 5 ft (1.5 m) from the duct entrance. A sprinkler(s) or an automatic spray nozzle(s) in exhaust ducts subject to freezing shall be properly protected against freezing by approved means. *(See 5-14.3.1.)*

Exception No. 1: Sprinklers or automatic spray nozzles shall not be required in a vertical riser located outside of a building, provided the riser does not expose combustible material or provided the interior of the building and the horizontal distance between the hood outlet and the vertical riser is at least 25 ft (7.6 m).

Exception No. 2: Sprinklers or automatic spray nozzles shall not be required where the entire exhaust duct is connected to a listed exhaust hood incorporating a specific duct collar and

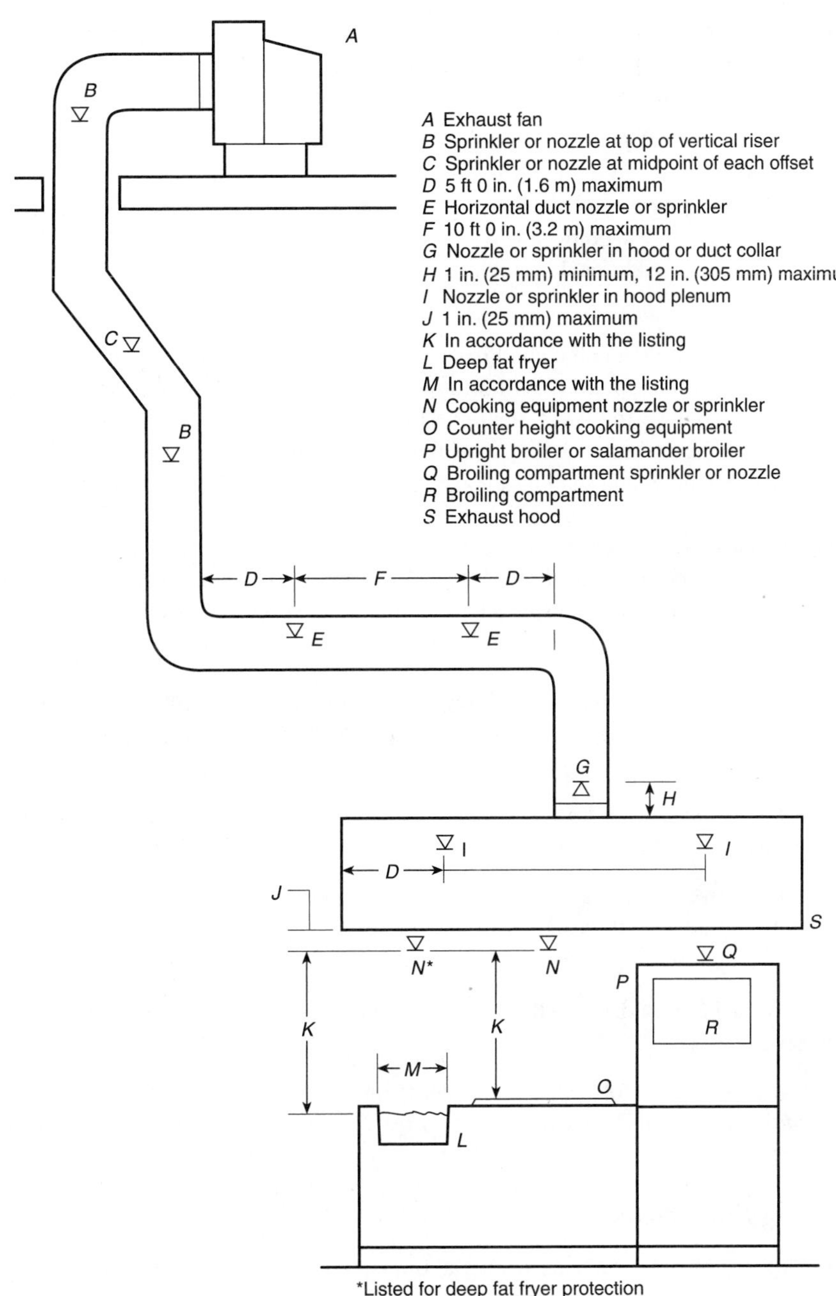

Figure A-4-9.2 *Typical installation showing automatic sprinklers or automatic nozzles being used for the protection of commercial cooking equipment and ventilation systems.*

sprinkler (or automatic spray nozzle) assembly that has been investigated and been shown to protect an unlimited length of duct in accordance with UL 300, Standard for Safety Fire Testing of Fire Extinguishing Systems for Protection of Restaurant Cooking Areas.

UL 300, *Standard for Safety Fire Testing of Fire Extinguishing Systems for Protection of Restaurant Cooking Areas*, is currently used to determine adequate protection for ducts of unlimited length used for cooking applications. If a manufacturer successfully passes these tests with a specific sprinkler or spray nozzle assembly in a specific listed hood and duct collar, additional sprinklers or spray nozzles in the duct are not needed.

4-9.4 Each hood exhaust duct collar shall have one sprinkler or automatic spray nozzle located 1 in. minimum to 12 in. maximum (25.4 mm minimum to 305 mm maximum) above the point of duct collar connection in the hood plenum. Hoods that have listed fire dampers located in the duct collar shall be protected with a sprinkler or automatic spray nozzle located on the discharge side of the damper and shall be so positioned as not to interfere with damper operation.

4-9.5 Hood exhaust plenum chambers shall have one sprinkler or automatic spray nozzle centered in each chamber not exceeding 10 ft (3 m) in length. Plenum chambers greater than 10 ft (3 m) in length shall have two sprinklers or automatic spray nozzles evenly spaced, with the maximum distance between the two sprinklers not to exceed 10 ft (3 m).

4-9.6 Sprinklers or automatic spray nozzles being used in duct, duct collar, and plenum areas shall be of the extra-high temperature classification [325°F to 375°F (163°C to 191°C)] and shall have orifice sizes not less than ¼ in. (6.4 mm) and not more than ½ in. (13 mm).

Exception: When use of a temperature-measuring device indicates temperatures above 300°F (149°C), a sprinkler or automatic spray nozzle of higher classification shall be used.

The sprinklers over the cooking surface must operate first so that water from the sprinklers in the plenum do not cool them and prevent their operation.

4-9.7 Access must be provided to all sprinklers or automatic spray nozzles for examination and replacement.

The access should not jeopardize the integrity of the hood or duct. Refer to NFPA 96, *Standard for Ventilation Control and Fire Protection of Commercial Cooking Operations*.

4-9.8 Cooking Equipment.

4-9.8.1 Cooking equipment (such as deep fat fryers, ranges, griddles, and broilers) that is considered to be a source of ignition shall be protected in accordance with the provisions of 4-9.1.

4-9.8.2 A sprinkler or automatic spray nozzle used for protection of deep fat fryers shall be listed for that application. The position, arrangement, location, and water supply for each sprinkler or automatic spray nozzle shall be in accordance with its listing.

As of early 1999, no listed sprinklers for the protection of deep fat fryers are available. The sprinklers that were previously listed for this function lost their listing following changes to UL 300. The remaining cooking equipment can still be protected with standard sprinklers.

4-9.8.3 The operation of any cooking equipment sprinkler or automatic spray nozzle shall automatically shut off all sources of fuel and heat to all equipment requiring protection. Any gas appliance not requiring protection but located under ventilating equipment shall also be shut off. All shutdown devices shall be of the type that requires manual resetting prior to fuel or power being restored.

4-9.9 A listed indicating valve shall be installed in the water supply line to the sprinklers and spray nozzles protecting the cooking and ventilating system.

4-9.10 A listed line strainer shall be installed in the main water supply preceding sprinklers or automatic spray nozzles having nominal K-factors smaller than 2.8 (4.0).

4-9.11 A system test connection shall be provided to verify proper operation of equipment specified in 4-9.8.3.

4-9.12 Sprinklers and automatic spray nozzles used for protecting commercial-type cooking equipment and ventilating systems shall be replaced annually.

Exception: Where automatic bulb-type sprinklers or spray nozzles are used and annual examination shows no buildup of grease or other material on the sprinklers or spray nozzles.

REFERENCES CITED IN COMMENTARY

National Fire Protection Association, 1 Batterymarch Park, P.O. Box 9101, Quincy, MA 02269-9101.

NFPA 25, *Standard for the Inspection, Testing, and Maintenance of Water-Based Fire Protection Systems,* 1998 edition.
NFPA 72, *National Fire Alarm Code®,* 1999 edition.
NFPA 80A, *Recommended Practice for Protection of Buildings from Exterior Fire Exposures,* 1996 edition.
NFPA 96, *Standard for Ventilation Control and Fire Protection of Commercial Cooking Operations,* 1998 edition.
NFPA 409, *Standard on Aircraft Hangars,* 1995 edition.
The SFPE Handbook of Fire Protection Engineering, second edition, 1995.

Underwriters Laboratories Inc., 333 Pfingsten Road, Northbrook, IL 60062.

UL 300, *Standard for Safety Fire Testing of Fire Extinguishing Systems for Protection of Restaurant Cooking Areas,* 1996 edition.

CHAPTER 5

Installation Requirements

5-1* Basic Requirements

A-5-1 The installation requirements are specific for the normal arrangement of structural members. There will be arrangements of structural members not specifically detailed by the requirements. By applying the basic principles, layouts for such construction can vary from specific illustrations, provided the maximum specified for the spacing and location of sprinklers (Section 5-4) are not exceeded.

Where buildings or portions of buildings are of combustible construction or contain combustible material, standard fire barriers should be provided to separate the areas that are sprinkler protected from adjoining unsprinklered areas. All openings should be protected in accordance with applicable standards, and no sprinkler piping should be placed in an unsprinklered area unless the area is permitted to be unsprinklered by this standard.

Water supplies for partial systems should be designed with consideration to the fact that in a partial system more sprinklers might be opened in a fire that originates in an unprotected area and spreads to the sprinklered area than would be the case in a completely protected building. Fire originating in a nonsprinklered area might overpower the partial sprinkler system.

Where sprinklers are installed in corridors only, sprinklers should be spaced up to the maximum of 15 ft (4.5 m) along the corridor, with one sprinkler opposite the center of any door or pair of adjacent doors opening onto the corridor, and with an additional sprinkler installed inside each adjacent room above the door opening. Where the sprinkler in the adjacent room provides full protection for that space, an additional sprinkler is not required in the corridor adjacent to the door.

Numerous factors, such as the available water supply, type of sprinkler, building construction features, and the anticipated fire hazards, must be considered in the design of automatic sprinkler systems. Chapter 5 deals exclusively with installation details, such as positioning, and location of system components, such as sprinklers, pipe hangers, and valves. The requirements in this chapter identify the location and positioning rules for system components and consider the impact of building construction and equipment on satisfactory sprinkler system performance.

For example, sprinklers must be positioned with regard to other sprinklers, structural and architectural features at the ceiling, and other building elements such as ductwork and lighting. Failure to do so greatly increases the likelihood of delayed sprinkler activation and of skewed or obstructed sprinkler spray patterns. Limiting distances between sprinklers allows for their timely activation and results in the intended level of fire control or fire suppression. NFPA 13 also recognizes that building features that block the sprinkler discharge can negatively impact spray distribution patterns.

NFPA 13 requires that sprinklers be provided throughout the premises. This requirement is further reinforced through Formal Interpretation 78-6 and the Formal Interpretations that follow 5-1.1. However, certain exceptions permit sprinklers to be omitted from certain spaces where specific conditions are satisfied. These spaces include concealed spaces (see 5-13.1 and 5-13.7), vertical shafts (see 5-13.2), bathrooms and clothes closets in dwellings units (see 5-13.9.1 and 5-13.9.2), elevator shafts (see 5-13.6), and electrical equipment rooms (see 5-13.11). For the most part, these spaces must include noncombustible or limited-combustible construction where the introduction of combustible contents and storage is precluded by building usage.

The installation of sprinklers throughout the building ensures that the sprinkler system's effectiveness is not compromised by a fire originating in a noncompliant unsprinklered space. Fires originating in most unsprinklered areas are likely to grow to a size where sprinklers in adjacent spaces have limited success in controlling them.

Formal Interpretation 78-6

Reference: 5-1

Question: Is it the intent of 5-1 to require sprinkler protection in walk-in type coolers and freezers in fully sprinklered buildings?

Answer: Yes.

Issue Edition: 1978

Reference: 4-4

Date: May 1979

5-1.1* The requirements for spacing, location, and position of sprinklers shall be based on the following principles:

(1) Sprinklers installed throughout the premises
(2) Sprinklers located so as not to exceed maximum protection area per sprinkler
(3) Sprinklers positioned and located so as to provide satisfactory performance with respect to activation time and distribution

Exception No. 1: For locations permitting omission of sprinklers, see 5-13.1, 5-13.2, and 5-13.9.

Exception No. 2: When sprinklers are specifically tested and test results demonstrate that deviations from clearance requirements to structural members do not impair the ability of the sprinkler to control or suppress a fire, their positioning and locating in accordance with the test results shall be permitted.

Exception No. 3: Clearance between sprinklers and ceilings exceeding the maximum specified in 5-6.4.1, 5-7.4.1, 5-8.4.1, 5-9.4.1, 5-10.4.1, and 5-11.4.1 shall be permitted provided that tests or calculations demonstrate comparable sensitivity and performance of the sprinklers to those installed in conformance with these sections.

A-5-1.1 This standard contemplates full sprinkler protection for all areas. Other NFPA standards that mandate sprinkler installation might not require sprinklers in certain areas. The requirements of this standard should be used insofar as they are applicable. The authority having jurisdiction should be consulted in each case.

Unless NFPA 13 specifically permits the omission of sprinklers in a certain area, sprinklers are required. The four Formal Interpretations in this commentary section further stress the requirement for locating sprinklers throughout the premises. In some situations, this requirement can result in conflicts with other regulations. For example, NFPA *101*®, *Life Safety Code*®, allows sprinklers to be omitted from certain size closets in any type of residential occupancy. NFPA 13 restricts this omission to closets in hotels and motels only. Fires originating in these size closets in other types of residential occupancies can result in substantial property damage (see 5-13.9.2). The very small number of fires originating in these size closets that result in one or more fatalities justifies NFPA *101*'s broader exception. In other words, NFPA *101* addresses only life safety, while NFPA 13 addresses both life safety and property protection.

NFPA 13 does not support the use of partial sprinkler systems. However, NFPA 13 recognizes that other regulations permit the use of partial systems (see 1-6.2). In such cases, NFPA 13 requirements should be followed to the extent possible. When considering partial sprinkler protection, it is important to recognize that a fire originating in an unsprinklered space is likely to grow and spread to the sprinklered area unless some other precautions are taken. Once the fire reaches the sprinklered space, it is likely to be of a size that cannot be controlled or suppressed by the partial system. Sprinkler systems designed and installed in accordance with NFPA 13 are intended to control or suppress fires during their early stages of development. When property protection is a primary concern and differences exist between NFPA 13 and other standards, the requirements of NFPA 13 should take precedence and sprinklers should be provided accordingly.

Proper placement of automatic sprinklers is necessary to ensure that sprinklers operate in a timely manner and achieve effective fire control. To accomplish the basic goals indicated in 5-1.1(2) and (3), specific NFPA 13 requirements on spacing, position, coverage, and obstructions to sprinkler discharge must be followed (see Section 5-5). For example, depending on the type of sprinkler, specific dimensions on deflector positioning with respect to various ceiling features and arrangements and various types of obstructions are provided. In some cases, such as for unusual ceiling arrangements that contain deep pockets or where heat and smoke vents are used, additional provisions could be necessary. Additionally, where tests or calculations indicate that other arrangements provide at least the same level of performance prescribed by this standard, then these other arrangements are permitted to be used in accordance with Exception Nos. 2 and 3 of 5-1.1.

Formal Interpretation 80-29(A)

Reference: 5-1.1, 5-13.1.3

Background: Construction—Structural steel frame with concrete floors. The floor-to-floor height is approximately 15½ feet. Approximately 9 feet above each floor, a gypsum deck has been poured to form an "interstitial space." Non-combustible ductwork, plumbing, electrical conduit, sprinkler piping and open cable trays run in the interstitial space. Access to the space is by five stairways, with access doors each being approximately 3 feet by 3 feet.

Question: Is it the intent of 5-1.1 of NFPA 13 that sprinkler protection of the interstitial space described above be provided to consider the building completely protected by automatic sprinklers?

Answer: No. The space described is essentially non-combustible with no occupancy and no combustible services [with the possible exception of the cable trays which could be protected in accordance with 5-5.1.3 if necessary] and so could be treated as a non-combustible concealed space. Use of the space for storage or the introduction of combustibles would require provision of sprinklers to maintain classification as completely protected by automatic sprinklers.

Issue Edition: 1980

Reference: 3-10.3, 4-1

Date: October 1982

Formal Interpretation

Reference: 5-1.1

Question: Is it the intent of 5-1.1 to require the installation of a sprinkler in every room in a building including (a) Shower Rooms, (b) Clothes Closets?

Answer: Yes. The intent is to require sprinklers in every area except where specifically excluded or where the authority having jurisdiction permits omission of the sprinklers.

Issue Edition: 1973

Reference: 4-1.1.1

Date: January 1978

Formal Interpretation 83-16

Reference: 5-1.1

Question: Does 5-1.1 require that a fire pump room containing a fire pump for a sprinkler system for a completely sprinklered building be protected by automatic sprinklers even if the fire pump is in a separate building?

Answer: Yes. Wherever the fire pump is located, it would be considered a part of the building and therefore required to be protected by automatic sprinklers.

Issue Edition: 1983

Reference: 4-1.1.1 & 3-10

Date: April 1984

Formal Interpretation 87-2

Reference: 5-1.1

Question: Is it the intent of this section that the premises be considered sprinklered throughout [see 5-1.1(a)] when sprinkler systems are installed in accordance with this standard and include the omission of sprinklers in certain areas permitted by this standard (e.g. Exception 1 to 5-1.1 and 5-13.1.1)?

Answer: Yes.

Issue Edition: 1987

Reference: 4-1.1.1

Date: February 1988

5-1.2* System valves and gauges shall be accessible for operation, inspection, tests, and maintenance.

A-5-1.2 The components need not be open or exposed. Doors, removable panels, or valve pits can satisfy this need. Such equipment should not be obstructed by such permanent features as walls, ducts, columns, or direct burial.

Access to critical system components is necessary to allow for emergency shutdown after a fire and to check the operation of the system during a fire. Proper access also facilitates inspection, testing, and maintenance of the system in accordance with NFPA 25, *Standard for the Inspection, Testing, and Maintenance of Water-Based Fire Protection Systems.* While valves and gauges are not required to be visible—that is, they can be located in valve pits or behind doors or removable panels—3-8.3 requires appropriate signage for valves, and 5-14.1.1.1 requires that the valves be in an accessible location. Additionally, 5-14.1.1.4 requires that where control valves are installed overhead, their indicating feature must be visible from the floor below.

5-2 System Protection Area Limitations

The maximum floor area on any one floor to be protected by sprinklers supplied by any one sprinkler system riser or combined system riser shall be as follows:

Light hazard—52,000 ft^2 (4831 m^2)

Ordinary hazard—52,000 ft^2 (4831 m^2)

Extra hazard—

Pipe schedule—25,000 ft^2 (2323 m^2)

Hydraulically calculated—40,000 ft^2 (3716 m^2)

Storage—High-piled storage (as defined in 1-4.2) and storage covered by other NFPA standards—40,000 ft^2 (3716 m^2)

Exception No. 1: The floor area occupied by mezzanines shall not be included in the above area.

Exception No. 2: Where single systems protect extra hazard, high-piled storage, or storage covered by other NFPA standards, and ordinary or light hazard areas, the extra hazard or storage area coverage shall not exceed the floor area specified for that hazard and the total area coverage shall not exceed 52,000 ft^2 (4831 m^2).

The specified maximum coverage area is the maximum floor area per system on any one floor in a given building. The maximum areas of coverage are not to be extended to other buildings as indicated in the Formal Interpretation to Section 5-2. This Formal Interpretation implies that each building is to have its own sprinkler system.

Formal Interpretation

Reference: 5-2

Question: Would the 52,000 sq ft maximum area on one floor apply to several buildings being supplied by one riser assembly? Can the 52,000 sq ft be exceeded if the individual buildings are not in excess of 52,000 sq ft and are separated by a clear space of 20 ft or more?

Answer: No. The 52,000 sq ft area limitation and the 20 ft separating distance are not relevant to the problem. Each building should have its individual system riser.

Issue Edition: 1976

Reference: 3-3.1

Date: August 1977

NFPA 13 does not limit the number of floors that can be protected by a single riser, because each floor is considered a separate fire area. However, limitations are placed on the area of each floor. Vertical openings are assumed to be protected as outlined in 5-13.4. In a single-story, 312,000-ft^2 (28,986-m^2) building of ordinary or light hazard occupancies, at least six systems would be required. If the building were six stories high with the floor area of each story not exceeding 52,000 ft^2 (4831m^2), the building could be protected by a single system riser. If each story consisted of a floor area in excess of 52,000 ft^2 (4831 m^2), then more than one riser would be needed. Floor control valves, along with waterflow alarms, are often provided on each floor of a high-rise building to limit the area affected by a single impairment and to identify the floor or zone in which sprinkler operation occurred. Floor control valves and waterflow devices are not specifically required for each floor in a multistory building. See 5-13.2.2 and 5-15.1.6.

Where multiple risers are necessary to meet the system area limitations, a manifold riser arrangement can be considered as shown in Exhibits 5.1 and 5.2. For a building containing only light hazard or ordinary hazard occupancies, each of the three risers in Exhibits 5.1 and 5.2 could protect up to 52,000 ft^2 (4831 m^2) in a single-story, 156,000-ft^2 (14,470-m^2) building. If the building contained only extra hazard occupancies that were hydraulically designed or high-piled storage, each riser could protect up to 40,000 ft^2 (3716 m^2) in a single-story, 120,000-ft^2 (11,148-m^2) building.

NFPA 13 formerly limited system size by the number of sprinklers on a system—for example, 400 sprinklers for ordinary hazard areas. The 52,000-ft^2 (4831-m^2) limitation is determined by multiplying the 400 sprinklers by the 130-ft^2 (12-m^2) maximum spacing. These area limitations are not related to the hydraulics or the operating characteristics of the system. Rather, the area limitations are based on judgmental factors concerning the maximum area within a single, vertical fire division that should be protected by a single system or that could be out of service at any given time.

Exception No. 1 was added in 1991. Some confusion can result when tabulating the areas noted for each occupancy category. This exception does not require the

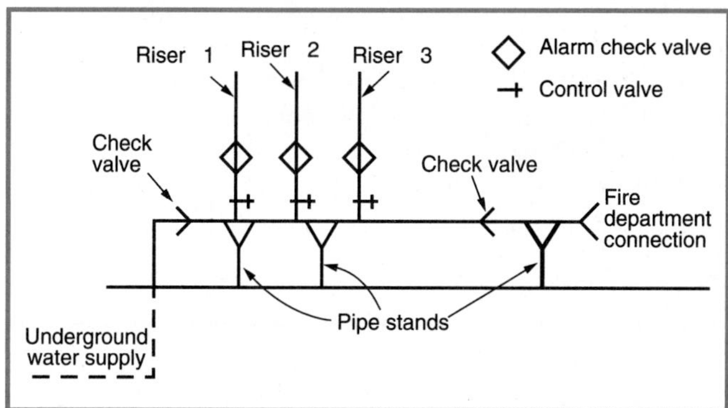

Exhibit 5.1 *Manifold riser arrangement consisting of three wet pipe systems.*

Exhibit 5.2 Manifold riser arrangement consisting of two wet pipe systems and one dry pipe system.

additional area occupied by the mezzanine to be counted against the area limitation. Exhibits 5.3 and 5.4 show examples of the intent of this exception. For high-piled storage, the area protected by one system riser cannot exceed 40,000 ft^2 (3716 m^2). When determining the area, the mezzanine area does not need to be added to the actual floor area.

A single system protecting both ordinary or light hazard areas and solid-piled, palletized, or rack storage greater than 12 ft (3.7 m) high, or protecting extra hazard areas, can have a coverage area of up to 52,000ft^2 (4831 m^2) as indicated by Exception No. 2. However, not more than 40,000 ft^2 (3716 m^2) of that coverage area can be high-piled storage or hydraulically designed for extra hazard occupancies. The 40,000-ft^2 (3716-m^2) maximum extra hazard coverage area for hydraulically designed systems is consistent with the requirements for storage areas that have similar fire loading.

5-3 Use of Sprinklers

Although numerous types of sprinklers are currently available, sprinkler technology continues to evolve, especially with regard to thermal sensitivity, spray pattern distribu-

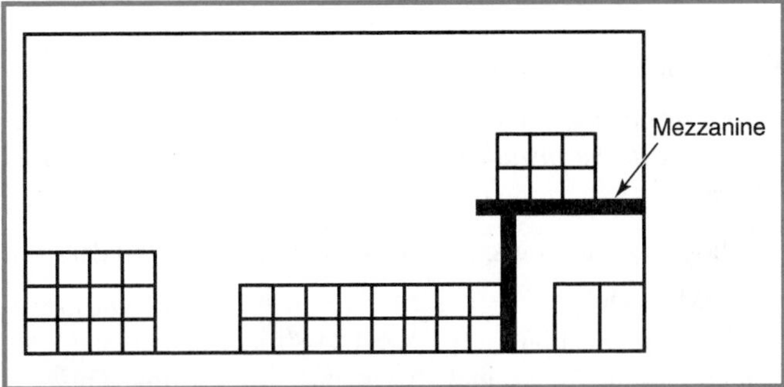

Exhibit 5.3 *A mezzanine used for storage.*

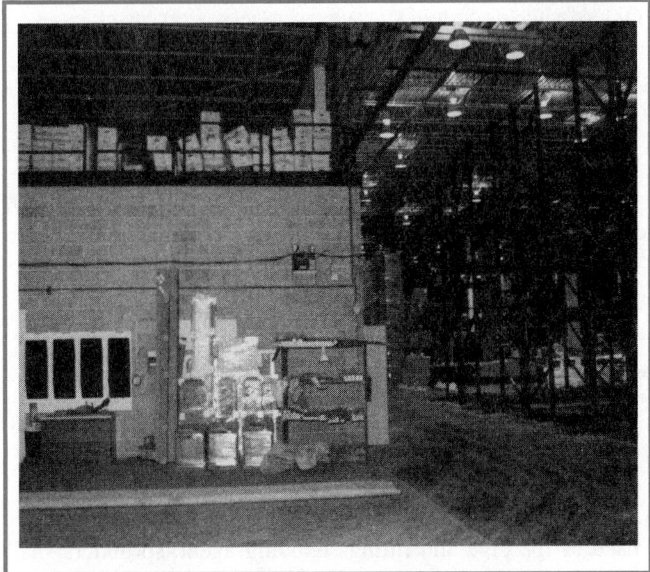

Exhibit 5.4 *Mezzanine storage area in a warehouse.*

tion characteristics, and droplet size. As a result, new types and styles of sprinklers with specific applications and installation requirements continue to be developed.

Beginning with the 1996 edition of NFPA 13, the requirements in Chapter 5 were reorganized by sprinkler types to better address their use and installation requirements. This standard provides use and installation requirements for the most common types of sprinklers, including standard spray upright and pendent, sidewall, residential, quick-response, extended coverage, large drop, early suppression fast-response (ESFR), and in-rack.

5-3.1 General.

5-3.1.1* Sprinklers shall be installed in accordance with their listing.

Exception: Where construction features or other special situations require unusual water distribution, listed sprinklers shall be permitted to be installed in positions other than anticipated by their listing to achieve specific results.

A-5-3.1.1 The evaluation for usage should be based upon a review of available technical data.

Any limitations placed on a sprinkler, such as the hazard classification, location in which it can be used, restrictions on its position with respect to structural elements, or limitations placed on its temperature rating, are included as part of its listing. This statement is to preclude the use of sprinklers for other than their intended use. For example, some architects have insisted on the installation of sprinklers within a structure to achieve symmetry. These sprinklers are not connected to a water supply and are merely being used as a decorative feature—clearly not an intended use.

Some specially listed sprinklers, such as certain storage sprinklers, attic sprinklers, window sprinklers, on-off sprinklers, and institutional sprinklers, have unique use and installation requirements. These and other special sprinklers must be installed in accordance not only with NFPA 13 but also with any additional requirements specified by their listing. Subsection 5-4.9 addresses special sprinklers.

5-3.1.2* Upright sprinklers shall be installed with the frame arms parallel to the branch line.

A-5-3.1.2 The purpose of this requirement is to minimize the obstruction of the discharge pattern.

The sprinkler's frame arm and the piping below the sprinkler both represent potential obstructions to the water distribution pattern and can prevent a uniform spray pattern. Even though the frame arm is designed to minimize this obstruction, the potential for obstruction cannot be completely eliminated. Installation of an upright sprinkler with its frame arm parallel to the branch line minimizes the likelihood of the water distribution pattern being obstructed by the frame arm. Exhibit 5.19 illustrates an upright sprinkler installed with its frame arm installed in the proper orientation.

5-3.1.3 Where solvent cement is used as the pipe and fittings bonding agent, sprinklers shall not be installed in the fittings prior to the fittings being cemented in place.

The solvent cement used to join some types of sprinkler piping, especially if used excessively, can drip onto the sprinkler and potentially block or plug the sprinkler orifice or otherwise prevent the sprinkler from operating properly. To prevent dripping of excess solvent, the sprinkler should not be installed in the fitting until the fitting is connected to the pipe and any excess solvent is removed.

5-3.1.4 Temperature Ratings.

Temperature ratings of a sprinkler are used as follows:

(1) To keep sprinklers from operating accidentally if installed in a high ambient temperature area
(2) To control the number of sprinklers operating in the design area

5-3.1.4.1* Ordinary-temperature-rated sprinklers shall be used throughout buildings.

Exception No. 1: Where maximum ceiling temperatures exceed 100°F (38°C), sprinklers with temperature ratings in accordance with the maximum ceiling temperatures of Table 3-2.5.1 shall be used.

Exception No. 2: Intermediate- and high-temperature sprinklers shall be permitted to be used throughout ordinary and extra hazard occupancies.

Exception No. 3: Sprinklers of intermediate- and high-temperature classifications shall be installed in specific locations as required by 5-3.1.4.2.

A-5-3.1.4.1 For protection of baled cotton, limited tests and actual fire experience indicate an initial low heat release; thus, sprinklers in the ordinary temperature range should offer some advantage by opening faster than those of intermediate- or high-temperature classifications under similar conditions.

Higher temperature classification sprinklers are preferable for some types of fast-developing fires. The use of ordinary temperature classification sprinklers to protect against fast-developing high heat release rate fires tends to result in the operation of sprinklers beyond the fire area. Sprinkler operation outside the fire area reduces the water discharge density available to the sprinklers directly adjacent to the fire, which reduces their effectiveness. In some high heat release fires with high thermal updrafts, the water discharged from sprinklers is carried back toward the ceiling as steam where it condenses on ordinary temperature sprinklers and causes them to operate. This phenomenon is one of the reasons considered to be responsible for sprinklers operating beyond the fire area.

Extra-high-temperature, very-extra-high-temperature, and ultra-high-temperature sprinklers identified in Table 3-2.5.1 can be utilized in areas immediately above equipment that produce large amounts of heat and high temperatures such as industrial and baking ovens, furnaces, and boilers.

5-3.1.4.2 The following practices shall be observed to provide sprinklers of other than ordinary temperature classification unless other temperatures are determined or unless high-temperature sprinklers are used throughout *[see Tables 5-3.1.4.2(a) and (b) and Figure 5-3.1.4.2]*.

(1) Sprinklers in the high-temperature zone shall be of the high-temperature classification, and sprinklers in the intermediate-temperature zone shall be of the intermediate-temperature classification.
(2) Sprinklers located within 12 in. (305 mm) to one side or 30 in. (762 mm) above an uncovered steam main, heating coil, or radiator shall be of the intermediate-temperature classification.
(3) Sprinklers within 7 ft (2.1 m) of a low-pressure blowoff valve that discharges free in a large room shall be of the high-temperature classification.
(4) Sprinklers under glass or plastic skylights exposed to the direct rays of the sun shall be of the intermediate-temperature classification.
(5) Sprinklers in an unventilated, concealed space, under an uninsulated roof, or in an unventilated attic shall be of the intermediate-temperature classification.

(6) Sprinklers in unventilated show windows having high-powered electric lights near the ceiling shall be of the intermediate-temperature classification.

(7) Sprinklers protecting commercial-type cooking equipment and ventilation systems shall be of the high- or extra-high-temperature classification as determined by use of a temperature-measuring device. *(See 4-9.6.)*

The requirements specified in 5-3.1.4.2 relate to the use of sprinklers discussed in Table 3-2.5.1. The temperature rating criteria allows the sprinklers to operate in response to a fire rather than to high ambient temperatures.

Table 5-3.1.4.2(a) Temperature Ratings of Sprinklers Based on Distance from Heat Sources

Type of Heat Condition	Ordinary Degree Rating	Intermediate Degree Rating	High Degree Rating
(a) Heating ducts			
1. Above	More than 2 ft 6 in.	2 ft 6 in. or less	
2. Side and below	More than 1 ft 0 in.	1 ft 0 in. or less	
3. Diffuser	Any distance except as shown under Intermediate Degree Rating column	*Downward discharge:* Cylinder with 1 ft 0 in. radius from edge extending 1 ft 0 in. below and 2 ft 6 in. above *Horizontal discharge:* Semi-cylinder with 2 ft 6 in. radius in direction of flow extending 1 ft 0 in. below and 2 ft 6 in. above	
(b) Unit heater			
1. Horizontal discharge		*Discharge side:* 7 ft 0 in. to 20 ft 0 in. radius pie-shaped cylinder *[see Figure 5-3.1.4.2]* extending 7 ft 0 in. above and 2 ft 0 in. below heater; also 7 ft 0 in. radius cylinder more than 7 ft 0 in. above unit heater	7 ft 0 in. radius cylinder extending 7 ft 0 in. above and 2 ft 0 in. below unit heater
2. Vertical downward discharge *(For sprinklers below unit heater, see Figure 5-3.1.4.2)*		7 ft 0 in. radius cylinder extending upward from an elevation 7 ft 0 in. above unit heater	7 ft 0 in. radius cylinder extending from the top of the unit heater to an elevation 7 ft 0 in. above unit heater
(c) Steam mains (uncovered)			
1. Above	More than 2 ft 6 in.	2 ft 6 in. or less	
2. Side and below	More than 1 ft 0 in.	1 ft 0 in. or less	
3. Blowoff valve	More than 7 ft 0 in.		7 ft 0 in. or less

For SI units, 1 in. = 25.4 mm; 1 ft = 0.3048 m.

Table 5-3.1.4.2(b) Ratings of Sprinklers in Specified Locations

Location	Ordinary Degree Rating	Intermediate Degree Rating	High Degree Rating
Skylights		Glass or plastic	
Attics	Ventilated	Unventilated	
Peaked roof: metal or thin boards, concealed or not concealed, insulated or uninsulated	Ventilated	Unventilated	
Flat roof: metal, not concealed	Ventilated or unventilated	Note: For uninsulated roof, climate and insulated or uninsulated occupancy can necessitate intermediate sprinklers. Check on job.	
Flat roof: metal, concealed, insulated or uninsulated	Ventilated	Unventilated	
Show windows	Ventilated	Unventilated	

Note: A check of job condition by means of thermometers may be necessary.

5-3.1.4.3 In case of occupancy change involving temperature change, the sprinklers shall be changed accordingly.

5-3.1.4.4* The minimum temperature rating of ceiling sprinklers in general storage, rack storage, rubber tire storage, roll paper storage, and baled cotton storage applications shall be 150°F.

A-5-3.1.4.4 Where high temperature–rated sprinklers are installed at the ceiling, high temperature–rated sprinklers also should extend beyond storage in accordance with Table A-5-3.1.4.4.

5-3.1.5 Thermal Sensitivity.

5-3.1.5.1* Sprinklers in light hazard occupancies shall be of the quick-response type as defined in 1-4.5.2.

Exception No. 1: Residential sprinklers shall be permitted in accordance with 5-4.5.

Exception No. 2: For modifications or additions to existing systems equipped with standard response sprinklers, standard response sprinklers shall be permitted to be used.

Exception No. 3: When individual standard response sprinklers are replaced in existing systems, standard response sprinklers shall be permitted to be used.

A-5-3.1.5.1 When renovations occur in an existing building and no changes are made in the existing sprinkler system, this section is not intended to require the replacement of existing standard sprinklers with quick-response sprinklers.

The use of quick-response sprinklers has been an option within NFPA 13 since the 1980 edition. Although quick-response sprinklers tend to enhance property protection and life safety, no requirements or incentives for their use were provided until the 1996

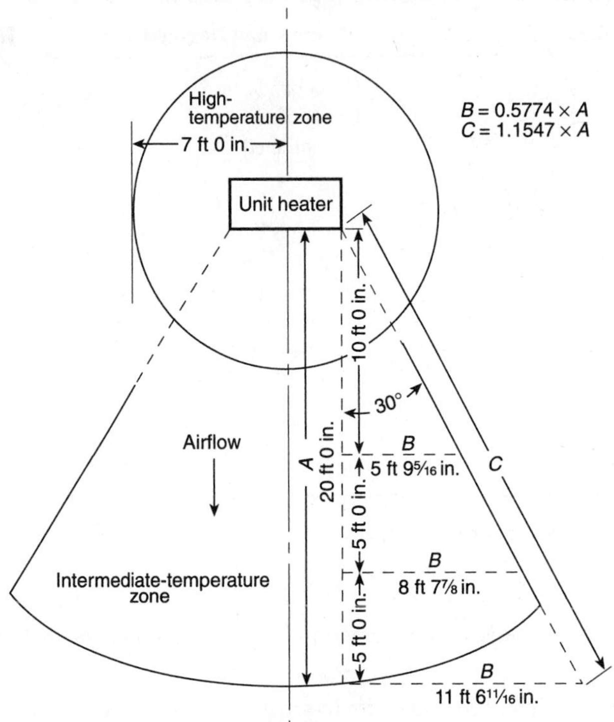

Figure 5-3.1.4.2 *High-temperature and intermediate-temperature zones at unit heaters.*

Table A-5-3.1.4.4 Distance Beyond Perimeter of Storage for High-Hazard Occupancies Protected with High Temperature–Rated Sprinklers

Design Area		Distance	
ft^2	m^2	ft	m
2000	186.0	30	9.1
3000	278.7	40	12.0
4000	371.6	45	13.7
5000	464.5	50	15.2
6000	557.4	55	16.7

edition. The requirement that quick-response sprinklers be used in all light hazard occupancies, with appropriate exceptions to address existing systems, raises the baseline level of system performance. Existing systems are not required to be upgraded when renovations occur.

Quick-response sprinklers are considered necessary because of their life safety benefits. The evidence clearly indicates that sprinklers also limit fire damage. Most of the sprinkler systems installed in light hazard occupancies, such as hospitals, hotels, and apartments, are installed for life safety purposes. However, even in those occupancies where life safety is not the primary reason for sprinkler system installation, such as in offices and restaurants, quick-response sprinklers are still considered important because they limit fire damage and the potential for injury or death of occupants and fire fighters. Given the current level of knowledge concerning the performance of quick-response sprinklers, the use of standard response technology instead of quick-response technology in light hazard occupancies is considered inappropriate. Like quick-response sprinklers, residential sprinklers have fast-response operating elements and can be used within some light hazard occupancies in accordance with 5-4.5.

5-3.1.5.2 When existing light hazard systems are converted to use quick-response or residential sprinklers, all sprinklers in a compartmented space shall be changed.

The mixing of quick-response or residential sprinklers with standard response sprinklers in a light hazard occupancy can cause more sprinklers to operate than necessary and change the order in which sprinklers operate—that is, those sprinklers further away from the fire could operate first. Where the sprinklers in a light hazard occupancy are converted to quick-response or residential sprinklers, all sprinklers in the space must be of the fast-response type.

5-4* Application of Sprinkler Types

Sprinklers shall be selected for use as indicated in this section. Sprinklers shall be positioned and spaced as described in Section 5-5.

A-5-4 The selection of a sprinkler type will vary by occupancy. Where more than one type of sprinkler is used within a compartment, sprinklers with similar response characteristics should be used (i.e., standard or quick response). However, some hazards might benefit from designs that include the use of both standard and quick-response sprinklers. Examples include rack storage protected by standard-response ceiling sprinklers and quick-response in-rack sprinklers. Another case might include opening protection using closely spaced quick-response sprinklers with standard-response sprinklers in the adjoining areas. Other designs can be compromised when sprinklers of differing sensitivity are mixed. An example is a system utilizing ESFR sprinklers adjacent to a system using high-temperature standard-response sprinklers as might be found in a warehouse. In this case, a fire occurring near the boundary might open ESFR sprinklers, which would not be contemplated in the standard-response system design.

Today, many sprinklers from which to choose are available when designing a system. Chapter 5 covers many of the different types of sprinklers that are currently on the market. Criteria have been established to provide a degree of order over the types of sprinklers that can be developed and listed in the future. NFPA 13 categorizes sprinklers into several types. These categories include upright and pendent spray, sidewall spray, extended coverage, open, residential, early suppression fast-response, large drop, in-rack, and special sprinklers. Section 5-4 provides general information on where to use each type of sprinkler and outlines any restrictions in occupancy and construction that apply to each type of sprinkler.

Upright and pendent spray sprinklers are the most common sprinkler type and can be used in all occupancies and construction types. Other types of sprinklers have more specific limitations associated with them. For example, extended coverage and sidewall sprinklers can only be installed under smooth, flat ceilings. Another example is that sidewall spray sprinklers are limited to light hazard occupancies unless specifically listed for ordinary hazard occupancies. One of the most common design errors is the assumption that a sprinkler can be used in all building construction types for the protection of all occupancy hazards types.

5-4.1 Standard Upright and Pendent Spray Sprinklers.

5-4.1.1 General Applications. Upright and pendent spray sprinklers shall be permitted in all occupancy hazard classifications and building construction types.

Exception: Quick-response sprinklers shall not be permitted for use in extra hazard occupancies under the area-density design method. (See 7-2.3.2.3, Exception No. 1.)

Quick-response sprinklers operate faster and react to smaller fires than standard response sprinklers. This faster response often results in improved sprinkler system performance where all other factors remain constant. Within extra hazard occupancies, fast-developing fires with large heat release rates can open a large number of quick-response sprinklers before the sprinklers have time to control the fire. The operation of a large number of sprinklers in a relatively short time period has the potential to overtax the system. For this reason, quick-response sprinklers using the area design method of Chapter 7 are not allowed for the protection of extra hazard occupancies.

5-4.1.2 Storage. For general storage, rack storage, rubber tire storage, roll paper storage, and baled cotton storage being protected with spray sprinklers with required densities of 0.34 gpm/ft^2 (13.9 mm/ min) or less, standard response sprinklers with a nominal K-factor of 8.0 or larger shall be used. For required densities greater than 0.34 gpm/ft^2 (13.9 mm/min), standard response spray sprinklers with a K-factor of 11.2 or larger that are listed for storage applications shall be used.

Exception No. 1: For densities of 0.20 gpm/ft^2 (8.2 mm/min) or less, standard response sprinklers with a K-factor of 5.6 shall be permitted.

Exception No. 2: For modifications to existing systems, sprinklers with K-factors of 8.0 or less shall be permitted.

Exception No. 3: The use of quick-response spray sprinklers shall be permitted when listed for such use.

Fire tests have shown that sprinklers with larger orifices—that is, K-factors of 8 or larger—perform better during storage fires with strong fire plumes than sprinklers with smaller orifices at the same design density. The lower water velocity through the sprinkler orifice generally results in larger water droplets that more readily penetrate the fire plume and allow more water to reach the fire's seat. At the higher pressures that are likely to be available when the first few sprinklers operate, sprinklers with K-factors of 5.6 can produce smaller water droplets that are unable to penetrate the fire plume and attack the fire.

5-4.2 Sidewall Spray Sprinklers.

Sidewall sprinklers shall be installed only in light hazard occupancies with smooth, flat ceilings.

Exception No. 1: Sidewall sprinklers shall be permitted to be used in ordinary hazard occupancies with smooth, flat ceilings where specifically listed for such use.

Exception No. 2: Sidewall sprinklers shall be permitted to be used to protect areas below overhead doors.

The discharge characteristics of sidewall sprinklers are not as effective as those of upright and pendent sprinklers for all applications. Accordingly, sidewall sprinklers are limited to light hazard occupancies unless the sprinkler is specifically listed for use in ordinary hazard occupancies. Several horizontal sidewall sprinklers are now listed for use in ordinary hazard occupancies.

In addition to being limited by the occupancy hazard classification, sidewall sprinklers are also limited to the ceiling configurations listed in this subsection. The ceilings under which sidewall sprinklers are installed cannot be sloped.

Sidewall sprinklers are often used in retrofit situations where access to ceilings can be limited. System piping can be installed along the intersection of a wall and ceiling and covered with a built-up soffit with the sidewall sprinkler protruding through the soffit.

5-4.3 Extended Coverage Sprinklers.

Extended coverage sprinklers shall be limited to a type of unobstructed construction consisting of flat, smooth ceilings with a slope not exceeding a pitch of one in six (a rise of two units in a run of 12 units, a roof slope of 16.7 percent).

Exception No. 1: Where sprinklers are specifically listed for unobstructed or noncombustible obstructed construction, they shall be permitted for such use.

Exception No. 2: Extended coverage upright and pendent spray sprinklers shall be permitted within trusses or bar joists having web members not greater than 1 in. (25.4 mm) maximum dimension or where trusses are spaced greater than 7½ ft (2.3 m) on center.

Exception No. 3: Where extended coverage sprinklers are specifically listed for use under smooth, flat ceilings that have slopes not exceeding a pitch of one in three (a rise of four units in a run of 12 units, a roof slope of 33.3 percent) they shall be permitted.

Extended coverage (EC) sprinklers are specifically listed as such. EC sprinklers cover an area that is larger than the area permitted for other types of sprinklers such as standard spray and sidewall. The extended coverage allows fewer sprinklers to be used. However, certain limitations are associated with the use of EC sprinklers.

The discharge from an EC sprinkler generally has a longer, flatter throw than a standard coverage sprinkler such as the standard spray upright. This flatter discharge pattern is affected by obstructions and the pitch of a ceiling more than other types of sprinklers. NFPA 13 limits the use of EC sprinklers to smooth, flat ceilings with relatively small slopes unless specific testing demonstrates their acceptability under greater slopes. Obstructions within bar joists and trusses are also limited to ensure proper distribution throughout the sprinkler's entire protection area.

5-4.4 Open Sprinklers.

Open sprinklers shall be permitted to be used in deluge systems to protect special hazards or exposures, or in other special locations. Open sprinklers shall be installed in accordance with all applicable requirements of this standard for their automatic counterpart.

Open sprinklers are spray sprinklers with their operating elements removed. For information on open window or cornice sprinklers designed for outside exposure fire protection usage, see Chapters 4 and 7. Open sprinklers are also used in deluge systems as discussed in Chapter 3. On a larger scale, NFPA 409, *Standard on Aircraft Hangars*, requires foam-water deluge systems, which utilize open sprinklers, as the predominant means of fire protection.

5-4.5 Residential Sprinklers.

Residential sprinklers are tested and listed in accordance with UL 1626, *Residential Sprinklers for Fire Protection Service* (see Exhibit 1.2). Because residential sprinklers are tested and listed using a residential fire scenario, they are permitted only in residential portions of all occupancies. In other portions of such occupancies, sprinklers must be listed for general usage and must be installed in accordance with the requirements of NFPA 13.

NFPA 13D, *Standard for the Installation of Sprinkler Systems in One- and Two-Family Dwellings and Manufactured Homes*, covers sprinkler system design in this specific category only. NFPA 13R, *Standard for the Installation of Sprinkler Systems in Residential Occupancies up to and Including Four Stories in Height*, covers sprinkler system design in certain low-rise residential facilities.

5-4.5.1* Residential sprinklers shall be permitted in dwelling units and their adjoining corridors provided they are installed in conformance with their listing and the positioning requirements of NFPA 13D, *Standard for the Installation of Sprinkler Systems in One-and Two-Family*

Dwellings and Manufactured Homes, or NFPA 13R, Standard for the Installation of Sprinkler Systems in Residential Occupancies up to and Including Four Stories in Height.

A

A-5-4.5.1 The response and water distribution pattern of listed residential sprinklers have been shown by extensive fire testing to provide better control than spray sprinklers in residential occupancies. These sprinklers are intended to prevent flashover in the room of fire origin, thus improving the chance for occupants to escape or be evacuated.

The protection area for residential sprinklers is defined in the listing of the sprinkler as a maximum square or rectangular area. Listing information is presented in even 2-ft (0.65-m) increments from 12 ft to 20 ft (3.9 m to 6.5 m). When a sprinkler is selected for an application, its area of coverage must be equal to or greater than both the length and width of the hazard area. For example, if the hazard to be protected is a room 13 ft 6 in. (4.4 m) wide and 17 ft 6 in. (5.6 m) long, a sprinkler that is listed to protect a rectangular area of 14 ft × 18 ft (4.5 m × 5.8 m) or a square area of 18 ft × 18 ft (5.8 m × 5.8 m) must be selected. The flow used in the calculations is then selected as the flow required by the listing for the selected coverage.

NFPA 13D and NFPA 13R are referenced in 5-4.5.1 as the source for positioning requirements when installing residential sprinklers. The term *positioning* includes spacing of the sprinklers and placement of the sprinklers' deflectors in relation to the ceiling. Positioning does not include the location of the sprinklers. Location refers to the spaces where sprinklers must be installed, such as in a bedroom, in a hallway, in kitchens, or in concealed spaces. Location of residential sprinklers is strictly within the jurisdiction of NFPA 13. The location requirement is further reinforced through Formal Interpretation 87-4.

Formal Interpretation 87-4

Reference: 5-4.5.1, 5-4.5.3, 7-9.2.2

Question 1: Is it the intent of 5-4.5.1 and 7-9.2.2 to allow residential sprinklers to be installed in corridors adjoining dwelling units?

Answer: Yes.

Question 2: Can the corridor also serve elevator lobbies, vending machine alcoves, linen closets, janitor closets and other spaces incidental to dwelling unit areas?

Answer: Yes.

Question 3: If corridors serve adjoining spaces other than dwelling units and their incidental service areas, may residential sprinklers be used in the corridors?

Answer: No.

Issue Edition: 1987

Reference: 3-16.2.9.1, 3-16.2.9.2, 7-4.4, and 7-4.4.3

Date: June 1988

5-4.5.2 Residential sprinklers shall be used only in wet systems.

Exception: Residential sprinklers shall be permitted for use in dry systems or preaction systems if specifically listed for such service.

The performance of a residential sprinkler, which is primarily intended to be a life safety device, depends on the prompt operation provided by the combination of a fast-response operating element and the location and spacing restrictions placed on these sprinklers. In wet pipe systems, the piping is filled with water under pressure so water is applied with minimal delay. Dry pipe systems are associated with a delay in water delivery time. Because a larger fire is anticipated due to such a delay, residential sprinklers may not perform as intended if installed on a dry pipe system. At this time, no listed residential sprinklers are acceptable for use in dry pipe or preaction sprinkler systems.

5-4.5.3 Where residential sprinklers are installed in a compartment as defined in 1-4.2, all sprinklers within the compartment shall be of the fast-response type that meets the criteria of 1-4.5.1(a)1.

When mixing sprinklers that are not of the same make, caution is needed to avoid the possibility of reverse operating order or spot operation of sprinklers. It is important that all residential sprinklers be of similar thermal sensitivity when they are installed in a compartment. The use of bulb-type and solder-type sprinklers or residential and other types of sprinklers is acceptable as long as all of the sprinklers in the compartment are fast response as defined in 1-4.5.1(a)1.

5-4.5.4 Residential sprinklers installed in conformance with this standard shall follow the sprinkler obstruction rules of 5-8.5 or 5-9.5 as appropriate for their installation orientation (upright, pendent, or sidewall and the obstruction criteria specified in the manufacturer's installation instructions).

5-4.6 Early Suppression Fast-Response (ESFR) Sprinklers.

5-4.6.1 ESFR sprinklers shall be used only in wet pipe systems.

Exception: ESFR sprinklers shall be permitted for use in dry systems if specifically listed for such service.

ESFR sprinklers use fast-response operating elements and are designed to respond very quickly to a fire. The design criteria specified in Chapter 7 are based on the assumption that a fire will not exceed a certain size when the sprinkler operates. Because fires in the spaces typically protected with ESFR sprinklers can grow extremely fast with large increases in heat release rates over a short period of time, the quickness with which water is applied is critical. Delayed application of water causes the fire to continue to grow, reaching a size at which the sprinkler system may no longer be able to suppress the fire. Once the fire has exceeded a critical size, many more sprinklers can be expected to operate and overtax the water supply.

Calculations have shown that the time delays inherent in delivering water to sprinklers in both dry pipe and preaction systems cause many more ESFR sprinklers to operate than are contemplated by NFPA 13. Therefore, ESFR sprinklers are not allowed in dry pipe systems and will not be allowed until they are specifically tested and listed for such application.

5-4.6.2 ESFR sprinklers shall be installed only in buildings where roof or ceiling slope above the sprinklers does not exceed a pitch of one in six (a rise of two units in a run of 12 units, a roof slope of 16.7 percent).

Roof slope requirements consider the effect of slope on sprinkler operating patterns. Sloped roofs tend to cause a skewed distribution of heat from a fire burning beneath the roof. With steep roof slopes, heat is likely to travel up the slope away from the sprinklers closest to the fire. This heat travel pattern delays operation of sprinklers near the fire and activates those sprinklers further from the fire, thus preventing early fire suppression.

5-4.6.3* ESFR sprinklers shall be permitted for use only in buildings with the following types of construction:

(1) Smooth ceiling, joists consisting of steel truss-shaped members, or wood truss-shaped members that consist of wood top or bottom chord members not exceeding 4 in. (102 mm) in depth with steel tube or bar web
(2) Wood beams of 4 in. by 4 in. (102 mm by 102 mm) or greater nominal dimension, concrete or steel beams spaced 3½ to 7½ ft (0.9 m to 2.3 m) on centers and either supported on or framed into girders
[Paragraphs (1) and (2) shall apply to construction with noncombustible or combustible roof or decks.]
(3) Construction with ceiling panels formed by members capable of trapping heat to aid the operation of sprinklers with members spaced greater than 7½ ft (2.3 m) and limited to a maximum of 300 ft^2 (27.9 m^2) in area

A-5-4.6.3 Storage in single-story or multistory buildings can be permitted, provided the maximum ceiling/roof height as specified in Table 5-11.2.2 is satisfied for each storage area.

Formal Interpretation 96-1

Reference: 5-4.6.3(2)

Question: Is it the intent of section 5-4.6.3(2) to prohibit the installation of ESFR sprinkler under ceiling construction consisting of concrete beams spaced 3.5 ft to 7.5 ft apart, that are framed into either concrete cross beams or the building walls?

Answer: No.

Issue Edition: 1996

Reference: 4-4.6.3(b)

Issue Date: December 14, 1998

Effective Date: January 3, 1999

5-4.6.4 Where ESFR sprinkler systems are installed adjacent to sprinkler systems with standard response sprinklers, a draft curtain of noncombustible construction and at least 2 ft (0.6 m) in depth shall be required to separate the two areas. A clear aisle of at least 4 ft (1.2 m) centered below the draft curtain shall be maintained for separation.

Where an ESFR sprinkler system is located adjacent to a standard response sprinkler system without any separation, it is possible for ESFR sprinklers to operate before the standard response sprinklers when the fire originates below the system using standard response sprinklers. The best way to prevent this situation is to separate the two systems by a wall. NFPA 13 recognizes that this type of separation is not always possible. In these cases, a draft curtain and an aisle can be used to separate the two systems. An aisle is needed to eliminate the potential for a fire to start under or near the draft curtain, operating sprinklers on both sides.

5-4.6.5 Sprinkler temperature ratings for ESFR sprinklers shall be ordinary.

Exception: Sprinklers of intermediate- and high-temperature ratings shall be installed in locations as required by Section 5-3.1.4.1.

5-4.7 Large Drop Sprinklers.

5-4.7.1 Large drop sprinklers shall be permitted to be used in wet, dry, or preaction systems.

The fires that have been used in testing large drop sprinklers (power law fires or t^2 fires) grow in an exponential manner. Therefore, a substantial escalation in heat release rates and thermal velocities would be expected in relatively short time periods. Testing large drop sprinklers installed on dry pipe and preaction systems has shown that fire control can be achieved if modifications to the number of design sprinklers and the minimum operating pressures are considered. Tables 7-9.4.1.1 and 7-9.4.1.2 reflect the design changes necessary when using large drop sprinklers on a dry pipe system. All of the supplemental provisions placed on preaction and dry pipe systems in Chapter 4 remain applicable when large drop sprinklers are used on these types of systems.

5-4.7.2* Where steel pipe is used in preaction and dry pipe systems, piping materials shall be limited to internally galvanized steel.

Exception: Nongalvanized fittings shall be permitted.

A-5-4.7.2 The purpose of this requirement is to avoid scale accumulation.

The use of internally galvanized steel piping materials applies only where large drop sprinklers are used. Dry pipe systems using other types of sprinklers are not limited to a particular piping material but must comply with the requirements of Sections 3-3 and 3-5.

5-4.7.3 Sprinkler temperature ratings shall be the same as those indicated in Tables 5-3.1.4.2 (a) and (b) or those used in large-scale fire testing to determine the protection requirements for the hazard involved.

Exception No. 1: Sprinklers of intermediate- and high-temperature ratings shall be installed in specific locations as required by 5-3.1.4.

Exception No. 2: In storage occupancies, ordinary, intermediate, or high temperature–rated sprinklers shall be used for wet pipe systems.

Exception No. 3: In storage occupancies, high temperature–rated sprinklers shall be used for dry pipe systems.

5-4.8 QRES. (Reserved)

This subsection is one of the several reserved in the 1991 edition for the quick-response early suppression (QRES) sprinkler, which still might be developed. Like other sprinklers, QRES devices will require specific requirements for their placement and temperature ratings. If the development of the QRES program is pursued, provisions for these and other parameters are likely to be added to NFPA 13.

5-4.9 Special Sprinklers.

5-4.9.1* Special sprinklers that are intended for the protection of specific hazards or construction features shall be permitted where such devices have been evaluated and listed for performance under the following conditions:

(1) Fire tests related to the intended hazard
(2) Distribution of the spray pattern with respect to wetting of floors and walls
(3) Distribution of the spray pattern with respect to obstructions
(4) Evaluation of the thermal sensitivity of the sprinkler
(5) Performance under horizontal or sloped ceilings
(6) Area of design

A-5-4.9.1 Tests of standard sprinklers by approved laboratories have traditionally encompassed a fire test using a 350-lb (160-kg) wood crib and water distribution tests in which water is collected in pans from several arrangements of sprinklers to evaluate distribution under non-fire conditions.

Tests of special sprinklers are customized to evaluate responsiveness, distribution, and other unique characteristics of the sprinkler to control or extinguish. These tests include variables such as the following:

(1) The location of the fire relative to the sprinklers (i.e., below one sprinkler or between two, four, or six sprinklers)
(2) Fire conditions that encompass a variety of fire growth rates representative of anticipated conditions of use
(3) Tests of room areas where sprinklers are expected to function in multiple arrays
(4) Adverse conditions of use (i.e., pipe shadows or other obstructions to discharge)

(5) Effect of a fire plume on water distribution and discharge under a variety of heat release rates

The 1996 edition eliminated EC sprinklers from the special sprinkler category. This previously largest segment of special sprinklers was made a general sprinkler type and requirements for their installation and design were included. Special sprinklers are intended to protect specific hazards or construction features. Special sprinklers must be listed, and their performance must be evaluated. The evaluation criteria includes fire tests related to the specific hazard, spray pattern distribution with respect to both the wetting of walls and floors and to obstructions, thermal sensitivity of the operating element, and the performance of the sprinkler under horizontal or sloped ceilings. The testing results are used to ensure adequate performance and identify special design and installation criteria to be included as part of the listing.

5-4.9.2 Special sprinklers shall maintain the following characteristics:

(1) Orifice size shall be in accordance with 3-2.3.
(2) Temperature ratings shall be in accordance with Table 3-2.5.1.
(3) The protection area of coverage shall not exceed 400 ft^2 (36 m^2) for light hazard and ordinary hazard occupancies.
(4) The protection area of coverage shall not exceed 196 ft^2 (17 m^2) for extra hazard and high-pile storage occupancies.

While a good deal of latitude is given to special sprinklers, the orifice size and temperature rating must be in compliance with the provisions of Chapter 3, and the maximum area of coverage cannot exceed 400 ft^2 (36 m^2) for light and ordinary hazard occupancies and 196 ft^2 (17 m^2) for extra hazard occupancies and high-piled storage. These requirements apply because of concerns for interchangeability and overall system reliability.

5-5 Position, Location, Spacing, and Use of Sprinklers

Proper positioning and spacing of sprinklers are important to ensure that sprinklers operate promptly and that obstructions to sprinkler distribution patterns do not adversely affect the performance of the sprinkler.

5-5.1 General.

Sprinklers shall be located, spaced, and positioned in accordance with the requirements of this section. Sprinklers shall be positioned to provide protection of the area consistent with the overall objectives of this standard by controlling the positioning and allowable area of coverage for each sprinkler. The requirements of 5-5.2 through 5-5.6 shall apply to all sprinkler types unless modified by more restrictive rules in Sections 5-6 through 5-11.

The general requirements with which all sprinklers must comply are covered in Section 5-5. Section 5-5 addresses protection area per sprinkler, sprinkler spacing, deflector position, obstruction to discharge, and clearance to storage. This generic information provides limiting minimum or maximum dimensions and guidance on how to determine these measurements.

Sections 5-6 through 5-12 provide specific requirements for standard upright and pendent spray sprinklers, standard sidewall spray sprinklers, EC upright and pendent spray sprinklers, EC sidewall sprinklers, large drop sprinklers, ESFR sprinklers, and in-rack sprinklers.

5-5.2 Protection Areas per Sprinkler.

5-5.2.1 Determination of the Protection Area of Coverage. The protection area of coverage per sprinkler (A_s) shall be determined as follows:

(a) *Along Branch Lines.* Determine distance between sprinklers (or to wall or obstruction in the case of the end sprinkler on the branch line) upstream and downstream. Choose the larger of either twice the distance to the wall or the distance to the next sprinkler. This dimension will be defined as *S*.

(b) *Between Branch Lines.* Determine perpendicular distance to the sprinkler on the adjacent branch line (or to a wall or obstruction in the case of the last branch line) on each side of the branch line on which the subject sprinkler is positioned. Choose the larger of either twice the distance to the wall or obstruction or the distance to the next sprinkler. This dimension will be defined as *L*.

5-5.2.1.1 The protection area of coverage of the sprinkler shall be established by multiplying the S dimension by the L dimension, as follows: $A_s = S \times L$

Exhibit 5.5 illustrates how to determine the *S* and *L* dimensions, and Exhibit 5.6 provides a specific example. The maximum dimensions for *S* and *L*, which are the distance from a wall and the area of coverage per sprinkler, are specified for specific types of sprinklers. See Sections 5-6 through 5-11.

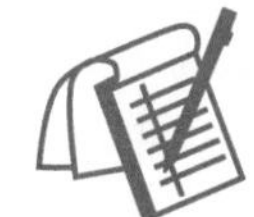

5-5.2.2 Maximum Protection Area of Coverage. The maximum allowable protection area of coverage for a sprinkler (A_s) shall be in accordance with the value indicated in the section for each type or style of sprinkler. The maximum area of coverage of any sprinkler shall not exceed 400 ft^2 (36 m^2).

The maximum allowable area of coverage for a specific sprinkler depends on the type of sprinkler being considered and the construction features and the occupancy hazard of the space in which the sprinkler is to be installed. The combustibility of the ceiling, the ceiling's effect on the flow of heat, and the ceiling's potential to obstruct the sprinkler's discharge pattern can affect overall sprinkler performance. Likewise, the actual distribution pattern—that is, the shape and throw—for a specific type of sprinkler also has an impact. The maximum protection area of 400 ft^2 (36 m^2) for any type of sprinkler limits the area that is permitted to be unprotected if the discharge from a single sprinkler is obstructed or if the sprinkler fails to operate. This maximum area of coverage is reduced for some sprinkler types as indicated in Sections 5-6 through 5-11.

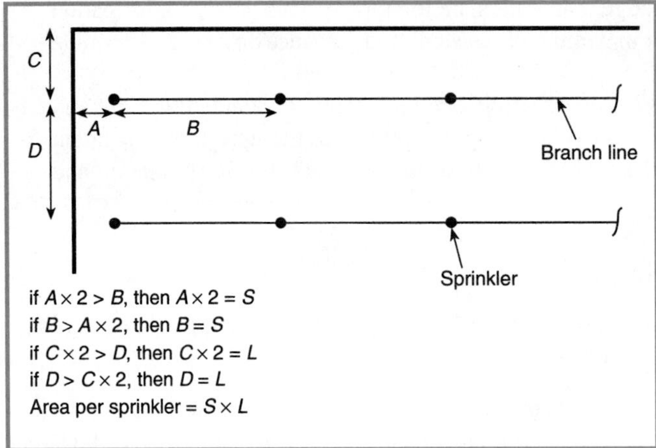

Exhibit 5.5 *Illustration of how to determine the area of coverage for a sprinkler.*

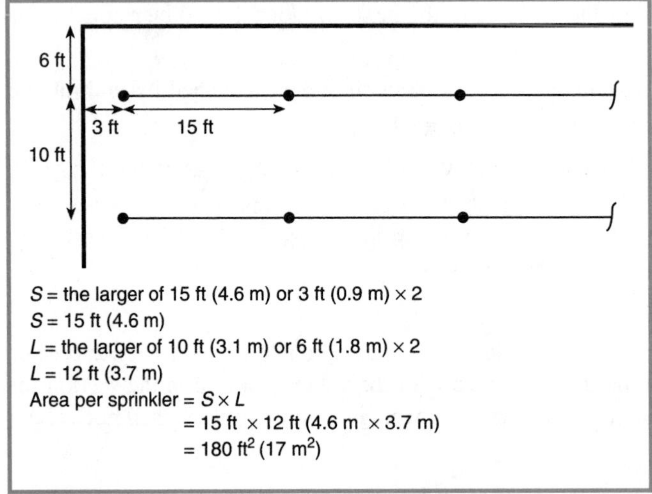

6 ft
3 ft
15 ft
10 ft
S = the larger of 15 ft (4.6 m) or 3 ft (0.9 m) × 2
S = 15 ft (4.6 m)
L = the larger of 10 ft (3.1 m) or 6 ft (1.8 m) × 2
L = 12 ft (3.7 m)
Area per sprinkler = S × L
= 15 ft × 12 ft (4.6 m × 3.7 m)
= 180 ft² (17 m²)

Exhibit 5.6 *Illustration of how to determine the area of coverage for a sprinkler.*

5-5.3 Sprinkler Spacing.

5-5.3.1 Maximum Distance Between Sprinklers. The maximum distance permitted between sprinklers shall be based on the centerline distance between sprinklers on the branch line or on adjacent branch lines. The maximum distance shall be measured along the slope of the ceiling.

The maximum distance permitted between sprinklers shall comply with the value indicated in the section for each type or style of sprinkler.

The spacing of sprinklers along branch lines depends on the maximum distance permitted between the sprinklers on branch lines and the maximum area of coverage permitted per sprinkler. To minimize the amount of piping used, branch lines are usually spaced as far apart as possible while maintaining even spacing within a ceiling bay or room. If the spacing of branch lines is uneven, the greatest distance between branch lines should be used to determine how far apart sprinklers can be spaced from each other along the branch lines.

The spacing of sprinklers is to be measured on the slope of the roof, and the maximum distances and areas given in Sections 5-6 through 5-11 are applied to the distance measured along the slope. For curved surfaces, distance should be measured along the slope projected between the two sprinklers as indicated in Exhibit 5.7.

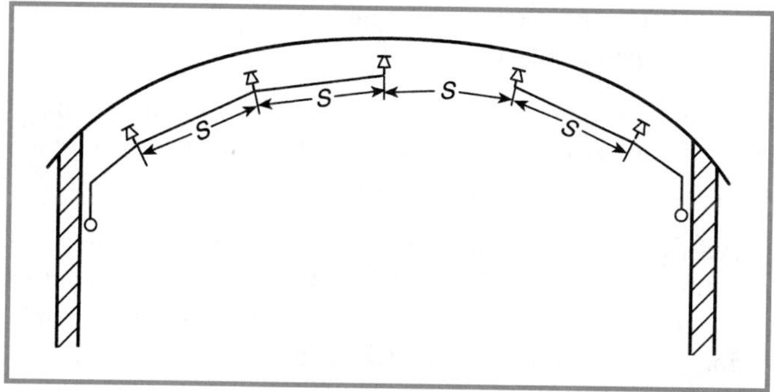

Exhibit 5.7 *Sprinklers installed along a curved ceiling.*

Question 1 of Formal Interpretation 83-5 applies with regard to the second sentence of 5-5.3.1, which requires that the distance be measured along the slope of the ceiling. Question 2 of Formal Interpretation 83-5 pertains to curved roofs.

Formal Interpretation 83-5

Reference: 5-5.3.1, 5-6.4.1.3

Question 1: Is it the intent of 5-5.3.1 and 5-6.4.1.3 to:

(a) Define the area of protection covered by a sprinkler as the distance between sprinklers and branch lines as measured on the slope, or

(b) To establish the location of the sprinkler with respect to the peak of a pitched roof building with the area of protection covered being determined by the spacing being projected to the horizontal plane of the floor?

Answer: The intent is expressed by (a).

(continues)

Question 2: If the answer to the above is (a), why would it be allowed by 5-5.3.1 and 5-6.3.2.1 to project the spacing to the horizontal plane of the floor for curved roof buildings?

Answer: Paragraphs 5-5.3.1 and 5-6.3.2.1 does not permit the spacing of sprinklers under a curved roof to be determined by their projected distance on the floor. Section 5-6.3.2.1 does however make an exception with respect to the sprinkler nearest a side wall when the roof curves down to the floor line.

Issue Edition: 1983

Reference: 4-4.5

Date: April 1983

5-5.3.2 Maximum Distance From Walls. The distance from sprinklers to walls shall not exceed one-half of the allowable maximum distance between sprinklers. The distance from the wall to the sprinkler shall be measured perpendicular to the wall.

5-5.3.3 Minimum Distance from Walls. The minimum distance permitted between a sprinkler and the wall shall comply with the value indicated in the section for each type or style of sprinkler. The distance from the wall to the sprinkler shall be measured perpendicular to the wall.

5-5.3.4 Minimum Distance Between Sprinklers. A minimum distance shall be maintained between sprinklers to prevent operating sprinklers from wetting adjacent sprinklers and to prevent skipping of sprinklers. The minimum distance permitted between sprinklers shall comply with the value indicated in the section for each type or style of sprinkler.

5-5.4 Deflector Position.

5-5.4.1* Distance Below Ceilings. The distances between the sprinkler deflector and the ceiling above shall be selected based on the type of sprinkler and the type of construction.

A-5-5.4.1 Batt insulation creates an effective thermal barrier and can be considered the ceiling/roof deck when determining distances between deflector and ceiling. The insulation needs to be installed in each pocket (not just above the sprinkler) and attached to the ceiling/roof in such a manner that it will not fall out during a fire prior to sprinkler activation.

In general, sprinklers should be located near the ceiling, because the ceiling is where heat from a fire typically collects. When the sprinkler is located further down from the ceiling, response time generally increases unless the sprinkler is located within the fire plume. Operation of sprinklers located very close to the ceiling can also be delayed if they are located in the dead-air space that develops under some ceilings. Obstructed construction requires that sprinklers be located further below the ceiling to allow the sprinkler discharge pattern to develop. This arrangement results in a slower sprinkler response time.

Ideally, for measurements affecting sprinkler sensitivity, the distance below the ceiling should be measured to the centerline of the thermal element rather than the deflector, because the relationship between the thermal element and the deflector varies with different sprinklers. In some cases, the centerline of the sprinkler's thermal element is not easily determined. Because the distance to the deflector is more readily attainable, NFPA 13 uses this method to measure the distance from the sprinkler to ceilings and obstructions. See Exhibit 5.8.

The Appendix A material was added to the 1999 edition to identify the conditions under which batt insulation can be treated as the ceiling for the purpose of determining

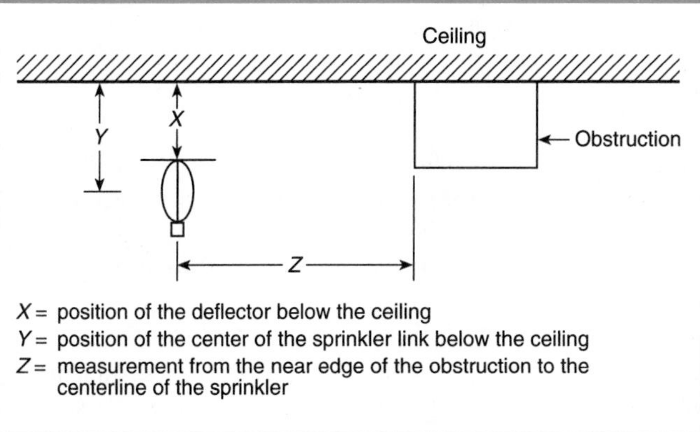

Exhibit 5.8 *Sprinkler deflector location.*

sprinkler placement. The ceiling provides a means of heat collection to aid in the sprinkler's activation. The use of insulation as a thermal barrier is readily accepted by NFPA 13 as indicated by the exceptions to 5-13.1.1 for the protection of concealed spaces. Therefore, heat from a fire starts to collect at the insulation rather than at the decking above the insulation. This concept is supported by Formal Interpretation 83-3. With regard to Formal Interpretation 83-3, it is important to recognize that the thermal barrier is created by the insulation and not by the aluminum sheathing.

Formal Interpretation 83-3

Reference: 5-5.4.1

Question: Under so-called Berkley Construction with a plywood deck on 2 in. × 6 in. wood stiffeners at 2 ft-0 in. centers framed into 3 in. × 14 in. purlins at 8 ft-0 in. centers framed into Glu-lam beams, with fiber glass insulation batts between the stiffeners with aluminum foil stapled to the lower edge of the stiffeners supporting the insulation, should sprinkler deflectors be positioned as for open wood joist construction or for panel construction with the aluminum sheathing considered as the "ceiling"?

Answer: The sheathing could be considered the ceiling in (panel) obstructed construction.

Issue Edition: 1983

Reference: 4-3.4

Date: May 1983

5-5.4.2 Deflector Orientation. Deflectors of sprinklers shall be aligned parallel to ceilings, roofs, or the incline of stairs.

Maintaining the deflector parallel to the ceiling results in minimum obstructions to discharge and a more effective discharge pattern.

5-5.5 Obstructions to Sprinkler Discharge.

In the 1991 edition, the sprinkler obstruction requirements were rewritten so that they no longer referred to specific construction features such as beams and bar joist webs. However, the revised method of categorizing obstructions as either horizontal or vertical was confusing.

For the 1996 edition, a new approach was introduced for dealing with obstructions to sprinkler discharge. Performance objectives were added, and the positioning rules were revised to specifically achieve these objectives. Subsection 5-5.5 provides general requirements, while Sections 5-6 through 5-11 provide requirements for specific types of sprinklers.

NFPA 13 addresses three general areas of concern with regard to obstructions. The first is the overall objective addressed by 5-5.5.1, which is to ensure that a sufficient amount of water from the sprinkler reaches the hazard. This concern is largely addressed by using the 1991 edition's requirements for horizontal obstructions, which dealt largely with continuous obstructions such as beams, top chord members, and ducts that are tight to or very near the ceiling and in close proximity to the sprinkler.

The second concern, addressed by 5-5.5.2 deals with the obstruction to sprinkler discharge pattern development. This concern addresses continuous and noncontinuous obstructions such as piping, light fixtures, truss webs, or building columns located within the first 18 in. (457 mm) of the sprinkler deflector. See Figure A-5-5.5.1. Obstructions located in this zone prevent the proper sprinkler discharge pattern from developing. As a result, sprinklers are required to be positioned so that they are located a certain distance away from the obstruction. The correct position is addressed within the section for each type of sprinkler. This requirement does not apply to continuous obstructions that are tight to the ceiling. Even though obstructions less than 18 in. (457 mm) below the sprinklers can prevent the proper discharge pattern from developing, the overall objective discussed in the previous paragraph provides for adequate sprinkler placement.

The third area of concern deals with obstructions that prevent the sprinkler discharge from reaching the hazard. These obstructions typically consist of continuous and noncontinuous obstructions that interrupt the water spray pattern once it is below the 18-in. (457-mm) discharge pattern development zone. These types of obstructions can include overhead doors, ducts, or decks. When these obstructions exceed a certain dimension, sprinklers are required to be beneath them.

Another category of obstruction that is applicable to certain types of sprinklers includes those obstructions that are suspended from the ceiling, such as privacy curtains in a hospital, or floor-mounted obstructions that do not reach the ceiling, such as partitions in an office cubicle. In both cases, the sprinklers must be positioned so that a sufficient portion of the discharge pattern can extend over the obstruction.

While these three objectives apply to all sprinklers, certain modifications are made to the objectives and positioning rules for specific types of sprinklers. See 5-6.5, 5-7.5, 5-8.5, 5-9.5, 5-10.5, and 5-11.5.

5-5.5.1* Performance Objective. Sprinklers shall be located so as to minimize obstructions to discharge as defined in 5-5.5.2 and 5-5.5.3, or additional sprinklers shall be provided to ensure adequate coverage of the hazard. *(See Figure A-5-5.5.1.)*

A-5-5.5.1 See Figure A-5-5.5.1.

Locating sprinklers so as to minimize the impact of obstructions to the sprinkler discharge is the preferred approach. The use of additional sprinklers to compensate for obstructions increases the water demand and the cost of the sprinkler system and may not provide the same level of protection that would be afforded if the effects of obstructions were minimized.

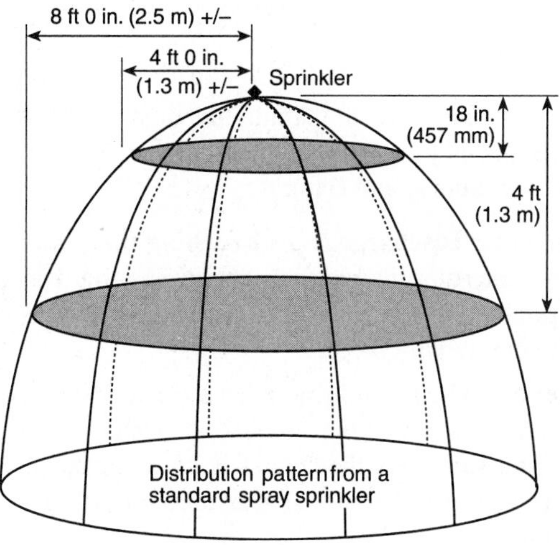

Figure A-5-5.5.1 *Obstructions to sprinkler discharge pattern development for standard upright or pendent spray sprinklers.*

5-5.5.2* Obstructions to Sprinkler Discharge Pattern Development.

A-5-5.5.2 Where of a depth that will obstruct the spray discharge pattern, girders, beams, or trusses forming narrow pockets of combustible construction along walls can require additional sprinklers.

Obstructions that affect sprinkler discharge pattern development are located within the sprinkler discharge pattern as shown in Figure A-5-5.5.1. The obstructions can be continuous, such as large beams; girders; top and bottom truss chord members; soffits; long horizontal light fixtures; heating, ventilation, and air conditioning (HVAC) ductwork; and similar items, or can be smaller noncontinuous obstructions, such as building columns, bar joist and truss webs, and certain light fixtures. The terms *horizontal* and *vertical*, which were used in prior editions, are no longer used to refer to these types of obstructions.

NFPA 13 requires a minimum distance to be maintained between the sprinkler and smaller noncontinuous obstructions. A sprinkler must be positioned so that the discharge pattern is below the solid continuous obstructions.

5-5.5.2.1 Continuous or noncontinuous obstructions less than or equal to 18 in. (457 mm) below the sprinkler deflector that prevent the pattern from fully developing shall comply with 5-5.5.2.

The discharge pattern for most sprinklers is fully developed at about a 4-ft (1.2-m) distance below the sprinkler (see Figure A-5-5.5.1). However, avoiding all obstructions in this space is not practical. Most of the pattern development occurs within a zone beginning at the sprinkler deflector and extending 18 in. to 36 in. (457 mm to 914 mm) below the sprinkler. The obstructions to sprinkler discharge pattern development are a concern within this zone. The minimum distance of 18 in. (457 mm) is modified in subsequent sections for ESFR and large drop sprinklers, which require a larger space for pattern development.

5-5.5.2.2 Sprinklers shall be positioned in accordance with the minimum distances and special exceptions of Sections 5-6 through 5-11 so that they are located sufficiently away from obstructions such as truss webs and chords, pipes, columns, and fixtures.

5-5.5.3* Obstructions that Prevent Sprinkler Discharge from Reaching the Hazard. Continuous or noncontinuous obstructions that interrupt the water discharge in a horizontal plane more than 18 in. (457 mm) below the sprinkler deflector in a manner to limit the distribution from reaching the protected hazard shall comply with 5-5.5.3.

A-5-5.5.3 Frequently, additional sprinkler equipment can be avoided by reducing the width of decks or galleries and providing proper clearances. Slatting of decks or walkways or the use of open grating as a substitute for automatic sprinklers thereunder is not acceptable. The use of cloth or paper dust tops for rooms forms obstruction to water distribution. If dust tops are used, the area below should be sprinklered.

Once the sprinkler discharge pattern is developed, obstructions in the horizontal plane can prevent the sprinkler discharge from reaching the protected hazard. In some situations, the obstruction cannot be avoided, and additional sprinklers are necessary to compensate for areas under the obstruction that would not receive adequate coverage.

5-5.5.3.1 Sprinklers shall be installed under fixed obstructions over 4 ft (1.2 m) wide such as ducts, decks, open grate flooring, cutting tables, and overhead doors.

Exception: Obstructions that are not fixed in place such as conference tables.

The size at which obstructions become too large to ignore is typically 4 ft (1.2 m). This width is reduced in Sections 5-10 and 5-11 for large drop and ESFR sprinklers. The exception recognizes that some obstructions larger than 4 ft (1.2 m), such as the example of a conference table, are often moved and are not likely to have any significant combustibles located beneath them. In these cases, additional sprinklers do not need to be placed below the obstruction. It also important to stress that overhead doors are considered obstructions. Although the overhead door is not considered an obstruction when in the closed position as indicated in Exhibit 5.9, the discharge from the sprinkler over the door will be obstructed when the door is in the open position. Sprinklers need to be positioned so that their discharge can adequately reach under the overhead door. The 1999 edition was revised to specifically allow the use of sidewall sprinklers for these applications. See Exception No. 2 to 5-4.2.

Exhibit 5.9 *Overhead door that will obstruct sprinkler discharge when in the open position.*

5-5.5.3.2 Sprinklers installed under open gratings shall be of the intermediate level/rack storage type or otherwise shielded from the discharge of overhead sprinklers.

Although open grating allows heat from a fire to pass through it and reach a sprinkler, the openings are not adequate to compensate for obstructions to the sprinkler spray pattern. Thus, supplemental sprinklers under the open grating deck become necessary. Gratings or slatted decks and walkways are frequently covered with goods in storage or by a light surface dust stop. Sprinklers are required under such gratings and walkways. Sprinklers must be provided with a water shield or be of the in-rack type to prevent water from operating sprinklers above from wetting the thermal element and delaying sprinkler operation.

5-5.6* Clearance to Storage.

The clearance between the deflector and the top of storage shall be 18 in. (457 mm) or greater.

Exception No. 1: Where other standards specify greater minimums, they shall be followed.

Exception No. 2: A minimum clearance of 36 in. (0.91 m) shall be permitted for special sprinklers.

Exception No. 3: A minimum clearance of less than 18 in. (457 mm) between the top of storage and ceiling sprinkler deflectors shall be permitted where proven by successful large-scale fire tests for the particular hazard.

Exception No. 4: The clearance from the top of storage to sprinkler deflectors shall be not less than 3 ft (0.9 m) where rubber tires are stored.*

A-5-5.6 The fire protection system design should consider the maximum storage height. For new sprinkler installations, maximum storage height is the usable height at which commodities can be stored above the floor while the minimum required unobstructed space below sprinklers is maintained. Where evaluating existing situations, maximum storage height is the maximum existing storage height if space between the sprinklers and storage is equal to or greater than that required.

Building heights where baled cotton is stored should allow for proper clearance between the pile height and sprinkler deflectors. Fire tests of high-piled storage have shown that sprinklers are generally more effective if located 1½ ft to 4½ ft (0.45 m to 1.4 m) above the storage height.

5-6 Standard Pendent and Upright Spray Sprinklers

5-6.1 General.

All requirements of Section 5-5 shall apply to standard pendent and upright spray sprinklers except as modified below.

5-6.2 Protection Areas per Sprinkler (Standard Pendent and Upright Spray Sprinklers).

5-6.2.1 Determination of the Protection Area of Coverage. The protection area of coverage per sprinkler (A_s) shall be determined in accordance with 5-5.2.1.

Exception: In a small room as defined in 1-4.2, the protection area of coverage for each sprinkler in the small room shall be the area of the room divided by the number of sprinklers in the room.

This exception applies to only small rooms in light hazard occupancies when the room meets the size and construction limitations as defined in 1-4.2.

5-6.2.2 Maximum Protection Area of Coverage. The maximum allowable protection area of coverage for a sprinkler *(A_s)* shall be in accordance with the value indicated in Tables 5-6.2.2(a) through 5-6.2.2(d). In any case, the maximum area of coverage of a sprinkler shall not exceed 225 ft^2 (21 m^2).

In previous editions of NFPA 13, the protection area per sprinkler was a function of occupancy hazard classification and specific ceiling construction type. This arrangement created difficulties when a type of ceiling construction was not specifically discussed in the standard. Additionally, no guidance was provided for certain combinations of construction type and occupancy hazard. The construction classification approach currently employed—that is, obstructed and unobstructed—minimizes that difficulty by dealing with only with those constructions features that can affect sprinkler performance.

For light hazard occupancies, NFPA 13 recognizes that more efficient and effective designs can result from the application of hydraulic calculation techniques. Therefore, for unobstructed construction, an increase in protection area to 225 ft^2 (21 m^2) per sprinkler is allowed.

NFPA 13 also recognizes that combustible construction can increase the associated fire hazards. Many types of combustible obstructed construction, including wood joist, pre-engineered wood I-beams, and lightweight wood construction, such as trusses, can have certain characteristics such as numerous small pockets, a tendency to obstruct heat flow, a greater likelihood for obstructing sprinkler discharge, and increased susceptibility to fire damage. The protection area per sprinkler is reduced to 130 ft^2 (12.1 m^2) where the distance between structural members is less than 3 ft (0.91 m) apart for these particular types of combustible ceiling construction. Effective sprinkler protection for the areas under these types of combustible ceiling construction is difficult, but not impossible, to provide.

The placement of sprinklers under combustible obstructed construction requires more attention, because obstructed construction by its very nature increases the possibility of sprinkler discharge pattern disruption, and the combustible materials can increase the overall fire hazard. If a fire is obstructed from sprinkler discharge and cannot be quickly controlled, the ceiling construction might become part of the fuel package. Once the combustible construction above the sprinkler is involved in the fire, controlling or suppressing the fire with sprinklers is very difficult, if not impossible.

Extra hazard occupancies contain large fuel loads that have the potential for developing into fires with very fast and high rates of heat re lease. Decreasing allowable areas of coverage to 90 ft^2 (8.4 m^2) for pipe schedule systems and 100 ft^2 (9.3 m^2) for hydraulically designed systems is intended to compensate for the larger, faster developing fires that are typical of extra hazard occupancies. For hydraulically designed systems below densities of 0.25 gpm/ft^2 (10.2 mm/min), areas of coverage per sprinkler are permitted to be extended to 130 ft^2 (12.1 m^2).

For high-piled storage, tests conducted on various commodities can indicate that smaller protection areas per sprinkler are required. Any such criteria identified as part of the design criteria outlined in Chapter 7 of this standard or other appropriate NFPA standards need to be followed.

Table 5-6.2.2(c) also recognizes that in buildings containing extra hazard occupancies or high-piled storage with 25-ft (7.6-m) wide ceiling bays, maximum sprinkler spacing of 12 ft 6 in (3.8 m) can be employed without compromising the system design integrity. In addition, a 15-ft (4.6-m) sprinkler spacing is still acceptable in extra hazard occupancies and high-piled storage for relatively low discharge densities of less than 0.25 gpm/ft^2 (10.2 mm/min),

5-6.3 Sprinkler Spacing (Standard Pendent and Upright Spray Sprinklers).

5-6.3.1 Maximum Distance Between Sprinklers. The maximum distance permitted between sprinklers shall comply with Tables 5-6.2.2(a) through (d).

5-6.3.2 Maximum Distance from Walls.

5-6.3.2.1* The distance from sprinklers to walls shall not exceed one-half of the allowable distance between sprinklers as indicated in Tables 5-6.2.2(a) through (d). The distance from the wall to the sprinkler shall be measured perpendicular to the wall. Where walls are angled or irregular, the maximum horizontal distance between a sprinkler and any point of floor area protected by that sprinkler shall not exceed 0.75 times the allowable distance permitted between sprinklers, provided the maximum perpendicular distance is not exceeded.

Exception: Within small rooms as defined in 1-4.2, sprinklers shall be permitted to be located not more than 9 ft (2.7 m) from any single wall. Sprinkler spacing limitations of 5-6.3 and area limitations of Table 5-6.2.2(a) shall not be exceeded.*

Table 5-6.2.2(a) Protection Areas and Maximum Spacing (Standard Spray Upright/Standard Spray Pendent) for Light Hazard

		Protection Area		Spacing (maximum)	
Construction Type	**System Type**	**ft^2**	**m^2**	**ft**	**m**
Noncombustible obstructed and unobstructed and combustible unobstructed	Pipe schedule	200	18.6	15	4.6
	Hydraulically calculated	225	20.9	15	4.6
Combustible obstructed	All	168	15.6	15	4.6
Combustible with members less than 3 ft on center	All	130	12.1	15	4.6

Table 5-6.2.2(b) Protection Areas and Maximum Spacing (Standard Spray Upright/Standard Spray Pendent) for Ordinary Hazard

		Protection Area		Spacing (maximum)	
Construction Type	**System Type**	**ft^2**	**m^2**	**ft**	**m**
All	All	130	12.1	15	4.6

Table 5-6.2.2(c) Protection Areas and Maximum Spacing (Standard Spray Upright/Standard Spray Pendent) for Extra Hazard

Construction Type	System Type	Protection Area		Spacing (maximum)	
		ft²	m²	ft	m
All	Pipe schedule	90	8.4	12	3.7
				[In buildings with storage bays 25 ft (7.6 m) wide, 12 ft 6 in. (3.8 m) shall be permitted]	
All	Hydraulically calculated with density ≥0.25	100	9.3	12	3.7
				[In buildings with storage bays 25 ft (7.6 m) wide, 12 ft 6 in. (3.8 m) shall be permitted]	
All	Hydraulically calculated with density <0.25	130	12.1	15	4.6

Table 5-6.2.2(d) Protection Areas and Maximum Spacing (Standard Spray Upright/Standard Spray Pendent) for High-Piled Storage

Construction Type	System Type	Protection Area		Spacing (maximum)	
		ft²	m²	ft	m
All	Hydraulically calculated with density ≥0.25	100	9.3	12	3.7
				[In buildings with storage bays 25 ft (7.6 m) wide, 12 ft 6 in. (3.8 m) shall be permitted]	
All	Hydraulically calculated with density <0.25	130	12.1	15	4.6

A-5-6.3.2.1 See Figure A-5-6.3.2.1(a).

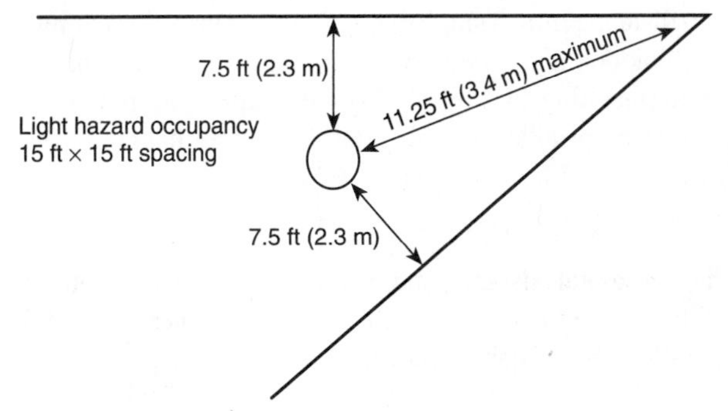

Figure A-5-6.3.2.1(a) *Maximum distance from walls.*

A-5-6.3.2.1 Exception. An example of sprinklers in small rooms for hydraulically designed and pipe schedule systems is shown in Figure A-5-6.3.2.1(b), and examples for hydraulically designed systems only are shown in Figures A-5-6.3.2.1(c), (d), and (e).

The examples illustrated in Exhibits 5.10 and 5.11 demonstrate the requirements for positioning a sprinkler from a wall. If the distance from a branch line to a wall exceeds one-half of the distance between the branch lines, then twice the distance between the branch line closest to the wall and the wall should be used in verifying if the distance between sprinklers on branch lines is in accordance with NFPA 13. More specifically, if the distance from the branch line to the wall is 7 ft (2.1 m), and the distance between branch lines is 13 ft (4.0 m), then twice the distance from the branch line to the wall would be 14 ft (4.3 m). The 14-ft (4.3-m) dimension needs to be used to determine the appropriate distance between sprinklers on the branch lines for the first branch line parallel to the wall. Other lines could use the 13-ft or 14-ft (4.0-m or 4.3-m) dimension, depending on whether symmetry was desired. If an ordinary hazard occupancy system with a protection area of 130 ft^2 (12 m^2) per sprinkler were being considered, the maximum distance between sprinklers on the branch lines could be as follows:

$$130 \text{ ft}^2 \ (12.1 \text{ m}^2) \div 14 \text{ ft} \ (4.3 \text{ m}) = 9.3 \text{ ft} \ (2.8 \text{ m})$$

Angled and other irregular walls and corners with less than 90-degree angles can cause situations where additional sprinklers are required to protect small spaces within a larger area. A change first introduced in the 1996 edition allowed the maximum distance from a sprinkler to a corner or wall to be slightly increased eliminating the need to install an additional sprinkler in these small spaces. For example, with a light hazard occupancy, the distance to the corner in Figure A-5-6.3.2.1(a) could be increased to 11.25 ft (3.4 m). Another common situation occurs when a window is recessed from the interior wall. This section allows the sprinkler to be further away from the window as long as the allowable distance from the wall is not exceeded.

The exception only pertains to small rooms (see definition of *Small Room* in 1-4.2). To accommodate ceiling patterns, lighting fixtures, diffusers, and so forth, sprinklers are permitted to be moved a maximum of 1½ ft (0.5 m) in one direction beyond the usual requirements so that the maximum distance from a single wall is 9 ft (2.7 m), eliminating the need for adding additional sprinklers. This requirement assumes that only those sprinklers in the room will operate, resulting in a higher available operating pressure and, thus, greater discharge from each sprinkler and an improved spray pattern. See Figures A-5-6.3.2.1(a) through (d).

5-6.3.2.2 Under curved surfaces, the horizontal distance shall be measured at the floor level from the wall, or the intersection of the curved surface and the floor to the nearest sprinkler shall not be greater than one-half the allowable distance between sprinklers.

5-6.3.3 Minimum Distance from Walls. Sprinklers shall be located a minimum of 4 in. (102 mm) from a wall.

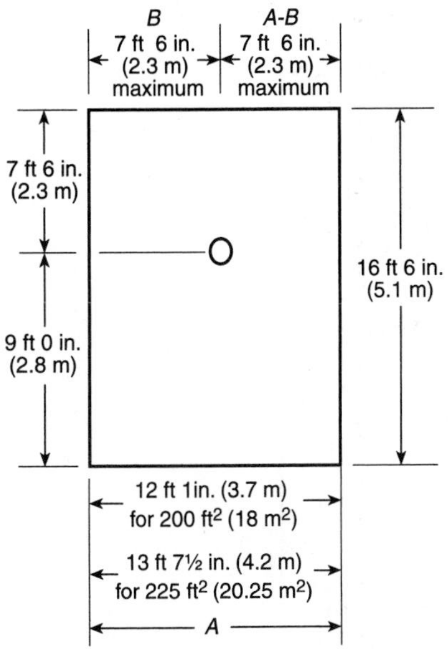

Figure A-5-6.3.2.1(b) *Small room provision—one sprinkler.*

30 ft 0 in. (9.3 m)
6 ft 0 in. (1.9 m)
9 ft 0 in. (2.8 m)
7 ft 6 in. (2.32 m)
15 ft 0 in. (4.6 m)
7 ft 6 in. (2.32 m)
15 ft 0 in. (4.65 m)

Figure A-5-6.3.2.1(c) *Small room provision—two sprinklers centered between sidewalls.*

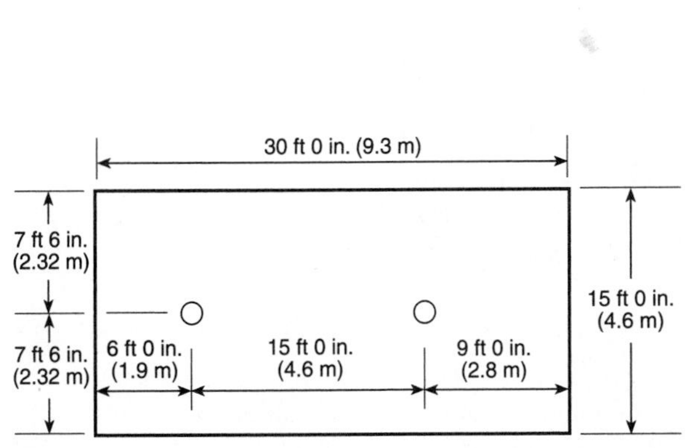

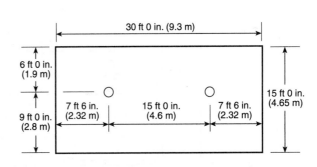

Figure A-5-6.3.2.1(d) *Small room provision—two sprinklers centered between top and bottom walls.*

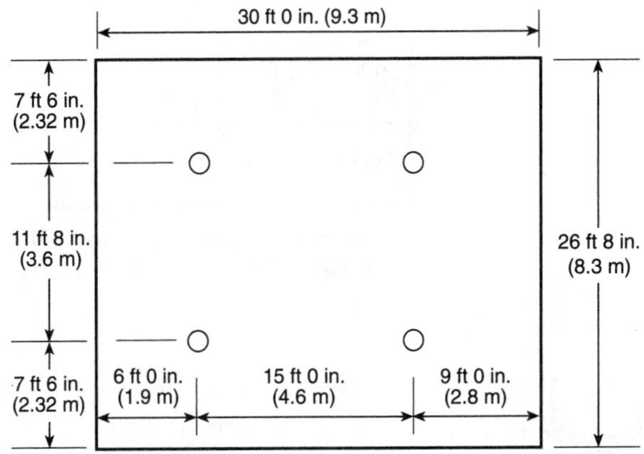

Figure A-5-6.3.2.1(e) *Small room provision—four sprinklers.*

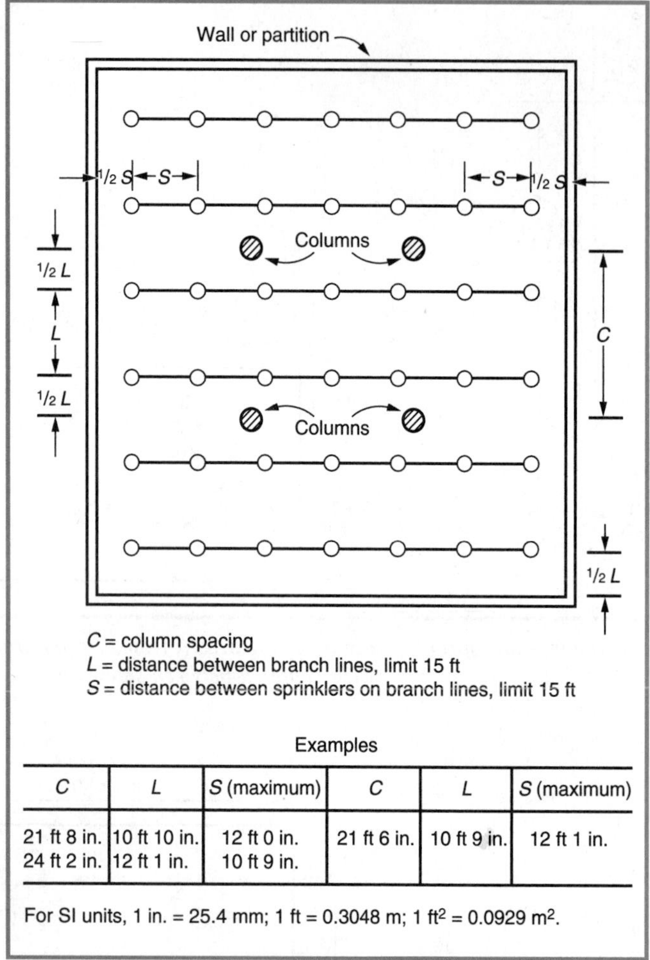

C	L	S (maximum)	C	L	S (maximum)
21 ft 8 in. 24 ft 2 in.	10 ft 10 in. 12 ft 1 in.	12 ft 0 in. 10 ft 9 in.	21 ft 6 in.	10 ft 9 in.	12 ft 1 in.

For SI units, 1 in. = 25.4 mm; 1 ft = 0.3048 m; 1 ft^2 = 0.0929 m^2.

Exhibit 5.10 *Positioning of standard spray sprinkler under a flat slab concrete ceiling in an ordinary hazard occupancy.*

Dead-air spaces in corners would increase a sprinkler's operation time when installed near the corner. The 4-in. (102-mm) minimum location from a wall ensures that the sprinkler will operate properly. NFPA 72, *National Fire Alarm Code®*, provides more discussion on this phenomenon. See also commentary to 5-7.3.3.

5-6.3.4 Minimum Distance Between Sprinklers. Sprinklers shall be spaced not less than 6 ft (1.8 m) on center.

Exception No. 1: Sprinklers shall be permitted to be placed less than 6 ft (1.8 m) on center where the following conditions are satisfied:

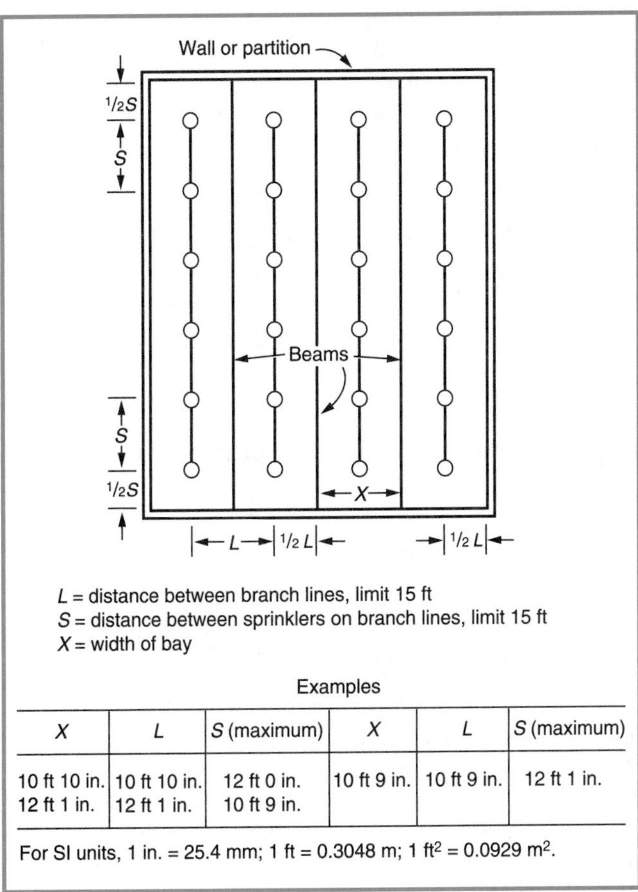

L = distance between branch lines, limit 15 ft
S = distance between sprinklers on branch lines, limit 15 ft
X = width of bay

Examples

X	L	S (maximum)	X	L	S (maximum)
10 ft 10 in.	10 ft 10 in.	12 ft 0 in.	10 ft 9 in.	10 ft 9 in.	12 ft 1 in.
12 ft 1 in.	12 ft 1 in.	10 ft 9 in.			

For SI units, 1 in. = 25.4 mm; 1 ft = 0.3048 m; 1 ft^2 = 0.0929 m^2.

Exhibit 5.11 *Positioning of standard spray sprinkler under a ceiling consisting of continuous smooth bays with beams supported on columns in an ordinary hazard occupancy.*

(a) Baffles shall be installed and located midway between sprinklers and arranged to protect the actuating elements.

(b) Baffles shall be of noncombustible or limited-combustible material that will stay in place before and during sprinkler operation.

(c) Baffles shall be not less than 8 in. (203 mm) wide and 6 in. (152 mm) high. The tops of baffles shall extend between 2 in. and 3 in. (51 mm and 76 mm) above the deflectors of upright sprinklers. The bottoms of baffles shall extend downward to a level at least even with the deflectors of pendent sprinklers.

Exception No. 2: In-rack sprinklers shall be permitted to be placed less than 6 ft (1.8 m) on center.

Exception No. 3: Old-style sprinklers protecting fur storage vaults shall be permitted to be placed less than 6 ft (1.8 m) on center.

Baffles minimize the possibility that discharge from one sprinkler will strike an adjacent sprinkler's actuating elements and delay or prevent the operation of the adjacent sprinklers. Because recessed and concealed sprinklers have their operating elements at or above the ceiling, they are less likely to be effected by the operation of adjacent sprinklers. However, NFPA 13 does not provide any exception that would allow recessed or concealed sprinklers to be located closer than 6 ft (1.8 m) apart.

Exception No. 2 pertains to sprinklers located within storage racks. In-rack sprinklers are not required to be separated by baffles when installed closer than 6 ft (1.8 m) apart, because they were tested for these configurations. Because in-rack sprinklers are located in or adjacent to the fire and not at the ceiling where the heat will collect, the lower sprinklers normally operate first and are not usually subjected to this phenomenon.

5-6.4 Deflector Position (Standard Pendent and Upright Spray Sprinklers).

5-6.4.1 Distance Below Ceilings.

The requirements for the position of sprinklers with respect to the floor or ceiling are established in 5-6.4.1. In all cases, a minimum of 1 in. (25.4 mm) of clearance is required to allow for the replacement of upright sprinklers. When applying the rules for positioning sprinklers, the dimensions are measured to the sprinkler deflector.

5-6.4.1.1 Under unobstructed construction, the distance between the sprinkler deflector and the ceiling shall be a minimum of 1 in. (25.4 mm) and a maximum of 12 in. (305 mm).

Exception: Ceiling-type sprinklers (concealed, recessed, and flush types) shall be permitted to have the operating element above the ceiling and the deflector located nearer to the ceiling where installed in accordance with their listing.

Formal Interpretations 78-4 and 80-4 provide some insight regarding the positioning of sprinklers underneath skylights and other types of irregular ceiling pockets. The concern with deep skylights and ceiling pockets is that unless a sprinkler is placed in the skylight or pocket, hot gases from a fire can collect in these spaces and delay the activation of the sprinkler system. The rules for positioning sprinklers below ceilings of obstructed, unobstructed, and sloped construction should be applied to the extent possible. In some cases an engineering analysis might be necessary to determine proper placement.

Exhibit 1.23 shows a concealed sprinkler. This type of sprinkler would have the deflector positioned at a point even with or slightly above the ceiling. Concealed sprinklers are specifically evaluated under the condition stated in the exception.

5-6.4.1.2 Under obstructed construction, the sprinkler deflector shall be located within the horizontal planes of 1 in. to 6 in. (25.4 mm to 152 mm) below the structural members and a maximum distance of 22 in. (559 mm) below the ceiling/roof deck.

Formal Interpretation 78-4

Reference: 5-6.4.1.1

Question: Is it the intent of 5-6.4.1.1 to require sprinklers to be installed in (4′ × 8′) skylights having 55 flame spread, when they do not support combustion? Additionally the skylights would have to be heated to over 600°F to start melting out.

Answer: No. It is not the intent of the standard to require the installation of sprinklers in pockets formed by 4′ × 8′ skylights as such areas would not significantly retard the operation of sprinklers.

Issue Edition: 1978

Reference: 4-3.1.2

Date: Reprinted April 1987

Formal Interpretation 80-4

Reference: 5-6.4.1.1

Question: Is it the intent of 5-6.4.1.1 to require sprinklers to be installed in (4′ × 8′) plastic skylights fitted with aluminum ventilation louvers approximately one foot high supporting the skylight?

Answer: No. It is not the intent of the standard to require the installation of sprinklers in pockets formed by plastic skylights as such areas would not significantly retard the operation of sprinklers.

Issue Edition: 1980

Reference: 4-3.1.2

Date: Reprinted April 1987

Exception No. 1 Sprinklers shall be permitted to be installed with the deflector at or above the bottom of the structural member to a maximum of 22 in. (559 mm) below the ceiling/roof deck where the sprinkler is installed in conformance with 5-6.5.1.2.

Exception No. 2: Where sprinklers are installed in each bay of obstructed construction, deflectors shall be permitted to be a minimum of 1 in. (25.4 mm) and a maximum of 12 in. (305 mm) below the ceiling.

Exception No. 3: Sprinkler deflectors shall be permitted to be 1 in. to 6 in. below composite wood joists to a maximum distance of 22 in. below the ceiling/roof deck only where joist channels are fire-stopped to the full depth of the joists with material equivalent to the web construction so that individual channel areas do not exceed 300 ft² (27.9 m²).

Exception No. 4: Deflectors of sprinklers under concrete tee construction with stems spaced less than 7½ ft (2.3 m) but more than 3 ft (0.9 m) on centers shall, regardless of the depth of the tee, be permitted to be located at or above a horizontal plane 1 in. (25.4 mm) below the bottom of the stems of the tees and shall comply with Table 5-6.5.1.2.*

A-5-6.4.1.2 Exception No. 4. For concrete joists spaced less than 3 ft (0.91 m) on center, the rules for obstructed construction shown in 5-6.4.1.2 apply. *(See Figure A-5-6.4.1.2.)*

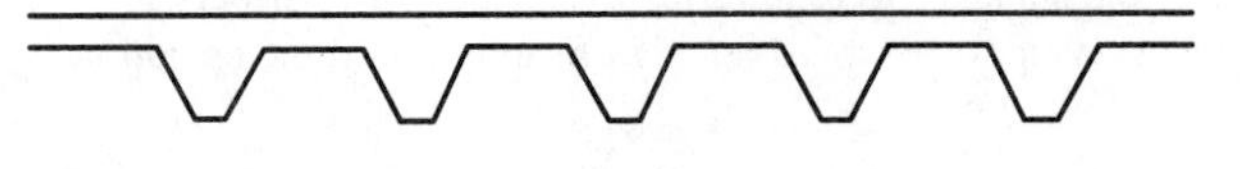

Figure A-5-6.4.1.2 *Typical concrete joist construction.*

The position of sprinklers under obstructed construction varies widely. A basic series of requirements is set to allow the deflector to be positioned in a number of ways, such as below the structural member in some circumstances, beside the members, in between all the members, or, for concrete joist construction, entirely below the members.

The 1991 edition resulted in substantive changes that permitted the sprinkler deflector to be as much as 22 in. (559 mm) below the ceiling for obstructed construction. A 22-in. (559-mm) maximum deflector position for other than concrete tee construction is set in 5-6.4.1.2 (see Exception No. 4). Positioning the sprinklers more than 22 in. (559 mm) below the ceiling negatively impacts sprinkler activation.

Exhibit 5.12 illustrates the placement of upright sprinklers underneath obstructed construction consisting of I-beams in accordance with the base rule of 5-6.4.1.2. If the I-beams had a depth of 10 in. (254 mm), then the sprinkler could be placed 1 in. to 6 in. (25.4 mm to 152 mm) below the bottom of the I-beam but no more than 11 in. to 16 in. (280 mm to 410 mm) below the ceiling. If the I-beam had a depth of 18 in. (0.46 m), another rule influences sprinkler placement. Because the sprinkler cannot be placed more than 22 in. (559 mm) from the ceiling, the sprinkler placement is limited to 1 in. to 4 in. (25.4 mm to 101.6 mm) below the I-beam and 19 in. to 22 in. (482.6 mm to 559 mm) below the ceiling. If the I-beams have a depth of 22 in. (559 mm) or more, then the sprinklers will need to be positioned in accordance with either Exception No. 1 or Exception No. 2 to 5-6.4.1.2.

Exhibit 5.13 shows the placement of sprinklers under solid wood joist construction in which the joists have a depth of 10 in. (254 mm). The sprinkler could be placed up to 6 in. (152 mm) below the joists, positioning it 16 in. (406 mm) from the ceiling or roof deck. Because sprinklers cannot be positioned more than 1 in. to 6 in. (25.4 mm to 152 mm) below the structural members, the maximum permitted dimension of 22 in. (559 mm) from the ceiling cannot be applied.

Exception No. 1 allows sprinkler deflectors to be positioned at or above the bottom of the structural member, provided the distance from the ceiling or roof deck to the deflectors does not exceed 22 in. (559 mm) and the sprinklers are positioned in accordance with the obstruction rules of 5-6.5.1.2, not including exceptions. Exhibit

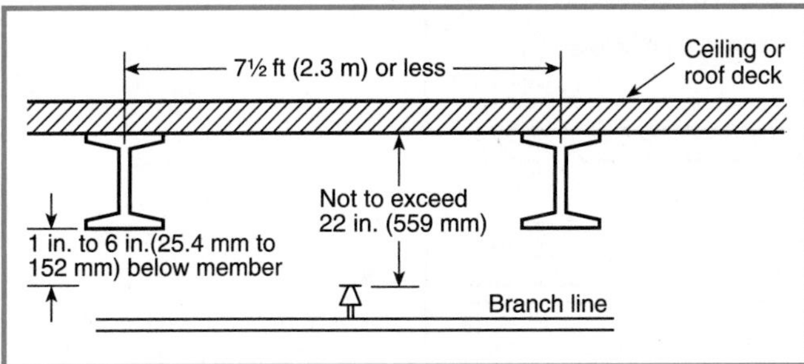

Exhibit 5.12 Placement of standard spray upright sprinklers underneath obstructed construction.

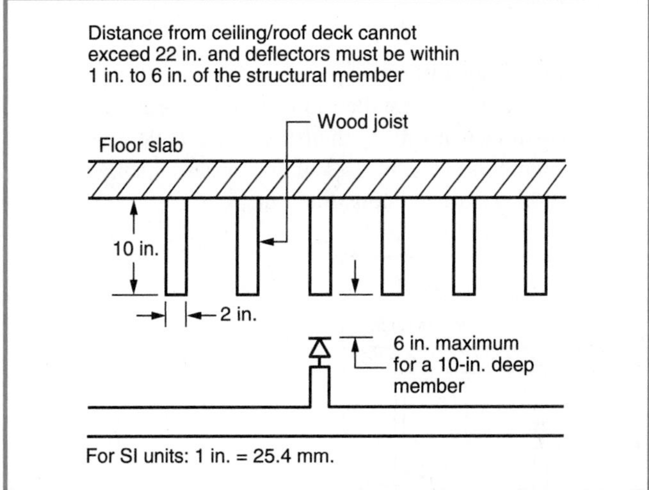

Exhibit 5.13 Placement of standard spray upright sprinklers underneath solid wood joist construction.

5.14 depicts one such arrangement and illustrates that the sprinkler is positioned within 22 in. (559 mm) of the ceiling so as to not negatively impact sprinkler activation and so that the sprinkler discharge will adequately clear the structural members. This rule is typically employed where sprinklers are not installed in each bay of obstructed construction. Exception No. 2 applies where sprinklers need to be installed in each bay of obstructed construction.

Where the depths of the structural members are of such a dimension that compliance with the obstruction rules of Table 5-6.5.1.2 is not possible, sprinklers would need to be installed in each bay, and the requirements of Exception No. 2 are applicable. The standard permits the sprinklers to be located 1 in. to 12 in. (25.4 mm to 305 mm) below the ceiling, as they are for unobstructed construction, but they can be located

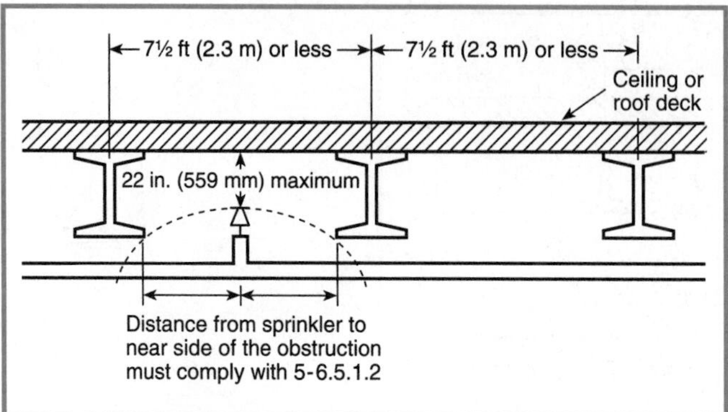

Exhibit 5.14 Placement of standard spray upright sprinklers underneath obstructed construction with the sprinkler deflector positioned above the bottom of the structural member.

up to 22 in. (559 mm) below the ceiling. The closer the sprinkler is to the ceiling, the faster its response to a fire. Where sprinklers are installed in each bay, the sprinklers should be spaced so that they are positioned no more than their allowable spacing from the near side of the structural members. Exhibit 5.15 shows an arrangement of upright sprinklers located in each bay of obstructed construction.

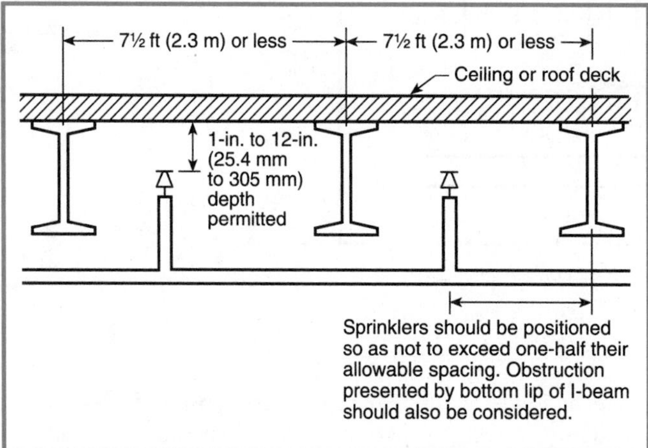

Exhibit 5.15 Placement of standard spray upright sprinklers in each bay of obstructed construction.

Exception No. 3 identifies the only conditions under which sprinklers can be placed under composite wood joist construction. This exception was revised for the 1999 edition to clarify that the distance between the sprinkler deflector and the ceiling cannot exceed 22 in. (559 mm) and that the maximum distance between the sprinkler deflector and the bottom of the joist cannot exceed 6 in. (152 mm).

Exception No. 4 applies only to specific types of concrete tee construction. The sprinkler is to be located as high as possible, taking into account the obstruction that could be created by the leg of the concrete tee. The allowance for this special condition is based on sensitivity tests conducted by the sprinkler industry. Concrete has a comparatively high threshold to withstand heat from fire. The concrete also has the capacity to act as a heat sink, collecting heat from the fire and creating a ceiling effect. This reaction ensures timely operation of the sprinkler.

5-6.4.1.3* Sprinklers under or near the peak of a roof or ceiling shall have deflectors located not more than 3 ft (0.9 m) vertically down from the peak. *[See Figures 5-6.4.1.3(a) and 5-6.4.1.3(b).]*

Exception No. 1: Under saw-toothed roofs, sprinklers at the highest elevation shall not exceed a distance of 3 ft (0.9 m) measured down the slope from the peak.

Exception No. 2: Under a steeply pitched surface, the distance from the peak to the deflectors shall be permitted to be increased to maintain a horizontal clearance of not less than 2 ft (0.6 m) from other structural members. [See Figure 5-6.4.1.3(c).]

A-5-6.4.1.3 Saw-toothed roofs have regularly spaced monitors of saw tooth shape, with the nearly vertical side glazed and usually arranged for venting. Sprinkler placement is limited to

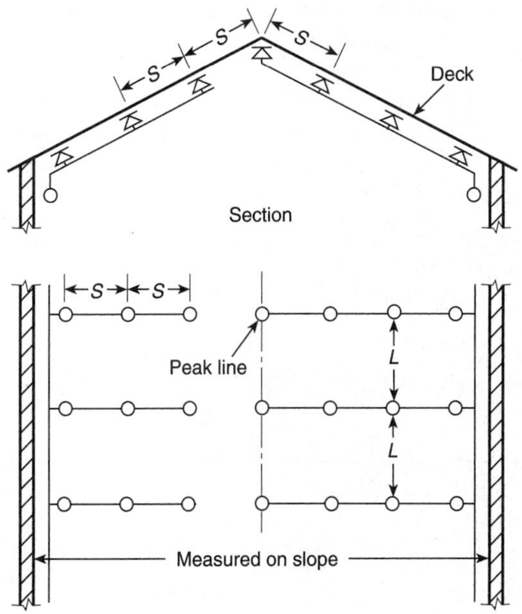

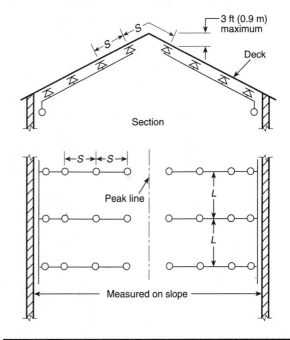

Figure 5-6.4.1.3(a) *Sprinklers under pitched roofs with sprinkler directly under peak; branch lines run up the slope.*

Figure 5-6.4.1.3(b) *Sprinklers at pitched roofs; branch lines run up the slope.*

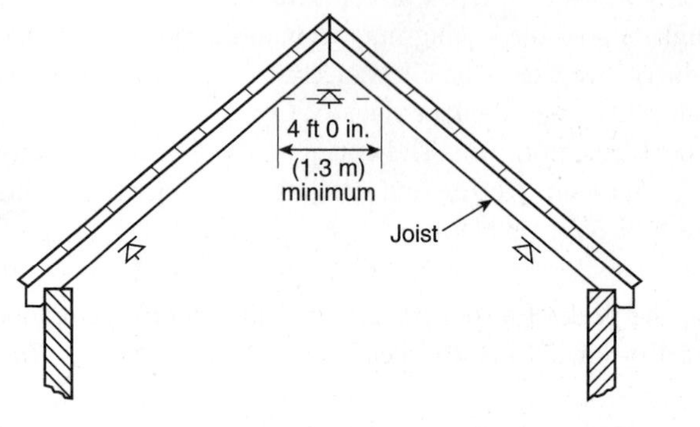

Figure 5-6.4.1.3(c) *Horizontal clearance for sprinklers at peak of pitched roof.*

a maximum of 3 ft (0.91 m) down the slope from the peak because of the effect of venting on sprinkler sensitivity.

Sprinklers are required to be within 3 ft (0.9 m) vertically of the peak but must also meet the position requirements of 5-6.4.1. The 3-ft (0.9-m) limitation addresses the concern that heat from a fire tends to collect at the peak. Locating sprinklers too far below the peak can delay their activation time. If the roof is steeply pitched, raising the sprinkler to improve responsiveness compromises the sprinkler spray pattern. A minimum 2-ft (0.6-m) horizontal clearance is required in this case.

5-6.4.1.4 Double Joist Obstructions. Where there are two sets of joists under a roof or ceiling, and there is no flooring over the lower set, sprinklers shall be installed above and below the lower set of joists where there is a clearance of 6 in. (152 mm) or more between the top of the lower joist and the bottom of the upper joist. *(See Figure 5-6.4.1.4.)*

Exception: Sprinklers are permitted to be omitted from below the lower set of joists where at least 18 in. (0.46 m) is maintained between the sprinkler deflector and the top of the lower joist.

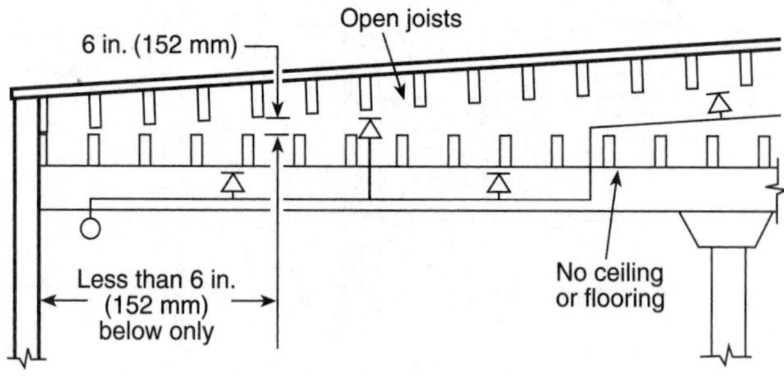

Figure 5-6.4.1.4 *Arrangement of sprinklers under two sets of open joists—no sheathing on lower joists.*

Formal Interpretation 80-16 provides additional information concerning double joist obstructions.

Formal Interpretation 80-16

Reference: 5-6.4.1.4

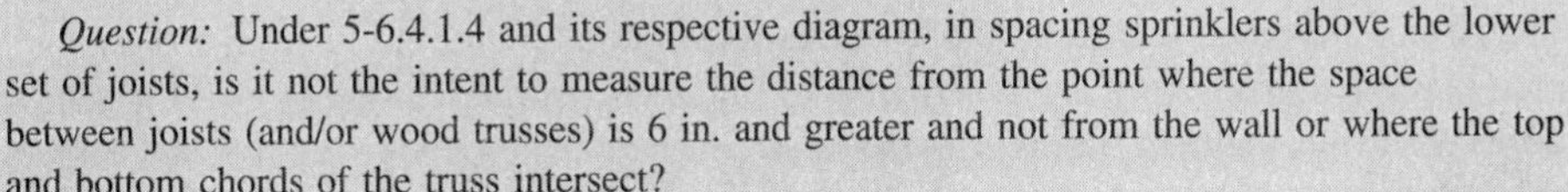

Question: Under 5-6.4.1.4 and its respective diagram, in spacing sprinklers above the lower set of joists, is it not the intent to measure the distance from the point where the space between joists (and/or wood trusses) is 6 in. and greater and not from the wall or where the top and bottom chords of the truss intersect?

Answer: No. This paragraph addresses only the special conditions stated in the beginning of the paragraph and does not apply to wood trusses. Under the conditions stated in 5-6.4.1.4, the first sprinkler between the two sets of joists should be located at the point where the clear space between the two sets of joists is six inches. In wood truss construction, sprinklers should be spaced from the end of the truss.

Issue Edition: 1980

Reference: 4-2.3.2

Date: July 1982

5-6.4.2* Deflector Orientation. Deflectors of sprinklers shall be aligned parallel to ceilings, roofs, or the incline of stairs.

Exception No. 1: Where sprinklers are installed in the peak below a sloped ceiling or roof surface, the sprinkler shall be installed with the deflector horizontal.

Exception No. 2: Pitched roofs having slopes not exceeding a pitch of one in six (a rise of two units in a run of 12 units, a roof slope of 16.7 percent) are considered level in the application of this rule, and sprinklers shall be permitted to be installed with deflectors horizontal.

A-5-6.4.2 On sprinkler lines larger than 2 in. (51 mm), consideration should be given to the distribution interference caused by the pipe, which can be minimized by installing sprinklers on riser nipples or installing sprinklers in the pendent position.

5-6.5 Obstructions to Sprinkler Discharge (Standard Pendent and Upright Spray Sprinklers).

5-6.5.1 Performance Objective.

5-6.5.1.1 Sprinklers shall be located so as to minimize obstructions to discharge as defined in 5-6.5.2 and 5-6.5.3, or additional sprinklers shall be provided to ensure adequate coverage of the hazard.

5-6.5.1.2 Sprinklers shall be arranged to comply with 5-5.5.2, Table 5-6.5.1.2, and Figure 5-6.5.1.2(a).

Exception No. 1: Sprinklers shall be permitted to be spaced on opposite sides of obstructions not exceeding 4 ft (1.2 m) in width provided the distance from the centerline of the obstruction to the sprinklers does not exceed one-half the allowable distance permitted between sprinklers.

Exception No. 2: Obstructions located against the wall and that are not over 30 in. (762 mm) in width shall be permitted to be protected in accordance with Figure 5-6.5.1.2(b).

In past editions, the provision addressed by 5-6.5.1.2 was commonly referred to as the Beam Rule for standard spray sprinklers. The dimensions specified in Table 5-6.5.1.2 outline the discharge pattern of the sprinkler and define how far away from a building element a sprinkler must be positioned to allow the sprinkler discharge to extend underneath the building element rather than to hit it. After reviewing the discharge patterns for sprinklers at pressures from 15 psi to 100 psi (1 bar to 7 bar), Table 5-6.5.1.2 was adjusted in 1996 to better reflect the patterns of typical standard spray upright and pendent sprinklers.

When applying Exception No. 1, the maximum spacing permitted between branch lines and between sprinklers on branch lines must still be followed. Therefore, the distance between sprinklers on opposite sides of the obstruction is measured from the centerline of the obstruction. The maximum allowable distances above the obstruction (see Table 5-6.5.1.2) can be exceeded if the trajectory and resultant discharge pattern of a particular sprinkler are verified by test.

Table 5-6.5.1.2 Positioning of Sprinklers to Avoid Obstructions to Discharge (SSU/SSP)

Distance from Sprinklers to Side of Obstruction *(A)*	Maximum Allowable Distance of Deflector above Bottom of Obstruction (in.) *(B)*
Less than 1 ft	0
1 ft to less than 1 ft 6 in.	2½
1 ft 6 in. to less than 2 ft	3½
2 ft to less than 2 ft 6 in.	5½
2 ft 6 in. to less than 3 ft	7½
3 ft to less than 3 ft 6 in.	9½
3 ft 6 in. to less than 4 ft	12
4 ft to less than 4 ft 6 in.	14
4 ft 6 in. to less than 5 ft	16½
5 ft and greater	18

For SI units, 1 in. = 25.4 mm; 1 ft = 0.3048 m.

Note: For *(A)* and *(B)*, refer to Figure 5-6.5.1.2(a).

The answer given in Formal Interpretation 76-11 relates to Exception No. 1 to 5-6.5.1.2. The exception cites permission to treat the centerline of the obstruction as a wall and to place sprinklers on either side of the obstruction. The spacing requirements of 5-6.3 are still applicable when applying this exception.

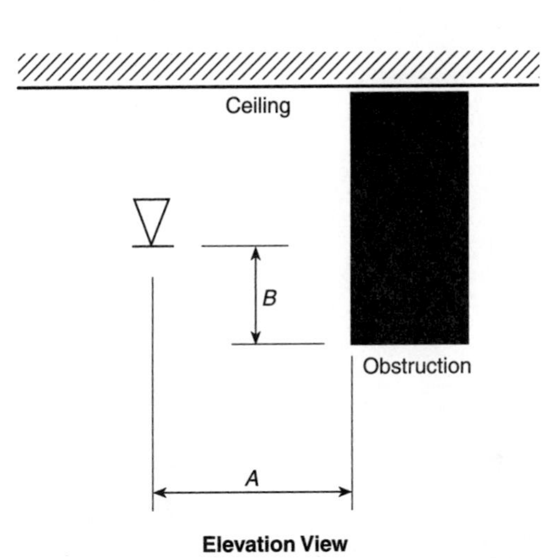

Figure 5-6.5.1.2(a) Positioning of sprinklers to avoid obstructions to discharge (SSU/SSP).

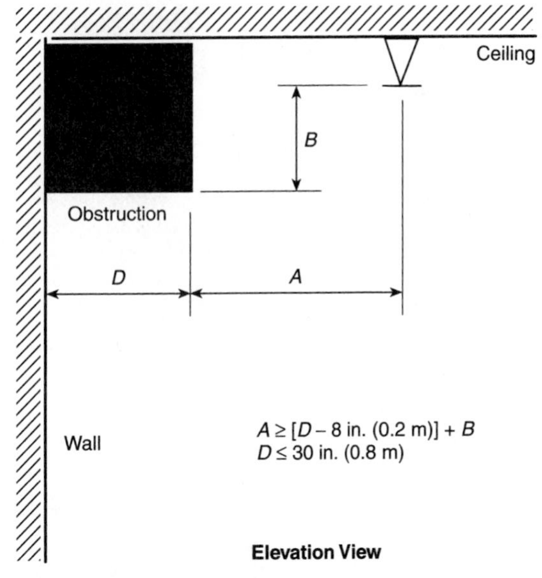

Figure 5-6.5.1.2(b) Obstructions against walls (SSU/ SSP).

Exception No. 2 and Figure 5-6.5.1.2(b) identify the conditions under which sprinklers are not required underneath soffits and other similar building features located against a wall. In other words, Exception No. 2 identifies when soffits do not present an obstruction to sprinkler discharge. If the obstruction is not over 30 in. (762 mm) in width and if the sprinklers are positioned in front of the obstruction as shown, then additional sprinklers under the soffit can be omitted. The exception is also intended to apply to building elements other than soffits located against the wall, such as cable trays, ductwork, cabinets, or similar small projections attached to a wall.

5-6.5.2 Obstructions to Sprinkler Discharge Pattern Development.

5-6.5.2.1* Continuous or noncontinuous obstructions less than or equal to 18 in. (457 mm) below the sprinkler deflector that prevent the pattern from fully developing shall comply with this section. Regardless of the rules of this section, solid continuous obstructions shall meet the requirements of 5-6.5.1.2.

A-5-6.5.2.1 The rules of 5-6.5.2.2 (known as the "Three Times Rule") have been written to apply to obstructions where the sprinkler can be expected to get water to both sides of the obstruction without allowing a significant dry shadow on the other side of the obstruction. This works for small noncontinuous obstructions and for continuous obstructions where the sprinkler can throw water over and under the obstruction, such as the bottom chord of an open truss or joist. For solid continuous obstructions, such as a beam, the Three Times Rule is ineffective since the sprinkler cannot throw water over and under the obstruction. Sufficient water must

Formal Interpretation 76-11

Reference: 5-6.5.1.2

Question: Does Table 5-6.5.1.2 apply to the layout shown in Figure 1?

Answer: No. Paragraph 5-6.5.1.2 Exception No. 1 offers the option as an alternate to Table 5-6.5.1.2 of spacing the sprinklers from the obstruction at a distance of not more than one-half the allowable distance between sprinklers. Under those conditions, the sprinklers may be positioned other than as specified in Table 5-6.5.1.2.

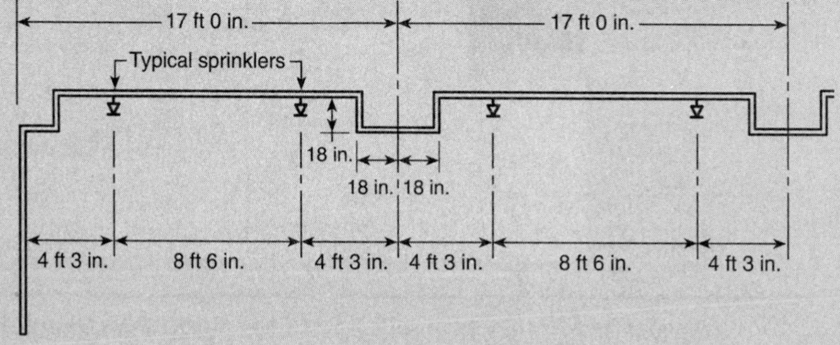

Figure 1 *Typical Ceiling Cross Section.*

Issue Edition: 1976

Reference: 4-2.4.6

Date: June 1978

be thrown under the obstruction to adequately cover the floor area on the other side of the obstruction. To ensure this, compliance with the rules of 5-6.5.1.2 is necessary.

Large solid continuous obstructions were referred to as horizontal obstructions prior to the 1996 edition. These obstructions, such as deep beams or a large duct mounted on or very near the ceiling, have a significant impact on discharge patterns and are required to be located above the sprinkler discharge pattern in accordance with 5-6.5.1.2. Smaller continuous obstructions, such as pipes and bottom members of trusses, and noncontinuous obstructions, such as the web members of bar joists located within 18 in. (457 mm) below the sprinkler, should be in compliance with the requirements of 5-6.5.2.1.

5-6.5.2.2 Sprinklers shall be positioned such that they are located at a distance three times greater than the maximum dimension of an obstruction up to a maximum of 24 in. (609 mm) (e.g., structural members, pipe, columns, and fixtures). *(See Figure 5-6.5.2.2.)*

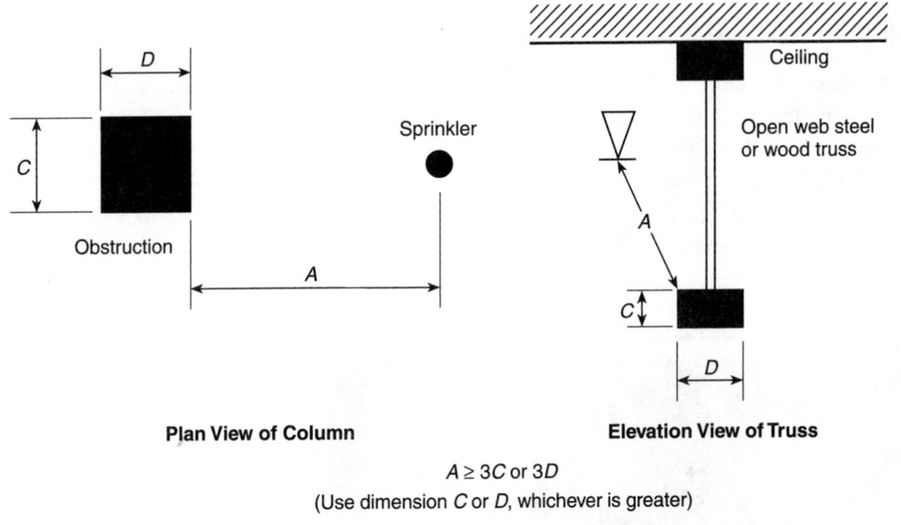

Figure 5-6.5.2.2 *Minimum distance from obstruction (SSU/SSP).*

Exception No. 1: For light and ordinary hazard occupancies, structural members only shall be considered.

Exception No. 2: Sprinklers shall be permitted to be spaced on opposite sides of the obstruction provided the distance from the centerline of the obstruction to the sprinklers does not exceed one-half the allowable distance between sprinklers.

Exception No. 3: Where the obstruction consists of open trusses 20 in. (0.51 m) or greater apart [24 in. (0.61 m) on center], sprinklers shall be permitted to be located one-half the distance between the obstruction created by the truss provided that all truss members are not greater than 4 in. (102 mm) (nominal) in width.

Exception No. 4: Sprinklers shall be permitted to be installed on the centerline of a truss, bar joist, or directly above a beam provided that the truss chord or beam dimension is not more than 8 in. (203 mm) and the sprinkler deflector is located at least 6 in. (152 mm) above the structural member. The sprinkler shall be positioned at a distance three times greater than the maximum dimension of the web members away from the web members.

Exception No. 5: Piping to which an upright sprinkler is directly attached less than 3 in. (76 mm) in diameter.

Exception No. 6: Piping to which pendent sprinklers are directly attached.

Exception No. 7: Sprinklers positioned with respect to obstructions in accordance with 5-6.5.1.2.

When installing sprinklers in unobstructed construction containing noncontinuous obstructions, such as building columns, bar joists, or some truss configurations, adequate horizontal and vertical clearance from the obstruction are critical to maintain proper sprinkler discharge patterns. This clearance is especially true where the obstructions are structural members that affect each line of sprinklers. Bar joists, trusses, and similar construction types do not impede heat flow, but they are capable of obstructing the sprinkler's spray pattern once it is in operation. Exhibit 5.16 shows a standard spray upright sprinkler positioned near bar joist construction. The sprinkler in Exhibit 5.16 needs to be positioned an adequate distance from the horizontal and vertical members of the bar joist.

Exhibit 5.16 Standard spray upright sprinkler positioned near bar joist construction.

The impact of an obstruction on sprinkler performance depends on many factors. One factor is the severity of the hazard being protected. Historically, NFPA 13 considered building structural elements and other large building elements, such as ducts, as being potential obstructions. In 1991, the standard was revised to consider other types of smaller building elements, such as piping and light fixtures, that could also obstruct sprinkler discharge patterns. Beginning with the 1999 edition, NFPA 13 (see Exception No. 1) acknowledges the difficulty in considering all nonstructural building elements as potential obstructions when designing the system. Additionally, NFPA 13 takes into account the fact that more obstructions can be tolerated in light and ordinary hazard occupancies when spray sprinklers are used. However, nonstructural building elements must be considered when designing sprinkler systems in extra hazard and storage occupancies.

Structural members that are obstructions in light and ordinary hazard occupancies must still be considered because they are easier to locate on building plans than lights and pipes, and because they tend to be continuous, thereby affecting more than one sprinkler in a symmetrical arrangement.

Many but not all of the obstruction problems are addressed in 5-6.5.2.2. When open truss members are closely spaced, it is virtually impossible to locate the sprinkler far enough away from all truss elements that could cause an obstruction. Exception No. 3 addresses this specific problem by permitting the installation of sprinklers in between the members without requiring adherence to the "Three Times Rule" as long as the joists are spaced 24 in. (0.61 m) or more apart. Exhibit 5.17 illustrates such an arrangement.

Exception No. 4 identifies the conditions under which standard spray upright and pendent sprinklers can be located directly over the centerline of the bottom chord of an open truss, bar joist, or beam without compliance with the Three Times Rule. Under

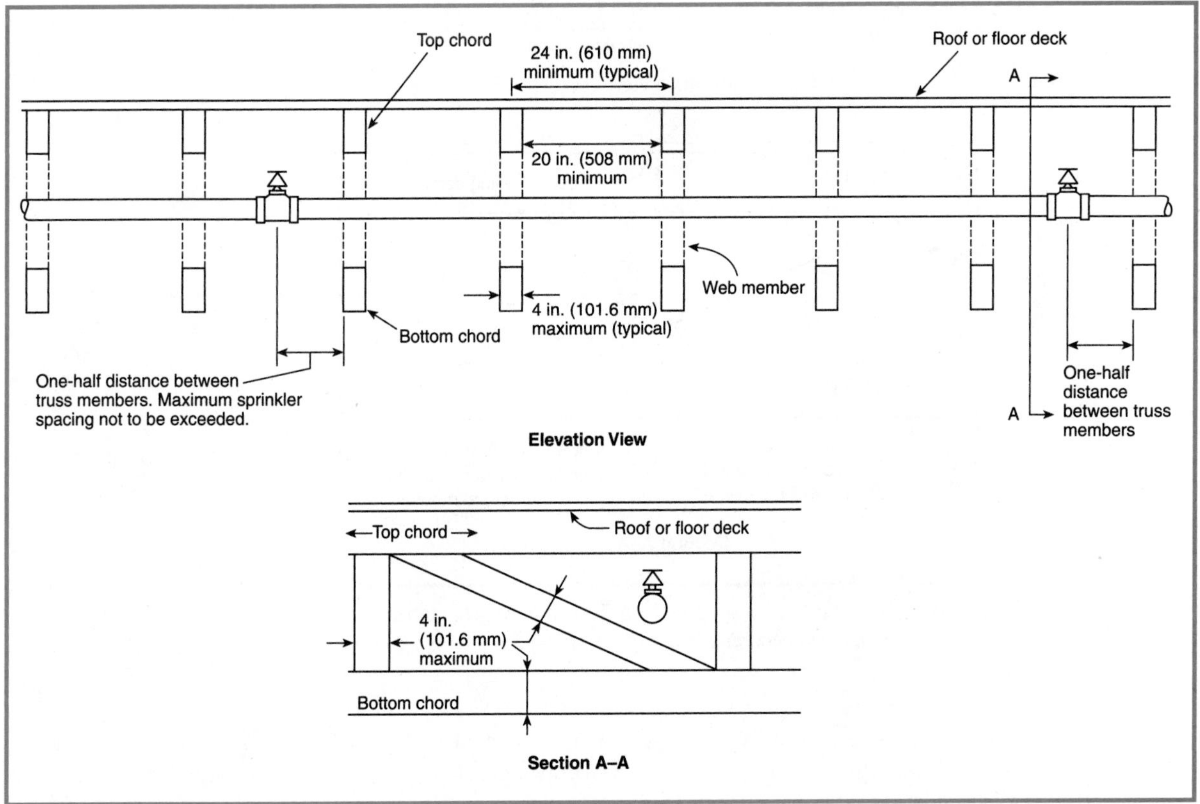

Exhibit 5.17 *Location of sprinklers in open truss construction.*

the conditions specified by Exception No. 4 and illustrated in Exhibit 5.18, the bottom chord and beams do not present a significant obstruction to sprinkler discharge. However, the vertical and diagonal truss and web members must comply with the Three Times Rule.

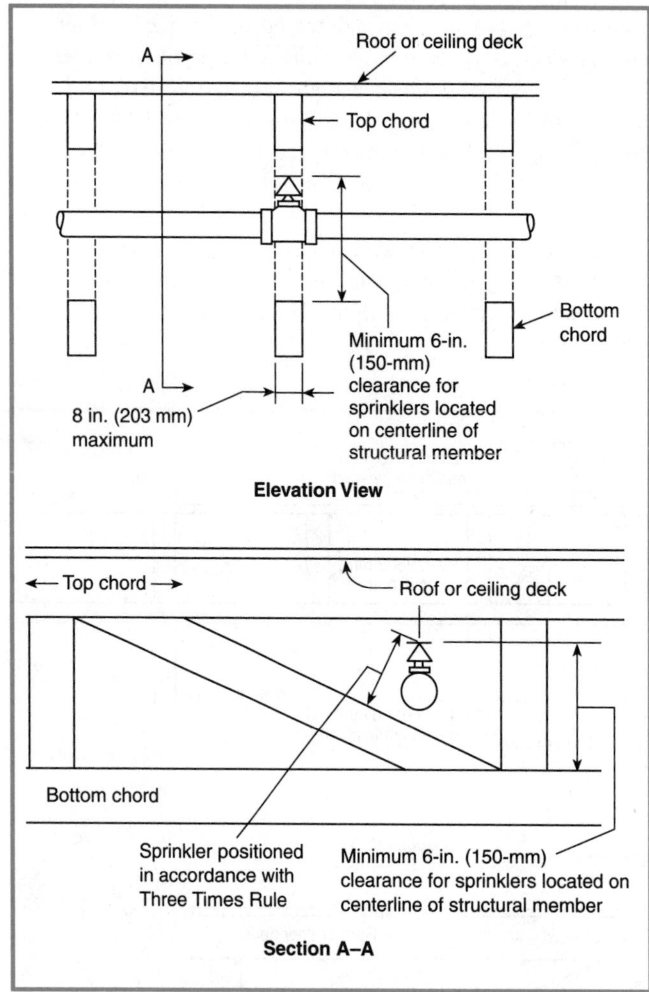

Exhibit 5.18 Sprinkler placed over the centerline of the bottom chord of a wood truss in accordance with Exception No. 4 to 5-6.5.2.2.

The pipe on which a sprinkler is located is not considered a major obstruction unless the piping is 3 in. (76 mm) or larger. Upright sprinklers on large branch lines or mains must be placed on riser nipples or offset from the branch line to eliminate the obstruction that is created directly below the sprinkler. The elevation of the sprinkler eliminates problems that can occur when a fire is located directly below a sprinkler or directly under the branch line between two sprinklers. The obstruction can prevent the discharge from reaching the fire and can also eliminate dead-air spots that might delay sprinkler actuation, such as when an upright sprinkler is located directly on a 6-in. (152-mm) pipe.

Other rules apply in addition to the obstruction rules of 5-6.5. The sprinkler's frame arm is required to be parallel to the branch line (see 5-3.1.2), and the distance between the centerline of an upright sprinkler and a hanger is required to be a minimum of 3 in. (76 mm) (see 6-2.3.2). Exhibit 5.19 shows an upright sprinkler attached directly to a branch line. The sprinkler is free of obstructions from structural ceiling elements, the branch line is less than 3 in. (76 mm) in diameter, the frame arm is parallel to the branch line, and the hanger is located at least 3 in. (76 mm) from the centerline of the sprinkler. However, verification that the sprinkler is properly positioned from the ceiling in accordance with 5-6.4 is needed, depending on whether the ceiling construction is obstructed or unobstructed.

Exhibit 5.19 *Upright sprinkler installation.*

5-6.5.2.3* Suspended or Floor-Mounted Vertical Obstructions. The distance from sprinklers to privacy curtains, free standing partitions, room dividers, and similar obstructions in light hazard occupancies shall be in accordance with Table 5-6.5.2.3 and Figure 5-6.5.2.3.

A-5-6.5.2.3 The distances given in Table 5-6.5.2.3 were determined through tests in which privacy curtains with either a solid fabric or close mesh [¼ in. (6.4 mm)] top panel were installed. For broader-mesh top panels—for example, ½ in. (13 mm) or greater measured on the diagonal—the obstruction of the sprinkler spray is not likely to be severe and the authority having jurisdiction might not need to apply the requirements in 5-6.5.2.3.

The distances specified in Table 5-6.5.2.3 were derived from NBSIR 80-2097, *Full-Scale Fire Tests with Automatic Sprinklers in a Patient Room.* The distance requirements

Table 5-6.5.2.3 Suspended or Floor-Mounted Obstructions (SSU/SSP)

Horizontal Distance *(A)*	Minimum Vertical Distance below Deflector (in.) *(B)*
6 in. or less	3
More than 6 in. to 9 in.	4
More than 9 in. to 12 in.	6
More than 12 in. to 15 in.	8
More than 15 in. to 18 in.	9½
More than 18 in to 24 in.	12½
More than 24 in. to 30 in.	15½
More than 30 in.	18

For SI units, 1 in. = 25.4 mm.

Note: For *(A)* and *(B)*, refer to Figure 5-6.5.2.3.

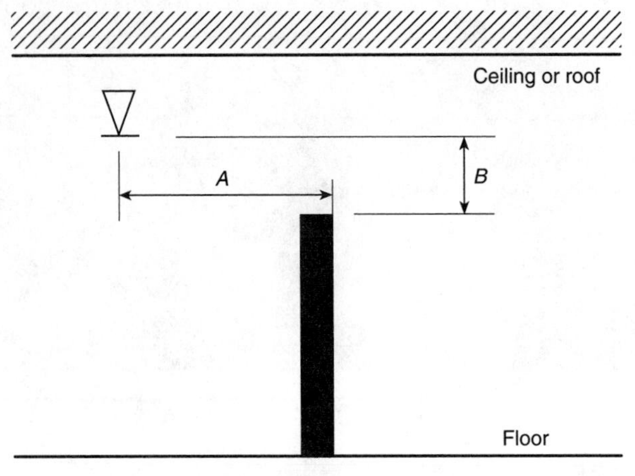

Figure 5-6.5.2.3 *Suspended or floor-mounted obstructions (SSU/SSP).*

should not be applied beyond light hazard occupancies, because the testing only evaluated sprinkler performance in a light hazard environment.

5-6.5.3* Obstructions that Prevent Sprinkler Discharge from Reaching the Hazard. Continuous or noncontinuous obstructions that interrupt the water discharge in a horizontal plane more than 18 in. (457 mm) below the sprinkler deflector in a manner to limit the distribution from reaching the protected hazard shall comply with this section.

Exception: The requirements of this section shall also apply to obstructions 18 in. or less below the sprinkler for light and ordinary hazard occupancies.

A-5-6.5.3 See A-5-5.5.3.

Once the sprinkler spray pattern has developed, small obstructions have a limited impact on sprinkler performance. However, larger obstructions located 18 in. (457 mm) or more below the sprinkler deflector prevent the sprinkler discharge from reaching combustibles located below them. This information identifies when additional upright or pendent sprinklers are required to compensate for these types of obstructions.

Because only structural members within 18 in. (457 mm) of the sprinkler are now considered by 5-6.5.2 for light and ordinary hazard occupancies, an exception was added to the 1999 edition. The exception to 5-6.5.3 requires that any continuous or noncontinuous obstructions wider than 4 ft (1.2 m) and located within 18 in. (457 mm) of the sprinkler deflector in light and ordinary hazard occupancies meet the requirements of 5-6.5.3.1 and 5-6.5.3.2.

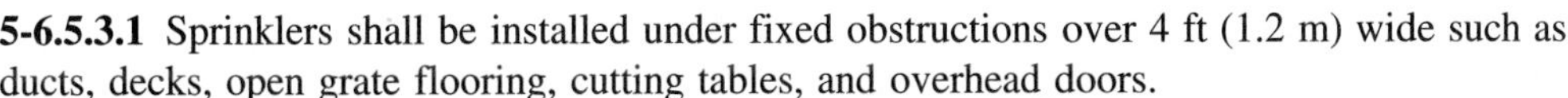

5-6.5.3.1 Sprinklers shall be installed under fixed obstructions over 4 ft (1.2 m) wide such as ducts, decks, open grate flooring, cutting tables, and overhead doors.

Exception: Obstructions that are not fixed in place, such as conference tables.

5-6.5.3.2 Sprinklers installed under open gratings shall be of the intermediate level/rack storage type or otherwise shielded from the discharge of overhead sprinklers.

As noted in 5-5.5.3.2, open gratings are often covered by storage or dust stops and require sprinklers below them. Sprinklers must be of the in-rack type or be provided with shields to prevent wetting of the thermal element by sprinkler discharge from above, which would delay the operation of the sprinkler.

5-6.6* Clearance to Storage (Standard Pendent and Upright Spray Sprinklers).

The clearance between the deflector and the top of storage shall be 18 in. (457 mm) or greater.

Exception: Where other standards specify greater minimums, they shall be followed.

A-5-6.6 The 18-in. (457-mm) dimension is not intended to limit the height of shelving on a wall or shelving against a wall in accordance with 5-6.6. Where shelving is installed on a wall and is not directly below sprinklers, the shelves, including storage thereon, can extend above the level of a plane located 18 in. (457 mm) below ceiling sprinkler deflectors. Shelving, and any storage thereon, directly below the sprinklers cannot extend above a plane located 18 in. (457 mm) below the ceiling sprinkler deflectors.

The discharge pattern of upright and pendent spray sprinklers takes a parabolic shape as indicated in Figure A-5-5.5.1. To ensure the distribution of water over the area that the sprinkler was designed to protect, the spray pattern must encounter minimal obstructions. Rack storage fire tests, other tests with solid-piled storage, and field experience have shown that standard spray sprinklers are effective with a minimum 18-in. (457-mm) clearance.

Sprinklers installed near wall-mounted shelves or piled storage against a wall are not intended to be governed by this requirement. The clear space beneath the sprinkler is needed for the spray pattern to fully develop to allow proper wetting of the floor and not the wall.

The exception to 5-6.6 identifies the need to observe the clearance provisions that can be needed for special sprinklers or special design features. One example concerns the design provisions for miscellaneous storage found in Table 7-2.3.2.2 for Group A plastics. The design criteria vary as the clearance increases.

5-7 Sidewall Standard Spray Sprinklers

Rules for horizontal sidewall, pendent sidewall, and upright sidewall sprinklers are covered by Section 5-7. EC sidewall sprinklers are covered in Section 5-9. In addition to the provisions of section 5-7, sidewall sprinklers must also comply with all of the general requirements in Section 5-5.

5-7.1 General.

All requirements of Section 5-5 shall apply to sidewall standard spray sprinklers except as modified below.

5-7.2 Protection Areas per Sprinkler (Standard Sidewall Spray Sprinklers).

5-7.2.1 Determination of the Protection Area of Coverage.

5-7.2.1.1 The protection area of coverage per sprinkler *(A_s)* shall be determined as follows:

(a) *Along the Wall.* Determine the distance between sprinklers along the wall (or to the end wall or obstruction in the case of the end sprinkler on the branch line) upstream and downstream. Choose the larger of either twice the distance to the end wall or the distance to the next sprinkler. This dimension will be defined as *S*.

(b) *Across the Room.* Determine the distance from the sprinkler to the wall opposite the sprinklers or to the midpoint of the room where sprinklers are installed on two opposite walls *(see 5-7.3.1).* This dimension will be defined as *L*.

5-7.2.1.2 The protection area of the sprinkler shall be established by multiplying the S dimension by the L dimension, as follows: $A_s = S \times L$

5-7.2.2 Maximum Protection Area of Coverage. The maximum allowable protection area of coverage for a sprinkler *(A_s)* shall be in accordance with the value indicated in Table 5-7.2.2. In any case, the maximum area of coverage of a sprinkler shall not exceed 196 ft^2 (59.7 m^2).

Protection areas for sidewall sprinklers differ from those of standard spray upright and pendent sprinklers. Because of the characteristic horizontal or sideways travel of the sprinkler discharge, the water spray pattern of sidewall sprinklers has a greater tendency of being obstructed or otherwise impacted by furniture and other objects near the floor. The discharge of standard spray sprinklers is not normally as adversely effected by furniture and other objects on the floor. In addition, the discharge patterns of sidewall spray sprinklers are less uniform than those of standard spray sprinklers.

Table 5-7.2.2 indicates the maximum allowable protection areas. Beginning with the 1996 edition, the table was revised to indicate that sprinkler protection areas are not effected by the building construction above the ceiling. The sprinkler system is intended to control the fire below the ceiling if properly designed and installed. Therefore, NFPA 13 only considers the combustibility of the exposed surfaces when determining the maximum allowable distance between sprinklers and the maximum sprinkler protection area. Section 5-13 identifies the concealed spaces above ceilings that are not required to be protected with sprinklers.

Table 5-7.2.2 Protection Areas and Maximum Spacing (Standard Sidewall Spray Sprinkler)

	Light Hazard		Ordinary Hazard	
	Combustible Finish	**Noncombustible or Limited-Combustible Finish**	**Combustible Finish**	**Noncombustible or Limited-Combustible Finish**
Maximum distance along the wall *(S)*	14 ft	14 ft	10 ft	10 ft
Maximum room width *(L)*	12 ft	14 ft	10 ft	10 ft
Maximum protection area	120 ft^2	196 ft^2	80 ft^2	100 ft^2

For SI units, 1 ft = 0.3048 m; 1 ft^2 = 0.0929 m^2.

5-7.3 Sprinkler Spacing (Standard Sidewall Spray Sprinklers).

5-7.3.1 Maximum Distance Between Sprinklers.

5-7.3.1.1 The maximum distance permitted between sprinklers shall be based on the centerline distance between sprinklers on the branch line. The maximum distance shall be measured along the slope of the ceiling.

5-7.3.1.2 Sidewall spray sprinklers shall be installed along the length of a single wall of rooms or bays in accordance with the maximum spacing provisions of Table 5-7.2.2.

Exception No. 1: Sidewall sprinklers shall not be installed back-to-back without being separated by a continuous lintel or soffit.

Exception No. 2: Where the width of the room or bay exceeds the maximum allowed, up to 24 ft (7.32 m) for light hazard occupancy or 20 ft (6.1 m) for ordinary hazard occupancy sidewall sprinklers shall be provided on two opposite walls or sides of bays with spacing as required by Table 5-7.2.2.

Exception No. 3: Sidewall sprinklers shall be permitted to be installed on opposing or adjacent walls provided no sprinkler is located within the maximum protection area of another sprinkler.

The use of a lintel or soffit (see Exception No. 1) is necessary to ensure that the sprinkler closest to the fire operates and to minimize the possibility of sprinklers on both sides of the lintel operating unnecessarily. The baffle also serves to prevent the discharge pattern of the operating sprinkler from impinging on the operating element of an adjacent sprinkler. The baffle is an important feature, because sidewall sprinklers are actually designed to discharge a portion of their spray pattern behind the sprinkler.

Due to the spacing limitations of sidewall sprinklers, only certain size rooms can be effectively protected. Exception No. 2 permits sidewall sprinklers to be used in larger rooms under certain conditions, including a maximum room width dimension and the use of two lines of sprinklers on opposite walls. The maximum room width allowed for light hazard occupancies when standard sidewall sprinklers are located on opposite walls was reduced from 30 ft (9.1 m) to 24 ft (7.32 m) beginning with the 1996 edition. EC sidewall sprinklers provide better coverage for wide rooms and should be used for rooms over 24 ft (7.32 m) wide.

Testing has determined the minimum required distance to prevent cold soldering from an adjacent sidewall sprinkler when both sprinklers are located along the same wall. However, no testing has been done to determine the minimum distance when sprinklers are located on opposite or adjacent walls. Placing sidewall sprinklers outside the protection area of the adjacent sprinklers is probably conservative but eliminates the potential for the discharge from one sprinkler wetting the thermal element of another.

5-7.3.2 Maximum Distance from Walls. The distance from sprinklers to the end walls shall not exceed one-half of the allowable distance permitted between sprinklers as indicated in Table 5-7.2.2.

5-7.3.3 Minimum Distance from Walls. Sprinklers shall be located a minimum of 4 in. (102 mm) from an end wall. The distance from the wall to the sprinkler shall be measured perpendicular to the wall.

The dead-air space described in the commentary to 5-6.3.3 can also have a detrimental effect on sidewall sprinklers. Exhibit 5.20 indicates the application of 5-7.3.3.

5-7.3.4 Minimum Distance Between Sprinklers. Sprinklers shall be spaced not less than 6 ft (1.8 m) on center.

5-7.4 Deflector Position from Ceilings and Walls (Standard Sidewall Spray Sprinklers).

5-7.4.1 Distance Below Ceilings and from Walls.

5-7.4.1.1 Sidewall sprinkler deflectors shall be located not more than 6 in. (152 mm) or less than 4 in. (102 mm) from ceilings.

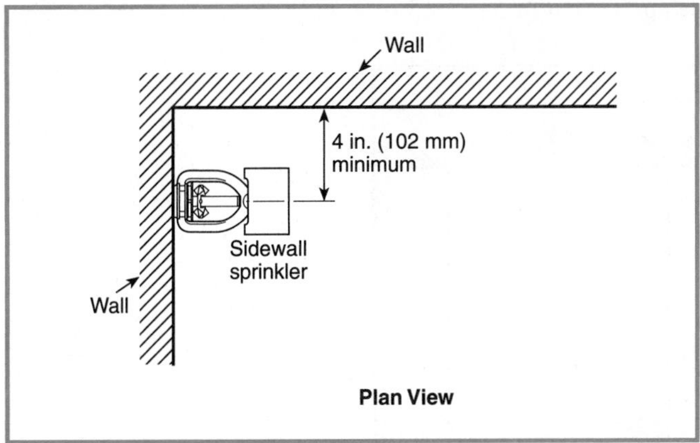

Exhibit 5.20 *Placement of sidewall sprinkler adjacent to an end wall.*

Exception: Horizontal sidewall sprinklers shall be permitted to be located in a zone 6 in. to 12 in. (152 mm to 305 mm) or 12 in. to 18 in. (305 mm to 457 mm) below noncombustible and limited-combustible ceilings where listed for such use.

Sidewall sprinklers should not be located in the dead-air space at the junction of the ceiling and the wall. The desirable sprinkler location is 4 in. to 6 in. (102 mm to 152 mm) from the ceiling to optimize performance. As the sprinkler is lowered to more than 6 in. (152 mm) from the ceiling, it is likely to take longer to operate. In addition, the ceiling has an impact on the discharge of a sidewall sprinkler, causing the spray pattern to change as the sprinkler is lowered. The sprinkler is allowed to be lowered as much as 18 in. (457 mm) when specifically tested and listed for such use.

5-7.4.1.2 Sidewall sprinkler deflectors shall be located not more than 6 in. (152 mm) or less than 4 in. (102 mm) from walls to which they are mounted.

Exception: Horizontal sidewall sprinklers are permitted to be located with their deflectors less than 4 in. (102 mm) from the wall on which they are mounted.

This exception allows horizontal sidewall sprinklers to be installed less than 4 in. (102 mm) from the wall to which they are mounted. This exception is similar to the exception to 5-6.4.1.1 for flush and concealed pendent sprinklers with respect to the position relative to the ceiling and allows the sprinkler to be mounted on the surface of a wall or soffit.

5-7.4.1.3 Sidewall sprinklers shall only be installed along walls, lintels, or soffits where the distance from the ceiling to the bottom of the lintel or soffit is at least 2 in. (51 mm) greater than the distances from the ceiling to sidewall sprinkler deflectors.

Exhibit 5.21 indicates the application of 5-7.4.1.3.

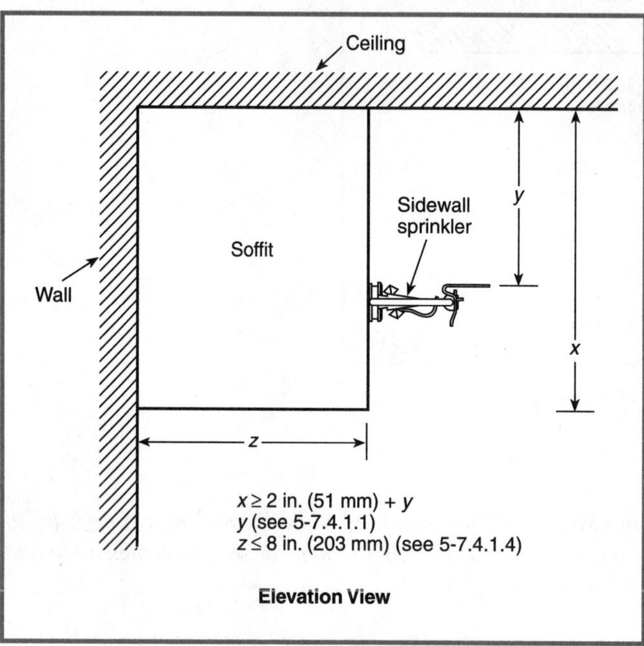

Exhibit 5.21 *Installation of sidewall sprinkler along a soffit.*

5-7.4.1.4 Where soffits are used for the installation of sidewall sprinklers, they shall not exceed 8 in. (203 mm) in width or projection from the wall.

Exception: Soffits shall be permitted to exceed 8 in. (203 mm) where additional sprinklers are installed below the soffit.

The evaluation of a sidewall sprinkler includes testing of its ability to project water to the back wall area of a soffit and wall. This ability decreases as the depth of the soffit becomes greater; thus, the 8-in. (203-mm) maximum soffit is viewed as a maximum practical width. See Exhibit 5.21. When installing sidewall sprinklers in a back-to-back configuration as described in 5-7.3.1.1, the width of the soffit should not exceed 16 in. (406 mm).

5-7.4.2 Deflector Orientation.

5-7.4.2.1 Deflectors of sprinklers shall be aligned parallel to ceilings or roofs.

5-7.4.2.2 Sidewall sprinklers, where installed under a sloped ceiling, shall be located at the high point of the slope and positioned to discharge downward along the slope.

When the ceiling is sloped, the sprinkler must be located on the high point of the slope, which is where heat from the fire collects. If sprinklers were located on any of the other three walls, a delay in sprinkler operation would occur.

5-7.5 Obstructions to Sprinkler Discharge (Standard Sidewall Spray Sprinklers).

5-7.5.1 Performance Objective.

Sidewall sprinklers are intended to be used in unobstructed construction under smooth, flat horizontal ceilings. The sprinklers are very sensitive to obstructions, such as fluorescent light fixtures or beams located near or along the ceiling, that would impair the horizontal throw of water needed to provide proper coverage. However, obstructions cannot be avoided in all cases, and provisions are given for locating sidewall sprinklers so as to reduce the effect of obstructions.

5-7.5.1.1 Sprinklers shall be located so as to minimize obstructions to discharge as defined in 5-5.5.2 and 5-5.5.3, or additional sprinklers shall be provided to ensure adequate coverage of the hazard.

5-7.5.1.2 Sidewall sprinklers shall be installed no closer than 4 ft (1.2 m) from light fixtures or similar obstructions. The distance between light fixtures or similar obstructions located more than 4 ft (1.2 m) from the sprinkler shall be in conformity with Table 5-7.5.1.2 and Figure 5-7.5.1.2.

This paragraph is analogous to the provisions for solid continuous obstructions for upright and pendent spray sprinklers (see 5-6.5.1.2) and is intended to ensure the sprinkler discharge falls below the obstruction. Table 5-7.5.1.2 is based on the discharge pattern falling away from the ceiling in a descending trajectory pattern. The farther the ceiling obstruction is from the wall, the higher the deflector can be positioned with respect to the ceiling.

Table 5-7.5.1.2 Positioning of Sprinklers to Avoid Obstructions (Standard Sidewall Spray Sprinklers)

Distance from Sidewall Sprinkler to Side of Obstruction *(A)*	Maximum Allowable Distance of Deflector above Bottom of Obstruction (in.) *(B)*
Less than 4 ft	0
4 ft to less than 5 ft	1
5 ft to less than 5 ft 6 in.	2
5 ft 6 in. to less than 6 ft	3
6 ft to less than 6 ft 6 in.	4
6 ft 6 in. to less than 7 ft	6
7 ft to less than 7 ft 6 in.	7
7 ft 6 in. to less than 8 ft	9
8 ft to less than 8 ft 6 in.	11
8 ft 6 in. or greater	14

For SI units, 1 in. = 25.4 mm; 1 ft = 0.3048 m.

Note: For *(A)* and *(B)*, refer to Figure 5-7.5.1.2.

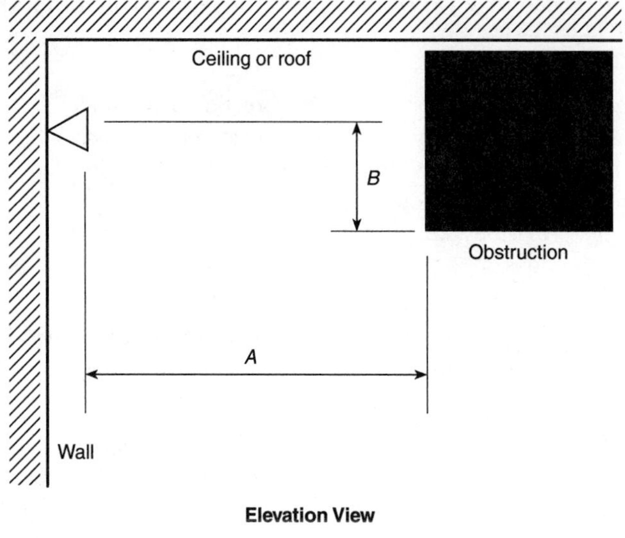

Figure 5-7.5.1.2 Positioning of sprinklers to avoid obstructions (standard sidewall spray sprinklers).

5-7.5.1.3 Obstructions projecting from the same wall as the one on which the sidewall sprinkler is mounted shall be in accordance with Table 5-7.5.1.3 and Figure 5-7.5.1.3.

Sidewall sprinkler spray distribution is adversely affected by obstructions projecting out from the same wall on which the sprinkler is mounted. The effect is similar to the problems caused by a beam parallel to the wall on which the sprinkler is mounted (see 5-7.5.1.2). The obstruction prevents the sprinkler discharge from reaching the hazard on the other side of the obstruction. Because the sprinkler throw along the wall on which it is mounted is shorter than the throw outward, the sprinkler can be located closer to the obstruction.

5-7.5.2 Obstructions to Sprinkler Discharge Pattern Development.

Sidewall sprinklers, like upright and pendent sprinklers, must be located to avoid obstructions that affect the discharge pattern development or prevent the discharge from reaching the protected hazard. Paragraphs 5-7.5.2 through 5-7.5.3 on obstructions are similar in content to those for upright and pendent sprinklers in 5-6.5.2 through 5-6.5.3. Rather than assume that these types of obstructions do not exist, these requirements were added to NFPA 13 to address what to do when these obstructions are encountered. Other types of sprinklers such as upright or pendent should be used if the obstruction cannot be relocated or if the sidewall sprinkler cannot be repositioned along the wall to avoid the obstruction.

Table 5-7.5.1.3 Positioning of Sprinklers to Avoid Obstructions Along the Wall (Standard Sidewall Spray Sprinklers)

Distance from Sidewall Sprinkler to Side of Obstruction *(A)*	Maximum Allowable Distance of Deflector above Bottom of Obstruction (in.) *(B)*
Less than 6 in.	1
6 in. to less than 1 ft	2
1 ft to less than 1 ft 6 in.	3
1 ft 6 in. to less than 2 ft	4½
2 ft to less than 2 ft 6 in.	5¾
2 ft 6 in. to less than 3 ft	7
3 ft to less than 3 ft 6 in.	8
3 ft 6 in. to less than 4 ft	9¼
4 ft to less than 4 ft 6 in.	10
4 ft 6 in. to less than 5 ft	11½
5 ft to less than 5 ft 6 in.	12¾
5 ft 6 in. to less than 6 ft	14
6 ft to less than 6 ft 6 in.	15
6 ft 6 in. to less than 7 ft	16¼
7 ft to less than 7 ft 6 in.	17½

For SI units, 1 in. = 25.4 mm; 1 ft = 0.3048 m.

Note: For *(A)* and *(B)*, refer to Figure 5-7.5.1.3.

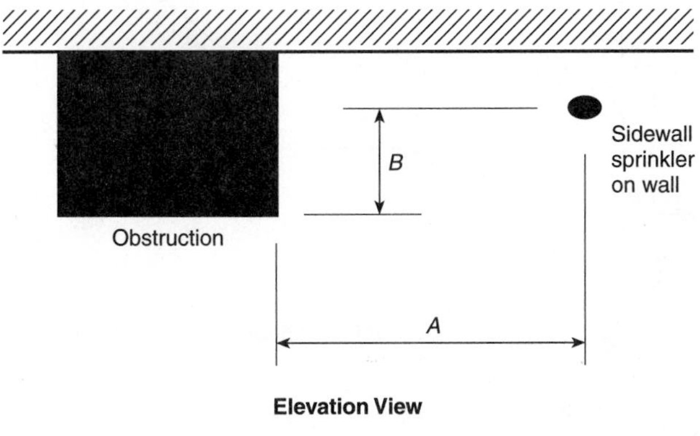

Figure 5-7.5.1.3 *Positioning of sprinklers to avoid obstructions along the wall (standard sidewall spray sprinklers).*

5-7.5.2.1* Continuous or noncontinuous obstructions less than or equal to 18 in. (457 mm) below the sprinkler deflector that prevent the pattern from fully developing shall comply with this section. Regardless of the rules of this section, solid continuous obstructions shall meet the requirements of 5-7.5.1.2.

A-5-7.5.2.1 The rules of 5-7.5.2.2 (known as the "Three Times Rule") have been written to apply to obstructions where the sprinkler can be expected to get water to both sides of the obstruction without allowing a significant dry shadow on the other side of the obstruction. This works for small noncontinuous obstructions and for continuous obstructions where the sprinkler can throw water over and under the obstruction, such as the bottom chord of an open truss or joist. For solid continuous obstructions, such as a beam, the Three Times Rule is ineffective since the sprinkler cannot throw water over and under the obstruction. Sufficient water must be thrown under the obstruction to adequately cover the floor area on the other side of the obstruction. To ensure this, compliance with the rules of 5-7.5.1.2 is necessary.

5-7.5.2.2 Sprinklers shall be positioned such that they are located at a distance three times greater than the maximum dimension of an obstruction up to a maximum of 24 in. (609 mm) (e.g., truss webs and chords, pipe, columns, and fixtures). Sidewall sprinklers shall be positioned in accordance with Figure 5-7.5.2.2 where obstructions are present.

Exception No. 1: Piping to which sidewall sprinklers are directly attached.

Exception No. 2: Sprinklers positioned with respect to obstructions in accordance with 5-7.5.1.2 and 5-7.5.1.3.

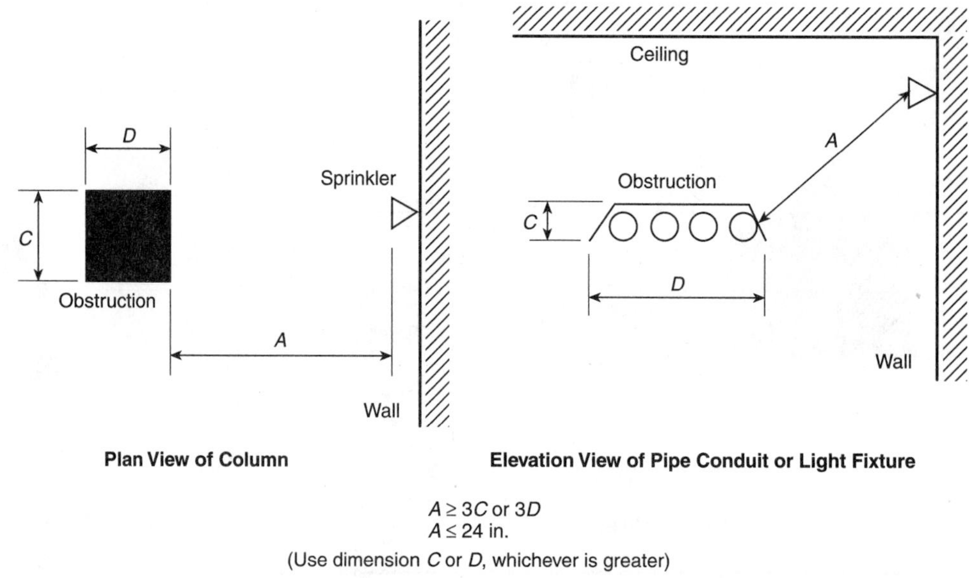

Figure 5-7.5.2.2 *Minimum distance from obstruction (standard sidewall spray sprinkler).*

5-7.5.2.3 Suspended or Floor-Mounted Vertical Obstructions. The distance from sprinklers to privacy curtains, free-standing partitions, room dividers, and similar obstructions in light hazard occupancies shall be in accordance with Table 5-7.5.2.3 and Figure 5-7.5.2.3.

Table 5-7.5.2.3 Suspended or Floor-Mounted Obstructions (Standard Sidewall Spray Sprinklers)

Horizontal Distance *(A)*	**Minimum Vertical Distance below Deflector (in.) *(B)***
6 in. or less	3
More than 6 in. to 9 in.	4
More than 9 in. to 12 in.	6
More than 12 in. to 15 in.	8
More than 15 in. to 18 in.	9½
More than 18 in. to 24 in.	12½
More than 24 in. to 30 in.	15½
More than 30 in.	18

For SI units, 1 in. = 25.4 mm.

Note: For *(A)* and *(B)*, refer to Figure 5-7.5.2.3.

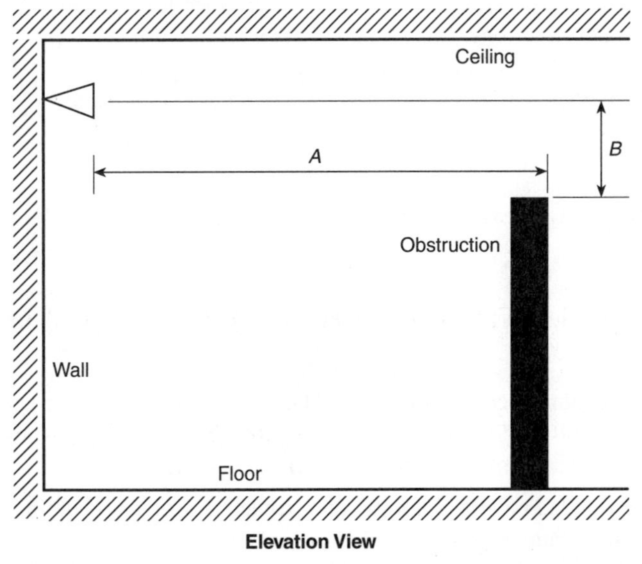

***Figure 5-7.5.2.3** Suspended or floor-mounted obstructions (standard sidewall spray sprinklers).*

5-7.5.3* Obstructions that Prevent Sprinkler Discharge from Reaching the Hazard.

A-5-7.5.3 See A-5-5.5.3.

5-7.5.3.1 Continuous or noncontinuous obstructions that interrupt the water discharge in a horizontal plane more than 18 in. (457 mm) below the sprinkler deflector in a manner to limit the distribution from reaching the protected hazard shall comply with this section.

5-7.5.3.2 Sprinklers shall be installed under fixed obstructions over 4 ft (1.2 m) wide such as ducts, decks, open grate flooring cutting tables, and overhead doors.

Exception: Obstructions that are not fixed in place such as conference tables.

5-7.6 Clearance to Storage (Standard Sidewall Spray Sprinklers).

The clearance between the deflector and the top of storage shall be 18 in. (457 mm) or greater.

5-8 Extended Coverage Upright and Pendent Spray Sprinklers

In recent years, the number of EC sprinklers has increased dramatically. Most manufacturers have models suitable for ordinary and light hazard occupancies. EC sprinklers may also become available for extra hazard occupancies and areas of high-piled storage. Many of the provisions in Section 5-8 are similar to those found in Section 5-6. The main difference is that EC sprinklers have larger protection areas and flatter distribution patterns than do standard spray pendent and upright sprinklers and, therefore, require greater separation distances from obstructions. Unless the rationale for the EC sprinkler requirements differs from those for standard upright and pendent sprinklers, commentary regarding related items in Section 5-6 is not repeated in Section 5-8 commentary.

5-8.1 General.

All requirements of Section 5-5 shall apply to extended coverage upright and pendent sprinklers except as modified below.

5-8.2 Protection Areas per Sprinkler (Extended Coverage Upright and Pendent Spray Sprinklers).

5-8.2.1* Determination of the Protection Area of Coverage. The protection area of coverage (A_s) for extended coverage sprinklers shall be not less than that prescribed by the listing. Listing dimensions shall be even-numbered square protection areas as shown in Table 5-8.2.1.

A-5-8.2.1 The protection area for extended coverage upright and pendent sprinklers is defined in the listing of the sprinkler as a maximum square area. Listing information is presented in even 2-ft (0.6-m) increments up to 20 ft (6.1 m). When a sprinkler is selected for an application, its area of coverage must be equal to or greater than both the length and width of the hazard area. For example, if the hazard to be protected is a room 13½ ft (4.1 m) wide and 17½ ft (5.3 m) long as indicated in Figure A-5-8.2.1, a sprinkler that is listed to protect an area of

Table 5-8.2.1 Protection Areas and Maximum Spacing (Extended Coverage Upright and Pendent Spray Sprinklers)

	Light Hazard		Ordinary Hazard		Extra Hazard		High-Pile Storage	
Construction Type	Protection Area (ft^2)	Spacing (ft)	Protection Area (ft^2)	Spacing (ft)	Protection Area (ft^2)	Spacing (ft)	Protection Area (ft^2)	Spacing (ft)
Unobstructed	400	20	400	20	—	—	—	—
	324	18	324	18	—	—	—	—
	256	16	256	16	—	—	—	—
listed for such use)	—	—	196	14	196	14	196	14
	—	—	144	12	144	12	144	12
Obstructed non-	400	20	400	20	—	—	—	—
combustible	324	18	324	18	—	—	—	—
(when specifically	256	16	256	16	—	—	—	—
	—	—	196	14	196	14	196	14
	—	—	144	12	144	12	144	12
Obstructed combustible	N/A	N/A	N/A	N/A	N/A	N/A	N/A	N/A

For SI units, 1 ft = 0.3048 m; 1 ft^2 = 0.0929 m^2.

18 ft × 18 ft (5.5 m × 5.5 m) must be selected. The flow used in the calculations is then selected as the flow required by the listing for the selected coverage.

Restrictions are placed on the allowable spacing of upright and pendent EC sprinklers to promote interchangeability. The protection area must be uniform—that is, a square pattern—so that the sprinkler can be installed without regard to the orientation of the frame arms. Limiting the spacing to even-numbered increments allows a variety of protection areas and promotes interchangeability among models from different manufacturers.

EC sprinklers are not allowed under combustible obstructed construction. Testing has shown that extending the sprinkler spacing under this type of construction allows the fire to ignite the ceiling, increasing the fire challenge and the potential for system failure.

5-8.2.2 Maximum Protection Area of Coverage. The maximum allowable area of coverage for a sprinkler (A_s) shall be in accordance with the value indicated in Table 5-8.2.1. In any case, the maximum area of coverage of a sprinkler shall not exceed 400 ft^2 (37.1 m^2).

The maximum area of coverage for an EC sprinkler limits the area that would be unprotected should a sprinkler be severely obstructed or fail to operate. The limit of 196 ft^2 (18.2 m^2) for extra hazard and high-piled storage considers that fires in these

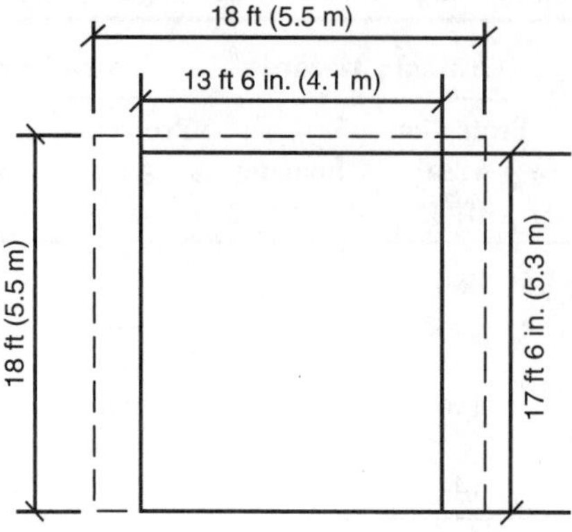

Figure A-5-8.2.1 Determination of protection area of coverage for extended coverage upright and pendent sprinklers.

occupancies develop much more rapidly than fires in light and ordinary hazard occupancies and can overtax adjacent sprinklers should one sprinkler not operate as anticipated.

5-8.3 Sprinkler Spacing (Extended Coverage Upright and Pendent Spray Sprinklers).

5-8.3.1 Maximum Distance Between Sprinklers. The maximum distance permitted between sprinklers shall be based on the centerline distance between sprinklers on the branch line or on adjacent branch lines. The maximum distance shall be measured along the slope of the ceiling. The maximum distance permitted between sprinklers shall comply with Table 5-8.2.1.

5-8.3.2 Maximum Distance from Walls. The distance from sprinklers to walls shall not exceed one-half of the allowable distance permitted between sprinklers as indicated in Table 5-8.2.1. The distance from the wall to the sprinkler shall be measured perpendicular to the wall. Where walls are angled or irregular, the maximum horizontal distance between a sprinkler and any point of floor area protected by that sprinkler shall not exceed 0.75 times the allowable distance permitted between sprinklers.

5-8.3.3 Minimum Distance from Walls. Sprinklers shall be located a minimum of 4 in. (102 mm) from a wall.

Exception: Where sprinklers have been listed for distances less than 4 in. (102 mm) from a wall, they shall be permitted.

5-8.3.4 Minimum Distance Between Sprinklers. Sprinklers shall be spaced not less than 8 ft (2.4 m) on center.

Exception: Sprinklers shall be permitted to be placed less than 8 ft (2.4 m) on center where the following conditions are satisfied:

(a) Baffles shall be installed and located midway between sprinklers and arranged to protect the actuating elements.

(b) Baffles shall be of noncombustible or limited-combustible material that will stay in place before and during sprinkler operation.

(c) Baffles shall be not less than 8 in. (203 mm) wide and 6 in. (152 mm) high. The tops of baffles shall extend between 2 in. and 3 in. (51 mm and 76 mm) above the deflectors of upright sprinklers. The bottoms of baffles shall extend downward to a level at least even with the deflectors of pendent sprinklers. (See A-5-13.4.)

5-8.4 Deflector Position (Extended Coverage Upright and Pendent Spray Sprinklers).

5-8.4.1 Distance Below Ceilings.

5-8.4.1.1 Under unobstructed construction, the distance between the sprinkler deflector and the ceiling shall be a minimum of 1 in. (25.4 mm) and a maximum of 12 in. (305 mm).

Exception No. 1: Ceiling-type sprinklers (concealed, recessed, and flush types) shall be permitted to have the operating element above the ceiling and the deflector located nearer to the ceiling where installed in accordance with their listing.

Exception No. 2: Where sprinklers are listed for use under other ceiling construction features or for different distances, they shall be permitted to be installed in accordance with their listing.

5-8.4.1.2 Under obstructed construction, the sprinkler deflector shall be located 1 in. to 6 in. (25.4 mm to 152 mm) below the structural members and a maximum distance of 22 in. (559 mm) below the ceiling/ roof deck.

Exception No. 1: Sprinklers shall be permitted to be installed with the deflector at or above the bottom of the structural member to a maximum of 22 in. (559 mm) below the ceiling/roof deck where the sprinkler is installed in conformance with 5-6.5.1.2.

Exception No. 2: Where sprinklers are installed in each bay of obstructed construction, deflectors shall be a minimum of 1 in. (25.4 mm) and a maximum of 12 in. (305 mm) below the ceiling.

Exception No. 3: Where sprinklers are listed for use under other ceiling construction features or for different distances, they shall be permitted to be installed in accordance with their listing.

5-8.4.1.3* Sprinklers under or near the peak of a roof or ceiling shall have deflectors located not more than 3 ft (0.9 m) vertically down from the peak. *[See Figures 5-6.4.1.3(a) and 5-6.4.1.3(b).]*

A-5-8.4.1.3 Saw-toothed roofs have regularly spaced monitors of saw tooth shape, with the nearly vertical side glazed and usually arranged for venting. Sprinkler placement is limited to a maximum of 3 ft (0.91 m) down the slope from the peak because of the effect of venting on sprinkler sensitivity.

The installation of EC sprinklers under peaked roofs rarely occurs due to the limitations on ceiling slope outlined in 5-4.3. When EC sprinklers are specifically listed for slopes of up to 4 in./ft, 5-8.4.1.3 may come into play if the maximum allowable sprinkler spacing is utilized.

5-8.4.2 Deflector Orientation. Deflectors of sprinklers shall be aligned parallel to ceilings or roofs.

5-8.5 Obstructions to Sprinkler Discharge (Extended Coverage Upright and Pendent Spray Sprinklers).

The provisions covering the location and placement of EC sprinklers are similar to those for standard spray sprinklers. Because their spray pattern is flatter and their throw is longer than standard spray sprinklers, EC sprinklers must be placed further away from obstructions than standard spray sprinklers. In addition, all obstructions must be considered, regardless of the occupancy hazard. The obstruction of the bottom chord must be considered even when the sprinkler is located in the centerline of a joist or truss. Also see commentary to Section 5-6.

5-8.5.1 Performance Objective.

5-8.5.1.1 Sprinklers shall be located so as to minimize obstructions to discharge as defined in 5-8.5.2 and 5-8.5.3, or additional sprinklers shall be provided to ensure adequate coverage of the hazard.

5-8.5.1.2 Sprinklers shall be arranged to comply with 5-5.5.2, Table 5-8.5.1.2, and Figure 5-8.5.1.2(a).

Exception No. 1: Sprinklers shall be permitted to be spaced on opposite sides of obstructions not exceeding 4 ft (1.2 m) in width provided the distance from the centerline of the obstruction to the sprinklers does not exceed one-half the allowable distance permitted between sprinklers.

Exception No. 2: Obstructions located against the wall and that are not over 30 in. (762 mm) in width shall be permitted to be protected in accordance with Figure 5-8.5.1.2(b).

5-8.5.2 Obstructions to Sprinkler Discharge Pattern Development.

5-8.5.2.1* Continuous or noncontinuous obstructions less than or equal to 18 in. (457 mm) below the sprinkler deflector that prevent the pattern from fully developing shall comply with 5-8.5.2. Regardless of the rules of this section, solid continuous obstructions shall meet the requirements of 5-8.5.1.2.

A-5-8.5.2.1 The rules of 5-8.5.2.2 (known as the "Four Times Rule") have been written to apply to obstructions where the sprinkler can be expected to get water to both sides of the obstruction without allowing a significant dry shadow on the other side of the obstruction. This works for small noncontinuous obstructions and for continuous obstructions where the sprinkler can throw water over and under the obstruction, such as the bottom chord of an open truss or joist. For solid continuous obstructions, such as a beam, the Four Times Rule is ineffective since the sprinkler cannot throw water over and under the obstruction. Sufficient water must be thrown under the obstruction to adequately cover the floor area on the other side of the obstruction. To ensure this, compliance with the rules of 5-8.5.1.2 is necessary.

Table 5-8.5.1.2 Position of Sprinklers to Avoid Obstructions to Discharge (Extended Coverage Upright and Pendent Spray Sprinklers)

Distance from Sprinklers to Side of Obstruction *(A)*	Maximum Allowable Distance of Deflector above Bottom of Obstruction (in.) *(B)*
Less than 1 ft	0
1 ft to less than 1 ft 6 in.	0
1 ft 6 in. to less than 2 ft	1
2 ft to less than 2 ft 6 in.	1
2 ft 6 in. to less than 3 ft	1
3 ft to less than 3 ft 6 in.	3
3 ft 6 in. to less than 4 ft	3
4 ft to less than 4 ft 6 in.	5
4 ft 6 in. to less than 5 ft	7
5 ft to less than 5 ft 6 in.	7
5 ft 6 in. to less than 6 ft	7
6 ft to less than 6 ft 6 in.	9
6 ft 6 in. to less than 7 ft	11
7 ft and greater	14

For SI units, 1 in. = 25.4 mm; 1 ft = 0.3048 m.

Note: For *(A)* and *(B)*, refer to Figure 5-8.5.1.2(a).

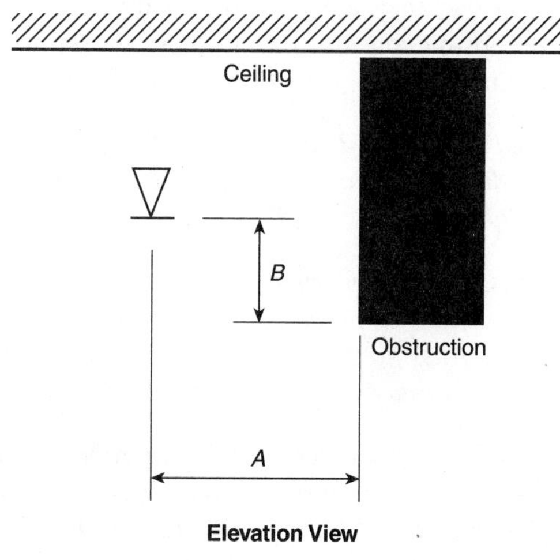

Figure 5-8.5.1.2(a) *Position of sprinklers to avoid obstructions to discharge (extended coverage upright and pendent spray sprinklers).*

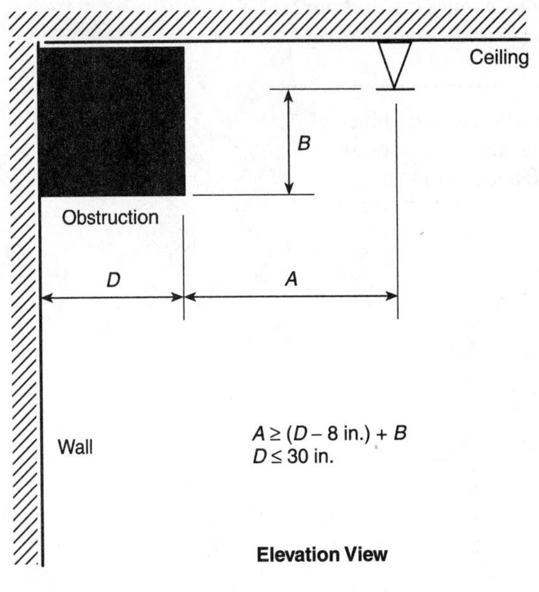

Figure 5-8.5.1.2(b) Obstructions against walls (extended coverage upright and pendent spray sprinklers).

5-8.5.2.2 Sprinklers shall be positioned such that they are located at a distance four times greater than the maximum dimension of an obstruction up to a maximum of 36 in. (.91 m) (e.g., truss webs and chords, pipe, columns, and fixtures). *(See Figure 5-8.5.2.2.)*

Exception No. 1: Sprinklers shall be permitted to be spaced on opposite sides of the obstruction provided the distance from the centerline of the obstruction to the sprinklers does not exceed one-half the allowable distance between sprinklers.

Exception No. 2: Where the obstruction consists of open trusses 20 in. (0.51 m) or greater apart [24 in. (0.61 m) on center], sprinklers shall be permitted to be located one-half the distance between the obstruction created by the truss provided that truss chords do not exceed 4 in. (101 mm) in width and web members do not exceed 1 in. (25.4 mm) in width.

Exception No. 3: Sprinklers shall be permitted to be installed on the centerline of a truss or bar joist or directly above a beam provided that the truss chord or beam dimension is not more than 8 in. (203 mm) and the sprinkler deflector is located at least 6 in. (152 mm) above the structural member. The sprinkler shall be positioned at a distance four times greater than the maximum dimension of the web members away from the web members.

Exception No. 4: Piping to which an upright sprinkler is directly attached less than 3 in. (75 mm) in diameter.

Exception No. 5: Piping to which pendent and sidewall sprinklers are directly attached.

Exception No. 6: Sprinklers positioned with respect to obstructions in accordance with 5-8.5.1.2.

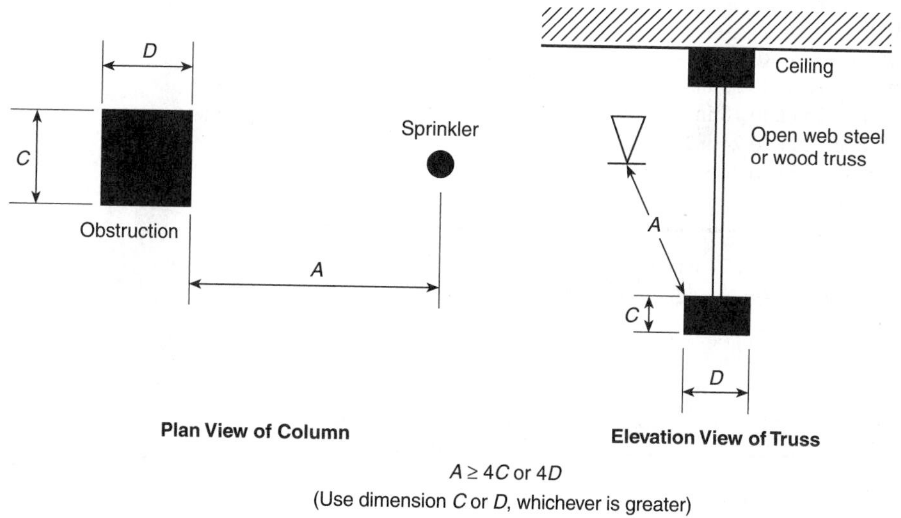

Figure 5-8.5.2.2 *Minimum distance from obstruction (extended coverage upright and pendent spray sprinklers).*

5-8.5.2.3 Suspended or Floor-Mounted Vertical Obstructions. The distance from sprinklers to privacy curtains, free-standing partitions, room dividers, and similar obstructions in light hazard occupancies shall be in accordance with Table 5-8.5.2.3 and Figure 5-8.5.2.3.

5-8.5.3* Obstructions that Prevent Sprinkler Discharge from Reaching the Hazard. Continuous or noncontinuous obstructions that interrupt the water discharge in a horizontal plane more than 18 in. (457 mm) below the sprinkler deflector in a manner to limit the distribution from reaching the protected hazard shall comply with 5-8.5.3.

A-5-8.5.3 See A-5-5.5.3.

5-8.5.3.1 Sprinklers shall be installed under fixed obstructions over 4 ft (1.2 m) wide such as ducts, decks, open grate flooring, cutting tables, and overhead doors.

Exception: Obstructions that are not fixed in place such as conference tables.

5-8.5.3.2 Sprinklers installed under open gratings shall be of the intermediate level/rack storage type or otherwise shielded from the discharge of overhead sprinklers.

Table 5-8.5.2.3 Suspended or Floor-Mounted Obstructions (Extended Coverage Upright and Pendent Spray Sprinklers)

Horizontal Distance *(A)*	**Minimum Vertical Distance below Deflector (in.) *(B)***
6 in. or less	3
More than 6 in. to 9 in.	4
More than 9 in. to 12 in.	6
More than 12 in. to 15 in.	8
More than 15 in. to 18 in.	9½
More than 18 in. to 24 in.	12½
More than 24 in. to 30 in.	15½
More than 30 in.	18

For SI units, 1 in. = 25.4 mm.

Note: For *(A)* and *(B)*, refer to Figure 5-8.5.2.3.

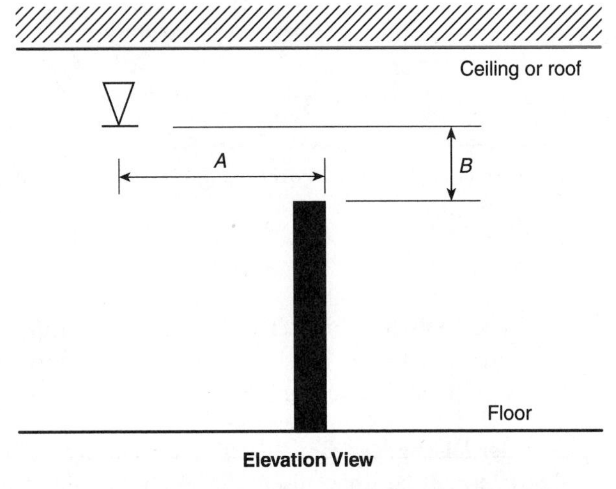

Figure 5-8.5.2.3 *Suspended or floor-mounted obstructions (extended coverage upright and pendent spray sprinklers).*

5-8.6 Clearance to Storage (Extended Coverage Upright and Pendent Spray Sprinklers).

The clearance between the deflector and the top of storage shall be 18 in. (457 mm) or greater.

Exception: Where other standards specify greater minimums, they shall be followed.

5-9 Extended Coverage Sidewall Spray Sprinklers

Most of the provisions in Section 5-9 are similar to those found in Section 5-7. The main difference is that EC sidewall sprinklers have larger protection areas and flatter distribution patterns than standard spray sidewall sprinklers and, therefore, require greater separation distances from obstructions. Unless the rationale for the EC sprinkler requirements differs from those for standard sidewall sprinklers, commentary regarding related items in Section 5-7 is not repeated in Section 5-9 commentary.

5-9.1 General.

All requirements of Section 5-5 shall apply to extended coverage sidewall spray sprinklers except as modified below.

5-9.2 Protection Areas per Sprinkler (Extended Coverage Sidewall Spray Sprinklers).

5-9.2.1* Determination of the Protection Area of Coverage. The protection area of coverage per sprinkler (A_s) for extended coverage sidewall sprinklers shall be not less than that prescribed by the listing. Listing dimensions shall be in 2-ft (0.61-m) increments up to 28 ft (8.5 m).

A-5-9.2.1 The protection area for extended coverage sidewall spray sprinklers is defined in the listing of the sprinkler as a maximum square or rectangular area. Listing information is presented in even 2-ft (0.65-m) increments up to 28 ft (9 m) for extended coverage sidewall spray sprinklers. When a sprinkler is selected for an application, its area of coverage must be equal to or greater than both the length and width of the hazard area. For example, if the hazard to be protected is a room 14½ ft (4.4 m) wide and 20⅔ ft (6.3 m) long as indicated in Figure A-5-9.2.1, a sprinkler that is listed to protect an area of 16 ft × 22 ft (4.9 m × 6.7 m) must be selected. The flow used in the calculations is then selected as the flow required by the listing for the selected coverage.

Restrictions are placed on the allowable spacing of EC sidewall sprinklers to maintain a degree of order and to promote interchangeability. Limiting the spacing to even-numbered increments allows a variety of protection areas and promotes interchangeability between models from different manufacturers.

5-9.2.2 Maximum Protection Area of Coverage. The maximum allowable protection area of coverage for a sprinkler (A_s) shall be in accordance with the value indicated in Table 5-9.2.2. In any case, the maximum area of coverage of a sprinkler shall not exceed 400 ft^2 (37.1 m^2).

5-9.3 Sprinkler Spacing (Extended Coverage Sidewall Spray Sprinklers).

5-9.3.1 Maximum Distance Between Sprinklers.

5-9.3.1.1 The maximum distance permitted between sprinklers shall be based on the centerline distance between sprinklers on the branch line along the wall.

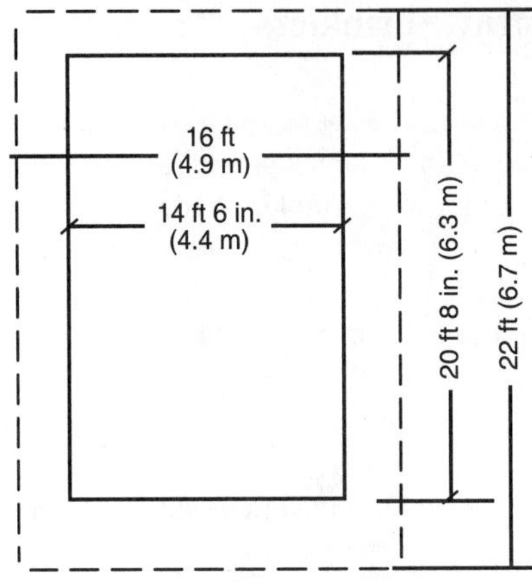

Figure A-5-9.2.1 Determination of protection area of coverage for extended coverage sidewall sprinklers.

Table 5-9.2.2 Protection Area and Maximum Spacing for Extended Coverage Sidewall Sprinklers

	Light Hazard				Ordinary Hazard			
	Protection Area		Spacing		Protection Area		Spacing	
Construction Type	ft²	m²	ft	m	ft²	m²	ft	m
Unobstructed, smooth, flat	400	37.2	28	8.5	400	37.2	24	7.3

5-9.3.1.2 Sidewall spray sprinklers shall be installed along the length of a single wall of a room.

Exception No. 1: Sidewall sprinklers shall not be installed back-to-back without being separated by a continuous lintel soffit or baffle.

Exception No. 2: Sidewall sprinklers shall be permitted to be installed on opposing or adjacent walls provided no sprinkler is located within the maximum protection area of another sprinkler.

5-9.3.2 Maximum Distance from Walls. The distance from sprinklers to the end walls shall not exceed one-half of the allowable distance permitted between sprinklers as indicated in Table 5-9.2.2.

5-9.3.3 Minimum Distance from Walls. Sprinklers shall be located a minimum of 4 in. (102 mm) from an end wall. The distance from the wall to the sprinkler shall be measured perpendicular to the wall.

5-9.3.4 Minimum Distance Between Sprinklers. No sprinklers shall be located within the maximum protection area of any other sprinkler.

5-9.4 Deflector Position from Ceilings and Walls (Extended Coverage Sidewall Spray Sprinklers).

5-9.4.1 Distance Below Ceilings and from Walls to Which Sprinklers are Mounted.

5-9.4.1.1 Sidewall sprinkler deflectors shall be located not more than 6 in. (152 mm) nor less than 4 in. (102 mm) from ceilings.

Exception: Horizontal sidewall sprinklers are permitted to be located in a zone 6 in. to 12 in. (152 mm to 305 mm) or 12 in. to 18 in. (305 mm to 457 mm) below noncombustible or limited-combustible ceilings where listed for such use.

5-9.4.1.2 Sidewall sprinkler deflectors shall be located not more than 6 in. (229 mm) or less than 4 in. (102 mm) from walls on which they are mounted.

Exception: Horizontal sidewall sprinklers shall be permitted to be located with their deflectors less than 4 in. (102 mm) from the wall on which they are mounted.

5-9.4.1.3 Sidewall sprinklers shall only be installed along walls, lintels, or soffits where the distance from the ceiling to the bottom of the lintel or soffit is at least 2 in. (51 mm) greater than the distances from the ceiling to sidewall sprinkler deflectors.

5-9.4.1.4 Where soffits are used for the installation of sidewall sprinklers, they shall not exceed 8 in. (203 mm) in width or projection from the wall.

Exception: Soffits shall be permitted to exceed 8 in. (203 mm) where additional sprinklers are installed below the soffit.

5-9.4.2 Deflector Orientation. Deflectors of sprinklers shall be aligned parallel to ceilings or roofs.

5-9.4.2.1 Sidewall sprinklers, where installed under a sloped ceiling, shall be located at the high point of the slope and positioned to discharge downward along the slope.

Exception: Unless specifically listed for other ceiling configurations.

The exception was added to the 1999 edition because listings have been achieved for sidewall sprinklers located along the sloping wall of a sloped ceiling room.

5-9.5 Obstructions to Sprinkler Discharge (Extended Coverage Sidewall Spray Sprinklers).

5-9.5.1 Performance Objective.

5-9.5.1.1 Sprinklers shall be located so as to minimize obstructions to discharge as defined in 5-5.5.2 and 5-5.5.3, or additional sprinklers shall be provided to ensure adequate coverage of the hazard.

5-9.5.1.2 Sidewall sprinklers shall be installed no closer than 8 ft (2.4 m) from light fixtures or similar obstructions. The distance between light fixtures or similar obstructions located more than 8 ft (2.4 m) from the sprinkler shall be in conformity with Table 5-9.5.1.2 and Figure 5-9.5.1.2.

Table 5-9.5.1.2 Positioning of Sprinklers to Avoid Obstructions (Extended Coverage Sidewall Sprinklers)

Distance from Sidewall Sprinkler to Side of Obstruction *(A)*	Maximum Allowable Distance of Deflector above Bottom of Obstruction (in.) *(B)*
8 ft to less than 10 ft	1
10 ft to less than 11 ft	2
11 ft to less than 12 ft	3
12 ft to less than 13 ft	4
13 ft to less than 14 ft	6
14 ft to less than 15 ft	7
15 ft to less than 16 ft	9
16 ft to less than 17 ft	11
17 ft or greater	14

For SI units, 1 in. = 25.4 mm; 1 ft = 0.3048 m.
Note: For *(A)* and *(B)*, refer to Figure 5-9.5.1.2.

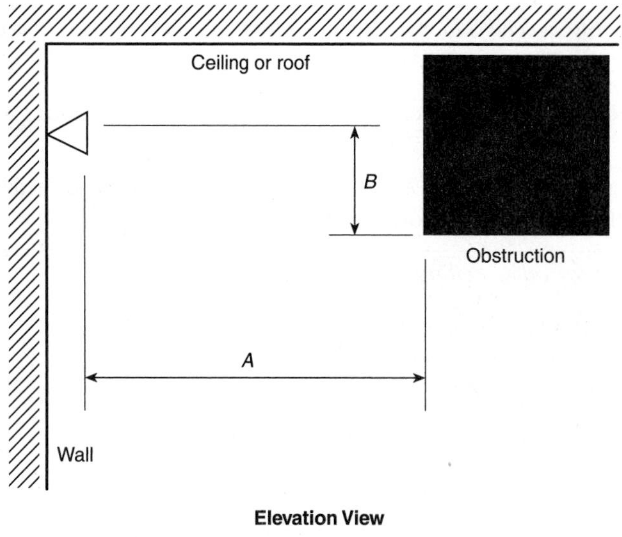

***Figure 5-9.5.1.2** Positioning of sprinklers to avoid obstructions (extended coverage sidewall sprinklers).*

5-9.5.2 Obstructions to Sprinkler Discharge Pattern Development.

5-9.5.2.1* Continuous or noncontinuous obstructions less than or equal to 18 in. (457 mm) below the sprinkler deflector that prevent the pattern from fully developing shall comply with 5-9.5.2. Regardless of the rules of this section, solid continuous obstructions shall meet the requirements of 5-9.5.1.2.

A-5-9.5.2.1 The rules of 5-9.5.2.2 (known as the "Four Times Rule") have been written to apply to obstructions where the sprinkler can be expected to get water to both sides of the obstruction without allowing a significant dry shadow on the other side of the obstruction. This works for small noncontinuous obstructions and for continuous obstructions where the sprinkler can throw water over and under the obstruction, such as the bottom chord of an open truss or joist. For solid continuous obstructions, such as a beam, the Four Times Rule is ineffective since the sprinkler cannot throw water over and under the obstruction. Sufficient water must be thrown under the obstruction to adequately cover the floor area on the other side of the obstruction. To ensure this, compliance with the rules of 5-9.5.1.2 is necessary.

A

5-9.5.2.2 Sprinklers shall be positioned such that they are located at a distance four times greater than the maximum dimension of the obstruction to a maximum of 36 in. (0.91 m) from the sprinkler (e.g., truss webs and chords, pipe, columns, and fixtures). Sidewall sprinklers shall be positioned in accordance with Figure 5-9.5.2.2 when obstructions are present.

Exception: Sprinklers positioned with respect to obstructions in accordance with 5-9.5.1.2.

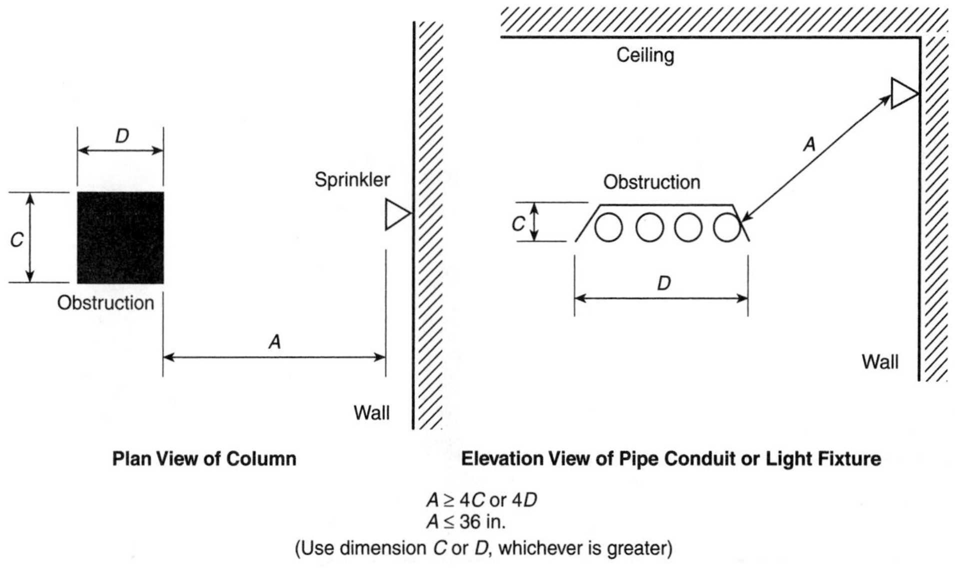

Figure 5-9.5.2.2 *Minimum distance from obstruction (extended coverage sidewall).*

5-9.5.2.3 Suspended or Floor-Mounted Vertical Obstructions. The distance from sprinklers to privacy curtains, free-standing partitions, room dividers, and similar obstructions in light hazard occupancies shall be in accordance with Table 5-9.5.2.3 and Figure 5-9.5.2.3.

Table 5-9.5.2.3 Suspended or Floor-Mounted Obstructions (Extended Coverage Sidewall Sprinklers)

Horizontal Distance *(A)*	Minimum Allowable Distance below Deflector (in.) *(B)*
6 in. or less	3
More than 6 in. to 9 in.	4
More than 9 in. to 12 in.	6
More than 12 in. to 15 in.	8
More than 15 in. to 18 in.	9½
More than 18 in. to 24 in.	12½
More than 24 in. to 30 in.	15½
More than 30 in.	18

For SI units, 1 in. = 25.4 mm.

Note: For *(A)* and *(B),* refer to Figure 5-9.5.2.3.

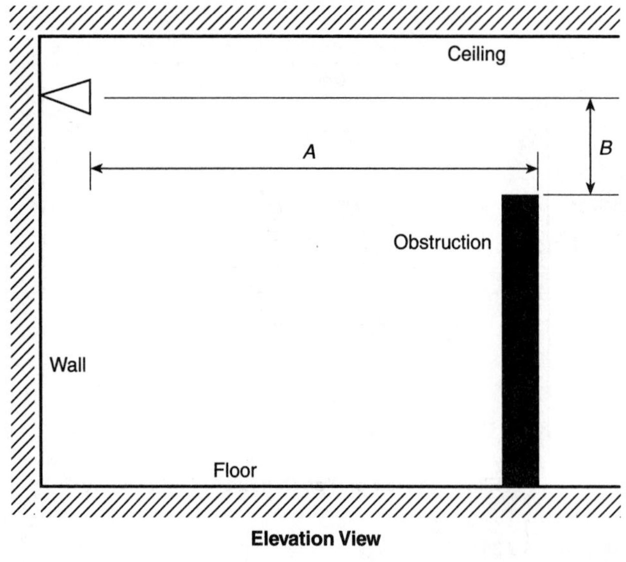

Figure 5-9.5.2.3 *Suspended or floor-mounted obstructions (extended coverage sidewall sprinklers).*

5-9.5.3* Obstructions that Prevent Sprinkler Discharge from Reaching the Hazard. Continuous or noncontinuous obstructions that interrupt the water discharge in a horizontal plane more than 18 in. (457 mm) below the sprinkler deflector in a manner to limit the distribution from reaching the protected hazard shall comply with 5-9.5.3.

A-5-9.5.3 See A-5-5.5.3.

5-9.5.3.1 Sprinklers shall be installed under fixed obstructions over 4 ft (1.2 m) wide such as ducts, decks, open grate flooring, cutting tables, and overhead doors.

Exception: Obstructions that are not fixed in place such as conference tables.

5-10 Large Drop Sprinklers

Many of the installation requirements for large drop sprinklers are similar to those found in Section 5-6. Unless the rationale for large drop requirements differs from those for standard upright and pendent sprinklers, commentary from related items in Section 5-6 is not repeated in Section 5-10 commentary.

5-10.1 General.

All requirements of Section 5-5 shall apply to large drop sprinklers except as modified below.

5-10.2* Protection Areas per Sprinkler (Large Drop Sprinklers).

A-5-10.2 Tests involving areas of coverage over 100 ft^2 (9.3 m^2) for large drop sprinklers are limited in number, and use of areas of coverage over 100 ft^2 (9.3 m^2) should be carefully considered.

5-10.2.1 Determination of the Protection Area of Coverage. The protection area of coverage per sprinkler *(A_s)* shall be determined in accordance with 5-5.2.1.

5-10.2.2 Maximum Protection Area of Coverage. The maximum allowable protection area of coverage for a sprinkler *(A_s)* shall be in accordance with the value indicated in Table 5-10.2.2. In any case, the maximum area of coverage of any sprinkler shall not exceed 130 ft^2 (12.9 m^2).

Tests involving areas of coverage over 100 ft^2 (9.3 m^2) for large drop sprinklers are limited in number. Limiting the area of coverage for large drop sprinklers to 100 ft^2 (9.3 m^2) for all constructions types is common practice. Careful consideration should be given when designing large-drop sprinkler installations with protection areas exceeding 100 ft^2 (9.3 m^2).

5-10.2.3 Minimum Protection Area of Coverage. The minimum allowable protection area of coverage for a sprinkler *(A_s)* shall be not less than 80 ft^2 (7.4 m^2).

The protection area for large drop sprinklers is limited to 80 ft^2 (7.4 m^2) to prevent "skipping." Skipping occurs when the sprinklers do not operate in a normal pattern

Table 5-10.2.2 Protection Areas and Maximum Spacing for Large Drop Sprinklers

Construction Type	Protection Area		Maximum Spacing	
	ft^2	m^2	ft	m
Noncombustible unobstructed	130	12.1	12	3.7
Noncombustible obstructed	130	12.1	12	3.7
Combustible unobstructed	130	12.1	12	3.7
Combustible obstructed	100	9.3	10	3.1
Rack storage applications	100	9.3	10	3.1

and when sprinklers some distance from the fire operate before those sprinklers closest to the fire. Air movement and general cooling of temperatures at the ceiling cause the skipping phenomenon. When skipping occurs, the fire continues to grow, resulting in a larger, more difficult fire to control. Limiting the sprinkler protection area, as well as the minimum distance between sprinklers set forth in 5-10.3.4, minimizes the possibility of skipping.

5-10.3 Sprinkler Spacing (Large Drop Sprinklers).

Spacing requirements were determined by testing at various sprinkler spacings and discharge pressures on the basis of the following parameters:

(1) Do not exceed the maximum allowable distance between sprinklers to ensure sufficient overlapping of sprinkler discharge patterns for adequate penetration of the fire plume.
(2) Maintain sufficient separation of sprinklers to prevent sprinkler skipping.
(3) Maintain sufficient cooling at ceiling level.

The importance of satisfying these three parameters cannot be overstated. These criteria are so closely intertwined that evaluating one without considering the effect on the others is difficult. Certain combinations of spacing dimensions, when used with the discharge pressures and fire hazards contemplated in 5-10.3, result in skipping, loss of sprinkler water penetration, and excessively high ceiling temperatures. If any of these adverse conditions occur, the adequacy of the sprinkler protection can be seriously compromised.

5-10.3.1* Maximum Distance Between Sprinklers. The distance between sprinklers shall be limited to not more than 12 ft (3.7 m) between sprinklers, as shown in Table 5-10.2.2.

Exception: Under obstructed combustible construction, the maximum distance shall be limited to 10 ft (3 m).

A-5-10.3.1 It is important that sprinklers in the immediate vicinity of the fire center not skip, and this requirement imposes certain restrictions on the spacing.

5-10.3.2 Maximum Distance from Walls. The distance from sprinklers to walls shall not exceed one-half of the allowable distance permitted between sprinklers as indicated in Table 5-10.2.2.

5-10.3.3 Minimum Distance from Walls. Sprinklers shall be located a minimum of 4 in. (102 mm) from a wall.

5-10.3.4 Minimum Distance Between Sprinklers. Sprinklers shall be spaced not less than 8 ft (2.4 m) on center.

5-10.4 Deflector Position (Large Drop Sprinklers).

5-10.4.1* Distance Below Ceilings.

A-5-10.4.1 If all other factors are held constant, the operating time of the first sprinkler will vary exponentially with the distance between the ceiling and deflector. At distances greater than 7 in. (178 mm), for other than open wood joist construction, the delayed operating time will permit the fire to gain headway, with the result that substantially more sprinklers operate. At distances less than 7 in. (178 mm), other effects occur. Changes in distribution, penetration, and cooling nullify the advantage gained by faster operation. The net result again is increased fire damage accompanied by an increase in the number of sprinklers operated. The optimum clearance between deflectors and ceiling is, therefore, 7 in. (178 mm). For open wood joist construction, the optimum clearance between deflectors and the bottom of joists is 3½ in. (89 mm).

5-10.4.1.1 Under unobstructed construction, the distance between the sprinkler deflector and the ceiling shall be a minimum of 6 in. (152 mm) and a maximum of 8 in. (203 mm).

5-10.4.1.2 Under obstructed construction, the distance between the sprinkler deflector and the ceiling shall be a minimum of 6 in. (152 mm) and a maximum of 12 in. (305 mm).

Exception No. 1: Under wood joist or composite wood joist construction, the sprinklers shall be located 1 in. to 6 in. (25.4 mm to 152 mm) below the structural members to a maximum distance of 22 in. (559 mm) below the ceiling/roof or deck.

Exception No. 2: Deflectors of sprinklers under concrete tee construction with stems spaces less than 7½ ft (2.3 m) but more than 3 ft (0.9 m) on centers shall, regardless of the depth of the tee, be permitted to be located at or above a horizontal plane 1 in. (25.4 mm) below the bottom of the stems of the tees and shall comply with Table 5-10.5.1.2.

Exception No. 2, which has been a long-standing exception for standard spray sprinklers, was added to the 1999 edition for large drop sprinklers, because a large drop sprinkler is a standard response sprinkler with operating features similar to those of standard spray sprinklers. Therefore, the positioning rules pertaining to sprinkler activation should remain uniform.

5-10.4.2 Deflector Orientation. Deflectors of sprinklers shall be aligned parallel to ceilings or roofs.

5-10.5* Obstructions to Sprinkler Discharge (Large Drop Sprinklers).

A-5-10.5 To a great extent, large drop sprinklers rely on direct attack to gain rapid control of both the burning fuel and ceiling temperatures. Therefore, interference with the discharge pattern and obstructions to the distribution should be avoided.

Large drop sprinklers are more sensitive to obstructions than standard spray pendent and upright sprinklers. Large drop sprinklers "attack" the fire source and rely less on prewetting of combustibles than do standard spray pendent and upright sprinklers. The discharge from large drop sprinklers contains larger water droplets. These droplets have greater momentum, which allows them to penetrate the fire plumes typically generated by the kinds of fires experienced in warehouse and storage operations. Any obstructions that interrupt the discharge reduce the amount of water that reaches the seat of the fire, limiting the sprinkler's effectiveness and allowing the fire to grow larger and become more difficult to control. As a result, the obstruction provisions for large drop sprinklers are more restrictive than for standard spray pendent and upright sprinklers.

5-10.5.1 Performance Objective.

5-10.5.1.1 Sprinklers shall be located so as to minimize obstructions to discharge as defined in 5-5.5.2 and 5-5.5.3, or additional sprinklers shall be provided to ensure adequate coverage of the hazard.

5-10.5.1.2 Sprinklers shall be arranged to comply with 5-5.5.2, Table 5-10.5.1.2, and Figure 5-10.5.1.2.

Exception: Where positioned on apposite sides of the obstruction.

The exception was added to the 1999 edition to formally permit the installation of large drop sprinklers on opposite sides of an obstruction.

5-10.5.2 Obstructions to Sprinkler Discharge Pattern Development.

5-10.5.2.1* Continuous or noncontinuous obstructions less than or equal to 36 in. (914 mm) below the sprinkler deflector that prevent the pattern from fully developing shall comply with 5-10.5.2. Regardless of the rules of this section, solid continuous obstructions shall meet the requirements of 5-10.5.1.2.

A-5-10.5.2.1 The rules of 5-10.5.2.2 (known as the "Three Times Rule") have been written to apply to obstructions where the sprinkler can be expected to get water to both sides of the obstruction without allowing a significant dry shadow on the other side of the obstruction. This works for small noncontinuous obstructions and for continuous obstructions where the sprinkler can throw water over and under the obstruction, such as the bottom chord of an open truss or joist. For solid continuous obstructions, such as a beam, the Three Times Rule is ineffective since the sprinkler cannot throw water over and under the obstruction. Sufficient water must be thrown under the obstruction to adequately cover the floor area on the other side of the obstruction. To ensure this, compliance with the rules of 5-10.5.1.2 is necessary.

Table 5-10.5.1.2 Positioning of Sprinklers to Avoid Obstructions to Discharge (Large Drop Sprinkler)

Distance from Sprinkler to Side of Obstruction *(A)*	Maximum Allowable Distance of Deflector above Bottom of Obstruction (in.) *(B)*
Less than 1 ft	0
1 ft to less than 1 ft 6 in.	1½
1 ft 6 in. to less than 2 ft	3
2 ft to less than 2 ft 6 in.	5½
2 ft 6 in. to less than 3 ft	8
3 ft to less than 3 ft 6 in.	10
3 ft 6 in. to less than 4 ft	12
4 ft to less than 4 ft 6 in.	15
4 ft 6 in. to less than 5 ft	18
5 ft to less than 5 ft 6 in.	22
5 ft 6 in. to less than 6 ft	26
6 ft	31

For SI units, 1 in. = 25.4 mm; 1 ft = 0.3048 m.
Note: For *(A)* and *(B)*, refer to Figure 5-10.5.1.2.

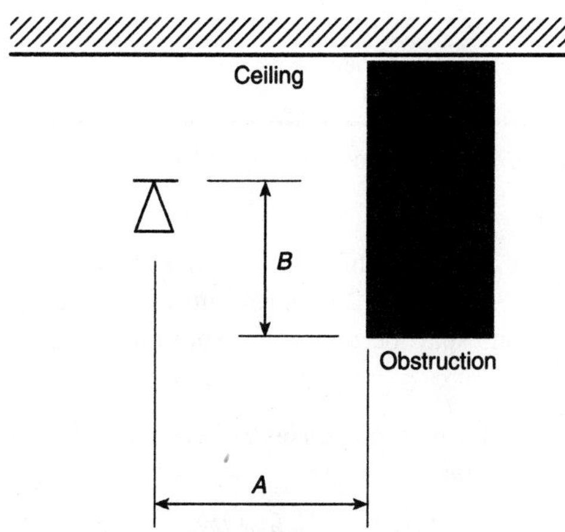

***Figure 5-10.5.1.2** Positioning of sprinklers to avoid obstructions to discharge (large drop sprinkler).*

The discharge pattern of a large drop sprinkler is more narrow than that of a standard spray pendent and upright sprinkler. To allow the discharge pattern to develop and the sprinkler patterns to overlap, a 36-in. (914-mm) zone below the sprinkler must be evaluated for potential obstructions to discharge pattern development.

5-10.5.2.2 For obstructions 8 in. (203 mm) or less in width, sprinklers shall be positioned such that they are located at least a distance three times greater than the maximum dimension of the obstruction from the sprinkler (e.g., webs and chord members, pipe, columns, and fixtures). Sprinklers shall be positioned in accordance with Figure 5-10.5.2.2 where obstructions are present.

Exception: Sprinklers positioned with respect to obstructions in accordance with 5-10.5.1.2.

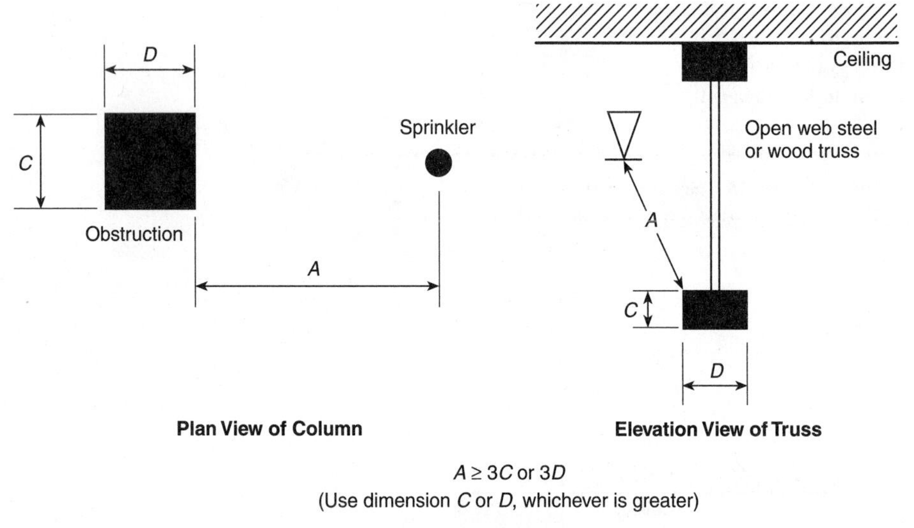

Figure 5-10.5.2.2 Minimum distance from obstruction (large drop sprinkler).

5-10.5.2.3 Where branch lines are larger than 2 in. (51 mm), the sprinkler shall be supplied by a riser nipple to elevate the sprinkler 13 in. (330 mm) for 2½ in. (64-mm) pipe and 15 in. (380 mm) for 3-in. (76-mm) pipe. These dimensions shall be measured from the centerline of the pipe to the deflector.

Exception No. 1: This provision shall not apply where the sprinklers are offset horizontally a minimum of 12 in. (305 mm) from the pipe.

Exception No. 2: Piping to which the sprinkler is directly attached less than 2 in. (51 mm) in diameter.

When an upright sprinkler is installed directly on the branch line, the branch line becomes an obstruction. As the branch line pipe size is increased, the "pipe shadow," or area below the sprinkler that is affected, increases. Because large drop sprinklers attack the fire, this obstruction can have a significant impact if the fire is located directly below the sprinkler and is a concern when the branch line piping is larger than 2 in. (51 mm).

Many systems require larger piping to meet the hydraulic design requirements of the large drop sprinkler. Raising the sprinkler on a riser nipple reduces the angle of obstruction by limiting the area affected by the obstruction to no more than would be created by a smaller branch line.

Another approach to the problem is taken in Exception No. 1 to 5-10.5.2.3. If the sprinkler is offset at least 12 in. (305 mm), the discharge from the sprinkler can flow under the branch line. This position eliminates the obstruction or limits the impact of the obstruction caused by the branch line so that it does not pose a significant problem.

5-10.5.3* Obstructions that Prevent Sprinkler Discharge from Reaching the Hazard. Continuous or noncontinuous obstructions that interrupt the water discharge in a horizontal plane below the sprinkler deflector in a manner to limit the distribution from reaching the protected hazard shall comply with 5-10.5.3.

A-5-10.5.3 See A-5-5.5.3.

5-10.5.3.1 Sprinklers shall be positioned with respect to fluorescent lighting fixtures, ducts, and obstructions more than 24 in. (610 mm) wide and located entirely below the sprinklers so that the minimum horizontal distance from the near side of the obstruction to the center of the sprinkler is not less than the value specified in Table 5-10.5.3.1. *(See Figure 5-10.5.3.1.)*

Because large drop sprinklers are more sensitive to obstructions, any obstruction more than 24 in. (610 mm) wide and less than 24 in. (610 mm) below the sprinkler must be located so that the sprinkler discharge pattern falls under it. The intent is to minimize obstruction to the lateral distribution from individual sprinklers.

Table 5-10.5.3.1 Obstruction Entirely Below the Sprinkler (Large Drop Sprinkler)

Distance of Deflector above Bottom of Obstruction *(B)*	Minimum Distance to Side of Obstruction (ft) *(A)*
Less than 6 in.	1½
6 in. to less than 12 in.	3
12 in. to less than 18 in.	4
18 in. to less than 24 in.	5
24 in. to less than 30 in.	5½
30 in. less than 36 in.	6

For SI units, 1 in. = 25.4 mm; 1 ft = 0.3048 m.
Note: For *(A)* and *(B)*, refer to Figure 5-10.5.3.1.

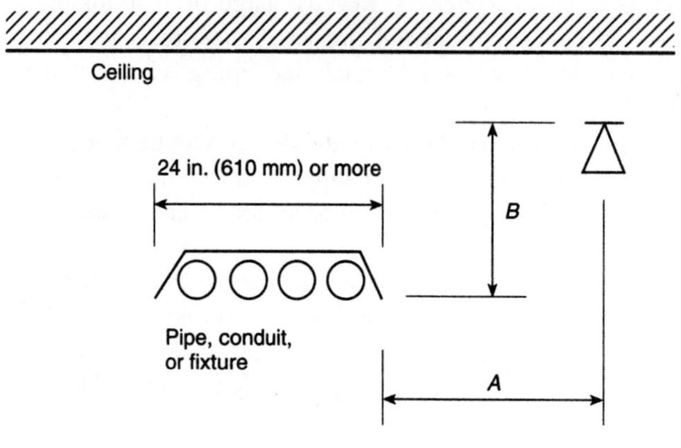

Figure 5-10.5.3.1 Obstruction entirely below the sprinkler (large drop sprinkler).

5-10.5.3.2 Sprinklers installed under open gratings shall be shielded from the discharge of overhead sprinklers.

5-10.5.3.3 Where the bottom of the obstruction is located 24 in. (610 mm) or more below the sprinkler deflectors, the following shall occur:

(1) Sprinklers shall be positioned so that the obstruction is centered between adjacent sprinklers. *(See Figure 5-10.5.3.3.)*

(2) The obstruction shall be limited to a maximum width of 24 in. (610 mm). *(See Figure 5-10.5.3.3.)*

Exception: Where the obstruction is greater than 24 in. (610 mm) wide, one or more lines of sprinklers shall be installed below the obstruction.

(3) The obstruction shall not extend more than 12 in. (305 mm) to either side of the midpoint between sprinklers. *(See Figure 5-10.5.3.3.)*

Exception: Where the extensions of the obstruction exceed 12 in. (305 mm), one or more lines of sprinklers shall be installed below the obstruction.

(4) At least 18 in. (457 mm) clearance shall be maintained between the top of storage and the bottom of the obstruction. *(See Figure 5-10.5.3.3.)*

At distances more than 24 in. (610 mm) below the sprinklers, obstructions interfere with the distribution from more than one sprinkler due to characteristics of the sprinkler's discharge pattern. These provisions provide for sufficient overlap of discharge patterns

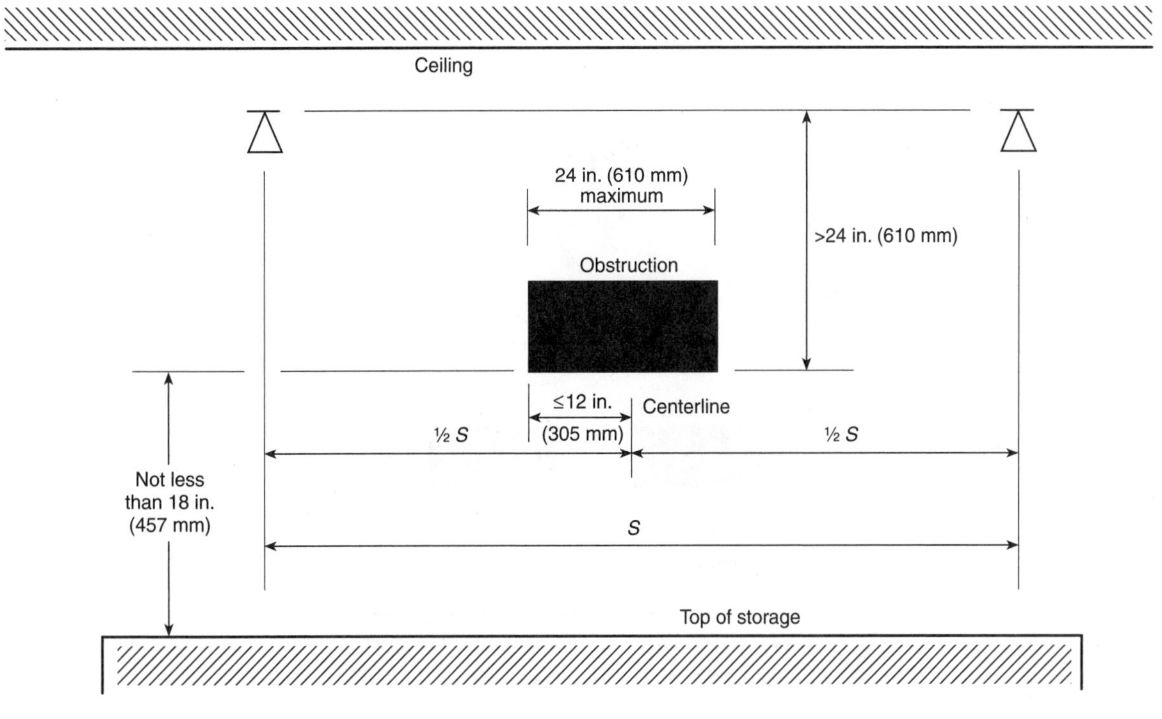

Figure 5-10.5.3.3 *Obstruction more than 24 in. (610 mm) below the sprinkler (large drop sprinkler).*

from adjacent sprinklers, which ensures that adequate water distribution is available to penetrate the fire plume.

5-10.5.3.4 In the special case of an obstruction running parallel to and directly below a branch line, the following shall occur:

(1) The sprinkler shall be located at least 36 in. (914 mm) above the top of the obstruction. *(See Figure 5-10.5.3.4.)*
(2) The obstruction shall be limited to a maximum width of 12 in. (305 mm). *(See Figure 5-10.5.3.4.)*
(3) The obstruction shall be limited to a maximum extension of 6 in. (152 mm) to either side of the centerline of the branch line. *(See Figure 5-10.5.3.4.)*

The obstruction to distribution as expected under the conditions specified in 5-10.5.3.4 would not be significantly greater than that which would occur due to pipe shadow. It is important to follow all of these obstruction provisions, because relatively minor obstructions can reduce the effectiveness of a large drop sprinkler. The use of large drop sprinklers can provide a more effective means of protecting high-challenge storage fire hazards, often without the use of in-rack sprinklers. However, large drop sprinklers are obstruction sensitive and require special attention to be paid to the details of these installation requirements.

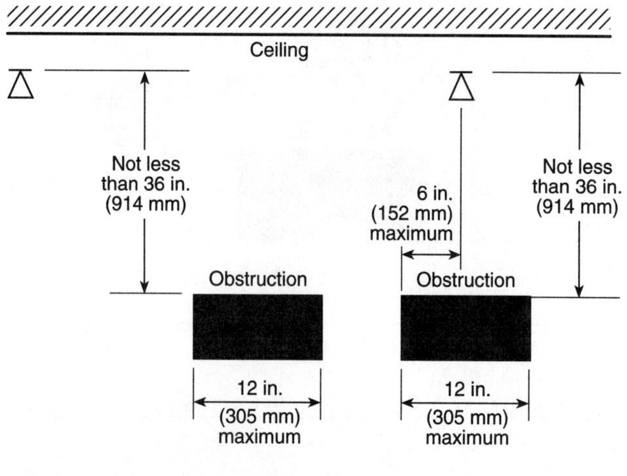

Figure 5-10.5.3.4 *Obstruction more than 36 in. (914 mm) below the sprinkler (large drop sprinkler).*

5-10.6 Clearance to Storage (Large Drop Sprinklers).

The clearance between the deflector and the top of storage shall be 36 in. (914 mm) or greater.

At clearances of less than 36 in. (914 mm), large drop sprinklers installed within the allowable spacing guidelines do not provide sufficient overlapping of sprinkler discharge patterns between adjacent sprinklers.

5-11 Early Suppression Fast-Response Sprinklers

5-11.1 General.

All requirements of Section 5-5 shall apply except as modified below.

The spacing and location requirements of ESFR sprinklers are designed to ensure that the sprinklers operate while the fire size is sufficiently small so that the fire can be suppressed. ESFR sprinklers are also designed to minimize the impact of obstructions so that the sprinkler discharge can reach the seat of the fire. Spacing requirements are determined on the basis of the following parameters:

(a) Maintain sufficient actual delivered density (see Chapter 1) for various sprinkler spacings and clearances between sprinklers and the top of storage (Scenarios consisting of a fire ignition source centered among four sprinklers, centered between two sprinklers, and located directly under one sprinkler are evaluated to ensure that early fire suppression will be achieved. Ideally, an unshielded fire should be completely extinguished. In the worst case, where the seat of the fire is shielded from direct sprinkler discharge, the fire's heat release rate should be severely diminished with the likelihood of the fire rekindling or spreading nearly nonexistent.)

(b) Maintain sufficient separation of sprinklers to prevent skipping

(c) Maintain sufficient cooling at ceiling level

5-11.2 Protection Areas per Sprinkler (Early Suppression Fast-Response Sprinklers).

5-11.2.1 Determination of the Protection Area of Coverage. The protection area of coverage per sprinkler (A_s) shall be determined in accordance with 5-5.2.1.

5-11.2.2 Maximum Protection Area of Coverage. The maximum allowable protection area of coverage for a sprinkler (A_s) shall be in accordance with the value indicated in Table 5-11.2.2. In any case, the maximum area of coverage of any sprinkler shall not exceed 100 ft^2 (9.3 m^2).

Exception: It shall be permitted to deviate from the maximum sprinkler spacing to eliminate obstructions created by trusses and bar joists by moving a sprinkler along the branch line a maximum of 1 ft (0.31 m) from its allowable spacing provided coverage for that sprinkler does not exceed 110 ft^2 (10.2 m^2) and the average spacing for the moved sprinkler and the adjacent sprinkler do not exceed 100 ft^2 (9.3 m^2). Adjacent branch lines shall maintain the same pattern. In no case shall the distance between sprinklers exceed 12 ft (3.7 m).*

Table 5-11.2.2 Protection Areas and Maximum Spacing of ESFR Sprinklers

	Ceiling/Roof Heights Up To 30 ft (9.1 m)				Ceiling/Roof Heights Over 30 ft (9.1 m)			
	Protection Area		Spacing		Protection Area		Spacing	
Construction Type	ft^2	m	ft	m	ft^2	m	ft	m
Noncombustible unobstructed	100	9.3	12	3.7	100	9.3	10	3.1
Noncombustible obstructed	100	9.3	12	3.7	100	9.3	10	3.1
Combustible unobstructed	100	9.3	12	3.7	100	9.3	10	3.1
Combustible obstructed	N/A		N/A		N/A		N/A	

Table 5-11.2.2 indicates that ESFR sprinklers cannot be used under obstructed combustible construction. However, 5-4.6.3 identifies construction that can be considered in this category.

The type of construction documented in 5-4.6.3(2) is identified in A-1-4.6(a) under obstructed construction as beam and girder construction. When the beams of this type of construction are made of wood, with or without a wood plank or other wood ceiling above, 5-4.6.3(2) allows ESFR sprinklers to be installed under such construction. Where used under this type of construction, the maximum protection area for ESFR sprinklers should not exceed 100 ft^2 (9.3 m^2).

The type of construction documented in 5-4.6.3(3) is identified in A-1-4.6(c) under obstructed construction as panel construction. Because panel construction is specifically allowed, the maximum protection area should not exceed 100 ft^2 (9.3 m^2). No other forms of obstructed combustible construction are allowed.

A-5-11.2.2 Exception. See Figure A-5-11.2.2.

ESFR sprinklers are required to be located a minimum distance from the nearest edge of any bottom chord of a bar joist or truss, as indicated in 5-11.5.2 and 5-11.5.3. If the spacing of sprinklers is not an even multiple of the chord spacing, then the standard sprinkler spacing will not meet this requirement. The exception to 5-11.2.2 allows that line of sprinklers to be moved 1 ft (0.31 m) to eliminate the obstruction caused by the bottom chord as long as the adjacent sprinkler on the branch line is not moved and the protection area restrictions outlined in the exception are met.

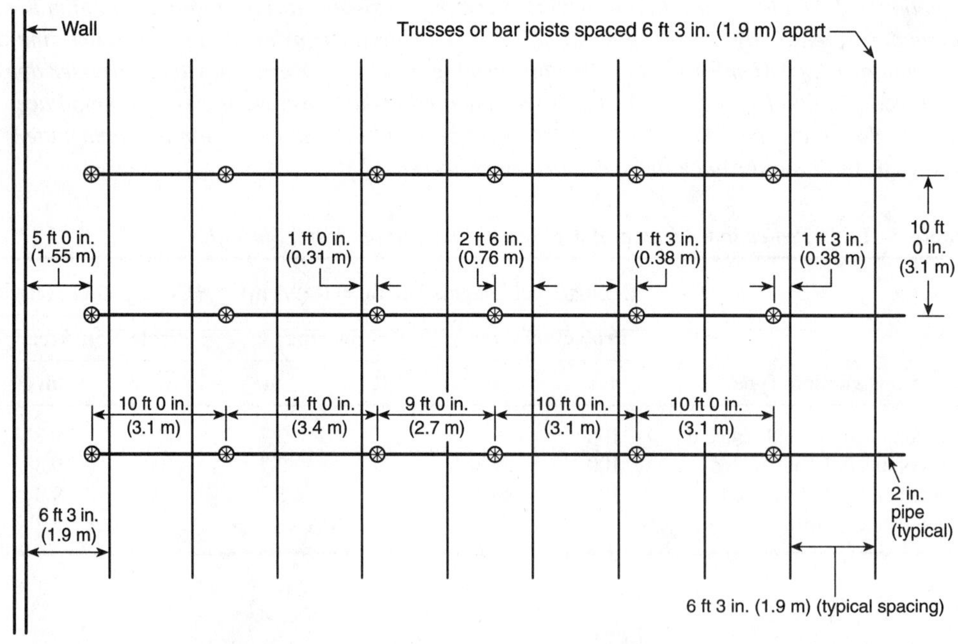

***Figure A-5-11.2.2** ESFR sprinkler spacing within trusses and bar joists.*

5-11.2.3 Minimum Protection Area of Coverage. The minimum allowable protection area of coverage for a sprinkler *(A_s)* shall be not less than 80 ft^2 (7.4 m^2).

The protection area for ESFR sprinklers is limited to 80 ft^2 (7.4 m^2) to prevent the possibility of skipping. (See commentary following 5-10.2.3 for an explanation of skipping.) Limiting the minimum distance between sprinklers as required in 5-11.3.4 also minimizes the potential for skipping.

The direct relationship between the spacing and the skipping of ESFR sprinklers is not known. With ESFR sprinkler installations, the first few sprinklers should suppress the fire and limit the potential for skipping. However, no tests have been conducted using ESFR sprinklers with less than 80-ft^2 (7.4-m^2) protection areas. If skipping should occur, the likely result is an overtaxing of the ESFR design and a failure of the sprinklers to suppress the fire.

5-11.3 Sprinkler Spacing (Early Suppression Fast-Response Sprinklers).

5-11.3.1 Maximum Distance Between Sprinklers. The distance between sprinklers shall be limited to not more than 12 ft (3.7 m) between sprinklers as shown in Table 5-11.2.2.

Exception No. 1: ESFR sprinklers used in buildings with storage heights greater than 25 ft (7.6 m) and ceiling heights greater than 30 ft (9.1 m) shall not be spaced more than 10 ft (3 m) between sprinklers.

Exception No. 2: It shall be permitted to deviate from the maximum sprinkler spacing to eliminate obstructions created by trusses and bar joists by moving a sprinkler along the branch line a maximum of 1 ft (0.31 m) from its allowable spacing provided coverage for that sprinkler does not exceed 110 ft^2 (10.2 m^2) and the average spacing for the moved sprinkler and the adjacent sprinkler do not exceed 100 ft^2 (9.3 m^2). Adjacent branch lines shall maintain the same pattern. In no case shall the distance between sprinklers exceed 12 ft (3.7 m).*

A-5-11.3.1 Exception No. 2. See Figure A-5-11.2.2.

ESFR sprinklers must be spaced so that patterns have sufficient overlap. As the storage height increases, so does the strength of the fire plume. As indicated by Exception No. 1, additional pattern overlap for storage heights over 25 ft (7.6 m) and ceiling heights over 30 ft (9.1 m) is required for suppression.

Exception No. 2 is similar to the exception to 5-11.2.2 concerning protection area limitations. The exception allows a line of sprinklers to be moved an additional 1 ft (0.31 m) to eliminate the obstruction caused by the bottom chord as long as the adjacent line is not moved and the maximum spacing between sprinklers does not exceed 12 ft (3.7 m).

5-11.3.2 Maximum Distance from Walls. The distance from sprinklers to walls shall not exceed one-half of the allowable distance permitted between sprinklers as indicated in Table 5-11.2.2.

5-11.3.3 Minimum Distance from Walls. Sprinklers shall be located a minimum of 4 in. (102 mm) from a wall.

5-11.3.4 Minimum Distance Between Sprinklers. Sprinklers shall be spaced not less than 8 ft (2.4 m) on center.

5-11.4 Deflector Position (Early Suppression Fast-Response Sprinklers).

5-11.4.1 Distance Below Ceilings. Pendent sprinklers with a nominal K-factor of 14 shall be positioned so that deflectors are a maximum 14 in. (356 mm) and a minimum 6 in. (152 mm) below the ceiling. Pendent sprinklers with a nominal K-factor of 25.2 shall be positioned so that deflectors are a maximum 18 in. (457 mm) and a minimum 6 in. (152 mm) below the ceiling. Upright sprinklers shall be positioned so that the deflector is 3 in. to 5 in. (76 mm to 127 mm) below the ceiling. With obstructed construction, the branch lines shall be permitted to be installed across the beams, but sprinklers shall be located in the bays and not under the beams.

The requirements for deflector positioning are based on the use of upright- or pendent-type sprinklers. Positioning the sprinklers below the ceiling at distances greater than that permitted will result in a delayed sprinkler response and allow the fire to grow larger in size. When the sprinklers do finally activate, the fire is likely to be of a size that does not allow early suppression.

The testing for pendent ESFR sprinklers with a nominal K factor of 25.2 demonstrated that a distance of 18 in. (457 mm) could be tolerated. All of the testing for the upright sprinkler was done with the sprinkler deflector placed 4 in. (102 mm) from the ceiling. Consequently, the impact of any delayed activation that would accompany increased sprinkler-to-ceiling distances is not known. As a result, the allowable distances below the ceiling are more restrictive for an upright ESFR sprinkler than for a pendent ESFR sprinkler.

5-11.4.2 Deflector Orientation. Deflectors of sprinklers shall be aligned parallel to ceilings or roofs.

5-11.5 Obstructions to Sprinkler Discharge (Early Suppression Fast-Response Sprinklers).

The effects of sprinkler discharge pattern interruption due to an obstruction must be minimized using the provisions contained in this subsection. Shadow areas in a high-piled storage occupancy provide enough of a fuel supply to a growing fire that the effect is not considered minimal. Thus, great care in positioning ESFR sprinklers with respect to lights, structural members, and similar obstructions must be exercised. Many obstructions that can be tolerated for other types of sprinklers are unacceptable where ESFR sprinklers are used.

Because of the sensitivity of ESFR sprinklers to obstructions, some types of buildings can have structural members located in positions where all of the ESFR obstruction requirements cannot be met. ESFR sprinklers should not be used in these cases. See 5-4.6.3.

Where possible, ESFR sprinklers should be positioned or obstructions should be relocated so that any obstructions are above the discharge pattern of the sprinklers.

All large continuous obstructions, including beams and ductwork, must be located so as to adhere to Table 5-11.5.1. Smaller isolated obstructions, such as light fixtures, can be handled by following 5-11.5.2. Continuous obstructions affecting the discharge pattern of more than one sprinkler can have a significant impact on sprinkler performance and must be located so as to comply with 5-11.5.3. Overall, these provisions are much more restrictive than those for other types of sprinklers but are necessary for successful performance of the ESFR sprinkler system.

5-11.5.1 Obstructions at or Near the Ceiling. Sprinklers shall be arranged to comply with Table 5-11.5.1 and Figure 5-11.5.1 for obstructions at the ceiling such as beams, ducts, lights, and top chords of trusses and bar joists.

Exception: Sprinklers shall be permitted to be spaced on opposite sides of obstructions provided the distance from the centerline on the obstructions to the sprinklers does not exceed one-half the allowable distance between sprinklers.

Table 5-11.5.1 Positioning of Sprinklers to Avoid Obstructions to Discharge (ESFR Sprinkler)

Distance from Sprinkler to Side of Obstruction *(A)*	Maximum Allowable Distance of Deflector above Bottom of Obstruction (in.) *(B)*
Less than 1 ft	0
1 ft to less than 1 ft 6 in.	1½
1 ft 6 in. to less than 2 ft	3
2 ft to less than 2 ft 6 in.	5½
2 ft 6 in. to less than 3 ft	8
3 ft to less than 3 ft 6 in.	10
3 ft 6 in. to less than 4 ft	12
4 ft to less than 4 ft 6 in.	15
4 ft 6 in. to less than 5 ft	18
5 ft to less than 5 ft 6 in.	22
5 ft 6 in. to less than 6 ft	26
6 ft	31

For SI units, 1 in. = 25.4 mm; 1 ft = 0.3048 m.

Note: For *(A)* and *(B)*, refer to Figure 5-11.5.1.

5-11.5.2* Isolated Obstructions Below the Elevation of Sprinklers. Sprinklers shall be installed below isolated noncontinuous obstructions that restrict only one sprinkler and are located below the elevation of sprinklers such as light fixtures and unit heaters.

Exception No. 1: Where the obstruction is 2 ft (0.6 m) or less in width and the sprinkler is located horizontally 1 ft (0.3 m) or greater from the nearest edge of the obstruction.

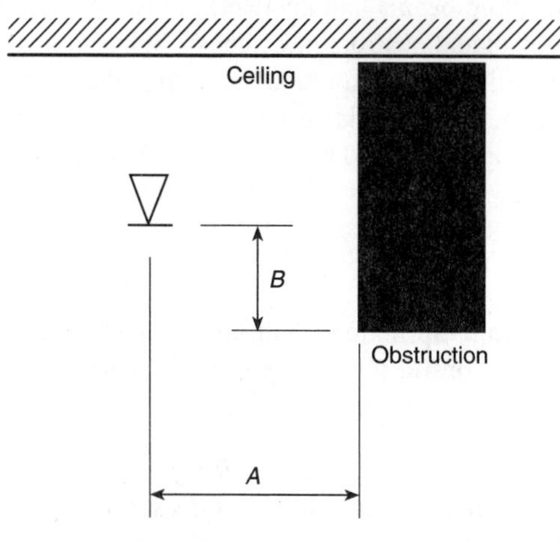

Figure 5-11.5.1 *Positioning of sprinklers to avoid obstructions to discharge (ESFR sprinkler).*

Exception No. 2: Where sprinklers are positioned with respect to the bottom of obstructions in accordance with 5-11.5.1.

Exception No. 3: If the obstruction is 2 in. (51 mm) or less in width and is located a minimum of 2 ft (0.6 m) below the elevation of the sprinkler deflector or is positioned a minimum of 1 ft (0.3 m) horizontally from the sprinkler.

A-5-11.5.2 Isolated obstructions that block adjacent sprinklers in a similar manner should be treated as a continuous obstruction.

5-11.5.3 Continuous Obstructions Below the Sprinklers. Sprinklers shall be arranged to comply with Table 5-11.5.1 for horizontal obstructions entirely below the elevation of sprinklers that restrict sprinkler discharge pattern for two or more adjacent sprinklers such as ducts, lights, pipes, and conveyors.

Exception No. 1: If the obstruction is 2 in. (51 mm) or less in width and is located a minimum of 2 ft (0.6 m) below the elevation of the sprinkler deflector or is positioned a minimum of 1 ft (0.3 m) horizontally from the sprinkler.

Exception No. 2: If the obstruction is 1 ft (0.3 m) or less in width and located a minimum of 1 ft (0.3 m) horizontally from the sprinkler.

Exception No. 3: If the obstruction is 2 ft (0.6 m) or less in width and located a minimum of 2 ft (0.6 m) horizontally from the sprinkler.

5-11.5.3.1 Upright sprinklers shall be installed on sprigs arranged so that the deflector is a minimum of 7 in. (178 mm) above the top of the sprinkler pipe.

5-11.5.3.2 ESFR sprinklers shall be positioned a minimum of 1 ft (0.3 m) horizontally from the nearest edge to any bottom chord of a bar joist or open truss.

5-11.5.3.3 Sprinklers installed under open gratings shall be of the intermediate level/rack storage type or otherwise shielded from the discharge of overhead sprinklers.

5-11.6 Clearance to Storage (Early Suppression Fast-Response Sprinklers).

The clearance between the deflector and the top of storage shall be 36 in. (914 mm) or greater.

At clearances of less than 36 in. (914 mm), ESFR sprinklers installed within the allowable spacing guidelines cannot provide sufficient overlapping of adjacent sprinkler discharge patterns.

5-12 In-Rack Sprinklers

A number of conditions in Table 7-2.3.2.2 require the installation of in-rack sprinklers to protect miscellaneous storage. Numerous protection schemes in Section 7-4 also depend on in-rack sprinklers. Section 5-12 identifies the installation criteria for in-rack sprinklers. Specific spacing and location criteria provided in Section 7-4 for in-rack sprinklers take precedence over the requirements in this section, which provides general requirements for in-rack installation.

5-12.1 System Size.

The area protected by a single system of sprinklers in racks shall not exceed 40,000 ft^2 (3716 m^2) of floor area occupied by the racks, including aisles, regardless of the number of levels of in-rack sprinklers.

5-12.2 Type of In-Rack Sprinklers.

Sprinklers in racks shall be ordinary temperature standard response classification with a nominal K-factor of 5.6 or 8.0, pendent or upright. Sprinklers with intermediate- and high-temperature ratings shall be used near heat sources as required by 5-3.1.3.

Exception: Quick-response sprinklers shall be permitted to be installed in racks.

Theoretically, the use of quick-response in-rack sprinklers can speed up sprinkler operation, resulting in the application of water to a smaller fire, thereby improving sprinkler performance. Because the in-rack sprinkler does not activate until it is immersed in the fire plume—that is, at very high temperatures—the difference between the size of the fire when a quick-response in-rack sprinkler opens and the size of a fire when a standard response in-rack sprinkler opens, however, is negligible. Therefore, the design requirements when using quick-response in-rack sprinklers remain the same as the design requirements when using standard response in-rack sprinklers.

5-12.3 In-Rack Sprinkler Water Shields.

Where there is more than one level of in-rack sprinklers and horizontal barriers are not used, water shields over the lower level in-rack sprinklers are required to minimize the likelihood of discharge from upper level in-rack sprinklers cold soldering the lower level in-rack sprinklers. As an exception to water shields, listed intermediate level/rack storage sprinklers are permitted. Exhibit 5.22 shows an in-rack sprinkler equipped with a water shield and a guard.

Exhibit 5.22 *In-rack sprinkler equipped with a water shield and guard.*

5-12.3.1† In-Rack Sprinkler Water Shields for Storage of Class I through IV Commodities. Water shields shall be provided directly above in-rack sprinklers, or listed intermediate level/rack storage sprinklers shall be used where there is more than one level, if not shielded by horizontal barriers.

5-12.3.2 In-Rack Sprinkler Water Shields for Plastic Storage. Where in-rack sprinklers are not shielded by horizontal barriers, water shields shall be provided above the sprinklers or listed intermediate level/rack storage sprinklers shall be used.

5-12.4 Location, Position, and Spacing of In-Rack Sprinklers.

(See Section 7-4).

In-rack sprinklers usually operate because they are located within the developing fire plume. Water distribution is not as critical because the pattern does not need to penetrate the fire plume. Obstructions to the sprinkler discharge patterns are much less critical and, therefore, the general requirements for obstructions do not apply to in-rack sprinklers.

5-12.5 Obstructions to In-Rack Sprinkler Discharge.

In-rack sprinklers shall not be required to meet the obstruction criteria and clearance from storage requirements of Section 5-5.

5-13 Special Situations

As indicated in 5-1.1(1), sprinklers are required throughout the premises. Under certain conditions, however, the omission of sprinklers in certain areas and spaces within a building is permitted. Section 5-13 identifies these spaces and conditions.

5-13.1 Concealed Spaces.

5-13.1.1* All concealed spaces enclosed wholly or partly by exposed combustible construction shall be protected by sprinklers.

Exception No. 1: Concealed spaces formed by studs or joists with less than 6 in. (152 mm) between the inside or near edges of the studs or joists. (See Figure 5-6.4.1.4.)

Exception No. 2: Concealed spaces formed by bar joists with less than 6 in. (152 mm) between the roof or floor deck and ceiling.

Exception No. 3: Concealed spaces formed by ceilings attached directly to or within 6 in. (152 mm) of wood joist construction.

Exception No. 4: Concealed spaces formed by ceilings attached directly to the underside of composite wood joist construction, provided the joist channels are firestopped into volumes each not exceeding 160 ft^3 (4.53 m^3) using materials equivalent to the web construction.

Exception No. 5: Concealed spaces entirely filled with noncombustible insulation.

Exception No. 6: Concealed spaces within wood joist construction and composite wood joist construction having noncombustible insulation filling the space from the ceiling up to the bottom edge of the joist of the roof or floor deck, provided that in composite wood joist construction the joist channels are firestopped into volumes each not exceeding 160 ft^3 (4.53 m^3) to the full depth of the joist with material equivalent to the web construction.

Exception No. 7: Concealed spaces over isolated small rooms not exceeding 55 ft^2 (4.6 m^2) in area.

Exception No. 8: Where rigid materials are used and the exposed surfaces have a flame spread rating of 25 or less and the materials have been demonstrated not to propagate fire in the form in which they are installed in the space.

Exception No. 9: Concealed spaces in which the exposed materials are constructed entirely of fire-retardant treated wood as defined by NFPA 703, Standard for Fire Retardant Impregnated Wood and Fire Retardant Coatings for Building Materials.

Exception No. 10: Noncombustib le concealed spaces having exposed combustible insulation where the heat content of the facing and substrate of the insulation material does not exceed 1000 Btu/ft² (11,356 kJ/m²).

Exception No. 11: Sprinklers shall not be required in the space below insulation that is laid directly on top of or within the ceiling joists in an otherwise sprinklered attic.

Exception No. 12: Pipe chases under 10 ft² (0.93 m²) formed by studs or wood joists, provided that in multifloor buildings the chases are firestopped at each floor using materials equivalent to the floor construction. Such pipe chases shall contain no sources of ignition, piping shall be noncombustible, and pipe penetrations at each floor shall be properly sealed.

A-5-13.1.1 Exception Nos. 1, 2, and 3 do not require sprinkler protection because it is not physically practical to install sprinklers in the types of concealed spaces discussed in these three exceptions. To reduce the possibility of uncontrolled fire spread, consideration should be given in these unsprinklered concealed space situations to using Exception Nos. 5, 8, and 10.

Concealed spaces can provide an unabated passage for firespread throughout a building unless protected. This paragraph applies to those portions of a building that have construction or finish materials of a combustible nature, are used for the storage of combustible materials, and can contain combustibles associated with building system features such as computer wiring or large quantities of nonmetallic piping.

Any of these scenarios could be found in a concealed space. It is important to recognize that concealed spaces are not exclusively limited to areas above ceilings but can also be found in walls and in spaces beneath the floor. A concealed space example is a raised floor in a computer room. If none of the three prescribed conditions exists, the space is defined as a concealed, noncombustible space with respect to combustible objects and requires no additional sprinkler protection.

Some minor quantities of combustible materials, such as communication wiring, can be present in some concealed spaces but should not typically be viewed as requiring sprinklers (see 5-13.1.1). The threshold value at which sprinklers become necessary in the concealed space is not defined. For example, the usual amounts of data or telephone wiring found above a ceiling would not typically constitute a threat. If bundles of unsheathed computer wiring are installed above the ceiling or beneath the floor in a manner where fire propagation in all directions is likely, then the concealed space should be treated the same as a combustible space, thereby requiring appropriate sprinkler protection. If some other protection measure is provided, such as a CO_2 system, then the concealed space is considered to be protected, and sprinklers are not required. The three Formal Interpretations regarding 5-13.1.1 identify some of the conditions under which concealed spaces are required to be sprinklered.

The exceptions indicate conditions where sprinklers are not required in spaces that would normally require sprinklers. Sometimes the building is modified to meet one of these exceptions to avoid installing sprinklers in the space. Not all of these exceptions provide an equivalent level of fire safety, because some of the exceptions simply recognize that the installation of sprinklers in the space is not physically practical. This difference in level of fire safety is especially true of Exception Nos. 1, 2, and 3, which

Formal Interpretation

Reference: 5-13.1.1

Question: Is it the intent of 5-13.1.1, of NFPA 13 to require sprinklers in the exterior canopy space of a shopping center, the interior of which is fully sprinklered and which is separated from the exterior canopy space by a rated partition in compliance with the National Building Code? The soffit of said canopy space is accessible by the removal of non-combustible 2 ft × 4 ft panels in a lay-in system with hold-down clips.

Answer: Yes. The only exception is where the space is less than 55 sq ft in area as covered in 5-13.1.1 Exception No. 7.

Issue Edition: 1975

Reference: 4-4.4

Date: September 1977

Formal Interpretation 80-14

Reference: 5-13.1.1

Question 1: Paragraph 5-13.1.1 of NFPA 13 deals with the installation of sprinklers in concealed spaces. Does 5-13.1.1 require the installation of sprinklers in a concealed space formed between a noncombustible roof or floor assembly and a suspended noncombustible lay-in tile ceiling system with enclosing end walls of masonry construction if the only combustibles present are fire retardant treated wood studs and top caps forming the top of one hour-rated interior partitions? The partitions consist of gypsumboard and noncombustible insulation fill terminating immediately above the ceiling and the studs with top caps extending another four to six inches above the gypsumboard but not to the structure above. The entire building is provided with an automatic sprinkler system below the ceiling.

Answer: No. The Committee's intent is to permit limited combustibles if the exposed surfaces have been demonstrated not to propagate fire in the form in which they are installed in the space.

Question 2: Does the response to No. 1 depend on whether or not the floor or roof/ceiling design has an hourly fire rating?

Answer: No.

Issue Edition: 1980

Reference: 4-4.4

Date: July 1982

Formal Interpretation

Reference: 5-13.1.1

Question: Would partitions constructed of 2 × 4 wood studs with ⅝-in. sheet rock on each side extending from floor through a 1 hour-rated drop ceiling and up to a concrete slab above be considered "combustible construction" as contemplated under 5-13.1.1?

Answer: No. It is the opinion of the Committee that where a combustible member is protected on both sides by noncombustible material such as sheet rock or metal-lath and plaster, the partition is not considered as "exposed combustible construction" in the intent of 5-13.1.1.

Issue Edition: 1975

Reference: 4-4.4.1

Date: October–November 1976

primarily exist to cover situations where sprinklers are retrofitted into existing buildings.

Exception No. 3 applies to sheathed joist construction or similar narrow spaces formed by attachment of a ceiling to or within 6 in. (152 mm) of joists. This exception is intended to apply only to joists with no openings in the members and up to a nominal depth of 14 in. (356 mm).

In some cases, filling an unsprinklered combustible concealed space with noncombustible insulation, as suggested by Exception No. 5, might be more economically advantageous than installing sprinklers.

Exception No. 8 permits the use of limited-combustible materials as a substitute for sprinkler protection. When considering these materials, it is important to verify that the testing used to determine the material's combustibility was conducted with the material arranged in the position in which it will be installed. Changes in the orientation or arrangement of the material can significantly change the flamespread characteristics and the combustibility of the material. Additionally, the materials are required to be rigid because experience indicates that nonrigid materials do not demonstrate the same characteristics during a fire.

Exception No. 10 allows the use of paper-coated insulation material in a space that is otherwise defined as a noncombustible space.

Exception No. 11 was added to the 1999 edition to indicate that sprinklers are not required in the space between the insulation in an attic and the ceiling sheathing. The sprinklers in the attic are anticipated to provide sufficient protection.

Exception No. 12 was added to the 1999 edition due to the impracticality of installing sprinklers in the small spaces that are usually behind the walls of bathrooms and kitchens in residential facilities. This exception supports the premise that sprinklers can only be omitted in concealed spaces where the installation of sprinklers is absolutely impractical, such as those spaces identified by Exception Nos. 1, 2, and 3 or where combustibles or ignition sources will not be present.

5-13.1.2 Sprinklers in concealed spaces having no access for storage or other use shall be installed in accordance with the requirements for light hazard occupancy.

This sprinkler requirement applies to a concealed combustible space that requires sprinkler protection because of 5-13.1.1. The reference to light hazard occupancy is for the purpose of selecting a density or the appropriate pipe schedule table (if allowed in Chapter 7). The type of construction must be evaluated to determine what area of coverage is needed for the sprinklers.

The intention of 5-13.1.2 is not to require sprinklers in a concealed noncombustible space that contains mechanical equipment made of noncombustible materials. Nor is the intent to require sprinklers simply because an access panel is installed to gain entry to the equipment. The existence of an access panel to a concealed space could lead to the space being used for storage or other purposes. However, if the size or arrangement of the access is such that the space cannot or will not be utilized and the combustibility is limited to the combustible construction, sprinklers can then be installed according to the requirements for a light hazard occupancy. If sprinklers are installed in a concealed space following the rules of light hazard occupancies, then they are required to be quick-response sprinklers.

5-13.1.3 Where heat-producing devices such as furnaces or process equipment are located in the joist channels above a ceiling attached directly to the underside of composite wood joist construction that would not otherwise require sprinkler protection of the spaces, the joist channel containing the heat-producing devices shall be sprinklered by installing sprinklers in each joist channel, on each side, adjacent to the heat-producing device.

5-13.1.4 In concealed spaces having exposed combustible construction, or containing exposed combustibles, in localized areas, the combustibles shall be protected as follows:

(a) If the exposed combustibles are in the vertical partitions or walls around all or a portion of the enclosure, a single row of sprinklers spaced not over 12 ft (3.7 m) apart nor more than 6 ft (1.8 m) from the inside of the partition shall be permitted to protect the surface. The first and last sprinklers in such a row shall not be over 5 ft (1.5 m) from the ends of the partitions.

(b) If the exposed combustibles are in the horizontal plane, the area of the combustibles shall be permitted to be protected with sprinklers on a light hazard spacing. Additional sprinklers shall be installed no more than 6 ft (1.8 m) outside the outline of the area and not more than 12 ft (3.7 m) on center along the outline. When the outline returns to a wall or other obstruction, the last sprinkler shall not be more than 6 ft (1.8 m) from the wall or obstruction.

Two situations where isolated combustibles are in an otherwise noncombustible concealed space are covered in 5-13.1.3 and 5-13.1.4. Sprinklers are provided to protect the combustible material and limit the potential for firespread. Where numerous areas contain combustibles, consideration should be given to providing sprinklers throughout the space.

5-13.2 Vertical Shafts.

5-13.2.1 One sprinkler shall be installed at the top of shafts.

Exception No. 1: Noncombustible or limited-combustible, nonaccessible vertical duct shafts.

Exception No. 2: Noncombustible or limited-combustible, nonaccessible vertical electrical or mechanical shafts.

Sprinklers are to be provided at the top of all shafts used for stairs or other shafts open to more than one floor. Previously, this requirement applied to elevator shafts, which is no longer the case. Elevator shafts now have a distinct set of requirements in 5-13.6 that addresses the needs of the elevator industry. Concealed combustible shafts must be sprinklered. Concealed shafts of noncombustible or limited-combustible construction and contents in a suitably rated enclosure do not require sprinklers.

5-13.2.2* Where vertical shafts have combustible surfaces, one sprinkler shall be installed at each alternate floor level. Where a shaft having combustible surfaces is trapped, an additional sprinkler shall be installed at the top of each trapped section.

The additional sprinklers for shafts with combustible sides must be placed to effectively wet the combustible surfaces. Where the shaft changes direction to form a trapped section, sprinklers are required at the top of each trapped section.

A-5-13.2.2 Where practicable, sprinklers should be staggered at the alternate floor levels, particularly where only one sprinkler is installed at each floor level.

5-13.2.3 Where accessible vertical shafts have noncombustible surfaces, one sprinkler shall be installed near the bottom.

Where shafts are accessible, trash and other material can potentially collect at the bottom of the shaft. In these cases, a sprinkler is required at the bottom of the shaft even if the shaft is of noncombustible construction.

5-13.3 Stairways.

5-13.3.1 Sprinklers shall be installed beneath all stairways of combustible construction.

5-13.3.2 In noncombustible stair shafts with noncombustible stairs, sprinklers shall be installed at the top of the shaft and under the first landing above the bottom of the shaft.

Exception: Sprinklers shall be installed beneath landings or stairways where the area beneath is used for storage.

The storage of materials in stairwells obstructs the egress route and is usually prohibited. When storage does occur, it is often at the top landing or under the first landing. As a result, sprinklers are required at these locations and at any other area in the stair shaft where storage can occur.

5-13.3.3* Sprinklers shall be installed in the stair shaft at each floor landing where two or more doors open from that landing into separate fire divisions.

A-5-13.3.3 See Figures A-5-13.3.3(a) and (b). Sprinklers would be required in the case shown in Figure A-5-13.3.3(a) but not in the case shown in Figure A-5-13.3.3(b).

Sprinklers are required at each floor landing where a noncombustible stair shaft serves two fire-separated buildings or two fire sections of one building as shown in Figure A-5-13.3.3(a) or where the stair landing serves as a horizontal exit. If the stair serves only one fire section, then sprinklers are required only at the roof and under the lowest landing. See Figures A-5-13.3.3(a) and A-5-13.3.3(b).

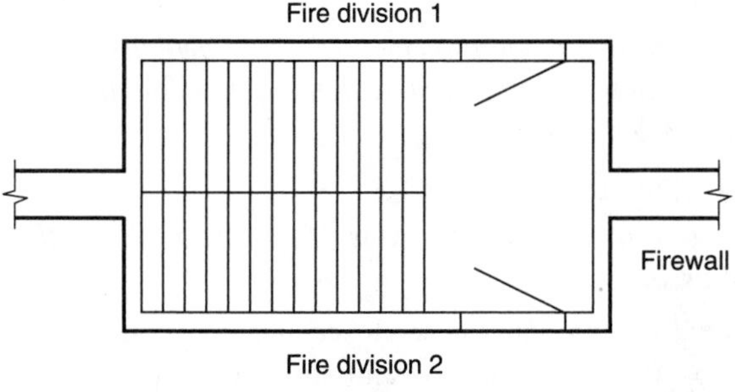

Figure A-5-13.3.3(a) *Noncombustible stair shaft serving two fire sections.*

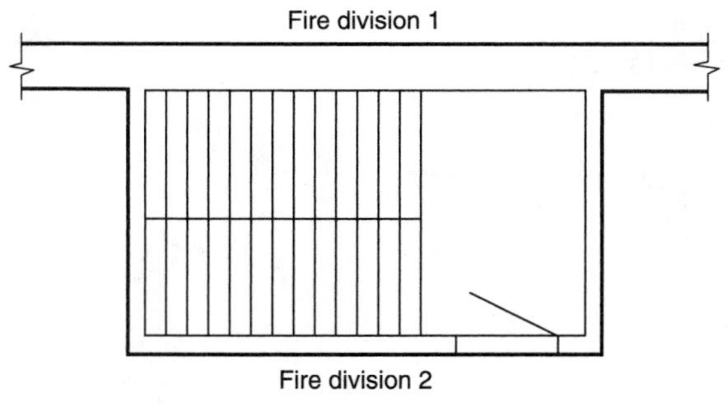

Figure A-5-13.3.3(b) *Noncombustible stair shaft serving one fire section.*

5-13.4* Vertical Openings.

Where moving stairways, staircases, or similar floor openings are unenclosed, the floor openings involved shall be protected by closely spaced sprinklers in combination with draft stops.

Exception No. 1: Closely spaced sprinklers and draft stops are not required around large openings such as those found in shopping malls, atrium buildings, and similar structures where all adjoining levels and spaces are protected by automatic sprinklers in accordance with this standard and where the openings have all horizontal dimensions between opposite edges of 20 ft (6 m) or greater and an area of 1000 ft² (93 m²) or greater.

Exception No. 2: Draft stops and closely spaced sprinklers are not required for convenience openings within individual dwelling units that meet all of the following criteria:

(a) Such openings shall connect a maximum of two adjacent stories (pierce one floor only).

(b) Such openings shall be separated from unprotected vertical openings serving other floors by a barrier with a fire resistance rating equal to that required for enclosure of floor openings by NFPA 101®, Life Safety Code®.*

(c) Such openings shall be separate from corridors.

(d) Such openings shall not serve as a required means of egress, although they can serve as a required means of escape.

A-5-13.4 Where sprinklers in the normal ceiling pattern are closer than 6 ft (1.8 m) from the water curtain, it might be preferable to locate the water curtain sprinklers in recessed baffle pockets. *(See Figure A-5-13.4.)*

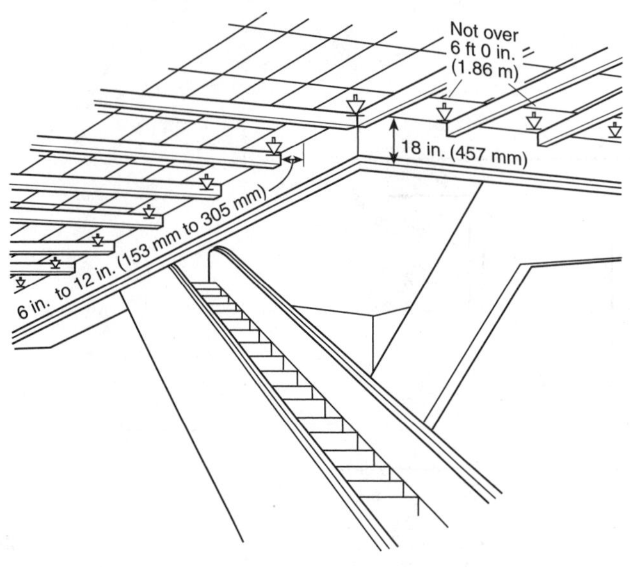

Figure A-5-13.4 *Sprinklers around escalators.*

The draft stops shall be located immediately adjacent to the opening, shall be at least 18 in. (457 mm) deep, and shall be of noncombustible or limited-combustible material that will stay in place before and during sprinkler operation. Sprinklers shall be spaced not more than 6 ft (1.8 m) apart and placed 6 in. to 12 in. (152 mm to 305 mm) from the draft stop on the side away from the opening. Where sprinklers are closer than 6 ft (1.8 m), cross baffles shall be provided in accordance with 5-6.3.4.

A-5-13.4 Exception No. 2(b). Subsection 6-2.4.4 of the 1997 edition of NFPA *101,*® *Life Safety Code*® requires a 2-hour separation for enclosures connecting four stories or more in new construction, a 1-hour separation for other enclosures in new construction, and a 30-minute separation for existing buildings. Special rules for residential construction exist in Chapters 16–20 of NFPA *101.*

By placing sprinklers close to a ceiling opening, the floor area under the opening can be protected. (Small openings are discussed in 5-13.2, and larger openings are addressed in 5-13.4.)

Subsection 5-13.4 is limited to openings that do not meet the definition of an *atrium.* These smaller openings tend to behave much the same as a chimney, allowing rapid vertical movement of the hot gases from the fire. This phenomenon is practically nonexistent in larger openings. Closely spaced sprinklers in conjunction with draft stops are an effective method of gaining control of the fire in these smaller sized openings and in preventing sprinklers from operating on the upper levels by cooling the convective air stream.

When the features of the opening warrant sprinklers spaced at intervals of less than 6 ft (1.8 m), the installation of baffles between adjacent sprinklers prevents the occurrence of the cold solder effect. The use of such devices is discussed in the commentary to 5-6.3.4 for standard spray sprinklers.

Sprinklers used to achieve the protection outlined in 5-13.4 can be the open or automatic type. Privately conducted tests using closed sprinklers have indicated their effectiveness. The use of deluge-type water curtains has become quite rare since the early 1960s. Accidental discharge of deluge-type water curtains results in considerable water damage as well as potential injury to persons on escalators when such false actuation occurs.

Openings are routinely provided between the floors of dwelling units, and Exception No. 2 addresses these arrangements. These openings do not need to be protected if they connect only two floors, are separated from other vertical openings by fire barriers, are not corridors, and are not required for egress by NFPA *101*. A typical example of this situation would be a stairway in a two-level apartment located in a high-rise building.

5-13.5* Building Service Chutes.

Building service chutes (e.g., linen, rubbish) shall be protected internally by automatic sprinklers. A sprinkler shall be provided above the top service opening of the chute, above the lowest service opening, and above service openings at alternate levels in buildings over two stories in height. The room or area into which the chute discharges shall also be protected by automatic sprinklers.

A-5-13.5 The installation of sprinklers at floor levels should be arranged so as to protect the sprinklers from mechanical injury and falling materials and not cause obstruction within the chute. This installation usually can be accomplished by recessing the sprinkler in the wall of the chute or by providing a protective deflector canopy over the sprinkler. Sprinklers should be placed so that there will be minimum interference of the discharge from the sprinklers. Sprinklers with special directional discharge characteristics might be advantageous. *(See Figure A-5-13.5.)*

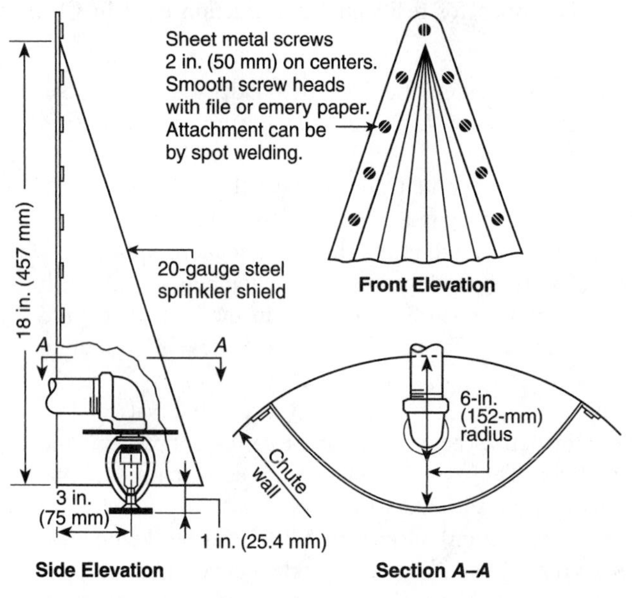

Figure A-5-13.5 *Canopy for protecting sprinklers in building service chutes.*

5-13.6 Elevator Hoistways and Machine Rooms.

Codes that cover elevator design, such as ASME A17.1, *Safety Code for Elevators and Escalators*, do not permit water discharge in elevator shafts until electrical power to the elevator cab has been shut down. This situation necessitates some special arrangement, such as a preaction system, to make sure that water does not flow in the elevator shaft until power shutdown has occurred. The additional cost of a special installation and the benefits returned for the protection must be weighed against the small number of fires in elevator shafts.

Several papers on this topic were presented at a symposium in February 1991 in Baltimore, MD.[1] Subjects presented at this symposium ranged from elevator safety in general to the potential problems associated with premature discharge of water onto

[1]Proceedings of the Symposium on Elevators and Fire, American Society of Mechanical Engineers, Three Park Avenue, New York, NY 10016-5990, February 1991.

elevator control elements. Following this symposium, representatives from the building code organizations and the American Society of Mechanical Engineers (ASME) worked to resolve the problem of providing proper fire protection without sacrificing any of the inherent safety features of the sprinkler system or elevator and its associated equipment. The result of this cooperation was the development of 5-13.6 to specifically address the installation of sprinklers in elevator shafts and equipment rooms.

5-13.6.1* Sidewall spray sprinklers shall be installed at the bottom of each elevator hoistway not more than 2 ft (0.61 m) above the floor of the pit.

Exception: For enclosed, noncombustible elevator shafts that do not contain combustible hydraulic fluids, the sprinklers at the bottom of the shaft are not required.

A-5-13.6.1 The sprinklers in the pit are intended to protect against fires cause by debris, which can accumulate over time. Ideally, the sprinklers should be located near the side of the pit below the elevator doors, where most debris accumulates. However, care should be taken that the sprinkler location does not interfere with the elevator toe guard, which extends below the face of the door opening.

ASME A17.1, *Safety Code for Elevators and Escalators,* allows the sprinklers within 2 ft (0.65 m) of the bottom of the pit to be exempted from the special arrangements of inhibiting waterflow until elevator recall has occurred.

Refuse and residual hydraulic fluids tend to collect at the bottom of shafts. A properly located sprinkler can control a fire of such material. Conventional requirements regarding the placement of the deflector and clear space below the sprinkler cannot always be adhered to in this area. These issues are not critical, however, because the sprinkler would be physically close to any point where a fire could originate, still allowing the sprinkler to control the fire. Because the sprinkler at the bottom of the shaft cannot discharge onto the elevator or other operating components of the elevator, ASME A17.1 no longer requires that the sprinkler discharge be delayed until power shutdown has occurred. The sprinkler at the bottom of the shaft, where installed, is allowed to be part of the normal building sprinkler system and is not required to be part of the special system used to protect the rest of the elevator equipment.

5-13.6.2* Automatic sprinklers in elevator machine rooms or at the tops of hoistways shall be of ordinary- or intermediate-temperature rating.

A-5-13.6.2 ASME A17.1, *Safety Code for Elevators and Escalators,* requires the shutdown of power to the elevator upon or prior to the application of water in elevator machine rooms or hoistways. This shutdown can be accomplished by a detection system with sufficient sensitivity that operates prior to the activation of the sprinklers *(see also NFPA 72, National Fire Alarm Code®).* As an alternative, the system can be arranged using devices or sprinklers capable of effecting power shutdown immediately upon sprinkler activation, such as a waterflow switch without a time delay. This alternative arrangement is intended to interrupt power before significant sprinkler discharge.

5-13.6.3* Upright or pendent spray sprinklers shall be installed at the top of elevator hoistways.

Exception: Sprinklers are not required at the tops of noncombustible hoistways of passenger elevators with car enclosure materials that meet the requirements of ASME A17.1, Safety Code for Elevators and Escalators.

A-5-13.6.3 Passenger elevator cars that have been constructed in accordance with ASME A17.1, *Safety Code for Elevators and Escalators,* Rule 204.2a (under A17.1a-1985 and later editions of the code) have limited combustibility. Materials exposed to the interior of the car and the hoistway, in their end-use composition, are limited to a flame spread rating of 0 to 75 and a smoke development rating of 0 to 450.

5-13.7 Spaces Under Ground Floors, Exterior Docks, and Platforms.

Sprinklers shall be installed in spaces under all combustible ground floors, exterior docks, and platforms.

Exception: Sprinklers shall be permitted to be omitted where all of the following conditions prevail:

(a) The space is not accessible for storage purposes and is protected against accumulation of wind-borne debris.

(b) The space contains no equipment such as conveyors or fuel-fired heating units.

(c) The floor over the space is of tight construction.

(d) No combustible or flammable liquids or materials that under fire conditions would convert into combustible or flammable liquids are processed, handled, or stored on the floor above the space.

The exception to 5-13.7 is another exception to the general requirement that all combustible spaces must be sprinklered. The exception recognizes the possibility of maintenance problems, such as lack of access and danger of freezing with piping under a floor. The conditions intend to eliminate sources of ignition from the concealed space and prevent combustibles from being stored or trash from accumulating, thereby limiting combustibles to the floor materials only.

5-13.8* Exterior Roofs or Canopies.

A-5-13.8 Small loading docks, covered platforms, ducts, or similar small unheated areas can be protected by dry-pendent sprinklers extending through the wall from wet sprinkler piping in an adjacent heated area. Where protecting covered platforms, loading docks, and similar areas, a dry-pendent sprinkler should extend down at a 45-degree angle. The width of the area to be protected should not exceed 7½ ft (2.3 m). Sprinklers should be spaced not over 12 ft (3.7 m) apart. *(See Figure A-5-13.8.)*

5-13.8.1 Sprinklers shall be installed under exterior roofs or canopies exceeding 4 ft (1.2 m) in width.

Exception: Sprinklers are permitted to be omitted where the canopy or roof is of noncombustible or limited combustible construction.

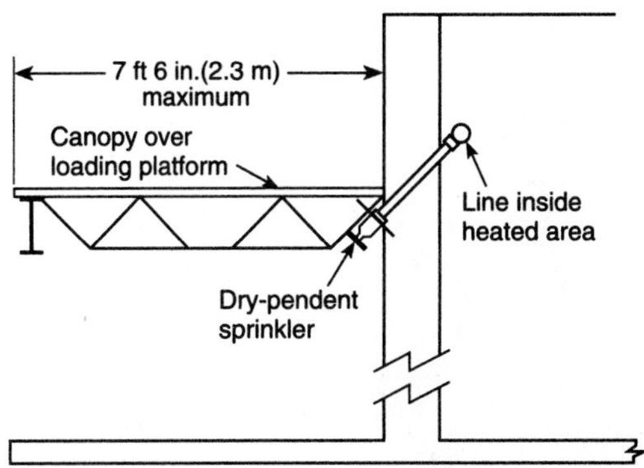

Figure A-5-13.8 *Dry-pendent sprinklers for protection of covered platforms, loading docks, and similar areas.*

5-13.8.2* Sprinklers shall be installed under roofs or canopies over areas where combustibles are stored and handled.

A-5-13.8.2 Short-term transient storage, such as that for delivered packages, and the presence of planters, newspaper machines, and so forth, should not be considered storage or handling of combustibles.

Exterior canopies exceeding 4 ft (1.2 m) in width are required to be sprinklered if they are of combustible construction regardless of whether or not combustible goods are stored or handled underneath them. Canopies less than 4 ft (1.2 m) in width do not need to be sprinklered regardless of construction type provided no combustibles are stored beneath them. The reference to noncombustible and limited-combustible construction applies to the entire canopy assembly and not just the exposed surface.

Sprinklers can be omitted if the canopy construction assembly is comprised totally of noncombustible or limited-combustible materials and the area underneath is essentially restricted to pedestrian use. The roof canopy typically found on strip shopping malls, in which the area under the canopy is limited to pedestrians, is one example of this condition. Automobiles stopping briefly to pick up or drop off passengers are not considered storage. Areas located at drive-in bank windows or *porte cocheres* at motels and hotels normally do not require sprinklers. However, the area under the exterior ceiling shown in Exhibit 5.23 requires sprinkler protection. The space is used primarily for parking vehicles, and the remainder of the building is sprinklered.

Figure A-5-13.8 shows one method of protecting areas under roofs or canopies up to 7½ ft (2.3 m) wide. Sprinklers are required under all such coverings where combustible goods are stored or handled.

5-13.9 Dwelling Units.

The risk associated with not providing sprinkler protection in the spaces described in 5-13.9 is considered acceptable, because the incidence of fires, property damage, and

Exhibit 5.23 *Parking area under an exterior ceiling that is required to be sprinklered.*

fatalities that occur in these spaces is relatively small. A specific limit, however, is imposed on the size of the space, its location, and the type of occupancy in which it is located.

5-13.9.1* Sprinklers are not required in bathrooms that are located within dwelling units, that do not exceed 55 ft^2 (5.1 m^2) in area, and that have walls and ceilings of noncombustible or limited-combustible materials with a 15-minute thermal barrier rating including the walls and ceilings behind fixtures. The area occupied by a noncombustible full height shower/bathtub enclosure shall not be required to be added to the floor area when determining the area of the bathroom.

Exception: Sprinklers are required in bathrooms of nursing homes and in bathrooms opening directly onto public corridors or exitways.

A-5-13.9.1 Fiberglass units are only considered noncombustible where indicated as such by testing.

5-13.9.2* Sprinklers are not required in clothes closets, linen closets, and pantries within dwelling units in hotels and motels where the area of the space does not exceed 24 ft^2 (2.2 m^2), the least dimension does not exceed 3 ft (0.9 m), and the walls and ceilings are surfaced with noncombustible or limited-combustible materials.

A-5-13.9.2 Portable wardrobe units, such as those typically used in nursing homes and mounted to the wall, do not require sprinklers to be installed in them. Although the units are attached to the finished structure, this standard views those units as pieces of furniture rather than as a part of the structure; thus, sprinklers are not required.

This provision, which permits sprinklers to be omitted in closets, is limited to clothes, linen, and pantry-type closets in hotel and motel guest rooms. This provision does not extend to any other types of dwelling units, such as apartments, nursing homes, dormitories, condominiums, or other residential properties that are not defined as a hotel or motel. Closets in hotels and motels are not expected to contain the types and quantities of combustibles typically found in the closets of residences.

5-13.10 Library Stack Rooms.

Sprinklers shall be installed in every aisle and at every tier of stacks with distance between sprinklers along aisles not to exceed 12 ft (3.7 m). *[See Figure 5-13.10(a).]*

Exception No. 1: Where vertical shelf dividers are incomplete and allow water distribution to adjacent aisles, sprinklers are permitted to be omitted in alternate aisles on each tier. Where ventilation openings are also provided in tier floors, sprinklers shall be staggered vertically. [See Figure 5-13.10(b).]

Exception No. 2: Sprinklers are permitted to be installed without regard to aisles where there is 18 in. (457 mm) or more clearance between sprinkler deflectors and tops of racks.

Library stack rooms are generally thought of as high-density book storage on shelves and, in some ways, can be thought of as a type of rack storage. To overcome the obstruction caused by the bookshelves, sprinklers are to be installed at every tier and aisle, unless the exceptions can be satisfied.

Exception No. 2 departs somewhat from the typical sprinkler clearance requirements, because 5-10.13 assumes that the 18 in. (457 mm) of clear space below the sprinkler deflector will not be maintained. In order to compensate for this closer sprinkler placement, sprinklers are required in each aisle unless the 18-in. (457-mm) clear space is available.

Although not explicitly covered by NFPA 13, the same logic with respect to high-density mobile shelving can be applied when the 18-in. (457-mm) clear space is not provided. If a determination can be made concerning the various combinations of aisles, then a row of sprinklers can be placed along the centerline to compensate for the lack of clearance. This approach frequently results in the need to install baffles, because the line spacing decreases to less than 6 ft (1.8 m) apart.

5-13.11 Electrical Equipment.

Sprinkler protection shall be required in electrical equipment rooms. Hoods or shields installed to protect important electrical equipment from sprinkler discharge shall be noncombustible.

Exception: Sprinklers shall not be required where all of the following conditions are met:

(a) The room is dedicated to electrical equipment only.

(b) Only dry-type electrical equipment is used.

(c) Equipment is installed in a 2-hour fire-rated enclosure including protection for penetrations.

(d) No combustible storage is permitted to be stored in the room.

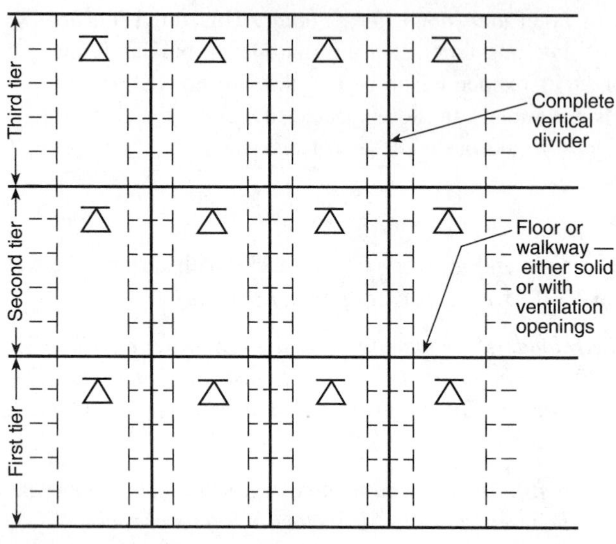

Figure 5-13.10(a) *Sprinklers in multitier library bookstacks with complete vertical dividers.*

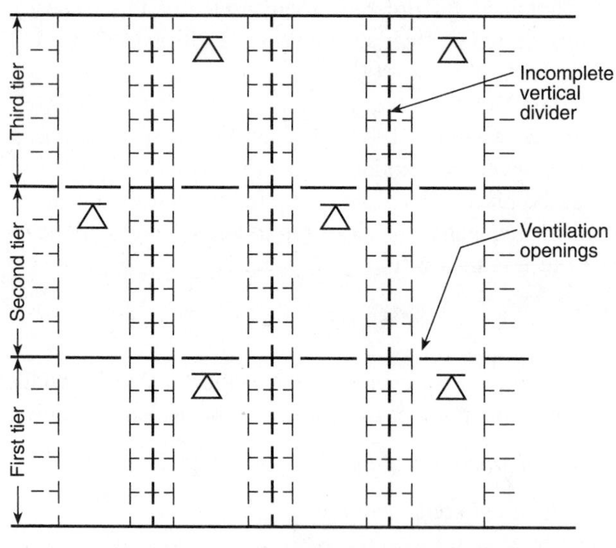

Figure 5-13.10(b) *Sprinklers in multitier library bookstacks with incomplete vertical dividers.*

Sprinkler protection in electrical equipment rooms is required by 5-1.1(1) and 5-13.11. Water shields of noncombustible material can be used to prevent direct contact between the electrical equipment and the discharging water.

Although sprinkler systems have been successfully installed in rooms containing electrical equipment for 100 years with no documented instances of a problem, NFPA 13 identifies certain conditions that, if followed, will permit sprinklers to be omitted from electrical equipment rooms. The conditions identified are relatively easy to determine; however, building owners need to police the areas. The building owner must control the access to all such electrical equipment rooms to reduce the likelihood that storage of any type occurs in these spaces.

5-13.12* Industrial Ovens and Furnaces.

A-5-13.12 The combustible materials present inside industrial ovens and furnaces can be protected by automatic sprinklers. Wet sprinkler systems are preferred. However, water-filled piping exposed to heat within an oven or furnace can incur deposition and buildup of minerals within the pipe. If the oven or furnace could be exposed to freezing temperatures, dry-pendent sprinklers are an alternative to wet pipe systems. Another option is to use a dry pipe system.

The preferred arrangement for piping is outside of the oven; the sprinkler should be installed in the pendent position. The sprinkler temperature rating should be at least 50°F (28°C) greater than the high-temperature limit setting of the oven or applicable zone. As a minimum, the sprinkler system inside the oven or furnace should be designed to provide 15 psi (1 bar) with all sprinklers operating inside the oven/furnace. Sprinkler spacing on each branch line should not exceed 12 ft (3.7 m).

5-13.13* Open-Grid Ceilings.

Open-grid ceilings shall not be installed beneath sprinklers.

Exception No. 1: Open-grid ceilings in which the openings are 1/4 in. (6.4 mm) or larger in the least dimension, where the thickness or depth of the material does not exceed the least dimension of the opening, and where such openings constitute 70 percent of the area of the ceiling material. The spacing of the sprinklers over the open-grid ceiling shall then comply with the following:

(a) In light hazard occupancies where sprinkler spacing (either spray or old-style sprinklers) is less than 10 ft × 10 ft (3 m × 3 m), a minimum clearance of at least 18 in. (457 mm) shall be provided between the sprinkler deflectors and the upper surface of the open-grid ceiling. Where spacing is greater than 10 ft × 10 ft (3 m × 3 m) but less than 10 ft × 12 ft (3 m × 3.7 m), a clearance of at least 24 in. (610 mm) shall be provided from spray sprinklers and at least 36 in. (914 mm) from old-style sprinklers. Where spacing is greater than 10 ft × 12 ft (3 m × 3.7 m), a clearance of at least 48 in. (1219 mm) shall be provided.

(b) In ordinary hazard occupancies, open-grid ceilings shall be permitted to be installed beneath spray sprinklers only. Where sprinkler spacing is less than 10 ft × 10 ft (3 m × 3 m), a minimum clearance of at least 24 in. (610 mm) shall be provided between the sprinkler deflectors and the upper surface of the open-grid ceiling. Where

spacing is greater than 10 ft × 10 ft (3 m × 3 m), a clearance of at least 36 in. (914 mm) shall be provided.

Exception No. 2: Other types of open-grid ceilings shall not be installed beneath sprinklers unless they are listed for such service and are installed in accordance with instructions contained in each package of ceiling material.

A-5-13.13 The installation of open-grid egg crate, louver, or honeycomb ceilings beneath sprinklers restricts the sideways travel of the sprinkler discharge and can change the character of discharge.

The requirements for clearance between the sprinkler deflectors and the top of the grid ceiling are to ensure that sprinkler discharge is not too severely impaired. The grid ceiling can obstruct the discharge pattern to some degree. In addition to the height above the grid ceiling, other obstructions, such as pipes or ducts, must also be considered.

5-13.14 Drop-Out Ceilings.

5-13.14.1 Drop-out ceilings shall be permitted to be installed beneath sprinklers where ceilings are listed for that service and are installed in accordance with their listings.

Exception: Special sprinklers shall not be installed above drop-out ceilings unless specifically listed for this purpose.

Drop-out ceilings are designed to shrink and fall when heated by a fire, permitting the activation of the sprinkler located above. Only ceiling panels that are listed and tested for this use are acceptable. The use of opaque ceilings, which make it difficult to determine if the tile has been painted or if material is stored above the ceiling, should be discouraged.

Quick-response and some EC sprinklers installed above drop-out ceiling panels can operate before all the ceiling panels drop out. Because both types of sprinklers have fast-acting heat-responsive elements, they should not be used above ceiling panels unless the panels have been specifically investigated for use with sprinklers containing fast-response elements.

5-13.14.2 Drop-out ceilings shall not be considered ceilings within the context of this standard.

Because the drop-out ceiling falls in the early stages of a fire, the ceiling does not need to be considered with respect to the positioning of sprinklers. The permanent ceiling/deck above the drop-out ceiling should be used as the point of reference for deflector placement.

5-13.14.3* Piping installed above drop-out ceilings shall not be considered concealed piping. *(See 3-6.4, Exception No. 2.)*

A-5-13.14.3 Drop-out ceilings do not provide the required protection for soft-soldered copper joints or other piping that requires protection.

5-13.14.4* Sprinklers shall not be installed beneath drop-out ceilings.

A-5-13.14.4 The ceiling tiles might drop before sprinkler operation. Delayed operation might occur because heat must then bank down from the deck above before sprinklers will operate.

The danger exists for the falling drop-out ceiling tiles to catch on a sprinkler located below the ceiling and interfere with that sprinkler's proper operation or distribution pattern.

5-13.15 Old-Style Sprinklers.

Old-style sprinklers shall not be used in a new installation.

Exception No. 1: Old-style sprinklers shall be installed in fur storage vaults. (See A-5-13.15, Exception No. 1.)*

Exception No. 2: Use of old-style sprinklers shall be permitted where construction features or other special situations require unique water distribution.

A-5-13.15 Exception No. 1. For tests of sprinkler performance in fur vaults see "Fact Finding Report on Automatic Sprinkler Protection for Fur Storage Vaults" of Underwriters Laboratories Inc., dated November 25, 1947.

Sprinklers should be listed old-style with orifice sizes selected to provide a flow rate as close as possible to, but not less than, 20 gpm (76 L/min) per sprinkler, for four sprinklers, based on the water pressure available.

Sprinklers in fur storage vaults should be located centrally over the aisles between racks and should be spaced not over 5 ft (1.5 m) apart along the aisles.

Where sprinklers are spaced 5 ft (1.5 m) apart along the sprinkler branch lines, pipe sizes should be in accordance with the following schedule:

1 in. (25.4 mm)—4 sprinklers

1¼ in. (31.7 mm)—6 sprinklers

1½ in. (38.1 mm)—10 sprinklers

2 in. (51 mm)—20 sprinklers

2½ in. (63.5 mm)—40 sprinklers

3 in. (76.2 mm)—80 sprinklers

Exception No. 1 identifies the only conditions under which NFPA 13 still requires the use of old-style sprinklers. Full-scale testing conducted in 1947 on fur storage vaults found that the optimum system design required the operation of four old-style sprinklers. The test results are reported in "Fact Finding Report on Automatic Sprinkler Protection for Fire Storage Vaults." Because no further testing was conducted for this storage configuration, the Technical Committee on Sprinkler System Installation Criteria has seen no basis for modifying this requirement. The pipe schedule identified in A-5-13.15 applies to piping running in the same direction as the aisles.

With regard to Exception No. 2, care should be taken when using old-style sprinklers to protect unusual construction features, especially when a heavy fire loading situation exists or the sprinklers are also protecting storage arrangements with designs that were originally tested with spray sprinklers.

5-13.16 Stages.

5-13.16.1 Sprinklers shall be installed under the roof at the ceiling, in spaces under the stage either containing combustible materials or constructed of combustible materials, and in all adjacent spaces and dressing rooms, storerooms, and workshops.

5-13.16.2 Where proscenium opening protection is required, a deluge system shall be provided with open sprinklers located not more than 3 ft (0.9 m) away from the stage side of the proscenium arch and spaced up to a maximum of 6 ft (1.8 m) on center. *(See Chapter 7 for design criteria.)*

The sprinkler requirements for proscenium opening protection are addressed in 5-13.16.2. Building codes and NFPA *101* describe a series of protection schemes to properly protect these openings. One method of protecting these openings is with a deluge-type system.

The primary purpose of the sprinklers is to protect the audience from a fire on the stage. When a curtain is provided, the sprinklers should be on the stage side of the curtain between the curtain and the fire, so that they wet the curtain when they operate.

5-13.17 Provision for Flushing Systems.

All sprinkler systems shall be arranged for flushing. Readily removable fittings shall be provided at the end of all cross mains. All cross mains shall terminate in 1¼ in. (31.8 mm) or larger pipe. All branch lines on gridded systems shall be arranged to facilitate flushing.

The readily removable fitting at the end of each cross main is not limited with regard to maximum pipe size. Branch lines in gridded systems must be installed in such a manner that the piping can be readily disconnected. Further requirements are provided in NFPA 25.

The system should provide for flushing, particularly where the water system is from a nonpotable water supply, such as a pond, stream, or lake, or where the system was changed from a wet system to a dry system or vice versa. In either of these cases, foreign material that blocks the system's individual sprinklers can be introduced. It is necessary to provide for flushing and removal of this foreign material.

5-13.18 Stair Towers.

Stairs, towers, or other construction with incomplete floors, if piped on independent risers, shall be treated as one area with reference to pipe sizes.

A fire in a stair or tower can open a large percentage of sprinklers in the area. An independent riser supplying a stair or tower must be sized to supply all the sprinklers.

5-13.19 Return Bends.

Return bends shall be used where pendent sprinklers are supplied from a raw water source, a mill pond, or open-top reservoirs. Return bends shall be connected to the top of branch lines in order to avoid accumulation of sediment in the drop nipples. *(See Figure 5-13.19.)*

Exception No. 1: Return bends shall not be required for deluge systems.

Exception No. 2: Return bends shall not be required where dry-pendent sprinklers are used.

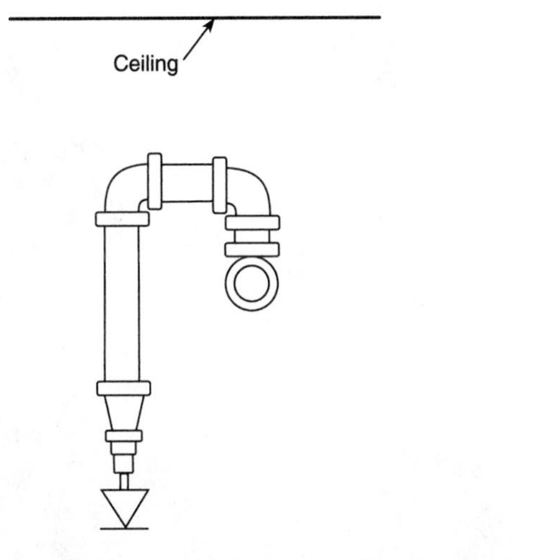

Figure 5-13.19 *Return bend arrangement.*

The purpose of return bends is to prevent collection of sediment in drop nipples. Return bends could be taken directly off the bottom of branch lines in wet pipe sprinkler systems, particularly where the systems are supplied from water sources that could contain excessive sediment. Return bends are not necessary on wet pipe systems that use a potable water supply. In this regard, water supplied from a swimming pool is not considered a raw source requiring return bends as indicated by Formal Interpretation 83-9B.

Return bends are also useful where centering of sprinklers in ceiling tile or exact positioning of sprinklers is desirable from an aesthetic or architectural viewpoint. The use of a return bend is shown in Exhibit 5.24. When standard pendent sprinklers, rather than dry pendent sprinklers, are installed on preaction and dry pipe systems, return bends are required per 4-2.2 Exception No. 2. Return bends are required due to the possibility of scale buildup obstructing the sprinkler orifice.

Formal Interpretation 83-9B

Reference: 5-13.19

Question: Would a filtered and chlorine-treated swimming pool be classified as a raw source of water thus requiring return bends to avoid accumulation of sediment in drop nipples?

Answer: No.

Issue Edition: 1983

Reference: 2-2.1.2.10, 3-8.4

Date: July 1983

Exhibit 5.24 *A return bend used for pendent sprinkler positioning.*

5-13.20 Piping to Sprinklers Below Ceilings.

5-13.20.1 In new installations expected to supply sprinklers below a ceiling, minimum 1-in. (25.4-mm) outlets shall be provided.

Exception: Hexagonal bushings shall be permitted to accommodate temporary sprinklers and shall be removed with the temporary sprinklers when the permanent ceiling sprinklers are installed.

Formal Interpretation 89-4 pertains to the application of 5-13.20.1.

Formal Interpretation 89-4

Reference: 5-13.20.1

Situation: Given the situation of a speculative type multi-purpose building with a roof elevation of 27 +/- feet and a tenant occupancy split into some office area with ceilings and some production, storage and processing areas without ceiling. The ceiling height in the office area is approximately ten feet protected by sprinklers and an overhead system in the entire space is installed at the 26 foot elevation.

Question 1: Is it the intent of 5-13.20.1 to supply the sprinklers on the overhead system, which protects the areas without ceilings, with 1" outlets?

Answer: No.

Question 2: If the answer to Question 1 is no, is 5-13.20.1 intended to cover situations in speculative office buildings, such as high rise buildings?

Answer: Yes.

Issue Edition: 1989

Reference: 3-3.5.1

Issue Date: March 23, 1990

Effective Date: April 12, 1990

5-13.20.2 When pipe schedule systems are revamped, a nipple not exceeding 4 in. (102 mm) in length shall be permitted to be installed in the branch line fitting. All other piping shall be 1 in. (25.4 mm) where it supplies a single sprinkler in an area. *[See Figure 5-13.20.2(a).]*

Exception No. 1: When it is necessary to pipe two new ceiling sprinklers from an existing outlet in an overhead system, the use of a nipple not exceeding 4 in. (102 mm) in length and of the same pipe thread size as the existing outlet shall be permitted, provided that a hydraulic calculation verifies that the design flow rate will be achieved. [See Figure 5-13.20.2(b).]

Exception No. 2: The use of pipe nipples less than 1 in. (25.4 mm) in diameter shall not be permitted in areas subject to earthquakes.

When a system is to be temporarily installed exposed and a false ceiling will eventually be added, the outlets for sprinklers on the pipes are to be 1 in. (25.4 mm) and reduced to the thread size of the sprinklers. A common example of the need to install temporary sprinklers is in an office building with some unfinished floors that are not yet occupied. The initial installation can involve upright sprinklers. If these sprinklers are installed in 1-in. (25.4-mm) outlets using bushings, the proper fitting will be available for the final sprinkler installation when the false ceiling is installed and the office space is finished.

Changing the existing line fitting is not required when revamping existing pipe schedule systems. Because a sprinkler retained in the concealed space and a sprinkler

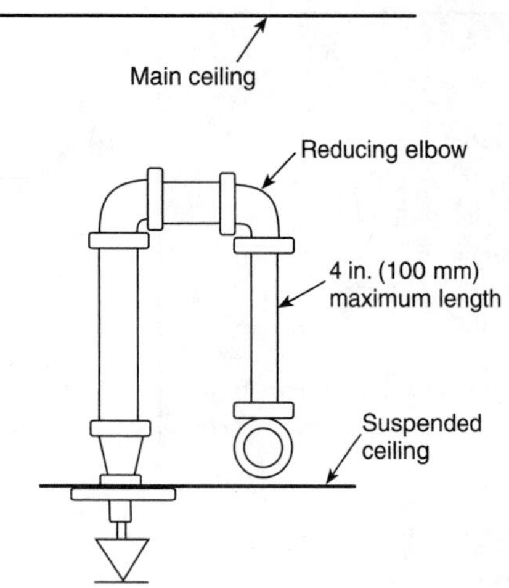

Figure 5-13.20.2(a) Nipple and reducing elbow supplying sprinkler below ceiling.

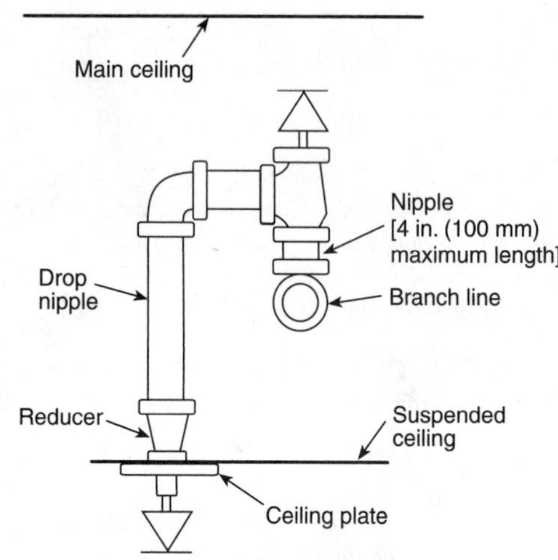

Figure 5-13.20.2(b) Sprinklers in concealed space and below ceiling.

below the ceiling are in separate areas, both can be supplied by a single connection. [See Figure 5-13.20.2(b).]

Exception No. 1 is critical when work on an existing pipe schedule system is contemplated. Without the exception, the in-place line tee would have to be removed, replaced with a 1-in. (25.4-mm) tee, and reassembled. The exception permits an existing tee with a ½-in. (12.7-mm) or ¾-in. (19-mm) outlet to remain in place to supply sprinklers below a newly installed ceiling. This arrangement must be verified with hydraulic calculations when two sprinklers are to be installed beneath the ceiling from the existing line tee.

Exception No. 2 addresses pipe nipples in areas subject to earthquakes. Piping with diameters less than 1 in. (25.4 mm) are likely to fail during an earthquake, especially when supplying sprinklers below a ceiling. To prevent the failures, a minimum 1-in. (25.4-mm) fitting is required.

5-13.20.3 When hydraulically designed systems are revamped, any existing bushing shall be removed and a nipple not exceeding 4 in. (102 mm) in length shall be permitted to be installed in the branch line fitting. Calculations shall be provided to verify that the system design flow rate will be achieved.

Exception No. 1: When it is necessary to pipe two new ceiling sprinklers from an existing outlet in an overhead system, any bushings shall be removed and the use of a nipple not exceeding 4 in. (102 mm) in length and of the same pipe thread size as the existing outlet shall be permitted, provided that a hydraulic calculation verifies that the design flow rate will be achieved.

Exception No. 2: The use of pipe nipples less than 1 in. (25.4 mm) in diameter is not permitted in areas subject to earthquakes.

When a hydraulically designed system is modified as indicated, new calculations must be provided to verify that the arrangement can provide the necessary flow and pressure. The calculation must be made in the case of one or two sprinklers being supplied from the existing line fitting.

5-13.21 Dry Pipe Underground.

Where necessary to place pipe that will be under air pressure underground, the pipe shall be protected against corrosion *(see 5-14.4.2).*

Exception: Unprotected cast- or ductile-iron pipe shall be permitted where joined with a gasketed joint listed for air service underground.

Corrosion that forms on the exterior of cast-iron or ductile-iron pipe insulates that pipe from further corrosion. Such pipe is normally used to transport water, which is less likely to leak than air. When this type of pipe is subject to air pressure, the gasketed joints must be specifically listed for use under air pressure to prevent leakage.

5-13.22* System Subdivision.

Where individual floor/zone control valves are not provided, a flanged joint or mechanical coupling shall be used at the riser at each floor for connections to piping serving floor areas in excess of 5000 ft^2 (465 m^2).

A-5-13.22 See Figure A-5-13.22.

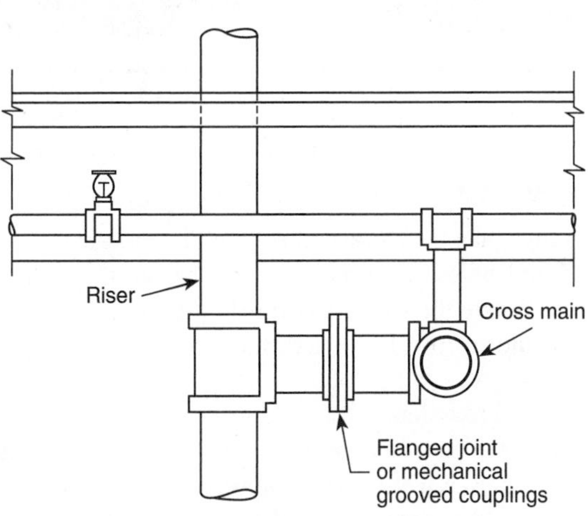

Figure A-5-13.22 *One arrangement of flanged joint at sprinkler riser.*

When a system protects multiple floors, each floor should be able to be isolated so that the entire system is not impaired when a repair or modification needs to be made on a given floor. Floor or zone control valves that are part of a floor control assembly (see Exhibit 5.25) are normally used for this purpose and are the preferred method although NFPA 13 does not require them. All NFPA 13 requires is a flanged joint or mechanical coupling, which enables the piping to be disconnected and a blank or cap to be inserted, allowing the remainder of the system to remain in service while repairs or modifications are being made on one floor.

Exhibit 5.25 *Floor control assembly.*

5-13.23 Spaces Above Nonstorage Areas.

Where nonstorage spaces have lower ceilings than the storage portion of the building, the space above this drop ceiling shall be sprinklered unless it complies with the rules of 5-13.1 for allowable unsprinklered concealed spaces. Where the space above a drop ceiling is sprinklered, the sprinkler system shall conform to the rules of 5-4.1.2 or its exceptions.

NFPA 13 considers sprinkler protection throughout the building. When sprinklers are provided in a warehouse or storage area and an office or other area with a drop ceiling occupies a portion of the space, the ceiling above the space must be sprinklered. The sprinkler design above the space should be the same as the remaining ceiling sprinklers in the storage area. The only exception is a case in which the wall around the space with the lower ceiling extends all the way to the roof and creates a concealed space that does not require sprinklers (see 5-13.1).

5-14 Piping Installation

The methods and procedures necessary for the proper installation of pipe and associated appurtenances are addressed in Section 5-14. Essentially, Section 5-14 describes the proper methods for assembling pipe, fittings, and associated equipment.

5-14.1 Valves.

5-14.1.1* Control Valves. See 3-8.1.

The control valve referred to in 5-14.1.1 can be any number of valve types. For example, a post-indicator valve located outside the building could be used to satisfy this requirement, as could a wall post-indicator valve.

A-5-14.1.1 See Figure A-5-14.1.1.

5-14.1.1.1* Each sprinkler system shall be provided with a listed indicating valve in an accessible location, so located as to control all automatic sources of water supply.

A-5-14.1.1.1 A water supply connection should not extend into a building or through a building wall unless such connection is under the control of an outside listed indicating valve or an inside listed indicating valve located near the outside wall of the building.

All valves controlling water supplies for sprinkler systems or portions thereof, including floor control valves, should be accessible to authorized persons during emergencies. Permanent ladders, clamped treads on risers, chain-operated hand wheels, or other accepted means should be provided where necessary.

Outside control valves are suggested in the following order of preference:

(1) Listed indicating valves at each connection into the building at least 40 ft (12.2 m) from buildings if space permits
(2) Control valves installed in a cutoff stair tower or valve room accessible from outside
(3) Valves located in risers with indicating posts arranged for outside operation.
(4) Key-operated valves in each connection into the building

Each source of water supply, with the exception of the fire department connection, is to be equipped with an indicating control valve to isolate each source of water supply. Formal Interpretation 83-15, which is located in 3-10.2.1, illustrates the application of the requirements of 5-14.1.1.3 for a sprinkler system loop on each floor of a high-rise building that is supplied by two risers. Each riser must be capable of being isolated.

5-14.1.1.2 At least one listed indicating valve shall be installed in each source of water supply.

Exception: There shall be no shutoff valve in the fire department connection.

Exhibit 5.26 illustrates a wet pipe sprinkler system riser with a fire department connection. The fire department connection does not have a control valve in accordance

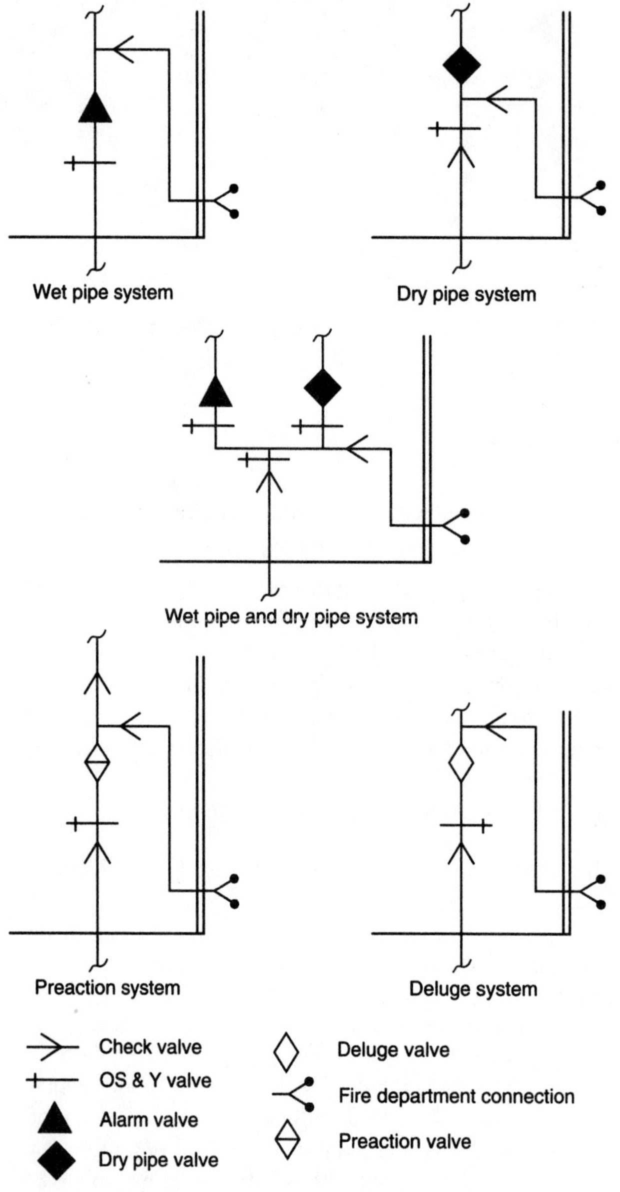

Figure A-5-14.1.1 *Examples of acceptable valve arrangements.*

with the exception. The Formal Interpretation to 5-14.1.1.2 also provides additional information about this requirement.

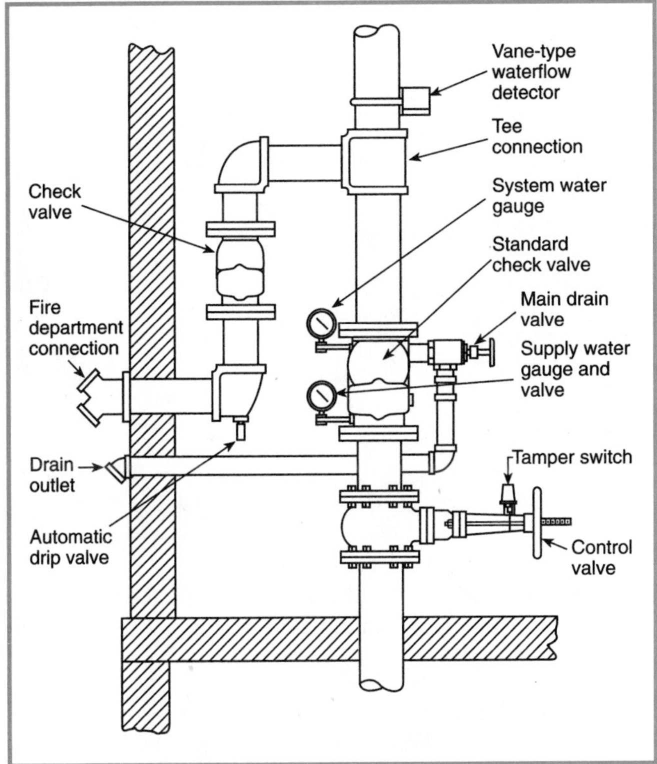

Exhibit 5.26 *Straight pipe riser.*

Formal Interpretation

Reference: 5-14.1.1.2

Question: For a high-rise building with two risers supplying both sprinklers and 2½-in. hose outlets with a connection from each of the risers to a sprinkler loop main on each floor, where each connection from the riser to the loop contains an indicating type pressure regulating control valve, waterflow switch, and OS and Y control valve—is the valve arrangement indicated below acceptable?

Answer: Yes. The Committee's answer is with respect to the valve arrangement only and excludes all other aspects of the system.

(continues)

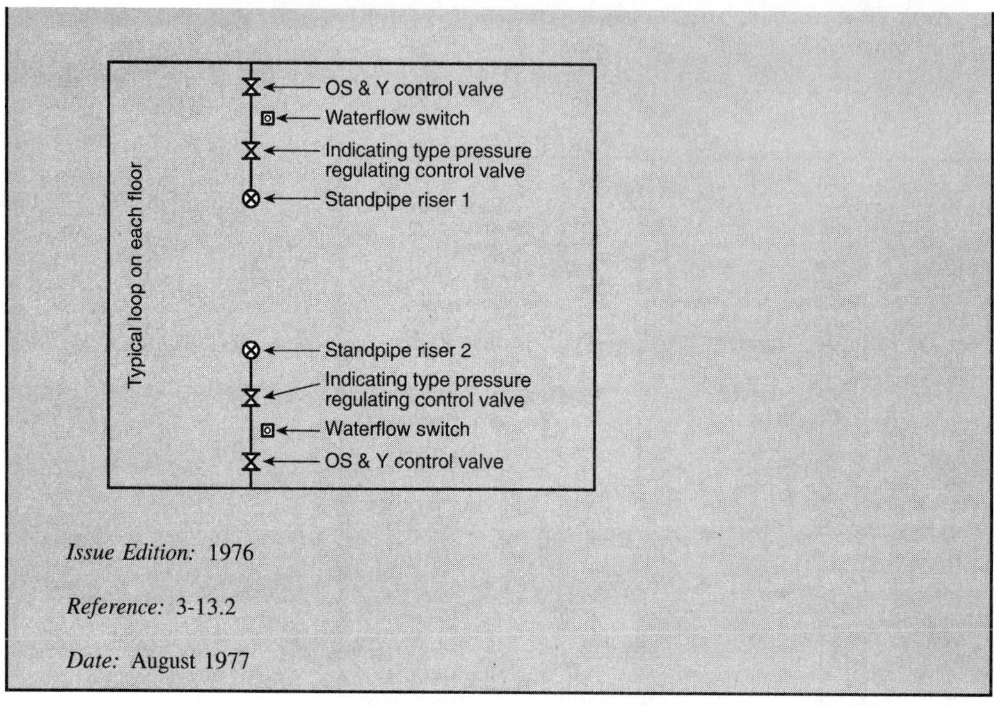

Issue Edition: 1976

Reference: 3-13.2

Date: August 1977

5-14.1.1.3* Valves on connections to water supplies, sectional control and isolation valves, and other valves in supply pipes to sprinklers and other fixed water-based fire suppression systems shall be supervised by one of the following methods:

(1) Central station, proprietary, or remote station signaling service
(2) Local signaling service that will cause the sounding of an audible signal at a constantly attended point
(3) Valves locked in the correct position
(4) Valves located within fenced enclosures under the control of the owner, sealed in the open position, and inspected weekly as part of an approved procedure

Floor control valves in high-rise buildings and valves controlling flow to sprinklers in circulating closed loop systems shall comply with 5-14.1.1.3(1) or (2).

Exception: Supervision of underground gate valves with roadway boxes shall not be required.

A-5-14.1.1.3 The management is responsible for the supervision of valves controlling water supply for fire protection and should exert every effort to see that the valves are maintained in the normally open position. This effort includes special precautions to ensure that protection is promptly restored by completely opening valves that are necessarily closed during repairs or alternations. The precautions apply equally to valves controlling sprinklers and other fixed water-based fire suppression systems, hydrants, tanks, standpipes, pumps, street connections, and sectional valves.

Either one or a combination of the methods of valve supervision described in the following list is considered essential to ensure that the valves controlling fire protection systems are in the normally open position. The methods described are intended as an aid to the person responsible for developing a systematic method of determining that the valves controlling sprinkler systems and other fire protection devices are open.

Continual vigilance is necessary if valves are to be kept in the open position. Responsible day and night employees should be familiar with the location of all valves and their proper use.

The authority having jurisdiction should be consulted as to the type of valve supervision required. Contracts for equipment should specify that all details are to be subject to the approval of the authority having jurisdiction.

(a) *Central Station Supervisory Service.* Central station supervisory service systems involve complete, constant, and automatic supervision of valves by electrically operated devices and circuits continually under test and operating through an approved outside central station, in compliance with NFPA 72, *National Fire Alarm Code.* It is understood that only such portions of NFPA 72 that relate to valve supervision should apply.

(b) *Proprietary Supervisory Service Systems.* Proprietary supervisory service systems include systems where the operation of a valve produces some form of signal and record at a common point by electrically operated devices and circuits continually under test and operating through a central supervising station at the property protected, all in compliance with the standards for the installation, maintenance, and use of local protective, auxiliary protective, remote station protective, and proprietary signaling systems. It is understood that only portions of the standards that relate to valve supervision should apply.

The standard method of locking, sealing, and tagging valves to prevent, so far as possible, their unnecessary closing, to obtain notification of such closing, and to aid in restoring the valve to normal condition is a satisfactory alternate to valve supervision. The authority having jurisdiction should be consulted as to details for specific cases.

Where electrical supervision is not provided, locks or seals should be provided on all valves and should be of a type acceptable to the authority having jurisdiction.

Seals can be marked to indicate the organization under whose jurisdiction the sealing is conducted. All seals should be attached to the valve in such a manner that the valves cannot be operated without breaking the seals. Seals should be of a character to prevent injury in handling and to prevent reassembly when broken. When seals are used, valves should be inspected weekly. The authority having jurisdiction can require a valve tag to be used in conjunction with the sealing.

A padlock, with a chain where necessary, is especially desirable to prevent unauthorized closing of valves in areas where valves are subject to tampering. When such locks are employed, valves should be inspected monthly.

If valves are locked, any distribution of keys should be restricted to only those directly responsible for the fire protection system. Multiple valves should not be locked together; they should be individually locked.

The individual performing the inspections should determine that each valve is in the normal position, properly locked or sealed, and so note on an appropriate record form while still at the valve. The authority having jurisdiction should be consulted for assistance in preparing a suitable report form for this activity.

Identification signs should be provided at each valve to indicate its function and what it controls.

The position of the spindle of OS&Y valves or the target on the indicator valves cannot be accepted as conclusive proof that the valve is fully open. The opening of the valve should be followed by a test to determine that the operating parts have functioned properly.

The test consists of opening the main drain valve and permitting free flow of water until the gauge reading becomes stationary. If the pressure drop is excessive for the water supply involved, the cause should be determined immediately and the proper remedies taken. When sectional valves or other special conditions are encountered, other methods of testing should be used.

If it becomes necessary to break a seal for emergency reasons, the valve, following the emergency, should be opened by the person responsible for the fire protection of the plant, or his or her designated representative, and this person should apply a seal at the time of the valve opening. This seal should be maintained in place until such time as the authority having jurisdiction can replace it with one of its own.

Seals or locks should not be applied to valves reopened after closure until such time as the inspection procedure is carried out.

Where water is shut off to the sprinkler or other fixed water-based fire suppression systems, a guard or other qualified person should be placed on duty and required to continuously patrol the affected sections of the premises until such time as protection is restored.

During specific critical situations, a person should be stationed at the valve so that the valve can be reopened promptly if necessary. It is the intent of this section that the person remain within sight of the valve and have no other duties beyond this responsibility. This procedure is considered imperative when fire protection is shut off immediately following a fire.

An inspection of all other fire protection equipment should be made prior to shutting off water in order to make sure it is in operative condition.

In case of changes to fire protection equipment, all possible work should be done in advance of shutting off the water so that final connections can be made quickly and protection restored promptly. Many times it will be found that by careful planning open outlets can be plugged and protection restored on a portion of the equipment while the alterations are being made.

Where changes are being made in underground piping, all possible piping should be laid before shutting off the water for final connections. Where possible, temporary feed lines, such as temporary piping for reconnection of risers by hose lines, and so forth, should be used to afford maximum protection. The plant, public fire department, and other authorities having jurisdiction should be notified of all impairments to fire protection equipment.

The four methods of supervision help to ensure that a common mode of sprinkler system failure is minimized. Approximately one-third of system failures are attributable to closed or partially closed sprinkler system control valves.[2] The means of supervision are listed in 5-14.1.1.3 in descending order of preference.

Central station, proprietary, or remote station supervision methods are preferred over the other methods listed. Supervisory signals received at such alarm facilities are

[2]Rohr, K.D., 1998, "U.S. Experience with Sprinklers: Who Has Them? How Well Do They Work?," Quincy, MA: National Fire Protection Association Fire Analysis and Research Division.

usually forwarded to trouble or maintenance crews responsible for the protected property. This method also allows for much quicker notification of the fire department. Early knowledge that a control valve has been closed for any reason can change the fire department response and fire-fighting tactics. The use of electrical supervision has not, however, been mandated due to the cost of such a system, particularly in existing, large, sprawling facilities. Exhibit 5.27 shows an outside stem and yoke (OS&Y) valve supervised both electronically with a tamper switch and manually with a lock and chain.

Exhibit 5.27 *OS&Y valve.*

5-14.1.1.4 Where control valves are installed overhead, they shall be positioned so that the indicating feature is visible from the floor below.

One example of this requirement would be the use of an OS&Y valve installed as a sectional control valve in a rack storage system. The valve should be installed so that the stem is in a horizontal position when viewed from the floor.

5-14.1.1.5 Where there is more than one source of water supply, a check valve shall be installed in each connection.

Exception: Where cushion tanks are used with automatic fire pumps, no check valve is required in the cushion tank connection.

Each source of supply must have a check valve to isolate the supplies from each other. In a fire situation, water flows from the highest pressure source. For example, consider a system supplied by a gravity tank, a city connection, and a fire department connection. The gravity tank has the highest pressure source. Initial flow would be from the gravity tank. If the tank's pressure dropped below that of the city connection, the tank's check valve would close and the supply would come from the city. When the fire department connected onto the system, the check valves in both automatic supplies would close and the supply would come through the fire department pumps via the fire department connection.

A cushion tank, when installed, is a part of the fire pump supply. The cushion tank is subject to the pump's check valve.

5-14.1.1.6 Check valves shall be installed in a vertical or horizontal position in accordance with their listing.

5-14.1.1.7* Where a single wet pipe sprinkler system is equipped with a fire department connection, the alarm valve is considered a check valve, and an additional check valve shall not be required.

A-5-14.1.1.7 Where a system having only one dry pipe valve is supplied with city water and a fire department connection, it will be satisfactory to install the main check valve in the water supply connection immediately inside of the building. In instances where there is no outside control valve, the system indicating valve should be placed at the service flange, on the supply side of all fittings.

An alarm valve is a check valve designed to activate waterflow alarms. Another check valve would be redundant and would increase the friction loss characteristics of the system, because an alarm check valve has the same checking ability as a gravity check valve. Exhibit 5.28 shows a wet pipe system employing an alarm valve that also functions as the check valve for the water supply source.

5-14.1.1.8* In a connection serving as one source of supply, listed indicating valves or indicator post valves shall be installed on both sides of all check valves required in 5-14.1.1.5.

Exception No. 1: There shall be no control valves in the fire department connection piping. (See 5-14.1.1.2.)

Exception No. 2: Where the city connection serves as the only automatic source of supply to a wet pipe sprinkler system, a control valve is not required on the system side of the check valve or the alarm check valve.

Exception No. 3: In the discharge pipe from a pressure tank or a gravity tank of less than 15,000 gal (56.78 m^3) capacity, no control valve need be installed on the tank side of the check valve.

A-5-14.1.1.8 See Figure A-5-14.1.1.8. For additional information on controlling valves, see NFPA 22, *Standard for Water Tanks for Private Fire Protection.*

The intent of requiring gate valves on both sides of a check valve is to isolate the check valve for maintenance or repair while permitting the system to remain in service. If the system has only one automatic supply, a gate valve on the system side of the check valve is not required. An alarm valve is a specialized device, and control valves are not required on the system side of alarm check valves.

Exhibit 5.28 Wet pipe system.

5-14.1.1.9* Where a gravity tank is located on a tower in the yard, the control valve on the tank side of the check valve shall be an outside screw and yoke or listed indicating valve; the other shall be either an outside screw and yoke, a listed indicating valve, or a listed valve having a post-type indicator. Where a gravity tank is located on a building, both control valves shall be outside screw and yoke or listed indicating valves and all fittings inside the building, except the drain tee and heater connections, shall be under the control of a listed valve.

A-5-14.1.1.9 For additional information on controlling valves, see NFPA 22, *Standard for Water Tanks for Private Fire Protection.*

5-14.1.1.10* When a pump is located in a combustible pump house or exposed to danger from fire or falling walls, or when a tank discharges into a private fire service main fed by another

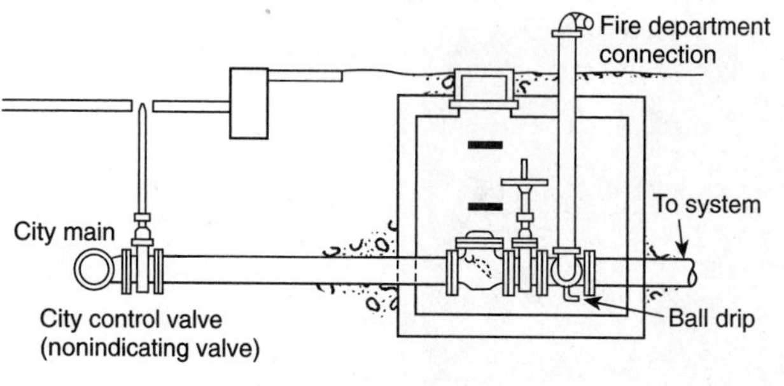

Figure A-5-14.1.1.8 *Pit for gate valve, check valve, and fire department connection.*

supply, either the check valve in the connection shall be located in a pit or the control valve shall be of the post-indicator type located a safe distance outside buildings.

A-5-14.1.1.10 Check valves on tank or pump connections, when located underground, can be placed inside of buildings and at a safe distance from the tank riser or pump, except in cases where the building is entirely of one fire area, when it is ordinarily considered satisfactory to locate the check valve overhead in the lowest level.

The intent of 5-14.1.1.10 is to ensure that the control valves for the fire pump discharge piping remain accessible during a fire. Fire pumps should not be located where they will be subject to falling walls or impaired during an emergency.

5-14.1.1.11* All control valves shall be located where readily accessible and free of obstructions.

A-5-14.1.1.11 It might be necessary to provide valves located in pits with an indicator post extending above grade or other means so that the valve can be operated without entering the pit.

5-14.1.1.12 Identification signs shall be provided at each valve to indicate its function and what it controls.

5-14.1.2 Pressure-Reducing Valves.

5-14.1.2.1 In portions of systems where all components are not listed for pressure greater than 175 psi (12.1 bar) and the potential exists for normal (nonfire condition) water pressure in excess of 175 psi (12.1 bar), a listed pressure-reducing valve shall be installed and set for an outlet pressure not exceeding 165 psi (2.4 bar) at the maximum inlet pressure.

Pressure-reducing and pressure control valves allow a predetermined quantity of water to pass through the system at some established pressure. These devices can be installed in a system riser that supports the entire system or at individual feed mains that supply a floor. The installation of these devices requires special knowledge. All provisions contained in the listing criteria as well as the manufacturer's instructions must be followed. These valves also require routine flow testing. Provisions for such testing should be provided as part of the initial design. Exhibit 5.29 shows a pressure-reducing valve installed on a sprinkler system riser. When the pressure on the system exceeds 165 psi, the valve will open and relieve pressure on the system. The relief valve is arranged so that waterflow due to excess pressure will drain through the system's main drain.

Exhibit 5.29 *Pressure-reducing valve installed on a wet pipe system riser.*

5-14.1.2.2 Pressure gauges shall be installed on the inlet and outlet sides of each pressure-reducing valve.

5-14.1.2.3* A relief valve of not less than ½ in. (13 mm) in size shall be provided on the discharge side of the pressure-reducing valve set to operate at a pressure not exceeding 175 psi (12.1 bar).

A-5-14.1.2.3 Where the relief valve operation would result in water being discharged onto interior walking or working surfaces, consideration should be given to piping the discharge from the valve to a drain connection or other safe location.

5-14.1.2.4 A listed indicating valve shall be provided on the inlet side of each pressure-reducing valve.

Exception: A listed indicating valve is not required where the pressure-reducing valve meets the listing requirements for use as an indicating valve.

5-14.1.2.5 Means shall be provided downstream of all pressure-reducing valves for flow tests at sprinkler system demand.

The requirement for a means of flow testing downstream of all pressure-reducing valves was added to the 1999 edition to facilitate their testing. NFPA 25 currently requires that these devices be tested every 5 years. In high-rise buildings, where pressure-reducing valves are typically used, a drain riser is normally provided and should now be sized to accommodate the testing of full sprinkler flow demand.

5-14.1.3* Post-Indicator Valves.

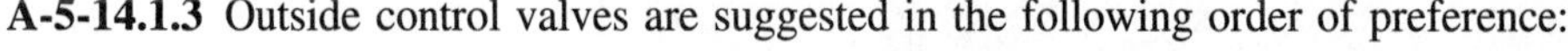

A-5-14.1.3 Outside control valves are suggested in the following order of preference:

(1) Listed indicating valves at each connection into the building at least 40 ft (12.2 m) from buildings if space permits
(2) Control valves installed in a cutoff stair tower or valve room accessible from outside
(3) Valves located in risers with indicating posts arranged for outside operation
(4) Key-operated valves in each connection into the building

Post-indicator valves should be located not less than 40 ft (12.2 m) from buildings. When post-indicator valves cannot be placed at this distance, they are permitted to be located closer, or wall post-indicator valves can be used, provided they are set in locations by blank walls where the possibility of injury by falling walls is unlikely and from which people are not likely to be driven by smoke or heat. Usually, in crowded plant yards, they can be placed beside low buildings, near brick stair towers, or at angles formed by substantial brick walls that are not likely to fall.

5-14.1.3.1 Post-indicator valves shall be set so that the top of the post will be 36 in. (0.9 m) above the final grade.

5-14.1.3.2 Post-indicator valves shall be properly protected against mechanical damage where needed.

Post-indicator valves should be located where they are not subject to mechanical damage. When post-indicator valves are located along roadways, near loading docks, or in other areas where they could be hit by vehicular traffic or otherwise damaged, they should be protected. One method is to provide concrete-filled pipes or other barriers for protection as indicated in Exhibit 5.30. Note that the conduit on the post-indicator valve, which is part of the means of valve supervision, is damaged and needs to be repaired. This condition would be in violation of NFPA 25. In addition to the post-indicator valve, Exhibit 5.30 also includes the outlet for the system main drain, the fire pump test header and a fire department connection. Exhibit 5.31 illustrates a wall post-indicator valve, which can also be used.

Exhibit 5.30 *Post-indicator valve with means of supervision damaged.*

5-14.1.4 Valves in Pits.

5-14.1.4.1 Where it is impractical to provide a post-indicator valve, valves shall be permitted to be placed in pits with permission of the authority having jurisdiction.

Whenever possible, valves that can be operated above grade should be used. Valve pits are considered confined spaces and can be a safety hazard. Valves in the pit may need to be arranged so they can be actuated remotely via an extension handle, allowing the valve to be operated in an emergency without the need to enter the pit.

5-14.1.4.2* When used, valve pits shall be of adequate size and readily accessible for inspection, operation, testing, maintenance, and removal of equipment contained therein. They shall be constructed and arranged to properly protect the installed equipment from movement of earth, freezing, and accumulation of water. Poured-in-place or precast concrete, with or without reinforcement, or brick (all depending upon soil conditions and size of pit) are appropriate materials for construction of valve pits. Other approved materials shall be permitted to be used. Where the water table is low and the soil is porous, crushed stone or gravel shall be permitted to be used for the floor of the pit. *[See Figure A-5-15.2(b) for a suggested arrangement.]*

Valve pits located at or near the base of the riser of an elevated tank shall be designed in accordance with Chapter 9 of NFPA 22, *Standard for Water Tanks for Private Fire Protection.*

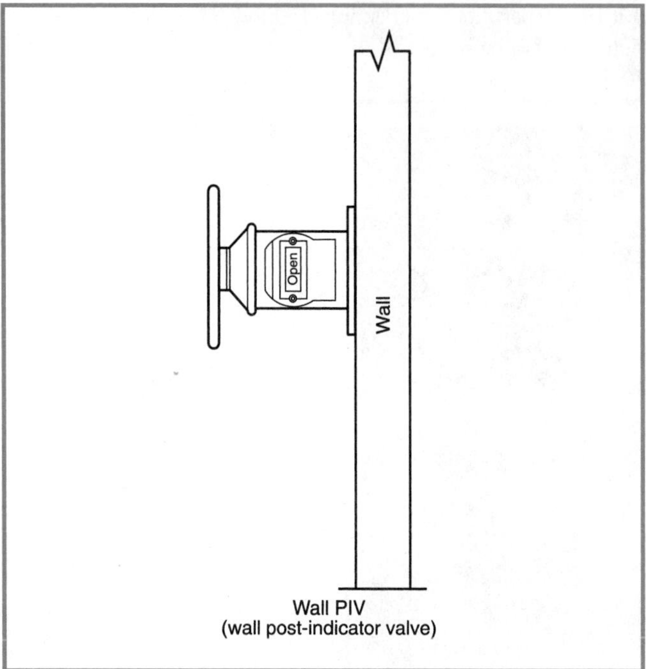

Exhibit 5.31 *Wall post-indicator valve.*

A-5-14.1.4.2 A valve wrench with a long handle should be provided at a convenient location on the premises.

5-14.1.4.3 The location of the valve shall be clearly marked, and the cover of the pit shall be kept free of obstructions.

In cold climates, the pit's cover can be covered with snow or ice, making the pit inaccessible and the cover difficult to find. To maintain pit accessibility, the locations must be clearly marked and snow or ice promptly removed.

5-14.1.5 Sectional Valves.

5-14.1.5.1 Large private fire service main systems shall have sectional controlling valves at appropriate points in order to permit sectionalizing the system in the event of a break or for the making of repairs or extensions.

Placing a sectional control valve between every four to six pieces of fire protection equipment, such as a sprinkler system or hydrant, is common practice. This spacing is so that only a small section of the underground main must be isolated when making repairs or system modifications.

5-14.1.5.2 A valve shall be provided on each bank where a main crosses water and outside the building foundation(s) where the main or section of main runs under a building. *(See 5-14.4.3.1.)*

5-14.1.6* In-Rack Sprinkler System Control Valves. Where sprinklers are installed in racks, separate indicating control valves and drains shall be provided and arranged so that ceiling and in-rack sprinklers can be controlled independently.

Exception No. 1: Installation of 20 or fewer in-rack sprinklers supplied by any one ceiling sprinkler system.

Exception No. 2: The separate indicating valves shall be permitted to be arranged as sectional control valves where the racks occupy only a portion of the area protected by the ceiling sprinklers.

A-5-14.1.6 In-rack sprinklers and ceiling sprinklers selected for protection should be controlled by at least two separate indicating valves and drains. In higher rack arrangements, consideration should be given to providing more than one in-rack control valve in order to limit the extent of any single impairment.

The control valves are intended to allow isolation of the in-rack system from the ceiling system as indicated in Exhibit 5.32. In-rack sprinklers are susceptible to damage due to their location. These valves are required to permit the ceiling system to remain in service while an in-rack sprinkler is replaced. Additionally, when the ceiling system is out of service, the valves are to be arranged so that the in-rack system remains operational.

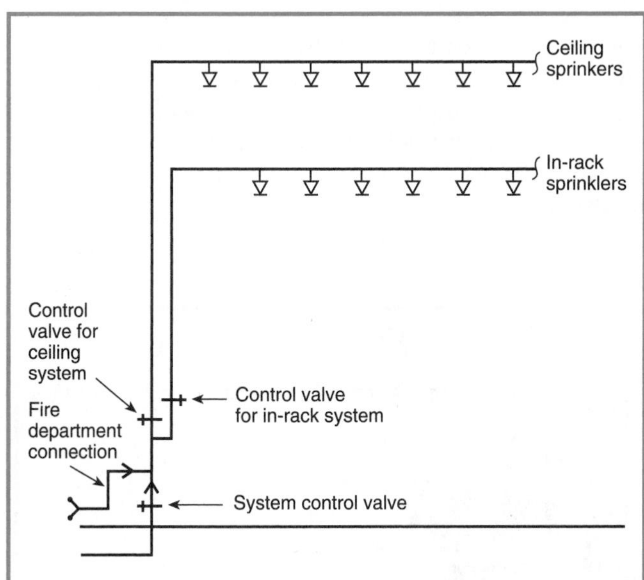

Exhibit 5.32 *Arrangement showing separately controlled ceiling and in-rack sprinklers.*

Exception No. 1 allows an in-rack sprinkler system consisting of up to 20 sprinklers to be supplied from the overhead system without a separate control valve (see Exhibit

5.33). The 20-sprinkler limit is the total number of in-rack sprinklers supplied from a given system riser.

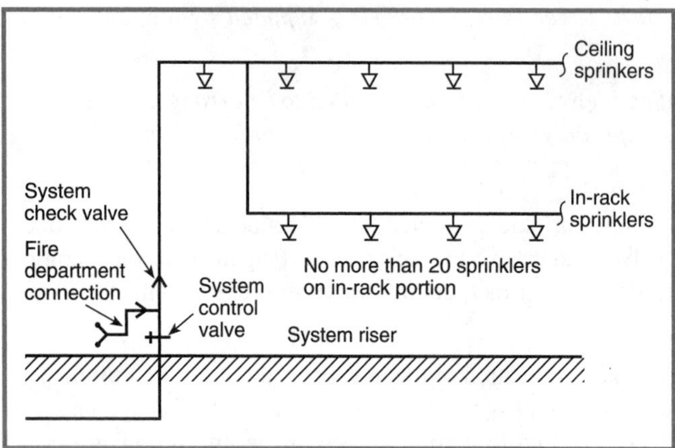

Exhibit 5.33 *In-rack sprinklers supplied directly from ceiling system without a separate control valve in accordance with Exception No. 1 to 5-14.1.6.*

Exception No. 2 allows the control valves used to isolate the overhead system from the in-rack system to be arranged as sectional valves so that portions rather than the entire overhead system can be isolated. This exception applies only where the in-rack system occupies a portion of the area protected by the overhead system. Exhibit 5.34 provides an example of this arrangement, and Exhibit 5.35 shows a sectional valve arrangement for an in-rack system.

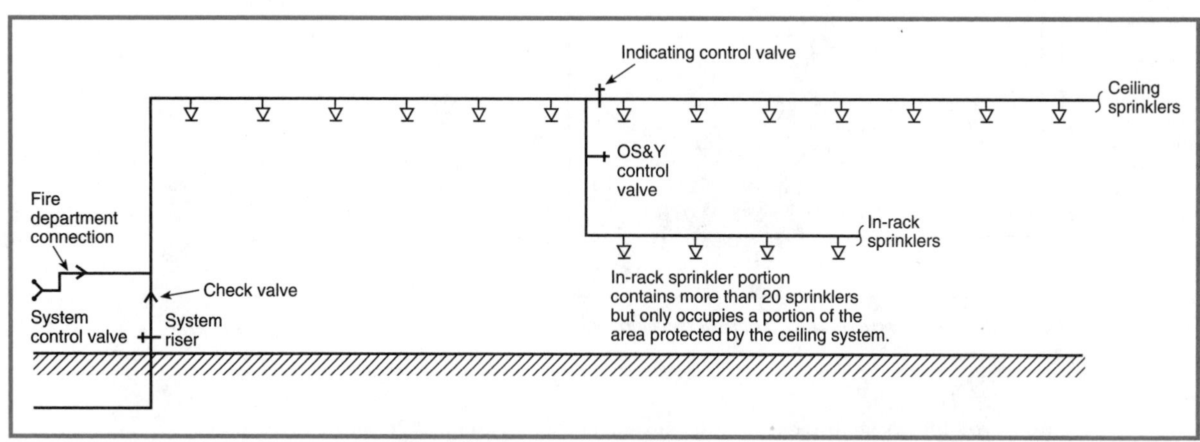

Exhibit 5.34 *In-rack sprinklers occupying only a portion of the area protected by ceiling sprinklers.*

Exhibit 5.35 Sectional control valve for in-rack sprinklers.

5-14.2 Drainage.

The requirements in the paragraphs that follow 5-14.2 regarding drains are necessary to facilitate system maintenance, repair, modifications, and additions. When systems are to be taken out of service for any reason, properly sized and arranged drains help minimize the amount of time the system is out of service. For dry pipe and preaction systems protecting cold environments, immediate drainage of any water that has entered the system is critical. Properly sized drains and the requirements in 5-14.2.3 for installing the pipe at a specific pitch serve this purpose. Additionally, after a system is involved in a fire and the fire is extinguished, the system needs to be shut down and drained so that the sprinklers that operated can be replaced.

Some of these drain features are also used to check the status of the water supply. NFPA 25 provides guidance on the main drain test, which is accomplished using the drain connections discussed in 5-14.2.4.

5-14.2.1* All sprinkler pipe and fittings shall be so installed that the system can be drained.

A-5-14.2.1 All piping should be arranged where practicable to drain to the main drain valve.

Ideally, the entire system should be arranged so that all water can be drained through the main drain. Where this cannot be accomplished, auxiliary drainage facilities per 5-14.2.5 and sectional drainage in accordance with 5-14.2.4.3 need to be sized and arranged with consideration for the type of system and the volume of trapped piping.

5-14.2.2 On wet pipe systems, sprinkler pipes shall be permitted to be installed level. Trapped piping shall be drained in accordance with 5-14.2.5.

In pipe schedule systems, the reduction in pipe size from the supply to the branch lines has the effect of a slight pitch when the piping is installed level. This effect is due to the slope of the reducing fittings necessitated by the reduction in pipe size.

5-14.2.3 In dry pipe systems and preaction systems branch lines shall be pitched at least ½ in. per 10 ft (4 mm/m) and mains shall be pitched at least ¼ in. per 10 ft (2 mm/m).

Because dry pipe systems and preaction systems are usually installed in areas subject to freezing, proper drainage is critical due to the potential freezing of a pipe network after a system trip. Water trapped in the pipe network can begin to freeze in a matter of minutes in systems installed in structures that maintain temperatures at or below 0°F (-18°C).

Exception No. 1: Mains shall be pitched at least ½ in. per 10 ft (4 mm/m) in refrigerated areas.

Exception No. 2: Preaction systems located entirely in areas not subject to freezing are not required to be pitched.

5-14.2.4 System, Main Drain, or Sectional Drain Connections. See Figures 5-14.2.4 and A-5-15.4.2(b).

Paragraphs 5-14.2.4 through 5-14.2.5.3.3 describe the facilities needed to drain any portion of a system in an efficient manner without undue risk of water damage. The arrangement of main, sectional, and auxiliary drains expedites the manner in which a system can be taken out of service for maintenance or repairs. See Exhibit 5.26 for an illustration of a main drain arrangement. Formal Interpretation 78-8 pertains to the application of Figure 5-14.2.4.

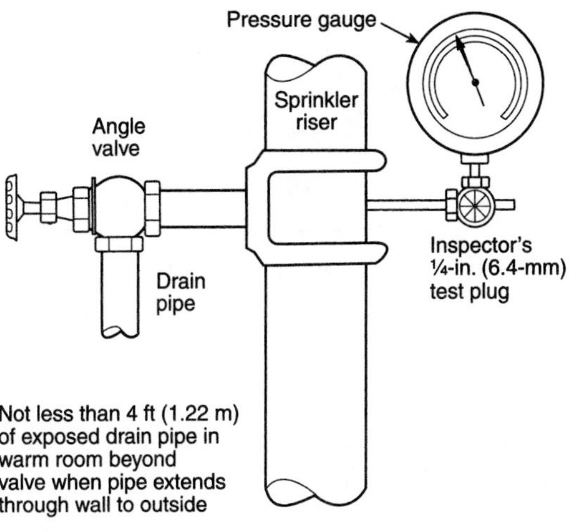

Figure 5-14.2.4 *Drain connection for system riser.*

Formal Interpretation 78-8

Reference: Figure 5-14.2.4

Question: Would the pressure gauge arrangement shown in Figure 1 be considered an acceptable alternate to that shown in Figure 5-14.2.4?

Answer: No. The pressure gauge location, as illustrated, will not give a true residual reading. It will indicate an excessive pressure drop.

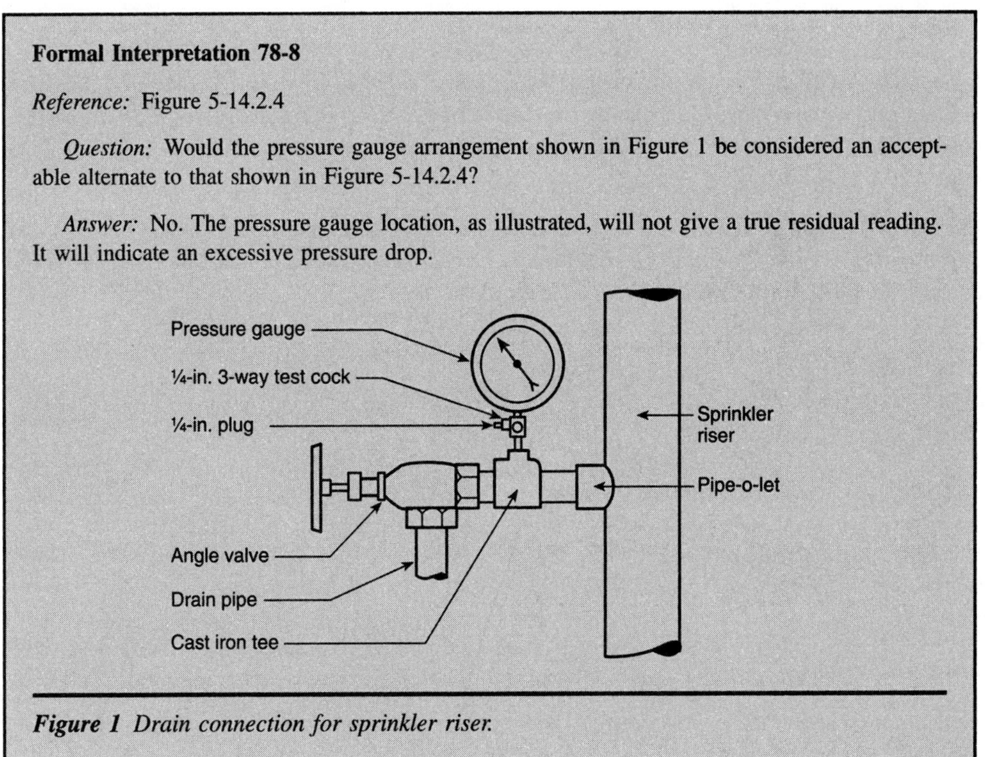

Figure 1 *Drain connection for sprinkler riser.*

(continues)

Issue Edition: 1978

Reference: 3-11.2

Date: August 1980

Reprinted for correction: January, 1993

5-14.2.4.1 Provisions shall be made to properly drain all parts of the system.

5-14.2.4.2 Drain connections for systems supply risers and mains shall be sized as shown in Table 5-14.2.4.2.

Table 5-14.2.4.2 Drain Size

Riser or Main Size	Size of Drain Connection
Up to 2 in.	¾ in. or larger
2½ in., 3 in., 3½ in.	1¼ in. or larger
4 in. and larger	2 in. only

For SI units, 1 in. = 25.4 mm.

5-14.2.4.3 Where an interior sectional or floor control valve(s) is provided, it shall be provided with a drain connection sized as shown in Table 5-14.2.4.2 to drain that portion of the system controlled by the sectional valve. Drains shall discharge outside or to a drain connection. *[See Figure A-5-15.4.2(b).]*

Exception: For those drains serving pressure-reducing valves, the drain, drain connection, and all other downstream drain piping shall be sized to permit a flow of at least the greatest system demand supplied by the pressure-reducing valve.

Piping controlled by sectional valves represents a significant segment of the sprinkler system. Sectional or floor control valves serve the same function for that portion of the system as the main drain does for the entire system. The use of express drains in high-rise buildings is one method of disposing of water from upper floors.

With regard to the exception, NFPA 25 requires annual flow testing of pressure-regulating devices. To facilitate flow testing, the drain connection serving such devices must be capable of handling the system flow rate and pressure for the area served by that pressure-reducing valve.

5-14.2.4.4 The test connections required by 5-15.4.1 shall be permitted to be used as main drain connections.

Exception: Where drain connections for floor control valves are tied into a common drain riser, the drain riser shall be one pipe size larger than the largest size drain connection tying into it.

5-14.2.5 Auxiliary Drains.

Drainage facilities are necessary for trapped piping. The type of facility, its size, and its arrangement are dependent on the type of system and the volume of trapped piping.

5-14.2.5.1 Auxiliary drains shall be provided where a change in piping direction prevents drainage of system piping through the main drain valve.

5-14.2.5.2 Auxiliary Drains for Wet Pipe Systems and Preaction Systems in Areas Not Subject to Freezing.

5-14.2.5.2.1 Where the capacity of trapped sections of pipes in wet systems is less than 5 gal (18.9 L), the auxiliary drain shall consist of a nipple and cap or plug not less than ½ in. (12 mm) in size.

Exception No. 1: Auxiliary drains are not required for system piping that can be drained by removing a single pendent sprinkler.

Exception No. 2: Where flexible couplings or other easily separated connections are used, the nipple and cap or plug shall be permitted to be omitted.

The drainage of trapped piping sections in wet pipe systems is not typically a common event. Therefore, a nipple and cap or a plug is adequate when the volume of the trapped section is 5 gal (18.9 L) or less. This small quantity of water can be emptied into a bucket. A nipple and cap are preferable to a plug because of the comparative ease with which they can be removed. For easier removal, the plug should be brass.

Exception No. 1 recognizes that the small volume of water trapped in one pendent sprinkler can be drained as readily through the sprinkler connection as through a plug. Requiring a drain is unrealistic where pendent sprinklers are installed below a ceiling and are supplied directly from branch lines concealed above the ceiling. Exception No. 2 recognizes that many of the fittings currently used can be separated easily and can be used to drain small trapped portions of a wet system with little inconvenience.

5-14.2.5.2.2 Where the capacity of isolated trapped sections of pipe is more than 5 gal (18.9 L) and less than 50 gal (189 L), the auxiliary drain shall consist of a valve ¾ in. (19 mm) or larger and a plug or a nipple and cap.

A drain valve is required when the trapped section in a wet pipe system exceeds 5 gal (18.9 L). A drain valve allows the trapped piping to be emptied in a reasonable amount of time. The valved outlet also permits the trapped portion to be drained without creating a nuisance water spill. Exhibit 5.36 shows an auxiliary drain consisting of a valve and a plug. Formal Interpretation 85-2 further illustrates the drain valve requirements.

5-14.2.5.2.3* Where the capacity of isolated trapped sections of pipe is 50 gal (189 L) or more, the auxiliary drain shall consist of a valve not smaller than 1 in. (25.4 mm), piped to an accessible location.

A-5-14.2.5.2.3 An example of an accessible location would be a valve located approximately 7 ft (2 m) above the floor level to which a hose could be connected to discharge the water in an acceptable manner.

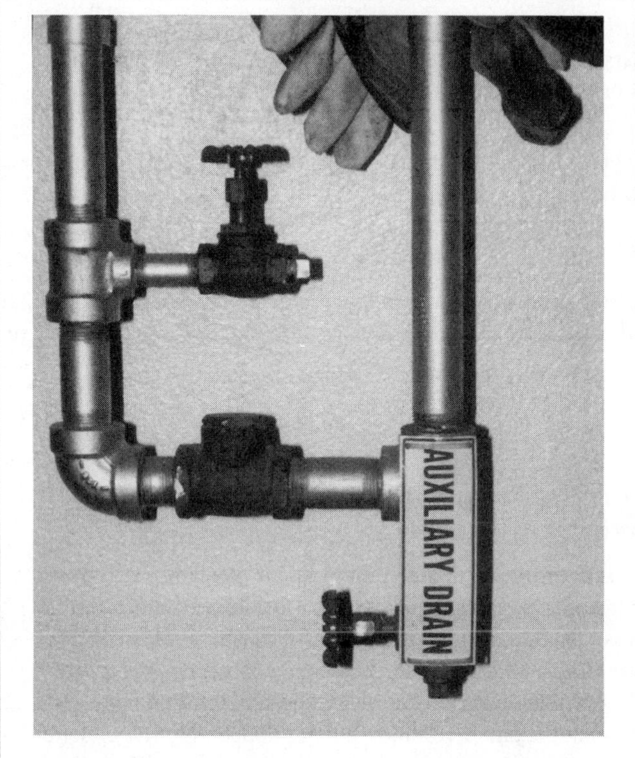

Exhibit 5.36 *Auxiliary drain consisting of a valve and a plug.*

Formal Interpretation 85-2

Reference: 5-14.2.5.2.2

Question 1: Is a ¾-in. auxiliary drain as described in 5-14.2.5.2.2 sufficient when a large portion of a main is trapped on a wet pipe sprinkler system?

Answer: Yes.

Question 1A: Must this drain discharge to the outside or to a drain connection?

Answer: No.

Issue Edition: 1985

Reference: 3-11.3.2.2

Date: February 1985

5-14.2.5.2.4 Tie-in drains are not required on wet pipe systems and preaction systems protecting non-freezing environments.

Tie-in drains are cross connections of the ends of trapped branch lines that are piped to a single drain valve. These drains are not required on wet pipe systems and preaction systems protecting non-freezing environments, because such piping is not drained at the same frequency as systems protecting cold environments.

5-14.2.5.3 Auxiliary Drains for Dry Pipe Systems and Preaction Systems in Areas Subject to Freezing.

5-14.2.5.3.1 Where the capacity of trapped sections of pipe is less than 5 gal (18.9 L), the auxiliary drain shall consist of a valve not smaller than ½ in. (12 mm) and a plug or a nipple and cap.

Exception: Auxiliary drains are not required for pipe drops supplying dry-pendent sprinklers installed in accordance with 4-2.2.

Because dry pipe systems and some preaction systems are installed in areas subject to freezing, small trapped sections of piping of 5 gal (18.9 L) or less must be provided with drain valves. Water that has entered the system, either because the dry pipe valve tripped or from condensation of moisture from the pressurized air in the system, must be easily drained to prevent freezing. An auxiliary drain is not required when pendent sprinklers in a heated area are supplied by piping that is also in a heated area.

5-14.2.5.3.2 Where the capacity of isolated trapped sections of system piping is more than 5 gal (18.9 L), the auxiliary drain shall consist of two 1-in. (25.4-mm) valves and one 2-in. × 12-in. (50-mm × 305-mm) condensate nipple or equivalent, accessibly located. *(See Figure 5-14.2.5.3.2.)*

The condensate nipple illustrated in Figure 5-14.2.5.3.2 or its equivalent is required for each section of trapped piping with more than 5 gal (18.9 L) capacity in a dry pipe system and a preaction system protecting cold environments. The condensate nipple is capable of collecting and removing moisture from the system while minimizing the potential for excessive loss of air pressure and possible unplanned operation of a dry pipe valve. The upper valve is normally open, allowing moisture to enter the chamber, while the lower valve is closed and sealed to avoid leaks. To drain the chamber, the upper valve is closed to temporarily isolate it from the system, and the lower valve is opened to remove the accumulated condensate. Where inside and outside temperatures are extreme, isolating the exterior drain pipe from the interior drain pipe with a threaded fitting can be desirable. This separation can have the effect of limiting moisture buildup by checking the transfer of a high exterior ambient temperature to the cooler interior temperature of the drain pipe. Also see the commentary to 5-14.2.5.3.3.

5-14.2.5.3.3 Tie-in drains shall be provided for multiple adjacent trapped branch pipes and shall be only 1 in. (25.4 mm). Tie-in drain lines shall be pitched a minimum of ½ in. per 10 ft (4 mm/m).

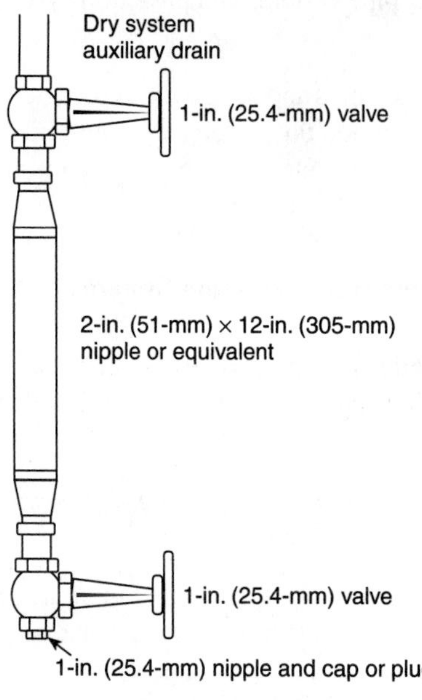

Figure 5-14.2.5.3.2 *Dry system auxiliary drain.*

When two or more adjacent branch lines are trapped in a dry pipe or preaction system, the ends of the branch lines must be piped together and run to a low-point drain at least 1 in. (25.4 mm) in size. The drain needs to be equipped with a valve and nipple and cap or a plug to facilitate moisture removal from the system.

Properly pitching tie-in drains and removing condensate from drains prior to freezing weather is extremely important. Typically, tie-in drains consist of smaller-sized piping connected to a rather large volume of piping, and, with these smaller drains, freeze-ups are possible.

Tie-in drains on multiple adjacent branch lines should be avoided if possible, because the use of tie-in drains in effect creates a gridded system. Gridded dry pipe systems and gridded double interlock preaction systems are prohibited by 4-2.3.2 and 4-3.2.5. In these types of systems, all of the air in the piping must be evacuated before a steady flow of water is discharged from the operated sprinklers. A gridded pipe arrangement is very likely to result in excessive time delays before effective sprinkler discharge occurs. Tie-in drains are considered necessary to allow rapid drainage of systems protecting cold environments. The restriction that tie-in drains be limited to 1 in. (25.4 mm) should help minimize any delays.

5-14.2.6 Discharge of Drain Valves.

5-14.2.6.1* Direct interconnections shall not be made between sprinkler drains and sewers. The drain discharge shall conform to any health or water department regulations.

A-5-14.2.6.1 Where possible, the main sprinkler riser drain should discharge outside the building at a point free from the possibility of causing water damage. Where it is not possible to discharge outside the building wall, the drain should be piped to a sump, which in turn should discharge by gravity or be pumped to a waste water drain or sewer. The main sprinkler riser drain connection should be of a size sufficient to carry off water from the fully open drain valve while it is discharging under normal water system pressures. Where this is not possible, a supplementary drain of equal size should be provided for test purposes with free discharge, located at or above grade.

The restriction on the connection of sprinkler drains and sewers is intended to prevent any harmful element from first entering the sprinkler system by way of the drain connection and then the public water system by way of the sprinkler system. An air gap is the method usually employed to avoid a direct connection and potential backflow. Exhibit 5.37 shows the main drain discharging to atmosphere.

The valve between the sprinkler system and the drain connection and the check valve between the sprinkler system and the public water supply provide sufficient protection against contamination. This additional regulation prohibiting direct interconnections is considered necessary to further enhance protection of the potable water.

The drain connections on risers must be capable of handling the water discharged under normal system pressure. In some situations, such as in high-rise buildings where the drain lines can be long, making the drain riser one size larger than the largest drain connection (see 5-14.2.4.2) may not be sufficient to prevent overpressurization of the normally atmospheric drain line from a high-pressure system.

5-14.2.6.2 Where drain pipes are buried underground, approved corrosion-resistant pipe shall be used.

5-14.2.6.3 Drain pipes shall not terminate in blind spaces under the building.

The drain discharge must be piped in a manner that allows the operator to ascertain that the drain is not obstructed and that it is not causing water damage.

5-14.2.6.4 Where exposed to the atmosphere, drain pipes shall be fitted with a turned-down elbow.

The turned-down elbow discourages the use of the drain piping as a refuse receptacle and minimizes the possibility of damage to property or of passersby getting wet.

5-14.2.6.5 Drain pipes shall be arranged to avoid exposing any part of the sprinkler system to freezing conditions.

When systems are installed where the exterior ambient temperatures are subject to freezing, at least 4 ft (1.2 m) of the drain line beyond the valve should be installed in a warm room. The recommended 4 ft (1.2 m) of pipe beyond the valve provides a frost break that prevents water in the system from freezing due to cold air from outside. Additionally, all moisture needs to be removed from trapped, unheated areas in dry pipe and preaction systems. (See NFPA 25.)

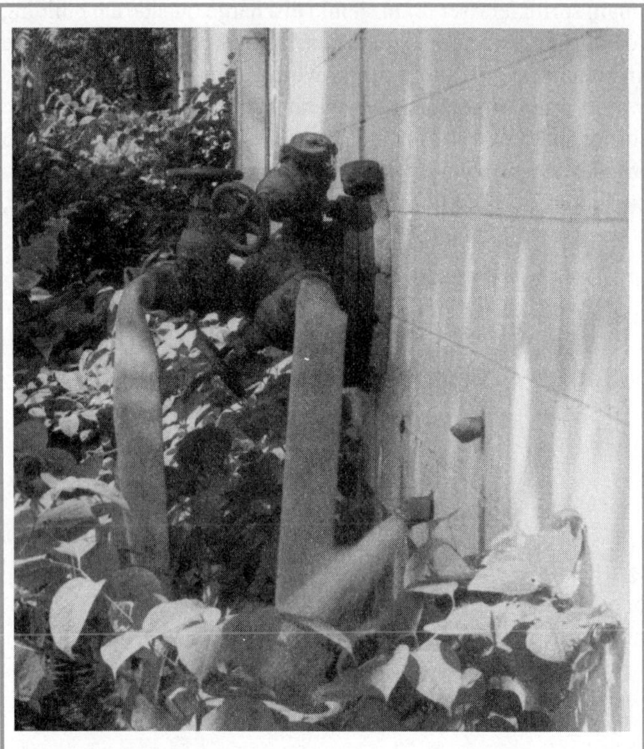

Exhibit 5.37 *System main drain discharging to atmosphere.*

5-14.3 Protection of Piping.

5-14.3.1 Protection of Piping Against Freezing.

5-14.3.1.1 Where portions of systems are subject to freezing and temperatures cannot reliably be maintained at or above 40°F (4°C), sprinklers shall be installed as a dry pipe or preaction system.

Exception: Small unheated areas are permitted to be protected by antifreeze systems or by other systems specifically listed for this purpose. (See 4-5.2.)

A wet pipe system should always be the first choice when selecting a system. Dry pipe systems should be considered when the minimum temperature of 40°F (4°C) cannot be maintained. The use of a dry pipe system for any unheated area should be limited to that area, and a separate wet pipe system should be used for the heated portions of a building. Dry pipe systems tend to have slower operating times, increased corrosion rates on the internal portions of the pipe, and greater maintenance requirements than wet pipe systems.

The exception to 5-14.3.1.1 permits an alternative to dry pipe systems such as antifreeze or other listed means of freeze protection. Other means of freeze protection include steam coils, heat tracing, or circulating hot water tubing that has been specifically investigated and listed for use with sprinkler systems. With any of those alternative approaches, caution must be used to verify that temperatures do not exceed the maximum ambient temperature allowed for the sprinkler's temperature rating. Excessive temperatures can be conducted through the piping and lead to unwanted sprinkler operation. In addition, the need for insulation in conjunction with an alternative approach system must also be evaluated. At least one electrical heating system is on the market that can be used to comply with the exception.

5-14.3.1.2 Where aboveground water-filled supply pipes, risers, system risers, or feed mains pass through open areas, cold rooms, passageways, or other areas exposed to freezing temperatures, the pipe shall be protected against freezing by insulating coverings, frostproof casings, or other reliable means capable of maintaining a minimum temperature between 40°F (4°C) and 120°F (48.9°C).

Although the response to Question 2 of Formal Interpretation 89-9 indicates that large portions of the system are not to be protected by heat tracing and insulation, the exception to 5-14.3.1.1 allows such an arrangement for small, unheated areas, provided the combination of heat tracing and insulation are listed for such use.

Formal Interpretation 89-9

Reference: 5-14.3.1.2

Question 1: Have branch lines been intentionally left out of this paragraph?

Answer: Yes.

Question 2: Is it considered an acceptable practice to heat trace and insulate branch lines with upright sprinklers installed on the line?

Answer: No.

Issue Edition: 1989

Reference: 3-5.1.2

Issue Date: August 3, 1990

Effective Date: August 23, 1990

5-14.3.2 Protection of Piping Against Corrosion.

5-14.3.2.1* Where corrosive conditions are known to exist due to moisture or fumes from corrosive chemicals or both, special types of fittings, pipes, and hangers that resist corrosion shall be used or a protective coating shall be applied to all unprotected exposed surfaces of the sprinkler system. *(See 3-2.6.)*

A-5-14.3.2.1 Types of locations where corrosive conditions can exist include bleacheries, dye houses, metal plating processes, animal pens, and certain chemical plants.

If corrosive conditions are not of great intensity and humidity is not abnormally high, good results can be obtained by a protective coating of red lead and varnish or by a good grade of commercial acid-resisting paint. The paint manufacturer's instructions should be followed in the preparation of the surface and in the method of application.

Where moisture conditions are severe but corrosive conditions are not of great intensity, copper tube or galvanized steel pipe, fittings, and hangers might be suitable. The exposed threads of steel pipe should be painted.

In instances where the piping is not readily accessible and where the exposure to corrosive fumes is severe, either a protective coating of high quality can be employed or some form of corrosion-resistant material used.

Piping installed in ocean-front facilities can also be subject to higher than normal rates of corrosion.

The referenced protective paints and coatings are not applied to the automatic sprinklers. Automatic sprinklers must be provided with corrosion-resistant coatings applied by the manufacturer in accordance with 3-2.6.

5-14.3.2.2 Where water supplies are known to have unusual corrosive properties and threaded or cut-groove steel pipe is to be used, wall thickness shall be in accordance with Schedule 30 [in sizes 8 in. (200 mm) or larger] or Schedule 40 [in sizes less than 8 in. (200 mm)].

Caution must be exercised when considering the inclusion of corrosion-control additives to sprinkler systems, and the authority having jurisdiction should be consulted. A remote possibility of contaminating the domestic supply through backflow exists. Some additives can solidify when exposed to air, causing obstruction to waterflow. Such chemicals have been used for poor quality water or as a corrosion inhibitor for the circulating closed-loop systems discussed in 4-6.1.

Although corrosion protection for exposed surfaces is independent of wall thickness and acceptable pipe-joining methods, the use of threaded thin-wall pipe must be avoided where subject internally to water having unusual corrosive properties. The thinner wall between the inside diameter of the pipe and the root diameter of the threads cannot withstand the same amount of corrosion and introduces a failure point for the piping system. If threaded pipe is used, it should be Schedule 30 or heavier for pipe sizes 8 in. (200 mm) or larger and Schedule 40 for pipe sizes less than 8 in. (200 mm). The use of welded or roll-grooved Schedule 10 pipe would also be satisfactory.

5-14.3.2.3 Where corrosive conditions exist or piping is exposed to the weather, corrosion-resistant types of pipe, fittings, and hangers or protective corrosion-resistant coatings shall be used.

5-14.3.2.4 Where steel pipe is used underground, the pipe shall be protected against corrosion.

5-14.3.3 Protection of Piping in Hazardous Areas. Private service main aboveground piping shall not pass through hazardous areas and shall be located so that it is protected from mechanical and fire damage.

Exception: Aboveground piping is permitted to be located in hazardous areas protected by an automatic sprinkler system.

Private fire service main piping should be located outside of buildings and buried. In some cases, running piping within the building may be necessary. The piping must not be exposed to hazardous areas because it would be subject to damage and not be able to provide water to other areas that may need protection. If automatic sprinkler protection designed for the hazard involved is provided, the potential for main damage is minimized.

5-14.4* Underground Private Fire Service Mains.

A-5-14.4 *Installation Standards.* The following documents apply to the installation of pipe and fittings:

AWWA C603, *Standard for the Installation of Asbestos-Cement Water Pipe*

AWWA C600, *Standard for the Installation of Ductile-Iron Water Mains and Their Appurtenances*

AWWA M11, *A Guide for Steel Pipe-Design and Installation*

AWWA M41, *Ductile Iron Pipe and Fittings*

Concrete Pipe Handbook, American Concrete Pipe Association

Handbook of PVC Pipe, Uni-Bell Plastic Pipe Association

Installation Guide for Ductile Iron Pipe, Ductile Iron Pipe Research Association

Thrust Restraint Design for Ductile Iron Pipe, Ductile Iron Pipe Research Association

5-14.4.1 Depth of Cover.

5-14.4.1.1* The depth of cover over water pipes shall be determined by the maximum depth of frost penetration in the locality where the pipe is laid. The top of the pipe shall be buried not less than 1 ft (0.3 m) below the frost line for the locality. In those locations where frost is not a factor, the depth of cover shall be not less than 2½ ft (0.8 m) to prevent mechanical damage. Pipe under driveways shall be buried a minimum of 3 ft (0.9 m) and under railroad tracks a minimum of 4 ft (1.2 m).

A-5-14.4.1.1 As there is normally no circulation of water in private fire mains, they require greater depth of covering than do public mains. Greater depth is required in a loose gravelly soil (or in rock) than in compact, clayey soil. Recommended depth of cover above the top of underground yard mains is shown in Figure A-5-14.4.1.1.

Buried piping must be located below the frost line to prevent freezing during the winter months. The piping must also be located deep enough to be protected from other surface loads that could cause mechanical damage. Piping located under driveways, roads, and railroad tracks must be buried deeper to prevent undue loads on the sprinkler piping. See 5-14.4.3.3.

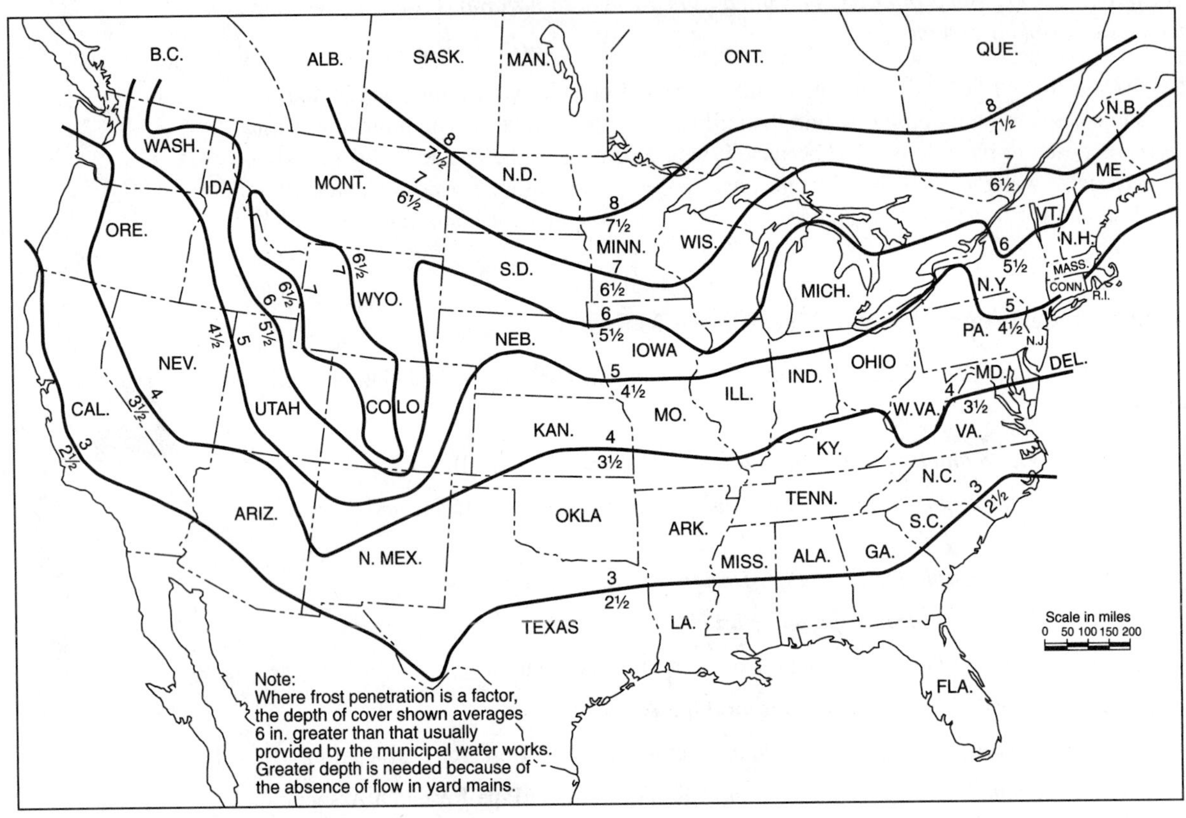

Figure A-5-14.4.1.1 Recommended depth of cover (in feet) above top of underground yard mains.

5-14.4.1.2 Depth of covering shall be measured from top of pipe to finished grade, and due consideration shall always be given to future or final grade and nature of soil.

5-14.4.2 Protection Against Freezing.

5-14.4.2.1 Where it is impracticable to bury pipe, it shall be permitted to be laid aboveground, provided the pipe is protected against freezing and mechanical damage.

5-14.4.2.2 Pipes shall not be placed over water raceways or near embankment walls without special attention being given to protection against frost.

5-14.4.2.3 Where pipe is laid in water raceways or shallow streams, care shall be taken that there will be sufficient depth of running water between the pipe and the frost line during all seasons of frost; a safer method is to bury the pipe 1 ft (0.3048 m) or more under the bed of the waterway. Care shall also be taken to keep the pipe back from the banks a sufficient distance to avoid any danger of freezing through the side of the bank above the water line. Pipe shall be buried below the frost line where entering the water.

5-14.4.3 Protection Against Damage.

5-14.4.3.1 Pipe shall not be run under buildings.

Exception No. 1: When absolutely necessary to run pipe under buildings, special precautions shall be taken that include arching the foundation walls over the pipe, running pipe in covered trenches, and providing valves to isolate sections of pipe under buildings. (See 5-14.5.2.)

Exception No. 2: Fire service mains shall be permitted to enter the building adjacent to the foundation.

Exception No. 3: Where adjacent structures or physical conditions make it impractical to locate risers immediately inside an exterior wall, such risers shall be permitted to be located as close as practical to exterior walls to minimize underground piping under the building.

Piping located under buildings is extremely difficult to repair. When buildings are built over existing underground piping, the piping should be rerouted around the new building. When piping under buildings requires repair, operations in the building must be curtailed, equipment may need to be moved, and the floor must be excavated. Leaks in the buried piping underneath the building could go undetected for long periods, and they can also undermine the building support. The location of system valves in the center of a building also places the system controls in the center of the building, which may not be desirable during a fire event.

The exceptions to 5-14.4.3.1 outline some of the measures that can be taken to minimize the exposure but do not completely eliminate it. Exception No. 2 permits the fire main to run underneath the building's footing or to penetrate the foundation wall if the main rises immediately inside the building adjacent to the exterior wall. Where the main penetrates the foundation wall, arching of the foundation wall over the main or other means of ensuring system and structural integrity need to be provided.

With regard to Exception No. 3, the fire main under the building that connects to the system riser needs to be accessible. Trench plates or other access means to the main running underneath the building are necessary.

5-14.4.3.2 Where a riser is close to building foundations, underground fittings of proper design and type shall be used to avoid pipe joints being located in or under the foundations.

5-14.4.3.3 Mains running under railroads carrying heavy trucking, under large piles of heavy commodities, or in areas that subject the main to heavy shock and vibrations shall be subjected to an evaluation of the specific loading conditions and suitably protected, if necessary. *(See 3-4.2.)*

5-14.4.3.4* When it is necessary to join metal pipe with pipe of dissimilar metal, the joint shall be insulated, by an approved method, against the passage of an electric current.

A-5-14.4.3.4 Gray cast iron is not considered galvanically dissimilar to ductile iron. Rubber gasket joints (unrestrained push-on or mechanical joints) are not considered connected electrically. Metal thickness should not be considered a protection against corrosive environments. In the case of cast-iron or ductile iron pipe for soil evaluation and external protection systems, see Appendix A of AWWA C105, *Polyethelene Encasement for Ductile Iron Pipe Systems.*

5-14.4.3.5 In no case shall the pipe be used for grounding of electrical services.

The use of sprinkler or underground piping for electrical grounding increases the potential for stray currents and increased galvanic corrosion. In addition, the use of nonconductive joints may not provide the expected ground. Electrical equipment should be grounded in accordance with NFPA 70, *National Electrical Code®*.

5-14.4.4 Care in Laying.

The precautions that must be taken to minimize damage to underground piping, eliminate stresses, and ensure a long service life are outlined in 5-14.4.4.

5-14.4.4.1 Pipes, valves, hydrants, and fittings shall be inspected for damage when received and shall be inspected prior to installation. Bolted joints shall be checked for proper torquing of bolts. Pipe, valves, hydrants, and fittings shall be clean inside. When work is stopped, open ends shall be plugged to prevent stones and foreign materials from entering.

Precautions must be taken to prevent rocks and other foreign materials from entering piping during installation. These materials can be carried into sprinkler and other fire protection piping and cause blockages and system failure during an emergency. Care in installation does not eliminate the need to flush the piping.

5-14.4.4.2 All pipe, fittings, valves, and hydrants shall be carefully lowered into the trench with suitable equipment. They shall be carefully examined for cracks or other defects while suspended above the trench immediately before installation. Plain ends shall be inspected with special attention, as these ends are the most susceptible to damage. Under no circumstances shall water main materials be dropped or dumped. Pipe shall not be rolled or skidded against other pipe materials.

5-14.4.4.3 Pipes shall bear throughout their full length and shall not be supported by the bell ends only or by blocks.

Exception: If ground is soft, or of a quicksand nature, special provisions shall be made for supporting pipe. For ordinary conditions of soft ground, longitudinal wooden stringers with cross ties will give good results.

5-14.4.4.4 Valves and fittings used with nonmetallic pipe shall be properly supported and restrained in accordance with the manufacturer's specifications.

5-14.4.5 Pipe Joint Assembly.

5-14.4.5.1 Joints shall be assembled by persons familiar with the particular materials being used and in accordance with the manufacturer's instructions and specifications.

5-14.4.5.2 All bolted joint accessories shall be cleaned and thoroughly coated with asphalt or other corrosion-retarding material after installation.

5-14.4.6 Backfilling.

5-14.4.6.1 Backfill shall be well tamped in layers under and around pipes (and puddled where possible) to prevent settlement or lateral movement and shall contain no ashes, cinders, refuse, organic matter, or other corrosive materials.

Improper backfill is a major cause of underground piping failures. Clean fill cushions the pipe and distributes the load to the surrounding earth. The presence of cinders, refuse, and other organic matter can create points of accelerated corrosion that can reduce the life of underground piping.

5-14.4.6.2 Rocks shall not be placed in trenches. Frozen earth shall not be used for backfilling.

5-14.4.6.3 In trenches cut through rock, tamped backfill shall be used for at least 6 in. (152 mm) under and around the pipe and for at least 2 ft (0.6 m) above the pipe.

5-15 System Attachments

5-15.1* Sprinkler Alarms/Waterflow Alarms.

A-5-15.1 Central station, auxiliary, remote station, or proprietary protective signaling systems are a highly desirable supplement to local alarms, especially from a safety to life standpoint. *(See 5-15.1.6.)*

Approved identification signs, as shown in Figure A-5-15.1, should be provided for outside alarm devices. The sign should be located near the device in a conspicuous position and should be worded as follows:

SPRINKLER FIRE ALARM—WHEN BELL RINGS
CALL FIRE DEPARTMENT OR POLICE.

Figure A-5-15.1 *Identification sign.*

5-15.1.1 Local waterflow alarms shall be provided on all sprinkler systems having more than 20 sprinklers.

NFPA 13 requires an audible waterflow alarm on the premises for all systems that have more than 20 sprinklers. The systems are not required to have supplemental alarm systems. The standard does require alarm supervision of valves controlling sprinkler systems with non-fire protection connections and recommends, but does not require, such supervision in other instances. The sprinkler alarms are waterflow alarms and are not building evacuation alarms. However, in some applications building owners can choose to react to waterflow alarms as evacuation alarms. The purpose for which the alarm is intended should be considered in the overall design of both actuating mechanisms and audible alarm devices.

5-15.1.2 On each alarm check valve used under conditions of variable water pressure, a retarding device shall be installed. Valves shall be provided in the connections to retarding devices to permit repair or removal without shutting off sprinklers; these valves shall be so arranged that they can be locked or sealed in the open position.

Retarding chambers (see Exhibit 5.28) are required when alarm devices connected to alarm valves would otherwise be subject to nuisance operation caused by surges from variable pressure supplies, such as city water or fire pumps. When the supply is a constant pressure source, such as a gravity tank or pressure tank, retarding chambers may not be necessary.

5-15.1.3 Alarm, dry pipe, preaction, and deluge valves shall be fitted with an alarm bypass test connection for an electric alarm switch, water motor gong, or both. This pipe connection shall be made on the water supply side of the system and provided with a control valve and drain for the alarm piping. A check valve shall be installed in the pipe connection from the intermediate chamber of a dry pipe valve.

Exception: The alarm test connection at the riser shall be permitted to be made on the system side of an alarm valve.

Local waterflow alarms should be tested at least quarterly. (See NFPA 25.) The required bypass from the water side of the system allows these tests to be conducted without tripping the system. Tripping the system in areas subject to freezing requires more care, as it could cause the system to freeze.

5-15.1.4 An indicating control valve shall be installed in the connection to pressure-type contactors or water motor–operated alarm devices. Such valves shall be locked or sealed in the open position. The control valve for the retarding chamber on alarm check valves shall be accepted as complying with this paragraph.

The control valve is to be located in the piping between the connection from the alarm valve, dry pipe valve, preaction valve, or deluge valve and the local waterflow alarm device(s). This piping is normally subject to only atmospheric air pressure, and lacking an indicating-type valve, no visual means exists to ascertain if the valve was open and if water could reach the alarm actuators. Exhibits 5.28 and 5.38 identify the bypass test connection along with the control valve.

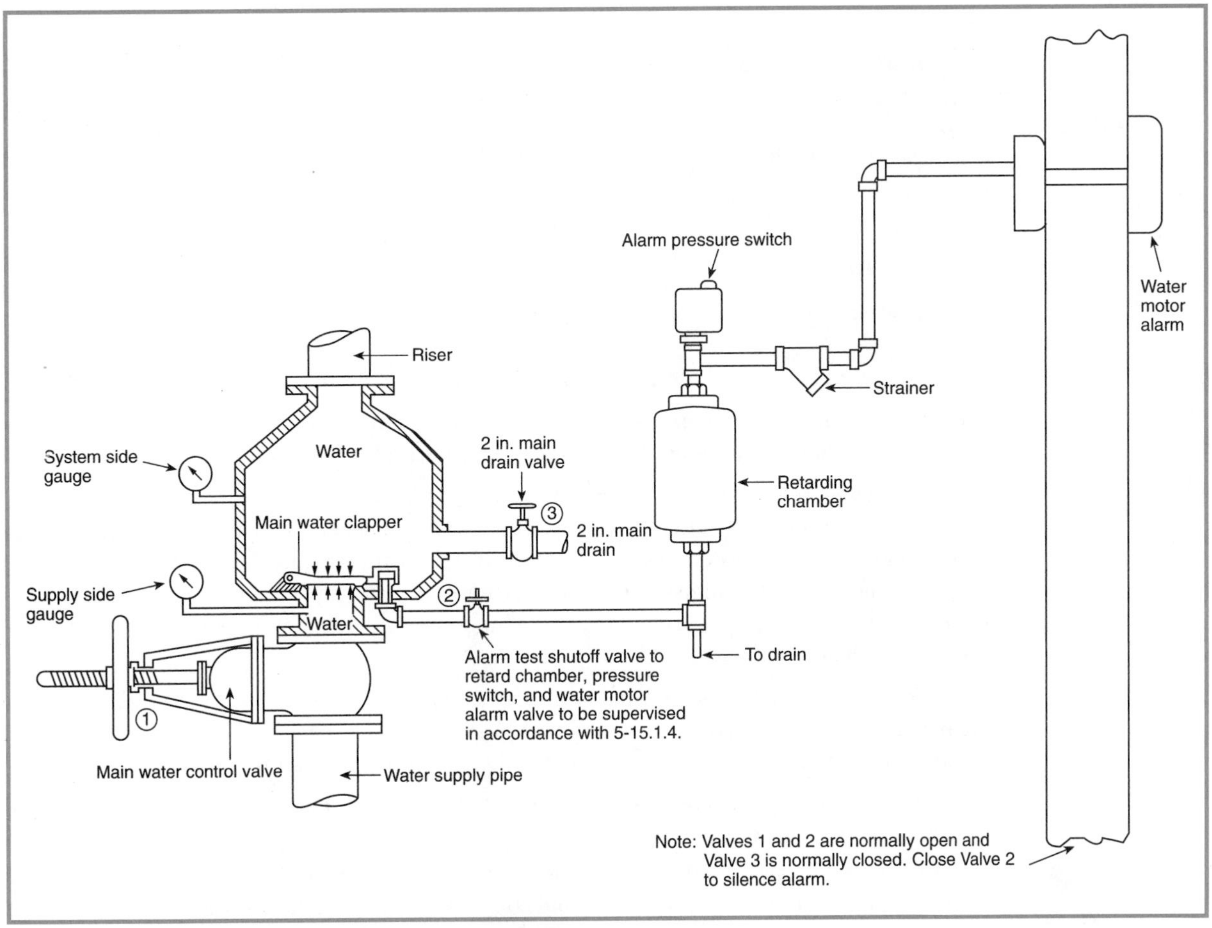

Exhibit 5.38 *Wet alarm valve with alarm test shutoff valve.*

5-15.1.5* Attachments—Mechanically Operated. For all types of sprinkler systems employing water motor–operated alarms, a listed ¾-in. (19-mm) strainer shall be installed at the alarm outlet of the waterflow detecting device.

Exception: Where a retarding chamber is used in connection with an alarm valve, the strainer shall be located at the outlet of the retarding chamber unless the retarding chamber is provided with an approved integral strainer in its outlet.

A-5-15.1.5 Water motor–operated devices should be located as near as practicable to the alarm valve, dry pipe valve, or other waterflow detecting device. The total length of the pipe to these devices should not exceed 75 ft (22.9 m), nor should the water motor–operated device be located over 20 ft (6.1 m) above the alarm device or dry pipe valve.

See commentary following 3-10.4, Exhibit 5.38, and Formal Interpretation 80-6. The strainer and the pipe of corrosion-resistant material are required to protect against obstruction of the small orifice through which water enters the water motor.

Formal Interpretation 80-6

Reference: 5-15.1.5, 3-9.6

Question: Is it the intent of 5-15.1.5 to require water motor operated alarm devices drain piping be of corrosion resistant material?

Answer: No.

Issue Edition: 1980

Reference: 3-17.5

Date: July 1981

5-15.1.6* Alarm Attachments—High-Rise Buildings. When a fire must be fought internally due to the height of a building, the following additional alarm apparatus shall be provided:

(a) Where each sprinkler system on each floor is equipped with a separate waterflow device, it shall be connected to an alarm system in such a manner that operation of one sprinkler will actuate the alarm system and the location of the operated flow device shall be indicated on an annunciator and/or register. The annunciator or register shall be located at grade level at the normal point of fire department access, at a constantly attended building security control center, or at both locations.

Exception: Where the location within the protected buildings where supervisory or alarm signals are received is not under constant supervision by qualified personnel in the employ of the owner, a connection shall be provided to transmit a signal to a remote central station.

(b) A distinct trouble signal shall be provided to indicate a condition that will impair the satisfactory operation of the sprinkler system.

A-5-15.1.6 Monitoring should include but not be limited to control valves, building temperatures, fire pump power supplies and running conditions, and water tank levels and temperatures. Pressure supervision should also be provided on pressure tanks.

Check valves can be required to prevent false waterflow signals on floors where sprinklers have not activated—for example, floor systems interconnected to two supply risers.

To improve the reliability of an automatic sprinkler system in a high-rise building, supervision of any portion of the system that could impair its operation is required. A sprinkler waterflow alarm annunciated by floor is not required, because some systems can take on configurations that would not lend themselves to this type of zoning. On the other hand, where sprinkler waterflow alarms are provided on each floor, they are intended to be annunciated at a point to allow rapid identification of the fire location by the fire department on arrival.

The overall reliability of the system is further improved by requiring remote monitoring of supervisory signals in the numerous cases where 24-hour surveillance of on-site supervisory equipment is not provided.

5-15.1.7* Alarm Service. A central station, auxiliary, remote station, or proprietary sprinkler waterflow alarm shall be provided for sprinkler systems protecting storage in accordance with Section 7-3. A local waterflow alarm shall be permitted where recorded guard service is provided.

A-5-15.1.7 For further information, see NFPA 72, *National Fire Alarm Code.*

5-15.1.8† Sprinkler Waterflow Alarm for In-Rack Sprinklers.

5-15.2* Fire Department Connections.

A-5-15.2 The fire department connection should be located not less than 18 in. (457 mm) and not more than 4 ft (1.2 m) above the level of the adjacent grade or access level.

Typical fire department connections are shown in Figures A-5-15.2(a) and A-5-15.2(b). See NFPA 13E, *Guide for Fire Department Operations in Properties Protected by Sprinkler and Standpipe Systems.*

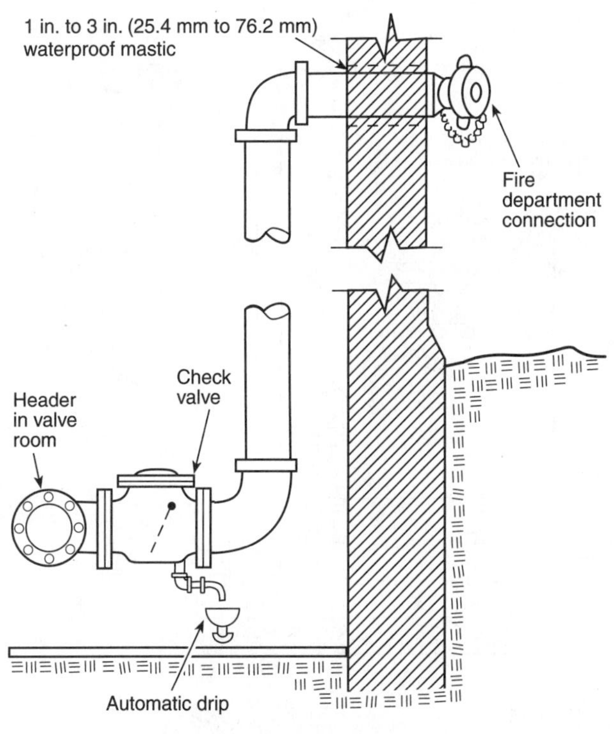

Figure A-5-15.2(a) *Fire department connection.*

5-15.2.1* A fire department connection shall be provided as described in 5-15.2. *(See Figure 5-15.2.1.)*

Exception No. 1: Buildings located in remote areas that are inaccessible for fire department support.

Exception No. 2: Large-capacity deluge systems exceeding the pumping capacity of the fire department.

Exception No. 3: Single-story buildings not exceeding 2000 ft^2 (186 m^2) in area.

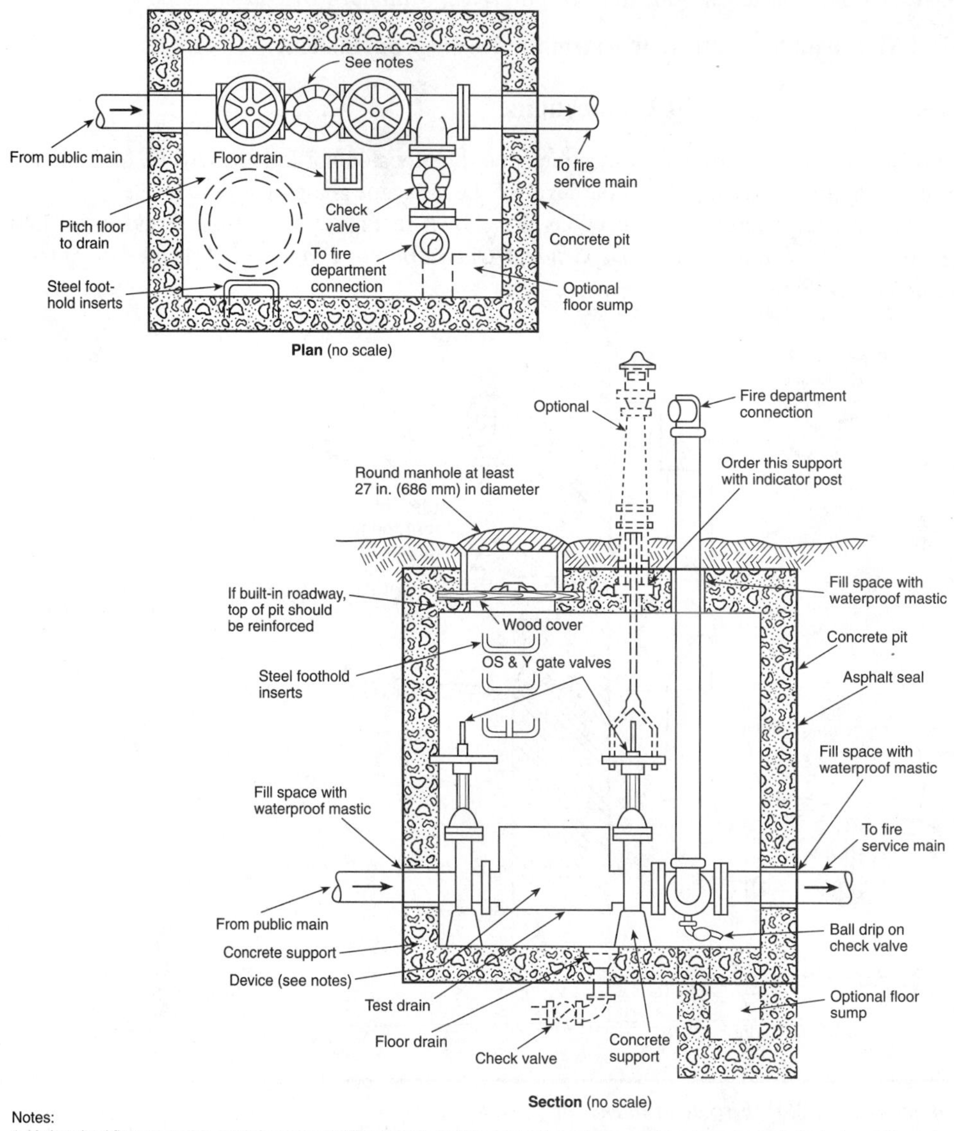

Notes:

1. Various backflow prevention regulations accept different devices at the connection between public water mains and private fire service mains.
2. The device shown in the pit could be any or a combination of the following:
   (a) Gravity check valve
   (b) Detector check valve
   (c) Double check valve assembly
   (d) Reduced pressure zone (RPZ) device
   (e) Vacuum breaker
3. Some backflow prevention regulations prohibit these devices from being installed in a pit.
4. In all cases, the device(s) in the pit should be approved or listed as necessary. The requirements of the local or municipal water department should be reviewed prior to design or installation of the connection.
5. Pressure drop should be considered prior to the installation of any backflow prevention devices.

Figure A-5-15.2(b) *Typical city water pit—valve arrangement.*

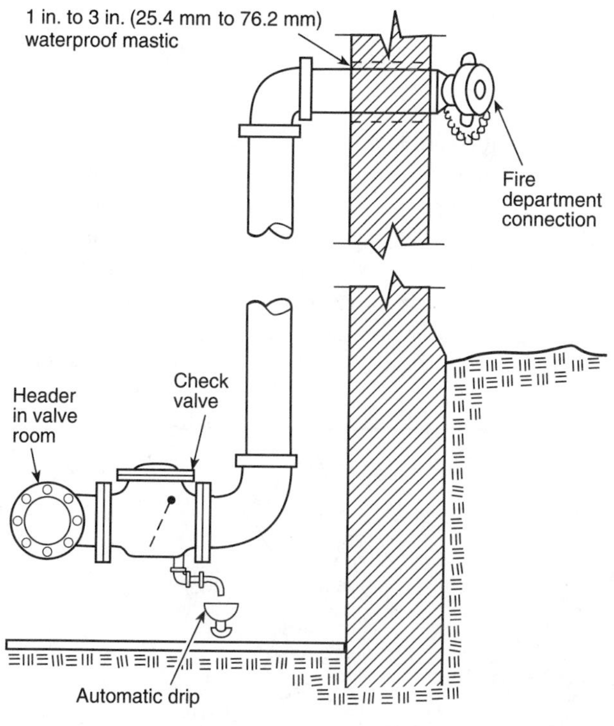

Figure 5-15.2.1 *Fire department connections.*

A-5-15.2.1 Fire department connections should be located and arranged so that hose lines can be readily and conveniently attached without interference from nearby objects including buildings, fences, posts, or other fire department connections. Where a hydrant is not available, other water supply sources such as a natural body of water, a tank, or reservoir should be utilized. The water authority should be consulted when a nonpotable water supply is proposed as a suction source for the fire department.

The three exceptions are examples of situations in which a fire department connection is not practical. Exception Nos. 1 and 2 are conditions in which the connection would be of little or no benefit, thus not to provide the connection is an economical decision. Exception No. 3 addresses a situation in which fire department access is readily available to all sides and areas of the structure; thus, the connection is likely to provide only a minimal benefit.

5-15.2.2 Size. Pipe size shall be 4 in. (102 mm) for fire engine connections and 6 in. (152 mm) for fire boat connections.

Exception No. 1: For hydraulically calculated systems, fire department connection pipe as small as the system riser shall be permitted where serving one system riser.

Exception No. 2: A single-outlet fire department connection shall be acceptable where piped to a 3-in. (76-mm) or smaller riser.

5-15.2.3* Arrangement. See Figure 5-15.2.1.

A-5-15.2.3 The check valve should be located to maximize accessibility and minimize freezing potential.

Each separate water supply connection for sprinkler systems, whether automatic or manual, should be made on the system side of a check valve. Only in this way can pressure differentials between municipal, fire pump, pressure tank, or fire department connections effectively supply the sprinklers. The highest pressure supply comes in first until its capacity is reached. As capacity is exceeded, the pressure drops until the next highest pressure supply comes in. The check valve prevents the higher pressure supply from backing up or circulating into lower pressure supplies. (See 5-14.1.1.5 and its associated commentary, and for examples, see Exhibits 5.39 and 5.40.)

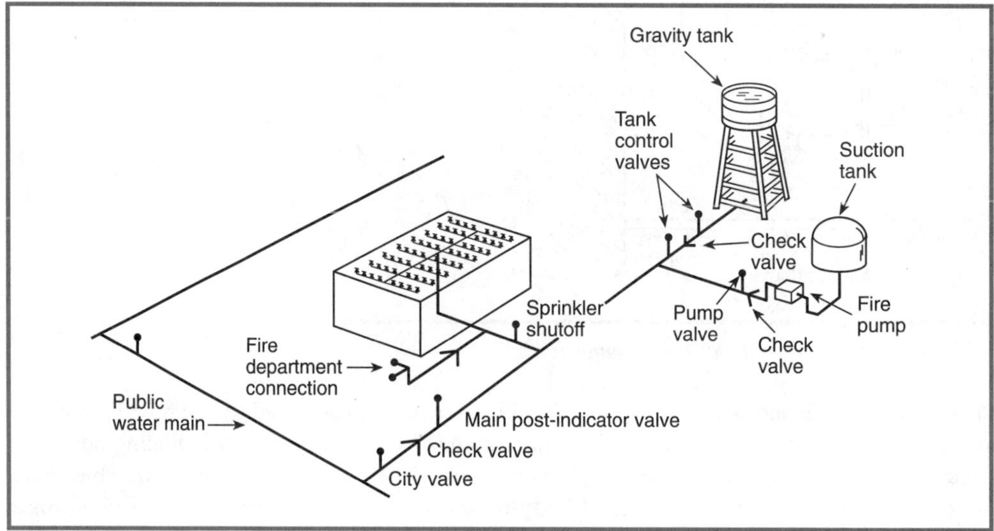

Exhibit 5.39 *Alternative water supplies to sprinkler system.*

Sprinklers in Exhibits 5.39 and 5.40 are normally supplied by a public main with a static pressure of 50 psi (3.4 bar). On arrival at a fire in this building, the fire officer would order a pumper to connect to the hydrant across the street and to pump at 150 psi (10.3 bar) into the yard fire department connection, adding supplementary pressure to the sprinkler system. If the connection were not on the system side of a check valve, the pumped water would circulate back into the same main to the suction side of the pumper, and no supplementary pressure could be added.

5-15.2.3.1 The fire department connection shall be on the system side of the water supply check valve.

See Exhibit 5.26 and Formal Interpretation 91-2.

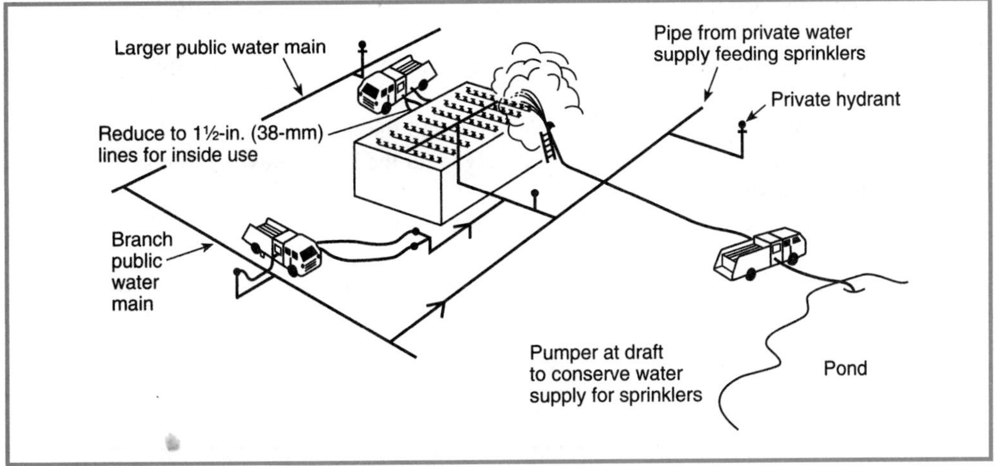

Exhibit 5.40 *Fire department water supply connections to sprinkler system.*

Formal Interpretation 91-2

Reference: 5-15.2.3.1, 5-15.2.3.2(1)

Background: Given a wet pipe sprinkler system with a single system riser.

Question 1: Can the fire department connection be connected to a system cross main?

Answer: Yes.

Question 2: Can the fire department connection be connected to a system feed main?

Answer: Yes.

Question 3: For gridded systems, can the fire department connection be connected to a system cross main?

Answer: Yes. The system designer must verify that the following conditions have been satisfied:

1. The system has been proven hydraulically based upon the use of the fire department connection to supply the necessary flow and pressure.
2. The point of attachment of the fire department connection to the system piping is at least as large as the fire department connection piping.
3. No supplemental or sectional control valves are installed in the feed main or cross main that could restrict delivery of water to the system.

Issue Edition: 1991

Reference: 4-6.2.3.1, 4-6.2.3.2(a)

Issue Date: January 11, 1993

Effective Date: January 30, 1993

5-15.2.3.2 For single systems, the fire department connection shall be installed as follows:

(1) *Wet System.* On the system side of system control, check, and alarm valves *(See Figure A-5-14.1.1.)*
(2) *Dry System.* Between the system control valve and the dry pipe valve
(3) *Preaction System.* Between the preaction valve and the check valve on the system side of the preaction valve
(4) *Deluge System.* On the system side of the deluge valve

Exception: Connection of the fire department connection to underground piping shall be permitted.

See Formal Interpretations 80-5 and 80-19.

Formal Interpretation 80-5

Reference: 5-15.2.3.2

Background: Paragraph 5-15.2.3.2 indicates that the fire department connection shall be made on the system side of an approved indicating valve, unless the sprinklers are supplied by fire department pumper connection in the yard.

Question: Is it the intent of 5-15.2.3.2 to permit a post indicator valve to be present, in line, between the fire department connection and the main riser? (Please refer to the following diagram for clarity on this matter):

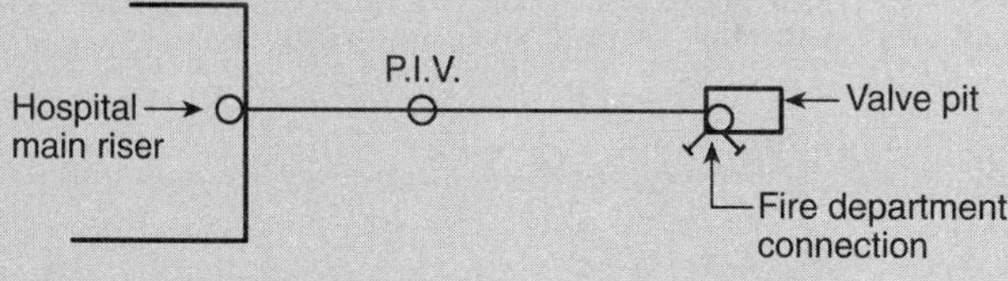

Answer: Yes.

Issue Edition: 1980

Reference: 2-7.3.2

Date: April 1981

5-15.2.3.3 For multiple systems, the fire department connection shall be connected between the supply control valves and the system control valves.

Exception: Connection of the fire department connection to underground piping shall be permitted.

Formal Interpretation 80-19

Reference: 5-15.2.3.2, 5-15.2.3.3

Question: In a sprinkler system having the fire department connection in the yard, may the system indicating valve be of the OS and Y type located on the system riser inside the building?

Answer: Yes.

Issue Edition: 1980

Reference: 2-7.3.2, 2-7.3.4

Date: September 1982

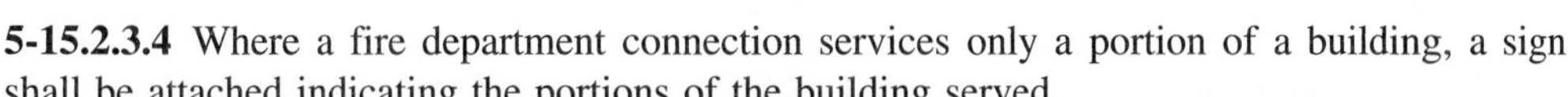

5-15.2.3.4 Where a fire department connection services only a portion of a building, a sign shall be attached indicating the portions of the building served.

Confusion can occur when there are multiple fire department connections or where the fire department connection serves only a portion of a building. If the fire department cannot determine which fire department connection to use, the proper connection or any connection at all may not be used. On other occasions, the fire department may pump into a fire department connection that does not serve the fire area. To avoid these types of problems, NFPA 13 requires that the fire department connection serve all of the sprinkler systems when multiple systems exist (see 5-15.2.3.3) and that connections that serve only a portion of the building be marked with the area they serve.

5-15.2.3.5 Fire department connections shall be on the street side of buildings and shall be located and arranged so that hose lines can be readily and conveniently attached to the inlets without interference from any nearby objects including buildings, fences, posts, or other fire department connections.

Each fire department connection to sprinkler systems shall be designated by a sign having raised or engraved letters at least 1 in. (25.4 mm) in height on plate or fitting reading service design—for example,

AUTOSPKR., OPEN SPKR. AND STANDPIPE

A sign shall also indicate the pressure required at the inlets to deliver the greatest system demand.

Exception: The sign is not required where the system demand pressure is less than 150 psi (10.3 bar).

According to 5-15.2.3 through 5-15.2.3.3, the fire department connection is made on the system side of a check valve. For installation on systems with fire pumps, the fire department connection needs to be made on the sprinkler system side of the discharge check valve for the fire pump. Full efficiency and reliability cannot be obtained for a suction-side connection because of losses through the pump and valves and because of the inability to supplement the system if the discharge-side control valves are closed. Pumping into the suction side of a fire pump or booster pump increases the discharge pressure of the pump, frequently beyond the pressure limits of the system. The fire department connection should be made on the system side of a pump and on the system side—that is, the discharge side—of the pump's discharge check valve and control valves.

The information provided on the sign at the fire department connection is directed at the engineer on the fire department pumper. Since fire department procedures normally require the pump operator to provide 150 psi (10.3 bar) at the connection, the sign is required to indicate that some pressure beyond 150 psi (10.3 bar) is necessary. (See NFPA 13E, *Guide for Fire Department Operations in Properties Protected by Sprinkler and Standpipe Systems*.)

5-15.2.3.6 Fire department connections shall not be connected on the suction side of fire pumps.

5-15.2.3.7 Fire department connections shall be properly supported.

5-15.2.4 Valves.

5-15.2.4.1 A listed check valve shall be installed in each fire department connection.

5-15.2.4.2 There shall be no shutoff valve in the fire department connection piping.

The installation of a shutoff valve on both sides of a check valve to facilitate repair is usually considered good practice. The check valve in the fire department connection is considered to be an exception to general good practice, and no shutoff valve is permitted at that point in the fire department connection. Control valves in a fire department connection serve no useful purpose and only provide a point of system failure. See Exhibit 5.26.

5-15.2.5 Drainage. The piping between the check valve and the outside hose coupling shall be equipped with an approved automatic drip.

Exception: An automatic drip is not required in areas not subject to freezing.

In the event that the check valve in the siamese connection leaks, the purpose of the automatic drip is to drain this water to a safe location and to maintain the piping between the check valve and the hose couplings free of water. Without the automatic drip, any collected water could freeze and prevent the fire department from pumping into the system under fire conditions. This automatic drip also facilitates any maintenance of the fire department connection piping, because this portion of the pipe is already free of water. Exhibits 5.26 and 5.41 show an automatic drip valve used on the fire department connections. The drain should be located at the lowest point of the fire department connection piping to allow complete drainage.

Exhibit 5.41 *Fire department connection with an automatic drain valve located on the outside of the building.*

5-15.3 Gauges.

5-15.3.1 A pressure gauge with a connection not smaller than ¼ in. (6.4 mm) shall be installed at the system main drain, at each main drain associated with a floor control valve, and on the inlet and outlet side of each pressure reducing valve. Each gauge connection shall be equipped with a shutoff valve and provisions for draining.

5-15.3.2 The required pressure gauges shall be listed and shall have a maximum limit not less than twice the normal system working pressure at the point where installed. They shall be installed to permit removal and shall be located where they will not be subject to freezing.

5-15.4 System Connections.

Paragraph 5-14.2.4 should be referenced to determine the relation between the test connections noted in 5-15.4 and the drain sizes previously discussed. A single drain normally serves both functions.

5-15.4.1* Main Drain Test Connections Main drain test connections shall be provided at locations that will permit flow tests of water supplies and connections. They shall be so installed that the valve can be opened wide for a sufficient time to assure a proper test without causing water damage. Main drain connections shall be sized in accordance with 5-14.2.4 and 5-14.2.6.

A-5-15.4.1 See Figure A-5-15.4.1.

Test connections or drain pipes are used to drain the system and to help evaluate the water supply. Test connections and drain pipes are used to drain systems when repairs are necessary or when dry pipe valves have tripped. They are also used to help confirm that the proper water supply is available at the system riser and on the system side of all check valves, control valves, and underground piping.

Test connections or drain pipes are normally smaller than all other water supply piping. Therefore, they are not capable of large-scale flow tests and should not be used as a means of evaluating the available water supply. Waterflow tests conducted through the main drain test do not provide an accurate reading of the water supply. One reason for an inaccurate reading is the unknown friction loss values of check valves under low flow conditions.

Two-in. (51-mm) main drain tests, as they are commonly called, normally consist of opening the 2-in. (51-mm) main drain wide until the pressure stabilizes. Records are kept of these tests to detect possible deterioration of water supplies or to detect valves in the overall water supply system that may have been closed. The 2-in. (51-mm) drain test on new systems necessitates the establishment of a benchmark when the system is first approved during the acceptance tests. See NFPA 25 for additional information on conducting 2-in. (51-mm) drain tests. See Exhibits 5.26, 5.30, and 5.37.

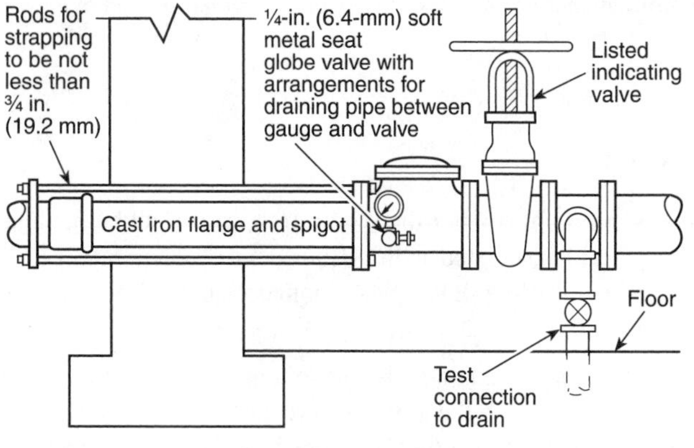

Figure A-5-15.4.1 Water supply connection with test connection.

5-15.4.2* Wet Pipe Systems An alarm test connection not less than 1 in. (25.4 mm) in diameter, terminating in a smooth bore corrosion-resistant orifice, giving a flow equivalent to one sprinkler of a type having the smallest orifice installed on the particular system, shall be provided to test each waterflow alarm device for each system. The test connection valve shall be readily accessible. The discharge shall be to the outside, to a drain connection capable of accepting full flow under system pressure, or to another location where water damage will not result.

A-5-15.4.2 This test connection should be in the upper story, and the connection preferably should be piped from the end of the most remote branch line. The discharge should be at a point where it can be readily observed. In locations where it is not practical to terminate the test connection outside the building, the test connection is permitted to terminate into a drain capable of accepting full flow under system pressure. In this event, the test connection should be made using an approved sight test connection containing a smooth bore corrosion-resistant orifice giving a flow equivalent to one sprinkler simulating the least flow from an individual sprinkler in the system. *[See Figures A-5-15.4.2(a) and A-5-15.4.2(b).]* The test valve should be located at an accessible point and preferably not over 7 ft (2.1 m) above the floor. The control valve on the test connection should be located at a point not exposed to freezing.

The primary function of the wet pipe system inspector's test is to verify the operation of the waterflow alarm device(s)—for example, water-motor gong, pressure switch, or flow switch—at a flow equivalent to that of one operating sprinkler.

The intent of 5-15.4.2 is to require inspector's test connections in a manner that provides for the testing of all waterflow alarm devices. Both an electrically operated alarm and a water-motor alarm supplied from an alarm valve are permitted to be tested through a single test connection.

Ideally the inspector's test connection is located at the end of the most remote branch line in the upper story. Such a location is not, however, required. For a wet pipe system, the exclusive purpose of the inspector's test is to verify operation of the alarm attachment, which is the reason that the test connection is not mandated at the remote portion of the system. The test connection is only required to be located downstream of the alarm device. Exhibit 5.42 shows a test connection located in the fire pump room near the sprinkler system alarm valve.

5-15.4.3* Dry Pipe Systems. A trip test connection not less than 1 in. (25.4 mm) in diameter, terminating in a smooth bore corrosion-resistant orifice, to provide a flow equivalent to one sprinkler of a type installed on the particular system, shall be installed on the end of the most distant sprinkler pipe in the upper story and shall be equipped with a readily accessible shutoff valve and plug not less than 1 in. (25.4 mm), at least one of which shall be brass. In lieu of a plug, a nipple and cap shall be acceptable.

The dry pipe system inspector's test connection is used to measure the approximate time from the opening of the most distant sprinkler in the system until water flows from that sprinkler. For that reason, the test connection must be located at the end of the farthest line in the top story of the protected occupancy. The valve must be sealed with a plug or nipple and cap when not in use to avoid leakage of air and to avoid accidental tripping of the dry pipe valve.

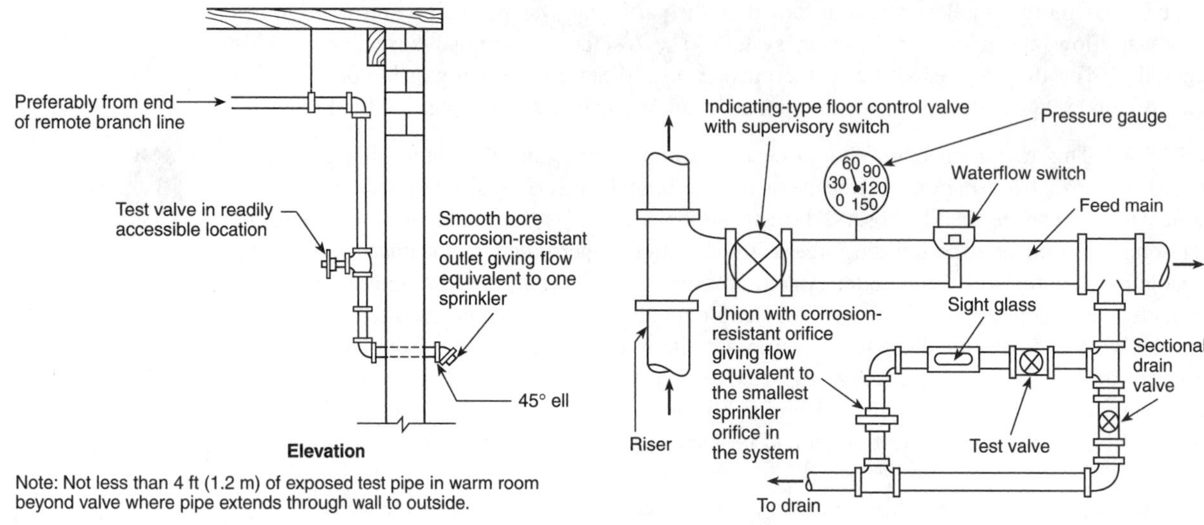

Figure A-5-15.4.2(a) System test connection on wet pipe system.

Figure A-5-15.4.2(b) Floor control valve.

The installation of the test valve on the drop pipe is permitted for the purpose of accessibility. In this case, the connection should be made to the top of the branch line to minimize condensation buildup. Moisture should be drained periodically as is done with low-point drains in areas subject to freezing. Manufacturer's instructions should be followed to avoid accidental operation of dry pipe valves equipped with quick-opening devices.

A-5-15.4.3 See Figure A-5-15.4.3.

5-15.4.4 Preaction Systems. A test connection shall be provided on a preaction system using supervisory air. The connection used to control the level of priming water shall be considered adequate to test the operation of the alarms monitoring the supervisory air pressure.

5-15.4.5 Deluge Systems. A test connection is not required on a deluge system.

5-15.4.6 Backflow Devices.

5-15.4.6.1* Backflow Prevention Valves. Means shall be provided downstream of all backflow prevention valves for flow tests at system demand.

A-5-15.4.6.1 The full flow test of the backflow prevention valve can be performed with a test header or other connection downstream of the valve. A bypass around the check valve in the fire department connector line with a control valve in the normally closed position can be an

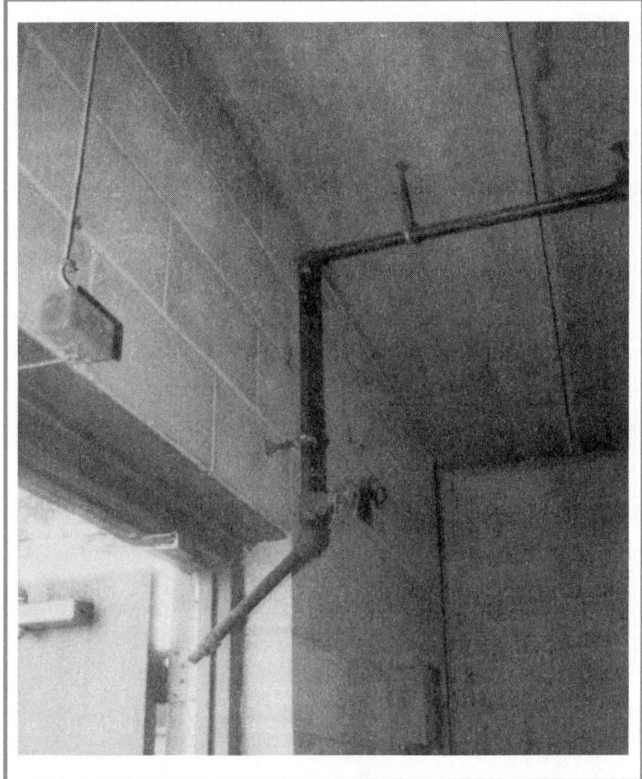

Exhibit 5.42 Inspector's test connection for a wet pipe system.

acceptable arrangement. When flow to a visible drain cannot be accomplished, closed loop flow can be acceptable if a flow meter or site glass is incorporated into the system to ensure flow.

5-15.4.6.2 When backflow prevention devices are to be retroactively installed on existing systems, a thorough hydraulic analysis, including revised hydraulic calculations, new fire flow data, and all necessary system modifications to accommodate the additional friction loss, shall be completed as a part of the installation.

A hydraulic analysis is required to determine and properly account for the impact of a retroactively installed backflow prevention device on the performance of an existing sprinkler system. Existing pipe schedule systems are not required to be recalculated in accordance with the hydraulic calculation methods of 7-2.3 when a backflow prevention device is retroactively installed. However, pipe schedule systems are required to meet certain pressure and flow requirements in accordance with 7-2.2.1. As with a hydraulically calculated sprinkler system, the water supply must be verified to still meet the demand of the pipe schedule system after the installation of a backflow prevention device. A backflow prevention device is shown in Exhibit 5.43.

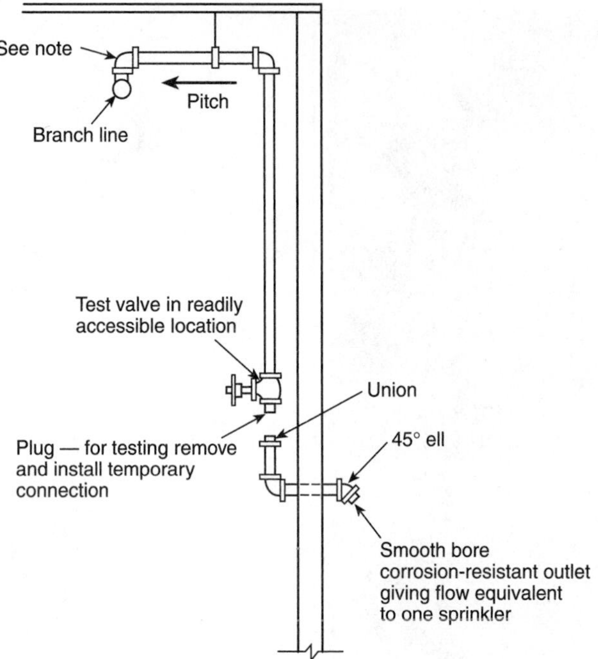

Figure A-5-15.4.3 *System test connection on dry pipe system.*

5-15.5 Hose Connections.

5-15.5.1† Small (1½-in.) Hose Connections.

5-15.5.1.1* Where required by Sections 7-3, 7-4, 7-6, 7-7, and 7-8, small (1½ in.) hose lines shall be available to reach all portions of the storage area. The hose connections shall not be required to meet the requirements of Class II hose systems defined by NFPA 14, *Standard for the Installation of Standpipe and Hose Systems.* Hose connections shall be supplied from one of the following:

(1) Outside hydrants
(2) A separate piping system for small hose stations
(3) Valved hose connections on sprinkler risers where such connections are made upstream of all sprinkler control valves
(4) Adjacent sprinkler systems
(5) In rack storage areas, the ceiling sprinkler system in the same area (as long as in-rack sprinklers are provided in the same area and are separately controlled)

Exhibit 5.43 *Backflow prevention device.*

A-5-15.5.1.1 In areas used to store baled cotton, due consideration to access aisle configuration should be given with maximum hose lengths not exceeding 100 ft (30.1 m). Additionally, in these areas, where a separate piping system is used to supply hose lines it should be in accordance with NFPA 14, *Standard for the Installation of Standpipe and Hose Systems.*

Storage fires normally require the use of small hose lines for complete extinguishment of fires within boxes or other areas that can be shielded from the sprinklers at the ceiling. These hose connections can be supplied from separate or adjacent systems or outside hydrants so that the ceiling sprinklers can be shut off during the clean-up operation. These hose connections are not intended to serve as standpipes in accordance with NFPA 14, *Standard for the Installation of Standpipe and Hose Systems*. Exhibit 5.44 shows a hose connection equipped with a hose line in a storage facility.

5-15.5.1.2* Hose used for fire purposes only shall be permitted to be connected to wet sprinkler systems only, subject to the following restrictions:

(1) Hose station's supply pipes shall not be connected to any pipe smaller than 2½ in. (64 mm).

Exception: For hydraulically designed loops and grids, the minimum size pipe between the hose station's supply pipe and the source shall be permitted to be 2 in. (51 mm).

Exhibit 5.44 *Sprinkler system equipped with a hose connection.*

(2) For piping serving a single hose station, pipe shall be minimum 1 in. (25.4 mm) for horizontal runs up to 20 ft (6.1 m), minimum 1¼ in. (33 mm) for the entire run for runs between 20 and 80 ft (6.1 and 24.4 m), and minimum 1½ in. (38 mm) for the entire run for runs greater than 80 ft (24.4 m). For piping serving multiple hose stations, runs shall be a minimum of 1½ in. (38 mm) throughout.
(3) Piping shall be at least 1 in. (25 mm) for vertical runs.
(4) When the pressure at any hose station outlet exceeds 100 psi (6.9 bar), an approved device shall be installed at the outlet to reduce the pressure at the outlet to 100 psi (6.9 bar).

A-5-15.5.1.2 This standard covers 1½-in. (38-mm) hose connections for use in storage occupancies and other locations where standpipe systems are not required. Where Class II standpipe systems are required, see the appropriate provisions of NFPA 14, *Standard for the Installation of Standpipe and Hose Systems,* with respect to hose stations and water supply for hose connections from sprinkler systems.

Hose connections of 1½ in. (38 mm) are restricted to wet pipe systems. The maintenance problems due to loss of air pressure through the hose valves in other types of systems would outweigh their value. These types of connections are convenience connections for first-aid fire-fighting conditions. These connections are not standpipe connections. Thus, the flows noted in Table 7-2.3.1.1 are all that need to be accounted for in this type of connection.

Hose connections of 1½ in. (38 mm) supplied from sprinkler systems have been successful in fire extinguishment and fire control for many years. The requirements for Class II standpipe systems in NFPA 14 do not apply to hose connections attached to sprinkler systems.

The pressure at the outlet is restricted to 100 psi (6.9 bar) due to the danger to the operator in using a hose subject to high pressure in either static or underflow conditions. Therefore, the maximum allowable pressure at the hose outlet under both conditions is 100 psi (6.9 bar).

5-15.5.2* Hose Connections for Fire Department Use. In buildings of light or ordinary hazard occupancy, 2½-in. (64-mm) hose valves for fire department use are permitted to be attached to wet pipe sprinkler system risers. *[See 7-2.3.1.3(d).]* The following restrictions shall apply:

(1) Sprinklers shall be under separate floor control valves.
(2) The minimum size of the riser shall be 4 in. (102 mm) unless hydraulic calculations indicate that a smaller size riser will satisfy sprinkler and hose stream demands.
(3) Each combined sprinkler and standpipe riser shall be equipped with a riser control valve to permit isolating a riser without interrupting the supply to other risers from the same source of supply.

(For fire department connections serving standpipe and sprinkler systems, refer to Section 3-9.)

A-5-15.5.2 Combined automatic sprinkler and standpipe risers should not be interconnected by sprinkler system piping.

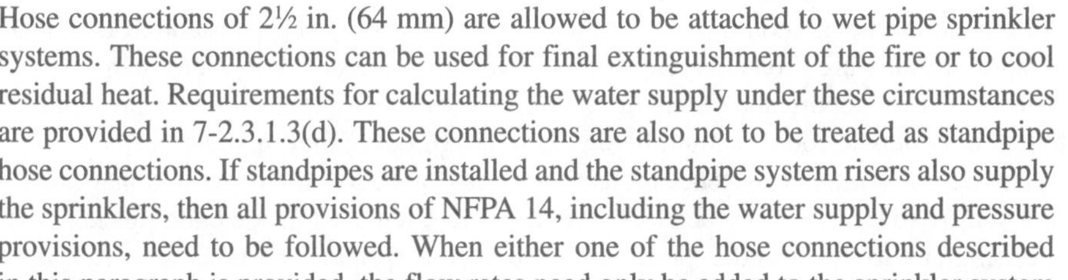

Hose connections of 2½ in. (64 mm) are allowed to be attached to wet pipe sprinkler systems. These connections can be used for final extinguishment of the fire or to cool residual heat. Requirements for calculating the water supply under these circumstances are provided in 7-2.3.1.3(d). These connections are also not to be treated as standpipe hose connections. If standpipes are installed and the standpipe system risers also supply the sprinklers, then all provisions of NFPA 14, including the water supply and pressure provisions, need to be followed. When either one of the hose connections described in this paragraph is provided, the flow rates need only be added to the sprinkler system at the design pressure available at the point of connection to the sprinkler system pipe.

When risers are used to supply combination sprinkler and standpipe systems, guidance is needed to avoid confusion where systems in remotely located areas are controlled by more than one valve. No appreciable improvement in reliability is realized by interconnection. In addition, if the risers are cross connected and the hydraulic calculations account for this feature, the sprinkler system might not perform as intended if one of the risers is removed from service.

5-16 Spray Application Using Flammable and Combustible Materials

The material in Sections 5-16 through 5-32 is extracted from other NFPA standards so that the sprinkler installation criteria are located in one place. To completely understand the criteria in these sections, the referenced standard should be consulted. In addition, the referenced NFPA document can provide options to standard sprinkler protection, such as water spray sprinklers, foam-water sprinklers, or another special extinguishing system, for the protection of those hazards.

5-16.1 For applicable terms not defined in Chapter 1, the terms defined in NFPA 33, *Standard for Spray Application Using Flammable or Combustible Materials,* shall be used.

5-16.2* The sprinklers for each spray area and mixing room shall be controlled by a separate, accessible, listed indicating valve. Sprinkler systems in stacks or ducts shall be automatic and of a type not subject to freezing. (**33:** 7-2.4)

A-5-16.2 Automatic sprinklers in spray areas, including the interior of spray booths and exhaust ducts, should be wet pipe, preaction, or deluge system in order that water can be placed on the fire in the shortest possible time. Automatic sprinklers in spray booths and exhaust ducts should be of the lowest practical temperature rating. Sprinklers outside the temperature rating. Sprinklers outside the booth at ceiling level should be high temperature–rated [286°F (141°C)]. The delay in application of water with ordinary dry pipe sprinklers can permit a fire to spread so rapidly that final extinguishment is difficult without large resulting damage.

The location of the sprinklers inside spray booths should be selected with care in order to avoid sprinklers being placed in the direct path of spray and yet afford protection for the entire booth interior. When in the direct path of spray even one day's operation can result in deposits on the sprinkler head that insulate the fusible link or choke open head orifices to the extent that sprinklers cannot operate efficiently.

Automatic sprinklers should also be located so that areas subject to substantial accumulations of overspray residue are protected. Generally, sprinklers are located no more than 4 ft (1.2 m) from side walls of booths and rooms and from dry overspray collectors (where applicable). Sprinklers in booths or rooms should be on extra hazard occupancy spacing of 90 ft^2 (9.4 m^2).

Sprinklers or sprinkler systems protecting stacks or ducts should be automatic and of a type not subject to freezing. Dry pendent sprinklers are often used inside buildings near exhaust duct penetrations to the outside. Nonfreeze or dry type sprinkler systems are often used for ducts outside buildings. Sprinklers should be spaced no more than 12 ft (3.7 m) apart in the duct for adequate protection.

All sprinklers in spray areas should be controlled by an accessible control valve, preferably an OS&Y valve. (**33:** A-7-2.4)

5-16.3 Sprinklers protecting spray areas and mixing rooms shall be protected against overspray residue so that they will operate quickly in event of fire. If covered, cellophane bags having a thickness of 0.003 in. (0.076 mm) or less, or thin paper bags shall be used. Coverings shall be replaced frequently so that heavy deposits of residue do not accumulate. Sprinklers that have been painted or coated, except by the sprinkler manufacturer, shall be replaced with new listed sprinklers having the same characteristics. (**33:** 7-2.5)

5-17 Storage and Handling of Cellulose Nitrate Motion Picture Film

5-17.1 For applicable terms not defined in Chapter 1, the terms defined in NFPA 40, *Standard for the Storage and Handling of Cellulose Nitrate Motion Picture Film,* shall be used.

5-17.2 In areas or room where nitrate film is handled, the area that is protected per sprinkler shall not exceed 64 ft^2 (6 m^2) with sprinklers and branch lines not being over 8 ft (2.4 m) apart. (**40:** 3-1.4)

5-17.3 Cabinet Protection.

(**40:** 4-2.5)

5-17.3.1 Where cabinets are required to be sprinklered, they shall be provided with at least one automatic sprinkler. (**40:** 4-2.5.1)

5-17.3.2 Where cans are stored on more than one shelf, as shown in Figure 5-17.3.2 and as described in 4-2.6.1 or 4-2.6.2 of NFPA 40, one sprinkler shall be provided for each shelf. (**40:** 4-2.5.2)

5-17.4 Vaults Other than Extended Term Storage Vaults.

(**40:** 4-3)

5-17.4.1 Sprinkler protection utilizing regular automatic sprinklers or open sprinklers shall be calculated on the basis of one sprinkler for each 62.5 ft^3 (1.8 m^3) of the interior vault volume. (**40:** 4-3.6.1)

5-17.4.2 The minimum number of sprinklers for a standard 750-ft^3 (21-m^3) vault shall be not less than 12. (**40:** 4-3.6.2)

5-17.5 Extended Term Storage Vaults.

5-17.5.1 Sprinklers shall be provided in a ratio of one sprinkler for each 62.5 ft^3 (1.8 m^3) of vault volume.

Exception: Sprinkler systems in existing extended term storage vaults that were in compliance with the provisions of NFPA 40 at the time of installation shall be permitted to be continued in use.

(**40:** 4-5.5.1)

5-17.5.2 The minimum number of sprinklers for a 1000-ft^3 (28-m^3) vault shall be 15 sprinklers.

Exception: Sprinkler systems in existing extended term storage vaults that were in compliance with the provisions of NFPA 40 at the time of installation shall be permitted to be continued in use.

(**40:** 4-5.5.2)

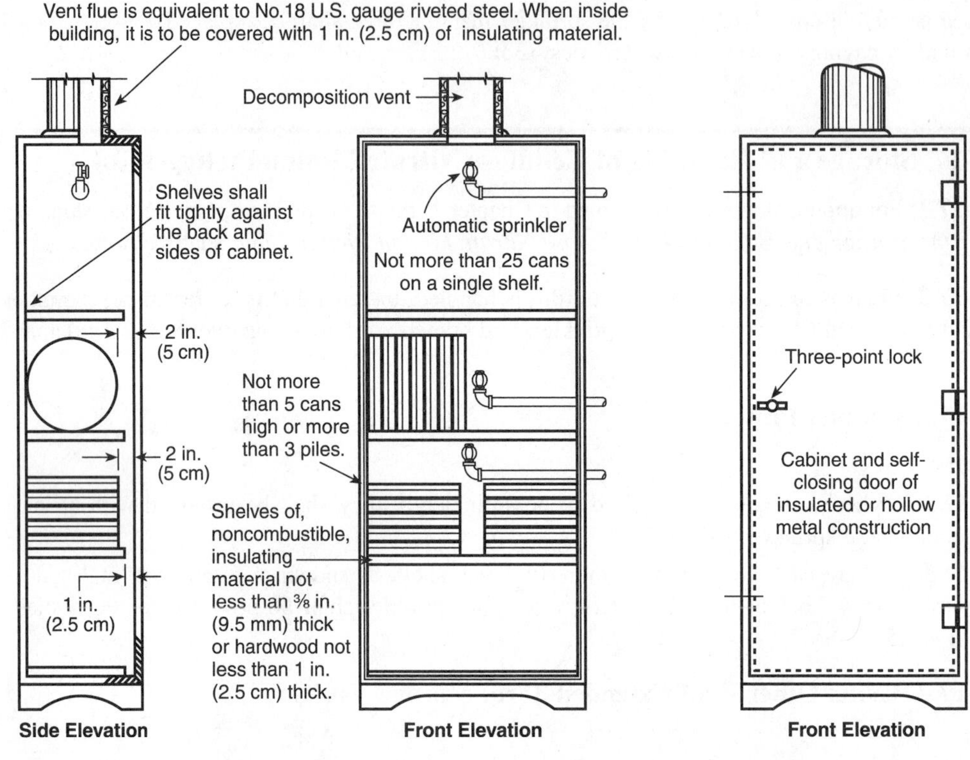

Figure 5-17.3.2 *Standard film cabinet for other than extended term storage film. (40: Figure 4-2)*

5-17.5.3 Directional sprinklers that will provide coverage into the face of the shelves shall be provided. (**40:** 4-5.5.3)

5-17.6 Motion Picture Film Laboratories.

In all cases, sprinklers shall be arranged so that not more than two machines are protected by any one sprinkler. (**40:** 7-2.5.2)

5-18 Storage of Pyroxylin Plastic

5-18.1 For applicable terms not defined in Chapter 1 the terms defined in NFPA 42, *Code for the Storage of Pyroxylin Plastic,* shall be used.

5-18.2 Where sprinkler systems are provided for isolated storage buildings per Section 3-4.3 of NFPA 42, sprinklers shall be spaced so that there is one sprinkler per 32 ft^2 (3 m^2). (**42:** 3-4.3)

5-18.3 Sprinklers in buildings used for storage of loose scrap shall be installed in the ratio of one sprinkler for each 1000 lb (454 kg) of storage.

Exception: The ratio in 5-18.3 shall need not apply if the scrap is in tanks or other receptacles kept filled with water.

(**42:** 3-4.4)

5-18.4 Where cabinets are required to be sprinklered, they shall have at least one automatic sprinkler in each compartment. (**42:** 4-2.10)

5-18.5* Vaults Containing Pyroxylin Plastic.

A-5-18.5 See Figures A-5-18.5(a) and (b).

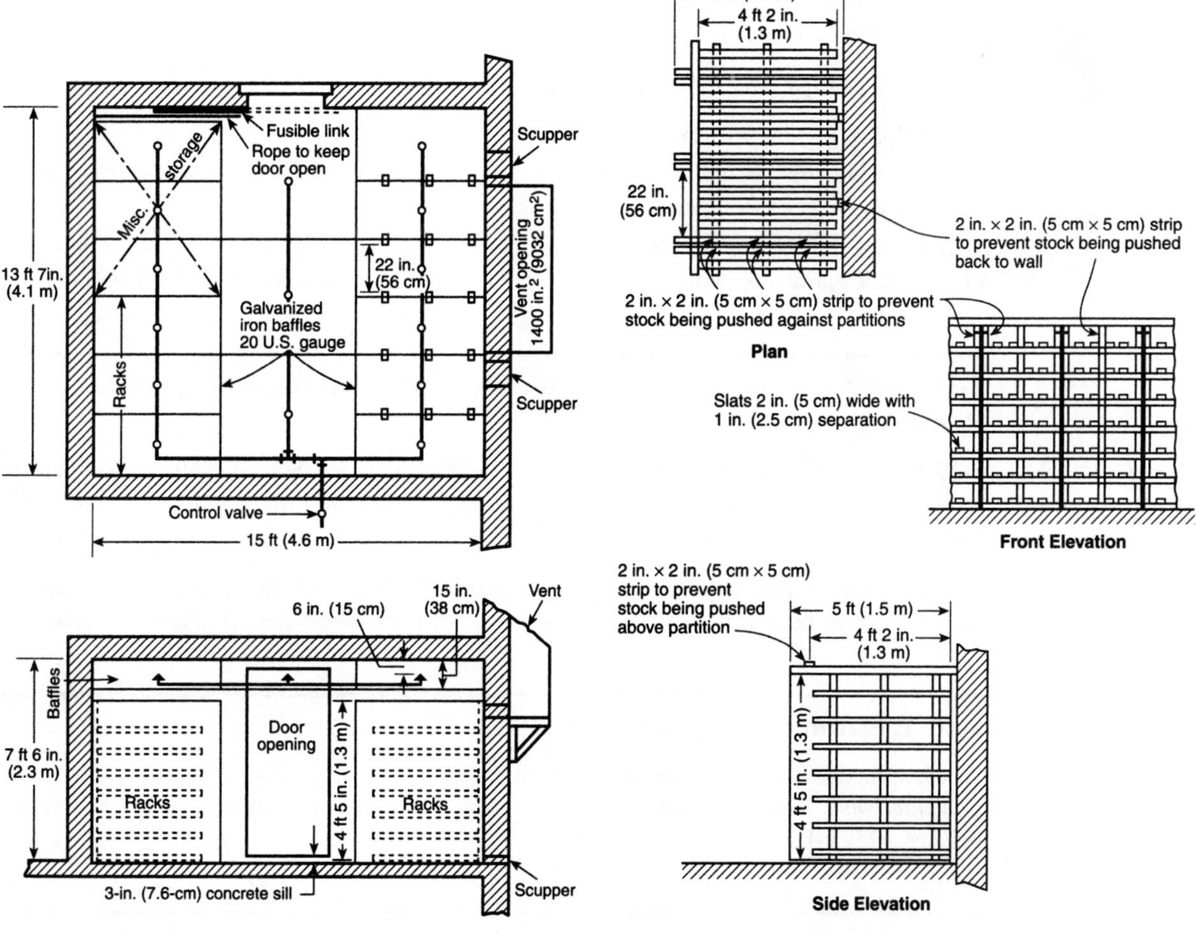

Figure A-5-18.5(a) *Raw stock storage vault, showing general arrangement of sprinklers, racks, and baffles. [42: Figure 4-3.3.7(a)]*

Figure A-5-18.5(b) *Details of storage racks in raw stock storage vault. [42: Figure 4-3.3.7(b)]*

5-18.5.1 Vaults shall be equipped with automatic sprinklers in a ratio of one sprinkler to each 834 lb (378 kg) of pyroxylin plastic or one sprinkler to each 125 ft^3 (3.5 m^3) of total vault space. (**42:** 4-4.1)

5-18.5.2 A vault that is divided into two or more sections shall have at least one automatic sprinkler in each section. (**42:** 4-4.2)

5-18.5.3 Sprinkler systems for vaults shall be equipped with a 1-in. (2.5-cm) drip line with a ½-in. (13-mm) outlet valve. (**42:** 4-4.5)

5-18.6* Tote-Box Storeroom for Pyroxylin Plastic.

Sprinkler protection provided for the tote box storeroom shall consist of one sprinkler in the center of the aisle immediately in front of the dividing partition between each pair of sections. Proper baffles shall be provided between heads. *[See Figure A-5-18.6(a).]* (**42:** 4-7.9)

A-5-18.6 See Figures A-5-18.6(a) and (b).

5-18.7* Finished Stock Storeroom for Pyroxylin Plastic.

A-5-18.7 See Figures A-5-18.7(a) and (b).

5-18.7.1 Automatic sprinklers shall be installed with proper baffles between sprinklers in the center of the aisle opposite each section. (**42:** 4-8.7)

5-18.7.2 Special Rooms for Stock in Shipping Cases. The special room shall be protected by automatic sprinklers, with at least one sprinkler for each 64 ft^2 (6 m^2). (**42:** 4-9.4)

5-19 Oxygen-Fuel Gas Systems for Welding, Cutting, and Allied Processes

5-19.1 For applicable terms not defined in Chapter 1, the terms defined in NFPA 51, Standard for the *Design and Installation of Oxygen-Fuel Gas Systems for Welding, Cutting, and Allied Processes,* shall be used.

5-19.2 Where sprinkler systems are provided per NFPA 51, 2-3.1, Exception No. 1, sprinklers shall be located not more than 20 ft (6 m) above the floor where the cylinders are stored.

5-20 Electronic Computer Systems

Where sprinkler systems are provided per NFPA 75, *Standard for the Protection of Electronic Computer/Data Processing Equipment,* they shall be valved separately from other sprinkler systems. (**75:** 6-1.3)

5-21 Incinerators, Systems and Equipment

5-21.1 Where sprinkler systems are provided per NFPA 82, *Standard on Incinerators and Waste and Linen Handling Systems and Equipment,* the following shall apply.

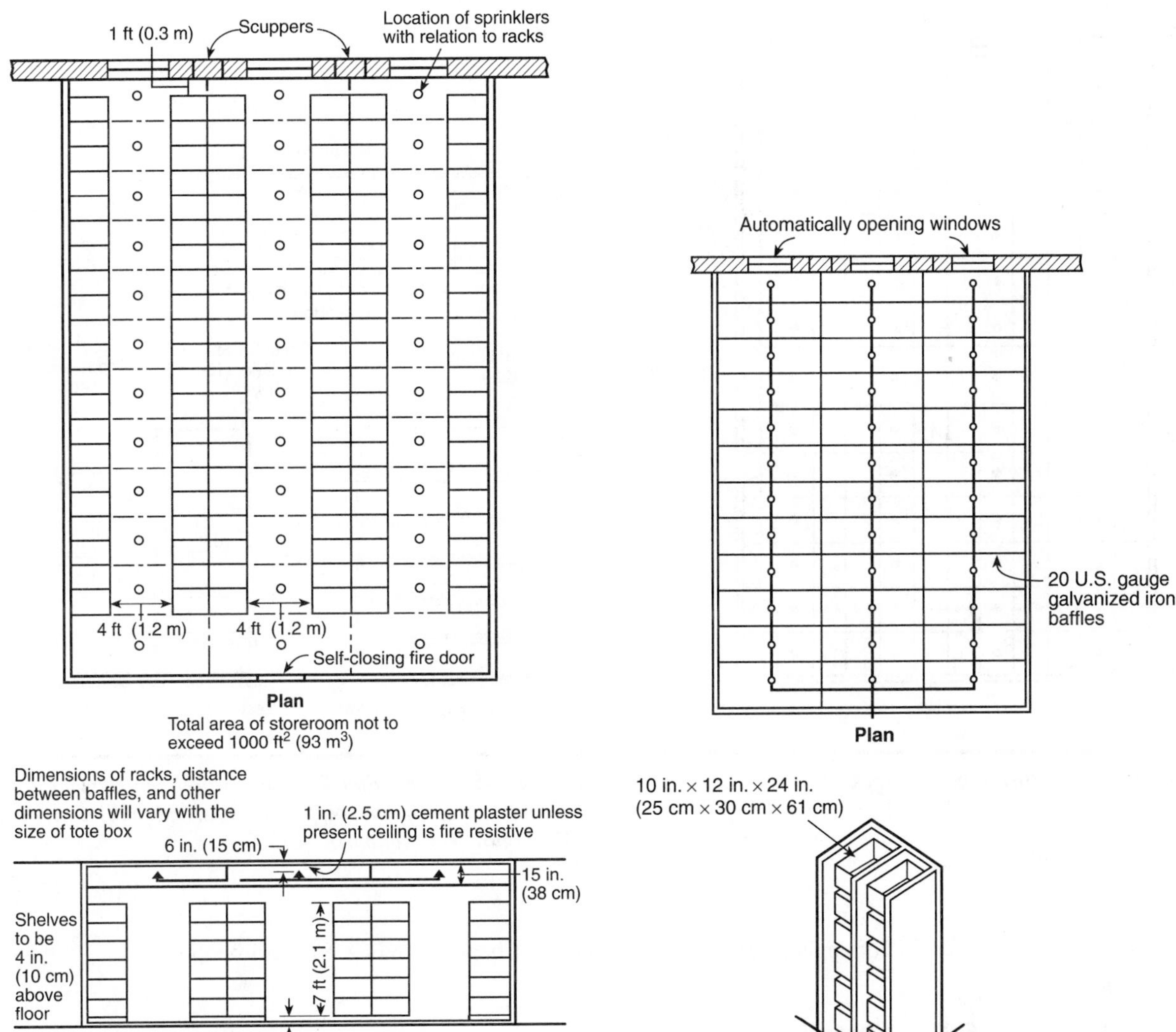

Figure A-5-18.6(a) *Tote-box storeroom showing general arrangement of racks and sprinklers. [42: Figure 4-7]*

Figure A-5-18.6(b) *Tote-box storeroom showing arrangement of sprinklers and baffles and section of tote-box storage rack. [42: Figure 4-7.7]*

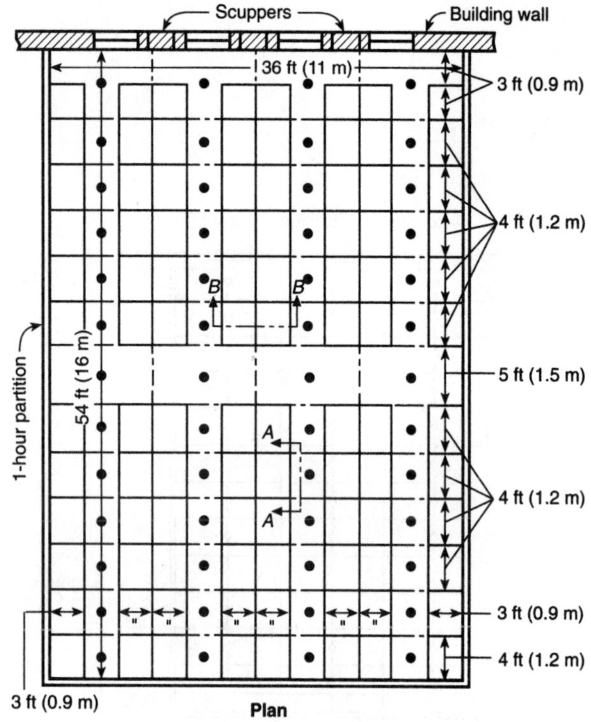

Figure A-5-18.7(a) *Finished-stock storeroom showing general arrangement of racks. [42: Figure 4-8 (top)]*

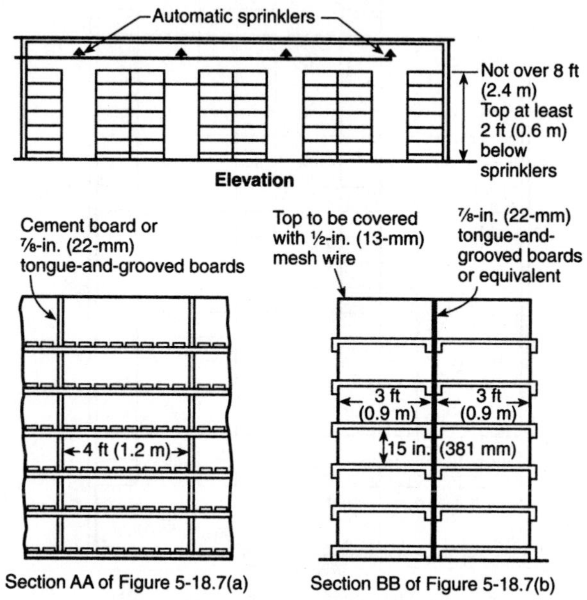

Figure A-5-18.7(b) *Finished-stock storeroom showing arrangement of sprinklers, baffles, shelves, and rack partitions. [42: Figure 4-8 (bottom)]*

5-21.2 For applicable terms not defined in Chapter 1, NFPA 82 shall be used.

5-21.3 Chute Automatic Sprinklers.

(**82:** 3-2.5)

5-21.3.1* Gravity Chute. Gravity chutes shall be protected internally by automatic sprinklers. This requires a sprinkler at or above the top service opening of the chute, and, in addition, a sprinkler shall be installed within the chute at alternate floor levels in buildings over two stories in height with a mandatory sprinkler located at the lowest service level. (**82:** 3-2.5.1)

A-5-21.3.1 See Figure A-5-21.3.1.

5-21.3.2 Chute Sprinkler Head Protection. Automatic sprinklers installed in gravity chute service openings shall be recessed out of the chute area through which the material travels. (**82:** 3-2.5.2)

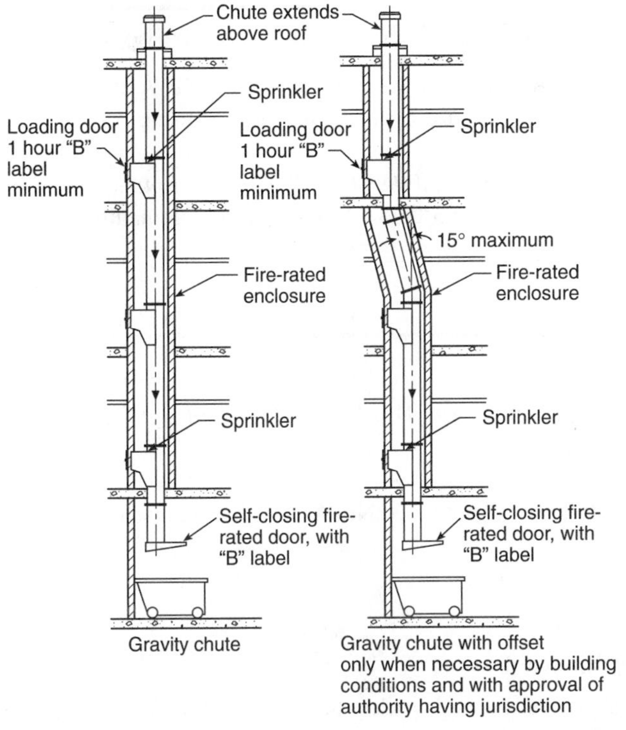

Figure A-5-21.3.1 *Gravity chute. [82: Figure 3-2.5.1]*

5-21.4 Automatic Sprinklers, Full Pneumatic Systems.

Full pneumatic-type risers shall be protected internally by automatic sprinklers. A sprinkler shall be required at or above the top loading station and at alternate floor levels in buildings over two stories in height, with a mandatory sprinkler located at the lowest loading station. Sprinklers shall be recessed out of the station area through which the material travels. (**82:** 3-3.4)

5-21.5 Commercial-Industrial Compactors.

All chute-fed compactors shall have an automatic special fine water spray sprinkler with a minimum ½-in. (13-mm) orifice installed in the hopper of the compactor. This sprinkler shall be an ordinary temperature–rated sprinkler. The sprinklers shall be supplied by a minimum 1-in. (25.4-mm) ferrous piping or ¾-in. (19-mm) copper tubing line from the domestic cold water supply.

The sprinkler shall provide a suitable spray into the hopper. A cycling (on-off), self-actuating, snap-action, heat-actuated sprinkler shall be permitted to be used, or the sprinkler shall be permitted to be controlled by a temperature sensor operating a solenoid valve. Sprinkler water piping shall be protected from freezing in outdoor installations. (**82:** 5-3, 5-3.1)

5-22 Industrial Furnaces Using a Special Processing Atmosphere

5-22.1 For applicable terms not defined in Chapter 1 the terms defined in NFPA 86C, *Standard for Industrial Furnaces Using a Special Processing Atmosphere,* shall be used.

5-22.2 Where sprinkler systems are provided per NFPA 86C, 18-1.2(b), sprinklers shall be of extra high-temperature rating [325°F to 650°F (163°C to 343°C] to avoid premature operation from localized flashing. [**86C:** 18-1.2(b)]

5-23 Water-Cooling Towers

5-23.1 Where sprinkler systems are provided per NFPA 214, *Standard on Water-Cooling Towers,* the following rules shall apply.

5-23.2 For applicable terms not defined in Chapter 13, NFPA 214 shall be used.

5-23.3* Counterflow Towers.

(**214:** 3-2.4.1)

A-5-23.3 See Figures A-5-23.3(a) through (d).

5-23.3.1 The discharge outlets shall be located under the fan deck and fan opening. (**214:** 3-2.4.1.1)

5-23.3.2 Except under the fan opening, all discharge outlets shall have deflector distances installed in accordance with Section 5-5. (**214:** 3-2.4.1.2)

Exception: Under fan openings.

5-23.3.3 Closed-head discharge outlets for dry-pipe and preaction systems shall be installed in the upright position only. (**214:** 3-2.4.1.3)

5-23.4* Crossflow Towers.

(**214:** 3-2.4.2)

A-5-23.4 See Figures A-5-23.4(a) through (d).

5-23.4.1 The discharge outlets protecting the plenum area shall be located under the fan deck and in the fan opening. (**214:** 3-2.4.2.1)

5-23.4.2 Discharge outlets protecting the fill shall be located under the distribution basin on either the louver or drift eliminator side, discharging horizontally through the joist channels. (**214:** 3-2.4.2.2)

5-23.4.3 Towers with a fill area longer than the maximum allowable for the discharge device being used shall have discharge devices placed on both sides of the fill area in each joist channel. The pressure at each discharge device shall be adequate to provide protection for half of the length of the fill area. (**214:** 3-2.4.2.3)

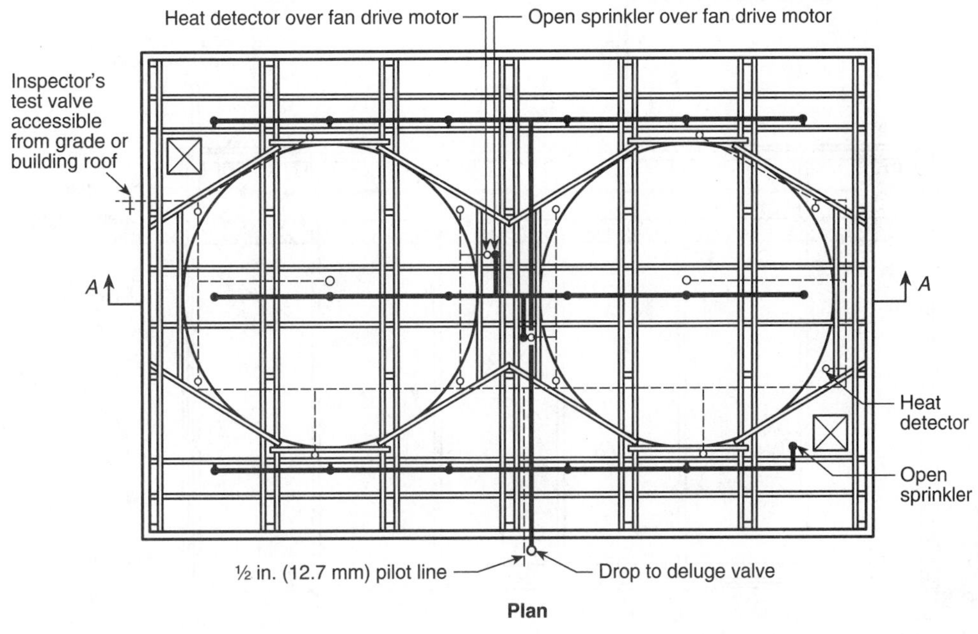

Figure A-5-23.3(a) Typical deluge fire protection arrangement for counterflow towers, illustration 1. [214: Figure A-3-2.4.1(a)]

5-23.4.4 Where joist channels are wider than 2 ft (0.6 m), more than one discharge device shall be required per joist channel.

Exception: If the discharge device being used is listed for the width of the joist channel being protected.

(**214:** 3-2.4.2.4)

5-23.5* On towers having extended fan decks that completely enclose the distribution basin, the discharge outlets protecting the fill area shall be located over the basin, under the extension of the fan deck. (**214:** 3-2.4.3)

A-5-23.5 Location of the nozzle relative to surfaces to be protected should be determined by the particular nozzle's discharge characteristics. Care should also be taken in the selection of nozzles to obtain waterways not easily obstructed by debris, sediment, sand, and so forth, in the water. *[See Figures A-5-23.5(a) and (b).]*

5-23.6 For deluge systems using directional spray nozzles in the pendent position, provisions shall be made to protect the underside of a combustible fan deck. (**214:** 3-2.4.4)

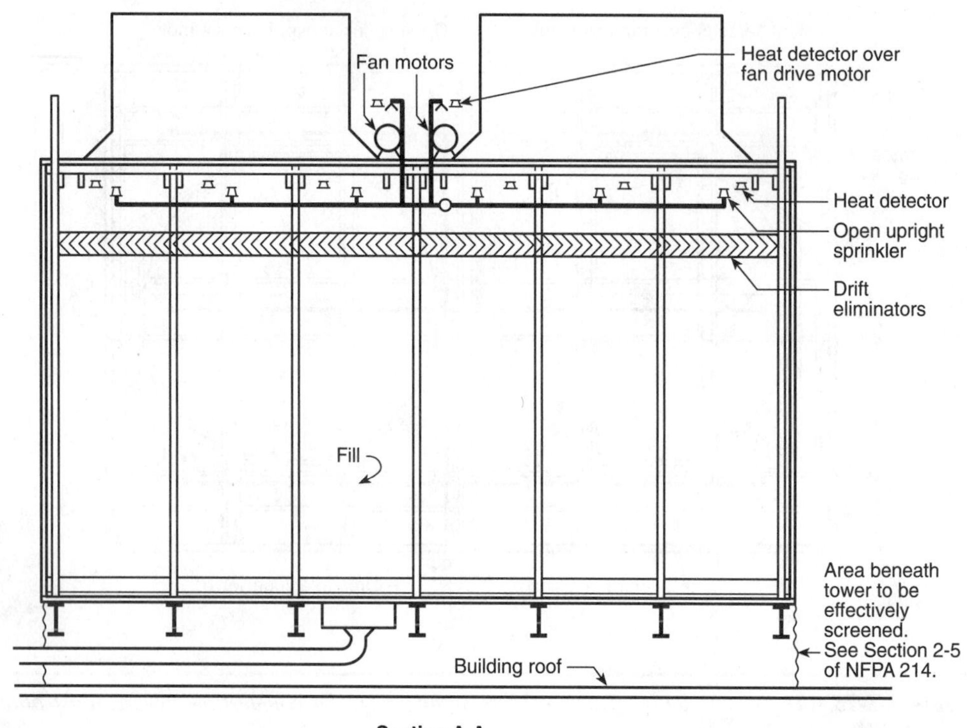

Figure A-5-23.3(b) Typical deluge fire protection arrangement for counterflow towers, illustration 2. [214: Figure A-3-2.4.1(b)]

5-23.6.1* On towers having basin covers that do not completely enclose the hot water basin, outlets protecting the fill shall be located under the distribution basin as set out in 5-23.4. (**214:** 3-2.4.5)

A-5-23.6.1 See Figure A-5-23.6.1.

5-23.7 Valves.

(**214:** 3-2.6)

5-23.7.1 General. Shutoff valves and automatically operated water control valves, if provided, shall be located:

(1) Outside the fire-exposed area;

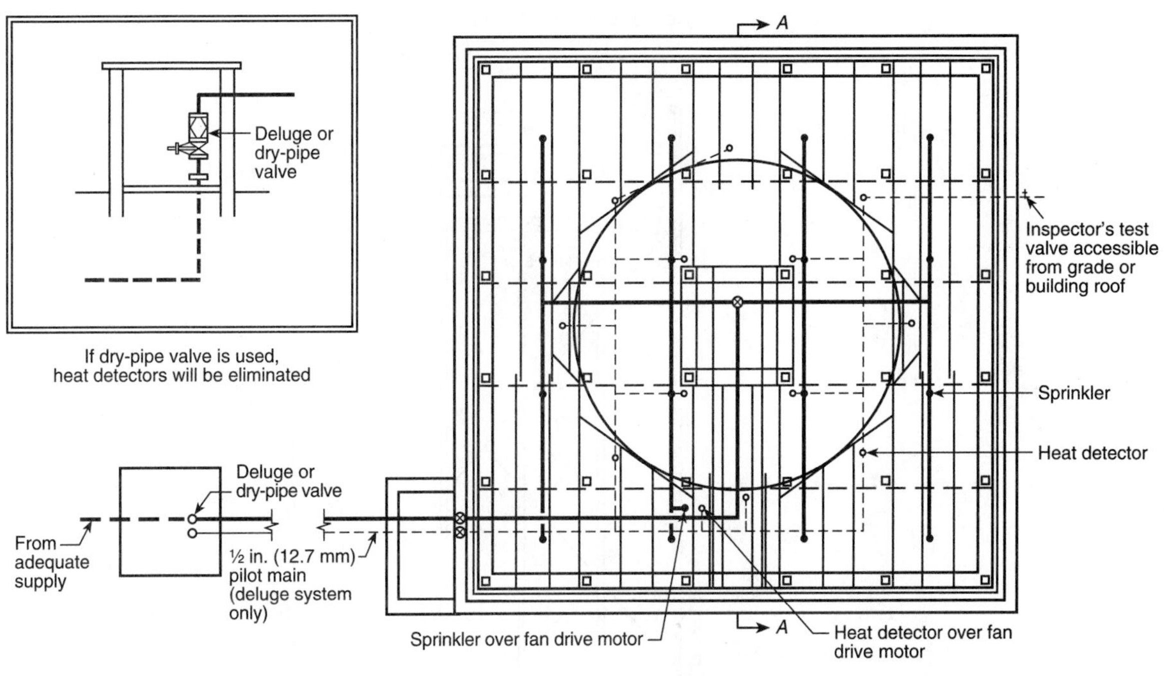

Figure A-5-23.3(c) *Typical deluge or dry pipe fire protection arrangement for counterflow towers, illustration 1. [214: Figure A-3-2.4.1(c)]*

(2) As close to the cooling tower as possible to minimize the amount of pipe to the discharge device; and
(3) Where they will be accessible during a fire emergency.

(**214:** 3-2.6.1)

5-23.7.2 Manual Release Valve. Remote manual release valves, where required, shall be conspicuously located and readily accessible during a fire emergency. If remote, manual release valves are not required, an inspector's test valve shall be provided for each pilot-head-operated system. (**214:** 3-2.6.2)

5-23.8 Strainers.

Strainers are required for systems utilizing discharge devices with waterways of less than 0.375-in. (9.5-mm) diameter. *(See NFPA 15, Standard for Water Spray Fixed Systems for Fire Protection, for further details.)* (**214:** 3-2.7)

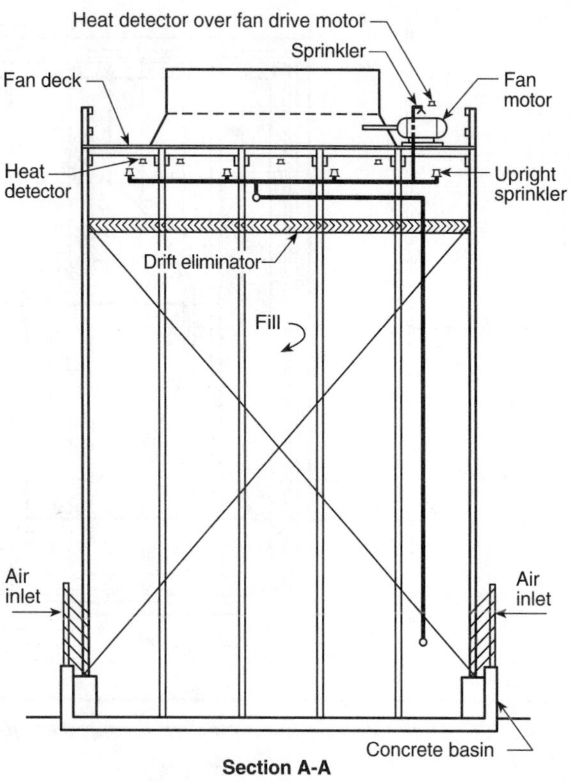

Figure A-5-23.3(d) Typical deluge or dry pipe fire protection arrangement for counterflow towers, illustration 2. [214: Figure A-3-2.4.1(d)]

5-23.9 Heat Detectors.

(**214:** 3-2.8)

5-23.9.1 Where deluge or preaction systems are used, heat detectors shall be installed in accordance with the applicable sections of NFPA 72, *National Fire Alarm Code®*. (**214:** 3-2.8.1)

5-23.9.2 In mechanical induced-draft towers, heat detectors shall be located under the fan deck at the circumference of the fan opening and under the fan opening where necessary to comply with the following spacing requirements. *(For extended fan decks, see 3-2.8.3 in NFPA 214.)* (**214:** 3-2.8.2)

5-23.9.2.1 Fixed-temperature detectors shall be spaced not more than 8 ft (2.4 m) apart in any direction including the fan opening. Temperature ratings shall be selected in accordance with operating conditions, but shall be no less than intermediate. (**214:** 3-2.8.2.1)

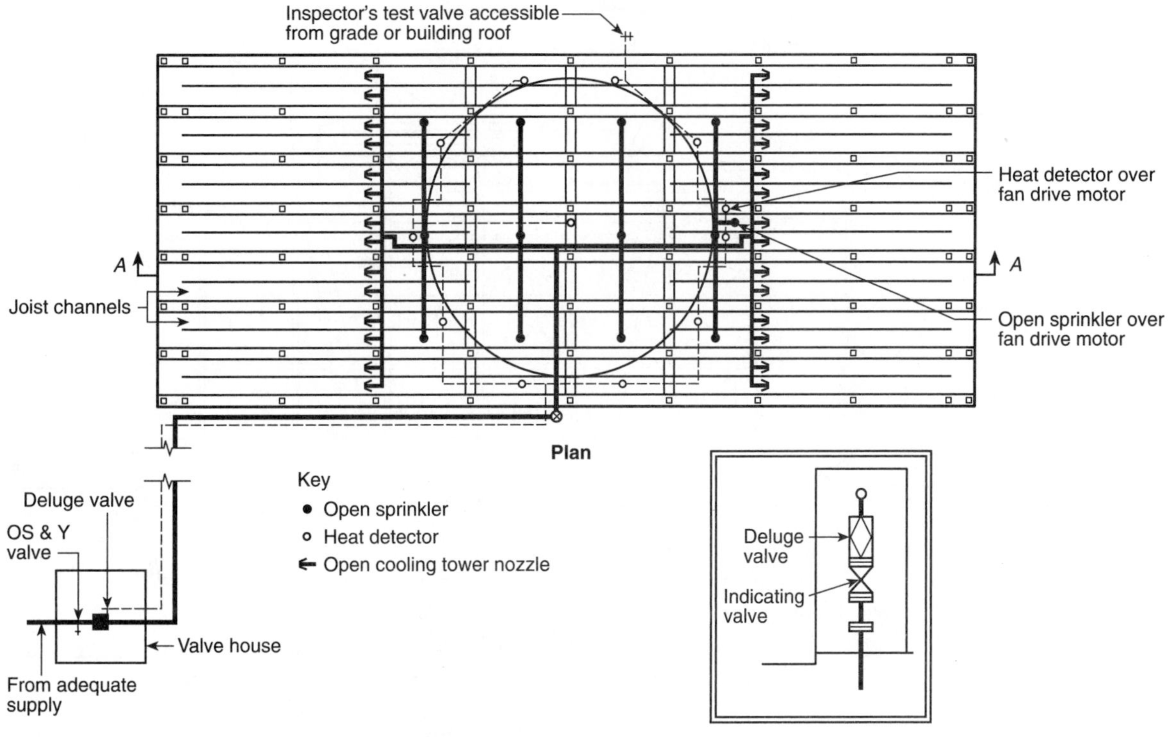

Figure A-5-23.4(a) Typical deluge fire protection arrangement for crossflow towers, illustration 1. [214: Figure A-3-2.4.2(a)]

5-23.9.2.2 Rate-of-rise detectors shall be spaced not more than 15 ft (4.6 m) apart in any direction. In pneumatic-type systems, for detectors inside the tower, there shall be no more than one detector for each mercury check in towers operating in cold climates, and two detectors for each mercury check in towers used during the warm months only or year-round in warm climates. There shall be no more than four detectors for each mercury check where the detectors are located outside the tower. (**214:** 3-2.8.2.2)

5-23.9.3 On towers having extended fan decks that completely enclose the distribution basin, detectors shall be located under the fan deck extension in accordance with standard, indoor-spacing rules for the type detectors used. *(See NFPA 72, National Fire Alarm Code.)*

Exception: Where the fan deck extension is 16 ft (4.9 m) or less and this dimension is the length of the joist channel, then only one row of detectors centered on and at right angles to the joist channels shall be required. Spacing between detectors shall be in accordance with NFPA 72, National Fire Alarm Code.

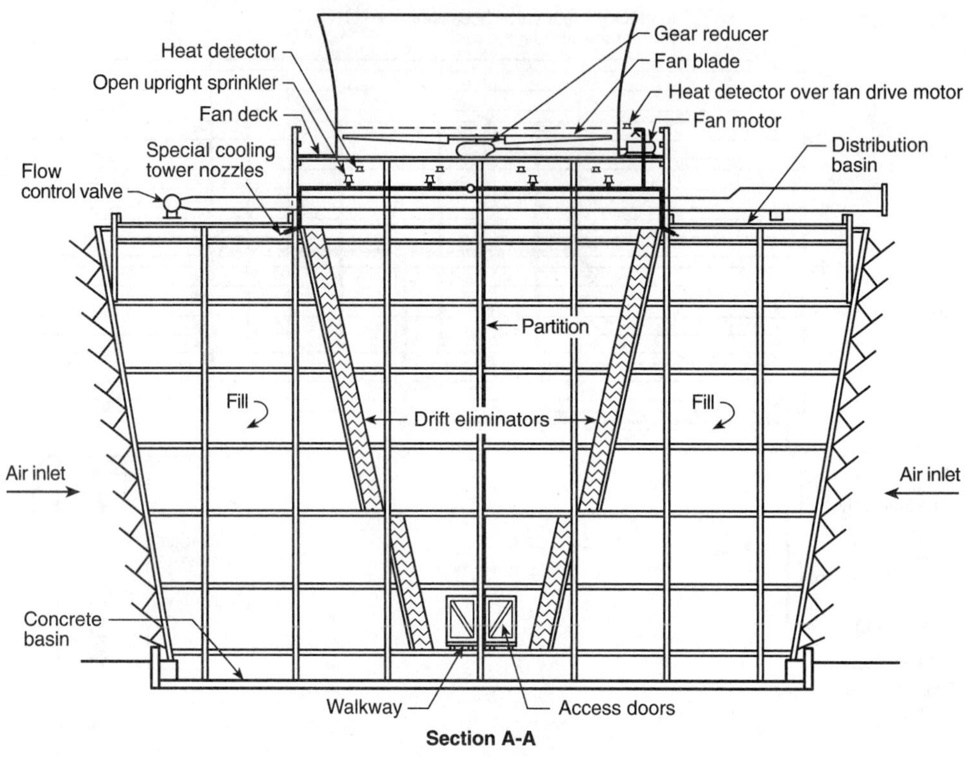

Figure A-5-23.4(b) *Typical deluge fire protection arrangement for crossflow towers, illustration 2. [214: Figure A-3-2.4.2(b)]*

On towers having extended fan decks that do not completely enclose the hot water basin, detectors shall not be required under the fan deck extension.

(**214:** 3-2.8.3)

5-23.9.4 Where the total number of deluge systems exceeds the number for which the water supply was designed, heat barriers shall be installed under the extended fan deck to separate the systems. Heat barriers shall extend from the fan deck structure to the distribution basin dividers. (**214:** 3-2.8.4)

5-23.9.5 Where heat detectors are inaccessible during tower operation, an accessible test detector shall be provided for each detection zone. (**214:** 3-2.8.5)

5-23.9.6 Heat detector components exposed to corrosive vapors or liquids shall be protected by materials of suitable construction or by suitable, protective coatings applied by the equipment manufacturer. (**214:** 3-2.8.6)

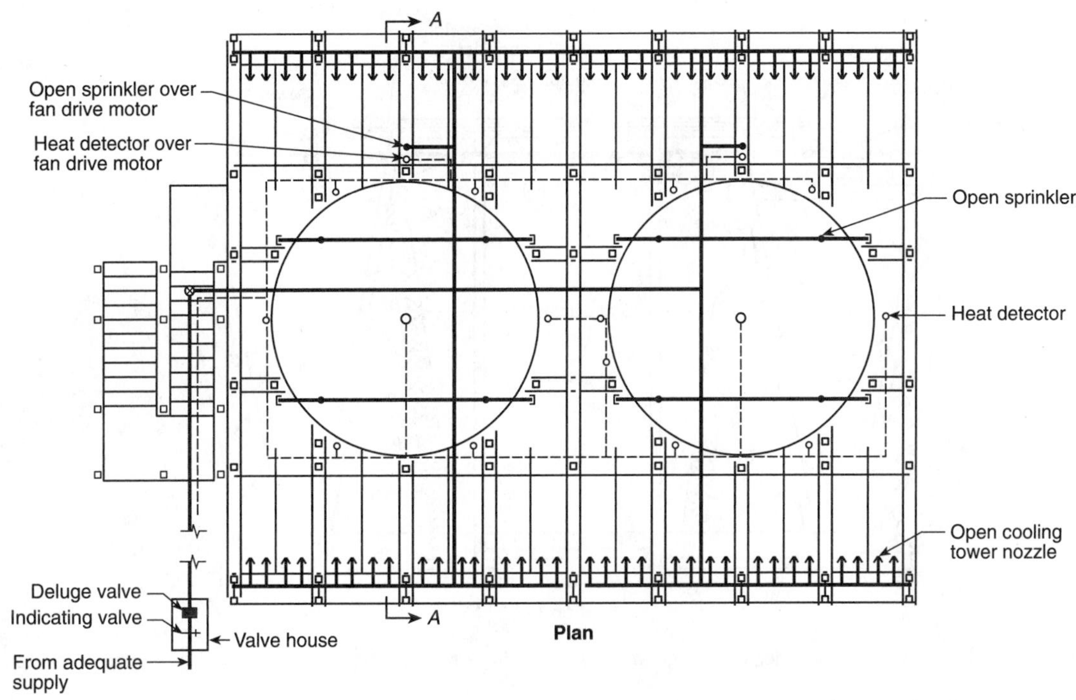

Figure A-5-23.4(c) *Typical deluge fire protection arrangement for crossflow towers, illustration 3. [214: Figure A-3-2.4.2(c)]*

5-23.10 Protection for Fan Drive Motor.

(**214:** 3-2.9)

5-23.10.1 A heat detector and water discharge outlet shall be provided over each fan drive motor when the motor is located so that it is not within the protected area of the tower. (**214:** 3-2.9.1)

5-23.10.2 Provision shall be made to interlock the fan motors with the fire protection system so that the cooling tower fan motors will be stopped in the cell(s) for which the system is actuated. Where the continued operation of the fans is vital to the process, a manual override switch may be provided to reactivate the fan when it is determined that there is no fire. (**214:** 3-2.9.2)

5-23.11 Corrosion Protection.

(**214:** 3-3)

5-23.11.1 Piping, fittings, hangers, braces, and attachment hardware including fasteners shall be hot-dip galvanized steel per ASTM A 153, *Standard Specification for Zinc Coating (Hot*

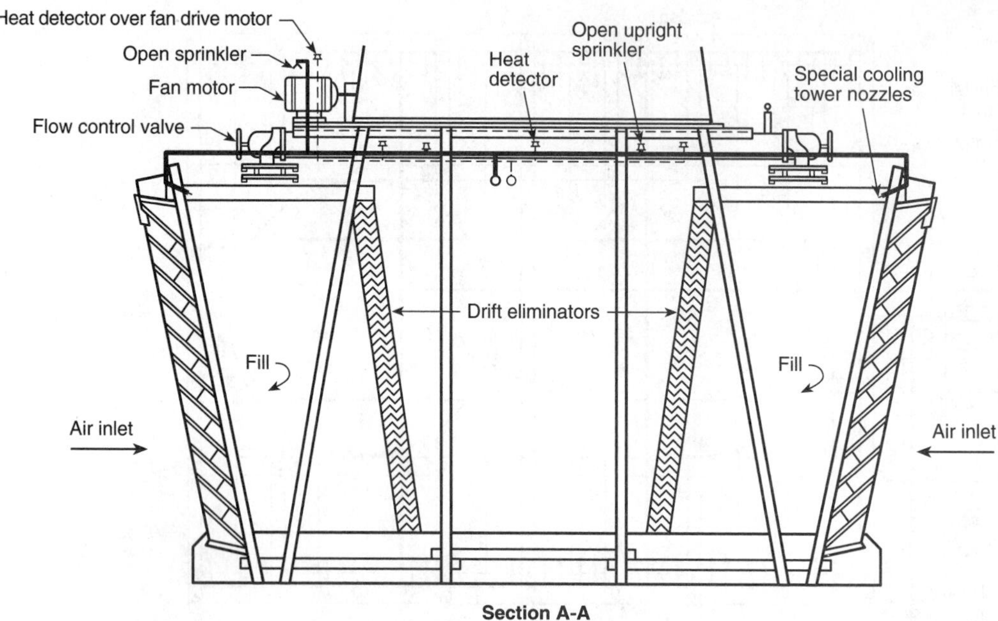

Figure A-5-23.4(d) *Typical deluge fire protection arrangement for crossflow towers, illustration 4. [214: Figure A-3-2.3.2(d)]*

Dip) on Iron and Steel Hardware, or other materials having a superior corrosion resistance. Exposed pipe threads and bolts on fittings shall be protected against corrosion. All other components shall be corrosion resistant or protected against corrosion by a suitable coating. (**214:** 3-3.1)

5-23.11.2 Approved sprinklers are made of nonferrous material and are corrosion resistant to normal atmospheres. Some atmospheres require special coatings on the discharge devices. Wax-type coatings shall not be used on devices without fusible elements. (**214:** 3-3.2)

5-23.11.3 Special care shall be taken in the handling and installation of wax-coated or similar sprinklers to avoid damaging the coating. Corrosion-resistant coatings shall not be applied to the sprinklers by anyone other than the manufacturer of the sprinklers, except that in all cases any damage to the protective coating occurring at the time of installation shall be repaired at once using only the coating of the manufacturer of the sprinkler in an approved manner so that no part of the sprinkler will be exposed after the installation has been completed. Otherwise, corrosion will attack the exposed metal and will, in time, creep under the wax coating. (**214:** 3-3.3)

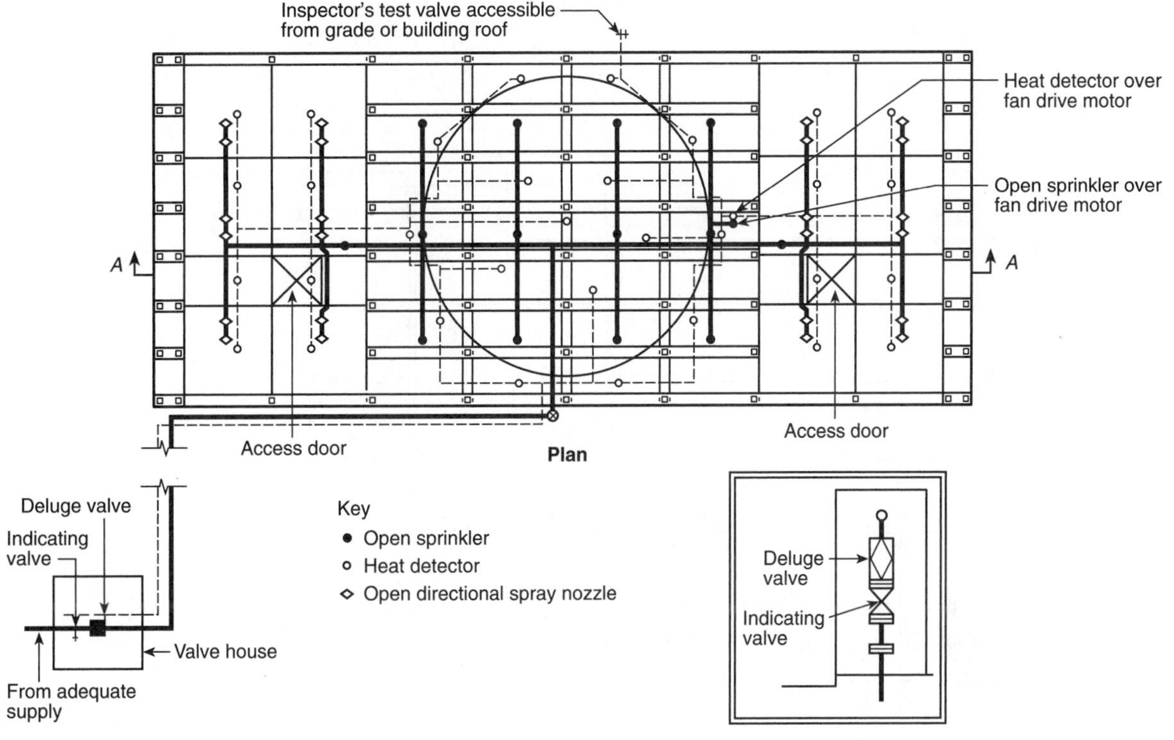

Figure A-5-23.5(a) Typical deluge fire protection arrangement for crossflow towers with completely enclosed distribution basins, illustration 1. [214: Figure A-3-2.4.3(a)]

5-24 Construction and Fire Protection of Marine Terminals, Piers, and Wharves

5-24.1 Where sprinkler systems are provided per NFPA 307, *Standard for the Construction and Fire Protection of Marine Terminals, Piers, and Wharves,* the following shall also apply.

5-24.2 For terms not defined in Chapter 1, NFPA 307 shall be used.

5-24.3 Where there is danger of damage to sprinkler equipment by floating objects, physical barriers shall be provided to exclude such objects. (**307:** 3-3.3.2)

5-24.4 The following installation criteria shall also apply.

(a) Where narrow horizontal channels or spaces are caused by caps, stringers, ties, and other structural members, the standard upright sprinkler might not project sufficient water

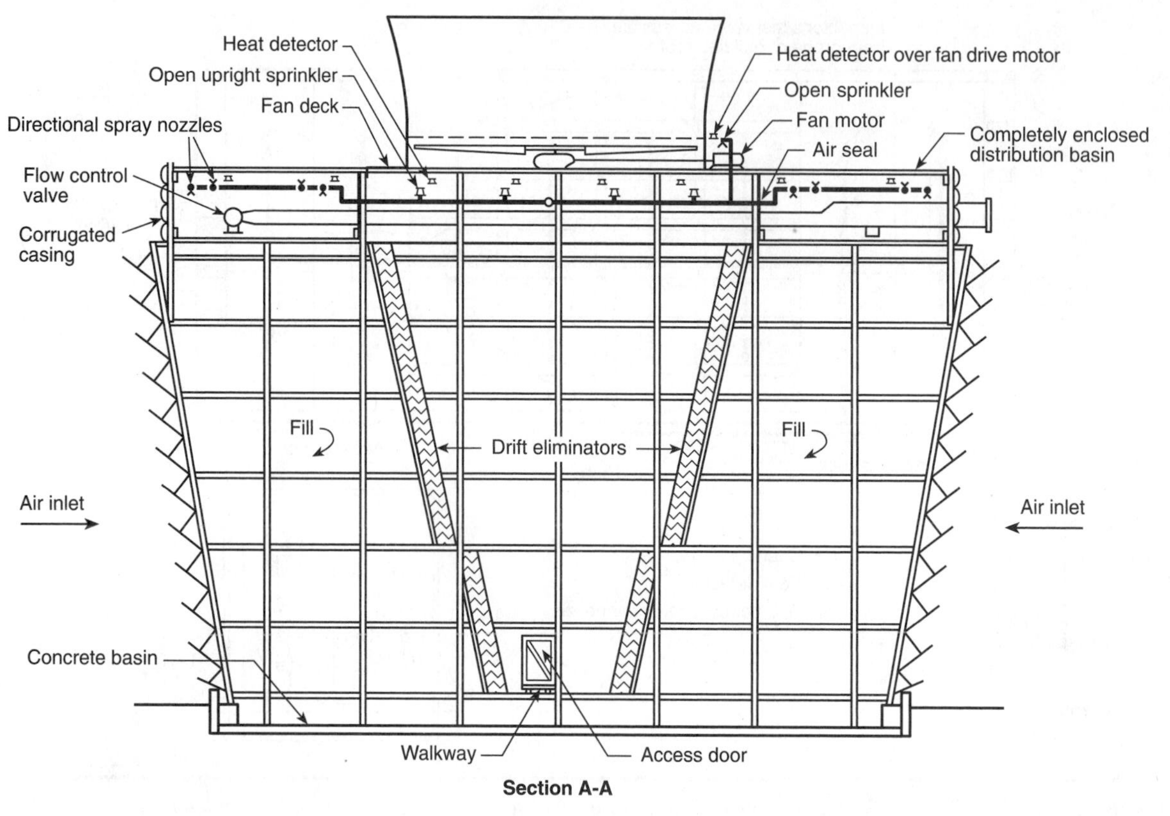

Figure A-5-23.5(b) *Typical deluge fire protection arrangement for crossflow towers with completely enclosed distribution basins, illustration 2. [214: Figure A-3-2.4.3(b)]*

upward to extinguish or control fires on the underside of the pier or wharf deck. In these cases, a sprinkler that projects water upward to wet the overhead, such as a pendent sprinkler installed in an upright position, or the old-style sprinkler shall be used. Location, spacing, and deflector position shall be governed by the discharge pattern of the sprinkler and the structure being protected. The following design and installation guides shall apply where pendent sprinklers in the upright position or old-style sprinklers are to be utilized:

(1) The maximum coverage per sprinkler head shall be limited to 7.5 m^2 (80 ft^2).
(2) Where spacing or arrangement of stringers constitutes typical open-joist construction directly supporting the deck, sprinkler branch lines shall be installed between the bents at right angles to the stringers. Spacing between branch lines shall not exceed 3 m (10 ft). Sprinklers on branch lines shall be staggered and spaced not to exceed 2.5 m (8 ft) on centers.

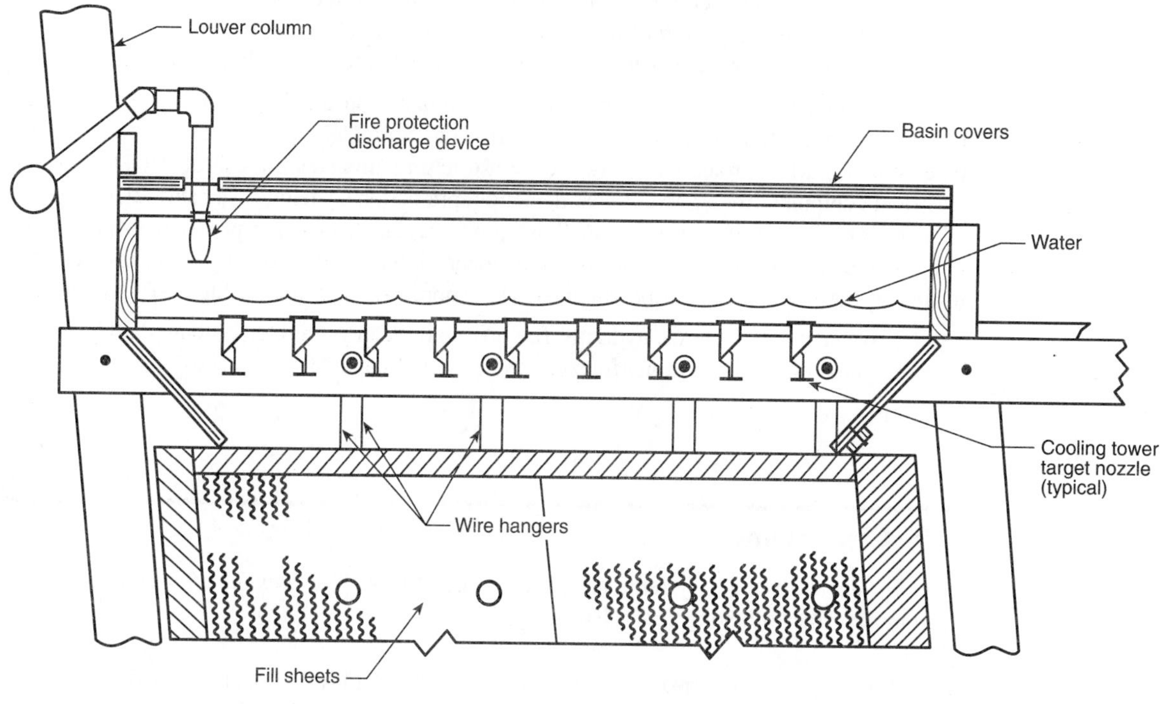

Figure A-5-23.6.1 *Typical deluge fire protection arrangement for crossflow towers with covers completely enclosing distribution basins. [214: Figure A-3-2.4.5]*

(3) Where crisscross construction (typically ties on stringers—see diagram in Appendix B) is involved, closer spacing of sprinklers shall be permitted as necessary to provide wetting of the entire structure.

(4) The deflectors of sprinklers on lines under stringers shall be located not less than 100 mm (4 in.) nor more than 250 mm (10 in.) below the bottom plane of the stringer, and not more than 450 mm (18 in.) below the underside of the pier or wharf deck.

(5) The temperature rating of the sprinkler shall not exceed 74°C (165°F).

(6) The maximum area to be protected by any one system shall be limited to 2325 m^2 (25,000 ft^2).

(b) Sprinklers designed and approved specifically for protection of combustible substructures shall be installed in conformity with their listing.

(c) The pipe hangers shall be placed in a location where they will be in the wetting pattern of the sprinkler to prevent the lag screws from burning or charring out, dropping sprinkler piping, and bleeding the system. The distance from the sprinkler to the hanger shall not exceed 460 mm (18 in.).

(d) Horizontal and vertical bracing shall be provided at not more than 6-m (20-ft) intervals on all sprinkler piping 76 mm (3 in.) or larger that is parallel to and within 15 m (50 ft) of the face of the pier or wharf and where it might be subjected to heavy fireboat nozzle streams.

(e) Sprinkler systems, including hanger assemblies and bracing, in underdeck areas shall be properly protected throughout against corrosion. Sprinklers shall be of corrosion-resistant type. When the fire protection design for substructures involves the use of detectors or other electrical equipment for smoke or heat detection, pre-action or deluge-type sprinkler protection, all detectors and wiring systems shall be moisture- and corrosion-proof to protect against unfavorable atmospheric conditions that exist beneath these structures. Frequent inspection and testing of these systems shall be conducted in accordance with applicable NFPA standards.

(f) Water supply systems, hydrants, fire hose valves, and sprinkler systems shall be installed with adequate protection against freezing and physical damage.

(**307:** 3-3.3.3)

5-25 Cleanrooms

5-25.1 Where sprinkler systems are provided per NFPA 318, *Standard for the Protection of Cleanrooms,* the following shall also apply.

5-25.2 For applicable terms not defined in Chapter 1, NFPA 318 shall be used.

5-25.3 Cleanrooms.

Wet pipe automatic sprinkler protection shall be provided throughout facilities containing cleanrooms and clean zones. (**318:** 2-1.1)

5-25.4* Quick response sprinklers shall be utilized for sprinkler installations within downflow airstreams in cleanrooms and clean zones. (**318:** 2-1.2.2)

A-5-25.4 The use of quick-response sprinklers, while still delayed in opening by the downward airflow, would respond to a smaller size fire quicker than conventional sprinklers. (Glass bulb–type quick response sprinklers might be preferable to other types of quick-response sprinklers.) (**318:** A-2-1.2.2)

5-25.5* Sprinklers installed in ductwork shall be spaced a maximum of 20 ft (6.1 m) apart horizontally and 12 ft (3.7 m) apart vertically. (**318:** 2-1.2.6.1)

A-5-25.5 Small orifice sprinklers, ⅜ in. (9.5 mm) or larger, can be used. (**318:** A-2-1.2.6.1)

5-25.6 A separate indicating control valve shall be provided for sprinklers installed in ductwork. (**318:** 2-1.2.6.2)

5-25.7 The sprinklers shall be accessible for periodic inspection and maintenance. (**318:** 2-1.2.6.5)

5-26 Aircraft Hangars

See NFPA 409, *Standard on Aircraft Hangars,* for sprinkler system installation criteria pertaining to the protection of aircraft hangars.

5-27* Liquid and Solid Oxidizers

See NFPA 430, *Code for the Storage of Liquid and Solid Oxidizers.*

A-5-27 At the time of the publication of this standard, the Technical Committee on Sprinkler System Installation Criteria is aware of ongoing research concerning the sprinkler protection of solid oxidizers. This research is being conducted by the National Fire Protection Research Foundation.

5-28 Organic Peroxide Formulations

5-28.1 For applicable terms not defined in Chapter 1, NFPA 432, *Code for the Storage of Organic Peroxide Formulations,* shall be used.

5-28.2 Where automatic sprinkler protection is provided for Class I organic peroxide formulations in quantities exceeding 2000 lb (907 kg), it shall be a deluge system. (**432:** 5-5.2)

5-29 Light Water Nuclear Power Plants

5-29.1 For applicable terms not defined in Chapter 1, NFPA 803, *Standard for Fire Protection for Light Water Nuclear Power Plants,* shall be used.

5-29.2 Each system shall have an independent connection to the plant yard main and be equipped with an approved indicating-type control or shutoff valve. (**803:** 10-2.2)

Multiple sprinkler and standpipe systems shall be supplied by interior headers or fire protection loops. When provided, such headers or loops are considered an extension of the yard main system and shall be provided with at least two connections to the yard main. The arrangement shall be supplied and valved so that no single impairment can affect sprinkler and hose protection at the same time.

5-29.3 The fire main system piping shall not serve service water system functions. (**803:** 12-3)

5-30 Advanced Light Water Reactor Electric Generating Plants

5-30.1 For applicable terms not defined in Chapter 1, NFPA 804, *Standard for Fire Protection for Advanced Light Water Reactor Electric Generating Plants,* shall be used.

5-30.2 Yard Mains, Hydrants, and Building Standpipes.

(**804:** 7-4)

5-30.2.1 Approved visually indicating sectional control valves such as post-indicator valves shall be provided to isolate portions of the main for maintenance or repair without simultaneously shutting off the supply to both primary and backup fire suppression systems. (**804:** 7-4.2)

5-30.2.2 A common yard fire main loop may serve multi-unit nuclear power plant sites if it is cross-connected between units. Sectional control valves shall permit maintaining independence of the individual loop around each unit. For such installations, common water supplies shall also be permitted to be utilized. For multiple-reactor sites with widely separated plants [approaching 1 mi. (1.6 km) or more], separate yard fire main loops shall be used. (**804:** 7-4.4)

5-30.2.3 Sprinkler systems and manual hose station standpipes shall have connections to the plant underground water main so that a single active failure or a crack in a moderate-energy line can be isolated so as not to impair both the primary and backup fire suppression systems. Alternatively, headers fed from each end are permitted inside buildings to supply both sprinkler and standpipe systems, provided steel piping and fittings meeting the requirements of ANSI B31.1, *Code for Power Piping,* are used for the headers (up to and including the first valve) supplying the sprinkler systems where such headers are part of the seismically analyzed hose standpipe system. Where provided, such headers are considered an extension of the yard main system. Each sprinkler and standpipe system shall be equipped with an outside screw and yoke (OS&Y) gate valve or other approved shutoff valve. (**804:** 7-4.7)

5-30.3 Cable Concentrations.

The location of sprinklers or spray nozzles shall consider cable tray arrangements and possible transient combustibles to ensure adequate water coverage for areas that could present exposure fire hazards to the cable raceways. (**804:** 8-4.2.2.2)

5-30.4 Turbine Building.

Deluge sprinkler systems or deluge spray systems shall be zoned to limit the area of protection to that which the drainage system can handle with any two adjacent systems actuated. The systems shall be hydraulically designed with each zone calculated with the largest adjacent zone flowing. (**804:** 8-4.2.2.3)

5-31* Electric Generating Plants and High Voltage Direct Current Converter Stations

For applicable terms not defined in Chapter 1, NFPA 850, *Recommended Practice for Fire Protection for Electric Generating Plants and High Voltage Direct Current Converter Stations,* shall be used.

A-5-31 Where an adequate and reliable water supply, such as a lake, cooling pond, river, or municipal water system, is unavailable, at least two separate water supplies should be provided for fire protection purposes with each supply capable of meeting the fire waterflow requirements determined by 4-2.1 of NFPA 850. (**850:** 4-2.2)

Each water supply should be connected to the yard main by separate connections arranged and valve controlled to minimize the possibility of multiple supplies being impaired simultaneously. (**850:** 4-2.3)

Indicator control valves should be installed to provide adequate sectional control of the fire main loop to minimize plant protection impairments. (**850:** 4-4.1.4)

Each hydrant should be equipped with a separate shutoff valve located on the branch connection to the supply main. (**850:** 4-4.1.5)

Interior fire protection loops are considered an extension of the yard main and should be provided with at least two valved connections to the yard main with appropriate sectional control valves on the interior loop. (**850:** 4-4.1.6)

If a sprinkler system is used to protect the coal conveyor, particular care should be exercised in locating closed sprinkler so that they will be in the path of the heat produced by the fire and still be in a position to provide good coverage of all belt surfaces along the conveyor. (**850:** 5-4.6.2.1)

Protection inside dust collectors should include the clean air plenum and the bag section. If the hopper is shielded from water discharge, sprinklers also should be provided in the hopper section.

All areas beneath the turbine-generator operating floor that are subject to oil flow, oil spray, or oil accumulation should be protected by an automatic sprinkler or foam-water sprinkler system. This coverage normally includes all areas beneath the operating floor in the turbine building. (**850:** 5-7.4.1.1)

Lubricating oil lines above the turbine operating floor should be protected with an automatic sprinkler system covering those areas subject to oil accumulation including the area within the turbine lagging (skirt). (**850:** 5-7.4.1.2)

Turbine-generator bearings should be protected with a manually or automatically operated closed-head sprinkler system utilizing directional nozzles. (**850:** 5-7.4.2.1)

Due to the large quantity of platforms, equipment, and walkways, care should be taken to include coverage under all obstructions greater than 4 ft (1.2 m) wide. (**850:** 7-4.4.8)

5-32* Hydroelectric Generating Plants

See NFPA 851, *Recommended Practice for Fire Protection for Hydroelectric Generating Plants,* for applicable terms not defined in Chapter 1.

A-5-32 Upstream water is frequently the fire protection water supply. Water for fire suppression should not be taken downstream from any closure device in a penstock, flume, or forebay. (**851:** 4-2.6)

Fire extinguishing systems, where installed for lube oil systems employing combustible-type oil, should include protection for the reservoirs, pumps, and all oil lines, especially where unions exist on piping and beneath any shielded area where flowing oil can collect. Facilities not provided with curbs or drains should extend coverage for a distance of 20 ft (6.1 m) from the oil lines, when measured from the outermost oil line. (**851:** 5-2.7)

REFERENCES CITED IN COMMENTARY

National Fire Protection Association, 1 Batterymarch Park, P.O. Box 9101, Quincy, MA 02269-9101.

NFPA 13D, *Standard for the Installation of Sprinkler Systems in One- and Two-Family Dwellings and Manufactured Homes*, 1999 edition.

NFPA 13E, *Guide for Fire Department Operations in Properties Protected by Sprinkler and Standpipe Systems*, 1995 edition.

NFPA 13R, *Standard for the Installation of Sprinkler Systems in Residential Occupancies up to and Including Four Stories in Height*, 1999 edition.

NFPA 14, *Standard for the Installation of Standpipe and Hose Systems*, 1996 edition.

NFPA 25, *Standard for the Inspection, Testing, and Maintenance of Water-Based Fire Protection Systems*, 1998 edition.

NFPA 70, *National Electrical Code®*, 1999 edition.

NFPA 72, *National Fire Alarm Code®*, 1999 edition.

NFPA *101®*, *Life Safety Code®*, 1997 edition.

NFPA 409, *Standard on Aircraft Hangars*, 1995 edition.

American Society of Mechanical Engineers, Three Park Avenue, New York, NY 10016-5990.

ASME A17.1, *Safety Code for Elevators and Escalators*, 1996 edition.

National Institute of Standards and Technology, Gaithersburg, MD 20899.

NBSIR 80-2097, *Full-Scale Fire Tests with Automatic Sprinklers in a Patient Room*, 1980.

Underwriters Laboratories Inc., 333 Pfingsten Road, Northbrook, IL 60062.

UL 1626, *Residential Sprinklers for Fire Protection Service*, 2nd edition, 1994.

"Fact Finding Report on Automatic Sprinkler Protection for Fire Storage Vaults," November 25, 1947.

CHAPTER 6

Hanging, Bracing, and Restraint of System Piping

Chapter 6 addresses the structural issues related to installation of fire protection piping systems. Section 6-1 contains regulations for acceptable types of hangers, Section 6-2 covers hanger installation, Section 6-3 addresses joint restraint for fire mains, and Section 6-4 deals with protection of piping where subject to earthquakes. The Technical Committee on Hanging and Bracing attempted to make Chapter 6 suitable for reference by other fire protection system documents. For example, NFPA 13 earthquake protection criteria have been referenced by other water-based fire protection system standards that use similar piping materials and configurations, such as NFPA 14, *Standard for the Installation of Standpipe and Hose Systems*, and NFPA 15, *Standard for Water Spray Fixed Systems for Fire Protection.*

6-1 Hangers

6-1.1* General.

Types of hangers shall be in accordance with the requirements of Section 6-1.

Exception No. 1: Hangers certified by a registered professional engineer to include all of the following shall be acceptable:

(a) Hangers shall be designed to support five times the weight of the water-filled pipe plus 250 lb (114 kg) at each point of piping support.

(b) These points of support shall be adequate to support the system.

(c) The spacing between hangers shall not exceed the value given for the type of pipe as indicated in Table 6-2.2.

(d) Hanger components shall be ferrous.

Detailed calculations shall be submitted, when required by the reviewing authority, showing stresses developed in hangers, piping, and fittings and safety factors allowed.

Exception No. 2 In areas subject to earthquake, hangers shall also meet the requirements of 6-4.7.

A-6-1.1 See Figure A-6-1.1.

Hangers can consist of one individual component such as a U-hook or up to three separate components as indicated in Exhibit 6.1. The first component is a building-attached component that attaches to the building structure. In Exhibit 6.1, the building-attached component includes a C-clamp attached to an I-beam. Fasteners used to attach the building-attached component to the building structure are considered part of the hanger and must meet the applicable requirements of NFPA 13.

The second possible component includes the pipe attachment component and consists of the portion of the hanger assembly that is attached to the sprinkler piping. Exhibit 6.1 shows the use of an adjustable swivel ring as the pipe attachment component.

The third possible component includes the connecting piece that joins the building attachment component with the pipe attachment component. An all-thread rod is used for this purpose in Exhibit 6.1.

Exception No. 1 permits a performance-based approach in place of the normal rules of 6-1.1 to accommodate unusual applications or situations. The performance-based approach requires the services of a professional engineer to certify that the four conditions specified in (a), (b), (c), and (d) of Exception No. 1 are satisfied. The safety factor of five times the weight of the water-filled pipe plus 250 lb (114 kg) is traditional for fire protection piping and includes the assumed weight of a sprinkler fitter with associated work equipment who could grasp the pipe to prevent falling from a ladder or platform. This exception is not intended to substitute for using listed hangers but rather to allow for the use of other methods where listed hangers cannot be used for a particular building arrangement. See also commentary to 6-2.1.3.

Since the issuance of Formal Interpretation 78-3, additional criteria need to be examined by the professional engineer. These criteria include a new item (c) to the exception to 6-1.1, which requires that the spacing between hangers not exceed the requirements of Table 6-2.2, and a clarification that detailed calculations include stresses developed in fittings as well as in the hangers and the piping.

Exception No. 2 was added to the 1999 edition because hangers can experience forces other than those resulting from gravity in areas subject to earthquakes. Additional measures as required in 6-4.7 are needed for certain hanger arrangements to guard against the effect of these forces.

6-1.1.1 The components of hanger assemblies that directly attach to the pipe or to the building structure shall be listed.

Exception No. 1: Mild steel hangers formed from rods shall be permitted to be not listed.

Exception No. 2: Fasteners as specifically identified in 6-1.5 shall be permitted to be not listed.

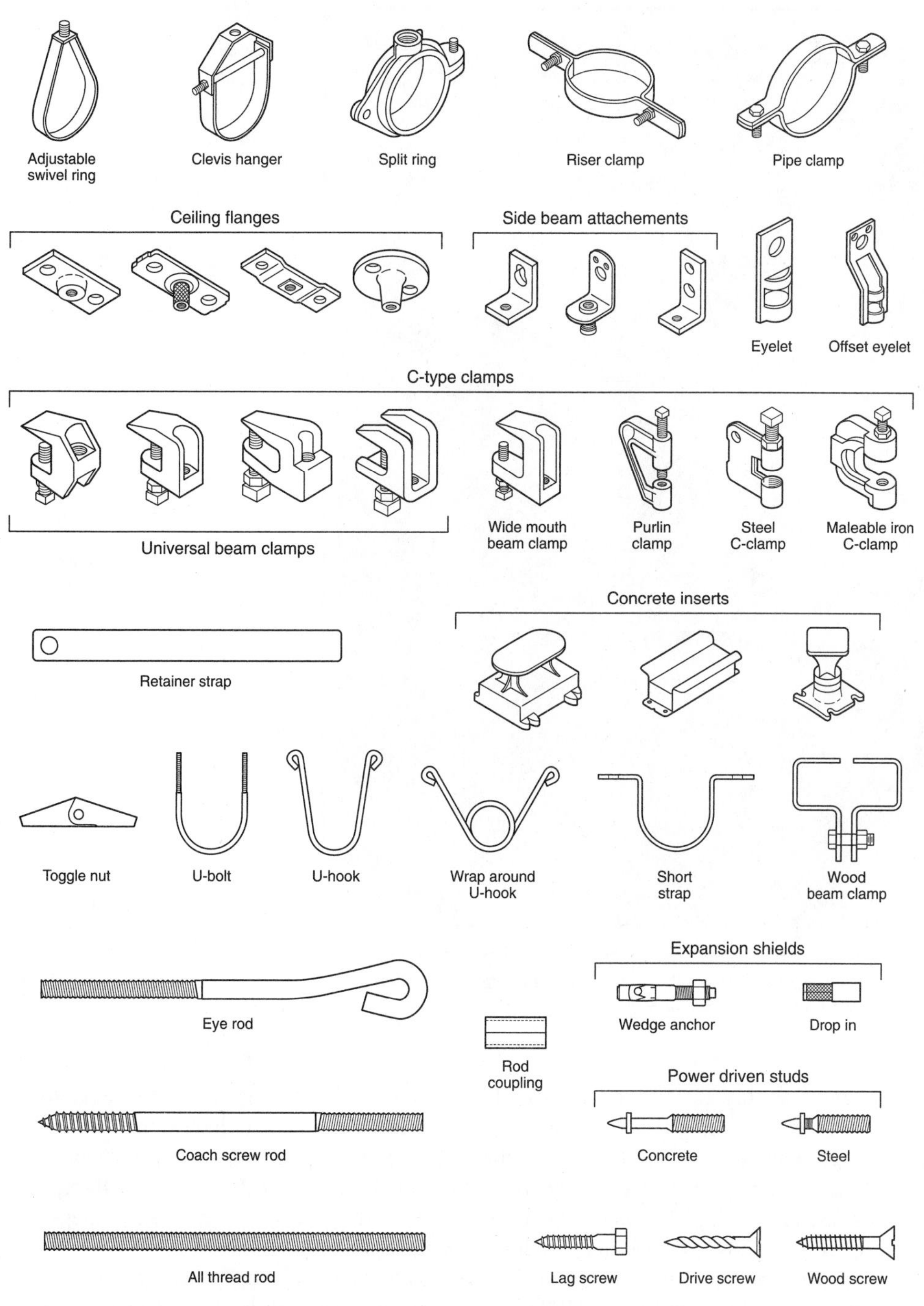

Figure A-6-1.1 *Common types of acceptable hangers.*

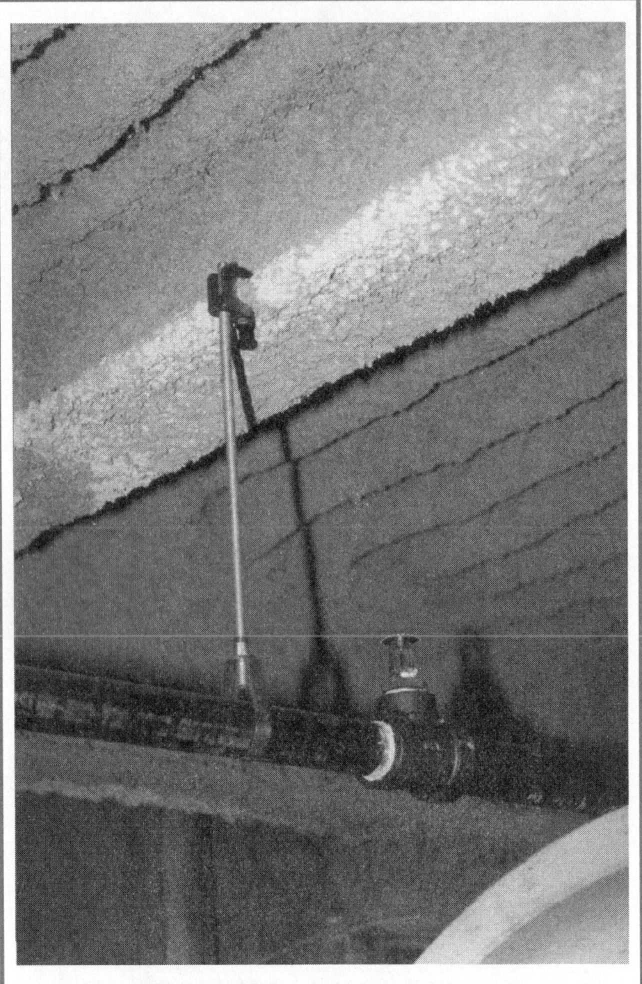

Exhibit 6.1 *Hanger assembly consisting of a C-clamp, an all-thread rod, and an adjustable swivel ring.*

Hanger assembly components must be listed. Mild steel rod hangers are not required to be listed according to Exception No. 1. Mild steel rod hangers have demonstrated through standard practice that rods in the sizes required in Table 6-1.4.1 can support, at maximum permitted spacing, five times the weight of water-filled Schedule 40 pipe plus 250 lb (114 kg). U-bolts and eye rods are considered a type of mild steel hanger formed from rods and meet the conditions of Exception No. 1.

Exception No. 2 was added to the 1999 edition to specifically identify that the screws and bolts identified in NFPA 13 are not considered to be part of the hanger assembly requiring listing. Other fastener assemblies not specifically identified in 6-1.5 are required to be listed.

Formal Interpretation 78-3

Reference: 6-1.1

Question: Does the method of hanging sprinkler piping as shown in Figure 1 comply with Section 6-1.1? It is capable of supporting five times the weight of water-filled piping plus 250 lb and has been certified by a registered professional engineer.

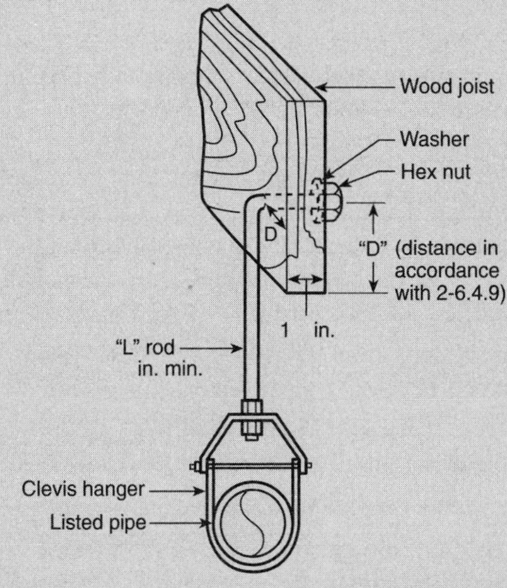

Figure 1

Answer: Yes. Certification by a registered professional engineer also includes compliance with other requirements of 6-1.1, Exception, namely:

"(b) These points of support are adequate to support the sprinkler system.

"(d) Ferrous materials are used for hanger components.

"Detailed calculations shall be submitted, when required by the reviewing authority, showing stresses developed both in hangers and piping and safety factors allowed."

Issue Edition: 1978

Reference: 3-15

Date: January 1979

6-1.1.2 Hangers and their components shall be ferrous.

Exception No. 1: Nonferrous components that have been proven by fire tests to be adequate for the hazard application, that are listed for this purpose, and that are in compliance with the other requirements of this section shall be acceptable.

Exception No. 2: Holes through solid structural members shall be permitted to serve as hangers for the support of system piping provided such holes are permitted by applicable building codes and the spacing and support provisions for hangers of this standard are satisfied.

Exception No. 2 recognizes that holes through structural members are occasionally used for piping support but requires that structural strength considerations be given priority. To avoid weakening a beam or joist, most codes prohibit the cutting of a hole in the top or bottom third of the depth and anywhere within twice the depth from either end support.

6-1.1.3* For trapeze hangers, the minimum size of steel angle or pipe span between purlins or joists shall be such that the available section modulus of the trapeze member from Table 6-1.1.3(b) equals or exceeds the section modulus required in Table 6-1.1.3(a).

Any other sizes or shapes giving equal or greater section modulus shall be acceptable. All angles shall be used with the longer leg vertical. The trapeze member shall be secured to prevent slippage. Where a pipe is suspended from a pipe trapeze of a diameter less than the diameter of the pipe being supported, ring, strap, or clevis hangers of the size corresponding to the suspended pipe shall be used on both ends.

A-6-1.1.3 Table 6-1.1.3(a) assumes that the load from 15 ft (5 m) of water-filled pipe, plus 250 lb (114 kg), is located at the midpoint of the span of the trapeze member, with a maximum allowable bending stress of 15 ksi (111 kg). If the load is applied at other than the midpoint, for the purpose of sizing the trapeze member, an equivalent length of trapeze can be used, derived from the following formula:

$$L = \frac{4ab}{a + b}$$

where:

L = equivalent length

a = distance from one support to the load

b = distance from the other support to the load

Where multiple mains are to be supported or multiple trapeze hangers are provided in parallel, the required or available section modulus can be added.

A trapeze hanger generally consists of a length of steel pipe or an angle iron that serves as a trapeze bar between the building structural members, along with other conventional hanger components. Conventional hangers are used to support the sprinkler system piping from the trapeze and can also be used to support the trapeze from the adjacent structural members. In any event, the hanger requirements of NFPA 13 apply to the hangers used to support the sprinkler pipe from the trapeze and any hangers used to support the trapeze from the building structure.

The selection of the minimum size of pipe or angle that can serve as the bar of the trapeze is made on the basis of required section modulus. The section modulus relates to the ability of a structural member, including a length of pipe or angle used as a trapeze, to withstand flexure. The top numbers in Table 6-1.1.3(a) provide the

Table 6-1.1.3(a) Section Modulus Required for Trapeze Members (in.^{3})

Span of Trapeze	1 in.	1¼ in.	1½ in.	2 in.	2½ in.	3 in.	3½ in.	4 in.	5 in.	6 in.	8 in.	10 in.
1 ft 6 in.	0.08	0.09	0.09	0.09	0.10	0.11	0.12	0.13	0.15	0.18	0.24	0.32
	0.08	0.09	0.09	0.10	0.11	0.12	0.13	0.15	0.18	0.22	0.30	0.41
2 ft 0 in.	0.11	0.12	0.12	0.13	0.13	0.15	0.16	0.17	0.20	0.24	0.32	0.43
	0.11	0.12	0.12	0.13	0.15	0.16	0.18	0.20	0.24	0.29	0.40	0.55
2 ft 6 in.	0.14	0.14	0.15	0.16	0.17	0.18	0.20	0.21	0.25	0.30	0.40	0.54
	0.14	0.15	0.15	0.16	0.18	0.21	0.22	0.25	0.30	0.36	0.50	0.68
3 ft 0 in.	0.17	0.17	0.18	0.19	0.20	0.22	0.24	0.26	0.31	0.36	0.48	0.65
	0.17	0.18	0.18	0.20	0.22	0.25	0.27	0.30	0.36	0.43	0.60	0.82
4 ft 0 in.	0.22	0.23	0.24	0.25	0.27	0.29	0.32	0.34	0.41	0.48	0.64	0.87
	0.22	0.24	0.24	0.26	0.29	0.33	0.36	0.40	0.48	0.58	0.80	1.09
5 ft 0 in.	0.28	0.29	0.30	0.31	0.34	0.37	0.40	0.43	0.51	0.59	0.80	1.08
	0.28	0.29	0.30	0.33	0.37	0.41	0.45	0.49	0.60	0.72	1.00	1.37
6 ft 0 in.	0.33	0.35	0.36	0.38	0.41	0.44	0.48	0.51	0.61	0.71	0.97	1.30
	0.34	0.35	0.36	0.39	0.44	0.49	0.54	0.59	0.72	0.87	1.20	1.64
7 ft 0 in.	0.39	0.40	0.41	0.44	0.47	0.52	0.55	0.60	0.71	0.83	1.13	1.52
	0.39	0.41	0.43	0.46	0.51	0.58	0.63	0.69	0.84	1.01	1.41	1.92
8 ft 0 in.	0.44	0.46	0.47	0.50	0.54	0.59	0.63	0.68	0.81	0.95	1.29	1.73
	0.45	0.47	0.49	0.52	0.59	0.66	0.72	0.79	0.96	1.16	1.61	2.19
9 ft 0 in.	0.50	0.52	0.53	0.56	0.61	0.66	0.71	0.77	0.92	1.07	1.45	1.95
	0.50	0.53	0.55	0.59	0.66	0.74	0.81	0.89	1.08	1.30	1.81	2.46
10 ft 0 in.	0.56	0.58	0.59	0.63	0.68	0.74	0.79	0.85	1.02	1.19	1.61	2.17
	0.56	0.59	0.61	0.65	0.74	0.82	0.90	0.99	1.20	1.44	2.01	2.74

For SI units, 1 in. = 25.4 mm; 1 ft = 0.3048 m.

Notes:

1. Top values are for Scheule 10 pipe; bottom values are for Schedule 40 pipe.
2. The table is based on a maximum allowable bending stress of 15 ksi and a midspan concentrated load from 15 ft (4.6 m) of water-filled pipe, plus 250 lb (114 kg).

required section modulus for support of Schedule 10 steel pipe, and the lower numbers show the required section modulus for support of the heavier Schedule 40 steel pipe. Table 6-1.1.3(b) provides the available section moduli for various diameters of Schedule 10 pipe, Schedule 40 pipe, and shapes of angles commonly used as trapeze bars. The section modulus for each member is denoted as Z in the following example.

The following example illustrates the intended use of Tables 6-1.1.3(a) and 6-1.1.3(b).

Example 6.1

An installation involves the support of 4-in. (102-mm) Schedule 10 pipe. The span, generally the horizontal distance between the building structural members that support the trapeze bar, is 6 ft (1.8 m). (See Exhibit 6.2.) The trapeze consists of an angle iron.

Table 6-1.1.3(b) Available Section Moduli of Common Trapeze Hangers (in.^{3})

Pipe (in.)	Modulus	Angles	Modulus
Schedule 10			
1	0.12	1½ × 1½ × 3⁄16	0.10
1¼	0.19	2 × 2 × ⅛	0.13
1½	0.26	2 × 1½ × 3⁄16	0.18
2	0.42	2 × 2 × 3⁄16	0.19
2½	0.69	2 × 2 × ¼	0.25
3	1.04	2½ × 1½ × 3⁄16	0.28
3½	1.38	2½ × 2 × 3⁄16	0.29
4	1.76	2 × 2 × 5⁄16	0.30
5	3.03	2½ × 2½ × 3⁄16	0.30
6	4.35	2 × 2 × ⅜	0.35
		2½ × 2½ × ¼	0.39
		3 × 2 × 3⁄16	0.41
Schedule 40			
1	0.13	3 × 2½ × 3⁄16	0.43
1¼	0.23	3 × 3 × 3⁄16	0.44
1½	0.33	2½ × 2½ × 5⁄16	0.48
2	0.56	3 × 2 × ¼	0.54
2½	1.06	2½ × 2 × ⅜	0.55
3	1.72	2½ × 2½ × ⅜	0.57
3½	2.39	3 × 3 × ¼	0.58
4	3.21	3 × 3 × 5⁄16	0.71
5	5.45	2½ × 2½ × ½	0.72
6	8.50	3½ × 2½ × ¼	0.75
		3 × 2½ × ⅜	0.81
		3 × 3 × ⅜	0.83
		3½ × 2½ × 5⁄16	0.93
		3 × 3 × 7⁄16	0.95
		4 × 4 × ¼	1.05
		3 × 3 × ½	1.07
		4 × 3 × 5⁄16	1.23
		4 × 4 × 5⁄16	1.29
		4 × 3 × ⅜	1.46
		4 × 4 × ⅜	1.52
		5 × 3½ × 5⁄16	1.94
		4 × 4 × ½	1.97
		4 × 4 × ⅝	2.40
		4 × 4 × ¾	2.81
		6 × 4 × ⅜	3.32
		6 × 4 × ½	4.33
		6 × 4 × ¾	6.25
		6 × 6 × 1	8.57

For SI units, 1 in. = 25.4 mm; 1 ft = 0.3048 m.

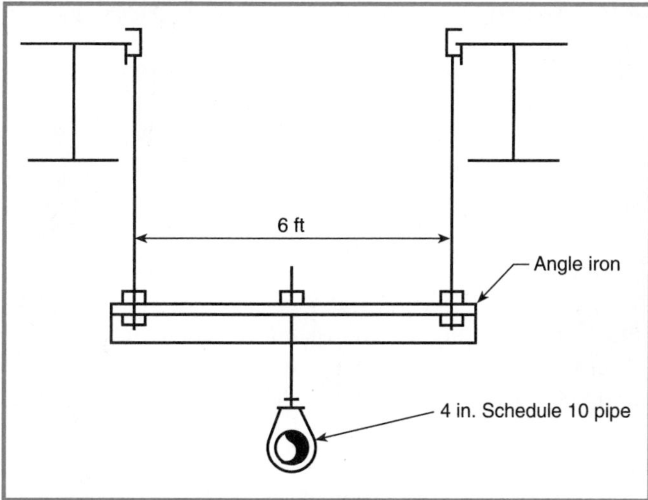

Exhibit 6.2 *Diagram of a trapeze installation.*

Determine acceptable size of trapeze.

Solution

From Table 6-1.1.3(a), the required section modulus is 0.51 in.^{3}. From Table 6-1.1.3(b), an angle iron with dimensions of 3 in. × 2 in. × ¼ in. (Z = 0.54 in.^{3}) provides a section modulus of 0.51 in.^{3} or more:

If Schedule 10 pipe were used as the trapeze, it would require a nominal diameter of 2 ½ in. (Z = 0.69 in.^{3}). If Schedule 40 pipe were used, it would require a nominal diameter of 2 in. (Z= 0.56 in.^{3}). Other members from Table 6-1.1.3(b) and other steel shapes that meet or exceed the predetermined Z of 0.51 in.^{3} are also acceptable.

Where materials other than steel are proposed for use as trapeze members, an engineering analysis must be completed to determine the material's capability. Comparable beam strength and appropriate fire endurance characteristics are required.

The phrase "on both ends" in the last sentence of 6-1.1.3 refers to the top and bottom ends of the rod supporting the main. The phrase does not mean both ends of the trapeze. The top hanger ring, strap, or clevis is being used in a position that is opposite its normal use, and it must have strength appropriate to the larger pipe size.

The following formula merits some explanation:

$$L = \frac{4ab}{a + b}$$

The section moduli given in Table 6-1.1.3(a) are based on the assumption that the pipe to be supported will be located in the exact center of the span and, thus, will put maximum stress on the trapeze bar. The nearer the load is to one side of the span, the less stress it exerts. If the load is sufficiently off center, a trapeze bar of smaller size may possibly be used to support it.

In Example 6.1, to determine acceptable trapeze members, assume that the 4-in. (102-mm) Schedule 10 pipe is to be located 1 ft (0.3 m) from one end of the span. Using the following formula, the equivalent span was determined to be 3.33 ft (10.2 m). Because Table 6-1.1.3(a) provides the required section modulus for 1-ft (0.3-m) increments of trapeze length, the equivalent span was rounded up to 4 ft (1.2 m).

$$L = \frac{(4)(1\text{ft})(5\text{ft})}{1\text{ft} + 5\text{ft}} = \frac{20\text{ft}^2}{6\text{ft}} = 3.33\text{ft}$$

Referring to Table 6-1.1.3(a), the section modulus needed for 4-in. (102-mm) Schedule 10 pipe with a 4-ft (1.22-m) span is 0.34. As Table 6-1.1.3(b) indicates, the trapeze bar could be either 2-in. (51-mm) Schedule 10 or Schedule 40 pipe or a 2-in. × 2-in. × 3⁄8-in. angle iron.

6-1.1.4 The size of hanger rods and fasteners required to support the steel angle iron or pipe indicated in Table 6-1.1.3(b) shall comply with 6-1.4. Holes for bolts shall not exceed 1⁄16 in. greater than the diameter of the bolt. Bolts shall be provided with a flat washer and nut.

Each of the rods used to support the trapeze needs to be sized according to the size of the sprinkler pipe being supported by the trapeze in accordance with Table 6-1.4.1. Where hanger rods are inserted through holes drilled into the trapeze member, the requirements for bolts also apply to threaded rods. This requirement includes tolerances as well as washers and nuts.

6-1.1.5* Sprinkler piping or hangers shall not be used to support nonsystem components.

A-6-1.1.5 The rules covering the hanging of sprinkler piping take into consideration the weight of water-filled pipe plus a safety factor. No allowance has been made for the hanging of nonsystem components from sprinkler piping.

In many cases, sprinkler pipe and even the sprinklers become convenient, accessible, and tempting hangers for signs and decorations. Such use of sprinkler system components is strictly prohibited. The detectors and wiring associated with a preaction or deluge system are permitted to be supported by the pipe and hangers because they are system components. However, detection systems not associated with the sprinkler system cannot be attached to the pipe. The arrangement shown in Exhibit 6.3 is not in compliance with 6-1.1.5. In Exhibit 6.3, the electrical conduit is draped over and is being supported by the sprinkler piping.

6-1.2 Concrete Inserts and Expansion Shields.

6-1.2.1 The use of listed inserts set in concrete and listed expansion shields to support hangers shall be permitted for mains and branch lines.

Exception No. 1: Expansion shields shall not be used in cinder concrete, except for branchlines where the expansion shields are alternated with through-bolts or hangers attached to beams.

Exhibit 6.3 Sprinkler pipe inappropriately used to support electrical conduit.

Exception No. 2: Expansion shields shall not be used in ceilings of gypsum or other similar soft material.

The effectiveness of expansion shields depends on the gripping power of the material with which they are used. Expansion shields pull out of soft materials such as gypsum.

6-1.2.2 Expansion shields shall be installed in a horizontal position in the sides of concrete beams.

Exception: Expansion shields shall be permitted to be installed in the vertical position under the following conditions:

(a) When used in concrete having gravel or crushed stone aggregate to support pipes 4 in. (102 mm) or less in diameter.

(b) When expansion shields are alternated with hangers connected directly to the structural members, such as trusses and girders, or to the sides of concrete beams [to support pipe 5 in. (127 mm) or larger].

(c) When expansion shields are spaced not over 10 ft (3 m) apart [to support pipe 4 in. (102 m) or larger].

Expansion shields installed in a horizontal position are more secure because the associated hanger assembly is not subject to a direct downward pulling force as indicated in Exhibit 6.4. However, under the conditions specified in the exception, expansion shields can be installed in a vertical position for certain sizes of pipe (see Exhibit 6.5).

6-1.2.3 Holes for expansion shields in the side of beams shall be above the centerline of the beam or above the bottom reinforcement steel rods.

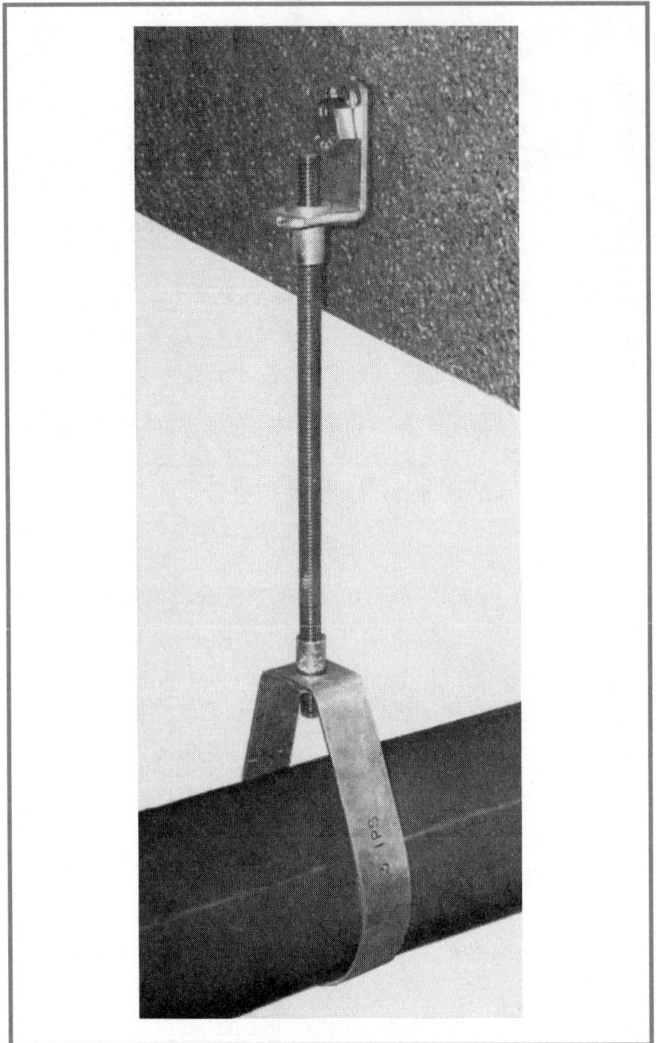

Exhibit 6.4 *Hanger assembly attached to concrete that uses an expansion shield in the horizontal position.*

Shields must be located where they provide maximum strength. If a reinforcing rod is contacted when drilling for a shield, a new hole above the bottom reinforcing rod must be made to the required depth.

6-1.2.4 Holes for expansion shields used in the vertical position shall be drilled to provide uniform contact with the shield over its entire circumference. The depth of the hole shall not be less than specified for the type of shield used.

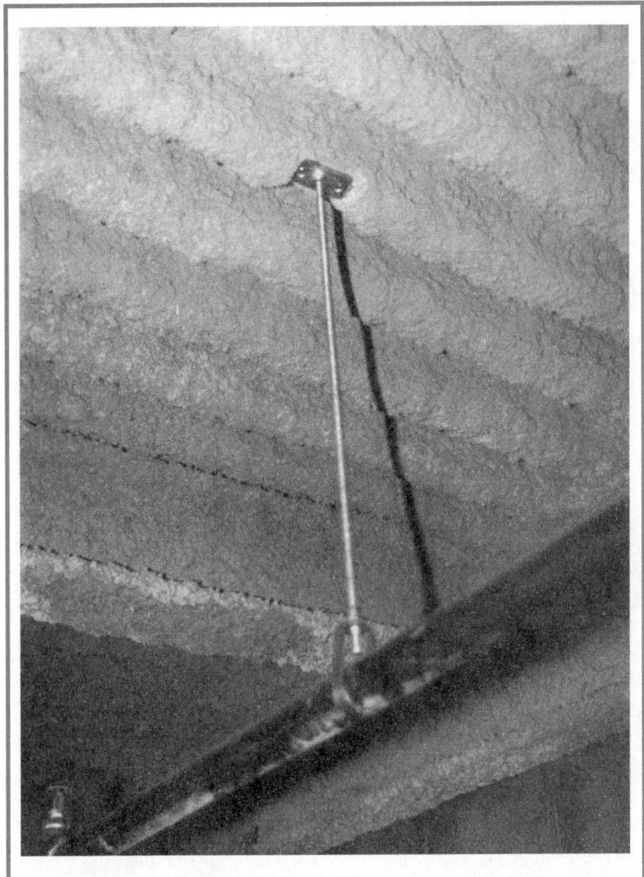

Exhibit 6.5 Hanger assembly attached to concrete decking that uses expansion shields in the vertical position.

6-1.3 Powder-Driven Studs and Welding Studs.

6-1.3.1* Powder-driven studs, welding studs, and the tools used for installing these devices shall be listed. Pipe size, installation position, and construction material into which they are installed shall be in accordance with individual listings.

A-6-1.3.1 Powder-driven studs should not be used in steel of less than $3/16$ in. (4.8 mm) total thickness.

The limitations placed on powder-actuated tools and stud welders vary depending on the individual tool. Limitations of each device and the components of its installation unit are specified in its individual listing.

6-1.3.2* Representative samples of concrete into which studs are to be driven shall be tested to determine that the studs will hold a minimum load of 750 lb (341 kg) for 2-in. (51-mm) or smaller pipe, 1000 lb (454 kg) for 2½-, 3-, or 3½-in. (64-, 76-, or 89-mm) pipe, and 1200 lb (545 kg) for 4- or 5-in. (102- or 127-mm) pipe.

A-6-1.3.2 The ability of concrete to hold the studs varies widely according to type of aggregate, quality of concrete, and proper installation.

The concrete in the facility where the system is being installed must be tested to ascertain that powder-driven studs do not have less than the indicated holding power. The strength of the concrete varies depending on the type of cement and aggregate used and the needed cure time for the concrete. The safety factor imposed in 6-2.1.3 must be increased as specified in 6-1.3.2 when powder-driven studs are installed in concrete, but no additional safety factors need to be prescribed. Formal Interpretation 87-6 provides more information on the use of powder-driven studs.

Formal Interpretation 87-6

Reference: 6-1.1, 6-2.1.3, 6-1.3, 6-1.3.1, 6-1.3.2

Question 1: Do the design requirements set forth in 6-1.1 apply to powder driven studs?

Answer: Yes.

Question 2: Does 6-1.1 apply to powder-driven studs when used in a manner other than that listed (such as limited penetration)?

Answer: No. 6-1.3.2 is imposed on powder driven studs and they must be used per their listing.

Question 3: Does 6-2.1.3 relate directly to 6-1.3.2 in that "Point of Hanging" equates to the "Ability of concrete to hold the stud?"

Answer: Yes.

Question 4: Are the minimum loads given in 6-1.3.2 test loads as opposed to actual working loads?

Answer: Yes.

Question 5: With due consideration for industry standards, is it safe to assume that a stud that passes a tension pull test (for example, of 750 lbs. for 2 in. pipe) is safe to carry a working load of 326.62 lbs. or 383.1 lbs., whichever is applicable?

Answer: Yes.

Question 6: Does 6-1.3.1 limit the loads (working) put on studs to that given in their listing?

Answer: No. The loads applied to the stud must meet the listing and NFPA 13 requirements.

(continues)

Question 7: If "Yes" to Question 6, would that limit the engineer to the listed load?

Answer: N/A.

Question 8: If "No" to Question 6, is approval then subject only to the acceptance of the Authority Having Jurisdiction.

Answer: No. Must comply with 6-1.3.

Question 9: Is the bottom line here the intent to require the hangar and its parts to meet at least 6-1.1 and that, if a powder driven stud in concrete or steel is used, that it test under tension to carry a load equal to the requirement of 6-2.1.3 except that it shall not pull out at less than 750 lbs. tension for 2 in. or smaller etc? Test under tension as used here would mean a load inclusive of the manufacturer's safety factor.

Answer: No. It is not intended to require safety factors beyond those in 6-1.3.2.

Issue Edition: 1987

Reference: 3-10.1, 3-10.3, 3-10.3.1, 3-10.3.2, 3-10.1.4

Issue Date: July 5, 1988

Effective Date: July 25, 1988

(Published August 1989)

6-1.3.3 Increaser couplings shall be attached directly to the powder-driven studs or welding studs.

Listings commonly permit studs with a smaller diameter than that required by Table 6-1.4.1 for rods. Hanger rods that comply with the dimensions given in the table must be connected directly to the studs with increaser couplings.

6-1.3.4 Welding studs or other hanger parts shall not be attached by welding to steel less than U.S. Standard, 12 gauge.

U.S. Standard, 12 gauge is 0.081 in. (2.06 mm) thick. Steel thinner than this lacks the necessary strength for welding.

6-1.4 Rods and U-Hooks.

6-1.4.1 Rods and Coach Screw Rods. Hanger rod size shall be the same as that approved for use with the hanger assembly, and the size of rods shall not be less than that given in Table 6-1.4.1.

Exception: Rods of smaller diameter shall be permitted where the hanger assembly has been tested and listed by a testing laboratory and installed within the limits of pipe sizes expressed in individual listings.

Table 6-1.4.1 Hanger Rod Sizes

Pipe Size	Diameter of Rod in.	Diameter of Rod mm
Up to and including 4 in.	⅜	9.5
5, 6, and 8 in.	½	12.7
10 and 12 in.	⅝	15.9

Table 6-1.4.1 indicates the smallest rod sizes that can be used as part of a hanger assembly. The exception to 6-1.4.1 allows for smaller diameters if proven satisfactory through testing and listing.

6-1.4.2 U-Hooks. The size of the rod material of U-hooks shall not be less than that given in Table 6-1.4.2.

Table 6-1.4.2 U-Hook Rod Sizes

Pipe Size	Hook Material Diameter in.	Hook Material Diameter mm
Up to 2 in.	5/16	7.9
2½ to 6 in.	⅜	9.5
8 in.	½	12.7

6-1.4.3 Eye Rods. The size of the rod material for eye rods shall not be less than specified in Table 6-1.4.3. Eye rods shall be secured with lock washers to prevent lateral motion. Where eye rods are fastened to wood structural members, the eye rod shall be backed with a large flat washer bearing directly against the structural member, in addition to the lock washer.

6-1.4.4 Threaded sections of rods shall not be formed or bent.

Bending the threaded section of hanger rods could result in cracks at the thread root. Such cracks affect the structural ability of the member and can lead to failure of the rod.

This requirement is also applicable to the rod once it is installed. Cracks can also result if the rod is arranged in such a manner that it has a lateral load applied to it. In other words, the rod must be in a true vertical position once it is connected between the building structure and the pipe. Exhibit 6.6 illustrates an unacceptable condition,

Table 6-1.4.3 Eye Rod Sizes

	Diameter of Rod			
	With Bent Eye		With Welded Eye	
Pipe Size	in.	mm	in.	mm
Up to 4 in.	⅜	9.5	⅜	9.5
5 to 6 in.	½	12.7	½	12.7
8 in.	¾	19.1	½	12.7

because the threaded portion of the hanger rod is bent. Also note that the bag placed over the sprinkler to protect the sprinkler from paint overspray has not been removed.

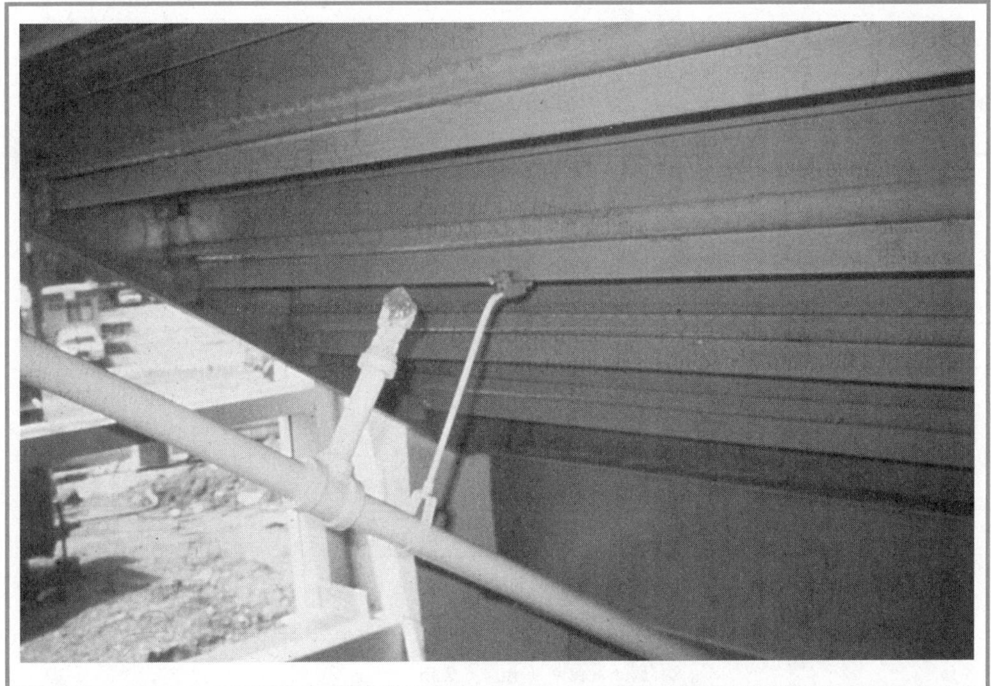

Exhibit 6.6 Bent section of threaded hanger rod, which is not permitted by NFPA 13.

6-1.5 Screws and Bolts.

6-1.5.1 Drive screws shall be used only in a horizontal position as in the side of a beam and only for 2-in. pipe or smaller. Drive screws shall only be used in conjunction with hangers that require two points of attachments.

These conditions are the only ones under which drive screws can be used. Table 6-1.5.2 limits the use of drive screws to U-hooks supporting pipe with diameters of 2 in. (51 mm) or less. The acceptable level of performance with drive screws is restricted to those cases where the weight requirements are relatively small and where two drive screws are installed horizontally. However, specially listed hanger assemblies with two points of attachment intended to be used with drive screws are available. See Figures A-6-1.1 and A-6-2.3.3(c) for illustrations of a drive screw and a U-hook.

6-1.5.2 For ceiling flanges and U-hooks, screw dimensions shall not be less than those given in Table 6-1.5.2.

Exception: When the thickness of planking and thickness of flange do not permit the use of screws 2 in. (51 mm) long, screws 1¾ in. (44 mm) long shall be permitted with hangers spaced not over 10 ft (3 m) apart. When the thickness of beams or joists does not permit the use of screws 2½ in. (64 mm) long, screws 2 in. (51 mm) long shall be permitted with hangers spaced not over 10 ft (3 m) apart.

Table 6-1.5.2 Screw Dimensions for Ceiling Flanges and U-Hooks

Pipe Size	Two Screw Ceiling Flanges
Up to 2 in.	Wood screw No. 18 × 1½ in. or Lag screw 5⁄16 in. × 1½ in.
Pipe Size	**Three Screw Ceiling Flanges**
Up to 2 in.	Wood screw No.18 × 1½ in.
2½ in., 3 in., 3½ in.	Lag screw ⅜ in. × 2 in.
4 in., 5 in., 6 in.	Lag screw ½ in. × 2 in.
8 in.	Lag screw ⅝ in. × 2 in.
Pipe Size	**Four Screw Ceiling Flanges**
Up to 2 in.	Wood screw No. 18 × 1½ in.
2½ in., 3 in., 3½ in.	Lag screw ⅜ in. × 1½ in.
4 in., 5 in., 6 in.	Lag screw ½ in. × 2 in.
8 in.	Lag screw ⅝ in. × 2 in.
Pipe Size	**U-Hooks**
Up to 2 in.	Drive screw No. 16 × 2 in.
2½ in., 3 in., 3½ in.	Lag screw ⅜ in. × 2½ in.
4 in., 5 in., 6 in.	Lag screw ½ in. × 3 in.
8 in.	Lag screw ⅝ in. × 3 in.

For SI units, 1 in. = 25.4 mm.

The types of screws that can be used to fasten ceiling flanges and U-hooks to the building structure are identified in 6-1.5.2. Where ceiling flanges are used, the screws are installed in a vertical position. For U-hook applications, the screws are installed in a horizontal position [see Figure A-6-2.3.3(c)]. As indicated in Table 6-1.5.2, only lag screws and wood screws are to be used to fasten ceiling flanges to the building structure. Other types of screws are not permitted. Drive screws, which are intended to be installed with a hammer, do not have the holding capacity of wood screws and lag screws. Drive screws can only be installed horizontally in conjunction with U-hooks to support 2-in. (51-mm) or smaller pipe.

The screw dimensions in Table 6-1.5.2 are intended for use with hangers spaced in accordance with Table 6-2.2 for steel pipe. The exception recognizes that it is not always possible to use screws with the lengths specified in the table and allows for use of shorter screws with closer hanger spacing.

6-1.5.3 The size of bolt or lag screw used with a hanger and installed on the side of the beam shall not be less than specified in Table 6-1.5.3. All holes for lag screws shall be pre-drilled ⅛ in. (3.2 mm) less in diameter than the maximum root diameter of the lag screw thread. Holes for bolts shall not exceed 1⁄16 in. (1.6 mm) greater than the diameter of the bolt. Bolts shall be provided with a flat washer and nut.

Exception: Where the thickness of beams or joists does not permit the use of screws 2½ in. (64 mm) long, screws 2 in. (51 mm) long shall be permitted with hangers spaced not over 10 ft (3 m) apart.

Table 6-1.5.3 Minimum Bolt or Lag Screw Sizes for Side of Beam Installation

	Size of Bolt or Lag Screw		Length of Lag Screw Used with Wood Beams	
Pipe Size	**in.**	**mm**	**in.**	**mm**
Up to and including 2 in.	⅜	9.5	2½	64
2½ to 6 in. (inclusive)	½	12.7	3	76
8 in.	⅝	15.9	3	76

The minimum sizes of lag screws and bolts that can be used to fasten hanger components other than U-hooks to the sides of beams are identified in 6-1.5.3. Pre-drilling holes ⅛ in. (3.2 mm) smaller than the root diameter of the lag screw thread combines ease of installation with the assurance of the needed gripping capacity of the lag screw into the wood.

6-1.5.4 Wood screws shall be installed with a screwdriver.

Installing wood screws with a screwdriver should not be violated in the interest of saving time. Wood screws and wood lag bolts can only provide the proper holding strength when installed using the proper tools. Hammering these devices in place can reduce their holding power and result in unacceptable installations.

6-1.5.5 Nails are not acceptable for fastening hangers.

6-1.5.6 Screws in the side of a timber or joist shall be not less than 2½ in. (64 mm) from the lower edge where supporting branch lines and not less than 3 in. (76 mm) where supporting main lines.

Exception: This requirement shall not apply to 2-in. (51-mm) or thicker nailing strips resting on top of steel beams.

6-1.5.7 The minimum plank thickness and the minimum width of the lower face of beams or joists in which coach screw rods are used shall be not less than that specified in Table 6-1.5.7.

The reference to nominal wood dimensions recognizes that nominal members such as 2 × 6 joists can only have an actual thickness of 1½ in. (38 mm). Exhibit 6.7 illustrates the installation of a coach screw rod in the bottom of a beam or joist. Exhibit 6.8 illustrates the installation of coach screw rods in wood ceiling planking.

Table 6-1.5.7 Minimum Plank Thicknesses and Beam or Joist Widths

	Nominal Plank Thickness		Nominal Width of Beam or Joist Face	
Pipe Size	**in.**	**mm**	**in.**	**mm**
Up to 2 in.	3	76	2	51
2½ to 3½ in.	4	102	2	51
4 in. and 5 in.	4	102	3	76
6 in.	4	102	4	102

6-1.5.8 Coach screw rods shall not be used for support of pipes larger than 6 in. (152 mm) in diameter. All holes for coach screw rods shall be predrilled ⅛ in. (3.2 mm) less in diameter than the maximum root diameter of the wood screw thread.

Predrilling holes ⅛ in. (3.2 mm) smaller than the root diameter of the coach screw thread combines ease of installation with assurance of the needed gripping power.

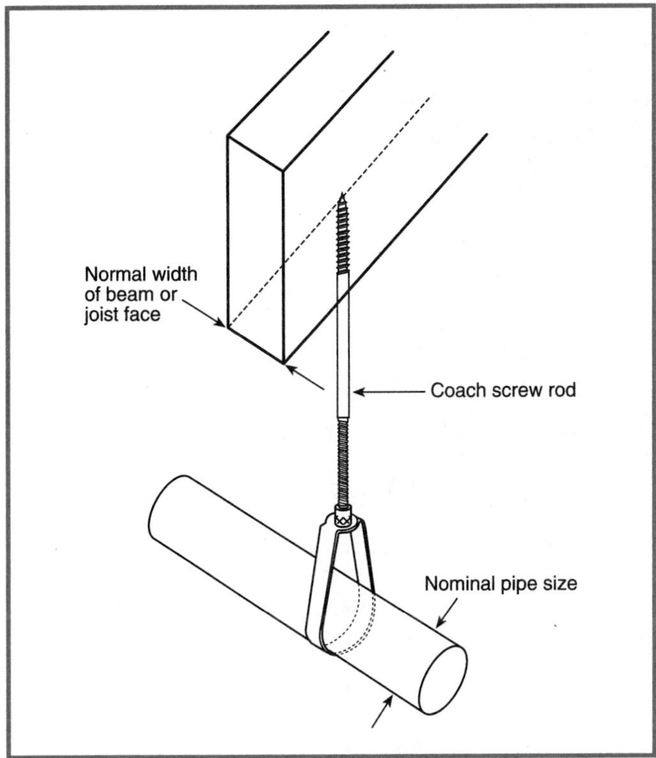

Exhibit 6.7 *Installation of a coach screw rod in the bottom of a beam or joist.*

6-2 Installation of Pipe Hangers

6-2.1 General.

6-2.1.1 Sprinkler piping shall be supported independently of the ceiling sheathing.

Exception: Toggle hangers shall be permitted only for the support of pipe 1½ in. (38 mm) or smaller in size under ceilings of hollow tile or metal lath and plaster.

Toggle hangers are not permitted for use with gypsum wallboard or other less substantial types of ceiling materials, including metal deck roofs. Sprinkler piping must be supported from the building structure, with the exception of 1½-in. (38-mm) or smaller piping, which can be supported under ceilings of hollow tile or metal lath and plaster with toggle hangers. (See Exhibit 6.9.)

6-2.1.2 Where sprinkler piping is installed in storage racks, piping shall be supported from the storage rack structure or building in accordance with all applicable provisions of Sections 6-2 and 6-3.

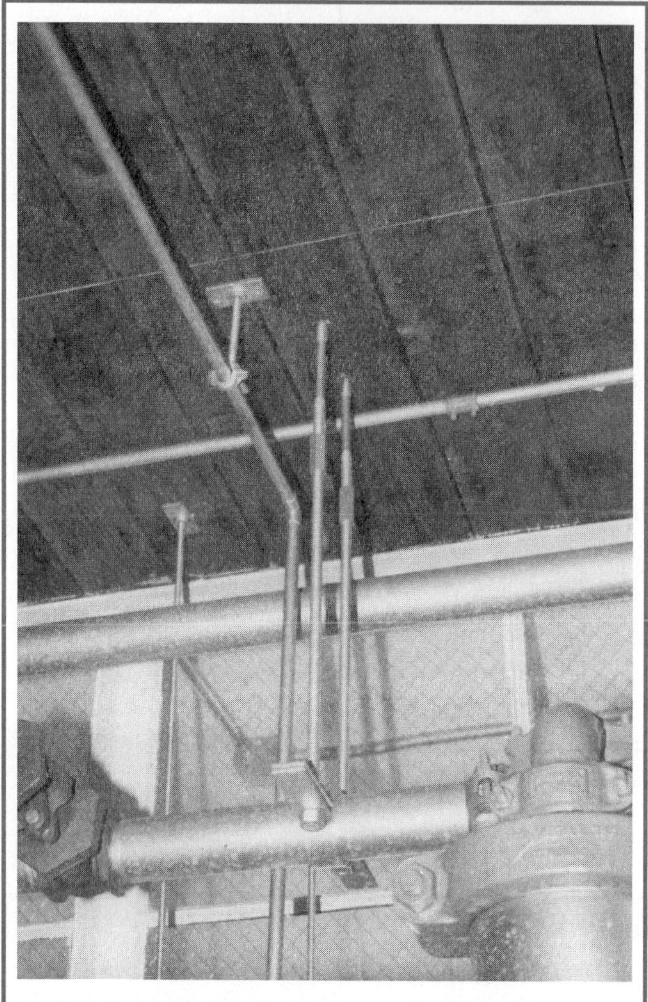

Exhibit 6.8 *Installation of coach screw rods in wood ceiling planking.*

6-2.1.3* Sprinkler piping shall be substantially supported from the building structure, which must support the added load of the water-filled pipe plus a minimum of 250 lb (114 kg) applied at the point of hanging. Trapeze hangers shall be used where necessary to transfer loads to appropriate structural members.

Exception: Branch line hangers under metal deck shall be permitted only for the support of pipe 1 in. (25.4 mm) or smaller in size, by drilling or punching vertical members and using through bolts. The distance from the bottom of the bolt hole to the bottom of the vertical member shall be not less than ⅜ in. (9.5 mm).

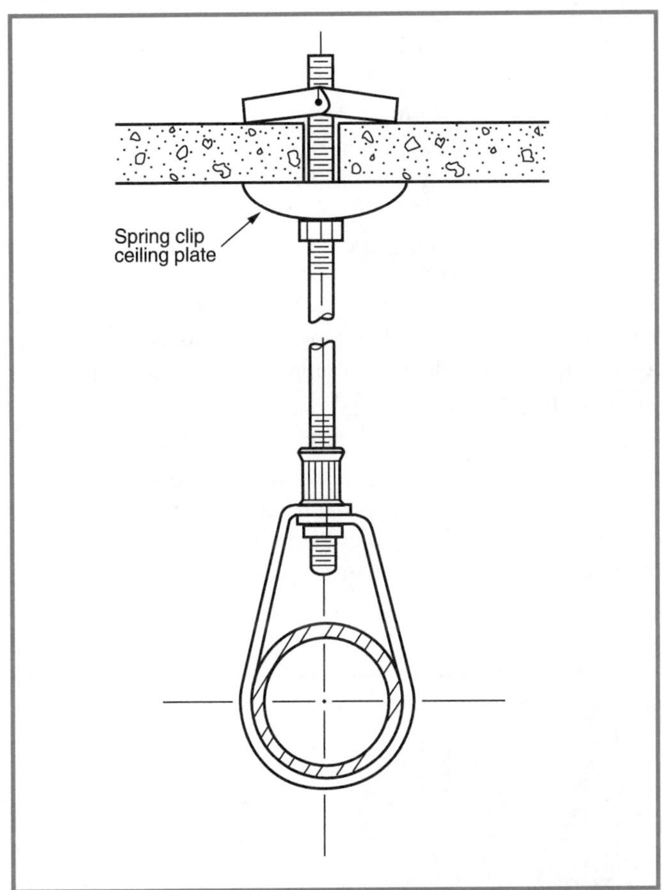

Exhibit 6.9 *Toggle hanger for the support of 1½-in. (38-mm) or smaller pipe.*

A-6-2.1.3 The method used to attach the hanger to the structure and the load placed on the hanger should take into account any limits imposed by the structure. Design manual information for pre-engineered structures or other specialty construction materials should be consulted, if appropriate.

System mains hung to a single beam, truss, or purlin can affect the structural integrity of the building by introducing excessive loads not anticipated in the building design. Also, special conditions such as collateral and concentrated load limits, type or method of attachment to the structural components, or location of attachment to the structural components might need to be observed when hanging system piping in pre-engineered metal buildings or buildings using other specialty structural components such as composite wood joists or combination wood and tubular metal joists.

Each individual point where the hanger is attached to the structure must be able to support the weight of the water-filled pipe plus 250 lb (114 kg). This requirement does not intend to add 250 lb (114 kg) concurrently for each hanger when determining the minimum strength of the building structure. As mentioned in the commentary to 6-1.1, 250 lb (114 kg) represents the assumed weight of a sprinkler pipe fitter with associated work equipment who grasps the pipe during a mishap. The safety factor for the structure is slightly different than the safety factor for the hanger, because it includes only one times, not five times, the weight of the water-filled pipe. The hangers are held to a higher minimum requirement.

The exception allows only very small diameter pipe to be attached to metal decks with through bolts.

6-2.1.4 Where sprinkler piping is installed below ductwork, piping shall be supported from the building structure or from the ductwork supports, provided such supports are capable of handling both the load of the ductwork and the load specified in 6-2.1.3.

If the ductwork supports are not capable of supporting both the ductwork load and the weight of the water-filled pipe plus 250 lb (114 kg), then either the ductwork supports must be increased in size or another method of support must be utilized. Additionally, where the ductwork is to be supported by the building structure rather than the ductwork supports, the sprinkler piping hanger assembly is not to penetrate the ductwork as shown in Exhibit 6.10.

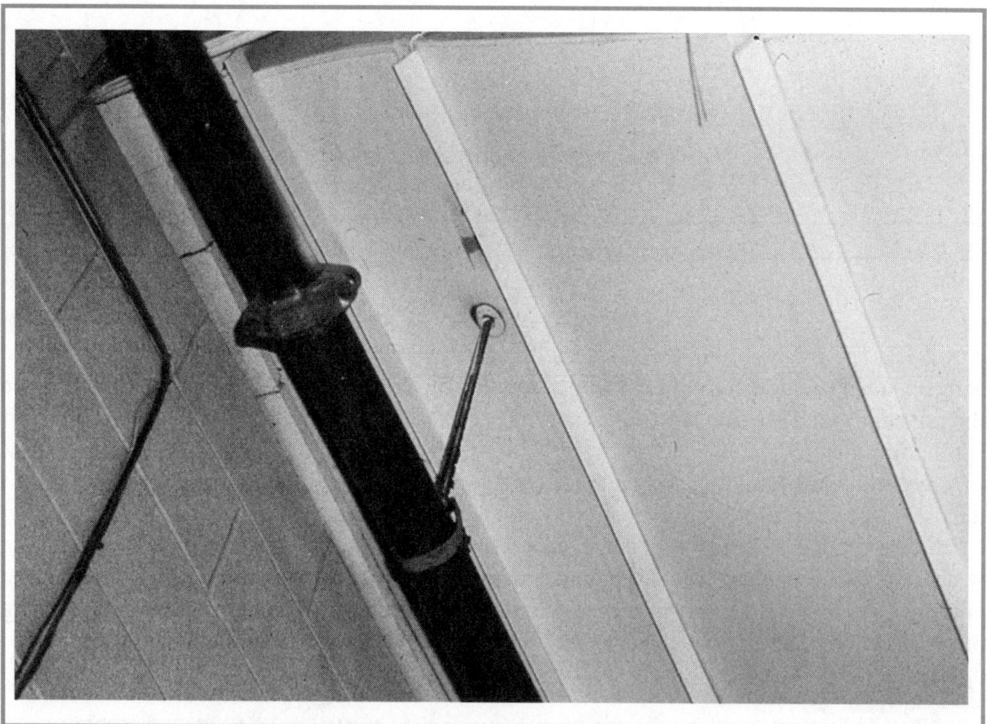

Exhibit 6.10 *Hanger assembly improperly penetrating ductwork.*

6-2.2* Maximum Distance between Hangers.

The maximum distance between hangers shall not exceed that specified in Table 6-2.2.

Exception: The maximum distance between hangers for listed nonmetallic pipe shall be modified as specified in the individual product listings.

Table 6-2.2 Maximum Distance Between Hangers (ft-in.)

Nominal Pipe Size (in.)	¾	1	1¼	1½	2	2½	3	3½	4	5	6	8
Steel pipe except threaded lightwall	N/A	12-0	12-0	15-0	15-0	15-0	15-0	15-0	15-0	15-0	15-0	15-0
Threaded lightwall steel pipe	N/A	12-0	12-0	12-0	12-0	12-0	12-0	N/A	N/A	N/A	N/A	N/A
Copper tube	8-0	8-0	10-0	10-0	12-0	12-0	12-0	15-0	15-0	15-0	15-0	15-0
CPVC	5-6	6-0	6-6	7-0	8-0	9-0	10-0	N/A	N/A	N/A	N/A	N/A
Polybutylene (IPS)	N/A	3-9	4-7	5-0	5-11	N/A	N/A	N/A	N/A	N/A	N/A	N/A
Polybutylene (CTS)	2-11	3-4	3-11	4-5	5-5	N/A	N/A	N/A	N/A	N/A	N/A	N/A
Ductile Iron Pipe	N/A	N/A	N/A	N/A	N/A	N/A	15-0	N/A	15-0	N/A	15-0	15-0

For SI units, 1 in. = 25.4 mm; 1 ft = 0.3048 m.

Note: IPS—iron pipe size; CTS—copper tube size

A-6-2.2 Where copper tube is to be installed in moist areas or other environments conducive to galvanic corrosion, copper hangers or ferrous hangers with an insulating material should be used.

The values given in Table 6-2.2 are based on the use of materials and component sizes given in Section 6-1. When pipe hangers are designed in accordance with Exception No. 1 to 6-1.1, the values given in Table 6-2.2 cannot be modified.

Because of the lack of material between the inside diameter and the root diameter of the thread, hangers are required to be spaced closer for threaded thin-wall pipe.

6-2.3 Location of Hangers on Branch Lines.

This subsection applies to the support of steel pipe or copper tube as specified in 3-3.1 and subject to the provisions of 6-2.2.

6-2.3.1 There shall be not less than one hanger for each section of pipe.

Exception No. 1: Where sprinklers are spaced less than 6 ft (1.8 m) apart, hangers spaced up to a maximum of 12 ft (3.7 m) shall be permitted.*

Exception No. 2: Starter lengths less than 6 ft (1.8 m) shall not require a hanger, unless on the end line of a sidefeed system or where an intermediate cross main hanger has been omitted.

A-6-2.3.1 Exception No. 1. See Figure A-6-2.3.1.

Because of the unique characteristics of some types of fittings, certain manufacturers recommend hanger spacings less than the maximums given in Table 6-2.2. In such circumstances, the manufacturer's recommendations should be followed.

Unless unique characteristics of the fitting make it appropriate to provide support near each fitting, spacing hangers up to 12 ft (3.7 m) apart is expected to adequately support the branch line piping when sprinklers are less than 6 ft (1.8 m) apart. Exception No. 1 can also be applied when multiple pipe fittings are installed at dimensions not exceeding 6 ft (1.8 m) from one another.

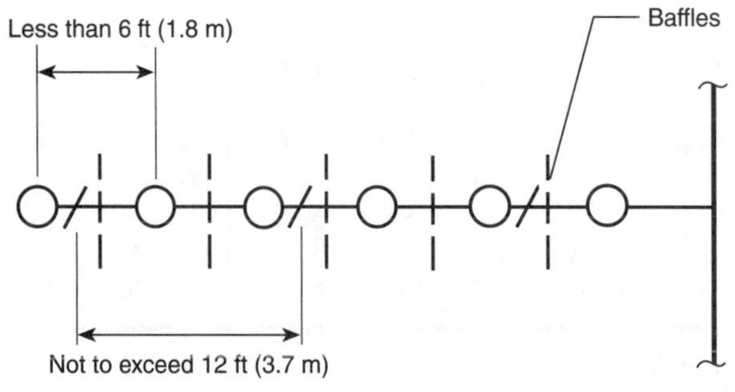

Figure A-6-2.3.1 *Distance between hangers.*

6-2.3.2 The distance between a hanger and the centerline of an upright sprinkler shall not be less than 3 in. (76 mm).

Guidance for deflector positioning with respect to the hanger is intended to ensure that the hanger does not become an obstruction to discharge. Where hangers, sway braces, or other pipe supports have an obstructing dimension in excess of ½ in. (12.7 mm), the obstruction rules in Sections 5-6 through 5-11 for the appropriate type of sprinkler should be followed.

6-2.3.3* The unsupported length between the end sprinkler and the last hanger on the line shall not be greater than 36 in. (0.9m) for 1-in. pipe, 48 in. (1.2 m) for 1¼-in. pipe, and 60 in. (1.5 m) for 1½-in. or larger pipe. Where any of these limits are exceeded, the pipe shall be extended beyond the end sprinkler and shall be supported by an additional hanger.

Exception: When the maximum pressure at the sprinkler exceeds 100 psi (6.9 bar) and a branch line above a ceiling supplies sprinklers in a pendent position below the ceiling, the hanger assembly supporting the pipe supplying an end sprinkler in a pendent position shall be of a type that prevents upward movement of the pipe.*

The unsupported length between the end sprinkler in a pendent position or drop nipple and the last hanger on the branch line shall not be greater than 12 in. (305 mm) for steel pipe or 6 in. (152 mm) for copper pipe. When this limit is exceeded, the pipe shall be extended beyond the end sprinkler and supported by an additional hanger. The hanger closest to the sprinkler shall be of a type that prevents upward movement of the piping.

A-6-2.3.3 Sprinkler piping should be adequately secured to restrict the movement of piping upon sprinkler operation. The reaction forces caused by the flow of water through the sprinkler could result in displacement of the sprinkler, thereby adversely affecting sprinkler discharge. Listed CPVC pipe and listed polybutylene pipe have specific requirements for piping support to include additional pipe bracing of sprinklers. See Figure A-6-2.3.3(a).

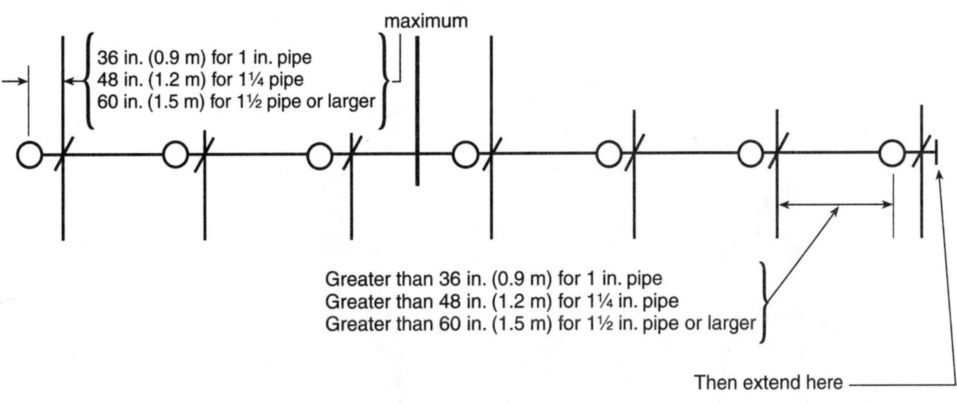

Figure A-6-2.3.3(a) *Distance from sprinkler to hanger.*

A-6-2.3.3 Exception. See Figures A-6-2.3.3(b) and A-6-2.3.3(c).

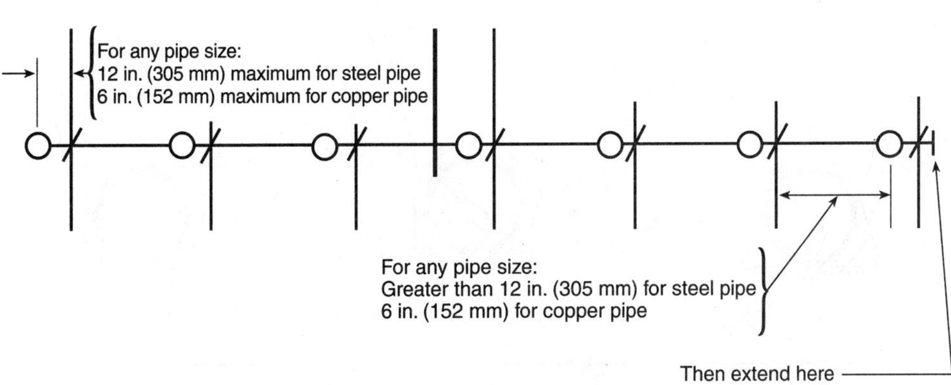

Figure A-6-2.3.3(b) *Distance from sprinkler to hanger where maximum pressure exceeds 100 psi (6.9 bar) and a branch line above a ceiling supplies pendent sprinklers below the ceiling.*

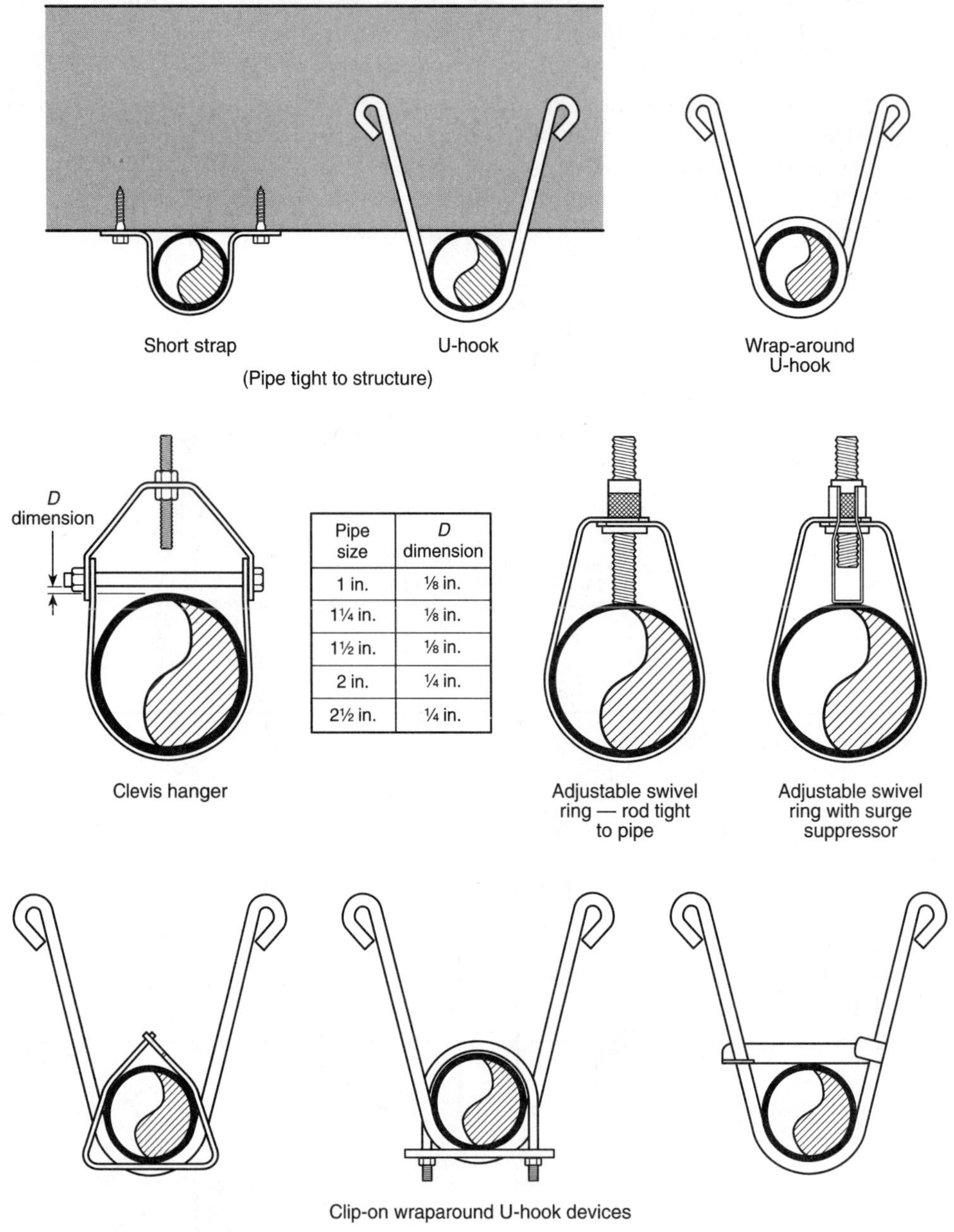

Pipe size	D dimension
1 in.	⅛ in.
1¼ in.	⅛ in.
1½ in.	⅛ in.
2 in.	¼ in.
2½ in.	¼ in.

Figure A-6-2.3.3(c) *Examples of acceptable hangers for end-of-line (or armover) pendent sprinklers.*

Formal Interpretation 89-5 provides more information on this requirement.

Formal Interpretation 89-5

Reference: 6-2.3.3

Question: Is it the intent of the exception to 6-2.3.3 to apply the end sprinkler support provisions to every system in which the maximum pressure at the sprinkler can exceed 100 psi (6.9 bars) when the fire department supports the system through the fire department connection?

Answer: No.

Issue Edition: 1989

Reference: 3-10.5.5

Issue Date: March 23, 1990

Effective Date: April 13, 1990

As Figure A-6-2.3.3(a) illustrates, extending the branch line to the next structural member prevents an excessive cantilever of the piping. The 36-in. (0.91-m) overhang limitation is also applicable to ¾-in. (19-mm) copper pipe.

Data have shown that unrestrained pendent sprinklers operating at high pressures can create a thrust force that lifts the branch line piping. At the end sprinkler and at armovers, the amount of thrust can be sufficient to lift a pendent sprinkler out of the ceiling tile and up into the concealed space.

Figure A-6-2.3.3(b) illustrates the exception to 6-2.3.3. Figure A-6-2.3.3(c) provides examples of acceptable types of hangers for end of line sprinklers or armover sprinklers.

6-2.3.4* The cumulative horizontal length of an unsupported armover to a sprinkler, sprinkler drop, or sprig-up shall not exceed 24 in. (610 mm) for steel pipe or 12 in. (305 mm) for copper tube.

Exception: Where the maximum pressure at the sprinkler exceeds 100 psi (6.9 bar) and a branch line above a ceiling supplies sprinklers in a pendent position below the ceiling, the cumulative horizontal length of an unsupported armover to a sprinkler or sprinkler drop shall not exceed 12 in. (305 mm) for steel pipe and 6 in. (152 mm) for copper tube. The hanger closest to the sprinkler shall be of a type that prevents upward movement of the piping.*

A-6-2.3.4 See Figure A-6-2.3.4(a).

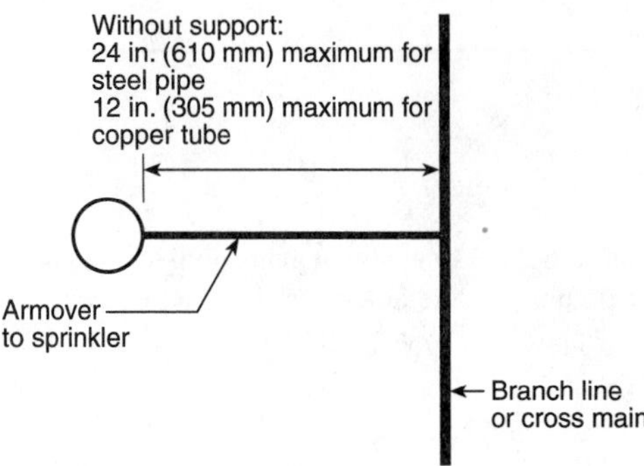

Figure A-6-2.3.4(a) *Maximum length for unsupported armover.*

A-6-2.3.4 Exception. See Figure A-6-2.3.4(b).

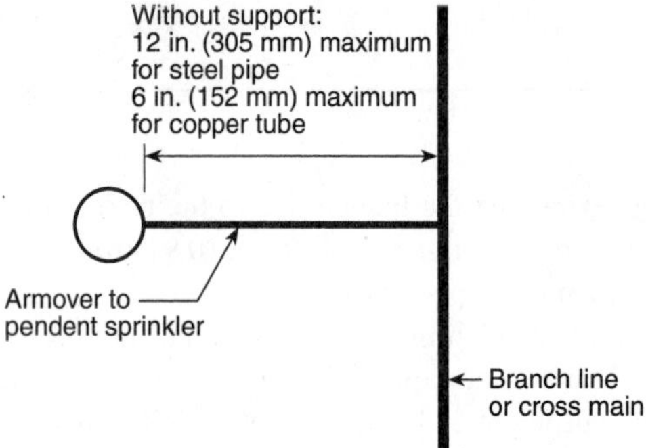

Note: The pendent sprinkler can be installed either directly in the fitting at the end of the armover or in a fitting at the bottom of a drop nipple.

Figure A-6-2.3.4(b) *Maximum length of unsupported armover where the maximum pressure exceeds 100 psi (6.9 bar) and a branch line above a ceiling supplies pendent sprinklers below the ceiling.*

Exhibit 6.11 illustrates an unsupported armover for an upright sprinkler. Special listed fittings that tend to rotate unless restrained either incorporate a mechanism to make rotation impossible or have, as part of their listing, a requirement for a hanger on any armover regardless of length. Incorporation of these requirements maintains system effectiveness even at high operating pressures.

Exhibit 6.11 *Unsupported armover to an upright sprinkler.*

6-2.3.5 Wall-mounted sidewall sprinklers shall be restrained to prevent movement.

6-2.4 Location of Hangers on Mains.

6-2.4.1 Hangers for mains shall be in accordance with 6-2.2 or between each branch line, whichever is the lesser dimension.

Exception No. 1: For cross mains in steel pipe systems in bays having two branch lines, the intermediate hanger shall be permitted to be omitted provided that a hanger attached to a purlin is installed on each branch line located as near to the cross main as the location of the purlin permits. Remaining branch line hangers shall be installed in accordance with 6-2.3.

Exception No. 2: For cross mains in steel pipe systems only in bays having three branch lines, either side or center feed, one (only) intermediate hanger shall be permitted to be omitted provided that a hanger attached to a purlin is installed on each branch line located as near to the cross main as the location of the purlin permits. Remaining branch line hangers shall be installed in accordance with 6-2.3.

Exception No. 3: For cross mains in steel pipe systems only in bays having four or more branch lines, either side or center feed, two intermediate hangers shall be permitted to be omitted provided the maximum distance between hangers does not exceed the distances specified in 6-2.2 and a hanger attached to a purlin on each branch line is located as near to the cross main as the purlin permits.

Cross mains of materials other than steel must have at least one hanger between every two branch lines. With steel pipe cross mains, the designer can either install a hanger between every two branch lines or install hangers on each branch line as near as possible to the cross main and omit one intermediate cross main hanger in each bay. This approach can reduce or eliminate the need for trapeze hangers.

The option to omit the intermediate cross main hanger in accordance with the exceptions to 6-2.4.1 is applicable to the last piece of cross main only if the main is extended to the next framing member and a hanger is also installed at that point. Formal Interpretation 85-3 provides more information pertaining to the hanger requirements for cross mains.

Formal Interpretation 85-3

Reference: 6-2.4.1, 6-2.2

Question: Is it permissible to hang each length with two hangers spaced in accordance with 6-2.2 when sprinkler mains are fabricated using full lengths of pipe (20 to 24 ft) and branch line connections are made with welded or saddle-type fittings?

Answer: No. Crossmain hangers must be installed in accordance with 6-2.4.1.

Issue Edition: 1985

Reference: 3-15.6.1

Date: February 1985

6-2.4.2 At the end of the main, intermediate trapeze hangers shall be installed unless the main is extended to the next framing member with a hanger installed at this point, in which event an intermediate hanger shall be permitted to be omitted in accordance with 6-2.4.1, Exceptions No. 1, 2, and 3.

6-2.4.5 Support of Risers.

6-2.5.1 Risers shall be supported by pipe clamps or by hangers located on the horizontal connections within 24 in. of the centerline of the riser.

Typically, risers are supported by a friction-type clamp that rests on or is secured to the floor slab. However, risers can also be supported using riser clamps that are fastened to the building structure (see Exhibit 6.12) or through the use of hangers that support the horizontal piping at the top of the riser.

6-2.5.2 Pipe clamps supporting risers by means of set screws shall not be used.

To adequately support a riser, the full surface of a clamp must bear against the surface of the pipe. See Exhibit 6.12.

6-2.5.3 In multistory buildings, riser supports shall be provided at the lowest level, at each alternate level above, above and below offsets, and at the top of the riser. Supports above the

Exhibit 6.12 Riser supported with a pipe clamp fastened to the building structure.

lowest level shall also restrain the pipe to prevent movement by an upward thrust where flexible fittings are used. Where risers are supported from the ground, the ground support constitutes the first level of riser support. Where risers are offset or do not rise from the ground, the first ceiling level above the offset constitutes the first level of riser support.

Flexible couplings and fittings allow limited axial movement. In the case of risers in multistory buildings, the axial expansion resulting from system pressurization can be significant unless the movement from upward thrust is prevented.

Risers with horizontal offsets can be supported at both the top and bottom in accordance with 6-2.5.1. A horizontal offset exceeding 6 ft (1.8 m) in length would be required to be supported with its own hanger in accordance with other provisions of NFPA 13.

6-2.5.4 Distance between supports for risers in vertical shafts or high bay areas shall not exceed 25 ft (7.6 m).

A maximum distance between supports is necessary to distribute the load.

6-3 Joint Restraint for Fire Mains

Most of the criteria for joint restraint of fire mains were formerly found within NFPA 24. The material is oriented primarily toward the installation of underground mains, although an effort has been made to generalize the material so that it applies to all buried or aboveground mains. Where mains are not buried, consideration must be given to their proper hanging in accordance with other sections of this chapter. Consideration also must be given to the lateral support of mains not provided with fused or welded joints or those with piping not joined in accordance with Section 3-6. The rules of this chapter address the need to provide thrust blocks only at turns in piping, because the soil surrounding underground pipe is generally satisfactory to hold the piping in place along straight runs. In an aboveground application, means must be provided to ensure similar lateral support.

6-3.1 General.

6-3.1.1* All tees, plugs, caps, bends, reducers, valves, and hydrant branches shall be restrained against movement by utilizing thrust blocks in accordance with 6-3.2 or restrained joint systems in accordance with 6-3.3. Piping with fused or welded joints and piping joined in accordance with Section 3-6 shall not require additional restraining.

A-6-3.1.1 It is a fundamental design principle of fluid mechanics that dynamic and static pressures, acting at changes in size or direction of a pipe, produce unbalanced thrust forces at bends, tees, wyes, deadends, reducers offsets, and so forth. This procedure includes consideration of lateral soil pressure and pipe/soil friction, variables that can be reliably determined using present-day soils engineering knowledge. Refer to A-3-4.1 for a list of references for use in calculating and determining joint restraint systems.

Except for the case of welded joints and approved special restrained joints, such as provided by approved mechanical joint retainer glands or locked mechanical and push-on joints, the usual joints for underground pipe are expected to be held in place by the soil in which the pipe is buried. Gasketed push-on and mechanical joints without special locking devices have limited ability to resist separation due to movement of the pipe.

6-3.1.2 On steep grades, mains shall be properly restrained to prevent slipping. The pipe shall be restrained at the bottom of a hill and at any turns (lateral or vertical). The restraining shall be done either to natural rock or by means of suitable piers built on the downhill side of the bell. Bell ends shall be installed facing uphill. Straight runs on hills shall be restrained as determined by the design engineer.

6-3.2* Thrust Blocks.

Thrust blocks shall be considered satisfactory where soil is suitable for their use.

A-6-3.2 Concrete thrust blocks are one of the most common methods of restraint now in use, provided stable soil conditions prevail and space requirements permit placement. Successful blocking is dependent upon factors such as location, availability and placement of concrete, and possibility of disturbance by future excavations.

Resistance is provided by transferring the thrust force to the soil through the larger bearing area of the block such that the resultant pressure against the soil does not exceed the horizontal bearing strength of the soil. Design of thrust blocks consists of determining the appropriate bearing area of the block for a particular set of conditions. The parameters involved in the design include pipe size, design pressure, angle of the bend (or configuration of the fitting involved), and the horizontal bearing strength of the soil.

Table A-6-3.2(a) gives the nominal thrust at fittings for various sizes of ductile iron and PVC piping. Figure A-6-3.2(a) shows an example of how thrust forces act on a piping bend. Figure A-6-3.2(b) shows an example of a typical connection to a fire protection systems riser.

Table A-6-3.2(a) Thrust at Fittings at 100 psi (6.9 bar) Water Pressure for Ductile Iron and PVC Pipe

Nominal Pipe Diameter (in.)	Total Pounds					
	Dead End	90-Degree Bend	45-Degree Bend	22½-Degree Bend	11¼-Degree Bend	5⅝-Degree Bend
4	1,810	2,559	1,385	706	355	162
6	3,739	5,288	2,862	1,459	733	334
8	6,433	9,097	4,923	2,510	1,261	575
10	9,677	13,685	7,406	3,776	1,897	865
12	13,685	19,353	10,474	5,340	2,683	1,224
14	18,385	26,001	14,072	7,174	3,604	1,644
16	23,779	33,628	18,199	9,278	4,661	2,126
18	29,865	42,235	22,858	11,653	5,855	2,670
20	36,644	51,822	28,046	14,298	7,183	3,277
24	52,279	73,934	40,013	20,398	10,249	4,675
30	80,425	113,738	61,554	31,380	15,766	7,191
36	115,209	162,931	88,177	44,952	22,585	10,302
42	155,528	219,950	119,036	60,684	30,489	13,907
48	202,683	286,637	155,127	79,083	39,733	18,124

Note: To determine thrust at pressure other than 100 psi (6.9 bar), multiply the thrust obtained in the table by the ratio of the pressure to 100 psi (6.9 bar). For example, the thrust on a 12-in., 90-degree bend at 125 psi (8.6 bar) is 19,353 × 125/100 = 24,191 pounds.

Thrust blocks are generally categorized into two groups—bearing and gravity blocks. Figure A-6-3.2(c) depicts a typical bearing thrust block on a horizontal bend.

The following are general criteria for bearing block design.

(a) Bearing surface should, where possible, be placed against undisturbed soil. Where it is not possible, the fill between the bearing surface and undisturbed soil must be compacted to at least 90 percent Standard Proctor density.

(b) Block height *(h)* should be equal to or less than one-half the total depth to the bottom of the block (H_t) but not less than the pipe diameter *(D)*.

(c) Block height *(h)* should be chosen such that the calculated block width *(b)* varies between one and two times the height.

The required block area is as follows:

$$A_b = (h)(b) = \frac{T(S_f)}{S_b}$$

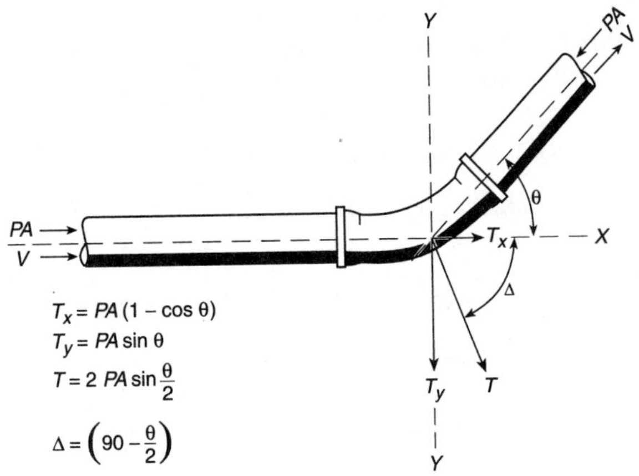

T Thrust force resulting from change in direction of flow
T_x Component of the thrust force acting parallel to the original direction of flow
T_y Component of the thrust force acting perpendicular to the original direction of flow
P Water pressure
A Cross-sectional area of the pipe interior
V Velocity in direction of flow

Figure A-6-3.2(a) *Thrust forces acting on a bend.*

System riser
Cast iron flange and spigot piece
Fire service main
Cast iron bell and spigot pipe
Rods

Figure A-6-3.2(b) *Typical connection to a fire protection system riser.*

Then, for a horizontal bend, the following formula is used:

$$b = \frac{2(S_f)(P)(A_b)\sin(\theta/2)}{(h)(S_b)}$$

where S_f is a safety factor (usually 1.5 thrust block design). A similar approach can be used to design bearing blocks to resist the thrust forces at tees, dead ends, and so forth. Typical values for conservative horizontal bearing strengths of various soil types are listed in Table A-6-3.2(b).

In lieu of the values for soil bearing strength shown in Table A-6-3.2(b), a designer might choose to use calculated Rankine passive pressure (P_p) or other determination of soil bearing strength based on actual soil properties.

It can be easily be shown that T_y = PA sin θ. The required volume of the block is as follows:

$$V_g = \frac{S_f PA\sin\theta}{W_m}$$

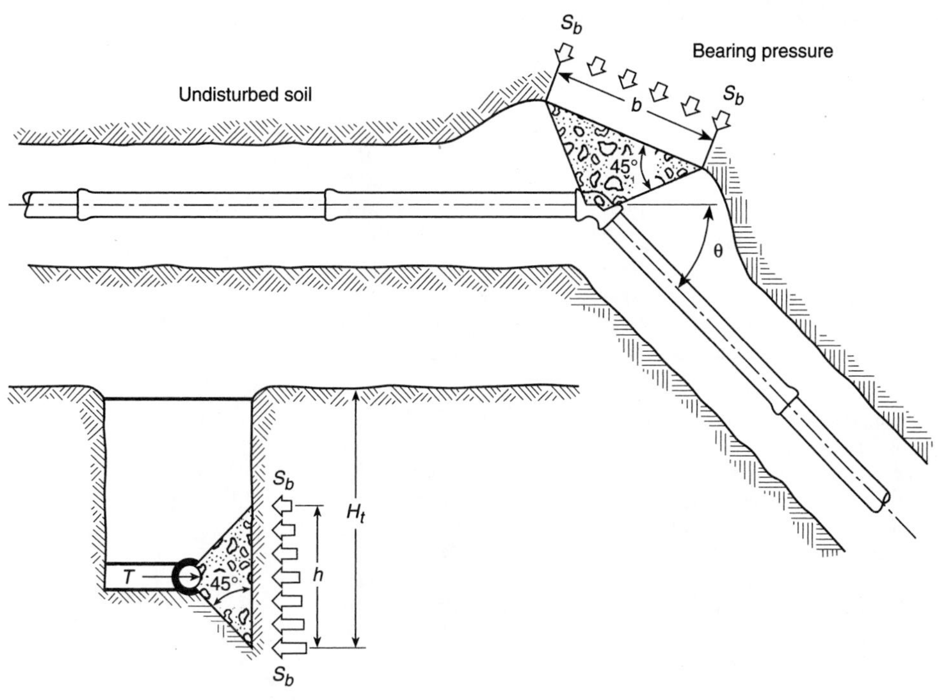

T Thrust force resulting from the change in direction of flow
S_b Horizontal bearing strength of the soil

Figure A-6-3.2(c) *Bearing thrust block.*

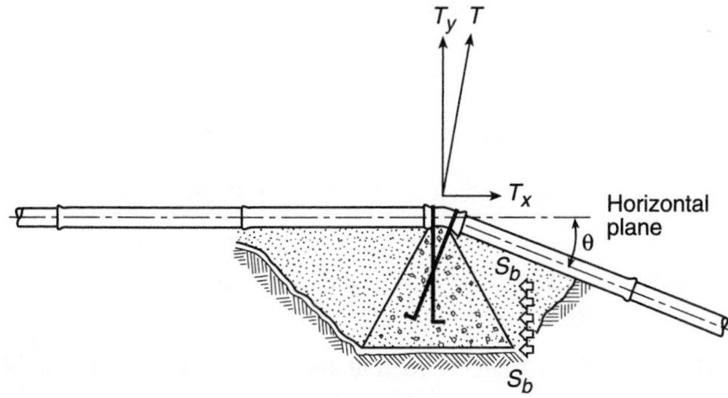

T Thrust force resulting from the change of direction of flow
T_x Horizontal component of the thrust force
T_y Vertical component of the thrust force
S_b Horizontal bearing strength of the soil

Figure A-6-3.2(d) *Gravity thrust block.*

Table A-6-3.2(b) Horizontal Bearing Strengths

Soil	Bearing Strength, S_b lb/ft^2	Bearing Strength, S_b KN/m^2
Muck	0	0
Soft clay	1000	47.9
Silt	1500	71.8
Sandy silt	3000	143.6
Sand	4000	191.5
Sandy clay	6000	287.3
Hard clay	9000	430.9

Notes:
1. Although the bearing strength values in this table have been used successfully in the design of thrust blocks and are considered to be conservative, their accuracy is totally dependent on accurate soil identification and evaluation. The ultimate responsibility for selecting the proper bearing strength of a particular soil type must rest with the design engineer.
2. Gravity thrust blocks can be used to resist thrust at vertical down bends. In a gravity thrust block, the weight of the block is the force providing equilibrium with the thrust force. The design problem is then to calculate the required volume of the thrust force in Figure A-6-3.2(d) is balanced by the weight of the block.

where W_m is the density of the block material. In a case such as the one shown, the horizontal component of thrust force, which is calculated as follows,

$$T_x = PA(1 - \cos \theta)$$

must be resisted by the bearing of the right side of the block against the soil. Analysis of this aspect will follow the same principles as the previous section on bearing blocks.

Exhibit 6.13 illustrates a typical thrust block arrangement.

6-3.2.1 The thrust blocks shall be of a concrete mix not leaner than one part cement, two and one-half parts sand, and five parts stone.

6-3.2.2 Thrust blocks shall be placed between undisturbed earth and the fitting to be restrained and shall be of such bearing to ensure adequate resistance to the thrust to be encountered.

6-3.2.3 Whenever possible, thrust blocks shall be placed so that the joints will be accessible for repair.

6-3.3 Restrained Joint Systems.

Fire mains utilizing restrained joint systems shall include locking mechanical or push-on joints, mechanical joints utilizing setscrew retainer glands, bolted flange joints, heat-fused or welded joints, pipe clamps and tie rods, or other approved methods or devices.

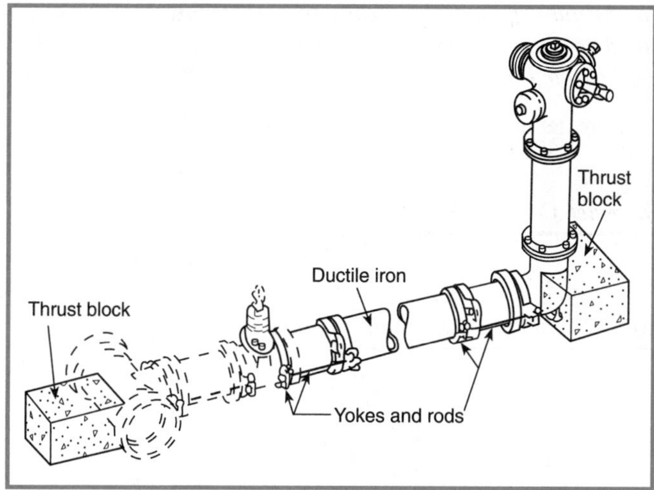

Exhibit 6.13 Typical thrust block arrangement. (Courtesy of the Los Angeles Department of Water and Power.)

Flange adapter fittings and other mechanical joints do not necessarily provide joint restraint. For example, flange adapter fittings are listed by Underwriters Laboratories as "fittings, retainer type" in UL 194, *Gasketed Joints for Ductile-Iron Pipe and Fittings for Fire Protection Service*. UL 194 further categorizes such fittings as either "gasketed joints with self-restraining feature" or "gasketed joints without self-restraining feature." The latter includes gasketed joints consisting of merely a pipe or fitting bell and a spigot end without a feature for joint restraint. Under the provisions of UL 194, all gasketed joints are provided with a pressure test at twice the rated working pressure with samples deflected to the maximum angle specified by the manufacturer. Joints for which the manufacturer claims a self-restraining feature are not provided with external restraint during the test. Manufacturer's literature should specify whether external restraint is required.

6-3.3.1 Sizing the Clamps, Rods, Bolts, and Washers.

6-3.3.1.1 Clamps shall be ½ in. × 2 in. (12.7 mm × 50.8 mm) for pipe 4 in. to 6 in., ⅝ in. × 2½ in. (15.9 mm × 63.5 mm) for pipe 8 in. to 10 in., and ⅝ in. × 3 in. (15.9 mm × 76.2 mm) for pipe 12 in. Bolt holes shall be 1/16 in. (1.6 mm) diameter larger than bolts.

6-3.3.1.2 Minimum rod size shall be ⅝ in. (15.9 mm) diameter. Table 6-3.3.1.2 gives numbers of various diameter rods required for a given pipe size. When using bolting rods, the diameter of mechanical joint bolts limits the size of rods to ¾ in. (19.1 mm).

Threaded sections of rods shall not be formed or bent. When using clamps, rods shall be used in pairs, two to a clamp.

Exception: Assemblies in which a restraint is made by means of two clamps canted on the barrel of the pipe shall be permitted to use one rod per clamp if approved for the specific installation by the authority having jurisdiction.

When using combinations of rods greater in number than two, the rods shall be symmetrically spaced.

Table 6-3.3.1.2 Rod Number—Diameter Combinations

Nominal Pipe Size (in.)	⅝ in. (15.9 mm)	¾ in. (19.1 mm)	⅞ in. (22.2 mm)	1 in. (25.4 mm)
4	2	—	—	—
6	2	—	—	—
8	3	2	—	—
10	4	3	2	—
12	6	4	3	2
14	8	5	4	3
16	10	7	5	4

Note: This table has been derived using pressure of 225 psi (15.5 bar) and design stress of 25,000 psi (172.4 MPa).

6-3.3.1.3 Clamp bolts shall be ⅝ in. (15.9 mm) diameter for pipe 4 in., 6 in., and 8 in., ¾ in. (19.1 mm) diameter for pipe 10 in., and ⅞ in. (22.2 mm) diameter for pipe 12 in.

6-3.3.1.4 Washers can be cast iron or steel, round or square. Dimensions for cast-iron washers shall be ⅝ in. × 3 in. (15.9 mm × 76.2 mm) for pipe 4 in., 6 in., 8 in., and 10 in. and ¾ in. × 3½ in. (19.1 mm × 88.9 mm) for pipe 12 in. Dimensions for steel washers shall be ½ in. × 3 in. (12.7 mm × 76.2 mm) for pipe 4 in., 6 in., 8 in., and 10 in. and ½ in. × 3½ in. (12.7 mm × 88.9 mm) for pipe 12 in. Holes shall be ⅛ in. (3.2 mm) larger than rods.

6-3.3.2 Sizes of Restraint Straps for Tees. Straps shall be ⅝ in. (15.9 mm) thick and 2½ in. (63.5 mm) wide for pipe 4 in., 6 in., 8 in., and 10 in. and ⅝ in. (15.9 mm) thick and 3 in. (76.2 mm) wide for pipe 12 in. Rod holes shall be ¹⁄₁₆ in. (1.6 mm) larger than rods. Dimensions in inches (mm) for straps are suitable either for mechanical or push-on joint tee fittings. Figure 6-3.3.2 and Table 6-3.3.2 shall be used in sizing the restraint straps.

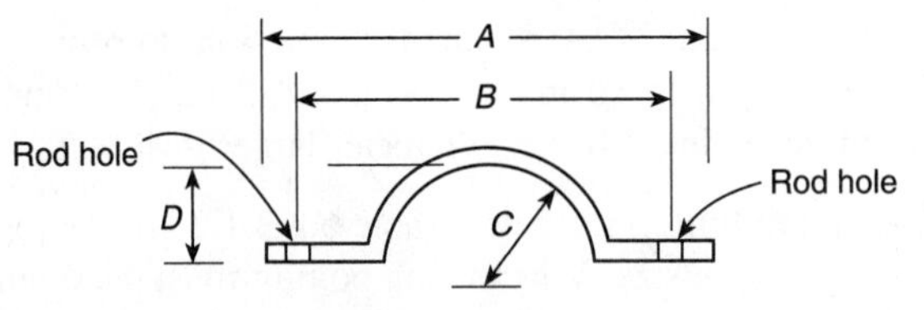

Figure 6-3.3.2 *Restraint straps for tees.*

Table 6-3.3.2 Restraint Straps for Tees

Nominal Pipe Size (in.)	A		B		C		D	
	in.	mm	in.	mm	in.	mm	in.	mm
4	12½	318	10⅛	257	2½	64	1¾	44
6	14½	368	12⅛	308	3 9/16	90	2 13/16	71
8	16¾	425	14⅜	365	4 21/32	118	3 29/32	99
10	19 1/16	484	16 11/16	424	5¾	146	5	127
12	22 5/16	567	19 3/16	487	6¾	171	5⅞	149

6-3.3.3 Sizes of Plug Strap for Bell End of Pipe. Strap shall be ¾ in. (19.1 mm) thick, 2½ in. (63.5 mm) wide. Strap length is the same as dimension A for tee straps given in Figure 6-3.3.2; distance between centers of rod holes is the same as dimension B for tee straps.

6-3.3.4 Material used for clamps, rods, rod couplings or turnbuckles, bolts, washers, restraint straps, and plug straps shall be of material having physical and chemical characteristics such that its deterioration under stress can be predicted with reliability.

6-3.3.5* After installation, rods, nuts, bolts, washers, clamps, and other restraining devices shall be cleaned and thoroughly coated with a bituminous or other acceptable corrosion-retarding material.

A-6-3.3.5 Examples of materials and the standards covering these materials are as follows:

(1) Clamps, steel *(see Note)*
(2) Rods, steel *(see Note)*
(3) Bolts, steel (ASTM A 307, *Standard Specification for Carbon Steel Bolts and Studs*)
(4) Washers, steel *(see Note);* cast iron (Class A cast iron as defined by ASTM A 126, *Standard Specification for Gray Iron Casting for Valves, Flanges and Pipe Fittings*)
(5) Anchor straps and plug straps, steel *(see Note)*
(6) Rod couplings or turnbuckles, malleable iron (ASTM A 197, *Standard Specification for Cupola Malleable Iron*)

Steel of modified range merchant quality as defined in U.S. Federal Standard No. 66C, *Standard for Steel Chemical Composition and Harden Ability,* April 18, 1967, change notice No. 2, April 16, 1970, as promulgated by the U.S. Federal Government General Services Administration.

The materials specified in A-6-3.3.5(1) through (6) do not preclude the use of other materials that will also satisfy the requirements of this section.

6-4 Protection of Piping Against Damage Where Subject to Earthquakes

Section 6-4 describes how to protect sprinkler systems against earthquake damage but not geographically where to provide such protection. The decision to require protection of sprinkler systems or other mechanical systems against earthquake forces is generally stated within building codes or determined by other authorities having jurisdiction. The decision is based on the past seismic activity of an area and its potential for future earthquakes.

Earthquake design criteria for sprinkler systems have been included in NFPA 13 since 1947. The performance of sprinkler systems in earthquakes has generally been quite good. Following the San Fernando, CA, earthquake in 1971 (6.7 moment magnitude), the Pacific Fire Rating Bureau surveyed 973 sprinklered buildings and reported that "if a sprinklered building fared well, so did the sprinkler system." Studies of fire sprinkler performance in more recent earthquakes, such as the Loma Prieta, CA, earthquake of 1989 (6.9 moment magnitude) and the Northridge, CA, earthquake of 1994 (6.7 moment magnitude), have reinforced this observation. Knowledge gained from past earthquakes, however, has led to modifications of the requirements of NFPA 13 over the years. The modifications have been aimed at avoiding significant damage to the sprinkler systems and permitting sprinkler systems to remain functional following an earthquake.

The protection criteria do not specifically address water supply components such as pumps and tanks, because installation of these components is outside the scope of NFPA 13. Nevertheless, pumps and tanks should be braced against horizontal accelerations in a fashion similar to system piping. The fastener load criteria of this standard can be applied.

6-4.1* General.

When sprinkler systems or aboveground fire service mains are to be protected against damage from earthquakes, the requirements of Section 6-4 shall apply.

Exception: Alternative methods of providing earthquake protection of sprinkler systems based on a dynamic seismic analysis certified by a registered professional engineer such that system performance will be at least equal to that of the building structure under expected seismic forces shall be permitted.

A-6-4.1 Sprinkler systems are protected against earthquake damage by means of the following:

(1) Stresses that would develop in the piping due to differential building movement are minimized through the use of flexible joints or clearances.
(2) Bracing is used to keep the piping fairly rigid when supported from a building component expected to move as a unit, such as a ceiling.

Areas known to have a potential for earthquakes have been identified in building code and insurance maps. Examples of two such maps are shown in Figures A-6-4.1(a) and A-6-4.1(b).

The earthquake protection criteria involve requirements aimed at avoiding piping stress through the use of flexible fittings and clearances and requirements to maintain system alignment and to prevent the development of damage-inducing momentum through the use of sway bracing. Subsection 6-4.1 references those criteria but also permits an exception based on a dynamic seismic analysis certified by a registered professional engineer. This exception provides a waiver of NFPA 13 criteria for special building designs, such as those buildings equipped with base isolation systems, to prevent the transfer of earthquake accelerations to the building and its mechanical systems.

NFPA 13 earthquake protection criteria have traditionally been based on an assumed horizontal acceleration of 0.5 g (A_h = 0.5 g). This acceleration results in a horizontal force factor of 0.5 W_p or a horizontal force equivalent to half the weight of the water-filled piping. This acceleration was judged to be suitably conservative for even highly active earthquake areas, and it was not considered economically justified to fine-tune system bracing loads to accommodate lower expected earthquake forces in some localities. The 0.5-g value continues to serve as the basis of the bracing criteria. Since the 1994 edition, NFPA 13 has permitted the use of a multiplier to accommodate code requirements that specify the use of higher or lower horizontal acceleration factors. Use of a multiplier is becoming common for the following two reasons:

(1) Recent earthquakes in California have demonstrated horizontal accelerations in excess of 0.5 g.
(2) Earthquake protection of sprinkler systems is now mandated in many parts of the country where expected accelerations can be less than 0.5 g.

Where earthquake protection of the system is required but other horizontal acceleration factors are not specified to the satisfaction of the authority having jurisdiction, the A_h = 0.5 g default value should continue to be used.

The maps in Figures A-6-4.1(a) and A-6-4.1(b) serve as examples only. The maps are typical of the maps used within building codes to identify areas in which buildings are required to be specially designed to resist earthquake forces.

In general, if the building is specially designed to resist earthquakes, the sprinkler system should also be protected. However, some codes exempt sprinkler and other mechanical systems from protection criteria in low-risk seismic zones. Using Figure A-6-4.1(b) as an example, most model building codes do not require protection of the sprinkler system in Zone 0. In Zone 1, only sprinkler systems located in essential facilities such as hospitals and police stations would be required to be protected.

6-4.2* Couplings.

Listed flexible pipe couplings joining grooved end pipe shall be provided as flexure joints to allow individual sections of piping 2½ in. (64 mm) or larger to move differentially with the individual sections of the building to which it is attached. Couplings shall be arranged to coincide with structural separations within a building. Systems having more flexible couplings than required here shall be provided with additional sway bracing as required in 6-4.5.3, Exception No. 4. The flexible couplings shall be installed as follows:

(1) Within 24 in. (610 mm) of the top and bottom of all risers.

Exception No. 1: In risers less than 3 ft (0.9 m) in length, flexible couplings are permitted to be omitted.

Exception No. 2: In risers 3 to 7 ft (0.9 to 2.1 m) in length, one flexible coupling is adequate.

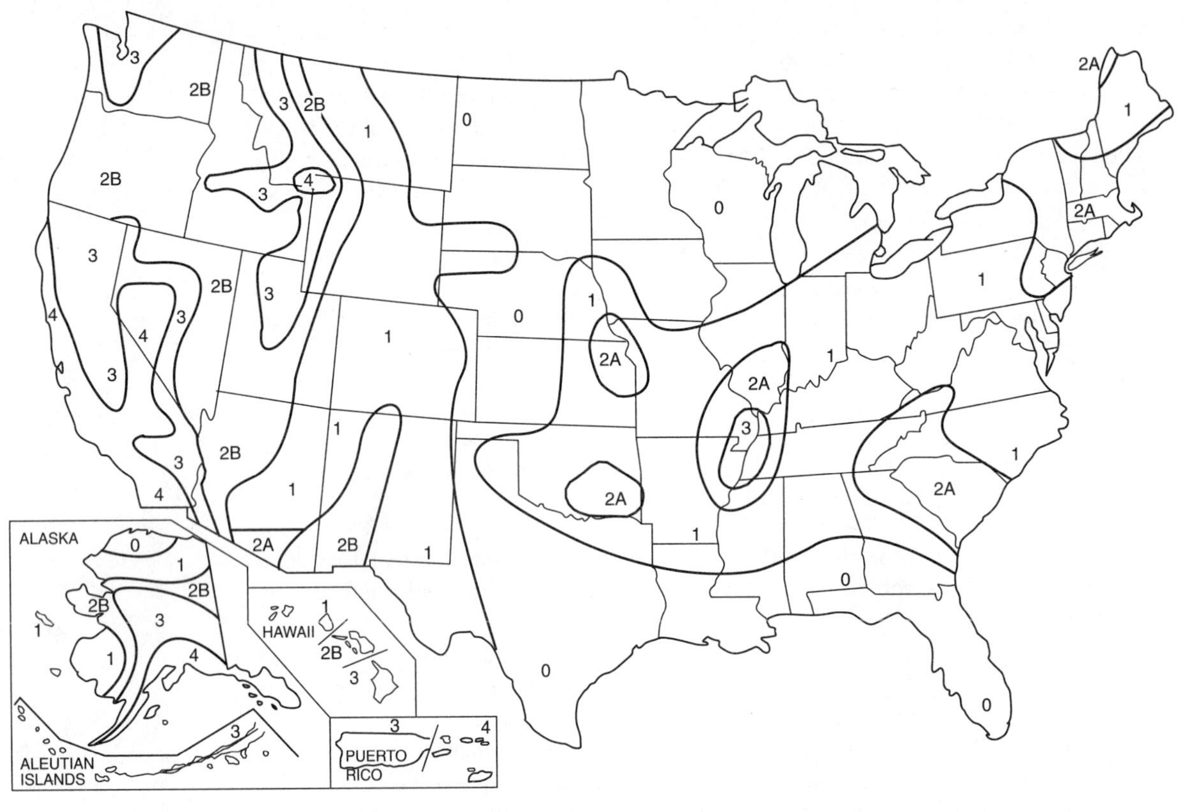

Figure A-6-4.1(a) *Seismic zone map of the United States.*

(2)* Within 12 in. (305 mm) above and within 24 in. below the floor in multistory buildings. When the flexible coupling below the floor is above the tie-in main to the main supplying that floor, a flexible coupling shall be provided on the vertical portion of the tie-in piping.
(3) On both sides of concrete or masonry walls within 1 ft of the wall surface.

Exception: Flexible pipe couplings are not required where clearance around the pipe is provided in accordance with 6-4.4.

(4)* Within 24 in. (610 mm) of building expansion joints.
(5) Within 24 in. (610 mm) of the top and bottom of drops to hose lines, rack sprinklers, and mezzanines, regardless of pipe size.
(6) Within 24 in. (610 mm) of the top of drops exceeding 15 ft (4.6 m) in length to portions of systems supplying more than one sprinkler, regardless of pipe size.
(7) Above and below any intermediate points of support for a riser or other vertical pipe.

Legend

X Seismic zone

X.XX Effective peak velocity–related acceleration (A_V)

$A_V \geq 0.40$

* Represents the maximum value in the center of the zone

0 100 200 300 400 500 Miles

Scale 1:20,000,000

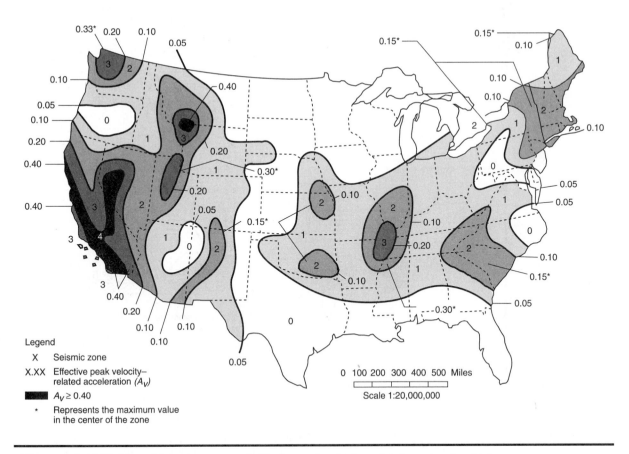

Figure A-6-4.1(b) *Map of seismic zones and effective peak velocity–related acceleration (A_v) for the contiguous 48 states; linear interpolation between contours is acceptable.*

A-6-4.2 Strains on sprinkler piping can be greatly lessened and, in many cases, damage prevented by increasing the flexibility between major parts of the sprinkler system. One part of the piping should never be held rigidly and another part allowed to move freely without provision for relieving the strain. Flexibility can be provided by using listed flexible couplings, by joining grooved end pipe at critical points, and by allowing clearances at walls and floors.

Tank or pump risers should be treated the same as sprinkler risers for their portion within a building. The discharge pipe of tanks on buildings should have a control valve above the roof line so any pipe break within the building can be controlled.

Piping 2 in. (51 mm) or smaller in size is pliable enough so that flexible couplings are not usually necessary. "Rigid-type" couplings that permit less than 1 degree of angular movement at the grooved connections are not considered to be flexible couplings. *[See Figures A-6-4.2(a) and (b).]*

Flexible listed pipe couplings are defined in 1-4.4 as couplings or fittings that allow axial displacement, rotation, and at least 1 degree of angular movement of the pipe without inducing harm on the pipe. Couplings or fittings that do not permit such movement are considered "rigid-type" mechanical couplings. Flexible couplings are required within the system at critical points to minimize stresses on the piping. Flexible special listed pipe should be considered to satisfy the requirements for flexible couplings only if equivalent levels of flexibility can be achieved without harm to the pipe. See Exhibit 6.14.

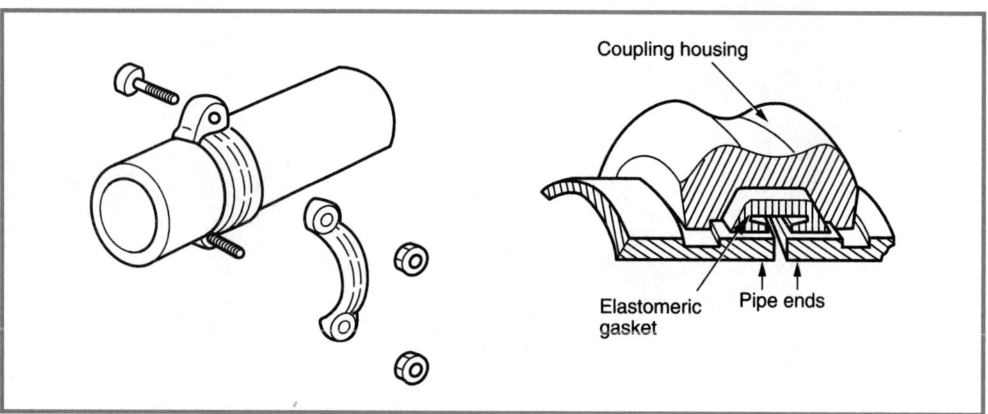

Exhibit 6.14 *Flexible pipe coupling.*

Since the 1994 edition, two flexible couplings per floor are required on the riser of a multistory building as indicated in 6-4.2(2). The couplings are intended to provide 2 degrees of angular displacement per story without damaging the riser. One flexible coupling is to be located within 12 in. (305 mm) above the floor. The other coupling is to be located just below the point of connection of the riser to the main supplying the system on that floor, but within 24 in. (610 mm) of the ceiling. In the 1999 edition, the requirement was modified, and Figure A-6-4.2(b) was added. This change requires an additional coupling on the piping serving the floor main if it is taken off the riser at an elevation lower than 24 in. (610 mm) below the ceiling.

With regard to 6-4.2(5), where racks are freestanding—that is, independent of the building structure—their movement relative to the ceiling can be considerably more than can be accommodated by a single flexible coupling. In cases where ceiling sprinkler piping drops to the top of a freestanding rack, a swing joint comprised of multiple flexible elbows can be needed. Earthquake experts have suggested that differential lateral movement of 5 percent of the rack height can be expected.

Drops to a single sprinkler are not required to be equipped with a flexible coupling because of the concern regarding the difficulty of sprinkler installation and replacement [see 6-4.2(6)]. The rotation inherent in a flexible coupling can prevent the development of the torque necessary to properly install the sprinkler.

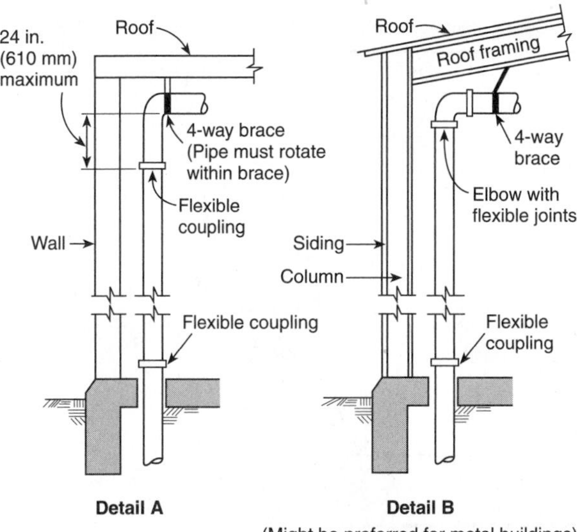

Note to Detail A: The four-way brace should be attached above the upper flexible coupling required for the riser and preferably to the roof structure if suitable. The brace should not be attached directly to a plywood or metal deck.

Figure A-6-4.2(a) Riser details.

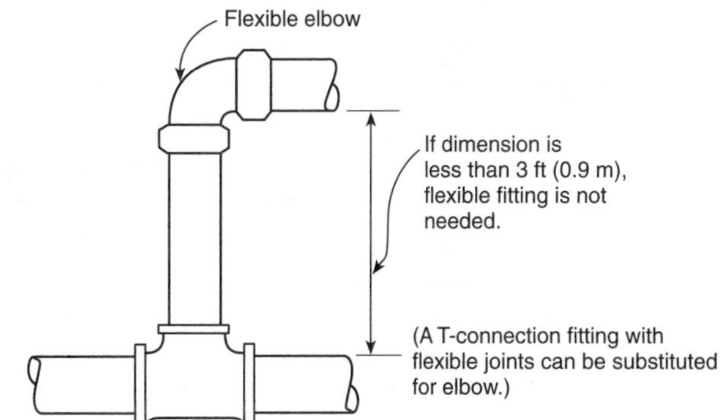

Figure A-6-4.2(b) Detail at short riser.

A-6-4.2(2) The flexible coupling should be at the same elevation as the flexible coupling on the main riser. See Figure A-6-4.2(2).

A-6-4.2(4) A building expansion joint is usually a bituminous fiber strip used to separate blocks or units of concrete to prevent cracking due to expansion as a result of temperature changes. Where building expansion joints are used, the flexible coupling is required on one side of the joint by 6-4.2(4).

For seismic separation joints, considerably more flexibility is needed, particularly for piping above the first floor. Figure A-6-4.3 shows a method of providing additional flexibility through the use of swing joints.

6-4.3* Seismic Separation Assembly.

Seismic separation assemblies with flexible fittings shall be installed where sprinkler piping, regardless of size, crosses building seismic separation joints above ground level.

A-6-4.3 Plan and elevation views of a seismic separation assembly assembled with flexible elbows are shown in Figure A-6-4.3.

A seismic separation assembly is considered to be an assembly of fittings, pipe, and couplings or an assembly of pipe and couplings that permits movement in all directions. The extent of permitted movement should be sufficient to accommodate calculated differential

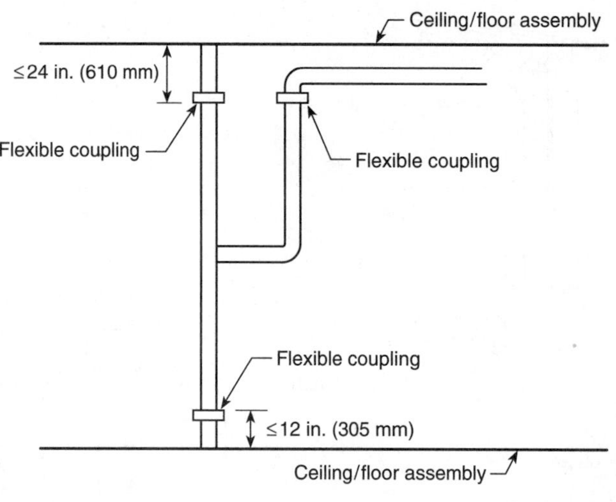

Figure A-6-4.2(2) Flexible coupling on main riser and branch line riser.

motions during earthquakes. In lieu of calculations, permitted movement can be made at least twice the actual separations, at right angles to the separation as well as parallel to it.

The seismic separation assembly (see Figure A-6-4.3) is intended to provide sufficient flexibility to accommodate the substantial relative motion that can be expected at building seismic separation joints. In many cases, providing each building section with its own riser can be more economical. The seismic separation assembly is not required on piping located below ground level. The seismic separation assembly is required for branch lines, as well as mains that cross seismic separation joints above ground level, without regard for pipe size.

6-4.4* Clearance.

Clearance shall be provided around all piping extending through walls, floors, platforms, and foundations, including drains, fire department connections, and other auxiliary piping.

A-6-4.4 While clearances are necessary around the sprinkler piping to prevent breakage due to building movement, suitable provision should also be made to prevent passage of water, smoke, or fire.

Drains, fire department connections, and other auxiliary piping connected to risers should not be cemented into walls or floors; similarly, pipes that pass horizontally through walls or foundations should not be cemented solidly or strains will accumulate at such points.

Where risers or lengths of pipe extend through suspended ceilings, they should not be fastened to the ceiling framing members.

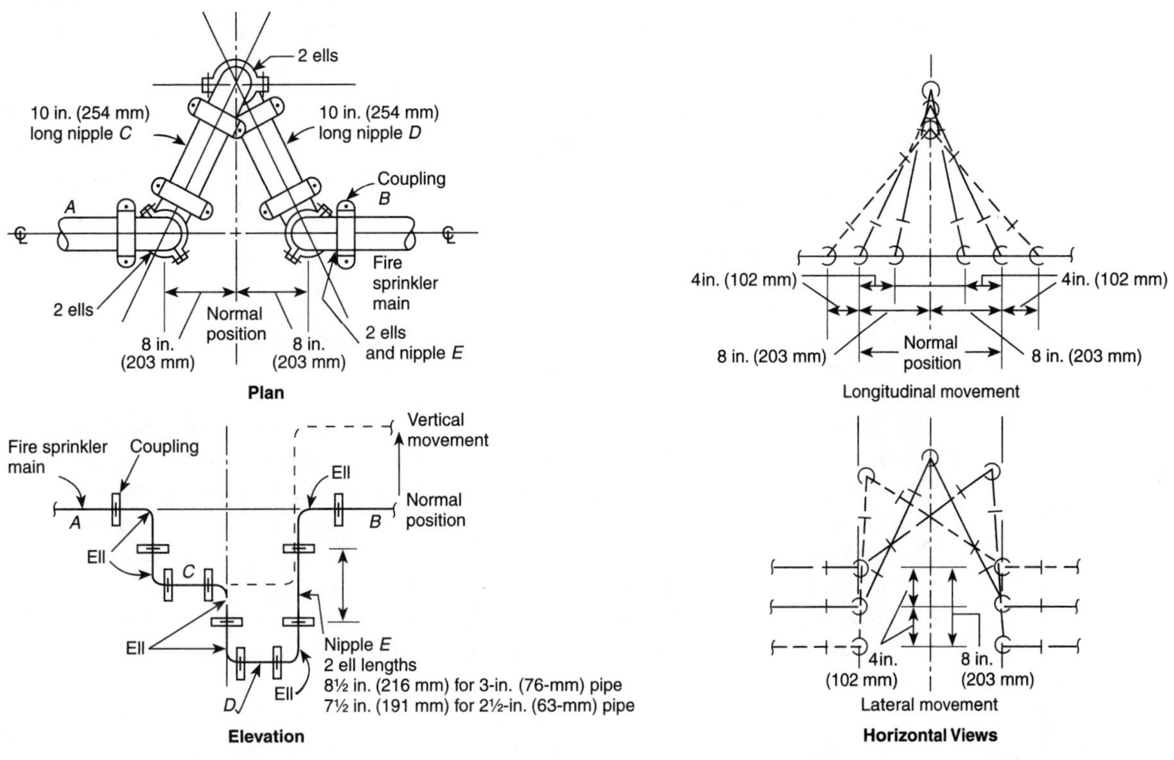

Figure A-6-4.3 Seismic separation assembly. Shown are an 8-in. (203-mm) separation crossed by pipes up to 4 in. (102 mm) in nominal diameter. For other separation distances and pipe sizes, lengths and distances should be modified proportionally.

Clearance is not required around pipes passing through successive floor joists or closely spaced beams expected to move as a unit, such as a part of a ceiling/roof deck. Clearance is required to be provided around pipes passing through beams that are considered primary structural members. Large openings can reduce the strength of structural members, and openings in beams should generally be located only in the center one-third of their depth and not close to either end of their span.

Where holes through primary structural members substitute for hangers as permitted in Exception No. 2 to 6-1.1.2, the intent of 6-4.4 is to provide total clearance, but the intent is not to require that the pipe be maintained directly in the center of the clearance opening.

6-4.4.1 Where pipe passes through holes in platforms, foundations, walls, or floors, the holes shall be sized such that the diameter of the holes is 2 in. (51 mm) larger than the pipe for 1 in. (25.4 mm) nominal to 3½ in. (89 mm) nominal and 4 in. (102 mm) larger than the pipe for pipe 4 in. (102 mm) nominal and larger. Clearance from structural members not penetrated or used, collectively or independently, to support the piping shall be at least 2 in. (51 mm).

Exception No. 1: Where clearance is provided by a pipe sleeve, a nominal diameter 2 in. (51 mm) larger than the nominal diameter of the pipe is acceptable for pipe sizes 1 in. (25.4 mm) through 3½ in. (89 mm), and the clearance provided by a pipe sleeve of nominal diameter 4 in. (102 mm) larger than the nominal diameter of the pipe is acceptable for pipe sizes 4 in. (102 mm) and larger.

Exception No. 2: No clearance is necessary for piping passing through gypsum board or equally frangible construction that is not required to have a fire resistance rating.

Exception No. 3: No clearance is necessary if flexible couplings are located within 1 ft (0.31 m) of each side of a wall, floor, platform, or foundation.

The clearance around the pipe does not need to be annular—that is, the pipe does not need to be placed in the exact center of the opening. Although the pipe penetrates frangible construction, the wall penetrated encloses an exit stair and is required to carry a fire resistance rating. Clearance must be provided according to Exception No. 2. Additionally, the clearance needs to be filled with flexible material that maintains the integrity of the fire resistance rating.

The clearance from members not penetrated or used for support is intended to avoid stresses within the piping system resulting from differential movement of such members.

As with walls, flexible couplings located within 1 ft (0.31 m) on each side of beams are permitted to be substituted for the required clearances.

6-4.4.2 Where required, the clearance shall be filled with a flexible material such as mastic.

Most building codes now require the annular space around piping penetrations through fire-rated assemblies, other than sprinkler penetrations of ceiling membranes, to be filled with an appropriate material. Listed penetration sealants are available that are tested in accordance with ASTM E 814, *Test Method for Fire Tests of Through-Penetration Fire Stops*, and that are flexible as installed.

6-4.5* Sway Bracing.

A-6-4.5 Figures A-6-4.5(a) and A-6-4.5(b) are examples of forms used to aid in the preparation of bracing calculations.

Sway bracing is provided to prevent excessive movement of system piping. Shifting of large pipe as a result of earthquake motion has led to the pull-out of hangers and fracture of fittings. With some exceptions as specifically stated, bracing is required for the following:

(1) The top of the system riser
(2) All feed and cross mains, regardless of size
(3) Branch lines 2½ in. (64 mm) in diameter and larger

Seismic Bracing Calculations

Sheet ________ of ________

Project: ______________________________

Address: ______________________________

____

Contractor: ______________________________

Address: ______________________________

____

Telephone: ______________________________

Fax: ______________________________

Brace Information

Length of brace: ____________

Diameter of brace: ____________

Type of brace: ____________

Angle of brace: ____________

Least radius of gyration:* ____________

L/R value:* ____________

Maximum horizontal load: ____________

Fastener Information

Orientation of connecting surface: ____________

Fastener:

Type: ____________

Diameter: ____________

Length (in wood): ____________

Maximum load: ____________

Seismic Brace Attachments

Structure attachment fitting or tension-only bracing system:

Make: ____________ Model: ____________

Listed load rating: ______ Adjusted load rating per 6-4.5.7: ______

Sway brace (pipe attachment) fitting:

Make: ____________ Model: ____________

Listed load rating: ______ Adjusted load rating per 6-4.5.7: ______

Seismic Brace Assembly Detail

(Provide detail on plans)

Brace identification no.
(to be used on plans) ____________

☐ Lateral brace ☐ Longitudinal brace

Sprinkler System Load Calculation

Diameter	Type	Length (ft)	Total (ft)	½ Weight per ft	½ Total Weight
				lb/ft	lb
				lb/ft	lb
				lb/ft	lb
				lb/ft	lb
				lb/ft	lb
				lb/ft	lb
				Total ½ weight of water-filled pipe	lb

* Excludes tension-only bracing systems

Figure A-6-4.5(a) *Seismic bracing calculation form.*

Seismic Bracing Calculations

Sheet ______ of ______

Project: Acme Warehouse
Address: 321 First Street
Any City, Any State

Contractor: Smith Sprinkler Company
Address: 123 Main Street
Any City, Any State
Telephone: (555) 555-1234
Fax: (555) 555-4321

Brace Information

Length of brace: 8 ft 8 in.
Diameter of brace: 1 in.
Type of brace: Schedule 40
Angle of brace: 30° to 45°
Least radius of gyration:* 0.42
L/R value:* 200
Maximum horizontal load: 1767 lb

Fastener Information

Orientation of connecting surface: "D"

Fastener:
Type: Through bolt
Diameter: 5/8 in.
Length (in wood): 4 in.
Maximum load: 491 lb

Seismic Brace Attachments

Structure attachment fitting or tension-only bracing system:
Make: Acme Model: 123
Listed load rating: ______ Adjusted load rating per 6-4.5.7: ______
Sway brace (pipe attachment) fitting:
Make: Acme Model: 321
Listed load rating: ______ Adjusted load rating per 6-4.5.7: ______

Seismic Brace Assembly Detail

(Provide detail on plans)

5/8 in. x 6 in. through bolt with nut and washer
Acme 123
1/2 beam depth minimum
30° minimum
4 in. x 12 in. beam
1 in. Schedule 40
Acme 321

Brace identification no. (to be used on plans) SB-1

☒ Lateral brace ☐ Longitudinal brace

Sprinkler System Load Calculation

Diameter	Type	Length (ft)	Total (ft)	½ Weight per ft	½ Total Weight
1 in.	Sch 40	15 ft + 25 ft + 8 ft + 22 ft	70 ft	1.03 lb/ft	72.1 lb
1¼ in.	Sch 40	25 ft + 33 ft + 18 ft	76 ft	1.47 lb/ft	111.7 lb
1½ in.	Sch 40	8 ft + 8 ft + 10 ft + 10 ft	36 ft	1.81 lb/ft	65.2 lb
2 in.	Sch 40	20 ft	20 ft	2.57 lb/ft	51.4 lb
4 in.	Sch 10	20 ft	20 ft	5.89 lb/ft	117.8 lb
				lb/ft	lb
				Total ½ weight of water-filled pipe	418.2 lb

* Excludes tension-only bracing systems

A-6-4.5(b) Sample seismic bracing calculation.

Branch line piping 2 in. (51 mm) in diameter and smaller is considered capable of considerable movement without damage.

A lesser degree of support than bracing is referred to as restraint and is separately addressed in 6-4.6.

NFPA 13 provides protection against an assumed horizontal force equal to one-half the weight of the water-filled piping ($F_p = 0.5\ W_p$). A multiplier permits the use of the bracing provisions with lower or higher horizontal accelerations, as could be specified in building codes.

NFPA 13 contains requirements for both lateral (perpendicular to the piping) and longitudinal (parallel to the piping) horizontal braces. A lateral sway brace is shown in Exhibit 6.15, and a longitudinal brace is shown in Exhibit 6.16. The normal maximum spacing of lateral braces [40 ft (12.2 m)] is based on the strength of the piping as a beam under the uniform load of its expected horizontal "weight." Longitudinal braces are required at a maximum spacing of 80 ft (24.4 m).

Lateral and longitudinal braces are "two-way" braces—that is, they prevent piping from moving back and forth in a single direction. Exhibits 6.15 and 6.16 illustrate two-way braces. "Four-way" bracing requires the simultaneous effect of lateral and longitudinal bracing. Exhibit 6.17 illustrates a four-way brace.

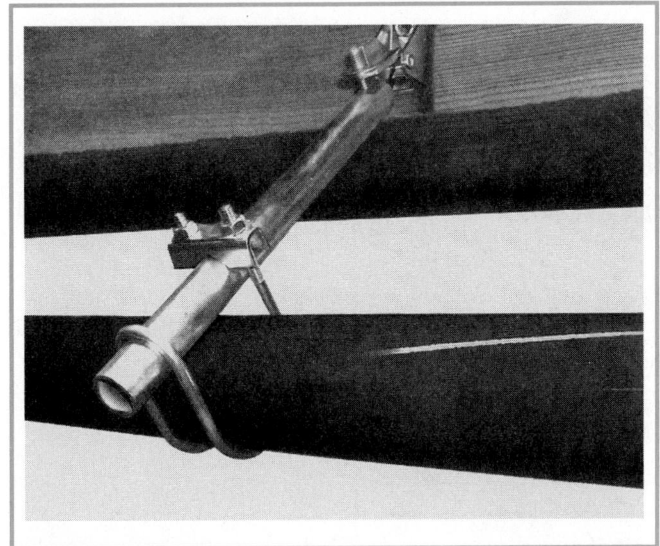

Exhibit 6.15 *Lateral sway brace attached wood. (Courtesy of Tolco.)*

As with system hangers, sway braces must be attached directly to the building structure. Fasteners and structural elements at the points of connection must be adequate to handle the expected loads.

6-4.5.1 The system piping shall be braced to resist both lateral and longitudinal horizontal seismic loads and to prevent vertical motion resulting from seismic loads. The structural compo-

Exhibit 6.16 *Longitudinal sway brace attached steel bar joist. (Courtesy of Tolco.)*

Exhibit 6.17 *Four-way brace. (Courtesy of AFCON.)*

nents to which bracing is attached shall be determined to be capable of carrying the added applied seismic loads.

The forces applied to piping during an earthquake can be vertical as well as horizontal. Minor vertical forces are expected to be offset by gravity. The rigid horizontal braces contemplated by NFPA 13 are also expected to assist the system hangers in resisting these vertical loads, especially because the braces are normally installed at an angle to the vertical.

6-4.5.2 Sway braces shall be designed to withstand forces in tension and compression.

Exception: Tension only bracing systems shall be permitted for use where listed for this service and where installed in accordance with their listing limitations, including installation instructions.*

A-6-4.5.2 Exception. The investigation of tension-only bracing using materials, connection methods, or both, other than those described in Table 6-4.5.8, should involve consideration of the following:

(a) Corrosion resistance.

(b) Prestretching to eliminate permanent construction stretch and to obtain a verifiable modulus of elasticity.

(c) Color coding or other verifiable marking of each different size cable for field verification.

(d) The capacity of all components of the brace assemblies, including the field connections, to maintain the manufacturer's minimum certified break strength.

(e) Manufacturer's published design data sheets/manual showing product design guidelines, including connection details, load calculation procedures for sizing of braces, and the maximum recommended horizontal load-carrying capacity of the brace assemblies including the associated fasteners as described in Table 6-4.5.9. The maximum allowable horizontal loads shall not exceed the manufacturer's minimum certified break strength of the brace assemblies, excluding fasteners, after taking a safety factor of 1.5 and then adjusting for the brace angle.

(f) Brace product shipments accompanied by the manufacturer's certification of the minimum break strength and prestretching and installation instructions.

(g) The manufacturer's literature, including any special tools or precautions required to ensure proper installation.

(h) A means to prevent vertical motion due to seismic forces when required.

Table A-6-4.5.2 identifies some specially listed tension-only bracing systems.

Table A-6-4.5.2 Specially Listed Tension-Only Seismic Bracing

Materials and Dimensions	Standard
Manual for Structural Application of Steel Cables	ASCE 19-96
Wire Rope Users Manual of the Wire Rope Technical Board	ASCE 19-96
Mechanical Strength Requirements	ASTM A 603
Breaking Strength Failure Testing	ASTM E 8

The exception that permits tension only bracing systems was added to the 1996 edition. As with nonmetallic pipe, these systems are acceptable for use by NFPA 13 only if specially investigated and listed for this service and installed in conformance with all requirements of the listing. Tension only bracing systems are not to be confused with the wire restraint method permitted by 6-4.6.1. A tension only bracing system is illustrated in Exhibit 6.18. The two AISI documents referenced in Appendix A of the 1996 edition have since been replaced by ASCE 19, *Structural Applications of Steel Cable for Buildings.*

Exhibit 6.18 *Tension only bracing system.*

6-4.5.3 Lateral sway bracing spaced at a maximum interval of 40 ft (12.2 m) on center shall be provided on all feed and cross mains regardless of size and all branch lines and other piping with a diameter of 2½ in. (63.5 mm) and larger. The last length of pipe at the end of a feed or cross main shall be provided with a lateral brace. Lateral braces shall be allowed to act as longitudinal braces if they are within 24 in. (610 mm) of the centerline of the piping braced longitudinally for lines that are 2½ in. (63.5 mm) and greater in diameter. The distance between the last brace and the end of the pipe shall not exceed 20 ft (6.1 m). This requirement shall not preclude the use of a lateral brace serving as a longitudinal brace as described in this paragraph.

Exception No. 1: Where the spacing of lateral braces is permitted to be up to 50 ft (15.2 m), the distance between the last brace and the end of the pipe is permitted to be extended to 25 ft (7.6 m).

Exception No. 2: Lateral sway bracing shall not be required on pipes individually supported by rods less than 6 in. (152 mm) long measured between the top of the pipe and the point of attachment to the building structure.

Exception No. 3: U-type hooks of the wraparound type or those U-type hooks arranged to keep the pipe tight to the underside of the structural element shall be permitted to be used to satisfy the requirements for lateral sway bracing provided the legs are bent out at least 30 degrees from the vertical and the maximum length of each leg and the rod size satisfies the conditions of Table 6-4.5.8.

Exception No. 4: Where flexible couplings are installed on mains other than as required in 6-4.2, a lateral brace shall be provided within 24 in. (610 mm) of every other coupling, but not more than 40 ft (12.2 m) on center.

Exception No. 5: Where building primary structural members exceed 40 ft (12.2 m) on center, lateral braces shall be permitted to be spaced up to 50 ft (15.2 m) on center.

Bracing is required for cross mains and feed mains regardless of size. For branch line and other piping smaller than 2½ in. (63.5 mm), bracing is not required. Branch lines smaller than 2 in. (51 mm) in diameter are considered to have sufficient flexibility to avoid damage under their own momentum. Mains, however, can also carry the loads of adjacent branch lines and must be braced regardless of size.

Prior to the 1996 edition of NFPA 13, the last lateral brace on a main could be located anywhere along the last length of pipe. The distance from the end is now limited to 20 ft (6.1 m).

According to Exception No. 2, lateral sway bracing is permitted to be omitted for pipes supported by rods less than 6 in. (152 mm) in length. Short hangers are expected to limit movement that could result in damage-inducing momentum. The types of hangers used should be able to withstand horizontal earthquake loads and should not be brittle or otherwise highly susceptible to fatigue failure. In general, the 6-in. (152-mm) length is measured between the points of support of the rod at the building structure and the pipe ring or clamp. Oversized pipe rings are not permitted as a means of extending the allowable distance between the structure and the pipe under this exception. A hole in a beam that is used to support the pipe gives the same intended result as either Exception No. 2 or Exception No. 3.

The intent of Exception No. 4 is to prevent excessive movement of the mains, possibly resulting in "bellows" or "accordion" effects. Bracing is not required near every flexible coupling but near every other such coupling. This additional bracing is not required where rigid-type mechanical couplings are used. Rigid-type mechanical couplings are those not meeting the definition of *flexible listed pipe coupling* in 1-4.4.

Exception No. 4 is applicable only to horizontal mains and branch lines required to be braced as mains. The exception is not intended to apply to other branch lines or to vertical piping such as risers.

6-4.5.4 Longitudinal sway bracing spaced at a maximum of 80 ft (24.4 m) on center shall be provided for feed and cross mains. Longitudinal braces shall be permitted to serve as lateral braces where they are installed within 24 in. (610 mm) of the piping that is braced laterally. The distance between the last brace and the end of the pipe shall not exceed 40 ft (12.2 m).

Although the 1994 edition of NFPA 13 permitted omission of longitudinal sway bracing for mains supported by rods less than 6 in. (152 mm) in length, experience in the 1994 Northridge, CA, earthquake showed that such omission is inappropriate, and, therefore, longitudinal sway bracing for mains supported by rods less than 6 in. (152 mm) in length is now required.

6-4.5.5* Tops of risers shall be secured against drifting in any direction, utilizing a four-way sway brace.

A-6-4.5.5 The four-way brace provided at the riser can also provide longitudinal and lateral bracing for adjacent mains.

For risers that extend above a roof to serve a standpipe outlet, the bracing can be located at the underside of the roof as indicated in Formal Interpretation 80-29. Although limited clearance around a riser extending through a substantial floor or roof in accordance with 6-4.4 can also limit movement and maintain alignment, Formal Interpretation 80-29 clarified that a four-way brace is required at the top of all risers.

Formal Interpretation 80-29

Reference: 6-4.5.5

Question 1: Is it the intent of 6-4.5.5 of NFPA 13 to require a four-way earthquake brace only at the top of risers in a multistory building?

Answer: A four-way brace is required at the top of the riser in all buildings subject to earthquake damage. Four-way bracing is not required for intermediate floors in multistory buildings.

Question 2: Is it the intent of 6-4.5.5 of NFPA 13 to require a four-way earthquake brace above the roof line when the sprinkler riser is used to supply 2½-inch hose outlets on each floor and above the roof?

Answer: No. The four-way brace just below the roof slab is adequate.

Issue Edition: 1980

Reference: 3-10.3, 4-1

Date: October 1982

6-4.5.6* Horizontal loads for braces shall be determined by analysis based on a horizontal force of $F_p = 0.5\ W_p$, where F_p is the horizontal force factor and W_p is the weight of the water-filled piping. For lateral braces, the load shall include all branch lines and mains, unless the branch lines are provided with longitudinal bracing, within the zone of influence of the brace. For longitudinal braces, the load shall include all mains within the zone of influence of the brace.

Exception: Where the use of other horizontal force factors is required or permitted by the authority having jurisdiction, they shall take precedence.

A-6-4.5.6 *Location of Sway Bracing.* Two-way braces are either longitudinal or lateral depending on their orientation with the axis of the piping. *[See Figures A-6-4.5.6(a), (b), (c), and (d).]* The simplest form of two-way brace is a piece of steel pipe or angle. Because the brace must act in both compression and tension, it is necessary to size the brace to prevent buckling.

An important aspect of sway bracing is its location. In Building 1 of Figure A-6-4.5.6(a), the relatively heavy main will pull on the branch lines when shaking occurs. If the branch lines are held rigidly to the roof or floor above, the fittings can fracture due to the induced stresses.

Bracing should be on the main as indicated at Location B. With shaking in the direction of the arrows, the light branch lines will be held at the fittings. Where necessary, a lateral brace or other restraint should be installed to prevent a branch line from striking against building components or equipment.

A four-way brace is indicated at Location A. This keeps the riser and main lined up and also prevents the main from shifting.

In Building 1, the branch lines are flexible in a direction parallel to the main, regardless of building movement. The heavy main cannot shift under the roof or floor, and it also steadies the branch lines. While the main is braced, the flexible couplings on the riser allow the sprinkler system to move with the floor or roof above, relative to the floor below.

Figures A-6-4.5.6(b), (c), and (d) show typical locations of sway bracing.

For all threaded connections, sight holes or other means should be provided to permit indication that sufficient thread is engaged.

To properly size and space braces, it is necessary to employ the following steps:

(a) Based on the distance of mains from the structural members that will support the braces, choose brace shapes and sizes from Table 6-4.5.8 such that the maximum slenderness ratios,

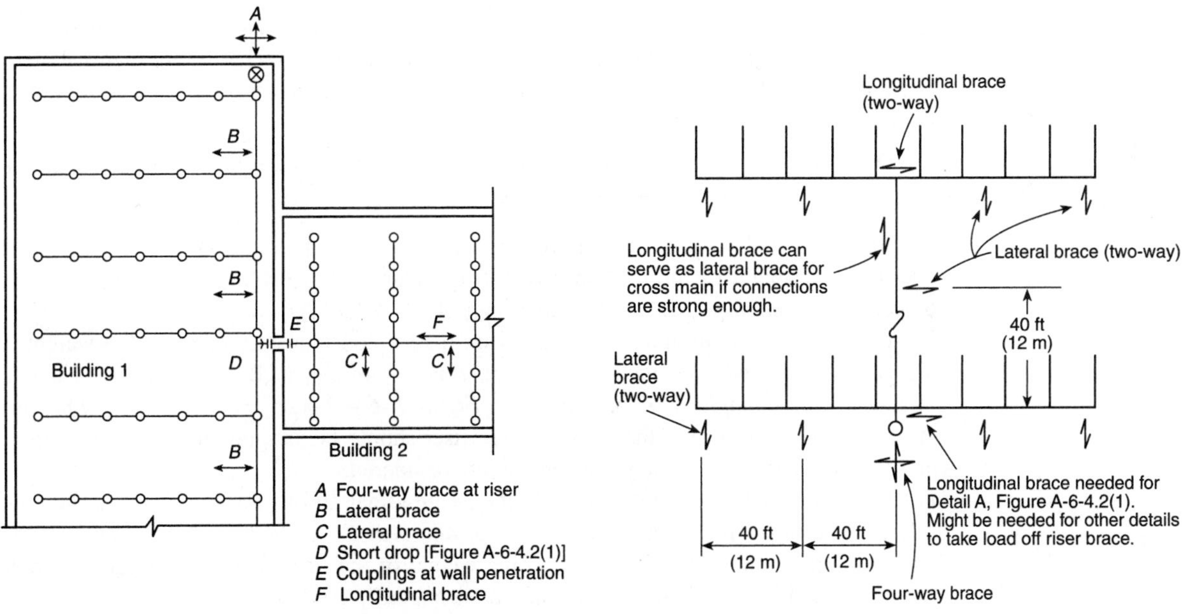

Figure A-6-4.5.6(a) *Earthquake protection for sprinkler piping.*

Figure A-6-4.5.6(b) *Typical location of bracing on a tree system.*

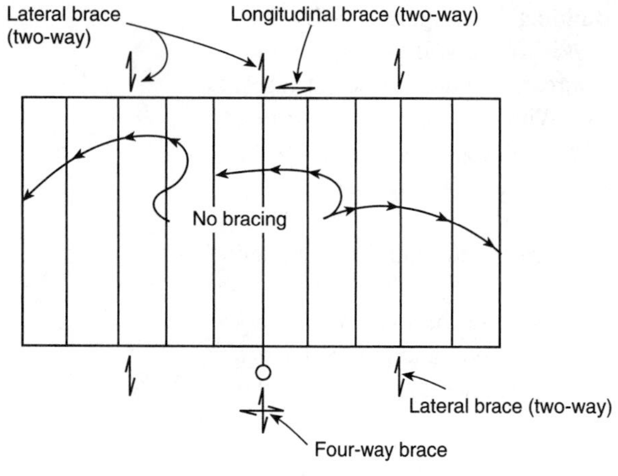

Figure A-6-4.5.6(c) *Typical location of bracing on a gridded system.*

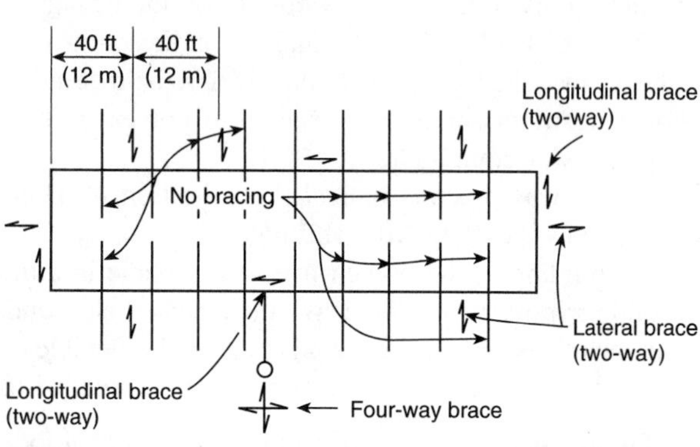

Figure A-6-4.5.6(d) *Typical location of bracing on a looped system.*

l/r, do not exceed 300. The angle of the braces from the vertical should be at least 30 degrees and preferably 45 degrees or more.

(b) Tentatively space lateral braces at 40-ft (12-m) maximum distances along mains and tentatively space longitudinal braces at 80-ft (24-m) maximum distances along mains. Lateral braces should meet the piping at right angles, and longitudinal braces should be aligned with the piping.

(c) Determine the total load tentatively applied to each brace in accordance with the examples shown in Figure A-6-4.5.6(e) and the following:

(1) For the loads on lateral braces on cross mains, add one-half the weight of the branch to one-half the weight of the portion of the cross main within the zone of influence of the brace. *[See examples 1, 3, 6, and 7 in Figure A-6-4.5.6(e).]*

(2) For the loads on longitudinal braces on cross mains, consider only one-half the weight of the cross mains and feed mains within the zone of influence. Branch lines need not be included. *[See examples 2, 4, 5, 7, and 8 in Figure A-6-4.5.6(e).]* For the four-way bracing at the top of the riser, half the weight of the riser should be assigned to both of the lateral and longitudinal loads as they are separately considered.

(3) For the four-way brace at the riser, add the longitudinal and lateral loads within the zone of influence of the brace *[see examples 2, 3, and 5 in Figure A-6-4.5.6(e)]*. For the four-way bracing at the top of the riser, half the weight of the riser should be assigned to both the lateral and longitudinal loads as they are separately considered.

(d) If the total expected loads are less than the maximums permitted in Table 6-4.5.8 for the particular brace and orientation, go on to step (e). If not, add additional braces to reduce the zones of influence of overloaded braces.

(e) Check that fasteners connecting the braces to structural supporting members are adequate to support the expected loads on the braces in accordance with Table 6-4.5.8. If not, again add additional braces or additional means of support.

Use the information on weights of water-filled piping contained within Table A-6-4.5.6.

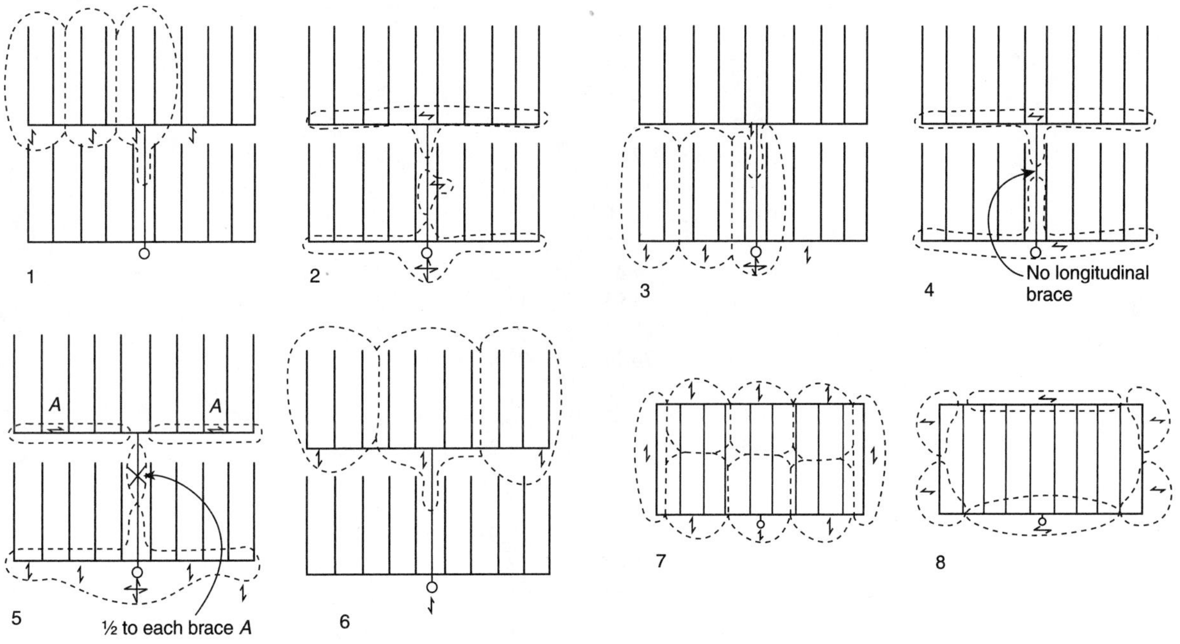

Figure A-6-4.5.6(e) *Examples of load distribution to bracing.*

Prior to the 1999 edition, the following two methods were permitted by NFPA 13 for determining the loads to be carried by earthquake braces:

(1) The assigned load method
(2) The "zone of influence" method

Using the assigned loads of Table A-6-4.5.6, loads could be assumed for various sizes of pipe. This table was developed based on conservative worst case assumptions relative to the loads contributed by branch lines. In almost all cases, loads could be reduced through the use of the zone of influence option, which assigns loads based on the particular configuration of the piping. Use of the zone of influence method is now required.

Under the zone of influence method, both branch lines and mains are considered to contribute to the loads on lateral braces, but only mains are considered to contribute to the loads in longitudinal braces. The longitudinal braces are not expected to be

Table A-6-4.5.6 Piping Weights for Determining Horizontal Load

Nominal Dimensions	Weight of Water-Filled Pipe		One-Half Weight of Water-Filled Pipe	
	lb/ft	kg/m	lb/ft	kg/m
Schedule 40 Pipe (in.)				
1	2.05	3.05	1.03	1.5
1¼	2.93	4.36	1.47	2.2
1½	3.61	5.37	1.81	2.7
2	5.13	7.63	2.57	3.8
2½	7.89	11.74	3.95	5.9
3	10.82	16.10	5.41	8.1
3½	13.48	20.06	6.74	10.0
4	16.40	24.41	8.20	12.2
5	23.47	34.93	11.74	17.5
6	31.69	47.16	15.85	23.6
8*	47.70	70.99	23.85	35.5
Schedule 10 Pipe (in.)				
1	1.81	2.69	0.91	1.4
1¼	2.52	3.75	1.26	1.9
1½	3.04	4.52	1.52	2.3
2	4.22	6.28	2.11	3.1
2½	5.89	8.77	2.95	4.4
3	7.94	11.82	3.97	5.9
3½	9.78	14.55	4.89	7.3
4	11.78	17.53	5.89	8.8
5	17.30	25.75	8.65	12.9
6	23.03	34.27	11.52	17.1
8	40.08	59.65	20.04	29.8

*Schedule 30.

impacted by branch line loads, because these forces are not uniformly transferred during earthquake motion. The exception permits the loads within the tables to be increased or decreased based on the use of a multiplier as required or permitted by the authority having jurisdiction.

Figures A-6-4.5.6(a), (b), (c), and (d) show typical locations of sway bracing on various configurations of systems. Figure A-6-4.5.6(c) has no lateral bracing on the outside lines because the outside lines are typical branch lines of the grid—that is, not oversized—and, therefore, are not considered mains.

6-4.5.7 Where the horizontal force factors used exceed 0.5 W_p and the brace angle is less than 45 degrees from vertical or where the horizontal force factor exceeds 1.0 W_p and the brace angle is less than 60 degrees from vertical, the braces shall be arranged to resist the net vertical reaction produced by the horizontal load.

First introduced in the 1996 edition, 6-4.5.7 addresses vertical movement of the pipe as a result of a strong horizontal load during a seismic event. Previous editions did not specifically address vertical movement.

6-4.5.8* Sway bracing shall be tight. For individual braces, the slenderness ratio *(l/r)* shall not exceed 300 where *l* is the length of the brace and *r* is the least radius of gyration. Where threaded pipe is used as part of a sway brace assembly, it shall not be less than Schedule 30. All parts and fittings of a brace shall lie in a straight line to avoid eccentric loadings on fittings and fasteners. For longitudinal braces only, the brace shall be permitted to be connected to a tab welded to the pipe in conformance with 3-6.2. For individual braces, the slenderness ratio, *l/r*, shall not exceed 300 where *l* is the length of the brace and *r* is the least radius of gyration. For tension-only braces, two tension-only brace components opposing each other must be installed at each lateral or longitudinal brace location. For all braces, whether or not listed, the maximum allowable horizontal load shall be based on the weakest component of the brace with safety factors. The loads determined in 6-4.5.6 shall not exceed the lesser of the maximum allowable loads provided in Table 6-4.5.8 or the manufacturer's certified maximum allowable horizontal loads for 30- to 44-degree, 45- to 59-degree, 60- to 89-degree, and 90-degree brace angles. These certified allowable horizontal loads must include a minimum safety factor of 1.5 against the ultimate break strength of the brace components and then be further reduced according to the brace angles.

Exception: Other pipe schedules and materials not specifically included in Table 6-4.5.8 shall be permitted to be used if certified by a registered professional engineer to support the loads determined in accordance with the above criteria. Calculations shall be submitted where required by the authority having jurisdiction.

A-6-4.5.8 Sway brace members should be continuous. Where necessary, splices in sway bracing members should be designed and constructed to insure that brace integrity is maintained.

The slenderness ratio of a brace *(l/r)* is determined by dividing the length of the brace *(l)* by its least radius of gyration *(l/r)*. The least radius of gyration is a mathematical abstraction related to the ability of a shape to resist buckling. When determining the least radius of gyration of a shape that is nonsymmetrical along its axis, the least stable axis must be considered, because that is where buckling is expected to occur. The slenderness ratio is limited to a maximum of 300 because most braces are expected to act as long columns when they are resisting horizontal and vertical loads in compression. Long columns can help absorb some of the energy of an earthquake, but excessive length reduces the column's ability to resist buckling and invites fatigue failure over time. Prior to the 1994 edition of NFPA 13, the maximum slenderness ratio was limited to 200.

Table 6-4.5.8 indicates the formulas used to determine the least radius of gyration for common brace shapes and the maximum lengths of braces that qualify for slenderness ratios of 100, 200, and 300.

Table 6-4.5.8 Maximum Horizontal Loads for Sway Braces

			Maximum Horizontal Load (lb)		
Shape and Size	Least Radius of Gyration	Maximum Length for:	30°–44° Angle from Vertical	45°–59° Angle from Vertical	60°–90° Angle from Vertical
Pipe (Schedule 40)	$= \frac{\sqrt{r_o^2 + r_i^2}}{2}$	$l/r = 200$			
1 in.	0.42	7 ft 0 in.	1,767	2,500	3,061
1¾ in.	0.54	9 ft 0 in.	2,392	3,385	4,145
1½ in.	0.54	10 ft 4 in.	2,858	4,043	4,955
2 in.	0.787	13 ft 1 in.	3,828	5,414	6,630
Angles		$l/r = 200$			
1½ × 1½ × ¾ in.	0.292	4 ft 10 in.	2,461	3,481	4,263
2 × 2 × ¼ in.	0.391	6 ft 6 in.	3,356	4,746	5,813
2½ × 2 × ¼ in.	0.424	7 ft 0 in.	3,792	5,363	6,569
2½ × 2½ × ¼ in.	0.491	8 ft 2 in.	4,257	6,021	7,374
3 × 2½ × ¼ in.	0.528	8 ft 10 in.	4,687	6,628	8,118
3 × 3 × ¼ in.	0.592	9 ft 10 in.	5,152	7,286	8,923
Rods	$= \frac{r}{2}$	$l/r = 200$			
⅜ in.	0.094	1 ft 6 in.	395	559	685
½ in.	0.125	2 ft 6 in.	702	993	1,217
⅝ in.	0.156	2 ft 7 in.	1,087	1,537	1,883
¾ in.	0.188	3 ft 1 in.	1,580	2,235	2,737
⅞ in.	0.219	3 ft 7 in.	2,151	3,043	3,726
Flats	$= 0.29h$ (where h is smaller of two side dimensions)	$l/r = 200$			
1½ × ¼ in.	0.0725	1 ft 2 in.	1,118	1,581	1,936
2 × ¼ in.	0.0725	1 ft 2 in.	1,789	2,530	3,098
2 × ⅜ in.	0.109	1 ft 9 in.	2,683	3,795	4,648
Pipe (Schedule 40)	$= \frac{\sqrt{r_0^2 + r_i^2}}{2}$	$l/r = 100$			
1 in.	0.42	3 ft 6 in.	7,068	9,996	12,242
1¼ in.	0.54	4 ft 6 in.	9,567	13,530	16,570
1½ in.	0.623	5 ft 2 in.	11,441	16,181	19,817
2 in.	0.787	6 ft 6 in.	15,377	21,746	26,634

Table 6-4.5.8 Maximum Horizontal Loads for Sway Braces (Continued)

Shape and Size	Least Radius of Gyration	Maximum Length for:	Maximum Horizontal Load (lb) 30°–44° Angle from Vertical	45°–59° Angle from Vertical	60°–90° Angle from Vertical
Rods	$= \frac{r}{2}$	$l/r = 100$			
⅜ in.	0.094	0 ft 9 in.	1,580	2,234	2,737
½ in.	0.125	1 ft 0 in.	2,809	3,972	4,865
⅝ in.	0.156	1 ft 3 in.	4,390	6,209	7,605
¾ in.	0.188	1 ft 6 in.	6,322	8,941	10,951
⅞ in.	0.219	1 ft 9 in.	8,675	12,169	14,904
Pipe (Schedule 40)	$= \frac{\sqrt{r_o^2 + r_i^2}}{2}$	$l/r = 300$			
1 in.	0.42	10 ft 6 in.	786	1,111	1,360
1½ in.	0.54	13 ft 6 in.	1,063	1,503	1,841
1½ in.	0.623	15 ft 7 in.	1,272	1,798	2,202
2 in.	0.787	19 ft 8 in.	1,666	2,355	2,885
Rods	$= \frac{r}{2}$	$l/r = 300$			
⅜ in.	0.094	2 ft 4 in.	176	248	304
½ in.	0.125	3 ft 1 in.	312	441	540
⅝ in.	0.156	3 ft 11 in.	488	690	845
¾ in.	0.219	5 ft 6 in.	956	1,352	1,656

The load values in Table 6-4.5.8 are calculated based on determining critical loads for buckling in columns in accordance with Euler's formula. Euler's formula is as follows:

$$P = \frac{AE\pi^2}{l/r^2}$$

where:

P = maximum axial load

A = cross-sectional area of brace

E = modulus of elasticity for steel equal to 29,000,000 psi (1,998,100 bar)

l/r = slenderness ratio

The table provides load values for slenderness ratios of 100, 200, and 300.

The following illustrates how the maximum horizontal load is determined for 1 in. Schedule 40 steel pipe where the maximum slenderness ratio of the braces is

300. The cross-sectional area of the pipe is calculated using the following formula and Table A-3-3.2.

$$A = \pi(d_o^2 - d_i^2)$$

$$4 = 0.494 \text{ in.}$$

where:

d_o = outside diameter

d_i = inside diameter

The maximum axial load is found as follows:

$$P = \frac{0.494 \text{ in.}^2 \text{ (29,000,000 psi)}\pi^2}{300^2} = 1571 \text{ lb}$$

The three values for maximum horizontal load that appear in Table 6-4.5.8 were determined by multiplying this maximum axial load by the sine of the angle made with the vertical plane. The determination is based on the worst case angle within each range of angles. This adjustment is needed because the axial load in the brace is higher than the horizontal load when the brace is at some angle from the horizontal. For the 1-in. Schedule 40 steel pipe with a *(l/r)* ratio of 300, the maximum horizontal load at each angle is as follows:

1571 lb (sin 30°) = 1571(0.5) = 786 lb
1571 lb (sin 45°) = 1571(0.707) = 1111 lb
1571 lb (sin 60°) = 1571(0.867) = 1360 lb

Therefore, maximum allowable loads for each type of brace in Table 6-4.5.8, are stated as a function of the angle of the brace from vertical. The allowable loads increase as the brace moves toward a horizontal position, where it can most efficiently handle lateral loads.

Exhibit 6.19 illustrates an unacceptable and an acceptable bracing arrangement. The arrangement on the left in Exhibit 6.20 is unacceptable because the hanger ring is not snug and the brace is not perpendicular to the pipe. The bracing is required to be tight. In addition, the bracing on all parts and fittings of the brace are to lie in a straight line to avoid an eccentric loading on fittings and fasteners.

6-4.5.9* For individual fasteners, the loads determined in 6-4.5.6 shall not exceed the allowable loads provided in Figure 6-4.5.9.

The type of fasteners used to secure the bracing assembly to the structure shall be limited to those shown in Figure 6-4.5.9. For connections to wood, through bolts with washers on each end shall be used. Holes for through bolts shall be 1⁄16 in. (1.6 mm) greater than the diameter of the bolt.

Exception No. 1: Where it is not practical to install through bolts due to the thickness of the member or inaccessibility, lag screws shall be permitted. Holes shall be pre-drilled 1⁄8 in. (3.2 mm) smaller than the maximum root diameter of the lag screw.

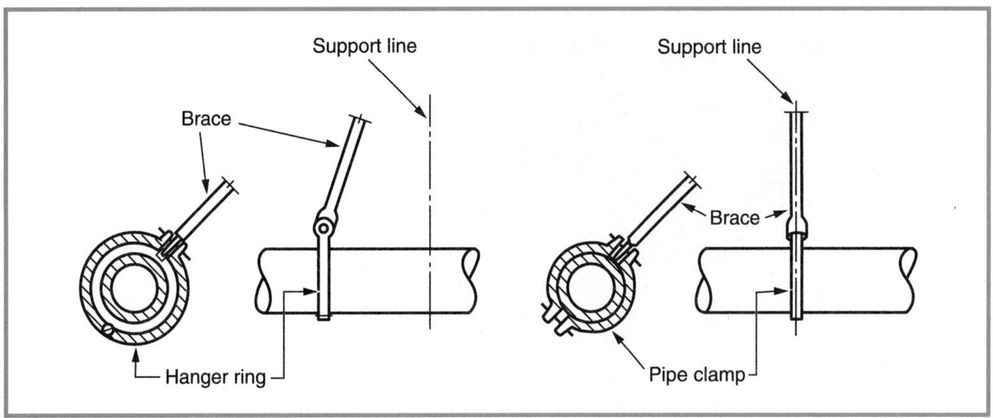

Exhibit 6.19 An unacceptable connection to brace pipe on the left and an acceptable connection to brace pipe on the right.

Exception No. 2: Other fastening methods are acceptable for use if certified by a registered professional engineer to support the loads determined in accordance with the criteria in 6-4.5.9. Calculations shall be permitted where required by the authority having jurisdiction.

A-6-4.5.9 The criteria in Table 6-4.5.8 are based upon the use of a shield-type expansion anchor. Use of other anchors in concrete should be in accordance with the listing provisions of the anchor.

Current fasteners for anchoring to concrete are referred to as expansion anchors. Expansion anchors come in two types. Deformation-controlled anchors are set by driving a plug into the expansion port in the anchor or driving the anchor over a plug that expands the end of the anchor into the concrete. Torque-controlled expansion anchors are set by applying a torque to the anchor, usually to a nut, which causes the expansion sleeves to be pressed against the wall of the drilled hole.

Consideration should be given with respect to the position near the edge of the concrete and to the type of bolts used in conjunction with the anchors.

For connections to wood components, the fastener load values within Figure 6-4.5.9 are based on tests conducted by the Forest Products Laboratory in Madison, WI. However, the values are increased 80 percent to adapt to anchorage design, the yield point of lag screws and bolts subject to a force of at least 45,000 psi (3103 bar), and basic stress of wood subject to a force of 1300 psi (90 bar) parallel to the grain and 275 psi (19 bar) perpendicular to the grain. For lag screws, depth of penetration, including shank, is assumed to be at least 8 diameters. For through bolts in wood, the allowable load increases with the bolt length up to a point, because more area is available for bearing. From that point, the allowable load decreases with the increasing length due to bending in the bolt.

The allowable loads on unfinished steel bolts (see ASTM A 307, *Standard Specifications for Carbon Steel Bolts and Studs*) are based on AISC N690, *Specification for*

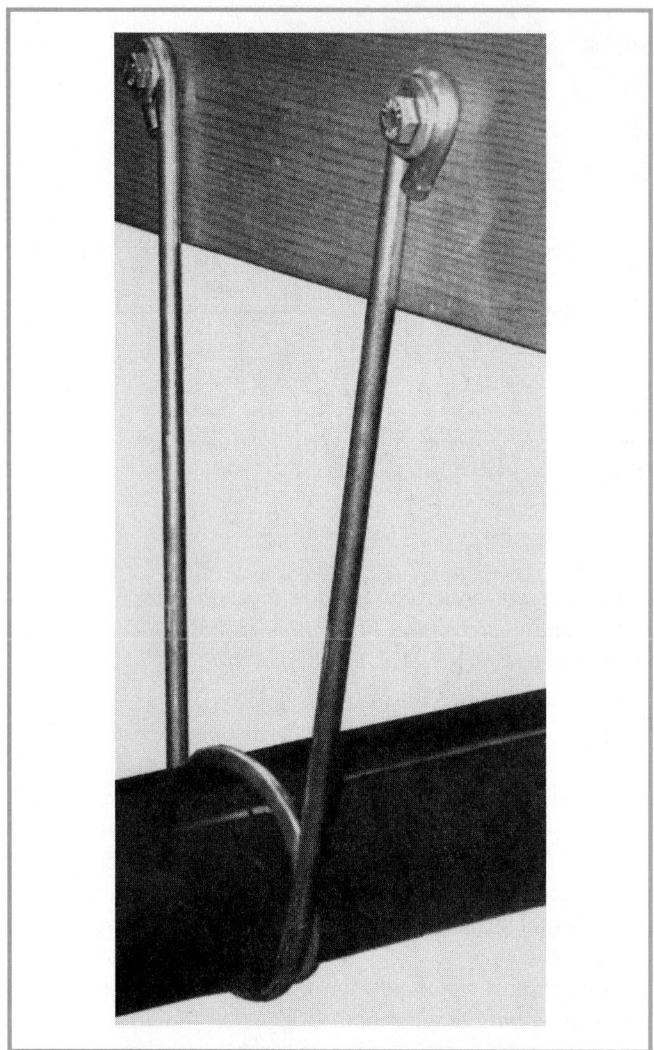

Exhibit 6.20 *Restraint of branch line with a wraparound U-hook. (Courtesy of Tolco.)*

the Design Fabrication and Erection of Structural Steel for Buildings, and are increased 33⅓ percent for earthquake loading.

For lag screws, expansion shields, and steel bolts, the maximum fastener loads are adjusted based on the brace angle with the structural base. The adjustments are based on a combination of shear and pull-out strength assuming the worst possible angle within the ranges that are shown in Figure 6-4.5.9, Examples *A* through *I*. The loads are further adjusted to account for the difference in axial loading of the brace as compared to the horizontal load as in Table 6-4.5.8. With a through bolt, pull-out is considered to be resisted by the washer on the far side, and only shear loads are considered.

Exception No. 2 allows for the use of fastening methods other than those specified by NFPA 13 to be used to fasten sway bracing provided the other methods have been evaluated by a registered professional engineer.

6-4.5.10 Sway bracing assemblies shall be listed for a maximum load rating. The loads shall be reduced as shown in Table 6-4.5.10 for loads that are less than 90 degrees from vertical.

Table 6-4.5.10 Allowable Horizontal Load on Brace Assemblies Based on the Weakest Component of the Brace Assembly

Brace Angle	Allowable Horizontal Load
30–40 degrees from vertical	Listed load rating divided by 2.000
45–59 degrees from vertical	Listed load rating divided by 1.414
60–89 degrees from vertical	Listed load rating divided by 1.155
90 degrees from vertical	Listed load rating

Exception: Where sway bracing utilizing pipe, angles, flats, or rods as shown in Table 6-4.5.8 is used, the components do not require listing. Bracing fittings and connections used with those specific materials shall be listed.

Since the 1996 edition, sway brace components other than pipes, angles, flats, and rods must be listed. These components include the brace fittings that attach to the pipe and to the building structure. Listings of the brace components include a maximum horizontal load. The listed loads are to be reduced in accordance with Table 6-4.5.10 for braces configured at less than 90 degrees from vertical. This reduction is consistent with the reductions built into Table 6-4.5.8 and Figure 6-4.5.9 for brace members and fasteners. The listing requirement ensures that the fitting does not compromise the integrity of the bracing assembly.

The most common cause of earthquake bracing failure is the overloading of fasteners at the point of connection to the building structure. Brace fitting manufacturers generally make available special brackets that allow multiple fasteners to be used to connect the end of the brace to the building structure. These brackets can distribute the load, which helps to reduce the load on each fastener to levels allowed by Figure 6-4.5.9. If the load cannot be reduced in this manner, additional braces should be added to reduce the load on each brace.

6-4.5.11 Bracing shall be attached directly to feed and cross mains. Each run of pipe between changes in direction shall be provided with both lateral and longitudinal bracing.

Exception: Pipe runs less than 12 ft (3.6 m) in length shall be permitted to be supported by the braces on adjacent runs of pipe.

Bracing of feed and cross mains through the branch lines is not permitted. The exception, added to the 1996 edition, clarifies that short runs of main do not require bracing if their loads can be assumed by adjacent piping runs.

6-4.5.12 A length of pipe shall not be braced to sections of the building that will move differentially.

This requirement is also intended to apply to rigidly connected portions of systems, because they are also susceptible to damage from differential movement. In the 1994 Northridge, CA, earthquake, for example, fracture of a large threaded elbow took place near the intersection of a wall and the roof of a mercantile occupancy. One section of piping extending from the elbow was secured to the wall, and the other section was secured to the roof. Following the earthquake, a roll-groove elbow was used to repair the main, incorporating two flexible couplings to handle future differential movement.

6-4.6 Restraint of Branch Lines.

Restraint is considered a way of holding piping components in place but to a lesser degree than bracing. Most of the restraint requirements were added to NFPA 13 based on the performance of sprinkler systems in the 1989 Loma Prieta, CA, earthquake, and were reinforced by the 1994 Northridge, CA, earthquake. Strong vertical movement of unrestrained branch lines pulled sprinklers upward then back down through the ceilings. Where lateral movement took place simultaneously, the operating mechanisms of flush-type sprinklers were susceptible to damage as the sprinklers broke through ceiling tiles. Ordinary gypsum board ceilings were found to shear concealed sprinklers where considerable lateral branch line movement took place. In evaluating the potential for damage to sprinklers, consideration should be given to the degree of lateral or upward movement permitted by the particular hanging arrangement. Exhibit 6.20 shows restraint of a branch line using a wraparound U-hook.

6-4.6.1* Restraint is considered a lesser degree of resisting loads than bracing and shall be provided by use of one of the following:

(1) A listed sway brace assembly
(2) A wraparound U-hook satisfying the requirements of 6-4.5.3, Exception No. 3
(3) No. 12, 440-lb (200-kg) wire installed at least 45 degrees from the vertical plane and anchored on both sides of the pipe
(4) Other approved means

Wire used for restraint shall be located within 2 ft (610 mm) of a hanger. The hanger closest to a wire restraint shall be of a type that resists upward movement of a branch line.

Note: Loads (given in pounds) are keyed to vertical angles of braces and orientation of connecting surface. These values are based on concentric loadings of the fastener. Use figures to determine proper reference within table. For angles between those shown, use most restrictive case. Braces should not be attached to light structure members.

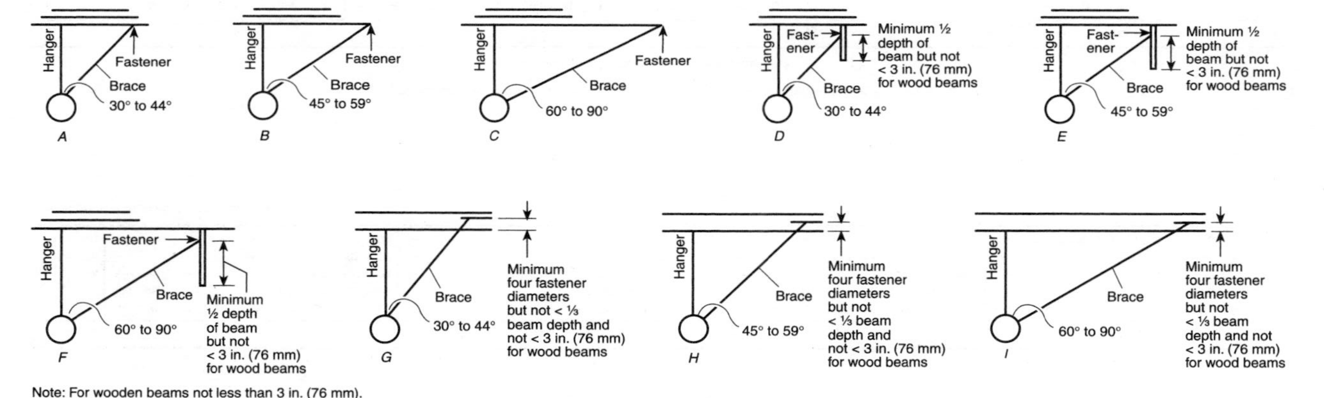

Note: For wooden beams not less than 3 in. (76 mm).

Lag Screws and Lag Bolts in Wood (Load Perpendicular to Grain — Holes Predrilled Using Good Practice)

Shank Diameter of Lag (in.)

Length under head (in.)	⅜ A	⅜ B	⅜ C	⅜ D	⅜ E	⅜ F	⅜ G	⅜ H	⅜ I
3	304	325	292	168	325	526	230	324	400
4	392	354	317	183	354	678	250	352	435
5	476	375	336	194	375	824	265	373	461
6	564	382	342	196	382	976	270	380	470
8	—	—	—	—	—	—	—	—	—

Length under head (in.)	½ A	½ B	½ C	½ D	½ E	½ F	½ G	½ H	½ I
3	366	—	—	—	—	632	—	—	—
4	473	509	456	264	509	818	360	507	626
5	582	545	488	282	545	1008	385	542	670
6	689	559	501	288	559	1192	395	556	687
8	905	573	513	296	573	1586	405	570	704

Length under head (in.)	⅝ A	⅝ B	⅝ C	⅝ D	⅝ E	⅝ F	⅝ G	⅝ H	⅝ I
3	410	—	—	—	—	716	—	—	—
4	538	—	—	—	—	929	—	—	—
5	687	728	653	277	728	1154	515	725	896
6	791	778	697	403	778	1360	550	775	957
8	1044	806	723	416	806	1807	570	803	991

Length under head (in.)	⅞ A	⅞ B	⅞ C	⅞ D	⅞ E	⅞ F	⅞ G	⅞ H	⅞ I
3	487	—	—	—	—	843	—	—	—
4	548	—	—	—	—	1122	—	—	—
5	813	—	—	—	—	1407	—	—	—
6	971	—	—	—	—	1630	—	—	—
8	1297	1365	1223	685	1365	2244	965	1359	1678

For SI units, 1 in. = 25.4 mm; 1 ft = 0.3048 m; 1 lb = 0.45 kg.

Figure 6-4.5.9 *Maximum loads for various types of structure and maximum loads for various types of fasteners to structure.*

Through Bolts in Wood (Load Perpendicular to Grain)

Diameter of Bolt (in.)

Length of Bolt in Timber (in.)	⅜						½					
	ABCE	D	F	G	H	I	ABCE	D	F	G	H	I
1½	300	173	519	150	211	261	340	197	589	170	239	296
2	370	214	641	185	261	322	420	243	727	210	296	365
2½	460	266	796	230	324	400	550	318	952	275	387	478
3	480	277	831	240	338	417	630	364	1091	315	444	548
3⅝	460	268	797	230	324	400	720	416	1247	360	507	626
5½	—	—	—	—	—	—	680	393	1177	340	479	591

Length of Bolt in Timber (in.)	⅝						⅞					
	ABCE	D	F	G	H	I	ABCE	D	F	G	H	I
1½	390	225	675	195	275	339	470	272	614	235	331	409
2	470	272	814	235	331	409	580	335	1004	290	408	504
2½	620	358	1074	310	437	539	760	439	1316	380	535	661
3	710	410	1229	355	500	617	870	503	1506	435	613	757
3⅝	850	491	1472	425	599	739	1050	607	1818	525	739	913
5½	1020	590	1766	510	718	887	1580	913	2736	790	1113	1374

For SI units, 1 in. = 25.4 mm; 1 lb = 0.45 kg.

Expansion Shields in Concrete

Diameter of Bolt (in.)

Min. Depth of Hole (in.)	⅜									½								
	A	B	C	D	E	F	G	H	I	A	B	C	D	E	F	G	H	I
2½	498	962	1173	678	668	860	925	1303	1609	—	—	—	—	—	—	1638	2306	2848
3¼	—	—	—	—	—	—	925	1303	1609	923	1782	2076	1200	1782	1597	1638	2306	2848
3¾	—	—	—	—	—	—	925	1303	1609	—	—	—	—	—	—	1638	2306	2848
4½	—	—	—	—	—	—	925	1303	1609	—	—	—	—	—	—	1638	2306	2848

Min. Depth of Hole (in.)	⅝									⅞								
	A	B	C	D	E	F	G	H	I	A	B	C	D	E	F	G	H	I
2½	—	—	—	—	—	—	2080	2930	3617	—	—	—	—	—	—	2470	4113	5078
3¼	—	—	—	—	—	—	2080	2930	3617	—	—	—	—	—	—	2970	4113	5078
3¾	1480	2857	2637	1524	2857	2581	2080	2930	3617	—	—	—	—	—	—	2970	4113	5078
4½	—	—	—	—	—	—	2080	2930	3617	3070	4130	3702	2139	4130	5312	2970	4113	5078

For SI units, 1 in. = 25.4 mm; 1 lb = 0.45 kg.

Connections to Steel (Values Assume Bolt Perpendicular to Mounting Surface)

Diameter of Unfinished Steel Bolt (in.)

¼									⅜								
A	B	C	D	E	F	G	H	I	A	B	C	D	E	F	G	H	I
400	500	600	300	500	650	325	458	565	900	1200	1400	800	1200	1550	735	1035	1278

½									⅝								
A	B	C	D	E	F	G	H	I	A	B	C	D	E	F	G	H	I
1600	2050	2550	1450	2050	2850	1300	1830	2260	2500	3300	3950	2250	3300	4400	2045	2880	3557

For SI units, 1 in. = 25.4 mm; 1 ft = 0.3048 m; 1 lb = 0.45 kg.

Figure 6-4.5.9 (Continued)

A-6-4.6.1 Wires used for piping restraints should be attached to the branch line with two tight turns around the pipe and fastened with four tight turns within 1½ in. (38 mm), and should be attached to the structure in accordance with the details shown in Figures A-6-4.6.1(a) through (d) or other approved method.

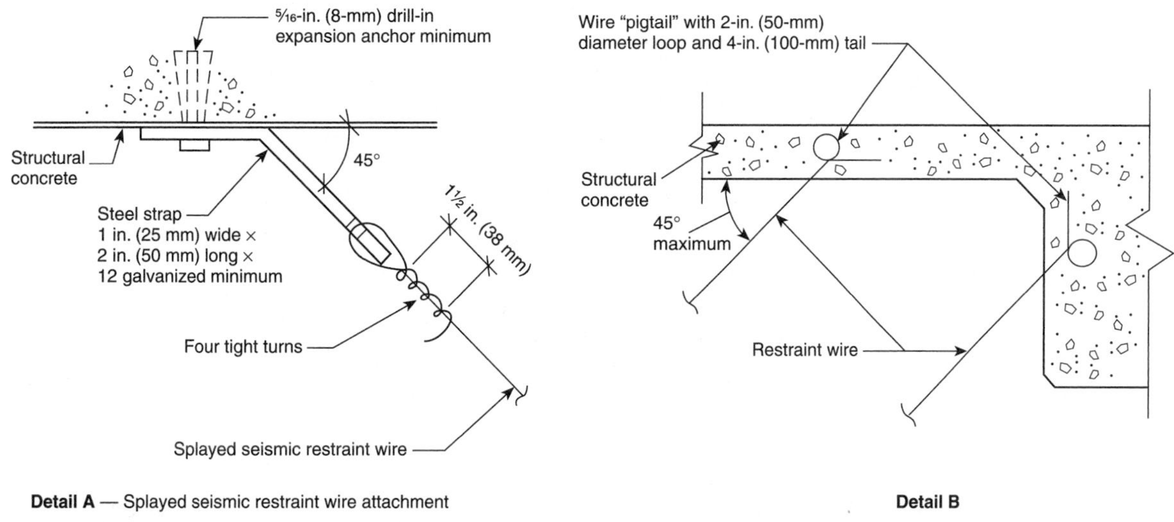

Figure A-6-4.6.1(a) Wire attachment to cast-in-place concrete.

Restraint for branch lines must be in the lateral as well as the vertical direction. Simple clips used to provide vertical restraint to meet the requirements of the exceptions to 6-2.3.3 and 6-2.3.4 can provide adequate restraint if used in combination with a substantial ceiling that provides lateral restraint. The necessary restraint can also be provided by additional hangers brought at an angle, wraparound U-hooks, lateral earthquake braces, or the wire restraint method detailed in A-6-4.6.1.

6-4.6.2 The end sprinkler on a line shall be restrained against excessive vertical and lateral movement.

A wraparound U-hook or other restraint at the end of each branch line is intended to prevent the piping from whipping or bouncing out of hangers, which is a possibility even with conventional U-hooks.

6-4.6.3* Where upward or lateral movement would result in an impact against the building structure, equipment, or finish materials, branch lines shall be restrained at intervals not exceeding 30 ft (9 m).

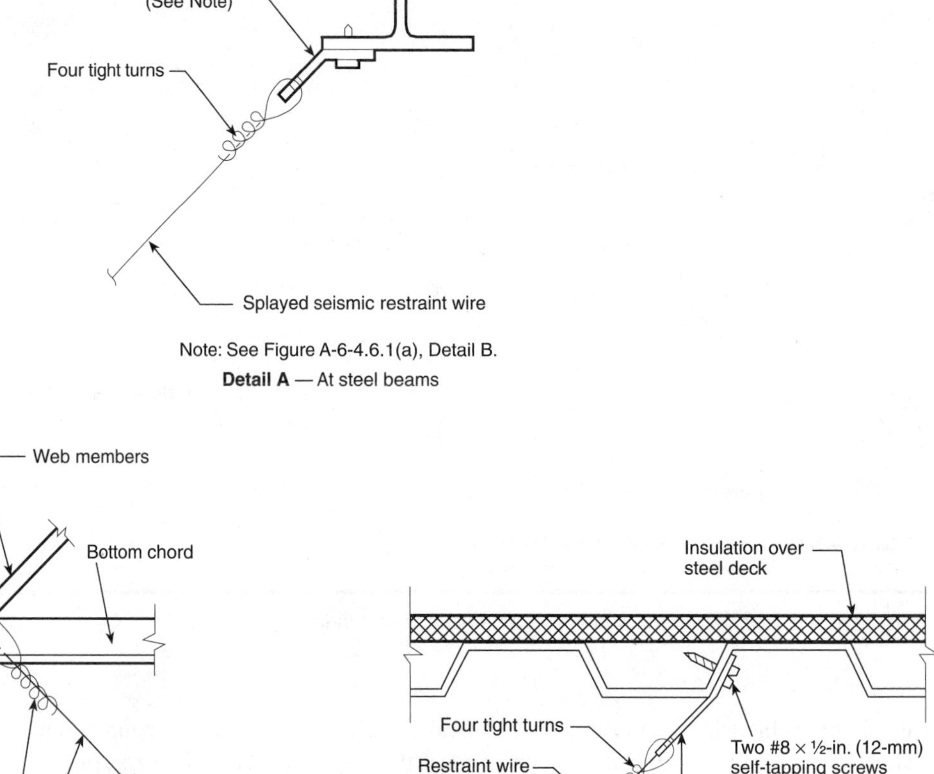

Figure A-6-4.6.1(b) *Acceptable details—wire connections to steel framing.*

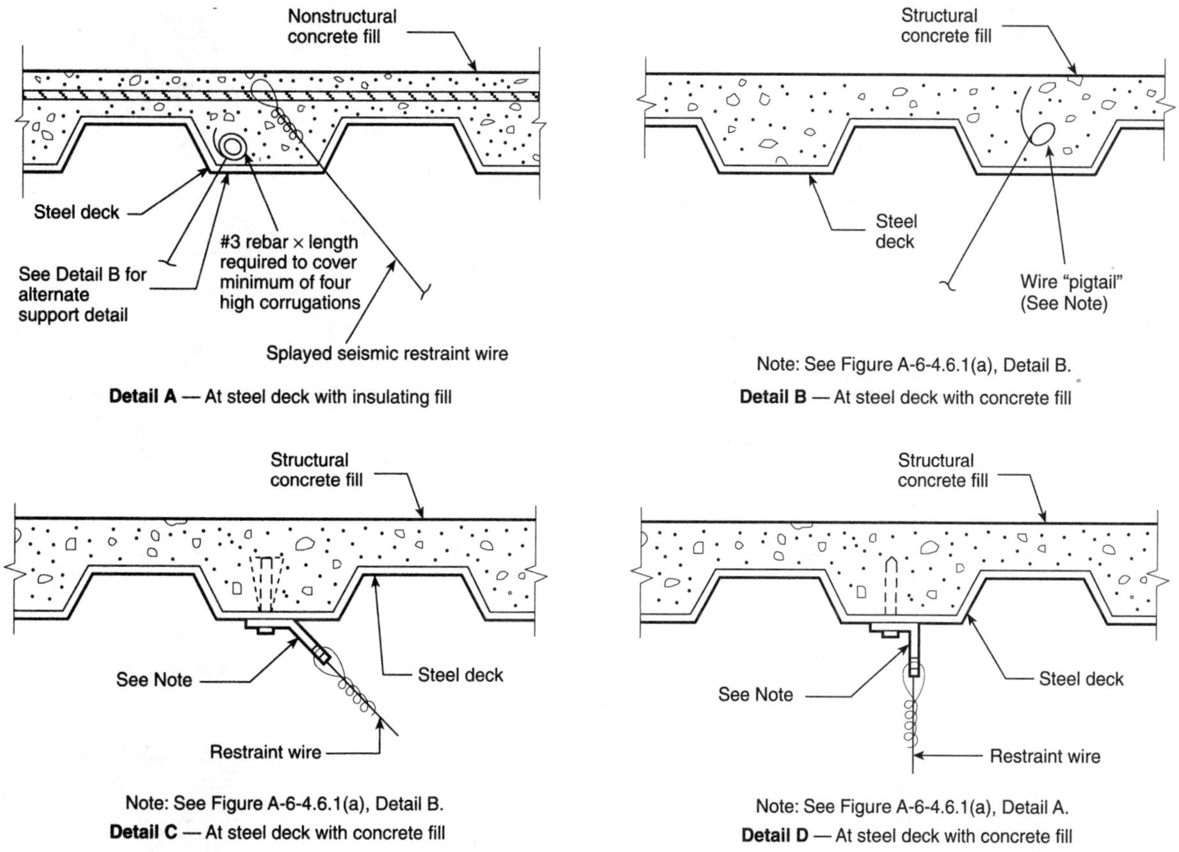

Nonstructural concrete fill
Steel deck
See Detail B for alternate support detail
#3 rebar × length required to cover minimum of four high corrugations
Splayed seismic restraint wire

Detail A — At steel deck with insulating fill

Structural concrete fill
Steel deck
Wire "pigtail" (See Note)

Note: See Figure A-6-4.6.1(a), Detail B.

Detail B — At steel deck with concrete fill

Structural concrete fill
See Note
Steel deck
Restraint wire

Note: See Figure A-6-4.6.1(a), Detail B.

Detail C — At steel deck with concrete fill

Structural concrete fill
See Note
Steel deck
Restraint wire

Note: See Figure A-6-4.6.1(a), Detail A.

Detail D — At steel deck with concrete fill

For SI units, 1 in. = 25.4 mm.
Note: If self-tapping screws are used with concrete fill, set screws before placing concrete.

Figure A-6-4.6.1(c) *Acceptable details—wire connections to steel decking with fill.*

A-6-4.6.3 The restraining wire should be provided as close as possible to the hanger.

Restraint is required at 30-ft (9-m) intervals for branch lines if branch line movement could result in damage to the sprinklers from contact with structural members, mechanical equipment, or finish materials such as rigid ceilings. While lay-in acoustical tile would not be considered a rigid ceiling, gypsum board or metal ceilings would, because they can damage most sprinklers. The exception to 6-4.6.3 would be pendent sprinklers with deep escutcheons such that the interface with the ceiling consisted of the penetration of a 1-in. (25.4-mm) drop not susceptible to damage.

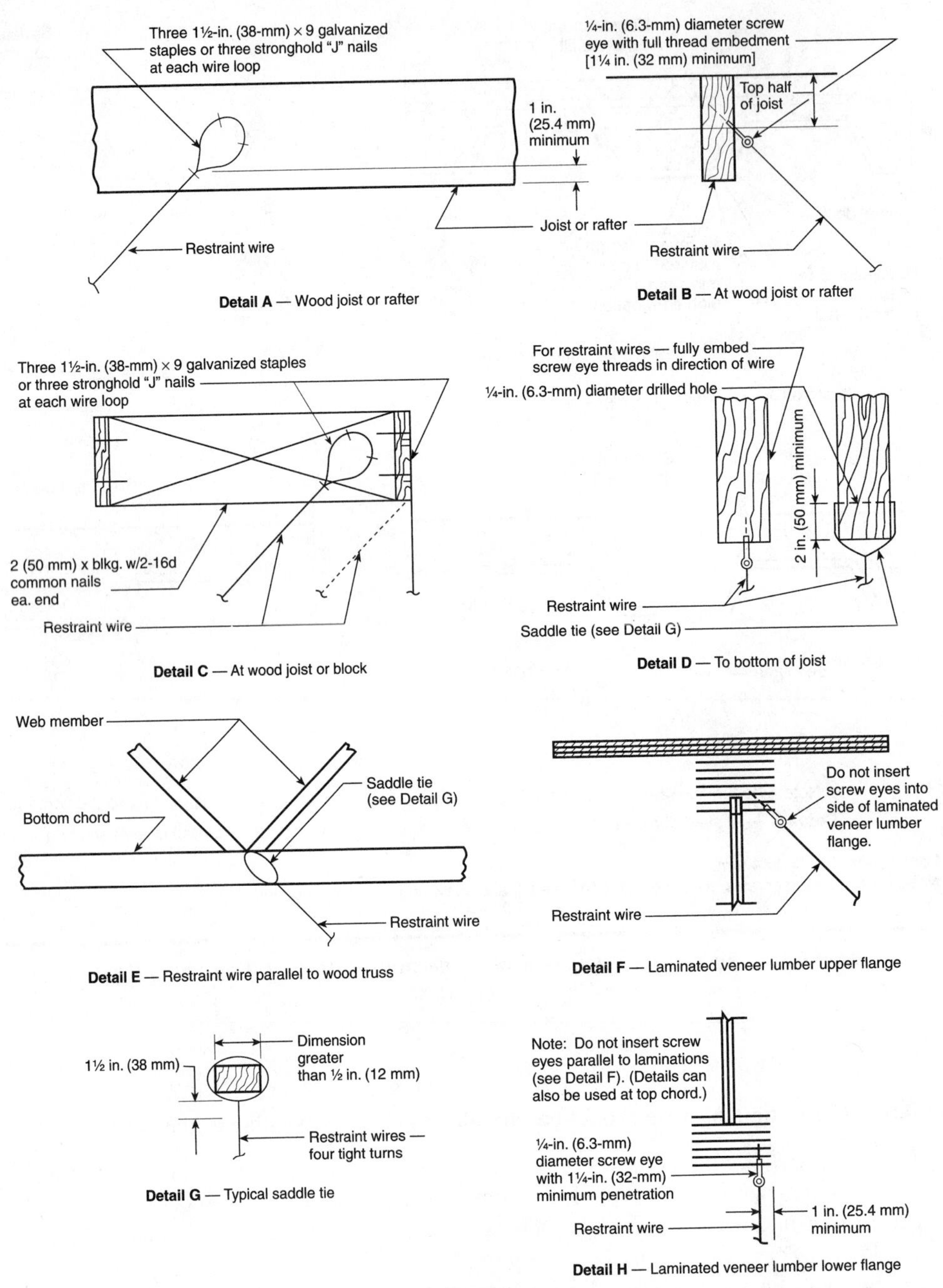

Figure A-6-4.6.1(d) Acceptable details—wire connections to wood framing.

6-4.6.4* Sprig-ups 4 ft (1.2 m) or longer shall be restrained against lateral movement.

A vertical sprig, or a sprig-up, is considered the opposite of a pipe drop. A vertical sprig is a pipe rising from the branch line to serve an individual upright sprinkler. They are susceptible to rotation during relatively minor seismic events. The requirement for restraint of sprigs was originally instituted for sprigs more than 8 ft (2.4 m) in length but was changed in the 1996 edition to apply to sprigs exceeding 4 ft (1.2 m) in length. The change was based on observations of rotated sprigs following the Northridge, CA, earthquake. The additional stability gained from this requirement results in time and money saved because sprigs do not have to be manually repositioned after the event. The restraint provisions of 6-4.6.4 do not apply to drops to individual sprinklers, because gravity acts to pull them back to center.

A-6-4.6.4 Such restraint can be provided by using the restraining wire discussed in 6-4.6.1.

6-4.7 Hangers and Fasteners Subject to Earthquakes.

6-4.7.1 C-type clamps (including beam and large flange clamps) used to attach hangers to the building structure in areas subject to earthquakes shall be equipped with a restraining strap. The restraining strap shall be listed for use with a C-type clamp or shall be a steel strap of not less than 16 gauge thickness and not less than 1 in. (25.4 mm) wide for pipe diameters 8 in. (203 mm) or less and 14 gauge thickness and not less than 1¼ in. (31.7 mm) wide for pipe diameters greater than 8 in. (203 mm). The restraining strap shall wrap around the beam flange not less than 1 in. (25.4 mm). A lock nut on a C-type clamp shall not be used as a method of restraint. A lip on a "C" or "Z" purlin shall not be used as a method of restraint.

Where purlins or beams do not provide an adequate lip to be secured by a restraining strap, the strap shall be through-bolted or secured by a self-tapping screw.

Beginning with the 1996 edition, restraining straps were required on all C-type clamps used for system hangers in areas subject to earthquakes. Previously, NFPA 13 permitted any other means the authority having jurisdiction determined acceptable to prevent C-type clamps from sliding off beam flanges during earthquake movement. Specific criteria for the restraining strap is provided in 6-4.7.1. Exhibits 6.21 and 6.22 and illustrate the use of a C-clamp with a restraining strap.

6-4.7.2 C-type clamps (including beam and large flange clamps), with or without restraining straps, shall not be used to attach braces to the building structure.

6-4.7.3 Powder-driven fasteners shall not be used to attach braces to the building structure.

Exception: Powder-driven fasteners shall be permitted where they are specifically listed for service in resisting lateral loads in areas subject to earthquakes.

Reports of powder-driven fasteners pulling out during earthquakes have led to the prohibition of their use in attaching braces, unless the fasteners are specifically listed for such use.

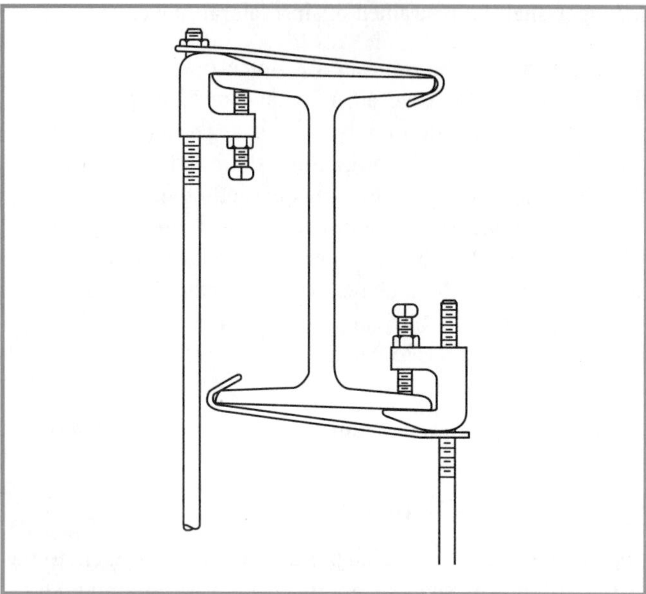

Exhibit 6.21 *Application of a C-clamp with a restraining strap on an I-beam. (Courtesy of AFCON.)*

Exhibit 6.22 *C-clamp with a restraining strap.*

6-4.7.4 Powder-driven fasteners shall not be used to attach hangers to the building structure where the systems are required to be protected against earthquakes using a horizontal force factor exceeding 0.50 W_p, where W_p is the weight of the water-filled pipe.

Exception: Powder-driven fasteners shall be permitted where they are specifically listed for horizontal force factors in excess of 0.50 W_p.

Powder-driven fasteners are prohibited as a means of attaching system hangers in strong motion earthquake areas where the authority having jurisdiction designates a horizontal force factor in excess of 0.50 W_p or one-half the weight of the water-filled pipe, unless the fasteners have been specifically investigated and listed for such loads.

References Cited in Commentary

National Fire Protection Association, 1 Batterymarch Park, P.O. Box 9101, Quincy, MA 02269-9101.

NFPA 14, *Standard for the Installation of Standpipe and Hose Systems,* 1996 edition.
NFPA 15, *Standard for Water Spray Fixed Systems for Fire Protection,* 1996 edition.

American Institute of Steel Construction, 1 East Wacker Drive, Suite 3100, Chicago, IL 60601

AISC N690, *Specification for the Design Fabrication and Erection of Structural Steel for Buildings,* 1984 edition.

American Society of Civil Engineers, 1801 Alexander Bell Drive, Alexandria, VA 20191-4400.

ASCE 19, *Structural Applications of Steel Cable for Buildings,* 1996 edition.

American Society for Testing and Materials, 100 Barr Harbor Drive, West Conshohocken, PA 19428-2959.

ASTM A 307, *Standard Specifications for Carbon Steel Bolts and Studs,* 1993 edition.
ASTM E 814, *Test Methods for Fire Tests of Through-Penetration Fire Stops,* 1988 edition.

Underwriters Laboratories Inc., 333 Pfingsten Road, Northbrook, IL 60062.

UL 194, *Gasketed Joints for Ductile-Iron Pipe and Fittings for Fire Protection Service,* 5th edition, 1996.

CHAPTER 7

Design Approaches

This chapter sets forth the minimum requirements concerning the amount of water—that is, sprinkler system discharge criteria—needed to effectively control or suppress a fire. Chapter 7 was expanded to address storage occupancies that were previously covered by the NFPA 231 document series (see Sections 7-3 through 7-8). Additionally, discharge criteria for special hazards addressed by other NFPA documents were either extracted or referenced in Section 7-10.

Chapter 7 recognizes occupancy hazard and, now, commodity fire control in combination with the area/density method as the traditional, more common design approach to determining water demand for sprinkler systems (see Section 7-2). However, other approaches for specific types of sprinklers, such as residential, large drop, and early suppression fast-response (ESFR), have been successfully developed and are also addressed.

The various design approaches are stand-alone requirements. For instance, when applying the criteria of Section 7-4 for rack storage, the criteria of Section 7-2 are not applicable (see exception to 7-2.3.2.1). Additionally, Chapter 7 also includes requirements for high-expansion foam used in combination with sprinkler systems.

7-1 General

7-1.1 Water demand requirements shall be determined from the occupancy hazard fire control approach of Section 7-2.

Exception: Special design approaches as permitted in Section 7-9.

Although not specifically identified in 7-1.1, the fire control approach for the protection of certain storage occupancies as specified in Sections 7-3 through 7-8 is considered an occupancy hazard fire control approach. The exception to 7-2.3.2.1 permits the use of the sprinkler demand in Sections 7-3 through 7-8 for storage occupancies. The sprinkler system discharge criteria in Section 7-2 are not intended to apply to storage occupancies where the storage exceeds 12 ft (3.7 m) in height. Additionally, the discharge criteria for special hazards in Section 7-10 are also considered an occupancy hazard fire control approach (see 7-2.1.2).

7-1.2 For buildings with two or more adjacent occupancies that are not physically separated by a barrier or partition capable of delaying heat from a fire in one area from fusing sprinklers in the adjacent area, the required sprinkler protection for the more demanding occupancy shall extend 15 ft (4.6 m) beyond its perimeter.

This requirement is new to the 1999 edition and is a carryover from the storage documents. It is not unusual for buildings to have various fire hazards within them resulting in different occupancy or commodity classifications. Because the discharge criteria for all fire hazards are not the same, this requirement's intent is to ensure that all fire hazards are adequately protected.

7-2 Occupancy Hazard Fire Control Approach

7-2.1 Occupancy Classifications.

7-2.1.1 Occupancy classifications for this standard relate to sprinkler installations and their water supplies only. They shall not be used as a general classification of occupancy hazards.

7-2.1.2 Occupancies or portions of occupancies shall be classified according to the quantity and combustibility of contents, the expected rates of heat release, the total potential for energy release, the heights of stockpiles, and the presence of flammable and combustible liquids, using the definitions contained in Section 1-4. Classifications are as follows:

Light hazard

Ordinary hazard (Groups 1 and 2)

Extra hazard (Groups 1 and 2)

Special occupancy hazard *(see Section 7-10)*

This is the only section that specifically identifies that a facility can contain multiple types of fire hazards or occupancies. Each hazard can be protected with the appropriate discharge criteria or the most demanding criteria can be used throughout the facility. Prior to the 1991 edition of NFPA 13, the ordinary hazard classification was divided into three groups. In 1991, the former ordinary hazard (Group 2) and ordinary hazard (Group 3) classifications were essentially merged into a single ordinary hazard (Group 2) classification. The impact of this change was considered minimal because the level of protection provided by the associated design densities was essentially the same. In the 1999 edition, discharge criteria for special occupancy hazards are specifically identified in Section 7-10. Exhibit 7.1 compares the previous area/density curves with those now in use. The solid lines in Exhibit 7.1 represent curves based on the 1989 edition of NFPA 13, and the dotted lines represent curves based on the 1999 edition.

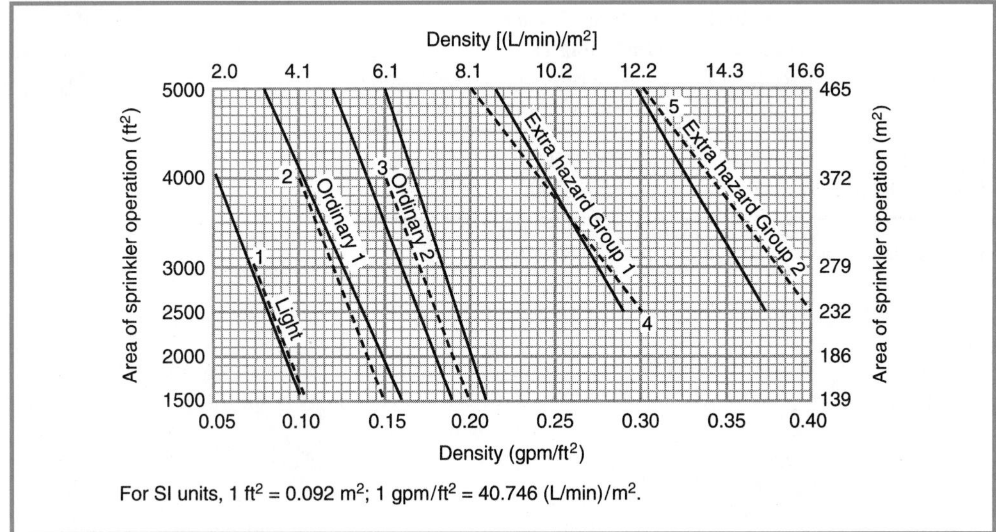

Exhibit 7.1 *Comparison of area/density curves.*

7-2.2 Water Demand Requirements—Pipe Schedule Method.

The pipe schedule method is the older means of determining the sprinkler system water demand under the occupancy hazard fire control approach. The pipe schedule method is still an acceptable method in many cases, although its application is not that common. The hydraulic method typically results in more economical systems. The use of pipe schedule systems was restricted to avoid their use under marginal conditions of water supply pressure and flow.

7-2.2.1 Table 7-2.2.1 shall be used in determining the minimum water supply requirements for light and ordinary hazard occupancies protected by systems with pipe sized according to the pipe schedules of Section 8-5. Pressure and flow requirements for extra hazard occupancies shall be based on the hydraulic calculation methods of 7-2.3. The pipe schedule method shall be permitted only for new installations of 5000 ft^2 (465 m^2) or less or for additions or modifications to existing pipe schedule systems sized according to the pipe schedules of Section 8-5. Table 7-2.2.1 shall be used in determining the minimum water supply requirements.

Exception No. 1: The pipe schedule method shall be permitted for use in systems exceeding 5000 ft^2 (465 m^2) where the flows required in Table 7-2.2.1 are available at a minimum residual pressure of 50 psi (3.4 bar) at the highest elevation of sprinkler.

Exception No. 2: The pipe schedule method shall be permitted for additions or modifications to existing extra hazard pipe schedule systems.

Table 7-2.2.1 Water Supply Requirements for Pipe Schedule Sprinkler Systems

Occupancy Classification	Minimum Residual Pressure Required (psi)	Acceptable Flow at Base of Riser (Including Hose Stream Allowance) (gpm)	Duration (minutes)
Light hazard	15	500–750	30–60
Ordinary hazard	20	850–1500	60–90

For SI units, 1 gpm = 3.785 L/min; 1 psi = 0.0689 bar.

A change to the 1996 edition clarified the sprinkler committee's intent that existing pipe schedule systems still must comply with the provisions of Section 8-5 if they are to be expanded.

Many mistakenly believe pipe schedule systems are restricted to small buildings of less than 5000 ft^2 (465 m^2). The size of the building (see Exception No. 1) only affects the required residual pressure (see Table 7-2.2.1) not whether the pipe schedule method can be used.

In 1991, NFPA 13 prohibited the use of the pipe schedule method for new extra hazard occupancies. However, Exception No. 2 permits the use of pipe schedule systems for additions and modifications to existing extra hazard systems.

7-2.2.2 The lower duration value of Table 7-2.2.1 shall be acceptable only where remote station or central station waterflow alarm service is provided.

The lower duration value is permitted where remote or central station waterflow alarm service is provided. The fire department or fire brigade is expected to be made aware of the fire more quickly than if only a local alarm were employed. Although remote or central station alarm monitoring does not have any impact on sprinkler system performance, it does allow the fire service to arrive at the scene sooner, when the fire is smaller, to initiate their fire attack.

7-2.2.3* The residual pressure requirement of Table 7-2.2.1 shall be met at the elevation of the highest sprinkler. *(See the Exceptions to 7-2.2.1).*

A-7-2.2.3 The additional pressure that is needed at the level of the water supply to account for sprinkler elevation is 0.433 psi/ft (0.098 bar/m) of elevation above the water supply. When backflow prevention valves are installed on pipe schedule systems, the friction losses of the device must be accounted for when determining acceptable residual pressure at the top level of sprinklers. The friction loss [in psi (bar)] should be added to the elevation loss and the residual pressure at the top row of sprinklers to determine the total pressure needed at the water supply.

The residual pressure requirement is to be applied at the elevation of the highest sprinkler. The process of calculating the available pressure starts at the connection to the water supply, which is typically at the street but can be at the base of the riser. The available residual pressure of the water supply at the acceptable flow identified in Table 7-2.2.1 must meet or exceed the minimum residual pressure required, which is also identified in Table 7-2.2.1. For installations larger than 5000 ft^2 (465 m^2), the 50 psi (3.4 bar) required by Exception No. 1 of 7-2.2.1 supersedes Table 7-2.2.1. Elevation pressure [see 8-4.4.5(a)] is the product of the elevation in feet (meters) multiplied by 0.433 psi/ft (0.096 bar/m). This requirement for elevation pressure demonstrates that some calculation is now necessary with pipe schedule systems. Traditional pipe schedule approaches required no additional hydraulic calculations.

The installation of backflow prevention devices on pipe schedule systems is quite common. Because of the high friction losses associated with backflow devices, the effect of backflow devices on system performance must be considered. This appendix material was added in 1996 to provide guidance in this regard. The friction loss for the backflow device must be obtained from the device manufacturer and based on the acceptable flow identified in Table 7-2.2. Also see commentary to 5-14.4.6.2.

7-2.2.4 The lower flow figure of Table 7-2.2.1 shall be permitted only where the building is of noncombustible construction or the potential areas of fire are limited by building size or compartmentation such that no open areas exceed 3000 ft^2 (279 m^2) for light hazard or 4000 ft^2 (372 m^2) for ordinary hazard.

7-2.3 Water Demand Requirements—Hydraulic Calculation Methods.

The following two hydraulic calculation approaches are permitted for determining the water demand for hydraulically calculated systems under the occupancy hazard fire control approach:

(1) The area/density method
(2) The room design method

Even though the subsection does not address the sprinkler discharge criteria for storage application (see Sections 7-3 through 7-8), it does provide hose stream demands for general and rack storage.

7-2.3.1 General.

7-2.3.1.1* The minimum water supply requirements for a hydraulically designed occupancy hazard fire control sprinkler system shall be determined by adding the hose stream demand from Table 7-2.3.1.1 to the water supply for sprinklers determined in 7-2.3.1.2. This supply shall be available for the minimum duration specified in Table 7-2.3.1.1.

Exception No. 1: An allowance for inside and outside hose shall not be required where tanks supply sprinklers only.

Exception No. 2: Where pumps taking suction from a private fire service main supply sprinklers only, the pump need not be sized to accommodate inside and outside hose. Such hose allowance shall be considered in evaluating the available water supplies.

Table 7-2.3.1.1† Hose Stream Demand and Water Supply Duration Requirements for Hydraulically Calculated Systems

Occupancy or Commodity Classification	Inside Hose (gpm)	Total Combined Inside and Outside Hose (gpm)	Duration (minutes)
Light hazard	0, 50, or 100	100	30
Ordinary hazard	0, 50, or 100	250	60–90
Extra hazard	0, 50, or 100	500	90–120
Rack storage, Class I, II, and III commodities up to 12 ft (3.7 m) in height	0, 50, or 100	250	90
Rack storage, Class IV commodities up to 10 ft (3.1 m) in height	0, 50, or 100	250	90
Rack storage, Class IV commodities up to 12 ft (3.7 m) in height	0, 50, or 100	500	90
Rack storage, Class I, II, and III commodities over 12 ft (3.7 m) in height	0, 50, or 100	500	90
Rack storage, Class IV commodities over 12 ft (3.7 m) in height and plastic commodities	0, 50, or 100	500	120
General storage, Class I, II, and III commodities over 12 ft (3.7 m) up to 20 ft (6.1 m)	0, 50, or 100	500	90
General storage, Class IV commodities over 12 ft (3.7 m) up to 20 ft (6.1 m)	0, 50, or 100	500	120
General storage, Class I, II, and III commodities over 20 ft (6.1 m) up to 30 ft (9.1 m)	0, 50, or 100	500	120
General storage, Class IV commodities over 20 ft (6.1 m) up to 30 ft (9.1 m)	0, 50, or 100	500	150
General storage, Group A plastics ≤5 ft (1.5 m)	0, 50, or 100	250	90
General storage, Group A plastics over 5 ft (1.5 m) up to 20 ft (6.1 m)	0, 50, or 100	500	120
General storage, Group A plastics over 20 ft (6.1 m) up to 25 ft (7.6 m)	0, 50, or 100	500	150

For SI units, 1 gpm = 3.785 L/min.

A-7-2.3.1.1 Appropriate area/density, other design criteria, and water supply requirements should be based on scientifically based engineering analyses that can include submitted fire testing, calculations, or results from appropriate computational models.

Recommended water supplies anticipate successful sprinkler operation. Because of the small but still significant number of uncontrolled fires in sprinklered properties, which have various causes, there should be an adequate water supply available for fire department use.

The discharge criteria for the specific special occupancy hazards identified in Section 7-10 might differ from the requirements of Table 7-2.3.1.1. For example, NFPA 415, *Standard on Airport Terminal Buildings, Fueling Ramp Drainage, and Loading Walkways,* classifies passenger handling areas as ordinary hazard (Group 1) occupancies.

Therefore, the discharge criteria for the sprinklers would be in accordance with the ordinary hazard (Group 1) area/density curve in Figure 7-2.3.1.2. However, NFPA 415 requires a hose stream demand of 500 gpm (1893 L/min), which exceeds the 250-gpm (946-L/min) hose stream allowance required by Table 7-2.3.1.1 for ordinary hazard occupancies. See 7-10.21.3.

7-2.3.1.2 The water supply for sprinklers only shall be determined either from the area/density curves of Figure 7-2.3.1.2 in accordance with the method of 7-2.3.2 or be based upon the room design method in accordance with 7-2.3.3, at the discretion of the designer. For special areas under consideration, as described in 7-2.3.4, separate hydraulic calculations shall be required in addition to those required by 7-2.3.2 or 7-2.3.3.

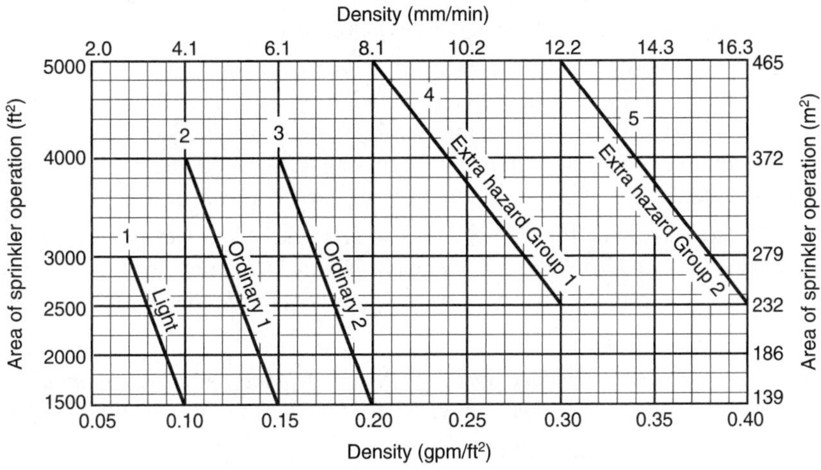

Figure 7-2.3.1.2 Area/density curves.

The sprinkler discharge criteria are permitted to be determined through hydraulic calculations in accordance with either the area/density method of 7-2.3.2, the room design method of 7-2.3.3, or the special design methods of 7-2.3.4. The special design methods mainly apply to building service chutes and spaces protected by a single line of sprinklers such as a corridor.

7-2.3.1.3 Regardless of which of the two methods is used, the following restrictions shall apply:

(a) For areas of sprinkler operation less than 1500 ft^2 (139 m^2) used for light and ordinary hazard occupancies, the density for 1500 ft^2 (139 m^2) shall be used. For areas of sprinkler operation less than 2500 ft^2 (232 m^2) for extra hazard occupancies, the density for 2500 ft^2 (232 m^2) shall be used.

(b)* For buildings having unsprinklered combustible concealed spaces (as described in 5-13.1.1 and 5-13.7), the minimum area of sprinkler operation shall be 3000 ft^2 (279 m^2).

Exception No. 1: Combustible concealed spaces filled entirely with noncombustible insulation.

Exception No. 2: Light or ordinary hazard occupancies where noncombustible or limited combustible ceilings are directly attached to the bottom of solid wood joists so as to create enclosed joist spaces 160 ft^3 (4.8 m^3) or less in volume.*

Exception No. 3: Concealed spaces where the exposed surfaces have a flame spread rating of 25 or less and the materials have been demonstrated to not propagate fire in the form in which they are installed in the space.*

(c) Water demand of sprinklers installed in racks or water curtains shall be added to the ceiling sprinkler water demand at the point of connection. Demands shall be balanced to the higher pressure. *(See Chapter 8.)*

(d) Water demand of sprinklers installed in concealed spaces or under obstructions such as ducts and cutting tables need not be added to ceiling demand.

(e) Where inside hose stations are planned or are required, a total water allowance of 50 gpm (189 L/min) for a single hose station installation or 100 gpm (378 L/ min) for a multiple hose station installation shall be added to the sprinkler requirements. The water allowance shall be added in 50-gpm (189-L/min) increments beginning at the most remote hose station, with each increment added at the pressure required by the sprinkler system design at that point.

(f) When hose valves for fire department use are attached to wet pipe sprinkler system risers in accordance with 5-15.5.2, the water supply shall not be required to be added to standpipe demand as determined from NFPA 14, *Standard for the Installation of Standpipe and Hose Systems.*

Exception No. 1: Where the combined sprinkler system demand and hose stream allowance of Table 7-2.3.1.1 exceeds the requirements of NFPA 14, Standard for the Installation of Standpipe and Hose Systems, this higher demand shall be used.

Exception No. 2: For partially sprinklered buildings, the sprinkler demand, not including hose stream allowance, as indicated in Table 7-2.3.1.1 shall be added to the requirements given in NFPA 14, Standard for the Installation of Standpipe and Hose Systems.

(g) Water allowance for outside hose shall be added to the sprinkler and inside hose requirement at the connection to the city water main or a yard hydrant, whichever is closer to the system riser.

(h) The lower duration values in Table 7-2.3.1.1 shall be permitted where remote station or central station waterflow alarm service is provided.

(i) Where pumps, gravity tanks, or pressure tanks supply sprinklers only, requirements for inside and outside hose need not be considered in determining the size of such pumps or tanks.

A-7-2.3.1.3(b) This section is included to compensate for possible delay in operation of sprinklers from fires in combustible concealed spaces found in wood frame, brick veneer, and ordinary construction.

A-7-2.3.1.3(b) Exception No. 2. Composite wood joists are not considered solid wood joists for the purposes of this section. Their web members are too thin and easily penetrated to adequately compartment a fire in an unsprinklered space.

A-7-2.3.1.3(b) Exception No. 3. This exception is intended to apply only when the exposed materials in the space are limited combustible materials or fire retardant–treated wood as defined in NFPA 703, *Standard for Fire Retardant Impregnated Wood and Fire Retardant Coatings for Building Materials.*

The sprinkler system is expected to control the fire when the proper discharge density is applied, assuming the system is properly designed, installed, and maintained. Although a sprinkler system can control or suppress a fire with fewer sprinklers than are required in the design area, restrictions for minimum design areas in accordance with 7-2.3.1.3(a) provide a safety factor. A number of factors such as ventilation and ceiling configuration can lead to an inconsistent sprinkler operating sequence—that is, skipping of sprinklers. A shielded fire can also require the use of most or all sprinklers in the design area.

The 3000-ft^2 (279-m^2) requirement of 7-2.3.1.3(b) also applies to the room design method and supersedes limiting the remote area based only on room size. In addition, this requirement has been further clarified by 7-2.3.2.8 whereby the 3000 ft^2 (279 m^2) is applied after all other area modifications have been made.

The intention of 7-2.3.1.3(b) is not to permit the omission of sprinklers from concealed combustible spaces but rather to compensate for the anticipated delay of sprinkler operation for a fire originating in an unprotected, concealed combustible space. As noted, 5-13.1.1 describes the conditions under which sprinklers do not have to be installed in concealed spaces.

The exceptions to 7-2.3.1.3(b) clarify that the 3000-ft^2 (279-m^2) design area does not apply where an unsprinklered concealed space that is entirely filled with noncombustible insulation is enclosed with limited-combustible construction materials or of a certain volume. Formal Interpretation 80-9 provides additional insight on the application of this requirement.

Formal Interpretation 80-9

Reference: 7-2.3.1.3(b)

Question 1: Is it the intent of 7-2.3.1.3(b) to impose the 3,000-square foot design area where sprinklers have been omitted in accordance with 5-13.1.1, regardless of the solid joist or beam depth; or, are these areas regarded as joist space voids which do not require 3,000 square feet to be used?

Answer: Yes, regardless of the solid joist or beam depth. (*See also exceptions to 7-2.3.1.3(b).*

Question 2: Would this also apply in part to walls?

Answer: No.

Issue Edition: 1980

Reference: 2-2.1.2.9

Date: March 1982

Also, if residential sprinklers are used in a dwelling unit and an unprotected, combustible concealed space exists, the area limitations of 7-2.3.1.3(b) do not apply because the discharge criteria for residential sprinklers are not based on the area/density method.

Sprinklers both at the ceiling and under an obstruction or in a concealed space as identified in 7-2.3.1.3(c) and (d) are not anticipated to operate during a fire when using the area/density method. This provision also applies to those situations in which a sprinkler is installed under a temporary obstruction such as an overhead garage door.

The size and type of hose connections contemplated by 7-2.3.1.3(e) are basically limited to those necessary for first-aid fire fighting or final extinguishment. Even though more than two such hose stations could be connected to the system, operation of more than two stations simultaneously during an incident is considered unlikely. Thus, only a maximum of two hose stations, simultaneously operating, are to be included in the total water supply. The demand of only one hose station is permitted to be included where only one hose station is needed. Table 7-2.3.1.1 highlights this point because the inside hose demand is presented as either 0 gpm, 50 gpm, or 100 gpm. Accounting for this potential use of water permits the sprinkler system to operate as intended while at the same time allowing some form of manual first-aid fire fighting. Hose connection requirements are identified in 5-15.5.

The hose connections contemplated in 7-2.3.1.3(f) are not standpipe connections as defined by NFPA 14, *Standard for the Installation of Standpipe and Hose Systems.* NFPA 14 provides specific requirements for Class II standpipe systems that are intended for building occupant use. These requirements are considerably more demanding in terms of flow and pressure for the standpipe system than the values given in NFPA 13. The 1996 edition of NFPA 14 requires that Class II standpipes provide 100 gpm (378 L/min) at 65 psi (4.5 bar). The distinction between hose connections and standpipes is important. NFPA 13 provides requirements for convenience connections, and NFPA 14 provides requirements for standpipe system connections. Although both have 1½-in. (40-mm) outlets, Class II standpipe outlets have 1½-in. (40-mm) inlets, and hose connections to sprinkler systems typically have 1-in. (25.4-mm) inlets.

For all practical purposes, the inside hose connections to sprinkler systems should be viewed as large sprinklers with no pressure requirements of their own but which are to flow the required amount of water at the pressure available at the hose outlet. The required flow at the hose outlets is to be carried through the calculations back to the system riser.

For a completely sprinklered building utilizing a combined sprinkler and standpipe riser, the pressure and flow demand for the standpipe system can, in most cases, exceed the demand of the sprinkler system. This results in exclusive adherence to NFPA 14 for determination of total water demand. In those instances where the sprinkler system demand is determined to be more demanding than the standpipe demand, the sprinkler system demand is to be used in accordance with Exception No. 1 of 7-2.3.1.3(f).

If a building is only partially sprinklered, which is strongly discouraged by 5-1.1, the sprinkler demand from NFPA 13 and hose demand from NFPA 14 must be added to determine the discharge criteria for the combined system in accordance with Exception No. 2 of 7-2.3.1.3(f). This combined demand is necessary because fires can originate in the nonsprinklered area and grow to the point that they activate nearby sprinklers while still requiring full standpipe water supplies to be available for manual suppression.

With regard to 7-2.3.1.3(g), if no inside hose connections are provided, the outside hose requirement is the value in the third column from the left in Table 7-2.3.1.1. This relatively small quantity of water is apportioned to the fire department to allow for final and complete extinguishment of the fire. The flow rate established in Table 7-2.3.1.1 can be added into the hydraulic calculations at the connection to the city water main or yard hydrant at the prevailing system pressure. The outside hose stream water typically receives its pressure from the fire department apparatus.

The lower durations indicated by 7-2.3.1.3(h) are permitted where remote or central station waterflow alarm service is provided. This reduction is permitted because the fire department or fire brigade is expected to be made aware of the fire much sooner than if only a local alarm were employed. Although remote or central station alarm monitoring does not have any impact on sprinkler system performance, it does allow the fire service to arrive at the scene more quickly to initiate its fire attack. Because systems designed in accordance with the area/density method are to control the fire until the fire department arrives, the earlier the fire department arrives, the smaller the amount of time needed for the sprinkler system to control the fire.

With regard to 7-2.3.1.3(i), where pumps or tanks supply sprinklers only, the hose stream requirements need not be added when sizing the pump or tank. For example, if the pump or tank supplies a system that is required to be equipped with small hose stations, the demand of the hose stations needs to be included when sizing the pump or tank.

7-2.3.1.4 Total system water supply requirements shall be determined in accordance with the hydraulic calculation procedures of Section 8-4.

7-2.3.2 Area/Density Method.

7-2.3.2.1 The water supply requirement for sprinklers only shall be calculated from the area/density curves in Figure 7-2.3.1.2 or from Section 7-10 where area/density criteria is specified for special occupancy hazards. When using Figure 7-2.3.1.2, the calculations shall satisfy any single point on the appropriate area/density curve as follows:

(1) Light hazard area/density curve 1
(2) Ordinary hazard (Group 1) area/density curve 2
(3) Ordinary hazard (Group 2) area/density curve 3
(4) Extra hazard (Group 1) area/ density curve 4
(5) Extra hazard (Group 2) area/ density curve 5

It shall not be necessary to meet all points on the selected curve.

Exception: Sprinkler demand for storage occupancies as determined in Sections 7-3 through 7-8.

The design approaches of Sections 7-3 through 7-8 also use an area/density approach to determine sprinkler discharge criteria. The modifiers in 7-2.3.2 do not apply to Sections 7-3 through 7-8.

Design of a sprinkler system in accordance with the area/density method requires that only a single point is met on an area/density curve. This requirement applies to the curves in Figure 7-2.3.1.2 and the area/ density curves identified in Sections 7-3 through 7-8 for the protection of storage occupancies. Selecting a point on the low end—that is, smaller operating area—of an area/density curve can result in a somewhat higher density but a lower total water demand. The higher density also requires a higher pressure but is generally considered superior in terms of fire control and is expected to confine the fire to a smaller area, reducing the total number of operating sprinklers. Selection of a point on the high end—that is, larger operating area—of the curve allows for a lower density, and, therefore, lower pressure, but a higher total water demand for the system, as well as a larger fire. Therefore, more sprinklers would be expected to operate.

The area of operation shown on the vertical axis of Figure 7-2.3.1.2 and the area/density curves in Sections 7-3 through 7-8 are independent of the building's size. Selection of an area/density combination can limit the number of sprinklers expected to operate to the area selected from the area/density curve.

The exception is new to the 1999 edition because the protection of storage occupancies is now within the scope of NFPA 13. Additionally, the area/density curve in Figure 7-2.3.1.2 is not intended to apply to storage in excess of 12 ft (3.7 m) in height.

The situation frequently arises where a small area of a higher hazard is surrounded by a lesser hazard. For example, consider a 600-ft^2 area consisting of 10-ft high on-floor storage of cartoned solid plastic commodities surrounded by a plaster injection molding operation in a 15-ft high building. In accordance with Table 7-2.3.2.2, the density required for the plastic storage must meet the requirements for extra hazard (Group 1) occupancies. The injection molding operation should be considered an ordinary hazard (Group 2) occupancy. In accordance with Figure 7-2.3.1.2, the corresponding discharge densities should be 0.3 gpm/ft^2 over 2500 ft^2 the storage and 0.2 gpm/ft^2 over 1500 ft^2 for the remainder of the area. (Also see 7-2.3.1.3 for the required minimum areas of operation.)

If the storage area is not separated from the surrounding area by a wall or partition (see 7-1.2), then the size of the operating area is determined by the higher hazard storage.

For example, the operating area is 2500 ft^2. The system must be able to provide the 0.30-gpm/ft^2 density over the storage area and 15 ft beyond. If part of the remote area is outside the 600 ft^2 plus the 15-ft overlap, then only 0.2 gpm/ft^2 is needed for that portion.

If the storage is separated from the surrounding area by a floor-to-ceiling/roof partition that is capable of preventing heat from a fire on one side from fusing sprinklers on the other side, then the size of the operating area is determined by the occupancy of the surrounding area. In this example, the design area is 1500 ft^2. A 0.30-gpm/ft^2 density is needed within the separated area with 0.15 gpm/ft^2 in the remainder of the remote area.

7-2.3.2.2 For protection of miscellaneous storage, miscellaneous tire storage, and storage up to 12 ft (3.7 m) in height, the discharge criteria in Table 7-2.3.2.2 shall apply.

Storage occupancies present unique fire protection challenges because of the fire load associated with the type, quantity, and arrangement of products present in these spaces. If storage exceeds 12 ft (3.7 m) in height, these storage occupancies require special fire protection features not contained within Section 7-2. Table 7-2.3.2.2 addresses storage not exceeding 12 ft (3.7 m) in height and the miscellaneous storage of certain commodities.

The miscellaneous storage of car or truck tires in facilities such as automobile service centers presents significant fire safety concerns. Burning tires tend to create

Table 7-2.3.2.2 Discharge Criteria for Miscellaneous Storage and Storage 12 ft (3.7 m) or Less in Height,[1] Commodity Classes I through IV

Commodity Classification	Palletized and Bin Box	Rack
I	OH-1	OH-1
II up to 8 ft (2.4 m)	OH-1	OH-1
II over 8 ft (2.4 m) up to 12 ft (3.6 m)	OH-2	OH-2
III	OH-2	OH-2
IV up to 10 ft (3 m)	OH-2	OH-2
IV over 10 ft (3 m) to 12 ft (3.6 m)	OH-2	EH-1

Group A Plastics Stored on Racks

Storage Height	Maximum Building Height	Cartoned		Exposed	
		Solid	Expanded	Solid	Expanded
Up to 5 ft	No limit	OH-2	OH-2	OH-2	OH-2
Over 5 ft to 10 ft	15 ft	EH-1	EH-1	EH-2	EH-2
Over 5 ft to 10 ft	20 ft	EH-2	EH-2	EH-2	OH-2 +1 level in-rack[5]
Over 10 ft to 12 ft	17 ft	EH-2[2]	EH-2[2]	EH-2[2]	EH-2[2]
Over 10 ft to 12 ft	No limit	OH-2 +1 level in-rack[5]	OH-2 +1 level in-rack[5]	OH-2 +1 level in-rack[5]	OH-2 +1 level in-rack[5]

Group A Plastics Solid-Piled, Palletized, Bin-Box, or Shelf Storage

Storage Height	Maximum Building Height	Cartoned		Exposed	
		Solid	Expanded	Solid	Expanded
Up to 5 ft	No limit	OH-2	OH-2	OH-2	OH-2
Over 5 ft to 10 ft	15 ft	EH-1	EH-1	EH-2	EH-2
Over 5 ft to 10 ft	20 ft	EH-2	EH-2	EH-2	—
Over 5 ft to 8 ft	No limit	—	—	—	EH-2
Over 10 ft to 12 ft	17 ft	EH-2	EH-2	EH-2	EH-2
Over 10 ft to 12 ft	27 ft	EH-2	EH-2	—	—

(continues)

Table 7-2.3.2.2 Discharge Criteria for Miscellaneous Storage and Storage 12 ft (3.7 m) or Less in Height,[1] Commodity Classes I through IV (Continued)

Miscellaneous Tire Storage[3]

Piling Methods	Height of Storage	Occupancy Group
On floor, on side	5 ft to 12 ft	EH-1
On floor, on tread or on side	To 5 ft	OH-2
Single-, double-, or multiple-row racks on tread or on side	To 5 ft	OH-2
Single-row rack, portable, on tread or on side	5 ft to 12 ft	EH-1
Single-row rack, fixed, on tread or on side	5 ft to 12 ft	EH-1 or OH-2 plus one level of in-rack sprinklers
Rolled Paper Stored on End	**Height of Storage**	**Occupancy Group**
Heavy and medium weight	To 8 ft	OH-2
On floor, on tread or on side	Over 8 ft to 12 ft	EH-1
Tissue	To 10 ft	EH-1
Idle Pallet Storage[4]	**Height of Storage**	**Occupancy Group**
Single row rack, fixed	To 6 ft wooden To 4 ft plastic	OH-2 OH-2

For SI units, 1 ft = 0.3048 in.

[1]The design of the sprinkler system shall be based on the conditions that will routinely or periodically exist in the building creating the greatest water demand, including pile height and clearance.

[2]For rack storage, OH-2 + 1 level in rack shall also be permitted.

[3]The discharge criteria for the storage in this table shall only apply to miscellaneous tire storage as defined in 1-4.10.

[4]The discharge criteria for pallets shall apply only to the storage of wooden pallets stored up to 6 ft (1.8 m) in height or plastic pallets up to 4 ft (1.2 m) in height with not over four stacks of wooden pallets or two stacks of plastic pallets separated from other stacks by at least an 8-ft (2.7-m) aisle. (For heights or quantities exceeding these limits see Section 7-5).

[5]See Section 7-11 for in-rack sprinkler discharge criteria.

very hot and smoky fires that are difficult to control and extinguish. Because of their structure, tires possess inherent flue spaces that allow for an unobstructed air supply during a fire. Additionally, fires burning on the interior surfaces of tires are often obstructed from sprinkler system discharge. Therefore, sprinkler activation during the incipient stages of a tire fire is critical.

7-2.3.2.3 The densities and areas provided in Figure 7-2.3.1.2 are for use only with spray sprinklers. For use with other types of sprinklers, see Section 7-9.

Exception No. 1: Quick-response sprinklers shall not be permitted for use with area/density curves 4 and 5 (extra hazard).*

Exception No. 2: Sidewall spray sprinklers shall be permitted for use with area/density curve 1 (light hazard) and, if specifically listed, with area/density curves 2 or 3 (ordinary hazard).

Exception No. 3: For extended coverage sprinklers, the minimum design area shall be that corresponding to the maximum density for the hazard in Figure 7-2.3.1.2 or the area protected by five sprinklers, whichever is greater. Extended coverage sprinklers shall be listed with and designed for the minimum flow corresponding to the density for the smallest area of operation for the hazard as specified in Figure 7-2.3.1.2.

A-7-2.3.2.3 Exception No. 1. It is not the intent of this exception to restrict the use of quick-response sprinklers in extra hazard occupancies but rather to indicate that the areas and densities shown in Figure 7-2.3.1.2 might not be appropriate for use with quick-response sprinklers in those environments due to a concern with water supplies.

Exception No. 1 imposes a limit on the use of quick-response sprinklers in extra hazard occupancies due to a lack of extensive full-scale testing. Because of their mode of operation, sidewall sprinklers must be specifically listed for ordinary hazard occupancies in accordance with Exception No. 2 if intended for such use.

Exception No. 3 clarifies that extended coverage (EC) sprinklers have a minimum design area, namely the area that comprises not less than five sprinklers or 1500 ft^2 (135 m^2). For an EC sprinkler listed for use in light hazard occupancies and a protection area per sprinkler of 20 ft × 20 ft (6.1 m × 6.1 m), the design area would be a 2000-ft^2 (186-m^2) area—that is, 400 ft^2 (36 m^2) per sprinkler multiplied by five sprinklers. If an EC sprinkler was listed with a protection area of 16 ft × 16 ft (5 m × 5 m), the design area would be based on a 1500-ft^2 (139-m^2) area of operation. In this case, the sprinkler is listed to cover 256 ft^2 (23 m^2). Multiplying the minimum five-sprinkler design area in Exception No. 3 by 256 ft^2 (23 m^2) results in an area of operation of 1280 ft^2 (115 m^2). Exception No. 3 requires the larger of the 1500-ft^2 (139-m^2) area from the area/ density curve or the five-sprinkler design area.

7-2.3.2.4 Where listed quick-response sprinklers are used throughout a system or portion of a system having the same hydraulic design basis, the system area of operation shall be permitted to be reduced without revising the density as indicated in Figure 7-2.3.2.4 when all of the following conditions are satisfied:

(1) Wet pipe system
(2) Light hazard or ordinary hazard occupancy
(3) 20-ft (6.1-m) maximum ceiling height

The number of sprinklers in the design area shall never be less than five. Where quick-response sprinklers are used on a sloped ceiling, the maximum ceiling height shall be used for determining the percent reduction in design area. Where quick-response sprinklers are installed, all sprinklers within a compartment shall be of the quick response type.

Exception: Where circumstances require the use of other than ordinary temperature–rated sprinklers, standard response sprinklers shall be permitted to be used.

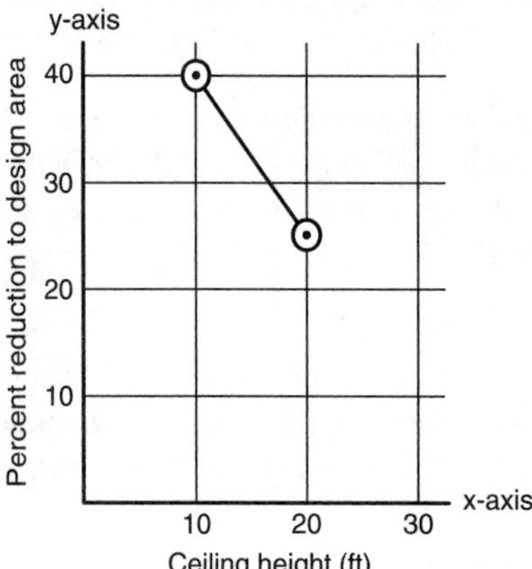

Note: $y = \frac{-3x}{2} + 55$

For ceiling height ≥ 10 ft and ≤ 20 ft, $y = \frac{-3x}{2} + 55$

For ceiling height < 10 ft, $y = 40$

For ceiling height > 20, $y = 0$

For SI units, 1 ft = 0.31 m.

Figure 7-2.3.2.4 *Design area reduction for quick-response sprinklers.*

The reduction in design area for quick-response sprinklers ranges from a maximum of 40 percent for ceiling heights of 10 ft (3 m) and less to 25 percent for ceiling heights of 20 ft (6.1 m). The design area reduction is based on comparative tests of quick-response and standard response spray sprinklers that demonstrated the fire can be controlled to a smaller size the earlier water is applied. The smaller fire size in turn results in the operation of fewer sprinklers.

The Technical Committee on Sprinkler System Discharge Criteria recognized that the time of sprinkler activation is the result of a number of factors in addition to sprinkler thermal sensitivity. These factors include temperature rating of sprinklers, spacing of sprinklers, distance of the sprinklers below the ceiling, ceiling height, and the fire itself. With the exception of ceiling height, however, most of these factors are controlled within specified ranges by NFPA 13 provisions for a particular occupancy class. The reduction in operating area provides a valid credit for the increased effectiveness of quick-response sprinklers with regard to ceiling heights.

The 1999 edition clarifies two items. The first item is that reduction for sloped ceilings is based on the maximum ceiling height. The second item is that quick-response sprinklers are not required to be used throughout the entire building. It is recognized that different portions of the building, or of a single floor, can have a different design basis. The portion of the building for which the pipe size is based on a reduced remote area must use quick-response sprinklers.

7-2.3.2.5 The system area of operation shall be increased by 30 percent without revising the density when the following types of sprinklers are used on sloped ceilings with a pitch exceeding one in six (a rise of two units in a run of 12 units, a roof slope of 16.7 percent):

(1) Spray sprinklers, including extended coverage sprinklers listed in accordance with 5-4.3, Exception No. 3, and quick-response sprinklers
(2) Large drop sprinklers

The design area increase of 30 percent for sloped ceilings with a pitch exceeding 2 in. in 12 in. (50.8 mm in 304.8 mm) (16.7 percent) is based on the test work and modeling conducted at the Factory Mutual Research Corporation. In those tests, slopes greater than 16.7 percent resulted in erratic sprinkler operating patterns that could adversely effect operating sprinklers outside of the design area. Heat from the fire is likely to accumulate near the peak of the slope and to activate sprinklers that are some distance from the fire. Subparagraph 7-2.3.2.5 compensates for the fact that more sprinklers are expected to operate under steeply pitched surfaces.

The 1999 edition clarifies that the modifier applies to all sprinklers allowed under a sloped ceiling including quick-response sprinklers.

The 30-percent area increase does not apply to those types of sprinklers that utilize methods other than the area/density method to determine their discharge criteria, such as residential sprinklers, or to those types of sprinklers that are limited to slopes of 2 in. in 12 in. (50.8 mm in 304.8 mm), such as extended coverage and ESFR sprinklers. (See Section 5-4.)

The design area increase is intended to apply only to continuously sloped ceilings or roofs such as those in large attic spaces. The increase is not intended to apply to sawtooth roofs or other similar situations in which the extent of the slope is small relative to the sprinkler design area.

7-2.3.2.6 For dry pipe systems and double interlock preaction systems, the area of sprinkler operation shall be increased by 30 percent without revising the density.

The delivery of water to operated sprinklers in dry pipe and double interlock preaction sprinkler systems takes longer than it does for wet pipe systems and non-interlock or single interlock preaction systems. Because of this delay, system size in terms of volume is limited (see 4-3.2.2). To compensate for this delay, a 30 percent increase in the design area is mandated. This increase is necessary because more sprinklers in the fire area are expected to operate while the pipe is being charged with water and the fire will be larger when water reaches the operating sprinklers. The 30 percent increase is applied as follows:

If the density and area for an ordinary hazard (Group 2) occupancy are selected at 0.20 gpm/ft^2 [8.1 (L/min)/m^2)] and 1500 ft^2 (139 m^2) (see Figure 7-2.3.1.2), this density would be adjusted for a dry pipe system as follows:

$$(30 \text{ percent} \times 1500 \text{ ft}^2) + 1500 \text{ ft}^2 = 1950 \text{ ft}^2 \ (182 \text{ m}^2)$$

The density of 0.20 gpm/ft^2 [8.1 (L/min)/m^2)] remains unchanged.

The 30 percent area increase rule does not apply to preaction systems other than to double interlock systems. Even though the pipe has no water in it, the delay in operation of non-interlock and single interlock systems is minimal because the detection device typically responds to the fire sooner than the sprinkler. (See the commentary to 4-3.2.1.)

Formal Interpretation 76-12 provides additional information on the application of the 30 percent increase requirement for dry pipe systems.

Formal Interpretation 76-12

Reference: 7-2.3.2.6, 7-2.3.1.3(b)

Question: For a dry pipe sprinkler system in an ordinary hazard occupancy having unsprinklered combustible concealed spaces, would 7-2.3.2.6 and 7-2.3.1.3(b) permit a density based on less than 3000 sq ft area of sprinkler operation provided the 30 percent increase in the area of sprinkler operation required by 7-2.3.2.6 brings the area of sprinkler operation to over 3000 sq ft as required by 7-2.3.1.3(b)?

Answer: Yes.

Issue Edition: 1976

Reference: Notes 5 & 6 Table 2-2.1(b)

Date: June 1978

7-2.3.2.7 Where high-temperature sprinklers are used for extra hazard occupancies, the area of sprinkler operation shall be permitted to be reduced by 25 percent without revising the density, but not to less than 2000 ft^2 (186 m^2).

Fire tests for extra hazard and storage occupancies using sprinklers with nominal K-factors of 5.6 and 8.0 indicate that fewer high-temperature sprinklers operate when compared to ordinary temperature–rated sprinklers. Because fewer high-temperature sprinklers activate, Section 7-3.2.2.2.2 permits reduced design areas if high-temperature sprinklers are used for the protection of Class I through Class IV commodities. However, this reduction in operating area has not been verified with sprinklers with larger nominal K-factors. Use of high-temperature sprinklers for light and ordinary hazard applications is nevertheless restricted to those conditions defined in 5-3.1.4.

7-2.3.2.8* Where multiple adjustments to the area of operation are required to be made in accordance with 7-2.3.2.4, 7-2.3.2.5, 7-2.3.2.6, or 7-2.3.2.7, these adjustments shall be compounded based on the area of operation originally selected from Figure 7-2.3.1.2. If the building has unsprinklered combustible concealed spaces, the rules of 7-2.3.1.3 shall be applied after all other modifications have been made.

A-7-2.3.2.8 *Example 1.* A dry pipe sprinkler system (OH$_2$) in a building with a ceiling slope exceeding two in 12 in. (16.6 percent slope). The initial area must be increased 30 percent for the dry pipe system and the resulting area an additional 30 percent for the roof slope. If the point 0.2 gpm/ft^2 (8.2 mm/min) over 1500 ft^2 (139 m^2) is chosen from Figure 7-2.3.1.2, the 1500-ft^2 (139-m^2) area is increased 450 ft^2 (42 m^2) to 1950 ft^2 (181 m^2), which is then further increased 585 ft^2 (54 m^2). The final discharge criteria is then 0.2 gpm/ft^2 (8.2 mm/min) over 2535 ft^2 (236 m^2).

Example 2. A wet pipe sprinkler system (light hazard) in a building with a ceiling slope exceeding two in 12 in. (16.6 percent slope) employs quick-response sprinklers qualifying for the 30 percent reduction permitted by 7-2.3.2.4. The initial area must be increased 30 percent for the ceiling slope and the resulting area decreased 30 percent for quick-response sprinklers. It does not matter if the reduction is applied first. If a discharge density of 0.1 gpm/ft^2 (4.1 mm/min) over 1500 ft^2 (139 m^2) is chosen from Figure 7-2.3.1.2, the 1500 ft^2 (139 m^2) is increased 450 ft^2 (42 m^2), resulting in 1950 ft^2 (181 m^2), which is then decreased 585 ft^2 (54 m^2). The final design is 0.1 gpm/ft^2 (4.1 mm/min) over 1365 ft^2 (126.8 m^2).

This requirement is new to the 1999 edition. The committee is not concerned with the order in which simultaneous modifications for dry pipe system quick-response sprinklers, sloped ceilings, or high-temperature sprinklers are applied to the area of system operation, provided that the 3000-ft^2 (270-m^2) minimum area limitation for unsprinklered concealed spaces in 7-2.3.1.3(b) is applied last. Additionally, where simultaneous modifications are made, the adjustments must be compounded based on the initial operating area selected from the area/density curve. Formal Interpretation 76-12 to 7-2.3.2.6 provides further guidance regarding the installation criteria committee's intent on overlapping design area modification.

7-2.3.3 Room Design Method.

7-2.3.3.1* The water supply requirements for sprinklers only shall be based upon the room that creates the greatest demand. The density selected shall be that from Figure 7-2.3.1.2 corresponding to the room size. To utilize this method, all rooms shall be enclosed with walls having a fire-resistance rating equal to the water supply duration indicated in Table 7-2.3.1.1.

A-7-2.3.3.1 This subsection allows for calculation of the sprinklers in the largest room, so long as the calculation produces the greatest hydraulic demand among selection of rooms and communicating spaces. For example, in a case where the largest room has four sprinklers and a smaller room has two sprinklers but communicates through unprotected openings with three other rooms, each having two sprinklers, the smaller room and group of communicating spaces should also be calculated.

Corridors are rooms and should be considered as such.

Walls can terminate at a substantial suspended ceiling and need not be extended to a rated floor slab above for this section to be applied.

This method has also been referred to as the largest room design method and is applicable to rooms of light hazard, ordinary hazard, and extra hazard occupancies. When opting for this approach, the designer must use the room that is also the most hydraulically demanding in terms of water supply and pressure. Enclosing walls of such a room must provide a fire resistance rating equal to the water supply duration requirements of Table 7-2.3.1.1 for the applicable occupancy hazard. Where two values are given for an occupancy, the lower value can be used where remote station or central station waterflow alarm service is provided.

Any communicating openings into such a room are to be protected in the same manner as any opening in a fire-resistive partition, with the exception that any self-closing or automatic doors satisfy the protection requirement for light hazard occupancies.

When use is made of the room design method, the various design area increases and decreases of 7-2.3.2 do not apply.

7-2.3.3.2 If the room is smaller than the smallest area shown in the applicable curve in Figure 7-2.3.1.2, the provisions of 7-2.3.1.3(a) shall apply.

7-2.3.3.3 Minimum protection of openings shall be as follows:

(1) Light hazard—automatic or self-closing doors

Exception: Where openings are not protected, calculations shall include the sprinklers in the room plus two sprinklers in the communicating space nearest each such unprotected opening unless the communicating space has only one sprinkler, in which case calculations shall be extended to the operation of that sprinkler. The selection of the room and communicating space sprinklers to be calculated shall be that which produces the greatest hydraulic demand.

(2) Ordinary and extra hazard—automatic or self-closing doors with appropriate fire-resistance ratings for the enclosure

7-2.3.4 Special Design Methods.

7-2.3.4.1 Where the design area consists of a building service chute supplied by a separate riser, the maximum number of sprinklers that needs to be calculated is three.

7-2.3.4.2 Where the room design method is used and the area under consideration is a corridor protected by one row of sprinklers, the maximum number of sprinklers that needs to be calculated is five. *(See 7-2.3.1.)*

Exception: Where the area under consideration is a corridor protected by a single row of sprinklers and the openings are not protected, the design area shall include all sprinklers in the corridor to a maximum of seven.

Where a corridor is present, the designer would have to calculate the largest room as well as the five sprinklers in the corridor to determine if the largest room or corridor requires the higher water supply. Where a corridor is not provided with self-closing or automatic doors, this exception recognizes that a seven-sprinkler design area is sufficient to control a fire in the corridor.

Due to the anticipated limited fuel load in the corridor, it is not necessary to add additional sprinklers from the communicating rooms to the seven-sprinkler design. It is necessary, however, to add the additional sprinklers required by the exception to 7-2.3.3.3(1) when evaluating the largest room and the communicating space is a corridor.

7-2.3.4.3 Where an area is to be protected by a single line of sprinklers, the design area shall include all sprinklers on the line up to a maximum of seven.

This provision was added to the 1999 edition to address all areas protected by a single line of sprinklers. Where protecting a long, narrow space with one line of sprinklers consisting of more than seven sprinklers, it is unlikely that all sprinklers in the space will activate unless the sprinkler system is overwhelmed.

7-3 Fire Control Approach for the Protection of Commodities That Are Stored Palletized, Solid Piled, in Bin Boxes, or in Shelves

7-3.1 General.

This section shall apply to a broad range of combustibles, including plastics that are stored palletized, solid piled, in bin boxes, or in shelves above 12 ft (3.7 m) high and using standard spray sprinklers.

Through experience and full-scale testing, the fire protection industry has determined that for Class I through Class IV commodities, the protection requirements in Section 7-3 are more a function of the class of commodity stored rather than the manner in which the commodity is stored. For example, the protection of Class III commodities is the same regardless of whether the commodity is stored palletized, in a solid pile, in bin boxes, or in shelves. Protection criteria beyond certain storage heights are not provided in 7-3.2.1.1.

The 12-ft (3.7-m) height limitation is based primarily on experience. For the broad range of Class I through Class IV commodities stored up to 12 ft (3.7 m) in height, the protection specified for light, ordinary, and extra hazard occupancies is adequate as stipulated in Table 7-2.3.2.2. This protection can be provided by a pipe schedule or hydraulically calculated sprinkler system. See 7-2.2 for limitations concerning pipe schedule systems.

As storage heights increase above 12 ft (3.7 m), the heat release rate and, therefore, protection requirements increase exponentially requiring sprinkler systems hydraulically designed to supply the sprinkler densities given in Figures 7-3.2.2.2.1 and 7-3.2.2.2.2.

7-3.1.1 For first-aid fire-fighting and mop-up operations, small hose [1½ in. (38 mm)] shall be provided in accordance with 5-15.5.

Exception: Hose connections shall not be required for the protection of Class I, II, III, and IV commodities stored 12 ft (3.7 m) or less in height.

Hose stations 1½ in. (38 mm) in size are considered indispensable for first-aid fire-fighting and final extinguishment (mop-up) operations in storage and warehouse facilities. The effectiveness of hose stations is dependent on the training of the plant fire brigade or plant emergency organization (PEO) with regard to fire-fighting tactics. Sounding the alarm, notifying the fire department, and evacuating all non-fire-fighting personnel are the initial steps before and during fire-fighting operations. Experience shows that many fires are extinguished in the early stages when trained plant personnel use interior hose stations. Experience also shows that maintenance of the hose and hose stations is critical.

Most warehouse fires, if not extinguished in the early stages with the use of hose stations, are controlled but not entirely extinguished by a properly designed, installed, and maintained sprinkler system. Generally, interior hose stations are needed for final extinguishment (mop-up) during salvage and overhaul operations.

The exception permits the omission of hose stations for certain storage arrangements and recognizes the cost of installing hose stations in any storage facility. Hose stations would still be beneficial, but the sprinkler system in combination with fire department response is considered adequate to guard against the associated fire risk. Whenever possible, hose stations should be considered in all storage facilities regardless of storage heights because their benefits generally outweigh their expense.

7-3.1.2 Minimum System Discharge Requirements.

7-3.1.2.1 The design density shall not be less than 0.15 gpm/ft^2 (6.1 mm/min), and the design area shall not be less than 2000 ft^2 (186 m^2) for wet systems or 2600 ft^2 (242 m^2) for dry systems for any commodity, class, or group.

Numerous factors should be considered when designing sprinkler protection for storage facilities. A limitation is placed on the discharge criteria for storage facilities because a number of these factors are not well understood. A sprinkler design less than the minimum specified is unlikely to control a fire in any combustible storage facility.

7-3.1.2.2 The sprinkler design density for any given area of operation for a Class III or Class IV commodity, calculated in accordance with 7-3.2, shall not be less than the density for the corresponding area of operation for ordinary hazard Group 2.

This requirement was added after miscellaneous storage of Class III and Class IV commodities was determined to need protection in accordance with the discharge criteria for ordinary hazard (Group 2) occupancies. Because miscellaneous storage is considered to present the least hazardous storage arrangement, the sprinkler protection of larger amounts of storage should not be any less than that required for miscellaneous storage.

7-3.1.2.3 The water supply requirements for sprinklers only shall be based on the actual calculated demand for the hazard in accordance with 7-3.2, 7-3.3, 7-9.4, or 7-9.5, depending on the type of sprinkler selected and the commodity being protected.

7-3.1.2.4 In buildings occupied in part for storage, within the scope of this standard, the required sprinkler protection shall extend 15 ft (4.6 m) beyond the perimeter of the storage area.

Exception: This requirement shall not apply where separated by a barrier partition that is capable of preventing heat from a fire in the storage area from fusing sprinklers in the nonstorage area.

In storage fires, sprinklers operate well beyond the actual fire unless a barrier as described by the exception is present. Typically, the sprinkler operating area is 10 times or more the actual fire area. Therefore, for a fire starting near the edge of the storage, sprinklers beyond the storage area will operate. The higher the storage and the more severe the commodity classification, the greater the likelihood that sprinklers further from the fire will operate. Where storage heights exceed 20 ft (6.1 m) for any commodity, especially Class IV and Group A plastics, extension of the designed sprinkler protection beyond 15 ft (4.6 m) should be considered.

7-3.1.3 The sprinkler system criteria specified in Section 7-3 is intended to apply to buildings with ceiling slopes not exceeding two in 12 (16.7 percent).

All full-scale and small-scale tests were conducted in laboratories using flat roofs/ceilings. A ⅙-scale model of a full-scale test facility was used to investigate the effects of varying ceiling slopes. These investigations indicated that sprinkler operating patterns and sequences will not be adversely affected when slopes do not exceed to 2 in. in 12 in. (16.7 percent).

Where the slope exceeds 16.7 percent, the sprinkler operating area and sequence can be skewed. Unless the fire occurs directly under a row of sprinklers, sprinklers nearest the fire may not operate. Heat from the fire will collect at the peak and result in the operation of an excessive number of sprinklers and loss of fire control.

A 30 percent increase in the area of sprinkler operation for light hazard, ordinary hazard, and extra hazard occupancies is required where roof/ceiling slopes exceed 16.7 percent (see 7-2.3.2.5). This requirement has not been tested and verified for warehouse occupancies.

7-3.2 Water Demand Requirements of Class I through IV Requirements.

7-3.2.1* General.

A-7-3.2.1 The following procedure should be followed in determining the proper density and area as specified in 7-3.2:

(1) Determine the commodity class.
(2) Select the density and area of application from Figure 7-3.2.2.2.1 or Figure 7-3.2.2.2.2.
(3) Adjust the required density for storage height in accordance with Figure 7-3.2.2.2.3.
(4) Increase the operating area by 30 percent in accordance with 7-3.2.2.2.4 where a dry pipe system is used.
(5) Satisfy the minimum densities and areas as indicated in 7-3.1.2.1 and 7-3.1.2.2.

Example:

Storage—greeting cards in boxes in cartons on pallets
Height—22 ft (6.7 m)
Clearance—6 ft (1.8 m)
Sprinklers—ordinary temperature
System type—dry

a. Classification—Class III
b. Selection of density/area—0.225 gpm/ft^2 (9.2 mm/min) over 3000 ft^2 (279 m^2) from Figure 7-3.2.2.2.1
c. Adjustment for height of storage using Figure 7-3.2.2.2.3—1.15 $\times$ 0.225 gpm/ft^3 = 0.259 gpm/ft^2 (10.553 mm/min), rounded up to 0.26 gpm/ft^2 (10.6 mm/min)

d. Adjustment of area of operation for dry system—1.3 × 3000 ft² = 3900 ft² (363 m²)
e. Confirmation that minimum densities and areas have been achieved

In 7-3.1.2.1, the minimum design density for a dry sprinkler system is 0.15 gpm/ft² over 2600 ft² (6.1 mm/min over 242 m²) [satisfied in A-7-3.2.1(5)b] for Class III.

Paragraph 7-3.1.2.2 refers to ordinary hazard, Group 2. The corresponding minimum density at 3000 ft² (279 m²) is 0.17 gpm/ft² (6.9 mm/min) (satisfied); 1.3 × 3000 ft² = 3900 ft² (363 m²), 0.17 gpm/ft² (6.9 mm/min) over 3900 ft² (363 m²).

The design density and area of application equals 0.26 gpm/ft² over 3900 ft² (10.6 mm/min over 363 m²).

7-3.2.1.1 Protection for Class I through Class IV commodities in the following configurations shall be provided in accordance with this chapter:

(1) Nonencapsulated commodities that are solid pile, palletized, or bin box storage up to 30 ft (9.1 m) in height

(2) Nonencapsulated commodities on shelf storage up to 15 ft (4.6 m) in height

(3)* Encapsulated commodities that are solid pile, palletized, bin box, or shelf storage up to 15 ft (4.6 m) in height

A-7-3.2.1.1(3) Full-scale tests show no appreciable difference in the number of sprinklers that open for either nonencapsulated or encapsulated products up to 15 ft (4.6 m) high. Test data is not available for encapsulated products stored higher than 15 ft (4.6 m). However, in rack storage tests involving encapsulated storage 20 ft (6.1 m) high, increased protection was needed over that for nonencapsulated storage.

The protection specified in Chapter 6 contemplates a maximum of 10-ft (3-m) clearances from top of storage to sprinkler deflectors for storage heights of 15 ft (4.6 m) and higher.

The only available test data on shelf storage is limited and was conducted at a storage height of 15 ft (4.6 m). In actual practice, shelf storage is rarely higher than 15 ft (4.6 m). Shelf storage exceeding 15 ft (4.6 m) can be protected in accordance with the criteria for rack storage in Section 7-4. If the shelf storage is considered record storage as described in NFPA 232, *Standard for the Protection of Records,* it should be protected in accordance with the guidelines of that standard.

7-3.2.1.2 Bin box and shelf storage that is over 12 ft (3.7 m) but not in excess of the height limits of 7-3.2.1.1 and that is provided with walkways at vertical intervals of not over 12 ft (3.7 m) shall be protected with automatic sprinklers under the walkway(s). Protection shall be as follows:

(a) Ceiling design density shall be based on the total height of storage within the building.

(b) Automatic sprinklers under walkways shall be designed to maintain a minimum discharge pressure of 15 psi (1 bar) for the most hydraulically demanding six sprinklers on each level. Walkway sprinkler demand shall not be required to be added to the ceiling sprinkler demand. Sprinklers under walkways shall not be spaced more than 8 ft (2.4 m) apart horizontally.

With regard to 7-3.2.1.2(a), sprinklers should be designed to protect the worst-case combination of storage height and clearance.

An operating pressure of at least 15 psi (1 bar) is necessary for sprinklers located underneath the walkway so that a sufficient amount of the sprinklers' discharge reaches into the bin boxes and storage shelves. The purpose of the sprinklers underneath the walkways is as follows:

(1) To reduce the likelihood of horizontal firespread under the walkway
(2) To protect any storage in the walkway
(3) To help control the fire in the shelves or bin boxes

Sprinklers under walkways should have water shields in accordance with A-3-2.8 to protect them from sprinkler discharge from above. Listed in-rack sprinklers also meet this requirement. Unless the walkways are solid, sprinklers over the walkway will not operate until flames directly strike them or unless they are located directly over the fire. Covering the entire walkway with plywood or metal eliminates the delayed operation of the sprinklers and is a better alternative for protecting the under-walkway sprinkler from ceiling sprinkler discharge. Warehouse managers often cover walkways to prevent debris or product from falling through grated walkways and causing damage or injuries below. However, not every warehouse does this, so this situation should not be anticipated.

7-3.2.2 Protection Criteria.

7-3.2.2.1 The water supply shall be capable of providing the sprinkler system demand determined in accordance with 7-3.2.2.3, including the hose stream demand and duration requirements of Table 7-2.3.1.1.

7-3.2.2.2 The area and density for the hydraulically remote area shall be determined as specified in 7-2-3.2.2 for storage under 12 ft (3.7 m) and 7-3.2.2.2.1 through 7-3.2.2.2.6 for storage over 12 ft (3.7 m).

7-3.2.2.2.1 Where using ordinary temperature–rated sprinklers, a single point shall be selected from the appropriate commodity curve on Figure 7-3.2.2.2.1.

The curves shown in Figure 7-3.2.2.2.1 were developed from tests conducted at Factory Mutual Research Corporation's test facilities. The curves are terminated at 2000 ft^2 (186 m^2) because control-mode protection with standard sprinklers cannot reliably limit sprinkler operating areas below 2000 ft^2 (186 m^2), even with increased discharge densities.

7-3.2.2.2.2 Where using high temperature–rated sprinklers, a single point shall be selected from the appropriate commodity curve on Figure 7-3.2.2.2.2.

The great majority of the testing on which protection requirements for storage of Class I through Class IV commodities are based was done with 165°F (74°C) sprinklers. At the time of the tests, the belief was that fewer 286°F (141°C) sprinklers than 165°F (74°C) sprinklers would operate in any storage fire. Therefore, the curves in figure 7-3.2.2.2.2 for protection with 286°F (141°C) sprinklers reflect that actual 165°F (74°C) sprinkler test results with no added safety factor. A safety factor was added to the test results to produce the 165°F (74°C) sprinkler curves.

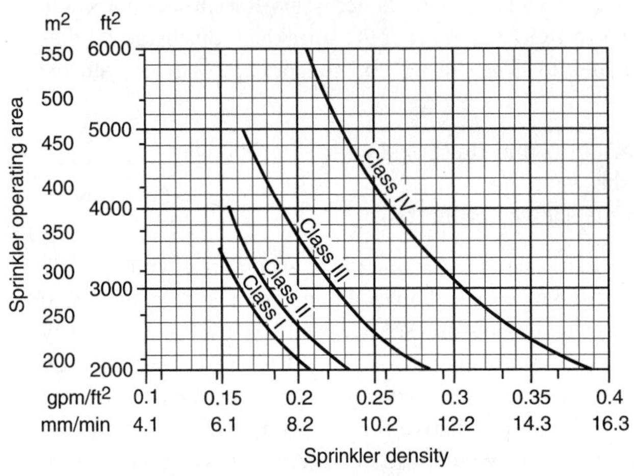

Figure 7-3.2.2.2.1 *Sprinkler system design curves, 20-ft (6.1-m) high storage—ordinary temperature–rated sprinklers.*

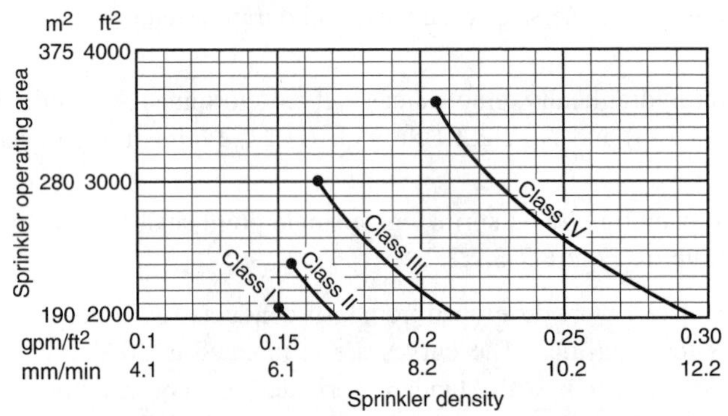

Figure 7-3.2.2.2.2 *Sprinkler system design curves, 20-ft (6.1-m) high storage—high temperature–rated sprinklers.*

Subsequent experience has shown that fewer 286°F (141°C) sprinklers than 165°F (74°C) sprinklers will operate in virtually all control-mode fire scenarios. When sprinkler protection is marginally adequate, significantly fewer 286°F (141°C) sprinklers will operate.

Concern is often expressed that the use of 286°F (141°C) sprinklers causes a delay in sprinkler operation, reducing the effectiveness of sprinklers. This concern is unfounded for two reasons. First, sprinkler sensitivity, as expressed in its response time index, has far more impact on the speed of operation of a sprinkler than its temperature rating. Second, under fire conditions, ceiling temperatures directly above the fire are increasing so rapidly that the difference in time required to operate a 286°F (141°C) sprinkler as compared to a 165°F (74°C) sprinkler is negligible.

Criteria for the protection of plastic commodities and for newer generation sprinklers, such as large drop, K-14 ESFR, K-17, and K-25 ESFR, are based on more recent testing and reflect actual performance at the specified temperature rating.

7-3.2.2.2.3 The densities selected in accordance with 7-3.2.2.2.1 or 7-3.2.2.2.2 shall be modified in accordance with Figure 7-3.2.2.2.3 without revising the design area.

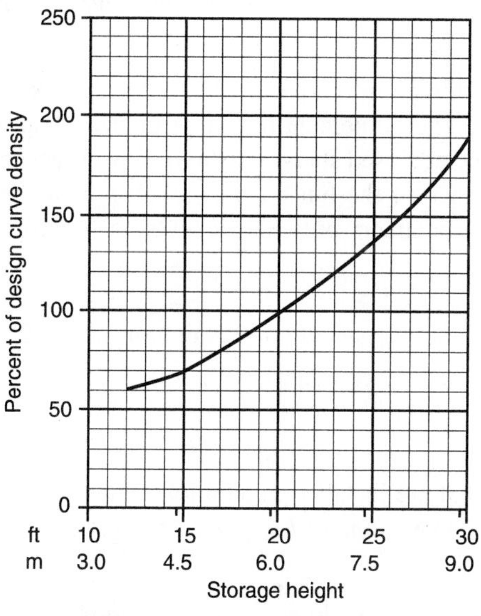

Figure 7-3.2.2.2.3 *Ceiling sprinkler density vs. storage height.*

The majority of testing that led to the development of the discharge criteria in Figures 7-3.2.2.2.1 and 7-3.2.2.2.2 is based on 20-ft (6.1-m) storage heights. Data from the tests in combination with field experience were extrapolated to produce discharge criteria for other storage heights. The values in Figure 7-3.2.2.2.3 have proven to be accurate when compared to actual fire events.

7-3.2.2.2.4 Where dry pipe systems are used, the areas of operation indicated in the design curves shall be increased by 30 percent.

Because no water immediately flows from the actuated sprinklers in a dry pipe system when the dry pipe valve operates, the fire grows unimpeded for a period of time before the water actually arrives at the open sprinklers in a properly designed system. During this water transit time, sprinklers continue to operate increasing the actual operating area. A review of fire loss data by Factory Mutual Research Corporation showed that about 30 percent more sprinklers operate with a dry pipe system than with a wet pipe system. Although the 30 percent penalty for dry pipe systems is not necessarily appropriate because of the many uncontrollable variables that affect the number of sprinkler operations in a dry pipe system, the 30 percent penalty was retained to allow for the likelihood that more sprinklers would operate in a dry pipe system than in a comparable wet pipe system. Also see 7-2.3.2.6.

7-3.2.2.2.5 In the case of metal bin boxes with face areas not exceeding 16 ft^2 (1.5 m^2) and metal closed shelves with face areas not exceeding 16 ft^2 (1.5 m^2), the area of application shall be permitted to be reduced by 50 percent, provided the minimum requirements of 7-3.1 through 7-3.1.3 are met.

The Technical Committee on General Storage developed 7-3.2.2.2.5 from a series of test results that were presented to them. The 50 percent reduction in area of sprinkler system operation for the protection of metal closed shelves and bin boxes is acceptable as long as the exposed storage on the face of the bin boxes or shelves is no greater than 16 ft^2 (1.5 m^2). Storage in metal containers presents as severe a hazard as the same storage in cardboard containers.

7-3.2.2.2.6 The final area and density shall not be less than the minimum specified in 7-3.1.2.

7-3.2.2.3 Given the area and density determined in accordance with 7-3.2.2.2, the fire sprinkler system shall be hydraulically calculated.

7-3.2.2.4 High-Expansion Foam Systems.

7-3.2.2.4.1 High-expansion foam systems that are installed in addition to automatic sprinklers shall be installed in accordance with NFPA 11A, *Standard for Medium- and High-Expansion Foam Systems.*

Exception: This requirement shall not apply where modified by this standard.

7-3.2.2.4.2 High-expansion foam used to protect the idle pallet shall have a maximum fill time of 4 minutes.

7-3.2.2.4.3 High-expansion foam systems shall be automatic in operation.

7-3.2.2.4.4 Detectors for high-expansion foam systems shall be listed and shall be installed at no more than one-half the listed spacing.

Quick detection of a fire in a warehouse protected by high-expansion foam is essential. The foam requires a period of time to cover the burning stock. The faster the foam is discharged onto the fire, the less the amount of damage that will occur. Reducing the detector spacing at the ceiling by one-half of what is typically required helps assure faster response of the foam system.

7-3.2.2.4.5 Detection systems, concentrate pumps, generators, and other system components that are essential to the operation of the system shall have an approved standby power source.

High-expansion foam systems also need electrical power to operate. Thus, a reliable standby power supply is needed.

7-3.2.2.4.6 A reduction in ceiling density to one-half that required for Class I through Class IV commodities, idle pallets, or plastics shall be permitted without revising the design area, but the density shall be no less than 0.15 gpm/ ft^2 (6.1 mm/min).

Ceiling sprinklers are necessary when protecting storage with high-expansion foam. The sprinklers are needed to help maintain the structural integrity of the building columns and roof structure by keeping them relatively cool while the foam covers the fire. Sprinkler discharge breaks down the foam, which needs to be taken into account when designing the system. Very few tests have been conducted using high-expansion foam in conjunction with sprinkler protection. The ceiling density of one-half that required for sprinkler protection is based largely on experience and judgement.

7-3.3 Water Demand Requirements for Plastic and Rubber Commodities.

7-3.3.1* General. See Appendix C.

A-7-3.3.1 The densities and area of application have been developed from fire test data. Most of these tests were conducted with K-8 orifice sprinklers and 80-ft^2 or 100-ft^2 (7.4-ft^2 or 9.3-m^2) sprinkler spacing. These and other tests have indicated that, with densities of 0.4 gpm/ ft^2 (16.3 mm/min) and higher, better results are obtained with K-8 orifice and 70-ft^2 to 100-ft^2 (7.4-m^2 to 9.3-m^2) sprinkler spacing than where using K-5.6 orifice sprinklers at 50-ft^2 (4.6-m^2) spacing. A discharge pressure of 100 psi (6.9 bar) was used as a starting point on one of the fire tests. It was successful, but has a 1½-ft (0.5-m) clearance between the top of storage and ceiling sprinklers. A clearance of 10 ft (3 m) could have produced a different result due to the tendency of the higher pressure to atomize the water and the greater distance that the fine water droplets had to travel to the burning fuel.

Table A-7-3.3.1 explains and provides an example of the method and procedure to follow in using this standard to determine proper protection for Group A plastics.

Table A-7-3.3.1 Metric Conversion Factors for Examples

To Convert from	to	Multiply by
feet (ft)	meters (m)	0.3048
square feet (ft^2)	square meters (m^2)	0.0920
gallons/minute (gpm)	liters/second (L/sec)	0.0631
gallons per minute per square foot (gpm/ft^2)	millimeters per minute (same as liters per minute per square meter) (mm/min)	40.746

Example 1. Storage is expanded, cartoned, stable, 15 ft (4.6 m) high in a 20-ft (6.1-m) building.

Answer 1. Column E—design density is 0.45 gpm/ft^2 (18.3 mm/min).

Example 2. Storage is nonexpanded, unstable, 15 ft (4.6 m) high in a 20-ft (6.1-m) building.

Answer 2. Column A—design density is listed as 0.25 gpm/ft^2 (10.2 mm/min), however, it is also possible that the storage can be 12 ft (3.66 m) in this 20-ft (6.1-m) building, which would require a design density of 0.3 (12.2 mm/min). Unless the owner can guarantee that the storage will always be 15 ft (4.6 m), the design density = 0.3 gpm/ft^2 (12.2 mm/ min).

Example 3. Storage is a nonexpanded, stable 15-ft (4.6-m) fixed-height unit load, one high, in an 18-ft (5.5-m) building.

Answer 3. Column A — design density is 0.25 gpm/ft^2 (10.2 mm/min). Note that this design density does not increase to 0.3 gpm/ft^2 (12.2 mm/min) as in the previous example because of the use of a fixed-height unit load. The storage height will never be 12 ft (3.66 m). It will always be 15 ft (4.6 m).

Example 4. Storage is expanded, exposed, unstable, 20 ft (6.1 m) high in a 27-ft (8.2-m) building.

Answer 4. Column C—design density is 0.7 gpm/ft^2 (28.5 mm/min). Note that other lower storage heights should also be checked, but they reveal the same, or lower, densities (0.7 gpm/ft^2 and 0.6 gpm/ft^2) (28.5 mm/min and 24.5 mm/min), so the design density remains at 0.7 gpm/ft^2 (28.5 mm/min).

Example 5. Storage is expanded, cartoned, unstable, 17 ft (5.2 m) high in 32-ft (9.75-m) building.

Answer 5. Column D—15-ft (4.6 m) storage in a 32-ft (9.75-m) building would be 0.55 gpm/ft^2 (22.4 mm/min); 20-ft (6.1-m) storage in a 32-ft (9.75-m) building would be 0.7 gpm/ft^2 (28.5 mm/min). Interpolation for 17-ft (5.2-m) storage is as follows:

$$0.7 - 0.55 = 0.15$$

$$\frac{0.15}{(20 - 15)} = 0.03$$

$$0.03 \times (17 - 15) = 0.06$$

$$0.55 + 0.06 = 0.61$$

Design density = 0.61 gpm/ft^2 (24.9 mm/ min)

Example 6. Storage is expanded, exposed, stable, 22 ft (6.71 m) high in a 23½-ft (7.16-m) building.

Answer 6. Column B—could interpolate between 0.6 gpm/ft^2 and 0.75 gpm/ft^2 (24.5 mm/min and 30.6 mm/min); however, this would be a moot point since the density for 15-ft (4.6-m) storage in this 23½-ft (7.16-m) building would be 0.8 gpm/ft^2 (32.6 mm/min). Unless the owner can guarantee 22-ft (6.71-m) storage, the design density is 0.8 gpm/ft^2 (32.6 mm/ min). If the owner can, in a manner acceptable to the authority having jurisdiction, guarantee 22-ft (6.71-m) storage, the interpolation would yield a design density of 0.66 gpm/ft^2 (26.9 mm/ min).

Example 7. Storage is nonexpanded, stable, exposed, 13½ ft (4.1 m) high in a 15-ft (4.6-m) building.

Answer 7. Column E—12-ft (3.66-m) storage in a 15-ft (4.6-m) building would be extra hazard, Group 2 (0.4 gpm/ft^2 over 2500 ft^2) (16.3 mm/min over 230 m^2).

Storage 15 ft (4.6 m) high in a 15-ft (4.6-m) building would be 0.45 gpm/ft^2 (18.3 mm/min). Interpolation for 13½-ft (4.1-m) storage is as follows:

$$0.45 - 0.4 = 0.05$$

$$\frac{0.05}{(15 - 12)} = 0.017$$

$$0.017 \times (13.5 - 12) = 0.026$$

$$0.4 + 0.026 = 0.426$$

Design density $= 0.426$ gpm/ft^2 (17.4 mm/ min)

7-3.3.1.1* Plastics stored up to 25 ft (7.62 m) in height protected by spray sprinklers shall be in accordance with this chapter. The decision tree shown in Figure 7-3.3.1.1 shall be used to determine the protection in each specific situation.

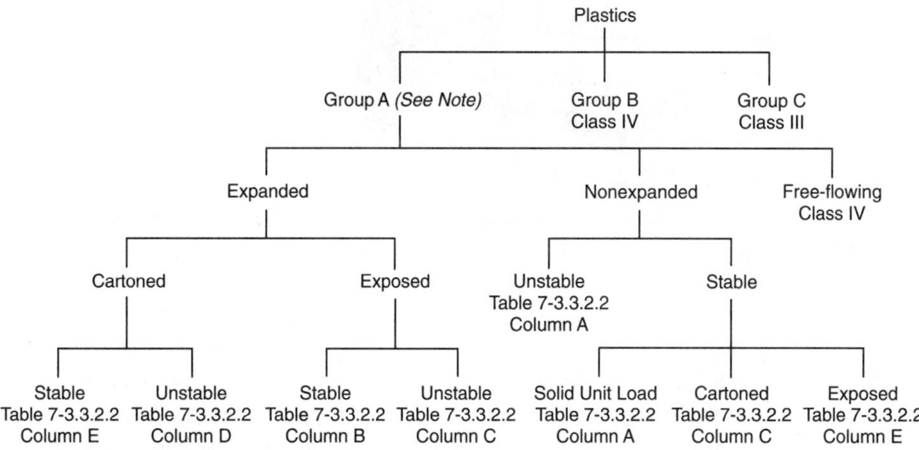

Note: Cartons that contain Group A plastic material shall be permitted to be treated as Class IV commodiites under the following conditions:

(a) There shall be multiple layers of corrugation or equivalent outer material that would significantly delay fire involvement of the Group A plastic

(b) The amount and arrangement of Group A plastic material within an ordinary carton would not be expected to significantly increase the fire hazard

Figure 7-3.3.1.1 *Decision tree.*

A-7-3.3.1.1 Two direct comparisons between ordinary temperature– and high temperature–rated sprinklers are possible, as follows:

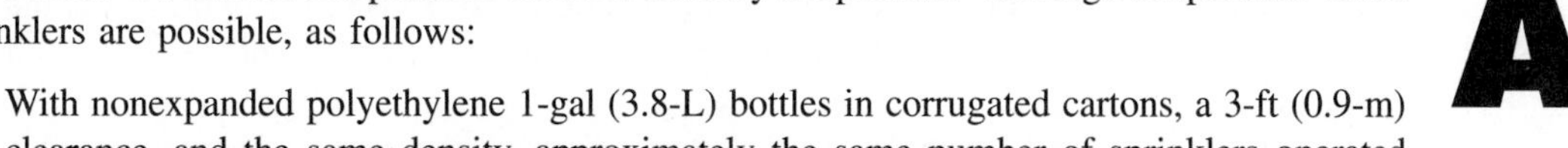

(1) With nonexpanded polyethylene 1-gal (3.8-L) bottles in corrugated cartons, a 3-ft (0.9-m) clearance, and the same density, approximately the same number of sprinklers operated (nine at high temperature versus seven at ordinary temperature).

(2) With exposed, expanded polystyrene meat trays, a 9.5-ft (1.9-m) clearance, and the same density, three times as many ordinary temperature–rated sprinklers operated as did high temperature–rated sprinklers (11 at high temperature versus 33 at ordinary temperature).

The cartoned plastics requirements of this standard are based to a great extent on test work that used a specific commodity—16-oz (0.473-L) polystyrene plastic jars individually separated by thin carton stock within a large corrugated carton [3½ ft^2 (0.32 m^2)]. *[See Figure A-7-3.3.1.1(a).]*

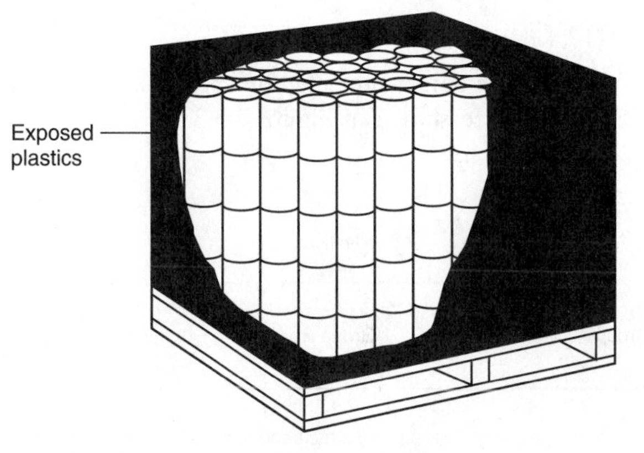

Figure A-7-3.3.1.1(a) *Corrugated carton containing individually separated plastic jars.*

Other Group A plastic commodities can be arranged in cartons so that they are separated by multiple thicknesses of carton material. In such arrangements, less plastic becomes involved in the fire at any one time. This could result in a less vigorous fire that can be controlled by Class IV commodity protection.

Other situations exist in which the plastics component is surrounded by several layers of less hazardous material and is therefore temporarily protected or insulated from a fire involving adjacent plastic products. Such conditions also could produce a less vigorous fire and be successfully handled by Class IV protection. *[See Figure A-7-3.3.1.1(b).]*

The decision to protect as a Class IV commodity, however, should be made only based on experienced judgment and only with an understanding of the consequences of underprotecting the storage segment.

The height limitation of 25 ft (7.6 m) is based on the fact that 25 ft (7.6 m) was the maximum height tested using sprinklers with nominal K-factors of 5-7, 8-0, and 11-2.

Figure 7-3.3.1.1 illustrates the differences in protection criteria for various types of plastic commodities (Groups A, B, and C). These three categories were developed using full-scale tests (as large as 30 ft × 30 ft × 25 ft high), small-scale sprinklered

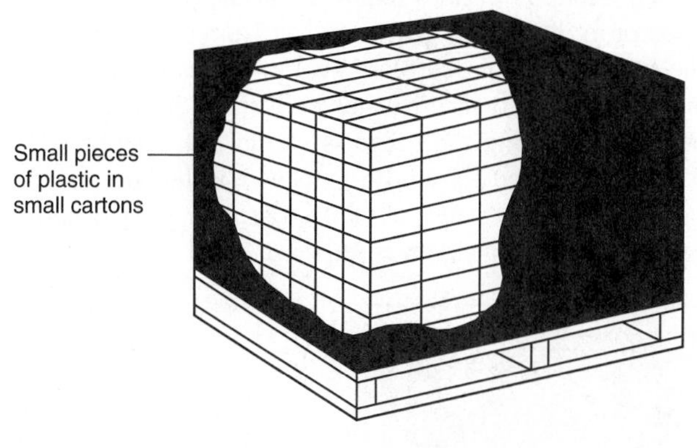

Figure A-7-3.3.1.1(b) *Corrugated carton containing plastic pieces individually separated by carton material.*

tests (usually 8 ft × 8 ft × 10 ft high), and "bench" tests to evaluate heat content and burning rate. After all the comparison tests were conducted, the three groups of plastic commodities were established.

The possible fire severity within any one of the broad categories of plastic is different. For example, polystyrene will normally burn much more severely (faster) than even the most highly plasticized polyvinyl chloride (PVC). However, the protection suggested by the decision tree should control a fire in the polystyrene product and do even a better job of controlling a fire in the highly plasticized PVC product.

The next level in the decision tree requires that Group A plastics be identified as either expanded, nonexpanded, or free-flowing. Expanded plastics have a cellular structure containing pockets of a blowing agent that is usually air but that could also be a flammable gas. Due the surface-to-air ratio inherent with expanded plastics, when they burn they do so much more rapidly than nonexpanded plastics. Virtually all expanded plastics are Group A plastics and, therefore, most of the testing of expanded plastics has been conducted on Group A plastics. However, it should be anticipated that Groups B and C expanded plastics would burn much more rapidly than nonexpanded Groups B and C plastics.

The sprinkler protection for free-flowing plastics is based on commodity tests with resins in large corrugated cartons. The appendix describes randomly packed, small objects as another example of a free-flowing product, but they cannot be expected to flow freely if they melt at a low temperature. Also, randomly packed, small objects can catch on each other and not flow freely from the carton. If the product does not flow freely very early in the fire—that is, before or very soon after sprinklers operate—the fire's intensity will not be reduced. A rule of thumb is that a product can be expected to flow freely from a box in a fire if it will flow through the hole created by cutting off a 2-in. × 2-in. × 2-in. (50-mm × 50-mm × 50-mm) corner from the container's bottom.

The next level under expanded Group A plastics requires that the commodity be identified as being stored exposed or in cartons. In general, water from sprinklers and hose streams does not affect the burning of exposed plastic commodities to the same degree as if it the plastic were in cartons. Cartons absorb water, prewetting the commodities ahead of the fire and slowing or stopping the progress of the fire. Exhibit 7.2 shows a cartoned nonexpanded plastic. Exposed plastics do not absorb water but instead allow the sprinkler discharge to flow over the plastics, virtually eliminating any benefit of prewetting. As a result, greater sprinkler protection is required.

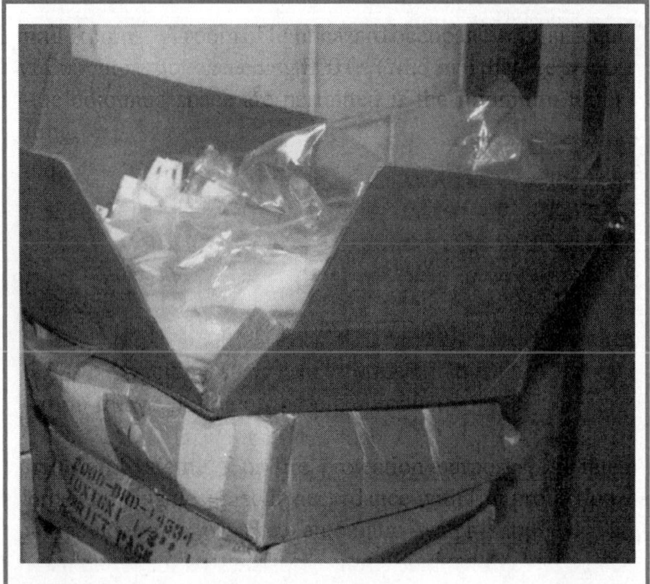

Exhibit 7.2 *Cartoned nonexpanded plastic.*

Pile stability is also a key factor in determining the proper protection of plastic commodities. The differences in the burning characteristics of unstable piles versus stable piles is described in A-7-3.3.1.1. Fundamentally, a pile can be considered unstable only if it can be expected to collapse within the first few minutes of a fire. In actual practice, a pile is rarely unstable. Predicting when, if ever, a pile will collapse during a fire is virtually impossible. The concept of unstable piles was added because in some, but not all, tests of expanded polystyrene meat trays in polyethylene bags on pallets in which the ignition source was close to the outside of the pile, some stacks collapsed and the fire was more easily controlled. However, other tests with the same commodity produced different results. Because of a different ignition location, the tendency of the stacks to lean in a different way, or variable air movement in the building, the fire spread into the pile, no collapse occurred, and a more severe fire challenge resulted. Variables such as these are generally beyond the control of the property owner or occupant, and a sprinkler protection scheme designed to protect against unstable fires would in all likelihood be ineffective against a fire in stable piles.

7-3.3.1.2* Factors affecting protection requirements such as closed/open array, clearance between storage and sprinklers, and stable/unstable piles shall be applicable only to storage of Group A plastics. This decision tree also shall be used to determine protection for commodities that are not wholly Group A plastics but contain such quantities and arrangements of the same that they are deemed more hazardous than Class IV commodities.

A-7-3.3.1.2 There are few storage facilities in which the commodity mix or storage arrangement remains constant, and a designer should be aware that the introduction of different materials can change protection requirements considerably. Design should be based on higher densities and areas of application, and the various reductions allowed should be applied cautiously. For evaluation of existing situations, however, the allowances can be quite helpful.

In a closed array, the stacks of material are butted together in one direction, and there are 6-in. (152-mm) flues in the other direction. Fire tests of closed arrays burned less severely on the interior of the piles. The limited flues restricted the airflow into the flues, significantly reducing the burning rate. In an open array, the stacks of material are butted together in one direction, and there are 12-in. (304-mm) flues in the other direction. By contrast, fire tests had more than enough airflow to sustain a high burning rate. The larger flues of the open stacks allowed adequate oxygen to the fire, and the plastic burned unimpeded. No fire tests were conducted with the flue space greater than 6 in. (152 mm) but less than 12 in. (304 mm). Considering any flue space greater than 6 in. (152 mm), an open array is, therefore, a conservative approach.

The size of the flue space is hard to police, and it is rare that a true closed array would always be present. Closed array protection should only be used when there is complete confidence in warehouse management and employees to maintain the closed condition at all times. If the protection is designed for a closed array, but open array conditions exist at the time of a fire, the sprinkler system would not be adequate and a major loss could occur. The closed array protection cannot control a fire in an open array.

The clearance between the storage and the roof/ceiling is an important factor as indicated in Table 7-3.3.2.2. High clearance affects sprinkler performance in two ways. First, as the clearance increases, the size of the fire before sprinklers operate also increases. Second, as the clearance increases, the fire plume that the sprinkler discharge must penetrate to reach the burning materials also increases. These two factors together significantly reduce the effectiveness of sprinklers. Clearances greater than 20 ft (6.1 m) are beyond the scope of testing used to develop the protection requirements in this standard, except for ESFR sprinklers.

Implicit in storage protection requirements is that protection for a given storage height in a building of a given height must also be adequate to protect any lesser storage height in that building. This requirement becomes necessary because warehouses are never completely full, and storage heights can fluctuate widely over both the short and long term. Therefore, protection adequate for a storage height that results in a clearance of less than 20 ft (6.1 m) will be adequate for lesser storage heights, despite the fact that clearance exceeds 20 ft (6.1 m).

Large drop and ESFR sprinklers have their own discharge criteria as indicated in 7-9.4 and 7-9.5 and should not be used with Table 7-3.3.2.2.

7-3.3.1.3 Group B plastics and free-flowing Group A plastics shall be protected in the same manner as a Class IV commodity. See 7-3.2 for protection of these storage commodities with spray sprinklers.

Small- and bench-scale testing of Group B plastics and large-scale testing of free-flowing Group A plastics show that Class IV commodity protection is adequate for this storage.

7-3.3.1.4 Group C plastics shall be protected in the same manner as a Class III commodity. See 7-3.2 for protection of these storage commodities with spray sprinklers.

Bench-scale tests with Group C plastics and Class III commodities produced results similar to those identified in the commentary to 7-3.3.1.3. The results indicate that Class III protection is adequate.

7-3.3.2 Protection Criteria.

7-3.3.2.1* The design of the sprinkler system shall be based on those conditions that routinely or periodically exist in a building that create the greatest water demand. These conditions include the following:

(1) Pile height
(2) Clearance
(3) Pile stability
(4) Array

Where the distance between roof/ceiling height and top of storage exceeds 20 ft (6.1 m), protection shall be provided for the storage height that would result in a 20-ft (6.1-m) distance between the roof/ceiling height and top of storage.

A-7-3.3.2.1 An evaluation for each field situation should be made to determine the worst applicable height-clearance relationship that can be expected to appear in a particular case. Fire tests have shown that considerably greater demands occur where clearance is 10 ft (3.1 m) as compared to 3 ft (0.9 m) and where a pile is stable as compared to an unstable pile. Since a system is designed for a particular clearance, the system could be inadequate when significant areas do not have piling to the design height and larger clearances exist between stock and sprinklers. This can also be true where the packaging or arrangement is changed so that stable piling is created where unstable piling existed. Recognition of these conditions is essential to avoid installation of protection that is inadequate or becomes inadequate because of changes.

No tests were conducted simulating a peaked roof configuration. However, it is expected that the principles of Section 7-3 still apply. The worst applicable height-clearance relationship that can be expected to occur should be found, and protection should be designed for it. If storage is all at the same height, the worst height-clearance relationship creating the greatest water demand would occur under the peak. If commodities are stored higher under the peak, the various height-clearance relationships should be tried and the one creating the greatest water demand used for designing protection.

7-3.3.2.2* Design areas and densities for the appropriate storage configuration shall be selected from Table 7-3.3.2.2. The columns A, B, C, D, and E correspond to the protection required by the decision tree shown in Figure 7-3.3.1.1.

Table 7-3.3.2.2 Design Densities for Plastic and Rubber Commodities

Storage Height		Roof/Ceiling Height		Density									
				A		B		C		D		E	
ft	m	ft	m	gpm/ft²	mm/min	gpm/ft²	mm/min	gpm/ft²	mm/min	gpm/ft²	mm/min	gpm/ft²	mm/min
≤5	1.52	up to 25	up to 7.62	OH-2	OH-2	0.2	8.2	OH-2	OH-2	OH-2	OH-2	OH-2	OH-2
≤12	3.66	up to 15	up to 4.57	0.2	8.2	EH-2	EH-2	0.3	12.2	EH-1	EH-1	EH-2	EH-2
		>15 to 20	>4.57 to 6.1	0.3	12.2	0.6	24.5	0.5	20.4	EH-2	EH-2	EH-2	EH-2
		>20 to 32	>6.1 to 9.75	0.4	16.3	0.8	32.6	0.6	24.5	0.45	18.3	0.7	28.5
15	4.5	up to 20	up to 6.1	0.25	10.2	0.5	20.4	0.4	16.3	0.3	12.2	0.45	18.3
		>20 to 25	>6.1 to 7.62	0.4	16.3	0.8	32.6	0.6	24.5	0.45	18.3	0.7	28.5
		>25 to 35	>7.62 to 10.67	0.45	18.3	0.9	36.7	0.7	28.5	0.55	22.4	0.85	34.6
20	6.1	up to 25	up to 7.62	0.3	12.2	0.6	24.5	0.45	18.3	0.35	14.3	0.55	22.4
		>25 to 30	>7.62 to 9.14	0.45	18.3	0.9	36.7	0.7	28.5	0.55	22.4	0.85	34.6
		>30 to 35	>9.14 to 10.67	0.6	24.5	1.2	48.9	0.85	34.6	0.7	28.5	1.1	44.8
25	7.62	up to 30	up to 9.14	0.4	16.3	0.75	30.6	0.55	22.4	0.45	18.3	0.7	28.5
		>30 to 35	>9.14 to 10.67	0.6	24.5	1.2	48.9	0.85	34.6	0.7	28.5	1.1	44.8

Notes:

1. Minimum clearance between sprinkler deflector and top of storage shall be maintained as required.
2. Column designations correspond to the configuration of plastics storage as follows:
   A: (1) Nonexpanded, unstable
   (2) Nonexpanded, stable, solid unit load
   B: Expanded, exposed, stable
   C: (1) Expanded, exposed, unstable
   (2) Nonexpanded, stable, cartoned
   D: Expanded, cartoned, unstable
   E: (1) Expanded, cartoned, stable
   (2) Nonexpanded, stable, exposed
3. OH-2 = Desity required for ordinary hazard Group 2 occupancies
   EH-1 = Density required for extra hazard Group 1 occupancies
   EH-2 = Density required for extra hazard Group 2 occupancies
4. Hose streams shall be provided in accordance with Table 7-2.3.1.1.

A-7-3.3.2.2 Test data is not available for all combinations of commodities, storage heights, and clearances. Some of the protection criteria in this standard are based on extrapolations of test data for other commodities and storage configurations, as well as available loss data.

For example, there is very limited test data for storage of expanded plastics higher than 20 ft (6.1 m). The protection criteria in this standard for expanded plastics higher than 20 ft (6.1 m) are extrapolated from test data for expanded plastics storage 20 ft (6.1 m) and less in height and test data for unexpanded plastics above 20 ft (6.1 m).

Further examples can be found in the protection criteria for clearances up to 15 ft (4.6 m). Test data is limited for clearances greater than 10 ft (3.1 m). It should be assumed that, if protection is adequate for a given storage height in a building of a given height, the same protection will protect storage of any lesser height in the same building. For example, protection adequate for 20-ft (6.1-m) storage in a 30-ft (9.1-m) building [10-ft (3.1-m) clearance] would also protect 15-ft (4.6-m) storage in a 30-ft (9.1-m) building [15-ft (4.6-m) clearance]. Therefore, the protection criteria in Table 7-3.3.2.2 for 15-ft (4.6-m) clearance are based on the protection criteria for storage 5 ft (1.5 m) higher than the indicated height with 10-ft (3.1-m) clearance.

Table 7-3.3.2.2 is based on tests that were conducted primarily with high temperature–rated, K-8 orifice sprinklers. Other tests have demonstrated that, where sprinklers are used with orifices greater than K-8, ordinary-temperature sprinklers are acceptable.

The use of sprinklers with a nominal K-factor of 11.2 to produce the densities in Table 7-3.3.2.2 will normally allow larger droplets, greater initial densities at lower pressures, and more efficient use of water affecting the fire, and the impact of large clearances is lessened.

7-3.3.2.2.1 For Table 7-3.3.2.2, the design areas are a minimum of 2500 ft^2 (232 m^2).

Exception No. 1: Where Table 7-3.3.2.2 allows densities and areas to be selected in accordance with 7-2.3, for ordinary hazard Group 2 occupancies, any area/density from that curve shall be acceptable.

Exception No. 2: For closed arrays, the area shall be permitted to be reduced to 2000 ft^2 (186 m^2).

7-3.3.2.2.2 Interpolation of densities between storage heights shall be permitted. Densities shall be based upon the 2500 ft^2 (232 m^2) design area. The "up to" in the table is intended to aid in the interpolation of densities between storage heights. Interpolation of ceiling/roof heights shall not be permitted.

7-3.3.2.3* Where dry pipe systems are used for Group A plastics, the operating area shall be increased by 30 percent without revising the density.

A-7-3.3.2.3 Wet systems are recommended for storage occupancies. Dry pipe systems should be permitted only where it is impractical to provide heat.

7-3.3.2.4 High-Expansion Foam Systems. Where high-expansion foam is used, a reduction in ceiling density to one-half that required for plastics shall be permitted without revising the design area but shall be not less than 0.15 gpm/ft^2 (6.1 mm/min).

See 7-3.2.2.4.

7-4† Fire Control Approach for the Protection of Commodities Stored on Racks

7-4.1 Protection Criteria—General.

7-4.1.1 This section shall apply to storage of materials representing the broad range of combustibles stored in racks using standard spray sprinklers.

7-4.1.2 The sprinkler system criteria of Section 7-4 is intended to apply to buildings with ceiling slopes not exceeding two in 12 (16.7 percent).

All of the testing used to develop storage protection requirements was done under flat roofs. Research done at Factory Mutual using modeling techniques indicates that roof slopes greater than 2 in. in 12 in. (16.7 percent) can result in skewed operating patterns and unpredictable results.

A ⅙-scale model of Factory Mutual's West Gloucester, RI, Test Center was used to conduct tests with varying ceiling slopes. The results indicated slopes of up to 16.7 percent did not appreciably change the results from a flat roof. No full-scale testing was conducted to verify the model testing.

7-4.1.3* Sprinkler protection criteria for the storage of materials on racks shall be in accordance with 7-4.1 and either 7-4.2 for storage up to 25 ft (7.6 m), 7-4.3 for storage over 25 ft (7.6 m), or 7-4.4 for plastics storage as appropriate.

Exception: Protection criteria, for Group A plastics as indicated in 7-4.4 shall be permitted for the protection of the same storage height and configuration of Class I, II, III, and IV commodities.*

A-7-4.1.3 The fire protection system design should consider the maximum storage height. For new sprinkler installations, maximum storage height is the usable height at which commodities can be stored above the floor while the minimum required unobstructed space below sprinklers is maintained. Where evaluating existing situations, maximum storage height is the maximum existing storage height if space between the sprinklers and storage is equal to or greater than that required.

A-7-4.1.3 Exception. Information for the protection of Classes I, II, III, and IV commodities was extrapolated from full-scale fire tests that were performed at different times than the tests that were used to develop the protection for plastic commodities. It is possible that, by selecting certain points from the curves shown in Figures 7-4.2.2.1.1(a) through (g) (and after applying the appropriate modifications), the protection specified by 7-4.2 exceeds the requirements of 7-4.4. In such situations, the protection specified for plastics, although less than that required by the curves shown in Figures 7-4.2.2.1.1(a) through (g), can adequately protect Classes I, II, III, and IV commodities.

This section also allows storage areas that are designed to protect plastics to store Classes I, II, III, and IV commodities without a reevaluation of fire protection systems.

Protection that is adequate for a higher commodity classification that presents a greater fire hazard, such as Class IV commodities, is also adequate for a lower commodity classification, such as Class I. Future changes regarding the classes of commodities to be stored at the premises should be anticipated to the extent possible, and protection for the highest anticipated commodity class likely to be stored should be provided. Designing for the most hazardous commodity anticipated when the sprinkler system is initially installed is often more cost effective in the long run.

7-4.1.3.1† Sprinkler protection criteria is based on the assumption that roof vents and draft curtains are not being used.

7-4.1.4 Flue Space.

Flue spaces exert two primary influences on fires in racks. Flues allow a fire to spread vertically through a rack, thereby providing a means for heat to rise up through the racks and activate the sprinklers. Flues also provide the path through which water from the sprinklers reaches a fire. Where they exist, flue spaces are to extend vertically through the entire height of the rack.

Where no flue spaces exist, fire development cannot occur up through the racks. The fire can only develop and spread horizontally down the length of the rack and up along the rack face in the aisles. This type of fire development severely delays the activation of sprinklers, and when sprinklers do activate, their discharge cannot penetrate the rack and reach the fire even though a greater number of sprinklers ultimately operate. A similar phenomenon also occurs where flue spaces exist but are obstructed, limited in number or size, or do not extend vertically through the entire height of the rack. The final outcome can often result in a very significant loss or complete destruction of the storage facility.

Clear transverse flue spaces between pallet loads that line up vertically for the full height of the rack are required for all single-, double-, and multiple-row racks regardless of overall storage heights. Longitudinal flues are only required for double-row racks where storage heights exceed 25 ft (7.6 m). However, where longitudinal flues are not provided, it is assumed that the pallet loads abut each other as indicated in Exhibit 7.3. Longitudinal flues are not required for multirow racks because it is assumed that the pallet loads abut each other in one direction. If pallet loads do not abut, an unobstructed 6-in. longitudinal flue space needs to be maintained through the rack. Where in-rack sprinklers are used, they are to be installed in the transverse flues or in the intersection of the transverse and longitudinal flues where longitudinal flues are used.

Rack storage tests of 20-ft (6.1-m) high double-row racks indicated that where pallet loads are pushed together in the back-to-back space, thus eliminating the longitudinal flues as indicated in Exhibit 7.3, virtually no difference was observed in results from those tests that had clear longitudinal flues.

Where a rack storage configuration is determined to be a multiple-row rack arrangement because the aisle widths are less than 3 ft (0.9 m), the aisles are not to be considered flue spaces. Flue spaces must have a nominal width of 6 in. (152 mm). Spaces larger than the nominal 6 in. (152 mm) rising up through the rack are not to be considered flues. Exhibit 7.4 illustrates the flue spaces required for a multiple-row rack arrangement consisting of aisles of less than 3 ft (.9 m).

In-rack sprinklers are to be installed in the transverse flues or in the intersection of the transverse and longitudinal flues where longitudinal flues are used in a multiple-row storage rack.

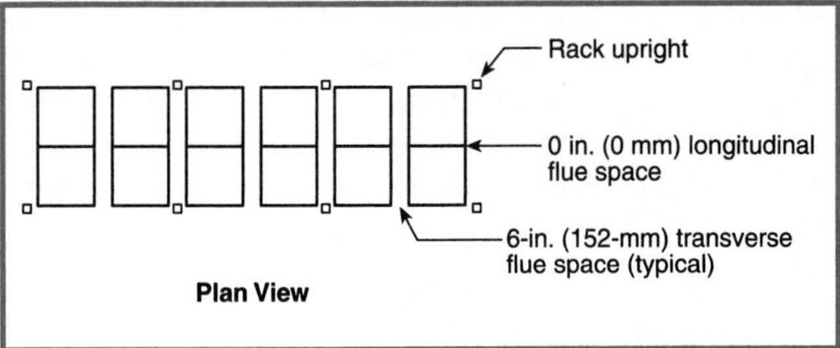

Exhibit 7.3 *Plan view of double-row rack without a longitudinal flue space.*

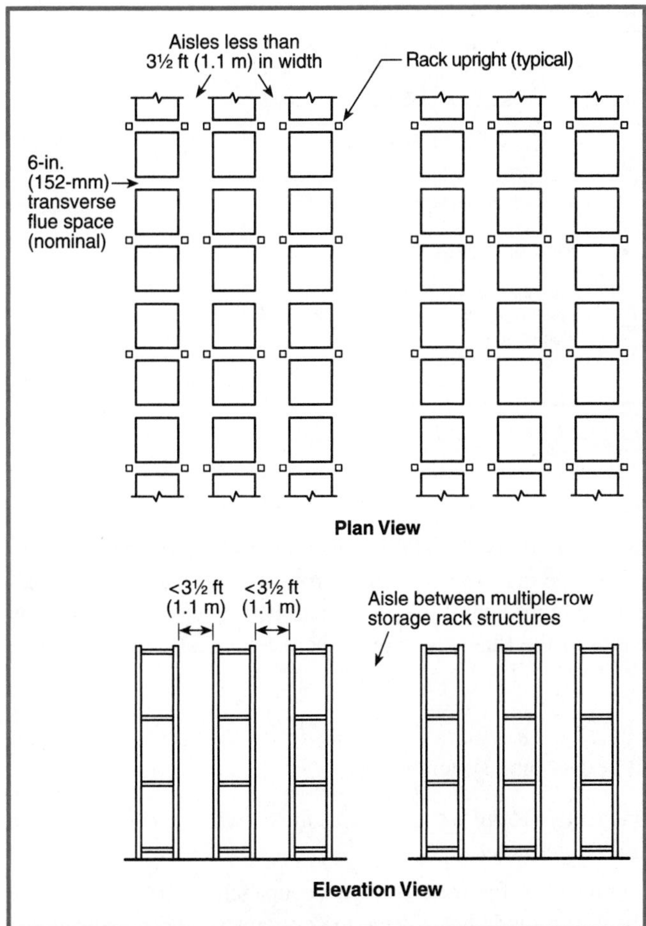

Exhibit 7.4 *Multiple-row rack storage arrangement consisting of aisles of less than 3 ft (0.9 m).*

7-4.1.4.1† Storage Up to and Including 25 ft (7.6 m). In double-row and multiple-row racks without solid shelves, a longitudinal (back-to-back clearance between loads) flue space shall not be required. Nominal 6-in. (152.4-mm) transverse flue spaces between loads and at-rack uprights shall be maintained in single-row, double-row, and multiple-row racks. Random variations in the width of flue spaces or in their vertical alignment shall be permitted. *(See Figure 7-4.1.4.1.)*

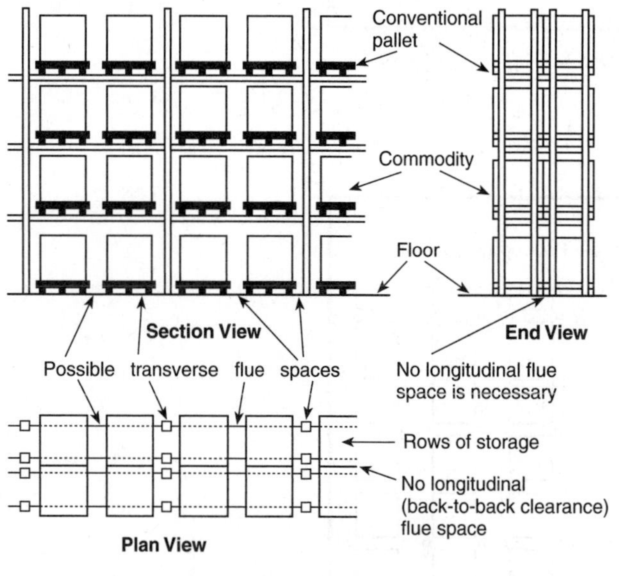

Figure 7-4.1.4.1 *Typical double-row rack with back-to-back loads.*

7-4.1.4.2 Storage Height Over 25 ft (7.6 m). Nominal 6-in. (152.4-mm) transverse flue spaces between loads and at-rack uprights shall be maintained in single-row, double-row, and multiple-row racks. Nominal 6-in. (152.4-mm) longitudinal flue spaces shall be provided in double-row racks. Random variations in the width of the flue spaces or in their vertical alignment shall be permitted.

7-4.1.5 Protection Systems.

7-4.1.5.1* Sprinkler systems shall be wet pipe systems.

Exception: In areas that are subject to freezing or where special conditions exist, dry-pipe systems and preaction systems shall be permitted.

A-7-4.1.5.1 Wet systems are recommended for rack storage occupancies. Dry systems are permitted only where it is impractical to provide heat. Preaction systems should be considered for rack storage occupancies that are unheated, particularly where in-rack sprinklers are installed or for those occupancies that are highly susceptible to water damage.

7-4.1.5.2 Where dry pipe systems are used, the ceiling sprinkler areas of operation shall be increased 30 percent over the areas specified by 7-4.2, 7-4.3, and 7-4.4. Densities and areas shall be selected so that the final area of operation after the 30 percent increase is not greater than 6000 ft^2 (557.4 m^2).

See 7-2.3.2.6 for a discussion of dry pipe systems.

7-4.1.5.3 Where preaction systems are used, preaction systems shall be treated as dry pipe systems.

Exception: This requirement shall not apply where it can be demonstrated that the detection system that activates the preaction system causes water to be discharged from sprinklers as quickly as the discharge from a wet pipe system.

Regardless of the detection system, gridded preaction systems, unless extremely small, cannot deliver water to sprinklers as quickly as a wet system. An amount of air greater than that expected in a wet pipe system will always be trapped in the piping and delay the uniform distribution of water from the activated sprinklers.

7-4.1.5.4 Detectors for preaction systems shall be installed in accordance with 7-4.1.12.3.

7-4.1.5.5* In buildings that are used in part for rack storage of commodities, the design of the ceiling sprinkler system that is within 15 ft (4.6 m) of the racks shall be the same as that provided for the rack storage area.

Exception: Where separated by a barrier partition that is capable of preventing heat from a fire in the rack storage area from fusing sprinklers in the nonrack storage area.

See commentary to 7-3.1.2.4.

A-7-4.1.5.5 Where high temperature–rated sprinklers are installed at the ceiling, high temperature–rated sprinklers also should extend beyond storage in accordance with Table A-7-4.1.5.5.

Table A-7-4.1.5.5 Extension of Installation of High-Temperature Sprinklers over Storage

Design Area for High Temperature–Rated Sprinklers		Distance Beyond Perimeter of High-Hazard Occupancy for High Temperature–Rated Sprinklers	
ft^2	m^2	ft	m
2000	185.8	30	9.14
3000	278.7	40	12.2
4000	371.6	45	13.72
5000	464.5	50	15.24
6000	557.4	55	16.76

7-4.1.6 Hose Connections. For first-aid fire-fighting and mop-up operations, small hose [1½ in. (38 mm)] shall be provided in accordance with 5-15.5.

See commentary to 7-3.1.1.

Exception: Hose connections shall not be required for the protection of Class I, II, III, and IV commodities stored 12 ft (3.7 m) or less in height.

7-4.1.7 Solid and Slatted Shelves.

7-4.1.7.1*† Slatted shelves shall be considered equivalent to solid shelves.

Exception: A wet pipe system that is designed to provide a minimum of 0.6 gpm/ft² (24.5 mm/min) density over a minimum area of 2000 ft² (186 m²) or K-14 ESFR sprinklers operating at a minimum of 50 psi (3.5 bar) shall be permitted to protect single-row and double-row slatted-shelf racks where all of the following conditions are met:

(a) Sprinklers shall be K-11.2 orifice spray sprinklers with a temperature rating of ordinary, intermediate, or high and shall be listed for storage occupancies or shall be K-14 ESFR.

(b) The protected commodities shall be limited to Class I–IV, Group B plastics, Group C plastics, cartoned (expanded and unexpanded) Group A plastics, and exposed (unexpanded) Group A plastics.

(c) Shelves shall be slatted using a minimum nominal 2-in. (51-mm) thick by maximum nominal 6-in. (152-mm) wide slat held in place by spacers that maintain a minimum 2-in. (51-mm) opening between each slat.

(d) Where K-11.2 orifice sprinklers are used, there shall be no slatted shelf levels in the rack above 12 ft (3.7 m). Wire mesh (greater than 50 percent opening) shall be permitted for shelf levels above 12 ft (3.7 m).

(e) Transverse flue spaces at least 3 in. (76 mm) wide shall be provided at least every 10 ft (3.1 m) horizontally.

(f) Longitudinal flue spaces at least 6 in. (152 mm) wide shall be provided for double-row racks.

(g) The aisle widths shall be at least 7½ ft (2.3 m).

(h) The maximum roof height shall be 27 ft (8.2 m).

(i) The maximum storage height shall be 20 ft (6.1 m).

(j) Solid plywood or similar materials shall not be placed on the slatted shelves so that they block the 2-in. (51-mm) spaces between slats, nor shall they be placed on the wire mesh shelves.

A-7-4.1.7.1 Slatting of decks or walkways or the use of open grating as a substitute for automatic sprinkler thereunder is not acceptable.

In addition, where shelving of any type is employed, it is for the basic purpose of providing an intermediate support between the structural members of the rack. As a result, it becomes almost impossible to define and maintain transverse flue spaces across the rack as required in 7-4.1.4 and illustrated in Figure 7-4.1.4.1.

7-4.1.7.2† Sprinklers shall be installed at the ceiling and beneath each shelf in single-, double-, or multiple-row racks with solid shelves that obstruct both longitudinal and transverse flue spaces. Design criteria for combined ceiling and in-rack sprinklers shall be used with this storage configuration.

This requirement should not be interpreted to mean that racks with solid shelves that have only their required longitudinal flues or only their transverse flues obstructed have adequate flue spaces. The requirements of 7-4.1.4 still apply.

Although in-rack sprinklers are specifically required to be installed underneath each shelf that obstructs both required flue spaces, no other criteria have been presented to the Technical Committee on Sprinkler System Discharge Criteria for those conditions in which only one of the required flue spaces is obstructed or if an inadequate number of flue spaces exists. An inadequate number of flues results where transverse flues are only provided at the rack uprights and not between pallet loads. Lacking other protection criteria, the provisions of 7-4.1.7.2 or some other verified means of protecting racks with obstructed flues needs to be considered where flue spaces cannot be provided in accordance with 7-4.1.4.

Exhibit 7.5 shows a double-row rack storage arrangement with inadequate flue spaces. Note that the solid shelves eliminate the required flues between loads [flues every 4 ft to 5 ft (1.2 m to 1.5 m) down the rack are considered adequate] and create a very narrow flue [less than 6 in. (152 mm)] at the rack uprights. NFPA 13 provides no criteria in which ceiling sprinklers alone would be effective in protecting this storage arrangement.

7-4.1.8† Open-Top Combustible Containers.

7-4.1.9 Movable Racks. Rack storage in movable racks shall be protected in the same manner as multiple-row racks.

7-4.1.10 In-Rack Sprinklers. The number of sprinklers and the pipe sizing on a line of sprinklers in racks shall be restricted only by hydraulic calculations and not by any piping schedule.

7-4.1.11* Horizontal Barriers and In-Rack Sprinklers. Horizontal barriers used in conjunction with in-rack sprinklers to impede vertical fire development shall be constructed of sheet metal, wood, or similar material and shall extend the full length and width of the rack. Barriers shall be fitted within 2 in. (51 mm) horizontally around rack uprights. *[See Table 7-4.3.1.5.1, Figures 7-4.3.1.5.1(a), (g), and (j), and Figures 7-4.3.1.5.3(c) and (e).]*

A-7-4.1.11 Where the ceiling is more than 10 ft (3.1 m) above the maximum height of storage, a horizontal barrier should be installed above storage with one line of sprinklers under the barrier for Classes I, II, and III commodities and two lines of sprinklers under the barrier for Class IV commodities. In-rack sprinkler arrays should be installed as indicated in Table 7-4.3.1.5.1 and Figures 7-4.3.1.5.1(a) through (j).

Barriers should be of sufficient strength to avoid sagging that interferes with loading and unloading operations.

Exhibit 7.5 *Double-row storage racks consisting of solid shelves and inadequate flue spaces.*

Horizontal barriers are not required to be provided above a Class I or Class II commodity with in-rack sprinkler arrays in accordance with Figures 7-4.3.1.5.1(a) and (b), provided one line of in-rack sprinklers is installed above the top tier of storage.

7-4.1.12 High-Expansion Foam.

See 7-3.2.2.4.

7-4.1.12.1* Where high-expansion foam systems are installed, they shall be in accordance with NFPA 11A, *Standard for Medium- and High-Expansion Foam Systems,* and they shall be automatic in operation.

Exception: This requirement shall not apply where modified by this standard.

A-7-4.1.12.1 Detection systems, concentrate pumps, generators, and other system components that are essential to the operation of the system should have an approved standby power source.

7-4.1.12.2 In-rack sprinklers shall not be required where high-expansion foam systems are used in combination with ceiling sprinklers.

7-4.1.12.3 Detectors for High-Expansion Foam Systems. Detectors shall be listed and shall be installed in one of the following configurations:

(1) At the ceiling only where installed at one-half the listed linear spacing [e.g., 15 ft × 15 ft (4.6 m × 4.6 m) rather than at 30 ft × 30 ft (9.1 m × 9.1 m)]

Exception: Ceiling detectors alone shall not be used where the ceiling/roof clearance from the top of the storage exceeds 10 ft (3.1 m) or the height of the storage exceeds 25 ft (7.6 m).

(2) At the ceiling at the listed spacing and in racks at alternate levels
(3) Where listed for rack storage installation and installed in accordance with the listing to provide response within 1 minute after ignition using an ignition source that is equivalent to that used in a rack storage testing program

7-4.1.12.4 High-Expansion Foam Submergence.

7-4.1.12.4.1 Storage of Classes I, II, III, and IV Commodities Up to and Including 25 ft (7.6 m) in Height.

7-4.1.12.4.1.1* Where high-expansion foam systems are used without sprinklers, the maximum submergence time shall be 5 minutes for Class I, Class II, or Class III commodities and 4 minutes for Class IV commodities.

A-7-4.1.12.4.1.1 Where high-expansion foam is contemplated as the protection media, consideration should be given to possible damage to the commodity from soaking and corrosion. Consideration also should be given to the problems associated with the removal of the foam after discharge.

7-4.1.12.4.1.2 Where high-expansion foam systems are used in combination with ceiling sprinklers, the maximum submergence time shall be 7 minutes for Class I, Class II, or Class III commodities and 5 minutes for Class IV commodities.

7-4.1.12.4.1.3 High-Expansion Foam Ceiling Sprinkler Density. Where high-expansion foam systems are used in combination with ceiling sprinklers, the minimum ceiling sprinkler design density shall be 0.2 gpm/ft^2 (8.2 mm/min) for Class I, Class II, or Class III commodities or 0.25 gpm/ft^2 (10.2 mm/min) for Class IV commodities for the most hydraulically remote 2000-ft^2 (186-m^2) operating area.

7-4.1.12.4.2 Storage of Classes I, II, III, and IV Commodities Over 25 ft (7.6 m) in Height.

7-4.1.12.4.2.1 Where high-expansion foam systems are used for storage over 25 ft (7.6 m) high up to and including 35 ft (10.7 m) high, they shall be used in combination with ceiling sprinklers. The maximum submergence time for the high-expansion foam shall be 5 minutes for Class I, Class II, or Class III commodities and 4 minutes for Class IV commodities.

7-4.1.12.4.2.2 Where high-expansion foam is used in combination with ceiling sprinklers, the minimum ceiling sprinkler design density shall be 0.2 gpm/ft^2 (8.2 mm/min) for Class I, Class II, or Class III commodities and 0.25 gpm/ft^2 (10.2 mm/min) for Class IV commodities for the most hydraulically remote 2000-ft^2 (186-m^2) operating area.

7-4.2 Spray Sprinkler Protection Criteria for Class I through Class IV Commodities Stored Up to and Including 25 ft (7.6 m) in Height.

7-4.2.1 In-Rack Sprinklers.

7-4.2.1.1 In-Rack Sprinkler Location.

In the testing on which this standard is based (see Appendix C), where in-rack sprinklers were used, the sprinklers were typically installed above the top of a storage tier at transverse flues. Where multiple levels of in-rack sprinklers were used, in-rack sprinklers were usually staggered among transverse flues, except when barriers were provided.

7-4.2.1.1.1*† The elevation of in-rack sprinkler deflectors with respect to storage shall not be a consideration in single- or double-row rack storage up to and including 20 ft (6.1 m) high.

A-7-4.2.1.1.1 Where possible, it is recommended that in-rack sprinkler deflectors be located at least 6 in. (152.4 mm) above pallet loads.

Where in-rack sprinklers are not located above the top of a storage tier, it is important that they be located at transverse flues. Tests were conducted with in-rack sprinklers located one-half the way up the load and between transverse flues. Although sprinkler discharge on top of a pallet load is more advantageous, water distributed to the side of a load would limit fire development up through the flue once the sprinklers activate.

7-4.2.1.1.2* In single- or double-row racks without solid shelves with storage over 20 ft (6.1 m) high, or in multiple-row racks, or in single- or double-row racks with solid shelves and storage height up to and including 25 ft (7.6 m), a minimum of 6-in. (152.4-mm) vertical clear space shall be maintained between the in-rack sprinkler deflectors and the top of a tier of storage. Sprinkler discharge shall not be obstructed by horizontal rack members.

A-7-4.2.1.1.2 Where possible, it is recommended that in-rack sprinklers be located away from rack uprights.

Formal Interpretation 86-1, originally issued for NFPA 231C, provides more guidance on the placement of in-rack sprinklers.

Formal Interpretation 86-1

Reference: 7-4.2.1.1.2

Question 1: Is it the intent of 7-4.2.1.1.2 to require the placement of in-rack sprinklers within rack flue spaces so that the discharge from the sprinklers will wet the storage when storage consists of double row racks without solid shelves with height of storage over 20 feet?

Answer: Yes.

Question 2: If the answer to Question 1 is "yes," is it the intent of 7-4.2.1.1.2 to prevent sprinklers from being located behind horizontal rack structural members that would obstruct the sprinkler spray onto the storage when the storage consists of double row racks without solid shelves with height of storage over 20 feet?

Answer: Yes.

Issue Edition: 1986 of NFPA 231C

Reference: 6-4.2

Issue Date: October 1987

Reissued to correct error: August 1995

7-4.2.1.1.3 In-rack sprinklers at one level only for storage up to and including 25 ft (7.6 m) high shall be located at the first tier level at or above one-half of the storage height.

Exhibit 7.6 illustrates the application of this placement requirement.

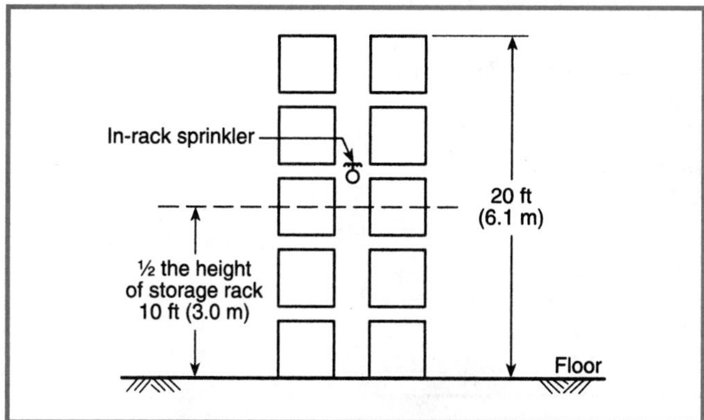

Exhibit 7.6 Positioning of one level of in-rack sprinklers for storage heights under 25 ft (7.6 m).

7-4.2.1.1.4 In-rack sprinklers at two levels only for storage up to and including 25 ft (7.6 m) high shall be located at the first tier level at or above one-third and two-thirds of the storage height.

Exhibit 7.7 illustrates the application of this placement requirement.

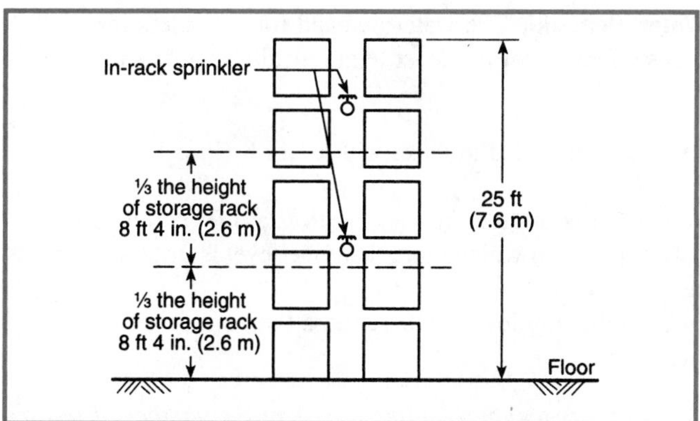

Exhibit 7.7 Positioning of two levels of in-rack sprinklers for storage heights under 25 ft (7.6 m).

7-4.2.1.2 In-Rack Sprinkler Spacing.

7-4.2.1.2.1* Maximum horizontal spacing of in-rack sprinklers in single- or double-row racks with nonencapsulated storage up to and including 25 ft (7.6 m) in height shall be in accordance with Table 7-4.2.1.2.1.

For encapsulated storage, maximum horizontal spacing shall be 8 ft (2.44 m).

Table 7-4.2.1.2.1 In-Rack Sprinkler Spacing for Class I, II, III, and IV Commodities Stored up to 25 ft in Height

Aisle Widths		Commodity Class					
		I and II		III		IV	
ft	m	ft	m	ft	m	ft	m
8	2.4	12	3.7	12	3.7	8	2.4
4	1.2	12	3.7	8	2.4	8	2.4

A-7-4.2.1.2.1 Spacing of sprinklers on branch lines in racks in the various tests demonstrates that maximum spacing as specified is proper.

7-4.2.1.2.2† Sprinklers installed in racks shall be spaced without regard to rack uprights.

Although NFPA 13 does not prohibit in-rack sprinklers from being positioned near rack uprights, A-7-4.2.1.1.2 recommends that in-rack sprinklers be located away from rack uprights.

7-4.2.1.3† In-Rack Sprinkler Discharge Pressure. Sprinklers in racks shall discharge at not less than 15 psi (1 bar) for all classes of commodities.

7-4.2.1.4† In-Rack Sprinkler Water Demand. The water demand for sprinklers installed in racks shall be based on simultaneous operation of the most hydraulically remote sprinklers as follows:

(1) Six sprinklers where only one level is installed in racks with Class I, Class II, or Class III commodities
(2) Eight sprinklers where only one level is installed in racks with Class IV commodities
(3) Ten sprinklers (five on each two top levels) where more than one level is installed in racks with Class I, Class II, or Class III commodities
(4) Fourteen sprinklers (seven on each two top levels) where more than one level is installed in racks with Class IV commodities

Exception: Where a storage rack, due to its length, requires less than the number of in-rack sprinklers specified, only those in-rack sprinklers in a single rack need to be included in the calculation.

7-4.2.1.5 In-Rack Sprinkler Location—Single- and Double-Row Racks. In single- or double-row racks without solid shelves, in-rack sprinklers shall be installed in accordance with Table 7-4.2.1.5.

This section is intended to apply only to those rack storage arrangements that contain flue spaces in accordance with 7-4.1.4. Where an inadequate number or type of flue spaces are provided or where required flue spaces are obstructed, the criteria provided by 7-4.1.4 are unlikely to provide adequate protection and the provisions of 7-4.1.7 should be considered.

The term *solid shelves* is used to indicate conditions where inadequate flues are provided. The provisions of 7-4.2.1.5 section can be used where solid shelving materials such as slave pallets are used, provided the required number and type of flue spaces are provided. Where slave pallets or other solid materials are used in racks up to 25 ft (7.6 m) in height and only ceiling sprinklers are used, the ceiling density must be increased 20 percent. See 7-4.2.2.1.11 for additional information.

7-4.2.1.6† In-Rack Sprinkler Location—Multiple-Row Racks.

The aisle width indicated in Tables 7-4.2.1.6.1 and 7-4.2.1.6.2 refers to the aisles between rack structures and not the aisles between individual racks. See Exhibit 7.4.

7-4.2.1.6.1 For encapsulated or nonencapsulated storage in multiple-row racks no deeper than 16 ft (4.9 m) with aisles 8 ft (2.4 m) or wider, in-rack sprinklers shall be installed in accordance with Table 7-4.2.1.6.1.

7-4.2.1.6.2 For encapsulated or nonencapsulated storage in multiple-row racks deeper than 16 ft (4.9 m) or with aisles less than 8 ft (2.4 m) wide, in-rack sprinklers shall be installed in accordance with Table 7-4.2.1.6.2.

7-4.2.1.6.3* Maximum horizontal spacing of in-rack sprinklers on branch lines, in multiple-row racks with encapsulated or nonencapsulated storage up to and including 25 ft (7.6 m) in height, shall not exceed 12 ft (3.7 m) for Class I, II, or III commodities and 8 ft (2.4 m) for Class IV commodities, with area limitations of 100 ft^2 (9.3 m^2) per sprinkler for Class I, II, or III commodities and 80 ft^2 (7.4 m^2) per sprinkler for Class IV commodities. The rack plan view shall be considered in determining the area covered by each sprinkler. The aisles shall not be included in area calculations.

A-7-4.2.1.6.3 In-rack sprinklers at one level only for storage up to and including 25 ft (7.6 m) in multiple-row racks should be located at the tier level nearest one-half to two-thirds of the storage height.

7-4.2.1.6.4 A minimum of 6 in. (152.4 mm) shall be maintained between the in-rack sprinkler deflector and the top of a tier of storage.

7-4.2.2† Ceiling Sprinkler Discharge Criteria—Area Density Method.

7-4.2.2.1 General.

7-4.2.2.1.1*† Ceiling Sprinkler Water Demand. Ceiling sprinkler water demand shall be determined in accordance with 7-4.2.2.2 for single- and double-row racks or 7-4.2.3 for multiple-

Table 7-4.2.1.5 Single- or Double-Row Racks—Storage Height Up to and Including 25 ft (7.6 m) Without Solid Shelves

			Aisles*			Ceiling Sprinkler Water Demand					
						With In-Rack Sprinklers			Without In-Rack Sprinklers		
Height	Commodity Class	Encap-sulated	ft	m	Sprinklers Mandatory In-Racks	Figure	Curves	Apply Figure 7-4.2.2.1.3	Figure	Curves	Apply Figure 7-4.2.2.1.3
Over 12 ft (3.7 m), up to and including 20 ft (6.1 m)	I	No	4	1.2	No	7-4.2.2.1.1(a)	C and D	Yes	7-4.2.2.1.1(a)	F and H	Yes
			8	2.4			A and B			E and G	
		Yes	4	1.2	No	7-4.2.2.1.1(e)	C and D		7-4.2.2.1.1(e)	G and H	Yes
			8	2.4			A and B			E and F	
	II	No	4	1.2	No	7-4.2.2.1.1(b)	C and D		7-4.2.2.1.1(b)	G and H	Yes
			8	2.4			A and B			E and F	
		Yes	4	1.2	No	7-4.2.2.1.1(e)	C and D		7-4.2.2.1.1(e)	G and H	Yes
			8	2.4			A and B			E and F	
	III	No	4	1.2	No	7-4.2.2.1.1(c)	C and D		7-4.2.2.1.1(c)	G and H	Yes
			8	2.4			A and B			E and F	
		Yes	4	1.2	1 level	7-4.2.2.1.1(f)	C and D		—	—	—
			8	2.4			A and B				
	IV	No	4	1.2	No	7-4.2.2.1.1(d)	C and D		7-4.2.2.1.1(d)	G and H	Yes
			8	2.4			A and B			E and F	
		Yes	4	1.2	1 level	7-4.2.2.1.1(g)	C and D		—	—	—
			8	2.4			A and B				
Over 20 ft (6.1 m), up to and including 22 ft (6.7 m)	I	No	4	1.2	No	7-4.2.2.1.1(a)	C and D	No	7-4.2.2.1.1(a)	F and H	Yes
			8	2.4			A and B			E and G	
		Yes	4	1.2	1 level	7-4.2.2.1.1(e)	C and D		—	—	—
			8	2.4			A and B				
	II	No	4	1.2	No	7-4.2.2.1.1(b)	C and D		7-4.2.2.1.1(b)	G and H	Yes
			8	2.4			A and B			E and F	
		Yes	4	1.2	1 level	7-4.2.2.1.1(e)	C and D		—	—	—
			8	2.4			A and B				
	III	No	4	1.2	No	7-4.2.2.1.1(c)	C and D		7-4.2.2.1.1(c)	G and H	Yes
			8	2.4			A and B			E and F	
		Yes	4	1.2	1 level	7-4.2.2.1.1(f)	C and D		—	—	—
			8	2.4			A and B				
	IV	No	4	1.2	No	7-4.2.2.1.1(d)	C and D		7-4.2.2.1.1(d)	G and H	Yes
			8	2.4			A and B			E and F	
		Yes	4	1.2	1 level	7-4.2.2.1.1(g)	C and D		—	—	—
			8	2.4			A and B				

Table 7-4.2.1.5 Single- or Double-Row Racks—Storage Height Up to and Including 25 ft (7.6 m) Without Solid Shelves (Continued)

Height	Commodity Class	Encapsulated	Aisles* ft	Aisles* m	Sprinklers Mandatory In-Racks	Ceiling Sprinkler Water Demand: With In-Rack Sprinklers: Figure	With In-Rack Sprinklers: Curves	With In-Rack Sprinklers: Apply Figure 7-4.2.2.1.3	Without In-Rack Sprinklers: Figure	Without In-Rack Sprinklers: Curves	Without In-Rack Sprinklers: Apply Figure 7-4.2.2.1.3
Over 22 ft (6.7 m), up to and including 25 ft (7.6 m)	I	No	4	1.2	No	7-4.2.2.1.1(a)	C and D	No	7-4.2.2.1.1(a)	F and H	Yes
			8	2.4			A and B			E and G	
		Yes	4	1.2	1 level	7-4.2.2.1.1(e)	C and D		—	—	—
			8	2.4			A and B				
	II	No	4	1.2	No	7-4.2.2.1.1(b)	C and D		7-4.2.2.1.1(b)	G and H	Yes
			8	2.4			A and B			E and F	
		Yes	4	1.2	1 level	7-4.2.2.1.1(e)	C and D		—	—	—
			8	2.4			A and B				
	III	No	4	1.2	No	7-4.2.2.1.1(c)	C and D		7-4.2.2.1.1(c)	G and H	Yes
			8	2.4			A and B			E and F	
		Yes	4	1.2	1 level	7-4.2.2.1.1(f)	C and D		—	—	—
			8	2.4			A and B				
	IV	No	4	1.2	1 level	7-4.2.2.1.1(d)	C and D		—	—	—
			8	2.4			A and B				
		Yes	4	1.2		7-4.2.2.1.1(g)	C and D		—	—	—
			8	2.4			A and B				

*See 7-4.2.2.2.2 for interpolation of aisle widths.

row racks. The design curves in Figures 7-4.2.2.1.1(a) through (g) shall apply to nominal 20 ft (6.1 m) height of storage.

A-7-4.2.2.1.1 Bulkheads are not a substitute for sprinklers in racks. Their installation does not justify reduction in sprinkler densities or design operating areas as specified in the design curves.

See commentary to 7-4.2.1.5.

7-4.2.2.1.2 The design curves indicate water demands for ordinary temperature–rated and nominal high temperature–rated sprinklers at the ceiling. The ordinary-temperature design curves corresponding to ordinary temperature–rated sprinklers shall be used for sprinklers with ordinary- and intermediate-temperature classification. The high-temperature design curve corresponding to high temperature–rated sprinklers shall be used for sprinklers having a high-temperature rating.

7-4.2.2.1.3 For storage height up to and including 25 ft (7.6 m) protected with ceiling sprinklers only and for storage height up to and including 20 ft (6.1 m) protected with ceiling sprinklers

Table 7-4.2.1.6.1 Multiple-Row Racks—Rack Depth Up to and Including 16 ft (4.9 m), Aisles 8 ft (2.4 m) or Wider, Storage Height Up to 25 ft (7.6 m)

Height	Commodity Class	Encapsulated	Sprinklers Mandatory In-Racks	Ceiling Sprinkler Water Demand							
				With In-Rack Sprinklers				Without In-Rack Sprinklers			
				Figure	Curves	Apply Figure 7-4.2.2.1.3	1.25 × Density	Figure	Curves	Apply Figure 7-4.2.2.1.3	1.25 × Density
Over 12 ft (3.7 m), up to and including 15 ft (4.6 m)	I	No	No	7-4.2.2.1.1(a)	C and D	Yes	No	7-4.2.2.1.1(a)	I and J	Yes	No
		Yes		7-4.2.2.1.1(a)			Yes	7-4.2.2.1.1(a)	I and J		Yes
	II	No		7-4.2.2.1.1(b)			No	7-4.2.2.1.1(b)	I and J	Yes	No
		Yes		7-4.2.2.1.1(b)			Yes	7-4.2.2.1.1(b)	I and J		Yes
	III	No	No	7-4.2.2.1.1(c)			No	7-4.2.2.1.1(c)	I and J	Yes	No
		Yes	1 level	7-4.2.2.1.1(c)			Yes		NA	NA	NA
	IV	No	No	7-4.2.2.1.1(d)			No	7-4.2.2.1.1(d)	C and D	No	No
		Yes	1 level	7-4.2.2.1.1(d)	A and B		1.50 × density		NA	NA	NA
Over 15 ft (4.6 m), up to and including 20 ft (6.1 m)	I	No	No	7-4.2.2.1.1(a)	C and D	Yes	No	7-4.2.2.1.1(a)	I and J	Yes	No
		Yes		7-4.2.2.1.1(a)			Yes	7-4.2.2.1.1(a)	I and J		Yes
	II	No		7-4.2.2.1.1(b)			No	7-4.2.2.1.1(b)	I and J	Yes	No
		Yes		7-4.2.2.1.1(b)			Yes	7-4.2.2.1.1(b)	I and J		Yes
	III	No	No	7-4.2.2.1.1(c)			No	7-4.2.2.1.1(c)	I and J	Yes	No
		Yes	1 level	7-4.2.2.1.1(c)			Yes	NA	NA	NA	NA
	IV	No	1 level	7-4.2.2.1.1(d)			No				
		Yes		7-4.2.2.1.1(d)	A and B		1.50 × density				
Over 20 ft (6.1 m), up to and including 25 ft (7.6 m)	I	No	No	7-4.2.2.1.1(a)	C and D	No	No	7-4.2.2.1.1(a)	I and J	Yes	No
		Yes	1 level	7-4.2.2.1.1(a)			Yes	NA	NA	NA	NA
	II	No	1 level	7-4.2.2.1.1(b)			No				
		Yes		7-4.2.2.1.1(b)			Yes				
	III	No		7-4.2.2.1.1(c)			No				
		Yes		7-4.2.2.1.1(c)			Yes				
	IV	No	2 levels	7-4.2.2.1.1(d)			No				
		Yes		7-4.2.2.1.1(d)	A and B		1.50 × density				

Table 7-4.2.1.6.2 Multiple-Row Racks—Rack Depth Over 16 ft (4.9 m) or Aisles Narrower than 8 ft (2.4 m), Storage Height Up to and Including 25 ft (7.6 m)

				Ceiling Sprinkler Water Demand							
				With In-Rack Sprinklers				Without In-Rack Sprinklers			
Height	Commodity Class	Encap-sulated	Sprinklers Mandatory In-Racks	Figure	Curves	Apply Figure 7-4.2.2.1.3	1.25 × Density	Figure	Curves	Apply Figure 7-4.2.2.1.3	1.25 × Density
Over 12 ft (3.7 m), up to and including 15 ft (4.6 m)	I	No	No	7-4.2.2.1.1(a)	C and D	Yes	No	7-4.2.2.1.1(a)	I and J	Yes	No
		Yes		7-4.2.2.1.1(a)			Yes	7-4.2.2.1.1(a)	I and J		Yes
	II	No		7-4.2.2.1.1(b)			No	7-4.2.2.1.1(b)	I and J	Yes	No
		Yes		7-4.2.2.1.1(b)			Yes	7-4.2.2.1.1(b)	I and J		Yes
	III	No		7-4.2.2.1.1(c)			No	7-4.2.2.1.1(c)	I and J	Yes	No
		Yes	1 level	7-4.2.2.1.1(c)			Yes				
	IV	No	No	7-4.2.2.1.1(d)			No	7-4.2.2.1.1(d)	C and D	No	No
		Yes	1 level	7-4.2.2.1.1(d)			1.50 × density				
Over 15 ft (4.6 m), up to and including 20 ft (6.1 m)	I	No	1 level	7-4.2.2.1.1(a)	C and D	Yes	No	NA	NA	NA	NA
		Yes		7-4.2.2.1.1(a)			Yes				
	II	No		7-4.2.2.1.1(b)			No				
		Yes		7-4.2.2.1.1(b)			Yes				
	III	No		7-4.2.2.1.1(c)			No				
		Yes		7-4.2.2.1.1(c)			Yes				
	IV	No		7-4.2.2.1.1(d)			No				
		Yes		7-4.2.2.1.1(d)			1.50 × density				
Over 20 ft (6.1 m), up to and including 25 ft (7.6 m)	I	No	1 level	7-4.2.2.1.1(a)	C and D	No	No	NA	NA	NA	NA
		Yes		7-4.2.2.1.1(a)			Yes				
	II	No		7-4.2.2.1.1(b)			No				
		Yes		7-4.2.2.1.1(b)			Yes				
	III	No		7-4.2.2.1.1(c)			No				
		Yes		7-4.2.2.1.1(c)			Yes				
	IV	No	2 levels	7-4.2.2.1.1(d)			No				
		Yes		7-4.2.2.1.1(d)			1.50 × density				

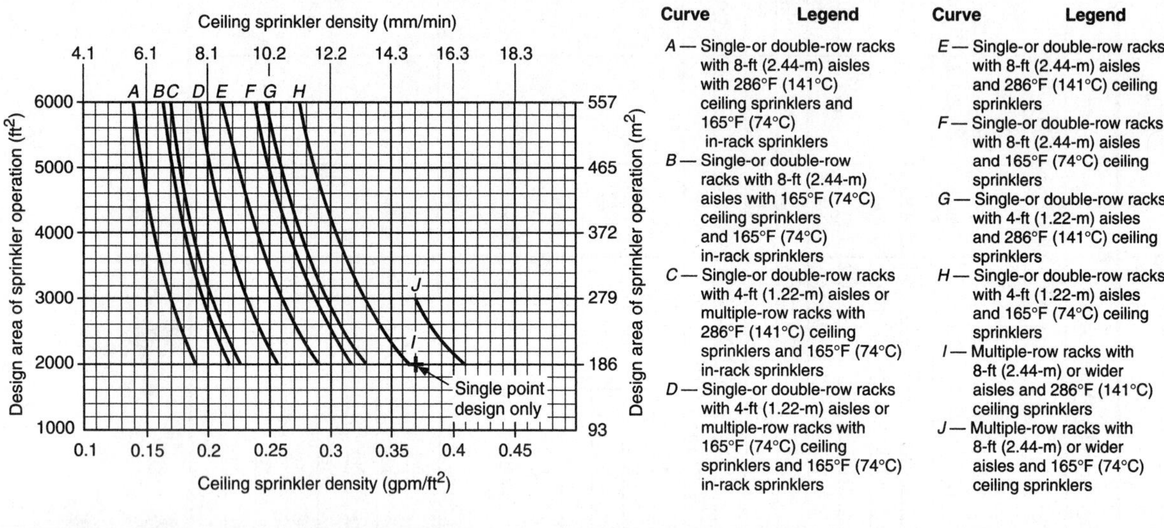

Figure 7-4.2.2.1.1(a) *Sprinkler system design curves—20-ft (6.1-m) high rack storage—Class I nonencapsulated commodities—conventional pallets.*

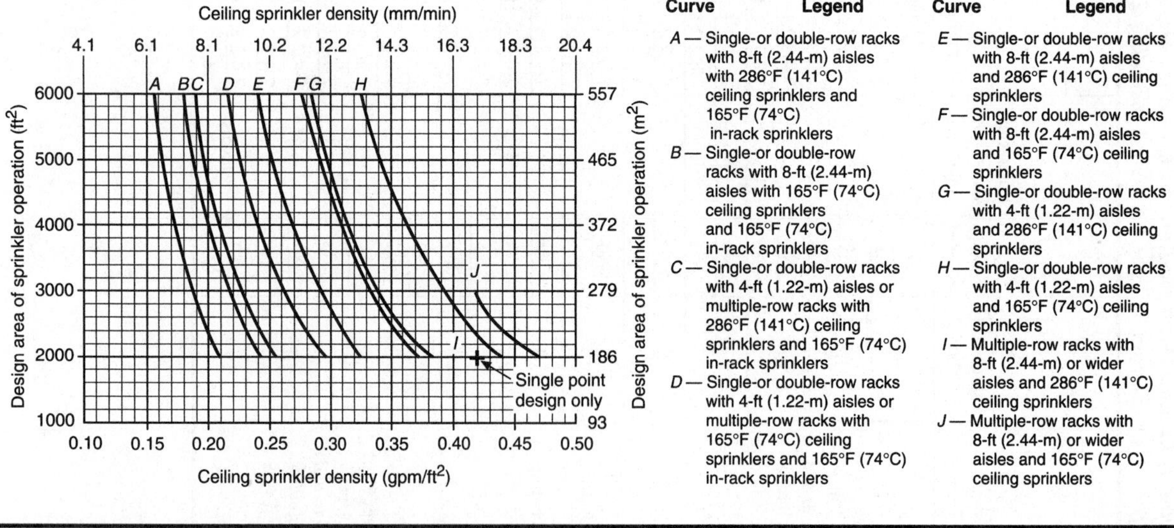

Figure 7-4.2.2.1.1(b) *Sprinkler system design curves—20-ft (6.1-m) high rack storage—Class II nonencapsulated commodities—conventional pallets.*

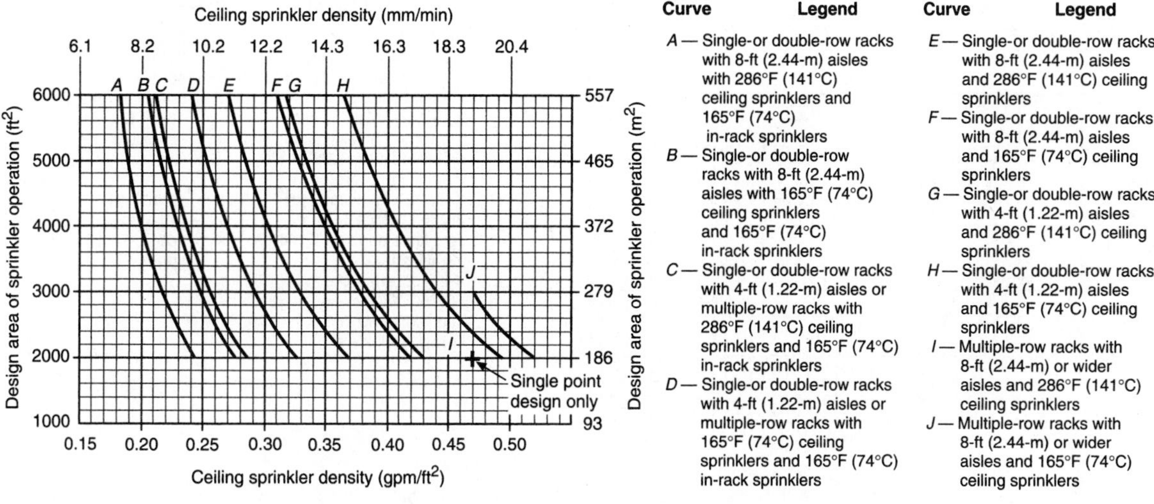

Figure 7-4.2.2.1.1(c) *Sprinkler system design curves—20-ft (6.1-m) high rack storage—Class III nonencapsulated commodities—conventional pallets.*

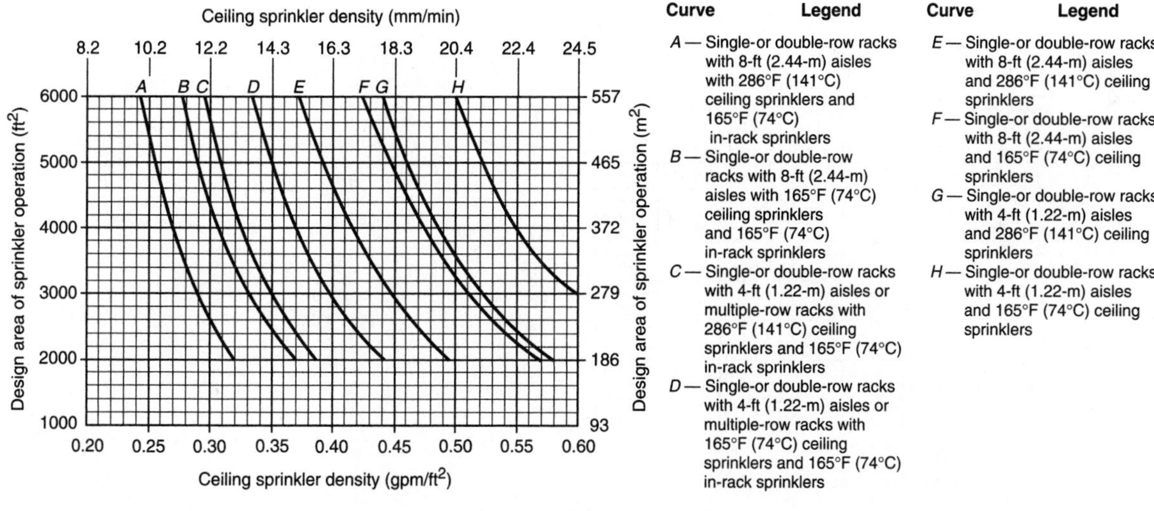

Figure 7-4.2.2.1.1(d) *Sprinkler system design curves—20-ft (6.1-m) high rack storage—Class IV nonencapsulated commodities—conventional pallets.*

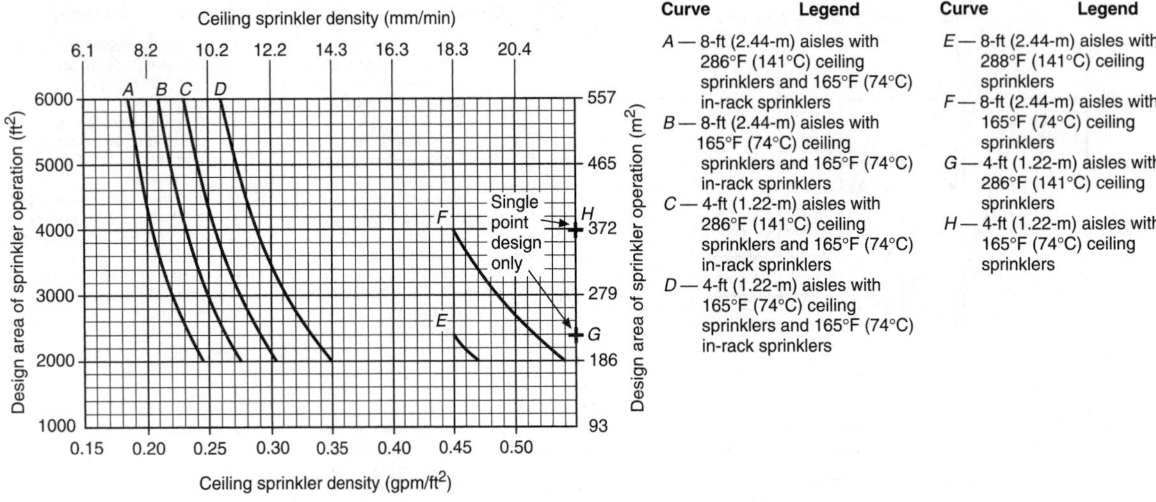

Figure 7-4.2.2.1.1(e) *Single- or double-row racks—20-ft (6.1-m) high rack storage—sprinkler system design curves—Class I and II encapsulated commodities—conventional pallets.*

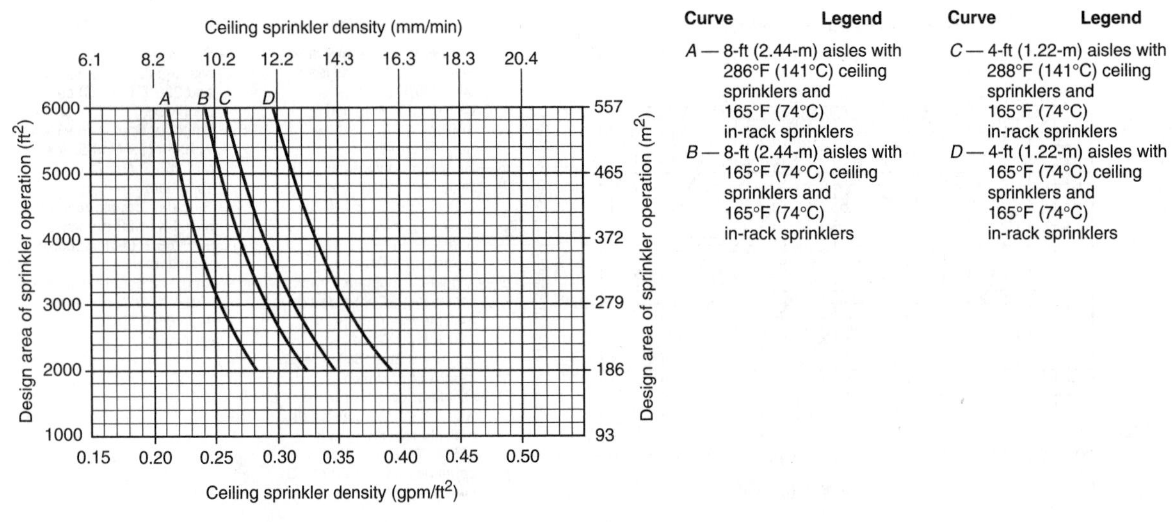

Figure 7-4.2.2.1.1(f) *Single- or double-row racks—20-ft (6.1-m) high rack storage—sprinkler system design curves—Class III encapsulated commodities—conventional pallets.*

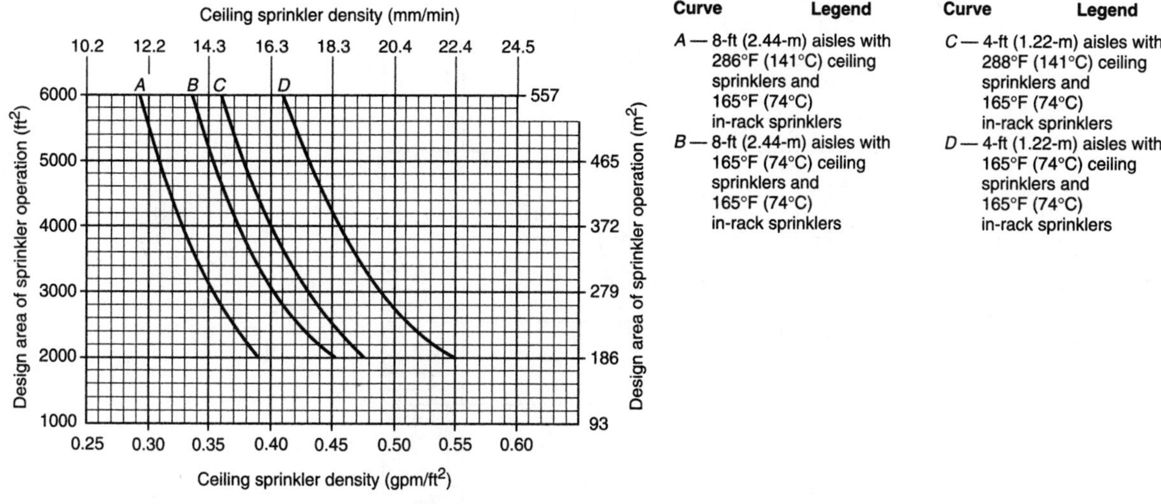

***Figure 7-4.2.2.1.1(g)** Single- or double-row racks—20-ft (6.1-m) high rack storage—sprinkler system design curves—Class IV encapsulated commodities—conventional pallets.*

and minimum required in-rack sprinklers, densities obtained from design curves shall be adjusted in accordance with Figure 7-4.2.2.1.3.

7-4.2.2.1.4 For storage height over 20 ft (6.1 m) up to and including 25 ft (7.6) protected with ceiling sprinklers and minimum required in-rack sprinklers, densities obtained from design curves shall be used. Densities shall not be adjusted in accordance with Figure 7-4.2.2.1.3.

7-4.2.2.1.5 For storage height up to and including 20 ft (6.1 m) protected with ceiling sprinklers and with more than one level of in-rack sprinklers, but not in every tier, densities obtained from design curves and adjusted in accordance with Figure 7-4.2.2.1.3 shall be permitted to be reduced an additional 20 percent, as indicated in Table 7-4.2.2.1.5.

Table 7-4.2.2.1.5 summarizes the adjustments to ceiling sprinkler density for storage height and where a level of in-rack sprinklers in excess of that specified by Table 7-4.2.1.5 is installed. The height adjustment is not dependent on the presence of in-rack sprinklers. With regard to the adjustment for in-rack sprinklers, if a nonencapsulated Class III commodity is stored to a height of 20 ft (6.1 m) in double-row racks with 4-ft (1.2-m) aisles with high-temperature sprinklers at the ceiling, in-rack sprinklers are not required per Table 7-4.2.1.5. However, the designer has the option of using in-rack sprinklers. Where one level of in-rack sprinklers is installed, the ceiling sprinkler density is chosen from curve C of Figure 7-4.2.2.1.1(c). An appropriate density would be 0.285 gpm/ft^2 (11.6 mm/min) over 2000 ft^2 (186 m^2). According to Table 7-4.2.2.1.5, no additional adjustment is permitted. If a second level of in-rack sprinklers were installed, the ceiling sprinkler density could be reduced by 20 percent, which would

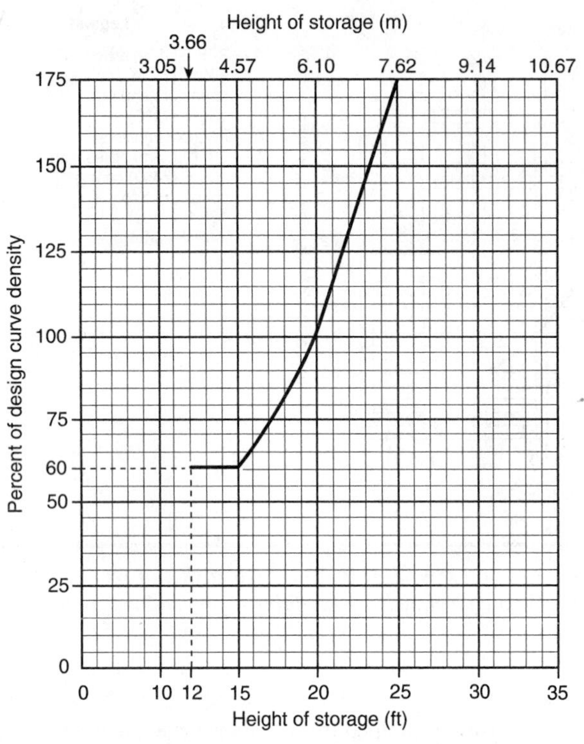

Figure 7-4.2.2.1.3 *Ceiling sprinkler density vs. storage height.*

result in a required density of 0.23 gpm/ft² (9.4 mm/min) over 2000 ft² (186 m²). If no in-rack sprinklers are installed, the density would be obtained from curve G of Figure 7-4.2.2.1.1(c). An appropriate density would be 0.43 gpm/ft² (17.5 mm/min) over 2000 ft² (186 m²).

7-4.2.2.1.6 For storage height over 20 ft (6.1 m) up to and including 25 ft (7.6 m) protected with ceiling sprinklers and with more than the minimum required level of in-rack sprinklers, but not in every tier, densities obtained from design curves shall be permitted to be reduced 20 percent as indicated in Table 7-4.2.2.1.5. Densities shall not be adjusted in accordance with Figure 7-4.2.2.1.3 for storage height.

7-4.2.2.1.7 For storage height up to and including 20 ft (6.1 m) protected with ceiling sprinklers and in-rack sprinklers at each tier, densities obtained from design curves and adjusted in accordance with Figure 7-4.2.2.1.3 shall be permitted to be reduced an additional 40 percent, as indicated in Table 7-4.2.2.1.5.

7-4.2.2.1.8 For storage height over 20 ft (6.1 m) up to and including 25 ft (7.6 m) protected with ceiling sprinklers and in-rack sprinklers at each tier, densities obtained from design curves

Table 7-4.2.2.1.5 Adjustment to Ceiling Sprinkler Density for Storage Height and In-Rack Sprinklers

Storage Height	In-Rack Sprinklers	Apply Figure 7-4.2.2.1.3 for Storage Height Adjustment	Permitted Ceiling Sprinklers Density Adjustments Where In-rack Sprinklers Are Installed
Over 12 ft (3.7 m) through 25 ft (7.6 m)	None	Yes	None
Over 12 ft (3.7 m) through 20 ft (6.1 m)	Minimum required	Yes	None
	More than minimum, but not in every tier	Yes	Reduce density 20% from that of minimum in-rack sprinklers
	In every tier	Yes	Reduce density 40% from that of minimum in-rack sprinklers
Over 20 ft (6.1 m) through 25 ft (7.6 m)	Minimum required	No	None
	More than minimum, but not in every tier	No	Reduce density 20% from that of minimum in-rack sprinklers
	In every tier	No	Reduce density 40% from that of minimum in-rack sprinklers

shall be permitted to be reduced 40 percent, as indicated in Table 7-4.2.2.1.5. Densities shall not be adjusted in accordance with Figure 7-4.2.2.1.3 for storage height.

7-4.2.2.1.9† Where clearance from ceiling to top of storage is less than 4½ ft (1.37 m), the sprinkler operating area indicated in curves E, F, G, and H in Figures 7-4.2.2.1.1(a) through (e) shall be permitted to be reduced as indicated in Figure 7-4.2.2.1.9 but shall not be reduced to less than 2000 ft^2 (185.8 m^2). *(See 7-4.2.2.1.10.)*

7-4.2.2.1.10 Where clearance from ceiling to top of Class I or Class II encapsulated storage is 1½ ft to 3 ft (0.46 m to 0.91 m), the sprinkler operating area indicated in curve F only of Figure 7-4.2.2.1.1(e) shall be permitted to be reduced by 50 percent but shall not be reduced to less than 2000 ft^2 (186 m^2).

7-4.2.2.1.11 Where solid, flat-bottom, combustible pallets are used with storage height up to and including 25 ft (7.6 m), the densities that are indicated in the design curves shown in Figures 7-4.2.2.1.1(a) through (g), based on conventional pallets, shall be increased 20 percent for the given area. The percentage shall be applied to the density determined in accordance with Figure 7-4.2.2.1.3. The increase in density shall not apply where in-rack sprinklers are installed.

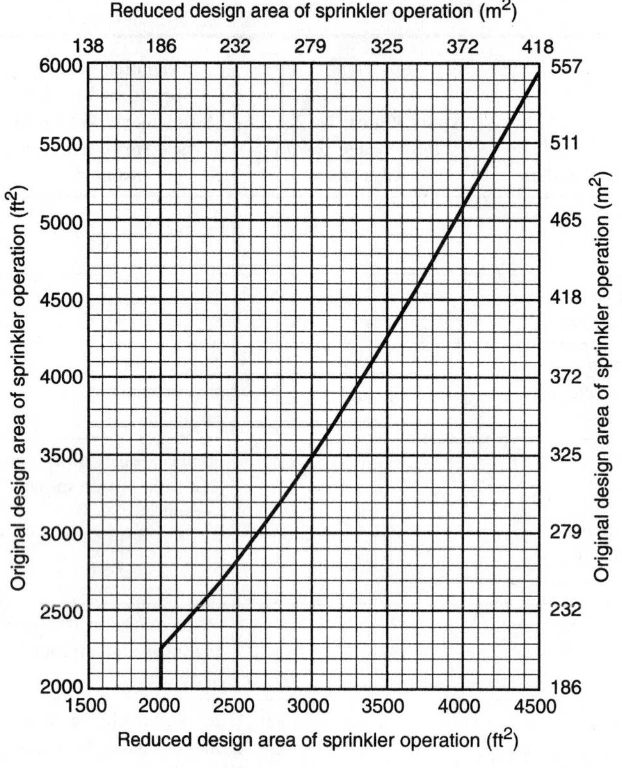

Figure 7-4.2.2.1.9 *Adjustment of design area of sprinkler operation for clearance from top of storage to ceiling.*

The most common examples of this type pallet are slave pallets as indicated in Chapter 1. See Figure A-1-4.9(b).

7-4.2.2.2* Ceiling Sprinkler Water Demand—Double- and Single-Row Racks.

A-7-4.2.2.2 Data indicates that the sprinkler protection criteria in Table 7-4.2.1.5 is ineffective, by itself, for rack storage with solid shelves, if the required flue spaces are not maintained. Use of Table 7-4.2.1.5 and the figures it references, along with the additional provisions that are required by this standard, can provide acceptable protection.

7-4.2.2.2.1 For Class I, Class II, Class III, or Class IV commodities, encapsulated or nonencapsulated in single- or double-row racks, ceiling sprinkler water demand in terms of density (gpm/ft^2) (mm/min) and area of sprinkler operation [ft^2 (m^2) of ceiling or roof] shall be selected from the curves in Figures 7-4.2.2.1.1-1.6(a) through (g) as directed by Table 7-4.2.1.5. The curves in Figures 7-4.2.2.1.1(a) through (g) also shall apply to portable racks arranged in the same manner as single-, double, or multiple-row racks. The design shall be sufficient to satisfy a single point on the appropriate curve related to the storage configuration and commodity class. It shall not be required to meet all points on the selected curve. Figure 7-4.2.2.1.3 shall be used to adjust the density for storage height unless otherwise specified.

In many cases, selecting the design point on the curve corresponding to the minimum design area results in the most economical and most effective protection. The following three examples and solutions illustrate how sprinkler protection is determined for double-row racks.

Example 7.1

A nonencapsulated Class II commodity is stored to a height of 17 ft (5.2 m) in double-row racks separated with 4-ft (1.2-m) aisles with ordinary temperature–rated sprinklers at the ceiling. Conventional pallets are used, and the clearance between the top of storage and the sprinkler deflectors will be maintained at 5 ft (1.5 m).

Solution

Per Table 7-4.2.1.5, in-rack sprinklers are not mandatory. The ceiling sprinkler density without in-rack sprinklers would be chosen from curve H of Figure 7-4.2.2.1.1(b). An appropriate density from this curve would be 0.44 gpm/ft^2 (17.9 mm/min) over 2000 ft^2 (186 m^2).This density needs to be adjusted for storage height according to 7-4.2.2.1.3. For 17-ft (5.2-m) storage height, the adjustment factor from Figure 7-4.2.2.1.3 is 75 percent, which results in a density of 0.33 gpm/ft^2 (13.5 mm/min) over 2000 ft^2 (186 m^2). Because the clearance exceeds 4 ft (1.2 m), the adjustment per 7-4.2.2.1.9 is not permitted.

Example 7.2

The same criteria as in example (a) are used with the exception that in-rack sprinklers will be installed in each level of storage. What is an acceptable discharge criteria for the ceiling system?

Solution

Although in-rack sprinklers are not mandated by Table 7-4.2.1.5, they are permitted as a design option. The ceiling sprinkler density with in-rack sprinklers would be chosen from curve D of Figure 7-4.2.2.1.1(b). An appropriate density from this curve would be 0.295 gpm/ft^2 (12.0 mm/min) over 2000 ft^2 (186 m^2). In accordance with 7-4.2.2.1.7, this density needs to be adjusted for storage height and for in-rack sprinklers at each tier. For a 17-ft (5.2-m) storage height, the adjustment factor from Figure 7-4.2.2.1.3 is 75 percent, which results in a density of 0.22 gpm/ft^2 (9.0 mm/min) over 2000 ft^2 (186 m^2). This density is reduced an additional 40 percent for in-rack sprinklers in every tier, which results in a density of 0.13 gpm/ft^2 (5.3 mm/min) over 2000 ft^2 (186 m^2). A clearance adjustment is not permitted where in-rack sprinklers are used in accordance with 7-4.2.2.1.9.

Example 7.3

A nonencapsulated Class IV commodity is stored to a height of 22 ft (6.7 m) in double-row racks separated with 5-ft (1.5-m) aisles with high temperature–rated sprinklers at

the ceiling. A solid-shelf material is used but 6-in. (152-mm) transverse flue spaces are maintained at both the rack uprights and between loads in accordance with 7-4.1.4.1. The clearance between the top of storage and the ceiling will be maintained at 3 ft (0.9 m).

Solution

In accordance with Table 7-4.2.1.5, in-rack sprinklers are not mandatory, so the solution will be based on in-rack sprinklers not being used. Because densities are not provided for 5-ft (1.5-m) aisles, the density needs to be obtained by interpolating the densities for 4-ft (1.2-m) and 8-ft (2.4-m) aisles in accordance with 7-4.2.2.2.2. The ceiling sprinkler densities would be chosen from Figure 7-4.2.2.1.1(d). Curve G would be used for 4-ft (1.2-m) aisles and curve E would be used for 8-ft (2.4-m) aisles. Appropriate densities would be 0.58 gpm/ft^2 (23.6 mm/min) over 2000 ft^2 (186 m^2) for 4-ft (1.2-m) aisles and 0.495 gpm/ft^2 (20.2 mm/min) over 2000 ft^2 (186 m^2) for 8-ft (2.4-m) aisles. In accordance with 7-4.2.2.1.3, these densities need to be adjusted for storage height. For a 22-ft^2 (6.7-m) storage height, the adjustment factor from Figure 7-4.2.2.1.3 is 130 percent, which results in densities of 0.754 gpm/ft^2 (30.7 mm/min) over 2000 ft^2 (186 m^2) and 0.644 gpm/ft^2 (26.2 mm/min) over 2000 ft^2 (186 m^2). Interpolating for the 5-ft (1.5-m) aisle width results in a density of 0.7265 gpm/ft^2 (29.6 mm/min) over 2000 ft^2 (186 m^2). In accordance with 7-4.2.2.1.9, the sprinkler operating area is permitted to be reduced because the clearance is less than 4 ft (1.2 m). However, the operating area cannot be reduced to less than 2000 ft^2 (186 m^2), so a reduction cannot be taken because the density is based on a 2000-ft^2 (186-m^2) operating area. If the initial densities were selected for a larger operating area such as 3000 ft^2 (279 m^2), a reduction would have been permitted. In accordance with 7-4.2.2.1.11, where solid-shelving materials are used, the density needs to be increased by 20 percent. The result is a final density of 0.87 gpm/ft^2 (35.5 mm/min) over 2000 ft^2 (186 m^2).

7-4.2.2.2.2*† Design curves for single- and double-row racks shall be selected to correspond to aisle width. For aisle widths between 4 ft (1.2 m) and 8 ft (2.4 m), a direct linear interpolation between curves shall be made. The density given for 8-ft (2.4-m) wide aisles shall be applied to aisles wider than 8 ft (2.4 m). The density given for 4-ft (1.2-m) wide aisles shall be applied to aisles narrower than 4 ft (1.2 m) down to 3½ ft (1.07 m). Where aisles are narrower than 3½ ft (1.07 m), racks shall be considered to be multiple-row racks.

A-7-4.2.2.2.2 The aisle width and the depth of racks are determined by material-handling methods. The widths of aisles should be considered in the design of the protection system. Storage in aisles can render protection ineffective and should be discouraged.

Sprinkler protection requirements do not continue to decrease as aisle width increases beyond 8 ft (2.4 m). For aisle widths less than 8 ft (2.4 m), a higher density is needed to prevent fire from jumping the aisle. For aisles greater than 8 ft (2.4 m), aisle jump is no longer a controlling factor in protection requirements.

7-4.2.3 Ceiling Sprinkler Water Demand—Multiple-Row Racks.

7-4.2.3.1 For nonencapsulated Class I, Class II, Class III, or Class IV commodities, ceiling sprinkler water demand in terms of density (gpm/ft^2) (mm/min) and area of sprinkler operation [ft^2 (m^2) of ceiling or roof] shall be selected from the curves in Figures 7-4.2.2.1.1(a) through (d). The curves in Figures 7-4.2.2.1.1(a) through (d) also shall apply to portable racks arranged in the same manner as single-, double-, or multiple-row racks. The design shall be sufficient to satisfy a single point on the appropriate curve related to the storage configuration and commodity class. It shall not be required to meet all points on the selected curve. Figure 7-4.2.2.1.3 shall be used to adjust density for storage height unless otherwise specified. *(See A-7-4.2.1.2.1.)*

In many cases, selecting the design point on the curve corresponding to the minimum design area results in the most economical and most effective protection.

The following example illustrates how to determine sprinkler protection for multiple-row racks with storage under 25 ft (7.6 m).

Example 7.4

An encapsulated Class III commodity is stored up to 24 ft (7.3 m) in height in a 25-ft (7.6-m) wide multiple-row rack structure with 10-ft (3.1-m) wide aisles between rack structures. Ordinary temperature–rated ceiling sprinklers are used.

Solution:

Table 7-4.2.1.6.2 would be used to determine the sprinkler protection because the rack depth exceeds 16 ft (4.9 m). Although the aisle width between rack structures exceeds 8 ft (2.4 m), limitations for both rack depth and aisle width need to be satisfied simultaneously in order to use Table 7-4.2.1.6.1. For encapsulated storage, one level of in-rack sprinklers is required. The ceiling density would be obtained from curve D of Figure 7-4.2.2.1.1(c). The result is a density of 0.325 gpm/ft^2 (13.2 mm/min) over 2000 ft^2 (186 m^2) The density is not adjusted for storage height as indicated Table 7-4.2.1.6.2. However, in accordance with 7-4.2.3.2, the density needs to be increased by 25 percent for encapsulated Class III commodities. The final density is 0.41 gpm/ft^2 (16.3 mm/min) over 2000 ft^2 (186 m^2) with one level of in-rack sprinklers.

7-4.2.3.2 For encapsulated Class I, Class II, or Class III commodities with storage height up to and including 25 ft (7.6 m) on multiple-row racks, ceiling sprinkler density shall be 25 percent greater than for nonencapsulated commodities on multiple-row racks.

7-4.2.3.3 For encapsulated Class IV commodities with storage height up to and including 25 ft (7.6 m) on multiple-row racks, ceiling sprinkler density shall be 50 percent greater than for nonencapsulated commodities on double-row racks.

7-4.3 Spray Sprinkler Protection Criteria for Class I through Class IV Commodities Stored Over 25 ft (7.6 m) in Height.

7-4.3.1 General.

Location and stagger of in-rack sprinklers are critical to the successful protection of rack storage above 25 ft (7.6 m). Fire must be confined to a small area within the rack or it will not be controlled at all. Increased ceiling sprinkler densities for these rack storage heights can almost never compensate for deficiencies in in-rack protection.

7-4.3.1.1 In-Rack Sprinkler Spacing. In-rack sprinklers shall be staggered horizontally and vertically where installed in accordance with Table 7-4.3.1.5.1, Figures 7-4.3.1.5.1(a) through (j), and Figures 7-4.3.1.5.3(a) through (e).

7-4.3.1.2 In-Rack Sprinkler Location. In single-row, double-row, or multiple-row racks, a minimum 6-in. (152.4-mm) vertical clear space shall be maintained between the sprinkler deflectors and the top of a tier of storage. Face sprinklers in such racks shall be located a minimum of 3 in. (76 mm) from rack uprights and no more than 18 in. (460 mm) from the aisle face of storage. Longitudinal flue in-rack sprinklers shall be located at the intersection with the transverse flue space and with the deflector located at or below the bottom of horizontal load beams or above or below other adjacent horizontal rack members. Such in-rack sprinklers shall be a minimum of 3 in. (76 mm) radially from the side of the rack uprights.

Face sprinklers must be located within racks and not over aisles. A fire spreads vertically through the flues or on the face of the pallet loads along the aisles depending on the location of the ignition source. If the face sprinklers are located in the aisle, their activation will be significantly delayed if it occurs at all. Because face sprinklers incorrectly positioned in the aisle are likely to operate only if they are subject to direct flame impingement, it can be assumed that the fire is overwhelming the sprinkler system design and that success is not likely.

7-4.3.1.3 In-Rack Sprinkler Discharge. Sprinklers in racks shall discharge at a rate not less than 30 gpm (113.6 L/min) for all classes of commodities. *(See C-7-4.2.1.4.)*

7-4.3.1.4 In-Rack Sprinkler Water Demand. The water demand for sprinklers installed in racks shall be based on simultaneous operation of the most hydraulically remote sprinklers as follows:

(1) Six sprinklers where only one level is installed in racks with Class I, Class II, or Class III commodities
(2) Eight sprinklers where only one level is installed in racks with Class IV commodities
(3) Ten sprinklers (five on each two top levels) where more than one level is installed in racks with Class I, Class II, or Class III commodities
(4) Fourteen sprinklers (seven on each two top levels) where more than one level is installed in racks with Class IV commodities

7-4.3.1.5 In-Rack Sprinkler Location—Double- and Single-Row Racks.

7-4.3.1.5.1* In double-row racks without solid shelves and with a maximum of 10 ft (3.1 m) between the top of storage and the ceiling, in-rack sprinklers shall be installed in accordance with Table 7-4.3.1.5.1 and Figures 7-4.3.1.5.1(a) through (j). The highest level of in-rack sprinklers shall be not more than 10 ft (3.1 m) below the top of storage. *(See 7-4.1.9.)*

Where a single-row rack is mixed with double-row racks, Table 7-4.3.1.5.1 and Figures 7-4.3.1.5.1(a) through 7-4.3.1.5.1(j) shall be used.

Exception: Figures 7-4.3.1.5.3(a) through (c) shall be permitted to be used for the protection of the single-row racks.

Table 7-4.3.1.5.1 Double-Row Racks without Solid Shelves—Storage Higher than 25 ft (7.6 m), Aisles 4 ft (1.2 m) or Wider

Commodity Class	In-Rack Sprinklers—Approximate Vertical Spacing at Tier Nearest the Vertical Distance and Maximum Horizontal Spacing[1,2,3]		Figure	Maximum Storage Height	Stagger	Ceiling Sprinkler Operating Area		Ceiling Sprinkler Density Clearance up to 10 ft (3.1 m)[4,5,6]			
								Ordinary Temperature		High Temperature	
	Longitudinal Flue[7]	Face[8,9]				ft^2	m^2	gpm/ft^2	mm/min	gpm/ft^2	mm/min
I	Vertical 20 ft (6.1 m) Horizontal 10 ft (3.1 m) under horizontal barriers	None	7-4.3.1.5.1(a)	30 ft (9.1 m)	No			0.25	10.2	0.35	14.3
	Vertical 20 ft (6.1 m) Horizontal 10 ft (3.1 m)	Vertical 20 ft (6.1 m) Horizontal 10 ft (3.1 m)	7-4.3.1.5.1(b)	Higher than 25 ft (7.6 m)	Yes			0.25	10.2	0.35	14.3
I, II, III	Vertical 10 ft (3.1 m) or at 15 ft (4.6 m) and 25 ft (7.6 m) Horizontal 10 ft (3.1 m)	None	7-4.3.1.5.1(c)	30 ft (9.1 m)	Yes	2000	186	0.3	12.2	0.4	16.3
	Vertical 10 ft (3.1 m) Horizontal 10 ft (3.1 m)	Vertical 30 ft (9.1 m) Horizontal 10 ft (3.1 m)	7-4.3.1.5.1(d)	Higher than 25 ft (7.6 m)	Yes			0.3	12.2	0.4	16.3
	Vertical 20 ft (6.1 m) Horizontal 10 ft (3.1 m)	Vertical 20 ft (6.1 m) Horizontal 5 ft (1.5 m)	7-4.3.1.5.1(e)		Yes			0.3	12.2	0.4	16.3
	Vertical 25 ft (7.6 m) Horizontal 5 ft (1.5 m)	Vertical 25 ft (7.6 m) Horizontal 5 ft (1.5 m)	7-4.3.1.5.1(f)		No			0.3	12.2	0.4	16.3
	Horizontal barriers at 20 ft (6.1 m) Vertical intervals—two lines of sprinklers under barriers—maximum horizontal spacing 10 ft (3.1 m), staggered		7-4.3.1.5.1(g)		Yes			0.3	12.2	0.4	16.3

(continues)

Table 7-4.3.1.5.1 Double-Row Racks without Solid Shelves—Storage Higher than 25 ft (7.6 m), Aisles 4 ft (1.2 m) or Wider (Continued)

	In-Rack Sprinklers—Approximate Vertical Spacing at Tier Nearest the Vertical Distance and Maximum Horizontal Spacing[1,2,3]					Ceiling Sprinkler Operating Area		Ceiling Sprinkler Density Clearance up to 10 ft (3.1 m)[4,5,6]			
								Ordinary Temperature		High Temperature	
Commodity Class	**Longitudinal Flue[7]**	**Face[8,9]**	**Figure**	**Maximum Storage Height**	**Stagger**	**ft²**	**m²**	**gpm/ft²**	**mm/min**	**gpm/ft²**	**mm/min**
I, II, III, IV	Vertical 15 ft (4.6 m) Horizontal 10 ft (3.1 m)	Vertical 20 ft (6.1 m) Horizontal 10 ft (3.1 m)	7-4.3.1.5.1(h)	Higher than 25 ft (7.6 m)	Yes	2000	186	0.35	14.3	0.45	18.3
	Vertical 20 ft (6.1 m) Horizontal 5 ft (1.5 m)	Vertical 20 ft (6.1 m) Horizontal 5 ft (1.5 m)	7-4.3.1.5.1(i)		No			0.35	14.3	0.45	18.3
	Horizontal barriers at 15 ft (4.6 m) Vertical intervals—two lines of sprinklers under barriers—maximum horizontal spacing 10 ft (3.1 m), staggered		7-4.3.1.5.1(j)		Yes			0.35	14.3	0.45	18.3

[1]Minimum in-rack sprinkler discharge, 30 gpm (114 L/min) *(see 7-4.3.1.3).*

[2]Water shields required *(see 5-12.3.1).*

[3]All in-rack sprinkler spacing dimensions start from the floor.

[4]For encapsulated commodity, increase density 25 percent *(see 7-4.3.2.2).*

[5]Clearance is distance between top of storage and ceiling.

[6]See A-7-4.3.1.5.3, A-7-4.1.11, and A-7-4.3.2.1 for protection recommendations where clearance is greater than 10 ft (3.1 m).

[7]Install sprinklers at least 3 in. (76.2 mm) from uprights *(see 7-4.3.1.2).*

[8]Face sprinklers shall not be required for a Class I commodity consisting of noncombustible products on wood pallets (without combustible containers), except for arrays shown in Figures 7-4.3.1.5.1(g) and 7-4.3.1.5.1(j).

[9]In Figures 7-4.3.1.5.1(a) through 7-4.3.1.5.1(j), each square represents a storage cube that measures 4 ft to 5 ft (1.2 m to 1.5 m) on a side. Actual load heights can vary from approximately 18 in. to 10 ft (0.46 m to 3.1 m). Therefore, there can be one load to six or seven loads between in-rack sprinklers that are spaced 10 ft (3.1 m) apart vertically.

A-7-4.3.1.5.1 Where storage tiers are not the same size on each side of the longitudinal flue, one side of the flue should be protected with sprinklers at the proper elevation above the load. The next level of sprinklers should protect the other side of the flue with the sprinklers at the proper elevation above that load as indicated in Figure A-7-4.3.1.5.1. The vertical spacing requirements for in-rack sprinklers specified in Tables 7-4-3.1.5.1 and 7-4.3.1.5.4 and 7-4.4 for plastics should be followed.

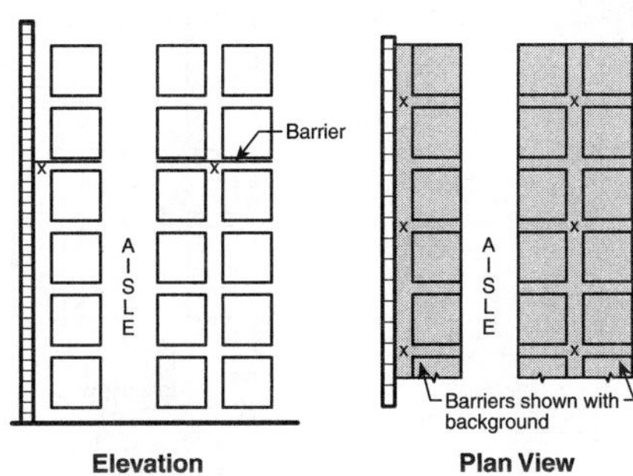

Note: Symbol x indicates in-rack sprinklers.

Figure 7-4.3.1.5.1(a) In-rack sprinkler arrangement, Class I commodities, storage height 25 ft to maximum 30 ft (7.6 m to maximum 9.1 m).

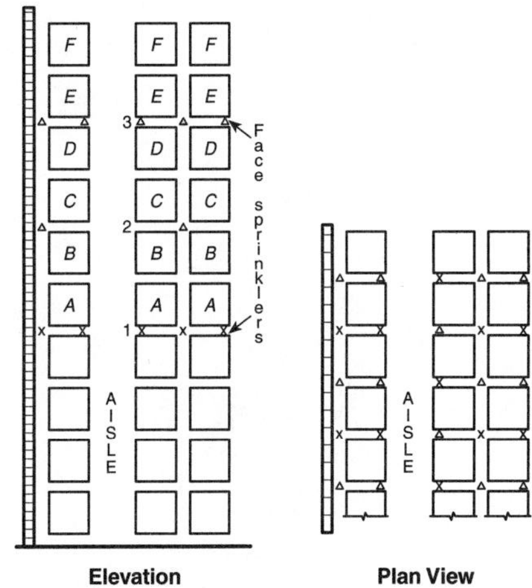

Notes:

1. Sprinklers labeled 1 (the selected array from Table 7-4.3.1.5.1) shall be required where loads labeled *A* or *B* represent top of storage.
2. Sprinklers labeled 1 and 2 shall be required where loads labeled *C* or *D* represent top of storage.
3. Sprinklers labeled 1 and 3 shall be required where loads labeled *E* or *F* represent top of storage.
4. For storage higher than represented by loads labeled *F*, the cycle defined by Notes 2 and 3 is repeated, with stagger as indicated.
5. Symbol Δ or x indicates sprinklers on vertical or horizontal stagger.

Figure 7-4.3.1.5.1(b) In-rack sprinkler arrangement, Class I commodities, storage height over 25 ft (7.6 m).

7-4.3.1.5.2 In-rack sprinklers for storage higher than 25 ft (7.6 m) in double-row racks shall be spaced horizontally and located in the horizontal space nearest the vertical intervals specified in Table 7-4.3.1.5.1 and Figures 7-4.3.1.5.1(a) through (j).

7-4.3.1.5.3* In single-row racks without solid shelves with storage height over 25 ft (7.6 m) and a maximum of 10 ft (3.1 m) between the top of storage and the ceiling, sprinklers shall be installed in accordance with Figures 7-4-3.1.5.3(a) through (e).

In single-row racks, where figures show in-rack sprinklers in transverse flue spaces centered between the rack faces, it shall be permitted to position these in-rack sprinklers in the transverse flue at any point between the load faces.

A-7-4.3.1.5.3 In single-row racks with more than 10 ft (3.1 m) between the top of storage and the ceiling, a horizontal barrier should be installed above storage with one line of sprinklers under the barrier.

See commentary to 7-4.2.1.5.

7-4.3.1.5.4* In-Rack Sprinkler Location—Multiple-Row Racks. In multiple-row racks with a maximum of 10 ft (3.1 m) between the top of storage and the ceiling, protection shall be in

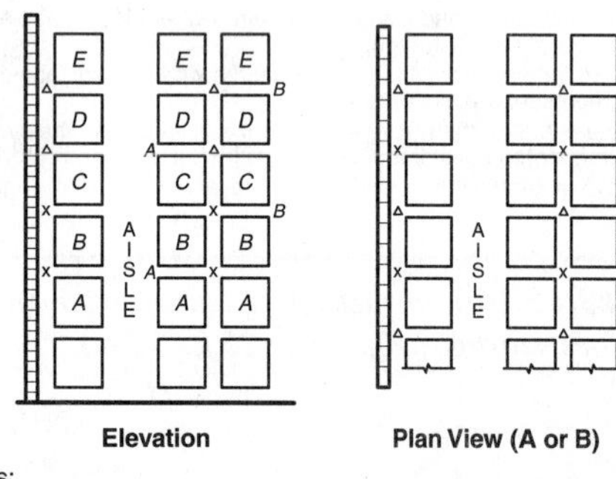

Notes:

1. Alternate location of in-rack sprinklers. Sprinklers shall be permitted to be installed above loads *A* and *C* or above loads *B* and *D*.
2. Symbol Δ or x indicates sprinklers on vertical or horizontal stagger.

Figure 7-4.3.1.5.1(c) *In-rack sprinkler arrangement, Class I, Class II, or Class III commodities, storage height 25 ft to maximum 30 ft (7.6 m to maximum 9.1 m).*

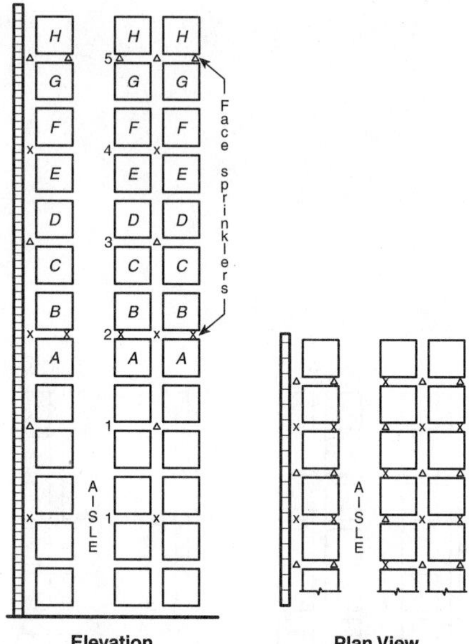

Notes:

1. Sprinklers labeled 1 shall be required where loads labeled *A* represent the top of storage.
2. Sprinklers labeled 1 and 2 shall be required where loads labeled *B* or *C* represent top of storage.
3. Sprinklers labeled 1, 2, and 3 shall be required where loads labeled *D* or *E* represent top of storage.
4. Sprinklers labeled 1, 2, 3, and 4 shall be required where loads labeled *F* or *G* represent top of storage.
5. Sprinklers labeled 1, 2, 3, 4, and 5 shall be required where loads labeled *H* represent top of storage.
6. For storage higher than represented by loads labeled *H*, the cycle defined by Notes 3, 4, and 5 is repeated with stagger as indicated.
7. The indicated face sprinklers shall be permitted to be omitted where commodity consists of unwrapped or unpackaged metal parts on wood pallets.
8. Symbol Δ or x indicates sprinklers on vertical or horizontal stagger.

Figure 7-4.3.1.5.1(d) *In-rack sprinkler arrangement, Class I, Class II, or Class III commodities, storage height over 25 ft (7.6 m)—option 1.*

accordance with Table 7-4.3.1.5.4 and in-rack sprinklers shall be installed as indicated in Figures 7-4.3.1.5.4(a), (b), and (c). The highest level of in-rack sprinklers shall be not more than 10 ft (3.1 m) below maximum storage height for Class I, Class II, or Class III commodities or 5 ft (1.5 m) below the top of storage for Class IV commodities.

A-7-4.3.1.5.4 In multiple-row racks with more than 10 ft (3.1 m) between the maximum height of storage and ceiling, a horizontal barrier should be installed above storage with a level of sprinklers, spaced as stipulated for in-rack sprinklers, installed directly beneath the barrier. In-rack sprinklers should be installed as indicated in Figures 7-4.3.1.5.4(a) through (c).

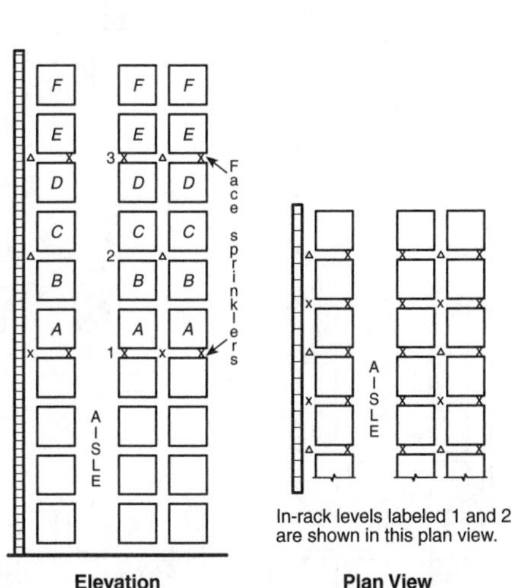

Notes:

1. Sprinklers labeled 1 (the selected array from Table 7-4.3.1.5.1) shall be required where loads labeled *A* or *B* represent top of storage.
2. Sprinklers labeled 1 and 2 shall be required where loads labeled *C* or *D* represent top of storage.
3. Sprinklers labeled 1 and 3 shall be required where loads labeled *E* or *F* represent top of storage.
4. For storage higher than represented by loads labeled *F*, the cycle defined by Notes 2 and 3 is repeated, with stagger as indicated.
5. Symbol Δ or x indicates sprinklers on vertical or horizontal stagger.

Figure 7-4.3.1.5.1(e) *In-rack sprinkler arrangement, Class I, Class II, or Class III commodities, storage height over 25 ft (7.6 m)—option 2.*

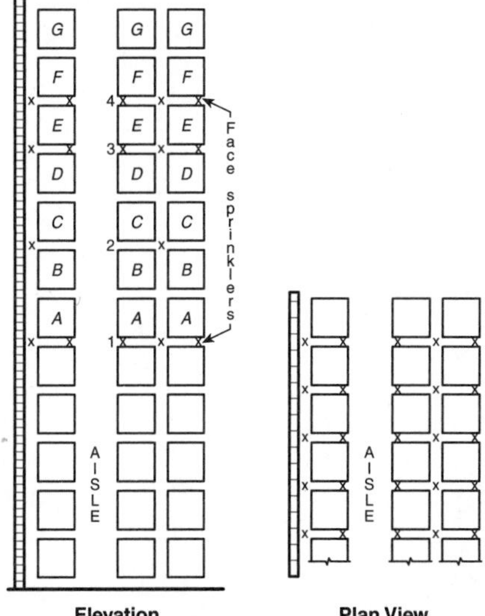

Notes:

1. Sprinklers labeled 1 (the selected array from Table 7-4.3.1.5.1) shall be required where loads labeled *A* or *B* represent top of storage.
2. Sprinklers labeled 1 and 2 shall be required where loads labeled *C* or *D* represent top of storage.
3. Sprinklers labeled 1 and 3 shall be required where loads labeled *E* represent top of storage.
4. Sprinklers labeled 1 and 4 shall be required where loads labeled *F* or *G* represent top of storage.
5. For storage higher than represented by loads labeled *G*, the cycle defined by Notes 2, 3, and 4 is repeated.
6. Symbol x indicates face and in-rack sprinklers.

Figure 7-4.3.1.5.1(f) *In-rack sprinkler arrangement, Class I, Class II, or Class III commodities, storage height over 25 ft (7.6 m)—option 3.*

Data indicate that the sprinkler protection criteria in 7-4.3.1.5.4 is ineffective, by itself, for rack storage with solid shelves, if the required flue spaces are not maintained. Use of Table 7-4.3.1.5.4, along with the additional provisions that are required by this standard, can provide acceptable protection.

7-4.3.1.5.5 In-Rack Sprinkler Spacing. Maximum horizontal spacing of sprinklers in multiple-row racks with storage higher than 25 ft (7.6 m) shall be in accordance with Figures 7-4.3.1.5.4(a), (b), and (c).

7-4.3.2 Ceiling Sprinklers for Single- and Double-Row Racks—Area Density Method.

7-4.3.2.1*† The water demand for nonencapsulated storage on racks without solid shelves separated by aisles at least 4 ft (1.2 m) wide and with not more than 10 ft (3.1 m) between the top of storage and the sprinklers shall be based on sprinklers in a 2000-ft^2 (186-m^2) operating

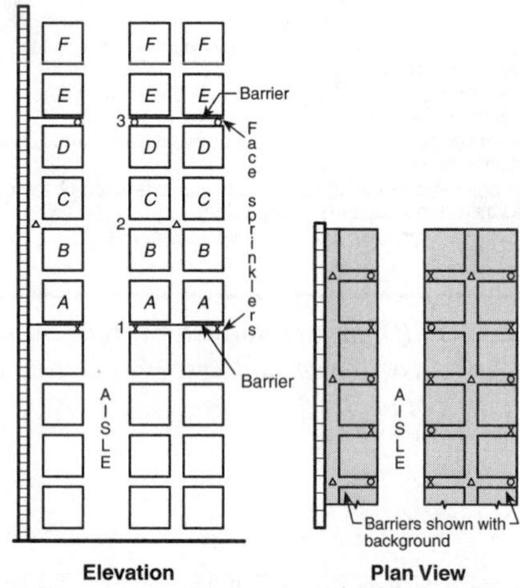

Notes:

1. Sprinklers labeled 1 (the selected array from Table 7-4.3.1.5.1) shall be required where loads labeled *A* or *B* represent top of storage.
2. Sprinklers labeled 1 and 2 shall be required where loads labeled *C* or *D* represent top of storage.
3. Sprinklers labeled 1 and 3 shall be required where loads labeled *E* or *F* represent top of storage.
4. For storage higher than represented by loads labeled *F*, the cycle defined by Notes 2 and 3 is repeated.
5. Symbols o, Δ, and x indicate sprinklers on vertical or horizontal stagger.

Figure 7-4.3.1.5.1(g) *In-rack sprinkler arrangement, Class I, Class II, or Class III commodities, storage height over 25 ft (7.6 m)—option 4.*

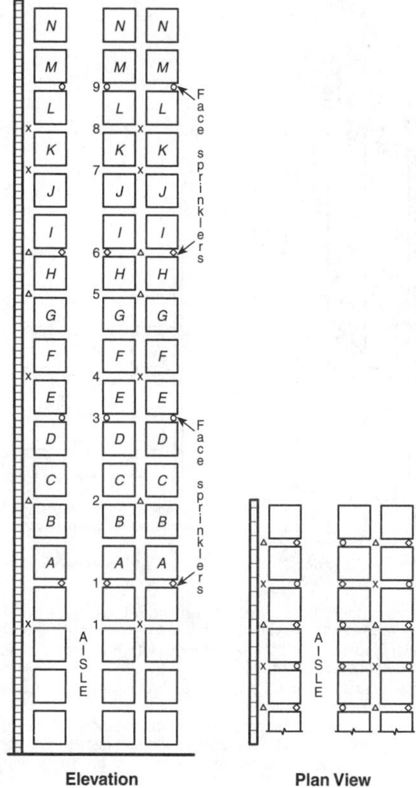

Notes:

1. Sprinklers labeled 1 (the selected array from Table 7-4.3.1.5.1) shall be required where loads labeled *A* or *B* represent top of storage.
2. Sprinklers labeled 1 and 2 shall be required where loads labeled *C* or *D* represent top of storage.
3. Sprinklers labeled 1, 2, and 3 shall be required where loads labeled *E* or *F* represent top of storage.
4. Sprinklers labeled 1, 2, 3, and 4 shall be required where loads labeled *G* represent top of storage.
5. Sprinklers labeled 1, 2, 3, 4, and 5 shall be required where loads labeled *H* represent top of storage.
6. Sprinklers labeled 1, 2, 3, 4, and 6 (not 5) shall be required where loads labeled *I* or *J* represent top of storage.
7. Sprinklers labeled 1, 2, 3, 4, 6, and 7 shall be required where loads labeled *K* represent top of storage.
8. Sprinklers labeled 1, 2, 3, 4, 6, and 8 shall be required where loads labeled *L* represent top of storage.
9. Sprinklers labeled 1, 2, 3, 4, 6, 8, and 9 shall be required where loads labeled *M* or *N* represent top of storage.
10. For storage higher than represented by loads labeled *N*, the cycle defined by Notes 1 through 9 is repeated, with stagger as indicated. In the cycle, loads labeled *M* are equivalent to loads labeled *A*.
11. Symbols o, x, and Δ indicate sprinklers on vertical or horizontal stagger.

Figure 7-4.3.1.5.1(h) *In-rack sprinkler arrangement, Class I, Class II, Class III, or Class IV commodities, storage height over 25 ft (7.6 m)—option 1.*

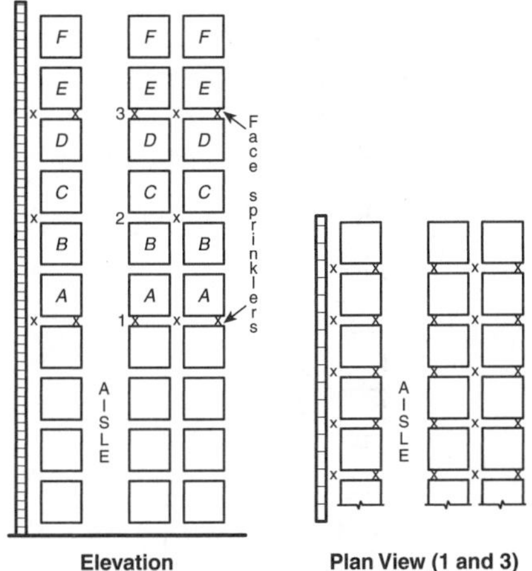

Notes:

1. Sprinklers labeled 1 (the selected array from Table 7-4.3.1.5.1) shall be required where loads labeled *A* or *B* represent top of storage.
2. Sprinklers labeled 1 and 2 shall be required where loads labeled *C* or *D* represent top of storage.
3. Sprinklers labeled 1 and 3 shall be required where loads labeled *E* or *F* represent top of storage.
4. For storage higher than represented by loads labeled *F*, the cycle defined by Notes 2 and 3 is repeated.
5. Symbol x indicates face and in-rack sprinklers.

Figure 7-4.3.1.5.1(i) *In-rack sprinkler arrangement, Class I, Class II, Class III, or Class IV commodities, storage height over 25 ft (7.6 m)—option 2.*

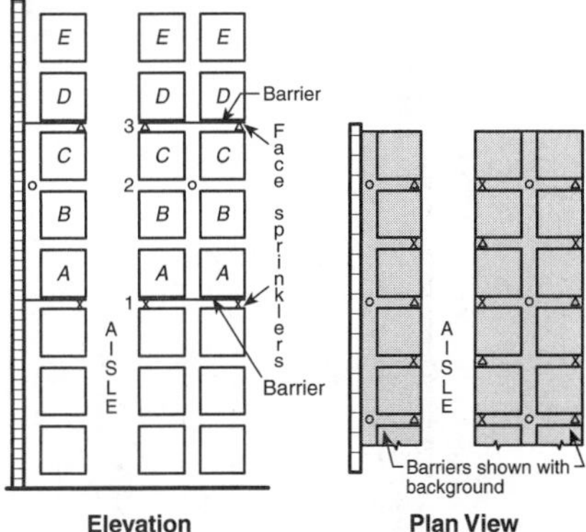

Notes:

1. Sprinklers labeled 1 (the selected array from Table 7-4.3.1.5.1) shall be required where loads labeled *A* or *B* represent top of storage.
2. Sprinklers labeled 1 and 2 and barrier labeled 1 shall be required where loads labeled *C* represent top of storage.
3. Sprinklers and barriers labeled 1 and 3, shall be required where loads labeled *D* or *E* represent top of storage.
4. For storage higher than represented by loads labeled *E*, the cycle defined by Notes 2 and 3 is repeated.
5. Symbol Δ or x indicates sprinklers on vertical or horizontal stagger.
6. Symbol o indicates longitudinal flue space sprinklers.

Figure 7-4.3.1.5.1(j) *In-rack sprinkler arrangement, Class I, Class II, Class III, or Class IV commodities, storage height over 25 ft (7.6 m)—option 3.*

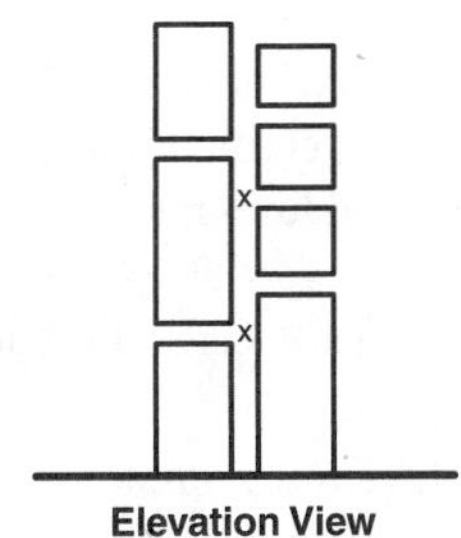

Figure A-7-4.3.1.5.1 *Placement of in-rack sprinkler where rack levels have varying heights.*

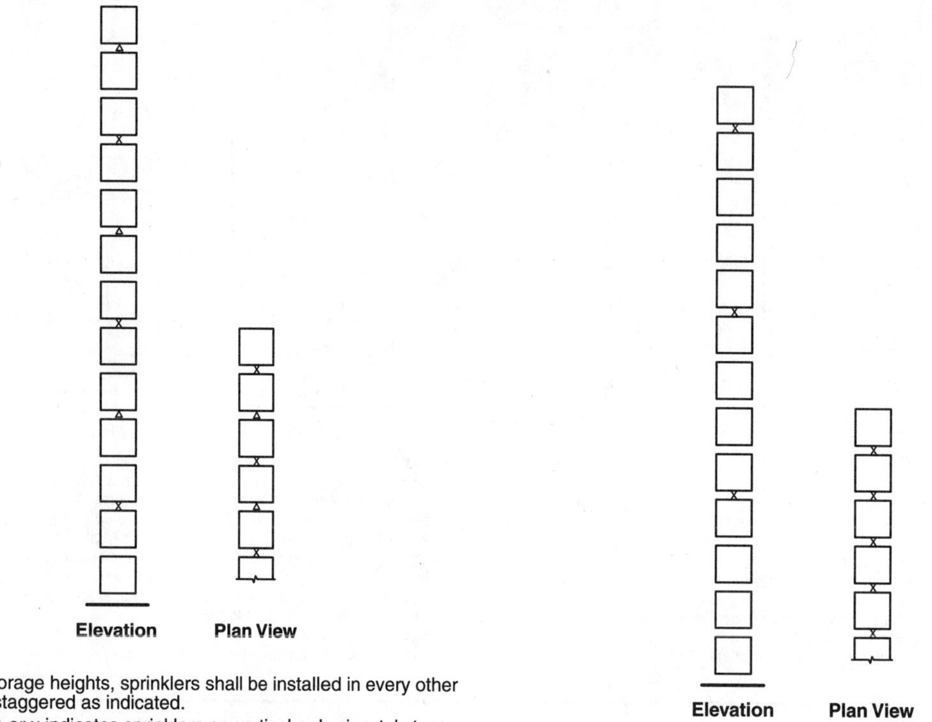

Notes:

1. For all storage heights, sprinklers shall be installed in every other tier and staggered as indicated.
2. Symbol Δ or x indicates sprinklers on vertical or horizontal stagger.
3. Each square in the figure represents a storage cube measuring 4 ft to 5 ft (1.25 m to 1.56 m) on a side.

Figure 7-4.3.1.5.3(a) *Class I, Class II, Class III, or Class IV commodities, in-rack sprinkler arrangement, single-row racks, storage height over 25 ft (7.6 m)—option 1.*

Note: Each square in the figure represents a storage cube measuring 4 ft to 5 ft (1.25 m to 1.56 m) on a side.

Figure 7-4.3.1.5.3(b) *Class I, Class II, or Class III commodities, in-rack sprinkler arrangement, single-row racks, storage height over 25 ft (7.6 m)—option 1.*

area, discharging a minimum of 0.25 gpm/ft^2 (10.2 mm/min) for Class I commodities, 0.3 gpm/ft^2 (12.2 mm/min) for Classes II and III commodities, and 0.35 gpm/ft^2 (14.3 mm/min) for Class IV commodities for ordinary temperature–rated sprinklers or a minimum of 0.35 gpm/ft^2 (14.3 mm/min) for Class I commodities, 0.4 gpm/ft^2 (16.3 mm/min) for Classes II and III commodities, and 0.45 gpm/ft^2 (18.3 mm/min)] for Class IV commodities for high temperature–rated sprinklers. *(See Table 7-4.3.1.5.1.)*

A-7-4.3.2.1 Water demand for storage height over 25 ft (7.6 m) on racks without solid shelves separated by aisles at least 4 ft (1.2 m) wide and with more than 10 ft (3.1 m) between the top of storage and the sprinklers should be based on sprinklers in a 2000-ft^2 (186-m^2) operating area for double-row racks and a 3000-ft^2 (278.7-m^2) operating area for multiple-row racks discharging a minimum of 0.18 gpm/ft^2 (7.33 mm/min) for Class I commodities, 0.21 gpm/ft^2 (8.56 mm/min) for Classes II and III commodities, and 0.25 gpm/ft^2 (10.2 mm/min) for Class IV commodities for ordinary temperature–rated sprinklers or a minimum of 0.25 gpm/ft^2 (10.2 mm/min) for Class I commodities, 0.28 gpm/ft^2 (11.41 mm/min) for Classes II and III

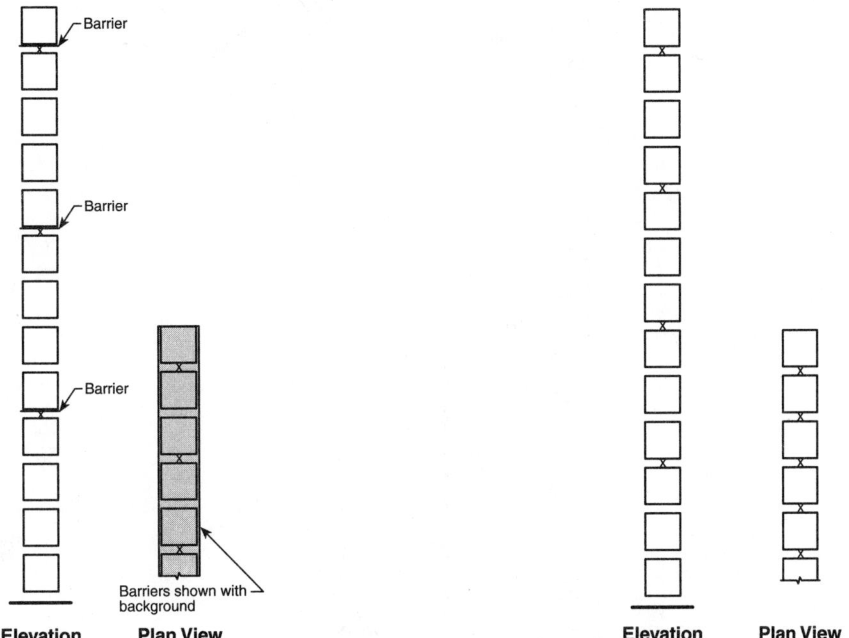

Note: Each square in the figure represents a storage cube measuring 4 ft to 5 ft (1.25 m to 1.56 m) on a side.

Figure 7-4.3.1.5.3(c) *Class I, Class II, or Class III commodities, in-rack sprinkler arrangement, single-row racks, storage height over 25 ft (7.6 m)—option 2.*

Note: Each square in the figure represents a storage cube measuring 4 ft to 5 ft (1.25 m to 1.56 m) on a side.

Figure 7-4.3.1.5.3(d) *Class I, Class II, Class III, or Class IV commodities, in-rack sprinkler arrangement, single-row racks, storage height over 25 ft (7.6 m)—option 2.*

commodities, and 0.32 gpm/ft^2 (13.04 mm/min) for Class IV commodities for high temperature–rated sprinklers. *(See A-7-4.1.11 and A-7-4.3.1.5.4.)*

Where such storage is encapsulated, ceiling sprinkler density should be 25 percent greater than for nonencapsulated storage.

Data indicate that the sprinkler protection criteria in 7-4.3.2.1 is ineffective, by itself, for rack storage with solid shelves, if the required flue spaces are not maintained. Use of 7-4.3.2.1, along with the additional provisions that are required by this standard, can provide acceptable protection.

See commentary to A-7-4.3.1.5.3.

7-4.3.2.2 Where storage as described in 7-4.3.2.1 is encapsulated, ceiling sprinkler density shall be 25 percent greater than for nonencapsulated storage.

7-4.3.3 Ceiling Sprinklers for Multiple-Row Racks—Area Density Method.

7-4.3.3.1 The water demand for nonencapsulated storage on racks without solid shelves separated by aisles at least 4 ft (1.2 m) wide and with not more than 10 ft (3.1 m) between the top

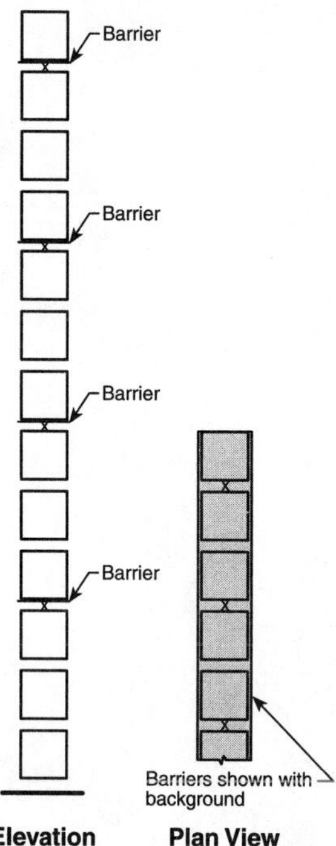

Note: Each square in the figure represents a storage cube measuring 4 ft to 5 ft (1.25 m to 1.56 m) on a side.

Figure 7-4.3.1.5.3(e) *Class I, Class II, Class III, or Class IV commodities, in-rack sprinkler arrangement, single-row racks, storage height over 25 ft (7.6 m)—option 3.*

of storage and the sprinklers shall be based on sprinklers in a 2000-ft^2 (186-m^2) operating area for multiple-row racks, discharging a minimum of 0.25 gpm/ft^2 (10.2 mm/min) for Class I commodities, 0.3 gpm/ ft^2 (12.2 mm/min) for Classes II and III commodities, and 0.35 gpm/ft^2 (14.3 mm/min) for Class IV commodities for ordinary temperature–rated sprinklers or a minimum of 0.35 gpm/ft^2 (14.3 mm/min) for Class I commodities, 0.4 gpm/ft^2 (16.3 mm/min) for Classes II and III commodities, and 0.45 gpm/ft^2 (18.3 mm/min) for Class IV commodities for high temperature–rated sprinklers. *(See Table 7-4.3.1.5.4.)*

7-4.3.3.2 Where such storage is encapsulated, ceiling sprinkler density shall be 25 percent greater than for nonencapsulated storage.

7-4.3.4 High-Expansion Foam Systems. Where high-expansion foam is used in combination with ceiling sprinklers, the minimum ceiling sprinkler design density shall be 0.2 gpm/ft^2 (8.2 mm/min) for Class I, Class II, or Class III commodities and 0.25 gpm/ft^2 (10.2 mm/min) for Class IV commodities for the most hydraulically remote 2000-ft^2 (186-m^2) area.

See 7-3.2.2.4.

Table 7-4.3.1.5.4 Multiple-Row Racks, Storage Heights over 25 ft (7.6 m)

		In-Rack Sprinklers[1,2,3]													Ceiling Sprinklers Density			
		Approximate Vertical Spacing		Maximum Horizontal Spacing in a Flue		Maximum Horizontal Spacing across Flue					Maximum Spacing from Top of Storage to Highest In-Rack Sprinklers		Ceiling Sprinkler Operating Area		165° Rating		286° Rating	
Commodity Class	Encap-sulated	ft	m	ft	m	ft	m	Height Limit ft	Stagger	Figure	ft	m	ft²	m²	gpm/ft²	mm/min	gpm/ft²	mm/min
I	No	20	6.1	12	3.7	10	3.1	None	Between adjacent flues	7-4.3.1.5.4(a)	10	3.1	2000	186	0.25	10.2	0.35	14.3
	Yes														0.31		0.44	
I, II, and III	No	15	4.6	10	3.1	10	3.1			7-4.3.1.5.4(b)	10	3.1			0.30	12.2	0.40	16.3
	Yes														0.37		0.50	20.4
I, II, III, and IV	No	10	3.1	10	3.1	10	3.1			7-4.3.1.5.4(c)	5	1.5			0.35	14.3	0.45	18.3
	Yes														0.44		0.56	

For SI units, °C = 5/9(°F − 32); 1 gpm/ft2 = 40.746 mm/min.

[1]All four rack faces shall be protected by sprinklers located within 18 in. (0.46 m) of the faces, as indicated in Figures 7-4.3.1.5.4(a), (b), and (c). It shall not be required for each sprinkler level to protect all faces. *(See A-7-4.3.1.5.4.)*

[2]All in-rack sprinkler spacing dimensions start from the floor.

[3]In Figures 7-4.3.1.5.4(a) through 7-4.3.1.5.4(c), each square represents a storage cube measuring 4 ft to 5 ft (1.2 m to 1.5 m) on a side. Actual load heights can vary from approximately 18 in. to 10 ft (0.46 m to 1 m). Therefore, there could be as few as one load or as many as six or seven loads between in-rack sprinklers that are spaced 10 ft (3.1 m) apart vertically.

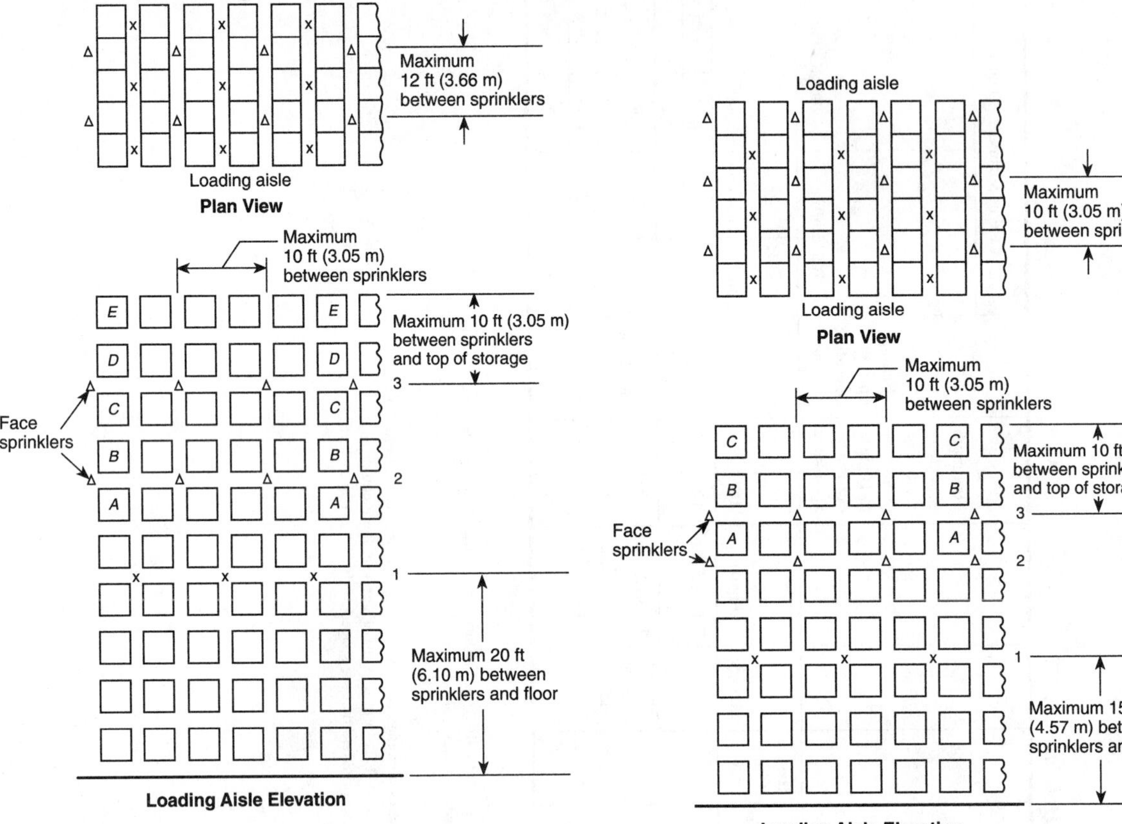

Notes:
1. Sprinklers labeled 1 shall be required if loads labeled *A* represent top of storage.
2. Sprinklers labeled 1 and 2 shall be required if loads labeled *B* or *C* represent top of storage.
3. Sprinklers labeled 1 and 3 shall be required if loads labeled *D* or *E* represent top of storage.
4. For storage higher than represented by loads labeled *E*, the cycle defined by Notes 2 and 3 is repeated, with stagger as indicated.
5. Symbol Δ or x indicates sprinklers on vertical or horizontal stagger.

Figure 7-4.3.1.5.4(a) *In-rack sprinkler arrangement—multiple-row racks, Class I commodities, storage height over 25 ft (7.6 m).*

Notes:
1. Sprinklers labeled 1 and 2 shall be required if loads labeled *A* represent top of storage.
2. Sprinklers labeled 1 and 3 shall be required if loads labeled *B* or *C* represent top of storage.
3. For storage higher than represented by loads labeled *C*, the cycle defined by Notes 2 and 3 is repeated, with stagger as indicated.
4. Symbol Δ or x indicates sprinklers on vertical or horizontal stagger.

Figure 7-4.3.1.5.4(b) *In-rack sprinkler arrangement—multiple-row racks, Class I, Class II, or Class III commodities, storage height over 25 ft (7.6 m).*

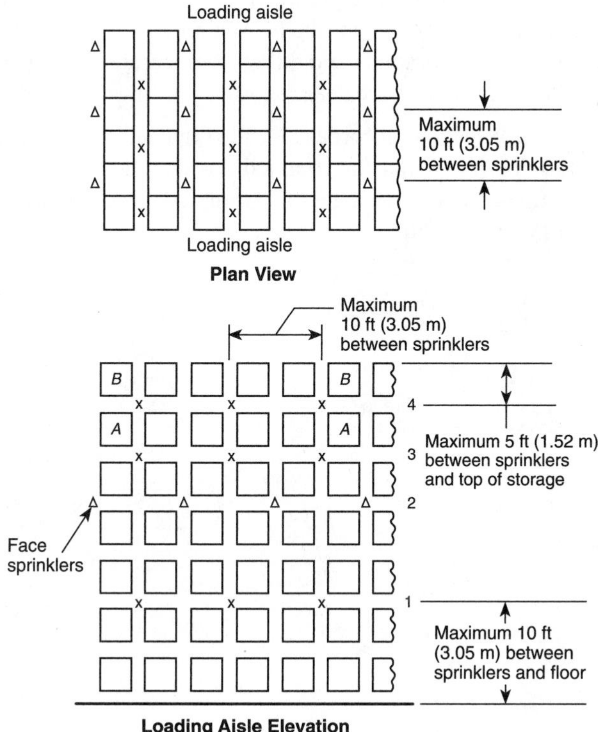
Loading aisle
Maximum
10 ft (3.05 m)
between sprinklers
Loading aisle
Plan View
Maximum
10 ft (3.05 m)
between sprinklers
B
A
4
3
2
1
Maximum 5 ft (1.52 m)
between sprinklers
and top of storage
Face
sprinklers
Maximum 10 ft
(3.05 m) between
sprinklers and floor
Loading Aisle Elevation
Notes:
1. Sprinklers labeled 1, 2, and 3 shall be required if loads labeled A represent top of storage.
2. Sprinklers labeled 1, 2, and 4 shall be required if loads labeled B represent top of storage.
3. For storage higher than represented by loads labeled B, the cycle defined by Notes 1 and 2 is repeated, with stagger as indicated.
4. Symbol Δ or x indicates sprinklers on vertical or horizontal stagger.

Figure 7-4.3.1.5.4(c) *In-rack sprinkler arrangement, Class I, Class II, Class III, or Class IV commodities—multiple-row racks, storage height over 25 ft (7.6 m).*

7-4.4 Spray Sprinkler Protection Criteria for Plastics Commodities.

See commentary to 7-3.3.1.1 on the protection of plastics.

7-4.4.1* General. For the storage of Group A plastics stored 5 ft (1.5 m) or less in height, the sprinkler design criteria for miscellaneous storage specified in 7-2.3.2.2 shall be used.

A-7-4.4.1 All rack fire tests of plastics were run with an approximate 10-ft (3.1-m) maximum clearance between the top of the storage and the ceiling sprinklers. Within 30-ft (9.1-m) high buildings, greater clearances above storage configurations should be compensated for by the addition of more in-rack sprinklers or the provision of greater areas of application, or both.

7-4.4.1.1† Plastic commodities shall be protected in accordance with Figure 7-4.4.1.1. This decision tree also shall be used to determine protection for commodities that are not entirely Group A plastics but contain such quantities and arrangements of Group A plastics that they are deemed more hazardous than Class IV commodities.

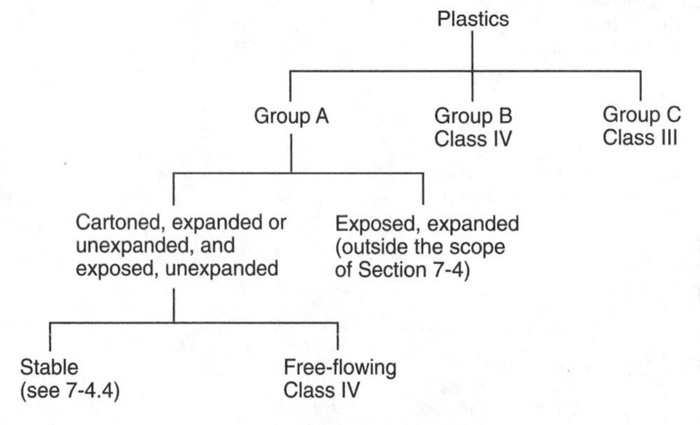

Figure 7-4.4.1.1 *Decision tree.*

7-4.4.1.2 Group B plastics and free-flowing Group A plastics shall be protected the same as Class IV commodities.

7-4.4.1.3 Group C plastics shall be protected the same as Class III commodities.

7-4.4.1.4† Ceiling sprinklers shall have a temperature rating of ordinary, intermediate, or high temperature.

Exception: High temperature–rated sprinklers shall be used where required by 5-3.1.3.

7-4.4.2 In-Rack Sprinklers.

7-4.4.2.1 In-Rack Sprinkler Clearance. The minimum of 6-in. (152.4-mm) vertical clear space shall be maintained between the sprinkler deflectors and the top of a tier of storage.

7-4.4.2.2 In-Rack Sprinkler Water Demand. The water demand for sprinklers installed in racks shall be based on simultaneous operation of the most hydraulically remote sprinklers as follows:

(1) Eight sprinklers where only one level is installed in racks
(2) Fourteen sprinklers (seven on each top two levels) where more than one level is installed in racks

7-4.4.2.3† In-Rack Sprinklers—Storage Up to and Including 25 ft (7.6 m). In-rack sprinklers shall be installed in accordance with Figures 7-4.4.2.3(a) through (g).

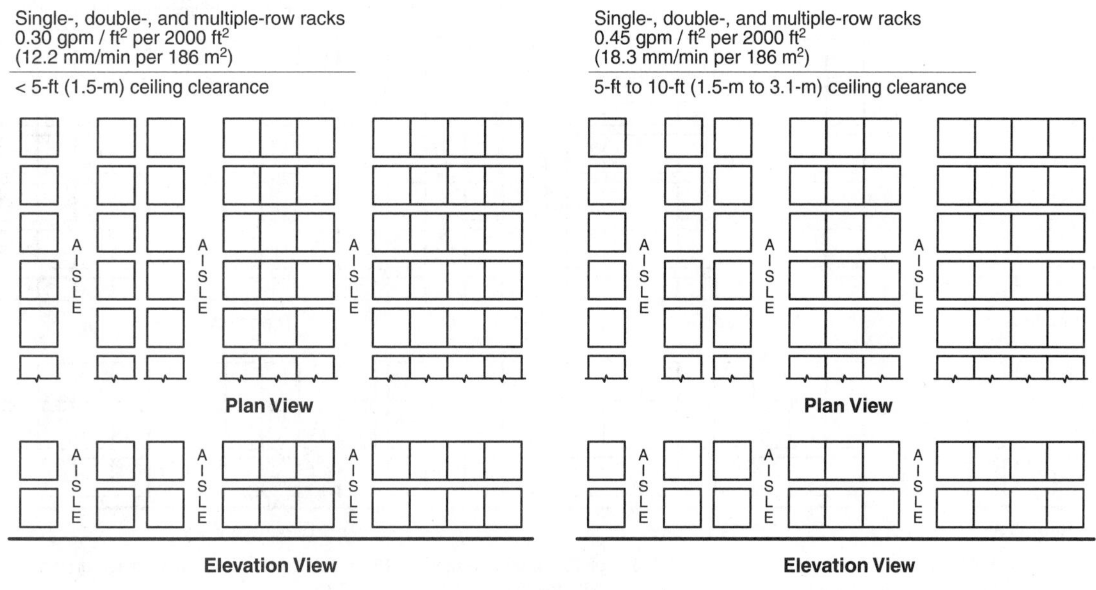

Figure 7-4.4.2.3(a) *5-ft to 10-ft (1.5-m to 3-m) storage.*

7-4.4.2.4 In-Rack Sprinkler Location—Storage Over 25 ft (7.6 m) in Height.

7-4.4.2.4.1 In double-row racks without solid shelves and with a maximum of 10 ft (3.1 m) between the top of storage and the ceiling, in-rack sprinklers shall be installed in accordance with Figure 7-4.4.2.4.1(a) or (b). The highest level of in-rack sprinklers shall be not more than 10 ft (3.1 m) below the top of storage.

See commentary to 7-4.2.1.5.

7-4.4.2.4.2 In-rack sprinklers for storage higher than 25 ft (7.6 m) in double-row racks shall be spaced horizontally and shall be located in the horizontal space nearest the vertical intervals specified in Figure 7-4.4.2.4.1(a) or (b).

7-4.4.2.4.3 In single-row racks without solid shelves with storage height over 25 ft (7.6 m) and a maximum of 10 ft (3.1 m) between the top of storage and the ceiling, sprinklers shall be installed as indicated in Figure 7-4.4.2.4.3(a), (b), or (c).

See commentary to 7-4.4.2.4.1.

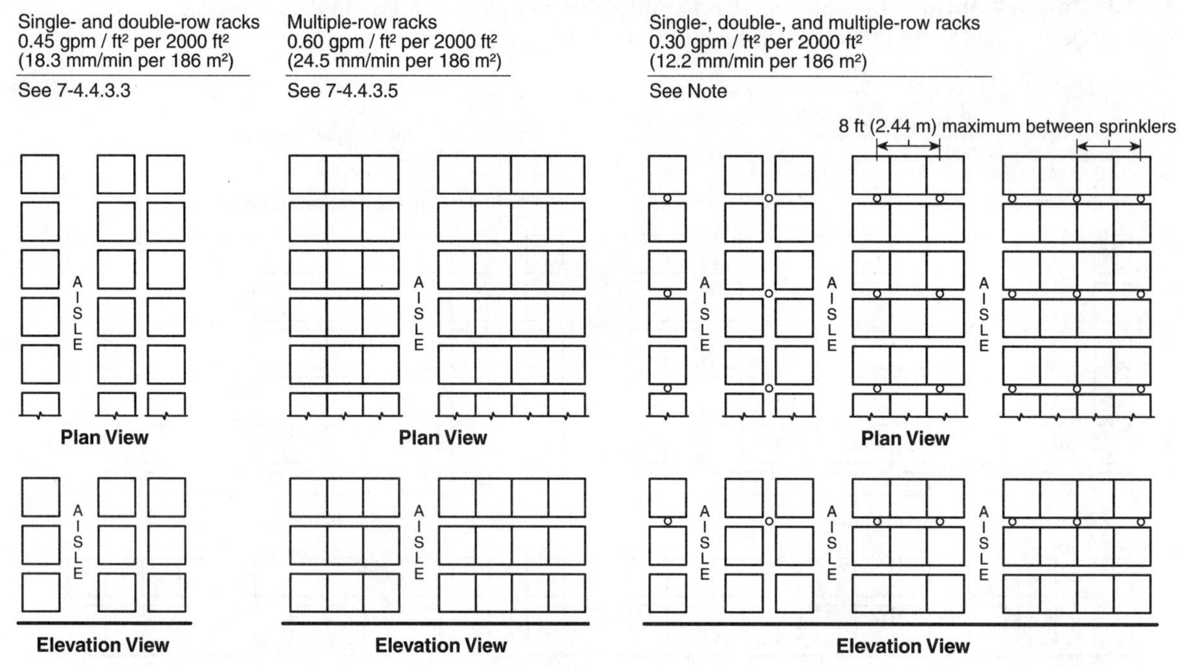

Figure 7-4.4.2.3(b) *15-ft (4.6-m) storage; <5-ft (1.5-m) clearance.*

7-4.4.2.4.4 In multiple-row racks without solid shelves with storage height over 25 ft (7.6 m) and a maximum of 10 ft (3.1 m) between the top of storage and the roof/ceiling, in-rack sprinklers shall be installed as indicated in Figures 7-4.4.2.4.4(a) through 7-4.4.2.4.4(f).

See commentary to 7-4.4.2.4.1.

A-7-4.4.2.4.4 Figure (a). The protection area per sprinkler under barriers should be no greater than 80 ft^2 (7.44 m^2).

A-7-4.4.2.4.4 Figure (b). The protection area per sprinkler under barriers should be no greater than 80 ft^2 (7.44 m^2).

A-7-4.4.2.4.4 Figure (c). The protection area per sprinkler under barriers should be no greater than 50 ft^2 (4.65 m^2).

A-7-4.4.2.4.4 Figure (d). The protection area per sprinkler under barriers should be no greater than 50 ft^2 (4.65 m^2).

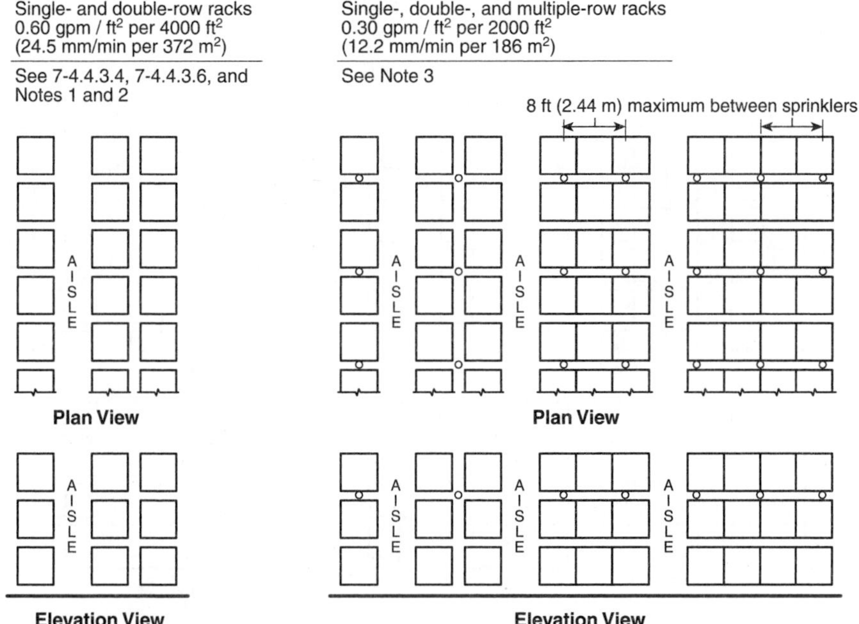

Notes:

1. † Where ⅝-in. (15.9-mm) orifice sprinklers listed for storage use are installed at the ceiling only, the ceiling sprinkler discharge criteria shall be permitted to be reduced to 0.6 gpm / ft^2 per 2000 ft^2 (24.5 mm/min per 186 m^2).
2. † Where ⅝-in. (15.9-mm) orifice sprinklers listed for storage use are installed at the ceiling only and the ceiling height in the protected area does not exceed 22 ft (6.7 m), the ceiling sprinkler discharge criteria shall be permitted to be reduced 0.45 gpm / ft^2 per 2000 ft^2 (18.3 mm/min per 186 m^2).
3. Single level of in-rack sprinklers [½ in. or 17⁄32 in. (12.7 mm or 13.5 mm) operating at 15 psi (1.03 bar) minimum] installed on 8 ft to 10 ft (2.5 m to 3.12 m) spacings located, as indicated, in the transverse flue spaces.

Figure 7-4.4.2.3(c) *15-ft (4.6-m) storage; 5-ft to 10-ft (1.5-m to 3.1-m) ceiling clearance.*

A-7-4.4.2.4.4 Figure (e). The protection area per sprinkler under barriers should be no greater than 50 ft^2 (4.65 m^2).

A-7-4.4.2.4.4 Figure (f). The protection area per sprinkler under barriers should be no greater than 50 ft^2 (4.65 m^2).

7-4.4.2.4.5 In-Rack Sprinkler Discharge Pressure. Sprinklers in racks shall discharge at not less than 30 gpm (113.6 L/min).

7-4.4.3 Ceiling Sprinklers for Single-, Double-, and Multiple-Row Racks—Storage Height Up to 25 ft (7.6 m)—Clearances Up to and Including 10 ft (3.1 m)—Area Density Method.

7-4.4.3.1 Ceiling Sprinkler Water Demand. For Group A plastic commodities in cartons, encapsulated or nonencapsulated in single-, double-, and multiple-row racks, ceiling sprinkler water demand in terms of density [gpm/ ft^2 (mm/min)] and area of operation [ft^2 (m^2)] shall be selected from Figures 7-4.4.2.3(a) through (g). Linear interpolation of design densities and

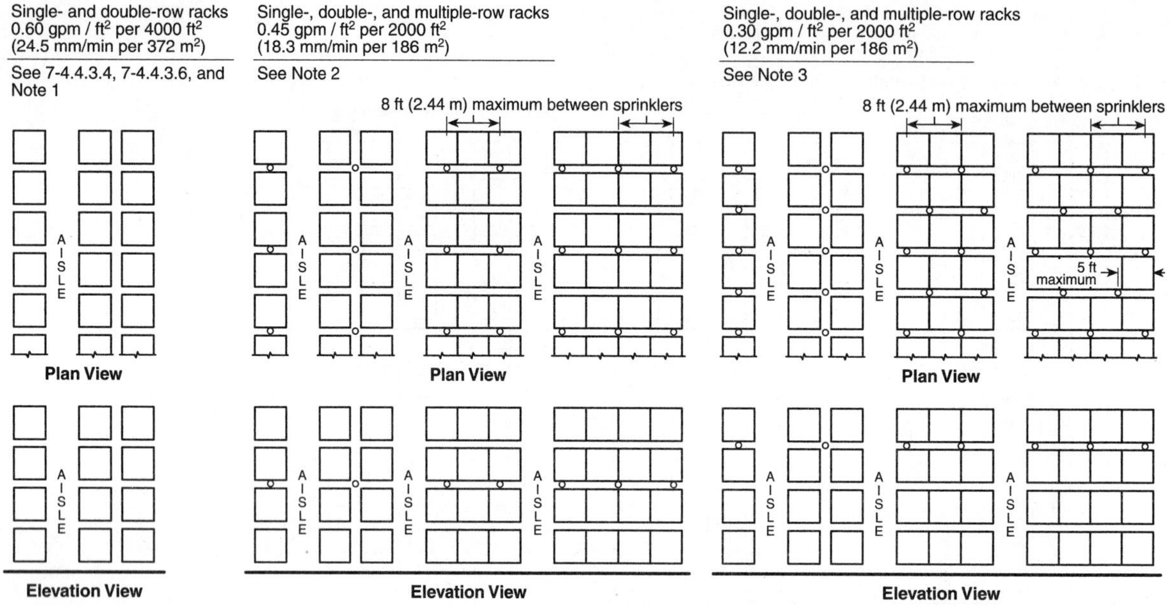

Single- and double-row racks
0.60 gpm / ft^2 per 4000 ft^2
(24.5 mm/min per 372 m^2)
See 7-4.4.3.4, 7-4.4.3.6, and Note 1

Single-, double-, and multiple-row racks
0.45 gpm / ft^2 per 2000 ft^2
(18.3 mm/min per 186 m^2)
See Note 2

Single-, double-, and multiple-row racks
0.30 gpm / ft^2 per 2000 ft^2
(12.2 mm/min per 186 m^2)
See Note 3

8 ft (2.44 m) maximum between sprinklers

8 ft (2.44 m) maximum between sprinklers

AISLE

5 ft maximum

Plan View

Elevation View

Notes:

1. † Where ⅝-in. (15.9-mm) orifice sprinklers listed for storage use are installed at the ceiling only, the ceiling sprinkler discharge criteria shall be permitted to be reduced to 0.6 gpm / ft^2 per 2000 ft^2 (24.5 mm/min per 186 m^2).
2. Single level of in-rack sprinklers [½ in. or $^{17}/_{32}$ in. (12.7 mm or 13.5 mm) operating at 15 psi (1.03 bar) minimum] installed on 8 ft to 10 ft (2.5 m to 3.12 m) spacings located, as indicated, in the transverse flue spaces.
3. Single level of in-rack sprinklers [$^{17}/_{32}$ in. (13.5 mm) operating at 15 psi (1.03 bar) minimum or ½ in. (12.7 mm) operating at 30 psi (2.07 bar) minimum] installed on 4 ft to 5 ft (1.25 m to 1.56 m) spacings located, as indicated, in the longitudinal flue space at the intersection of every transverse flue space.

Figure 7-4.4.2.3(d) *20-ft (6.1-m) storage; <5-ft (1.5-m) ceiling clearance.*

areas of application shall be permitted between storage heights with the same clearances. No interpolation between clearances shall be permitted.

7-4.4.3.2 Single-, Double-, and Multiple-Row Racks Up to 10-ft (3.1-m) Storage with Up to 10-ft (3.1-m) Clearance. The protection strategies utilizing only ceiling sprinklers, as shown in Figure 7-4.4.2.3(a), shall be acceptable for single-, double-, and multiple-row rack storage.

7-4.4.3.3 Single- and Double-Row Rack Storage Greater than 10 ft (3.1 m) Up to 15 ft (4.6 m) with Less than 5-ft (1.25-m) Clearance. The protection strategy utilizing only ceiling sprinklers, as shown in Figure 7-4.4.2.3(b), shall be acceptable only for single- and double-row rack storage.

7-4.4.3.4 Single- and Double-Row Rack Storage Greater than 10 ft (3.1 m) Up to 15 ft (4.6 m) with Clearance from 5 ft to 10 ft (1.5 m to 3.1 m), and Single- and Double-Row Rack Storage Up to 20 ft (6.1 m) with Less than 5-ft (1.5-m) Clearance. The protection strategies utilizing only ceiling sprinklers, as shown in Figures 7-4.4.2.3(c) and (d), shall be acceptable only for single- and double-row rack storage.

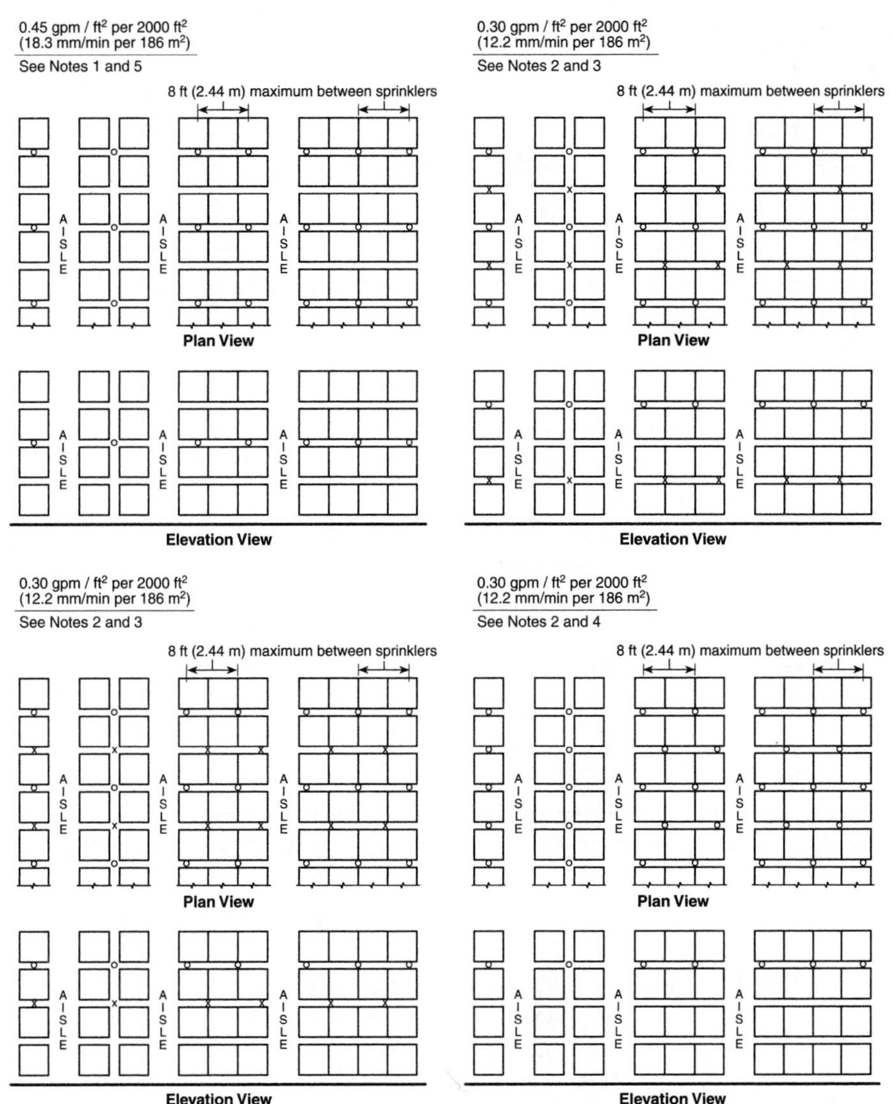

Notes:

1. Single level of in-rack sprinklers [½ in. or 17/32 in. (12.7 mm or 13.5 mm) operating at 15 psi (1.03 bar) minimum] installed on 8 ft to 10 ft (2.5 m to 3.12 m) spacings located, as indicated, in the transverse flue spaces.
2. Ceiling-only protection shall not be permitted for this storage configuration.
3. Two levels of in-rack sprinklers [½ in. or 17/32 in. (12.7 mm or 13.5 mm) operating at 15 psi (1.03 bar) minimum] installed on 8 ft to 10 ft (2.5 m to 3.12 m) spacings located as indicated and staggered in the transverse flue space.
4. Single level of in-rack sprinklers [17/32 in. (13.5 mm) operating at 15 psi (1.03 bar) minimum or ½ in. (12.7 mm) operating at 30 psi (2.07 bar) minimum] installed on 4 ft to 5 ft (1.25 m to 1.56 m) spacings located, as indicated, in the longitudinal flue space at the intersection of every transverse flue space.
5. Where ⅝-in. (15.9-mm) orifice sprinklers listed for storage use are installed at the ceiling, the in-rack sprinklers shall not be required, provided the ceiling sprinkler discharge criteria is increased to 0.6 gpm per ft²/2000 ft² (24 L/ min per m²/186m²) and the ceiling height in the protected area does not exceed 27 ft (8.2 m).

Figure 7-4.4.2.3(e) *20-ft (6.1-m) storage; 5-ft to 10-ft (1.5-m to 3.1-m) ceiling clearance.*

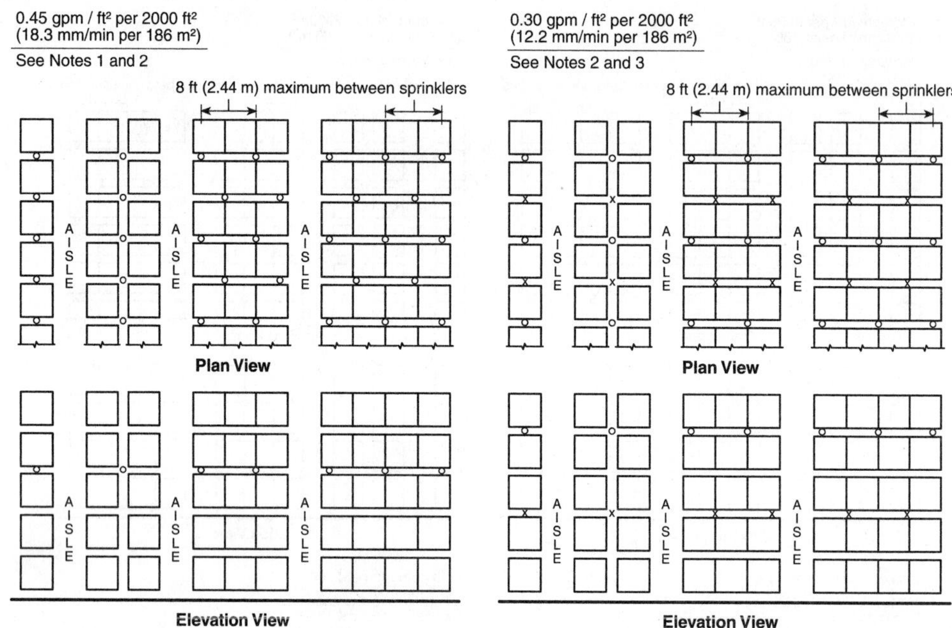

Notes:

1. Single level of in-rack sprinklers [17/32 in. (13.5 mm) operating at 15 psi (1.03 bar) minimum or ½ in. (12.7 mm) operating at 30 psi (2.07 bar) minimum] installed on 4 ft to 5 ft (1.25 m to 1.56 m) spacings located, as indicated, in the longitudinal flue space at the intersection of every transverse flue space.
2. Ceiling-only protection shall not be permitted for this storage configuration.
3. Two levels of in-rack sprinklers [½ in. or 17/32 in. (12.7 mm or 13.5 mm) operating at 15 psi (1.03 bar) minimum] installed on 8 ft to 10 ft (2.5 m to 3.12 m) spacings located as indicated and staggered in the transverse flue space.

Figure 7-4.4.2.3(f) 25-ft (7.6-m) storage; <5-ft (1.5-m) ceiling clearance. (See Note 2.)

7-4.4.3.5 Multiple-Row Racks—15-ft (4.6-m) Storage with Less than 5-ft (1.5-m) Clearance. Where using the protection strategy utilizing only ceiling sprinklers, as shown in Figure 7-4.4.2.3(b), for multiple-row rack storage, the density to be used shall be 0.6 gpm/ft^2 (24.5 mm/min) over 2000 ft^2 (186 m^2). The combination of ceiling and in-rack sprinklers specified in Figure 7-4.4.2.3(b) shall be permitted as an alternative.

7-4.4.3.6 Multiple-Row Racks—15-ft (4.6-m) Storage with 10-ft (3.1-m) Clearance, and 20-ft (6.1-m) Storage with Less than 5-ft (1.5-m) Clearance. The protection strategies utilizing only ceiling sprinklers, as shown in Figures 7-4.4.2.3(c) and (d), shall not be permitted for multiple-row rack storage. Only the specified combinations of ceiling and in-rack sprinklers shall be used.

7-4.4.4 Ceiling Sprinklers for Single-, Double-, and Multiple-Row Racks—Storage Over 25 ft (7.1 m) in Height—Area Density Method.

0.30 gpm / ft² per 2000 ft²
(12.2 mm/min per 186 m²)

See Notes 1 and 2

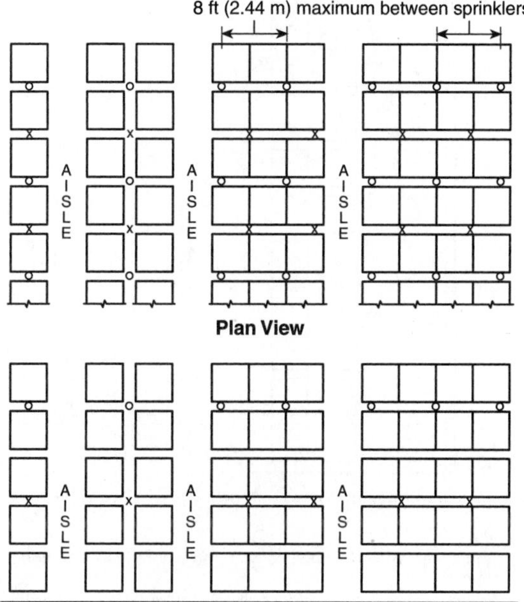

Notes:

1. Two levels of in-rack sprinklers [½ in. or 17⁄32 in. (12.7 mm or 13.5 mm) operating at 15 psi (1.03 bar) minimum] installed on 8 ft to 10 ft (2.5 m to 3.12 m) spacings located as indicated and staggered in the transverse flue space.
2. Ceiling-only protection shall not be permitted for this storage configuration.

Figure 7-4.4.2.3(g) *25-ft (7.6-m) storage; 5-ft to 10-ft (1.5-m to 3.1-m) ceiling clearance. (See Note 2.)*

7-4.4.4.1 Ceiling Sprinkler Water Demand. For Group A plastic commodities in cartons, encapsulated or nonencapsulated, ceiling sprinkler water demand in terms of density [gpm/ft^2 (mm/min)] and area of operation [ft^2 (m^2)] shall be selected from Table 7-4.4.4.1.

7-4.4.4.2 Where a single-row rack is mixed with double-row racks, either Figures 7-4.4.2.3(a) through (g) or Figures 7-4.4.2.4.1(a) and (b) shall be used in accordance with the corresponding storage height.

Exception: Figures 7-4.4.2.4.3(a) through 7-4.4.2.4.3(c) shall be permitted to be used for the protection of the single-row racks.

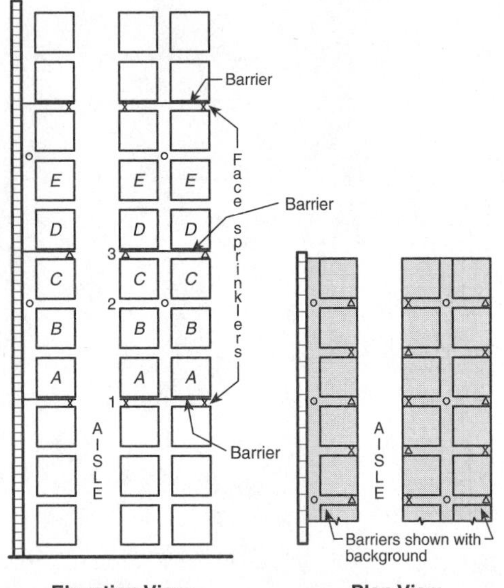

Elevation View **Plan View**

Notes:

1. Sprinklers and barriers labeled 1 shall be required where loads labeled *A* or *B* represent top of storage.
2. Sprinklers labeled 1 and 2 and barriers labeled 1 shall be required where loads labeled *C* represent top of storage.
3. Sprinklers and barriers labeled 1 and 3 shall be required where loads labeled *D* or *E* represent top of storage.
4. For storage higher than represented by loads labeled *E*, the cycle defined by Notes 2 and 3 is repeated.
5. Symbol Δ or x indicates face sprinklers on vertical or horizontal stagger.
6. Symbol o indicates longitudinal flue space sprinklers.
7. Each square represents a storage cube measuring 4 ft to 5 ft (1.22 m to 1.53 m) on a side. Actual load heights can vary from approximately 18 in. (0.46 m) up to 10 ft (3.05 m). Therefore, there could be as few as one load to as many as six or seven loads between in-rack sprinklers that are spaced 10 ft (3.05 m) apart vertically.

Figure 7-4.4.2.4.1(a) *In-rack sprinkler arrangement, Group A plastic commodities, storage height over 25 ft (7.6 m)—option 1.*

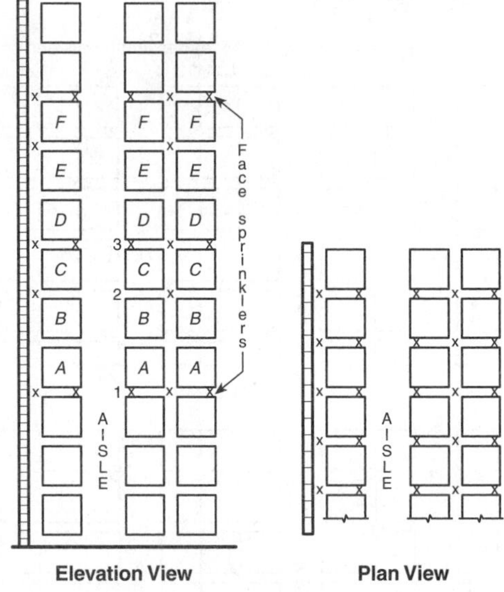

Elevation View **Plan View**

Notes:

1. Sprinklers labeled 1 shall be required where loads labeled *A* or *B* represent top of storage.
2. Sprinklers labeled 1 and 2 shall be required where loads labeled *C* represent top of storage.
3. Sprinklers labeled 1 and 3 shall be required where loads labeled *D* or *E* represent top of storage.
4. For storage higher than loads labeled *F*, the cycle defined by Notes 2 and 3 is repeated.
5. Symbol x indicates face and in-rack sprinklers.
6. Each square represents a storage cube measuring 4 ft to 5 ft (1.22 m to 1.53 m) on a side. Actual load heights can vary from approximately 18 in. (0.46 m) up to 10 ft (3.05 m). Therefore, there could be as few as one load to as many as six or seven loads between in-rack sprinklers that are spaced 10 ft (3.05 m) apart vertically.

Figure 7-4.4.2.4.1(b) *In-rack sprinkler arrangement, Group A plastic commodities, storage height over 25 ft (7.6 m)—option 2.*

7-5* Protection of Idle Pallets

A-7-5 Idle pallet storage introduces a severe fire condition. Stacking idle pallets in piles is the best arrangement of combustibles to promote rapid spread of fire, heat release, and complete combustion. After pallets are used for a short time in warehouses, they dry out and edges become frayed and splintered. In this condition they are subject to easy ignition from a small

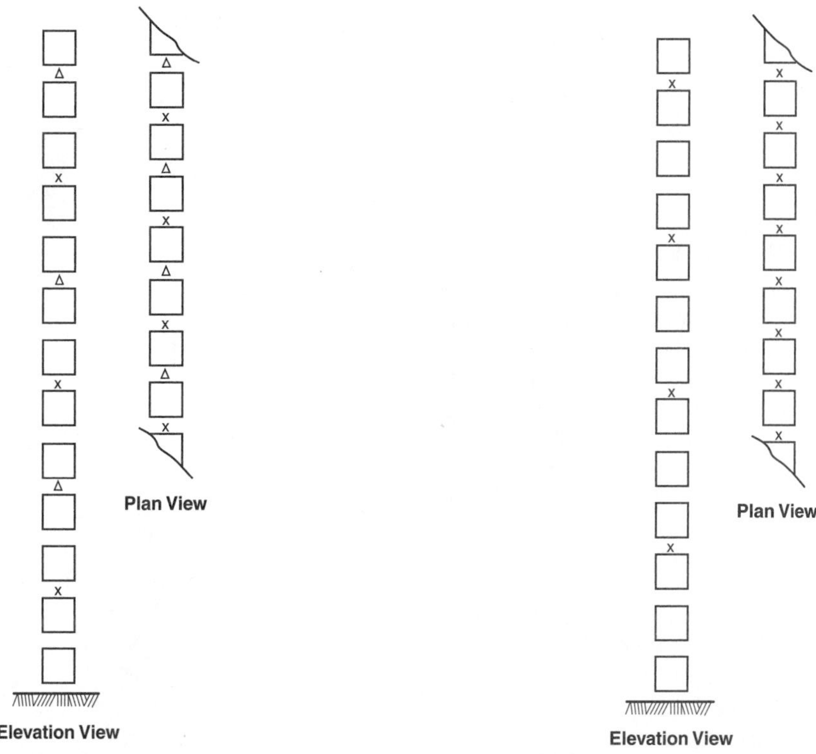

Note: Each square represents a storage cube measuring 4 ft to 5 ft (1.22 m to 1.53 m) on a side. Actual load heights can vary from approximately 18 in. (0.46 m) up to 10 ft (3.05 m). Therefore, there could be as few as one load to as many as six or seven loads between in-rack sprinklers that are spaced 10 ft (3.05 m) apart vertically.

Figure 7-4.4.2.4.3(a) *In-rack sprinkler arrangement, Group A plastic commodities, single-row racks, storage height over 25 ft (7.6 m)—option 1.*

Note: Each square represents a storage cube measuring 4 ft to 5 ft (1.22 m to 1.53 m) on a side. Actual load heights can vary from approximately 18 in. (0.46 m) up to 10 ft (3.05 m). Therefore, there could be as few as one load to as many as six or seven loads between in-rack sprinklers that are spaced 10 ft (3.05 m) apart vertically.

Figure 7-4.4.2.4.3(b) *In-rack sprinkler arrangement, Group A plastic commodities, single-row racks, storage height over 25 ft (7.6 m)—option 2.*

ignition source. Again, high piling increases considerably both the challenge to sprinklers and the probability of involving a large number of pallets when fire occurs. Therefore, it is preferable to store pallets outdoors where possible.

A fire in stacks of idle plastic or wooden pallets is one of the greatest challenges to sprinklers. The undersides of the pallets create a dry area on which a fire can grow and expand to other dry or partially wet areas. This process of jumping to other dry, closely located, parallel, combustible surfaces continues until the fire bursts through the top of the stack. Once this happens, very little water is able to reach the base of the fire. The only practical method of stopping a fire in a large concentration of pallets with ceiling sprinklers is by means of prewetting. In high stacks, this cannot be done without abnormally high water supplies. The storage of

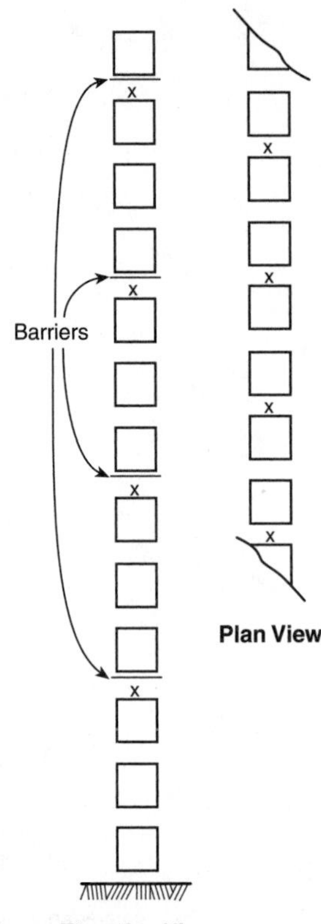

Note: Each square represents a storage cube measuring 4 ft to 5 ft (1.22 m to 1.53 m) on a side. Actual load heights can vary from approximately 18 in. (0.46 m) up to 10 ft (3.05 m). Therefore, there could be as few as one load to as many as six or seven loads between in-rack sprinklers that are spaced 10 ft (3.05 m) apart vertically.

Figure 7-4.4.2.4.3(c) *In-rack sprinkler arrangement, Group A plastic commodities, single-row racks, storage height over 25 ft (7.6 m)—option 3.*

empty wood pallets should not be permitted in an unsprinklered warehouse containing other storage.

Wooden pallets present a unique challenge for sprinkler protection. Idle pallets create an ideal configuration for efficient combustion by presenting many surfaces for burning and many openings to provide an almost unlimited source of air. At the same time, the configuration shields much of the burning surfaces from sprinkler discharge. In addition, pallets are subject to easy ignition due to their frayed, splintered edges and typical dried-out condition.

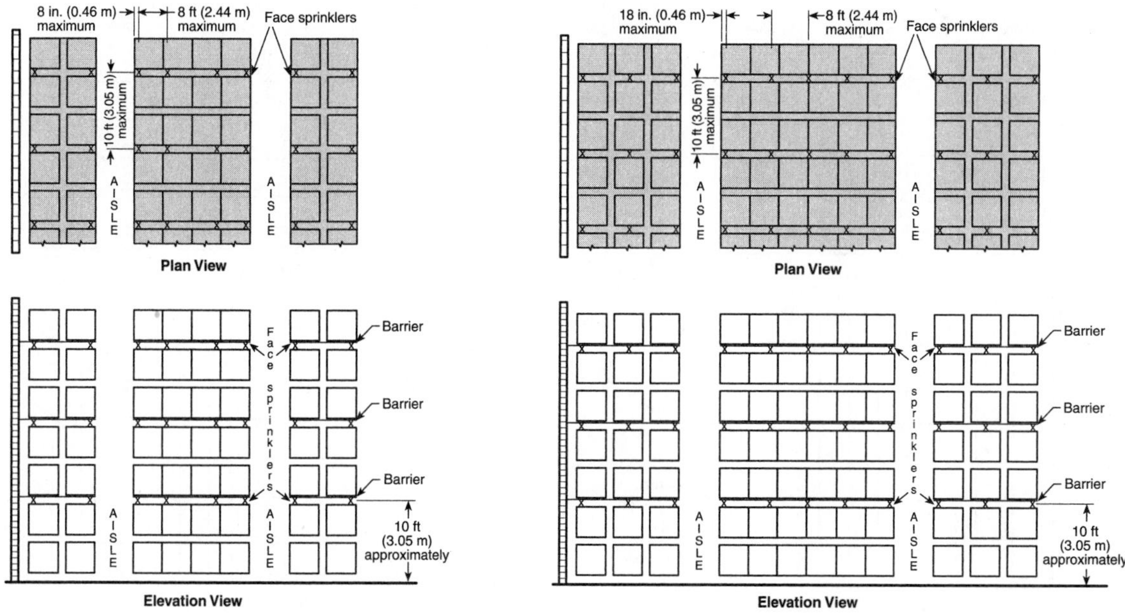

Figure 7-4.4.2.4.4(a) In-rack sprinkler arrangement, cartoned plastic and uncartoned unexpanded plastic, multiple-row racks, storage height over 25 ft (7.6 m)—option 1.*

Figure 7-4.4.2.4.4(b) In-rack sprinkler arrangement, cartoned plastic and uncartoned unexpanded plastic, multiple-row racks, storage height over 25 ft (7.6 m)—option 2.*

7-5.1 General.

7-5.1.1 The following criteria is intended to apply to buildings with ceiling slopes not exceeding 2 in 12 (16.7 percent).

See commentary to 7-2.3.2.5 and 7-3.1.3.

7-5.1.2 Where dry pipe systems are used, the sprinkler system design area shall be increased by 30 percent.

See commentary to 7-2.3.2.5 and 7-3.1.2.1.

7-5.2 Wood Pallets.

7-5.2.1* Pallets shall be stored outside or in a detached structure.

Exception: Indoor pallet storage shall be permitted in accordance with 7-5.2.2.

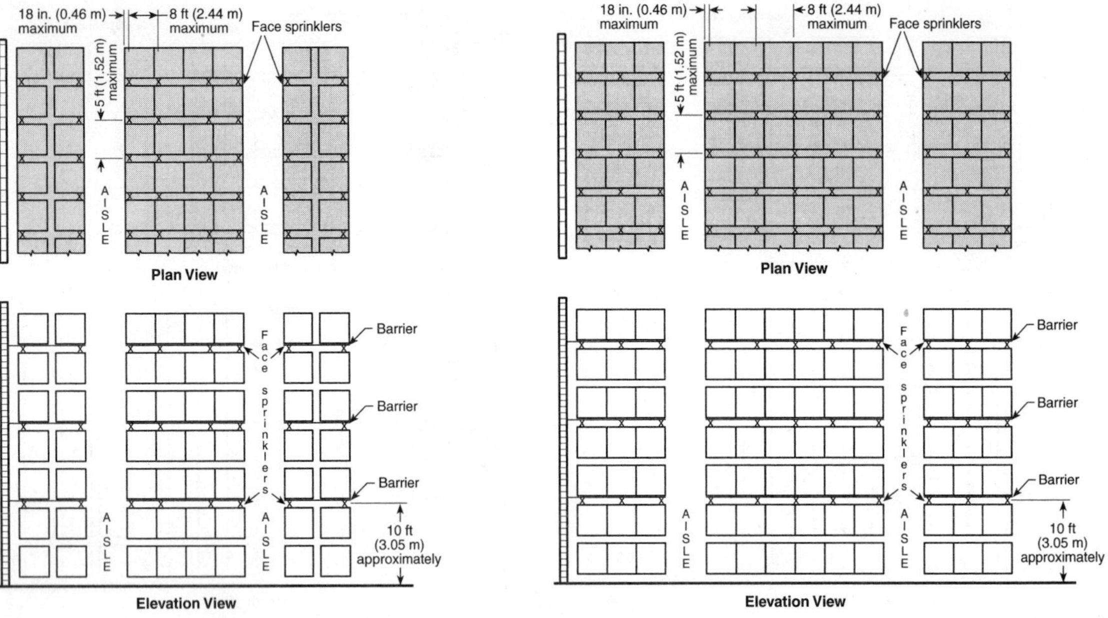

Figure 7-4.4.2.4.4(c)* *In-rack sprinkler arrangement, cartoned plastic, and uncartoned unexpanded plastic, multiple-row racks, storage height over 25 ft (7.6 m)—option 1.*

Figure 7-4.4.2.4.4(d)* *In-rack sprinkler arrangement, cartoned plastic, and uncartoned unexpanded plastic, multiple-row racks, storage height over 25 ft (7.6 m)—option 2.*

A-7-5.2.1 See Table A-7-5.2.1.

7-5.2.2* Pallets, where stored indoors, shall be protected as indicated in Table 7-5.2.2 using standard spray sprinklers, Tables 7-9.4.1.1 and 7-9.4.1.2 using large drop sprinklers, or Tables 7-9.5.1.1 and 7-9.5.1.2 using ESFR sprinklers, unless the following conditions are met:

(1) Pallets shall be stored no higher than 6 ft (1.8 m).
(2) Each pallet pile of no more than four stacks shall be separated from other pallet piles by at least 8 ft (1.4 m) of clear space or 25 ft (7.6 m) of commodity.

A-7-5.2.2 No additional protection is necessary, provided the requirements of 7-5.2.2(a) and (b) are met.

Where idle wood pallets are stored in accordance with the two conditions of 7-5.2.2, experience and judgement by the Technical Committee on Sprinkler System Discharge Criteria indicate that most sprinklered buildings will normally withstand the heat produced as long as it is not prolonged. Thus, some idle pallet storage without extra protection is considered acceptable. Exhibit 7.8 shows such an arrangement.

The two conditions identified in 7-5.2.2 are often difficult to enforce. Designing the sprinkler protection for the idle pallet storage heights anticipated is usually the preferred approach.

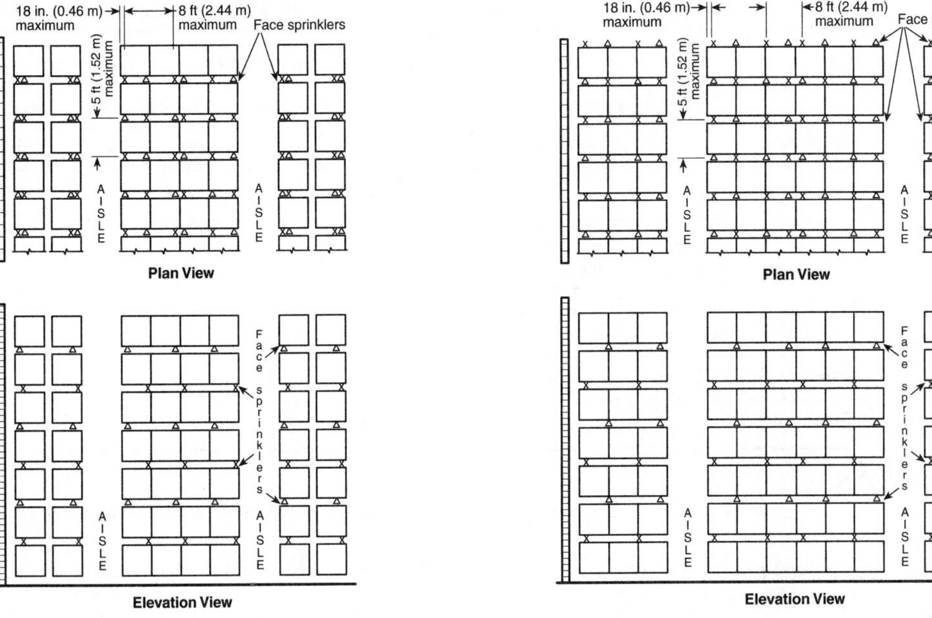

Figure 7-4.4.2.4.4(e)* *In-rack sprinkler arrangement, cartoned plastic, and uncartoned unexpanded plastic, multiple-row racks, storage height over 25 ft (7.6 m)—option 3.*

Figure 7-4.4.2.4.4(f)* *In-rack sprinkler arrangement, cartoned plastic, and uncartoned unexpanded plastic, multiple-row racks, storage height over 25 ft (7.6 m)—option 4.*

Table 7-4.4.4.1 Ceiling Sprinkler Discharge Criteria for Single-, Double-, and Multiple-Row Racks with Storage Over 25 ft (7.6 m) in Height

Storage Height above Top Level In-Rack Sprinklers	Ceiling Sprinklers Density (gpm/ft^2)/Area of Application (ft^2)
5 ft or less	0.30/2000
Over 5 ft up to 10 ft	0.45/2000

For SI units, 1 ft = 0.3048 m; 1 gpm/ft^2 = 40.746 mm/min; 1 ft^2 = 0.0929 m^2.

Note: Provide in-rack sprinkler protection per Figures 7-4.4.2.4.1(a) and (b) and Figures 7-4.4.2.4.3(a) through (c).

Table A-7-5.2.1 Recommended Clearance Between Outside Idle Pallet Storage and Building

Wall Construction		Minimum Distance of Wall from Storage of					
		Under 50 Pallets		50 to 200 Pallets		Over 200 Pallets	
Wall Type	Openings	ft	m	ft	m	ft	m
Masonry	None	0	0	0	0	0	0
	Wired glass with outside sprinklers and 1-hour doors	0	0	10	3.1	20	6.1
	Wired or plain glass with outside sprinklers and ¾-hour doors	10	3.1	20	6.1	30	9.1
Wood or metal with outside sprinklers		10	3.1	20	6.1	30	9.1
Wood, metal, or other		20	6.1	30	9.1	50	15.2

Notes:
1. Fire-resistive protection comparable to that of the wall also should be provided for combustible eaves lines, vent openings, and so forth.
2. Where pallets are stored close to a building, the height of storage should be restricted to prevent burning pallets from falling on the building.
3. Manual outside open sprinklers generally are not a reliable means of protection unless property is attended to at all times by plant emergency personnel.
4. Open sprinklers controlled by a deluge valve are preferred.

Table 7-5.2.2 Protection for Indoor Storage of Idle Wood Pallets

Height of Pallet Storage		Sprinkler Density Requirements		Area of Sprinkler Demand			
				High Temperature		Ordinary Temperature	
ft	m	gpm/ft^2	mm/min	ft^2	m^2	ft^2	m^2
Up to 6	Up to 1.8	0.2	8.2	2000	186	3000	279
6–8	1.8–2.4	0.3	12.2	2500	232	4000	372
8–12	2.4–3.7	0.6	24.5	3500	325	6000	557
12–20	3.7–6.1	0.6	24.5	4500	418	—	—

Where floor storage of idle pallets does not meet the conditions of 7-5.2.2, standard spray sprinklers in accordance with Table 7-5.2.2, large drop sprinklers, or ESFR sprinklers need to be considered.

7-5.2.3† Idle wood pallets shall not be stored in racks.

Exception: Idle wooden pallets shall be permitted to be stored in racks when protected in accordance with the appropriate provisions of 7-9.5.1.2.

Exhibit 7.8 *Idle pallet storage arrangement that meets the conditions of 7-5.2.2.*

Only ESFR sprinklers can be used to protect idle wood pallets stored in racks as indicated in 7-9.5. No protection schemes using standard spray sprinklers at the ceiling along with in-rack sprinklers have been tested to confirm effective protection of idle pallet storage in racks.

7-5.3 Plastic Pallets.

7-5.3.1 Plastic pallets shall be stored outside or in a detached structure.

Exception No. 1: Indoor plastic pallet storage shall be permitted in accordance with 7-5.3.2.

Exception No. 2: Indoor storage of nonexpanded polyethylene solid deck pallets shall be permitted to be protected in accordance with 7-5.2.2.

Exception No. 3: Indoor storage of plastic pallets shall be permitted to be protected in accordance with Table 7-9.5.1.1 or *Table 7-9.5.1.2.*

Exception No. 4: Indoor storage of non-wood pallets having a demonstrated fire hazard that is equal to or less than idle wood pallets and is listed for such equivalency shall be permitted to be protected in accordance with 7-5.2.2.

Underwriters Laboratories Inc. has recently introduced a new testing standard for non-wood pallets that would be permitted to be considered under Exception No. 4.

7-5.3.2 Plastic pallets where stored indoors shall be protected as specified in 7-5.3.2(1) and (2):

(1) Where stored in cutoff rooms the following shall apply:

a. The cutoff rooms shall have at least one exterior wall.

b. The plastic pallet storage shall be separated from the remainder of the building by 3-hour-rated fire walls.

c. The storage shall be protected by sprinklers designed to deliver 0.6 gpm/ft^2 (24.5 mm/min) for the entire room or by high-expansion foam and sprinklers as indicated in 7-3.2.2.4.

d. The storage shall be piled no higher than 12 ft (3.7 m).

e. Any steel columns shall be protected by 1-hour fireproofing or a sidewall sprinkler directed to one side of the column at the top or at the 15-ft (4.6-m) level, whichever is lower. Flow from these sprinklers shall be permitted to be omitted from the sprinkler system demand for hydraulic calculations.

(2) Where stored without cutoffs from other storage the following shall apply:

a. Plastic pallet storage shall be piled no higher than 4 ft (1.2 m).

b. Sprinkler protection shall employ high temperature–rated sprinklers.

c. Each pallet pile of no more than two stacks shall be separated from other pallet piles by at least 8 ft (2.4 m) of clear space or 25 ft (7.6 m) of stored commodity.

With regard to item 7-5.3.2(1), for those plastic pallets not listed as burning like a four-way wood pallet (or less) in accordance with Exception No. 4 to 7-5.3.1, a cutoff room designed as required in 7-5.3.2(1)a should control the fire until the fire department arrives.

7-5.3.3 Idle plastic pallets shall not be stored in racks.

Exception: Idle plastic pallets shall be permitted to be stored in racks when protected in accordance with the appropriate provisions of 7-9.5.1.2.

Certain arrangements of idle plastic pallets stored in racks can be protected using ESFR sprinklers.

7-6 Protection of Rubber Tire Storage

The sprinkler protection of rubber tires presents unique challenges. Because of their structure, tires possess inherent air spaces that provide a sufficient amount of air for combustion. Additionally, fires burning on the interior surface of a tire are usually shielded from sprinkler discharge.

The selection of protection criteria from Tables 7-6.2.1(a), 7-6.2.1(b), or 7-6.2.1(c) requires identification of the specific tire storage configuration to be protected. See 1-4.10 and 1-4.10.1 for identification of tire storage configurations.

7-6.1 General.

The sprinkler system criteria of Section 7-6 shall apply to buildings with ceiling slopes not exceeding 2 in 12 (16.7 percent).

Although no fire tests have been conducted with rubber tires as the commodity where the ceiling slope has exceeded 2 in. in 12 in. (16.7 percent), the technical committee believes that slopes in excess of 16.7 percent require additional protection in terms of sprinkler system discharge criteria. Also see commentary to 7-2.3.2.5 and 7-3.1.3.

7-6.2 Ceiling Systems.

7-6.2.1* Sprinkler discharge and area of application shall be in accordance with Table 7-6.2.1(a) and Figure 7-6.2.1 for standard spray sprinklers. See 5-4.1.2 for appropriate orifice sizes. Large drop and ESFR sprinklers shall be in accordance with Tables 7-6.2.1(b) and (c), respectively.

Figure 7-6.2.1 shall be used as follows.

(1) Note the example indicated by the broken line.
(2) Read across the graph to the storage height of 14 ft (4.3 m) until the storage height intersects the storage height curve at a sprinkler density of 0.45 gpm/ft^2 (18.3 mm/min).
(3) Read down until the sprinkler density intersects the sprinkler operating area curves at 3200 ft^2 (397 m^2) for ordinary-temperature sprinklers and 2000 ft^2 (186 m^2) for high-temperature sprinklers.

Table 7-6.2.1(a) Protection Criteria for Rubber Tire Storage Using Standard Spray Sprinklers

Piling Method	Piling Height (ft)	Sprinkler Discharge Density *(See Note 1.)* (gpm/ft^2)	Areas of Application (ft^2) *(See Note 1.)* Ordinary Temperature	High Temperature
(1) On-floor storage	Up to 5	0.19	2000	2000
a. Pyramid piles, on-side	Over 5 to 12	0.30	2500	2500
b. Other arrangements such that no horizontal channels are formed *(See Note 2.)*	Over 12 to 18	0.60	Not allowed	2500
(2) On-floor storage	Up to 5	0.19	2000	2000
Tires on-tread	Over 5 to 12	0.30	2500	2500
(3) Palletized portable rack storage	Up to 5	0.19	2000	2000
On-side or on-tread	Over 5 to 20	See Figure 7-6.2.1	—	—
	Over 20 to 30	0.30 plus high-expansion foam	3000	3000

(continues)

Table 7-6.2.1(a) Protection Criteria for Rubber Tire Storage Using Standard Spray Sprinklers (Continued)

Piling Method	Piling Height (ft)	Sprinkler Discharge Density *(See Note 1.)* (gpm/ft^2)	Areas of Application (ft^2) *(See Note 1.)* Ordinary Temperature	High Temperature
(4) Palletized portable rack storage	Up to 5	0.19	2000	2000
On-side	Over 5 to 20	See Figure 7-6.2.1	—	—
	20 to 25	0.60 and	Not allowed	5000
		0.90 *(see Note 3)*;	Not allowed	3000
		or		
		0.75 with 1-hour fire-resistive rating of roof and ceiling assembly	Not allowed	4000
(5) Open portable rack storage, on-side or on-tread	Up to 5	0.19	2000	2000
	Over 5 to 12	0.60	5000	3000
	Over 12 to 20	0.60 and	Not allowed	5000
		0.90 *(see Note 3)*;	Not allowed	3000
		or		
		0.30 plus high-expansion foam	3000	3000
(6) Single-, double-, and multiple-row fixed rack storage on pallets, on-side or on-tread	Up to 5	0.19	2000	2000
	Over 5 to 20	See Figure 7-6.2.1;		
		or		
		0.40 plus one level in-rack sprinklers;	3000	3000
		or		
		0.30 plus high-expansion foam	3000	3000
	Over 20 to 30	0.30 plus high-expansion foam	Not allowed	3000
(7) Single-, double-, and multiple-row fixed rack storage without pallets or shelves, on-side or on-tread	Up to 5	0.19	2000	2000
	Over 5 to 12	0.60	5000	3000
	Over 12 to 20	0.60 and	Not allowed	5000
		0.90 *(see Note 3)*;	Not allowed	3000
		or		
		0.40 plus one level in-rack sprinklers;	3000	3000
		or		
		0.30 plus high-expansion foam	3000	3000
	Over 20 to 30	0.30 plus high-expansion foam	Not allowed	3000

For SI units, 1 ft = 0.3048 m; 1 ft^2 = 0.0929 m^2; 1 gpm/ft^2 = 40.746 mm/min.

Notes:

1. Sprinkler discharge densities and areas of application are based on a maximum clearance of 10 ft (3.1 m) between sprinkler deflectors and the maximum available height of storage. The maximum clearance is noted from actual testing and is not a definitive measurement.

2.* Laced tires on-floor, vertical stacking on-side (typical truck tires), and off-road tires.

3. Water supply shall fulfill both requirements.

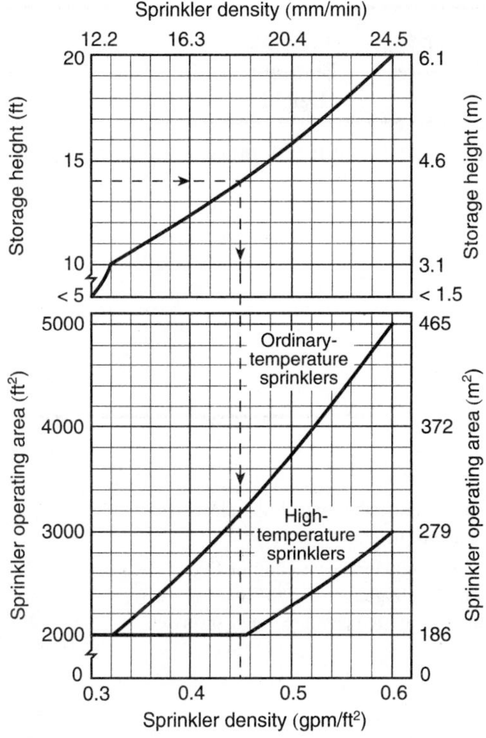

Figure 7-6.2.1 *Sprinkler system design curves for palletized portable rack storage and fixed rack storage with pallets over 5 ft to 20 ft (1.5 m to 6.1 m) in height.*

A-7-6.2.1 The protection criteria in Tables 7-6.2.1(a), (b), and (c) have been developed from fire test data. Protection requirements for other storage methods are beyond the scope of this standard at the present time. From fire testing with densities of 0.45 gpm/ft^2 (18.3 mm/min) and higher, there have been indications that large-orifice sprinklers at greater than 50-ft^2 (4.6-m^2) spacing produce better results than the ½-in. (12.7-mm) orifice sprinklers at 50-ft^2 (4.6-m^2) spacing.

Tables 7-6.2.1(a) and (b) are based on operation of standard sprinklers. Use of quick-response or other special sprinklers should be based on appropriate tests as approved by the authority having jurisdiction.

The current changes to Tables 7-6.2.1(a), (b), and (c) represent test results from rubber tire fire tests performed at the Factory Mutual Research Center.

Storage heights and configurations, or both, [e.g., automated material-handling systems above 30 ft (9.1 m)] beyond those indicated in the table have not had sufficient test data

Table 7-6.2.1(b) Large Drop Sprinkler Protection for Rubber Tires (See Note 1.)

Piling Method	Pile Height	Number of Sprinklers and Minimum Operating Pressures *(See Note 2.)*	Maximum Building Height	Duration (hr)	Hose Demand
Rubber tire storage, on-side or on-tread, in palletized portable racks, or open portable racks, or fixed racks without solid shelves	Up to 25 ft (7.6 m)	15 sprinklers @ 75 psi (5.2 bar) *(See Note 3.)*	32 ft (9.8 m)	3	500 gpm (1893 L/min)

Notes:

1. Wet systems only.
2. Sprinkler operating pressures and number of sprinklers in the design are based on tests in which the clearance was 5 ft to 7 ft (1.5 m to 2.1 m) between the sprinkler deflector and the maximum height of storage.
3. The design area shall consist of the most hydraulically demanding area of 15 sprinklers, consisting of five sprinklers on each of three branch lines. The design shall include a minimum operating area of 1200 ft^2 (112 m^2) and a maximum operating area of 1500 ft^2 (139 m^2) and shall utilize a high temperature–rated sprinkler.

developed to establish recommended criteria. Detailed engineering reviews of the protection should be conducted and approved by the authority having jurisdiction.

In the 1974 tentative standard of 231D, *Standard for Storage of Rubber Tires,* only the following two piling methods were listed:

(1) On-floor storage (pyramid piles, on-tread, off-road tires, and arrangements with no horizontal channels)
(2) Palletized rack storage (on-side and on-tread, bundled and compressed)

A third option, on-tread storage in portable metal racks, was in the appendix.

The 1974 appendix of NFPA 231D states that only on-floor and palletized rack storage had actually been tested. No other storage arrangements had been tested at that time; the appendix item for the portable metal racks was intended to be good engineering judgement.

In the 1975 edition of NFPA 231D, the sprinkler protection criteria had been expanded to include open portable racks, double- and multi-row fixed rack storage on pallets, and double- and multi-row fixed rack storage without pallets or shelves as well as the two piling methods listed in the previous paragraph. All three of these added piling methods included tires on-side and on-tread. The height limits in NFPA 231D were 12 ft (3.7 m) for the on-floor storage, 30 ft (9.1 m) for the palletized storage, but only with high-expansion foam aiding the sprinkler protection, and 20 ft (6.1 m)

Table 7-6.2.1(c) ESFR Sprinklers for Protection of Rubber Tires (See Note 1.)

Piling Method	Pile Height	Maximum Building Height		Nominal K-factor	Number of Sprinklers *(See Note 2.)*	Minimum Operating Pressure *(See Note 2.)*		Duration (hours)	Hose Demand	
		ft	m			psi	bar		gpm	L/min
Rubber tire storage, on-side or on-tread, in palletized portable racks, open portable racks, or fixed racks without solid shelves	Up to 25 ft (7.6 m)	30	9.1	13.5–14.5	12 *(See Note 3.)*	50	3.5	1	250	946
				23.9–26.5	12 *(See Note 3.)*	20	1.4	1	250	946
Rubber tire storage, on-side, in palletized portable racks, open portable racks, or fixed racks without solid shelves	Up to 25 ft (7.6 m)	35	10.7	13.5–14.5	12 *(See Note 3.)*	75	5.2	1	250	946
				23.9–26.5	12 *(See Note 3.)*	30	2.1	1	250	946
Laced tires in open portable steel racks *[See Figure A-1-4.10.1(g).]*	Up to 25 ft (7.6 m)	30	9.1	13.5–14.5	20 *(See Notes 4 and 5.)*	75	5.2	3	500	1892

Notes:

1. Wet systems only.
2. Sprinkler operating pressures and number of sprinklers in the design are based on tests in which the clearance was 5 ft to 7 ft (1.5 m to 2.1 m) between the sprinkler deflector and the maximum height of storage.
3. The shape of the design area shall be in accordance with 7-9.5.2.2.
4. Where used in this application, ESFR protection is expected to control rather than to suppress the fire.
5. The design area shall consist of the most hydraulically demanding area of 20 sprinklers, consisting of 5 sprinklers on each of 4 branch lines. The design shall include a minimum operating area of 1600 ft^2 (149 m^2).

for each of three added piling methods, but with high-expansion foam aiding the sprinkler protection.

By the 1986 edition, NFPA 231D increased the storage heights in piling methods for double- and multi-row fixed racks, on-tread or on-side, with or without pallets or shelves, to 30 ft (9.1 m) as long as high-expansion foam aided the sprinkler protection.

Also in 1986, the appendix reminded the user that the protection in the standard was based on standard response sprinklers. The use of quick-response sprinklers or other special sprinklers was outside the scope of the standard. If special sprinklers or quick-response sprinklers were to be used, testing needed to be conducted to establish protection criteria. This comment is still in the appendix of the 1998 edition of NFPA 231D and applies to two of the three protection tables—the third protection table is for ESFR sprinklers.

In the early 1990s, the tire industry proposed two major changes to NFPA 231D. The industry wanted to increase the storage height in several areas and to introduce laced tires as a piling or storage method. Tests were conducted at the Factory Mutual Research Center to determine if existing, accepted protection options were applicable to this higher and unique storage. This testing led to the following changes to NFPA 231D:

(a) Since 1974, on-floor storage had been restricted to 12 ft (3.7 m) in height. However, testing in 1993 showed that some on-floor storage, such as pyramid piles and arrangements that had no horizontal channels, could be stored to 18 ft (5.5 m) when protected by a density of 0.60 gpm/ft^2 (24.5 mm/min).

(b) Since 1974, palletized portable rack storage had been allowed to exceed 20 ft (6.1m) up to 30 ft (9.1 m) but only when high-expansion foam aided the sprinkler protection. The 1993 testing showed that this storage could extend to 25 ft (7.6 m) and be protected with a density of 0.60 gpm/ft^2 (24.5 mm/min) when the ceiling was provided with 1-hour fireproofing. In the 1998 edition of 231D, the density was changed to 0.75 gpm/ft^2 to be more conservative.

(c) Laced tire storage was not a warehouse concern in the United States until the early 1990s. Lacing of tires was often used while shipping the tires in a tractor trailer, but no manufacturer had warehoused tires in this fashion. About this time, a manufacturer introduced this storage method to its warehouses, storing tires up to 25 ft (7.6 m) high. A test was conducted to determine if a density of 0.60 gpm/ ft^2 (24.5 mm/min) would provide adequate protection. In the test, 77 sprinklers operated and the steel temperatures exceeded 1500°F (816°C), forcing NFPA 231D to indicate that storage of this type did not have "sufficient test data developed to establish recommended criteria."

In 1996, another test series was conducted to introduce new technology to NFPA 231D. These tests resulted in the addition of large drop and ESFR sprinklers to the protection criteria as follows:

(1) Table 7-6.2.1(a) applies to sprinklers ½-in., 17⁄32-in., ⅝-in., and 0.70-in. size orifices. A version of these criteria has existed in NFPA 231D over the years.
(2) Table 7-6.2.1(b) applies to large drop sprinklers protecting a range of piling methods except for laced tires.
(3) Table 7-6.2.1(c) applies to ESFR sprinklers protecting a range of piling methods, including laced tires.

Table 7-6.2.1(b) is based on tests conducted at the Factory Mutual Research Center. One test was for tires stored on-tread in portable racks up to 25 ft (7.6 m). The second test was for tires stored on-side on palletized portable racks up to 25 ft (7.6 m). The sprinklers operated at 75 psi (5.2 bar). In the former test, eight sprinklers controlled the fire and, in the latter test, six sprinklers controlled the fire.

Table 7-6.2.1(c) is based on tests conducted at the Factory Mutual Research Center. Initially only the K-14 ESFR sprinkler was permitted. Testing using plastic commodities resulted in the addition of the K-25 ESFR sprinkler when the table was transferred to NFPA 13. Rubber tires have not been used as the commodity in tests using K-25 ESFR sprinklers.

A-7-6.2.1 Table (a), Note 2. Laced tires are not stored to a significant height by this method due to the damage inflicted on the tire (i.e., bead).

7-6.2.2 In buildings used in part for tire storage, for the purposes of this standard, the required sprinkler protection shall extend 15 ft (4.6 m) beyond the perimeter of the tire storage area.

This requirement for sprinkler protection 15 ft (4.6 m) beyond the storage perimeter is necessary for the following reasons:

(1) The building and the area adjoining the tire storage area need to be protected should a fire start at or spread to the perimeter of the tire storage area.
(2) Storage often extends beyond the designated normal storage area for a variety of reasons.
(3) Sprinklers will normally operate beyond the designated storage area as the heat rises and spreads along the ceiling.

7-6.2.3 Where high-expansion foam systems are installed in accordance with NFPA 11A, *Standard for Medium- and High-Expansion Foam Systems,* a reduction in sprinkler discharge density to one-half the density specified in Table 7-6.2.1(a) or 0.24 gpm/ft^2 (9.78 mm/min), whichever is higher, shall be permitted.

Ceiling sprinklers are necessary when protecting storage with high-expansion foam. The sprinklers are primarily needed to maintain the integrity of the columns and ceiling structure for the time it takes to cover the storage commodity with foam. Sprinkler discharge also breaks down the foam, and the designer must take this into account when determining the amount of foam required.

This requirement is not intended to apply when protection criteria of sprinklers plus high-expansion foam have been selected from Table 7-6.2.1(a) as the discharge criteria.

In existing buildings used for tire storage, high-expansion foam can be used to augment an existing sprinkler system whose calculated density is below that required for the proposed storage height. For example, an existing system calculated to provide 0.25 gpm/ft^2 (10.2 mm/min) can be used for storage requiring up to 0.50 gpm/ft^2 (20.4 mm/min) where high-expansion foam is used. Reinforcement or redesign of the sprinkler system is an alternative.

7-6.3 In-Rack Sprinkler System Requirements.

7-6.3.1 In-rack sprinklers, where provided, shall be installed in accordance with 7-4, except as modified by 7-6.3.2 through 7-6.3.4.

7-6.3.2 The maximum horizontal spacing of sprinklers in rack shall be 8 ft (2.4 m).

7-6.3.3 Sprinklers in racks shall discharge at not less than 30 psi (2.1 bar).

7-6.3.4 Water demand for sprinklers installed in racks shall be based on simultaneous operation of the most hydraulically remote 12 sprinklers where only one level is installed in racks.

7-6.4 Water Supplies.

7-6.4.1 The rate of water supply shall be sufficient to provide the required sprinkler discharge density over the required area of application plus provision for generation of high-expansion foam and in-rack sprinklers where used.

7-6.4.2 Total water supplies shall include provision for not less than 750 gpm (2835 L/ min) for hose streams in addition to that required for automatic sprinklers and foam systems. Water supplies shall be capable of supplying the demand for sprinkler systems and hose streams for not less than 3 hours.

Exception No. 1: For on-floor storage up to and including 5 ft (1.5 m) in height, hose stream requirements shall be permitted to be 250 gpm (946 L/min) with a water supply duration of not less than 2 hours.

Exception No. 2: For ESFR and large drop sprinkler systems approved for rubber tire storage, duration and hose demand shall be in accordance with Tables 7-6.2.1(b) and 7-6.2.1(c).

7-6.4.3* Where dry pipe systems are used, the area of sprinkler application shall be increased by not less than 30 percent.

A-7-6.4.3 Wet systems are recommended for tire storage occupancies. Dry systems are permitted only where it is impracticable to provide heat.

Because no water initially flows from the sprinklers in a dry pipe system, the fire grows unimpeded by the sprinklers for a period of time before the dry pipe valve operates and water reaches the operated sprinklers. During this delay, sprinklers continue to operate. It has been agreed that about 30 percent more sprinklers operate with a dry pipe system than with a wet pipe system.

This requirement does not apply to large drop sprinkler protection of rubber tires, and ESFR sprinklers are not listed for use on dry pipe systems.

7-6.5 Miscellaneous Tire Storage.

Miscellaneous tire storage shall be protected in accordance with 7-2.3.2.2.

Miscellaneous tire storage was defined in recognition of that storage found in conjunction with another operation. The definition of *miscellaneous tire storage* can be found in 1-4.10 and A-1-4.10 and also in NFPA 230, *Standard for the Fire Protection of Storage.*

7-6.6 Small Hose.

For first-aid fire-fighting and mop-up operations, small hose [1½ in. (38 mm)] shall be provided in accordance with 5-15.5.

Hose stations of 1½ in. are indispensable for first-aid fire-fighting and mop-up operations in warehouse facilities.

The training of the plant fire brigade or plant emergency organization in fire-fighting tactics largely determines the effectiveness of hose stations. Experience shows that many fires are suppressed or extinguished in the early stages when trained plant personnel use interior hand hose stations in conjunction with a properly designed, installed, and maintained sprinkler system.

Most rubber tire fires are controlled but not entirely extinguished by the sprinkler systems. Generally, interior hose stations are needed for final extinguishment or mop-up during salvage and overhaul operations.

7-7 Protection of Baled Cotton Storage

7-7.1 General.

The sprinkler system criteria in Section 7-7 shall apply to buildings with ceiling slopes not exceeding 2 in 12 (16.7 percent).

The 1999 edition marks the first time that NFPA 13 covers the protection of baled cotton as a requirement and not just as recommended good practice. Fire test data for fires in baled cotton is very limited. Protection criteria are largely based on empirical data gathered and observations made over the course of several years by members of the baled cotton subcommittees and task groups. With baled cotton fires, sprinkler protection should provide undelayed response over large design areas due to the possibility of rapidly spreading flameover-type fires and should provide sustained quantities of water to fight deep-seated fires. The cotton fiber itself contains enough entrained oxygen to sustain a fire deep within a compressed bale. Such a fire will slowly develop until it "breaks out" and exposes nearby bales. Flameover fires, or flashover, as it is known to baled cotton industries, can result from any number of sources, burning across the surface of bales involving areas such as exposed sample holes or fan-head bale ends where cotton ties have come loose. Current bale-wrapping practices lessen the chance of flameover, but the possibility still exists. Sustained application of water over large design areas is necessary due both to the possibility of rapid firespread and the large fuel loading (Btu content) present in baled cotton storage. Recent developments in quick-response and larger orifice sprinkler heads should aid in control of the rapidly spreading nature of some baled cotton fires and in control of more deeply seated fires.

7-7.2 Sprinkler Systems.

7-7.2.1 For tiered or rack storage up to a nominal 15 ft (4.6 m) in height, sprinkler discharge densities and areas of application shall be in accordance with Figure 7-7.2.1. The density provided for the area of operation can be taken from any point on the selected curve. It is not necessary to meet more than one point on the selected curve.

Curve A of Figure 7-7.2.1 is based on the criteria for protection of non-rack storage of Class IV commodities. Curve C is an extrapolation between curve A and the criteria for protection of Class IV commodities in racks. Curves B and D represent a 30 percent increase in area for use of dry pipe systems.

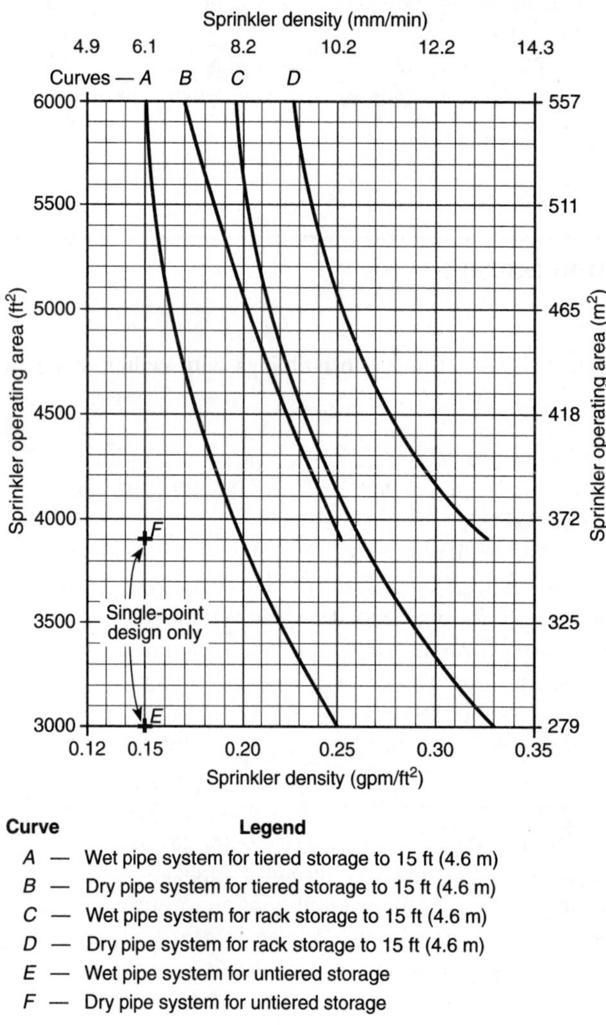

Figure 7-7.2.1 *Sprinkler system design curves.*

7-7.2.2 Where roof or ceiling heights would prohibit storage above a nominal 10 ft (3.1 m), the sprinkler discharge density shall be permitted to be reduced by 20 percent of that indicated in Figure 7-7.2.1 but shall not be reduced to less than 0.15 gpm/ft^2 (6.1 mm/min).

Low roof or ceiling heights that do not permit storage above 10 ft (3 m) effectively limit storage of baled cotton to no more than two bales high on ends or five bales high on sides. Limited numbers of bales reduce the potential Btu content available to a fire.

7-7.2.3* Baled storage that is not tiered can be based on the single-point design curve E for wet pipe systems and curve F for dry pipe systems.

A-7-7.2.3 This untiered design density limits storage to the height of one bale, on side or on end, and would likely prohibit any future tiering without redesign of the sprinkler system.

Baled cotton that is not tiered is stacked one bale high on end. Curve E is a single-point design curve that allows the minimum density over the minimum design area. Curve F represents a 30 percent increase in design area for use of dry pipe systems.

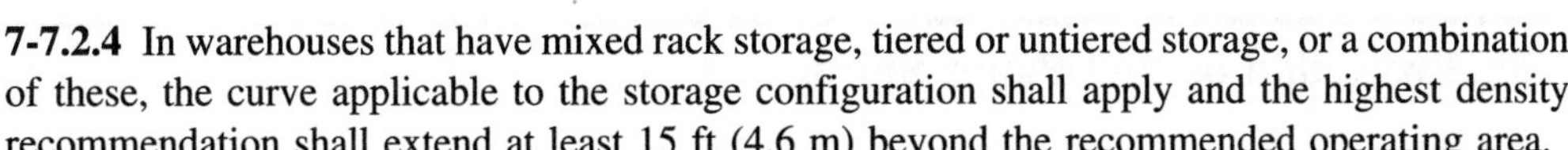

7-7.2.4 In warehouses that have mixed rack storage, tiered or untiered storage, or a combination of these, the curve applicable to the storage configuration shall apply and the highest density recommendation shall extend at least 15 ft (4.6 m) beyond the recommended operating area.

The user is required to design for protection of the most demanding storage configuration. The design is required to be extended at least 15 ft (4.6 m) beyond the area actually occupied by that configuration.

7-7.2.5 Minimum sprinkler operating areas shall be 3000 ft^2 (279 m^2) for wet pipe systems and 3900 ft^2 (363 m^2) for dry pipe systems. The maximum operating area shall not exceed 6000 ft^2 (557 m^2). No area credit is recommended for the use of high-temperature sprinklers.

The minimum design area of 3000 ft^2 (279 m^2) for wet pipe systems and 3900 ft^2 (363 m^2) for dry pipe systems acknowledges the high Btu content of baled cotton, which is 6800 Btu per pound or 3,400,000 Btu per bale. The propensity for rapid firespread through initial flameover across exposed surfaces of cotton is also acknowledged. The use of ordinary temperature–rated sprinklers minimizes the time for the first sprinklers to operate. Use of quick-response spray sprinklers or larger orifice sprinkler heads should enhance control of fires in baled cotton.

7-7.3 Water Supplies.

7-7.3.1 The total water supply available shall be sufficient to provide the recommended sprinkler discharge density over the area to be protected, plus a minimum of 500 gpm (1893 L/min) for hose streams.

A hose stream allowance of 500 gpm (1893 L/min) allows for application of two hose streams through warehouse doors located on the leeward side of warehouses. The leeward side is the side protected from the wind. Warehouse doors are recommended to be kept closed as much as possible to limit oxygen supply to the involved interior of the warehouse. Hard streams directly applied to bale surfaces should be used cautiously as cotton fire can thus actually be spread to uninvolved bales.

7-7.3.2 Water supplies shall be capable of supplying the total demand for sprinklers and hose streams for not less than 2 hours.

Fires involving baled cotton can be deep seated and require considerable water for final extinguishment before salvage operations.

7-7.4 Small Hose.

For first-aid fire-fighting and mop-up operations, small hose [1½ in. (38 mm)] shall be provided in accordance with 5-15.5.

Bales involved in fires have the initial appearance of being totally unusable. However, water does not damage cotton. The availability of small hose for mop-up and final extinguishment is essential to optimize salvage of fire-damaged cotton.

7-8* Protection of Roll Paper Storage

A-7-8 This section provides a summary of the data developed from the tissue test series of full-scale roll paper tests conducted at the Factory Mutual Research Center in West Gloucester, RI.

The test building is approximately 200 ft × 250 ft [50,000 ft^2 (4.65 km^2)] in area, of fire-resistive construction, and has a volume of approximately 2.25 million ft^3 (63,761.86 m^3), the equivalent of a 100,000-ft^2 (9.29-km^2) building 22.5 ft (6.86 m) high. The test building has two primary heights beneath a single large ceiling. The east section is 30 ft (9.1 m) high and the west section is 60 ft (18.29 m) high.

The tissue test series was conducted in the 30-ft (9.1-m) section, with clearances from the top of storage to the ceiling nominally 10 ft (3.1 m).

Figure A-7-8 illustrates a typical storage array used in the tissue series of tests.

The basic criteria used in judging test failure included one or more of the following:

(1) Firespread to the north end of the storage array
(2) Gas temperatures near the ceiling maintained at high levels for a time judged to be sufficient to endanger exposed structural steel
(3) Fire reaching the target stacks

Table A-7-8 outlines the tissue test results.

Fire tests have been conducted on 20-ft (6.1-m) and 25-ft (7.6-m) high vertical storage of tissue with 10-ft (3.1-m) and 5-ft (1.5-m) clear space to the ceiling in piles extending up to seven columns in one direction and six columns in the other direction. In these tests, target columns of tissue were located directly across an 8-ft (2.4-m) aisle from the main pile. Three tests were conducted using 17⁄32-in. (13.5-mm) 286°F (141°C) high-temperature sprinklers on a 100-ft^2 (9.3-m^2) spacing and at constant pressures of 14 psi, 60 psi, and 95 psi (1 bar, 4.1 bar, and 6.6 bar), respectively. One test was run using 0.64-in. (16.3-mm) 286°F (141°C) high-temperature sprinklers on a 100-ft^2 (9.3-m^2) spacing at a constant pressure of 50 psi (3.5 bar). Two tests were conducted following a scheduled decay from an initial pressure of 138 psi (9.5 bar) to a design point of 59 psi (4.1 bar) if 40 sprinklers opened. The significant characteristic of these fire tests was the rapid initial firespread across the surface of the rolls. Ceiling temperatures were controlled during the decaying pressure tests and during the higher constant pressure

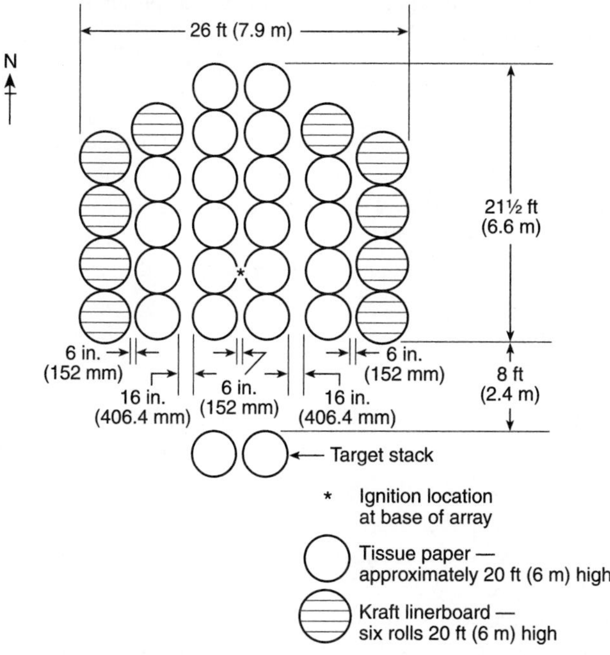

Figure A-7-8 *Plan view of typical tissue storage array.*

tests. With the exception of the 20-ft (6.1-m) high decaying pressure test, the extent of firespread within the pile could not be clearly established. Aisle jump was experienced, except at the 95-psi (6.6-bar) constant pressure, 20-ft (6.1-m) high decaying pressure, and large drop test. Water absorption and pile instability caused pile collapse in all tests. This characteristic should be considered where manually attacking a fire in tissue storage occupancies.

Available fire experience in roll tissue storage occupancies does not correlate well with the constant pressure full-scale fire tests with respect to the number of sprinklers operating and the extent of firespread. Better correlation is noted with the decaying pressure tests. Thirteen fires reported in storage occupancies with storage piles ranging from 10 ft to 20 ft (3.1 m to 6.1 m) high and protected by wet-pipe sprinkler systems ranging from ordinary hazard design densities to design densities of 0.6 gpm/ft^2 (24.5 mm/min) were controlled with an average of 17 sprinklers. The maximum number of wet pipe sprinklers that opened was 45 and the minimum number was five, versus 88 and 26, respectively, in the constant pressure tests. Seventeen sprinklers opened in the 20-ft (6.1-m) high decaying pressure test. One actual fire in tissue storage provided with a dry pipe system opened 143 sprinklers but was reported as controlled.

One fire test was conducted with plastic-wrapped rolls of heavyweight kraft paper. The on-end storage was in a standard configuration, 20 ft (6.1 m) high with 9½-ft (2.9-m) clearance to ceiling sprinklers. The prescribed 0.3-gpm/ft^2 (12.2-mm/min) density controlled the firespread, but protection to roof steel was marginal to the point where light beams and joists might be expected to distort. A lower moisture content in the paper as a result of the protective plastic wrapping was considered to be the reason for the higher temperatures in this test as compared to a similar test where the rolls were not wrapped.

The appendix data and discussion of the full-scale roll paper tissue fire tests, which presented the highest fire challenge tested, are indicative of the overall roll paper fire test program that was conducted over a period of 3 years. Roll tissue presented the highest challenge to sprinklers in the tests because of the rapid flamespread across the surface of the rolls. A summary of the 3-year roll paper test program for tissue is shown in Table A-7-8. A summary of the test program for kraft linerboard and newsprint is shown in Table 7.1.

7-8.1 General.

7-8.1.1 The sprinkler system criteria in Section 7-8 is intended to apply to buildings with ceiling slopes not exceeding 2 in 12 (16.7 percent).

The requirements for sprinkler protection of roll paper were developed from data obtained from full-scale fire tests of roll paper storage that were conducted at the Factory Mutual Test Center in West Gloucester, RI, under flat ceilings. These sprinkler protection criteria were previously located in NFPA 231F.

Although no tests have been conducted on roll paper in buildings with sloped ceilings, a design area increase of 50 percent was applied to the criteria of NFPA 231F as a safety factor for those types of sprinklers that employ the area/density design method. This safety factor was added to address potential adverse conditions that were not accounted for in the test program. Whether the safety factor adequately addresses roof slopes in excess of 16.7 percent is uncertain.

7-8.1.2 Where buildings are occupied in part for vertical roll paper storage and only a portion of the sprinkler system is hydraulically designed, the design area shall extend not less than 20 ft (6.1 m) beyond the area occupied by the roll paper storage.

In storage fires, sprinklers can operate well beyond the actual fire. The sprinkler operating area can be up as much as 10 or more times the actual fire area. Therefore, for a fire starting near the edge of the storage, sprinklers beyond the storage area will operate. The more severe the hazard presented by the type of the commodity stored, the greater the likelihood that sprinklers further from the fire will operate.

7-8.1.3 Wet pipe systems shall be used in tissue storage areas.

Fire tests conducted on roll paper storage demonstrate that the weight and surface characteristics of the paper determine the rate of flamespread over the surface of the paper. Lightweight and thin paper sheets burn much faster than heavyweight thick paper sheets. The surface characteristics of a paper sheet or coatings on the paper can also influence the rate of flamespread over the surface. A "fuzzy" soft sheet of paper will burn faster than paper of the same weight that has a polished surface (calandered). The rate of flamespread across the exposed surface or outer ply of a roll of paper is the most critical factor in designing a sprinkler system that will control a roll paper

Table A-7-8 Summary of Roll Paper Tissue Tests

Test Specifications	Test Number					
	B1[a]	B2	B3	B4	B5[b]	B6[b]
Test date	10/4/79	7/23/80	7/30/80	10/15/80	7/28/82	8/5/82
Paper type	Tissue	Tissue	Tissue	Tissue	Tissue	Tissue
Stack height [ft–in. (m)]	21–10 (6.66)	20–0 (6.1)	21–8 (21.60)	18–6 (6.64)	19–10 (6.05)	25–3 (7.69)
Paper, banded	No	No	No	No	No	No
Paper, wrapped	No	No	No	No	No	No
Fuel array	Standard	Standard	Standard	Standard	Standard	Standard
Clearance to ceiling [ft–in. (m)]	8–2 (2.49)	10–0 (3.05)	8–4 (2.54)	11–6 (3.51)	5–2 (1.58)	4–9 (1.45)
Clearance to sprinklers [ft–in. (m)]	7–7 (2.31)	9–5 (2.87)	7–9 (2.36)	10–9 (3.28)	4–7 (1.40)	4–2 (1.27)
Sprinkler orifice [in. (mm)]	17⁄32 (13.5)	17⁄32 (13.5)	17⁄32 (13.5)	0.64 (16.33)	17⁄32 (13.5)	17⁄32 (13.5)
Sprinkler temp. rating [°F (°C)]	280 (138)	280 (138)	280 (138)	280 (138)	280 (138)	280 (138)
Sprinkler spacing [ft × ft (m × m)]	10 × 10 (3.05 × 3.05)	10 × 10 (3.05 × 3.05)	10 × 10 (3.05 × 3.05)	10 × 10 (3.05 × 3.05)	10 × 10 (3.05 × 3.05)	10 × 10 3.05 × 3.05)
Water pressure [psi (bar)]	14 (0.9)[c]	60 (4.1)	95 (6.6)	50 (3.4)	138 (9.5) initial 102 (7.0) final	138 (9.5) initial 88 (6.1) final
Moisture content of paper (%)	9.3	9.3	10.2	6.0	8.2	9.2
First sprinkler operation (min:sec)	0:43	0:32	0:38	0:31	0:28	0:22
Total sprinklers open	88	33	26	64	17	29
Final flow [gpm (L/min)]	2575 (9746)[c]	1992 (7540)	1993 (7544)	4907 (18573)	1363 (5159)	2156 (8161)
Sprinkler demand area [ft^2 (m^2)]	8800 (817.5)	3300 (306.6)	2600 (241.5)	6400 (595)	1700 (158)	2900 (269)
Average discharge density [gpm/ft^2 (mm/min)]	0.29 (11.8)[c]	0.60 (24.4)	0.77 (31.4)	—	0.92 (37.5) initial 0.80 (32.6) final	0.96 (39.1) initial 0.74 (30.2) final
Maximum 1-minute average gas temperature over ignition [°F (°C)]	1680 (916)[c]	1463 (795)	1634 (890)	1519 (826)	[d]	[e]
Duration of high temperature within acceptable limits	No	Yes	Yes	Marginal	Yes	Yes
Maximum 1-minute average fire plume gas velocity over ignition [ft/sec (m/sec)]	—	40.7 (12.4)	50.2 (15.3)	47.8 (14.6)	—	—
Target ignited	Yes	Yes	No	No	No	Briefly
Extent of fire damage within acceptable limits	No	No	Marginal	Marginal	Yes	Marginal
Test duration (min)	17.4	20	20	25.5	45	45

[a]Phase I test.

[b]Phase III tests decaying pressure.

[c]Pressure increased to 50 psi (3.5 bar) at 10 minutes.

[d]Maximum steel temperature over ignition 341°F (172°C).

[e]Maximum steel temperature over ignition 132°F (56°C).

Table 7.1 Summary of Roll Paper Fire Test Program for Kraft Linerboard and Newsprint

	Test Number										
	A1[a]	A2[a]	A3[a]	A4[a]	A5	A6	A7	C1	C2	C3	C4
Test date	9/18/79	9/25/79	10/1/79	10/9/79	7/9/80	7/16/80	8/14/80	8/16/80	8/20/80	11/19/80	12/17/80
Paper type	Kraft linerboard	Kraft linerboard	Kraft linerboard	Kraft linerboard	Kraft linerboard	Kraft linerboard	Kraft linerboard	Newsprint	Newsprint	Newsprint	Newsprint
Stack height [ft–in. (m)]	20–0 (6.1)	20–0 (6.1)	20–0 (6.1)	25–4 (7.7)	20–0 (6.1)	26–0 (7.9)	26–0 (7.9)	19–0 (5.6)	19–0 (5.6)	19–0 (5.6)	19–0 (5.6)
Paper, banded	Yes	Yes	No	Yes	No	Yes	Yes	No	No	No	No
Paper, wrapped	No	No	No	No	No	No	No	Yes	No	No	No
Fuel array	Std.	Std.	Std.	Std.	Open	Open	Open	Closed	Open	Std.	Std.
Clearance to ceiling [ft–in. (m)]	5–0 (1.5)	10–0 (3.05)	10–0 (3.05)	4–8 (1.4)	10–0 (3.05)	34–0 (10.4)	34–0 (10.4)	11–0 (3.4)	11–0 (3.4)	11–0 (3.4)	11–0 (3.4)
Clearance to sprinklers [ft–in. (m)]	4–5 (1.4)	9–5 (2.87)	9–5 (2.87)	4–1 (1.3)	9–5 (2.87)	33–5 (10.2)	33–5 (10.2)	10.5 (3.2)	10.5 (3.2)	10.5 (3.2)	10.5 (3.2)
Sprinkler orifice [in. (mm)]	17/32 (13.5)	17/32 (13.5)	17/32 (13.5)	17/32 (13.5)	17/32 (13.5)	17/32 (13.5)	0.64 (16.33)	17/32 (13.5)	17/32 (13.5)	1/2 (12.7)	17/32 (13.5)
Sprinkler temp. rating [°F (°C)]	280 (138)	280 (138)	280 (138)	280 (138)	280 (138)	280 (138)	280 (138)	280 (138)	280 (138)	280 (138)	280 (138)
Sprinkler spacing [ft × ft (m × m)]	10 × 10 (3.05 × 3.05)	10 × 10 (3.05 × 3.05)	10 × 10 (3.05 × 3.05)	10 × 10 (3.05 × 3.05)	10 × 10 (3.05 × 3.05)	10 × 10 (3.05 × 3.05)	10 × 10 (3.05 × 3.05)	10 × 10 (3.05 × 3.05)	10 × 10 (3.05 × 3.05)	10 × 10 (3.05 × 3.05)	10 × 10 (3.05 × 3.05)
Water pressure [psi (bar)]	14 (0.9)	14 (0.9)	14 (0.9)	14 (0.9)	34 (2.3)	60 (4.1)	60 (4.1)	15 (1.0)	34 (2.3)	34 (2.3)	34 (2.3)
Moisture content of paper (%)	8.4	9.9	9.2	7.8	9.2	9.4	9.6	10.6/10.9[c]	9.5	7.5	7.2
First sprinkler operation (min:sec)	1:52	1:30	1:34	1:39	1:36	2:03	2:09	2:31	0:54	0:39	0:41
Total sprinklers open	12	18	20	22	28	39	7	4	31	45	34
Final flow [gpm (L/min)]	350 (1325)	520 (1968)	575 (2176)	630 (2385)	1282 (4852)	2378 (9001)	601 (2275)	121 (458)	1414 (5352)	1538 (5821)	1550 (5867)
Sprinkler demand area [ft² (m²)]	1200 (112)	1800 (167)	2000 (186)	2200 (204)	2800 (260)	3900 (362)	700 (65)	400 (37)	3100 (288)	4500 (418)	3400 (316)
Average discharge density [gpm/ft² (mm/min)]	0.29 (11.8)	0.29 (11.8)	0.29 (11.8)	0.29 (11.8)	0.46 (18.7)	0.61 (24.9)	—	0.30 (12.2)	0.46 (18.7)	0.34 (13.9)	0.46 (18.7)
Maximum 1-minute average gas temperature over ignition [°F (°C)]	1480 (804)	1640 (893)	1550 (843)	1470 (799)	1386 (752)	1654 (901)	949 (509)	1311 (711)	1512 (822)	1851 (1011)	1646 (897)
Duration of high temperature within acceptable limits	Yes	Yes	Yes	Marginal	Yes	No	Yes	Yes	Yes	No	Marginal
Maximum 1-minute average fire plume gas velocity over ignition [ft/sec (m/sec)]	—	—	—	—	47.4 (14.4)	80.7 (24.6)	36.4 (11.1)	34.6 (10.5)	49.2 (15.0)	53.2 (16.2)	48.5 (14.8)
Target ignited	No	No	No	No	No	Yes	No	No	Yes	Yes	Yes
Extent of fire damage within acceptable limits	Yes	Yes	Yes	Yes	Yes	No	Yes	Yes	No	No	No
Test duration (min)	12.3	12.9	30.2	25.1	30	20	30	25.1	25.2	25	25.4

[a]Phase I Test.

[b]Pressure increased to 50 psi at 10 minutes.

[c]Wrapper—10.6 percent; Newsprint—10.9 percent.

fire. Lightweight tissue paper exhibits a particularly rapid surface flamespread that can open an excessive number of sprinklers in a relatively short period of time and make it very difficult for a dry pipe sprinkler system to establish a controlling sprinkler density. Appendix C contains a summary of the full-scale roll paper tissue tests. The classification of roll paper by basis weight can be found in 2-2.5.

7-8.1.4 Horizontal storage of heavyweight or mediumweight paper shall be protected as a closed array.

Fire tests conducted on vertical storage of roll paper in a closed array indicated a relatively slow developing fire that required a relatively small sprinkler system discharge density and area of operation to achieve fire control. A closed paper array is one in which the distance between paper columns in all directions is small—that is, not more than 2 in. (50 mm)—as shown in Exhibit 7.9. The slow fire development is due to the limited size of the flue spaces between paper rolls. Horizontal (nested) storage of paper rolls shown in Exhibit 7.10 also minimizes the size of the flue spaces in the horizontal direction, which tends to limit fire growth. See definitions of *array* in 1-4.12 for further details.

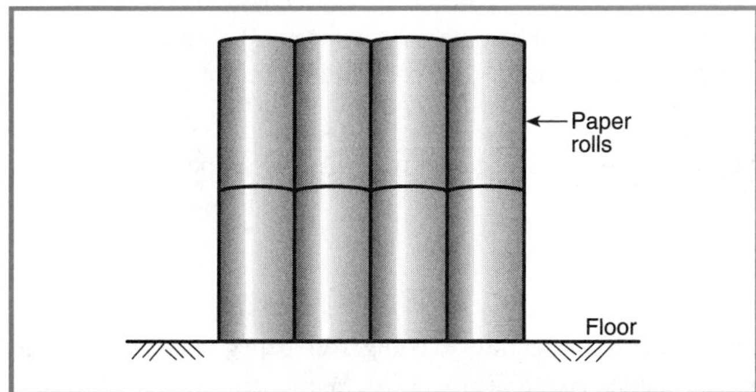

Exhibit 7.9. *Closed paper roll vertical storage array.*

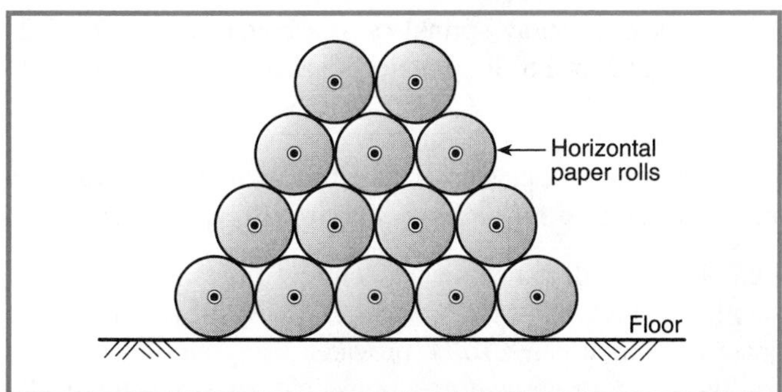

Exhibit 7.10 *Horizontal (nested) storage of paper rolls.*

7-8.1.5 Mediumweight paper shall be permitted to be protected as heavyweight paper where wrapped completely on the sides and both ends, or where wrapped on the sides only with steel bands.

Wrapping material shall be either a single layer of heavyweight paper with a basis weight of 40 lb (18.1 kg), or two layers of heavyweight paper with a basis weight of less than 40 lb (18.1 kg).

7-8.1.6 Lightweight paper or tissue paper shall be permitted to be protected as mediumweight paper where wrapped completely on the sides and both ends, or where wrapped on the sides only with steel bands.

Wrapping material shall be either a single layer of heavyweight paper with a basis weight of 40 lb (18.1 kg), or two layers of heavyweight paper with a basis weight of less than 40 lb (18.1 kg).

Fire tests conducted on roll paper storage demonstrate that the flamespread characteristics of the exposed outer sheets of the paper rolls are the critical factor in determining the sprinkler system design required to achieve fire control. Heavyweight paper used as a protective wrapper on mediumweight newsprint and writing papers has a relatively low flamespread rating. When mediumweight papers were wrapped in heavyweight paper for testing, sprinkler systems designed to protect heavyweight paper successfully controlled the fire. See 2-2.5 for a description of paper classifications. Further, when heavyweight wrappers were used on lightweight tissue rolls, successful control of the fire was achieved by sprinkler systems designed to protect mediumweight paper.

7-8.1.7 For purposes of sprinkler system design criteria, lightweight class paper shall be protected as tissue.

7-8.2* Protection Criteria.

A-7-8.2 *Existing Systems.* Sprinkler systems protecting existing roll paper storage facilities should be evaluated in accordance with Tables A-7-8.2(a) and (b). While fire can be controlled by the protection shown in Tables A-7-8.2(a) and (b), greater damage can occur when the densities in Tables A-7-8.2(a) and (b) are used rather than those specified in Tables 7-8.2.2.3(a) and (b).

7-8.2.1 Roll paper storage shall be protected by spray sprinklers in accordance with 7-8.2.2, large drop sprinklers in accordance with 7-8.2.3, or ESFR sprinklers in accordance with 7-8.2.4.

7-8.2.2 Spray Sprinklers.

7-8.2.2.1 Storage of heavyweight or mediumweight classes of rolled paper up to 10 ft (3.1 m) in height shall be protected by sprinklers designed for Ordinary Hazard, Group 2 densities.

Testing was used to establish the discharge densities for the control of fires in heavyweight and mediumweight paper stored up to 20 ft (6.1 m) and 25 ft (7.6 m) in height. The tests indicated that adequate protection can also be provided using an ordinary hazard (Group 2) density where the heavyweight and mediumweight paper are stored no higher than 10 ft (3.1 m).

Table A-7-8.2(a) Automatic Sprinkler System Design Criteria—Spray Sprinklers for Existing Storage Facilities (Discharge densities are gpm/ft² over ft².)

		Heavyweight					Mediumweight			
		Closed Array	Standard Array		Open Array		Closed Array	Standard Array		Open Array
Storage Height (ft)	Clearance (ft)	Banded or Unbanded	Banded	Unbanded	Banded	Unbanded	Banded or Unbanded	Banded	Unbanded	Banded or Unbanded
10	≤5	0.2/2000	0.2/2000	0.2/2000	0.25/2000	0.25/2000	0.2/2000	0.25/2000	0.3/2000	0.3/2000
10	>5	0.2/2000	0.2/2000	0.2/2000	0.25/2500	0.25/2500	0.2/2000	0.25/2000	0.3/2000	0.3/2000
15	≤5	0.25/2000	0.25/2000	0.25/2500	0.3/2500	0.3/3000	0.25/2000	0.3/2000	0.45/2500	0.45/2500
15	>5	0.25/2000	0.25/2000	0.25/2500	0.3/3000	0.3/3500	0.25/2000	0.3/2500	0.45/3000	0.45/3000
20	≤5	0.3/2000	0.3/2000	0.3/2500	0.45/3000	0.45/3500	0.3/2000	0.45/2500	0.6/2500	0.6/2500
20	>5	0.3/2000	0.3/2500	0.3/3000	0.45/3500	0.45/4000	0.3/2500	0.45/3000	0.6/3000	0.6/3000
25	≤5	0.45/2500	0.45/3000	0.45/3500	0.6/2500	0.6/3000	0.45/3000	0.6/3000	0.75/2500	0.75/2500
25	>5	0.45/3000	0.45/3500	0.45/4000	0.6/3000	0.6/3500	0.45/3500	0.6/3500	0.75/3000	0.75/3000
30	≤5	0.6/2500	0.6/3000	0.6/3000	0.75/2500	0.75/3000	0.6/4000	0.75/3000	0.75/3500	0.75/3500

Note: Densities or areas, or both, can be interpolated between any 5-ft storage height increment.

Table A-7-8.2(b) Automatic Sprinkler System Design Criteria—Spray Sprinklers for Existing Storage Facilities (Discharge densities are mm/min over m^2.)

		Heavyweight					Mediumweight			
		Closed Array	Standard Array		Open Array		Closed Array	Standard Array		Open Array
Storage Height (m)	Clearance (m)	Banded or Unbanded	Banded	Unbanded	Banded	Unbanded	Banded or Unbanded	Banded	Unbanded	Banded or Unbanded
3.1	≤1.5	0.76/ 185.8	0.76/ 185.8	0.76/ 185.8	0.95/ 185.8	0.95/ 185.8	0.76/ 185.8	0.95/ 185.8	12.2/ 185.8	12.2/ 185.8
3.1	>1.5	0.76/ 185.8	0.76/ 185.8	0.76/ 185.8	0.95/ 232.3	0.95/ 232.3	0.76/ 185.8	0.95/ 185.8	12.2/ 185.8	12.2/ 185.8
4.6	≤1.5	0.95/ 185.8	0.95/ 185.8	0.95/ 232.3	12.2/ 232.3	12.2/ 278.7	0.95/ 185.8	12.2/ 185.8	18.3/ 232.3	18.3/ 232.3
4.6	>1.5	0.95/ 185.8	0.95/ 185.8	0.95/ 232.3	12.2/ 278.7	12.2/ 325.2	0.95/ 185.8	12.2/ 232.3	18.3/ 278.7	18.3/ 278.7
6.1	≤1.5	12.2/ 185.8	12.2/ 185.8	12.2/ 232.3	18.3/ 278.7	18.3/ 325.2	12.2/ 185.8	18.3/ 232.3	24.5/ 232.3	24.5/ 232.3
6.1	>1.5	12.2/ 185.8	12.2/ 232.3	12.2/ 278.7	18.3/ 325.2	18.3/ 371.6	12.2/ 232.3	18.3/ 278.7	24.5/ 278.7	24.5/ 278.7
7.6	≤1.5	18.3/ 232.3	18.3/ 278.7	18.3/ 325.2	24.5/ 232.3	24.5/ 278.7	18.3/ 278.7	24.5/ 278.7	30.6/ 232.3	30.6/ 232.3
7.6	>1.5	183/ 278.7	18.3/ 325.2	18.3/ 371.6	24.5/ 278.7	24.5/ 325.2	18.3/ 325.2	24.5/ 325.2	30.6/ 278.7	30.6/ 278.7
9.1	≤1.5	24.5/ 232.3	24.5/ 278.7	24.5/ 278.7	30.6/ 232.3	30.6/ 278.7	24.5/ 371.6	30.6/ 278.7	30.6/ 325.2	30.6/ 325.2

Note: Densities or areas, or both, can be interpolated between any 1.5-m storage height increment.

7-8.2.2.2 Storage of tissue and lightweight classes of paper up to 10 ft (3.1 m) in height shall be protected by sprinklers in accordance with Extra Hazard, Group 1 densities.

Because of the rapid firespread across the surface of paper rolls classified as lightweight or tissue, protection criteria must be increased to extra hazard (Group 1).

7-8.2.2.3 Sprinkler design criteria for storage of roll paper 10 ft (3.1 m) high and higher in buildings or structures with roof or ceilings up to 30 ft (9.1 m) shall be in accordance with Tables 7-8.2.2.3(a) and (b).

Tables 7-8.2.2.3 (a) and 7-8.2.2.3 (b) are based on data from a series of full-scale roll paper fire tests conducted at the Factory Mutual Research Center at West Gloucester, RI. See Appendix C and commentary to Table A-7-8.

7-8.2.2.4* Where dry pipe systems are used in heavyweight class or mediumweight class storage areas, the areas of operation indicated by Tables 7-8.2.2.3(a) and (b) shall be increased by 30 percent.

A-7-8.2.2.4 In a dry pipe system, the area increase of 30 percent should be compounded [e.g., 2000 ft^2 (186 m^2) (1.67 for low-temperature sprinklers and 1.3 for dry pipe systems) = 4343 ft^2 (403.5 m^2) total area]. Where dry pipe systems are used in existing installations, the areas of operation indicated by Tables 7-8.2.2.3(a) and (b) should be increased by 30 percent.

See commentary to 7-2.3.2.6 for a general discussion on the effects of dry pipe systems.

7-8.2.2.5* High-temperature sprinklers shall be used for installations protecting roll paper stored 15 ft (4.6 m) or higher.

A-7-8.2.2.5 Generally, more sprinklers open in fires involving roll paper storage protected by sprinklers rated below the high-temperature range. An increase of 67 percent in the design area should be considered.

High-temperature sprinklers were used in all of the fire tests conducted on roll paper that provided the data for developing the area/ density design criteria. Large drop sprinklers used in the fire tests on roll paper also had high-temperature ratings. The ESFR design criteria are based on EFSR technology and tests conducted by Factory Mutual Research Corporation. ESFR sprinklers are not subject to 7-8.2.2.5.

7-8.2.2.6 The protection area per sprinkler shall not exceed 100 ft^2 (9.3 m^2) or be less than 70 ft^2 (6.5 m^2).

The protection area per sprinkler used in the roll paper fire test program described in A-7-8 was 100 ft^2 (9.3 m^2).

Table 7-8.2.2.3(a) Automatic Sprinkler System Design Criteria—Spray Sprinklers for Buildings or Structures with Roof or Ceilings up to 30 ft (Discharge densities are gpm/ft² over ft².)

		Heavyweight					Mediumweight				Tissues
		Closed Array	Standard Array		Open Array		Closed Array	Standard Array		Open Array	
Storage Height (ft)	Clearance (ft)	Banded or Unbanded	Banded	Unbanded	Banded	Unbanded	Banded or Unbanded	Banded	Unbanded	Banded or Unbanded	All Storage Array
10	≤5	0.3/ 2000	0.3/ 2000	0.3/ 2000	0.3/ 2000	0.3/ 2000	0.3/ 2000	0.3/ 2000	0.3/ 2000	0.3/ 2000	0.45/ 2000
10	>5	0.3/ 2000	0.3/ 2000	0.3/ 2000	0.3/ 2000	0.3/ 2000	0.3/ 2000	0.3/ 2000	0.3/ 2000	0.3/ 2000	0.45/ 2500
15	≤5	0.3/ 2000	0.3/ 2000	0.3/ 2000	0.3/ 2500	0.3/ 3000	0.3/ 2000	0.3/ 2000	0.45/ 2500	0.45/ 2500	0.60/ 2000
15	>5	0.3/ 2000	0.3/ 2000	0.3/ 2000	0.3/ 3000	0.3/ 3500	0.3/ 2000	0.3/ 2500	0.45/ 3000	0.45/ 3000	0.60/ 3000
20	≤5	0.3/ 2000	0.3/ 2000	0.3/ 2500	0.45/ 3000	0.45/ 3500	0.3/ 2000	0.45/ 2500	0.6/ 2500	0.6/ 2500	0.75/ 2500
20	>5	0.3/ 2000	0.3/ 2500	0.3/ 3000	0.45/ 3500	0.45/ 4000	0.3/ 2500	0.45/ 3000	0.6/ 3000	0.6/ 3000	0.75/ 3000
25	≤5	0.45/ 2500	0.45/ 3000	0.45/ 3500	0.6/ 2500	0.6/ 3000	0.45/ 3000	0.6/ 3000	0.75/ 2500	0.75/ 2500	See Note 1

Notes:

1. Sprinkler protection requirements for tissue stored above 20 ft have not been determined.
2. Densities or areas, or both, shall be permitted to be interpolated between any 5-ft storage height increment.

Table 7-8.2.2.3(b) Automatic Sprinkler System Design Criteria—Spray Sprinklers for Buildings or Structures with Roof or Ceilings Up to 9.1 m (Discharge densities are mm/min over m^2.)

		Heavyweight					Mediumweight				
		Closed Array	Standard Array		Open Array		Closed Array	Standard Array		Open Array	Tissues
Storage Height (ft)	Clearance (ft)	Banded or Unbanded	Banded	Unbanded	Banded	Unbanded	Banded or Unbanded	Banded	Unbanded	Banded or Unbanded	All Storage Array
3.1	≤1.5	12.2/ 185.8	12.2/ 185.8	12.2/ 185.8	12.2/ 185.8	12.2/ 185.8	12.2/ 185.8	12.2/ 185.8	12.2/ 185.8	12.2/ 185.8	18.3/ 185.8
3.1	>1.5	12.2/ 185.8	12.2/ 185.8	12.2/ 185.8	12.2/ 185.8	12.2/ 185.8	12.2/ 185.8	12.2/ 185.8	12.2/ 185.8	12.2/ 185.8	18.3/ 232.3
4.6	≤1.5	12.2/ 185.8	12.2/ 185.8	12.2/ 185.8	12.2/ 232.3	12.2/ 278.7	12.2/ 185.8	12.2/ 185.8	18.3/ 232.3	18.3/ 232.3	24.5/ 185.8
4.6	>1.5	12.2/ 185.8	12.2/ 185.8	12.2/ 185.8	12.2/ 278.7	12.2/ 322.2	12.2/ 185.8	12.2/ 232.3	18.3/ 278.7	18.3/ 278.7	24.5/ 278.7
6.1	≤1.5	12.2/ 185.8	12.2/ 185.8	12.2/ 232.3	18.3/ 278.7	18.3/ 325.2	12.2/ 185.8	18.3/ 232.3	24.5/ 232.3	24.5/ 232.3	30.6/ 232.3
6.1	>1.5	12.2/ 185.8	12.2/ 232.3	12.2/ 278.7	18.3/ 325.2	18.3/ 371.6	12.2/ 232.3	18.3/ 278.7	24.5/ 278.7	24.5/ 278.7	30.6/ 278.7
7.6	≤1.5	18.3/ 232.3	18.3/ 278.7	18.3/ 325.2	24.5/ 232.3	24.5/ 278.7	18.3/ 278.7	24.5/ 278.7	30.6/ 232.3	30.6/ 232.3	See Note 1

Notes:

1. Sprinkler protection requirements for tissue stored above 6.1 m have not been determined.
2. Densities or areas, or both, shall be permitted to be interpolated between any 1.5-m storage height increment.

7-8.2.2.7 Where high-expansion foam systems are installed in heavyweight class and mediumweight class storage areas, sprinkler discharge design densities can be reduced to not less than 0.24 gpm/ft^2 (9.8 mm/min) with a minimum operating area of 2000 ft^2 (186 m^2).

The roll paper fire test program conducted to provide data for automatic sprinkler design criteria did not include tests using high-expansion foam. However, the effect on high-expansion foam has been demonstrated on other storage commodities. See 7-3.2.2.4 for more information on high-expansion foam.

7-8.2.2.8 Where high-expansion foam systems are installed in tissue storage areas, sprinkler discharge densities and areas of application shall not be reduced below those provided in Tables 7-8.2.2.3(a) and (b).

7-8.2.3 Large Drop Sprinklers. Where automatic sprinkler system protection utilizes large drop sprinklers, hydraulic design criteria shall be as specified in Table 7-8.2.3. Design discharge pressure shall be 50 psi (3.4 bar). The number of sprinklers to be calculated is indicated based on storage height, clearance, and system type.

7-8.2.4 ESFR Sprinklers. Where automatic sprinkler system protection utilizes ESFR sprinklers, hydraulic design criteria shall be as specified in Table 7-8.2.4. Design discharge pressure shall be applied to 12 operating sprinklers.

7-8.3 Water Supplies.

7-8.3.1 The water supply system for automatic fire protection systems shall be designed for a minimum duration of 2 hours.

Exception: For ESFR sprinklers, the water supply duration shall be 1 hour.

7-8.3.2 At least 500 gpm (1893 lpm) shall be added to the sprinkler demand for large and small hose stream demand.

Exception: For ESFR sprinklers, the hose stream allowance shall be for 250 gpm (947 lpm).

7-8.3.3 The water supply design shall include the demand of the automatic sprinkler system plus the hose stream demand plus, where provided, the high-expansion foam system.

7-8.4 Small Hose.

For first-aid fire-fighting and mop-up operations, small hose [1½ in. (38 mm)] shall be provided in accordance with Section 5-15.5.

7-9 Special Design Approaches

7-9.1 General.

All special design approaches utilize the hydraulic calculation procedures of Section 8-4, except as specified.

Table 7-8.2.3 Automatic Sprinkler System Design Criteria—Large Drop Sprinklers (number of sprinklers to be calculated)

Storage Height		Clearance			Heavyweight					Mediumweight					Tissues
					Closed Array	Standard Array		Open Array		Closed Array	Standard Array		Open Array		All Storage Array
ft	m	ft	m	System Type	Banded or Unbanded	Banded	Unbanded	Banded	Unbanded	Banded or Unbanded	Banded	Unbanded	Banded	Unbanded	
20	6.1	<10	<3.1	W	15	15	15	15	NA	15	15	15	NA	NA	See Note 3
20	6.1	<10	<3.1	D	25	25	25	NA	NA	25	25	25	NA	NA	NA
26	7.9	<34	<10.4	W	15	15	15	15	NA	NA	NA	NA	NA	NA	NA
26	7.9	<34	<10.4	D	NA	NA	NA	NA	NA	NA	NA	NA	NA	NA	NA

Notes:

1. W = wet; D = dry; NA = not applicable.
2. For definition of storage height, see Section 1-4.
3. Twenty-five large drop sprinklers @ 75 psi (5.2 bar) for closed or standard array; other arrays NA.

Table 7-8.2.4 ***Rolled Paper Storage—Automatic Sprinkler Design Criteria—ESFR Sprinklers (maximum height of storage permitted)***

ESFR	System	Pressure		Building Height		Heavyweight						Mediumweight						Tissue
						Closed		Standard		Open		Closed		Standard		Open		All
K-factor	Type	psi	bar	ft	m	ft	m	ft	m	ft	m	ft	m	ft	m	ft	m	Arrays
11.0–11.5	Wet	50	3.4	25	7.6	20	6.1	20	6.1	20	6.1	20	6.1	20	6.1	20	6.1	NA
13.5–14.5	Wet	50	3.4	30	9.1	25	7.6	25	7.6	25	7.6	25	7.6	25	7.6	25	7.6	NA
13.5–14.5	Wet	75	5.2	40	12.2	30	9.1	30	9.1	30	9.1	NA		NA		NA		NA
24.2–26.2	Wet	20	1.4	30	9.1	25	7.6	25	7.6	25	7.6	25	7.6	25	7.6	25	7.6	NA
24.2–26.2	Wet	40	2.8	40	12.2	30	9.1	30	9.1	30	9.1	NA		NA		NA		NA

Note: NA = not applicable.

7-9.2 Residential Sprinklers.

The spaces in which residential sprinklers can be used within the scope of NFPA 13 are identified in 5-4.5. Because residential sprinklers were developed to protect against residential fire scenarios, they are permitted in residential portions (primarily dwelling units) of all buildings regardless of size or height. For example, NFPA 13 allows the use of residential sprinklers in high-rise apartment buildings. In accordance with 7-9.2.3, nonresidential spaces (those primarily outside of the dwelling unit) must be protected with other types of sprinklers appropriate for the hazard.

NFPA 13R, *Standard for the Installation of Sprinkler Systems in Residential Occupancies up to and Including Four Stories in Height,* is available as an alternative to NFPA 13 for residential occupancies up to and including four stories in height. NFPA 13R does not address all types of residential facilities. Where sprinkler systems are being considered for single-family homes, NFPA 13D, *Standard for the Installation of Sprinkler Systems in One- and Two-Family Dwellings and Manufactured Homes,* should be used. NFPA provides three different design and installation standards for residential sprinklers. However, the differences in scope between NFPA 13, NFPA 13R, and NFPA 13D need to be understood, because the design and performance objectives of these documents are not identical.

7-9.2.1* Sprinkler discharge rates shall be provided in accordance with minimum flow rates indicated in individual residential sprinkler listings, both for the single sprinkler discharge and the multiple sprinkler discharge of the design sprinklers.

A-7-9.2.1 The protection area for residential sprinklers with extended coverage areas is defined in the listing of the sprinkler as a maximum square or rectangular area. Listing information is presented in even 2-ft (0.61-m) increments from 12 ft to 20 ft (3.6 m to 6.1 m) for residential sprinklers. When a sprinkler is selected for an application, its area of coverage must be equal to or greater than both the length and width of the hazard area. For example, if the hazard to be protected is a room 14 ft 6 in. (4.3 m) wide and 20 ft 8 in. (6.2 m) long, a sprinkler that is listed to protect an area of 16 ft × 22 ft (4.9 m × 6.8 m) must be selected. The flow used in the calculations is then selected as the flow required by the listing for the selected coverage. *(See Figure A-7-9.2.1.)*

Listed residential sprinklers are provided with single-sprinkler and multiple-sprinkler minimum flows and pressures. A minimum operating pressure of 7 psi (0.5 bar) is required for all sprinklers (see 8-4.4.8). As a result, the listings of some residential sprinklers were revised to require higher multiple-sprinkler flows corresponding to this minimum pressure.

7-9.2.2* The design area shall be that area that includes the four hydraulically most demanding sprinklers. Calculations shall be provided to verify the single (one) operating sprinkler criteria and the multiple (four) operating sprinkler criteria.

A-7-9.2.2 In Figure A-7-9.2.2, calculate the area indicated by the heavy outline and X. The circle indicates sprinklers.

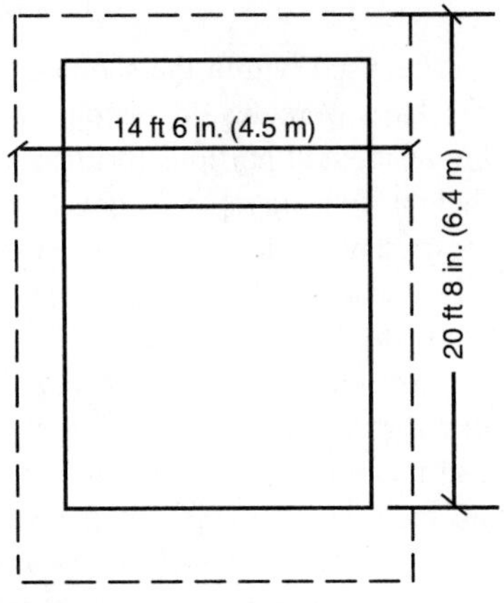

Figure A-7-9.2.1 Determination of protection area of coverage for residential sprinklers.

NFPA 13D requires the sprinkler system design to be based on a maximum of two sprinklers operating. Two sprinklers are considered satisfactory for a one- or two-family dwelling, but the four-sprinkler design area required for an NFPA 13 system is intended to compensate for conditions that could occur in a larger residential occupancy, such as an apartment or hotel. Unlike the four-sprinkler design basis of NFPA 13R, the four-sprinkler design area in NFPA 13 is required to be determined without regard to compartment boundaries. The NFPA 13 residential design area consists of the four contiguous sprinklers that together produce the greatest demand, even if they are located in four individual adjacent compartments.

At least two sets of calculations need to be completed when using residential sprinklers in accordance with NFPA 13. Additional sets of calculations can be required for variations in sprinkler models or maximum protection areas per sprinkler.

7-9.2.3 Where areas such as attics, basements, or other types of occupancies are outside of dwelling units but within the same structure, these areas shall be protected in accordance with the provisions of this standard, including appropriate design criteria of 7-2.3.

7-9.2.4 Hose stream demand and water supply duration requirements shall be in accordance with those for light hazard occupancies in Table 7-2.3.1.1.

7-9.3 Quick-Response Early Suppression (QRES) Sprinklers.

(Reserved) *(See 1-4.5.2 and A-1-4.5.2.)*

7-9.4 Large Drop Sprinklers.

Applications of large drop sprinklers are described in 5-4.7. Chapter 3 requires these sprinklers to have a minimum nominal K-factor of 11.2. Listing of large drop sprinklers

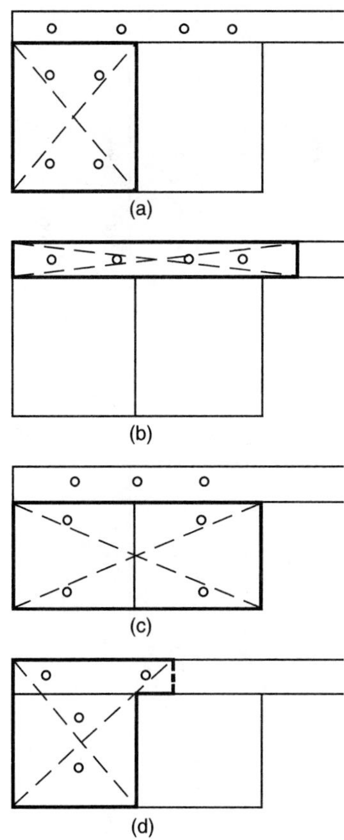

Figure A-7-9.2.2 Examples of design area for dwelling units.

takes into consideration the ability of the sprinkler discharge pattern to penetrate a high-velocity fire plume. Although listed extra-large-orifice spray sprinklers and one category of ESFR sprinklers have similar orifice sizes, only listed large drop sprinklers are to be used in accordance with the special design approach of 7-9.4.

7-9.4.1 Large drop sprinklers shall be permitted to protect ordinary hazard, storage of Class I through Class IV commodities, plastic commodities, miscellaneous storage, and other storage as specified in Sections 7-5, 7-6, and 7-8 or by other NFPA standards.

The discharge criteria for large drop sprinklers formerly found in NFPA 231 and NFPA 231C have been consolidated and moved to 7-9.4 of NFPA 13.

7-9.4.1.1 Protection of palletized and solid-piled storage of Class I through Class IV, unexpanded plastic and expanded plastic commodities shall be in accordance with Table 7-9.4.1.1.

Table 7-9.4.1.1 Large Drop Sprinkler Design Criteria for Palletized and Solid-Piled Storage

		Maximum Storage Height		Maximum Building Height			Number of Design Sprinklers by Minimum Operating Pressure			Hose Stream Demand		Water Supply Duration (hours)
Configuration	Commodity Class	ft	m	ft	m	Type of System	25 psi (1.7 bar)	50 psi (3.4 bar)	75 psi (5.2 bar)	gpm	L/min	
Palletized	I, II, or III	25	7.6	35	10.7	Wet	15	15	15	500	1900	2
						Dry	25	25	25	—	—	—
Palletized	IV	20	6.1	30	9.1	Wet	20	15	15	500	1900	2
						Dry	NA	NA	NA	—	—	—
Palletized	Cartoned or exposed unexpanded plastics	20	6.1	30	9.1	Wet	25	15	15	500	1900	2
						Dry	NA	NA	NA	—	—	—
Palletized	Cartoned or exposed expanded plastics	18	5.5	26	7.9	Wet	NA	15	15	500	1900	2
						Dry	NA	NA	NA	—	—	—
Palletized	Idle wood pallets	20	6.1	30	9.1	Wet	15	15	15	500	1900	1½
						Dry	25	25	25	—	—	—
Solid pile	I, II, or III	20	6.1	30	9.1	Wet	15	15	15	500	1900	1½
						Dry	25	25	25	—	—	—
Solid pile	IV	20	6.1	30	9.1	Wet	NA	15	15	500	1900	1½
						Dry	NA	NA	NA	—	—	—
Solid pile	Cartoned or exposed unexpanded plastics	20	6.1	30	9.1	Wet	NA	15	15	500	1900	1½
						Dry	NA	NA	NA	—	—	—

Note: NA = not allowed.

7-9.4.1.2 Protection of single-, double-, and multiple-row rack storage without solid shelves for Classes I through IV unexpanded plastic commodities shall be in accordance with Table 7-9.4.1.2.

7-9.4.1.2.1 For rack storage, a minimum of 6-in. (152.4-mm) longitudinal flue spaces shall be maintained in addition to transverse flue spaces.

Where large drop sprinklers are used to protect rack storage, longitudinal as well as transverse flue spaces are required regardless of storage height. Additionally, the transverse flues are to be located at the rack uprights and between pallet loads as described in 7-4.1.4.1.

7-9.4.1.2.2 Where in-rack sprinklers are required by Table 7-9.4.1.2, in-rack sprinkler spacing, design pressure, and hydraulic calculation criteria shall be in accordance with the requirements of Section 7-4 as applicable for the commodity.

7-9.4.2 Protection Criteria.

7-9.4.2.1 Protection shall be provided as specified in Tables 7-9.4.1.1 and 7-9.4.1.2, Section 7-6, Section 7-8, or appropriate NFPA standards in terms of minimum operating pressure and the number of sprinklers to be included in the design area.

Protection criteria for Class I through Class IV commodities and plastic commodities, whether stored on the floor or in racks, are addressed in Tables 7-9.4.1.1 and 7-9.4.1.2. Protection criteria for rubber tires and roll paper using large drop sprinklers are addressed by Sections 7-6 and 7-8, respectively.

7-9.4.2.2 The minimum number of design sprinklers for ordinary hazard and miscellaneous storage in accordance with this standard shall be 15 for wet pipe and preaction systems and 25 for double interlock preaction systems and dry pipe systems. For other storage configurations, the number of design sprinklers shall be in accordance with Section 7-6, Section 7-8, or other NFPA standards.

The criteria of 7-9.4.2.2 are intended to apply primarily to the protection of ordinary hazard and miscellaneous storage using large drop sprinklers.

7-9.4.3 Large drop sprinkler systems shall be designed such that the minimum operating pressure is not less than 25 psi (1.7 bar).

7-9.4.4 For design purposes, 95 psi (6.6 bar) shall be the maximum discharge pressure at the hydraulically most remote sprinkler.

Discharge pressures in excess of 95 psi (6.6 bar) negatively impact the sprinkler's discharge pattern and droplet size.

7-9.4.5 The design area shall be a rectangular area having a dimension parallel to the branch lines at least 1.2 times the square root of the area protected by the number of sprinklers to be included in the design area. Any fractional sprinkler shall be included in the design area.

Table 7-9.4.1.2 Large Drop Sprinkler Design Criteria for Single-, Double-, and Multiple-Row Racks without Solid Shelves

Commodity Class	Maximum Storage Height		Maximum Ceiling/ Roof Height		Type of System	Number of Design Sprinklers by Minimum Operating Pressure			Hose Stream Demand		Water Supply Duration
	ft	m	ft	m		25 psi (1.7 bar)	50 psi (3.4 bar)	75 psi (5.2 bar)	gpm	L/min	(hours)
I, II	25	7.6	30	9.1	Wet	20	20	20	500	1900	1½
					Dry	30	30	30	—	—	—
I, II	30	9.1	35	10.7	Wet	20 plus one level of in-rack sprinklers	20 plus one level of in-rack sprinklers	20 plus one level of in-rack sprinklers	500 —	1900 —	1½ —
					Dry	30 plus one level of in-rack sprinklers	30 plus one level of in-rack sprinklers	30 plus one level of in-rack sprinklers	500 —	1900 —	1½ —
I, II, III	20	6.1	30	9.1	Wet	15	15	15	500	1900	1½
					Dry	25	25	25	—	—	—
I, II, III	25	7.6	35	10.7	Wet	15 plus one level of in-rack sprinklers	15 plus one level of in-rack sprinklers	15 plus one level of in-rack sprinklers	500 —	1900 —	1½ —
					Dry	25 plus one level of in-rack sprinklers	25 plus one level of in-rack sprinklers	25 plus one level of in-rack sprinklers	500 —	1900 —	1½ —
IV	20	6.1	30	9.1	Wet	NA	20	15	500	1900	2
					Dry	NA	NA	NA			
IV	25	7.6	35	10.7	Wet	NA	20 plus one level of in-rack sprinklers	15 plus one level of in-rack sprinklers	500	1900	2
					Dry	NA	NA	NA			
Cartoned or exposed unexpanded plastics	20	6.1	30	9.1	Wet	NA	30	20	500	1900	2
					Dry	NA	NA	NA	—	—	—
IV	20	6.1	25	7.6	Wet	NA	15	15	500	1900	2
					Dry	NA	NA	NA	—	—	—
Cartoned or exposed unexpanded plastics	25	7.6	35	10.7	Wet	NA	30 plus one level of in-rack sprinklers	20 plus one level of in-rack sprinklers	500	1900	2
IV	25	7.6	30	9.1	Wet	NA	15 plus one level of in-rack sprinklers	15 plus one level of in-rack sprinklers	500	1900	2
					Dry	NA	NA	NA	—	—	—
Cartoned or exposed unexpanded plastics	20	6.1	25	7.6	Wet	NA	15	15	500	1900	2
					Dry	NA	NA	NA	—	—	—
Cartoned or exposed unexpanded plastics	25	7.6	30	9.1	Wet	NA	15 plus one level of in-rack sprinklers	15 plus one level of in-rack sprinklers	500	1900	2
	25	7.6	36	9.1	Dry	NA	NA	NA	—	—	—

Note: NA = not allowed.

Starting with the 1996 edition, NFPA 13 provided more information about the shape of the design area where large drop sprinklers are used. Although earlier editions implied that the design area was to be a rectangular shape having a dimension parallel to the branch lines of at least 1.2 times the square root of the area protected by the number of sprinklers, it was not explicitly stated.

7-9.4.6 The nominal diameter of branch line pipes (including riser nipples) shall be not less than 1¼ in. (33 mm) nor greater than 2 in. (51 mm).

Exception No. 1: Starter pieces shall be permitted to be 2½-in. (64 mm).

Exception No. 2: Where branch lines are larger than 2 in. (51 mm), the sprinkler shall be supplied by a riser nipple to elevate the sprinkler 13 in. (330 mm) for 2½-in. (64-mm) pipe and 15 in. (380 mm) for 3-in. (76-mm) pipe. These dimensions are measured from the centerline of the pipe to the deflector. In lieu of this, sprinklers shall be permitted to be offset horizontally a minimum of 12 in. (305 mm).

The purpose of limiting the maximum diameter of branch lines or riser nipples is to minimize the effect of pipe shadow below the sprinkler. Pipe shadow can result in significant shielded, or dry, areas with materially reduced water application and loss of sprinkler water penetration.

7-9.4.7 Hose stream demand and water supply duration requirements shall be in accordance with those for extra hazard occupancies in Table 7-2.3.1.1 or Table 7-9.4.1.1 and Table 7-9.4.1.2, whichever applies.

7-9.4.8 Where large drop sprinklers are installed under open wood joist construction, their minimum operating pressure shall be 50 psi (3.4 bar).

Exception: Where each joist channel of open, wood joist construction is fully fire stopped to its full depth at intervals not exceeding 20 ft (6.1 m), the lower pressures specified in Table 7-9.4.1.2 shall be permitted to be used.

7-9.4.9 For the purpose of using Table 7-9.4.1.1 and Table 7-9.4.1.2, preaction systems shall be classified as dry pipe systems.

Exception: Where it can be demonstrated that the detection system activating the preaction system will cause water to be at the sprinklers when they operate, preaction systems shall be permitted to be treated as wet pipe systems.

7-9.4.10 Building steel shall not require special protection where Table 7-9.4.1.1 or Table 7-9.4.1.2 is applied as appropriate for the storage configuration.

7-9.5* Early Suppression Fast-Response (ESFR) Sprinklers.

A-7-9.5 ESFR sprinklers are designed to respond quickly to growing fires and deliver heavy discharge to suppress fires rather than to control them. ESFR sprinklers should not be relied on to provide suppression if they are used outside the design parameters.

While these sprinklers are intended primarily for use in high-pile storage situations, this section permits their use and extension into adjacent portions of an occupancy that might have a lesser classification.

Applications of ESFR sprinklers are described in 5-4.6. The special design approach for ESFR sprinklers is based on fire suppression by the first operating ring of sprinklers. In order to allow for conditions that can deter the ease of suppression, the design area must include 12 ESFR sprinklers flowing simultaneously.

Unlike large drop sprinklers, no special provisions for the use of ESFR sprinklers are available in dry pipe or preaction systems.

7-9.5.1 ESFR sprinklers shall be permitted to protect ordinary hazard, storage of Class I through Class IV commodities, plastic commodities, miscellaneous storage, and other storage as specified in Sections 7-5, 7-6, and 7-8 or by other NFPA standards.

The discharge criteria for ESFR sprinklers formerly found in NFPA 231 and NFPA 231C have been consolidated and moved into 7-9.5 of NFPA 13.

7-9.5.1.1* Protection of palletized and solid pile storage of Classes I through IV, cartoned or uncartoned unexpanded plastic, cartoned expanded plastic, and idle wood or plastic pallets shall be in accordance with Table 7-9.5.1.1.

A-7-9.5.1.1 Storage in single-story or multistory buildings can be permitted, provided the maximum ceiling/roof height as specified in Table 7-9.5.1.1 is satisfied for each storage area.

ESFR protection criteria are provided in Table 7-9.5.1.1, which addresses protection of palletized and solid-piled storage, and Table 7-9.5.1.2, which addresses rack storage without solid shelves. ESFR sprinklers can also be used for protection of ordinary hazards and miscellaneous storage.

The criteria in Tables 7-9.5.1.1 and 7-9.5.1.2 have been expanded to permit broader applications of ESFR sprinkler technology, based on large-scale test results reviewed by the Technical Committee on Sprinkler System Discharge Criteria. Design criteria for K-11 and K-25 ESFR sprinkler technology have been incorporated into the tables where test data supports such inclusion.

7-9.5.1.2 Protection of single-, double-, and multiple-row rack storage of Classes I through IV, cartoned or uncartoned unexpanded plastic, cartoned expanded plastic, and idle wood or plastic pallets shall be in accordance with Table 7-9.5.1.2.

Exception: ESFR protection as defined shall not apply to the following:

(a) Rack storage involving solid shelves

See commentary to 7-4.1.7.2 and 7-9.5.1.2.2.

(b) Rack storage involving combustible, open-top cartons or containers

7-9.5.1.2.1 Where required by Table 7-9.5.1.2, one level of K-8.0 quick-response, ordinary-temperature in-rack sprinklers shall be installed at the tier level closest to but not exceeding ½

Table 7-9.5.1.1 ESFR Protection of Palletized and Solid-Pile Storage

Type of Storage	Commodity	Maximum Height of Storage		Maximum Ceiling/ Roof Height of Building		Nominal K-factor	Sprinkler Design Pressure		Hose Stream Demand		Duration (hours)
		ft	m	ft	m		psi	bar	gpm	L/min	
Palletized and solid-pile storage (no open-top containers or solid shelves)	1. Cartoned unexpanded plastic 2. Cartoned expanded plastic 3. Uncartoned unexpanded plastic 4. Class I, Class II, Class III, or Class IV commodities, encapsulated or unencapsulated 5. Idle wood or plastic pallets	25	7.6	30	9.1	14	50	3.4	250	946	1
	1. Cartoned or uncartoned unexpanded plastic 2. Class I, Class II, Class III, or Class IV commodities, encapsulated or unencapsulated 3. Idle wood or plastic pallets	35	10.7	40	12.2	14	75	5.2	250	946	1
	1. Cartoned or uncartoned unexpanded plastic 2. Class I, Class II, Class III, or Class IV commodities, encapsulated or unencapsulated	35	10.7	45	13.7	14	90	6.2	250	946	1
	1. Cartoned unexpanded plastic 2. Class I, Class II, Class III, or Class IV commodities, encapsulated or unencapsulated	20 25 30 35 40	6.1 7.6 9.1 10.7 12.2	25 30 35 40 45	7.6 9.1 10.7 12.2 13.7	11.2 25.2 25.2 25.2 25.2	50 20 30 40 50	3.4 1.4 2.1 2.8 3.4	250 250 250 250 250	946 946 946 946 946	1 1 1 1 1

of the maximum storage height. In-rack sprinkler hydraulic design criteria shall be the most hydraulically remote eight sprinklers at 50 psi (3.4 bar). In-rack sprinklers shall be located at the intersection of the longitudinal and transverse flue space. Horizontal spacing shall not be permitted to exceed 5-ft (1.5-m) intervals.

7-9.5.1.2.2 Where ESFR sprinklers are installed, see 7-4.1.4 for special requirements for longitudinal flue spaces in double-row racks.

With any rack storage sprinkler protection, especially using ESFR sprinklers, it is critical that nominal 6-in. (152-mm) flue spaces up through the rack be maintained. Transverse flues as described in 7-4.1.4 are required at rack uprights and between pallet loads regardless of rack arrangement and storage height. Additionally, where longitudinal flues exist, they must be maintained vertically through the rack. See commentary to 7-4.1.4 for additional information.

Table 7-9.5.1.2 ESFR Protection of Rack Storage without Solid Shelves

Type of Storage*	Commodity	Maximum Height of Storage		Maximum Ceiling/ Roof Height of Building		Nominal K-factor	Sprinkler Design Pressure		In-Rack Sprinkler Requirements	Hose Stream Demand		Duration (hours)
		ft	m	ft	m		psi	bar		gpm	L/min	
Single-row rack storage, double-row rack storage, multiple-row rack storage	1. Cartoned unexpanded plastic 2. Cartoned expanded plastic 3. Exposed unexpanded plastic 4. Classes I, II, III, and IV commodities, encapsulated or unencapsulated 5. Idle wood and plastic pallets	25	7.6	30	9.1	14	50	3.4	No	250	946	1
	1. Cartoned or exposed unexpanded plastic 2. Classes I, II, III, and IV commodities, encapsulated or unencapsulated 3. Idle wood and plastic pallets	35	10.7	40	12.2	14	75	5.2	No	250	946	1
	1. Cartoned or exposed unexpanded plastic 2. Classes I, II, III, and IV commodities, encapsulated or unencapsulated	35 40	10.7 12.2	45 45	13.7 13.7	14 14	90 90	6.2 6.2	No Yes	250 250	946 946	1 1
	1. Cartoned unexpanded plastic 2. Classes I, II, III, and IV commodities, encapsulated or unencapsulated	20 25 30 35 40	6.1 7.6 9.1 10.7 12.2	25 30 35 40 45	7.6 9.1 10.7 12.2 13.7	11.2 25.2 25.2 25.2 25.2	50 20 30 40 50	3.4 1.4 2.1 2.8 3.4	No No No No No	250 250 250 250 250	946 946 946 946 946	1 1 1 1 1

*No open-top containers

7-9.5.1.3 ESFR sprinklers shall be permitted to be used in other specific hazard classifications and configurations only when proven by large-scale or other suitable fire testing.

7-9.5.2* General Protection Criteria.

A-7-9.5.2 Design parameters were determined from a series of full-scale fire tests that were conducted as a joint effort between Factory Mutual Research Corporation and the National Fire Protection Research Foundation. (Copies of the test reports are available from the NFPRF.)

7-9.5.2.1 ESFR sprinkler systems shall be designed such that the minimum operating pressure is not less than that indicated in Table 7-9.5.1.1, Table 7-9.5.1.2, Section 7-6, or Section 7-8 for type of storage, commodity, storage height, and building height involved.

7-9.5.2.2 The design area shall consist of the most hydraulically demanding area of 12 sprinklers, consisting of four sprinklers on each of three branch lines. The design shall include a minimum of 960 ft^2 (89 m^2).

7-9.5.2.3 The maximum building height shall be measured to the underside of the roof deck or ceiling.

7-9.5.2.4 ESFR sprinklers shall be limited to wet pipe systems.

7-9.5.2.5 Early suppression fast-response (ESFR) sprinklers shall be used only in buildings equal to, or less than, the height of the building for which they have been listed.

7-9.6 Exposure Protection.

7-9.6.1* Piping shall be hydraulically calculated in accordance with Section 8-4 to furnish a minimum of 7 psi (0.5 bar) at any sprinkler with all sprinklers facing the exposure operating.

A-7-9.6.1 If the system is a deluge type, then all the sprinklers need to be calculated even if they are located on different building faces.

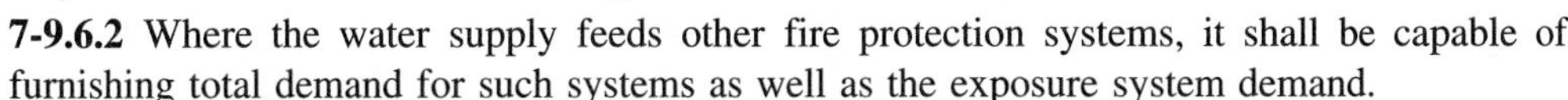

7-9.6.2 Where the water supply feeds other fire protection systems, it shall be capable of furnishing total demand for such systems as well as the exposure system demand.

7-9.7 Water Curtains.

Sprinklers in a water curtain such as described in 5-13.4 shall be hydraulically designed to provide a discharge of 3 gpm per lineal foot (37 L/min per lineal meter) of water curtain, with no sprinklers discharging less than 15 gpm (56.8 L/min). For water curtains employing automatic sprinklers, the number of sprinklers calculated in this water curtain shall be the number in the length corresponding to the length parallel to the branch lines in the area determined by 8-4.4.1(a). If a single fire can be expected to operate sprinklers within the water curtain and within the design area of a hydraulically calculated system, the water supply to the water curtain shall be added to the water demand of the hydraulic calculations and shall be balanced to the calculated area demand. Hydraulic design calculations shall include a design area selected to include ceiling sprinklers adjacent to the water curtain. *(See 5-13.4.)*

7-9.8 Fire Protection of Steel Columns.

Protection of steel columns is necessary primarily to ensure that the structural integrity of the building is maintained during sprinkler system operation. The criteria in this subsection were previously located in NFPA 231C and NFPA 231D, *Standard for Storage of Rubber Tires.*

7-9.8.1† Columns within Storage Racks of Class I through Class IV and Plastic Commodities. Where sprinkler protection of building columns within the rack structure or vertical rack members supporting the building are required in lieu of fireproofing, sprinkler protection in accordance with one of the following shall be provided:

(1) Sidewall sprinklers at the 15-ft (4.6-m) elevation, pointed toward one side of the steel column
(2) Provision of ceiling sprinkler density for a minimum of 2000 ft^2 (186 m^2) with 165°F (74°C) or 286°F (141°C) rated sprinklers as shown in Table 7-9.8.1 for storage heights above 15 ft (4.6 m), up to and including 20 ft (6.1 m)
(3) Provision of large drop or ESFR ceiling sprinkler protection in accordance with 7-9.4 and 7-9.5, respectively.

Table 7-9.8.1 Ceiling Sprinkler Densities for Protection of Steel Building Columns

	Aisle Width	
Commodity Class	**4 ft (1.2 m) gpm/ft^2**	**8 ft (2.4 m) mm/min**
I	0.37	15.1
II	0.44	17.9
III	0.49	20.0
IV and plastics	0.68	27.7

Note: For aisle widths of 4 ft to 8 ft (1.2 m to 2.4 m), a direct linear interpolation between densities can be made.

As indicated by Formal Interpretation 75-4, which was originally issued for NFPA 231C, additional protection of steel columns is only required where the columns are located within the rack structure.

Formal Interpretation 75-4

Reference: 7-9.8.1(1)

Question: Does 7-9.8.1(1) apply when columns occur adjacent to single row racks (whether at building wall or otherwise)?

Answer: No.

Issue Edition: 1975 of NFPA 231C

Reference: 3-2.3(b)

Date: November 1979

Reprinted to correct error: January 1989

7-9.8.2 Columns within Rubber Tire Storage. Where fireproofing is not provided, steel columns shall be protected as follows:

(1) Storage exceeding 15 ft through 20 ft (4.6 m through 6 m) in height—one sidewall sprinkler directed to one side of the column at a 15-ft (4.6-m) level
(2) Storage exceeding 20 ft (6.1 m) in height—two sidewall sprinklers, one at the top of the column and the other at a 15-ft (4.6-m) level, both directed to the side of the column

Exception: The protection specified in 7-9.8.1(1) and (2) shall not be required where storage in fixed racks is protected by in-rack sprinklers.

7-10 Sprinkler System Discharge Criteria for Special Occupancy Hazards

7-10.1 Flammable and Combustible Liquids.

See NFPA 30, *Flammable and Combustible Liquids Code,* for sprinkler system discharge criteria pertaining to the protection of flammable and combustible liquids.

7-10.2 Aerosol Products.

See NFPA 30B, *Code for the Manufacture and Storage of Aerosol Products,* for sprinkler system discharge criteria pertaining to the protection of aerosol products.

7-10.3 Spray Application Using Flammable or Combustible Materials.

7-10.3.1 See NFPA 33, *Standard for Spray Application Using Flammable or Combustible Materials,* for applicable terms not defined in Chapter 1.

7-10.3.2* Spray Areas and Mixing Rooms. The automatic sprinkler system in spray areas and mixing rooms shall be designed for Extra Hazard (Group 2) occupancies. (**33:** 7-2.1)

A-7-10.3.2 Spray application operations should only be located in buildings that are completely protected by an approved system of automatic sprinklers. If located in unsprinklered buildings, sprinklers should be installed to protect spray application processes where practical. Because of the rapidity and intensity of fires that involve spray operations, the available water should be ample to simultaneously supply all sprinklers likely to open in one fire without depleting the available water for use by hose streams. Noncombustible draft curtains can be used to limit the number of sprinklers that will open.

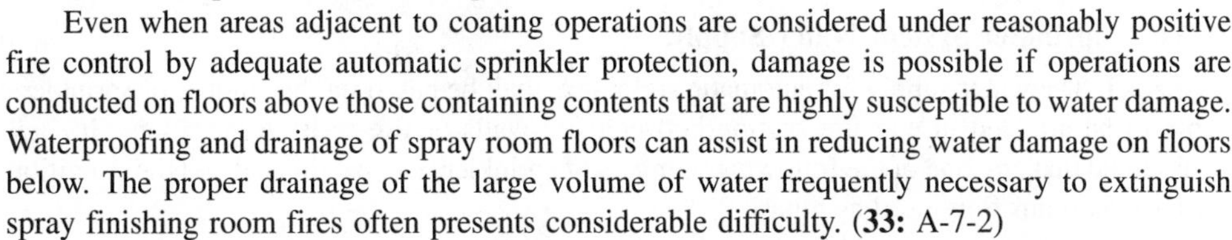

Even when areas adjacent to coating operations are considered under reasonably positive fire control by adequate automatic sprinkler protection, damage is possible if operations are conducted on floors above those containing contents that are highly susceptible to water damage. Waterproofing and drainage of spray room floors can assist in reducing water damage on floors below. The proper drainage of the large volume of water frequently necessary to extinguish spray finishing room fires often presents considerable difficulty. (**33:** A-7-2)

7-10.3.3 Water supply for sprinklers shall be sufficient to supply all sprinklers likely to open in any one fire incident without depleting the available water for use in hose streams. Where sprinklers are installed to protect spray areas and mixing rooms only, water shall be permitted to be furnished from the domestic supply, subject to the approval of the authority having jurisdiction and provided the domestic supply can meet the design criteria for extra hazard, Group 2 occupancies. (**33:** 7-2.3)

7-10.3.4 For spray application areas using styrene cross-linked thermoset resins (commonly known as glass fiber reinforced plastics), automatic sprinkler systems shall be designed and installed for at least ordinary hazard, Group 2 occupancies. (**33:** 15-3)

7-10.4 Solvent Extraction Plants.

7-10.4.1 See NFPA 36, *Standard for Solvent Extraction Plants,* for applicable terms not defined in Chapter 1.

7-10.4.2* Sprinkler Systems. (**36:** 2-9)

A-7-10.4.2 Water spray systems that are used to protect solvent extraction process equipment or structures should be designed to provide a density of not less than 0.25 gpm per ft^2 (10.2 mm/min) of protected surface area. Deluge systems that are used for the same purposes should be designed to provide a density of not less than 0.16 gpm per ft^2 (6.5 mm/min) of protected surface area. Preparation buildings should be protected with automatic sprinkler systems designed for ordinary hazard, Group 2. (**36:** A-2-9)

7-10.5 Nitrate Film.

7-10.5.1 See NFPA 40, *Standard for the Storage and Handling of Cellulose Nitrate Motion Picture Film,* for applicable terms not defined in Chapter 1.

7-10.5.2 Where rooms containing nitrate film are required to be sprinklered per NFPA 40, the sprinkler system shall be installed in accordance with the requirements for extra hazard occupancies. (**40:** 3-1.2)

7-10.5.3 Sprinkler System Water Supply. Water supplies for automatic sprinklers shall be based on 20 gpm (1.26 L/ sec) per sprinkler for 20 minutes for the total number of sprinklers in one vault, plus 25 percent of the sprinklers in the communicating fire area. (**40:** 3-2.2)

7-10.6 Storage of Pyroxylin Plastic.

7-10.6.1 See NFPA 42, *Code for the Storage of Pyroxylin Plastic,* for applicable terms not defined in Chapter 1.

7-10.6.2 Sprinkler System Water Supply.

7-10.6.2.1 The water supply for automatic sprinklers shall be based on the number of sprinklers liable to be affected in any fire section between fire walls or fire-resistive partitions. It shall be assumed that any one of the following numbers of sprinklers can be affected and the condition giving maximum flow used as a basis:

(1) All sprinklers in a vault
(2) All sprinklers in a tote box storeroom

(3) Three-fourths of the sprinklers in a finished-stock storeroom
(4) All sprinklers in a section of an isolated storage building (**42:** 2-4.3.1)

7-10.6.2.2 The water supply for an automatic sprinkler system shall be based on a flow of 20 gpm (76 Lpm) per sprinkler for 20 minutes, with a minimum rate of flow of 500 gpm (1900 Lpm). Such flow shall be with an effective pressure at the top line of sprinklers of not less than 40 psi (2.8 bar). (**42:** 2-4.3.2)

7-10.7 Laboratories Using Chemicals.

7-10.7.1 See NFPA 45, *Standard for Fire Protection for Laboratories Using Chemicals,* for applicable terms not defined in Chapter 1.

7-10.7.2 An automatic sprinkler system, where required by Table 3-1(a) of NFPA 45, depending on the construction of the building, the hazard class of the laboratory unit, the construction of the laboratory unit enclosure, and the area of the laboratory unit shall be in accordance with the following:

(1) Automatic sprinkler system protection for Class A and Class B laboratories shall be in accordance with ordinary hazard, Group 2 occupancies.
(2) Automatic sprinkler system protection for Class C and Class D laboratories shall be in accordance with ordinary hazard, Group 1 occupancies. (**45:** 4-2.1.1)

7-10.8 Oxygen-Fuel Gas Systems for Welding, Cutting, and Allied Processes.

7-10.8.1 See NFPA 51, *Standard for the Design and Installation of Oxygen-Fuel Gas Systems for Welding, Cutting, and Allied Processes,* for applicable terms not defined in Chapter 1.

7-10.8.2 Where sprinkler systems are required per 2-3.1 of NFPA 51, they shall provide a sprinkler discharge density of at least 0.25 gpm/ft^2 (10.2 mm/min) over a minimum operating area of at least 3000 ft^2 (88 m^2). (**51:** 2-3.1, Exception No. 1)

7-10.9 Acetylene Cylinder Charging Plants.

7-10.9.1 See NFPA 51A, *Standard for Acetylene Cylinder Charging Plants,* for applicable terms not defined in Chapter 1.

7-10.9.2 Where an automatic sprinkler system is required per NFPA 51A, *Standard for Acetylene Cylinder Charging Plants,* it shall be an extra hazard open or closed head sprinkler system. (**51A:** 9-2.2)

7-10.10 Storage, Use, and Handling of Compressed and Liquefied Gases in Portable Cylinders.

7-10.10.1 See NFPA 55, *Standard for the Storage, Use, and Handling of Compressed and Liquefied Gases in Portable Cylinders,* for applicable terms not defined in Chapter 1.

7-10.10.2 Where an automatic sprinkler system is required per NFPA 55, the sprinkler system protecting the gas cylinder storage, and for a distance of 25 ft (7.6 m) beyond in all directions, shall be capable of providing a sprinkler density of at least 0.3 gpm/ft^2 (12.2 mm/min) over the most hydraulically remote 2500 ft^2 (232.25 m^2). (**55:** 2-2.2.1)

7-10.10.3 Where sprinkler systems are provided per NFPA 55, 2-2.2.2, Exception No. 1, they shall be designed for ordinary hazard, Group 1 occupancies. (**55:** 2-2.2.2)

7-10.10.4 Where sprinkler systems are provided per NFPA 55, 2-2.2.2, Exception No. 2, they shall be designed for ordinary hazard, Group 1 occupancies. (**55:** 2-2.2.2)

7-10.10.5 Where sprinkler systems are required for gas cylinder storage rooms per NFPA 55, they shall be capable of providing a minimum density of 0.3 gpm/ft^2 (12.2 mm/min) over the most hydraulically remote 2500 ft^2 (232.25 m^2) or the entire room area, whichever is smaller. (**55:** 2-2.3.2)

7-10.11 Storage and Handling of Liquefied Petroleum Gases at Utility Gas Plants.

7-10.11.1 See NFPA 59, *Standard for the Storage and Handling of Liquefied Petroleum Gases at Utility Gas Plants,* for applicable terms not defined in Chapter 1.

7-10.11.2 Sprinkler System Water Supply. The design of fire water supply and distribution systems, if required by NFPA 59, shall provide for the simultaneous supply of those fixed fire protection systems, including monitor nozzles, at their design flow and pressure, involved in the maximum single incident expected in the plant. An additional supply of 1000 gpm (3785 L/min) shall be available for hand hose streams for a period of not less than 2 hours. Manually actuated monitors shall be permitted to be used to augment hand hose streams. (**59:** 10-5.2)

7-10.12 Production, Storage, and Handling of Liquefied Natural Gas (LNG).

7-10.12.1 See NFPA 59A, *Standard for the Production, Storage, and Handling of Liquefied Natural Gas (LNG),* for applicable terms not defined in Chapter 1.

7-10.12.2 The design of fire water supply and distribution systems, if required by NFPA 59A, shall provide for the simultaneous supply of those fixed fire protection systems, including monitor nozzles, at their design flow and pressure, involved in the maximum single incident expected in the plant plus an allowance of 1000 gpm (3785 L/min) for hand hose streams for not less than 2 hours. (**59A:** 9-5.2)

7-10.13 Ventilation Control and Fire Protection of Commercial Cooking Operations.

7-10.13.1 See NFPA 96, *Standard for Ventilation Control and Fire Protection of Commercial Cooking Operations,* for applicable terms not defined in Chapter 1.

7-10.13.2 Sprinkler System Water Supply. Solid fuel appliances with fire boxes exceeding 5 ft^3 (0.14 m^3) shall be provided with a fixed water pipe system with a hose in the immediate vicinity of the appliance. The system shall have a minimum operating pressure of 40 psi (2.8 bar) and shall provide a minimum of 5 gpm per cu ft (18.9 L/min per 0.03 m^3) of fire box volume. (**96:** 11-7.5)

7-10.14 Class A Hyperbaric Chambers.

7-10.14.1 See NFPA 99, *Standard for Health Care Facilities,* for applicable terms not defined in Chapter 1.

7-10.14.2 In chambers that consist of more than one chamber compartment (lock), the design of the deluge system shall ensure adequate operation when the chamber compartments are at different depths (pressures). The design shall also ensure the independent or simultaneous operation of deluge systems. (**99:** 19-2.5.2)

7-10.14.3 Water shall be delivered from the sprinklers as specified in 7-10.14.5 within 3 seconds of activation of any affiliated deluge control. (**99:** 19-2.5.2.2)

7-10.14.4* Where an automatic sprinkler system is provided per NFPA 99, the average spray density at floor level shall be not less than 2 gpm/ft^2 (81.5 mm/min) with no floor area larger than 1 m^2 receiving less than 1 gpm/ft^2 (40.8 mm/min). (**99:** 19-2.5.2.3)

A-7-10.14.4 Experience has shown that when water is discharged through conventional sprinklers into a hyperbaric atmosphere, the spray angle is reduced because of increased resistance to water droplet movement in the denser atmosphere. This is so even though the water pressure differential is maintained above chamber pressure. Therefore, it is necessary to compensate by increasing the number of sprinklers. It is recommended that spray coverage tests be conducted at maximum chamber pressure.

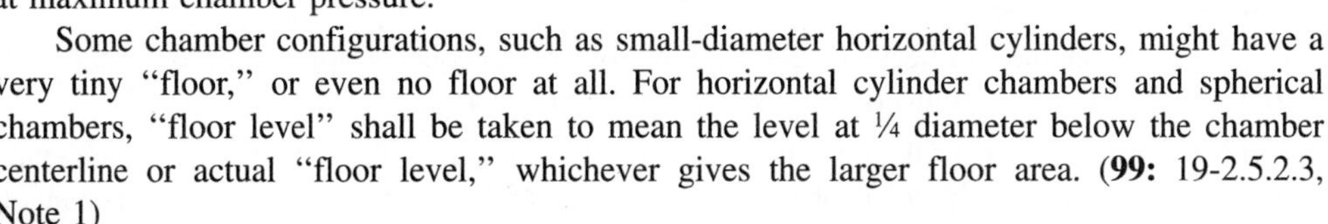

Some chamber configurations, such as small-diameter horizontal cylinders, might have a very tiny "floor," or even no floor at all. For horizontal cylinder chambers and spherical chambers, "floor level" shall be taken to mean the level at ¼ diameter below the chamber centerline or actual "floor level," whichever gives the larger floor area. (**99:** 19-2.5.2.3, Note 1)

7-10.14.5 There shall be sufficient water available in the deluge system to maintain the flow specified in NFPA 99, 19-2.5.2.3 simultaneously in each chamber compartment (lock) containing the deluge system for 1 minute. The limit on maximum extinguishment duration shall be governed by the chamber capacity (bilge capacity also, if so equipped) and/or its drainage system. (**99:** 19-2.5.2.4)

7-10.14.6 The deluge system shall have stored pressure to operate for at least 15 seconds without electrical branch power. (**99:** 19-2.5.2.5)

7-10.15 Fixed Guideway Transit Systems.

7-10.15.1 See NFPA 130, *Standard for Fixed Guideway Transit Systems,* for applicable terms not defined in Chapter 1.

7-10.15.2 Where an automatic sprinkler system is installed per NFPA 130, Section 6-4, it shall be of a closed-head type for ordinary hazard classification. (**130:** 6-4.1)

7-10.16 Race Track Stables.

7-10.16.1 See NFPA 150, *Standard on Fire Safety in Racetrack Stables,* for applicable terms not defined in Chapter 1.

7-10.16.2 Automatic sprinkler systems shall be designed in accordance with ordinary hazard, Group 2 classification. (**150:** 4-1.2)

7-10.17 Water-Cooling Towers.

7-10.17.1 See NFPA 214, *Standard on Water-Cooling Towers,* for applicable terms not defined in Chapter 1.

7-10.17.2 Types of Systems.

7-10.17.2.1 The counterflow tower design lends itself to either closed- or open-head systems. Therefore, wet pipe, dry pipe, preaction, or deluge systems shall be permitted to be used. A deluge system provides a higher degree of protection where water supplies are adequate. In climates that are subject to freezing temperatures, a deluge system minimizes the possibility of failure due to pipes freezing. (**214:** 3-2.2.1)

7-10.17.2.2 The crossflow design is such that it is difficult to locate sprinklers in the most desirable spots for both water distribution and heat detection. This situation can be solved by separating these two functions and using separate water discharge and detection systems. The open-head deluge system does this and, therefore, shall be used in crossflow towers. (**214:** 3-2.2.2)

7-10.17.3 Minimum Rate of Application.

7-10.17.3.1 Under the fan decks of counterflow towers, the rate of application of water shall be 0.5 gpm/ft^2 (20.4 mm/min) (including fan opening). (**214:** 3-2.3.1)

7-10.17.3.2 Towers with a fill area longer than the maximum allowable for the discharge device being used shall have discharge devices placed on both sides of the fill area in each joist channel. The pressure at each discharge device shall be adequate to provide protection for half of the length of the fill area. (**214:** 3-2.4.2.3)

7-10.17.3.3 Under the fan decks of crossflow towers, the rate of application of water shall be 0.33 gpm/ft^2 (13.5 mm/min) (including fan opening). (**214:** 3-2.3.2)

7-10.17.3.4 Over the fill areas of crossflow towers, the rate of application of water shall be 0.5 gpm/ft^2 (20.4 mm/min). (**214:** 3-2.3.3)

7-10.17.4 Discharge outlets [in crossflow towers] shall be open directional spray nozzles or other approved spray devices arranged to discharge 0.35 gpm/ft^2 (14.3 mm/min) directly on the distribution basin and 0.15 gpm/ft^2 (6.1 mm/min) on the underside of the fan deck extension. On towers having extended fan decks that do not completely enclose the hot water basin, outlets protecting the fill shall be located under the distribution basin as set out in 3-2.4.2 of NFPA 214. (**214:** 3-2.4.3)

7-10.17.5 For deluge systems [in crossflow towers] using directional spray nozzles in the pendent position, provisions shall be made to protect the underside of a combustible fan deck at a minimum of 0.15 gpm/ft^2 (6.1 mm/min), which shall be included as part of the application rate specified in 7-10.17.3 of NFPA 13. (**214:** 3-2.4.4)

7-10.17.6 On film-filled towers that have solid, hot-water basin covers over the complete basin, the discharge outlets protecting the fill area shall be permitted to be located under the basin covers. These discharge outlets shall be open directional spray nozzles or other approved devices arranged to discharge 0.35 gpm/ft^2 (14.3 mm/min) directly on the distribution basin, and 0.15 gpm/ft^2 (6.1 mm/min) on the underside of the water basin covers. (**214:** 3-2.4.5)

7-10.17.7 Exposure Protection. Systems for exterior protection shall be designed with the same attention and care as interior systems. Pipe sizing shall be based on hydraulic calculations. The water supply and discharge rate shall be based on a minimum 0.15 gpm/ft^2 (6.1 mm/min) for all surfaces being protected. (**214:** 3-2.10.2)

7-10.17.8 Sprinkler System Water Supply.

7-10.17.8.1 Deluge Systems.

7-10.17.8.1.1* Where all cells of a cooling tower are protected by a single deluge system, the water supply shall be adequate to supply all discharge outlets on that system. (**214:** 3-6.1.1)

A-7-10.17.8.1.1 Where a single deluge system protects an entire water-cooling tower, regardless of the number of cells, the water supply needs to be based on the entire deluge system coverage. Refer to Figure A-7-10.17.8.1.1. (**214:** A-3-6.1.1)

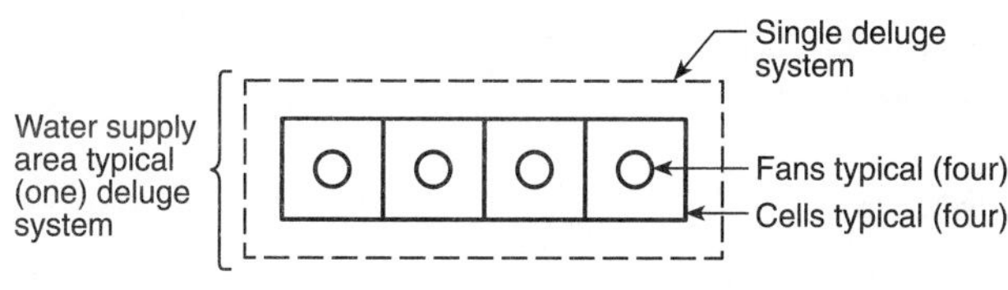

Figure A-7-10.17.8.1.1 *Single deluge system. (214: Figure A-3-6.1.1)*

7-10.17.8.1.2 Where two or more deluge systems are used to protect a cooling tower and fire-resistant partitions are not provided between the deluge systems, the water supply shall be adequate to supply all discharge outlets in the two most hydraulically demanding adjacent systems. (**214:** 3-6.1.2)

7-10.17.8.1.3* Where two or more deluge systems are separated by fire-resistant partitions, the water supply shall be adequate to supply all discharge outlets in the single most hydraulically demanding system. (**214:** 3-6.1.3)

A-7-10.17.8.1.3 Deluge systems separated by fire-resistant partitions can be treated independently as worst-case water supply situations. Refer to Figure A-7-10.17.8.1.3. (**214:** A-3-6.1.3)

7-10.17.8.2* Wet, Dry, and Preaction Systems.

A-7-10.17.8.2 Water-cooling towers with each cell separated by a fire-resistant partition and protected by wet, dry, or preaction system(s) should have the water supply based on the most demanding individual cell. Refer to Figure A-7-10.17.8.2. (**214:** A-3-6.2.1)

7-10.17.8.2.1 Where each cell of the cooling tower is separated by a fire-resistant partition, the water supply shall be adequate to supply all discharge outlets in the hydraulically most demanding single cell. (**214:** 3-6.2.1)

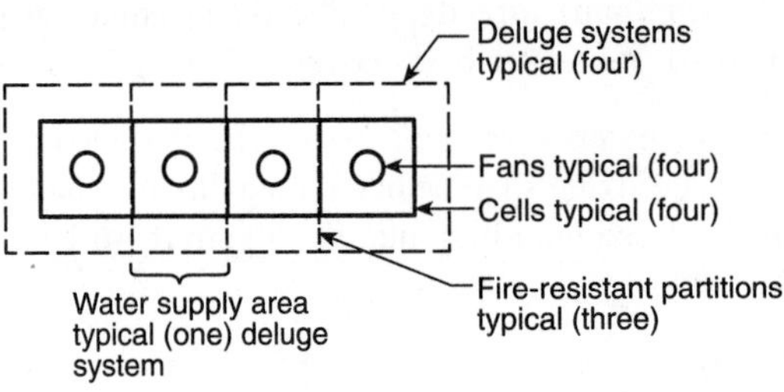

__Figure A-7-10.17.8.1.3__ Multiple deluge systems. (214: Figure A-3-6.1.3)

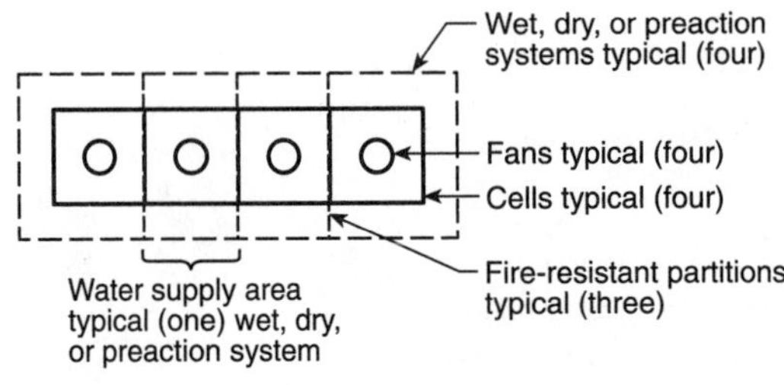

__Figure A-7-10.17.8.2__ Multiple wet, dry, or preaction systems with fire-resistant partitions. (214: Figure A-3-6.2.1)

7-10.17.8.2.2* Where fire-resistant partitions are not provided between each cell of a cooling tower, the water supply shall be adequate to supply all discharge outlets in the two most hydraulically demanding adjoining cells. (**214:** 3-6.2.2)

A-7-10.17.8.2.2 Without fire-resistant partitions between cells, the worst-case situation involves the most demanding adjoining cells. Refer to Figure A-7-10.17.8.2.2. (**214:** A-3-6.2.2)

7-10.17.8.3 Hose Streams. Water supplies shall be sufficient to include a minimum of 500 gpm (1892.5 L/min) for hose streams in addition to the sprinkler requirements. (**214:** 3-6.3)

7-10.17.8.4 Duration. An adequate water supply of at least 2-hour duration shall be provided for the combination of the water supply specified in 7-10.17.8.1 or 7-10.17.8.2 of NFPA 13, plus the hose stream demand specified in 7-10.17.8.3 of NFPA 13. (**214:** 3-6.4)

7-10.18 Piers, Terminals, and Wharves.

7-10.18.1 See NFPA 307, *Standard for the Construction and Fire Protection of Marine Terminals, Piers, and Wharves,* for applicable terms not defined in Chapter 1.

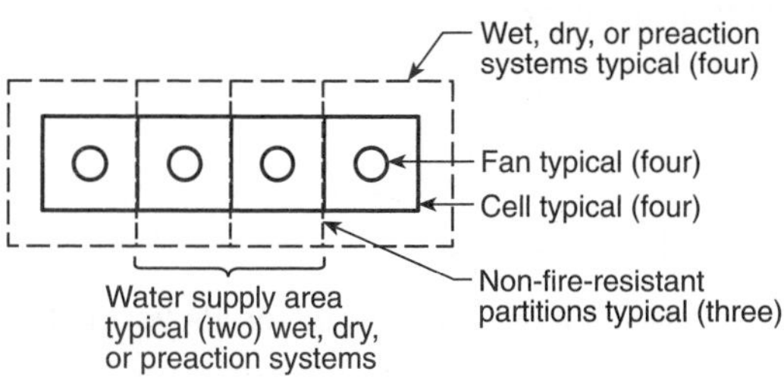

Figure A-7-10.17.8.2.2 Multiple wet, dry, or preaction systems with no fire-resistant partitions. (214: Figure A-3-6.2.2)

7-10.18.2* Piers and Wharves. Where sprinkler systems are required per NFPA 307, the sprinklers shall be K-5.6 and shall discharge at a minimum pressure of 12.5 psi (0.9 bar). The design area shall be based upon the largest area between firestops plus an additional area embracing at least two branch lines on opposite sides of the firestop. The minimum design area shall be not less than 5000 ft^2 (465 m^2). [**307:** 3-3.3.3(a)5]

A-7-10.18.2 The use of firestops for draft control to bank heat, facilitate the opening of sprinklers, and prevent the overtaxing of the sprinkler system is particularly important in the design of sprinkler protection for combustible substructures. The fire walls and firestops of 3-3.3.6 of NFPA 307, *Standard for the Construction and Fire Protection of Marine Terminals, Piers, and Wharves,* should be incorporated into the sprinkler system design for this purpose to the maximum extent practical; however, due to limitations in the size of the design area for the sprinkler system, additional firestops will normally be needed. These additional or supplemental firestops need only have limited fire resistance but should be as deep as possible and be of substantial construction, such as double 3-in. (76.2-mm) planking where exposed to the elements. Where not exposed to physical damage, ¾-in. (19.05-mm) treated plywood extending 48 in. (1219.2 mm) below stringers with solid blocking between stringers should provide adequate durability and reasonable effectiveness. [**307:** A-3-3.3.3(a)(5)]

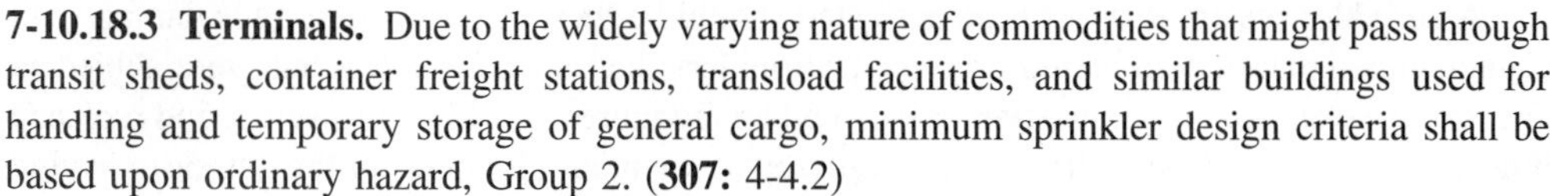

7-10.18.3 Terminals. Due to the widely varying nature of commodities that might pass through transit sheds, container freight stations, transload facilities, and similar buildings used for handling and temporary storage of general cargo, minimum sprinkler design criteria shall be based upon ordinary hazard, Group 2. (**307:** 4-4.2)

7-10.18.4 If the maximum storage height that the building will permit exceeds 12 ft (3.7 m), [the sprinkler system shall be designed for the requirements for] Class IV commodities piled to the maximum height permitted by building construction. *(See Sections 7-3 and 7-4 of NFPA 13.)* (**307:** 4-4.3)

7-10.18.5 If racks or shelving are present or likely to be present, [the sprinkler system shall be designed for the requirements for] Class IV commodities. Protection in warehouses for the long-term storage of specific commodities shall be designed for the specific use. *(See Sections 7-3 and 7-4 of NFPA 13.)*

Exception: Buildings not exceeding 5000 ft² (465 m²) total floor area.

(**307:** 4-4.4)

7-10.19 Cleanrooms.

7-10.19.1 See NFPA 318, *Standard for the Protection of Cleanrooms,* for applicable terms not defined in Chapter 1.

7-10.19.2* Automatic sprinklers for cleanrooms or clean zones shall be hydraulically designed for a density of 0.2 gpm/ft² (8.2 mm/min) over a design area of 3000 ft² (278.8 m²). (**318:** 2-1.2.2)

A-7-10.19.2 Typical configurations of cleanrooms and their chases and plenums create numerous areas that might be sheltered from sprinkler protection. These areas can include air-mixing boxes, catwalks, hoods, protruding lighting, open waffle slabs, equipment, piping, ducting, and cable trays. Care should be taken to relocate or supplement sprinkler protection to ensure that sprinkler discharge covers all parts of the occupancy. Care should also be taken to ensure that sprinklers are located where heat will be satisfactorily collected for reliable operation of the sprinkler.

Gaseous fire suppression systems are not substitutes for automatic sprinkler protection. The large number of air changes in cleanrooms can cause dilution or stratification of the gaseous agent.

It is recommended that sprinkler systems be inspected at least semiannually by a qualified inspection service *(see NFPA 25, Standard for the Inspection, Testing, and Maintenance of Water-Based Fire Protection Systems).* The length of time between such inspections can be decreased due to ambient atmosphere, water supply, or local requirements of the authority having jurisdiction.

Prior to taking a sprinkler system out of service, one should be certain to receive permission from all authorities having jurisdiction and notify all personnel who might be affected during system shutdown. A fire watch during maintenance periods is a recommended precaution. Any sprinkler system taken out of service for any reason should be returned to service as promptly as possible.

A sprinkler system that has been activated should be thoroughly inspected for damage and components replaced or repaired promptly. Sprinklers that did not operate but were subjected to corrosive elements of combustion or elevated temperatures should be inspected, and replaced if necessary, in accordance with the minimum replacement requirements of the authority having jurisdiction. Such sprinklers should be destroyed to prevent their reuse. (**318:** A-2-1.2.1)

7-10.19.3 Automatic sprinkler protection shall be designed and installed in the plenum and interstitial space above cleanrooms for a density of 0.2 gpm/ft² (8.2 mm/ min) over a design area of 3000 ft² (278.8 m²). (**318:** 2-1.2.6)

7-10.19.4* Sprinklers installed in duct systems shall be hydraulically designed to provide 0.5 gpm (1.9 L/min) over an area derived by multiplying the distance between the sprinklers in a horizontal duct by the width of the duct. Minimum discharge shall be 20 gpm (76 L/min) per sprinkler from the five hydraulically most remote sprinklers. (**318:** 2-1.2.7.1)

A-7-10.19.4 Small-orifice sprinklers, ⅜ in. (9.5 mm) or larger, can be used. (**318:** A-2-1.2.1.6.1)

7-10.20 Aircraft Hangars.

See NFPA 409, *Standard on Aircraft Hangars,* for sprinkler system discharge criteria pertaining to the protection of aircraft hangars.

7-10.21 Airport Terminal Buildings, Fueling Ramp Drainage, and Loading.

7-10.21.1 See NFPA 415, *Standard on Airport Terminal Buildings, Fueling Ramp Drainage, and Loading Walkways,* for applicable terms not defined in Chapter 1.

7-10.21.2 Sprinkler System Design.

7-10.21.2.1 Passenger handling areas [in airport terminal buildings] shall be classified as ordinary hazard, Group 1 occupancy for the purpose of sprinkler system design. (**415:** 2-5.1.1)

7-10.21.2.2* Other areas of the airport terminal building shall be classified in accordance with Chapter 2 of NFPA 13, based on the occupancy of the area. (**415:** 2-5.1.2)

A-7-10.21.2.2 The exposure to the airport terminal building from the airport ramp is significant. The number of building sprinklers operating from the exposure fire can be greater than from an internal ignition source. (**415:** A-2-5.1.2)

7-10.21.3 Sprinkler System Water Supply. Water supply from public or private sources shall be adequate to supply maximum calculated sprinkler demand plus a minimum of 500 gpm (1893 L/min) for hose streams. The supply shall be available at the rate specified for a period of at least 1 hour. (**415:** 2-5.5)

7-10.22 Aircraft Engine Test Facilities.

7-10.22.1 See NFPA 423, *Standard for Construction and Protection of Aircraft Engine Test Facilities,* for applicable terms not defined in Chapter 1.

7-10.22.2* In engine test cells, the minimum design discharge density shall be 0.5 gpm/ft^2 (20.4 mm/min) of protected area. (**423:** 5-6.3)

A-7-10.22.2 Because of the nature of the test cell fire potential, deluge systems are considered more appropriate than automatic sprinklers due to their speed of operation and simultaneous discharge of all nozzles; however, automatic sprinklers can be used under the following conditions.

(1) In small cells [600 ft^2 (56 m^2) or less] where it is likely that all sprinklers would fuse at the same time
(2) As a backup to a manual water spray or other manual system

(**423:** A-5-6.3)

7-10.22.3 In engine test cells, water supplies shall be capable of meeting the largest demand at the design rate plus hose stream demand for a period of 30 minutes. Hose stream demand shall be a minimum of 250 gpm (946 L/min). The hydraulic calculation and the water supply shall be based on the assumption that all sprinklers in the test cell are operating simultaneously. (**423:** 5-6.4)

7-10.23 Liquid and Solid Oxidizers.

7-10.23.1 See NFPA 430, *Code for the Storage of Liquid and Solid Oxidizers,* for applicable terms not defined in Chapter 1.

7-10.23.2 Only wet pipe sprinkler systems shall be employed for protection of buildings or areas containing Class 2 or Class 3 oxidizers. (**430:** 2-11.3)

7-10.23.3 Sprinkler System Water Supplies.

7-10.23.3.1 Water supplies shall be adequate for the protection of the oxidizer storage by hose streams and automatic sprinklers. The water system shall be capable of providing not less than 750 gpm (2840 L/min) where protection is by means of hose streams, or 500 gpm (1890 L/min) for hose streams in excess of the automatic sprinkler water demand. (**430:** 2-11.4.1)

7-10.23.3.2 Duration of the water supply shall be a minimum of 2 hours. (**430:** 2-11.4.2)

7-10.23.4 Class 1 Oxidizers. Class I oxidizers in noncombustible or combustible containers (paper bags or noncombustible containers with removable combustible liners) shall be designated as a Class 1 commodity; as a Class 2 commodity where contained in fiber packs or noncombustible containers in combustible packaging; and as a Class 3 commodity where contained in plastic containers. (**430:** 3-3.2)

7-10.23.5 Class 2 Oxidizers.

7-10.23.5.1* Sprinkler protection for Class 2 oxidizers shall be designed in accordance with Table 7-10.23.5.1. (**430:** 4-4.1)

A-7-10.23.5.1 For the purposes of Table 7-10.23.5.1, the fire hazard potential of Class 2 oxidizers has been considered as approximately equal to Group A plastic (non-expanded), stable cartoned. (**430:** A-4-4.1)

Table 7-10.23.5.1 Sprinkler Protection for Class 2 Oxidizers

	Ceiling Sprinklers						
	Storage Height		Density		Area of Application		
Type of Storage	ft	m	gpm/ft^2	mm/min	ft^2	m^2	In-Rack Sprinklers
Palletized	8	2.4	0.20	8.2	3750	348	—
Bulk	12	3.7	0.35	14.3	3750	348	—
Rack	12	3.7	0.20	8.2	3750	348	One line above each level of storage except the top level
	16	4.9	0.30	12.2	2000	186	

(**430:** Table 4-4.1)

7-10.23.5.2 Storage Protection with In-Rack Sprinklers. In-rack sprinklers shall be designed to provide 30 psi (2.1 bar) on the hydraulically most remote six sprinklers on each level. (**430:** 4-4.4, 4-4.4.1)

7-10.23.6 Class 3 Oxidizers.

7-10.23.6.1* Sprinkler protection for Class 3 oxidizers shall be designed in accordance with Table 7-10.23.6.1. (**430:** 5-4.1)

Table 7-10.23.6.1 Sprinkler Protection for Class 3 Oxidizers

Type of Storage	Storage Height		Density		Area of Application		In-Rack Sprinklers
	ft	m	gpm/ft^2	mm/min	ft^2	m^2	
Palletized	5	1.5	0.35	14.3	5000	465	—
Bulk	10	3	0.65	26.5	5000	465	—
Rack	10	3	0.35	14.3	5000	465	1 level at midpoint of rack

(**430:** Table 5-4.1)

A-7-10.23.6.1 For the purposes of Table 7-10.23.6.1, the sprinkler density has been derived from fire loss history. (**430:** A-5-4.1)

7-10.23.6.2 Storage Protection with In-Rack Sprinklers. In-rack sprinklers shall be designed to provide 30 psi (2.1 bar) on the hydraulically most remote six sprinklers on each level. (**430:** 5-4.4, 5-4.4.1)

7-10.23.7 Class 4 Oxidizers. Sprinkler protection for Class 4 oxidizers shall be installed on a deluge sprinkler system to provide water density of 0.35 gpm (14.3 mm/min) over the entire storage area. (**430:** 6-4.1)

7-10.24 Storage of Organic Peroxide Formulations.

7-10.24.1 See NFPA 432, *Code for the Storage of Organic Peroxide Formulations,* for applicable terms not defined in Chapter 1.

7-10.24.2 Where [automatic sprinkler systems are required per NFPA 432, *Code for the Storage of Organic Peroxide Formulations,* they] shall provide the following discharge densities:

Class I [organic peroxides]—0.5 gpm/ft^2 (20. 4 mm/min)

Class II [organic peroxides]—0.4 gpm/ft^2 (16.3 mm/min)

Class III [organic peroxides]—0.3 gpm/ft (12.2 mm/min)

Class IV [organic peroxides]—0.25 gpm/ft^2 (10.2 mm/min) (**432:** 2-8.2)

7-10.24.2.1 The system shall be designed to provide the required density over a 3000-ft^2 (279-m^2) area for areas protected by a wet pipe sprinkler system or 3900 ft^2 (363 m^2) for areas protected by a dry pipe sprinkler system. The entire area of any building of less than 3000 ft^2 (279 m^2) shall be used as the area of application. (**432:** 2-8.2.1)

7-10.24.3 Sprinkler System Water Supply. Water supplies for automatic sprinkler systems, fire hydrants, and so forth, shall be capable of supplying the anticipated demand for at least 90 minutes. (**432:** 2-8.3)

7-10.24.4 Where automatic sprinkler systems are required for Class I organic peroxide formulations in quantities exceeding 2000 lb (908 kg) in detached storage, per NFPA 432, *Code for the Storage of Organic Peroxide Formulations,* automatic sprinkler protection shall be of the deluge type. (**432:** 5-5.2)

7-10.25 Organic Peroxide Formulations.

7-10.25.1 See NFPA 432, *Code for the Storage of Organic Peroxide Formulations,* for applicable terms not defined in Chapter 1.

7-10.25.2 Automatic sprinkler systems for the protection of storage of organic peroxide formulations shall provide the following discharge densities:

Class I—0.5 gpm/ft^2 (20.4 mm/min)

Class II—0.4 gpm/ft^2 (16.3 mm/min)

Class III—0.3 gpm/ft^2 (12.2 mm/min)

Class IV—0.25 gpm/ft^2 (10.2 mm/min)

7-10.25.3 The system shall be designed to provide the required density over a 3000 ft^2 (280 m^2) area for areas protected by a wet pipe sprinkler system or 3900 ft^2 (360 m^2) for areas protected by a dry pipe sprinkler system. The entire area of any building of less than 3000 ft^2 (280 m^2) shall be used as the area of application.

7-10.26 Light Water Nuclear Power Plants.

7-10.26.1 See NFPA 803, *Standard for Fire Protection for Light Water Nuclear Power Plants,* for applicable terms not defined in Chapter 1.

7-10.26.2* The yard mains shall be looped and shall be of sufficient size to meet the flow requirements specified in Section 7-10.26.3 of NFPA 13. (**803:** 11-2)

A-7-10.26.2 Cement-lined pipe 12 in. (304.8 mm) in diameter is recommended. Main sizes should be designed to encompass any anticipated expansion.

The underground main should be arranged such that any one break will not put both a fixed water extinguishing system and hose lines protecting the same area out of service.

7-10.26.3 The water supply for the permanent fire protection installation shall be based on the maximum automatic sprinkler system demand, with simultaneous flow of 750 gpm at grade (2835 L/min) for hose streams and the shortest portion of the fire loop main out of service. (**803:** 12-4)

7-10.27 Advanced Light Water Reactor Electric Generating Plants.

7-10.27.1 See NFPA 804, *Standard for Fire Protection for Advanced Light Water Reactor Electric Generating Plants,* for applicable terms not defined in Chapter 1.

7-10.27.2* Sprinkler System Water Supply. The fire water supply shall be calculated on the basis of the largest expected flow rate for a period of 2 hours, but shall not be less than 300,000 gal (1,135,500 L). This flow rate shall be based on 500 gpm (1892.5 L/min) for manual hose streams plus the largest design demand of any sprinkler system. The fire water supply shall be capable of delivering this design demand with the hydraulically least demanding portion of fire main loop out of service. (**804:** 7-2.1)

A-7-10.27.2 The water supply for the permanent fire protection water system should be based on providing a 2-hour water supply for both items (1) and (2) as follows:

(1) Either of the following items, a or b, whichever is larger:

  a. The largest fixed fire suppression system demand

  b. Any fixed fire suppression system demand that could be reasonably expected to operate simultaneously during a single event (e.g., turbine underfloor protection in conjunction with other fire protection systems in the turbine area)

(2) The hose stream demand of not less than 500 gpm (1892.5 L/min).

  (**804:** A-7-2.1)

7-10.27.3 Yard Mains. The underground yard fire main loop shall be installed to furnish anticipated water requirements. The type of pipe and water treatment shall be design considerations, with tuberculation as one of the parameters. Means for inspecting and flushing the systems shall be provided. (**804:** 7-4.1)

7-10.27.4 Cable Tunnels. (**804:** 8-4.2)

7-10.27.4.1 Automatic sprinkler systems shall be designed for a density of 0.3 gpm/ft^2 (12.2 mm/min) for the most remote 100 linear feet (30.5 linear meters) of cable tunnel up to the most remote 2500 ft^2 (232.2 m^2). (**804:** 8-4.2.2.1)

7-10.27.4.2 Deluge sprinkler systems or deluge spray systems shall be zoned to limit the area of protection to that which the drainage system can handle with any two adjacent systems actuated. The systems shall be hydraulically designed with each zone calculated with the largest adjacent zone flowing. (**804:** 8-4.2.2.3)

7-10.27.5* Beneath Turbine Generator Operating Floor. [When automatic sprinkler systems are provided per NFPA 804] all areas beneath the turbine generator operating floor shall be protected by an automatic sprinkler or foam-water sprinkler system. The sprinkler system beneath the turbine generator shall take into consideration obstructions from structural members and piping and shall be designed to a minimum density of 0.3 gpm/ ft^2 (12.2 mm/min) over a minimum application of 5000 ft^2 (464.5 m^2). (**804:** 8-8.2, 8-8.2.1)

A-7-10.27.5 To avoid water application to hot parts or other water-sensitive areas and to provide adequate coverage, designs that incorporate items such as fusible element operated spray nozzles might be necessary. (**804:** A-8-8.2.1)

7-10.27.6* Turbine Generator Bearings. (**804:** 8-8.3)

A-7-10.27.6 Additional information concerning turbine generator fire protection can be found in EPRI Research Report 1843-2, "Turbine Generator Fire Protection by Sprinkler System," July 1985. (**804:** A-8-8.3)

7-10.27.6.1 Lubricating oil lines above the turbine operating floor shall be protected with an automatic sprinkler system covering those areas subject to oil accumulation, including the area within the turbine lagging (skirt). The automatic sprinkler system shall be designed to a minimum density of 0.30 gpm/ft^2 (12.2 mm/min). (**804:** 8-8.4)

7-10.27.6.2 If shaft-driven ventilation systems are used, an automatic preaction sprinkler system providing a density of 0.3 gpm/ft^2 (12.2 mm/min) over the entire area shall be provided. (**804:** 8-8.6)

7-10.27.7 Standby Emergency Diesel Generators and Combustion Turbines. Sprinkler and water spray protection systems shall be designed for a 0.25-gpm/ ft^2 (10.2-mm/min) density over the entire area. (**804:** 8-9.2)

7-10.27.8 Fire Pump Room/House. If sprinkler and water spray systems are provided for fire pump houses, they shall be designed for a minimum density of 0.25 gpm/ft^2 (10.2 mm/min) over the entire fire area. (**804:** 8-22)

7-10.27.9 Oil-Fired Boilers. Sprinkler and water spray systems shall be designed for a minimum density of 0.25 gpm/ ft^2 (10.2 mm/min) over the entire area. (**804:** 8-24.2)

7-10.28* Electric Generating Plants and High-Voltage Direct Current Converter Stations.

A-7-10.28 Sprinkler System Discharge Criteria for Electric Generating Plants and High-Voltage Direct Current Converter Stations. See NFPA 850, *Recommended Practice for Fire Protection for Electric Generating Plants and High Voltage Direct Current Converter Stations,* for applicable terms not defined in Chapter 1.

(a) *Sprinkler System Water Supply.* The water supply for the permanent fire protection installation should be based on providing a 2-hour supply for both items (1) and (2) as follows:

(1) Either of items a or b below, whichever is larger:

a. The largest fixed fire suppression system demand

b. Any fixed fire suppression system demands that could reasonably be expected to operate simultaneously during a single event [e.g., turbine under floor protection in conjunction with other fire protection system(s) in the turbine area; coal conveyor protection in conjunction with protection for related coal handling structures during a conveyor fire; adjacent transformers not adequately separated according to 3-1.3 of NFPA 850].

(2) The hose stream demand of not less than 500 gpm (31.5 L/sec).

(**850:** 4-2.1)

Where an adequate and reliable water supply, such as a lake, cooling pond, river, or municipal water system, is unavailable, at least two separate water supplies should be provided for fire protection purposes with each supply capable of meeting the fire waterflow requirements determined by 4-2.1 of NFPA 850. (**850:** 4-2.2)

(b) *Yard Mains.* The supply mains should be looped around the main power block and should be of sufficient size to supply the flow requirements determined by 4-2.1 of NFPA 850 to any point in the yard loop considering the most direct path to be out of service. Pipe sizes should be designed to encompass any anticipated expansion and future water demands. (**850:** 4-4.1.3)

(c) *Coal Handling Structures.* Sprinkler systems should be designed for a minimum of 0.25 gpm/ft^2 (10.2 mm/min) density over a 2500 ft^2 (232 m^2) area. (**850:** 5-4.6.1)

(d) *Coal Conveyors.* Sprinklers should be designed for a minimum of 0.25 gpm/ft^2 (10.2 mm/min) density over 2000 ft^2 (186 m^2) of enclosed area or the most remote 100 linear ft (30 m) of conveyor structure up to 2000 ft^2 (186 m^2). (**850:** 5-4.6.2)

(e) [In areas over conveyor belts and striker plates within the stacker reclaimer,] the water supply [should] be from a 3000-gal to 5000-gal (11,355-L to 18,925-L) capacity pressure tank located on-board. (**850:** 5-4.6.4)

(f) Sprinklers for bag-type dust collectors should be designed for ordinary hazard systems. Sprinkler systems should be designed for a density of 0.2 gpm (8.2 mm/ min) over the projected plan area of the dust collector. (**850:** 5-4.6.5.1)

(g) *Steam Generator.* Boiler front fire protection systems should be designed to cover the fuel oil burners and ignitors, adjacent fuel oil piping and cable, a 20 ft (6.1 m) distance from the burner and ignitor including structural members and walkways at these levels. Additional coverage should include areas where oil may collect. Sprinkler and water spray systems should be designed for a density of 0.25 gpm/ft^2 (10.2 mm/min) over the protected area. (**850:** 5-5.1.2)

(h) *Flue Gas Bag-Type Dust Collectors.* The design density should be 0.2 gpm/ft^2 (8.2 mm/min) over the plan area of the dust collector. (**850:** 5-6.3.3)

(i) *Electrostatic Precipitators.* If mineral oil insulating fluids are used, hydrants or standpipes should be located so that each transformer-rectifier set can be reached by at least one hose stream. In addition the following should be provided:

Automatic sprinkler protection. Automatic sprinkler systems should be designed for a density of 0.25 gpm/ft^2 (10.2 mm/min) over 3500 ft^2 (325 m^2). The drain system should be capable of handling oil spillage plus the largest design waterflow from the fire protection system. (**850:** 5-6.4.3)

(j) *Scrubber Buildings.* Where scrubbers have plastic or rubber linings, one of the following methods of protection for the building should be provided:

Automatic sprinkler protection at ceiling level sized to provide a density 0.2 gpm/ft^2 (8.2 mm/min). The area of operation should be the area of the building or 10,000 ft^2 (930 m^2). Where draft curtains are provided the area of operation can be reduced to the largest area subdivided by draft curtains.(**850:** 5-6.5.2.2)

(k) *Turbine-Generator Area.* The sprinkler system beneath the turbine-generator should take into consideration obstructions from structural members and piping and should be designed to a density of 0.3 gpm/ft^2 (12.2 mm/min) over a minimum application of 5000 ft^2 (464 m^2).

NOTE: To avoid water application to hot parts or other water sensitive areas and to provide adequate coverage, designs that incorporate items such as fusible element operated directional spray nozzles may be necessary. (**850:** 5-7.4.1, 5-7.4.1.1)

The automatic sprinkler system [protecting the lubricating oil lines above the turbine operating floor] should be designed to a density of 0.3 gpm/ft^2 (12.2 mm/min). (**850:** 5-7.4.1.2)

(l) *Turbine-Generator Bearings.* Fire protection systems for turbine-generator bearings should be designed for a density of 0.25 gpm/ft^2 (10.2 mm/min) over the protected area. (**850:** 5-7.4.2.1)

(m) *Cable Spreading Room and Cable Tunnels.* Automatic sprinkler systems should be designed for a density of 0.3 gpm/ ft^2 (12.2 mm/min) over 2500 ft^2 (232 m^2) or the most remote 100 linear ft (30 m) of cable tunnels up to 2500 ft^2 (232 m^2). (**850:** 5-8.2.1)

(n) *Emergency Generators.* Sprinkler systems should be designed for a 0.25 gpm/ft^2 (10.2 mm/min) density over the fire area. (**850:** 5-9.1.2.1)

(o) *Fire Pumps.* If sprinkler systems are provided for fire pump houses, they should be designed for a density of 0.25 gpm/ ft^2 (10.2 mm/min) over the fire area. (**850:** 5-9.4)

(p) *Oil- or Coal-Fueled Auxiliary Boilers.* If a sprinkler system is provided it should be designed for a density of 0.25 gpm/ ft^2 (10.2 mm/min) over the entire room.

(q) *Alternative Fuels.*

(1) *Hydraulic Equipment, Reservoirs, Coolers, and Associated Oil-Filled Equipment.* Sprinklers should be over oil-containing equipment and for 20 ft (6.1 m) beyond in all directions. A density of 0.25 gpm/ft^2 (10.2 mm/min) should be provided. (**850:** 7-3.4.3)
(2) *Tipping/Receiving Building.* Systems should be designed for a minimum of 0.25 gpm/ft^2 (10.2 mm/min) over the most remote 3000 ft^2 (279 m^2) (increase by 30 percent for dry pipe systems) of floor area with the protection area per sprinkler not to exceed 130 ft^2 (120 m^2). High temperature sprinklers [250°F to 300°F (121°C to 149°C)] should be used.

NOTE: The above requirements are based on storage heights not exceeding 20 ft (6.1 m). (**850:** 7-3.4.4)

(3) *The MSW Storage Pit, Charging Floor, and Grapple Laydown Areas.* Systems should be designed for a minimum of 0.2 gpm/ft^2 (8.2 mm/min) over the most remote 3000 ft^2 (279 m^2) (increase by 30 percent for dry pipe systems) of pit/floor area with the protection area per sprinkler not to exceed 100 ft^2 (9.3 m^2). High temperature sprinklers [250°F to 300°F (121°C to 149°C)] should be used. (**850:** 7-3.4.5.1)

(r) *Refuse Derived Fuels.*

(1) *Hydraulic Equipment, Reservoirs, Coolers, and Associated Oil-Filled Equipment.* Sprinklers should be over oil-containing equipment and for 20 ft (6.1 m) beyond in all directions. A density of 0.25 gpm/ft^2 (10.2 mm/min) should be provided. (**850:** 7-4.4.6)
(2) *Tipping/Receiving Building.* Systems should be designed for a minimum of 0.25 gpm/ft^2 (10.2 mm/min) over the most remote 3000 ft^2 (279 m^2) (increase by 30 percent for dry

pipe systems) of floor area with the protection area per sprinkler not to exceed 130 ft^2 (12.0 m^2). High temperature sprinklers [250°F to 300°F (121°C to 149°C)] should be used.

NOTE: The above requirements are based on storage heights not exceeding 20 ft (6.1 m).

(**850:** 7-4.4.7)

(3) *Processing Building.* Systems should be designed for a minimum of 0.25 gpm/ft^2 (10.2 mm/min) over the most remote 3000 ft^2 (279 m^2) (increase by 30 percent for dry pipe systems) of floor area with the protection area per sprinkler not to exceed 130 ft^2 (12.0 m^2). (**850:** 7-4.4.8)

(4) *RDF Storage Building.* Systems should be designed for a minimum of 0.35 gpm/ft^2 (14.3 mm/min) over the most remote 3000 ft^2 (279 m^2) (increase by 30 percent for dry pipe systems) of floor area with the protection area per sprinkler not to exceed 100 ft^2 (9.3 m^2). High temperature sprinklers [250°F to 300°F (121°C to 149°C)] should be used. Storage heights in excess of 20 ft (6.1 m) will require higher design densities. (**850:** 7-4.4.9)

(5) *RDF Boiler Feed System Area, Including Bins, Hoppers, Chutes, Conveyors, and So Forth.* Where provided, the systems should be designed for a minimum of 0.2 gpm/ft^2 (8.2 mm/min) over the most remote 2000 ft^2 (186 m^2) (increase by 30 percent for dry pipe systems) of floor area with the protection area per sprinkler not to exceed 130 ft^2 (12.0 m^2). Internal, as well as external, protection also should be considered depending upon specific equipment design, ceiling heights, and accessibility for manual fire fighting. (**850:** 7-4.4.10)

(6) *Shredder Enclosures.* Systems should be designed for a minimum of 0.25 gpm/ft^2 (10.2 mm/min) over the most remote 3000 ft^2 (279 m^2) (increase by 30 percent for dry pipe systems) of floor area with the protection area per sprinkler not to exceed 100 ft^2 (9.3 m^2). (**850:** 7-4.4.11)

(s) *Biomass Fuels.*

(1) *Biomass Storage Buildings.* Systems should be designed for a minimum of 0.25 gpm/ft^2 (10.2 mm/min) over the most remote 3000 ft^2 (279 m^2) (increase by 30 percent for dry pipe systems) of floor area with the protection area per sprinkler not to exceed 130 ft^2 (12.0 m^2).

NOTE: Biomass fuels exhibit a wide range of burning characteristics and upon evaluation can require increased levels of protection.

(**850:** 7-5.4.4)

(2) *Hydraulic Equipment, Reservoirs, Coolers, and Associated Oil-Filled Equipment.* Sprinklers or spray nozzles should be over oil-containing equipment and for 20 ft (6.1 m) beyond in all directions. A density of 0.25 gpm/ft^2 (10.2 mm/min) should be provided. (**850:** 7-5.4.6)

(t) *Rubber Tire Fuel—Hydraulic Equipment, Reservoirs, Coolers, and Associated Oil-Filled Equipment.* Sprinklers should be over oil-containing equipment and for 20 ft (6.1 m) beyond in all directions. A density of 0.25 gpm/ft^2 (10.2 mm/min) should be provided. (**850:** 7-6, 7-6.4.10)

7-10.29* Hydroelectric Generating Plants.

A-7-10.29 Sprinkler System Discharge Criteria for Hydroelectric Generating Plants. See NFPA 851, *Recommended Practice for Fire Protection for Hydroelectric Generating Plants,* for applicable terms not defined in Chapter 1.

(a) *Sprinkler Systems Water Supply.* The water supply for the permanent fire protection installation should be based on the largest fixed fire suppression system demand plus the maximum hose stream demand of not less than 500 gpm (31.5 L/sec) for a 2-hour duration. (**851:** 4-2.2)

(b) If a single water supply is utilized, two independent connections should be provided. If a situation can arise in which the primary water supply can become unavailable (e.g., dewatering of penstocks), an auxiliary supply should be provided. Each supply should be capable of meeting the requirements in 4-2.2 of NFPA 851. (**851:** 4-2.3)

(c) Fixed fire protection for this equipment, where provided, should be automatic wet pipe sprinkler systems utilizing a design density of 0.25 gpm/ft^2 (10.2 mm/ min) for the entire hazard area *(see 3-5.3 of NFPA 803).* (**851:** 5-2.4)

(d) Sprinkler or water spray systems should be designed for a density of 0.3 gpm/ft^2 (12.2 mm/min) over 2500 ft^2 (232 m^2). This coverage is for area protection. Individual cable tray tier coverage could be required based on the Fire Risk Evaluation. (**851:** 5-5.3)

(e) *Cable Tunnels.* Automatic sprinkler systems should be designed for a density of 0.3 gpm/ft^2 (12.2 mm/min) over 2500 ft^2 (232 m^2) or the most remote 100 linear ft (30 m) of cable tunnel up to 2500 ft^2 (232 m^2). (**851:** 5-6.1)

(f) *Emergency Generators.* Sprinkler and water spray protection systems should be designed for a 0.25 gpm/ft^2 (10.2 mm/ min) density over the fire area. (**851:** 5-11.2)

(g) *Air Compressors.* Automatic sprinkler protection, with a density of 0.25 gpm/ft^2 (10.2 mm/min) over the postulated oil spill, should be considered for air compressors containing a large quantity of oil. *(See 4-8.2 of NFPA 851.)* (**851:** 5-12)

(h) *Hydraulic Systems for Gate and Valve Operators.* Automatic sprinkler protection designed for a density of 0.25 gpm/ft^2 (10.2 mm/min) over the fire area should be considered for hydraulic systems not using a listed fire-resistant fluid. *(See 4-8.2 of NFPA 851.)* (**851:** 5-13)

(i) *Fire Pumps.* If sprinkler systems are provided they should be designed for a density of 0.25 gpm/ft^2 (10.2 mm/min) over the fire area. For automatic foam-water sprinkler systems, a density of 0.16 gpm/ft^2 (6.5 mm/min) should be provided. (**851:** 5-14)

7-10.30* Fire Protection in Places of Worship.

A-7-10.30 Sprinkler systems for specific areas associated with religious facilities should be designed as follows:

(1) All assembly areas, except state—Light Hazard
(2) Stages—Ordinary Hazard (Group 2)

(3) Kitchens—Ordinary Hazard (Group 1)
(4) Storage rooms—Ordinary Hazard (Group 2)
(5) Unused attics/lofts/steeples/ concealed spaces—Light Hazard
(6) Schools/day-care centers—Light Hazard
(7) Gift shops—Ordinary Hazard (Group 1)
(8) Special exhibit area—Ordinary Hazard (Group 2)
(9) Libraries—Ordinary Hazard (Group 2)
(10) Offices—Light Hazard

(**909:** A-10-4.2)

7-11 In-Rack Sprinklers

In-rack sprinklers mandated by this standard shall meet the requirements of this section.

7-11.1 In-rack sprinklers shall operate at a minimum of 15 psi (1 bar).

The hydraulic design criteria for in-rack sprinklers noted in this subsection are intended to apply to the protection of miscellaneous storage, miscellaneous tire storage, and storage under 12 ft (3.7 m) as described in 7-2.3.2.2. In-rack sprinklers with nominal K-factors of 5.6 or 8.0 are required in accordance with 5-12.2. A discharge pressure of 15 psi (1 bar) produces flows of 21 gpm (79 L/min) and 31 gpm (117 L/min), respectively.

7-11.2 Water Demand.

Where one level of in-rack sprinklers is installed for miscellaneous storage, water demand shall be based on simultaneous operation of the hydraulically most demanding four adjacent sprinklers.

References Cited in Commentary

National Fire Protection Association, 1 Batterymarch Park, P.O. Box 9101, Quincy, MA 02269-9101.

NFPA 13D, *Standard for the Installation of Sprinkler Systems in One- and Two-Family Dwellings and Manufactured Homes,* 1999 edition.
NFPA 13R, *Standard for the Installation of Sprinkler Systems in Residential Occupancies up to and Including Four tories in Height,* 1999 edition.
NFPA 14, *Standard for the Installation of Standpipe and Hose Systems,* 1996 edition.
NFPA 230, *Standard for the Fire Protection of Storage,* 1999 edition.
NFPA 231D, *Standard for Storage of Rubber Tires,* 1998 edition.
NFPA 232, *Standard for the Protection of Records,* 1995 edition.
NFPA 415, *Standard on Airport Terminal Buildings, Fueling Ramp Drainage, and Loading Walkways,* 1997 edition.

CHAPTER 8

Plans and Calculations

This chapter describes the procedures and information necessary to calculate the hydraulic demand of a sprinkler system and to verify that the system's overall layout is compliant with NFPA 13. Emphasis is placed on the methodology used to calculate and verify the system's hydraulic demand and available water supply.

8-1* Working Plans

A-8-1 Preliminary layouts should be submitted for review to the authority having jurisdiction before any equipment is installed or remodeled in order to avoid error or subsequent misunderstanding *(see Figure A-8-1)*. Any material deviation from approved plans will require permission of the authority having jurisdiction.

Preliminary layouts should show as much of the following information as is required to provide a clear representation of the system, hazard, and occupancy:

(1) Name of owner and occupant.
(2) Location, including street address.
(3) Point of compass.
(4) Construction and occupancy of each building. Data on special hazards should be submitted as they can require special rulings.
(5) Building height in feet.
(6) If it is proposed to use a city main as a supply, whether the main is dead end or circulating, size of main and pressure in psi, and, if dead end, direction and distance to nearest circulating main.
(7) Distance from nearest pumping station or reservoir.
(8) In cases where reliable, up-to-date information is not available, a waterflow test of the city main should be conducted in accordance with A-9-2.1. The preliminary plans should

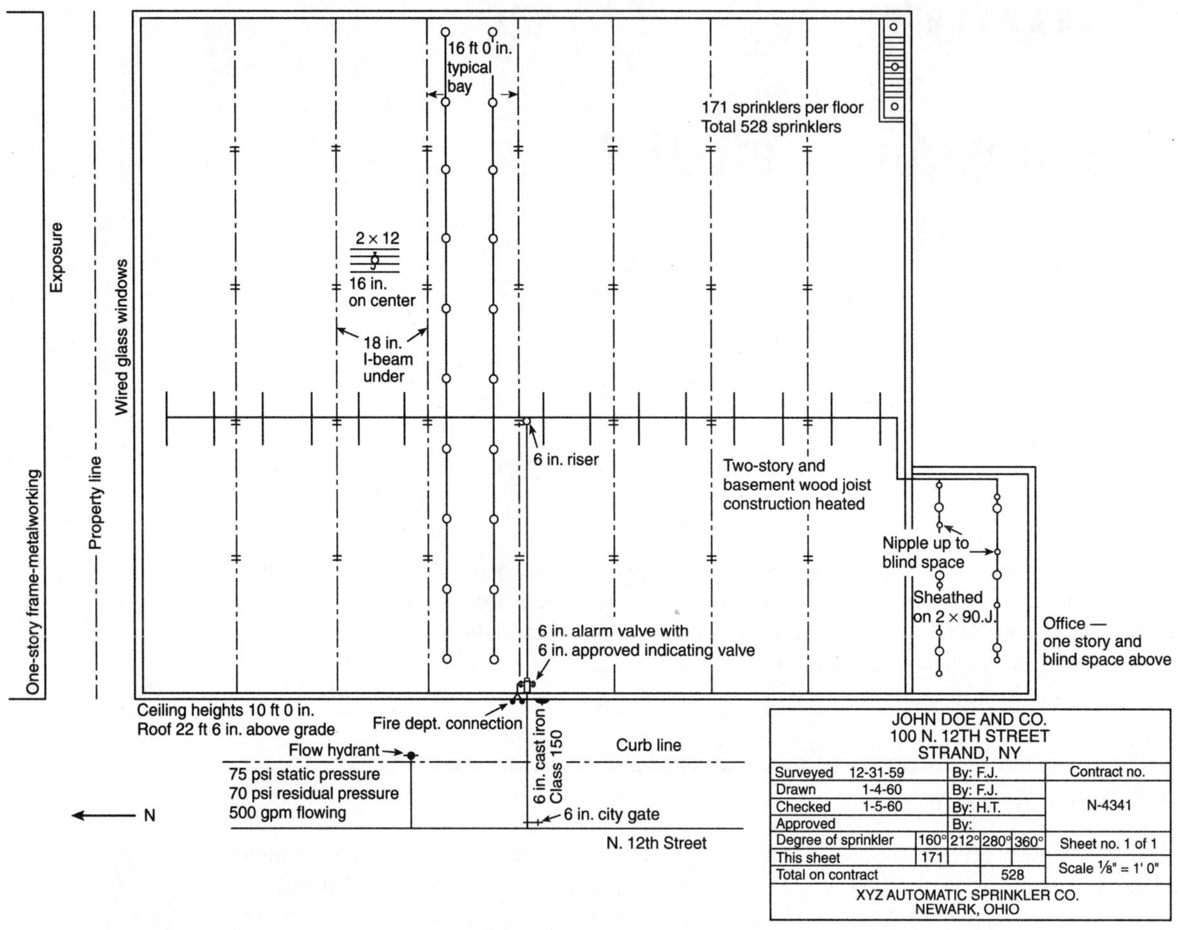

Figure A-8-1 *Typical preliminary plan.*

specify the person who conducted the test, date and time, the location of the hydrants where flow was taken and where static and residual pressure readings were recorded, the size of main supplying these hydrants, and the results of the test, giving size and number of open hydrant butts flowed. Also, data covering minimum pressure in the connection with the city main should be included.

(9) Data covering waterworks systems in small towns in order to expedite the review of plans.

(10) Fire walls, fire doors, unprotected window openings, large unprotected floor openings, and blind spaces.

(11) Distance to and construction and occupancy of exposing buildings—for example, lumber yards, brick mercantiles, and fire-resistive office buildings.

(12) Spacing of sprinklers, number of sprinklers in each story or fire area and total number of sprinklers, number of sprinklers on each riser and on each system by floors, total area protected by each system on each floor, total number of sprinklers on each dry pipe system or preaction or deluge system and if extension to present equipment, sprinklers already installed.
(13) Capacities of dry pipe systems with bulk pipe included *(see Table A-4-2.3)* and, if an extension is made to an existing dry pipe system, the total capacity of the existing and also the extended portion of the system.
(14) Weight or class, size, and material of any proposed underground pipe.
(15) Whether property is located in a flood or earthquake area requiring consideration in the design of sprinkler system.
(16) Name and address of party submitting the layout.

Working plans are prepared primarily for the qualified workers who conduct the actual system installation and for those responsible for verifying that the design complies with NFPA 13. The plans also serve to protect the interest of the owner, who usually is not knowledgeable about sprinkler systems and who usually relies on others to verify that the proposed design is acceptable. Several parties usually conduct a review of the proposed design and at various stages during the design process. These parties typically include the owner's representatives, such as the engineer or architect of record via the services of a registered fire protection engineer, and society's representatives, such as the fire department or building department. Representatives from the associated insurance companies will also most likely take part in the review process. Regardless of the type of building in which the system is installed, the success of the overall fire and life safety system is dependent on proper sprinkler system design and installation. Verification of compliance with NFPA 13 is critical.

With a speculative-type building in which the nature of the future tenants or their operations is uncertain during the building's design and construction stage, the difficulties associated with the verification process increase. Much of the installation criteria of NFPA 13 hinges on the determination of the building's occupancy hazard classification. This determination usually cannot be made without knowing the type of tenant and the tenant's intended operations. To address these concerns, some jurisdictions have implemented additional regulations for speculative-type buildings. One example is to require any speculative building to be designed based on the criteria for an ordinary hazard (Group 2) occupancy. This approach provides a sprinkler system capable of protecting a wide range of hazards, and it allows the building owner more latitude in attracting prospective tenants. However, the limitations of this approach must also be acknowledged. Tenants with operations that present a greater hazard than those addressed by an ordinary hazard (Group 2) classification will be inadequately protected by the sprinkler system. The owner needs to be aware of the limits of the sprinkler system.

The owner is required by NFPA 25, *Standard for the Inspection, Testing, and Maintenance of Water-Based Fire Protection Systems*, to retain the initial working plans and specifications because this facilitates the necessary inspection, testing, and

maintenance activities. Any future modifications or additions of the sprinkler system also become more workable.

Some of the symbols commonly used in plan drawings can be found in NFPA 170, *Standard for Fire Safety Symbols*. Any symbol used in the plan drawings that is not a standard symbol needs to be explained in the legend or some other obvious location on the working plans.

The plans describing the preliminary layout often do not contain all of the items identified in A-8-1. Preliminary plans are useful in assessing the available water supply system or assessing if the basis of overall design is flawed. For example, in A-8-1(8), an out-of-date water test would indicate that more current information is needed for an accurate evaluation.

Another example involves A-8-1(5). An inexperienced designer, architect, or engineer might specify a design approach based on an ordinary hazard (Group 2) occupancy classification for a rack storage operation in a proposed single-story, 35-ft (10.7-m) high building. A more accurate approach would consider a 30-ft to 34-ft (9.1-m to 10.4-m) storage height and the appropriate design criteria in accordance with Section 7-4 for the type of commodity in question. Although preliminary plans are not always submitted or required, they are very helpful during the design process and can eliminate costly problems and delays later in the design process.

8-1.1* Working plans shall be submitted for approval to the authority having jurisdiction before any equipment is installed or remodeled. Deviation from approved plans shall require permission of the authority having jurisdiction.

A-8-1.1 See Figure A-8-1.1. Underground mains should be designed so that the system can be extended with a minimum of expense. Possible future plant expansion should also be considered and the piping designed so that it will not be covered by buildings.

8-1.1.1 Working plans shall be drawn to an indicated scale, on sheets of uniform size, with a plan of each floor, and shall show those items from the following list that pertain to the design of the system.

(1) Name of owner and occupant.
(2) Location, including street address.
(3) Point of compass.
(4) Full height cross section, or schematic diagram, including structural member information if required for clarity and including ceiling construction and method of protection for nonmetallic piping.
(5) Location of partitions.
(6) Location of fire walls.
(7) Occupancy class of each area or room.
(8) Location and size of concealed spaces, closets, attics, and bathrooms.
(9) Any small enclosures in which no sprinklers are to be installed.
(10) Size of city main in street and whether dead end or circulating; if dead end, direction and distance to nearest circulating main; and city main test results and system elevation relative to test hydrant *(see A-9-2.1)*.
(11) Other sources of water supply, with pressure or elevation.

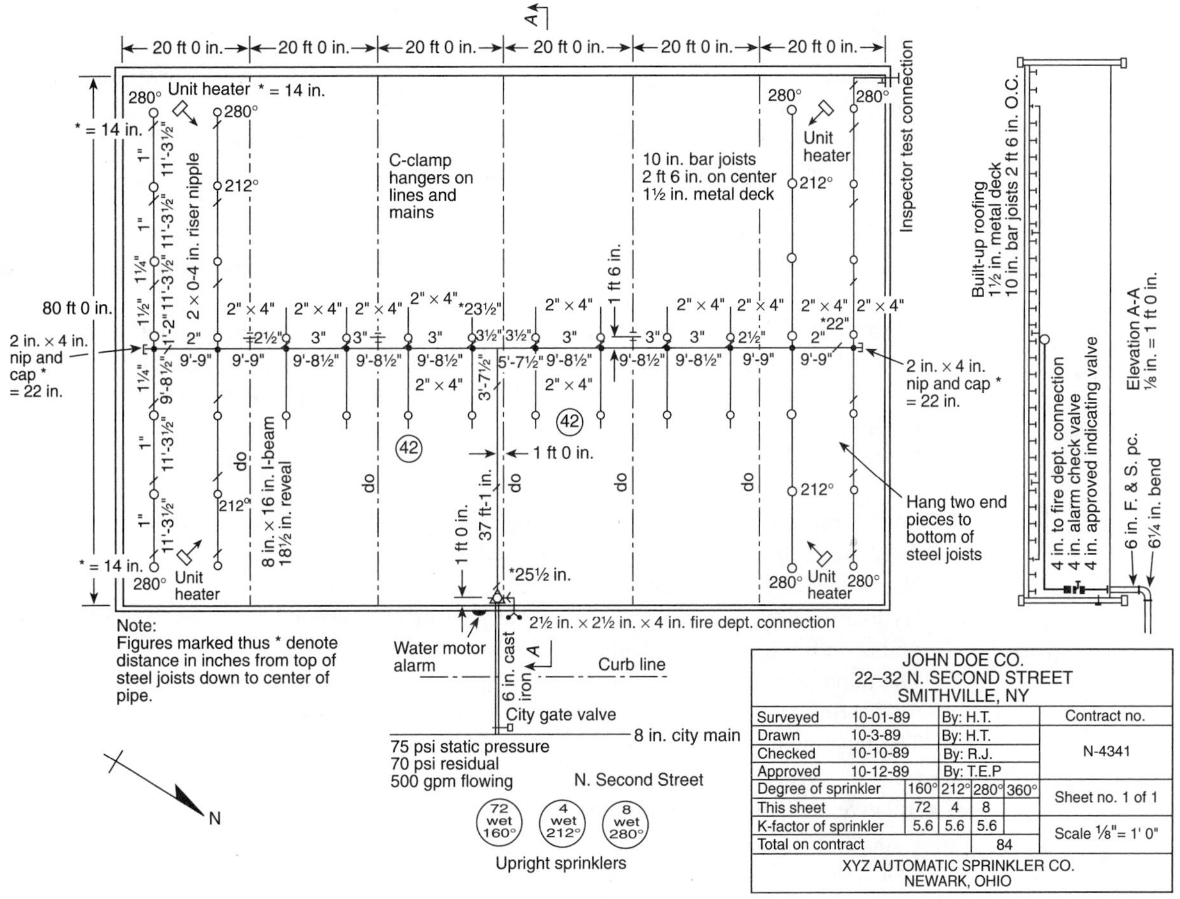

Figure A-8-1.1 *Typical working plans.*

(12) Make, type, model, and nominal K-factor of sprinklers.
(13) Temperature rating and location of high-temperature sprinklers.
(14) Total area protected by each system on each floor.
(15) Number of sprinklers on each riser per floor.
(16) Total number of sprinklers on each dry pipe system, preaction system, combined dry pipe-preaction system, or deluge system.
(17) Approximate capacity in gallons of each dry pipe system.
(18) Pipe type and schedule of wall thickness.
(19) Nominal pipe size and cutting lengths of pipe (or center-to-center dimensions). Where typical branch lines prevail, it shall be necessary to size only one typical line.
(20) Location and size of riser nipples.
(21) Type of fittings and joints and location of all welds and bends. The contractor shall specify

on drawing any sections to be shop welded and the type of fittings or formations to be used.

(22) Type and locations of hangers, sleeves, braces, and methods of securing sprinklers when applicable.

(23) All control valves, check valves, drain pipes, and test connections.

(24) Make, type, model, and size of alarm or dry pipe valve.

(25) Make, type, model, and size of preaction or deluge valve.

(26) Kind and location of alarm bells.

(27) Size and location of standpipe risers, hose outlets, hand hose, monitor nozzles, and related equipment.

(28) Private fire service main sizes, lengths, locations, weights, materials, point of connection to city main; the sizes, types and locations of valves, valve indicators, regulators, meters, and valve pits; and the depth that the top of the pipe is laid below grade.

(29) Piping provisions for flushing.

(30) Where the equipment is to be installed as an addition to an existing system, enough of the existing system indicated on the plans to make all conditions clear.

(31) For hydraulically designed systems, the information on the hydraulic data nameplate.

(32) A graphic representation of the scale used on all plans.

(33) Name and address of contractor.

(34) Hydraulic reference points shown on the plan that correspond with comparable reference points on the hydraulic calculation sheets.

(35) The minimum rate of water application (density), the design area of water application, in-rack sprinkler demand, and the water required for hose streams both inside and outside.

(36) The total quantity of water and the pressure required noted at a common reference point for each system.

(37) Relative elevations of sprinklers, junction points, and supply or reference points.

(38) If room design method is used, all unprotected wall openings throughout the floor protected.

(39) Calculation of loads for sizing and details of sway bracing.

(40) The setting for pressure-reducing valves.

(41) Information about backflow preventers (manufacturer, size, type).

(42) Information about antifreeze solution used (type and amount).

(43) Size and location of hydrants, showing size and number of outlets and if outlets are to be equipped with independent gate valves. Whether hose houses and equipment are to be provided, and by whom, shall be indicated. Static and residual hydrants that were used in flow tests shall be shown.

(44) Size, location, and piping arrangement of fire department connections.

Formal Interpretation 80-26 provides additional information on the application of 8-1.1.1(22).

With regard to 8-1.1.1(34), hydraulic reference points are important to both the designer and the plan reviewer because the points help ensure that all essential components of the system are included in the calculations. Hydraulic reference points serve to provide coordination between the plans and the hydraulic calculations. The reference points are especially important for more hydraulically complex systems such as those that are gridded.

Formal Interpretation 80-26

Reference: 8-1.1.1(22)

Question: Is it the intent of 8-1.1.1(22) that hanger locations be dimensioned?

Answer: No, but hanger locations are required to be in accordance with 4-14.2.3 and 4-14.2.4.

Issue Edition: 1980

Reference: 1-9.2(z)1

Date: October 1982

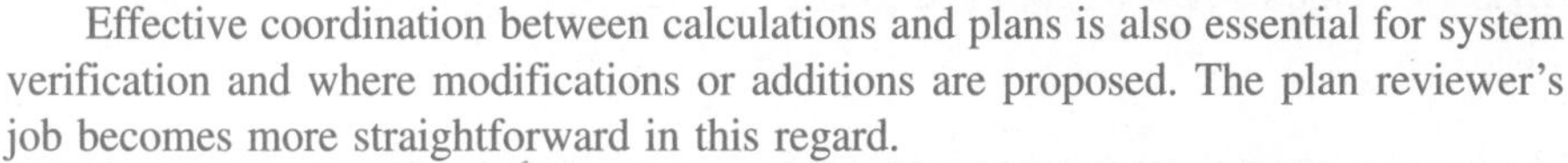

Effective coordination between calculations and plans is also essential for system verification and where modifications or additions are proposed. The plan reviewer's job becomes more straightforward in this regard.

The information described in 8-1.1.1(35) establishes the basis for the majority of the design options. If an incorrect discharge density was selected or if the area of system operation is not large enough for the occupancy hazard, the proposed system design is subject to rejection. The following information, which is associated with the water supply requirements, must be shown on the plans:

(1) Density
(2) Area of water application
(3) In-rack sprinkler demand, if applicable
(4) Inside hose demand, if applicable
(5) Outside hose demand

The sum of the inside and outside hose demands must always be equal to or greater than the values given in Table 7-2.3.1.1.

With regard to 8-1.1.1(36), the actual calculated demand is normally referenced at the base of the riser in buildings containing several systems. In single-system buildings, the reference point can be at the riser's base or the connection point to the water supply (city main).

With regard to 8-1.1.1(41), the friction loss through the backflow preventer must be accounted for in the calculations. Because each model device has different friction loss characteristics, the model on the plans, the model on the calculations, and the model installed all have to be the same. A change during the construction phase could cause an inadequate sprinkler system discharge.

8-1.1.2 The working plan submittal shall include the manufacturer's installation instructions for any specially listed equipment, including descriptions, applications, and limitations for any sprinklers, devices, piping, or fittings.

NFPA 13 does not address specially listed equipment and system components in detail. Therefore, when specially listed equipment is used in the system, a proper evaluation cannot be conducted using NFPA 13 alone. Specific instructions and information about the component must be obtained from the manufacturer or listing agency and included on the plans and hydraulic calculations. Examples of specially listed equipment include the use of special fittings whose equivalent lengths—that is, hydraulic characteristics—are not addressed in NFPA 13, and special sprinklers with minimum flows and pressures not addressed within NFPA 13.

8-1.1.3* Working Plans for Automatic Sprinkler Systems with Non-Fire Protection Connections. Special symbols shall be used and explained for auxiliary piping, pumps, heat exchangers, valves, strainers, and the like, clearly distinguishing these devices and piping runs from those of the sprinkler system. Model number, type, and manufacturer's name shall be identified for each piece of auxiliary equipment.

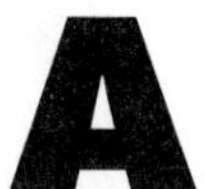

A-8-1.1.3 See Figures A-8-1.1.3(a) and (b).

8-2 Water Supply Information

8-2.1 Water Supply Capacity Information.

The following information shall be included:

(1) Location and elevation of static and residual test gauge with relation to the riser reference point
(2) Flow location
(3) Static pressure, psi (bar)
(4) Residual pressure, psi (bar)
(5) Flow, gpm (L/min)
(6) Date
(7) Time
(8) Test conducted by or information supplied by
(9) Other sources of water supply, with pressure or elevation

Waterflow test data must be current, analyzed during periods of peak usage, and handled in such a manner that it represents the supply's true orientation, or direction, to the system being designed. Determining when peak demands occur can be difficult, but water departments will usually have information readily available concerning the time of year and time of day when the highest distribution system demand occurs.

8-2.2 Water Supply Treatment Information.

The following information shall be included where required by 9-1.5:

(1) Type of condition that requires treatment
(2) Type of treatment needed to address the problem
(3) Details of treatment plan

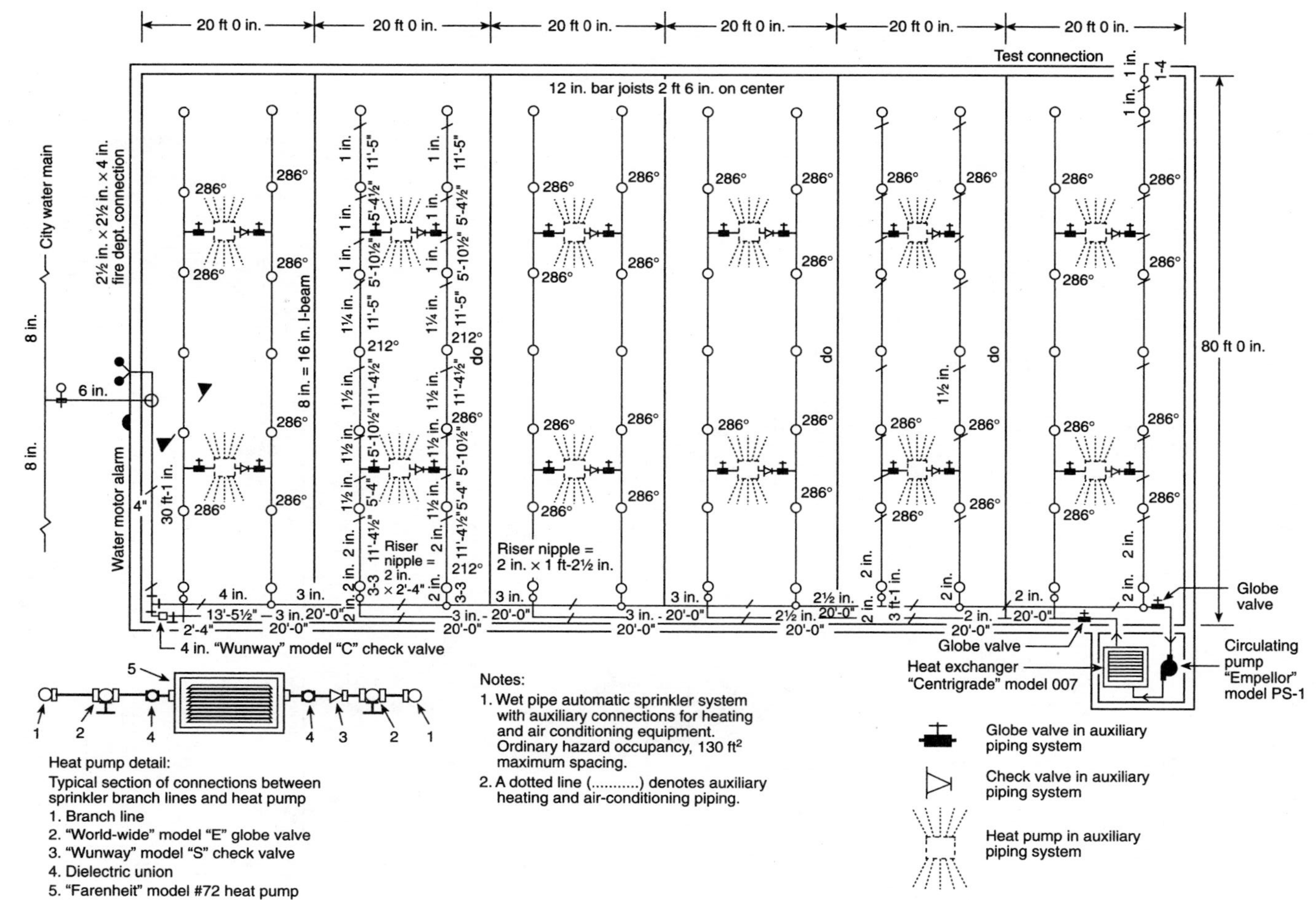

Figure A-8-1.1.3(a) Working plans for circulating closed-loop systems (example 1).

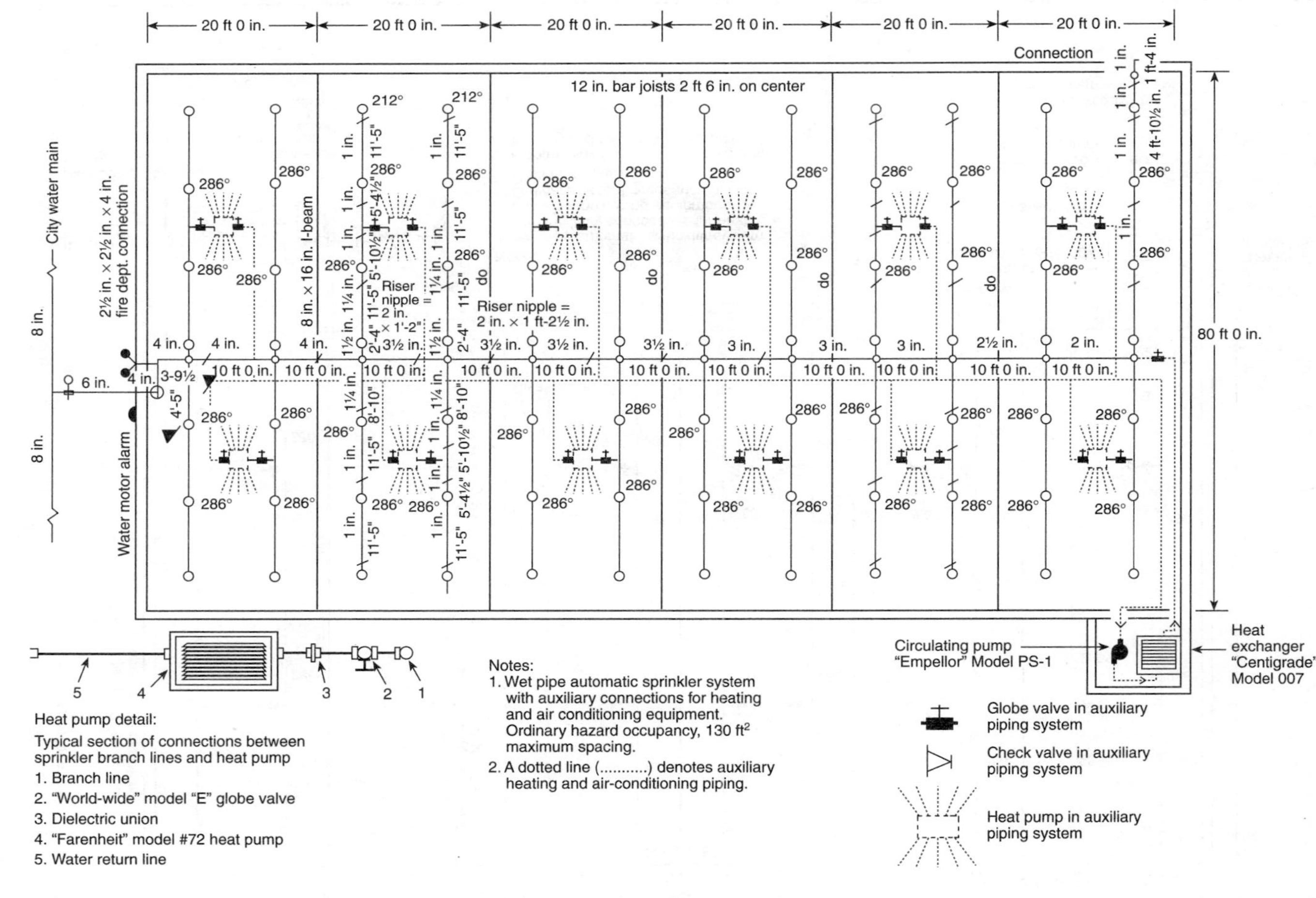

***Figure A-8-1.1.3(b)** Working plans for circulating closed-loop systems (example 2).*

Microbiologically induced corrosion (MIC) is a relatively rare but growing problem. Microbes in the water attach themselves to the pipe and then eat their way through, creating pinhole leaks in the pipe. Microbe byproducts can also build up and restrict the effective diameter of the pipe. Neither situation is beneficial for a sprinkler system.

MIC must be dealt with once it is known to be a problem. The affected pipes must be cleaned or replaced, and then the water entering the system needs to be treated with a biocide so that the microbes are destroyed.

Proper treatment depends on the type of microbes causing the corrosion or buildup. Different microbes respond differently to each kind of biocide. Each site where MIC is known to be a problem needs its own plan for how to deal with the particular microbes in question.

Chapter 10 of NFPA 25 provides more information on identification and treatment of systems with MIC. The NFSA report "Detection, Treatment and Prevention of Microbiologically Influenced Corrosion in Water-Based Fire Protection Systems," also provides information on the subject.

8-3 Hydraulic Calculation Forms

8-3.1 General.

Hydraulic calculations shall be prepared on form sheets that include a summary sheet, detailed worksheets, and a graph sheet. *[See copies of typical forms in Figures A-8-3.2(a), A-8-3.3, and A-8-3.4.]*

The calculation forms in Section 8-3 are tools that help monitor the waterflow through the system, summarize system design information, and allow for a quick evaluation of the system. As illustrated in Figure A-8-3.2(c), waterflow through the system is tracked in a direction opposite to that of the actual waterflow—that is, the first evaluation point begins at the hydraulically most demanding sprinkler, which many times is the furthest sprinkler from the system riser, and ends at the point of connection to the system water supply. These forms permit the designer or other persons reviewing the plans to account for individual pipe lengths and fittings and to keep track of changes in pressure and flow that result from friction loss and elevation changes.

8-3.2* Summary Sheet.

The summary sheet shall contain the following information, where applicable:

(1) Date
(2) Location
(3) Name of owner and occupant
(4) Building number or other identification
(5) Description of hazard
(6) Name and address of contractor or designer
(7) Name of approving agency

(8) System design requirements, as follows:
  a. Design area of water application, ft^2 (m^2)
  b. Minimum rate of water application (density), gpm/ft^2 (mm/min)
  c. Area per sprinkler, ft^2 (m^2)

(9) Total water requirements as calculated, including allowance for inside hose, outside hydrants, and water curtain and exposure sprinklers

(10) Allowance for in-rack sprinklers, gpm (L/min)

(11) Limitations (dimension, flow, and pressure) on extended coverage or other listed special sprinklers

A-8-3.2 See Figures A-8-3.2(a) through (d).

The summary sheet shown in Figure A-8-3.2(a) is normally used as the cover sheet for the calculations and contains the information required by 8-3.2 concerning the system design parameters. This information is useful in estimating the total amount of water needed for sprinkler system demand. For example, multiplying the value given in 8-3.2(8)a by the value given in 8-3.2(8)b will give an estimation of the minimum flow rate needed to supply the sprinklers, assuming no friction loss in the system. More specifically, for a light hazard occupancy with a density of 0.1 gpm/ft^2 operating over 1500 ft^2 (see Figure 7-2.3.1.2), the water from demand for the sprinklers cannot be less than 150 gpm (0.1 × 1500). Because no friction loss in the system occurs, the actual minimum flow rate will be greater than the value obtained by multiplying 8-3.2(8)a by 8-3.2(8)b. Experienced designers and plan reviewers can apply a correction factor for the pipe sizes and arrangement and make rough estimates regarding sprinkler demand.

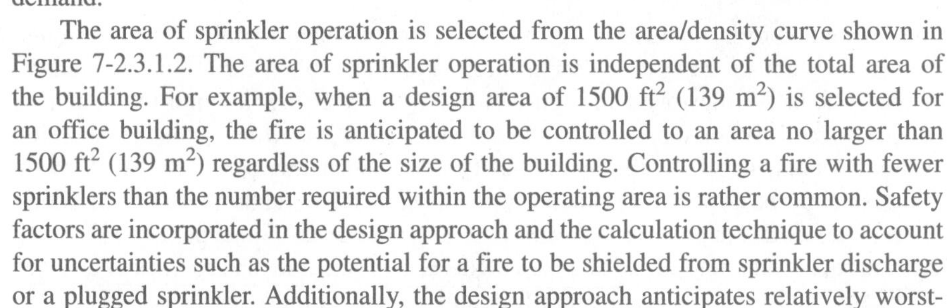

The area of sprinkler operation is selected from the area/density curve shown in Figure 7-2.3.1.2. The area of sprinkler operation is independent of the total area of the building. For example, when a design area of 1500 ft^2 (139 m^2) is selected for an office building, the fire is anticipated to be controlled to an area no larger than 1500 ft^2 (139 m^2) regardless of the size of the building. Controlling a fire with fewer sprinklers than the number required within the operating area is rather common. Safety factors are incorporated in the design approach and the calculation technique to account for uncertainties such as the potential for a fire to be shielded from sprinkler discharge or a plugged sprinkler. Additionally, the design approach anticipates relatively worst-case fire situations for each respective occupancy hazard.

Selecting the hydraulically most demanding area of operation ensures that an area located hydraulically closer to the water supply will also control the fire. In buildings containing two or more occupancy hazards, multiple sets of hydraulic calculations are usually necessary to verify that each hazard is adequately protected.

Figure A-8-3.2(b) shows the floor plan and elevation of a hydraulically designed sprinkler system in a small building that is 130 ft × 200 ft (39 m × 60 m) in area. The design density of 0.15 gpm/ft^2 (6.1 mm/min) over an area of operation of 1500 ft^2 (139 m^2) is obtained from Figure 7-2.3.1.2 and is sufficient for an ordinary hazard (Group 1) occupancy. This design approach requires that only a single point from the curve be satisfied. For example, a design density of 0.12 gpm/ft^2 (4.5 mm/min) operating

Hydraulic Calculations

for

ABC Company, employee garage

7499 Franklin Road

Charleston, SC

Contract No. 4001

Date 1-7-99

Design data:

Occupancy classification ORD. GR. 1

Density 0.15 gpm/ft2

Area of application 1500 ft2

Coverage per sprinkler 130 ft2

Special sprinklers —

No. of sprinklers calculated 12

In-rack demand —

Hose streams 250 gpm

Total water required 510.4 gpm
including hose streams

Name of contractor

Name of designer

Address

Authority having jurisdiction

Figure A-8-3.2(a) *Summary sheet.*

over an area of 3000 ft^2 (279 m^2) is also acceptable for protecting an ordinary hazard (Group 1) occupancy. The designer has the option to choose any point along the curve. However, many designers usually choose a higher density operating over a smaller area of operation. As a result, many designs correspond to the design densities and areas of operations found at the lower ends of these curves.

To determine the number of operating sprinklers for the design, the area of operation is divided by the area of coverage per sprinkler. Using the information provided in Figure A-8-3.2(b), 1500 ft^2 is divided by 130 ft^2 (139 m^2 is divided by 12 m^2), giving a value of 11.54 sprinklers. Because a fraction of a sprinkler is not possible, this value

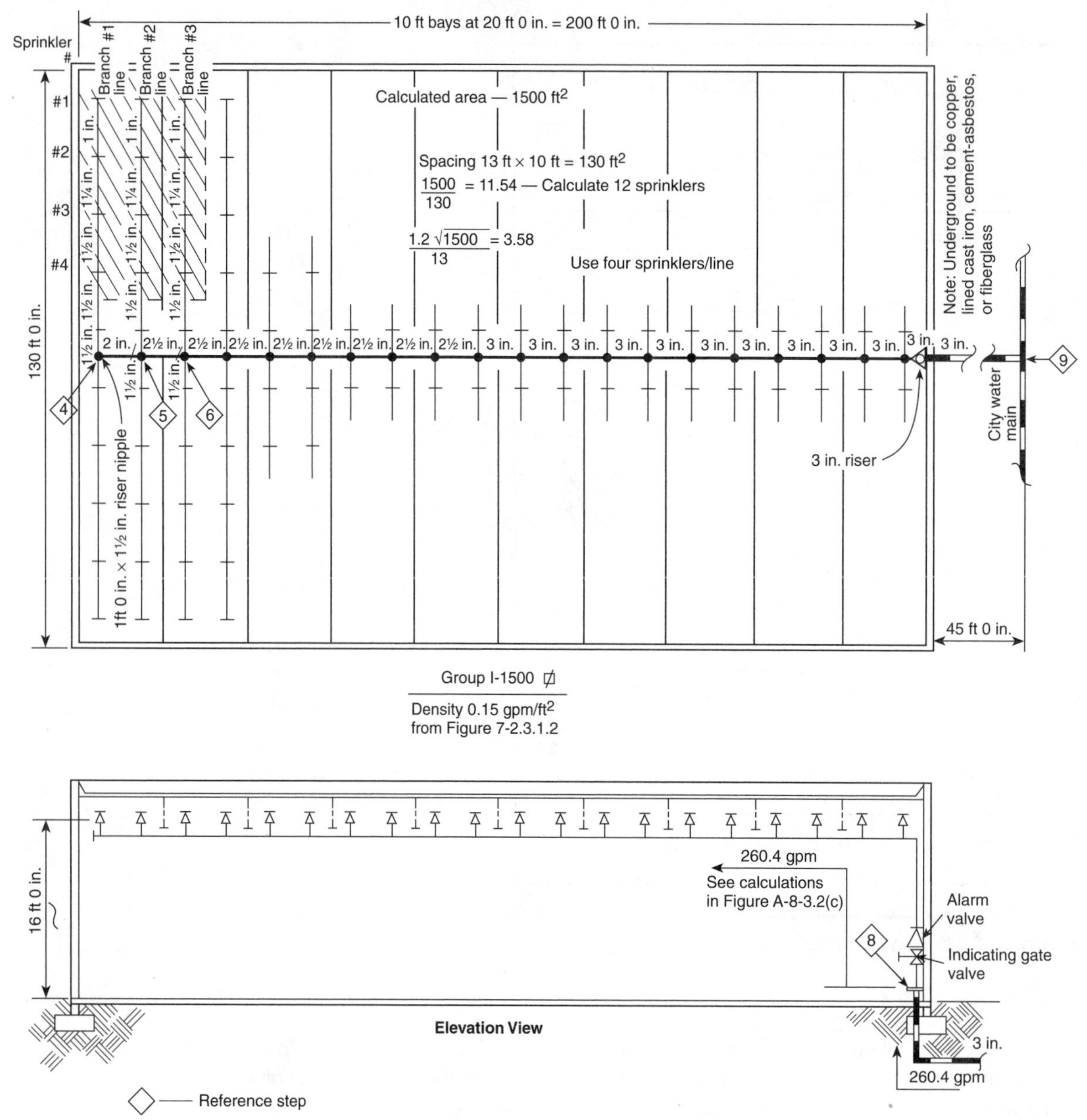

Figure A-8-3.2(b) Hydraulic calculation example (plan view and elevation view).

Contract Name GROUP I 1500 ft^2 Sheet 2 Of 3

Step No.	Nozzle Ident. and Location		Flow in gpm	Pipe Size	Pipe Fittings and Devices	Equiv. Pipe Length	Friction Loss psi Foot	Pressure Summary	Normal Pressure	D = 0.15 GPM/ ft^2 Notes K = 5.6	Ref. Step
1	1	BL-1	q	1		L 13.0	C=120	Pt 12.1	Pt		
						F		Pe	Pv	q = 130 x. 15 = 19.5	
			Q 19.5			T 13.0	0.124	Pf 1.6	Pn	$P = (19.5/5.6)^2 = 12.1$ psi	
2	2		q 20.7	$1^1/_4$		L 13.0		Pt 13.7	Pt		
						F		Pe	Pv	$q = 5.65\sqrt{13.7}$	
			Q 40.2			T 13.0	0.125	Pf 1.6	Pn		
3	3		q 21.9	$1^1/_2$		L 13.0		Pt 15.3	Pt		
						F		Pe	Pv	$q = 5.65\sqrt{15.3}$	4
			Q 62.1			T 13.0	0.131	Pf 1.7	Pn		
4	4	DN RN	q 23.1	$1^1/_2$	2T-16	L 20.5		Pt 17.0	Pt		
						F 16.0		Pe 0.4	Pv	$q = 5.65\sqrt{17}$	5
			Q 85.2			T 36.5	0.236	Pf 8.6	Pn	Pe = 1 x 0.433	
5		CM TO BL-2	q	2		L 10.0		Pt 26.0	Pt	$K = \frac{85.2}{\sqrt{26}}$	
						F		Pe	Pv		
			Q 85.2			T 10.0	0.07	Pf .7	Pn	K = 16.71	
6		BL-2 CM TO BL-3	q 86.3	$2^1/_2$		L 10.0		Pt 26.7	Pt		
						F		Pe	Pv	$q = 16.71\sqrt{26.1}$	6
			Q 171.5			T 10.0	0.107	Pf 1.1	Pn		
7		BL-3 CM	q 88.1	$2^1/_2$		L 70.0		Pt 27.8	Pt		
						F		Pe	Pv	$q = 16.7\sqrt{27.8}$	
			Q 259.6			T 70.0	0.231	Pf 16.2	Pn		
8		CM TO FIS	q	3	E5	L 119.0		Pt 44.0	Pt		
					AV15	F 21		Pe 6.5	Pv	Pe = 15 x 0.433	8
			Q 259.6		GV1	T140.0	.081	Pf 11.2	Pn		
9		THROUGH UNDER-GROUND TO CITY MAIN	q	3	E5	L 50.0	C=150	Pt 61.7	Pt	$F = F_{40} \times 1.51 \times F_c$	
					GV1	F 27.6	TYPE'M'	Pe	Pv	$F_c = [2.981/3.068]^{4.87} = 0.869$; $F = 21 \times 1.51 \times 0.869$	9
			Q 259.6		T15	T 77.6	0.061	Pf 4.7	Pn	F = 27.6	
			q			L		Pt 66.4	Pt		
						F		Pe	Pv		
			Q			T		Pf	Pn		
			q			L		Pt	Pt		
						F		Pe	Pv		
			Q			T		Pf	Pn		
								Pt			

Figure A-8-3.2(c) *Hydraulic calculations.*

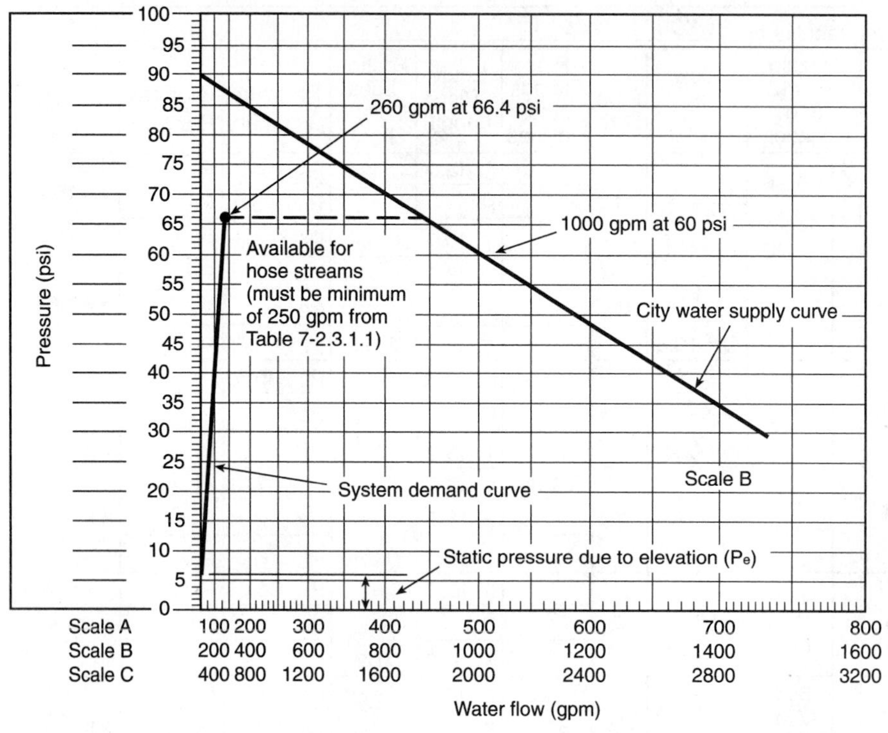

Figure A-8-3.2(d) *Hydraulic graph.*

is always rounded up to the next whole number. Therefore, the number of sprinklers is 12. The area of coverage per sprinkler is determined by the methodology discussed in 5-5.2.

The number of sprinklers required to operate on a branch line in the area of operation is determined by dividing 1.2 times $\sqrt{A}$ by the distance between sprinklers on the branch line. This calculation ensures that the area of operation takes the shape of a rectangle, with the longer dimension parallel to the direction of the branch lines. (see A-8-4.4 for more discussion on this procedure.) In this case, the number of sprinklers along the branch line is 4—that is, 1.2 ft^2 $\times$ 13 ft $=$ 3.58 sprinklers. The length of the rectangle's longer side in the design area is measured in terms of a number of sprinklers rather than an actual length dimension. Because 0.58 or any other fractional part of a sprinkler cannot operate, the resulting number must always be rounded up to the next whole number. Therefore, four sprinklers must be used. Any fractional number in this calculation technique is always rounded up to the next whole number, even if the calculation result has a fractional value less than 0.5—for example, 4.1 sprinklers would be rounded up to 5 sprinklers.

Once these values are determined, the actual design area can be laid out based on a relative minimum length of four sprinklers per branch line. Because the first calculation in this example indicated that at least 12 sprinklers must be considered, the shorter length of the design area—that is, the length parallel to the cross main—is determined by counting the number of branch lines along the cross main, which produces 12 sprinklers with 4 sprinklers on each branch line. In this case, three branch lines need to be included in the operation area.

The operating area is the hydraulically most demanding rectangle encompassing 4 sprinklers on each of 3 branch lines, which provides the required total of 12 sprinklers. This area is indicated by cross-hatching, and, because this system is symmetrical about the cross main, this area can be located as shown in Figure A-8-3.2(b), or it can be in the lower left corner of the building shown in Figure A-8-3.2(b).

In Step 1 of Figure A-8-3.2(c), sprinkler 1 and branch line 1 (BL-1) are identified in the Nozzle Identification and Location column. The flow for the first sprinkler is determined using the formula $Q = D \times A$ where the selected density of 0.15 gpm/ft^2 *(D)* is multiplied by the area of coverage per sprinkler, 130 ft^2 *(A)*. This results in a flow of 19.5 gpm from sprinkler 1. The final pipe size is determined through a combination of trial and error and experience. An initial pipe sizing can be based on the pipe schedule tables and then modified through hydraulic analysis to determine the most ideal pipe size. In this case, a 1-in. pipe has been selected. Note that the pipe size is given as a nominal dimension.

The fitting directly connected to a sprinkler is not usually included in the calculations because it is accounted for in the sprinkler's K-factor. [see 8-4.4.5(d) for more information.] As a result, no values are shown under the Pipe Fittings and Devices column in Figure A-8-3.2(c). The equivalent pipe length is the total center-to-center distance between sprinklers, which in this case is the actual pipe length of 13 ft (3.9 m) because no fittings were used. The friction loss (psi/ft) is determined by using the Hazen-Williams formula in 8-4.2.1. When using 1-in. Schedule 40 pipe with a *C* factor of 120, the formula becomes:

$$p = 5.10 \times 10^{-4} \times Q^{1.85}$$

(Note that the actual inside diameter of 1-in. Schedule 40 pipe is 1.049 in.)

When the flow *(Q)* is 19.5 gpm, as in this example

$$p = 0.124 \text{ psi/ft}$$

By multiplying 0.124 psi/ft by 13 ft, a friction loss of 1.6 psi (0.11 bar) from sprinkler 1 to sprinkler 2 exists.

The total pressure (P_t) at sprinkler 1 is determined by using the following formula:

$$P_t = \left(\frac{Q}{K}\right)^2$$

where:

P_t = pressure (psi)

Q = flow from sprinkler (gpm)

K = K-factor of sprinkler

The total pressure in the case of the example is as follows:

$$\left(\frac{19.5 \text{ gpm}}{5.6}\right)^2 = 12.1 \text{ psi}$$

The pressure at sprinkler 2 is determined by adding the pressure at sprinkler 1 plus the pressure drop caused by friction loss. In this case, the pressure at sprinkler 2 is 13.7 psi (0.93 bar) [12.1 psi (0.82 bar) at sprinkler 1 plus 1.6 psi (0.11 bar) for friction loss]. The flow *(Q)* from sprinkler 2 is determined using the formula $Q = K\sqrt{P}$ where *K* is the K-factor of sprinkler 2 (5.6) and *P* is the pressure at sprinkler 2 (13.7 psi). The flow at sprinkler 2 is therefore 20.7 gpm (78 L/min). This flow is added to sprinkler 1, and the same procedures for determining friction loss and flow for the pipe between sprinklers 1 and 2 are followed for determining the information for the pipe between sprinklers 2 and 3.

In Step 4 of Figure A-8-3.2(c), the four sprinklers have been calculated, and the pressure loss between sprinkler 4 and the cross main needs to be determined. The pipe length *(L)* included is 13.0 ft (4.0 m) between sprinklers plus 6.5 ft (2.0 m) for a starter piece plus 1.0 ft (0.3 m) for a riser nipple for a total of 20.5 ft (6.3 m). The tees at the top and bottom of the riser nipple are included in the analysis of the friction loss and are equal to 8 ft (2.4 m) of pipe each. (See Table 8-4.3.1.) The change in elevation [1 ft × 0.433 psi/ft (0.31 m × 0.0979 bar/m)] also needs to be taken into account at this point.

A K-factor is established for branch line 1 (BL-1) by dividing the flow *(Q)* by $\sqrt{P}$, which in this case is 85.2 divided by $\sqrt{26}$, or 16.71. This K-factor is used in predicting the flow in subsequent branch lines that are identical to BL-1. This K-factor is based on the same hydraulic principle as the K-factor assigned to a sprinkler—namely, that the K-factor for BL-1 describes the physical characteristics of the pipe opening in the cross main for the branch line and identifies the constant relationship between the flow and the pressure at this point. When a pressure is associated with the branch line K-factor, a corresponding flow can then be established for the branch line. If the system design area is of such a nature that it is not symmetrical and if sprinklers for the design area occur on both sides of the branch line, then two K-factors must be determined and the flow must be balanced to properly reflect the effect of the higher pressure.

In Step 8 of Figure A-8-3.2(c), the pressure due to elevation (P_e) of 6.5 psi (15 ft × 0.433) is added. The elevation pressure is considered here because the change in elevation occurs within the step with no additional water flowing from the system between the two points considered by this step. Any change in elevation needs to be taken into account where it occurs.

In Step 9 of Figure A-8-3.2(c), as indicated by the note, copper pipe is used, so the *C* factor is 150 (see Table 8-4.4.5). The fitting and device equivalent lengths are totaled and multiplied by 1.51 (see 8-4.3.2) due to the different *C* factor. The total of the equivalent lengths also needs to be multiplied by 0.869 due to the different internal pipe diameter. The factor 0.869 is derived by taking the ratio of the internal diameter of copper tube to the internal diameter of Schedule 40 steel and raising this ratio to the 4.87 power.

In Figure A-8-3.2(d), the water supply curve is based on a 90-psi (6.2-bar) static pressure and a 60-psi (4.1-bar) residual pressure with 1000 gpm (3785 L/min) flowing.

The system demand curve is started at the 6.5-psi (0.5-bar) point on the ordinate to consider the pressure due to elevation (P_e) and is continued to the demand point of 260 gpm (984 L/min) at 66.4 psi (4.5 bar). The system demand curve represents the friction loss in the system. The horizontal dashed line drawn from the above demand point to the city water supply curve represents the amount of water available for hose streams. In this case, about 610 gpm (2309 L/min) are available, but only 250 gpm (946 L/min) are required. The design demand in this example falls well below the water supply curve and, therefore, is acceptable.

Some authorities having jurisdiction will lower the water supply curve to take into account daily and seasonal fluctuations in available pressures and to prepare for future growth of areas to be developed. However, this adjustment is not required by NFPA 13.

Because the demand curve falls well below the water supply curve, some pipe sizes are allowed to be reduced to utilize the 19-psi (1.3-bar) excess pressure and reduce the cost of the sprinkler system. If the system demand point reached beyond the city water supply curve, some of the piping could be enlarged to decrease friction loss. Some adjustments can also be made to reduce the system demand to a point at or below the water supply curve. These adjustments include the use of sprinklers with larger K-factors (orifice sizes), selection of a lower discharge density with a larger area of operation, or a reduction in sprinkler spacing. If these actions do not correct the situation, then a more detailed analysis of the water supply is needed. If the system demand point cannot be reduced, then the water supply needs to be increased through the installation of a fire pump and water storage tank.

8-3.3* Detailed Worksheets.

Detailed worksheets or computer printout sheets shall contain the following information:

(1) Sheet number
(2) Sprinkler description and discharge constant (*K*)
(3) Hydraulic reference points
(4) Flow in gpm (L/min)
(5) Pipe size
(6) Pipe lengths, center-to-center of fittings
(7) Equivalent pipe lengths for fittings and devices
(8) Friction loss in psi/ft (bar/m) of pipe
(9) Total friction loss between reference points
(10) In-rack sprinkler demand balanced to ceiling demand
(11) Elevation head in psi (bar) between reference points
(12) Required pressure in psi (bar) at each reference point
(13) Velocity pressure and normal pressure if included in calculations

(14) Notes to indicate starting points or reference to other sheets or to clarify data shown
(15)* Diagram to accompany gridded system calculations to indicate flow quantities and directions for lines with sprinklers operating in the remote area
(16) Combined K-factor calculations for sprinklers on drops, armovers, or sprigs where calculations do not begin at the sprinkler

A-8-3.3 See Figure A-8-3.3.

A-8-3.3(15) See Figure A-8-3.3(15).

Figure A-8-3.3 illustrates just one of several worksheets used in the sprinkler industry today. Many of the computer printouts are similar. The form is used by the designer to track the waterflow and hydraulic characteristics of the system. The form allows the calculation procedure to be followed from the last sprinkler in the hydraulically most remote area to the point in the system where the water supply is connected. Pipe diameters, fittings, and friction loss values are tabulated based on the information retrieved from the sheet.

Figure A-8-3.3(15) depicts a gridded system that has been calculated. The dotted lines outline the calculated area, and the numbers in circles are reference points. The arrows on the pipe indicate the flow directions, and the numbers alongside indicate quantity in gpm (L/min). The flow directions are of particular importance in a gridded system due to the various flow paths that the water can take from the source to the operating sprinklers.

To facilitate the review of a proposed system, a diagram similar to Figure A-8-3.3(15) must accompany hydraulic calculations for a gridded system. Most of the commonly used software programs available today produce similar flow diagrams.

8-3.4* Graph Sheet.

A graphic representation of the complete hydraulic calculation shall be plotted on semiexponential graph paper ($Q^{1.85}$) and shall include the following:

(1) Water supply curve
(2) Sprinkler system demand
(3) Hose demand (where applicable)
(4) In-rack sprinkler demand (where applicable)

A-8-3.4 See Figure A-8-3.4.

The graph sheet in Figure A-8-3.4 is used to plot the water supply curve and the sprinkler system demand. The graph sheet is referred to as hydraulic graph paper. The scale along the abscissa (x-axis) must be to the 1.85 power because, in the Hazen-Williams formula, pressure *(p)* is proportional to the flow *(Q)* to the 1.85 power. Without this exponential adjustment, the water supply graph would be in the shape of a curve (similar to a fire pump curve) rather than a straight-line function. The straight line allows the water supply information to be more easily plotted using two data points.

The flow ($Q^1.85$) axis can be multiplied by any number to extend the range of the flow value. For example, if no adjustment is desired, the flow numbers are merely

Contract no. ____________________ Sheet no. ________ of ________

Name and location ____

Reference	Nozzle type and location	Flow in gpm (L/min)	Pipe size (in.)	Fitting and devices	Pipe equivalent length	Friction loss psi/ft (bar/m)	Required psi (bar)	Normal Pressure	Notes
		q			length		Pt	Pt	
					fitting		Pf	Pv	
		Q			total		Pe	Pn	
		q			length		Pt	Pt	
					fitting		Pf	Pv	
		Q			total		Pe	Pn	
		q			length		Pt	Pt	
					fitting		Pf	Pv	
		Q			total		Pe	Pn	
		q			length		Pt	Pt	
					fitting		Pf	Pv	
		Q			total		Pe	Pn	
		q			length		Pt	Pt	
					fitting		Pf	Pv	
		Q			total		Pe	Pn	
		q			length		Pt	Pt	
					fitting		Pf	Pv	
		Q			total		Pe	Pn	
		q			length		Pt	Pt	
					fitting		Pf	Pv	
		Q			total		Pe	Pn	
		q			length		Pt	Pt	
					fitting		Pf	Pv	
		Q			total		Pe	Pn	
		q			length		Pt	Pt	
					fitting		Pf	Pv	
		Q			total		Pe	Pn	
		q			length		Pt	Pt	
					fitting		Pf	Pv	
		Q			total		Pe	Pn	
		q			length		Pt	Pt	
					fitting		Pf	Pv	
		Q			total		Pe	Pn	
		q			length		Pt	Pt	
					fitting		Pf	Pv	
		Q			total		Pe	Pn	
		q			length		Pt	Pt	
					fitting		Pf	Pv	
		Q			total		Pe	Pn	
		q			length		Pt	Pt	
					fitting		Pf	Pv	
		Q			total		Pe	Pn	
		q			length		Pt	Pt	
					fitting		Pf	Pv	
		Q			total		Pe	Pn	
		q			length		Pt	Pt	
					fitting		Pf	Pv	
		Q			total		Pe	Pn	
		q			length		Pt	Pt	
					fitting		Pf	Pv	
		Q			total		Pe	Pn	
		q			length		Pt	Pt	
					fitting		Pf	Pv	
		Q			total		Pe	Pn	

Pt = total pressure; Pf = friction loss pressure; Pv = velocity pressure; Pe = elevation pressure

Figure A-8-3.3 *Sample worksheet.*

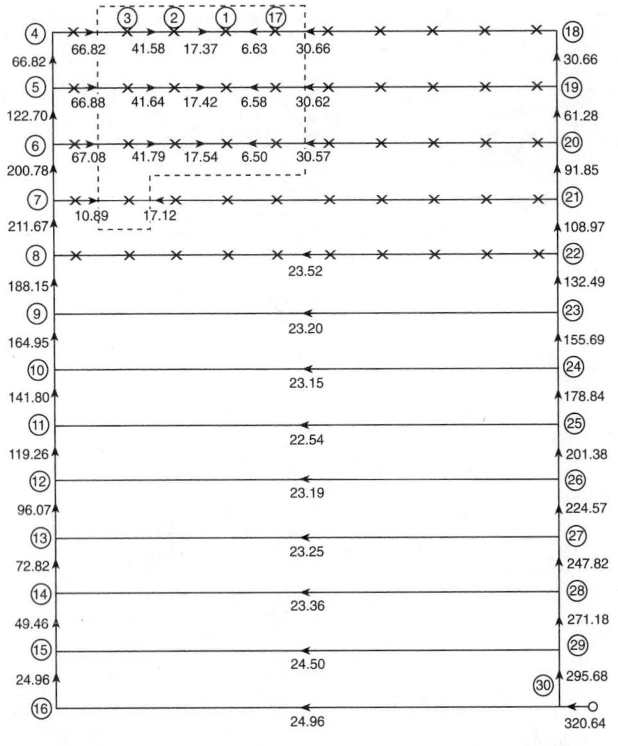

Figure A-8-3.3(15) *Example of hydraulically remote area—grid system.*

multiplied by 1. If a flow of 2500 gpm (9463 L/min) needs to be evaluated, the scale could be multiplied by 3, resulting in an extension of the scale to 3000 gpm (11,355 L/min).

8-4 Hydraulic Calculation Procedures

The procedure for hydraulically calculating a sprinkler system is different than that for other types of systems that utilize waterflow. A combination of factors must be integrated into the design, including the hazard, the spacing of sprinklers, the type of pipe, and the type of sprinkler system.

8-4.1* General.

A calculated system for a building, or a calculated addition to a system in an existing sprinklered building, shall supersede the rules in this standard governing pipe schedules, except that all

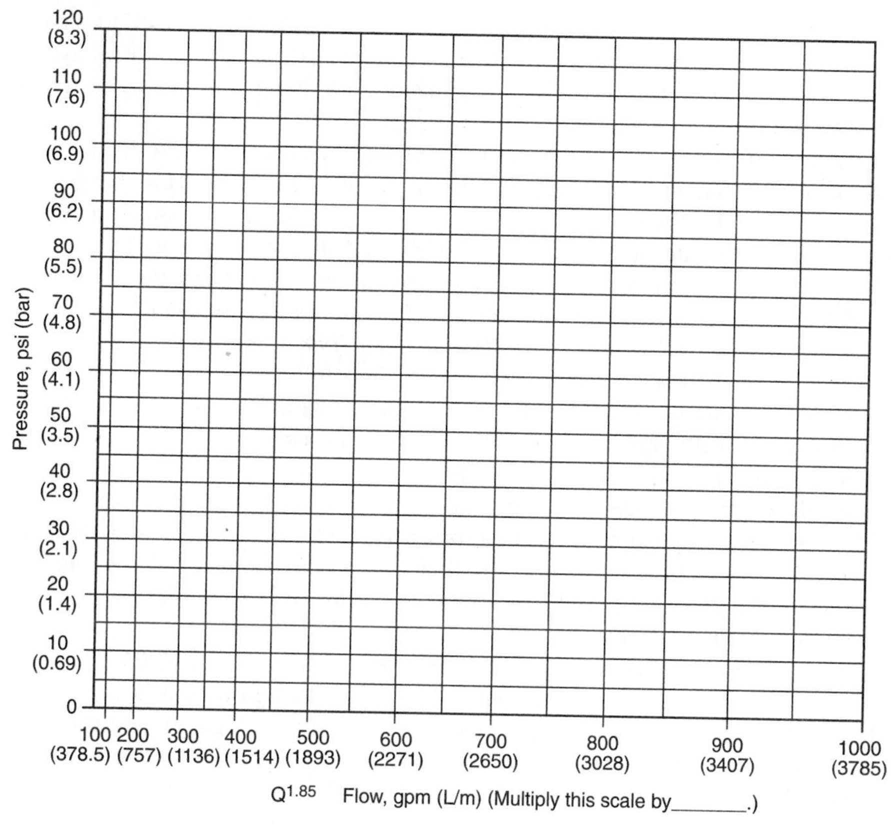

Figure A-8-3.4 *Sample graph sheet.*

systems shall continue to be limited by area and pipe sizes shall be no less than 1 in. (25.4 mm) nominal for ferrous piping and ¾ in. (19 mm) nominal for copper tubing or nonmetallic piping listed for fire sprinkler service. The size of pipe, number of sprinklers per branch line, and number of branch lines per cross main shall otherwise be limited only by the available water supply. However, sprinkler spacing and all other rules covered in this and other applicable standards shall be observed.

A-8-4.1 When additional sprinkler piping is added to an existing system, the existing piping does not have to be increased in size to compensate for the additional sprinklers, provided the new work is calculated and the calculations include that portion of the existing system that can be required to carry water to the new work. It is not necessary to restrict the water velocity when determining friction losses using the Hazen-Williams formula.

Under no circumstances can a new pipe schedule system (see Chapter 7 and Section 8-5) be connected to or supplied from an existing hydraulically designed system.

Extending an existing pipe schedule system with an existing hydraulically designed system can be acceptable in some cases where a careful analysis of the resultant hybrid system has been done. This analysis would include shifting the remote area design to include a part of the existing pipe schedule design.

The restriction on ¾-in. (19-mm) steel pipe was included in NFPA 13 in 1940, about 30 years before the advent of hydraulically calculated systems. This change was made to improve the discharge characteristics of the sprinkler and to reduce the likelihood of the smaller diameter line becoming clogged. It should be noted that ¾-in. (19-mm) nonmetallic piping and copper tube are currently listed and are acceptable. Additionally, NFPA 13D, *Standard for the Installation of Sprinkler Systems in One- and Two-Family Dwellings and Manufactured Homes*, allows the use of ½-in. (12.7-mm) nonmetallic pipe when specific criteria are followed.

A number of sprinkler systems are still in service that utilize ¾-in. (19-mm) steel pipe. No requirements are in NFPA 13 to replace all such pipe. During a routine obstruction investigation required by NFPA 25, if the smaller diameter pipe appears blocked, the obstruction needs to be cleared or the pipe needs to be replaced.

The use of the hydraulic calculation procedure, which is favored over the pipe schedule procedure, does not impose any limit on the number of sprinklers supplied by certain diameters of pipe. The determining factor for any such limit rests with the capacity of the water supply.

The last sentence of A-8-4.1 was added to state definitively that NFPA 13 has no limits on water velocity in any of the pipes. The impact of high water velocities tends to be self-correcting. When the velocity of water in a pipe is high, the pressure loss in that pipe is also high. If the pressure loss is too high, the pipe size will be increased to reduce the pressure loss and the velocity will correspondingly go down.

8-4.2 Formulas.

8-4.2.1 Friction Loss Formula. Pipe friction losses shall be determined on the basis of the Hazen-Williams formula, as follows:

$$p = \frac{4.52Q^{1.85}}{C^{1.85}d^{4.87}}$$

where:

p = frictional resistance in psi per foot of pipe

Q = flow in gpm

C = friction loss coefficient

d = actual internal diameter of pipe in inches

For SI units, the following equation shall be used:

$$p_m = 6.05\left(\frac{Q_m^{1.85}}{C^{1.85}d_m^{4.87}}\right)10^5$$

where:

p_m = frictional resistance in bar per meter of pipe

Q_m = flow in L/min

C = friction loss coefficient

d_m = actual internal diameter in mm

The Hazen-Williams formula is the most common of the empirical formulas used to determine relationships between flow, friction loss, and available pressure when designing a sprinkler system. The Darcy-Weisbach equation can provide a more precise method for establishing these relationships, but it is not commonly used for fire protection calculations due to its complexity.

The Hazen-Williams formula is dependent on the relationship between the pipe type (the *C* factor), the pipe diameter, and the flow through the pipe. The *C* factor describes the relative roughness of the pipe interior and is similar to the factor used in the Reynolds Number calculation in the Darcy-Weisbach equation. *C* factors for various pipe types are shown in Table 8-4.4.5.

Although NFPA 13 does not stipulate a maximum limit on velocities, some insurance companies modify this standard and incorporate such limits. Other mechanical piping standards impose limits on velocity because of concerns over noise and pipe erosion. These concerns are not relevant in fire protection pipe systems because water is not normally flowing through them. Automatic sprinkler systems only operate during a fire or during testing.

NFPA 13 does not require individual detailed calculations using the Hazen-Williams formula for individual pipe segments. As an alternative, friction loss tables or internal hydraulic calculations that are part of a software program are usually used to establish the friction loss. These tables and programs provide a set of accurate friction loss values for a combination of pipe types and flows, all of which are based on the Hazen-Williams formula.

8-4.2.2 Velocity Pressure Formula. Velocity pressure shall be determined on the basis of the following formula:

$$P_v = \frac{0.001123Q^2}{D^4}$$

where:

P_v = velocity pressure in psi

Q = flow in gpm

D = inside diameter in inches

For SI units, 1 in. = 25.4 mm; 1 gal = 3.785 L; 1 psi = 0.0689 bar.

The velocity pressure is a measure of the energy required to keep the water in a pipe in motion and is necessary for certain hydraulic calculations (see 8-4.2.3). The velocity pressure always acts in the direction of waterflow. The basic formula for velocity pressure is as follows:

$$P_v = \left(\frac{V^2}{2_g}\right) 0.433 \text{ psi/ft}$$

where:

P_v = velocity pressure in psi

V = velocity in ft/sec

g = 32.2 ft/sec^2

By substituting Q/A for V in the previous formula and using proper units, the more usable formula that appears in the text is found. In this approach, Q is the volumetric flow rate ft^3/sec, and A is the cross-sectional area (ft^2) of the pipe.

8-4.2.3 Normal Pressure Formula. Normal pressure (P_n) shall be determined on the basis of the following formula:

$$P_n = P_t - P_v$$

where:

P_n = normal pressure

P_t = total pressure in psi (bar)

P_v = velocity pressure in psi (bar)

The normal pressure is the pressure acting against, or perpendicular to, the pipe wall. Therefore, the normal pressure also acts perpendicular to the velocity pressure, because the general direction of waterflow is parallel to the pipe wall. When no flow occurs, the velocity pressure is zero, and the normal pressure is equal to the total pressure. See Exhibits 8.1 and 8.2.

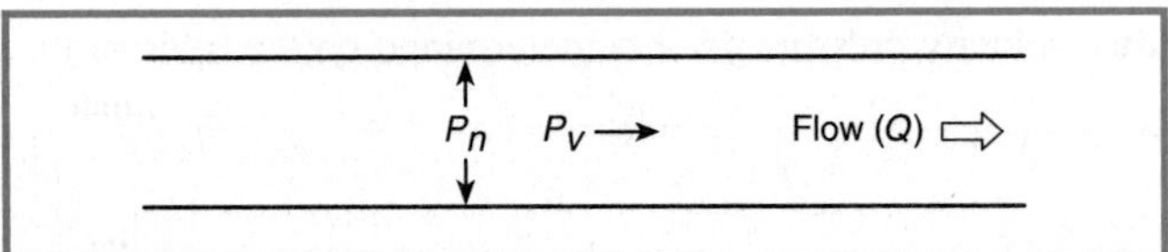

Exhibit 8.1 *Velocity and normal pressures when water flows through a pipe.*

NFPA 13 allows the user to choose between two methods of hydraulic calculations—the total pressure method and the velocity pressure method. Hydraulic calculations of sprinkler systems using the total pressure *(P_t)* are the most common. Use of the total pressure method simplifies the calculations and, in most cases, builds a safety factor into the calculations because the assumption is made that the total pressure is responsible for pushing the water out of the pipe through the sprinkler. This assumption

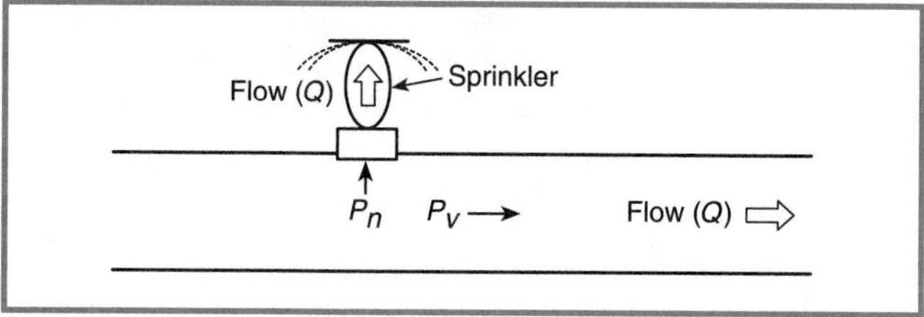

Exhibit 8.2 Normal pressure is responsible for flow from sprinklers where additional flow is downstream of the operating sprinkler.

is somewhat conservative because it is the normal pressure that is really responsible for pushing the water out of the pipe. The normal pressure is always less than the total pressure. By using a higher pressure to calculate the flow, the demand is overestimated, creating a safety factor.

The two places where the safety factor does not apply are at the last sprinkler on the branch line and the last branch line on a cross main. In these places, the total pressure overestimates the flow that will be achieved; the actual flow related to the normal pressure may not be sufficient to achieve the density chosen by the designer. When using the normal pressure method of calculations, the user must be careful that the end sprinklers and end branch lines have sufficient flow. Due to the additional safety factors built into the total pressure method of calculations, no additional considerations are necessary when using this calculation method.

NFPA 15, *Standard for Water Spray Fixed Systems for Fire Protection*, requires the use of the normal pressure method of hydraulic calculations because of the possibility of long runs of small diameter pipe to the end nozzle. These long runs cause a greater than acceptable disparity between the flow predicted by the total pressure method and the flow predicted by the normal pressure method. Because sprinkler systems do not generally have long runs of small diameter pipe to the end sprinkler, the same concerns are not shared by the Technical Committee on Sprinkler System Installation Criteria.

8-4.2.4 Hydraulic Junction Points. Pressures at hydraulic junction points shall balance within 0.5 psi (0.03 bar). The highest pressure at the junction point, and the total flows as adjusted, shall be carried into the calculations.

Calculations for loops and grids need to be done using an iterative technique in which an assumption is made and a series of calculations are made to test the assumption. If the calculated value is the same as the assumed value, the assumption is correct. Due to the number of significant figures used in the calculations and the method of rounding, getting exactly the same number between the assumed value and the calculation is not always possible. Therefore, a value needs to be selected to let the user know how close is close enough to continue with the calculations.

The value of 0.5 psi (0.03 bar) was selected because it provides a reasonable degree of accuracy for balancing at hydraulic junction points. This value is easily obtained by most computer programs without an excessive number of iterations. By requiring a balance at the hydraulic junction points, the designer will be assured that an excess system demand is not concealed in the calculations. Several software programs have the capability to carry the balance to four significant digits. The use of four significant digits does not necessarily mean that the design is better but, rather, that the computer can do a faster calculation with more significant digits to the right of the decimal point.

8-4.3 Equivalent Pipe Lengths of Valves and Fittings.

8-4.3.1 Table 8-4.3.1 shall be used to determine the equivalent length of pipe for fittings and devices unless manufacturer's test data indicate that other factors are appropriate. For saddle-type fittings having friction loss greater than that shown in Table 8-4.3.1, the increased friction loss shall be included in hydraulic calculations. For internal pipe diameters different from Schedule 40 steel pipe, the equivalent feet shown in Table 8-4.3.1 shall be multiplied by a factor derived from the following formula:

$$\left(\frac{\text{Actual inside diameter}}{\text{Schedule 40 steel pipe inside diameter}}\right)^{4.87} = \text{Factor}$$

The factor thus obtained shall be further modified as required by Table 8-4.3.1.

This table shall apply to other types of pipe listed in Table 8-4.3.1 only where modified by factors from 8-4.3.1 and 8-4.3.2.

Table 8-4.3.1 Equivalent Schedule 40 Steel Pipe Length Chart

	Fittings and Valves Expressed in Equivalent Feet of Pipe														
Fittings and Valves	**½ in.**	**¾ in.**	**1 in.**	**1¼ in.**	**1½ in.**	**2 in.**	**2½ in.**	**3 in.**	**3½ in.**	**4 in.**	**5 in.**	**6 in.**	**8 in.**	**10 in.**	**12 in.**
45° elbow	—	1	1	1	2	2	3	3	3	4	5	7	9	11	13
90° standard elbow	1	2	2	3	4	5	6	7	8	10	12	14	18	22	27
90° long-turn elbow	0.5	1	2	2	2	3	4	5	5	6	8	9	13	16	18
Tee or cross (flow turned 90°)	3	4	5	6	8	10	12	15	17	20	25	30	35	50	60
Butterfly valve	—	—	—	—	—	6	7	10	—	12	9	10	12	19	21
Gate valve	—	—	—	—	—	1	1	1	1	2	2	3	4	5	6
Swing check*	—	—	5	7	9	11	14	16	19	22	27	32	45	55	65

For SI units, 1 in. = 25.4 mm; 1 ft = 0.3048 m.

Notes:

1. This table applies to all types of pipe listed in Table 8-4.4.5.
2. Information on ½-in. pipe is included in this table only because it is allowed under 5-13.20.2 and 5-13.20.3.

*Due to the variations in design of swing check valves, the pipe equivalents indicated in this table are considered average.

The fitting and valve losses shown in Table 8-4.3.1 are calculated, experimentally derived values that have been rounded to a whole number for convenience. The exception is the swing check valve, in which several makes of valves were averaged to produce the numbers in the table. Where possible, the designer should use the friction loss value for the specific check valve that is to be installed. This information can usually be obtained from the manufacturer.

Where valves or fittings that are not in Table 8-4.3.1 are used in the system, the user must account for the friction loss by obtaining the equivalent length from the manufacturer or by obtaining the actual friction loss at the specific flow. This figure needs to be included in the calculations.

Some specially listed 1-in. (25.4-mm) saddle-type fittings have a friction loss equivalent to 21 ft (6.4 m) of pipe as opposed to the 5 ft (1.5 m) shown in Table 8-4.3.1. This difference in friction loss is another case where the listing information for a given product is of utmost importance. Ignoring this increased loss would introduce an appreciable error into the calculations.

To make the equivalent pipe length table even more accurate, the modification formula shown in 8-4.3.1 is intended to account for the difference in the internal diameter between Schedule 40 pipe and other pipe types.

Values for ½-in. (12.7-mm) and ¾-in. (19-mm) pipe appear in Table 8-4.3.1 because when hydraulic and pipe schedule systems are redesigned or modified, 5-13.20.2 and 5-13.20.3 allow the use of pipe nipples with a pipe diameter smaller than 1 in. (25.4 mm) under certain conditions.

8-4.3.2 Table 8-4.3.1 shall be used with a Hazen-Williams *C* factor of 120 only. For other values of C, the values in Table 8-4.3.1 shall be multiplied by the factors indicated in Table 8-4.3.2.

Table 8-4.3.2 C Value Multiplier

Value of *C*	100	130	140	150
Multiplying factor	0.713	1.16	1.33	1.51

Note: These factors are based upon the friction loss through the fitting being independent of the *C* factor available to the piping.

The method of hydraulic calculations used for sprinkler systems mathematically represents the friction loss for a valve or fitting as an extra length of pipe. This assumed extra length of pipe is referred to as the fitting's or valve's equivalent length. In an effort to make the calculations easier, the equivalent length of pipe is added to the actual length of pipe attached to the fitting. To correctly add the equivalent pipe length to the actual pipe length, the two pipes need to be of the same material, with the same internal diameter, and the same *C* factor.

Table 8-4.3.1 contains equivalent lengths for various types of fittings and valves for Schedule 40 pipe with a *C* factor of 120. Where other types of pipe are used, the formula in 8-4.3.1 is used to adjust for different pipe diameter (other schedules of pipe), and 8-4.3.2, including the table, is used to adjust for differing *C* factors.

The following example illustrates the effect of differing *C* factors. The base formula for friction loss through a given length of pipe using the Hazen-Williams approach is as follows:

$$P_f = L \times \frac{4.52\ Q^{1.85}}{(C^{1.85})(d^{4.87})}$$

where:

P_f = friction loss in psi

L = length in ft

(The other factors are as described in 8-4.2.1.)

The example illustrates why an adjustment is needed and considers a 2-in., Schedule 40, 10-ft length of steel pipe with a *C* factor of 100 when 135 gpm is flowing through the pipe. The actual internal diameter of the pipe must be used. Tables A-3-3.2 and A-3.3.4 provide information in this regard.

When the *C* factor is 100, the friction loss is as follows:

$$P_{f(100)} = 10\left[\frac{4.52\ (135^{1.85})}{(100^{1.85})(2.067^{4.87})}\right] = 2.7 \text{ psi}$$

From Table 8-4.3.2, the multiplier for pipe having a *C* value of 100 is 0.713. The correction to the friction loss value is as follows:

$$P_{f(100)} = 2.7 \text{ psi} \times 0.713 = 1.9 \text{ psi}$$

When the *C* factor is 120, the friction loss is as follows:

$$P_{f(120)} = 10\left[\frac{4.52\ (135^{1.85})}{(120^{1.85})(2.067^{4.87})}\right] = 1.9 \text{ psi}$$

Therefore, the values from Table 8-4.3.1 need to be adjusted accordingly when *C* factors of other than 120 are used. Another example specifically considers the adjustment needed for a fitting. Determine the equivalent length of a 1¼-in. (31.8-mm) 90-degree elbow used in a system employing Schedule 10 steel pipe on a dry pipe system.

In Step 1, determine the internal diameter of Schedule 10 and Schedule 40 steel pipe. From Table A-3-3.2, Schedule 10 is 1.442 in. (36.6 mm) and Schedule 40 is 1.380 in. (35.1 mm).

In Step 2, derive the factor using the following equation (see also 8-4.3.1):

$$factor = \left(\frac{1.442 \text{ in.}}{1.380 \text{ in.}}\right)^{4.87} = 1.2387$$

In Step 3, establish the equivalent length of a 1¼-in., Schedule 10, 90-degree elbow with a *C* value of 120. Multiply the equivalent length obtained from Table 8-4.3.1 for a 1¼-in., Schedule 40, 90-degree elbow by the factor determined in Step 2.

$$3 \text{ ft} \times 1.2387 = 3.716 \text{ ft } (1.13 \text{ m})$$

In Step 4, make a final adjustment for the different *C* factor for a dry pipe system. From Table 8-4.4.5, the *C* value for a dry system using black steel is 100. From Table 8-4.3.2, the multiplier for pipe with a *C* value of 100 is 0.713. Therefore, the equivalent length of a 1¼-in. 90-degree elbow used in a system employing Schedule 10 steel pipe on a dry pipe system is:

$$0.713 \times 3.716 = 2.6495 \text{ ft}$$

8-4.3.3 Specific friction loss values or equivalent pipe lengths for alarm valves, dry pipe valves, deluge valves, strainers, and other devices shall be made available to the authority having jurisdiction.

8-4.3.4 Specific friction loss values or equivalent pipe lengths for listed fittings not in Table 3-5.1 *(see 3-5.2)* shall be used in hydraulic calculations where these losses or equivalent pipe lengths are different from those shown in Table 8-4.3.1.

A number of special fittings that are listed for various pipe types are available on the market. These fittings often have friction loss characteristics that are different than the values given in Table 8-4.3.1. Although these fittings tend to facilitate the installation of a system, they do come with a pressure penalty that must be accounted for in the hydraulic calculations. As indicated in 8-1.1.2, the working plans are required to include manufacturer's instructions and information on these specially listed fittings.

8-4.4* Calculation Procedure.

A-8-4.4 See Figure A-8-4.4.

Details on how to properly calculate the water demand of a sprinkler system are provided in 8-4.4. The various concepts and equations described throughout Chapters 7 and 8 to hydraulically design a system are tied together in 8-4.4.

The concepts in Chapter 7 are used to determine the amount of water each sprinkler will need to discharge, the pressure necessary to achieve this flow, the number of sprinklers to calculate, the duration of the water supply, and the existence and demand of hose connections. Hydraulic calculations are then typically done by starting with the flow and pressure demand at the most hydraulically demanding single point in the system, which is usually the most distant sprinkler from the water supply. Working back towards the water supply, pressure losses and additional flows are added to determine the total water demand.

At some points within a system, the waterflow will split and travel in two or more directions. This flow split happens on cross mains with multiple branch lines, at connections to lines supplying in-rack sprinklers, and at connections to hose stations. When this type of flow split occurs, the demand for each leg of the split needs to be calculated independently. Because only one pressure reading can exist at a single point in the system, the more demanding pressure reading is used at the point of separation. Therefore, the flow through the line that requires less pressure needs to be adjusted to the higher pressure. This adjustment results in a greater waterflow because the greater the amount of pressure, the greater the amount of flow when all other variables remain unchanged.

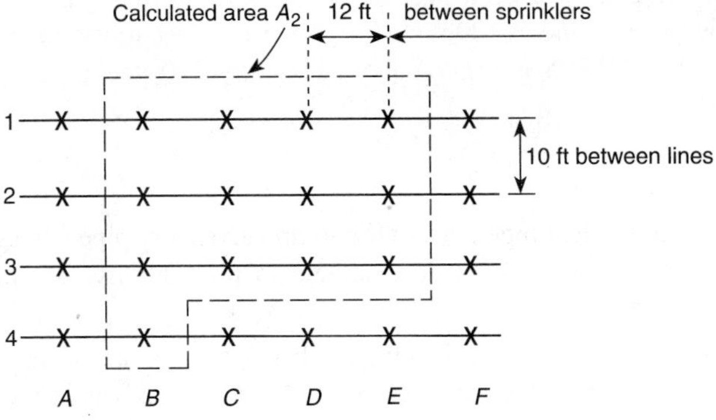

Notes:

1. For gridded systems, the extra sprinkler (or sprinklers) on branch line 4 can be placed in any adjacent location from *B* to *E* at the designer's option.
2. For tree and looped systems, the extra sprinkler on line 4 should be placed closest to the cross main.

Assume a remote area of 1500 ft² with sprinkler coverage of 120 ft²

$$\text{Total sprinklers to calculate} = \frac{\text{Design area}}{\text{Area per sprinkler}}$$

$$= \frac{1500}{120} = 12.5\text{, calculate 13}$$

$$\text{Number of sprinklers on branch line} = \frac{1.2\sqrt{A}}{S}$$

Where:
A = design area
S = distance between sprinklers on branch line

$$\text{Number of sprinklers on branch line} = \frac{1.2\sqrt{1500}}{12} = 3.87$$

For SI units, 1 ft = 0.3048 m; 1 ft² = 0.0929 m².

Figure A-8-4.4 *Example of determining the number of sprinklers to be calculated.*

For example, consider the portion of a sprinkler system shown in Exhibit 8.3. The flow through sprinklers 1 and 2 is calculated on the first branch line and the flow through sprinkler 3 is calculated on the second branch line. Hydraulic calculations for the first branch at point A indicate a demand of 37 gpm at 15 psi. This includes discharge through sprinklers 1 and 2 and the associated pressure and elevation losses. Hydraulic calculations for the second branch line when sprinkler 3 is operating indicate a demand of 18 gpm (68.1 L/min) at 12 psi (0.8 bar) at point A. Because only one pressure reading can exist at a single point in the system, the more demanding pressure of 15 psi (1.0 bar) will need to be provided at point A so that the proper discharge from sprinklers 1 and 2 is obtained. Supplying a pressure of 12 psi (0.8 bar) will result in less than 37 gpm (140.1 L/min) through the first branch line. At point A, 15 psi (1.0 bar) will create a waterflow in excess of 18 gpm (68.1 L/min) through the second branch line to sprinkler 3. The increased flow can be calculated by determining a K-factor at point A using the demand information of 18 gpm at 12 psi and rearranging the formula $Q = K\sqrt{P}$ to solve for K. This equation yields a K-factor of 5.2 (18 gpm) $\div \sqrt{12}$ psi. The K-factor is used to determine the flow at sprinkler 3 if the pressure is increased to 15 psi with the formula $Q = K\sqrt{P}$. In this case, $Q = 5.2\sqrt{15} = 20$ *gpm*. The total demand at point A becomes 57 gpm (37 + 20) at 15 psi.

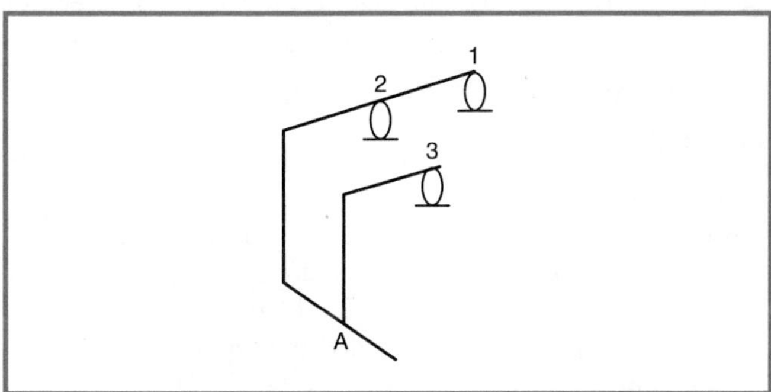

Exhibit 8.3 *Branch line schematic.*

Calculations at the point of connection of the line that supplies in-rack sprinklers are done as indicated in the preceding example. At the point of connection to the ceiling sprinkler system, the two demands are evaluated with the higher pressure demand selected and the flow at the lower demand adjusted to the higher pressure. For some specific storage commodities stored in racks, the protection criteria do not require the in-rack sprinkler water demand to be added to the ceiling system demand. One example includes combustible liquids in plastic containers protected in accordance with NFPA 30, *Flammable and Combustible Liquids Code*. The protection criteria for this commodity employ ceiling sprinklers, in-rack sprinklers, and solid barriers throughout the rack. In this circumstance, only the most demanding flow from among the ceiling system and in-rack lines needs to be met.

Hose connections to sprinkler systems that are a part of a standpipe system required to meet NFPA 14, *Standard for the Installation of Standpipe and Hose Systems*, also may need to be adjusted as described in the previous example. The requirements of NFPA 14 for combined systems provide more information on this piping arrangement.

Hose connections to sprinkler systems that are not a part of a standpipe system, such as those installed for first aid in warehouses, are not required to follow the adjustments described in the previous example. Instead, the flow demand is added to the sprinkler demand at the point of connection using the pressure available in the system at that point. In effect, no minimum pressure requirement exists for hose connections that are not part of a standpipe system.

Figure A-8-4.4 illustrates how to determine the number of sprinklers in the gridded system's area of operation and how to determine the number of sprinklers on each branch line in the area of operation. To apply to a tree or loop system to this figure, the branch line would terminate at the last sprinkler on the right in the calculated area and the cross main would be to the left. Also, Note 1 only applies to gridded systems. The extra sprinkler on branch line 4 would have to be *B* for tree and loop systems, because it is closest to the supply and would be the hydraulically most demanding on that branch line.

8-4.4.1* For all systems the design area shall be the hydraulically most demanding based on the criteria of 7-2.3.

Exception: Special design approaches in accordance with Section 7-9.

A

A-8-4.4.1 See Figures A-8-4.4.1(a) and (b).

The exception recognizes that the use of specially listed sprinklers or alternative design approaches can result in some design areas that are not rectangular. In a design involving listed residential sprinklers, the design area of operation can be a perfect square.

System *A* in Figure A-8-4.4.1(a) shows the design area for 12 sprinklers with 3 branch lines and 4 sprinklers per branch line. Systems *B* and *C* indicate the location of the extra sprinkler on the fourth branch line.

System *D* illustrates the exception to 8-4.4.1.1.

System *E* illustrates that sprinklers on both sides of the cross main can be required to fulfill the required design area and the 1.2 $\sqrt{A}$ restriction on the branch line sprinklers.

Figure A-8-4.4.1(b) illustrates the design area for looped systems with various riser locations and branch line configurations.

8-4.4.1.1 Where the design is based on area/density method, the design area shall be a rectangular area having a dimension parallel to the branch lines at least 1.2 times the square root of the area of sprinkler operation *(A)* used, which shall permit the inclusion of sprinklers on both sides of the cross main. Any fractional sprinkler shall be carried to the next higher whole sprinkler.

Exception: In systems having branch lines with an insufficient number of sprinklers to fulfill the 1.2 $\sqrt{A}$ requirement, the design area shall be extended to include sprinklers on adjacent branch lines supplied by the same cross main.

The 1966 edition of NFPA 13 was the first to include a chapter that dealt with hydraulic calculations. The following sentence is from that edition: "The design area shall be the hydraulically most remote area and shall include all sprinklers on both sides of the cross main." This sentence meant that if the system shown in Figure A-8-3.2(b) were to be calculated, the ten sprinklers on the end of the system (five on each side of the cross main) plus two on the next line would be calculated.

However, under certain conditions, such as where sprinklers were spaced 13 ft (3.96 m) apart on a branch line and the branch lines were spaced 10 ft (3.05 m) apart

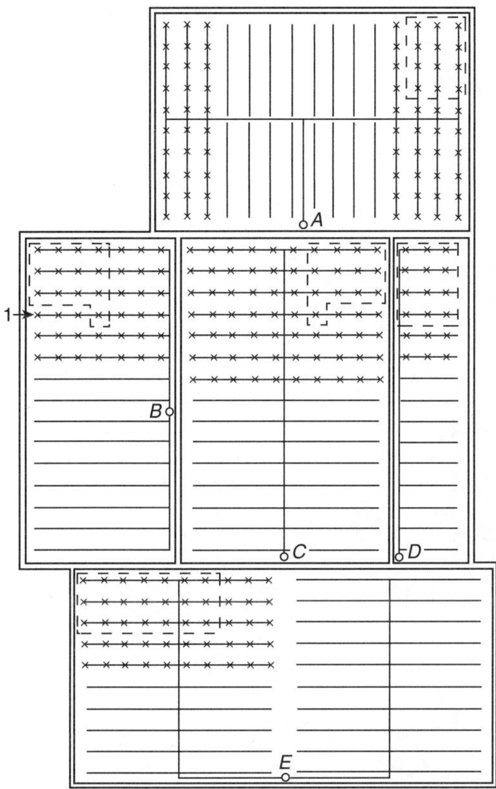

1 This sprinkler is not in the selected area of operation.

Figure A-8-4.4.1(a) *Example of hydraulically most demanding area.*

and where the design area included 12 sprinklers, some designers were calculating 3 sprinklers on the line (3 × 13 ft = 39 ft) over 4 branch lines (4 × 10 ft = 40 ft). This approach resulted in a design area that was almost a perfect square. The square shape for the design area would be inadequate if four sprinklers operated on the branch line. Therefore, the requirement that the area of operation take the shape of a rectangle with the length of the longer side having a dimension of 1.2 $\sqrt{A}$ was eventually added.

Gridded systems, along with the mention of a square-shaped design area, first appeared in the 1975 edition of NFPA 13 (7-4.3.1, Exception No. 2) as follows: "For gridded systems, the design area shall be the hydraulically most remote area which approaches a square." This approach was adopted because it did not seem logical to calculate all of the sprinklers on one line of a grid, which had been done on tree systems since 1966.

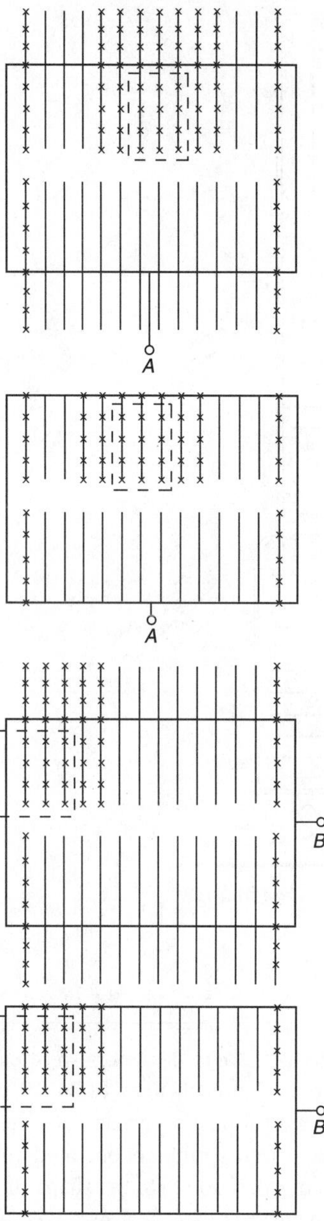

Figure A-8-4.4.1(b) *Example of hydraulically most demanding area.*

In later editions of NFPA 13, the characteristic square design area was changed to represent a rectangular area. The rectangle is required to have its longer side at least 1.2 times greater in length than its shorter side. The rectangle is required to be longer in the dimension parallel to the branch lines because this arrangement accounts for the possibility that a fire could spread in this direction and open multiple sprinklers on a single branch line before opening sprinklers on other branch lines. To require a smaller number of individual branch lines to flow more water through them is hydraulically more demanding than to require the same demand to be supplied by more branch lines with fewer sprinklers operating on them. Obviously, a worst-case scenario would be to make the designer calculate all of the sprinklers on the remote branch line. However, at a certain point in time the fire is much more likely to spread to areas under adjacent branch lines than to spread entirely under one branch line. Part of the rationale behind design areas is an attempt to predict where heat from a fire will travel. To provide an even greater factor of safety, some insurance companies require the longer side of the design area to have a dimension of $1.4\sqrt{A}$. This more conservative approach, however, is not considered necessary by NFPA 13.

The number of sprinklers to be calculated in the design area can be calculated using the area of coverage for the sprinklers in the design area. For example, with a design area of 1500 ft^2 and an area of coverage of 130 ft^2 per sprinkler, the total number of sprinklers to be considered in the design area is 1500 ft^2 divided by 130 ft^2, which gives a total of 12 sprinklers (the actual number of 11.5 needs to rounded up to the next whole number).

The number of sprinklers along a branch line can be determined by dividing the length of the design area's longer side by the spacing of the sprinklers along the branch line. For example, with a design area of 1500 ft^2 and a sprinkler spacing of 13 ft along the branch lines and 10 ft between branch lines, the number of sprinklers along the branch line is 4—that is, $(1.2\sqrt{1500})/13$. With a total of 12 sprinklers to be considered and 4 sprinklers required per branch line, a total of 3 branch lines must be considered in the design area—that is, 12 divided by 4 equals 3. This technique is only applicable where all sprinklers are uniformly spaced.

When determining the total number of sprinklers in the design area and along a branch line, the proper geometric area for system design, which is selected from the curves in Chapter 7, must be verified. Additionally, the number of sprinklers to be calculated must also be verified against the geometry of the design area. This requirement is frequently misinterpreted and misapplied. This situation usually arises when sprinklers are not symmetrically spaced and when they are not spaced exactly at one-half the allowable distance from the wall. In actuality, a large percentage of sprinkler systems fall into this situation.

Exhibit 8.4 illustrates one example of how the design area can incorrectly be calculated to an area less than that selected from Chapter 7. The area of coverage for the sprinklers in Exhibit 8.4 appears to be 104 ft^2 ($A_s = S \times L =$ 13 ft $\times$ 8 ft). Sprinklers labeled E, F, H, and I are covering this actual area. However, sprinklers labeled A, B, C, D, and G are covering a physical area of something less than 104 ft^2 because of their position adjacent to the walls. The actual areas of coverage for sprinklers

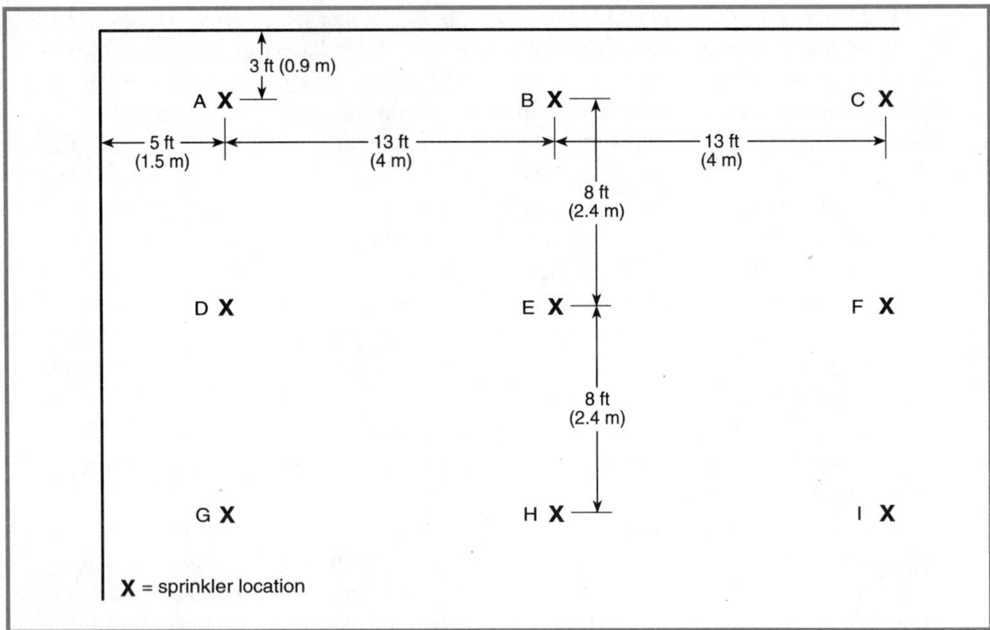

Exhibit 8.4 *Sprinkler spacing.*

A, B, C, D, and G are 80.5 ft^2 (11.5 ft × 7 ft), 91 ft^2 (7 ft × 13 ft), 91 ft^2 (7 ft × 13 ft), 92 ft^2 (11.5 ft × 8 ft), and 92 ft^2 (11.5 ft × 8 ft), respectively. If the determination of the total number of sprinklers is based on 104 ft^2, the actual design area will be less than 1500 ft^2, which is only permitted under certain conditions as specified in Chapter 7. To provide the required design area of 1500 ft^2, additional sprinklers must be included, which will increase the overall system demand in terms of flow and pressure.

The exception to 8-4.4.1.1 was added in 1980 for clarification and is illustrated by System *D* in Figure A-8-4.4.1(a). Note that the design area's longer dimension is along the cross main.

In all cases shown in Figure A-8-4.4.1(a), the area of sprinkler operation becomes a unit of measurement defined by the number of sprinklers rather than by a unit of length such as feet or meters. Because fractional parts of a sprinkler cannot operate, the conventional rules for rounding off numbers do not apply and the numbers are always rounded up. Thus, if the application of the rectangular area rule results in the need to calculate 5.07 sprinklers on a branch line, 6 sprinklers must be included.

8-4.4.1.2 Where the design is based on the room design method, the calculation shall be based on the room and communicating space, if any, that is hydraulically the most demanding. *(See 7-2.3.3.)*

8-4.4.2* For gridded systems, the designer shall verify that the hydraulically most demanding area is being used. A minimum of two additional sets of calculations shall be submitted to demonstrate peaking of demand area friction loss when compared to areas immediately adjacent on either side along the same branch lines.

Exception: Computer programs that show the peaking of the demand area friction loss shall be acceptable based on a single set of calculations.

A-8-4.4.2 See Figure A-8-4.4.2.

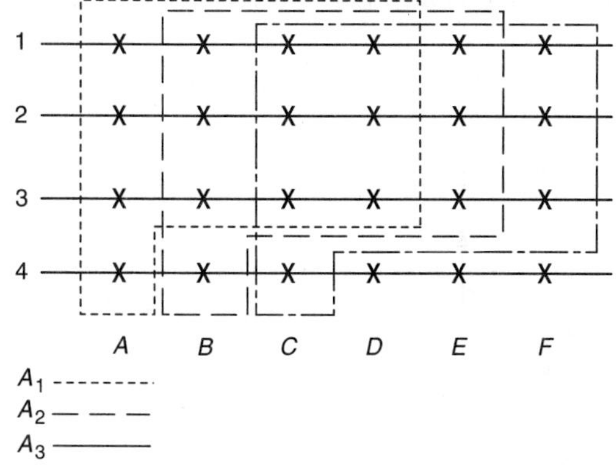

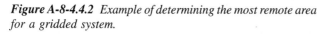

Figure A-8-4.4.2 Example of determining the most remote area for a gridded system.

In the 1975 edition of NFPA 13, the shape of the design area for a gridded system was shown as a square area located along the far cross main (not the supply cross main). However, this configuration did not necessarily represent the hydraulically most demanding area. Because of the complexity of the flow in a gridded system and the bidirectional flows that often occur within the same branch line, the hydraulically most demanding area is not readily obvious.

To determine the appropriate design area, 8-4.4.2 requires that an initial design area be selected and that two additional areas be considered on either side of the initially selected area. This approach results in three separate design areas that overlap each other as illustrated in Figure A-8-4.4.2. A_2 represents the initially chosen design area. Hydraulic calculations for each of the three design areas are required in order to determine which is the most demanding area. If the initially chosen area (A_2) is determined not to be the most demanding, then an additional area (A_4) adjacent to the area with the highest demand of the three (either A_1 or A_3) needs to be calculated to verify that A_4 is not more demanding. This technique is typically referred to as peaking the system.

Hydraulic calculations for a gridded system verify the hydraulically most demanding area and determine the system demand. These calculations are very tedious and time consuming. Conducting the calculations with a hydraulics computer program is most economical.

8-4.4.3 System piping shall be hydraulically designed using design densities and areas of operation in accordance with 7-2.3.2.1 or 7-2.3.2.2 as required for the occupancies or hazards involved.

In the 1999 edition of NFPA 13, design densities and areas of operation for hazards not presented by Figure 7-2.3.2.1 have been included.

8-4.4.3.1* The density shall be calculated on the basis of floor area of sprinkler operation. The area covered by any sprinkler used in hydraulic design and calculations shall be the horizontal distances measured between the sprinklers on the branch line and between the branch lines in accordance with 5-5.2.1.

A-8-4.4.3.1 See Figure A-8-4.4.3.1.

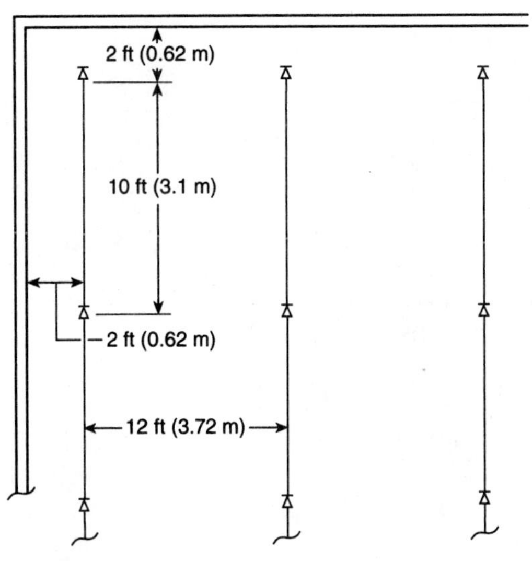

***Figure A-8-4.4.3.1** Sprinkler design area.*

The first sentence in 8-4.4.3.1 has been in NFPA 13 for several years and applies to cases where the sprinklers are installed under a sloping roof or ceiling. Under sloped ceilings, where the design area on the floor is 1500 ft^2 (139 m^2), the area projected along the ceiling will be larger than this depending on the slope of the ceiling.

The area of coverage per sprinkler is based on the area of coverage of the sprinkler as determined by 5-5.2.1—that is, along branch lines and between branch lines. Even though the branch line adjacent to the wall in Figure A-8-4.4.3.1 is only 2 ft (0.62 m) from the wall, the *L* dimension is considered to be 12 ft (3.72 m). Similarly, the last sprinkler on this branch line is also 2 ft (0.62 m) from the wall. However, the *S* dimension is considered to be 10 ft (3.1 m). Therefore, the area of coverage of sprinklers in this figure is 120 ft^2 (11.1 m^2) [10 ft × 12 ft (3.1 m × 3.72 m)].

Where small rooms as defined in 1-4.2 are considered, the area of sprinkler coverage is defined as the area of the room divided by the number of sprinklers in that room. See 5-6.3.2.1 for additional information regarding small rooms.

The piping layout of many systems are not as uniform as the one illustrated in Figure A-8-4.4.3.1. Due to building features, sprinkler placement and piping layouts can take on a more varied arrangement. This varied arrangement can result in differing areas of coverage for sprinklers in the design area and requires evaluation of several portions of a system to determine the hydraulically most demanding area so that the system demand is properly determined. Once the area of coverage per sprinkler and the appropriate dimensions for *S* and *L* are determined, other sprinklers are not required to be uniformly spaced in this arrangement. However, any other sprinkler layout that results in a greater hydraulic demand will need to be calculated separately to verify that it will work.

8-4.4.3.2* Where sprinklers are installed above and below a ceiling or in a case where more than two areas are supplied from a common set of branch lines, the branch lines and supplies shall be calculated to supply the largest water demand.

A-8-4.4.3.2 This subsection assumes a ceiling constructed so as to reasonably assure that a fire on one side of the ceiling will operate sprinklers on one side only. Where a ceiling is sufficiently open, or of such construction that operation of sprinklers above and below the ceiling can be anticipated, the operation of such additional sprinklers should be considered in the calculations.

8-4.4.4* Each sprinkler in the design area and the remainder of the hydraulically designed system shall discharge at a flow rate at least equal to the stipulated minimum water application rate (density) multiplied by the area of sprinkler operation. Calculations shall begin at the hydraulically most remote sprinkler. Discharge at each sprinkler shall be based on the calculated pressure at that sprinkler.

Exception No. 1: Where the area of application is equal to or greater than the minimum allowable area of Figure 7-2.3.1.2 for the appropriate hazard classification (including a 30 percent increase for dry pipe systems), sprinkler discharge in closets, washrooms, and similar small compartments requiring only one sprinkler shall be permitted to be omitted from hydraulic calculations within the area of application. Sprinklers in these small compartments shall, however, be capable of discharging minimum densities in accordance with Figure 7-2.3.1.2.

Exception No. 2: Where spray sprinklers and large drop sprinklers are provided above and below obstructions such as wide ducts or tables, the water supply for one of the levels of sprinklers shall be permitted to be omitted from the hydraulic ceiling design calculations within the area of application. Where ESFR sprinklers are installed above and below obstructions, the discharge for up to two sprinklers from one of the levels shall be included with those of the other level in the hydraulic calculation.

A-8-4.4.4 When it is not obvious by comparison that the design selected is the hydraulically most remote, additional calculations should be submitted. The most distant area is not necessarily the hydraulically most remote.

When applying velocity pressure, sometimes the discharge from the second sprinkler from the end of the branch line is less than the discharge from the end sprinkler. To correct this deficiency, either the pipe sizing should be changed or the end sprinkler discharge should be slightly increased.

The last sentence of 8-4.4.4 requires that the discharge from each sprinkler in the design area be calculated based on the pressure acting at that sprinkler. The intent of this statement is to prevent the calculation of the discharge from each sprinkler in the design area based on the discharge from the most remote sprinkler by manipulating the density/area requirements. For instance, if the hydraulically most demanding sprinkler produces a flow of 20.8 gpm (847 L/min) based on its area of coverage and design density, it is not correct to assume that the remaining sprinklers in the design area will also flow at 20.8 gpm (847 L/min). When water flows through a piping network, pressure changes result throughout the system due to friction, turbulence, and elevation changes. Therefore, the pressures acting at each sprinkler in the design area can vary, resulting in different amounts of flow at each sprinkler. These pressure changes must be accounted for, thereby increasing the complexity of the hydraulic calculations.

Exception No. 1 allows the sprinkler discharge in certain small compartments to be omitted from the hydraulic calculations if specific conditions are met. These conditions include a requirement for the design area to be equal to or greater than the minimum allowable area identified in Figure 7-2.3.1.2 for the appropriate hazard. This limitation includes the area increase for dry pipe systems. Therefore, when the room design method is used, this exception cannot be applied. Additionally, these spaces are to be of a size that only one sprinkler is required to protect them. The sprinkler in the small compartment must be verified as capable of producing the appropriate discharge density as identified in Figure 7-2.3.1.2. The intent of this provision is to prevent designers from falsely minimizing system demands by omitting sprinkler discharge from the compartments addressed by this exception.

It is compelling to note the types of areas described in Exception No. 1. Small compartments as addressed in this exception are intended not only to be physically small but also to contain a relatively small fuel load. For example, the size of a small office, because of its size, can require only one sprinkler. However, this exception is not intended to apply to small offices because the anticipated fuel load is greater than that expected in closets and washrooms.

In compartmented areas, such as the small spaces described by this exception, the simultaneous operation of all of the sprinklers in the design area is unlikely. On the other hand, the sprinklers in the foyer, closet, and bathroom of a hotel room could all operate in response to a fire in the bedroom.

The intent of Exception No. 1 is not to eliminate the rooms that constitute the primary occupancy of a building or space from the calculations. If sprinklers in these spaces are not included in the calculations, the proper water demand will not be determined and can cause unfavorable performance of the sprinkler system. The sprinkler discharge in small compartments can be omitted from the calculations, but the design area needs to include the square footage of the small compartments.

Small compartments, such as bathrooms, closets, or foyers in hotels, can be protected with small-orifice sprinklers—that is, those with nominal K-factors less than 5.6—even though the sprinklers in the bedrooms and hallways have a nominal K-factor of 5.6 or more. See Exception No. 1 to 8-4.4.6.

With regard to Exception No. 2, the presence of a permanent or temporary obstruction, such as an overhead garage door, is treated in a similar manner to a ceiling configuration that requires sprinklers above and below it where spray sprinklers and large drop sprinklers are installed. Simultaneous operation of sprinklers above and below the obstruction during a fire does little to improve the overall fire protection scheme and is not necessary. Although not specified by NFPA 13, good practice would include the more hydraulically demanding level of sprinklers.

Because ESFR sprinklers are designed to suppress rather than control a fire, it is critical that the sprinklers react promptly and that their discharge reach the fire. Therefore, up to two additional sprinklers need to be included in the design area where additional sprinklers are positioned around obstructions.

The appendix section warns that the most remote area is not necessarily the hydraulically most demanding and that a hydraulic evaluation would be necessary to verify the location of the area. Figures A-8-4.4.1(a) and A-8-4.4.1(b) provide examples of the location and the shape of the hydraulically most demanding areas for tree and looped systems. These examples illustrate uniform piping and sprinkler layouts in a space assumed to have the same occupancy classification throughout. However, in many situations, the system layout is not so uniform, more than one occupancy hazard exists, and the hydraulically most demanding area is not so obvious.

Occupancy hazard classification, piping configuration, and sprinkler spacing are factors that must be considered when determining the location of the design area. Exhibit 8.5 illustrates a building that has three different operations and hazards in it. Even though the office space, which is a light hazard occupancy, is the furthest from the riser, it may not represent the hydraulically most demanding area. The space used for the storage of finished product up to 14 ft (4.27 m) in height needs to be protected for the appropriate commodity classification in accordance with either Section 7-3 or Section 7-4. The flow-coating operation is considered an extra hazard (Group 2) occupancy. An evaluation of areas in each of the three spaces is needed to verify the most demanding area.

The difficulties associated with determining the hydraulically most demanding area for gridded systems is addressed in 8-4.4.2.

8-4.4.5 Pipe friction loss shall be calculated in accordance with the Hazen-Williams formula with *C* values from Table 8-4.4.5.

(a) Include pipe, fittings, and devices such as valves, meters, and strainers, and calculate elevation changes that affect the sprinkler discharge.

Exception: Tie-in drain piping shall not be included in the hydraulic calculations.

(b) Calculate the loss for a tee or a cross where flow direction change occurs based on the equivalent pipe length of the piping segment in which the fitting is included. The tee at the top of a riser nipple shall be included in the branch line, the tee at the base of a riser nipple shall be included in the riser nipple, and the tee or cross at a cross main-feed main junction shall be included in the cross main. Do not include fitting loss for straight-through flow in a tee or cross.

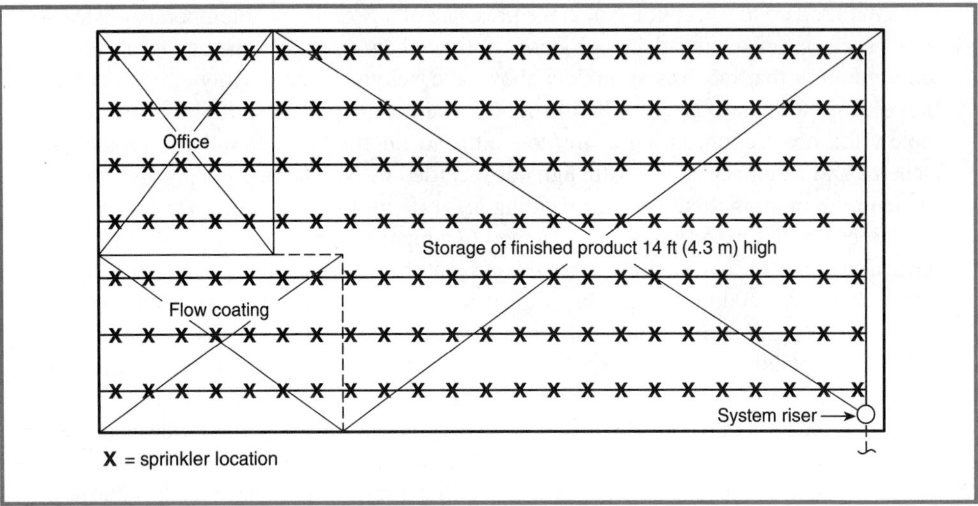

Exhibit 8.5 *Building containing various operations and fire hazards.*

(c) Calculate the loss of reducing elbows based on the equivalent feet value of the smallest outlet. Use the equivalent feet value for the standard elbow on any abrupt 90-degree turn, such as the screw-type pattern. Use the equivalent feet value for the long-turn elbow on any sweeping 90-degree turn, such as a flanged, welded, or mechanical joint-elbow type. *(See Table 8-4.3.1.)*

(d) Friction loss shall be excluded for the fitting directly connected to a sprinkler.

(e) Losses through a pressure-reducing valve shall be included based on the normal inlet pressure condition. Pressure loss data from the manufacturer's literature shall be used.

Table 8-4.4.5 Hazen-Williams C Values

Pipe or Tube	*C* Value*
Unlined cast or ductile iron	100
Black steel (dry systems including preaction)	100
Black steel (wet systems including deluge)	120
Galvanized (all)	120
Plastic (listed) all	150
Cement lined cast or ductile iron	140
Copper tube or stainless steel	150
Asbestos cement	140
Concrete	140

*The authority having jurisdiciton is permitted to consider other *C* values.

With regard to 8-4.4.5(a), water flowing through a pipe, a fitting, or any other attachment, such as a valve or strainer, results in a pressure loss due to friction and turbulent flow. Additionally, changes in elevation also cause pressure changes because the force of gravity either acts against or aids waterflow. Therefore, the type of pipe, any devices attached to the pipe, and changes in elevation must be accounted for in the hydraulic calculations. However, NFPA 13 does allow certain devices to be excluded because their effects are incorporated into other parts of the system—for example, the fitting addressed in 8-4.4.5(d).

With regard to 8-4.4.5(b), the evaluation of the waterflow from the supply to the last branch line in the system is illustrated in Exhibit 8.6. Tee A is included with the 1½-in. (38.1-mm) starting piece. Tee B is included with the 2-in. (50.8-mm) riser nipple. Tees C, D, and E are not included because the flow is straight through them. Tee F is included with the 3-in. (76.2-mm) piece of cross main.

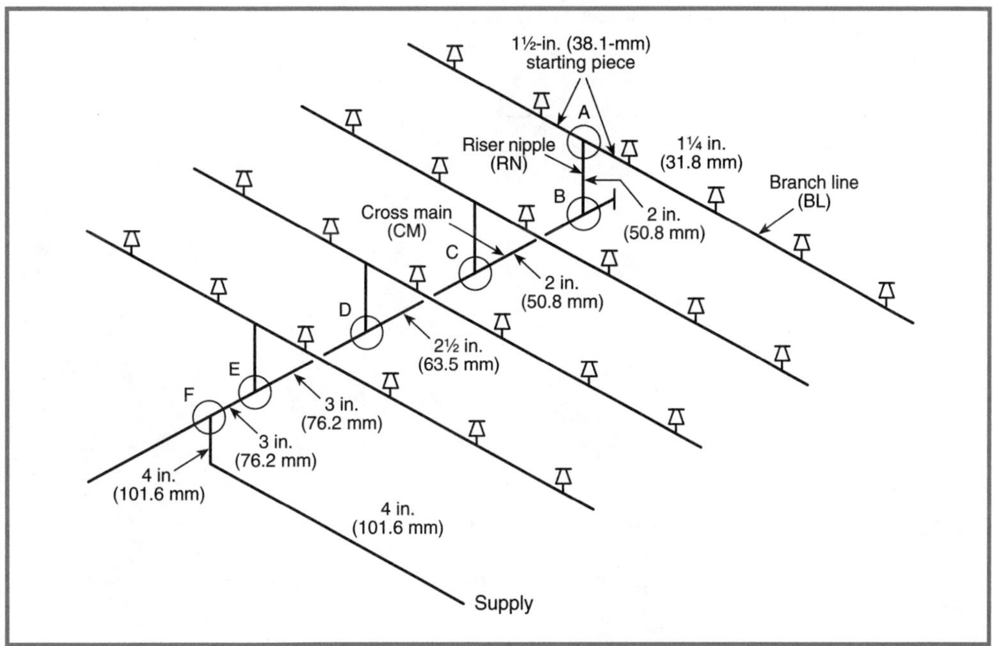

Exhibit 8.6 *Diagram of a typical sprinkler system used in making sample calculations.*

The Formal Interpretation provides additional information regarding 8-4.4.5(b).

With regard to 8-4.4.5(d), on branch lines with upright or pendent sprinklers screwed into the line tee, the fitting losses associated with the line tee are excluded from the calculations because it is assumed that one fitting is included in the approval tests of a sprinkler. The presence of this fitting is accounted for by the sprinkler's K-factor.

In cases where sprinklers are installed on sprigs or drops, the reducing coupling at the sprinkler is ignored, but the tee on the branch line must be included. In this situation, a new K-factor for the entire assembly, which includes the tee fitting, the

Formal Interpretation

Reference: 8-4.4.5(b)

Question: Regarding 8-4.4.5(b) in calculating the friction loss for a tee where the flow is turned 90°, is there any additional equivalent feet of pipe required for friction loss calculation because a bushing is used instead of a tee without a bushing? Referring to the sketch, would the equivalent feet of pipe be greater using a 1½ × 1½ × 2½ tee with 1½ × 1¼ bushing, as shown, than if a 1½ × 1¼ × 2½ tee were used?

Answer: The tee at the top of a riser nipple, where flow is turned 90°, should be considered part of the branch line segment. For example, if a 2½-in. riser supplies 1½-in. branch line piping on one side of the riser and 1½-in. branch line piping on the other side, the loss through the tee should be represented by equivalent feet of 1½-in. pipe in one direction and by 1¼-in. pipe in the other.

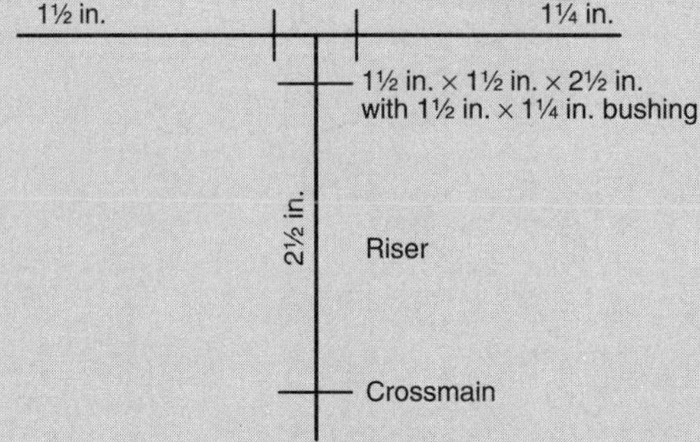

For SI Units: 1 in = 25.4 mm.

The use of bushings, as indicated by 3-5.5, is discouraged by the standard. However, when used in crosses or tees, the method should be no different from that described in 8-4.4.5(b). For a change in direction, the fitting friction loss equivalent length is sized according to the pipe supplied, not from any of the other outlet sizes.

Issue Edition: 1974

Reference: 7-4.3.1(f)

Date: October—November 1975

sprig or drop, and the sprinkler, could be determined. This K-factor allows individual calculations to be simplified for each junction between the branch line and the sprig.

With regard to 8-4.4.5(e), pressure-reducing devices would be installed where system pressures are in excess of the system component's pressure rating. Many components are limited to a maximum working pressure of 175 psi (12.1 bar). These devices act to intentionally reduce the available pressure so that system components will not be damaged by high pressures. The manufacturer's instructions must be closely followed to establish the correct outlet pressure and flow parameters.

NFPA 13 requires the use of the Hazen-Williams formula when determining friction losses—that is, pressure changes between pipe segments. (Refer to 8-4.2.1 for the formula.) One variable in this formula is represented by the letter *C*. This variable, often referred to as the *C* value, is a coefficient that reflects the roughness of the internal surface of the piping material.

C values for steel and iron pipe materials tend to decrease with time due to the buildup of scale and the effects of corrosion. The values shown in Table 8-4.4.5 include *C* values for aged pipe. If pipe could remain in its original condition after installation, the *C* factors would be somewhat higher than those shown in Table 8.1. A comparison between new pipe *C* factors and the design *C* factors required by NFPA 13 is shown in Table 8.1. For ferrous and cement-lined products, an adjustment is applied to compensate for the aging process of the pipe.

Table 8.1 Comparison of C Values for New Pipe

Pipe Type	*C* New	*C* Design	Upscale Factor (%)
Steel (welded or seamless)	140	120*	16
Unlined and ductile iron	130	100	23
Plastic	150	150	0
Copper	150	150	0
Cement-lined cast iron	150	140	7

*Wet pipe system.

Factors that can affect the *C* value for a particular piece of pipe include the pipe's manufacturing process, the rate of pipe corrosion, and the buildup of scale and other materials that decrease the inside diameter of the pipe. The synergistic effect of these contributing factors is accounted for by reducing the pipe's *C* value.

Pipe manufacturing techniques do not usually produce a perfectly round piece of pipe, and some inherent imperfections exist. However, these imperfections are usually very subtle and cannot be detected without taking precise measurements.

Corrosion of ferrous materials is very difficult, if not impossible, to prevent in sprinkler systems. Degradation of pipe in this manner can actually cause a slight increase in the internal diameter of the pipe. However, the effects of corrosion should not be considered as having a positive effect on system performance. Corrosion causes the internal surface of the pipe to become much rougher, thus increasing the associated pressure loss due to friction.

Tuberculation of pipe is a function of the naturally occurring elements and chemicals added to the water supply by the water purveyor. These waterborne chemicals can fall out over time and cause buildups on the internal portions of the pipe network. The result is a rougher internal pipe surface and a smaller internal diameter. Exhibit 8.7 illustrates a pipe that has experienced tuberculation.

Exhibit 8.7 *Tuberculated pipe.*

Table 8-4.4.5 includes the types of materials that are commonly used in the design and installation of sprinkler systems. If the water quality is poor and the pipe deteriorates faster than would be expected in most water supplies, further adjustments to the *C* value should then be made. Table 8.1 identifies the adjustments (safety factors) used for a particular type of pipe material. Because plastic and copper piping materials do not experience the same effects as ferrous pipe, their *C* values are not adjusted.

8-4.4.6* Orifice plates or sprinklers of different orifice sizes shall not be used for balancing the system.

Exception No. 1: Sprinklers with different orifice sizes shall be acceptable for special use such as exposure protection, small rooms or enclosures, or directional discharge. (See 1-4.2 for definition of small rooms.)

Exception No. 2: Extended-coverage sprinklers with a different orifice size shall be acceptable for part of the protection area where installed in accordance with their listing.

A-8-4.4.6 The use of sprinklers with differing orifice sizes in situations where different protection areas are needed is not considered balancing. An example would be a room that could be protected with sprinklers having different orifice size in closet, foyer, and room areas. However, this procedure introduces difficulties when restoring a system to service after operation since it is not always clear which sprinklers go where.

Orifice plates are prohibited because they can be removed while modifying a system and not be reinstalled. This removal would invalidate the hydraulic calculations supporting the initial system design. Also, if installed in a horizontal run of pipe, orifice plates are subject to the accumulation of debris on the surface facing the direction of waterflow. This accumulation would further impact the hydraulics as well as affect proper system drainage.

In general, the use of sprinklers with differing orifice sizes is prohibited. The concern is that if sprinklers with various orifice sizes are used and if some of those sprinklers have to be replaced at a future date, a sprinkler with a differing orifice size might be installed.

The two exceptions recognize the following:

(1) Certain arrangements can require a reduced sprinkler discharge, thus a sprinkler with a smaller orifice is permitted.
(2) The use of extended coverage sprinklers with a different orifice size than the other sprinklers in the room can be needed to provide proper coverage of a space.

An example of where Exception No. 2 applies is illustrated in Exhibit 8.8. The center portions of the room can be protected with extended coverage sprinklers that have a nominal K-factor of 8. However, the space under the 3-ft (0.91-m) wide soffit also needs to be protected. Because the areas under the soffits are so small, Exception No. 2 permits the use of a smaller orifice sprinkler.

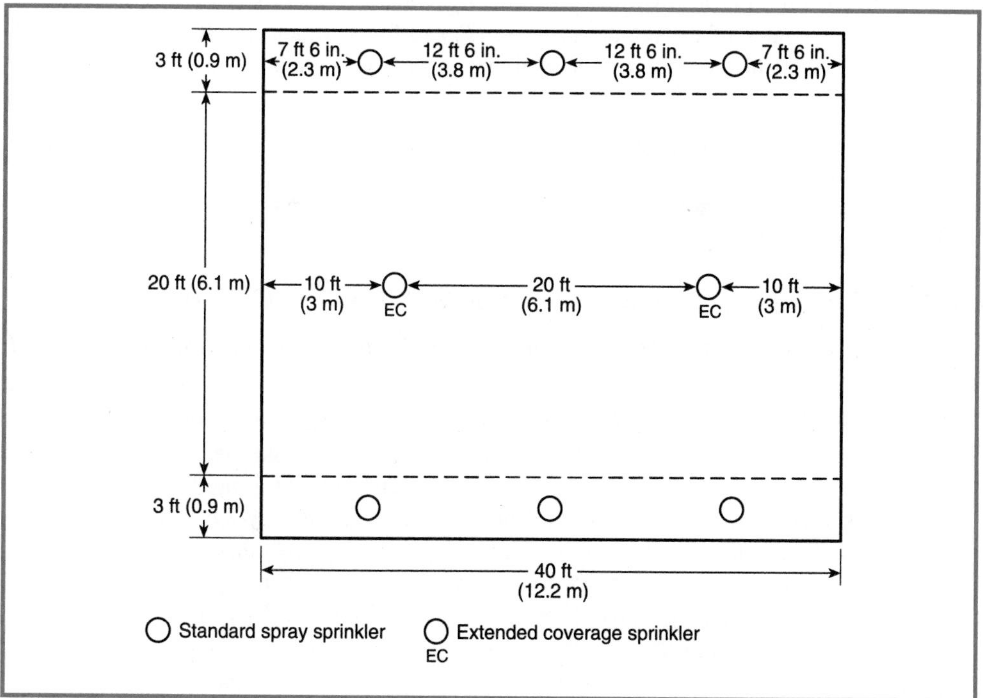

Exhibit 8.8 *Sprinkler layout using the multiple-orifice approach.*

8-4.4.7* When calculating flow from an orifice, the total pressure (P_t) shall be used. Flow from a sprinkler shall be calculated using the nominal K-factor.

Exception: Use of the normal pressure (P_n) calculated by subtracting the velocity pressure from the total pressure shall be permitted. Where the normal pressure is used, it shall be used on all branch lines and cross mains where applicable.

A-8-4.4.7 Where the normal pressure (P_n) is used to calculate the flow from an orifice, the following assumptions should be used.

(a) At any flowing outlet along a pipe, except the end outlet, only the normal pressure (P_n) can act on the outlet. At the end outlet the total pressure (P_t) can act. The following should be considered end outlets:

(1) The last flowing sprinkler on a dead-end branch line
(2) The last flowing branch line on a dead-end cross main
(3) Any sprinkler where a flow split occurs on a gridded branch line
(4) Any branch line where a flow split occurs on a looped system

(b) At any flowing outlet along a pipe, except the end outlet, the pressure acting to cause flow from the outlet is equal to the total pressure (P_t) minus the velocity pressure (P_v) on the upstream (supply) side.

(c) To find the normal pressure (P_n) at any flowing outlet, except the end outlet, assume a flow from the outlet in question and determine the velocity pressure (P_v) for the total flow on the upstream side. Because normal pressure (P_n) equals total pressure (P_t) minus velocity pressure (P_v), the value of the normal pressure (P_n) so found should result in an outlet flow approximately equal to the assumed flow; if not, a new value should be assumed, and the calculations should be repeated.

The total pressure acts on the remote outlet at the point where the flow enters the outlet from opposing directions. Velocity pressure is usually small compared to the total pressure and normally does not have a major impact on the outcome of the hydraulic calculation.

8-4.4.8 Minimum operating pressure of any sprinkler shall be 7 psi (0.5 bar).

Exception: Where higher minimum operating pressure for the desired application is specified in the listing of the sprinkler.

When sprinklers are submitted to testing laboratories for examination and listing, they are subjected to water distribution and fire tests. These tests are conducted at a minimum flow of 15 gpm (57 L/min) per sprinkler. The pressure required to produce this flow through sprinklers with a nominal K-factor of 5.6 is approximately 7 psi (0.5 bar), as illustrated in the equation that follows. The exception recognizes that specially listed sprinklers might have minimum operating pressures greater than 7 psi (0.5 bar). In some cases, the minimum operating pressure can be as high as 75 psi (1.93 bar).

$$p = \frac{Q^2}{K^2} = \frac{15^2}{5.6^2} = 7.17 \text{ psi}$$

In recent years, some sprinklers were listed with operating pressures less than 7 psi (0.5 bar). Such pressures are no longer permitted by NFPA 13. The concern was that, at these lower pressures, sprinklers that had been installed for an extended period of time would not operate properly. These lower pressures' ability to blow off the orifice cap on activation of the operating element was uncertain. Additional concerns were that sprinklers operating at pressures less than 7 psi (0.5 bar) could produce unsuitable discharge patterns.

8-5 Pipe Schedules

Pipe schedules shall not be used, except in existing systems and in new systems or extensions to existing systems described in Chapter 7. Water supplies shall conform to 7-2.2.

As noted in 7-2.2, the use of pipe schedule design approaches is restricted to rather small systems, unless relatively high residual pressures are available. Although the pipe schedule approach has previously served as an acceptable design option for a wider range of operations and fire hazards, its misuse has resulted in its restricted application.

8-5.1* General.

The pipe schedule sizing provisions shall not apply to hydraulically calculated systems. Sprinkler systems having sprinklers with orifices other than ½ in. (13 mm) nominal, listed piping material other than that covered in Table 3-3.1, extra hazard, Groups 1 and 2 systems, and exposure protection systems shall be hydraulically calculated.

A-8-5.1 The demonstrated effectiveness of pipe schedule systems is limited to their use with ½-in. (13-mm) orifice sprinklers. The use of other size orifices can require hydraulic calculations to prove their ability to deliver the required amount of water within the available water supply.

The numbers of sprinklers shown in the pipe schedule tables are maximum numbers for the minimum size pipes shown. The graduated type of system—that is, larger pipe serving increasingly smaller pipe—is not mandatory as long as the minimum pipe sizes established in the pipe schedule tables are met. Formal Interpretation 80-22 provides more information on this requirement.

8-5.1.1 The number of automatic sprinklers on a given pipe size on one floor shall not exceed the number given in 8-5.2, 8-5.3, or 8-5.4 for a given occupancy.

The number of sprinklers per branch line is limited (see 8-5.2.1, 8-5.3.1, and A-8-5.4) on pipe schedule systems, and, prior to 1966, the number of branch lines on a cross main was limited to 14. These rules do not apply to hydraulically designed systems.

Formal Interpretation 80-22

Reference: 8-5.1

Question 1: Referring to Figure 1, does 8-5.1 permit the piping supplying multiple floors to be sized for the maximum number of sprinklers on one floor or must it be sized for all sprinklers connected to it?

Answer: It may be sized to supply the maximum number of sprinklers on one floor. Referring to Figure 1, the pipe in question may be 1¼ in. for light hazard or ordinary hazard occupancies.

Question 2: Is the sizing of piping to supply sprinklers on any one floor restricted to system risers?

Answer: No.

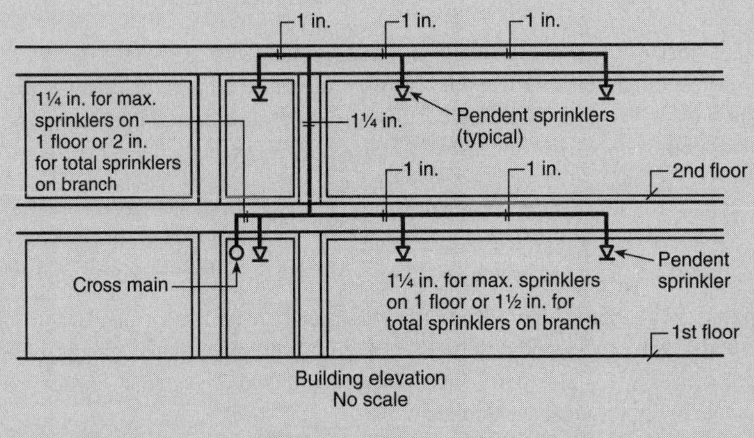

Figure 1

Issue Edition: 1980

Reference: 3-4.2

Date: October 1982

8-5.1.2* Size of Risers. Each system riser shall be sized to supply all sprinklers on the riser on any one floor as determined by the standard schedules of pipe sizes in 8-5.2, 8-5.3, or 8-5.4.

A-8-5.1.2 Where the construction or conditions introduce unusually long runs of pipe or many angles in risers or feed or cross mains, an increase in pipe size over that called for in the schedules can be required to compensate for increased friction losses.

8-5.1.3 Slatted Floors, Large Floor Openings, Mezzanines, and Large Platforms. Buildings having slatted floors or large unprotected floor openings without approved stops shall be treated as one area with reference to pipe sizes, and the feed mains or risers shall be of the size required for the total number of sprinklers.

Fire spreads readily through grated or slatted floors. Therefore, an area with such divisions must be treated as a single fire area when sizing pipe.

8-5.1.4 Stair Towers. Stairs, towers, or other construction with incomplete floors, if piped on independent risers, shall be treated as one area with reference to pipe sizes.

A fire in a stair or tower can open a large number of sprinklers in that area. An independent riser supplying such an area must be sized to supply all the sprinklers in the area.

8-5.2 Schedule for Light Hazard Occupancies.

8-5.2.1 Branch lines shall not exceed eight sprinklers on either side of a cross main.

Exception: Where more than eight sprinklers on a branch line are necessary, lines shall be permitted to be increased to nine sprinklers by making the two end lengths 1 in. (25.4 mm) and 1¼ in. (33 mm), respectively, and the sizes thereafter standard. Ten sprinklers shall be permitted to be placed on a branch line, making the two end lengths 1 in. (25.4 mm) and 1¼ in. (33 mm), respectively, and feeding the tenth sprinkler by a 2½-in. (64-mm) pipe.

The amount of pressure lost to friction in long runs of small-diameter pipe is significant. To ensure that pressure is not excessively reduced, the number of sprinklers on a branch line is limited to eight sprinklers in light hazard occupancies. The exception allows up to ten sprinklers on branch lines in light hazard occupancies, provided that pipe sizing is increased to reduce the pressure drop along the run of pipe.

8-5.2.2 Pipe sizes shall be in accordance with Table 8-5.2.2.

Exception: Each area requiring more sprinklers than the number specified for 3½-in. (89-mm) pipe in Table 8-5.2.2 and without subdividing partitions (not necessarily fire walls) shall be supplied by mains or risers sized for ordinary hazard occupancies.

Table 8-5.2.2 Light Hazard Pipe Schedules

Steel		Copper	
1 in.	2 sprinklers	1 in.	2 sprinklers
1¼ in.	3 sprinklers	1¼ in.	3 sprinklers
1½ in.	5 sprinklers	1½ in.	5 sprinklers
2 in.	10 sprinklers	2 in.	12 sprinklers
2½ in.	30 sprinklers	2½ in.	40 sprinklers
3 in.	60 sprinklers	3 in.	65 sprinklers
3½ in.	100 sprinklers	3½ in.	115 sprinklers
4 in.	See Section 5-2	4 in.	See Section 5-2

For SI units, 1 in. = 25.4 mm.

A system riser in a light hazard occupancy is limited to the protection area of 52,000 ft^2 (4831 m^2). Piping is sized in accordance with Table 8-5.2.2, unless more than 100 sprinklers are in an area without subdividing partitions. Piping supplying such sprinklers must be sized in accordance with 8-5.3 to minimize the friction loss.

8-5.2.3 Where sprinklers are installed above and below ceilings *[see Figures 8-5.2.3(a), (b), and (c)]* and such sprinklers are supplied from a common set of branch lines or separate branch lines from a common cross main, such branch lines shall not exceed eight sprinklers above and eight sprinklers below any ceiling on either side of the cross main. Pipe sizing up to and including 2½ in. (64 mm) shall be as shown in Table 8-5.2.3 utilizing the greatest number of sprinklers to be found on any two adjacent levels.

Exception: Branch lines and cross mains supplying sprinklers installed entirely above or entirely below ceilings shall be sized in accordance with Table 8-5.2.2.

Table 8-5.2.3 Number of Sprinklers above and below a Ceiling

Steel		Copper	
1 in.	2 sprinklers	1 in.	2 sprinklers
1¼ in.	4 sprinklers	1¼ in.	4 sprinklers
1½ in.	7 sprinklers	1½ in.	7 sprinklers
2 in.	15 sprinklers	2 in.	18 sprinklers
2½ in.	50 sprinklers	2½ in.	65 sprinklers

For SI units, 1 in. = 25.4 mm.

Formal Interpretations 80-1 and 91-1 pertain to 8-5.2.3.

Formal Interpretation 80-1

Reference: 8-5.2.3

Question: Is the ceiling, as referenced in 8-5.2.3, required to be a rated assembly (i.e., one-hour, two-hour, etc.)?

Answer: No.

Issue Edition: 1980

Reference: 3-5.3

Date: October 1980

Formal Interpretation 91-1

Reference: 8-5.2.3, 8-5.3.3

Background: An existing pipe schedule system is present in a structure. The original system was installed using NFPA 13, Table 8-5.2.2, 8-5.3.2(a), or 8-5.3.2(b) as appropriate for the hazard and conditions.

The existing pipe schedule system must be modified in order to accommodate a new ceiling. Combustible construction is present, thus sprinklers must be maintained in the concealed space as well as below the new ceiling. No consideration will be given to hydraulically calculating the modified system.

Question: Is there an exception for conditions as described above that would allow sprinklers to be placed on the branchlines above and below the ceiling but that would not require conformance with Table 8-5.2.3 or Table 8-5.3.3?

Answer: No.

Issue Edition: 1991

Reference: Tables 6-5.2.3, 6-5.3.3

Issue Date: July 7, 1992

Effective Date: July 27, 1992

A fire either above or below a ceiling in a light hazard occupancy will normally be contained in that area by operating sprinklers. In light hazard pipe schedule systems, NFPA 13 permits more sprinklers on a branch line in the above or below configuration than are permitted on a branch line protecting a single area. NFPA 13 also permits more total sprinklers to be supplied by a given pipe size. When determining if sprinkler protection is needed above the ceiling, 5-13.1.1 should be consulted.

The exception recognizes that piping dedicated to a given area—that is, above a ceiling or below a ceiling—can be sized to accommodate the sprinklers for that area only.

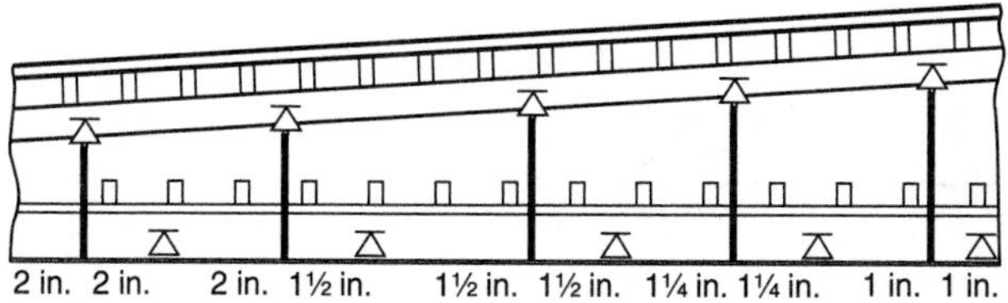

Figure 8-5.2.3(a) *Arrangement of branch lines supplying sprinklers above and below a ceiling.*

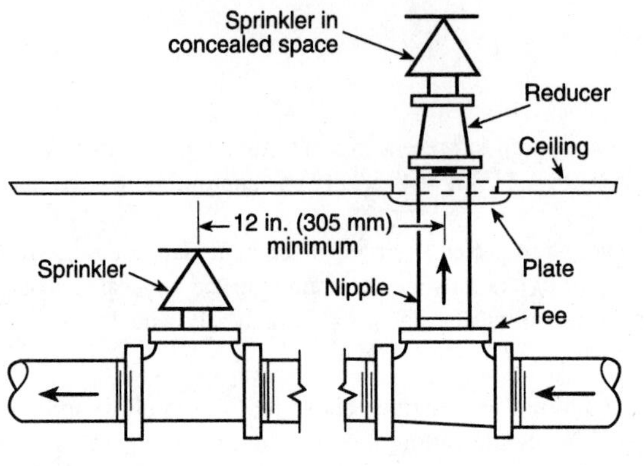

Figure 8-5.2.3(b) *Sprinkler on riser nipple from branch line in lower fire area.*

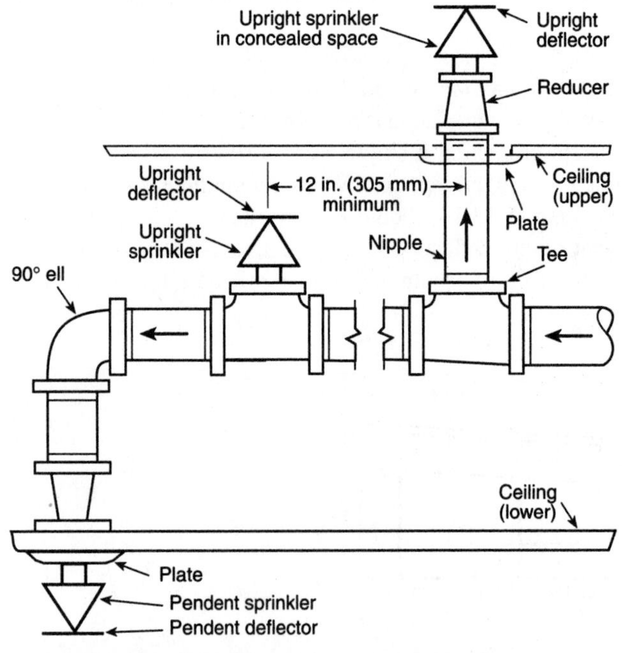

Figure 8-5.2.3(c) *Arrangement of branch lines supplying sprinklers above, in between, and below ceilings.*

8-5.2.3.1* Where the total number of sprinklers above and below a ceiling exceeds the number specified in Table 8-5.2.3 for 2½-in. (64-mm) pipe, the pipe supplying such sprinklers shall be increased to 3 in. (76 mm) and sized thereafter according to the schedule shown in Table 8-5.2.2 for the number of sprinklers above or below a ceiling, whichever is larger.

A-8-5.2.3.1 For example, a 2½-in. (64-mm) steel pipe, which is permitted to supply 30 sprinklers, can supply a total of 50 sprinklers where not more than 30 sprinklers are above or below a ceiling.

The probability of a fire involving the areas both above and below a ceiling in a light hazard occupancy and opening in excess of 50 sprinklers is extremely remote. When sizing piping for systems with more than 50 sprinklers, the pipe diameter is increased to 3 in. (76 mm) to reduce friction loss.

8-5.3 Schedule for Ordinary Hazard Occupancies.

8-5.3.1 Branch lines shall not exceed eight sprinklers on either side of a cross main.

Exception: Where more than eight sprinklers on a branch line are necessary, lines shall be permitted to be increased to nine sprinklers by making the two end lengths 1 in. (25.4 mm) and 1¼ in. (33 mm), respectively, and the sizes thereafter standard. Ten sprinklers shall be permitted to be placed on a branch line, making the two end lengths 1 in. (25.4 mm) and 1¼ in. (33 mm), respectively, and feeding the tenth sprinkler by a 2½-in. (64-mm) pipe.

8-5.3.2 Pipe sizes shall be in accordance with Table 8-5.3.2(a).

Exception: Where the distance between sprinklers on the branch line exceeds 12 ft (3.7 m) or the distance between the branch lines exceeds 12 ft (3.7 m), the number of sprinklers for a given pipe size shall be in accordance with Table 8-5.3.2(b).

This exception recognizes the increased friction loss when either the distance between sprinklers on branch lines or the distance between branch lines exceeds 12 ft (3.7 m). The exception also recognizes that the sprinkler's area of coverage needs to reach or overlap any adjacent sprinkler's area of coverage so that a uniform water discharge and density are achieved. Increasing the pipe sizing as required reduces pressure losses due to friction and results in a greater amount of pressure at the sprinkler. The higher pressure increases sprinkler discharge and improves water distribution.

The pipe sizes specified in Table 8-5.3.2(a) take into consideration normal water supplies and friction loss. The pipe sizes tend to be somewhat smaller for branch lines and somewhat larger for cross mains, feed mains, and system risers when compared to hydraulically designed systems.

8-5.3.3 Where sprinklers are installed above and below ceilings and such sprinklers are supplied from a common set of branch lines or separate branch lines supplied by a common cross main, such branch lines shall not exceed eight sprinklers above and eight sprinklers below any ceiling on either side of the cross main. Pipe sizing up to and including 3 in. (76 mm) shall be as shown in Table 8-5.3.3 *[see Figures 8-5.2.3(a), (b), and (c)]* utilizing the greatest number of sprinklers to be found on any two adjacent levels.

Table 8-5.3.2(a) Ordinary Hazard Pipe Schedule

Steel		Copper	
1 in.	2 sprinklers	1 in.	2 sprinklers
1¼ in.	3 sprinklers	1¼ in.	3 sprinklers
1½ in.	5 sprinklers	1½ in.	5 sprinklers
2 in.	10 sprinklers	2 in.	12 sprinklers
2½ in.	20 sprinklers	2½ in.	25 sprinklers
3 in.	40 sprinklers	3 in.	45 sprinklers
3½ in.	65 sprinklers	3½ in.	75 sprinklers
4 in.	100 sprinklers	4 in.	115 sprinklers
5 in.	160 sprinklers	5 in.	180 sprinklers
6 in.	275 sprinklers	6 in.	300 sprinklers
8 in.	See Section 5-2	8 in.	See Section 5-2

For SI units, 1 in. = 25.4 mm.

Table 8-5.3.2(b) Number of Sprinklers—Greater than 12-ft (3.7-m) Separations

Steel		Copper	
2½ in.	15 sprinklers	2½ in.	20 sprinklers
3 in.	30 sprinklers	3 in.	35 sprinklers
3½ in.	60 sprinklers	3½ in.	65 sprinklers

For SI units, 1 in. = 25.4 mm.

Note: For other pipe and tube sizes, see Table 8-5.3.2(a).

Exception: Branch lines and cross mains supplying sprinklers installed entirely above or entirely below ceilings shall be sized in accordance with Tables 8-5.3.2(a) or (b).

A fire either above or below a ceiling in an ordinary hazard occupancy will normally be contained in that area by operating sprinklers. In ordinary hazard pipe schedule systems, NFPA 13 permits more sprinklers on a branch line in the above or below configuration than are permitted on a branch line protecting a single area. NFPA 13 also permits more total sprinklers to be supplied by a given pipe size. When determining if sprinkler protection is needed above the ceiling, 5-13.1.1 should be consulted.

8-5.3.3.1* Where the total number of sprinklers above and below a ceiling exceeds the number specified in Table 8-5.3.3 for 3-in. (76-mm) pipe, the pipe supplying such sprinklers shall be increased to 3½ in. (89 mm) and sized thereafter according to the schedule shown in Table

Table 8-5.3.3 Number of Sprinklers above and below a Ceiling

Steel		Copper	
1 in.	2 sprinklers	1 in.	2 sprinklers
1¼ in.	4 sprinklers	1¼ in.	4 sprinklers
1½ in.	7 sprinklers	1½ in.	7 sprinklers
2 in.	15 sprinklers	2 in.	18 sprinklers
2½ in.	30 sprinklers	2½ in.	40 sprinklers
3 in.	60 sprinklers	3 in.	65 sprinklers

For SI units, 1 in. = 25.4 mm.

8-5.2.2 or Table 8-5.3.2(a) for the number of sprinklers above or below a ceiling, whichever is larger.

Exception: Where the distance between the sprinklers protecting the occupied area exceeds 12 ft (3.7 m) or the distance between the branch lines exceeds 12 ft (3.7 m), the branch lines shall be sized in accordance with either Table 8-5.3.2(b), taking into consideration the sprinklers protecting the occupied area only, or 8-5.3.3, whichever requires the greater size of pipe.

A-8-5.3.3.1 For example, a 3-in. (76-mm) steel pipe, which is permitted to supply 40 sprinklers in an ordinary hazard area, can supply a total of 60 sprinklers where not more than 40 sprinklers protect the occupied space below the ceiling.

The probability of a fire involving the areas both above and below a ceiling in an ordinary hazard occupancy and opening in excess of 60 sprinklers is extremely remote. When sizing piping or tubing for systems with more than 60 sprinklers with steel pipe and more than 65 sprinklers with copper tubing, the pipe or tubing diameter is increased to reduce friction loss.

8-5.4* Extra hazard occupancies shall be hydraulically calculated.

Exception: For existing systems, see A-8-5.4.

A-8-5.4 The piping schedule shown in Table A-8-5.4 is reprinted only as a guide for existing systems. New systems for extra hazard occupancies should be hydraulically calculated as required in 8-5.4.

A set of pipe schedule design tables and flow values were provided in NFPA 13 until 1991. Since then, concern about pipe schedule systems intended to protect facilities with fire hazards that are likely to operate a large number of sprinklers has placed a restriction on the pipe schedule method for extra hazard occupancies. Because many existing pipe schedule systems are currently in place, the selection table for extra hazard occupancies has been retained in Appendix A and appears as Table A-8-5.4.

Table A-8-5.4 Extra Hazard Pipe Schedule

Steel		Copper	
1 in.	1 sprinkler	1 in.	1 sprinkler
1¼ in.	2 sprinklers	1¼ in.	2 sprinklers
1½ in.	5 sprinklers	1½ in.	5 sprinklers
2 in.	8 sprinklers	2 in.	8 sprinklers
2½ in.	15 sprinklers	2½ in.	20 sprinklers
3 in.	27 sprinklers	3 in.	30 sprinklers
3½ in.	40 sprinklers	3½ in.	45 sprinklers
4 in.	55 sprinklers	4 in.	65 sprinklers
5 in.	90 sprinklers	5 in.	100 sprinklers
6 in.	150 sprinklers	6 in.	170 sprinklers

For SI units, 1 in. = 25.4 mm.

8-6 Deluge Systems

Open sprinkler and deluge systems shall be hydraulically calculated according to applicable standards.

Pipe schedule systems are designed with a safety margin that presumes that only some of the sprinklers will operate in a fire. In open and deluge systems, water discharges from all sprinklers, including those outside the fire area. Therefore, the system must be specifically designed to provide the density of coverage required to suppress or control the fire over the entire protected area.

8-7* Exposure Systems

Exposure sprinklers shall be hydraulically calculated using Table 8-7 and a relative classification of exposures guide number.

A-8-7 In the design of an exposure protection system, the flow rate from window and cornice sprinklers is shown in Table 8-7. The flow rates are based on the guide numbers selected from Table 2-3 of NFPA 80A, *Recommended Practice for Protection of Buildings from Exterior Fire Exposures.*

Section A of the table is for window sprinklers. The orifice size is selected according to the level on which the sprinkler is located. Section B of the table is for cornice sprinklers.

Recognizing that sprinkler systems for exposure protection require a unique design approach, a calculation procedure is included in Table 8-7 to show the technique to be followed. An analysis of the level of exposure protection needed for an adjacent structure must be completed. This study includes an evaluation of the building construction materials used and the area and height of the exposing structure. The resultant

Table 8-7 Exposure Protection

Section A—Window Sprinklers

Guide Number	Level of Window Sprinkler	Window Sprinkler Orifice Size	Discharge Coefficient (K-factor)	Flow Rate (Q) (gpm)	Application Rate Over 25 ft of Window Area (gpm/ft²)
1.50 or less	Top 2 levels	⅜ in. (9.5 mm)	2.8	7.4	0.30
	Next lower 2 levels	5⁄16 in. (7.9 mm)	1.9	5.0	0.20
	Next lower 2 levels	¼ in. (6.4 mm)	1.4	3.7	0.15
1.51–2.20	Top 2 levels	½ in. (12.7 mm)	5.6	14.8	0.59
	Next lower 2 levels	7⁄16 in. (11.1 mm)	4.2	11.1	0.44
	Next lower 2 levels	⅜ in. (9.5 mm)	2.8	7.4	0.30
2.21–13.15	Top 2 levels	⅝ in. (15.9 mm)	11.2	29.6	1.18
	Next lower 2 levels	17⁄32 in. (13.5 mm)	8.0	21.2	0.85
	Next lower 2 levels	½ in. (12.7 mm)	5.6	14.8	0.59

Section B—Cornice Sprinklers

Guide Number	Cornice Sprinkler Orifice Size	Application Rate per Lineal Foot (gpm)
1.50 or less	⅜ in. (9.5 mm)	0.75
1.51–2.20	½ in. (12.7 mm)	1.50
2.21–13.15	⅝ in. (15.9 mm)	3.00

For SI units, 1 in. = 25.4 mm; 1 gpm = 3.785 L/min; 1 gpm/ft² = 40.76 mm/min.

guide number is based on the values derived from NFPA 80A, *Recommended Practice for Protection of Buildings from Exterior Fire Exposures.*

8-8 In-Rack Sprinklers

8-8.1 Pipes to in-rack sprinklers shall be sized by hydraulic calculations.

8-8.2 Water demand of sprinklers installed in racks shall be added to ceiling sprinkler water demand over the same protected area at the point of connection. The demand shall be balanced to the higher pressure.

The flow and pressure demand associated with in-rack sprinklers must be hydraulically balanced to the higher pressure in the same manner that is required for branch lines on opposite sides of a cross main. See commentary to 8-4.4.

References Cited in Commentary

National Fire Protection Association, 1 Batterymarch Park, P.O. Box 9101, Quincy, MA 02269-9101.

NFPA 13D, *Standard for the Installation of Sprinkler Systems in One- and Two-Family Dwellings and Manufactured Homes*, 1999 edition.

NFPA 14, *Standard for the Installation of Standpipe and Hose Systems*, 1996 edition.

NFPA 15, *Standard for Water Spray Fixed Systems for Fire Protection*, 1996 edition.

NFPA 25, *Standard for the Inspection, Testing, and Maintenance of Water-Based Fire Protection Systems*, 1998 edition.

NFPA 30, *Flammable and Combustible Liquids Code*, 1996 edition.

NFPA 80A, *Recommended Practice for Protection of Buildings from Exterior Fire Exposures*, 1996 edition.

NFPA 170, *Standard for Fire Safety Symbols*, 1999 edition.

National Fire Sprinkler Association, Robin Hill Corporate Park, P.O. Box 1000, Patterson, NY 12563.

"Detection, Treatment and Prevention of Microbiologically Influenced Corrosion in Water-Based Fire Protection Systems," June 1998.

CHAPTER 9

Water Supplies

The effectiveness of even the most conservative sprinkler system design is dependent on its water supply. Unless the sprinkler system discharge criteria can be satisfied by the water supply in terms of pressure, flow, and duration, the system will not perform as intended. Sprinkler system design considerations such as sprinkler spacing, hazard classification, design density, and pipe sizes will have little or no bearing without sufficient water.

In many cases, efforts are made to design a sprinkler system so that an existing water supply can adequately supply the system's demand—that is, larger pipes or sprinklers with larger orifices are selected so that the system demand will not exceed the water supply curve of the city water main. If the water supply, which includes a pressure, flow, and capacity (duration times flow) component, is deficient, supplemental measures in increasing the water supply will be needed for proper system operation. In some situations, a fire pump can be needed to compensate for a lack of pressure and flow even if a sufficient volume of water is available. For other cases, such as a large industrial complex, the installation of a gravity tank could be needed to provide the necessary water supply capacity. Chapter 9 identifies water supply sources that are acceptable for sprinkler systems.

9-1 General

9-1.1 Number of Supplies.

Every automatic sprinkler system shall have at least one automatic water supply.

The requirement for each sprinkler system to have at least one automatic water supply does not mean that each sprinkler system (system riser) in a building must have its own dedicated water supply. NFPA 13 assumes that only one fire will occur in a building at any given time. For example, if a building's size requires it to have three separate sprinkler systems, each system is not required to be connected to its own water supply. Nor is it necessary for a single water supply source to be sized to account for the simultaneous operation of all three systems. Instead, the water supply must be capable of meeting the demands of the hydraulically most demanding areas of the three systems.

The term *automatic water supply* means that the activation of the water supply does not depend on human intervention. When a sprinkler activates due to heat from a fire, the waterflow out of the sprinkler causes a pressure drop in the sprinkler system. This drop in pressure initiates waterflow through the system to the open sprinkler. An automatic sprinkler system cannot be arranged in such a manner that the operation of a sprinkler sounds an alarm that notifies someone to either turn on a fire pump or open a control valve to the water supply.

9-1.2 Capacity.

Water supplies shall be capable of providing the required flow and pressure for the required duration as specified in Chapter 7.

As with all components critical to the effective operation of the sprinkler system, water supply sources must be reliable. The sources must be capable of meeting the system demand at all times. If the water supply is taken from a municipal water works system, potential future degradation of water supply conditions, as well as seasonal and daily fluctuations, must be considered. Additionally, waterflow tests should be conducted on a routine basis to monitor the condition of the water supply. Over time, water supply systems are likely to be degraded by pipe corrosion, scale buildup, and inadvertently closed valves. The impact of future users of the water supply must also be evaluated. Additionally, if water supplies employ fire pumps or water tanks, certain requirements are placed on this equipment to maintain their reliability. See NFPA 20, *Standard for the Installation of Stationary Pumps for Fire Protection,* and NFPA 22, *Standard for Water Tanks for Private Fire Protection,* for additional information. NFPA 25, *Standard for the Inspection, Testing, and Maintenance of Water-Based Fire Protection Systems,* further requires that specific inspection, testing, and maintenance activities be conducted to ensure the integrity of the water supply.

Where a waterworks system serves as the sprinkler system's supply, 7-2.3.1.3(g) requires that the supply be of sufficient capacity so that it meets the sprinkler system's discharge criteria, any inside hose demand, and a waterflow allowance for outside hose. The outside hose allowance is to be added at either the connection to the water main or a yard hydrant, whichever is closer.

Regardless of its arrangement, the underground supply must have sufficient capacity to satisfy the calculated system demand based on the selected design approach from Chapter 7. In all cases, this supply must require no human intervention to operate and must be capable of supplying the system for the necessary duration.

9-1.3 Size of Fire Mains.

No pipe smaller than 6 in. (152.4 mm) in diameter shall be installed as a private service main.

Exception: For mains that do not supply hydrants, sizes smaller than 6 in. (152.4 mm) shall be permitted to be used subject to the following restrictions:

(a) The main supplies only automatic sprinkler systems, open sprinkler systems, water spray fixed systems, foam systems, or Class II standpipe systems.

(b) Hydraulic calculations show that the main will supply the total demand at the appropriate pressure. Systems that are not hydraulically calculated shall have a main at least as large as the riser.

Private fire service mains must be at least 6 in. in diameter where they supply hydrants. In some cases, the use of a minimum 8-in. (203-mm) pipe for dead-end mains is advised. Where the fire service mains do not supply hydrants, they are permitted to be sized based on hydraulic calculations or to be as large the riser under certain conditions. Fire mains need sufficient supply to meet the demands of the sprinkler systems, inside hose, outside hose, and other anticipated water demands.

9-1.4 Underground Supply Pipe.

For pipe schedule systems, the underground supply pipe shall be at least as large as the system riser.

The minimum diameter of the underground pipe connecting the water supply with a pipe schedule sprinkler system, regardless of the type of water supply used, is specified in 9-1.4.

9-1.5 Water Supply Treatment.

In areas with water supplies known to have contributed to microbiologically influenced corrosion (MIC) of sprinkler system piping, water supplies shall be tested and appropriately treated prior to filling or testing of metallic piping systems.

Subsection 9-1.5 was added to the 1999 edition. Microbiologically influenced corrosion (MIC) is caused in part by microorganisms in the water supply that react with system piping. The overall effect can result in the formation of deposits and pitting of system piping in a relatively short period of time.

Over the past several years, additional data on the effects of MIC have become available. This subsection recognizes that MIC can cause severe problems with metallic pipe in areas with specific water supply and system factors. These provisions were introduced to guard against the effects of MIC on sprinkler system performance when the sprinkler system is installed. Because the effects of MIC are often not a one-time event, follow-up evaluation and treatment of the water supply is likely to be necessary once the system is in service. Additional information on MIC can be found in the NFSA report "Detection, Treatment and Prevention of Microbiologically Influenced Corrosion in Water-Based Fire Protection Systems."

9-1.6 Arrangement.

9-1.6.1 Connection Between Underground and Aboveground Piping. The connection between the system piping and underground piping shall be made with a suitable transition piece and shall be properly strapped or fastened by approved devices. The transition piece shall be protected against possible damage from corrosive agents, solvent attack, or mechanical damage.

The provisions of 9-1.6.1 imply that piping serving as the transition piece between the underground supply piping and the aboveground system piping at the base of the riser be of a metallic material. Many of the types of pipe acceptable for use as underground fire service mains are limited to conditions involving below-grade applications only. The use of certain pipe materials, such as plastic and concrete, for aboveground fire service mains is restricted by their listings. As indicated by 9-1.6.1, these transition pieces must be able to withstand damage from likely corrosive agents, solvent attack, and mechanical damage. As a result, the need to use steel, iron, or copper pipe materials at these transition areas is evident.

9-1.6.2* Connection Passing Through or Under Foundation Walls. When system piping pierces a foundation wall below grade or is located under the foundation wall, clearance shall be provided to prevent breakage of the piping due to building settlement.

A-9-1.6.2 Where the system riser is close to an outside wall, underground fittings of proper length should be used in order to avoid pipe joints located in or under the wall. Where the connection passes through the foundation wall below grade, a 1-in. to 3-in. (25-mm to 76-mm) clearance should be provided around the pipe and the clear space filled with asphalt mastic or similar flexible waterproofing material.

Without clearance around the wall, normal building movement and settlement can cause the failure of pipe that is rigidly held in place by foundation walls.

9-1.7* Meters.

Where meters are required by other authorities, they shall be listed.

A-9-1.7 Where water meters are in the supply lines to a sprinkler system, they should be rated to deliver the proper system demand. The amount of water supplied through a water meter varies with its size and type and might not provide the required demand, regardless of the water supply available.

Meters are not necessary for the proper performance of a sprinkler system. However, meters could be required by the local water authority. Because of the pressure loss they introduce, the use of meters on automatic sprinkler systems should be discouraged. Meters also introduce an additional expense and maintenance concern.

Figure B-1 of NFPA 13 provides guidance on possible arrangements for metering water that is used for purposes other than fire protection. Where a concern about possible leakage or loss from the fire protection piping exists, a detector check valve can be substituted for the check valve indicated.

9-1.8* Connection from Waterworks System.

Where connections are made from public waterworks systems, it might be necessary to guard against possible contamination of the public supply. The requirements of the public health authority having jurisdiction shall be determined and followed. Where equipment is installed to guard against possible contamination of the public water system, such equipment and devices shall be listed for fire protection service.

A-9-1.8 Where connections are made from public waterworks systems, such systems should be guarded against possible contamination as follows *(see AWWA M14, Backflow Prevention and Cross Connection Control).*

(a) For private fire service mains with direct connections from public waterworks mains only or with booster pumps installed in the connections from the street mains, no tanks or reservoirs, no physical connection from other water supplies, no antifreeze or other additives of any kind, and with all drains discharging to atmosphere, dry well, or other safe outlets, no backflow protection is recommended at the service connection.

(b) For private fire service mains with direct connection from the public water supply main plus one or more of the following: elevated storage tanks or fire pumps taking suction from aboveground covered reservoirs or tanks (all storage facilities are filled or connected to public water only and the water in the tanks is to be maintained in a potable condition), an approved double check valve assembly is recommended.

(c) For private fire service mains directly supplied from public mains with an auxiliary water supply such as a pond or river on or available to the premises and dedicated to fire department use; or for systems supplied from public mains and interconnected with auxiliary supplies, such as pumps taking suction from reservoirs exposed to contamination or rivers and ponds; driven wells, mills, or other industrial water systems; or for systems or portions of systems where antifreeze or other solutions are used, an approved reduced pressure zone–type backflow preventer is recommended.

Backflow prevention devices are not a requirement of NFPA 13 and serve no benefit to the sprinkler system. The devices are typically required by public health authorities for non-fire protection purposes. These devices typically impact the system's hydraulic characteristics and are a potential point of system impairment if performance is not properly monitored. Where installed, the pressure loss through a backflow prevention device must be accounted for and the valves on the device must be supervised as any other system control valve. Where a backflow prevention device is installed retroactively, the pressure loss through the additional pipe and fittings used in conjunction with the preventor must also be taken into account. See 5-15.4.6 for more details. Additionally, the backflow prevention device must be properly inspected, tested, and maintained to ensure proper operation.

9-2 Types

This section describes those sources of water that are acceptable for use as the permanent, automatic supply for the sprinkler system. Although not specifically mentioned, suction tanks, embankment tanks, wells, and ponds could also be considered as water supply sources for automatic sprinkler systems. Along with the other water supply sources identified, the sources must be reliable and have enough capacity to meet the sprinkler system demand at all times.

When water supplies other than circulating public waterworks systems are used, the proper installation of system components such as piping, pumps, or tanks should be verified. Even though private systems are designed to provide adequate water capacity, flow, and pressure, the reliability of these systems needs to be monitored through the implementation of a periodic inspection, testing, and maintenance program and through the proper supervision of certain system components. Public systems are tested and supervised continually by the daily demands placed on them. Private systems can remain idle for longer periods of time.

9-2.1* Connections to Water Works Systems.

A connection to a reliable water works system shall be an acceptable water supply source. The volume and pressure of a public water supply shall be determined from waterflow test data. An adjustment to the waterflow test data to account for daily and seasonal fluctuations, possible interruption by flood or ice conditions, large simultaneous industrial use, future demand on the water supply system, or any other condition that could affect the water supply shall be made as appropriate.

A-9-2.1 Care should be taken in making water tests to be used in designing or evaluating the capability of sprinkler systems. The water supply tested should be representative of the supply that might be available at the time of a fire. For example, testing of public water supplies should be done at times of normal demand on the system. Public water supplies are likely to fluctuate widely from season to season and even within a 24-hour period. Allowance should be made for seasonal or daily fluctuations, for drought conditions, for possibility of interruption by flood, or for ice conditions in winter. Testing of water supplies also normally used for industrial use should be done while water is being drawn for industrial use. The range of industrial-use demand should be taken into account. In special situations where the domestic water demand could significantly reduce the sprinkler water supply, an increase in the size of the pipe supplying both the domestic and sprinkler water can be justified.

Future changes in water supplies should be considered. For example, a large, established, urban supply is not likely to change greatly within a few years. However, the supply in a growing suburban industrial park might deteriorate quite rapidly as greater numbers of plants draw more water.

Dead-end mains should be avoided, if possible, by arranging for mains supplied from both directions. When private fire service mains are connected to dead-end public mains, each situation should be examined to determine if it is practical to request the water utility to loop the mains in order to obtain a more reliable supply.

Testing of Water Supply. To determine the value of public water as a supply for automatic sprinkler systems, it is generally necessary to make a flow test to determine how much water can be discharged at a residual pressure at a rate sufficient to give the required residual pressure under the roof (with the volume flow hydraulically translated to the base of the riser)—that is, a pressure head represented by the height of the building plus the required residual pressure.

The proper method of conducting this test is to use two hydrants in the vicinity of the property. The static pressure should be measured on the hydrant in front of or nearest to the property and the water allowed to flow from the hydrant next nearest the property, preferably the one farthest from the source of supply if the main is fed only one way. The residual pressure will be that indicated at the hydrant where water is not flowing.

Referring to Figure A-9-2.1, the method of conducting the flow tests is as follows:

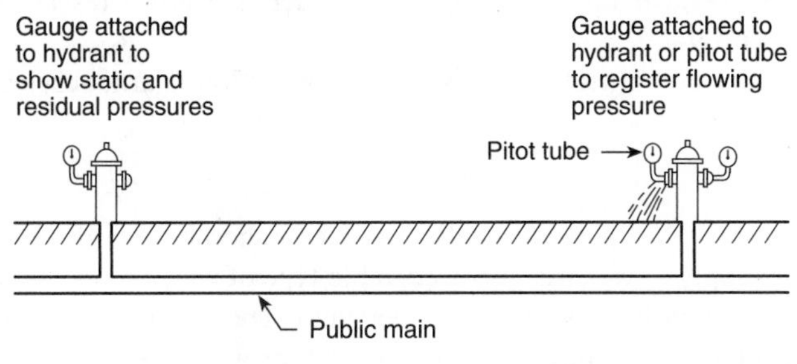

Figure A-9-2.1 *Method of conducting flow tests.*

(1) Attach the gauge to the hydrant *(A)* and obtain static pressure.
(2) Either attach a second gauge to the hydrant *(B)* or use the pitot tube at the outlet. Have hydrant *(B)* opened wide and read pressure at both hydrants.
(3) Use the pressure at *(B)* to compute the gallons flowing and read the gauge on *(A)* to determine the residual pressure or that which will be available on the top line of sprinklers in the property.

Water pressure in psi for a given height in feet equals height multiplied by 0.434.

In making flow tests, whether from hydrants or from nozzles attached to hose, always measure the size of the orifice. While hydrant outlets are usually 2½ in. (64 mm), they are sometimes smaller and occasionally larger. The Underwriters Laboratories play pipe is 1⅛ in. (29 mm) and 1¾ in. (44 mm) with the tip removed, but occasionally nozzles will be 1 in. (25.4 mm) or 1¼ in. (33 mm), and with the tip removed the opening can be only 1½ in. (38 mm).

The pitot tube should be held approximately one-half the diameter of the hydrant or nozzle opening away from the opening. It should be held in the center of the stream, except that in using hydrant outlets the stream should be explored to ascertain the average pressure.

For further information on water supply testing, see NFPA 291, *Recommended Practice for Fire Flow Testing and Marking of Hydrants.*

Connection to a public or private water works system is often the first choice when considering water supply options. Such systems are typically very reliable, accessible, and capable of accommodating the demand for a wide range of sprinkler systems.

If a waterworks system is contemplated for the supply source, the associated water supply test data must be current. Peak loads on the system, as well as any seasonal fluctuations and likely interruptions of waterflow, must be contemplated. When considering a waterworks system, the least amount of water in terms of pressure and flow available should be used as the water supply, because the sprinkler system demand must be met 24 hours a day, 7 days a week, 365 days a year.

9-2.2* Pumps.

A single automatically controlled fire pump installed in accordance with NFPA 20, *Standard for the Installation of Centrifugal Fire Pumps,* shall be an acceptable water supply source.

A-9-2.2 An automatically controlled vertical turbine pump taking suction from a reservoir, pond, lake, river, or well complies with 9-2.2.

See sections dealing with sprinkler equipment supervisory and waterflow alarm services in NFPA 72, *National Fire Alarm Code.*

Listed horizontal and vertical turbine fire pumps installed in accordance with NFPA 20 and connected to a sufficient supply of water are an acceptable type of water supply. Fire pumps cannot create water, which is important to understand. The pumps serve to increase the pressure and resultant flow from an existing supply of water such as a waterworks system. If a sufficient capacity of water does exist, means in addition to the fire pump, such as a tank, are necessary.

A vertical turbine pump is a type of centrifugal fire pump that operates in a vertical position. In that position, the first impeller, or bowl, of the pump is always submerged. Therefore, the first pump impeller has a positive suction pressure on it. This type of pump, with its first impeller submerged, does not actually take suction but, rather, pumps from a water source located below the pump. NFPA's *Fire Pump Handbook* provides additional information on fire pumps and their proper application.

To provide sufficient reliability of a single fire pump that can serve as the sole supply for a sprinkler system, supervisory service of pump conditions to a constantly attended location is required. Conditions to be supervised should include pump-running indication, pump power supply availability, and other conditions that could render the pump inoperative. These requirements, as well as additional details, are contained in NFPA 20.

9-2.3 Pressure Tanks.

Unlike gravity tanks or suction tanks, pressure tanks contain both water and air under pressure. A sufficient capacity of air must be available to discharge the water from the tank at the necessary rate. See 9-2.3.3 for details on water level and air pressure requirements.

9-2.3.1 Acceptability.

9-2.3.1.1 A pressure tank installed in accordance with NFPA 22, *Standard for Water Tanks for Private Fire Protection,* shall be an acceptable water supply source.

9-2.3.1.2 Pressure tanks shall be provided with an approved means for automatically maintaining the required air pressure. Where a pressure tank is the sole water supply, there shall also be provided an approved trouble alarm to indicate low air pressure and low water level with the alarm supplied from an electrical branch circuit independent of the air compressor.

As with fire pumps, certain features and conditions about the pressure tank should be supervised to increase the reliability of this sole source of water supply. Trouble alarms that indicate low air pressure and low water levels should be received at a constantly attended location. NFPA 72, *National Fire Alarm Code®*, provides additional details on acceptable supervisory methods.

9-2.3.1.3 Pressure tanks shall not be used to supply other than sprinklers and hand hose attached to sprinkler piping.

9-2.3.2 Capacity. In addition to the requirements of 9-1.2, the water capacity of a pressure tank shall include the extra capacity needed to fill dry pipe or preaction systems where installed. The total volume shall be based on the water capacity plus the air capacity required by 9-2.3.3.

Unlike wet pipe systems, dry pipe and preaction systems are filled with air immediately before a fire. (See Chapter 4 for requirements on dry pipe and preaction systems.) Therefore, where a pressure tank serves as the water supply, it must be sized so that it has sufficient capacity to fill the system piping in addition to the capacity needed to meet the systems discharge requirements.

9-2.3.3* Water Level and Air Pressure. Pressure tanks shall be kept with a sufficient supply of water to meet the demand of the fire protection system as calculated in Chapter 8 for the duration required by Chapter 7. The pressure shall be sufficient to push all of the water out of the tank while maintaining the necessary residual pressure (required by Chapter 8) at the top of the system.

A-9-2.3.3 For pipe schedule systems, the air pressure to be carried and the proper proportion of air in the tank can be determined from the following formulas where:

P = air pressure carried in pressure tank

A = proportion of air in tank

H = height of highest sprinkler above tank bottom

When the tank is placed above the highest sprinkler, use the following formula:

$$P = \frac{30}{A} - 15$$

If $A = \frac{1}{3}$, then $P = 90 - 15 = 75$ lb psi
If $A = \frac{1}{2}$, then $P = 60 - 15 = 45$ lb psi
If $A = \frac{2}{3}$, then $P = 45 - 15 = 30$ lb psi
When the tank is below the level of the highest sprinkler, use the following formula:

$$P = \frac{30}{A} - 15 + \frac{1.434H}{A}$$

If $A = \frac{1}{3}$, then $P = 75 + 1.30H$
If $A = \frac{1}{2}$, then $P = 45 + 0.87H$
If $A = \frac{2}{3}$, then $P = 30 + 0.65H$

The respective air pressures above are calculated to ensure that the last water will leave the tank at a pressure of 15 psi (1 bar) when the base of the tank is on a level with the highest sprinkler or at such additional pressure as is equivalent to a head corresponding to the distance between the base of the tank and the highest sprinkler when the latter is above the tank.

For hydraulically calculated systems, the following formula should be used to determine the tank pressure and ratio of air to water:

$$P_i \quad \frac{P_f \quad 15}{A} \quad 15$$

where:

P_i tank pressure

P_f pressure required from hydraulic calculations

A proportion of air

Example: Hydraulic calculations indicate 75 psi (5.2 bar) is required to supply the system. What tank pressure will be required?

$$P_i \quad \frac{75 \quad 15}{0.5} \quad 15$$

$$P_i \quad 180 \quad 15 \quad 165 \text{ psi}$$

For SI units, 1 ft 0.3048 m; 1 psi 0.0689 bar.

In this case, the tank would be filled with 50 percent air and 50 percent water, and the tank pressure would be 165 psi (11.4 bar). If the pressure is too high, the amount of air carried in the tank will have to be increased.

Pressure tanks should be located above the top level of sprinklers but can be located in the basement or elsewhere.

Table 7-2.2.1 requires that certain residual pressures be provided at the elevation of the highest sprinkler for pipe schedule systems. Table 7-2.2.1 also requires a minimum flow at the base of the riser. These criteria need to be considered when contemplating a pressure tank for pipe schedule systems.

9-2.4 Gravity Tanks.

An elevated tank installed in accordance with NFPA 22, *Standard for Water Tanks for Private Fire Protection,* shall be an acceptable water supply source.

Gravity tanks are not as common as they were years ago. The availability of a wider range of fire pump capacities in terms of flow–pressure combinations has made pumps a more desirable option in many cases.

Where gravity tanks are used, they must be of a capacity that satisfies the system discharge requirements of Chapter 7. If gravity tanks are also used to supply domestic or industrial needs, the arrangement must be such that the entire volume of water needed for fire protection is available during and after other demands have been placed on the tank.

9-2.5 Penstocks or Flumes, Rivers, or Lakes.

Water supply connections from penstocks, flumes, rivers, lakes, or reservoirs shall be arranged to avoid mud and sediment and shall be provided with approved double removable screens or approved strainers installed in an approved manner.

When contemplating the use of naturally occurring water supply sources, such as lakes, rivers, wells, and ponds, their reliability and ability to meet the system demand at all times must be verified. For example, seasonal fluctuations must be considered. Because of the degree of uncertainty associated with some naturally occurring water supplies, local regulations and insurance company guidelines can limit the use of these water supply sources.

References Cited in Commentary

National Fire Protection Association, 1 Batterymarch Park, P.O. Box 9101, Quincy, MA 02269-9101.

NFPA 20, *Standard for the Installation of Stationary Pumps for Fire Protection,* 1999 edition.
NFPA 22, *Standard for Water Tanks for Private Fire Protection,* 1998 edition.
NFPA 25, *Standard for the Inspection, Testing, and Maintenance of Water-Based Fire Protection Systems,* 1998 edition.
NFPA 72, *National Fire Alarm Code®,* 1999 edition.
Fire Pump Handbook, 1998 edition.

National Fire Sprinkler Association, Robin Hill Corporate Park, P.O. Box 1000, Patterson, NY 12563.

"Detection, Treatment and Prevention of Microbiologically Influenced Corrosion in Water-Based Fire Protection Systems," June 1998.

CHAPTER 10

Systems Acceptance

The approval process for system acceptance is important in verifying the design and installation of the system and should receive top priority. These final steps in the design and installation of a sprinkler system are to confirm that the basic requirements of NFPA 13 are satisfied, that the work was completed in an acceptable manner, and that the customer receives a system that performs as intended. Failure to comply with system acceptance requirements will likely delay the issuance of the building's certificate of occupancy. The remedy of any deficiency in the sprinkler system design or installation discovered after the building is occupied is likely to be extremely costly.

10-1 Approval of Sprinkler Systems and Private Fire Service Mains

The installing contractor shall do the following:

(1) Notify the authority having jurisdiction and owner's representative of the time and date testing will be performed
(2) Perform all required acceptance tests *(see Section 10-2)*
(3) Complete and sign the appropriate contractor's material and test certificate(s) *[see Figures 10-1(a) and 10-1(b)]*

Occasionally the authority having jurisdiction wants to be present during acceptance tests. The authority having jurisdiction is responsible for informing the building owner either directly or through the installing contractor or design engineer that he or she wants to be present. The building owner is obligated to comply with such a request and to give the authority having jurisdiction advance notice. Failure to do so can result in the need to repeat the tests.

Material and test certificates for aboveground and underground installations are shown in Figures 10-1(a) and 10-1(b). These certificates help in verifying that the materials used and the tests performed are in accordance with the requirements of

NFPA 13. The certificates also provide a record of the test results that can be used for comparison with any tests conducted in the future. It is important to complete all applicable portions of these certificates for each installed system.

10-2 Acceptance Requirements

10-2.1* Flushing of Piping.

Fire service mains (from the water supply to the system riser) and lead-in connections to system risers shall be completely flushed before connection is made to sprinkler piping. The flushing operation shall be continued for a sufficient time to ensure thorough cleaning. The minimum rate of flow shall be not less than one of the following:

(1) The hydraulically calculated water demand rate of the system including any hose requirements
(2) That flow necessary to provide a velocity of 10 ft/sec (3.1 m/sec) [see Table 10-2.1(2)]
(3) The maximum flow rate available to the system under fire conditions

A-10-2.1 Underground mains and lead-in connections to system risers should be flushed through hydrants at dead ends of the system or through accessible aboveground flushing outlets allowing the water to run until clear. Figure A-10-2.1 shows acceptable examples of flushing the system. If water is supplied from more than one source or from a looped system, divisional valves should be closed to produce a high-velocity flow through each single line. The flows specified in Table 10-2.1(2) will produce a velocity of at least 10 ft/sec (3 m/sec), which is necessary for cleaning the pipe and for lifting foreign material to an aboveground flushing outlet.

A quantity of water producing velocities of 10 ft/sec (3.1 m/sec) may not be available under all circumstances. If the velocity isn't available, then the system's anticipated actual demand should be simulated to determine if any foreign matter is obstructing the underground pipe network.

Stones, gravel, blocks of wood, bottles, work tools, work clothes, and other objects have been found in piping when flushing procedures were performed. Also, objects in underground piping that are quite remote from the sprinkler installation and that would otherwise remain stationary can sometimes be carried into sprinkler system piping when sprinkler systems operate. Sprinkler systems can draw greater flows than most domestic or process uses. Fire department pumpers taking suction from hydrants for pumping into sprinkler systems and normal fire-fighting operations further increase flow rates and velocities and can further dislodge other materials in the piping network, forcing them into sprinkler system piping.

Because of the inherent nature of sprinkler system design in which pipe sizes usually decrease beginning at the point of connection to the underground piping, objects that move from the underground piping into the sprinkler system can become lodged at a point in the system where they may obstruct the passage of water. Exhibit 10.1

Contractor's Material and Test Certificate for Aboveground Piping

PROCEDURE
Upon completion of work, inspection and tests shall be made by the contractor's representative and witnessed by an owner's representative. All defects shall be corrected and system left in service before contractor's personnel finally leave the job.

A certificate shall be filled out and signed by both representatives. Copies shall be prepared for approving authorities, owners, and contractor. It is understood the owner's representative's signature in no way prejudices any claim against contractor for faulty material, poor workmanship, or failure to comply with approving authority's requirements or local ordinances.

Property name	Date
Property address	

Plans	Accepted by approving authorities (names)		
	Address		
	Installation conforms to accepted plans	☐ Yes	☐ No
	Equipment used is approved If no, explain deviations	☐ Yes	☐ No
Instructions	Has person in charge of fire equipment been instructed as to location of control valves and care and maintenance of this new equipment? If no, explain?	☐ Yes	☐ No
	Have copies of the following been left on the premises?	☐ Yes	☐ No
	1. System components instructions	☐ Yes	☐ No
	2. Care and maintenance instructions	☐ Yes	☐ No
	3. NFPA 25	☐ Yes	☐ No
Location of system	Supplies buildings		

Sprinklers	Make	Model	Year of manufacture	Orifice size	Quantity	Temperature rating

Pipe and fittings	
	Type of pipe
	Type of fittings

Alarm valve or flow indicator	Alarm device			Maximum time to operate through test connection	
	Type	Make	Model	Minutes	Seconds

Dry pipe operating test	Dry valve			Q. O. D.		
	Make	Model	Serial no.	Make	Model	Serial no.

	Time to trip through test connection[1]		Water pressure	Air pressure	Trip point air pressure	Time water reached test outlet[1]		Alarm operated properly	
	Minutes	Seconds	psi	psi	psi	Minutes	Seconds	Yes	No
Without Q.O.D.									
With Q.O.D.									

If no, explain

[1] Measured from time inspector's test connection is opened

Figure 10-1(a) Contractor's material and test certificate for aboveground piping. (Continues)

<table>
<tr><td rowspan="6">Deluge and preaction valves</td><td colspan="9">Operation ☐ Pneumatic ☐ Electric ☐ Hydraulics</td></tr>
<tr><td colspan="4">Piping supervised ☐ Yes ☐ No</td><td colspan="5">Detecting media supervised ☐ Yes ☐ No</td></tr>
<tr><td colspan="9">Does valve operate from the manual trip, remote, or both control stations? ☐ Yes ☐ No</td></tr>
<tr><td colspan="5">Is there an accessible facility in each circuit for testing? ☐ Yes ☐ No</td><td colspan="4">If no, explain</td></tr>
<tr><td rowspan="2">Make</td><td rowspan="2">Model</td><td colspan="2">Does each circuit operate supervision loss alarm?</td><td colspan="2">Does each circuit operate valve release?</td><td colspan="2">Maximum time to operate release</td><td></td></tr>
<tr><td>Yes</td><td>No</td><td>Yes</td><td>No</td><td>Minutes</td><td>Seconds</td><td></td></tr>
<tr><td rowspan="3">Pressure reducing valve test</td><td rowspan="2">Location and floor</td><td rowspan="2">Make and model</td><td rowspan="2">Setting</td><td colspan="2">Static pressure</td><td colspan="2">Residual pressure (flowing)</td><td>Flow rate</td><td></td></tr>
<tr><td>Inlet (psi)</td><td>Outlet (psi)</td><td>Inlet (psi)</td><td>Outlet (psi)</td><td>Flow (gpm)</td><td></td></tr>
<tr><td></td><td></td><td></td><td></td><td></td><td></td><td></td><td></td><td></td></tr>
<tr><td>Test description</td><td colspan="9">Hydrostatic: Hydrostatic tests shall be made at not less than 200 psi (13.6 bar) for 2 hours or 50 psi (3.4 bar) above static pressure in excess of 150 psi (10.2 bar) for 2 hours. Differential dry-pipe valve clappers shall be left open during the test to prevent damage. All aboveground piping leakage shall be stopped.

Pneumatic: Establish 40 psi (2.7 bar) air pressure and measure drop, which shall not exceed 1½ psi (0.1 bar) in 24 hours. Test pressure tanks at normal water level and air pressure and measure air pressure drop, which shall not exceed 1½ psi (0.1 bar) in 24 hours.</td></tr>
<tr><td rowspan="5">Tests</td><td colspan="6">All piping hydrostatically tested at ____ psi (____ bar) for ____ hours
Dry piping pneumatically tested ☐ Yes ☐ No
Equipment operates properly ☐ Yes ☐ No</td><td colspan="3">If no, state reason</td></tr>
<tr><td colspan="9">Do you certify as the sprinkler contractor that additives and corrosive chemicals, sodium silicate or derivatives of sodium silicate, brine, or other corrosive chemicals were not used for testing systems or stopping leaks?
☐ Yes ☐ No</td></tr>
<tr><td>Drain test</td><td colspan="4">Reading of gauge located near water supply test connection: ____ psi (____ bar)</td><td colspan="4">Residual pressure with valve in test connection open wide: ____ psi (____ bar)</td></tr>
<tr><td colspan="6">Underground mains and lead in connections to system risers flushed before connection made to sprinkler piping
Verified by copy of the U Form No. 85B ☐ Yes ☐ No
flushed by installer of underground sprinkler piping ☐ Yes ☐ No</td><td colspan="3">Other Explain</td></tr>
<tr><td colspan="6">If powder-driven fasteners are used in concrete, has representative sample testing be satisfactorily completed? ☐ Yes ☐ No</td><td colspan="3">If no, explain</td></tr>
<tr><td>Blank testing gaskets</td><td colspan="2">Number used</td><td colspan="5">Locations</td><td colspan="2">Number removed</td></tr>
<tr><td rowspan="5">Welding</td><td colspan="9">Welding piping ☐ Yes ☐ No</td></tr>
<tr><td colspan="9">If yes. . .</td></tr>
<tr><td colspan="9">Do you certify as the sprinkler contractor that welding procedures comply with the requirements of at least AWS B2.1? ☐ Yes ☐ No</td></tr>
<tr><td colspan="9">Do you certify that the welding was performed by welders qualified in compliance with the requirements of at least AWS B2.1? ☐ Yes ☐ No</td></tr>
<tr><td colspan="9">Do you certify that the welding was carried out in compliance with a documented quality control procedure to ensure that all discs are retrieved, that openings in piping are smooth, that slag and other welding residue are removed, and that the internal diameters of piping are not penetrated? ☐ Yes ☐ No</td></tr>
<tr><td>Cutouts (discs)</td><td colspan="9">Do you certify that you have a control feature to ensure that all cutouts (discs) are retrieved? ☐ Yes ☐ No</td></tr>
</table>

Figure 10-1(a) *(Continued)*

Hydraulic data nameplate	Nameplate provided ☐ Yes ☐ No		If no, explain
Remarks	Date left in service with all control valves open		
Signatures	Name of sprinkler contractor		
	Tests witnessed by		
	For property owner (signed)	Title	Date
	For sprinkler contractor (signed)	Title	Date
Additional explanations and notes			

Figure 10-1(a) (Continued)

on page 735 illustrates an object that was carried from the underground piping into the sprinkler system. When the dry pipe valve was reset, the object was discovered.

Previous studies by Factory Mutual Research Corporation concluded that the size of particles that will move upward in piped water streams can be determined if the specific gravity of the particle and the velocity of the water stream are known. For example, granite that is 2 in. (51 mm) in diameter will move upward in piping if the water stream velocity is 5.64 ft/sec (1.72 m/sec), which is equivalent to the minimum flows that 10-2.1 previously required. In the 1991 edition, the flow rates were increased to reflect velocities of 10 ft/sec (3.1 m/sec), because this number is more in line with requirements of other accepted standards. The velocity of 10 ft/sec (3.1 m/sec) is also in agreement with the flushing requirements of NFPA 15, *Standard for Water Spray Fixed Systems for Fire Protection.*

In the field, actual flushing is normally done without measuring the flow rate. Flushing is normally accomplished at the maximum flow rate available from the water supply allowed by 10-2.1(3).

10-2.2 Hydrostatic Tests.

10-2.2.1* All piping and attached appurtenances subjected to system working pressure shall be hydrostatically tested at 200 psi (13.8 bar) and shall maintain that pressure without loss for 2 hours. Loss shall be determined by a drop in gauge pressure or visual leakage. The test pressure shall be read from a gauge located at the low elevation point of the system or portion being tested.

Contractor's Material and Test Certificate for Underground Piping

PROCEDURE

Upon completion of work, inspection and tests shall be made by the contractor's representative and witnessed by an owner's representative. All defects shall be corrected and system left in service before contractor's personnel finally leave the job.

A certificate shall be filled out and signed by both representatives. Copies shall be prepared for approving authorities, owners, and contractor. It is understood the owner's representative's signature in no way prejudices any claim against contractor for faulty material, poor workmanship, or failure to comply with approving authority's requirements or local ordinances.

Property name		Date
Property address		
Plans	Accepted by approving authorities (names)	
	Address	
	Installation conforms to accepted plans Equipment used is approved If no, state deviations	☐ Yes ☐ No ☐ Yes ☐ No
Instructions	Has person in charge of fire equipment been instructed as to location of control valves and care and maintenance of this new equipment? If no, explain	☐ Yes ☐ No
	Have copies of appropriate instructions and care and maintenance charts been left on premises? If no, explain	☐ Yes ☐ No
Location	Supplies buildings	
Underground pipes and joints	Pipe types and class	Type joint
	Pipe conforms to ________ standard Fittings conforms to ________ standard If no, explain	☐ Yes ☐ No ☐ Yes ☐ No
	Joints needed anchorage clamped, strapped, or blocked in accordance with ________ standard If no, explain	☐ Yes ☐ No
Test description	Flushing: Flow the required rate until water is clear as indicated by no collection of foreign material in burlap bags at outlets such as hydrants and blow-offs. Flush at flows not less than 390 gpm (1476 L/min) for 4-in. pipe, 880 gpm (3331 L/min) for 6-in. pipe, 1560 gpm (5905 L/min) for 8-in. pipe, 2440 gpm (9235 L/min) for 10-in. pipe, and 3520 gpm (13,323 L/min) for 12-in. pipe. When supply cannot produce stipulated flow rates, obtain maximum available. Hydrostatic: Hydrostatic tests shall be made at not less than 200 psi (13.8 bar) for 2 hours or 50 psi (3.4 bar) above static pressure in excess of 150 psi (10.3 bar) for 2 hours. Leakage: New pipe laid with rubber gasketed joints shall, if the workmanship is satisfactory, have little or no leakage at the joints. The amount of leakage at the joints shall not exceed 2 quarts per hour (1.89 L/hr) per 100 joints irrespective of pipe diameter. The leakage shall be distributed over all joints. If such leakage occurs at a few joints, the installation shall be considered unsatisfactory and necessary repairs made. The amount of allowable leakage specified above can be increased by 1 fluid ounce per inch valve diameter per hr. (30 mL/25 mm/hr) for each metal seated valve isolating the test section. If dry barrel hydrants are tested with the main valve open so the hydrants are under pressure, an additional 5 ounces per minute (150 mL/min) leakage is permitted for each hydrant.	
Flushing tests	New underground piping flushed according to ________ standard by (company) If no, explain	☐ Yes ☐ No
	How flushing flow was obtained ☐ Public water ☐ Tank or reservoir ☐ Fire pump	Through what type opening ☐ Hydrant butt ☐ Open pipe
	Lead-ins flushed according to ________ standard by (company) If no, explain	☐ Yes ☐ No
	How flushing flow was obtained ☐ Public water ☐ Tank or reservoir ☐ Fire pump	Through what type opening ☐ Y connection to flange and spigot ☐ Open pipe

Figure 10-1(b) *Contractor's material and test certificate for underground piping.*

Hydrostatic test	All new underground piping hydrostatically tested at ________ psi for ________ hours		Joints covered ☐ Yes ☐ No
Leakage test	Total amount of leakage measured ________ gallons ________ hours		
	Allowable leakage ________ gallons ________ hours		
Hydrants	Number installed	Type and make	All operate satisfactorily ☐ Yes ☐ No
Control valves	Water control valves left wide open If no, state reason		☐ Yes ☐ No
	Hose threads of fire department connections and hydrants interchangeable with those of fire department answering alarm		☐ Yes ☐ No
Remarks	Date left in service		
Signatures	Name of installing contractor		
	Tests witnessed by		
	For property owner (signed)	Title	Date
	For installing contractor (signed)	Title	Date
Additional explanation and notes			

Figure 10-1(b) *(Continued)*

Table 10-2.1(2) Flow Required to Produce a Velocity of 10 ft/sec (3 m/sec) in Pipes

Pipe Size		Flow Rate	
in.	**mm**	**gpm**	**L/min**
4	102	390	1476
6	152	880	3331
8	203	1560	5905
10	254	2440	9235
12	305	3520	13323

Exception No. 1: Portions of systems normally subjected to system working pressures in excess of 150 psi (10.4 bar) shall be tested as described in 10-2.2.1 at a pressure of 50 psi (3.5 bar) in excess of system working pressure.

Exception No. 2: Where cold weather will not permit testing with water, an interim air test shall be permitted to be conducted as described in 10-2.3.

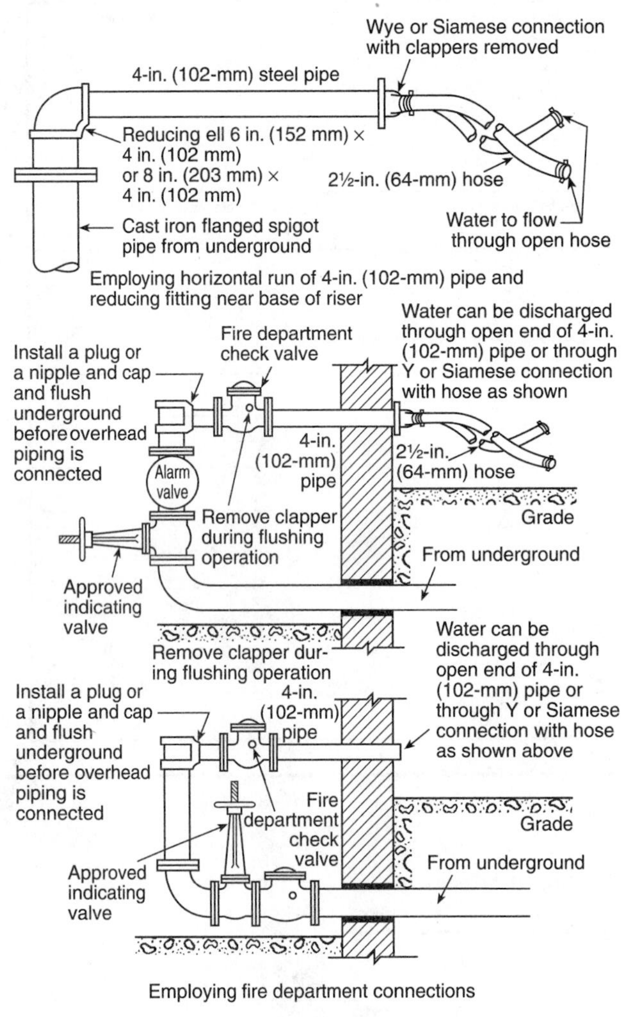

Figure A-10-2.1 *Methods of flushing water supply connections.*

Exception No. 3: Modifications affecting 20 or fewer sprinklers shall not require testing in excess of system working pressure.

Exception No. 4: Where addition or modification is made to an existing system affecting more than 20 sprinklers, the new portion shall be isolated and tested at not less than 200 psi (13.8 bar) for 2 hours.

Exception No. 5: Modifications that cannot be isolated, such as relocated drops, shall not require testing in excess of system working pressure.

Exhibit 10.1 *Object trapped in a dry pipe sprinkler system.*

Exception No. 6: In buried pipe, leakage shall be permitted as follows:

(a) The amount of leakage at the joints shall not exceed 2 qt/hr (1.89 L/hr) per 100 gaskets or joints, irrespective of pipe diameter.*

(b) The amount of allowable leakage specified in item (a) of this exception shall be permitted to be increased by 1 fluid ounce (30 ml) per inch valve diameter per hour for each metal seated valve isolating the test section.*

(c) If dry barrel hydrants are tested with the main valve open so the hydrants are under pressure, an additional 5 oz/min (150 ml/min) of leakage shall be permitted for each hydrant.

(d) The amount of leakage in buried piping shall be measured at the specified test pressure by pumping from a calibrated container.

A-10-2.2.1 A sprinkler system has for its water supply a connection to a public water service main. A 100-psi (6.9-bar) rated pump is installed in the connection. With a maximum normal public water supply of 70 psi (4.8 bar) at the low elevation point of the individual system or portion of the system being tested and a 120-psi (8.3-bar) pump (churn) pressure, the hydrostatic test pressure is 70 psi + 120 psi + 50 psi or 240 psi (16.5 bar).

To reduce the possibility of serious water damage in case of a break, pressure can be maintained by a small pump, the main controlling gate meanwhile being kept shut during the test.

Polybutylene pipe will undergo expansion during initial pressurization. In this case, a reduction in gauge pressure might not necessarily indicate a leak. The pressure reduction should not exceed the manufacturer's specifications and listing criteria.

When systems having rigid thermoplastic piping such as CPVC are pressure tested, the sprinkler system should be filled with water. The air should be bled from the highest and farthest sprinklers. Compressed air or compressed gas should never be used to test systems with rigid thermoplastic pipe.

A recommended test procedure is as follows: The water pressure is to be increased in 50-psi (3.4-bar) increments until the test pressure described in 10-2.2.1 is attained. After each increase in pressure, observations are to be made of the stability of the joints. These observations are to include such items as protrusion or extrusion of the gasket, leakage, or other factors likely to affect the continued use of a pipe in service. During the test, the pressure is not to be increased by the next increment until the joint has become stable. This applies particularly to movement of the gasket. After the pressure has been increased to the required maximum value and held for 1 hour, the pressure is to be decreased to 0 psi while observations are made for leakage. The pressure is again to be slowly increased to the value specified in 10-2.2.1 and held for 1 more hour while observations are made for leakage and the leakage measurement is made.

A-10-2.2.1 Exception No. 6(a). New pipe laid with rubber gasketed joints should, if the workmanship is satisfactory, have no leakage at the joints. Unsatisfactory amounts of leakage usually result from twisted, pinched, or cut gaskets. However, some leakage might result from small amounts of grit or small imperfections in the surfaces of the pipe joints.

A-10-2.2.1 Exception No. 6(b). The use of a blind flange or skillet is preferred for use when hydrostatically testing segments of new work. Metal seated valves are susceptible to developing slight imperfections during transport, installation, and operation and thus can be likely to leak more than 1 fluid ounce (30 ml) per inch of valve diameter per hour. For this reason, the blind flange should be used when hydrostatically testing.

All new systems are required to be tested hydrostatically at a pressure of at least 200 psi (13.8 bar). This value is set to ensure that pipe joints are made to withstand that pressure without coming apart or leaking. Although the hydrostatic test is primarily a workmanship test and not a materials performance test, damaged materials—for example, cracked fittings or leaky valves—are routinely discovered during the hydrostatic test. All materials used must be rated for a minimum working pressure of 175 psi (12.1 bar), which is the maximum pressure at which the system is normally maintained. The three Formal Interpretations related to 10-2.2 and 10-2.2.1 provide additional information on hydrostatic tests.

Formal Interpretation

Reference: 10-2.2

Question: Is hydropneumatic the substitute or equal of hydrostatic procedure according to Testing Laboratory or NFPA Standards?

Answer: No. NFPA 13, *Standard for Installation of Automatic Sprinkler Systems,* requires hydrostatic pressure testing of new systems under 10-2.2. During seasons in which freezing may occur, interim testing of dry pipe systems with air is required with the customary hydrostatic testing required when weather permits in accordance with 10-2.2.1.

Hydropneumatic procedures are associated with flushing of foreign materials that may obstruct waterways in piping and not with the pressure tests required for leakage testing in new systems.

Issue Edition: 1973

Reference: 1-11.3

Date: January 1977

Formal Interpretation 80-24

Reference: 10-2.2.1

Question: Is it the intent of 10-2.2.1 that the sprinkler piping at the top of a system fed by a riser 105 feet in height be pressurized to 200 psi for 2 hours when the sprinkler piping and riser are considered as "the individual system or zone"?

Answer: No.

Issue Edition: 1980

Reference: 1-11.3.1, 1-11.3.2

Date: October 1982

Formal Interpretation

Reference: 10-2.2.1

Question: In determining hydrostatic test pressure, when a booster pump is installed, should the shutoff pressure or normal operating pressure be considered in determining hydrostatic test pressure? If a relief valve is installed, should the relief valve setting be considered in determining the hydrostatic pressure test?

Answer: The hydrostatic pressure should, in the case of new installations, be based upon the maximum pressure exerted upon the new system piping and components. When a booster pump is provided, the maximum pressure is shutoff pressure or "churn" pressure without consideration of relief valve pressure settings.

Issue Edition: 1976

Reference: 1-11.3.1

Date: August 1977

The measures of success for inside sprinkler piping under hydrostatic test are no visible leakage and no loss of pressure in the 2-hour test period. Often a very small bead of water forms on a fitting during the test. Unless the bead continues to grow and drip, it is not considered visible leakage.

The hydrostatic test pressure is measured at the lowest elevation within the system or portion of the system being tested. Testing at the high point of the system, which, due to static head, would increase the test pressure significantly, is not considered necessary. The procedure is carried out in this way due to the fact that application of pressure typically occurs at the lower elevation, and these high pressures would not be anticipated at the higher elevations within the system.

When conducting a hydrostatic test, it is important to note that temperature changes affect the pressure readings of the tested systems. The expansion or contraction of the air in the system changing due to the temperature affects the pressure shown on the gauge. These fluctuations due to temperature apply to both hydrostatic and pneumatic tests because trapped air is in the system even in a hydrostatic test. This pressure reading is not an indication of a change in the amount of internal air or water or a leak but, rather, is an indication that the expansion or contraction of the air molecules has caused the pressure to fluctuate. For example, if the system were to be put under pressure when the temperature was 65°F (18°C) and the temperature dropped to 40°F (4°C) by the end of the test, the gauge would indicate less pressure, even if no leaks occurred. Conversely, if the temperature were to rise to 90°F (32°C), the pressure would increase, even though no additional air or water had been pumped into the system.

Where system pressures are in excess of 150 psi (10.4 bar), the hydrostatic test must be conducted at a pressure 50 psi (3.5 bar) in excess of the working pressure as indicated in Exception No. 1. This requirement is intended to ensure system integrity for these high-pressure systems.

The air test for dry pipe and double interlock systems in 10-2.3 is permitted by Exception No. 2 to 10-2.2.1 as an interim substitution for the hydrostatic test where a danger of water freezing in the system exists. The standard hydrostatic test must be made, however, when it is possible to do so without danger of freezing.

Exception Nos. 3, 4, and 5 address additions and modifications and specify when new piping installations are to be isolated and tested. In many cases, segregating new work from existing work is difficult. The Technical Committee on Sprinkler System Installation Criteria believes that to require the entire system to undergo another hydrostatic test when only relatively minor changes are made is unreasonable. When new portions of a system cannot be isolated, such as relocated drops, NFPA 13 now provides some flexibility and permits hydrostatic tests to be conducted at the system's normal static pressure. In general, existing portions of the system do not need to be subjected to a new hydrostatic test.

Exception No. 6 pertains to buried pipe only. A certain amount of leakage due to the nature of underground piping systems and the way they are installed is allowed.

10-2.2.2 Additives, corrosive chemicals such as sodium silicate or derivatives of sodium silicate, brine, or other chemicals shall not be used while hydrostatically testing systems or for stopping leaks.

Water additives, such as sodium silicate, that are intended to plug small system leaks during hydrostatic testing can clog small orifices, bind sprinkler parts, and prevent or delay operation. These chemicals tend to harden when exposed to air and can create significant obstructions later when they detach themselves from the internal portions of the pipe. Water additives for stopping leaks should never be used on sprinkler systems.

10-2.2.3 Piping between the exterior fire department connection and the check valve in the fire department inlet pipe shall be hydrostatically tested in the same manner as the balance of the system.

A reminder that the fire department connection must be tested is contained in 10-2.2.3. The fact that the piping from the check valve to the hose connection is not normally subjected to water pressure can lead the contractor to believe testing of that portion is not required. Other portions of pipe, such as the inspector's test connection drain and auxiliary drains, are not typically subject to high pressures and do not have to be tested in accordance with this requirement to ensure integrity of the entire system. Any buried piping connecting the fire department connection to the system can be tested in accordance with the requirements of Exception No. 6 to 10-2.2.1.

10-2.2.4 When deluge systems are being hydrostatically tested, plugs shall be installed in fittings and replaced with open sprinklers after the test is completed, or the operating elements of automatic sprinklers shall be removed after the test is completed.

10-2.2.5* The trench shall be backfilled between joints before testing to prevent movement of pipe.

At times contractors must leave trenches exposed for long periods of time while scheduling tests. Because of the inherent safety concerns with open trenches, the joints are permitted to be buried in accordance with the exception.

Exception: Where required for safety measures presented by the hazards of open trenches, the pipe and joints shall be permitted to be backfilled provided the installing contractor takes the responsibility for locating and correcting leakage in excess of that permitted in 10-2.2.1, Exception No. 6.

A-10-2.2.5 Hydrostatic tests should be made before the joints are covered so that any leaks can be readily detected. Thrust blocks should be sufficiently hardened before hydrostatic testing is begun. If the joints are covered with backfill prior to testing, the contractor remains responsible for locating and correcting any leakage in excess of that permitted in 10-2.2.1, Exception No. 6.

The pipeline should be prepared 24 hours prior to testing by filling it with water in such a manner as to remove all air. The test pressure should be applied to stabilize the system, which should minimize losses due to entrapped air, changes in water temperature, distention of components under pressure, movement of gaskets, and absorption of air by the water and water by the pipe wall.

10-2.2.6 Provision shall be made for the proper disposal of water used for flushing or testing.

10-2.2.7* Test blanks shall have painted lugs protruding in such a way as to clearly indicate their presence. The test blanks shall be numbered, and the installing contractor shall have a recordkeeping method ensuring their removal after work is completed.

A-10-2.2.7 Valves isolating the section to be tested might not be "drop-tight." When such leakage is suspected, test blanks of the type required in 10-2.2.7 should be used in a manner that includes the valve in the section being tested.

All test blanks must be removed after the work is completed. This procedure is extremely important. A test blank without a protruding lug is virtually impossible to detect by visual inspection. Test blanks are commonly referred to as frying pans because of their characteristic shape.

10-2.2.8 When subject to hydrostatic test pressures, the clapper of a differential-type valve shall be held off its seat to prevent damaging the valve.

The design of differential dry pipe valves incorporates flexible gaskets, which prevent water from entering the system. Pressures in excess of 50 psi (3.4 bar) can damage the mechanism of the valve.

10-2.3 Dry Pipe and Double Interlock System(s) Air Test.

In addition to the standard hydrostatic test, an air pressure leakage test at 40 psi (2.8 bar) shall be conducted for 24 hours. Any leakage that results in a loss of pressure in excess of 1½ psi (0.1 bar) for the 24 hours shall be corrected.

The air pressure leakage test verifies the integrity of the system, since both types of systems use pressurized air or nitrogen for system monitoring. Under non-fire conditions, dry pipe and double interlock preaction systems are subject to lower pressures than when the system is operating and is charged with water.

Because these systems are subjected to pressures from both air and water, they must be pressure tested using both air and water. Differences in physical properties between air and water make it impossible to judge system integrity based on only one of these tests. See commentary to 10-2.2.1 regarding the impact on temperature changes during the test.

10-2.3.1 Where systems are installed in spaces that are capable of being operated at temperatures below 32°F (0°C), air pressure leakage tests required in 10-2.3 shall be conducted at the lowest nominal temperature of the space.

10-2.4 System Operational Tests.

10-2.4.1 Waterflow detecting devices including the associated alarm circuits shall be flow tested through the inspector's test connection and shall result in an audible alarm on the premises within 5 minutes after such flow begins and until such flow stops.

An audible alarm is required on the premises when the inspector's test is operated. The alarm is to sound within 5 minutes after flow begins and must continue to sound until the flow stops. The 5-minute requirement in NFPA 13 was prompted by conditions created when a sprinkler system incorporates both a strong water supply and a backflow device or check valve.

Pressure differentials in water supplies can cause surges through the system. In strong water supplies, these surges can force water into the sprinkler system. Air pockets, which can be trapped in the system, become compressed and provide space for the added water. The presence of a backflow device or check valve does not allow the water to flow back out of the system.

Over a period of such occurrences, the system's static pressure continues to increase. As a result, when the inspector's test valve is opened, there can be enough pressure in the system to allow flow out of the test connection for a period of time, without requiring water from the water supply. If the built-up pressure in the system is large enough, the alarm device cannot operate until the system pressure is reduced sufficiently to initiate flow from the source of supply.

10-2.4.2 A working test of the dry pipe valve alone and with a quick-opening device, if installed, shall be made by opening the inspector's test connection. The test shall measure the time to trip the valve and the time for water to be discharged from the inspector's test connection. All times shall be measured from the time the inspector's test connection is completely opened. The results shall be recorded using the contractor's material and test certificate for aboveground piping.

Full operational tests of dry pipe valves are required. The Contractor's Material and Test Certificate includes provisions for recording the following information:

(1) Time it takes for the valve to trip after opening the system test valve
(2) Air and water pressures before the test
(3) Air pressure at trip point
(4) Delivery time for water to reach the test outlet
(5) Indication of whether or not alarms operated properly

The valve should operate in accordance with the tolerance specified by the manufacturer for the trip point. When this test is conducted, no maximum time limit is imposed by NFPA 13 as long as the system capacity is less than 750 gal (2850 L). For example, a 3-minute interval from the time the inspector's test valve is opened until the water is delivered to that inspector's test valve is an acceptable time lapse. The 60-second time limit imposed by 4-2.3.1 is only applicable to systems that exceed a capacity of 750 gal (2850 L).

10-2.4.3 The automatic operation of a deluge or preaction valve shall be tested in accordance with the manufacturer's instructions. The manual and remote control operation, where present, shall also be tested.

10-2.4.4 The main drain valve shall be opened and remain open until the system pressure stabilizes. The static and residual pressures shall be recorded on the contractor's test certificate.

The main drain test is crucial in that it establishes a baseline against which future tests can be compared. By recording available static and residual pressures during each test, a database can be established and used for tracking the relative condition of the water supply during the life of the system. Decreases in the static pressure, residual pressure,

or both could indicate a deteriorating water supply, an obstructed supply pipe, or a partially closed valve. Chapter 2 of NFPA 25, *Standard for the Inspection, Testing, and Maintenance of Water-Based Fire Protection Systems,* requires periodic tests of the main drain valve.

10-2.4.5 Operating Test.

10-2.4.5.1 Each hydrant shall be fully opened and closed under system water pressure, and dry barrel hydrants shall be checked for proper drainage. Where fire pumps are available, this check shall be done with the pumps running.

10-2.4.5.2 All control valves shall be fully closed and opened under system water pressure to ensure proper operation.

10-2.5 Each pressure-reducing valve shall be tested upon completion of installation to ensure proper operation under flow and no-flow conditions. Testing shall verify that the device properly regulates outlet pressure at both maximum and normal inlet pressure conditions. The results of the flow test of each pressure-reducing valve shall be recorded on the contractor's test certificate. The results shall include the static and residual inlet pressures, static and residual outlet pressures, and the flow rate.

If pressure-reducing valves are installed to control excessive pressures—for example, greater than 175 psi (12.1 bar) for most systems—they must be flow tested to ensure that they are correctly set. Pressure-reducing valves can also be required for combined sprinkler/ standpipe systems. The flow-testing requirements of these devices originated not from a potential functional failure of the device but to ensure that the installer properly followed the installation instructions. Following the manufacturer's instructions is very important when determining the appropriate pressure and flow values for these types of devices. An improperly set pressure-reducing device can have catastrophic results.

10-2.6 The backflow prevention assembly shall be forward flow tested to ensure proper operation. The minimum flow rate shall be the system demand, including hose stream demand where applicable.

Additional information on the testing of backflow devices is provided in 5-15.4.6. Due to their effect on sprinkler systems, backflow prevention devices must be accounted for through hydraulic analysis and flow testing.

10-2.7 Operating tests shall be made of exposure protection systems upon completion of the installation, where such tests do not risk water damage to the building on which they are installed or to adjacent buildings.

Acceptance tests should reveal that all surfaces to be protected from radiant and connective heat are uniformly wetted and that the most remote sprinkler has an adequate discharge pressure. The designer needs to consider the potential effects of wind and other drafts so that they do not substantially impact the sprinkler's intended discharge and spray pattern and do not prevent the proper wetting of the surfaces to be protected.

10-3 Circulating Closed Loop Systems

For sprinkler systems with non-fire protection connections, additional information shall be appended to the Contractor's Material and Test Certificate for Aboveground Piping shown in Figure 10-1(a) as follows:

(1) Certification that all auxiliary devices, such as heat pumps, circulating pumps, heat exchangers, radiators, and luminaries, if a part of the system, have a pressure rating of at least 175 psi or 300 psi (12.1 bar or 20.7 bar) if exposed to pressures greater than 175 psi (12.1 bar).
(2) All components of sprinkler system and auxiliary system have been pressure tested as a composite system in accordance with 10-2.2.
(3) Waterflow tests have been conducted and waterflow alarms have operated while auxiliary equipment is in each of the possible modes of operation.
(4) With auxiliary equipment tested in each possible mode of operation and with no flow from sprinklers or test connection, waterflow alarm signals did not operate.
(5) Excess temperature controls for shutting down the auxiliary system have been properly field tested.

Discharge tests of sprinkler systems with non-fire protection connections shall be conducted using system test connections described in 3-8.2. Pressure gauges shall be installed at critical points and readings shall be taken under various modes of auxiliary equipment operation. Waterflow alarm signals shall be responsive to discharge of water through system test pipes while auxiliary equipment is in each of the possible modes of operation.

10-4 Instructions

The installing contractor shall provide the owner with the following:

(1) All literature and instructions provided by the manufacturer describing proper operation and maintenance of any equipment and devices installed

(2) NFPA 25, *Standard for the Inspection, Testing, and Maintenance of Water-Based Fire Protection Systems*

Following the requirements of Section 10-4 is important so that the system is operable during its entire life expectancy. Many systems installed more than 50 years ago are still in service, are being properly maintained, and continue to protect lives and property.

NFPA 13 has previously provided a number of system maintenance provisions. In 1992, NFPA introduced NFPA 25. As a standard, NFPA 25 contains legally enforceable language. NFPA 25 superseded NFPA 13A, which was the maintenance document since 1939. NFPA 13A was withdrawn from the NFPA standards system in 1994, because all items relating to the periodic inspection, testing, and maintenance of sprinkler systems are contained within NFPA 25.

10-5* Hydraulic Design Information Sign

The installing contractor shall identify a hydraulically designed sprinkler system with a permanently marked weatherproof metal or rigid plastic sign secured with corrosion-resistant wire, chain, or other approved means. Such signs shall be placed at the alarm valve, dry pipe valve, preaction valve, or deluge valve supplying the corresponding hydraulically designed area. The sign shall include the following information:

(1) Location of the design area or areas
(2) Discharge densities over the design area or areas
(3) Required flow and residual pressure demand at the base of the riser
(4) Occupancy classification or commodity classification and maximum permitted storage height and configuration
(5) Hose stream demand included in addition to the sprinkler demand

In more cases than not, as-built drawings become lost or misplaced over time. By keeping a permanent record of the design parameters attached to the system riser, any future modifications or work on the system are much easier and less costly to perform. The information contained on the nameplate is also of vital importance in assessing the ability of the system to control fires as the building's occupancy changes or the water supply's strength deteriorates. Exhibit 10.2 illustrates a hydraulic nameplate.

Exhibit 10.2 *Example of a hydraulic nameplate.*

A-10-5 See Figure A-10-5.

This system as shown on .... company

print no .... dated ....

for ....

at .... contract no ....

is designed to discharge at a rate of .... gpm/ft^2

(L/min/m^2) of floor area over a maximum area of ....

ft^2 (m^2) when supplied with water at a rate of ....

gpm (L/min) at .... psi (bar) at the base of the riser.

Hose stream allowance of .... gpm (L/min)

is included in the above.

Occupancy classification ....

Commodity classification ....

Maximum storage height ....

Figure A-10-5 *Sample nameplate.*

References Cited in Commentary

National Fire Protection Association, 1 Batterymarch Park, P.O. Box 9101, Quincy, MA 02269-9101.

NFPA 15, *Standard for Water Spray Fixed Systems for Fire Protection,* 1996 edition.

NFPA 25, *Standard for the Inspection, Testing, and Maintenance of Water-Based Fire Protection Systems,* 1998 edition.

CHAPTER 11

Marine Systems

11-1 General

This chapter outlines the deletions, modifications, and additions that shall be required for marine application. All other requirements of this standard shall apply to merchant vessel systems except as modified by this chapter.

The chapter on marine systems first appeared in the 1996 edition of NFPA 13 as Chapter 9. Prior to the 1990s, relatively few automatic sprinkler systems were installed on commercial ships in the United States. Because sprinkler installations on ships were relatively rare, the sprinkler system design and installation requirements contained in the *Code of Federal Regulations* applicable to ships had not kept pace with modern technology.

However, changes in the demographics of the passenger vessel fleet in the United States — from ships with overnight accommodations that resembled hotel-type occupancies to vessels with no overnight accommodations that served as assembly occupancies — prompted changes in vessel construction and fire protection requirements. These changes favored sprinkler system installations.

Internationally, a number of fire casualties led the International Maritime Organization to require installation of sprinkler systems on all new passenger vessels and the retrofitting of sprinkler systems on existing ships. Similarly, NFPA 301, *Code for Safety to Life from Fire on Merchant Vessels*, which was first published in 1998, requires installation of sprinkler systems on most passenger vessels.

In Exhibit 11.1, the fire on the cruise ship Ecstasy is shown being extingished. More than 2500 people were aboard the cruise ship in 1998 when a fire began in the laundry room. Fire boats and on-board fire fighters were needed to extinguish the fire.

Instead of revising its regulations applicable to sprinkler systems, the U.S. Coast Guard, which is the authority having jurisdiction for commercial ships in the United States, looked to NFPA 13 for shipboard sprinkler system design and installation

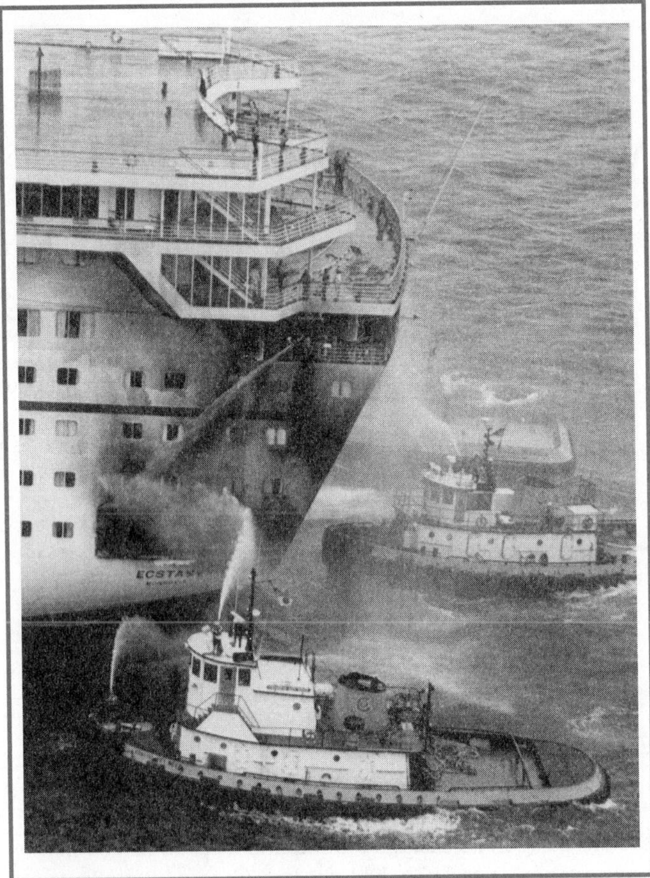

Exhibit 11.1 *Fire on cruise ship Ecstasy in 1998. (Courtesy of Reuters/Colin Bradley/Archive Photos.)*

requirements. Although the fire hazards and the purpose of a sprinkler system on marine vessels closely resemble those of shoreside structures, certain modifications to the requirements in Chapters 1 through 10 of NFPA 13 are necessary for marine applications.

As an initial step, the U.S. Coast Guard informally adopted the 1991 edition of NFPA 13 and published a set of changes to the standard that were necessary to make NFPA 13 applicable to shipboard installations. These changes were published in Navigation and Vessel Inspection Circular (NVIC) 10-93. The Technical Committee on Automatic Sprinklers subsequently reviewed NVIC 10-93 and included the necessary changes in Chapter 9 of the 1996 edition of NFPA 13. The U.S. Coast Guard subsequently replaced its sprinkler regulations by referencing the 1996 edition of NFPA 13 in the *Code of Federal Regulations.*

The changes in Chapter 11 primarily reflect unique shipboard factors such as operating environment, the unavailability of outside fire-fighting assistance, and egress philosophy. Because passengers and crew of a passenger ship cannot simply egress to safety, the overall shipboard fire protection system also demands a slightly higher level of protection from the sprinkler system. Although Chapters 1 through 10 are applicable to shipboard sprinkler systems, the requirements in Chapter 11 take precedence over those found in those chapters.

The marine chapter was revised from the 1996 edition to reflect experience gained through its application over the preceding 3 years.

11-1.1 The following definitions shall be applicable to this chapter.

A-Class Boundary. A boundary designed to resist the passage of smoke and flame for 1 hour when tested in accordance with ASTM E 119, *Standard Test Methods for Fire Tests of Building Construction and Materials.*

B-Class Boundary. A boundary designed to resist the passage of flame for ½ hour when tested in accordance with ASTM E 119, *Standard Test Methods for Fire Tests of Building Construction and Materials.*

Central Safety Station. A continuously manned control station from which all of the fire control equipment is monitored. If this station is not the bridge, direct communication with the bridge shall be provided by means other than the ship's service telephone.

The central safety station is typically the bridge.

Heat-Sensitive Material.* A material whose melting point is below 1700 F (926.7 C).

A-11-1.1 Heat-Sensitive Material. The backbone of the fire protection philosophy for U.S. flagged vessels and passenger vessels that trade internationally is limiting a fire to the compartment of origin by passive means. Materials that do not withstand a 1-hour fire exposure when tested in accordance with ASTM E 119, *Standard Test Methods for Fire Tests of Building Construction and Materials,* are considered "heat sensitive." *(See Figure A-11-1.1.)*

Ships are highly compartmentalized by passive fire barrier systems, such as bulkheads (walls) and decks (floors). Overall, shipboard fire protection places a greater emphasis on passive fire safety systems than on active fire suppression systems, such as automatic sprinklers. An assembly that is not expected to withstand fire exposure for 1 hour in accordance with the time–temperature profile of ASTM E 119, *Standard Test Methods for Fire Tests of Building Construction and Materials*, is considered heat sensitive. The fire endurance of materials, such as piping, that penetrate fire barriers on ships must be tested dry, so that a failure of the sprinkler system would not cause a failure of the passive system.

Heel. The inclination of a ship to one side.

Heel Angle. The angle defined by the intersection of a vertical line through the center of a vessel and a line perpendicular to the surface of the water.

Exhibit 11.2 illustrates measurement of the heel angle.

International Shore Connection

Threads to mate hydrants and hose at shore facilities

Threads to mate hydrants and hose on ship

in. (14 mm) minimum

2.75 in. (70 mm)

3.5 in. (89 mm)

1.25 in. (32 mm)

0.75 in. (19 mm)

Shore

Material: Any suitable for 150 psi (10.3 bar) service (shore)
Flange surface: Flat face
Gasket material: Any suitable for 150 psi (10.3 bar) service
Bolts: Four -in. (16-mm) minimum diameter, 2-in. (51-mm) long, threaded to within 1 in. (25.4 mm) of bolt head
Nuts: Four, to fit bolts
Washers: Four, to fit bolts

Ship

Material: Brass or bronze suitable for 150 psi (10.3 bar) service (ship)

Figure A-11-1.1 *International Shore Fire Connection.*

International Shore Connection. A universal connection complying with ASTM F 1121, *Standard Specification for International Shore Connections for Marine Fire Applications,* to which shoreside fire-fighting hose are to be connected.

Marine System.* A sprinkler system installed on a ship, boat, or other floating structure that takes its supply from the water on which the vessel floats.

A-11-1.1 Marine System. Some types of sprinkler systems can closely resemble marine systems, such as a system installed on a floating structure that has a permanent water supply connection to a public main. For these types of systems, judgment should be used in determining if certain aspects of Chapter 11 are applicable.

The definition of *marine system* is new to the 1999 edition. Although originally intended only for application to ships, other types of waterborne structures were recognized to more closely resemble ships than buildings. Examples include restaurants and casinos that are built on permanently moored barges. Where a sprinkler system on a floating

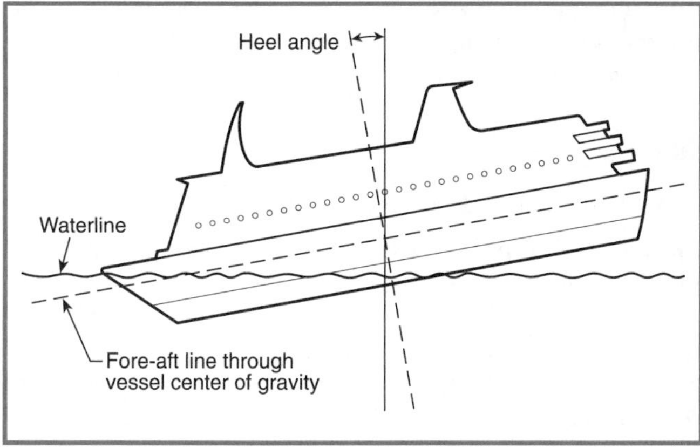

Exhibit 11.2 *Measurement of the heel angle. (Courtesy of Morgan Hurley.)*

structure takes its water supply from the water on which the structure floats, as opposed to from a permanent water main, the sprinkler system must comply with Chapter 11.

Marine Thermal Barrier.* An assembly that is constructed of noncombustible materials and made intact with the main structure of the vessel, such as shell, structural bulkheads, and decks. A marine thermal barrier shall meet the requirements of a B-Class boundary. In addition, a marine thermal barrier shall be insulated such that, if tested in accordance with ASTM E 119, *Standard Test Methods for Fire Tests of Building Construction and Materials,* for 15 minutes, the average temperature of the unexposed side does not rise more than 250 F (193 C) above the original temperature, nor does the temperature at any one point, including any joint, rise more than 405 F (225 C) above the original temperature.

A-11-1.1 Marine Thermal Barrier. A marine thermal barrier is typically referred to as a B-15 boundary.

Supervision. A visual and audible alarm signal given at the central safety station to indicate when the system is in operation or when a condition that would impair the satisfactory operation of the system exists. Supervisory alarms shall give a distinct indication for each individual system component that is monitored.

Survival Angle. The maximum angle to which a vessel is permitted to heel after the assumed damage required by stability regulations is imposed.

The survival angle for a ship is typically calculated when the ship is designed. A sprinkler designer would likely need to consult with a vessel's representative, such as the captain or naval architect, to determine the survival angle.

Type 1 Stair. A fully enclosed stair that serves all levels of a vessel in which persons can be employed.

Water Supply. The supply portion of the sprinkler system from the water pressure tank or the sea suction of the designated sprinkler system pump up to and including the valve that isolates the sprinkler system from these two water sources.

11-1.2* Occupancy Classifications.

Marine environment classifications shall be in accordance with Section 2-1.

A-11-1.2 In addition to the examples provided in A-2-1, Table A-11-1.2 provides additional examples of occupancy definitions of typical shipboard spaces.

Table A-11-1.2 Examples of Shipboard Space Occupancy Classification

	Space Types Included		
Occupancy Type	**CFR[1]**	**SOLAS[2]**	**Examples**
Light hazard	1[3], 2, 3, 4, 5, 6, 7, 8[4], 13	1[3], 2, 3, 4, 5, 6, 7, 8, 9	Accommodation spaces Small pantries
Ordinary hazard (Group 1)	8[4], 9[4]	12, 13[4]	Galleys Storage areas Sales shops Laundries Pantries with significant storage
Ordinary hazard (Group 2)	9[4], 11[4]	12[4], 13[4]	Sales shops Storage areas Stages (with sets) Machine shops
Extra hazard (Group 1)	1, 9[4], 10, 11[4]	1, 12[4], 13[4]	Auxiliary machinery—limited combustible liquids[5] Steering rooms—combustible hydraulic fluid in use[5]
Extra hazard (Group 2)	1, 9[4], 10, 11[4]	1, 12[4], 13[4]	Auxiliary machinery—with combustible liquids[5] Machinery spaces[5]

[1]Space type designations are given in 46 *CFR* 72.05-5.

[2]Space type designations are given in the *International Convention for the Safety of Life at Sea,* 1974 (SOLAS 74), as amended, regulations II-2/3 and II-2/26.

[3]Primarily for accommodation-type control stations, such as the wheel house, which would not include generator rooms or similar-type spaces.

[4]Depends on storage type, quantity, and height and distance below sprinkler.

[5]Automatic sprinklers typically will not be the primary means of protection in these areas; total flooding systems are usually used.

The classifications in Table A-11-1.2 are not meant to be applied without giving consideration to the definition of each occupancy hazard given in the standard. Table A-11-1.2 is general guidance for classification of typical spaces. Where a space is outfitted such that the occupancy definitions indicate that another classification would be more appropriate, the most representative and most demanding occupancy classification should be used. For example, it would certainly be possible to outfit a stateroom to require upgrading the occupancy to ordinary hazard, Group 1.

When a vessel undergoes modifications, alterations, or service changes that significantly affect the fire risk of the occupancy of one or more compartments, the occupancy classification should be reevaluated to determine if it has changed.

11-1.3* Partial installation of automatic sprinklers shall not be permitted.

Exception No. 1: Spaces shall be permitted to be protected with an alternative, approved fire suppression system where such areas are separated from the sprinklered areas with a 1-hour-rated assembly.

Exception No. 2: Where specific sections of this standard permit the omission of sprinklers.

A-11-1.3 Experience has shown that structures that are partially sprinklered can be overrun by well-developed fires originating in unsprinklered areas. Therefore, the entire vessel should be sprinklered whenever sprinkler systems are considered.

Although discussed in 1-6.2, partial sprinkler systems are not acceptable for shipboard fire protection. Exception No. 1 recognizes that some areas of a ship, such as the engine room, are typically not protected by an automatic sprinkler system but would be protected by another type of fire protection system such as a carbon dioxide system. However, these spaces need to be separated by a 1-hour-rated assembly. Exception No. 2 recognizes that sprinkler coverage in certain areas of a ship might not be necessary and maintains the exception for their omission under certain conditions.

11-2 System Components, Hardware, and Use

11-2.1* Sprinklers shall have a nominal discharge coefficient greater than 1.9.

A-11-2.1 Sprinklers with a nominal K-factor of 2.8 or less coupled with a system strainer minimize the potential for clogging.

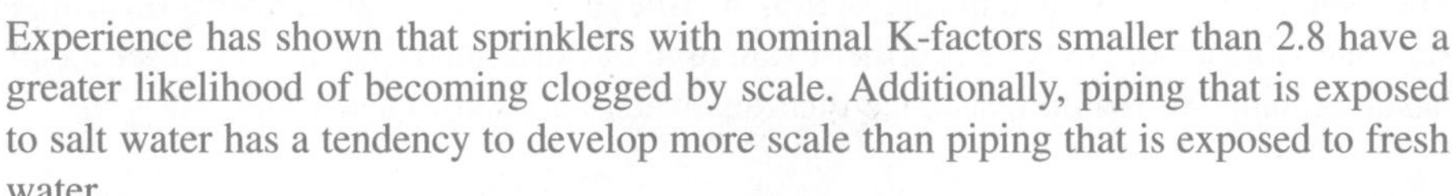

Experience has shown that sprinklers with nominal K-factors smaller than 2.8 have a greater likelihood of becoming clogged by scale. Additionally, piping that is exposed to salt water has a tendency to develop more scale than piping that is exposed to fresh water.

The 1996 edition required a minimum operating pressure of 10 psi (0.69 bar) to account for changes in elevation between sprinklers and their water supply as the vessel lists (leans to a side) or heels (points upwards or downwards). However, this 10-psi (0.69-bar) minimum was replaced with the more performance-oriented requirements contained in 11-6.3.

11-2.2* Sprinkler piping penetrations shall be designed to preserve the fire integrity of the ceiling or bulkhead penetrated.

A-11-2.2 Where a marine thermal barrier is penetrated, limiting the opening around the sprinkler pipe to 1⁄16 in. (1.6 mm) is considered as meeting this requirement.

Because of the importance of passive fire protection systems, piping that penetrates walls, floors, and ceilings must not compromise the integrity of the passive fire protection system. For an A-Class (1-hour) barrier, the penetration is usually protected with an approved fire stop or the piping is welded or flanged to the vessel structure at the penetration. A small 1⁄16-in. (1.6-mm) gap is considered acceptable for penetrations of any B-Class (30-minute) barrier, such as a marine thermal barrier.

11-2.3 Spare Sprinklers.

11-2.3.1 The required stock of spare sprinklers shall be carried for each type of sprinkler installed onboard the vessel. Where fewer than six sprinklers of a particular type are installed, 100 percent spares shall be kept in stock. Where applicable, at least one elastometric gasket shall be kept in the cabinet for each fire department connection that is installed onboard the vessel.

In the event of sprinkler system activation at sea, obtaining replacement sprinklers can be impossible for a long period of time. Additionally, because of the variety of spaces that are found on a large ship, it would not be unusual to find a wide variety of sprinkler types installed. The requirement for spare sprinklers in 3-2.9.1 is increased in 11-2.3.1 to ensure that if, for example, a sprinkler actuation occurs in a small compartment that is protected by sprinklers that do not match the majority found on the ship, it would be possible to replace the sprinklers at sea. The marine environment can also accelerate aging of the gaskets in fire department connections. Readily available spares are important.

11-2.3.2 The cabinet containing spare sprinklers, special wrenches, and elastometric gaskets shall be located in the same central safety station that contains the alarm annunciator panel(s) and supervisory indicators.

11-2.4 System Pipe and Fittings.

The 1996 edition prohibited the use of brazing filler metal because of concerns regarding saltwater corrosion. However, further evaluation has indicated that brazing filler metal has acceptable corrosion resistance qualities when exposed to salt water.

11-2.4.1* When ferrous materials are used for piping between the sea chest and zone control valves, these materials shall be protected against corrosion by hot dip galvanizing or by the use of Schedule 80 piping.

A-11-2.4.1 When nonferrous materials are used, consideration should be given to protecting against galvanic corrosion where the non-ferrous materials connect to steel pipe. Consideration should also be given to protection against galvanic corrosion from pipe hangers in areas of high humidity.

The piping between the sea chest and the sprinkler zone valves are likely to see the frequent flow of salt water when testing. Sprinkler zone piping will rarely, if ever, be exposed to salt water. In such an event, NFPA 25, *Standard for the Inspection, Testing, and Maintenance of Water-Based Fire Protection Systems,* requires flushing of the piping. Even if the piping is not flushed, the salt water will not be replenished and will lose oxygen content in fairly short order.

Even if galvanized, the failure from corrosion from the interior of the pipe is likely to be at all threaded connections, welded assembly connections, and where brass sprinklers thread into ferrous pipe. Only hot dipped galvanized after fabrication of assembly (as opposed to simply hot dipped galvanized pipe and fittings) will protect against some of those failures. Hot dipped galvanized after fabrication of assembly is practical from the sea chest to the sprinkler manifold where spaces are open and pipe is relatively large and uses flanged takedown joints instead of threaded unions. Hot dipped galvanized after fabrication of assembly is not practical in the sprinkler zone pipe where it is mainly field fit.

The additional requirements aimed at protecting ferrous materials reflect the concerns regarding the potential of corrosion damage caused by salt water. The 1996 edition required all sprinkler piping used in marine systems to be protected against corrosion by either hot dip galvanization or the use of extra heavy schedule material. However, this requirement was removed in the 1999 edition for all but the piping between the sea chest and sprinkler zone valves, because 11-7.4.3 requires charging of all other piping with fresh water, and NFPA 25, *Standard for the Inspection, Testing, and Maintenance of Water-Based Fire Protection Systems*, requires that piping be flushed following exposure to salt water.

Galvanic corrosion could occur where piping transitions from copper to steel or where a metal different from the piping is used for hangers.

11-2.4.2 Maximum design pressure for copper and brass pipe shall not exceed 250 psi (17.2 bar).

Vessel construction regulations typically do not permit copper or brass piping to be subjected to pressures in excess of 250 psi (17.2 bar).

11-2.5 Pipe Support.

The additional requirements for pipe support in this subsection are intended to protect sprinkler piping from the motions typically found on ships. When in doubt as to the magnitude of these motions, consultation with a naval architect can be necessary.

11-2.5.1* Pipe supports shall comply with the following:

(a) Pipe supports shall be designed to provide adequate lateral, longitudinal, and vertical sway bracing. The design shall account for the degree of bracing, which varies with the route and operation of the vessel. Bracing shall be designed to ensure the following:

(1) Slamming, heaving, and rolling will not shift sprinkler piping, potentially moving sprinklers above ceilings, bulkheads, or other obstructions.

(2) Piping and sprinklers will remain in place at a steady heel angle at least equal to the maximum required damaged survival angle.

(b) Pipe supports shall be welded to the structure. Hangers that can loosen during ship motion or vibration, such as screw-down-type hangers, shall not be permitted.

Exception: Hangers that are listed for seismic use shall be permitted to be used in accordance with their listing.

A-11-2.5.1 When designing supports, the selection and spacing of pipe supports should take into account the pipe dimensions, mechanical and physical properties of piping materials and supports, operating temperature, thermal expansion effects, external loads, thrust forces, vibration, maximum accelerations, differential motions to which the system might be subjected, and the type of support.

The route of the vessel is intended to be descriptive of its usual operating area. For example, expected motion of the system on an ocean vessel is expected to be considerably greater than the motion of a vessel that operates on a river. A vessel that operates within the confines of any of the Great Lakes is expected to subject the system pipe to greater motion than would a vessel that operates on a lake such as Lake Tahoe.

It is recommended that the designer review the requirements for automatic sprinkler systems that are subject to earthquakes. While it is obvious that shipboard motions and accelerations differ from those that occur during an earthquake, the general principle of protecting the piping system against damage applies. Individual hanger design, however, will be very similar. *(See Section 6-4.)*

Earthquake protection does not apply to ships; however, motions are similar to those that a ship will experience in a seaway. The design principles discussed in this section should be used as a guide for shipboard system design. See 6-1.1.

11-2.5.2 Sprinkler piping shall be supported by the primary structural members of the vessel such as beams, girders, and stiffeners.

11-2.5.3* The components of hanger assemblies that are welded directly to the ship structure shall not be required to be listed.

A-11-2.5.3 Use of heat-sensitive materials for pipe hangers and supports might be desirable in some cases. Where heat-sensitive materials are used, the hangers and supports should be adequately protected by either the direct application of insulation or installation behind a marine thermal barrier. Insulation materials applied directly to hangers should be insulated in accordance with the method provided in Society of Naval Architects and Marine Engineers Technical Research Bulletin 2-21, "Aluminum Fire Protection Guidelines."

11-2.5.4* U-hook sizes shall be no less than that specified in Table 6-1.4.1.

A-11-2.5.4 Consideration should be given to increasing the size of rods and U-hooks as necessary, to account for service and operational loading, including ship motion and vibrations.

11-2.6 Valves.

11-2.6.1* All indicating, supply, and zone control valves shall be supervised open from a central safety station.

A-11-2.6.1 Shipboard installations will normally require more than one valve per water supply. Locking valves in the open position is not an acceptable substitute for the requirement of 11-2.6.1 but can be done in addition to the supervision requirement.

By supervising these valves from the central safety station, any system impairment is intended to be quickly detected and corrected.

11-2.6.2 Drain and test valves shall meet the applicable requirements of 46 *CFR* 56.20 and 56.60.

A requirement of 46 *CFR* 56.20 is that the valve have limited leakage after removal of all resilient material. The removal of the resilient material simulates extended aging or exposure of the valve to a fire. The intention is to ensure that failure of a drain line does not allow large quantities of water to flow uncontrolled into the structure of a vessel.

A requirement of 46 *CFR* 56.60 is that valve castings meet ASTM A 47, *Standard Specification for Ferritic Malleable Iron Castings*, or ASTM A 197, *Standard Specification for Cupola Malleable Iron*, for malleable iron, ASTM A 126, *Standard Specification for Gray Iron Castings for Valves, Flanges and Pipe Fittings*, for gray iron, or ASTM A 395, *Standard Specification for Ferritic Ductile Iron Pressure-Retaining Castings for Use at Elevated Temperatures*, for ductile iron. Additionally, the valve itself is required by 46 *CFR* 56.60 to meet ANSI B16.34, *Valves — Flanged, Threaded, and Welding End.*

11-2.6.3 Valve markings shall include the information required by 46 *CFR* 56.20-5(a).

A requirement of 46 *CFR* 56.20-5(a) is that the valve be marked with the manufacturer's name and the service conditions for which it is intended.

11-2.7 Fire Department Connections and International Shore Connections.

11-2.7.1* A fire department connection and an International Shore Connection shall be installed.

Exception: Fire department connections shall not be required on vessels that operate primarily on international voyages.

A-11-2.7.1 International Shore Connections are portable universal couplings that permit connections of shipboard sprinkler or firemain systems between one ship and another or between a shore facility and a ship. Both the ship and the shore facility are expected to have an International Shore Connection fitting such that in an emergency they can be attached to their respective fire hoses and bolted together to permit charging the ship's system. It must be portable to accommodate hose-to-hose connection and allow assistance from any position.

Installation of an additional fire boat connection might be required on-board vessels whose route is such that regular access to fire boats is possible. An additional fire boat connection might not be necessary where fire boats are equipped to connect to the regular fire department connection. *(See A-11-2.7.3).*

11-2.7.2 Connections shall be located near the gangway or other shore access point so that they are readily accessible to the land-based fire department. Fire department and International Shore Connections shall be colored and marked so that the connections are easily located from

the shore access point (i.e., gangway location) and will not be confused with a firemain connection. An 18 in. 18 in. (0.46 m 0.46 m) sign displaying standard symbol 4-2.1 of NFPA 170, *Standard for Fire Safety Symbols,* shall be placed at the connection so that it is in plain sight from the shore access point. Connections on both sides of the vessel shall be provided where shore access arrangements make it necessary.

11-2.7.3* Fire department connection thread type shall be compatible with fire department equipment.

A-11-2.7.3 Selection of the pipe thread for the fire department connection should be done very carefully. It is recommended that a 2½-in. (63.5-mm) siamese connection with National Standard Hose Thread be used since a majority of fire department hose lines will be compatible with this thread. However, it must be noted that some fire jurisdictions might not be compatible with a connection of this type. Serious consideration should be given to the vessel's typical operating area. Precautions and planning should avert the possibility of the vessel being forced ashore by fire at a location where the fire department equipment is not compatible with this connection. Carriage of extra fittings and pre-voyage arrangements with all applicable jurisdictions should be considered. The International Shore Connection is required to ensure that all vessels fitted with sprinkler systems have at least one type of common connection.

11-3 System Requirements

11-3.1* Relief Valves.

Relief valves shall be provided on all wet pipe systems.

A-11-3.1 Special consideration should be given to the installation of relief valves in all wet pipe systems. Ambient ship temperatures can vary greatly depending on operating environment, duration of voyage, and failure of climate control systems.

Large ambient temperature variations occur on ships because of changes in climate or solar heating. Relief valves are required on all wet pipe systems to protect against piping failure caused by pressure buildup. This provision extends beyond that required for gridded systems in shoreside structures as indicated in 4-1.2.

11-3.2 Spare Detection Devices.

The number of spare detection devices or fusible elements used for protection systems that shall be carried per temperature rating is as follows:

(1) Vessels shall have two spare detection devices or fusible elements when operating voyages are normally less than 24 hours.
(2) Vessels shall have four spare detection devices or fusible elements when operating voyages are normally more than 24 hours.

The provisions of 4-3.1.4 are amplified in 11-3.2 by requiring additional detection devices for preaction and deluge systems for vessels with routes that routinely exceed 24 hours.

11-3.3 System Piping Supervision.

All preaction sprinkler systems shall be supervised regardless of the number of sprinklers supplied.

Because of the large number of small spaces on a ship and the level of reliability required of all shipboard fire protection systems, all preaction systems — not just those with more than 20 sprinklers — need to be supervised.

11-3.4 Circulating Closed Loop Systems.

Circulating closed loop systems shall not be permitted.

11-4 Installation Requirements

11-4.1 Temperature Zones.

Intermediate temperature–rated sprinklers shall be installed under a noninsulated steel deck that is exposed to sunlight.

To reduce the possibility of sprinkler operation caused by solar heating, intermediate temperature–rated sprinklers are required under uninsulated steel decks, such as ceilings, that are exposed to sunlight.

11-4.2* Residential Sprinklers.

Residential sprinklers shall be permitted for use only in sleeping accommodation areas.

A-11-4.2 Areas fitted primarily with multiple staterooms and corridors should be considered sleeping accommodation areas.

On ships, sleeping accommodation areas are the only spaces that resemble dwelling units.

11-4.3 Window Protection.

Where required, windows shall be protected by sprinklers installed at a distance not exceeding 1 ft (0.3 m) from the glazing at a spacing not exceeding 6 ft (1.8 m) such that the entire glazing surface is wetted at a linear density not less than 6 gpm/ft (75 mm/min).

Exception: Window sprinkler protection systems installed in accordance with their installation and testing criteria.

Vessel construction codes require protection of windows that face areas where passengers can be expected to gather in the event of a fire. The requirements for sprinkler spacing and design densities are based on research conducted by Richardson and Oleszkiewicz.[1] The exception recognizes that sprinklers specifically listed for window protection are available. It is noteworthy that the required density is based on protection of ordinary, non-fire-rated glazing.

11-4.4* Concealed Spaces.

Concealed spaces that are constructed of combustible materials, or materials with combustible finishes or that contain combustible materials, shall be sprinklered.

Exception: Spaces that contain only nonmetallic piping that is continuously filled with water are not required to be sprinklered.

A-11-4.4 If combustibles are present such that they constitute a threat, the space should be sprinklered. One example would be the presence of large bundles of unsheathed computer or electrical cable. Typical amounts of lighting or control cabling should not be considered to constitute a fire threat.

This subsection exceeds the requirements of 5-13.1.1. Ships, particularly passenger ships, typically have large concealed spaces, such as interstitial spaces above ceilings, that can contain large quantities of combustible piping, cabling, and so forth. The intent of this subsection is to protect against possible avenues where a fire could spread undetected.

11-4.5 Vertical Shafts.

11-4.5.1 Sprinklers are not required in vertical shafts used as duct, electrical, or pipe shafts that are nonaccessible, noncombustible, and enclosed in an A-Class-rated assembly.

The requirement that sprinklers are not required in vertical shafts echoes Exception Nos. 1 and 2 to 5-13.2.1 and allows for the omission of sprinklers in vertical shafts enclosed by A-Class (1-hour) assemblies. Many of the concealed spaces alluded to in 11-4.4 are enclosed by B-Class (30-minute) assemblies or non-rated assemblies.

11-4.5.2 Stairway enclosures shall be fully sprinklered.

The stairways on passenger ships typically contain more combustible materials, such as carpeting and small amounts of furniture, than their shoreside counterparts. Therefore, the requirement in 5-13.3.2 is expanded to require sprinklers in all levels of stairways.

11-4.6 Bath Modules.

Sprinklers shall be installed in bath modules (full room modules) constructed with combustible materials, regardless of room fire load.

Modern marine construction often utilizes bathrooms that are prefabricated from combustible composite materials. Because of concern over the potential combustible load

[1]Richardson and Oleszkiewicz, *Fire Technology*, Vol. 23, No. 2, May 1987, pp. 115–132.

that these materials would contribute to a fire, the provision in 5-13.9.1 that allows for the omission of sprinklers in certain bathrooms is not applicable to combustible bathroom modules on ships.

11-4.7 Ceiling Types.

Drop-out ceilings shall not be used in conjunction with sprinklers.

Because drop-out ceilings can affect sprinkler operation, they are not permitted on ships.

11-4.8 Return Bends.

To prevent sediment buildup, return bends shall be installed in all shipboard sprinkler systems where pendent-type or dry pendent–type sprinklers are used in wet systems *(see Figure 5-13.19).* Consideration shall be given concerning the intrusion of saltwater into the system. Specifically, sprinklers shall not be rendered ineffective by corrosion related to saltwater entrapment within the return bend.

All ships utilize a raw water source; therefore, return bends are required. The last sentence of this paragraph is in response to research that has indicated that sprinklers constructed using silicon brass alloy (UNS C87800) are susceptible to corrosion caused by salt water.

11-4.9 Hose Connections.

Sprinkler system piping shall not be used to supply hose connections or hose connections for fire department use.

Ships are generally required to have an independent fire main or standpipe system dedicated to manual fire-fighting operations. Although 5-15.5 permits hose connections to be supplied by sprinkler system piping in shoreside structures, hose connections for manual fire fighting are not permitted to be supplied by the ship's sprinkler system piping. Due to the limited stored water supply requirements discussed in Section 11-7, there is concern that operation of a hose connection supplied by the sprinkler system piping will very quickly deplete the stored water supply dedicated solely for the sprinkler system.

11-4.10 Heat-Sensitive Piping Materials.

This subsection is intended to ensure that where heat-sensitive piping materials, such as plastic pipe, are used, they will not be susceptible to failure caused by exposure to fire, even if the sprinkler system is impaired, such as by a closed valve.

11-4.10.1 Portions of the piping system constructed with a heat-sensitive material shall be subject to the following restrictions:

(1) Piping shall be of non-heat-sensitive type from the sea suction up through the penetration of the last A-Class barrier enclosing the space(s) in which the heat-sensitive piping is installed.
(2) B-Class draft stops shall be fitted not more than 45 ft (13.7 m) apart between the marine thermal barrier *(see definitions in 11-1.1)* and the deck or shell.
(3) Portions of a system that are constructed from heat-sensitive materials shall be installed behind a marine thermal barrier.

Exception: Piping materials with brazed joints shall not be required to be installed behind a marine thermal barrier, provided the following conditions are met:*

(a) The system is of the wet pipe type.

(b) The piping is not located in spaces containing boilers, internal combustion engines, or piping containing flammable or combustible liquids or gases under pressure, cargo holds, or vehicle decks.

(c) A relief valve in compliance with 4-1.2 is installed in each section of piping that is capable of being isolated by a valve(s).

(d) A valve(s) isolating the section of piping from the remainder of the system is installed in accordance with 11-4.10.2.

A-11-4.10.1 Exception to (3). Because of its melting point, brazing would be considered heat sensitive. The criteria of this paragraph is intended to permit brazed joints without requiring that they be installed behind a marine thermal barrier, while maintaining the fire endurance performance as stated in 11-4.10.1 under reasonably foreseeable failure modes.

The 1996 edition only permitted brazed copper piping when it was installed behind a marine thermal barrier. The brazing material would have been considered heat sensitive because it melts at a temperature less than 1700°F (926.7°C). This exception was added to the 1999 edition to permit brazing of copper tube, while maintaining the fire endurance performance as stated in 11-4.10.1.

The 1700°F (926.7°C) definition of heat-sensitive materials was derived from the endpoint temperature of a 1-hour ASTM E 119 exposure. Brazing filler metals have a solidous melting temperature between 1125°F and 1310°F (607°C and 710°C).[2] However, while dry piping could be expected to reach temperatures approaching 1700°F (926.7°C) after a 1-hour exposure, stagnant water–filled piping would likely not reach temperatures hot enough to melt brazing metals; hence, the first criterion of the exception. Soldered joints, which are permitted for non-marine systems under Exception No. 1 to 3-6.4, are not permitted for marine systems, because solders have melting temperatures ranging from 361°F to 464°F (183°C to 240°C).[3] Under foreseeable circumstances, such as exposure to a fire when a valve is shut, the solder could melt.

The second criterion of the exception recognizes that hydrocarbon fires or fires in cargo spaces or vehicle decks could easily reach temperatures that could jeopardize

[2] *Copper Tube Handbook*, Copper Development Association, 1995.
[3] ibid.

the integrity of brazed joints. A fuel spray fire on the USS Conyngham caused failure of a brazed joint in a fire main that was filled with stagnant water.

The third criterion of the exception is intended to allow for the expansion of boiling water in piping when there is no waterflow.

The fourth criterion of the exception is intended to ensure that valves that could prevent waterflow to a protected space are accessible in the event of a fire.

11-4.10.2 Each zone in which heat-sensitive piping is installed shall be fitted with a valve capable of segregating that zone from the remainder of the system. The valve shall be supervised and located outside of the zone controlled and within a readily accessible compartment having A-Class boundaries or within a Type 1 stair.

11-4.11 Discharge of Drain Lines.

11-4.11.1 Drain lines shall not be connected to housekeeping, sewage, or deck drains. Drains shall be permitted to be discharged to bilges. Overboard discharges shall meet the requirements of 46 *CFR* 56.50-95 and shall be corrosion resistant in accordance with 46 *CFR* 56.60. Systems that contain water additives that are not permitted to be discharged into the environment shall be specially designed to prevent such discharge.

Ships typically have two wastewater systems — gray and black. Black water includes piping from toilets, while gray water typically consists of drainage from sinks, deck drains, showers, and so forth. Gray water can often be discharged directly into the sea, but black water must be held or processed on the ship. Introduction of any pressurized water source into gray or black water piping could result in overflowing the system. Bilges are similar in function to sumps, and drainage from bilge pumps is always overboard.

11-4.11.2 Discharges shall be provided with a down-turned elbow.

The down-turned elbow serves to direct the water discharge in a downward direction and minimizes the likelihood of water spray hitting people or other objects on the deck and adjacent boats.

11-4.12 Alarm Signals and Devices.

11-4.12.1* A visual and audible alarm signal shall be given at the central safety station to indicate when the system is in operation or when a condition that would impair the satisfactory operation of the system exists. Alarm signals shall be provided for, but not limited to, each of the following: monitoring position of control valves, fire pump power supplies and operating condition, water tank levels and temperatures, zone waterflow alarms, pressure of tanks, and air pressure on dry pipe valves. Alarms shall give a distinct indication for each individual system component that is monitored. An audible alarm shall be given at the central safety station within 30 seconds of waterflow.

A-11-4.12.1 While not required, a dual annunciator alarm panel system is recommended. One panel should show the piping system layout and indicate status of zone valves, tank pressures, water supply valves, pump operation, and so forth. The second panel should show the vessel's general arrangement and indicate status of waterflow (i.e., fire location) alarms.

11-4.12.2 Waterflow alarms shall be installed for every zone of the sprinkler system. Sprinkler zones shall not encompass more than two adjacent decks or encompass more than one main vertical zone.

Many passenger vessels are subdivided into areas referred to as main vertical zones. These spaces have a function similar to that of horizontal exits in buildings. Main vertical zones are usually about 131 ft (40 m) in length. Sprinkler zones are limited to not more than two decks in any main vertical zone.

11-4.12.3 Electrically operated alarm attachments shall comply with, meet, and be installed in accordance with the requirements of 46 *CFR,* Subchapter J, "Electrical Engineering." All wiring shall be chosen and installed in accordance with IEEE 45, *Marine Supplement.*

NFPA 70, *National Electrical Code*®, is not fully applicable to shipboard electrical systems. The requirements for electric systems on ships are contained in 46 *CFR* 110-113, Subchapter J.

11-4.13 Test Connections.

Where test connections are below the bulkhead deck, they shall comply with the overboard discharge arrangements of 46 *CFR* 56.50-95.

The bulkhead deck is the highest deck that can be expected to be submerged when the vessel's hull is breached. Any penetration of the hull below the bulkhead deck is required to be protected so that it does not serve as a path for water ingress.

11-4.14 Copper tubing materials shall be protected against physical damage in areas where vehicles and stores handling equipment operate.

Copper tubing has a lower yield strength and thinner walls than steel pipe. Therefore, copper is more prone to damage such as pinching.

11-5 Design Approaches

11-5.1 Design Options.

Marine sprinkler systems shall be designed using the hydraulic calculation procedure of Chapter 7. The pipe schedule method shall not be used to determine the water demand requirements.

Marine sprinkler systems must be fully operational when the vessel is upright or inclined as a result of damage (see 11-6.3).

11-5.2* Window Protection.

Minimum water demand requirements shall include sprinklers that are installed for the protection of windows as described in 11-4.3.

A-11-5.2 For example, a design area of 1500 ft^2 (139.3 m^2) is used to design a sprinkler system for an unobstructed light hazard occupancy. In this case, the system must supply at least seven sprinklers that are installed within that area. If eight sprinklers are installed to protect windows within this design area, the water demand of these sprinklers is added to the total water demand. Thus, 15 sprinklers must be supplied by this system.

11-5.3* Hose Stream Allowance.

No allowance for hose stream use shall be required.

A-11-5.3 Hose stream flow need not be added to the water demand. The water supply for fire streams is supplied by separate fire pump(s) that supply the vessel's fire main.

As discussed in 11-4.9, sprinkler system piping is not permitted to be used to supply hose connections.

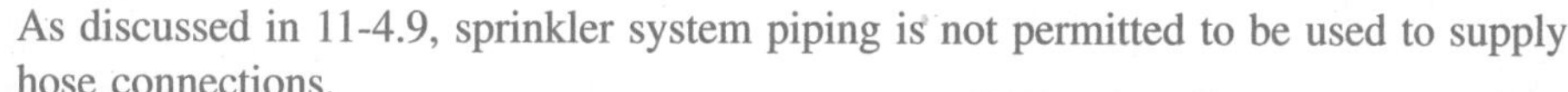

11-6 Plans and Calculations

11-6.1 Additional Information.

The pressure tank size, high pressure relief setting, high and low water alarm settings, low pressure alarm setting, and pump start pressure shall be provided.

11-6.2 Sprinklers specifically installed for the protection of windows under 11-4.3 are permitted to be of a different size from those protecting the remainder of the occupancy classification. All of the window sprinklers, however, shall be of the same size.

11-6.3* Marine sprinkler systems shall be designed and installed to be fully operational without a reduction in system performance when the vessel is upright and inclined at the angles of inclination specified in 46 *CFR* 58.01-40.

A-11-6.3 In vessels, the elevation of sprinklers with respect to the water supply varies as the vessel heels to either side or trims by the bow or stern. The water demand requirements can be increased or decreased under these conditions. This requirement aligns the operational parameters of this safety system with that required for other machinery vital to the safety of the vessel.

A list of 22.5 degrees and a trim of 7.5 degrees is specified in 46 *CFR* 58.01-40. Also, 46 *CFR* 58.01-40 contains flexibility where lesser angles can be accepted. For example, lesser angles are accepted on vessels that operate only in calm waters or do not get underway. Verification of this requirement requires hydraulic calculation of the hydraulically most remote area with the vessel in the following three conditions:

(1) Upright
(2) Listing to port
(3) Listing to starboard

11-7 Water Supplies

The water supply for a sprinkler system on a ship typically comes from two sources. The initial supply comes from a small limited water source stored on the vessel. This stored supply is sized to provide for 1 minute of system demand plus the volume of the piping if a dry pipe, preaction, or deluge system is used. During the discharge of the stored water source, a pump that takes suction from the sea activates. Because of concerns regarding drawing large quantities of mud into the fire pump suction when the vessel is close to shore or on an inland waterway, a single water supply that takes suction from the sea is not considered sufficiently reliable.

11-7.1 General.

The water supply requirements for marine applications shall be in accordance with Section 11-7.

11-7.2 Pressure Tank.

A pressure tank is the preferred source of initial water supply because, during a loss of power, it will still operate. As indicated in 11-7.2.1, the tank must be sized to provide the system demand for at least 1 minute plus the volume of any dry piping used.

11-7.2.1 A pressure tank shall be provided. The pressure tank shall be sized and constructed so that the following occurs:

(1) The tank shall contain a standing charge of fresh water equal to that specified by Table 11-7.2.1.
(2) The pressure tank shall be sized in accordance with 9-2.3.2.
(3) A glass gauge shall be provided to indicate the correct level of water within the pressure tank.
(4) Arrangements shall be provided for maintaining an air pressure in the tank such that, while the standing charge of water is being expended, the pressure will not be less than that necessary to provide the design pressure and flow of the hydraulically most remote design area.
(5) Suitable means of replenishing the air under pressure and the fresh water standing charge in the tank shall be provided.

(6) Tank construction shall be in accordance with the applicable requirements of 46 *CFR*, Subchapter F, "Marine Engineering."

Exception: In lieu of a pressure tank, a dedicated pump connected to a fresh water tank shall be permitted to be used, provided the following conditions are met:

(a) The pump is listed for marine use and is sized to meet the required system demand.

(b) The suction for the fire pump is located below the suction for the fresh water system so that there shall be a minimum water supply of at least 1 minute for the required system demand.

(c) Pressure switches are provided in the system and the controller for the pump that automatically start the pump within 10 seconds after detection of a pressure drop of more than 5 percent.

(d) There shall be a reduced pressure zone backflow preventer to prevent contamination of the potable water system by salt water.

(e) There are at least two sources of power for this pump. Where the sources of power are electrical, these shall be a main generator and an emergency source of power. One supply shall be taken from the main switchboard, by separate feeder reserved solely for that purpose. This feeder shall be run to an automatic change-over switch situated near the sprinkler unit and the switch shall normally be kept closed to the feeder from the emergency switchboard. The changeover switch shall be clearly labeled and no other switch shall be permitted in these feeders.

Table 11-7.2.1 Required Water Supply

System Type	Additional Water Volume
Wet pipe system	Flow requirement of the hydraulically most remote system demand for 1 minute
Dry pipe system Preaction system Deluge system	Flow requirement of the hydraulically most remote system demand for 1 minute of system demand plus the volume needed to fill all dry piping

The exception permits a dedicated pump that takes suction from a freshwater tank. Because of concerns regarding contamination of the vessel's potable water tank with seawater, a single pump that takes suction from both the sea and the potable water tank is not permitted. The suction from the pump must be located at a lower elevation from any other pumps, ensuring that the minimum required volume of water is always available. When the option permitted by the exception is used, both the pump dedicated to the potable water tank and the pump that takes suction from seawater must be powered by the emergency switchboard.

11-7.2.2 Relief valves shall be installed on the tank to avoid overpressurization and false actuation of any dry pipe valve. Relief valves shall comply with 46 *CFR* 54.15-10.

11-7.2.3 There shall be not less than two sources of power for the compressors that supply air to the pressure tank. Where the sources of power are electrical, these shall be a main generator and an emergency source of power. One supply shall be taken from the main switchboard, by separate feeders reserved solely for that purpose. Such feeders shall be run to a changeover switch situated near the air compressor, and the switch normally shall be kept closed to the feeder from the emergency switchboard. The changeover switch shall be clearly labeled, and no other switch shall be permitted in these feeders.

11-7.2.4 More than one pressure tank can be installed provided that each is treated as a single water source when determining valve arrangements. Check valves shall be installed to prohibit flow from tank to tank or from pump to tank.

Exception: Arrangements where a tank is designed to hold only pressurized air.

Where sufficient space on the vessel is not available to install a single pressure tank meeting the requirements of Table 11-7.2.1, several smaller tanks can be considered. Each tank must be fitted with valves to keep seawater out of the tank and keep water out of the air supply.

The exception permits the omission of check valves in piping between a supplemental pressurized air tank and a pressure tank containing water.

11-7.2.5 In systems subject to use with saltwater, valves shall be so arranged as to prohibit contamination of the pressure tank with saltwater.

11-7.2.6* Where applicable, a means shall be provided to restrict the amount of air that can enter the pressure tank from the air supply system. A means shall also be provided to prevent water from backflowing into the air supply system.

A-11-7.2.6 The purpose of this is to ensure that the pressure tank air supply will not keep the tank "fully" pressurized while water is expelled, thus preventing pump actuation.

11-7.3 Fire Pump.

11-7.3.1 A dedicated, automatically controlled pump that is listed for marine service, which takes suction from the sea, shall be provided to supply the sprinkler system. Where two pumps are required to ensure the reliability of the water supply, the pump that supplies the fire main shall be allowed to serve as the second fire pump.

The requirement for the installation of multiple fire pumps to increase system reliability would typically come from a vessel construction standard, such as NFPA 301.

11-7.3.2* The pump shall be sized to meet the water demand of the hydraulically most demanding area. Pumps shall be designed to not exceed 120 percent of the rated capacity of the pump.

A-11-7.3.2 NFPA 20, *Standard for the Installation of Centrifugal Fire Pumps,* requires that fire pumps furnish not less than 150 percent of their rated capacity at not less than 65 percent of their rated heat. The intention of the requirement of 11-7.3.2 is to limit designers to 120 percent of the rated capacity of the pump to provide an additional factor of safety for marine systems.

As a means of increasing the overall level of system reliability, pumps are not permitted to be sized in excess of 120 percent of their rated capacity.

11-7.3.3 The system shall be designed so that, before the supply falls below the design criteria, the fire pump shall be automatically started and shall supply water to the system until manually shut off.

Exception: Where pump and fresh water tank arrangement is used in lieu of the pressure tank, there must be a pressure switch that senses a system pressure drop of 25 percent, and the controller must automatically start the fire pump(s) if pressure is not restored within 20 seconds.

11-7.3.4 There shall be not less than two sources of power supply for the fire pumps. Where the sources of power are electrical, these shall be a main generator and an emergency source of power. One supply shall be taken from the main switchboard by separate feeders reserved solely for that purpose. Such feeders shall be run to a changeover switch situated near to the sprinkler unit, and the switch normally shall be kept closed to the feeder from the emergency switchboard. The changeover switch shall be clearly labeled and no other switch shall be permitted in these feeders.

11-7.3.5 A test valve(s) shall be installed on the discharge side of the pump with a short open-ended discharge pipe. The area of the pipe shall be adequate to permit the release of the required water output to supply the demand of the hydraulically most remote area.

11-7.3.6 Where two fire pumps are required to ensure the reliability of the water supply, each fire pump shall meet the requirements of 11-7.3.1 through 11-7.3.4. In addition, a system that is required to have more than one pump shall be designed to accommodate the following features:

(a)* Pump controls and system sensors shall be arranged such that the secondary pump will automatically operate if the primary pump fails to operate or deliver the required water pressure and flow. *[Figure A-11-7.3.6(a) is an example of an acceptable dual pump arrangement.]*

(b) Both pumps shall be served from normal and emergency power sources. However, where approved by the authority having jurisdiction, the secondary pump shall be permitted to be nonelectrically driven.

(c) Pump failure or operation shall be indicated at the central safety station.

A-11-7.3.6(a) Pumps should not be located within the same compartment. However, where this is not reasonable or practical, special attention should be given to protecting pumps such that a single failure will not render the sprinkler system inoperative. *[See Figure A-11-7.3.6(a).]*

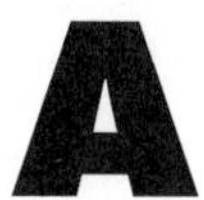

11-7.3.7* If not specifically prohibited, the fire pump that supplies the firemain is permitted to be used as the second pump, provided the following conditions are met:

(1) The pump is adequately sized to meet the required fire hose and sprinkler system pressure and flow demands simultaneously.
(2) The fire main system is segregated from the sprinkler system by a normally closed valve that is designed to automatically open upon failure of the designated fire pump.
(3) The fire pump that supplies the fire main is automatically started in the event of dedicated fire pump failure or loss of pressure in the sprinkler main. *(See Figure A-11-7.3.7.)*

A-11-7.3.7 See Figure A-11-7.3.7.

11-7.4 Water Supply Configurations.

11-7.4.1 The pressure tank and fire pump shall be located in a position reasonably remote from any machinery space of Category A.

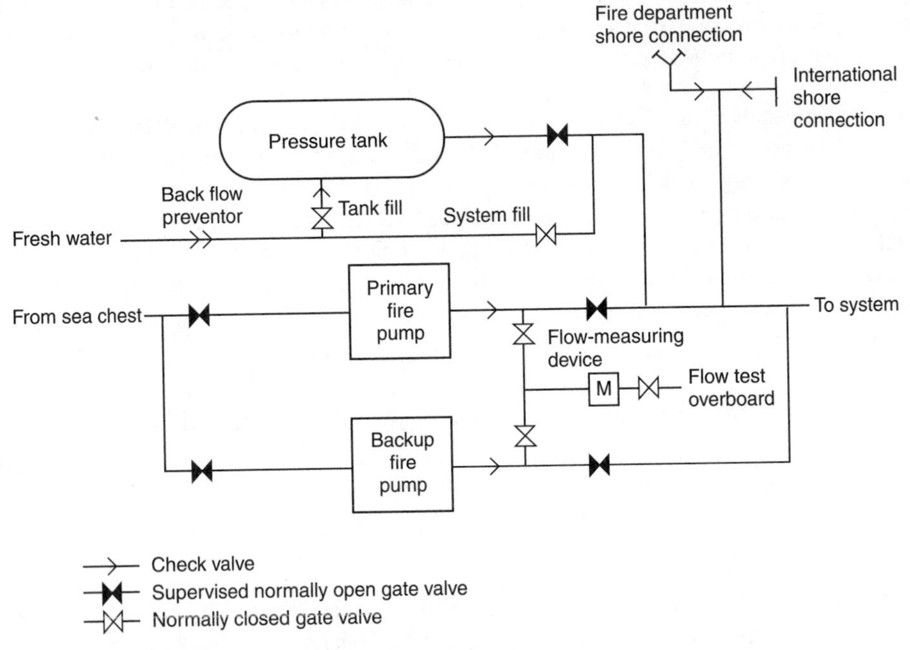

Figure A-11-7.3.6(a) Abbreviated example of a dual fire pump water supply.

Many shipboard fires start in machinery spaces. The intent of 11-7.4.1 is to locate the fire pumps and pressure tanks so that a fire in the machinery space does not cause an impairment of the sprinkler system.

11-7.4.2 All valves within the water supply piping system shall be supervised.

11-7.4.3 Only fresh water shall be used as the initial charge within the piping network.

Limiting the initial charge of water in the piping network to fresh water is intended to minimize the potential for corrosion. See 2-4.4 of NFPA 25 for information on flushing of piping following introduction of seawater.

11-7.4.4 The sprinkler system shall be cross-connected with the ship's fire main system and fitted with a lockable screw-down nonreturn valve such that backflow from the sprinkler system to the fire main is prevented.

The fire main (standpipe) used for manual fire-fighting operations also has one or more pumps that take suction from the sea. The cross-connection between these two systems minimizes the effects caused by a pump failure in either the sprinkler system piping or the fire main and increases the reliability of the overall fire protection system.

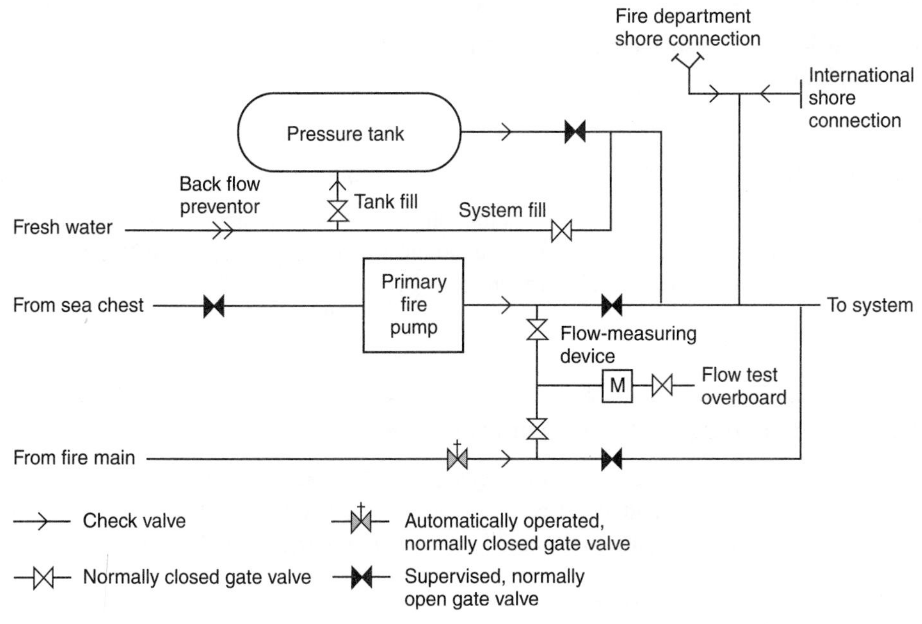

Figure A-11-7.3.7 *Abbreviated example of a water supply with fire pump backup.*

11-7.4.5 The piping, tanks, and pumps that make up the water supply shall be installed in accordance with the applicable requirements of 46 *CFR,* Subchapter F, "Marine Engineering."

11-7.4.6* When a shore water supply is to be used during extended dockside periods, the water supply shall be qualified in the manner described in 9-2.1. Tests shall be conducted in accordance with the requirements of the local shore-based authority having jurisdiction. The water supply information listed in Section 8-3 shall then be provided to the authority having jurisdiction.

A-11-7.4.6 This procedure should be used to qualify each water supply to which the vessel is to be attached. For example, this might require testing of multiple hydrants or connections in the same mooring area. The pressure loss effect of the hose or piping leading from the water supply to the ship should also be considered when qualifying each hydrant.

Some vessels, such as casino vessels, can operate at the dock for extended periods of time. These vessels can use a water supply independent of the vessel, such as a connection to a shoreside water main.

11-8 System Acceptance

11-8.1 Hydrostatic Tests.

In addition to the interior piping, the test required by 10-2.2.3 shall also be conducted on all external water supply connections including international shore and fireboat connections.

11-8.2 Alarm Test.

A waterflow test shall result in an alarm at the central safety station within 30 seconds after flow through the test connection begins.

11-8.3 Operational Tests.

Pressure tank and pump operation, valve actuation, and waterflow shall also be tested. Pump operation and performance shall be tested in accordance with Chapter 11 of NFPA 20, *Standard for the Installation of Centrifugal Fire Pumps.*

11-9 System Instructions and Maintenance

Instructions for operation, inspection, maintenance, and testing shall be kept on the vessel. Records of inspections, tests, and maintenance required by NFPA 25, *Standard for the Inspection, Testing, and Maintenance of Water-Based Fire Protection Systems,* shall also be kept on the vessel.

The requirements in NFPA 25 for the inspection, testing, and maintenance of sprinkler systems are applicable for marine purposes. NFPA 25 is referenced in Chapter 10. Chapter 9 expands on the requirements of NFPA 25 as indicated in Section 11-9.

Although NFPA 25 requires keeping records of inspections, tests, and maintenance, these records could be kept in a location other than on the vessel and still meet the requirements of NFPA 25. This section requires that instructions and records be kept on the vessel.

NFPA 25 requires flushing of piping following a system activation to limit corrosion caused by seawater ingress into the sprinkler system.

References Cited in Commentary

National Fire Protection Association, 1 Batterymarch Park, P.O. Box 9101, Quincy, MA 02269-9101.

NFPA 25, *Standard for the Inspection, Testing, and Maintenance of Water-Based Fire Protection Systems*, 1998 edition.
NFPA 70, *National Electrical Code®*, 1999 edition.
NFPA 301, *Code for Safety to Life from Fire on Merchant Vessels*, 1998 edition.

American National Standards Institute, Inc., 11 West 42nd Street, 13th floor, New York, NY 10036.

ANSI B16.34, Valves — Flanged, Threaded, and Welding End, 1988 edition.

American Society for Testing and Materials, 100 Barr Harbor Drive, West Conshohocken, PA 19428-2959.

ASTM A 47, *Standard Specification for Ferritic Malleable Iron Castings*, 1995 edition.

ASTM A 126, *Standard Specification for Gray Iron Castings for Valves, Flanges and Pipe Fittings*, 1995 edition.

ASTM A 197, *Standard Specification for Cupola Malleable Iron*, 1992 edition.

ASTM A 395, *Standard Specification for Ferritic Ductile Iron Pressure-Retaining Castings for Use at Elevated Temperatures*, 1998 edition.

ASTM E 119, *Standard Test Methods for Fire Tests of Building Construction and Material*, 1998 edition.

U.S. Government Printing Office, Washington, DC 20402.

Navigation and Vessel Inspection Circular 10-93, "Guide to the Acceptance of National Fire Protection Association Code No. 13 for Automatic Sprinkler System Design, Installation and Maintenance," U.S. Coast Guard, 1993.

Title 46, *Code of Federal Regulations*, Parts 56.20, 56.20-5(a), 56.60, and 58.01-40.

Title 46, *Code of Federal Regulations*, Parts 110–113, Subchapter J, "Electrical Engineering."

CHAPTER 12

System Inspection, Testing, and Maintenance

This chapter specifically requires that sprinkler systems be inspected, tested, and maintained in accordance with NFPA 25, *Standard for the Inspection, Testing, and Maintenance of Water-Based Fire Protection Systems*. Previous editions of NFPA 13 referred to NFPA 13A for inspection, testing, and maintenance activities. In 1992 the information from NFPA 13A was incorporated in NFPA 25, and NFPA 13A was withdrawn. Inspection and testing activities are critical in ensuring proper system operation. Additionally, many insurance companies insist that an ongoing program be in place. NFPA 25 identifies the activities that need to be conducted and the frequency at which they need to be conducted.

12-1* General

A sprinkler system installed in accordance with this standard shall be properly inspected, tested, and maintained in accordance with NFPA 25, *Standard for the Inspection, Testing, and Maintenance of Water-Based Fire Protection Systems,* to provide at least the same level of performance and protection as designed.

A-12-1 *Impairments.* Before shutting off a section of the fire service system to make sprinkler system connections, notify the authority having jurisdiction, plan the work carefully, and assemble all materials to enable completion in the shortest possible time. Work started on connections should be completed without interruption, and protection should be restored as promptly as possible. During the impairment, provide emergency hose lines and extinguishers and maintain extra watch service in the areas affected.

When changes involve shutting off water from any considerable number of sprinklers for more than a few hours, temporary water supply connections should be made to sprinkler systems so that reasonable protection can be maintained. In adding to old systems or revamping them, protection should be restored each night so far as possible. The members of the private fire brigade as well as public fire departments should be notified as to conditions.

Maintenance Schedule. The items shown in Table A-12-1 should be checked on a routine basis.

Table A-12-1 Maintenance Schedule

Parts	Activity	Frequency
Flushing piping	Test	5 years
Fire department connections	Inspection	Monthly
Control valves	Inspection	Weekly—sealed
	Inspection	Monthly—locked
	Inspection	Monthly—tamper switch
	Maintenance	Yearly
Main drain	Flow test	Quarterly—Annual
Open sprinklers	Test	Annual
Pressure gauge	Calibration test	
Sprinklers	Test	50 years
Sprinklers—high temperature	Test	5 years
Sprinklers—residential	Test	20 years
Waterflow alarms	Test	Quarterly
Preaction/deluge detection system	Test	Semiannually
Preaction/deluge systems	Test	Annually
Antifreeze solution	Test	Annually
Cold weather valves	Open and close valves	Fall, close; spring, open
Dry/preaction/deluge systems		
Air pressure and water pressure	Inspection	Weekly
Enclosure	Inspection	Daily—cold weather
Priming water level	Inspection	Quarterly
Low-point drains	Test	Fall
Dry pipe valves	Trip test	Annual—spring
Dry pipe valves	Full flow trip	3 years—spring
Quick-opening devices	Test	Semiannually

Inspection, testing, and maintenance issues are beyond the scope of NFPA 13, but they are critical for effective sprinkler system performance. These issues are especially true because of the inactive nature of a sprinkler system. Unlike other types of building systems that are used on a routine basis, the sprinkler system is only used during emergency situations. The system's proper operating condition is not verifiable through day-to-day operations, as are the heating, air-conditioning, and plumbing system. Therefore, the building owner or representative is required to employ and follow an inspection, testing, and maintenance program as specified by NFPA 25.

A system that does not comply with the requirements of NFPA 25 is not in compliance with NFPA 13. An improperly maintained system is likely to be impaired in some way and cannot be expected to meet its fire protection objectives. An impaired system is likely to have the same effect as no sprinkler protection at all.

It is critical that steps be taken to compensate for the lack of protection in a building when the sprinkler system is impaired. NFPA 25 provides information on dealing with system impairments. One step that can be taken regarding life safety is evacuation of the building. Provisions for emergency hose lines, extinguishers, and/or an approved fire watch are alternatives to an evacuation. See Chapter 11 of NFPA 25 for a more complete discussion of impairments and recommendations on how to handle the occupation of a building with an impaired system.

Table A-12-1 was added to NFPA 13 to give the user guidance on the kinds of activities that need to be performed and some general sense as to how frequently such activities should be carried out. The table was extracted from the 1995 edition of NFPA 25 and was not updated. As a result, some of the frequencies for activities no longer match the 1998 edition of NFPA 25. Because this table is in the appendix, and because the reference to NFPA 25 is a mandatory requirement of NFPA 13, the rules of NFPA 25 apply if a conflict arises.

For example, the table shows that a main drain test needs to be conducted on the sprinkler system at quarterly intervals, which was consistent with the 1995 edition of NFPA 25. However, the 1998 edition of NFPA 25 revised this requirement to annual intervals and after each time a valve is closed. Because the 1998 edition is the one that is referenced as the mandatory requirement of NFPA 13, the annual test and the test after each valve closure is the enforceable requirement.

Reference Cited in Commentary

National Fire Protection Association, 1 Batterymarch Park, P.O. Box 9101, Quincy, MA 02269-9101.

NFPA 25, *Standard for the Inspection, Testing, and Maintenance of Water-Based Fire Protection Systems*, 1998 edition.

CHAPTER 13

Referenced Publications

13-1

The following documents or portions thereof are referenced within this standard as mandatory requirements and shall be considered part of the requirements of this standard. The edition indicated for each referenced mandatory document is the current edition as of the date of the NFPA issuance of this standard. Some of these mandatory documents might also be referenced in this standard for specific informational purposes and, therefore, are also listed in Appendix E.

13-1.1 NFPA Publications.

National Fire Protection Association, 1 Batterymarch Park, P.O. Box 9101, Quincy, MA 02269-9101.

NFPA 11A, *Standard for Medium- and High-Expansion Foam Systems,* 1999 edition.

NFPA 13D, *Standard for the Installation of Sprinkler Systems in One- and Two-Family Dwellings and Manufactured Homes,* 1999 edition.

NFPA 13R, *Standard for the Installation of Sprinkler Systems in Residential Occupancies up to and Including Four Stories in Height,* 1999 edition.

NFPA 14, *Standard for the Installation of Standpipe and Hose Systems,* 1996 edition.

NFPA 15, *Standard for Water Spray Fixed Systems for Fire Protection,* 1996 edition.

NFPA 20, *Standard for the Installation of Centrifugal Fire Pumps,* 1999 edition.

NFPA 22, *Standard for Water Tanks for Private Fire Protection,* 1998 edition.

NFPA 25, *Standard for the Inspection, Testing, and Maintenance of Water-Based Fire Protection Systems,* 1998 edition.

NFPA 30, *Flammable and Combustible Liquids Code,* 1996 edition.

NFPA 30B, *Code for the Manufacture and Storage of Aerosol Products,* 1998 edition.

NFPA 33, *Standard for Spray Application Using Flammable or Combustible Materials,* 1995 edition.

NFPA 36, *Standard for Solvent Extraction Plants,* 1997 edition.

NFPA 40, *Standard for the Storage and Handling of Cellulose Nitrate Motion Picture Film,* 1997 edition.

NFPA 42, *Code for the Storage of Pyroxylin Plastic,* 1997 edition.

NFPA 45, *Standard on Fire Protection for Laboratories Using Chemicals,* 1996 edition.

NFPA 51, *Standard for the Design and Installation of Oxygen-Fuel Gas Systems for Welding, Cutting, and Allied Processes,* 1997 edition.

NFPA 51A, *Standard for Acetylene Cylinder Charging Plants,* 1996 edition.

NFPA 51B, *Standard for Fire Prevention During Welding, Cutting, and Other Hot Work,* 1999 edition.

NFPA 55, *Standard for the Storage, Use, and Handling of Compressed and Liquefied Gases in Portable Cylinders,* 1998 edition.

NFPA 59, *Standard for the Storage and Handling of Liquefied Petroleum Gases at Utility Gas Plants,* 1998 edition.

NFPA 59A, *Standard for the Production, Storage, and Handling of Liquefied Natural Gas (LNG),* 1996 edition.

NFPA 70, *National Electrical Code®,* 1999 edition.

NFPA 72, *National Fire Alarm Code®,* 1999 edition.

NFPA 75, *Standard for the Protection of Electronic Computer/Data Processing Equipment,* 1999 edition.

NFPA 82, *Standard on Incinerators and Waste and Linen Handling Systems and Equipment,* 1999 edition.

NFPA 86C, *Standard for Industrial Furnaces Using a Special Processing Atmosphere,* 1999 edition.

NFPA 96, *Standard for Ventilation Control and Fire Protection of Commercial Cooking Operations,* 1998 edition.

NFPA 99, *Standard for Health Care Facilities,* 1999 edition.

NFPA 101®, *Life Safety Code®,* 1997 edition.

NFPA 130, *Standard for Fixed Guideway Transit Systems,* 1997 edition.

NFPA 150, *Standard on Fire Safety in Racetrack Stables,* 1995 edition.

NFPA 170, *Standard for Fire Safety Symbols,* 1999 edition.

NFPA 214, *Standard on Water-Cooling Towers,* 1996 edition.

NFPA 220, *Standard on Types of Building Construction,* 1999 edition.

NFPA 231D, *Standard for Storage of Rubber Tires,* 1998 edition.

NFPA 251, *Standard Methods of Tests of Fire Endurance of Building Construction and Materials,* 1999 edition.

NFPA 307, *Standard for the Construction and Fire Protection of Marine Terminals, Piers, and Wharves,* 1995 edition.

NFPA 318, *Standard for the Protection of Cleanrooms,* 1998 edition.

NFPA 409, *Standard on Aircraft Hangars,* 1995 edition.

NFPA 415, *Standard on Airport Terminal Buildings, Fueling Ramp Drainage, and Loading Walkways,* 1997 edition.

NFPA 423, *Standard for Construction and Protection of Aircraft Engine Test Facilities,* 1999 edition.

NFPA 430, *Code for the Storage of Liquid and Solid Oxidizers,* 1995 edition.

NFPA 432, *Code for the Storage of Organic Peroxide Formulations,* 1997 edition.

NFPA 703, *Standard for Fire Retardant Impregnated Wood and Fire Retardant Coatings for Building Materials,* 1995 edition.

NFPA 803, *Standard for Fire Protection for Light Water Nuclear Power Plants,* 1998 edition.

NFPA 804, *Standard for Fire Protection for Advanced Light Water Reactor Electric Generating Plants,* 1995 edition.

NFPA 850, *Recommended Practice for Fire Protection for Electric Generating Plants and High Voltage Direct Current Converter Stations,* 1996 edition.

NFPA 851, *Recommended Practice for Fire Protection for Hydroelectric Generating Plants,* 1996 edition.

NFPA 1963, *Standard for Fire Hose Connections,* 1998 edition.

13-1.2 Other Publications.

13-1.2.1 ANSI Publications. American National Standards Institute, Inc., 11 West 42nd Street, 13th floor, New York, NY 10036.

ANSI B31.1, *Code for Power Piping.*

ANSI B36.10M, *Welded and Seamless Wrought Steel Pipe,* 1995.

13-1.2.2 ASME Publications. American Society of Mechanical Engineers, 345 East 47th Street, New York, NY 10017.

ASME A17.1, *Safety Code for Elevators and Escalators,* 1993.

ASME B1.20.1, *Pipe Threads, General Purpose (Inch),* 1983.

ASME B16.1, *Cast Iron Pipe Flanges and Flanged Fittings,* 1989.

ASME B16.3, *Malleable Iron Threaded Fittings,* 1992.

ASME B16.4, *Cast Iron Threaded Fittings,* 1992.

ASME B16.5, *Pipe Flanges and Flanged Fittings,* 1996.

ASME B16.9, *Factory-Made Wrought Steel Buttwelding Fittings,* 1993.

ASME B16.11, *Forged Steel Fittings, Socket-Welding and Threaded,* 1996.

ASME B16.18, *Cast Copper Alloy Solder Joint Pressure Fittings,* 1984.

ASME B16.22, *Wrought Copper and Copper Alloy Solder Joint Pressure Fittings,* 1995.

ASME B16.25, *Buttwelding Ends,* 1997.

13-1.2.3 ASTM Publications. American Society for Testing and Materials, 100 Barr Harbor Drive, West Conshohocken, PA 19428-2959.

ASTM A 53, *Standard Specification for Pipe, Steel, Black and Hot-Dipped, Zinc-Coated, Welded and Seamless,* 1995.

ASTM A 135, *Standard Specification for Electric-Resistance-Welded Steel Pipe,* 1993.

ASTM A 153, *Standard Specification for Zinc Coating (Hot Dip) on Iron and Steel Hardware,* 1998.

ASTM A 234/A 234M, *Standard Specification for Piping Fittings of Wrought-Carbon Steel and Alloy Steel for Moderate and Elevated Temperatures,* 1995.

ASTM A 795, *Standard Specification for Black and Hot-Dipped Zinc-Coated (Galvanized) Welded and Seamless Steel Pipe for Fire Protection Use,* 1995.

ASTM B 32, *Standard Specification for Solder Metal,* 1995.

ASTM B 75, *Standard Specification for Seamless Copper Tube,* 1995.

ASTM B 88, *Standard Specification for Seamless Copper Water Tube,* 1995.

ASTM B 251, *Standard Specification for General Requirements for Wrought Seamless Copper and Copper-Alloy Tube,* 1993.

ASTM B 446, *Standard Specification for Nickel-Chromium-Molybdenum-Columbium Alloy Rod and Bar,* 1993.

ASTM B 813, *Standard Specification for Liquid and Paste Fluxes for Soldering Applications of Copper and Copper-Alloy Tube,* 1993.

ASTM D 3309, *Standard Specification for Polybutylene (PB) Plastic Hot- and Cold-Water Distribution Systems,* 1995.

ASTM E 119, *Standard Test Methods for Fire Tests of Building Construction and Materials,* 1995.

ASTM E 136, *Standard Test Method for Behavior of Materials in a Vertical Tube Furnace at 750°C,* 1995.

ASTM F 437, *Standard Specification for Threaded Chlorinated Poly (Vinyl Chloride) (CPVC) Plastic Pipe Fittings, Schedule 80,* 1995.

ASTM F 438, *Standard Specification for Socket-Type Chlorinated Poly (Vinyl Chloride) (CPVC) Plastic Pipe Fittings, Schedule 40,* 1993.

ASTM F 439, *Standard Specification for Socket-Type Chlorinated Poly (Vinyl Chloride) (CPVC) Plastic Pipe Fittings, Schedule 80,* 1993.

ASTM F 442, *Standard Specification for Chlorinated Poly (Vinyl Chloride) (CPVC) Plastic Pipe (SDR-PR),* 1994.

ASTM F 1121, *Standard Specification for International Shore Connections for Marine Fire Applications,* 1987.

ASTM SI 10, *Standard for Use of the International System of Units (SI): the Modern Metric System,* 1997.

13-1.2.4 AWS Publications. American Welding Society, 550 N.W. LeJeune Road, Miami, FL 33126.

AWS A5.8, *Specification for Filler Metals for Brazing and Braze Welding,* 1992.

AWS B2.1, *Specification for Qualification of Welding Procedures and Welders for Piping and Tubing,* 1998.

AWS D10.9, *Specification for Qualification of Welding Procedures and Welders for Piping and Tubing,* 1980.

13-1.2.5 AWWA Publications. American Water Works Association, 6666 West Quincy Avenue, Denver, CO 80235.

AWWA C104, *Cement Mortar Lining for Ductile Iron Pipe and Fittings for Water,* 1995.
AWWA C105, *Polyethylene Encasement for Ductile Iron Pipe Systems,* 1993.
AWWA C110, *Ductile Iron and Gray Iron Fittings, 3-in. Through 48-in., for Water and Other Liquids,* 1993.
AWWA C111, *Rubber Gasket Joints for Ductile Iron Pressure Pipe and Fittings,* 1990.
AWWA C115, *Flanged Ductile Iron Pipe with Ductile Iron or Gray Iron Threaded Flanges,* 1994.
AWWA C150, *Thickness Design of Ductile Iron Pipe,* 1996.
AWWA C151, *Ductile Iron Pipe, Centrifugally Cast for Water,* 1996.
AWWA C200, *Steel Water Pipe 6 in. and Larger,* 1986.
AWWA C203, *Coal-Tar Protective Coatings and Linings for Steel Water Pipelines Enamel and Tape—Hot Applied,* 1997.
AWWA C205, *Cement-Mortar Protective Lining and Coating for Steel Water Pipe 4 in. and Larger—Shop Applied,* 1995.
AWWA C206, *Field Welding of Steel Water Pipe,* 1997.
AWWA C207, *Steel Pipe Flanges for Waterworks Service—Sizes 4 in. Through 144 in.,* 1978.
AWWA C208, *Dimensions for Fabricated Steel Water Pipe Fittings,* 1996.
AWWA C300, *Reinforced Concrete Pressure Pipe, Steel-Cylinder Type for Water and Other Liquids,* 1997.
AWWA C301, *Prestressed Concrete Pressure Pipe, Steel-Cylinder Type, for Water and Other Liquids,* 1992.
AWWA C302, *Reinforced Concrete Pressure Pipe, Non-Cylinder Type, for Water and Other Liquids,* 1995.
AWWA C303, *Reinforced Concrete Pressure Pipe, Steel-Cylinder Type, Pretensioned, for Water and Other Liquids,* 1995.
AWWA C400, *Standard for Asbestos-Cement Distribution Pipe, 4 in. Through 16 in., for Water and Other Liquids,* 1993.
AWWA C401, *Standard Practice for the Selection of Asbestos-Cement Water Pipe,* 1993.
AWWA C600, *Standard for the Installation of Ductile-Iron Water Mains and Their Appurtenances,* 1982.
AWWA C602, *Cement-Mortar Lining of Water Pipe Lines 4 in. and Larger—in Place,* 1995.
AWWA C603, *Standard for the Installation of Asbestos-Cement Water Pipe,* 1978.
AWWA C900, *Polyvinyl Chloride (PVC) Pressure Pipe, 4 in. Through 12 in., for Water and Other Liquids,* 1989.
AWWA M11, *A Guide for Steel Pipe-Design and Installation, 3rd edition,* 1989.

13-1.2.6 IEEE Publication. Institute of Electrical and Electronics Engineers, 445 Hoes Lane, P.O. Box 1331, Piscataway, NJ 08855-1331.

IEEE 45, *Marine Supplement.*

13-1.2.7 UL Publication. Underwriters Laboratories Inc., 333 Pfingsten Road, Northbrook, IL 60062.

UL 300, *Standard for Safety Fire Testing of Fire Extinguishing Systems for Protection of Restaurant Cooking Areas,* 1996.

13-1.2.8 U.S. Government Publications. U.S. Government Printing Office, Washington, DC 20402.

Title 46, *Code of Federal Regulations,* Parts 54.15-10, 56.20, 56.20-5(a), 56.50-95, 56.60, and 58.01-40.

Title 46, *Code of Federal Regulations,* Subchapter F, "Marine Engineering."

Title 46, *Code of Federal Regulations,* Subchapter J, "Electrical Engineering."

APPENDIX A

Explanatory Material

Appendix A is not a part of the requirements of this NFPA document but is included for informational purposes only. This appendix contains explanatory material, numbered to correspond with the applicable text paragraphs.

The material contained in Appendix A of NFPA 13 is included within the text of this handbook and, therefore, is not repeated here.

APPENDIX B

Miscellaneous Topics

This appendix is not a part of the requirements of this NFPA document but is included for informational purposes only.

B-1

Figure B-1 shows acceptable methods for interconnection of the fire protection and domestic water supply.

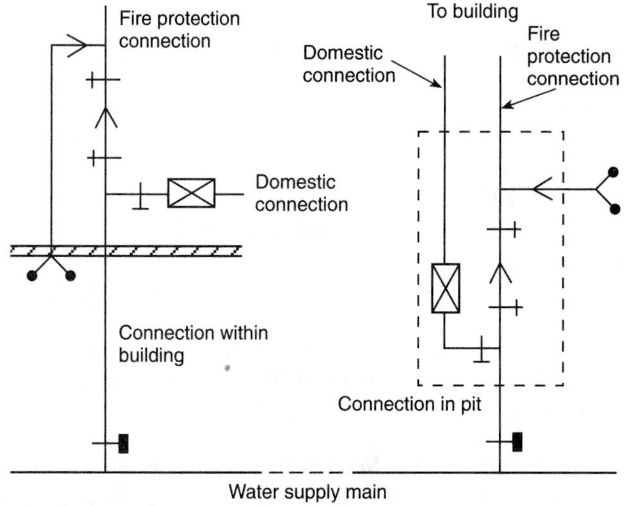

Figure B-1 *Permitted arrangements between the fire protection water supply and the domestic water supply.*

B-2 Sprinkler System Performance Criteria

B-2.1 Sprinkler system performance criteria have been based on test data. The factors of safety are generally small, are not definitive, and can depend on expected (but not guaranteed) inherent characteristics of the sprinkler systems involved. These inherent factors of safety consist of the following:

(1) The flow-declining pressure characteristic of sprinkler systems whereby the initial operating sprinklers discharge at a higher flow than with all sprinklers operating within the designated area.
(2) The flow-declining pressure characteristic of water supplies, which is particularly steep where fire pumps are the water source. This characteristic similarly produces higher than design discharge at the initially operating sprinklers.

The user of these standards can elect an additional factor of safety if the inherent factors are not considered adequate.

B-2.1.1 Performance-specified sprinkler systems, as opposed to scheduled systems, can be designed to take advantage of multiple loops or gridded configurations. Such configurations result in minimum line losses at expanded sprinkler spacing, in contrast to the older tree-type configurations, where advantage cannot be taken of multiple path flows.

Where the water supply characteristics are relatively flat with pressures being only slightly above the required sprinkler pressure at the spacing selected, gridded systems with piping designed for minimal economic line losses can all but eliminate the inherent flow-declining pressure characteristic generally assumed to exist in sprinkler systems. In contrast, the economic design of a tree-type system would likely favor a system design with closer sprinkler spacing and greater line losses, demonstrating the inherent flow-declining pressure characteristic of the piping system.

Elements that enter into the design of sprinkler systems include the following:

(1) Selection of density and area of application
(2) Geometry of the area of application (remote area)
(3) Permitted pressure range at sprinklers
(4) Determination of the water supply available
(5) Ability to predict expected performance from calculated performance
(6) Future upgrading of system performance
(7) Size of sprinkler systems

In developing sprinkler specifications, each of these elements needs to be considered individually. The most conservative design should be based on the application of the most stringent conditions for each of the elements.

B-2.1.2 Selection of Density and Area of Application. Specifications for density and area of application are developed from NFPA standards and other standards. It is desirable to specify densities rounded upward to the nearest 0.005 gpm/ft^2 (0.2 mm/min).

Prudent design should consider reasonable-to-expect variations in occupancy. This design would include not only variations in type of occupancy, but also, in the case of warehousing, the anticipated future range of materials to be stored, clearances, types of arrays, packaging, pile height, and pile stability, as well as other factors.

Design should also consider some degree of adversity at the time of a fire. To take this into account, the density and/or area of application can be increased. Another way is to use a dual-performance specification where, in addition to the normal primary specifications, a secondary density and area of application is specified. The objective of such a selection is to control the declining pressure-flow characteristic of the sprinkler system beyond the primary design flow.

A case can be made for designing feed and cross mains to lower velocities than branch lines to achieve the same result as specifying a second density and area of application.

B-2.1.3 Geometry of the Area of Application (Remote Area). It is expected that, over any portion of the sprinkler system equivalent in size to the area of application, the system will achieve the minimum specified density for each sprinkler within that area.

Where a system is computer-designed, ideally the program should verify the entire system by shifting the area of application the equivalent of one sprinkler at a time so as to cover all portions of the system. Such a complete computer verification of performance of the system is most desirable, but unfortunately not all available computer verification programs currently do this.

This selection of the proper Hazen-Williams coefficient is important. New unlined steel pipe has a Hazen-Williams coefficient close to 140. However, it quickly deteriorates to 130 and, after a few years of use, to 120. Hence, the basis for normal design is a Hazen-Williams coefficient of 120 for steel-piped wet systems. A Hazen-Williams coefficient of 100 is generally used for dry pipe systems because of the increased tendency for deposits and corrosion in these systems. However, it should be realized that a new system will have fewer line losses than calculated, and the distribution pattern will be affected accordingly.

Conservatism can also be built into systems by intentionally designing to a lower Hazen-Williams coefficient than that indicated.

B-2.1.4 Ability to Predict Expected Performance from Calculated Performance. Ability to accurately predict the performance of a complex array of sprinklers on piping is basically a function of the pipe line velocity. The greater the velocity, the greater is the impact on difficult-to-assess pressure losses. These pressure losses are presently determined by empirical means that lose validity as velocities increase. This is especially true for fittings with unequal and more than two flowing ports.

The inclusion of velocity pressures in hydraulic calculations improves the predictability of the actual sprinkler system performance. Calculations should come as close as practicable to predicting actual performance. Conservatism in design should be arrived at intentionally by known and deliberate means. It should not be left to chance.

B-2.1.5 Future Upgrading of System Performance. It is desirable in some cases to build into the system the capability to achieve a higher level of sprinkler performance than needed at present. If this is to be a consideration in conservatism, consideration needs to be given to maintaining sprinkler operating pressures on the lower side of the optimum operating range and/or designing for low pipe line velocities, particularly on feed and cross mains, to facilitate future reinforcement.

APPENDIX C

Explanation of Test Data and Procedures for Rack Storage

This appendix is not a part of the requirements of this NFPA document but is included for informational purposes only.

Appendix C provides an explanation of the test data and procedures that led to the development of sprinkler system discharge criteria for rack storage applications. The paragraphs are identified by the same number as the text in this standard to which they apply.

C-2-2 A review of full-scale fire tests run on the standard commodity (double tri-wall carton with metal liner), of Hallmark products and 3M products (e.g., abrasives, pressure-sensitive tapes of plastic fiber, and paper), and of the considerable number of commodity tests conducted provides a guide for commodity classifications. Such guidance is not related to any other method of classification of materials; therefore, sound engineering judgment and analysis of the commodity and the packaging should be used when selecting a commodity classification.

C-5-12.3.1 Tests 71, 73, 81, 83, 91, 92, 95, and 100 in the 20-ft (6.1-m) high array involving a single level of in-rack sprinklers were conducted without heat or water shields. Results were satisfactory.

Test 115 was conducted with two levels of sprinklers in racks with shields. Test 116, identical to Test 115 but without water shields, produced a lack of control. Visual observation of lower level in-rack sprinklers that did not operate although they were in the fire area indicated a need for water shields.

Tests 115 and 116 were conducted to investigate the necessity for water shields where multiple levels of in-rack sprinklers are installed.

Where water shields were not installed in Test 116, the fire jumped the aisle, and approximately 76 boxes were damaged. In Test 115 with water shields, the fire did not jump the aisle, and only 32 boxes were damaged. Water shields are, therefore, suggested wherever multiple levels of in-rack sprinklers are installed, except for installations with horizontal barriers or shelves that serve as water shields.

C-5-15.1.8 The time of operation of the first sprinkler varied from 52 seconds to 3 minutes and 55 seconds, with most tests under 3 minutes, except in Test 64 (Class III), where the first sprinkler operated in 7 minutes and 44 seconds. Fire detection more sensitive than waterflow is, therefore, considered necessary only in exceptional cases.

C-5-15.5.1 In most tests conducted, it was necessary to use small hose for mop-up operations. Small hose were not used in the high-expansion foam test.

Test 97 was conducted to evaluate the effect of dry pipe sprinkler operation. Test results were approximately the same as the base test with a wet pipe system. A study of NFPA records, however, indicates an increase in area of operation of 30 percent to be in order for dry pipe systems as compared with wet pipe systems.

C-7-2.3.1.1 Table. In all valid tests with double-row racks, sprinkler water supplies were shut off at approximately 60 minutes. In only one test did the last sprinkler operate in excess of 30 minutes after ignition; the last sprinkler operated in excess of 25 minutes in three tests, with the majority of tests involving the last sprinkler operating within 20 minutes.

C-7-4 The discharge criteria of Section 7-4 uses as a basis the large-scale fire test series conducted at the Factory Mutual Research Center, West Gloucester, Rhode Island.

The test building is approximately 200 ft × 250 ft (61 m × 76 m) [50,000 ft^2 (4.65 km^2) in area], of fire-resistive construction, and contains a volume of approximately 2.25 million ft^3 (63,761.86 m^3), the equivalent of a 100,000-ft^2 (9.29-km^2) building that is 22.5 ft (6.86 m) high. The test building has two primary heights beneath a single large ceiling. The east section is 30 ft (9.1 m) high, and the west section is 60 ft (18.29 m) high.

The test series for storage height of 20 ft (6.1 m) was conducted in the 30-ft (9.1-m) section with clearances from the top of storage to the ceiling nominally 10 ft (3.1 m).

Doors at the lower and intermediate levels and ventilation louvers at the tops of walls were kept closed during the majority of the fire tests, which minimized the effect of exterior conditions.

The entire test series was fully instrumented with thermocouples attached to rack members, simulated building columns, bar joists, and the ceiling.

Racks were constructed of steel vertical and horizontal members designed for 4000-lb (1814-kg) loads. Vertical members were 8 ft (2.4 m) on center for conventional racks and 4 ft (1.2 m) on center for simulated automated racks. Racks were 3½ ft (1.07 m) wide with 6-in. (152.4-mm) longitudinal flue space for an overall width of 7½ ft (2.29 m). Simulated automated racks and slave pallets were used in the main central rack in the 4-ft (1.2-m) aisle tests. Conventional racks and conventional pallets were used in the main central rack in the 8-ft (2.4-mm) aisle tests. The majority of the tests were conducted with 100-ft^2 (9.29-m^2) sprinkler spacing.

The test configuration for storage heights of 15 ft (4.6 m), 20 ft (6.1 m), and 25 ft (7.6 m) covered an 1800-ft^2 (167.2-m^2) floor area, including aisles between racks. Tests that were used in producing this standard limited fire damage to this area. The maximum water damage area anticipated in the standard is 6000 ft^2 (557.4 m^2), the upper limit of the design curves.

The test data shows that, as density is increased, both the extent of fire damage and sprinkler operation are reduced. The data also indicates that, with sprinklers installed in the racks, a reduction is gained in the area of fire damage and sprinkler operations (e.g., water damage).

Table C-7-4 illustrates these points. The information shown in the table is taken from the test series for storage height of 20 ft (6.1 m) using the standard commodity.

Table C-7-4 Summary of Relationship Between Sprinkler Discharge Density and the Extent of Fire Damage and Sprinkler Operation

Density gpm/ft²	Fire Damage in Test Array %	Fire Damage in Test Array ft²	Sprinkler Operation (165°F) Area (ft²)
0.30 (Ceiling only)	22	395	4500–4800
0.375 (Ceiling only)	17	306	1800
0.45 (Ceiling only)	9	162	700
0.20 (Ceiling only)	28–36	504–648	13,100–14,000
0.20 (Sprinklers at ceiling and in racks)	8	144	4100
0.30 (Sprinklers at ceiling and in racks)	7	126	700

For SI units: 1 ft = 0.3048 m; °C = 5⁄9 (°F-32); 1 gpm/ft² = 40.746 mm/min.

The fact that there is a reduction in both fire damage and area of water application as sprinkler densities are increased or where sprinklers are installed in racks should be considered carefully by those responsible for applying this standard to the rack storage situation.

In the test for storage height of 25 ft (7.6 m), a density of 0.55 gpm/ft² (22.4 mm/min) produced 42 percent, or 756 ft² (70.26 m²), fire damage in the test array and a sprinkler-wetted area of 1400 ft² (130.1 m²). Lesser densities would not be expected to achieve the same limited degree of control. Therefore, if the goal of smaller areas of fire damage is to be achieved, sprinklers in racks should be considered.

The test series for storage height over 25 ft (7.6 m) was conducted in the 60-ft (18.3-m) section of the test building with nominal clearances from the top of storage to the ceiling of either 30 ft (9.1 m) or 10 ft (3.1 m).

Doors at the lower and intermediate levels and ventilation louvers at the top of walls were kept closed during the fire tests, which minimized the effect of exterior wind conditions.

The purpose of the tests for storage height over 25 ft (7.6 m) was to accomplish the following:

(1) Determine the arrangement of in-rack sprinklers that can be repeated as pile height increases and that provide control of the fire
(2) Determine other protective arrangements, such as high-expansion foam, that provide control of the fire

Control was considered to have been accomplished if the fire was unlikely to spread from the rack of origin to adjacent racks or spread beyond the length of the 25-ft (7.6-m) test rack.

To aid in this judgment, control was considered to have been achieved if the fire failed to exhibit the following characteristics:

(1) Jump the 4-ft (1.2-m) aisles to adjoining racks
(2) Reach the end face of the end stacks (north or south ends) of the main rack

Control is defined as holding the fire in check through the extinguishing system until the commodities initially involved are consumed or until the fire is extinguished by the extinguishing system or manual aid.

The standard commodity as selected in the 20-ft (6.1-m) test series was used in the majority of tests for storage over 25 ft (7.6 m). Hallmark products and 3M products described in the 20-ft (6.1-m) test series report also were used as representative of Class III or Class IV commodities, or both, in several tests. The results of privately sponsored tests on Hallmark products and plastic encapsulated standard commodities also were made available to the committee.

A 25-ft (7.6-m) long test array was used for the majority of the tests for storage over 25 ft (7.6 m). The decision to use such an array was made because it was believed that a fire in racks over 25 ft (7.6 m) high that extended the full length of a 50-ft (15.24-m) long rack could not be considered controlled, particularly as storage heights increased.

One of the purposes of the tests was to determine arrangements of in-rack sprinklers that can be repeated as pile height increases and that provide control of the fire. The tests for storage height of 30 ft (9.1 m) explored the effect of such arrays. Many of these tests, however, produced appreciable fire spread in storage in tiers above the top level of protection within the racks. (In some cases, a total burnout of the top tiers of both the main rack and the target rack occurred.) In the case of the 30-ft (9.1-m) Hallmark Test 134 on the 60-ft (18.3-m) site, the material in the top tiers of storage burned vigorously, and the fire jumped the aisle above the fourth tier. The fire then burned downward into the south end of the fourth tier. In the test on the floor, a nominal 30-ft (9.1-m) clearance occurred between the top of storage and the ceiling sprinklers, whereas on the platform this clearance was reduced to nominal 10 ft (3.1 m). In most cases, the in-rack sprinklers were effective in controlling fire below the top level of protection within the racks. It has been assumed by the Test Planning Committee that, in an actual case with a clearance of 10 ft (3.1 m) or less above storage, ceiling sprinklers would be expected to control damage above the top level of protection within the racks. Tests have been planned to investigate lesser clearances.

Tests 114 and 128 explore the effect of changing the ignition point from the in-rack standard ignition point to a face ignition location. It should be noted, however, that both of these tests were conducted with 30-ft (9.1-m) clearance from the ceiling sprinklers to the top of storage and, as such, ceiling sprinklers had little effect on the fire in the top two tiers of storage. Fire spread in the three lower tiers is essentially the same. A similar change in the fire spread where the ignition point is changed was noted in Tests 126 and 127. Once again, 30-ft (9.1-m) clearance occurred between the top of storage and the ceiling sprinklers, and, as such, the ceiling sprinklers had little effect on the face fire. Comparisons of Tests 129, 130, and 131 in the test series for storage height of 50 ft (15.24 m) indicate little effect of point of ignition in the particular configuration tested.

Test 125, when compared with Test 133, indicates no significant difference in result between approved low-profile sprinklers and standard sprinklers in the racks.

C-7-4.1.3.1 Tests were conducted as a part of this program with eave line windows or louvers open to simulate smoke and heat venting. These tests opened 87.5 percent and 91 percent more sprinklers than did comparative tests without windows or louvers open. Venting tests that have been conducted in other programs were without the benefit of sprinkler protection and, as such, are not considered in this report, which covers only buildings protected by sprinklers. The design curves are based upon the absence of roof vents or draft curtains in the building. During mop-up operations, ventilating systems, where installed, should be capable of manual exhaust operations.

C-7-4.1.4.1 Test 80 was conducted to determine the effect of closing back-to-back longitudinal 6-in. (152.4-mm) flue spaces in conventional pallet racks. Test results indicated fewer sprinklers operating than with the flue space open, and, as such, no minimum back-to-back clearance is necessary if the transverse flue space is kept open.

Tests 145 and 146 were conducted to investigate the influence of longitudinal and transverse flue dimensions in double-row racks without solid shelves. Results were compared with Tests 65 and 66. Flue dimensions in Tests 65, 66, 145, and 146 were 6 in. (152.4 mm), 6 in. (152.4 mm), 3 in. (76.2 mm), and 12 in. (0.3 m), respectively. All other conditions were the same.

In Tests 65 and 66, 45 and 48 sprinklers operated compared with 59 and 58 for Tests 145 and 146, respectively. Fire damage in Tests 145 and 146 was somewhat less than in Tests 65 and 66; 2100 ft^3 (59.51 m^3) and 1800 ft^3 (51 m^3) in Tests 145 and 146, respectively, versus 2300 ft^3 (65.13 m^3) and 2300 ft^3 (65.13 m^3) in Tests 65 and 66, respectively, of combustible material were consumed.

Test results indicate narrow flue spaces of about 3 in. (76.2 mm) allow reasonable passage of sprinkler water down through the racks.

Tests 96 and 107, on multiple-row racks, used 6-in. (152.4-mm) transverse flue spaces. The water demand recommended in the standard is limited to those cases with nominal 6-in. (152.4-mm) transverse flues in vertical alignment.

C-7-4.1.7.1 A full-scale test program was conducted with various double-row rack storage arrangements of a cartoned Group A unexpanded plastic commodity at the Factory Mutual Research Corporation (FMRC) test facility. The series of nine tests included several variations, one of which involved the use of the following four distinct shelving arrangements: slatted wood, solid wood, wire mesh, and no shelving. The results of the testing program, specifically Tests 1, 2, 3, and 5, clearly demonstrate the acceptable performance of sprinkler systems protecting storage configurations that involve the use of slated shelving as described in 5-10.1. As a result of the test program, Factory Mutual has amended FM Loss Prevention Data Sheet 8-9 to allow slatted shelving to be protected in the same manner as an open rack arrangement.

Complete details of the test program are documented in the FMRC technical report FMRC J. I. 0X1R0.RR, "Large-Scale Fire Tests of Rack Storage Group A Plastics in Retail Operation Scenarios Protected by Extra Large Orifice (ELO) Sprinklers."

C-7-4.1.7.2 Test 98 with solid shelves 24 ft (7.3 m) long and 7½ ft (2.3 m) deep at each level produced total destruction of the commodity in the main rack and jumped the aisle. Density was 0.3 gpm/ft^2 (12.2 mm/min) from the ceiling sprinklers only. Test 108 with shelves 24 ft (7.3 m) long and 3½ ft (1.07 m) deep and with a 6-in. (152.4-mm) longitudinal flue space and one level of sprinklers in the rack resulted in damage to most of the commodity in the main rack but did not jump the aisle. Density from ceiling sprinklers was 0.375 gpm/ft^2 (15.3 mm/min), and rack sprinklers discharged at 15 psi (1 bar).

These tests did not yield sufficient information to develop a comprehensive protection standard for solid shelf racks. Items such as increased ceiling density, use of bulkheads, other configurations of sprinklers in racks, and limitation of shelf length and width should be considered.

Where such rack installations exist or are contemplated, the damage potential should be considered, and sound engineering judgment should be used in designing the protection system.

Test 98, with solid shelving obstructing both the longitudinal and transverse flue space, produced unsatisfactory results and indicates a need for sprinklers at each level in such a rack structure.

Test 147 was conducted with ceiling sprinklers only. Density was 0.45 gpm/ft^2 (18.3 mm/min) with a sprinkler spacing of 100 ft^2 (9.29 m^2). A total of 47 sprinklers opened, and 83 percent of the commodity was consumed. The fire jumped both aisles and spread to both ends of the main and target racks. The test was considered unsuccessful.

Test 148 was conducted with ceiling sprinklers and in-rack sprinklers. In-rack sprinklers were provided at each level (top of first, second, and third tiers) and were located in the longitudinal flue. They were directly above each other and 24 ft (7.3 m) on center or 22 ft (6.7 m) on each side of the ignition flue. Ceiling sprinkler discharge density was 0.375 gpm/ft^2 (15.3 mm/min). In-rack sprinkler discharge pressure was 30 psi (2.1 bar). A total of 46 ceiling sprinklers and three in-rack sprinklers opened, and 34 percent of the commodity was consumed. The fire consumed most of the material between the in-rack sprinklers and jumped both aisles.

C-7-4.1.8 Fire tests with open-top containers in the upper tier of storage and a portion of the third tier of storage produced an increase in sprinkler operation from 36 to 41 sprinklers and a more pronounced aisle jump and increase in fire spread in the main array. The smooth underside of the containers closely approximates fire behavior of slave pallets.

Installation of in-rack sprinklers or an increase in ceiling sprinkler density should be considered.

C-7-4.2.1.1.1 In one 20-ft (6.1-m) high test, sprinklers were buried in the flue space 1 ft (0.3 m) above the bottom of the pallet load, and results were satisfactory. Coverage of aisles by in-rack sprinklers is, therefore, not necessary, and distribution across the tops of pallet loads at any level is not necessary for the occupancy classes tested.

C-7-4.2.1.2.2 In all tests with in-rack sprinklers, obstructions measuring 3 in. × 3 ft (7.62 mm × 0.3 m) were introduced on each side of the sprinkler approximately 3 in. (76.2 mm) from the sprinkler to simulate rack structure member obstruction. This obstruction had no effect on sprinkler performance in the 20-ft (6.1-m) high tests.

Tests 103, 104, 105, and 109 in the 30-ft (9.1-m) high test with in-rack sprinklers obstructed by rack uprights produced unsatisfactory results. Tests 113, 114, 115, 117, 118, and 120 in the 30-ft (9.1-m) high test series with in-rack sprinklers located a minimum of 2 ft (0.61 m) from rack uprights produced improved results.

C-7-4.2.1.3 Operating pressures were 15 psi (1 bar) on all tests of sprinklers in racks with storage 20 ft (6.1 m) high and 30 psi (2.1 bar) for storage 30 ft (9.1 m) and 50 ft (15.24 m) high.

Tests 112 and 124 were conducted to compare the effect of increasing sprinkler discharge pressure at in-rack sprinklers from 30 psi to 75 psi (2.1 bar to 5.2 bar). With the higher discharge pressure, the fire did not jump the aisle, and damage below the top level of protection within the racks was somewhat better controlled by the higher discharge pressure of the in-rack sprinklers. A pressure of 15 psi (1 bar) was maintained on in-rack sprinklers in the first 30-ft (9.1-m) high tests (Tests 103 and 104). Pressure on in-rack sprinklers in subsequent tests was 30 psi (2.1 bar), except in Test 124, where it was 75 psi (5.2 bar).

C-7-4.2.1.4 In all except one case, using the standard commodity with one line of sprinklers installed in racks, only two sprinklers opened. In the one exception, two sprinklers opened in the main rack, and two sprinklers opened in the target rack.

C-7-4.2.1.6 Test 107, a multiple-row rack test conducted with pallet loads butted against each other, was 12 rows long. Each row was four boxes deep. With 0.45 gpm/ft^2 (18.3 mm/min) density from ceiling sprinklers only, fire spread to a depth of three rows on both sides of the ignition point. Fire damage, number of sprinklers open, and time rack steel temperature above 1000°F (538°C) were considerably greater than in comparable double-row rack Test 68. Temperatures at the ceiling did not reach dangerous limits. Fire intensity at the ends of rows was sufficiently intense to conclude that racks with deeper rows need additional protection.

C-7-4.2.2 Most tests for storage heights of 25 ft (7.6 m) and under were conducted with a clearance of 10 ft (3.1 m) from the top of storage to the sprinkler deflectors, and the basic design curves in Figures 7-4.2.2.1.1(a) through (g) reflect this condition.

Tests 140 and 141 were conducted with a 3-ft (0.9-m) clearance between the top of storage and the ceiling sprinkler deflectors. In Test 140, using a density of 0.3 gpm/ft^2 (12.2 mm/ min), 36 sprinklers operated compared with 45 and 48 sprinklers in Tests 65 and 66 with a 10-ft (3.1-m) clearance. In Test 141, 89 sprinklers operated compared with 140 sprinklers in Test 70 with a 10-ft (3.1-m) clearance. Firespread in Tests 140 and 141 was somewhat less than in Tests 65, 66, and 70.

Test 143 was conducted with an 18-in. (0.46-m) clearance between the top of storage and the ceiling sprinkler deflectors, and with a density of 0.3 gpm/ft^2 (12.2 mm/min). Thirty-seven sprinklers operated compared with 36 sprinklers in Test 140 with a 3-ft (0.9-m) clearance and 45 and 48 sprinklers in Tests 65 and 76 with a 10-ft (3.1-m) clearance. Firespread in Test 143 with an 18-in. (0.46-m) clearance was somewhat less than in Tests 65 and 66 with a 10-ft (3.1-m) clearance and Test 140 with a 3-ft (0.9-m) clearance.

Privately sponsored tests, using a 0.45 ceiling sprinkler density and an encapsulated commodity, indicated 40 sprinklers operating with a 10-ft (3.1-m) clearance, 11 sprinklers operating with a 3-ft (0.9-m) clearance, and 10 sprinklers operating with an 18-in. (0.46-m) clearance. Firespread was less in the test with the 18-in. (0.46-m) clearance than the 3-ft (0.9-m) clearance and also was less with the 3-ft (0.9-m) clearance than with the 10-ft (3.1-m) clearance.

C-7-4.2.2.1.1 Tests 65 and 66, compared with Test 69, and Test 93, compared with Test 94, indicated a reduction in areas of application of 44.5 percent and 45.5 percent, respectively, with high temperature–rated sprinklers as compared with ordinary temperature–rated sprinklers. Other extensive Factory Mutual tests produced an average reduction of 40 percent. Design curves are based on this area reduction. In constructing the design curves, the high-temperature curves above 3600 ft^2 (334.6 m^2) of application, therefore, represent 40 percent reductions in area of application of the ordinary-temperature curves in the 6000-ft^2 to 10,000-ft^2 (557.6-m^2 to 929.41-m^2) range.

Test 84 indicated the number of intermediate temperature–rated sprinklers operating is essentially the same as ordinary temperature–rated sprinklers.

C-7-4.2.2.1.9 Tests 77 and 95 were conducted to investigate protection needed on encapsulated commodities. The standard commodity [38 in. × 38 in. × 36 in. high (0.97 m × 0.97 m × 0.91 m high) sheet metal container inside a 42 in. × 42 in. × 42 in. (1.07 m × 1.07 m × 1.07 m) double tri-walled carton] was covered with a sheet of 4-mm to 6-mm thick polyethylene film stapled in place at the bottom. Test 77, at a density of 0.3 gpm/ft^2 (12.2 mm/min), with ceiling sprinklers only, went beyond the parameters for validity. Subsequent privately sponsored tests indicated control at a density of 0.45 gpm/ft^2 (18.3 mm/min). Test 95 indicated sprinklers at the ceiling and in racks adequately control this hazard. These test results were compared with Tests 65, 66, and 82 with comparable test configurations but without the plastic film covering.

A privately sponsored test was made with ceiling sprinklers only. At a density of 0.45 gpm/ft^2 (18.3 mm/min), 40 sprinklers operated. Firespread was slightly greater than in Test 65 with 0.3 gpm/ft^2 (12.2 mm/min) discharging from 45 sprinklers. Where the distance from the top of storage to the ceiling was reduced from 10 ft to 3 ft (3.1 m to 0.9 m) with 0.45 gpm/ft^2 (18.3 mm/min) density, 11 sprinklers operated. Firespread was less than in Test 65 or the previous privately sponsored test.

In order to evaluate the effect on plastic wrapping or encapsulation of pallet loads, Tests 77 and 95 were conducted as a part of the 20-ft (6.1-m) test series within the rack storage testing program, and Tests 1 and 2 were conducted as a part of privately sponsored Society of the Plastics Industry, Inc. (SPI) tests. Both SPI Tests 1 and 2 are considered valid and indicate that Classes I and II commodities can be protected by ceiling sprinklers only, using densities as indicated in design curves. Tests 1 and 2 also compare the results of a 3-ft (0.9-m) clearance from the top of storage to the sprinkler head deflectors with a 10-ft (3.1-m) clearance from the top of storage to the sprinkler head deflectors. A significant reduction in the number of sprinklers opening is indicated with the 3-ft (0.9-m) deflector clearance to the top of storage.

Subsequently, Tests 140 and 141 were made using the standard commodity. The distance from the top of storage to the sprinkler deflector was reduced to 3 ft (0.9 m). With 0.3 gpm/ ft^2 (12.2 mm/min) density, 36 sprinklers operated, and with 0.2 gpm/ft^2 (8.2 mm/min) density, 89 sprinklers operated. Firespread was somewhat less than in Tests 65 and 70 with a 10-ft (3.1-m) space between the top of storage and the ceiling.

C-7-4.2.2.2.2 Tests were not conducted with aisles wider than 8 ft (2.4 m) or narrower than 4 ft (1.2 m). It is, therefore, not possible to determine whether lower ceiling densities should be used for aisle widths greater than 8 ft (2.4 m) or if higher densities should be used for aisle widths less than 4 ft (1.2 m).

C-7-4.3.2.1 The recommended use of ordinary temperature–rated sprinklers at ceiling for storage higher than 25 ft (7.6 m) was determined by the results of fire test data. A test with high temperature–rated sprinklers and 0.45 gpm/ft^2 (18.3 mm/min) density resulted in fire damage in the two top tiers just within acceptable limits, with three ceiling sprinklers operating. A test with 0.45 gpm/ft^2 (18.3 mm/ min) density and ordinary temperature–rated sprinklers produced a dramatic reduction in fire damage with four ceiling sprinklers operating.

The four ordinary temperature–rated ceiling sprinklers operated before the first of the three high temperature–rated ceiling sprinklers. In both tests, two in-rack sprinklers at two levels operated at approximately the same time. The high temperature–rated sprinklers were at all times fighting a larger fire with less water than the ordinary temperature–rated ceiling sprinklers.

Tests 115 and 119 compare ceiling sprinkler density of 0.3 gpm/ft^2 (12.2 mm/min) with 0.45 gpm/ft^2 (18.3 mm/min). Damage patterns coupled with the number of boxes damaged in the main rack suggest that the increase in density produces improved control, particularly in the area above the top tier of in-rack sprinklers.

Tests 119 and 122 compare ceiling sprinkler temperature ratings of 286°F (141°C) and 165°F (74°C). A review of the number of boxes damaged and the firespread patterns indicates that the use of ordinary temperature–rated ceiling sprinklers on a rack configuration that incorporates in-rack sprinklers dramatically reduces the amount of firespread. Considering that in-rack sprinklers in the tests for storage over 25 ft (7.6 m) operated prior to ceiling sprinklers, it would seem that the installation of in-rack sprinklers converts an otherwise rapidly developing fire, from the standpoint of ceiling sprinklers, to a slower developing fire with a lower rate of heat release.

In the 20-ft (6.1-m) high test series, ceiling sprinklers operated before in-rack sprinklers. In the 30-ft (9.1-m) high series, ceiling sprinklers operated after in-rack sprinklers. The 50-ft (15.24-m) high test did not operate ceiling sprinklers. Ceiling sprinklers would, however, be needed if fire occurred in upper levels.

The results of these tests indicate the effect of in-rack sprinklers on storage higher than 25 ft (7.6 m). From the ceiling sprinkler operation standpoint, a fire with an expected high heat release rate was converted to a fire with a much lower heat release rate.

Since the fires developed slowly and opened sprinklers at two levels in the racks, only a few ceiling sprinklers were needed to establish control. Thus, the sprinkler operating area does not vary with height for storage over 25 ft (7.6 m) or for changes in sprinkler temperature rating and density.

All tests with sprinklers in racks were conducted using nominal ½-in. (12.7-mm) orifice size sprinklers of ordinary temperature.

C-7-4.4.1.1 In the RSP rack storage test series as well as the stored plastics program palletized test series, compartmented 16-oz (0.47-L) polystyrene jars were found to produce significantly higher protection requirements than the same commodity in a nested configuration. Polystyrene glasses and expanded polystyrene plates were comparable to the nested jars.

Different storage configurations within cartons or different products of the same basic plastic might, therefore, require reduced protection requirements.

In Test RSP-7, with nominal 15-ft (4.6-m) high storage with compartmented jars, a 0.6 gpm/ft^2 (24.5 mm/min) density, 8-ft (2.4-m) aisles, and a 10-ft (3.1-m) ceiling clearance, 29 sprinklers opened. In Tests RSP-4 with polystyrene glasses, RSP-5 with expanded polystyrene plates, and RSP-16 with nested polystyrene jars all stored at nominal 15-ft (4.6-m) height, 10-ft (3.1-m) ceiling clearance, 8-ft (2.4-m) aisles, and 0.6 gpm/ft^2 (24.5 mm/min) density, only four sprinklers opened.

However, Test RSP-11, with expanded polystyrene plates and 6-ft (1.8-m) aisles, demonstrated an increase in the number of operating sprinklers to 29. Test RSP-10 with expanded polystyrene plates, nominally 15 ft (4.6 m) high with a 10-ft (3.1-m) clearance and 8-ft (2.4-m) aisles, but protected only by 0.45 gpm/ft^2 (18.3 mm/min) density, opened 46 sprinklers and burned 100 percent of the plastic commodity.

At a nominal 20-ft (6.1-m) storage height with 8-ft (2.4-m) aisles, a 3-ft (0.9-m) ceiling clearance, and a 0.6 gpm/ft^2 (24.5 mm/min) density opened four sprinklers with polystyrene glasses in Test RSP-2 and 11 sprinklers with expanded polystyrene plates in Test RSP-6. In Test RSP-8, however, with the ceiling clearance increased to 10 ft (3.1 m) and other variables held constant, 51 sprinklers opened, and 100 percent of the plastic commodity burned.

Test RSP-3 with polystyrene glasses at a nominal height of 25 ft (7.6 m) with a 3-ft (0.9-m) ceiling clearance, 8-ft (2.4-m) aisles, and 0.6 gpm/ft^2 (24.5 mm/min) ceiling sprinkler density in combination with one level of in-rack sprinklers, resulted in four ceiling sprinklers and two in-rack sprinklers operating. Test RSP-9, with the same configuration but with polystyrene plates, opened 12 ceiling sprinklers and three in-rack sprinklers.

No tests were conducted with compartmented polystyrene jars at storage heights in excess of a nominal 15 ft (4.6 m) as a part of this program.

C-7-4.4.1.4 All tests in the RSP series were conducted utilizing ordinary temperature–rated sprinklers. However, after close review of all test data, the Technical Committee on Rack Storage believes that using intermediate or high temperature–rated sprinklers does not cause the demand areas to be any larger than those designated in Chapter 8; therefore, their use should be permitted.

C-7-4.4.2.3 Notes 1 and 2 to Figure 7-4.4.2.3(c) and Note 1 to Figure 7-4.4.2.3(d). The protection of Group A plastics by extra-large-orifice (ELO) sprinklers designed to provide 0.6 gpm/ft^2/2000 ft^2 (24.5 mm/min/186 m^2) or 0.45 gpm/ft^2/2000 ft^2 (18.3 mm/min/186 m^2) without the installation of in-rack sprinklers was developed from full-scale testing conducted with various double-row rack storage arrangements of a cartoned Group A unexpanded plastic commodity at the Factory Mutual Research Corporation (FMRC) test facility. The results of this test program are documented in the FMRC technical report, FMRC J.I. 0X1R0.RR, "Large Scale Fire Tests of Rack Stored Group A Plastics in Retail Operation Scenarios Protected by Extra Large Orifice (ELO) Sprinklers." The test program was initiated to address the fire protection issues presented by warehouse-type retail stores with regard to the display and storage of Group A plastic commodities including, but not limited to, acrylonitrile-butadiene-styrene copolymer (ABS) piping, polyvinyl chloride (PVC) hose and hose racks, tool boxes, polypropylene trash and storage containers, and patio furniture. Tests 1 and 2 of this series included protection of the Group A plastic commodity stored to 20 ft (6.1 m) under a 27-ft (8.2-m) ceiling by a design density of 0.6 gpm/ft^2 (24.5 mm/min) utilizing ELO sprinklers. The results of the testing program clearly demonstrate the acceptable performance of sprinkler systems that protect storage configurations involving Group A plastics up to 20 ft (6.1 m) in height under a 27-ft (8.2-m) ceiling where using ELO sprinklers to deliver a design density of 0.6 gpm/ft^2 (24.5 mm/min) and Group A plastics up to 14 ft (4.3 m) in height under a 22-ft (6.7-m) ceiling where using ELO sprinklers to deliver a design density of 0.45 gpm/ft^2 (18.3 mm/min). The tabulation of the pertinent tests shown in Table C-7-4.4.2.3 demonstrates acceptable performance.

C-7-5.2.3 No tests were conducted with idle pallets in racks using standard spray sprinklers. However, tests were conducted using ESFR and large drop sprinklers. Such storage conceivably

Table C-7-4.4.2.3 Summary of Test Results for Plastic Commodities Using ⅝-in. (15.9-mm) Orifice Sprinklers

	Date of Test						
Test Parameters	**8/20/93**	**8/25/93**	**9/2/93**	**10/7/93**	**2/17/94**	**2/25/94**	**4/27/94**
Type of shelving	Slatted wood	Slatted wood	Slatted wood	Slatted wood	Slatted wood	Slatted wood	Wire mesh
Other conditions/inclusions	—	—	—	—	Draft curtains	Draft curtains	—
Storage height (ft-in.)	19-11	19-11	15-4	15-4	19-11	19-11	13-11
Number of tiers	6[a]	6[a]	5[b]	5[b]	6[a]	6[b]	3
Clearance to ceiling/sprinklers (ft-in.)	6-10/ 6-3	6-10/ 6-3	11-5/ 10-10	11-5/ 10-10	6-10/ 6-3	6-10/ 6-3	8-4/ 7-9
Longitudinal/transverse flues (in.)	6/6 to 7½	6/6 to 7½	6/6 to 7	6/6 to 7½	6/6 to 7½	6/6 to 7½	6/3[c]
Aisle width (ft)	7½	7½	7½	7½	7½	7½	7½
Ignition centered below (number of sprinklers)	2	2	1	1	2	2	1
Sprinkler orifice size (in.)	0.64	0.64	0.64	0.64	0.64	0.64	0.64
Sprinkler temperature rating (°F)	165	286	286	165	165	286	286
Sprinkler RTI (ft-sec)$^{1/2}$	300	300	300	300	300	300	300
Sprinkler spacing (ft × ft)	8 × 10	8 × 10	8 × 10	8 × 10	8 × 10	8 × 10	10 × 10
Sprinkler identification	ELO-231	ELO-231	ELO-231	ELO-231	ELO-231	ELO-231	ELO-231
Constant water pressure (psi)	19	19	19	19	19	19	15.5
Minimum density (gpm/ft^2)	0.6	0.6	0.6	0.6	0.6	0.6	0.45
Test Results							
First sprinkler operation (min:sec)	2:03	2:25	1:12	0:44	1:25	0:52	0:49
Last sprinkler operation (min:sec)	2:12	15:19	6:34	7:34	15:54	14:08	10:58
Total sprinklers opened	4	9	7	13	35	18	12
Total sprinkler discharge (gpm)	205	450	363	613	1651	945	600
Average discharge per sprinkler (gpm)	51	50	52	47	47	52	50
Peak/maximum 1-min average gas temperature (°F)	1107/ 566	1412/ 868	965/308	662/184	1575/883	1162/767	1464/ 895
Peak/maximum 1-min average steel temperature (°F)	185/172	197/196	233/232	146/145	226/225	255/254	502/500
Peak/maximum 1-min average plume velocity (ft/sec)	27/15	25/18	18/15[d]	14/10[d]	26/23	20/18[d]	33/20
Peak/maximum 1-min heat flux (Btu/ ft^2/sec)	0.6/0.5	2.0/1.9	2.8/2.5	1.1/0.8	1.0/0.9	4.8/3.0	1.6/1.4
Aisle jump, east/west target ignition (min:sec)	None	8:24/ None	5:35/ 10:10	None	None	[e]/8:18	[e]/None
Equivalent number of pallet loads consumed	3	9	6	5	12	13	12
Test duration (min)	30	30	30	30	30	30	30
Results acceptable	Yes	Yes	Yes	Yes	No[f]	No[g]	Yes

For SI units, 1 ft = 0.305 m; 1 in. = 25.4 mm; °F = (1.8 × °C) + 32; °C = (°F − 32)/1.8; 1 psi = 0.069 bar; 1 gpm = 3.8 L/min; 1 ft/sec = 0.31 m/sec; 1 gpm/ft^2 = 40.746 mm/min.

[a]Main (ignition) racks divided into five or six tiers; bottom tiers each approximately 2 ft (0.6 m) high and upper tiers each about 5 ft (1.5 m) high; wood shelving below commodity at second through fifth tiers.

[b]Main (ignition) racks divided into five or six tiers; bottom tiers each approximately 2 ft (0.6 m) high and upper tiers each about 5 ft (1.5 m) high; wood shelving below commodity at second through fifth tiers; wire mesh shelving below commodity at sixth tier or below fifth (top) tier commodity.

[c]Transverse flues spaced 8 ft (2.4 m) apart [versus 3½ ft (1.1 m) apart in all other tests].

[d]Instrumentation located 5 ft (1.5 m) north of ignition.

[e]Minor surface damage to cartons.

[f]High water demand.

[g]Excessive firespread; marginally high water demand.

would introduce fire severity in excess of that contemplated by protection criteria for an individual commodity classification.

C-7-9.8.1 Temperatures in the test column were maintained below 1000°F (538°C) with densities, of roof ceiling sprinklers only, of 0.375 gpm/ft^2 (15.3 mm/min) with 8-ft (2.4-m) aisles and 0.45 gpm/ft^2 (18.3 mm/min) with 4-ft (1.2-m) aisles using the standard commodity.

APPENDIX D

Sprinkler System Information from the 1997 Edition of the *Life Safety Code*

This appendix is not a part of the requirements of this NFPA document but is included for informational purposes only.

D-1 Introduction

This appendix is provided as an aid to the user of NFPA 13 by identifying those portions of the 1997 edition of NFPA *101, Life Safety Code,* that pertain to sprinkler system design and installation. It is not intended that this appendix provide complete information regarding all aspects of fire protection addressed by NFPA *101.* It is important to note that this information was not copied from NFPA *101* using NFPA's extract policy and is not intended to be a part of the requirements of NFPA 13. While the 1997 edition of the *Life Safety Code* was the most current at the time of the publication of the 1999 edition of NFPA 13, a 2000 edition of the *Life Safety Code* is in preparation.

D-2 Definitions

See NFPA *101, Life Safety Code,* for terms not defined in Chapter 1.

D-3 Atriums

Glass walls and inoperable windows shall be permitted in lieu of the fire barriers where automatic sprinklers are spaced 6 ft (1.8 m) apart or less along both sides of the glass wall and inoperable windows, not more than 1 ft (0.3 m) from the glass, and with the automatic sprinklers located so that the entire surface of the glass is wet upon operation of the sprinklers. The glass shall be tempered, wired, or laminated glass held in place by a gasket system that permits the glass framing system to deflect without breaking (loading) the glass before the sprinklers operate. Automatic sprinklers shall not be required on the atrium side of the glass wall and inoperable windows where there is no walkway or other floor area on the atrium side above the main floor level. Doors in such walls shall be glass or other material that will resist the passage of smoke. Doors shall be self-closing or automatic-closing upon detection of smoke. [**101:** 6-2.4.6, Exception No. 2 to (a)]

D-4 Connection to Domestic Water Supply

Sprinkler piping serving not more than six sprinklers for any isolated hazardous area shall be permitted to be connected directly to a domestic water supply system having a capacity sufficient to provide 0.15 gpm/sq ft (6.1 L/min/sq m) of floor area throughout the entire enclosed area. An indicating shut-off valve shall be installed in an accessible location between the sprinklers and the connection to the domestic water supply. (**101:** 7-7.1.2)

D-5 Supervision

(**101:** 7-7.2)

D-5.1 Supervisory Signals.

Where supervised, automatic sprinkler systems are required by another section of NFPA *101,* supervisory attachments shall be installed and monitored for integrity in accordance with NFPA 72, *National Fire Alarm Code,* and a distinctive supervisory signal shall be provided to indicate a condition that would impair the satisfactory operation of the sprinkler system. This shall include, but not be limited to, monitoring of control valves, fire pump power supplies and running conditions, water tank levels and temperatures, pressure of tanks, and air pressure on dry-pipe valves. Supervisory signals shall sound and shall be displayed either at a location within the protected building that is constantly attended by qualified personnel or at an approved, remotely located receiving facility. (**101:** 7-7.2.1)

D-5.2 Alarm Signal Transmission.

Where supervision of automatic sprinkler systems is provided in accordance with another provision of NFPA *101,* waterflow alarms shall be transmitted to an approved, proprietary alarm receiving facility, a remote station, a central station, or the fire department. Such connection shall be installed in accordance with 7-6.1.4 of NFPA *101.* (**101:** 7-7.2.2)

D-6 Stages

D-6.1 Sprinklers shall not be required for stages 1000 sq ft (93 sq m) or less in area and 50 ft (15 m) or less in height where curtains, scenery, or other combustible hangings are not retractable vertically. Combustible hangings shall be limited to a single main curtain, borders, legs, and a single backdrop. (**101:** 8-4.5.10, Exception No. 1)

D-6.2 Sprinklers shall not be required under stage areas less than 4 ft (1.2 m) in clear height used exclusively for chair or table storage and lined on the inside with ⅝-in. (1.6-cm) Type X gypsum wallboard or the approved equivalent. (**101:** 8-4.5.10, Exception No. 2)

D-7 Exhibition Booths

The following shall be protected by automatic extinguishing systems:

(a) Single-level exhibit booths greater than 300 sq ft (27.9 sq m) and covered with a ceiling.

(b) Throughout each level of multilevel exhibit booths, including the uppermost level if the uppermost level is covered with a ceiling.

(c) A single exhibit or group of exhibits with ceilings that do not require sprinklers shall be separated by a minimum of 10 ft (3 m) where the aggregate ceiling exceeds 300 sq ft (27.9 sq m).

The water supply and piping for the sprinkler system shall be permitted to be of approved temporary means taken from an existing domestic water supply, an existing standpipe system, or an existing sprinkler system.

Exception No. 1: Ceilings that are constructed of open grate design or listed dropout ceilings in accordance with NFPA 13, Standard for the Installation of Sprinkler Systems, shall not be considered ceilings within the context of this section.

Exception No. 2: Vehicles, boats, and similar exhibited products having over 100 sq ft (9.3 sq m) of roofed area shall be provided with smoke detectors acceptable to the authority having jurisdiction.

Exception No. 3: Where fire protection of multilevel exhibit booths is consistent with the criteria developed through a life safety evaluation of the exhibition hall in accordance with 8-4.1 of NFPA 101, subject to approval of the authority having jurisdiction. (See A-8-2.3.2 of NFPA 101.)

(**101:** 8-7.5.3.7)

D-8 Proscenium Curtain

The proscenium opening of every legitimate stage shall be provided with a curtain constructed and mounted so as to intercept hot gases, flames, and smoke and to prevent flame from a fire on the stage from becoming visible from the auditorium side for a 5-min period where the curtain is of asbestos. Other materials shall be permitted if they have passed a 30-min fire test in a small scale 3 ft × 3 ft (0.9 m × 0.9 m) furnace with the sample mounted in the horizontal plane at the top of the furnace and subjected to the standard time-temperature curve.

The curtain shall be automatic-closing without the use of applied power.

All proscenium curtains shall be in the closed position except during performances, rehearsals, or similar activities.

Exception No. 1: In lieu of the protection required herein, all the following shall be provided:

(a) A noncombustible opaque fabric curtain shall be arranged so that it will close automatically; and

(b) An automatic, fixed waterspray deluge system shall be located on the auditorium side of the proscenium opening and shall be arranged so that the entire face of the curtain will be wetted. The system shall be activated by combination of rate-of-rise and fixed-temperature detectors located on the ceiling of the stage. Detectors shall be spaced in accordance with their listing. The water supply shall be controlled by a deluge valve and shall be sufficient to keep the curtain completely wet for 30 min or until the valve is closed by fire department personnel; and

(c) The curtain shall be automatically operated in case of fire by a combination of rate-of-rise and fixed-temperature detectors that also activates the deluge spray system. Stage sprinklers and vents shall be automatically operated by fusible elements in case of fire; and

(d) Operation of the stage sprinkler system or spray deluge valve shall automatically activate the emergency ventilating system and close the curtain; and

(e) The curtain, vents, and spray deluge system valve shall also be capable of manual operation.

Exception No. 2: Proscenium fire curtains or water curtains complying with 8-4.5.7 of NFPA 101.

(**101:** 9-4.5.7)

D-9

Listed quick response or listed residential sprinklers shall be used throughout smoke compartments containing patient sleeping rooms. (**101:** 12-3.5.2)

The requirements for use of quick response sprinklers intends that quick response sprinklers be the predominant type of sprinkler installed in the smoke compartment. It is recognized, however, that quick response sprinklers may not be approved for installation in all areas such as those where NFPA 13, *Standard for the Installation of Sprinkler Systems,* requires sprinklers of the intermediate- or high-temperature classification. It is not the intent of the 12-3.5.2 of NFPA *101* requirements to prohibit the use of standard sprinklers in limited areas of a smoke compartment where intermediate- or high-temperature sprinklers are required.

Where the installation of quick response sprinklers is impracticable in patient sleeping room areas, appropriate equivalent protection features acceptable to the authority having jurisdiction should be provided. It is recognized that the use of quick response sprinklers may be limited in facilities housing certain types of patients, or due to the installation limitations of quick response sprinklers. (**101:** A-12-3.5.2)

D-10

Where an automatic sprinkler system is installed, either for total or partial building coverage, the system shall be installed in accordance with Section 7-7 of NFPA *101.* In buildings up to and including four stories in height, systems installed in accordance with NFPA 13R, *Standard for the Installation of Sprinkler Systems in Residential Occupancies up to and Including Four Stories in Height,* shall be permitted.

Exception No. 1: In individual dwelling units, sprinkler installation shall not be required in closets not over 12 sq ft (1.1 sq m). Closets that contain equipment such as washers, dryers, furnaces, or water heaters shall be sprinklered regardless of size.

Exception No. 2: The draft stop and closely spaced sprinkler requirements of NFPA 13, Standard for the Installation of Sprinkler Systems, shall not be required for convenience openings complying with 6-2.4.8 of NFPA 101 where the convenience opening is within the dwelling unit.

(**101:** 18-3.5.1)

Where an automatic sprinkler system is installed, [illegible] for total or partial building coverage, the system shall be installed in accordance with Section [illegible]. In buildings up to and including four stories in height, systems installed in accordance with NFPA 13R, Standard for the Installation of Sprinkler Systems in Residential Occupancies up to and Including Four Stories in Height, shall be permitted.

[illegible]

Exception No. 2: [illegible]

[illegible]

APPENDIX E

Referenced Publications

E-1

The following documents or portions thereof are referenced within this standard for informational purposes only and are thus not considered part of the requirements of this standard unless also listed in Chapter 13. The edition indicated here for each reference is the current edition as of the date of the NFPA issuance of this standard.

E-1.1 NFPA Publications.

National Fire Protection Association, 1 Batterymarch Park, P.O. Box 9101, Quincy, MA 02269-9101.

NFPA 13E, *Guide for Fire Department Operations in Properties Protected by Sprinkler and Standpipe Systems,* 1995 edition.

NFPA 13R, *Standard for the Installation of Sprinkler Systems in Residential Occupancies up to and Including Four Stories in Height,* 1996 edition.

NFPA 14, *Standard for the Installation of Standpipe and Hose Systems,* 1996 edition.

NFPA 20, *Standard for the Installation of Centrifugal Fire Pumps,* 1999 edition.

NFPA 22, *Standard for Water Tanks for Private Fire Protection,* 1998 edition.

NFPA 25, *Standard for the Inspection, Testing, and Maintenance of Water-Based Fire Protection Systems,* 1998 edition.

NFPA 33, *Standard for Spray Application Using Flammable or Combustible Materials,* 1995 edition.

NFPA 36, *Standard for Solvent Extraction Plants,* 1997 edition.

NFPA 72, *National Fire Alarm Code®,* 1996 edition.

NFPA 80A, *Recommended Practice for Protection of Buildings from Exterior Fire Exposures,* 1996 edition.

NFPA 99, *Standard for Health Care Facilities,* 1999 edition.

NFPA *101®, Life Safety Code®,* 1997 edition.

NFPA 214, Standard on Water-Cooling Towers, 1996 edition.

NFPA 220, *Standard on Types of Building Construction,* 1999 edition.

NFPA 291, *Recommended Practice for Fire Flow Testing and Marking of Hydrants,* 1995 edition.

NFPA 307, *Standard for the Construction and Fire Protection of Marine Terminals, Piers, and Wharves,* 1995 edition.

NFPA 318, *Standard for the Protection of Cleanrooms,* 1998 edition.

NFPA 409, *Standard on Aircraft Hangars,* 1995 edition.

NFPA 415, *Standard on Airport Terminal Buildings, Fueling Ramp Drainage, and Loading Walkways,* 1997 edition.

NFPA 423, *Standard for Construction and Protection of Aircraft Engine Test Facilities,* 1999 edition.

NFPA 430, *Code for the Storage of Liquid and Solid Oxidizers,* 1995 edition.

NFPA 703, *Standard for Fire Retardant Impregnated Wood and Fire Retardant Coatings for Building Materials,* 1995 edition.

NFPA 803, *Standard for Fire Protection for Light Water Nuclear Power Plants,* 1998 edition.

NFPA 804, *Standard for Fire Protection for Advanced Light Water Reactor Generating Plants,* 1995 edition.

NFPA 850, *Recommended Practice for Fire Protection for Electric Generating Plants and High Voltage Direct Current Converter Stations,* 1996 edition.

NFPA 851, *Recommended Practice for Fire Protection for Hydroelectric Generating Plants,* 1996 edition.

NFPA 909, *Standard for the Protection of Cultural Resources, Including Museums, Libraries, Places of Worship, and Historic Properties,* 1997 edition.

E-1.2 Other Publications.

E-1.2.1 ACPA Publication. American Concrete Pipe Association, 222 W. Las Collinas Boulevard, Suite 641, Irving, TX 75039.

Concrete Pipe Handbook.

E-1.2.2 ASCE Publication. American Society of Civil Engineers, 1801 Alexander Bell Drive, Reston, VA 20191-4400.

Standard Guidelines for the Structural Applications of Steel Cables for Buildings, 1996.

E-1.2.3 ASME Publications. American Society of Mechanical Engineers, 345 East 47th Street, New York, NY 10017.

ASME B16.1, *Cast-Iron Pipe Flanges and Flanged Fittings for 25, 125, 250 and 800 lb,* 1989.

ASME A17.1, *Safety Code for Elevators and Escalators,* 1996.
ASME B1.20.1, *Pipe Threads, General Purpose (Inch),* 1983.

E-1.2.4 ASTM Publications. American Society for Testing and Materials, 100 Barr Harbor Drive, West Conshohocken, PA 19428-2959.

ASTM A 126, *Standard Specification for Gray Iron Casting for Valves, Flanges, and Pipe Fittings,* 1995.
ASTM A 135, *Standard Specification for Electric-Resistance-Welded Steel Pipe,* 1997.
ASTM A 197, *Standard Specification for Cupola Malleable Iron,* 1987.
ASTM A 307, *Standard Specification for Carbon Steel Bolts and Studs,* 1997.
ASTM D 3309, *Standard Specification for Polybutylene (PB) Plastic and Hot- and Cold-Water Distribution Systems,* 1996.
ASTM E 119, *Standard Test Methods for Fire Tests of Building Construction and Materials,* 1998.
ASTM F 437, *Standard Specification for Threaded Chlorinated Poly (Vinyl Chloride) (CPVC) Plastic Pipe Fittings, Schedule 80,* 1996.
ASTM F 438, *Standard Specification for Socket-Type Chlorinated Poly (Vinyl Chloride) (CPVC) Plastic Pipe Fittings, Schedule 40,* 1997.
ASTM F 439, *Standard Specification for Socket-Type Chlorinated Poly (Vinyl Chloride) (CPVC) Plastic Pipe Fittings, Schedule 80,* 1997.
ASTM F 442, *Standard Specification for Chlorinated Poly (Vinyl Chloride) (CPVC) Plastic Pipe (SDR-PR),* 1997.

E-1.2.5 AWWA Publications. American Water Works Association, 6666 West Quincy Avenue, Denver, CO 80235.

AWWA C104, *Cement Mortar Lining for Ductile Iron Pipe and Fittings for Water,* 1995.
AWWA C105, *Polyethylene Encasement for Ductile Iron Pipe Systems,* 1993.
AWWA C110, *Ductile Iron and Gray Iron Fittings, 3-in. Through 48-in., for Water and Other Liquids,* 1993.
AWWA C115, *Flanged Ductile Iron Pipe with Ductile Iron or Gray Iron Threaded Flanges,* 1994.
AWWA C150, *Thickness Design of Ductile Iron Pipe,* 1996.
AWWA C151, *Ductile Iron Pipe, Centrifugally Cast for Water,* 1996.
AWWA C153, *Ductile Iron Compact Fittings, 3 in. through 24 in. and 54 in. through 64 in. for Water Service,* 1994.
AWWA C203, *Coal-Tar Protective Coatings and Linings for Steel Water Pipelines Enamel and Tape — Hot Applied,* 1997.
AWWA C205, *Cement-Mortar Protective Lining and Coating for Steel Water Pipe 4 in. and Larger — Shop Applied,* 1995.
AWWA C206, *Field Welding of Steel Water Pipe,* 1997.
AWWA C208, *Dimensions for Fabricated Steel Water Pipe Fittings,* 1996.

AWWA C300, *Reinforced Concrete Pressure Pipe, Steel-Cylinder Type for Water and Other Liquids,* 1997.

AWWA C301, *Prestressed Concrete Pressure Pipe, Steel-Cylinder Type, for Water and Other Liquids,* 1992.

AWWA C302, *Reinforced Concrete Pressure Pipe, Non-Cylinder Type, for Water and Other Liquids,* 1995.

AWWA C303, *Reinforced Concrete Pressure Pipe, Steel-Cylinder Type, Pretensioned, for Water and Other Liquids,* 1995.

AWWA C400, *Standard for Asbestos-Cement Distribution Pipe, 4 in. Through 16 in., for Water and Other Liquids,* 1993.

AWWA C401, *Standard Practice for the Selection of Asbestos-Cement Water Pipe,* 1993.

AWWA C600, *Standard for the Installation of Ductile-Iron Water Mains and Their Appurtenances,* 1993.

AWWA C602, *Cement-Mortar Lining of Water Pipe Lines 4 in. and Larger — in Place,* 1995.

AWWA C603, *Standard for the Installation of Asbestos-Cement Water Pipe,* 1996.

AWWA C606, *Grooved and Shouldered Joints,* 1997.

AWWA C900, *Polyvinyl Chloride (PVC) Pressure Pipe, 4 in. Through 12 in., for Water and Other Liquids,* 1997.

AWWA M11, *A Guide for Steel Pipe-Design and Installation, 3rd edition,* 1989.

AWWA M14, *Recommended Practice for Backflow Prevention and Cross Connection Control, 2nd edition,* 1990.

AWWA M41, *Ductile Iron and Pipe Fittings.*

E-1.2.6 DIRPA Publication. Ductile Iron Pipe Research Association, 245 Riverchase Parkway, East, Suite 0, Birmingham, AL 35244.

Installation Guide for Ductile Iron Pipe.
Thrust Restraint Design for Ductile Iron Pipe.

E-1.2.7 EPRI Publication.

1843-2, "Turbine Generator Fire Protection by Sprinkler System," July 1985.

E-1.2.8 FMRC Publication. Factory Mutual Research Corporation, 1151 Boston-Providence Turnpike, Norwood, MA 02061.

FMRC J. I. 0X1R0.RR, "Large Scale Fire Tests of Rack Storage Group A Plastics in Retail Operation Scenarios Protected by Extra Large Orifice (ELO) Sprinklers."

E-1.2.9 IMO Publication. International Maritime Organization, 4 Albert Embankment, London, SEI 7SR, United Kingdom.

International Convention for the Safety of Life at Sea, 1974 (SOLAS 74), as amended, regulations II-2/3 and II-2/26.

E-1.2.10 SNAME Publication. Society of Naval Architects and Marine Engineers, 601 Pavonia Ave., Suite 400, Jersey City, NJ 07306.

Technical Research Bulletin 2-21, "Aluminum Fire Protection Guidelines."

E-1.2.11 UL Publication. Underwriters Laboratories Inc., 333 Pfingsten Road, Northbrook, IL 60062.

"Fact Finding Report on Automatic Sprinkler Protection for Fur Storage Vaults," November 25, 1947.

E-1.2.12 Uni-Bell Plastic Pipe Association.

Handbook of PVC Pipe.

E-1.2.13 U.S. Government Publications. U.S. Government Printing Office, Washington, DC 20402.

Title 46, *Code of Federal Regulations, Part 72.05-5.*

U.S. Federal Standard No. 66C, Standard for Steel Chemical Composition and Harden Ability, April 18, 1967, change notice No. 2, April 16, 1970.

Cross-References to Previous Editions

The 1999 edition of NFPA 13 is a consolidation of NFPA 13, NFPA 24, NFPA 231, NFPA 231C, NFPA 231D, NFPA 231E, and NFPA 231F.

1999 Edition	Previous Edition
1-1	13:1-1; 24:1-1; 231:1-1.1; 231C:1-1.1, 1-1.4
1-2	13:1-2; 24:1-2, 1-5; 231:1-1.4; 231C:1-1.4
1-2.1	13:1-2 Note 1
1-3 (N/C)	
1-3.1	13:1-3; 231:1-1.2; 231C:1-2, 1-4
1-4 (N/C)	
1-4.1	13:1-4.1
1-4.2	13:1-4.2
1-4.3	13:1-4.3
1-4.4	13:1-4.4
1-4.5	13:1-4.5
1-4.5.1	13:1-4.5.1
1-4.5.2	13:1-4.5.2
1-4.5.3	13:1-4.5.3
1-4.5.4	13:1-4.5.4
1-4.6	13:1-4.6
1-4.7	24:1-3
1-4.8	231:1-3
1-4.9	231C:1-3
1-4.10	231D:1-2
1-4.10 Misc. Storage	231D:3-3.1, 3-3.2
1-4.10.1	231D:1-3
1-4.11	231E:1-3
1-4.12	231F:1-4.11
1-5	13:1-5
1-6	13:1-6
1-6.1	13:1-6.1
1-6.2	13:1-6.2
1-7	24:1-6
1-7 Table	24:1-6
1-7.1	24:1-6.1
1-7.2	24:1-6.2
2-1	13:1-4.7
2-1.1	13:1-4.7.1
2-1.2	13:1-4.7.2
2-1.2.1	13:1-4.7.2.1
2-1.2.2	13:1-4.7.2.2
2-1.3	13:1-4.7.3
2-1.3.1 (R)	13:1-4.7.3.1
2-1.3.2 (R)	13:1-4.7.3.2
2-1.4 (R)	13:1-4.7.4, 1-4.7.4.1
2-2	231:2-1; 231C:2-1
2-2.1	231:2-1.1; 231C:2-1.1
2-2.1.1	231:2-1.1.1; 231C:2-1.1.1
2-2.1.2	231:2-1.1.2; 231C:2-1.1.2
2-2.2	231:2-1.2; 231C:2-1.2
2-2.3	231:2-1.3; 231C:2-1.3
2-2.3.1	231:2-1.3.1; 231C:2-1.3.1
2-2.3.2	231:2-1.3.2; 231C:2-1.3.2
2-2.3.3	231:2-1.3.3; 231C:2-1.3.3

1999 Edition	Previous Edition
2-2.3.4	231:2-1.3.4; 231C:2-1.3.4
2-2.4	231:2-1.4; 231C:2-3.1.4
2-2.4.1	231:2-1.4.1; 231C:2-3.1.4.1
2-2.4.2	231:2-1.4.2; 231C:2-3.1.4.2
2-2.4.3	231:2-1.4.3; 231C:2-3.1.4.3
2-2.5	231F:2-1
2-2.5.1	231F:2-1.1
2-2.5.2	231F:2-1.2
2-2.5.3	231F:2-1.3
2-2.5.4	231F:2-1.4
3-1	13:2-1
3-1.1	13:2-1.1; 24:3-1.1
3-1.2	13:2-1.2
3-2	13:2-2
3-2.1	13:2-2.1
3-2.2 (N)	
3-2.3	13:2-2.2
3-2.3.1 (R)	13:2-2.2.1
3-2.3.1 Table (R)	13:2-2.2
3-2.3.2	13:2-2.2.2
3-2.4	13:2-2.3
3-2.4.1	13:2-2.3.1
3-2.4.2	13:2-2.3.2
3-2.4.3	13:2-2.3.3
3-2.5	13:2-2.4
3-2.5.1	13:2-2.4.1
3-2.5.1 Table	13:2-2.4.1
3-2.5.2	13:2-2.4.2
3-2.6	13:2-2.5

1999 Edition	Previous Edition
3-2.6.1	13:2-2.5.1
3-2.6.2	13:2-2.5.2
3-2.6.3	13:2-2.5.3
3-2.6.4	13:2-2.5.4
3-2.6.5 (N)	33:7-2.5
3-2.7	13:2-2.6
3-2.7.1	13:2-2.6.1
3-2.7.2	13:2-2.6.2
3-2.8	13:2-2.7
3-2.9	13:2-2.8
3-2.9.1	13:2-2.8.1
3-2.9.2	13:2-2.8.2
3-2.9.3	13:2-2.8.3
3-3	13:2-3; 24:7-5.1
3-3.1 (R)	13:2-3.1
3-3.1 Table	13:2-3.1
3-3.2	13:2-3.2
3-3.3	13:2-3.3
3-3.4	13:2-3.4
3-3.5	13:2-3.5
3-3.5 Table	13:2-3.5
3-3.6	13:2-3.6
3-3.7	13:2-3.7
3-4	24:7-1
3-4.1 (R)	24:7-1.1
3-4.2	24:7-1.2
3-4.3	24:7-1.3
3-4.4 (R)	24:7-2
3-5	13:2-4; 24:7-5.1
3-5.1	13:2-4.1
3-5.1 Table	13:2-4.1
3-5.2	13:2-4.2

1999 Edition	Previous Edition	1999 Edition	Previous Edition
3-5.2 Table	13:2-4.2	3-6.7 (R)	24:7-3
3-5.3	13:2-4.3	3-7 (R)	13:2-6
3-5.4	13:2-4.4	3-8	13:2-7; 24:3-1
3-5.5	13:2-4.5	3-8.1	13:2-7.1
3-5.6	24:7-4	3-8.1.1	13:2-7.1.1; 24:3-1.2
3-6	13:2-5	3-8.1.2	13:2-7.1.2
3-6.1	13:2-5.1	3-8.1.3	13:2-7.1.3
3-6.1.1	13:2-5.1.1	3-8.2	13:2-7.2
3-6.1.2	13:2-5.1.2	3-8.3	13:2-7.3
3-6.1.3	13:2-5.1.3	3-9	13:2-8
3-6.2	13:2-5.2	3-9.1	13:2-8.1; 24:2-6.7
3-6.2.1 (R)	13:2-5.2.1	3-9.2	13:2-8.2; 24:2-6.8
3-6.2.2	13:2-5.2.2	3-9.3	24:2-6.6
3-6.2.3	13:2-5.2.3	3-10	13:2-9
3-6.2.4	13:2-5.2.4	3-10.1	13:2-9.1
3-6.2.5	13:2-5.2.5	3-10.2	13:2-9.2
3-6.2.6	13:2-5.2.6	3-10.2.1	13:2-9.2.1
3-6.2.7	13:2-5.2.7	3-10.2.2	13:2-9.2.2
3-6.2.8	13:2-5.2.8	3-10.2.3	13:2-9.2.3
3-6.2.8.1 (R)	13:2-5.2.8.1	3-10.2.4	13:2-9.2.4
3-6.2.8.2	13:2-5.2.8.2	3-10.3	13:2-9.3
3-6.2.9	13:2-5.2.9	3-10.3.1	13:2-9.3.1
3-6.2.9.1	13:2-5.2.9.1	3-10.3.2	13:2-9.3.2
3-6.2.9.2	13:2-5.2.9.2	3-10.4	13:2-9.4
3-6.3	13:2-5.3	3-10.5	13:2-9.5
3-6.3.1	13:2-5.3.1	3-10.5.1	13:2-9.5.1
3-6.3.2	13:2-5.3.2	3-10.5.2	13:2-9.5.2
3-6.4	13:2-5.4	3-10.6	13:2-9.6
3-6.4.1	13:2-5.4.1	4-1	13:3-1
3-6.5	13:2-5.5	4-1.1	13:3-1.1
3-6.6	13:2-5.6	4-1.2	13:3-1.2
3-6.6.1	13:2-5.6.1	4-1.3	13:3-1.3

1999 Edition	Previous Edition
4-2	13:3-2
4-2.1	13:3-2.1
4-2.2	13:3-2.2
4-2.3	13:3-2.3
4-2.3.1	13:3-2.3.1
4-2.3.2	13:3-2.3.2
4-2.4	13:3-2.4
4-2.4.1	13:3-2.4.1
4-2.4.2	13:3-2.4.2
4-2.4.3	13:3-2.4.3
4-2.4.4	13:3-2.4.4
4-2.4.5	13:3-2.4.5
4-2.5	13:3-2.5
4-2.5.1	13:3-2.5.1
4-2.5.2	13:3-2.5.2
4-2.5.3	13:3-2.5.3
4-2.5.4	13:3-2.5.4
4-2.6	13:3-2.6
4-2.6.1	13:3-2.6.1
4-2.6.2	13:3-2.6.2
4-2.6.3	13:3-2.6.3
4-2.6.4	13:3-2.6.4
4-2.6.5	13:3-2.6.5
4-2.6.5 Figure	13:3-2.6.5
4-2.6.6	13:3-2.6.6
4-2.6.7	13:3-2.6.7
4-2.6.8	13:3-2.6.8
4-3	13:3-3
4-3.1	13:3-3.1
4-3.1.1	13:3-3.1.1
4-3.1.2	13:3-3.1.2
4-3.1.3	13:3-3.1.3

1999 Edition	Previous Edition
4-3.1.4	13:3-3.1.4
4-3.1.5	13:3-3.1.5
4-3.1.6	13:3-3.1.6
4-3.1.7	13:3-3.1.7
4-3.1.7.1	13:3-3.1.7.1
4-3.1.7.2	13:3-3.1.7.2
4-3.1.8	13:3-3.1.8
4-3.1.8.1	13:3-3.1.8.1
4-3.1.8.2	13:3-3.1.8.2
4-3.2	13:3-3.2
4-3.2.1	13:3-3.2.1
4-3.2.2	13:3-3.2.2
4-3.2.3	13:3-3.2.3
4-3.2.4	13:3-3.2.4
4-3.2.5	13:3-3.2.5
4-3.3	13:3-3.3
4-3.3.1	13:3-3.3.1
4-3.3.2	13:3-3.3.2
4-4	13:3-4
4-4.1	13:3-4.1
4-4.1.1	13:3-4.1.1
4-4.1.2	13:3-4.1.2
4-4.1.3	13:3-4.1.3
4-4.1.4	13:3-4.1.4
4-4.2	13:3-4.2
4-4.2.1	13:3-4.2.1
4-4.2.1 Figure	13:3-4.2
4-4.2.2	13:3-4.2.2
4-4.2.3	13:3-4.2.3
4-4.2.4	13:3-4.2.4
4-4.3	13:3-4.3
4-4.4	13:3-4.4

1999 Edition	Previous Edition	1999 Edition	Previous Edition
4-4.4.1	13:3-4.4.1	4-6.1.3.2	13:3-6.1.3.2
4-4.4.2	13:3-4.4.2	4-6.1.4	13:3-6.1.4
4-4.4.3	13:3-4.4.3	4-6.1.5	13:3-6.1.5
4-4.5	13:3-4.5	4-6.1.6	13:3-6.1.6
4-4.6	13:3-4.6	4-6.1.7	13:3-6.1.7
4-5	13:3-5	4-6.1.7.1	13:3-6.1.7
4-5.1	13:3-5.1	4-6.1.7.2	13:3-6.1.7.1
4-5.2	13:3-5.2	4-7	13:3-7
4-5.2.1	13:3-5.2.1	4-7.1	13:3-7.1
4-5.2.1 Table	13:3-5.2.1 Table	4-7.2	13:3-7.2
4-5.2.2	13:3-5.2.2	4-7.2.1	13:3-7.2.1
4-5.2.2 Table (R)	13:3-5.2.2 Table	4-7.2.2	13:3-7.2.2
4-5.2.3	13:3-5.2.3	4-7.3	13:3-7.3
4-5.2.3(a) Figure	13:3-5.2.3(a)	4-7.3.1	13:3-7.3.1
4-5.2.3(b) Figure	13:3-5.2.3(b)	4-7.3.2	13:3-7.3.2
4-5.3 (N)		4-7.3.3	13:3-7.3.3
4-5.3.1 (R)	12:3-5.3.1	4-7.3.4	13:3-7.3.4
4-5.3.1 Figure	13:3-5.3(a) Figure	4-7.4	13:3-7.4
4-5.3.2 (R)	13:3-5.3.1 Ex., 3-5.3.2	4-7.4.1	13:3-7.4.1
4-5.3.2 Figure (R)	13:3-5.3(b) Figure	4-7.4.2	13:3-7.4.2
4-6	13:3-6	4-7.4.2(a) Figure	13:3-7.4.2
4-6.1	13:3-6.1	4-7.4.2(b) Figure	13:3-7.4.2
4-6.1.1	13:3-6.1.1	4-7.4.3	13:3-7.4.3
4-6.1.1.1	13:3-6.1.1.1	4-7.5	13:3-7.5
4-6.1.1.2	13:3-6.1.1.2	4-7.6	13:3-7.6
4-6.1.1.3	13:3-6.1.1.3	4-7.7	13:3-7.7
4-6.1.1.4	13:3-6.1.1.4	4-7.8	13:3-7.8
4-6.1.1.5	13:3-6.1.1.5	4-8	13:3-8
4-6.1.1.6	13:3-6.1.1.6	4-8.1	13:3-8.1
4-6.1.2	13:3-6.1.2	4-8.2	13:3-8.2
4-6.1.3	13:3-6.1.3	4-8.2.1	13:3-8.2.1
4-6.1.3.1	13:3-6.1.3.1	4-8.2.2	13:3-8.2.2

1999 Edition	Previous Edition
4-8.2.3	13:3-8.2.3
4-8.2.4 (R)	13:3-8.2.4
4-8.2.5	13:3-8.2.5
4-8.2.6	13:3-8.2.6
4-8.2.7	13:3-8.2.7
4-8.2.7 Figure (R)	13:3-8.2.7
4-9	13:3-9
4-9.1	13:3-9.1
4-9.2	13:3-9.2
4-9.3	13:3-9.3
4-9.4	13:3-9.4
4-9.5	13:3-9.5
4-9.6	13:3-9.6
4-9.7	13:3-9.7
4-9.8	13:3-9.8
4-9.8.1	13:3-9.8.1
4-9.8.2	13:3-9.8.2
4-9.8.3	13:3-9.8.3
4-9.9	13:3-9.9
4-9.10	13:3-9.10
4-9.11	13:3-9.11
4-9.12	13:3-9.12
5-1	13:4-1
5-1.1	13:4-1.1
5-1.2	13:4-1.2
5-2	13:4-2
5-3	13:4-3
5-3.1	13:4-3.1
5-3.1.1	13:4-3.1.1
5-3.1.2	13:4-3.1.2
5-3.1.3 (N)	
5-3.1.4	13:4-3.1.3

1999 Edition	Previous Edition
5-3.1.4.1	13:4-3.1.3.1; 231E: 4-1.3.5
5-3.1.4.2	13:4-3.1.3.2
5-3.1.4.2(a) Table	13:4-3.1.3.2(a)
5-3.1.4.2(b) Table	13:4-3.3.1.3.2(b)
5-3.1.4.2 Figure	13:4-3.1.3.2
5-3.1.4.3	13:4-3.1.3.3
5-3.1.4.4	231:5-1.2; 231C:5-2.3
5-3.1.5 (N)	
5-3.1.5.1	13:5-2.1.3
5-3.1.5.2	13:5-2.1.3.1
5-4	13:4-4
5-4.1	13:4-4.1; 231C:5-2
5-4.1.1	13:4-4.1
5-4.1.2 (R)	231:5-1.5, 7-2.2.3; 231C:5-2.2
5-4.2 (R)	13:4-4.2
5-4.3 (R)	13:4-4.3
5-4.4	13:4-4.4
5-4.5	13:4-4.5
5-4.5.1	13:4-4.5.1
5-4.5.2 (R)	13:4-4.5.2
5-4.5.3	13:4-4.5.3
5-4.5.4 (N)	
5-4.6	13:4-4.6
5-4.6.1	13:4-4.6.1; 231C:10-2.3
5-4.6.2	13:4-4.6.2; 231:9-1.4
5-4.6.3	13:4-4.6.3; 231:9-1.3
5-4.6.4 (N)	
5-4.6.5	13:4-4.6.4
5-4.7	13:4-4.7
5-4.7.1	13:4-4.7.1

1999 Edition	Previous Edition
5-4.7.2	13:4-4.7.2
5-4.7.3	13:4-4.7.3
5-4.7.3 Ex. 2	231:8-2.7; 231C:9-1.6
5-4.7.3 Ex. 3	231:8-2.7; 231C:9-1.6
5-4.8	13:4-4.8
5-4.9	13:4-4.9
5-4.9.1	13:4-4.9.1
5-4.9.2	13:4-4.9.2
5-5	13:4-5
5-5.1	13:4-5.1
5-5.2	13:4-5.2
5-5.2.1	13:4-5.2.1
5-5.2.1.1	13:4-5.2.1.1
5-5.2.2	13:4-5.2.2
5-5.3	13:4-5.3
5-5.3.1	13:4-5.3.1
5-5.3.2	13:4-5.3.2
5-5.3.3	13:4-5.3.3
5-5.3.4	13:4-5.3.4
5-5.4	13:4-5.4
5-5.4.1	13:4-5.4.1
5-5.4.2	13:4-5.4.2
5-5.5	13:4-5.5
5-5.5.1	13:4-5.5.1, 4-5.5.1.1
5-5.5.2	13:4-5.5.2
5-5.5.2.1 (R)	13:4-5.5.2.1
5-5.5.2.2 (R)	13:4-5.5.2.2
5-5.5.3	13:4-5.5.3
5-5.5.3.1 (R)	13:4-5.5.3.1
5-5.5.3.2	13:4-5.5.3.2
5-5.6	13:4-5.6; 231C:4-5; 231D:3-2.1; 231E:4-1.4
5-6	13:4-6
5-6.1	13:4-6.1
5-6.2	13:4-6.2
5-6.2.1	13:4-6.2.1
5-6.2.2	13:4-6.2.2
5-6.2.2(a) Table (R)	13:4-6.2 Table
5-6.2.2(b) Table (R)	13:4-6.2 Table
5-6.2.2(c) Table (R)	13:4-6.2 Table
5-6.2.2(d) Table (R)	13:4-6.2 Table
5-6.3	13:4-6.3
5-6.3.1	13:4-6.3.1
5-6.3.2	13:4-6.3.2
5-6.3.2.1	13:4-6.3.2
5-6.3.2.2	13:4-6.3.2.1
5-6.3.3	13:4-6.3.3
5-6.3.4	13:4-6.3.4
5-6.4	13:4-6.4
5-6.4.1	13:4-6.4.1
5-6.4.1.1	13:4-6.4.1.1
5-6.4.1.2 (R)	13:4-6.4.1.2
5-6.4.1.3	13:4-6.4.1.3
5-6.4.1.3(a) Figure	13:4-6.4.1.3(a)
5-6.4.1.3(b) Figure	13:4-6.4.1.3(b)
5-6.4.1.3(c) Figure	13:4-6.4.1.3(c)
5-6.4.1.4	13:4-6.4.1.4
5-6.4.1.4 Figure	13:4-6.4.1.4
5-6.4.2 (R)	13:4-6.4.2
5-6.5	13:4-6.5
5-6.5.1	13:4-6.5.1
5-6.5.1.1	13:4-6.5.1.1
5-6.5.1.2	13:4-6.5.1.2
5-6.5.1.2(a) Figure	13:4-6.5.1.2(a)

1999 Edition	Previous Edition
5-6.5.1.2(b) Figure	13:4-6.5.1.2(b)
5-6.5.2	13:4-6.5.2
5-6.5.2.1 (R)	13:4-6.5.2.1
5-6.5.2.2 (R)	13:4-6.5.2.2
5-6.5.2.2 Figure (R)	13:4-6.5.2.2 Figure
5-6.5.2.3	13:4-6.5.4
5-6.5.2.3 Figure	13:4-6.5.4 Figure
5-6.5.3 (R)	13:4-6.5.3
5-6.5.3.1 (R)	13:4-6.5.3.1
5-6.5.3.2	13:4-6.5.3.2
5-6.6	13:4-6.6
5-7	13:4-7
5-7.1	13:4-7.1
5-7.2	13:4-7.2
5-7.2.1	13:4-7.2.1
5-7.2.1.1	13:4-7.2.1
5-7.2.1.2	13:4-7.2.1.1
5-7.2.2	13:4-7.2.2
5-7.2.2 Table	13:4-7.2 Table
5-7.3	13:4-7.3
5-7.3.1	13:4-7.3.1
5-7.3.1.1	13:4-7.3.1
5-7.3.1.2	13:4-7.3.1.1
5-7.3.2	13:4-7.3.2
5-7.3.3	13:4-7.3.3
5-7.3.4	13:4-7.3.4
5-7.4	13:4-7.4
5-7.4.1	13:4-7.4.1
5-7.4.1.1 (R)	13:4-7.4.1.1
5-7.4.1.2 (R)	13:4-7.4.1.2
5-7.4.1.3	13:4-7.4.1.3
5-7.4.1.4	13:4-7.4.1.3.1

1999 Edition	Previous Edition
5-7.4.2	13:4-7.4.2
5-7.4.2.1	13:4-7.4.2
5-7.4.2.2	13:4-7.4.2.1
5-7.5	13:4-7.5
5-7.5.1	13:4-7.5.1
5-7.5.1.1	13:4-7.5.1.1
5-7.5.1.2	13:4-7.5.1.2
5-7.5.1.2 Table	13:4-7.5.1.2 Table
5-7.5.1.2 Figure	13:4-7.5.1.2 Figure
5-7.5.1.3	13:4-7.5.1.3
5-7.5.1.3 Table	13:4-7.5.1.3 Table
5-7.5.1.3 Figure	13:4-7.5.1.3 Figure
5-7.5.2	13:4-7.5.2
5-7.5.2.1 (R)	13:4-7.5.2.1
5-7.5.2.2 (R)	13:4-7.5.2.2
5-7.5.2.2 Figure (R)	13:4-7.5.2.2 Figure
5-7.5.2.3	13:4-7.5.4
5-7.5.2.3 Figure	13:4-7.5.4 Figure
5-7.5.3	13:4-7.5.3
5-7.5.3.1	13:4-7.5.3
5-7.5.3.2	13:4-7.5.3.1
5-7.6	13:4-7.6
5-8	13:4-8
5-8.1	13:4-8.1
5-8.2	13:4-8.2
5-8.2.1	13:4-8.2.1
5-8.2.1 Table	13:4-8.2
5-8.2.2	13:4-8.2.2
5-8.3	13:4-8.3
5-8.3.1	13:4-8.3.1
5-8.3.2	13:4-8.3.2
5-8.3.3	13:4-8.3.3

1999 Edition	Previous Edition
5-8.3.4	13:4-8.3.4
5-8.4	13:4-8.4
5-8.4.1	13:4-8.4.1
5-8.4.1.1	13:4-8.4.1.1
5-8.4.1.2	13:4-8.4.1.2
5-8.4.1.3	13:4-8.4.1.3
5-8.4.2	13:4-8.4.2
5-8.5	13:4-8.5
5-8.5.1	13:4-8.5.1
5-8.5.1.1	13:4-8.5.1.1
5-8.5.1.2	13:4-8.5.1.2
5-8.5.1.2 Table	13:4-8.5.1.2(a) Table
5-8.5.1.2(a) Figure	13:4-8.5.1.2(a) Figure
5-8.5.1.2(b) Figure	13:4-8.5.1.2(b) Figure
5-8.5.2	13:4-8.5.2
5-8.5.2.1 (R)	13:4-8.5.2.1
5-8.5.2.2 (R)	13:4-8.5.2.2
5-8.5.2.2 Figure (R)	13:4-8.5.2.2 Figure
5-8.5.2.3	13:4-8.5.4
5-8.5.2.3 Table	13:4-8.5.4 Table
5-8.5.2.3 Figure	13:4-8.5.4 Figure
5-8.5.3	13:4-8.5.3
5-8.5.3.1 (R)	13:4-8.5.3.1
5-8.5.3.2	13:4-8.5.3.2
5-8.6	13:4-8.6
5-9	13:4-9
5-9.1	13:4-9.1
5-9.2	13:4-9.2
5-9.2.1	13:4-9.2.1
5-9.2.2	13:4-9.2.2
5-9.2.2 Table (R)	13:4-9.2 Table
5-9.3	13:4-9.3

1999 Edition	Previous Edition
5-9.3.1	13:4-9.3.1
5-9.3.1.1	13:4-9.3.1
5-9.3.1.2	13:4-9.3.1.1
5-9.3.2	13:4-9.3.2
5-9.3.3	13:4-9.3.3
5-9.3.4	13:4-9.3.4
5-9.4	13:4-9.4
5-9.4.1	13:4-9.4.1
5-9.4.1.1 (R)	13:4-9.4.1.1
5-9.4.1.2 (R)	13:4-9.4.1.2
5-9.4.1.3	13:4-9.4.1.3
5-9.4.1.4	13:4-9.4.1.3.1
5-9.4.2	13:4-9.4.2
5-9.4.2.1 (R)	13:4-9.4.2.1
5-9.5	13:4-9.5
5-9.5.1	13:4-9.5.1
5-9.5.1.1	13:4-9.5.1.1
5-9.5.1.2	13:4-9.5.1.2
5-9.5.1.2 Table	13:4-9.5.1.2 Table
5-9.5.1.2 Figure	13:4-9.5.1.2 Figure
5-9.5.2	13:4-9.5.2
5-9.5.2.1 (R)	13:4-9.5.2.1
5-9.5.2.2 (R)	13:4-9.5.2.2
5-9.5.2.2 Figure (R)	13:4-9.5.2.2 Figure
5-9.5.2.3 (R)	13:4-9.5.4
5-9.5.2.3 Figure	13:4-9.5.4 Figure
5-9.5.2.3 Table	13:4-9.5.4 Table
5-9.5.3	13:4-9.5.3
5-9.5.3.1 (R)	13:4-9.5.3.1
5-10	13:4-10
5-10.1	13:4-10.1
5-10.2	13:4-10.2

1999 Edition	Previous Edition
5-10.2.1	13:4-10.2.1
5-10.2.2	13:4-10.2.2
5-10.2.2 Table	13:4-10.2; 231C:5-9.1.8
5-10.2.3	13:4-10.2.3; 231C:5-9.1.8
5-10.3	13:4-10.3
5-10.3.1	13:4-10.3.1
5-10.3.2	13:4-10.3.2
5-10.3.3	13:4-10.3.3
5-10.3.4	13:4-10.3.4
5-10.4	13:4-10.4
5-10.4.1	13:4-10.4.1
5-10.4.1.1	13:4-10.4.1.1
5-10.4.1.2 (R)	13:4-10.4.1.2
5-10.4.2	13:4-10.4.2
5-10.5	13:4-10.5
5-10.5.1	13:4-10.5.1
5-10.5.1.1	13:4-10.5.1.1
5-10.5.1.2 (R)	13:4-10.5.1.2
5-10.5.1.2 Figure	13:4-10.5.1.2
5-10.5.2	13:4-10.5.2
5-10.5.2.1 (R)	13:4-10.5.2.1
5-10.5.2.2 (R)	13:4-10.5.2.2
5-10.5.2.2 Figure (R)	13:4-10.5.2.2 Figure
5-10.5.2.3	13:4-10.5.2.3
5-10.5.3 (R)	13:4-10.5.3
5-10.5.3.1	13:4-10.5.3.1
5-10.5.3.1 Table	13:4-10.5.3.1
5-10.5.3.1 Figure	13:4-10.5.3.1
5-10.5.3.2 (R)	13:4-10.5.3.2
5-10.5.3.3	13:4-10.5.3.3
5-10.5.3.3 Figure	13:4-10.5.3.3
5-10.5.3.4	13:4-10.5.3.4
5-10.5.3.4 Figure	13:4-10.5.3.4
5-10.6	13:4-10.6
5-11	13:4-11
5-11.1	13:4-11.1
5-11.2	13:4-11.2
5-11.2.1	13:4-11.2.1
5-11.2.2 (R)	13:4-11.2.2; 231C:10-2.2
5-11.2.2 Table (R)	13:4-11.2.2 Table
5-11.2.3	13:4-11.2.3
5-11.3	13:4-11.3; 231C:10-2.2
5-11.3.1 (R)	13:4-11.3.1
5-11.3.2	13:4-11.3.2
5-11.3.3	13:4-11.3.3
5-11.3.4	13:4-11.3.4
5-11.4	13:4-11.4
5-11.4.1 (R)	13:4-11.4.1, 4-11.4.1.1
5-11.4.2	13:4-11.4.2
5-11.5	13:4-11.5
5-11.5.1 (R)	13:4-11.5.1
5-11.5.1 Table	13:4-11.5.1.2 Table
5-11.5.1 Figure	13:4-11.5.1.2 Figure
	13:4-11.5.1.1 (D)
	13:4-11.5.1.2 (D)
5-11.5.2 (R)	13:4-11.5.2
	13:4-11.5.2.1 (D)
	13:4-11.5.2.2 (D)
	13:4-11.5.2.2 Figure (D)
5-11.5.3 (R)	13:4-11.5.3
5-11.5.3.1 (R)	13:4-11.5.3.1

1999 Edition	Previous Edition
5-11.5.3.2 (N)	
	13:4-11.5.3.2 Figure (D)
5-11.5.3.3 (R)	13:4-11.5.3.2
5-11.6	13:4-11.6
5-12	13:4-12
5-12.1	13:4-12.1; 231C:5-3.1
5-12.2 (R)	13:4-12.3; 231C:6-2, 7-1, 8-1.5.1, 8-3.3
5-12.3 (N)	
5-12.3.1 (R)	231C:6-4, 7-4
5-12.3.2 (R)	231C:8-1.5.3
5-12.4 (R)	13:4-12.4, 4-12.4.1, 4-12.4.2, 4-12.4.3, 4-12.4.4, 4-12.4.5
5-12.5	231C:5-4.4
5-13	13:4-13
5-13.1	13:4-13.1
5-13.1.1 (R)	13:4-13.1.1
5-13.1.2	13:4-13.1.2
5-13.1.3	13:4-13.1.3
5-13.1.4 (N)	
5-13.2	13:4-13.2
5-13.2.1	13:4-13.2.1
5-13.2.2	13:4-13.2.2
5-13.2.3	13:4-13.2.3
5-13.3	13:4-13.3
5-13.3.1	13:4-13.3.1
5-13.3.2	13:4-13.3.2
5-13.3.3	13:4-13.3.3
5-13.4 (R)	13:4-13.2.4, 4-13.3.4
5-13.5	13:4-13.4
5-13.6	13:4-13.5

1999 Edition	Previous Edition
5-13.6.1 (R)	13:4-13.5.1
5-13.6.2	13:4-13.5.2
5-13.6.3	13:4-13.5.3
5-13.7	13:4-13.6
5-13.8 (R)	13:4-13.7
5-13.8.1 (R)	13:4-13.7.1
5-13.8.2 (R)	13:4-13.7.2
5-13.9	13:4-13.8
5-13.9.1 (R)	13:4-13.8.1
5-13.9.2	13:4-13.8.2
5-13.10	13:4-13.9
5-13.10(a) Figure	13:4-13.9(a)
5-13.10(b) Figure	13:4-13.9(b)
5-13.11	13:4-13.10
5-13.12 (N)	
5-13.13	13:4-13.11
5-13.14	13:4-13.12
5-13.14.1	13:4-13.12.1
5-13.14.2	13:4-13.12.2
5-13.14.3	13:4-13.12.3
5-13.14.4	13:4-13.12.4
5-13.15	13:4-13.13
5-13.16	13:4-13.14
5-13.16.1	13:4-13.14.1
5-13.16.2	13:4-13.14.2
5-13.17	13:4-13.15
5-13.18	13:4-13.16
5-13.19	13:4-13.17
5-13.19.1	13:4-13.17
5-13.19.1 Figure	13:4-13.17
5-13.20	13:4-13.18
5-13.20.1	13:4-13.18.1

1999 Edition	Previous Edition
5-13.20.2	13:4-13.18.2
5-13.20.2(a) Figure	13:4-13.18.2(a)
5-13.20.2(b) Figure	13:4-13.18.2(b)
5-13.20.3	13:4-13.18.3
5-13.21	13:4-13.19
5-13.22	13:4-13.22
5-13.23	231:5-1.8; 231C:5-2.5
5-14	13:4-14; 24:7-5.1
5-14.1	13:4-14.1
5-14.1.1 (R)	13:4-14.1.1
5-14.1.1.1	13:4-14.1.1.1
5-14.1.1.2	13:4-14.1.1.2; 24:3-2.1
5-14.1.1.3 (R)	13:4-14.1.1.3; 24:3-6.2
5-14.1.1.4	13:4-14.1.1.4
5-14.1.1.5	13:4-14.1.1.5; 24:3-2.2
5-14.1.1.6	13:4-14.1.1.6
5-14.1.1.7	13:4-14.1.1.7
5-14.1.1.8 (R)	13:4-14.1.1.8; 24:3-2.3
5-14.1.1.9	24:3-2.4
5-14.1.1.10	24:3-2.5
5-14.1.1.11	24:3-2.6
5-14.1.1.12	24:3-6.1
5-14.1.2	13:4-14.1.2
5-14.1.2.1	13:4-14.1.2.1
5-14.1.2.2	13:4-14.1.2.2
5-14.1.2.3	13:4-14.1.2.3
5-14.1.2.4	13:4-14.1.2.4
5-14.1.2.5 (N)	
5-14.1.3	24:3-3
5-14.1.3.1	24:3-3.3
5-14.1.3.2	24:3-3.4
5-14.1.4	24:3-4

1999 Edition	Previous Edition
5-14.1.4.1	24:3-4.1
5-14.1.4.2	24:3-4.2
5-14.1.4.3	24:3-4.3
5-14.1.5	24:3-5
5-14.1.5.1	24:3-5.1
5-14.1.5.2	24:3-5.2
5-14.1.6	13:4-12.2; 231C:5-4.2
5-14.2	13:4-14.3
5-14.2.1	13:4-14.3.1
5-14.2.2	13:4-14.3.2
5-14.2.3 (R)	13:4-14.3.3
5-14.2.4	13:4-14.3.4
5-14.2.4 Figure	13:4-14.3.4
5-14.2.4.1	13:4-14.3.4.1
5-14.2.4.2	13:4-14.3.4.2
5-14.2.4.2 Table	13:4-14.3.4.2
5-14.2.4.3	13:4-14.3.4.3
5-14.2.4.4	13:4-14.3.4.4
5-14.2.5	13:4-14.3.5
5-14.2.5.1	13:4-14.3.5.1
5-14.2.5.2	13:4-14.3.5.2
5-14.2.5.2.1	13:4-14.3.5.2.1
5-14.2.5.2.2	13:4-14.3.5.2.2
5-14.2.5.2.3	13:4-14.3.5.2.3
5-14.2.5.2.4	13:4-14.3.5.2.4
5-14.2.5.3	13:4-14.3.5.3
5-14.2.5.3.1	13:4-14.3.5.3.1
5-14.2.5.3.2	13:4-14.3.5.3.2
5-14.2.5.3.2 Figure	13:4-14.3.5.3.2
5-14.2.5.3.3	13:4-14.3.5.3.3
5-14.2.6	13:4-14.3.6
5-14.2.6.1	13:4-14.3.6.1

1999 Edition	Previous Edition
5-14.2.6.2	13:4-14.3.6.2
5-14.2.6.3	13:4-14.3.6.3
5-14.2.6.4	13:4-14.3.6.4
5-14.2.6.5	13:4-14.3.6.5
5-14.3	13:4-14.4
5-14.3.1	13:4-14.4.1
5-14.3.1.1	13:4-14.4.1.1
5-14.3.1.2 (R)	13:4-14.4.1.2; 24:7-5.2.2
5-14.3.2	13:4-14.4.2
5-14.3.2.1	13:4-14.4.2.1
5-14.3.2.2	13:4-14.4.2.2
5-14.3.2.3 (R)	13:4-14.4.2.3; 24:7-5.2.3
5-14.3.2.4	13:4-14.4.2.4
5-14.3.3	24:7-5.2.1
5-14.4 (N)	
5-14.4.1	24:8-1
5-14.4.1.1	24:8-1.1
5-14.4.1.2	24:8-1.2
5-14.4.2	24:8-2
5-14.4.2.1	24:8-2.1
5-14.4.2.2	24:8-2.2
5-14.4.2.3	24:8-2.3
5-14.4.3	24:8-3
5-14.4.3.1 (R)	24:8-3.1
5-14.4.3.2	24:8-3.2
5-14.4.3.3	24:8-3.3
5-14.4.3.4	24:8-3.4
5-14.4.3.5	24:8-3.5
5-14.4.4	24:8-4
5-14.4.4.1	24:8-4.1

1999 Edition	Previous Edition
5-14.4.4.2	24:8-4.2
5-14.4.4.3	24:8-4.3
5-14.4.4.4	24:8-4.4
5-14.4.5	24:8-5
5-14.4.5.1	24:8-5.1
5-14.4.5.2	24:8-5.2
5-14.4.6	24:8-7
5-14.4.6.1	24:8-7.1
5-14.4.6.2	24:8-7.2
5-14.4.6.3	24:8-7.3
5-15	13:4-15
5-15.1	13:4-15.1, 4-15.1.1
5-15.1.1	13:4-15.1.1.1
5-15.1.2	13:4-15.1.1.2
5-15.1.3	13:4-15.1.1.3
5-15.1.4	13:4-15.1.1.4
5-15.1.5	13:4-15.1.1.5
5-15.1.6	13:4-15.1.1.6
5-15.1.7	231:5-6
5-15.1.8	231C:5-5
5-15.2	13:4-15.2; 24:2-6
5-15.2.1 (R)	13:4-15.2.1; 24:2-6.1
5-15.2.1 Figure	13:4-15.2.1
5-15.2.2	13:4-15.2.2
5-15.2.3	13:4-15.2.3
5-15.2.3.1	13:4-15.2.3.1
5-15.2.3.2	13:4-15.2.3.2
5-15.2.3.3	13:4-15.2.3.3
5-15.2.3.4	13:4-15.2.3.4; 24:2-6.10.2
5-15.2.3.5	13:4-15.2.3.5; 24:2-6.9, 2-6.10.1

1999 Edition	Previous Edition
5-15.2.3.6	13:4-15.2.3.6
5-15.2.3.7	24:2-6.2
5-15.2.4	13:4-15.2.4
5-15.2.4.1	13:4-15.2.4.1; 24:2-6.4
5-15.2.4.2	13:4-15.2.4.2; 24:2-6.3
5-15.2.5	13:4-15.2.5; 24:2-6.5
5-15.3	13:4-15.3
5-15.3.1	13:4-15.3.1
5-15.3.2	13:4-15.3.2
5-15.4	13:4-15.4
5-15.4.1	13:4-15.4.1
5-15.4.2	13:4-15.4.2
5-15.4.3	13:4-15.4.3
5-15.4.4	13:4-15.4.4
5-15.4.5	13:4-15.4.5
5-15.4.6	13:4-15.4.6
5-15.4.6.1	13:4-15.4.6.1
5-15.4.6.2	13:4-15.4.6.2
5-15.5 (N)	
5-15.5.1 (N)	
5-15.5.1.1	231:5-3.1; 231C:5-6; 231D:4-4.1; 231E:4-4; 231F:5-4.1
5-15.5.1.2	13:4-13.20
5-15.5.2	13:4-13.21
5-16 (N)	
5-16.1 (N)	
5-16.2	(33:7-2.4)
5-16.3	(33:7-25)
5-17	(40:3-1.4)
5-17.1 (N)	
5-17.2	(40:3-1.4)

1999 Edition	Previous Edition
5-17.3	(40:4-2.5)
5-17.3.1	(40:4-2.5.1)
5-17.3.2	(40:4-2.5.2)
5-17.4	(40:4-3)
5-17.4.1	(40:4-3.6.1)
5-17.4.2	(40:4-3.6.2)
5-17.5	(40:7-2)
5-17.5.1	(40:4-5.5.1)
5-17.5.2	(40:4-5.5.2)
5-17.5.3	(40:4-5.5.3)
5-17.6	(40:7-2.5.2)
5-18 (N)	
5-18.1	(42:1-3)
5-18.2	(42:3-4.3)
5-18.3	(42:3-4.4)
5-18.4	(42:4-2.10)
5-18.5 (N)	
5-18.5.1	(42:4-4.1)
5-18.5.2	(42:4-4.2)
5-18.5.3	(42:4-4.5)
5-18.6	(42:4-7.9)
5-18.7 (N)	
5-18.7.1	(42:4-8.7)
5-18.7.2	(42:4-9.4)
5-19 (N)	
5-19.1 (N)	
5-19.2 (N)	
5-20 (N)	
5-20.1	(75:6-1.3)
5-21 (N)	
5-21.1 (N)	
5-21.2 (N)	

1999 Edition	Previous Edition	1999 Edition	Previous Edition
5-21.3	(82:3-2.5)	5-23.9.3	(214:3-2.8.3)
5-21.3.1	(82:3-2.5.1)	5-23.9.4	(214:3-2.8.4)
5-21.3.2	(82:3-2.5.2)	5-23.9.5	(214:3-2.8.5)
5-21.4	(82:3-3.4)	5-23.9.6	(214:3-2.8.6)
5-21.5	(82:5-3, 5-3.1)	5-23.10	(214:3-2.9)
5-22 (N)		5-23.10.1	(214:3-2.9.1)
5-22.1 (N)		5-23.10.2	(214:3-2.9.2)
5-22.2	[86C:18-1.2(b)]	5-23.11	(214:3-3)
5-23 (N)		5-23.11.1	(214:3-3.1)
5-23.1 (N)		5-23.11.2	(214:3-3.2)
5-23.2	[86C:18-1.2(b)]	5-23.11.3	(214:3-3.3)
5-23.3	(214:3-2.4.1)	5-24 (N)	
5-23.3.1	(214:3-2.4.1.1)	5-24.1 (N)	
5-23.3.2	(214:3-2.4.1.2)	5-24.2 (N)	
5-23.3.3	(214:3-2.4.1.3)	5-24.3	(307:3-3.3.2)
5-23.4	(214:3-2.4.2)	5-24.4	(307:3-3.3.3)
5-23.4.1	(214:3-2.4.2.1)	5-25 (N)	
5-23.4.2	(214:3-2.4.2.2)	5-25.1 (N)	
5-23.4.3	(214:3-2.4.2.3)	5-25.2 (N)	
5-23.4.4	(214:3-2.4.2.4)	5-25.3	(318:2-1.1)
5-23.5	(214:3-2.4.3)	5-25.4	(318:2-1.2.2)
5-23.6	(214:3-2.4.4)	5-25.5	(318:2-1.2.6.1)
5-23.6.1	(214:3-2.4.5)	5-25.6	(318:2-1.2.6.2)
5-23.7	(214:3-2.6)	5-25.7	(318:2-1.2.6.5)
5-23.7.1	(214:3-2.6.1)	5-26 (N)	
5-23.7.2	(214:3-2.6.2)	5-27 (N)	
5-23.8	(214:3-2.7)	5-28 (N)	
5-23.9	(214:3-2.8)	5-28.1 (N)	
5-23.9.1	(214:3-2.8.1)	5-28.2	(432:5-5.2)
5-23.9.2	(214:3-2.8.2)	5-29 (N)	
5-23.9.2.1	(214:3-2.8.2.1)	5-29.1 (N)	
5-23.9.2.2	(214:3-2.8.2.2)	5-29.2	(803:10-2.2)

1999 Edition	Previous Edition
5-29.3	(803:12-3)
5-30 (N)	
5-30.1 (N)	
5-30.2	(804:7-4)
5-30.2.1	(804:7-4.2)
5-30.2.2	(804:7-4.4)
5-30.2.3	(804:7-4.7)
5-30.3	(804:8-4.2.2.2)
5-30.4	(804:8-4.2.2.3)
5-31 (N)	
5-32 (N)	
6-1	13:2-6
6-1.1 (R)	13:2-6.1
6-1.1.1	13:2-6.1.1
6-1.1.2 (R)	13:2-6.1.2
6-1.1.3	13:2-6.1.5
6-1.1.3(a) Table	13:2-6.1.5(a)
6-1.1.3(b) Table	13:2-6.1.5(b)
6-1.1.4	13:2-6.1.6
6-1.1.5	13:2-6.1.7
6-1.2 (R)	13:2-6.2, 2-6.2.1, 2-6.2.2, 2-6.2.3, 2-6.2.4, 2-6.2.5, 2-6.2.6
6-1.2.1 (N)	
6-1.2.2 (N)	
6-1.2.3 (N)	
6-1.2.4 (N)	
6-1.3	13:2-6.3
6-1.3.1	13:2-6.3.1
6-1.3.2 (R)	13:2-6.3.3
6-1.3.3	13:2-6.3.4
6-1.3.4	13:2-6.3.5

1999 Edition	Previous Edition
6-1.4	13:2-6.4
6-1.4.1 (R)	13:2-6.4.1
6-1.4.1 Table	13:2-6.4.1
6-1.4.2 (R)	13:2-6.4.2
6-1.4.2 Table	13:2-6.4.2
6-1.4.3 (R)	13:2-6.4.3, 2-6.4.3.1
6-1.4.3 Table	13:2-6.4.3
6-1.4.4	13:2-6.4.4
6-1.5 (R)	13:2-6.4.5
6-1.5.1 (R)	13:2-6.4.2
6-1.5.2	13:2-6.4.5
6-1.5.2 Table	13:2-6.4.5
6-1.5.3 (R)	13:2-6.4.6
6-1.5.3 Table	13:2-6.4.6
6-1.5.4	13:2-6.4.7
6-1.5.5 (R)	13:2-6.4.7
6-1.5.6	13:2-6.4.8
6-1.5.7 (R)	13:2-6.4.9
6-1.5.7 Table	13:2-6.4.9
6-1.5.8 (R)	13:2-6.4.10
6-2	13:4-14.2
6-2.1	13:4-14.2.1
6-2.1.1	13:4-14.2.1.1
6-2.1.2 (R)	13:4-14.2.1.2
6-2.1.3 (R)	13:2-6.1.3
6-2.1.4	13:2-6.1.4
6-2.2 (R)	13:4-14.2.2, 4-14.2.2.1
6-2.2 Table	13:4-14.2.2.1
6-2.3	13:4-14.2.3
6-2.3.1	13:4-14.2.3.1
6-2.3.2	13:4-14.2.3.2
6-2.3.3 (R)	13:4-14.2.3.3

1999 Edition	Previous Edition
6-2.3.4 (R)	13:4-14.2.3.4
6-2.3.5	13:4-14.2.3.5
6-2.4 (R)	13:4-14.2.4
6-2.4.1 (R)	13:4-14.2.4.1, 4-14.2.4.2 (D)
6-2.4.2	13:4-14.2.4.3
6-2.5	13:4-14.2.5
6-2.5.1 (R)	13:4-14.2.5.1
6-2.5.2	13:4-14.2.5.2
6-2.5.3	13:4-14.2.5.3
6-2.5.4 (R)	13:4-14.2.5.4
6-3 (R)	24:8-6
6-3.1 (N)	
6-3.1.1 (R)	24:8-6.1
6-3.1.2 (R)	24:8-6.3.9
6-3.2	24:8-6.2.1
6-3.2.1	24:8-6.2.1
6-3.2.2	24:8-6.2.1
6-3.2.3	24:8-6.2.1
6-3.3 (R)	24:8-6.2.2
6-3.3.1	24:8-6.2.3
6-3.3.1.1	24:8-6.2.3(a)
6-3.3.1.2	24:8-6.2.3(b), 8-6.2.6
6-3.3.1.2 Table	24:8-6.2.3 Table
6-3.3.1.3	24:8-6.2.3(c)
6-3.3.1.4	24:8-6.2.3(d)
6-3.3.2 (R)	24:8-6.2.4
6-3.3.2 Figure	24:8-6.2.4 Figure
6-3.3.2 Table	24:8-6.2.4 Table
6-3.3.3	24:8-6.2.5
6-3.3.4	24:8-6.2.7
6-3.3.5 (R)	24:8-6.2.8
6-4	13:4-14.4.3
6-4.1	13:4-14.4.3.1; 24:7-5.2.4
6-4.2 (R)	13:4-14.4.3.2
6-4.3	13:4-14.4.3.3
6-4.4	13:4-14.4.3.4
6-4.4.1	13:4-14.4.3.4.1
6-4.4.2	13:4-14.4.3.4.2
6-4.5	13:4-14.4.3.5
6-4.5.1	13:4-14.4.3.5.1
6-4.5.2	13:4-14.4.3.5.2
6-4.5.3 (R)	13:4-14.4.3.5.10
6-4.5.3 Table (R)	13:4-14.4.3.5.3
6-4.5.3 Ex. 1 (D)	13:4-14.4.3.5.3
6-4.5.3 Ex. 3 (D)	13:4-14.4.3.5.3
6-4.5.4 (R)	13:4-14.4.3.5.8
6-4.5.5 (R)	13:4-14.4.3.5.9
6-4.5.6 (R)	13:4-14.4.3.5.3
6-4.5.7 (R)	13:4-14.4.3.5.4
6-4.5.8 (R)	13:4-14.4.3.5.5
6-4.5.8 Table	13:4-14.4.3.5.5
6-4.5.9	13:4-14.4.3.5.6
6-4.5.9 Ex. 2 (N)	
6-4.5.9 Figure (R)	13:4-14.4.3.5.6 Figure
6-4.5.10 (R)	13:4-14.4.3.5.7
6-4.5.10 Table (R)	13:4-14.4.3.5.7 Table
6-4-5.11	13:4-14.4.3.5.11
6-4.5.12	13:4-14.4.3.5.12
6-4.5.13.1 (D)	
6-4.5.13.2 (D)	
6-4.5.13.3 (D)	13:4-14.4.3.5.14
6-4.6 (R)	13:4-14.4.3.5.13

1999 Edition	Previous Edition
6-4.6.1 (R)	13:4-14.4.3.5.13 Ex. 1 and 2
6-4.6.2 (R)	13:4-14.4.3.5.13 Ex. 1
6-4.6.3 (R)	13:4-14.4.3.5.13 Ex. 1 and 2
6-4.6.4 (R)	13:4-14.4.3.5.14
6-4.7 (N)	
6-4.7.1 (R)	13:4-14.4.3.5.15
6-4.7.2 (R)	13:4-14.4.3.5.16
6-4.7.3 (R)	13:4-14.4.3.5.17
6-4.7.4 (R)	13:2-6.3.2
7-1	13:5-1
7-1.1 (R)	13:5-1
7-1.2 (N)	
7-2	13:5-2
7-2.1	13:5-2.1
7-2.1.1	13:5-2.1.1
7-2.1.2 (R)	13:5-2.1.2
7-2.2	13:5-2.2
7-2.2.1	13:5-2.2.1
7-2.2.1 Table (R)	13:5-2.2 Table
7-2.2.2	13:5-2.2.2
7-2.2.3	13:5-2.2.3
7-2.2.4	13:5-2.2.4
7-2.3	13:5-2.3
7-2.3.1	13:5-2.3.1
7-2.3.1.1	13:5-2.3.1.1
7-2.3.1.1 Table (R)	13:5-2.3; 231:6-2.4, 6-2.5, 7-2.3, 7-2.4; 231C:5-7, 5-7.1, 5-7.2, 5-8
7-2.3.1.2	13:5-2.3.1.2
7-2.3.1.2 Figure	13:5-2.3 Figure
7-2.3.1.3	13:5-2.3.1.3
7-2.3.1.3(c)	13:5-2.3.1.3(c); 231C:5-4.3
7-2.3.1.4	13:5-2.3.1.4
7-2.3.2	13:5-2.3.2
7-2.3.2.1 (R)	13:5-2.3.2.1
7-2.3.2.1 Ex. (N)	
7-2.3.2.2 (R)	13:1-4.7.4.2, 5-2.3.2.2; 231:6-2.2.1; 231C:6-1
7-2.3.2.2 Table (R)	13:1-4.7.4.2 Table, 5-2.3.2.2
7-2.3.2.3 (R)	13:5-2.3.2.3
7-2.3.2.4 (R)	13:5-2.3.2.4
7-2.3.2.4 Ex. (N)	
7-2.3.2.4 Figure	13:5-2.3.2.4
7-2.3.2.5 (R)	13:5-2.3.2.5
7-2.3.2.6	13:5-2.3.2.6
7-2.3.2.7	13:5-2.3.2.7
7-2.3.2.8 (N)	
7-2.3.3	13:7-5.2.3.3
7-2.3.3.1	13:5-2.3.3.1
7-2.3.3.2	13:5-2.3.3.2
7-2.3.3.3	13:5-2.3.3.3
7-2.3.4	13:5-2.3.4
7-2.3.4.1	13:5-2.3.4.1
7-2.3.4.2	13:5-2.3.4.2
7-2.3.4.2 Ex. (R)	13:7-2.3.4.3 Ex.
7-2.3.4.3 (N)	
7-3 (N)	
7-3.1 (R)	231:1-1.1
7-3.1.1 (R)	231:5-3.1
7-3.1.2	231:5-1
7-3.1.2.1	231:5-1.3

1999 Edition	Previous Edition
7-3.1.2.2	231:5-1.3.1
7-3.1.2.3	231:5-1.3.2
7-3.1.2.4	231:5-1.7
7-3.1.2.4 Ex. (N)	
7-3.1.3 (N)	
7-3.2	231:6-1
7-3.2.1	231:6-1
7-3.2.1.1	231:6-1.1
7-3.2.1.2	231:6-1.2
7-3.2.2	231:6-2
7-3.2.2.1	231:6-2.1
7-3.2.2.2	231:6-2.2
7-3.2.2.2.1	231:6-2.2.2
7-3.2.2.2.1 Figure	231:6-2.2.2
7-3.2.2.2.2	231:6-2.2.3
7-3.2.2.2.2 Figure	231:6-2.2.3
7-3.2.2.2.3	231:6-2.2.4
7-3.2.2.2.3 Figure	231:6-2.2.4
7-3.2.2.2.4	231:6-2.2.5
7-3.2.2.2.5	231:6-2.2.6
7-3.2.2.2.6	231:6-2.2.7
7-3.2.2.3	231:6-2.3
7-3.2.2.4	231:5-2.6
7-3.2.2.4.1	231:5-2.1
7-3.2.2.4.2	231:5-2.2
7-3.2.2.4.3	231:5-2.3
7-3.2.2.4.4	231:5-2.4
7-3.2.2.4.5	231:5-2.5
7-3.2.2.4.6	231:5-2.6
7-3.3 (N/C)	
7-3.3.1	231:7-1
7-3.3.1.1	231:7-1.1

1999 Edition	Previous Edition
7-3.3.1.1 Figure	231:7-1.1
7-3.3.1.2	231:7-1.2
7-3.3.1.3	231:7-1.3
7-3.3.1.4	231:7-1.4
7-3.3.2	231:7-2
7-3.3.2.1	231:7-2.1
7-3.3.2.2	231:7-2.2
7-3.3.2.2 Table	231:7-2.2 Table
7-3.3.2.2.1	231:7-2.2.1
7-3.3.2.2.2	231:7-2.2.2
7-3.3.2.3	231:7-2.5
7-3.3.2.4	231:5-2.6
7-4 (N/C)	
7-4.1 (N/C)	
7-4.1.1	231C:1-1
7-4.1.2 (N)	
7-4.1.3 (N)	
7-4.1.3 Ex.	231C:6-10
7-4.1.3.1	231C:C-3.3
7-4.1.4	231C:4-3
7-4.1.4.1	231C:4-3.1
7-4.1.4.1 Figure	231C:4-3.1
7-4.1.4.2	231C:4-3.2
7-4.1.5	231C:5-1
7-4.1.5.1	231C:5-1.1
7-4.1.5.1 Ex.	231C:5-1.1 Ex.
7-4.1.5.2	231C:5-1.2
7-4.1.5.3	231C:5-1.3
7-4.1.5.4	231C:5-1.4
7-4.1.5.5	231C:5-2.4
7-4.1.6	231C:5-6
7-4.1.7	231C:5-9

1999 Edition	Previous Edition
7-4.1.7.1	231C:5-10.1
7-4.1.7.1 Ex. (N)	
7-4.1.7.2	231C:5-10.2
7-4.1.8	231C:5-11
7-4.1.9	231C:5-12
7-4.1.10	231C:6-3, 7-3, 8-1.5.2
7-4.1.11	231C:7-9, 8-3.5
7-4.1.12	231C:5-9
7-4.1.12.1	231C:5-9.1
7-4.1.12.2	231C:5-9.2
7-4.1.12.3	231C:5-9.3
7-4.1.12.4	231C:5-9.4
7-4.1.12.4.1	231C:5-9.4.1
7-4.1.12.4.1.1	231C:5-9.4.1.1
7-4.1.12.4.1.2	231C:5-9.4.1.2
7-4.1.12.4.1.3	231C:5-9.4.1.3
7-4.1.12.4.2	231C:5-9.4.2
7-4.1.12.4.2.1	231C:5-9.4.2.1
7-4.1.12.4.2.2	231C:5-9.4.2.2
7-4.2 (N)	
7-4.2.1 (N)	
7-4.2.1.1	231C:6-5
7-4.2.1.1.1	231C:6-5.1
7-4.2.1.1.2	231C:6-5.2
7-4.2.1.1.3	231C:6-5.3
7-4.2.1.1.4	231C:6-5.4
7-4.2.1.2	231C:6-6
7-4.2.1.2.1	231C:6-6.1
7-4.2.1.2.1 Table	231C:6-6.1 Table
7-4.2.1.2.2	231C:6-6.2
7-4.2.1.3	231C:6-7
7-4.2.1.4	231C:6-8
7-4.2.1.5	231C:6-12
7-4.2.1.5 Table	231C:6-11 Table
7-4.2.1.6	231C:6-13
7-4.2.1.6.1	231C:6-13.1
7-4.2.1.6.1 Table	231C:6-13.1 Table
7-4.2.1.6.2	231C:6-13.2
7-4.2.1.6.2 Table	231C:6-13.2 Table
7-4.2.1.6.3	231C:6-13.3
7-4.2.1.6.4	231C:6-13.2
7-4.2.2 (N)	
7-4.2.2.1 (N)	
7-4.2.2.1.1	231C:6-9
7-4.2.2.1.1(a) Figure	231C:6-11(a) Figure
7-4.2.2.1.1(b) Figure	231C:6-11(b) Figure
7-4.2.2.1.1(c) Figure	231C:6-11(c) Figure
7-4.2.2.1.1(d) Figure	231C:6-11(d) Figure
7-4.2.2.1.1(e) Figure	231C:6-11(e) Figure
7-4.2.2.1.1(f) Figure	231C:6-11(f) Figure
7-4.2.2.1.1(g) Figure	231C:6-11(g) Figure
7-4.2.2.1.2	231C:6-9.1
7-4.2.2.1.3	231C:6-9.2
7-4.2.2.1.3 Figure	231C:6-9.2 Figure
7-4.2.2.1.4	231C:6-9.3
7-4.2.2.1.5	231C:6-9.4
7-4.2.2.1.5 Table	231C:6-9.2 Table
7-4.2.2.1.6	231C:6-9.5
7-4.2.2.1.7	231C:6-9.6
7-4.2.2.1.8	231C:6-9.7
7-4.2.2.1.9	231C:6-9.8
7-4.2.2.1.9 Figure	231C:6-9.8 Figure

1999 Edition	Previous Edition
7-4.2.2.1.10	231C:6-9.9
7-4.2.2.1.11	231C:6-9.10
7-4.2.2.2	231C:6-11
7-4.2.2.2.1	231C:6-11.1
7-4.2.2.2.2	231C:6-11.2
7-4.2.3	231C:6-14
7-4.2.3.1	231C:6-14.1
7-4.2.3.2	231C:6-14.2
7-4.2.3.3	231C:6-14.3
7-4.3 (N)	
7-4.3.1 (N)	
7-4.3.1.1	231C:7-2
7-4.3.1.2	231C:7-5
7-4.3.1.3	231C:7-6
7-4.3.1.4	231C:7-7, 7-7.1
7-4.3.1.5	231C:7-8
7-4.3.1.5.1	231C:7-8.1, 7-8.1.1
7-4.3.1.5.1 Table	231C:7-8.1 Table
7-4.3.1.5.1(a) Figure	231C:7-8.1(a) Figure
7-4.3.1.5.1(b) Figure	231C:7-8.1(b) Figure
7-4.3.1.5.1(c) Figure	231C:7-8.1(c) Figure
7-4.3.1.5.1(d) Figure	231C:7-8.1(d) Figure
7-4.3.1.5.1(e) Figure	231C:7-8.1(e) Figure
7-4.3.1.5.1(f) Figure	231C:7-8.1(f) Figure
7-4.3.1.5.1(g) Figure	231C:7-8.1(g) Figure
7-4.3.1.5.1(h) Figure	231C:7-8.1(h) Figure
7-4.3.1.5.1(i) Figure	231C:7-8.1(i) Figure
7-4.3.1.5.1(j) Figure	231C:7-8.1(j) Figure
7-4.3.1.5.2	231C:7-8.2

1999 Edition	Previous Edition
7-4.3.1.5.3	231C:7-8.3
7-4.3.1.5.3(a) Figure	231C:7-8.3(a) Figure
7-4.3.1.5.3(b) Figure	231C:7-8.3(b) Figure
7-4.3.1.5.3(c) Figure	231C:7-8.3(c) Figure
7-4.3.1.5.3(d) Figure	231C:7-8.3(d) Figure
7-4.3.1.5.3(e) Figure	231C:7-8.3(e) Figure
7-4.3.1.5.4	231C:7-11
7-4.3.1.5.4 Table	231C:7-11 Table
7-4.3.1.5.4(a) Figure	231C:7-11(a) Figure
7-4.3.1.5.4(b) Figure	231C:7-11(b) Figure
7-4.3.1.5.4(c) Figure	231C:7-11(c) Figure
7-4.3.1.5.5	231C:7-12
7-4.3.2 (N)	
7-4.3.2.1	231C:7-10.1
7-4.3.2.2	231C:7-10.2
7-4.3.3 (N)	
7-4.3.3.1	231C:7-13.1
7-4.3.3.2	231C:7-13.2
7-4.3.4	231C:5-9.4.2.2
7-4.4 (N)	
7-4.4.1	231C:8-1
7-4.4.1.1	231C:8-1.1
7-4.4.1.1 Figure	231C:8-1.1
7-4.4.1.2	231C:8-1.2
7-4.4.1.3	231C:8-1.3
7-4.4.1.4	231C:8-1.4
7-4.4.1.4 Ex.	231C:8-1.4 Ex. 2
7-4.4.2 (N)	
7-4.4.2.1	231C:8-1.5.4
7-4.4.2.2	231C:8-1.5.5

1999 Edition	Previous Edition
7-4.4.2.3	231C:8-2.2
7-4.4.2.3(a) Figure	231C:8-2 Figure
7-4.4.2.3(b) Figure	231C:8-2 Figure
7-4.4.2.3(c) Figure	231C:8-2 Figure
7-4.4.2.3(d) Figure	231C:8-2 Figure
7-4.4.2.3(e) Figure	231C:8-2 Figure
7-4.4.2.3(f) Figure	231C:8-2 Figure
7-4.4.2.3(g) Figure	231C:8-2 Figure
7-4.4.2.4	231C:8-3.2
7-4.4.2.4.1	231C:8-3.2.1
7-4.4.2.4.1(a) Figure	231C:8-3.2.1(a) Figure
7-4.4.2.4.1(b) Figure	231C:8-3.2.1(b) Figure
7-4.4.2.4.2	231C:8-3.2.2
7-4.4.2.4.3	231C:8-3.2.3
7-4.4.2.4.3(a) Figure	231C:8-3.2.3(a) Figure
7-4.4.2.4.3(b) Figure	231C:8-3.2.3(b) Figure
7-4.4.2.4.3(c) Figure	231C:8-3.2.3(c) Figure
7-4.4.2.4.4	231C:8-3.2.4
7-4.4.2.4.4(a) Figure	231C:8-3.2.4(a) Figure
7-4.4.2.4.4(b) Figure	231C:8-3.2.4(b) Figure
7-4.4.2.4.4(c) Figure	231C:8-3.2.4(c) Figure
7-4.4.2.4.4(d) Figure	231C:8-3.2.4(d) Figure
7-4.4.2.4.4(e) Figure	231C:8-3.2.4(e) Figure
7-4.4.2.4.4(f) Figure	231C:8-3.2.4(f) Figure
7-4.4.2.4.5	231C:8-3.4
7-4.4.3	231C:8-2
7-4.4.3.1	231C:8-2.1
7-4.4.3.2	231C:8-2.1.1
7-4.4.3.3	231C:8-2.1.2

1999 Edition	Previous Edition
7-4.4.3.4	231C:8-2.1.3
7-4.4.3.5	231C:8-2.1.4
7-4.4.3.6	231C:8-2.1.5
7-4.4.4 (N)	
7-4.4.4.1	231C:8-3.1.1
7-4.4.4.1 Table	231C:8-3.1 Table
7-4.4.4.2	231C:8-3.1.2
7-5	231:4-4; 231C:4-7, 4-7.1
7-5.1 (N)	
7-5.1.1 (N)	
7-5.1.2 (N)	
7-5.2	231:4-4.1
7-5.2.1	231:4-4.1.1
7-5.2.2	231:4-4.1.2
7-5.2.2 Table	231:4-4.1.2
7-5.2.3	231C:4-7.2
7-5.3	231:4-4.2
7-5.3.1	231:4-4.2.1
7-5.3.2	231:4-4.2.2
7-5.3.3	231C:4-7.2
7-6	231D
7-6.1 (N)	
7-6.2 (N)	
7-6.2.1	231D:4-1.2
7-6.2.1(a) Table	231D:4-1.2(a) Table
7-6.2.1(b) Table	231D:4-1.2(b) Table
7-6.2.1(c) Table (R)	231D:4-1.2(c) Table
7-6.2.1 Figure	231D:4-1.2 Figure
7-6.2.2	231D:4-1.3
7-6.2.3	231D:4-2.1
7-6.3	231D:4-1.4.1

1999 Edition	Previous Edition	1999 Edition	Previous Edition
7-6.3.1	231D:4-1.4.1	7-8.1.4	231F:5-1.1.3
7-6.3.2	231D:4-1.4.2	7-8.1.5	231F:5-1.1.4
7-6.3.3	231D:4-1.4.3	7-8.1.6	231F:5-1.1.5
7-6.3.4	231D:4-1.4.4	7-8.1.7	231F:5-1.1.6
7-6.4	231D:4-3	7-8.2 (N)	
7-6.4.1	231D:4-3.1	7-8.2.1 (N)	
7-6.4.2	231D:4-3.2	7-8.2.2	231F:5-1.3
7-6.4.2 Ex. 1	231D:4-3.2 Ex.	7-8.2.2.1	231F:5-1.2.1
7-6.4.2 Ex. 2 (N)		7-8.2.2.2	231F:5-1.2.2
7-6.4.3	231D:4-3,3	7-8.2.2.3	231F:5-1.3.1
7-6.5	231D:3-3	7-8.2.2.3(a) Table	13:7-5.4.1.3.1(a) Table
7-6.5.1.1 (D)	13:5-3.5.1.1	7-8.2.2.3(b) Table	13:7-5.4.1.3.1(b) Table
7-6.5.5 (D)	13:5-3.5.5	7-8.2.2.4	231F:5-1.3.3
7-6.6	231D:4-4.1	7-8.2.2.5	231F:5-1.3.5
7-7	231E	7-8.2.2.6	231F:5-1.3.6
7-7.1 (N)		7-8.2.2.7	231F:5-2.2
7-7.2	231E:4-1	7-8.2.2.8	231F:5-2.3
7-7.2.1	231E:4-1.3	7-8.2.3	231F:5-1.4
7-7.2.1 Figure	231E:4-1.3 Figure	7-8.2.3 Table	13:7-5.4.1.4
7-7.2.2	231E:4-1.3.1	7-8.2.4	231F:5-1.5
7-7.2.3	231E:4-1.3.2	7-8.2.4 Table	13:7-5.4.1.5
7-7.2.4	231E:4-1.3.3	7-8.3	231F:5-3
7-7.2.5	231E:4-1.3.4	7-8.3.1	231F:5-3.1
7-7.3	231E:4-2	7-8.3.2	231F:5-3.2
7-7.3.1	231E:4-2.1	7-8.3.3	231F:5-3.3
7-7.3.2	231E:4-2	7-8.4	231F:5-4.1
7-7.4	231E:4-4.1	7-9	13:5-3
7-8	231F	7-9.1	13:5-3.1
7-8.1 (N)		7-9.2	13:5-3.2
7-8.1.1 (N)		7-9.2.1	13:5-3.2.1
7-8.1.2	231F:5-1.1.1	7-9.2.2	13:5-3.2.2
7-8.1.3	231F:5-1.1.2	7-9.2.3	13:5-3.2.3

1999 Edition	Previous Edition
7-9.2.4	13:5-3.2.4
7-9.3	13:5-3.3
7-9.4	13:5-3.4
7-9.4.1	13:5-3.4.1
7-9.4.1.1	231:8-1.1
7-9.4.1.1 Table	231:8-1 Table
7-9.4.1.2	231C:9-1.1
7-9.4.1.2 Table (R)	231:9-1.1 Table
7-9.4.1.2.1	231C:9-1.5
7-9.4.1.2.2	231C:9-1.9
7-9.4.2	13:5-3.4.2; 231:8-2.1
7-9.4.2.1	13:5-3.4.2.1; 231:8-2.1
7-9.4.2.2	13:5-3.4.2.2
7-9.4.3	13:5-3.4.3
7-9.4.4	13:5-3.4.4
7-9.4.5	13:5-3.4.5
7-9.4.6	13:5-3.4.6
7-9.4.7	13:5-3.4.7; 231:8-2.2, 8-2.3
7-9.4.8	231:8-2.4; 231C:9-1.2
7-9.4.8 Ex.	13:A-5-3.4 Table, Note 1; 231C:9-1.2 Ex.
7-9.4.9	231:8-2.6; 231C:9-1.7
7-9.4.9 Ex.	231:8-2.6 Ex.; 231C:9-1.7 Ex.
7-9.4.10	231C:9-1.3
7-9.4.11	231C:9-1.5
7-9.5	13:5-3.5; 231:9; 231C:10
7-9.5.1	13:5-3.5.1; 231:9-1; 231C:10-1
7-9.5.1.1	231:9-1.1
7-9.5.1.1 Table	13:5-3.5.5; 231:9-1.1 Table, 9-2.2, 9-2.3

1999 Edition	Previous Edition
7-9.5.12	231:10-1.1
7-9.5.1.2 Table	13:5-3.5.5; 231C:10-1.1 Table, 10-3.1, 10-3.2
7-9.5.1.2.1	231C:10-2.5
7-9.5.1.2.2	231C:10-4.1
7-9.5.1.3	13:5-3.5.2
7-9.5.2	231:9-1, 9-2; 231C:10-2, 10-3
7-9.5.2.1	13:5-3.5.3
7-9.5.2.2	13:5-3.5.4; 231:9-2.1; 231C:10-2.1
7-9.5.2.3	231:9-1.2; 231C:10-1.2
7-9.5.2.4	231:9-2.4
7-9.5.2.5	231C:10-1
7-9.6	13:5-3.6
7-9.6.1	13:5-3.6.1
7-9.6.2	13:5-3.6.2
7-9.7	13:5-3.7
7-9.8 (N)	
7-9.8.1	231C:3-2.3(b), (c), and (d)
7-9.8.2	231D:2-1.2
7-10 (N)	
7-10.1 (N)	
7-10.2 (N)	
7-10.3 (N)	
7-10.3.1 (N)	
7-10.3.2	(33:7-2.1)
7-10.3.3	(33:7-2.3)
7-10.3.4	(33:15-3)
7-10.4 (N)	
7-10.4.1 (N)	
7-10.4.2	(36:2-9)

1999 Edition	Previous Edition
7-10.5 (N)	
7-10.5.1 (N)	
7-10.5.2	(40:3-1.2)
7-10.5.3	(40:3-2.2)
7-10.6 (N)	
7-10.6.1 (N)	
7-10.6.2 (N)	
7-10.6.2.1	(42:2-4.3.1)
7-10.6.2.2	(42:2-4.3.2)
7-10.7 (N)	
7-10.7.1	13:7-5.8.1
7-10.7.2	(45:5-2.1.1)
7-10.8 (N)	
7-10.8.1	13:7-5.9.1
7-10.8.2	(51:2-3.1 Ex. 1)
7-10.9 (N)	
7-10.9.1	13:7-5.10.1
7-10.9.2	(51A:9-2.2)
7-10.10 (N)	
7-10.10.1	13:7-5.11.1
7-10.10.2	(55:2-2.2.1)
7-10.10.3	(55:2-2.2.2)
7-10.10.4	(55:2-2.2.2)
7-10.10.5	(55:2-2.3.2)
7-10.11 (N)	
7-10.11.1 (N)	
7-10.11.2	(59:10-5.2)
7-10.12 (N)	
7-10.12.1 (N)	
7-10.12.2	(59A:9-5.2)
7-10.13 (N)	
7-10.13.1 (N)	

1999 Edition	Previous Edition
7-10.13.2	(96:11-7.5)
7-10.14 (N)	
7-10.14.1	13:7-5.12.1
7-10.14.2	(99:19-2.5.2)
7-10.14.3	(99:19-2.5.2.2)
7-10.14.4	(99:19-2.5.2.3)
7-10.14.5	(99:19-2.5.2.4)
7-10.14.6	(99:19-2.5.2.5)
7-10.15 (N)	
7-10.15.1	13:7-5.13.1
7-10.15.2	(130:6-4.1)
7-10.16 (N)	
7-10.16.1 (N)	
7-10.16.2	(150:4-1.2)
7-10.17 (N)	
7-10.17.1	13:7-5.15.1
7-10.17.2	(214:3-2.2)
7-10.17.2.1	(214:3-2.2.1)
7-10.17.2.2	(214:3-2.2.2)
7-10.17.3	(214:3-2.3)
7-10.17.3.1	(214:3-2.3.1)
7-10.17.3.2	(214:3-2.4.2.3)
7-10.17.3.3	(214:3-2.3.2)
7-10.17.3.4	(214:3-2.3.3)
7-10.17.4	(214:3-2.4.3)
7-10.17.5	(214:3-2.4.4)
7-10.17.6	(214:3-2.4.5)
7-10.17.7	(214:3-2.10.2)
7-10.17.8	(214:3-6)
7-10.17.8.1	(214:3-6.1)
7-10.17.8.1.1	(214:3-6.1.1)
7-10.17.8.1.2	(214:3-6.1.2)

1999 Edition	Previous Edition
7-10.17.8.1.3	(214:3-6.1.3)
7-10.17.8.2	(214:3-6.2)
7-10.17.8.2.1	(214:3-6.2.1)
7-10.17.8.2.2	(214:3-6.2.2)
7-10.17.8.3	(214:3-6.3)
7-10.17.8.4	(214:3-6.4)
7-10.18 (N)	
7-10.18.1 (N)	
7-10.18.2	[307:3-3.3.3(a)5]
7-10.18.3	(307:4-4.2)
7-10.18.4	(307:4-4.3)
7-10.18.5	(307:4-4.4)
7-10.19 (N)	
7-10.19.1 (N)	
7-10.19.2	(318:2-1.2.1)
7-10.19.3	(318:2-1.2.5)
7-10.19.4	(318:2-1.2.6.1)
7-10.20 (N)	
7-10.21 (N)	
7-10.21.1 (N)	
7-10.21.2	(415:2-5.1)
7-10.21.2.1	(415:2-5.1.1)
7-10.21.2.2	(415:2-5.1.2)
7-10.21.3	(415:2-5.5)
7-10.22 (N)	
7-10.22.1 (N)	
7-10.22.2	(423:5-6.3)
7-10.22.3	(423:5-6.4)
7-10.23 (N)	
7-10.23.1 (N)	
7-10.23.2	(430:2-11.3)
7-10.23.3	(430:2-11.4)

1999 Edition	Previous Edition
7-10.23.3.1	(430:2-11.4.1)
7-10.23.3.2	(430:2-11.4.2)
7-10.23.4	(430:3-3.2)
7-10.23.5 (N)	
7-10.23.5.1	(430:4-4.1)
7-10.23.5.1 Table	(430:4-4.1 Table)
7-10.23.5.2	(430:4-4.4, 4-4.4.1)
7-10.23.6 (N)	
7-10.23.6.1	(430:5-4.1)
7-10.23.6.1 Table	(430:5-4.1 Table)
7-10.23.6.2	(430:5-4.4, 5-4.4.1)
7-10.23.7	(430:6-4.1)
7-10.24 (N)	
7-10.24.1 (N)	
7-10.24.2	(432:2-8.2)
7-10.24.2.1	(432:2-8.2.1)
7-10.24.3	(432:2-8.3)
7-10.24.4	(432:5-5.2)
7-10.25 (N)	
7-10.25.1 (N)	
7-10.25.2 (N)	
7-10.25.3 (N)	
7-10.26 (N)	
7-10.26.1 (N)	
7-10.26.2	(803:11-2)
7-10.26.3	(803:12-4)
7-10.27 (N)	
7-10.27.1 (N)	
7-10.27.2	(804:4-7.2.1)
7-10.27.3	(804:7-4.1)
7-10.27.4	(804:8-4.2)
7-10.27.4.1	(804:8-4.2.2.1)

1999 Edition	Previous Edition
7-10.27.4.2	(804:8-4.2.2.3)
7-10.27.5	(804:8-8.2, 8-8.2.1)
7-10.27.6	(804:8-8.3)
7-10.27.6.1	(804:8-8.4)
7-10.27.6.2	(804:8-8.6)
7-10.27.7	(804:8-9.2)
7-10.27.8	(804:8-22)
7-10.27.9	(804:8-24.2)
7-10.28 (N)	
7-10.29 (N)	
7-10.30 (N)	
7-11	13:5-4
7-11.1	13:5-4.1
7-11.2	13:5-4.2, 5-4.2.1
8-1	13:6-1, 24:1-4
8-1.1 (R)	13:6-1.1, 24:1-4.1
8-1.1.1	13:6-1.1.1, 24:1-4.2
8-1.1.2	13:6-1.1.2
8-1.1.3	13:6-1.1.3
8-2 (R)	13:6-3
8-2.1	13:6-3
8-2.2 (N)	
8-3	13:6-2
8-3.1	13:6-2.1
8-3.2	13:6-2.2
8-3.3	13:6-2.3
8-3.4	13:6-2.4
8-4	13:6-4
8-4.1	13:6-4.1
8-4.2	13:6-4.2
8-4.2.1	13:6-4.2.1; 24:A-7-1.4
8-4.2.2	13:6-4.2.2

1999 Edition	Previous Edition
8-4.2.3	13:6-4.2.3
8-4.2.4	13:6-4.2.4
8-4.3	13:6-4.3
8-4.3.1	13:6-4.3.1
8-4.3.1 Table	13:6-4.3.1
8-4.3.2	13:6-4.3.2
8-4.3.3	13:6-4.3.3
8-4.3.4	13:6-4.3.4
8-4.4	13:6-4.4
8-4.4.1	13:6-4.4.1
8-4.4.1.1	13:6-4.4.1
8-4.4.1.2	13:6-4.4.1
8-4.4.2	13:6-4.4.2
8-4.4.3	13:6-4.4.3
8-4.4.3.1	13:6-4.4.3
8-4.4.3.2	13:6-4.4.3
8-4.4.4 (R)	13:6-4.4.4
8-4.4.5	13:6-4.4.5; 24:3-3.8.4, A-24-3.3.8.4
8-4.4.5 Table (R)	13:6-4.4.5, A-7-1.4 Table
8-4.4.6	13:6-4.4.6
8-4.4.7 (R)	13:6-4.4.7
8-4.4.8	13:6-4.4.8
8-5	13:6-5
8-5.1	13:6-5.1
8-5.1.1	13:6-5.1.1
8-5.1.2	13:6-5.1.2
8-5.1.3	13:6-5.1.3
8-5.1.4	13:6-5.1.4
8-5.2	13:6-5.2
8-5.2.1	13:6-5.2.1

1999 Edition	Previous Edition
8-5.2.2	13:6-5.2.2
8-5.2.2 Table	13:6-5.2.2
8-5.2.3	13:6-5.2.3
8-5.2.3 Table	13:6-5.2.3
8-5.2.3(a) Figure	13:6-5.2.3(a)
8-5.2.3(b) Figure	13:6-5.2.3(b)
8-5.2.3(c) Figure	13:6-5.2.3(c)
8-5.2.3.1	13:6-5.2.3.1
8-5.3	13:6-5.3
8-5.3.1	13:6-5.3.1
8-5.3.2	13:6-5.3.2
8-5.3.2(a) Table	13.6-5.3.2(a)
8-5.3.2(b) Table	13:6-5.3.3(b)
8-5.3.3	13:6-5.3.3
8-5.3.3.1	13:6-5.3.3.1
8-5.4	13:6-5.4
8-6	13:6-5.5
8-7	13:6-5.6
8-7 Table	13:6-5.6
8-8	13:6-6
8-8.1	13:6-6.1
8-8.2	13:6-6.2
9-1 (N/C)	
9-1.1	13:7-1
9-1.2	13:7-1.1
9-1.3	24:7-6.1
9-1.4	13:7-1.2.1
9-1.5 (N)	
9-1.6	13:7-1.2
9-1.6.1	13:7-1.2.2
9-1.6.2	13:7-1.2.3
9-1.7	13:7-1.3; 24:2-2.4

1999 Edition	Previous Edition
9-1.8	24:2-2.5
9-2	13:7-2
9-2.1	13:7-2.1; 24:2-2.1, 2-2.2
9-2.2	13:7-2.2, 7-2.2.1; 24:2-3
9-2.3	13:7-2.3
9-2.3.1	13:7-2.3.1
9-2.3.1.1	13:7-2.3.1.1; 24:2-4, A-2-4
9-2.3.1.2	13:7-2.3.1.2
9-2.3.1.3	13:7-2.3.1.3
9-2.3.2	13:7-2.3.2
9-2.3.3	13:7-2.3.3
9-2.4	13:7-2.4; 24:2-4
9-2.5	24:2-5
10-1	13:8-1; 24:9-1, 9-2.3.4
10-1(a) Figure	13:8-1(a)
10-1(b) Figure	13:8-1(b)
10-2	13:8-2; 24:9-2
10-2.1	13:8-2.1; 24:9-1.1, 9-2.1
10-2.1(b) Table	13:8-2.1(b)
10-2.2	13:8-2.2
10-2.2.1 (R)	13:8-2.2.1; 24:9-2.3.1
10-2.2.1 Ex. 3 (N)	
10-2.2.1 Ex. 4 (N)	
10-2.2.1 Ex. 6	24:9-2.3.2, 9-2.3.3
10-2.2.2	13:8-2.2.2; 24:9-2.3.5
10-2.2.3	13:8-2.2.3
10-2.2.4	13:8-2.2.4
10-2.2.5 (R)	13:8-2.2.5; 24:9-2.2
10-2.2.6	13:8-2.2.6
10-2.2.7	13:8-2.2.7
10-2.2.8	13:8-2.2.8

1999 Edition	Previous Edition
10-2.3	13:8-2.3
10-2.3.1	13:8-2.3.1
10-2.4	13:8-2.4
10-2.4.1	13:8-2.4.1
10-2.4.2	13:8-2.4.2
10-2.4.3	13:8-2.4.3
10-2.4.4	13:8-2.4.4
10-2.4.5	24:9-2.4
10-2.4.5.1	24:9-2.4.1
10-2.4.5.2	24:9-2.4.2
10-2.5	13:8-2.5
10-2.6	13:8-2.6
10-2.7	13:8-2.7
10-3	13:8-3, 8-6
10-4	13:8-4
10-4.1	13:8-4.1
10-5	13:8-5
11-1	13:9-1
11-1.1 (R)	13:9-1.1
11-1.2	13:9-1.2
11-1.3	13:9-1.3
11-2	13:9-2
11-2.1 (R)	13:9-2.1
11-2.2	13:9-2.2
11-2.3	13:9-2.3
11-2.3.1	13:9-2.3.1
11-2.3.2	13:9-2.3.2
11-2.4	13:9-2.4
11-2.4.1 (R)	13:9-2.4.1, 9-4.10
11-2.4.2 (R)	13:9-2.4.2
11-2.5	13:9-2.5
11-2.5.1	13:9-2.5.1

1999 Edition	Previous Edition
11-2.5.2	13:9-2.5.2
11-2.5.3	13:9-2.5.3
11-2.5.4	13:9-2.5.4
11-2.6	13:9-2.6
11-2.6.1	13:9-2.6.1
11-2.6.2	13:9-2.6.2
11-2.6.3	13:9-2.6.3
11-2.7	13:9-2.7
11-2.7.1	13:9-2.7.1
11-2.7.2	13:9-2.7.2
11-2.7.3	13:9-2.7.3
11-3	13:9-3
11-3.1	13:9-3.1
11-3.2	13:9-3.2
11-3.3	13:9-3.3
11-3.4	13:9-3.4
11-4	13:9-4
11-4.1	13:9-4.1
11-4.2	13:9-4.2
11-4.3	13:9-4.3
11-4.4	13:9-4.4
11-4.5	13:9-4.5
11-4.5.1	13:9-4.5.1
11-4.5.2	13:9-4.5.2
11-4.6	13:9-4.6
11-4.7	13:9-4.7
11-4.8	13:9-4.8
11-4.9	13:9-4.9
11-4.10 (N)	
11-4.10.1 (R)	13:9-4.10.1
11-4.10.2 (R)	13:9-4.10.2
11-4.11	13:9-4.11

1999 Edition	Previous Edition
11-4.11.1	13:9-4.11.1
11-4.11.2	13:9-4.11.2
11-4.12	13:9-4.12
11-4.12.1	13:9-4.12.1
11-4.12.2	13:9-4.12.2
11-4.12.3	13:9-4.12.3
11-4.13	13:9-4.13
11-4.14 (N)	
11-5	13:9-5
11-5.1	13:9-5.1
11-5.2	13:9-5.2
11-5.3	13:9-5.3
11-6	13:9-7
11-6.1	13:9-7.1
11-6.2	13:9-7.2
11-6.3 (N)	
11-7	13:9-6
11-7.1	13:9-6.1
11-7.2	13:9-6.2
11-7.2.1	13:9-6.2.1
11-7.2.1 Table	13:9-6.2.1
11-7.2.2	13:9-6.2.2
11-7.2.3	13:9-6.2.3
11-7.2.4	13:9-6.2.4
11-7.2.5	13:9-6.2.5
11-7.2.6	13:9-6.2.6
11-7.3	13:9-6.3
11-7.3.1	13:9-6.3.1
11-7.3.2	13:9-6.3.2
11-7.3.3	13:9-6.3.3
11-7.3.4	13:9-6.3.4
11-7.3.5	13:9-6.3.5

1999 Edition	Previous Edition
11-7.3.6 (R)	13:9-6.3.6
11-7.3.7	13:9-6.3.7
11-7.4	13:9-6.4
11-7.4.1	13:9-6.4.1
11-7.4.2	13:9-6.4.2
11-7.4.3 (N)	
11-7.4.4	13:9-6.4.3
11-7.4.5	13:9-6.4.4
11-7.4.6	13:9-6.4.5
11-8	13:9-8
11-8.1	13:9-8.1
11-8.2	13:9-8.2
11-8.3	13:9-8.3
11-9 (N)	
12-1	13:10-1, 10-1.1
A-1-2	13:1-2 Note 2
A-1-4.1 Approved	13:A-1-4.1
A-1-4.1 Authority Having Jurisdiction	13:A-1-4.1
A-1-4.1 Listed	13:A-1-4.1
A-1-4.2 Miscellaneous Storage	13:A-1-4.2
A-1-4.2 Sprinkler System	13:A-1-4.2
A-1-4.3 Gridded Sprinkler System	13:A-1-4.3
A-1-4.3(a) Figure	13:A-1-4.3(a)
A-1-4.3 Looped Sprinkler System	13:A-1-4.3
A-1-4.3(b) Figure	13:A-1-4.3(b)
A-1-4.3 Preaction Sprinkler System	13:1-4.3
A-1-4.4	13:A-1-4.4

1999 Edition	Previous Edition
A-1-4.4 Figure	13:A-1-4.4
A-1-4.5.1	13:A-1-4.5.1
A-1-4.5.2 ESFR Sprinkler	13:A-1-4.5.2
A-1-4.5.2 QRES Sprinkler	13:A-1-4.5.2
A-1-4.5.4 Dry Sprinkler	13:A-1-4.5.4
A-1-4.6 Obstructed Construction	13:A-1-4.6
A-1-4.6(a) Figure	13:A-1-4.6(a)(ii)
A-1-4.6 Unobstructed Construction	13:A-1-4.6
A-1-4.6(b) Figure	13:A-1-4.6(b)(i)1
A-1-4.6(c) Figure	13:A-1-4.6(b)(i)2
A-1-4.6(d) Figure	13:A-1-4.6(b)(v)
A-1-4.7 (N/C)	
A-1-4.7 Figure	24:A-1-3
A-1-4.8 Array, Open	231:A-1-3
A-1-4.8 Available Height for Storage	231:A-1-3
A-1-4.8 Compartmented	231:A-1-3
A-1-4.8 Container	231:A-1-3
A-1-4.8 Pile Stability, Stable Piles	231:A-1-3
A-1-4.8 Pile Stability, Unstable Piles	231:A-1-3
A-1-4.9 Aisle Width (N)	
A-1-4.9(a) Figure	231C:1-3(a)
A-1-4.9 Conventional Pallets (N)	
A-1-4.9(b) Figure	231C:1-3(b)
A-1-4.9 Longitudinal Flue Space (N)	
A-1-4.9(c) Figure	231C:1-3(c)
A-1-4.9 Rack	231C:A-1-2
A-1-4.9(d) Figure	231C:4-1(a)
A-1-4.9(e) Figure	231C:4-1(b)
A-1-4.9(f) Figure	231C:4-1(c)
A-1-4.9(g) Figure	231C:4-1(d)
A-1-4.9(h) Figure	231C:4-1(e)
A-1-4.9(i) Figure	231C:4-1(f)
A-1-4.9(j) Figure	231C:4-1(g)
A-1-4.9(k) Figure	231C:4-1(h)
A-1-4.9(l) Figure	231C:4-1(i)
A-1-4.9(m) Figure	231C:4-1(j)
A-1-4.9(n) Figure	231C:4-1(k)
A-1-4.10 Miscellaneous Tire Storage	231D:A-1-2
A-1-4.10.1	231D:A-1-3
A-1-4.10.1(a) Figure	231D:1-3(a)
A-1-4.10.1(b) Figure	231D:1-3(b)
A-1-4.10.1(c) Figure	231D:1-3(c)
A-1-4.10.1(d) Figure	231D:1-3(d)
A-1-4.10.1(e) Figure	231D:1-3(e)
A-1-4.10.1(f) Figure	231D:1-3(f)

1999 Edition	Previous Edition
A-1-4.10.1(g) Figure	231D:1-3(g)
A-1-4.11 Baled Cotton	231E:A-1-3
A-1-4.11 Table	231E:A-1-3 Table
A-1-12 Array, Standard	231F:A-1-4
A-1-12 Roll Paper Storage Height	231F:A-1-4
A-1-12 Roll Paper Storage Wrapped	231F:A-1-4
A-1-12 Figure	231F:A-1-4
A-2-1	13:A-1-4.7
A-2-1.1	13:A-1-4.7.1
A-2-1.2.1	13:A-1-4.7.2.1
A-2-1.2.2	13:A-1-4.7.2.2
A-2-1.3.1	13:A-1-4.7.3.1
A-2-1.3.2 (N/C)	
A-2-1.4	13:A-1-4.7.4.1
A-2-2	231:A-2-1; 231C:A-2-1
A-2-2 Table	231:A-2-1; 231C:A-2-1 Table
A-2-2.1.1	231:A-2-1.1.1; 231C:2-1.1.1
A-2-2.3	231:A-2-1.3; 231C:A-2-1.3
A-2-2.3 Table	231:A-2-1.3; 231C:A-2-1.3
A-2-2.3.1	231:A-2-1.3.1; 231C:A-2-1.3.1
A-2-2.3.1 Table	231:A-2-1.3.1; 231C:A-2-1.3.1
A-2-2.3.2	231:A-2-1.3.2; 231C:A-2-1.3.2
A-2-2.3.2 Table	231:A-2-1.3.2; 231C:A-2-1.3.2
A-2-2.3.3	231:A-2-1.3.3; 231C:A-2-1.3.3
A-2-2.3.3 Table	231:A-2-1.3.3; 231C:A-2-1.3.3
A-2-2.3.4	231:A-2-1.3.4; 231C:A-2-3.1.3.4
A-2-2.3.4 Table	231:A-2-1.3.4; 231C:A-2-3.1.3.4
A-2-2.4	231:2-1.4; 231C:2-3.1.4
A-2-2.4.1	231:A-2-1.4.1; 231C:A-2-3.1.4.1
A-2-2.4.1 Table	231:A-2-1.4.1; 231C:A-2-3.1.4.1
A-2-2.5	231F:A-2
A-2-2.5 Table	13:13-4
A-3-1.1	13:A-2-1.1
A-3-2.2 (N)	
A-3-2.3.1 (N)	
A-3-2.3.1 Table (N)	
A-3-2.5	13:A-2-2.4
A-3-2.6.1	13:A-2-2.5.1
A-3-2.6.2	13:A-2-2.5.2
A-3-2.6.3	13:A-2-2.5.3
A-3-2.7.2	13:A-2-2.6.2
A-3-2.8	13:A-2-2.7
A-3-3.2	13:A-2-3.2
A-3-3.2 Table	13:A-2-3.2
A-3-3.4	13:A-2-3.4
A-3-3.4 Table	13:A-2-3.4
A-3-3.5	13:A-2-3.5
A-3-4	24:A-7-1
A-3-4.1	24:A-7-1.1
A-3-4.2	24:A-7-1.2
A-3-4.4	24:A-7-2

1999 Edition	Previous Edition
A-3-5.2	13:A-2-4.2
A-3-5.3 (N)	
A-3-5.4	13:A-2-4.4
A-3-5.6	24:A-7-4
A-3-6.1.2	13:A-2-5.1.2
A-3-6.2	13:A-2-5.2
A-3-6.2(a) Figure	13:A-2-5.2(a)
A-3-6.2(b) Figure	13:A-2-5.2(b)
A-3-6.2.2	13:A-2-5.2.2
A-3-6.2.5(1)	13:A-2-5.2(a)
A-3-6.4	13:A-2-5.4
A-3-6.4.1	13:A-2-5.4.1
A-3-6.7	24:7-3
A-3-7	13:2-6
A-3-8.3	13:A-2-7.3
A-3-10.2.4	13:A-2-9.2.4
A-3-10.3.1	13:A-2-9.3.1
A-3-10.3.2	13:A-2-9.3.2
A-3-10.5	13:A-2-9.5
A-4-2	13:A-3-2
A-4-2.2	13:A-3-2.2 Ex. 1
A-4-2.3	13:A-3-2.3
A-4-2.3 Table	13:A-3-2.3
A-4-2.3.1	13:A-3-2.3.1
A-4-2.5	13:A-3-2.5
A-4-2.6.2	13:A-3-2.6.2
A-4-3.1	13:A-3-3.1
A-4-3.2.3 (N)	
A-4-3.2.4	13:A-3-3.2.4 Ex. 1
A-4-3.3	13:A-3-3.3
A-4-4.1	13:A-3-4.1
A-4-4.1.1	13:A-3-4.1.1

1999 Edition	Previous Edition
A-4-4.1.1 Figure	13:A-3-4.1.1
A-4-4.1.4	13:A-3-4.1.4 Ex. 1
A-4-4.3	13:A-3-4.3
A-4-4.3 Figure	13:A-3-4.3
A-4-5.1	13:A-3-5.1
A-4-5.2	13:A-3-5.2
A-4-5.2.3	13:A-3-5.2.3
A-4-5.2.3 Figure	13:A-3-5.2.3
A-4-5.3.1	13:A-3-5.3.1
A-4-5.3.2 (N)	
A-4-6.1.2	13:A-3-6.1.2
A-4-7.2.1	13:A-3-7.2.1
A-4-8	13:A-3-8
A-4-8.2	13:A-3-8.2
A-4-8.2.4	13:A-3-8.2.4
A-4-8.2.4 Figure	13:A-3-8.2.4
A-4-8.2.5	13:A-3-8.2.5
A-4-8.2.6	13:A-3-8.2.6
A-4-8.2.7	13:A-3-8.2.7
A-4-9.2	13:A-3-9.2
A-4-9.2 Figure	13:A-3-9.2
A-5-1	13:A-4-1
A-5-1.1	13:A-4-1.1
A-5-1.2	13:A-4-1.2
A-5-3.1.1	13:A-4-3.1.1
A-5-3.1.2	13:A-4-3.1.2
A-5-3.1.4.1	231E:A-4-1.3.5
A-5-3.1.4.4	231C:A-5-2.3
A-5-3.1.4.4 Table	231C:A-5-2.3
A-5-3.1.5.1	13:A-5-2.1.3
A-5-4	13:A-4-4
A-5-4.5.1	13:A-4-4.5.1

1999 Edition	Previous Edition
A-5-4.6.3	231:A-9-1.3
A-5-4.7.2	13:A-4-4.7.2
A-5-4.9.1	13:A-4-4.9.1
A-5-5.4.1 (N)	
A-5-5.5.1 (N)	
A-5-5.5.1 Figure	13:4-5.5.1.1
A-5-5.5.2	13:A-4-5.5.2
A-5-5.5.3	13:A-4-5.5.3
A-5-5.6	231C:A-4-5; 231E:4-1.4
A-5-6.3.2.1	13:A-4-6.3.2
A-5-6.3.2.1(a) Figure	13:A-4-6.3.2
A-5-6.3.2.1 Ex.	13:A-4-6-3.2.1 Ex.
A-5-6.3.2.1(b) Figure	13:A-4-6.3.2 Ex.(a)
A-5-6.3.2.1(c) Figure	13:A-4-6.3.2 Ex.(b)
A-5-6.3.2.1(d) Figure	13:A-4-6.3.2 Ex.(c)
A-5-6.3.2.1(e) Figure	13:A-4-6.3.2 Ex.(d)
A-5-6.4.1.2 Ex. 4	13:A-4-6.4.1.2 Ex. 4
A-5-6.4.1.3	13:A-4-6.4.1.3
A-5-6.4.2	13:A-4-6.4.2
A-5-6.5.2.1 (N)	
A-5-6.5.2.3	13:A-4-6.5.4
A-5-6.5.3	13:A-4-6.5.3
A-5-6.6	13:A-4-6.6
A-5-7.5.2.1 (N)	
A-5-7.5.3	13:A-4-7.5.3
A-5-8.2.1	13:A-4-8.2.1
A-5-8.2.1 Figure	13:A-4-8.2.1
A-5-8.4.1.3	13:A-4-8.4.1.3
A-5-8.5.2.1 (N)	
A-5-8.5.3	13:A-4-8.5.3
A-5-9.2.1	13:A-4-9.2.1
A-5-9.2.1 Figure	13:A-4-9.2.1
A-5-9.5.2.1 (N)	
A-5-9.5.3	13:A-4-9.5.3
A-5-10.2	13:A-4-10.2
A-5-10.3.1	13:A-4-10.3.1
A-5-10.4.1	13:A-4-10.4.1
A-5-10.5	13:A-4-10.5
A-5-10.5.2.1 (N)	
A-5-10.5.3	13:A-4-10.5.3
A-5-11.2.2 Ex. (N)	
A-5-11.2.2 Figure (N)	
A-5-11.3.1 Ex. 2 (N)	
A-5-11.5.2 (N)	
A-5-13.1.1	13:A-4-13.1.1
A-5-13.2.2	13:A-4-13.2.2
A-5-13.3.3	13:A-4-13.3.3
A-5-13.3.3(a) Figure	13:A-4-13.3.3(a)
A-5-13.3.3(b) Figure	13:A-4-13.3.3(b)
A-5-13.4	13:A-4-13.3.4
A-5-13.4 Figure	13:A-4-13.3.4
A-5-13.4 Ex. 2(b) (N)	
A-5-13.5	13:A-4-13.4
A-5-13.5 Figure	13:A-4-13.4
A-5-13.6.1 (R)	13:A-4-13.5.1
A-5-13.6.2	13:A-4-13.5.2
A-5-13.6.3	13:A-4-13.5.3

1999 Edition	Previous Edition
A-5-13.8	13:A-4-13.7
A-5-13.8 Figure	13:A-4-13.7
A-5-13.8.2 (N)	
A-5-13.9.1 (N)	
A-5-13.9.2	13:A-4-13.8.2
A-5-13.12 (N)	
A-5-13.13	13:A-4-13.11
A-5-13.14.3	13:A-4-13.12.3
A-5-13.14.4	13:A-4-13.12.4
A-5-13.15 Ex. 1	13:A-4-13.13 Ex. 1
A-5-13.22	13:A-4-13.22
A-5-13.22 Figure	13:A-4-13.22
A-5-14.1.1	13:A-4-14.1.1
A-5-14.1.1 Figure	13:A-4-14.1.1
A-5-14.1.1.1	13:A-4-14.1.1.1
A-5-14.1.1.3	24:A-3-6.2
A-5-14.1.1.7	13:A-4-14.1.1.7
A-5-14.1.1.8	13:A-4-14.1.1.8; 24:A-3-2.3
A-5-14.1.1.8 Figure	13:A-4-14.1.1.8
A-5-14.1.1.9	24:A-3-2.4
A-5-14.1.1.10	24:A-3-2.5
A-5-14.1.1.11	24:A-3-2.6
A-5-14.1.2.3	13:A-4-14.1.2.3
A-5-14.1.3	24:A-3-3.1, 3-3.2
A-5-14.1.4.2	24:A-3-4.2
A-5-14.1.6	231C:A-5-3.2
A-5-14.2.1	13:A-4-14.3.1
A-5-14.2.5.2.3	13:A-4-14.3.5.2.3
A-5-14.2.6.1	13:A-4-14.3.6.1
A-5-14.3.2.1	13:A-4-14.4.2.1
A-5-14.4	24:A-8
A-5-14.4.1.1	24:A-8-1.1
A-5-14.4.1.1 Figure	24:A-8-1.1
A-5-14.4.3.4	24:A-8-3.4
A-5-15.1	13:A-4-15.1.1
A-5-15.1 Figure	13:A-4-15.1.1
A-5-15.1.5	13:A-4-15.1.1.5
A-5-15.1.6	13:A-4-15.1.1.6
A-5-15.1.7	231:A-5-6
A-5-15.2	13:A-4-15.2; 24:A-2-6
A-5-15.2(a) Figure	24:A-2-6(a)
A-5-15.2(b) Figure	24:A-2-6(b)
A-5-15.2.1	13:A-4-15.2.1
A-5-15.2.3	13:A-4-15.2.3
A-5-15.4.1 (N)	
A-5-15.4.1 Figure (N)	
A-5-15.4.2	13:A-4-15.4.2
A-5-15.4.2(a) Figure	13:A-4-15.4.2(a)
A-5-15.4.2(b) Figure	13:A-5-15.4.2(b)
A-5-15.4.3	13:A-4-15.4.3
A-5-15.4.6.1	13:A-4-15.4.6.1
A-5-15.5.1.1	231E:4-4
A-5-15.5.1.2	13:A-4-13.20
A-5-15.5.2	13:A-4-13.21
A-5-16.2	(33:A-7-2.4)
A-5-18.5 (N)	
A-5-18.5(a) Figure (N)	
A-5-18.5(b) Figure (N/C)	
A-5-18.6	13:A-5-17.6.2

1999 Edition	Previous Edition
A-5-18.6(a) Figure	[42:4-7(a)]
A-5-18.6(b) Figure	[42:4-7(b)]
A-5-18.7 (N/C)	
A-5-18.7(a) Figure	[42:4-8(a)]
A-5-18.7(b) Figure	[42:4-8(b)]
A-5-21.3.1	(82:3-2.5.1)
A-5-21.3.1 Figure	(82:3-2.5.1)
A-5-23.3(a) Figure	[214:A-3-2.4.1(a)]
A-5-23.3(b) Figure	[214:A-3-2.4.1(b)]
A-5-23.3(c) Figure	[214:A-3-2.4.1(c)]
A-5-23.3(d) Figure	[214:A-3-2.4.1(d)]
A-5-23.4(a) Figure	[214:A-3-2.4.2(a)]
A-5-23.4(b) Figure	[214:A-3-2.4.2(b)]
A-5-23.4(c) Figure	[214:A-3-2.4.2(c)]
A-5-23.4(d) Figure	[214:A-3-2.4.2(d)]
A-5-23.5	(214:A-3-2.4.3)
A-5-23.5(a) Figure	[214:A-3-2.4.3(a)]
A-5-23.5(b) Figure	[214:A-3-2.4.3(b)]
A-5-23.6.1 Figure	(214:A-3-2.4.5)
A-5-25.4	(318:A-2-1.2.1)
A-5-25.5	(318:A-2-1.2.6.1)
A-5-27 (N)	
A-5-31	(850:4-2.2, 4-2.3, 4-4.1.4, 4-4.1.5, 4-4.1.6, 5-4.6.2.1, 5-7.4.1.1, 5-7.4.1.2, 5-7.4.2.1, 7-4.4.8)
A-5-32	(851:4-2.6, 5-2.7)
A-6-1.1	13:A-2-6.1
A-6-1.1 Figure	13:A-2-6.1
A-6-1.1.3	13:A-2-6.1.5
A-6-1.1.5	13:A-2-6.1.7
A-6-1.3.1	13:A-2-6.3.1

1999 Edition	Previous Edition
A-6-1.3.2	13:A-2-6.3.3
A-6-2.1.3 (N)	
A-6-2.2	13:A-4-14.2.2.1
A-6-2.3.1 Ex. 1	13:A-4-14.2.3.1 Ex. 1
A-6-2.3.1 Figure	13:A-4-14.2.3.1 Ex. 1
A-6-2.3.3	13:A-4-14.2.3.3
A-6-2.3.3(a) Figure	13:A-4-14.2.3.3
A-6-2.3.3 Ex.	13:A-4-14.2.3.3 Ex. 1 and 2
A-6-2.3.3(b) Figure	13:A-4-14.2.3.3 Ex. 1
A-6-2.3.3(c) Figure	13:A-4-14.2.3.3 Ex. 2
A-6-2.3.4	13:A-4-14.2.3.4
A-6-2.3.4(a) Figure	13:A-4-14.2.3.4
A-6-2.3.4 Ex.	13:A-4-14.2.3.4 Ex.
A-6-2.3.4(b) Figure	13:A-4-14.2.3.4 Ex.
A-6-3.1.1	24:A-8-6.1
A-6-3.2 (R)	24:A-8-6.2, A-8-6.2.1
A-6-3.2(a) Table	24:A-8-6.2
A-6-3.2(a) Figure	24:A-8-6.2(a)
A-6-3.2(b) Table	24:A-8-6.2.1
A-6-3.2(b) Figure	24:A-8-6.2(b)
A-6-3.2(c) Figure	24:A-8-6.2.1(a)
A-6-3.2(d) Figure	24:A-8-6.2.1(b)
A-6-3.3.5	24:A-8-6.2.8
A-6-4.1	13:A-4-14.4.3.1
A-6-4.1(a) Figure	13:A-4-14.4.3.1(a)
A-6-4.1(b) Figure	13:A-4-14.4.3.1(b)
A-6-4.2	13:A-4-14.4.3.2
A-6-4.2(a) Figure	13:A-4-14.4.3.2(a)
A-6-4.2(b) Figure	13:A-4-14.4.3.2(b)
A-6-4.2(2) (N)	

1999 Edition	Previous Edition
A-6-4.2(2) Figure (N)	
A-6-4.2(4)	13:A-4-14.4.3.2(d)
A-6-4.3	13:A-4-14.4.3.3
A-6-4.3 Figure	13:A-4-14.4.3.3
A-6-4.4	13:A-4-14.4.3.4
A-6-4.5 (N)	
A-6-4.5(a) Figure (N)	
A-6-4.5(b) Figure (N)	
A-6-4.5.2 Ex.	13:A-4-14.4.3.5.2 Ex.
A-6-4.5.2 Ex. Table	13:A-4-14.4.3.5.2 Ex.
A-6-4.5.5	13:A-4-14.4.3.5.9
A-6-4.5.6	13:A-4-14.4.3.5.3
A-6-4.5.6(a) Figure	13:A-4-14.4.3.5.3(a)
A-6-4.5.6(b) Figure	13:A-4-14.4.3.5.3(b)
A-6-4.5.6(c) Figure	13:A-4-14.4.3.5.3(c)
A-6-4.5.6(d) Figure	13:A-4-14.4.3.5.3(d)
A-6-4.5.6(e) Figure	13:A-4-14.4.3.5.3(h)
A-6-4.5.6 Table	13:A-4-14.4.3.5.3
A-6-4.5.8	13:A-4-14.4.3.5.5
A-6-4.5.9	13:A-4-14.4.3.5.6
A-6-4.6.1	13:A-4-14.4.3.5.13
A-6-4.6.1(a) Figure	13:A-4-14.4.3.5.13(a)
A-6-4.6.1(b) Figure	13:A-4-14.4.3.5.13(b)
A-6-4.6.1(c) Figure	13:A-4-14.4.3.5.13(c)
A-6-4.6.1(d) Figure	13:A-4-14.4.3.5.13(d)
A-6-4.6.3	13:A-4-14.4.3.5.13 Ex. 2
A-6-4.6.4	13:A-4-14.4.3.5.14
A-7-2.2.3	13:A-5-2.2.3
A-7-2.3.1.1	13:A-5-2.3.11, 231: A-6-2.4
A-7-2.3.1.3(b)	13:A-5-2.3.1.3(b)
A-7-2.3.1.3(b) Ex. 2	13:A-5-2.3.1.3(b) Ex. 2
A-7-2.3.1.3(b) Ex. 3	13:A-7-2.3.1.3(b) Ex. 3
A-7-2.3.2.3 Ex. 1	13:A-5-2.3.2.3 Ex. 1
A-7-2.3.2.8 (N)	
A-7-2.3.3.1	13:A-5-2.3.3.1
A-7-3.2.1	231:A-6-1
A-7-3.2.1.1(3)	231:A-6-1.1(c)
A-7-3.3.1	231:A-7-1, App. B
A-7-3.3.1 Table	231:A-7-1, App. B
A-7-3.3.1.1	231:A-7-1.1
A-7-3.3.1.1(a) Figure	231:A-7-1.1(a)
A-7-3.3.1.1(b) Figure	231:A-7-1.1(b)
A-7-3.3.1.2	231:A-7-1.2
A-7-3.3.2.1	231:A-7-2.1
A-7-3.3.2.2	231:A-7-2.2
A-7-3.3.2.3	231:A-7-2.5
A-7-4.1.3	231C:A-4-5
A-7-4.1.3 Ex.	231C:A-6-10
A-7-4.1.5.1	231C:A-5-1.1
A-7-4.1.5.5	231C:A-5-2.3
A-7-4.1.5.5 Table	231C:A-5-2.3
A-7-4.1.7.1	231C:A-5-10.1
A-7-4.1.11	231C:A-7-9, A-8-3.5
A-7-4.1.12.1	231C:A-5-9.1
A-7-4.1.12.4.1.1	231C:A-5-9.4.1.1
A-7-4.2.1.1.1	231C:A-6-5.1
A-7-4.2.1.1.2	231C:A-6-5.2
A-7-4.2.1.2.1	231C:A-6-6.1

1999 Edition	Previous Edition
A-7-4.2.1.6.3	231C:A-6-13.3
A-7-4.2.2.1.1	231C:A-6-9
A-7-4.2.2.2	231C:A-6-11
A-7-4.2.2.2.2	231C:4-4.1, A-4-4
A-7-4.3.1.5.1	231C:A-7-8.1
A-7-4.3.1.5.1 Figure	231C:A-7-8.1
A-7-4.3.1.5.3	231C:A-7-8.3
A-7-4.3.1.5.4	231C:A-7-11
A-7-4.3.2.1	231C:A-7-10.1
A-7-4.4.1	231C:A-8-1
A-7-4.4.2.4.4(a) Figure	231C:A-8-3.2.4, 8-3.2.4(a)
A-7-4.4.2.4.4(b) Figure	231C:A-8-3.2.4, 8-3.2.4(b)
A-7-4.4.2.4.4(c) Figure	231C:A-8-3.2.4, 8-3.2.4(c)
A-7-4.4.2.4.4(d) Figure	231C:A-8-3.2.4, 8-3.2.4(d)
A-7-4.4.2.4.4(e) Figure	231C:A-8-3.2.4, 8-3.2.4(e)
A-7-4.4.2.4.4(f) Figure	231C:A-8-3.2.4, 8-3.2.4(f)
A-7-5	231:A-4-4
A-7-5.2.1	231:A-4-4.1.1
A-7-5.2.1 Table	231:A-4-4.1.1
A-7-5.2.2	231:A-4-4.1.2
A-7-6.2.1	231D:A-4-1.2
A-7-6.2.1(a) Table Note 2	231D:A-4-1.2 Table Note 3
A-7-6.4 (D)	13:A-5-3.4
A-7-6.4 Table (D)	13:A-5-3.4
A-7-6.4.3	231D:A-4-3.3
A-7-6.5 (D)	13:A-5-3.5
A-7-6.5 Table (D)	13:A-5-3.5
A-7-7.2.3	231E:4-1.3.2
A-7-8	231F:App. B
A-7-8 Figure	231F:B-1
A-7-8 Table (N)	
A-7-8.2	231F:A-1-1.2
A-7-8.2(a) Table	231F:A-1-1.2(a)
A-7-8.2(b) Table	231F:A-1-1.2(b)
A-7-8.2.2.4	231F:A-5-1.3.3
A-7-8.2.2.5	231F:A-5-1.3.5
A-7-9.2.1	13:A-5-3.2.1
A-7-9.2.1 Figure	13:A-5-3.2.1
A-7-9.2.2	13:A-5-3.3.2
A-7-9.2.2 Figure	13:A-5-3.2.2
A-7-9.5	13:A-5-3.5.1.1; 231: A-9-1; 231C:A-10-1
A-7-9.5.1.1	231:A-9-1.3
A-7-9.5.2	231:A-9-2; 231C: A-10-2.1
A-7-9.6.1	13:A-5-3.6.1
A-7-10.3.2	(33:A-7-2)
A-7-10.4.2	(36:A-2-9)
A-7-10.14.4	(99:19-2.5.2.3 Note 1)
A-7-10.17.8.1.1	(214:A-3-6.1.1)
A-7-10.17.8.1.1 Figure	(214:A-3-6.1.1)
A-7-10.17.8.1.3	(214:A-3-6.1.3)
A-7-10.17.8.1.3 Figure	(214:A-3-6.1.3)
A-7-10.17.8.2	(214:A-3-6.2.1)
A-7-10.17.8.2 Figure	(214:A-3-6.2.1)
A-7-10.17.8.2.2	(214:A-3-6.2.2)
A-7-10.17.8.2.2 Figure	(214:A-3-6.2.2)

1999 Edition	Previous Edition
A-7-10.18.2	[307:A-3-3.3.3(a)(5)]
A-7-10.19.2	(318:A-2-1.2.1)
A-7-10.19.4	(318:A-2-1.6.1)
A-7-10.21.2.2	(415:A-2-5.1.2)
A-7-10.22.2	(423:A-5-6.3)
A-7-10.23.5.1	(430:A-4-4.1)
A-7-10.23.6.1	(430:A-5-4.1)
A-7-10.26.2	(803:11-2 Notes 1 and 2)
A-7-10.27.2	(804:A-7-2.1)
A-7-10.27.5	(804:A-8-8.2.1)
A-7-10.27.6	(804:A-8-8.3)
A-7-10.28	(850: 4-2.1, 4-2.2, 4-4.1.3, 5-4.6.1, 5-4.6.2, 5-4.6.4, 5-4.6.5.1, 5-5.1.2, 5-6.3.3, 5-6.4.3, 5-6.5.2.2, 5-7.4.1, 5-7.4.1.1, 5-7.4.1.2, 5-7.4.2.1, 5-8.2.1, 5-9.1.2.1, 5-9.4, 7-3.4.3, 7-3.4.4, 7-3.4.5.1, 7-4.4.6, 7-4.4.7, 7-4.4.8, 7-4.4.9, 7-4.4.10, 7-4.4.11, 7-5.4.4, 7-5.4.6, 7-6, 7-6.4.10)
A-7-10.29	(851:4-2.2, 4-2.3, 5-2.4, 5-5.3, 5-6.1, 5-11.2, 5-12, 5-13, 5-14; 909: A-10-4.2)
A-8-1	13:A-6-1; 24:A-1-4
A-8-1 Figure	13:A-6-1
A-8-1.1	13:A-6-1.1; 24:A-1-4
A-8-1.1 Figure	13:A-6-1.1
A-8-1.1.3	13:A-6-1.1.3
A-8-1.1.3(a) Figure	13:A-6-1.1.3(a)
A-8-1.1.3(b) Figure	13:A-6-1.1.3(b)
A-8-3.2	13:A-6-2.2

1999 Edition	Previous Edition
A-8-3.2(a) Figure	13:A-6-2.2(a)
A-8-3.2(b) Figure	13:A-6-2.2(b)
A-8-3.2(c) Figure	13:A-6-2.2(c)
A-8-3.2(d) Figure	13:A-6-2.2(d)
A-8-3.3	13:A-6-2.3
A-8-3.3 Figure	13:A-6-2.3
A-8-3.3(o)	13:A-6-2.3(o)
A-8-3.3(o) Figure	13:A-6-2.3(o)
A-8-3.4	13:A-6-2.4
A-8-3.4 Figure	13:A-6-2.4
A-8-4.1	13:A-6-4.1
A-8-4.4	13:A-6-4.4
A-8-4.4 Figure	13:A-6-4.4
A-8-4.4.1	13:A-6-4.4.1
A-8-4.4.1(a) Figure	13:A-6-4.4.1(a)
A-8-4.4.1(b) Figure	13:A-6-4.4.1(b)
A-8-4.4.2	13:A-6-4.4.2
A-8-4.4.2 Figure	13:A-6-4.4.2
A-8-4.4.3.1	13:A-6-4.4.3(a)
A-8-4.4.3.1 Figure	13:A-6-4.4.3(a)
A-8-4.4.3.2	13:A-6-4.4.3(b)
A-8-4.4.4	13:A-6-4.4.4
A-8-4.4.6	13:A-6-4.4.6
A-8-4.4.7	13:A-6-4.4.7
A-8-5.1	13:A-6-5.1
A-8-5.1.2	13:A-6-5.1.2
A-8-5.2.3.1	13:A-6-5.2.3.1
A-8-5.3.3.1	13:A-6-5.3.3.1
A-8-5.4	13:A-6-5.4
A-8-5.4 Table	13:A-6-5.4
A-8-7	13:A-6-5.6
A-9-1.3.2 (D)	24:7-6.2

1999 Edition	Previous Edition	1999 Edition	Previous Edition
A-9-1.6.2	13:A-7-1.2.3	A-11-2.4.1 (N)	
A-9-1.7 (N)		A-11-2.5.1	13:A-9-2.5.1
A-9-1.8	24:A-2-2.5	A-11-2.5.3	13:A-9-2.5.3
A-9-2.1	13:A-7-2.1; 24:A-2-2.3	A-11-2.5.4	13:A-9-2.5.4
A-9-2.1 Figure	13:A-7-2.1	A-11-2.6.1	13:A-9-2.6.1
A-9-2.2	13:A-7-2.2.1; 24:A-2-3	A-11-2.7.1	13:A-9-2.7.1
A-9-2.3.3	13:A-7-2.3.3	A-11-2.7.3	13:A-9-2.7.3
A-9-2.4.3(b) (D)		A-11-3.1	13:A-9-3.1
A-9-4.10 (D)		A-11-4.2	13:A-9-4.2
A-10-2.1	13:A-8-2.1	A-11-4.4	13:A-9-4.4
A-10-2.1 Figure	13:A-8-2.1	A-11-4.10.1 Ex. to (c) (N)	
A-10-2.2.1	13:A-8-2.2.1; 24:A-9-2.3.1	A-11-4.12.1	13:A-9-4.12.1
A-10-2.2.1 Ex. 6(a)	24:A-9-2.3.2	A-11-5.2	13:A-9-5.2
A-10-2.2.1 Ex. 6(b)	24:A-9-2.3.3	A-11-5.3	13:A-9-5.3
A-10-2.2.5	24:A-9-2.2	A-11-6.3 (N)	
A-10-2.2.7	13:A-8-2.2.7	A-11-7.2.6	13:A-9-6.2.6
A-10-5	13:A-8-5	A-11-7.3.2 (N)	
A-10-5 Figure	13:A-8-5	A-11-7.3.6(a)	13:A-9-6.3.6(a)
A-11-1.1 Heat-Sensitive Material	13:A-9-1.1	A-11-7.3.6(a) Figure	13:A-9-6.3.6(a)
A-11-1.1 Figure (N)		A-11-7.3.7 Figure	13:A-9-6.3.7
A-11-1.1 Marine System (N)		A-11-7.4.6	13:A-9-6.4.5
		A-12-1	13:A-10-1.1
A-11-1.1 Marine Thermal Barrier	13:A-9-1.1	A-12-1 Table	13:A-10-1.1
		Appendix B (N/C)	
A-11-1.2	13:A-9-1.2		
A-11-1.2 Table	13:A-9-1.2	Appendix C (N)	
		Appendix D (N)	
A-11-1.3	13:A-9-1.3		
A-11-2.1	13:A-9-2.1		
A-11-2.2	13:A-9-2.2		

( ) Extracted section; (N/C) no change; (D) deleted; (N) new; (R) revised

PART TWO

NFPA 13D, *Standard for the Installation of Sprinkler Systems in One- and Two-Family Dwellings and Manufactured Homes,* with Commentary

Part Two of this handbook includes the complete text and illustrations of the 1999 edition of NFPA 13D. The text and illustrations from the standard are printed in black and are the official requirements of NFPA 13D. Line drawings and photographs from the standard are labeled "Figures."

Paragraphs that begin with the letter "A" are extracted from Appendix A of the standard. Although printed in black ink, this nonmandatory material is purely explanatory in nature. For ease of use, this handbook places Appendix A material immediately after the standard paragraph to which it refers.

Part Two also includes Formal Interpretations on NFPA 13D; these apply to previous and subsequent editions for which the requirements remain substantially unchanged. Formal Interpretations are not part of the standard and are printed in shaded gray boxes.

In addition to standard text, appendixes, and Formal Interpretations, Part Two includes commentary that provides the history and other background information for specific paragraphs in the standard. This insightful commentary takes the reader behind the scenes, into the reasons underlying the requirements.

To make commentary material readily identifiable, commentary text, captions, and tables are all printed in green. So that the reader can easily distinguish between line drawings and photographs from the standard and from the commentary, line drawings and photographs in the commentary are labeled "Exhibits." The distinction between figures in the standard and exhibits in the commentary is new for this edition of the *Automatic Sprinkler Systems Handbook.*

Preface to 1999 Edition of NFPA 13D

It is intended that this standard provide a method for those individuals wishing to install a sprinkler system for additional life safety and property protection. It is not the purpose of this standard to require the installation of an automatic sprinkler system. This standard assumes that one or more smoke detectors will be installed in accordance with NFPA 72, *National Fire Alarm Code.*

CHAPTER 1

General Information

The 1973 publication of the report "America Burning" by the National Commission on Fire Prevention and Control focused national attention on the residential fire problem. This report, with its cover shown in Exhibit 1.1, indicated that the majority of fire deaths occurred in residential occupancies. As the magnitude of the U.S. residential fire problem became clear, the Technical Committee on Residential Sprinkler Systems expanded its focus to help address the residential fire problem.

In the summer of 1973, the Technical Committee on Residential Sprinkler Systems established the Subcommittee on Residential and Light Hazard Occupancies. This subcommittee was charged with developing a standard that would produce a reliable but inexpensive sprinkler system for these occupancies. In its first meeting, the subcommittee established the philosophies and principles of NFPA 13D. These philosophies and principles still apply to the current edition and include the following:

Cost is a major factor. A system that was slightly less reliable and had fewer operational features than those described in NFPA 13, *Standard for the Installation of Sprinkler Systems*, but that could be effective and installed at a substantially lower cost, was necessary to achieve acceptance of a residential system. This goal was a statement of "America Burning."

(a) Life safety is the primary goal of NFPA 13D with property protection a secondary goal.

(b) System design should be such that a fire could be controlled for sufficient time to enable people to escape—that is, a 10-minute stored water supply with an adequate local audible alarm.

(c) Piping arrangements, components, and hangers must be compatible with residential construction techniques. Combined sprinkler/plumbing systems are acceptable from a fire protection standpoint.

(d) The fire record in residential properties can reasonably serve as a baseline to permit omission of sprinklers in areas of low incidence of fire deaths, thus lowering

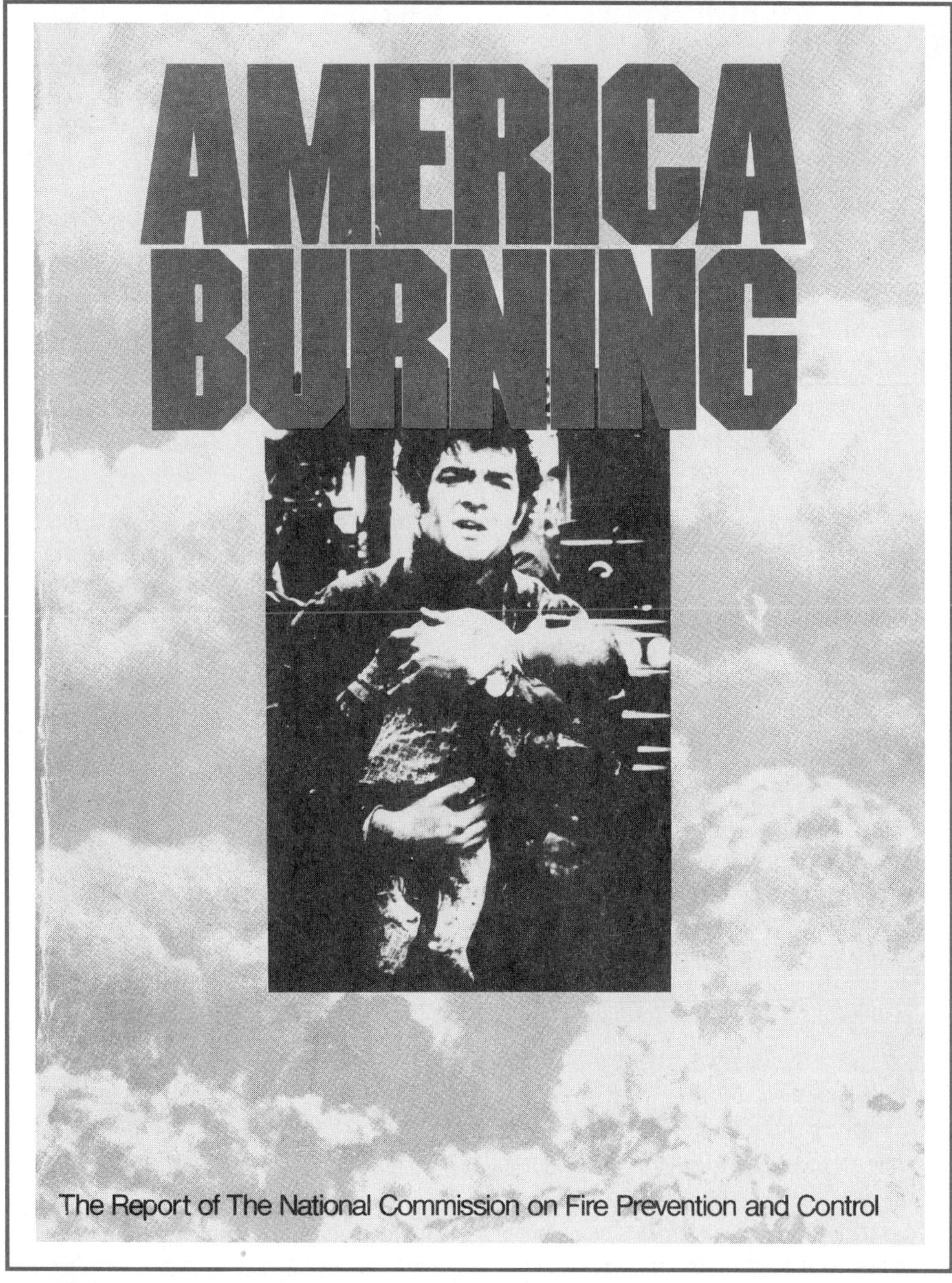

Exhibit 1.1 *"America Burning," the report of the National Commission on Fire Prevention and Control.*

costs. NFPA 13D gives permission to omit sprinklers in spaces where fires result in a small percentage of fatalities compared to other spaces in the dwelling.

The first draft document produced by the subcommittee addressed residential systems for one- and two-family dwellings, mobile homes, and multifamily housing up to four stories in height. However, when finally adopted in 1975, the multifamily housing portion had been removed because the Technical Committee on Residential Sprinkler Systems believed that such occupancies needed to include a system with more safety factors due to the greater number of people living there and the buildings' larger, more complex layout. As a result, multifamily buildings were required to be protected with sprinkler systems complying with NFPA 13. It would be another 13 years before sufficient understanding about fires and people's response to them in such occupancies allowed for the development of NFPA 13R, *Standard for the Installation of Sprinkler Systems in Residential Occupancies up to and Including Four Stories in Height*.

The first edition of NFPA 13D was developed with a relatively limited scientific understanding of residential fires and how sprinklers should protect against them. As a result, the first edition was based on technology that was largely applicable to property protection of commercial and industrial facilities such as factories and warehouses. In 1975, these types of buildings had a more prevalent use of sprinkler systems. Although the technical committee had good intentions, the first edition of NFPA 13D did not provide the type of cost-effective fire protection system recommended by the report "America Burning." However, the efforts of the technical committee clearly indicated that more scientific research would be necessary if the overall goals were to be realized.

In 1976, the National Fire Prevention and Control Administration (NFPCA), later renamed the U.S. Fire Administration (USFA), began to fund a number of research programs to scientifically evaluate the residential fire problem with the intent of developing a viable solution. The overall goals and objectives documented in the report "America Burning," which were embraced by the technical committee, centered on the system's ability to permit adequate time for occupants to escape from the dwelling. The NFPCA/USFA programs included studies that assessed the impact of using sprinklers on reducing the incidence of deaths and injuries in residential fires.[1] Other studies evaluated the practicality, design, installation, and user acceptance of residential systems while focusing on the reliability of these systems;[2] the evaluation of sprinkler discharge rates, spray patterns, response sensitivity, and overall design criteria;[3, 4] and the evaluation of prototype systems using full-scale tests.[5, 6, 7, 8]

Research and development programs also included evaluation of factors that contributed to the death of humans in a fire event—for example, the determination of threshold values of carbon monoxide, temperature, and oxygen depletion. These factors served as test criteria for evaluating whether a sprinkler could maintain occupant tenability in the room of fire origin. These research efforts collectively provided a greater understanding of the associated phenomenon of residential fires and created a new class of sprinkler—the residential sprinkler. This development resulted in a complete revision of NFPA 13D, published in 1980, at which time residential sprinkler technology was developed to a level where it could begin to be standardized and used commercially.

This part of the *Automatic Sprinkler Systems Handbook* addresses the 1999 edition of NFPA 13D. The commentary provides background information on the standard's requirements to the reader. Beginning with the 1999 edition, the Technical Committee on Residential Sprinkler Systems was established to specifically address NFPA 13D and NFPA 13R and to continue the efforts of increasing the effectiveness and appeal of residential sprinkler systems in terms of both their cost and their ability to save lives. Since 1980, a number of special situations such as the protection of piping in areas subject to freezing, placement of sprinklers beneath other than smooth, flat ceilings, the use of specially listed piping materials, alternative design options for limited area dwellings, and multipurpose piping systems have been addressed. For example, the 1996 edition was the first to allow the use of ½-in. nonmetallic pipe when certain conditions are met, and the 1999 edition provides for reduced system working pressures for multipurpose piping systems.

While many improvements have been incorporated into the standard over the years, associated efforts in terms of scientific research and overall acceptance of residential sprinklers are needed. One such effort includes the development of the Home Fire Sprinkler Coalition (HFSC) that was established in 1997 to make the general public more aware of the potential for fire in their homes and to promote the option of residential fire sprinklers. The HFSC acts independently from the technical committee and is a nonprofit organization made up of various interests. For more information on HFSC, call 1-888-635-7222.

Users of NFPA 13D are encouraged to review the publications presented in Chapter 6 to obtain additional information on residential sprinkler systems. Also, users should be aware that research efforts are ongoing in the rapidly changing field of residential fire protection and that more current information could become available.

A sprinkler system installed in accordance with NFPA 13D greatly enhances the chance for occupant escape. However, the residential sprinkler system is not considered a substitute for smoke detectors, which is indicated in the preface to NFPA 13D. The installation of smoke detectors is considered an important element in a balanced approach to residential fire safety.

1-1* Scope

This standard covers the design and installation of automatic sprinkler systems for protection against the fire hazards in one- and two-family dwellings and manufactured homes.

A-1-1 NFPA 13D is appropriate for protection against fire hazards only in one- and two-family dwellings and manufactured homes. Residential portions of any other type of building should be protected with residential sprinklers in accordance with 4-4.5 of NFPA 13, *Standard for the Installation of Sprinkler Systems.* Other portions of such buildings should be protected in accordance with NFPA 13.

The criteria in this standard are based on full-scale fire tests of rooms containing typical furnishings found in residential living rooms, kitchens, and bedrooms. The furnishings were arranged as typically found in dwelling units in a manner similar to that shown in Figures

A-1-1(a), A-1-1(b), and A-1-1(c). Sixty full-scale fire tests were conducted in a two-story dwelling in Los Angeles, California, and 16 tests were conducted in a 14-ft (4.3-m) wide mobile home in Charlotte, North Carolina.

Sprinkler systems designed and installed according to this standard are expected to prevent flashover within the compartment of origin where sprinklers are installed in the compartment. A sprinkler system designed and installed according to this standard cannot, however, be expected to completely control a fire involving fuel loads that are significantly higher than average for dwelling units [10 lb/ft^2 (49 kg/m^2)] and where the interior finish has an unusually high flame spread rating (greater than 225).

(For protection of multifamily dwellings, see NFPA 13, Standard for the Installation of Sprinkler Systems, or NFPA 13R, Standard for the Installation of Sprinkler Systems in Residential Occupancies up to and Including Four Stories in Height.)

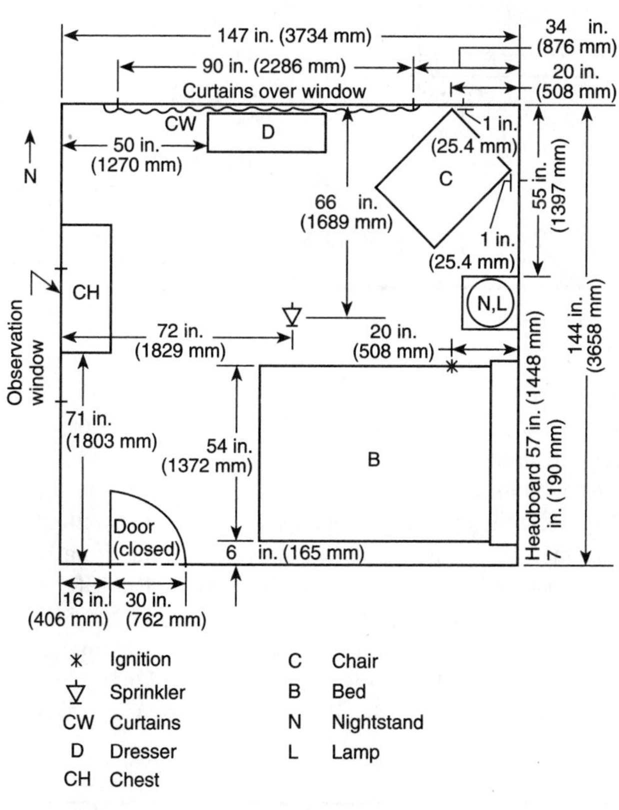

Figure A-1-1(a) *Bedroom.*

This standard applies only to one- and two-family dwellings and manufactured homes, formerly referred to as mobile homes, as defined in Section 1-3. Because of a number

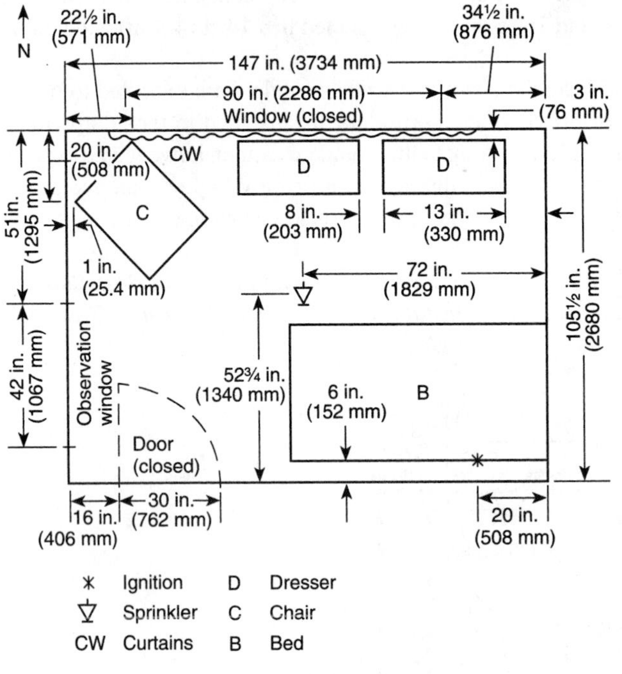

Figure A-1-1(b) Manufactured home bedroom.

of factors such as a two-sprinkler design discharge, the allowance for the omission of sprinklers in certain spaces, and a 10-minute water supply (7 minutes under specific conditions), a sprinkler system installed in accordance with NFPA 13D is inappropriate for multifamily occupancies. (A multifamily occupancy contains three or more dwelling units.) Sprinkler systems using residential sprinklers that are intended for multifamily facilities should be designed and installed in accordance with either NFPA 13R or NFPA 13. It should also be noted that the development of NFPA 13D did not consider the application of NFPA 13D for residential board and care facilities. Formal Interpretation 80-1 provides more insight on the scope of NFPA 13D. Since the Formal Interpretation was issued prior to the development of NFPA 13R, it was updated to include this new information.

Formal Interpretation 80-1

Reference: 1-1

Question: Is NFPA 13D appropriate for use in multiple (three or more) attached dwellings under any condition?

Answer: No. NFPA 13D is appropriate for use only in one- and two-family dwellings and mobile homes. Buildings which contain more than two dwelling units shall be protected in accordance with either NFPA 13R or NFPA 13 as appropriate. Paragraph 5-4.5 of NFPA 13 permits residential sprinklers to be used in residential portions of other buildings provided all other requirements of NFPA 13, including water supplies, are satisfied.

Note: Building codes may contain requirements such as 2-hour fire separations which would permit adjacent dwellings to be considered unattached. This FI has been editorially updated to include NFPA 13R since NFPA 13R was not available when the Technical Committee was balloted.

Issue Edition: 1980

Reference: 1-1

Date: June 1983

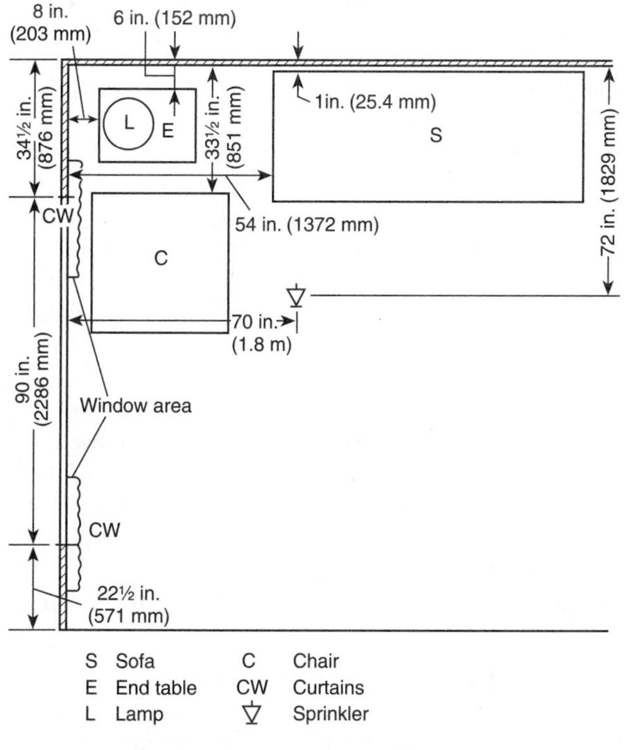

Figure A-1-1(c) *Living room.*

The requirements of NFPA 13D are based on full-scale fire tests conducted in dwellings in Los Angeles and Charlotte, NC, [6, 7, 8, 9] and earlier laboratory tests conducted by Factory Mutual Research Corporation and Battelle Columbus Laboratories.[3, 4] Some of these tests indicated that the associated life safety objectives could possibly be achieved with water application rates as low as 0.025 gpm/ft^2 [1.02 (L/min)/m^2]. However, these tests usually involved smoldering fires in typical residential room configurations and flaming fires generated by a gas burner rather than by a fuel package consisting of household combustibles such as furniture and draperies. Subsequent tests at Factory Mutual Research Corporation [5] and at the test sites in Los Angeles and Charlotte, NC, [6, 7, 8, 9] established that sprinkler discharge had to be in accordance with the design parameters as indicated in Section 4-1. Using these parameters, the basic design density acceptable is 0.09/gpm ft^2 [3.7 (L/ min)/m^2]. The likelihood of successfully achieving life safety and reducing injuries and property damage is reduced when smaller discharge densities are used. However, since April 1992, additional tests have been completed that have led to different design discharge rates for smaller dwellings as indicated in Chapter 5. These resultant discharge densities actually exceed the parameters found in the Los Angeles and Charlotte, NC, tests.

1-2* Purpose

The purpose of this standard is to provide a sprinkler system that aids in the detection and control of residential fires and thus provides improved protection against injury, life loss, and property damage. A sprinkler system designed and installed in accordance with this standard is expected to prevent flashover (total involvement) in the room of fire origin, where sprinklered, and to improve the chance for occupants to escape or be evacuated.

Guidelines have been established for the design and installation of sprinkler systems for one- and two-family dwellings and manufactured homes. Nothing in this standard is intended to restrict new technologies or alternative arrangements, provided that the level of safety prescribed by the standard is not reduced.

A-1-2 Various levels of fire safety are available to dwelling occupants to provide life safety and property protection.

This standard recommends, but does not require, sprinklering of all areas in a dwelling; it permits sprinklers to be omitted in certain areas. These areas have been proved by NFPA statistics *[see Tables A-1-2(a) and A-1-2(b)]* to be those where the incidence of life loss from fires in dwellings is low. Such an approach provides a reasonable degree of fire safety. Greater protection to both life and property is achieved by sprinklering all areas.

Guidance for the installation of smoke detectors and fire detection systems is found in NFPA 72, *National Fire Alarm Code.*

Both Underwriters Laboratories Inc. and Factory Mutual Research Corporation have developed test standards for the evaluation of residential sprinklers.[10, 11] These test standards include significantly different test parameters than those used to evaluate nonresidential sprinklers such as the standard spray sprinkler. The general criteria for residential sprinklers during a fire test are as follows:

(1) Limit the maximum ceiling air temperature to no greater than approximately 600°F (315°C)
(2) Limit the maximum temperature at 5 ft 3 in. (1.6 m) above floor to no greater than 200°F (93°C)
(3) Attain the preceding criteria with no more than two sprinklers operating

Compliance with the preceding criteria serves to provide an indication that flashover will not occur, the environment through which occupants must evacuate is tenable, and that both of these criteria can be accomplished with the water supply specified by NFPA 13D.

The basic laboratory test configuration is a 12 ft × 24 ft (3.7 m × 7.3 m) room with a combustible array simulating residential furnishings. The test facility is adjustable so that it can be configured to evaluate other sprinkler spacing parameters as permitted by 4-1.6.

The technical committee adopted the concept of "levels of protection" in order to achieve a reasonable degree of safety while controlling the cost of the system. Because sprinklers are permitted to be omitted from certain spaces in the dwelling as indicated by Sections 4-6 and 5-5, the level of protection is less than if those spaces were sprinklered. The areas where omission of sprinklers is permitted are those where NFPA fire statistics indicate a relatively low incidence of fire deaths. [See Table A-1-2(a).]

Dry pipe systems in dwellings are not desirable for life safety because of the delay in application of water from activated sprinklers. This delay was noted in early Factory Mutual Research Corporation tests and substantiated in the Los Angeles and Charlotte, NC, fire tests.[6, 7, 8, 9] The key to achieving control of a residential fire is the rapid application of water upon activation of the sprinkler. Additionally, residential systems must be simple and easy to maintain. Because dry pipe systems are more complex than wet pipe systems, their maintenance requires greater effort.

1-3 Definitions

Approved.* Acceptable to the authority having jurisdiction.

A-1-3 Approved. The National Fire Protection Association does not approve, inspect, or certify any installations, procedures, equipment, or materials; nor does it approve or evaluate testing laboratories. In determining the acceptability of installations, procedures, equipment, or materials, the authority having jurisdiction may base acceptance on compliance with NFPA or other appropriate standards. In the absence of such standards, said authority may require evidence of proper installation, procedure, or use. The authority having jurisdiction may also refer to the listings or labeling practices of an organization that is concerned with product evaluations and is thus in a position to determine compliance with appropriate standards for the current production of listed items.

The term *approved*, which means acceptable to the authority having jurisdiction, differs from the term *listed*. An item that is approved is not necessarily listed. Subsection

Table A-1-2(a) Casual Factors in One- and Two-Family Dwelling Fires that Caused One or More Deaths

Area of Origin (Based on 6066 incidents where area of origin was reported)	
Living room	41%
Bedroom	27%
Kitchen	15%
Storage area	4%
Heating equipment room	3%
Structural area	2%
Other areas	8%
Form of Materials Ignited (Based on 5080 incidents where form of material ignited was reported)	
Furniture	27%
Bedding	18%
Combustible liquid or gas	13%
Interior finish	9%
Structural member	9%
Waste, rubbish	4%
Clothing (on a person)	3%
Cooking materials	3%
Electrical insulation	2%
Curtains, draperies	2%
Other	10%
Form of Heat of Ignition (Based on 5016 incidents where form of heat of ignition was reported)	
Smoking materials	36%
Heat from fuel-fire or powered object	25%
Heat from miscellaneous open flame (including match)	15%
Heat from electrical equipment arcing or overload	14%
Hot objects, including properly operating electrical equipment	7%
Other	3%

Note: Total number of incidents reported: 10,194.

Source: FIDO Database 1973 to 1982, NFPA Fire Analysis Department.

Table A-1-2(b) Fires and Associated Deaths and Injuries in Dwellings, Duplexes, and Manufactured Homes by Area of Origin: Annual Average of 1986–1990 Structure Fires Reported to U.S. Fire Departments

Area of Origin	Civilian Deaths	Civilian Percent	Fires	Percent	Injuries	Percent
Living room, family room, or den	1,330	37.1	42,600	10.5	2,546	18.6
Bedroom	919	25.6	50,200	12.4	3,250	23.7
Kitchen	541	15.1	92,670	22.9	3,987	29.1
Dining room	83	2.3	3,780	0.9	189	1.4
Heating equipment room or area	62	1.7	15,130	3.7	374	2.7
Hallway or corridor	48	1.3	3,690	0.9	155	1.1
Laundry room or area	47	1.3	15,370	3.8	363	2.7
Garage or carport*	45	1.2	14,580	3.6	524	3.8
Bathroom	44	1.2	8,040	2.0	271	2.0
Unclassified structural area	43	1.2	4,530	1.1	104	0.8
Crawl space or substructure space	41	1.2	11,200	2.8	317	2.3
Multiple areas	41	1.1	3,350	0.8	96	0.7
Ceiling/floor assembly or concealed space	32	0.9	3,470	0.9	64	0.5
Wall assembly or concealed space	27	0.8	7,090	1.8	93	0.7
Closet	23	0.6	5,020	1.2	186	1.4
Exterior balcony or open porch	22	0.6	5,570	1.4	121	0.9
Exterior wall surface	22	0.6	14,620	3.6	118	0.9
Unclassified area	21	0.6	2,590	0.6	87	0.6
Attic or ceiling/roof assembly or concealed space	21	0.6	10,740	2.7	98	0.7
Tool room or other supply storage room or area	20	0.5	4,160	1.0	133	1.0
Lobby or entrance way	17	0.5	1,410	0.3	44	0.3
Interior stairway	17	0.5	1,100	0.3	41	0.3
Chimney	17	0.5	60,530	14.9	75	0.5
Unclassified function area	17	0.5	1,090	0.3	43	0.3
Unclassified storage area	14	0.4	2,460	0.6	80	0.6
Area not applicable	11	0.3	1,180	0.3	22	0.2
Exterior stairway	8	0.2	1,090	0.3	25	0.2
Lawn or field	7	0.2	1,670	0.4	24	0.2
Trash room or area	5	0.1	1,140	0.3	14	0.1
Product storage area	5	0.1	780	0.2	23	0.2
Unclassified means of egress	5	0.1	610	0.2	15	0.1
Unclassified service or equipment area	4	0.1	380	0.1	12	0.1
Library	3	0.1	180	0.0	11	0.0
Other known area	26	0.7	12,880	3.2	195	1.4
Total	3,589	100.0	404,900	100.0	13,691	100.0

Note: Fires and estimated to the nearest 10; civilian deaths and injuries are estimated to the nearest 1.

*Does not include dwelling garages coded as a separate property, which averaged 19 deaths, 259 injuries, and 21,170 fires per year.

Source: 1986–1990 NFIRS and NFPA survey.

1-5.2 requires that tanks, pumps, hangers, waterflow devices, and waterflow valves be approved. These system components are not required to be listed. Residential sprinklers, however, are required to be listed per 3-5.1 and 5-3.1, and certain types of pipe and fittings are also required to be listed in accordance with 3-3.2 and 3-3.7.

Authority Having Jurisdiction.* The organization, office, or individual responsible for approving equipment, materials, an installation, or a procedure.

A-1-3 Authority Having Jurisdiction. The phrase "authority having jurisdiction" is used in NFPA documents in a broad manner, since jurisdictions and approval agencies vary, as do their responsibilities. Where public safety is primary, the authority having jurisdiction may be a federal, state, local, or other regional department or individual such as a fire chief; fire marshal; chief of a fire prevention bureau, labor department, or health department; building official; electrical inspector; or others having statutory authority. For insurance purposes, an insurance inspection department, rating bureau, or other insurance company representative may be the authority having jurisdiction. In many circumstances, the property owner or his or her designated agent assumes the role of the authority having jurisdiction; at government installations, the commanding officer or departmental official may be the authority having jurisdiction.

Check Valve. A valve that allows flow in one direction only.

Control Valve.* A valve employed to control (shut) a supply of water to a sprinkler system.

A-1-3 Control Valve. System control valves should be of the indicating type, such as plug valves, ball valves, butterfly valves, or OS & Y gate valves.

The type of control valve used should be an indicating type. An indicating valve has some external means to indicate to the occupant that the valve is either in the open or closed position.

Design Discharge. The rate of water discharged by an automatic sprinkler expressed in gpm (L/min).

Dry System. A system employing automatic sprinklers attached to a piping system containing air under atmospheric or higher pressures. Loss of pressure from the opening of a sprinkler or detection of a fire condition causes the release of water into the piping system and out the opened sprinkler.

Because of the concern for delayed water delivery and the fact that there are no residential sprinklers listed for use on dry pipe systems, residential dry pipe systems are not allowed. (See Exception No. 1 to 3-5.1.) Furthermore, data from residential fire tests do not support the four-sprinkler design that was included in the 1980 edition of NFPA 13D. Rather than deleting all references to dry pipe systems in NFPA 13D, the technical committee left selective wording in place to allow for the development and listing of a residential sprinkler that can be used in a dry pipe system.

Areas subject to freezing can be protected by antifreeze systems or by the use of standard dry pendent or dry sidewall sprinklers. (See 4-3.2.) Additionally, piping passing through areas subject to freezing can be protected with insulation as indicated in A-4-3.1.

Dwelling. Any building that contains not more than one or two dwelling units intended to be used, rented, leased, let, or hired out to be occupied or that are occupied for habitation purposes.

Dwelling Unit. One or more rooms, arranged for the use of one or more individuals living together, as in a single housekeeping unit, that normally have cooking, living, sanitary, and sleeping facilities.

Labeled. Equipment or materials to which has been attached a label, symbol, or other identifying mark of an organization that is acceptable to the authority having jurisdiction and concerned with product evaluation, that maintains periodic inspection of production of labeled equipment or materials, and by whose labeling the manufacturer indicates compliance with appropriate standards or performance in a specified manner.

Listed.* Equipment, materials, or services included in a list published by an organization that is acceptable to the authority having jurisdiction and concerned with evaluation of products or services, that maintains periodic inspection of production of listed equipment or materials or periodic evaluation of services, and whose listing states that either the equipment, material, or service meets appropriate designated standards or has been tested and found suitable for a specified purpose.

A-1-3 Listed. The means for identifying listed equipment may vary for each organization concerned with product evaluation; some organizations do not recognize equipment as listed unless it is also labeled. The authority having jurisdiction should utilize the system employed by the listing organization to identify a listed product.

Manufactured Home.* A structure, transportable in one or more sections, that in the traveling mode is 8 body ft (2.4 m) or more in width and 40 body ft (12 m) or more in length or, where erected on-site, is 320 ft^2 ($28m^2$) or more, and that is built on a permanent chassis and designed to be used as a dwelling with or without a permanent foundation where connected to the required utilities, and includes the plumbing, heating, air conditioning, and electrical systems contained therein.

A-1-3 Manufactured Home. Manufactured homes were formerly referred to as "mobile homes" or "trailer coaches."

In the United States, the Department of Housing and Urban Development (HUD) regulates manufactured homes. The regulation of manufactured homes is unique when compared to modular housing units or stick-built homes. Both of these types of housing tend to be governed more by local codes. The previous reputation of manufactured homes as rural, low- to middle-income housing units is no longer applicable. Developments that are exclusively reserved for these units are widely used and accepted. In some southern and southwestern states, more manufactured homes are estimated to be brought in than are stick-built and modular homes to be constructed. The units' cost, functionality, and appearance make them attractive, viable, and economically feasible for many people.

Multipurpose Piping System. A piping system within dwellings and manufactured homes intended to serve both domestic and fire protection needs.

Preengineered System. A packaged sprinkler system including all components connected to the water supply and designed to be installed according to pretested limitations.

Pump. A mechanical device that transfers or raises, or transfers and raises, the pressure of a fluid (water).

Residential Sprinkler. A type of sprinkler that meets the definition of fast response as defined by NFPA 13, Standard for the Installation of Sprinkler Systems, and that has been specifically investigated for its ability to enhance survivability in the room of fire origin and that is listed for use in the protection of dwelling units.

Residential sprinklers are specifically listed by Underwriters Laboratories Inc. and Factory Mutual Research Corporation for residential service and have been tested for compliance with NFPA 13D.[10, 11] The primary purpose of residential sprinklers is to achieve life safety in residential occupancies. The sprinklers contain fast-response operating elements and can be of the upright, pendent, or sidewall configuration, although no upright devices are available at this time. The discussion on sprinkler sensitivity in the commentary to 1-4.5.1 of NFPA 13 provides additional information with regard to sprinklers with fast-response operating elements.

Shall. Indicates a mandatory requirement.

Should. Indicates a recommendation or that which is advised but not required.

Sprinkler, Automatic. A fire suppression or control device that operates automatically when its heat-actuated element is heated to its thermal rating or above, allowing water to discharge over a specific area.

Depending on the type used, automatic sprinklers are intended to achieve fire control or fire suppression. Additionally, automatic sprinklers can be used for property protection or life safety, or both. Residential sprinklers that are specifically listed automatic devices are intended to achieve life safety as their primary objective. Property protection is considered a secondary benefit. NFPA 13 provides more information with regard to the types and styles of automatic sprinklers.

Sprinkler System. An integrated system of piping, connected to a water supply, with listed sprinklers that automatically initiate water discharge over a fire area. Where required, the sprinkler system also includes a control valve and a device for actuating an alarm when the system operates.

The definition of *sprinkler system* includes water supplies and underground piping. These components are critical to proper sprinkler system performance and must be treated as part of the system.

Standard. A document, the main text of which contains only mandatory provisions using the word "shall" to indicate requirements and which is in a form generally suitable for mandatory reference by another standard or code or for adoption into law. Nonmandatory provisions shall be located in an appendix, footnote, or fine-print note and are not to be considered a part of the requirements of a standard.

Supply Pressure. The pressure within the supply (e.g., city or private supply water source).

Supply pressure is the pressure that can be expected from the municipal or stored water supply, which is discussed in A-4-4.3.

System Pressure. The pressure within the system (e.g., above the control valve).

System pressure is determined by subtracting friction losses between the street connection and the residence's water supply valve from the water pressure in the street after making any necessary adjustments for changes in friction loss in the pipe, fittings, appurtenances, and elevation. (See A-4-4.3.)

System Working Pressure. The maximum anticipated static (nonflowing) or flowing pressure applied to sprinkler system components exclusive of surge pressures.

The definition of *system working pressure* was added to the 1999 edition to clearly indicate that this pressure pertains to the installed system. The maximum static pressure could be the result of surge pressures held in a system by a check valve or backflow preventer and are not to be considered the system's working pressure.

Waterflow Alarm. A sounding device activated by a waterflow detector or alarm check valve and arranged to sound an alarm that is audible in all living areas over background noise levels with all intervening doors closed.

Waterflow Detector. An electric signaling indicator or alarm check valve actuated by waterflow in one direction only.

Wet System. A system employing automatic sprinklers that are attached to a piping system containing water and connected to a water supply so that water discharges immediately from sprinklers opened by a fire.

1-4* Maintenance

The owner is responsible for the condition of a sprinkler system and shall keep the system in normal operating condition.

A-1-4 The responsibility for properly maintaining a sprinkler system is that of the owner or manager, who should understand the sprinkler system operation. A minimum monthly maintenance program should include the following.

(1) Visual inspection of all sprinklers to ensure against obstruction of spray.
(2) Inspection of all valves to ensure that they are open.
(3) Testing of all waterflow devices.
(4) Testing of the alarm system, where installed.

NOTE: Where it appears likely that the test will result in a fire department response, notification to the fire department should be made prior to the test.

(5) Operation of pumps, where employed. (See NFPA 20, Standard for the Installation of Stationary Pumps for Fire Protection.)
(6) Checking of the pressure of air used with dry systems.
(7) Checking of water level in tanks.
(8) Special attention to ensure that sprinklers are not painted either at the time of installation or during subsequent redecoration. When sprinkler piping or areas next to sprinklers are being painted, the sprinklers should be protected by covering them with a bag, which should be removed immediately after painting is finished.

(For further information, see NFPA 25, Standard for the Inspection, Testing, and Maintenance of Water-Based Fire Protection Systems.)

An excellent resource for the maintenance of sprinkler systems is in NFPA 25, *Standard for the Inspection, Testing, and Maintenance of Water-Based Fire Protection Systems.* Further guidance on maintenance of the sprinkler system can be obtained from a local sprinkler contractor. Because of their simplicity, residential sprinkler systems generally need no more maintenance than residential plumbing systems. However, the owner can need to be educated on issues unique to sprinkler systems, such as obstructing discharge patterns, painting of sprinklers, hanging items from sprinklers, and maintaining control valves in the open position at all times. Sprinkler systems installed in accordance with NFPA 13D are generally exempted from the requirements contained within NFPA 25, which applies to sprinkler systems in commercial, industrial, and multifamily residential facilities.

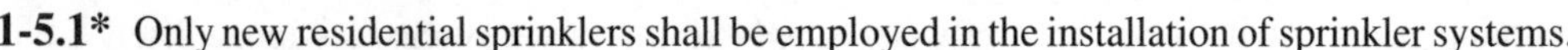

1-5 Devices, Materials, Design, and Installation

1-5.1* Only new residential sprinklers shall be employed in the installation of sprinkler systems.

A-1-5.1 At least one spare sprinkler of each type, temperature rating, and orifice size used in the system should be kept on the premises. Where fused sprinklers are replaced by the owner, fire department, or others, care should be taken to ensure that the replacement sprinkler has the same operating characteristics.

In agreement with NFPA 13 and NFPA 13R systems, only new sprinklers are permitted to be used in NFPA 13D systems. The performance and reliability of used sprinklers are not easily verified. The installation of new, listed residential sprinklers provides a much higher degree of reliability.

Sprinklers used in residential systems are a special type that have been tested for residential use.[11] Residential sprinklers are one-time operating devices that must be treated carefully during installation. When the system is originally installed, only new sprinklers should be used to ensure proper operating condition and type.

On those occasions when it is necessary to replace a sprinkler that has operated (fused) or has been damaged, the orifice size, the temperature rating, and the deflector configuration should be carefully checked so that the replacement sprinkler is the same as the original. Many sprinklers—for example, upright, pendent, or sidewall—have unique operating characteristics, and some have differing areas of coverage.

1-5.2 Only listed and approved devices and approved materials shall be used in sprinkler systems.

Exception: Listing shall be permitted to be waived for tanks, pumps, hangers, waterflow detection devices, and waterflow valves.

When the original edition of NFPA 13D was prepared, testing laboratories had not listed or evaluated tanks, pumps, hangers, waterflow devices, and water control valves in the sizes needed for residential use. Although some of these components are now listed, they are not required to be listed. One exception exists, however, involving the use of specially listed pipe materials. The exception to 3-4.1 requires the use of listed hangers for these materials.

The technical committee anticipated that the type of equipment used would be similar to that commonly found in residential plumbing systems. The economic impact of requiring listed tanks, pumps, and valves could increase the system cost by as much as 500 percent in some cases, which works against the low-cost objective established by USFA and adopted by the technical committee when they initiated their residential sprinkler program. Such a large cost increase would discourage installation of residential sprinkler systems. Although the use of listed components would increase the overall reliability of the system, the technical committee is of the opinion that the use of some nonlisted components still provides a sufficient level of reliability and performance.

1-5.3 Where listed, preengineered systems shall be installed within the limitations that have been established by the testing laboratories.

Approval laboratories have sometimes tested and listed complete systems that include all of the components necessary for an application. In this case, the system could include sprinklers, water supply tanks, valves, pumps, and special piping limitations. A manufacturer is permitted to develop such a system for application as a residential sprinkler system but does not require the use of such a packaged system.

1-5.4* All systems shall be tested for leakage at normal system operating water pressure.

Exception: Where a fire department pumper connection is provided, hydrostatic pressure tests shall be provided in accordance with NFPA 13, Standard for the Installation of Sprinkler Systems.

A-1-5.4 Testing of a system can be accomplished by filling the system with water and checking visually for leakage at each joint or coupling.

Fire department connections are not required for systems covered by this standard but can be installed at the discretion of the owner. In these cases, hydrostatic tests in accordance with NFPA 13, *Standard for the Installation of Sprinkler Systems,* are necessary.

Dry systems also should be tested by placing the system under air pressure. Any leak that results in a drop in system pressure greater than 2 psi (0.14 bar) in 24 hours should be corrected. Leaks should be identified using soapy water brushed on each joint or coupling. The presence of bubbles indicates a leak. This test should be made prior to concealing the piping.

Although A-1-5.4 describes the test procedures for a dry system, at this time, dry systems are not permitted. (See 3-5.1 and commentary to 4-1.2.1 and 4-3.2 Exception.)

Unlike the requirements of NFPA 13, and in some cases NFPA 13R, residential systems do not require a hydrostatic test unless the system is equipped with a fire department connection. If a hydrostatic test was required, it would complicate the installation and add significantly to the system's cost. Plumbing systems have been successfully installed in homes without special hydrostatic testing for years. Because the residential sprinkler system is intended to be similar to or integrated with the plumbing system, the technical committee concluded that a special hydrostatic test was not generally required.

Although a fire department connection to a residential system is not required, one can be installed. When a fire department connection is added, the system pressure can be increased using a fire department pumper. In such instances, the installer and owner should verify that the system has the integrity to withstand such pressures, and a hydrostatic test should be done in accordance with NFPA 13. Normally, the test would be a pressure of 200 psi (13.8 bar).

1-5.5 Where solvent cement is used as the pipe and fittings bonding agent, sprinklers shall not be installed in the fittings prior to the fittings being cemented in place.

In the 1999 edition, this installation requirement, which is often part of the installation instructions issued by the pipe manufacturer, was added to ensure that solvent cement will not drip into the piping drop and obstruct or clog the sprinkler.

1-6* Units

Metric units of measurement in this standard shall be in accordance with the modernized metric system known as the International System of Units (SI). The liter and bar units are outside of, but recognized by, SI and are commonly used in international fire protection. These units are provided in Table 1-6 with their conversion factors.

Table 1-6 Metric Conversions

Name of Unit	Unit Symbol	Conversion Factor
liter	L	1 gal = 3.785 L
pascal	Pa	1 psi = 6894.757 Pa
bar	bar	1 psi = 0.0689 bar
bar	bar	1 bar = 105 Pa

A-1-6 For additional conversions and information, see ASTM SI 10, *Standard for Use of the International System of Units (SI): the Modernized Metric System.*

1-6.1 Where a value for measurement as specified in this standard is followed by an equivalent value in other units, the first stated value shall be regarded as the requirement. A given equivalent value shall be considered to be approximate.

1-6.2 SI units have been converted by multiplying the quantity by the conversion factor and then rounding the result to the appropriate number of significant digits.

References Cited in Commentary

National Fire Protection Association, 1 Batterymarch Park, P. O. Box 9101, Quincy, MA 02269-9101.

NFPA 13, *Standard for the Installation of Sprinkler Systems*, 1999 edition.
NFPA 13R, *Standard for the Installation of Sprinkler Systems in Residential Occupancies up to and Including Four Stories in Height*, 1999 edition.
NFPA 25, *Standard for the Inspection, Testing, and Maintenance of Water-Based Fire Protection Systems*, 1998 edition.

U.S. Government Printing Office, Washington, DC 20402.

National Commission on Fire Prevention and Control, "America Burning," 1973.

Bibliography

1. Halpin, B.M., Dinan, J.J., and Deters, O.J., "Assessment of the Potential Impact of Fire Protection Systems on Actual Fire Incidents," Johns Hopkins University —Applied Physics Laboratory (JHU/APL), Laurel, MD, October 1978. (This study describes an in-depth analysis of fires involving fatalities and includes an assessment of how use of detectors, sprinklers, or remote alarms would have changed the results.)
2. Yurkonis, Peter, "Study to Establish the Existing Automatic Fire Suppression Technology for Use in Residential Occupancies," Rolf Jensen & Associates, Inc., Deerfield, IL, December 1978. (This study identifies suppression systems and evaluates design and cost factors affecting practical usage and user acceptance.)
3. Kung, H., Haines, D., and Green, R., Jr., "Development of Low-Cost Residential Sprinkler Protection," Factory Mutual Research Corporation, Norwood, MA, February 1978. (This study addresses development of low-cost residential sprinkler systems having minimal water discharge rates and providing adequate protection to life and property from smoldering fires.)
4. Henderson, N.C., Riegel, P.S., Patton, R.M., and Larcomb, D.B., "Investigation of Low-Cost Residential Sprinkler Systems," Battelle Columbus Laboratories (BCL), Columbus, OH, June 1978. (This study addresses fire tests of commercial nozzles and evaluates piping methods to achieve low-cost systems based on propane burner, wood crib, and furniture fire tests.)
5. Kung, H.C., Spaulding, R.D., and Hill, E.E., Jr., "Sprinkler Performance in Residential Fire Tests," Factory Mutual Research Corporation, Norwood, MA, Decem-

ber 1980. (This study includes fire tests on representative living room and bedroom residential configurations with combinations of furnishings, with open and closed windows, and with variations in sprinkler application rates and response sensitivity of sprinklers. This work has been closely evaluated by the NFPA 13D subcommittee during its conduct, and the subcommittee was responsible for directing many of the test conditions that were evaluated.)

6. Cote, A., and Moore, D., "Field Test and Evaluation of Residential Sprinkler Systems, Los Angeles Test Series," National Fire Protection Association, Boston, MA, April 1980. (This report was done for the NFPA 13D subcommittee and included a series of fire tests in actual dwellings with sprinkler systems installed in accordance with the data resulting from all prior test work and the criteria included in the proposed 1980 revisions to NFPA 13D so as to evaluate the effectiveness of the system under conditions approaching actual use.)
7. Moore, D., "Data Summary of the North Carolina Test Series of USFA Grant 79027 Field Test and Evaluation of Residential Sprinkler Systems," National Fire Protection Association, Boston, MA, September 1980. (This report was done for the NFPA 13D subcommittee and included a series of fire tests in actual mobile homes with sprinkler systems installed in accordance with the data resulting from prior test work and following criteria described in the 1980 edition of NFPA 13D in order to evaluate the effectiveness of the system under conditions approaching actual use.)
8. Kung, H.C., Spaulding, R.D., Hill, E.E., Jr., and Symonds, A.P., "Technical Report Field Evaluation of Residential Prototype Sprinkler, Los Angeles Fire Test Program," Factory Mutual Research Corporation, Norwood, MA, February 1982.
9. Cote, A.E., "Final Report on Field Test and Evaluation of Residential Sprinkler Systems," National Fire Protection Association, Quincy, MA, July 1982.
10. Factory Mutual Research Corporation/Underwriters Laboratories Inc., *Standard for Residential Automatic Sprinklers for Fire Protection Service*, Northbrook, IL, April 1980. (This study developed a performance standard for testing residential sprinklers under room fire conditions that conform with the requirements of NFPA 13D.)
11. Underwriters Laboratories Inc., "Proposed Standard for Residential Automatic Sprinklers for Fire Protection Service," UL 1626, Northbrook, IL, October 1980. (This standard describes the requirements for the testing of residential sprinklers by a testing laboratory. Factory Mutual Research Corporation has a similar standard.)

CHAPTER 2

Water Supply

2-1 General Provisions

Every automatic sprinkler system shall have at least one automatic water supply. Where stored water is used as the sole source of supply, the minimum quantity shall equal the water demand rate times 10 minutes. (See 4-1.3.)

Exception: For dwelling units that are one story in height and less than 2000 ft^2 (186 m^2) in area, the water supply shall be at least 7 minutes for the two-sprinkler demand.

A sprinkler system's effectiveness greatly depends on its water supply, which must be automatic and reliable. The adequacy of a public water supply is determined in the manner described in 4-4.3 and A-4-4.3 of NFPA 13D. When a stored water supply is used, it must provide the required amount of water and pressure for a minimum of 10 minutes. For dwellings that satisfy the limitations of the exception to Section 2-1 and 5-2.1, a 7-minute water supply can be used.

When considering a system design using the prototype residential sprinkler discussed in Chapter 4, the tank capacity must be at least 260 gal (984 L), which is based on two sprinklers operating at a total of 26 gpm (98 L/min). The tank would have sufficient water to control a fire while giving occupants enough time to evacuate the residence. The 10-minute supply includes an implied safety factor to account for a delay in the response of an occupant while the fire condition is recognized.

For a design based on the conditions noted in Chapter 5, tank capacity must be at least 91 gal (344 L), which is based on two sprinklers operating at a total of 13 gpm (49 L/min). The single sprinkler flows must also be determined and would result in a flow of 10 gpm (37.7 L/min) for a period of 10 minutes. This flow and duration requirement results in a tank capacity of at least 100 gal (378 L). These estimations for the size of the stored water supply do not take into account losses due to friction, turbulence, and elevation.

2-2* Water Supply Sources

The following water supply sources shall be considered to be acceptable by this standard:

(1) A connection to a reliable waterworks system with or without an automatically operated pump
(2) An elevated tank
(3) A pressure tank designed to American Society of Mechanical Engineers (ASME) standards for a pressure vessel with a reliable pressure source
(4) A stored water source with an automatically operated pump

A-2-2 The connection to city mains for fire protection is often subject to local regulation of metering and backflow prevention requirements. Preferred and acceptable water supply arrangements are shown in Figures A-2-2(a), A-2-2(b), and A-2-2(c). Where it is necessary to use a meter between the city water main and the sprinkler system supply, an acceptable arrangement as shown in Figure A-2-2(c) can be used. Under these circumstances, the flow characteristics of the meter are to be included in the hydraulic calculation of the system [see Table 4-4.3(e)]. Where a tank is used for both domestic and fire protection purposes, a low water alarm that actuates when the water level falls below 110 percent of the minimum quantity specified in Section 2-1 should be provided.

The effect of pressure-reducing valves on the system should be considered in the hydraulic calculation procedures.

The Technical Committee on Residential Sprinkler Systems' intent that pumps can be used on municipal supplies as well as on stored supplies is indicated in 2-2(1). If the municipal supply is not sufficient, the technical committee does not intend that a stored supply be used. A pump can be used to boost the pressure of a municipal supply provided a sufficient quantity of water is available. The commentary pertaining to 2-2(4) provides more discussion on pumps.

With regard to 2-2(2), a system can occasionally be supplied from an elevated tank. Where an elevated tank is used, it must have an elevation sufficient to provide adequate pressure to supply the system. Adequate pressure is determined by the methods described in 4-4.3 and 4-4.4.

Where a pressure tank is used for the water supply in accordance with 2-2(3), the amount of water and air in the tank is determined according to the information in 4-4.4. Subsection 9-2.3 of NFPA 13, *Standard for the Installation of Sprinkler Systems,* contains criteria for pressure tanks that are acceptable as a water source for NFPA 13 systems. Pressure tanks used as a water supply do not have to meet the requirements of NFPA 22, *Standard for Water Tanks for Private Fire Protection*. Tanks constructed in accordance with the ASME *Boiler and Pressure Vessel Code* can be acceptable, provided the authority having jurisdiction considers the air or nitrogen supply reliable.

With regard to 2-2(4), a water supply source, whether stored or from a municipal supply used in conjunction with an automatically operated pump, must be as follows:

(1) Have a pump with adequate capacity to supply the water demand as described in Section 4-1
(2) Have adequate pressure to overcome friction and elevation pressure losses

(3) Meet the operating pressure established by the testing laboratory for the sprinkler being used

NFPA 13D does not require that a fire pump conforming to NFPA 20, *Standard for the Installation of Stationary Pumps for Fire Protection*, be installed. The type of pump and the arrangement of power supplies to that pump only need to satisfy the system demand and to be electrically wired to conform to NFPA 70, *National Electrical Code®*.

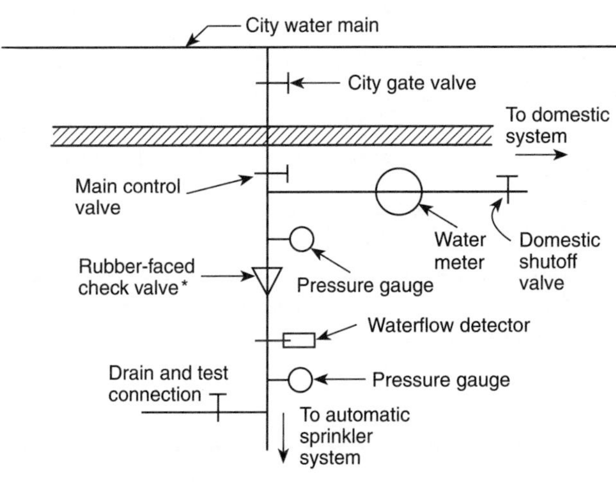

Figure A-2-2(a) *Preferable arrangement.*

Figures A-2-2(a), A-2-2(b), and A-2-2(c) show preferred and acceptable arrangements of connections to city mains and include appropriately located control valves, meters, domestic takeoffs, waterflow detectors, pressure gauges, and check valves. Where the local water authority requires backflow preventers, the devices are normally located in the position where the rubber-faced check valve is shown on the figures. The piping arrangement, including valves and fittings, must be taken into account when performing the calculations for system adequacy described in 4-4.3.

From a fire protection standpoint, meters and backflow preventers are undesirable because of the high friction loss characteristic of the devices, which must be included in the system's hydraulic calculation in accordance with 4-4.3. Unfortunately, many water authorities require metering of all water connections to a residence. In addition, clean water environmental concerns are leading more jurisdictions to require backflow preventers instead of traditional, less costly check valves. Check valves have performed satisfactorily, and no case histories involving backflow contamination from fire protection systems using check valves are available. Nonetheless, local jurisdictions often require the use of backflow devices.

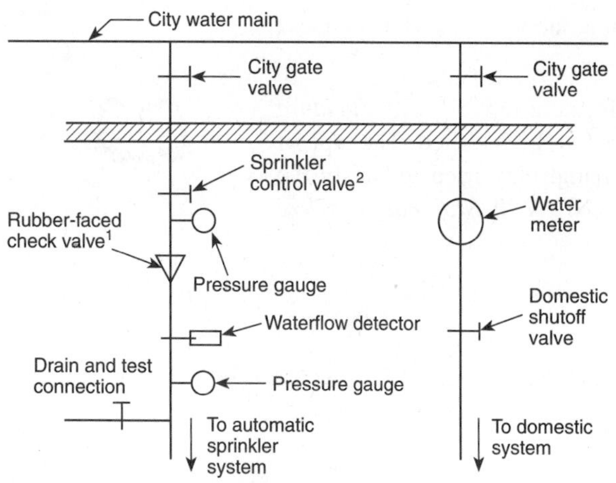

Figure A-2-2(b) Acceptable arrangement with valve supervision—Option 1.

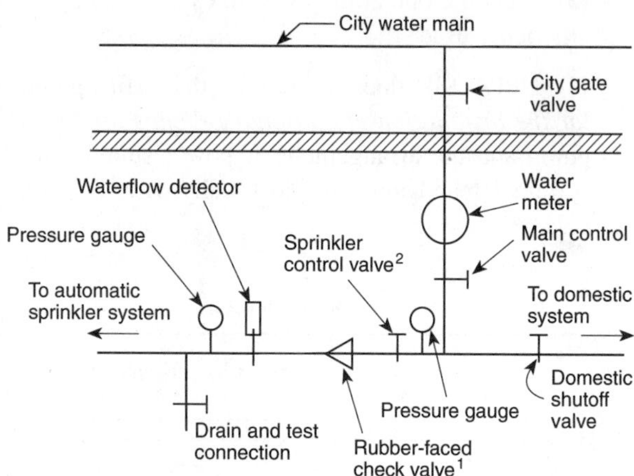

Figure A-2-2(c) Acceptable arrangement with valve supervision—Option 2.

2-3* Multipurpose Piping System

A piping system serving both sprinkler and domestic needs shall be considered to be acceptable by this standard where the following conditions are met.

(a)* In common water supply connections serving more than one dwelling unit, 5 gpm (19 L/min) are added to the sprinkler system demand to determine the size of common piping and the size of the total water supply requirements.

Exception: Domestic design demand shall not be required to be added where provision is made to prevent flow into the domestic water system upon operation of a sprinkler.

(b) Smoke detectors are provided in accordance with NFPA 72, *National Fire Alarm Code* .

(c) All piping in the system supplying sprinklers is listed and conforms to the piping specifications of this standard. Piping connected to this system and supplying plumbing fixtures only is required to comply with local plumbing and health authority requirements but is not required to be listed.

(d) Permitted by the local plumbing or health authority.

(e) A sign is affixed adjacent to the main shut-off valve that states in minimum ¼-inch letters, "Warning, the water system for this home supplies a fire sprinkler system that depends on certain flows and pressures being available to fight a fire. Devices that restrict the flow or decrease the pressure such as water softeners shall not be added to this system without a review of the fire sprinkler system by a fire protection specialist. Do not remove this sign."

A-2-3 Figures A-2-3(a), A-2-3(b), and A-2-3(c) illustrate acceptable water supply arrangements. When contemplating the use of multipurpose piping system, consideration should be given to possible future modifications for domestic purposes, such as the installation of a water softener or a water filtration system, which effect the performance of the sprinkler system.

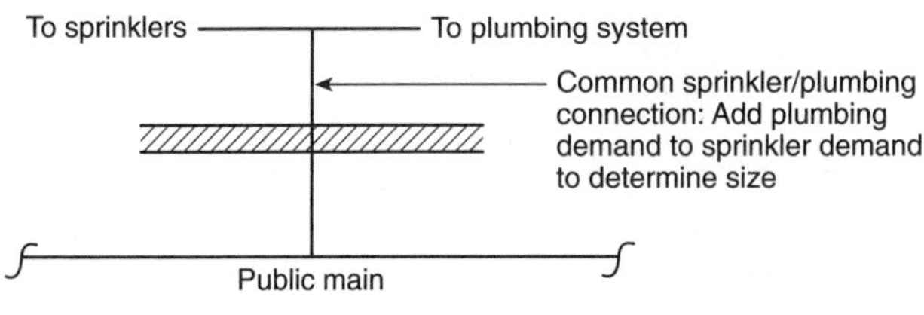

Figure A-2-3(a) *Multipurpose pipe system with separate supply.*

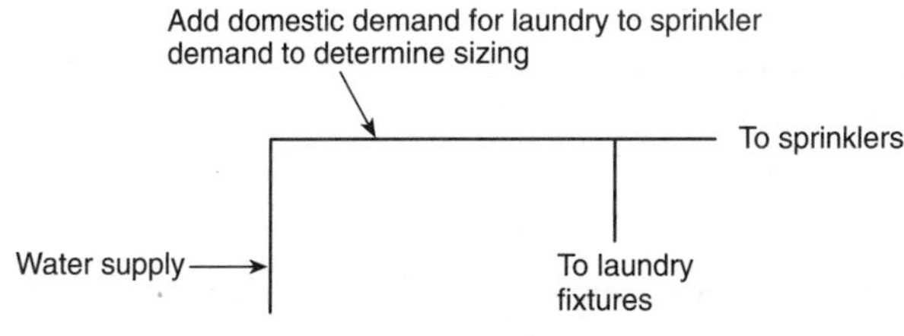

Figure A-2-3(b) *Multipurpose pipe system calculation procedure.*

A-2-3(a) In dwellings where long-term use of lawn sprinklers is common, provision should be made for such usage.

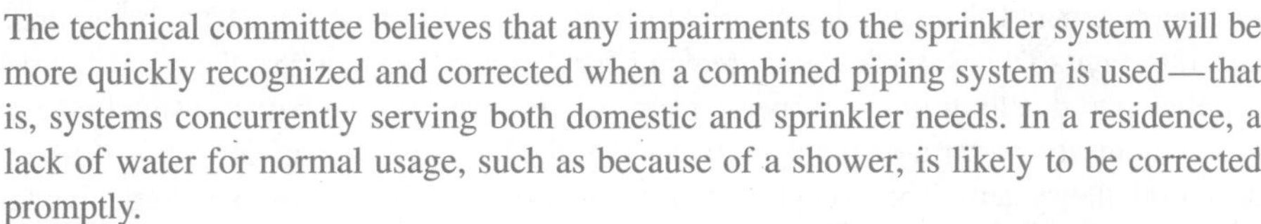
The technical committee believes that any impairments to the sprinkler system will be more quickly recognized and corrected when a combined piping system is used—that is, systems concurrently serving both domestic and sprinkler needs. In a residence, a lack of water for normal usage, such as because of a shower, is likely to be corrected promptly.

Figures A-2-3(a) and A-2-3(b) show two arrangements for a multipurpose piping system. Figure A-2-3(a) show a common point where domestic water and fire protection water are divided. A common piping line that is tapped to supply laundry fixtures is shown in Figure A-2-3(b). Figure A-2-3(c) shows an isometric view of a multilevel dwelling and one possible arrangement of the multipurpose piping. System design can be adjusted for simultaneous sprinkler flow and one or more domestic fixture outlets

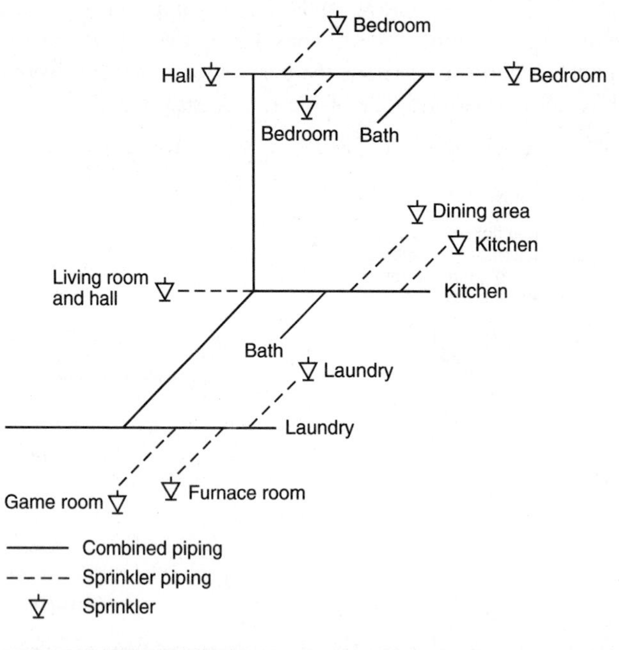

Figure A-2-3(c) *Multipurpose pipe system arrangement.*

by adding 5 gpm (19 L/min) to the sprinkler system demand where the supply serves a townhouse, for example.

A multipurpose piping system can sometimes be used to obtain a waiver for special backflow devices. Because water circulates through most portions of the system, the health concern alluded to in the commentary to A-2-2 is reduced.

Section 2-3 also clarifies under what circumstances the domestic use must be added to the sprinkler system demand.

With regard to 2-3(a), the technical committee recognized that periods of typical domestic water usage do not occur over great lengths of time except for an automatic long-term demand, such as a lawn sprinkler system. Although NFPA 13D only requires an adjustment to the waterflow requirements for domestic usage where the water source supplies two dwelling units, proper system performance can be better ensured where a peak domestic water demand is anticipated and added to the sprinkler demand when designing the system. The exception following Section 2-3(a) allows an alternative to considering the added demand by making an arrangement that shuts down the waterflow to the domestic demand. One way to accomplish this shutdown is through the use of an automatically controlled solenoid valve. On actuation of the sprinkler system waterflow switch, an electrical signal can be sent to the solenoid valve, which closes and shuts down the supply to the domestic demand or the individual appliance. The result is that no water is diverted from the sprinkler system.

Section 2-3(c) was revised for the 1999 edition to further promote the use of multipurpose residential sprinkler systems in one- and two-family dwellings. The 1999 edition permits the limited use of non-listed pipe in a sprinkler system other than that specified in Table 3-3.1.

Section 2-3(e) was added to the 1999 edition to make the homeowner aware of the impact of retroactively installing devices such as water softeners on the system. Unless properly accounted hydraulically, water softeners and other such devices can have an adverse effect on system performance by decreasing the available pressure and resultant waterflow to the sprinklers. This lower pressure ultimately affects the amount of water discharged from the sprinklers and the sprinklers' discharge pattern.

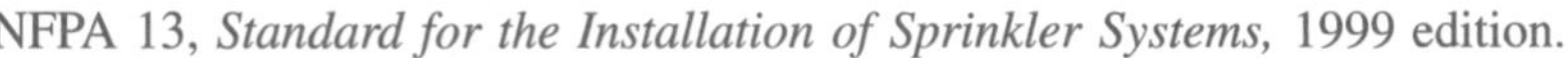

2-4 Manufactured Home Water Supply

A water supply for a sprinklered dwelling manufactured off-site shall not be less than that specified on the manufacturer's nameplate. *[See 4-4.3(k), Exception No. 2. See Chapter 5 for an alternative design approach for manufactured homes.]*

The piping for a sprinkler system installed in a manufactured home is typically installed by the home's manufacturer. NFPA 13D anticipates that the manufacturer will specify the water capacity and pressure needed to supply the system. The water supply that ultimately serves the manufactured home sprinkler system must meet the demand specified for the 7-minute or 10-minute duration, as appropriate.

References Cited in Commentary

National Fire Protection Association, 1 Batterymarch Park, P.O. Box 9101, Quincy, MA 02269-9101.

NFPA 13, *Standard for the Installation of Sprinkler Systems,* 1999 edition.
NFPA 20, *Standard for the Installation of Stationary Pumps for Fire Protection,* 1999 edition.
NFPA 22, *Standard for Water Tanks for Private Fire Protection,* 1998 edition.
NFPA 70, *National Electrical Code®*, 1999 edition.

American Society of Mechanical Engineers, Three Park Avenue, New York, NY 10016-5990.

ASME *Boiler and Pressure Vessel Code*, 1998 edition.

CHAPTER 3

System Components

3-1 Valves and Drains.

3-1.1 Each system shall have a single control valve arranged to shut off both the domestic system and the sprinkler system, and there shall be a separate shutoff valve for the domestic system only.

Exception No. 1: The sprinkler system piping shall be permitted to have a separate control valve where supervised by one of the following methods:

(a) Central station, proprietary, or remote station alarm service
(b) Local alarm service that causes the sounding of an audible signal at a constantly attended location.
(c) Valves that are locked open

Exception No. 2: A separate shutoff valve shall not be required for the domestic water supply in multipurpose piping systems.

A control valve permits anyone to shut off the water supply to the sprinkler system. That the control valve remains in the open position at all times is critical. The only exception should be for system maintenance or modification.

With only one control valve serving both the domestic and sprinkler water supplies, the possibility of inadvertently shutting off the sprinkler system is significantly reduced. The need for domestic water such as for a toilet would necessitate that water supply situation be promptly corrected. A separate control valve for the sprinkler system is acceptable only when the valve is supervised as described in Exception No. 1.

With regard to Exception No. 2, the status of the water supply on a multipurpose piping system, is essentially self-supervising. If the water supply is not available for the sprinkler system, it will also not be available for the toilets, sinks, and other domestic fixtures. This condition is likely to result in prompt corrective action regarding the water supply.

3-1.2 Each sprinkler system shall have a drain and/or test connection with a valve on the system side of the control valve.

This arrangement allows for the verification of waterflow and permits the system piping to be drained for maintenance or repairs. See 3-1.4 for requirements pertaining to waterflow alarms.

3-1.3 Additional drains shall be installed for each trapped portion of a dry system that is subject to freezing temperatures.

Exhibit 3.1 illustrates a trapped section of piping and how a drain is arranged to comply with 3-1.3. Such portions of the system are important to be carefully drained before freezing weather each year. If water from the system's operation or condensation collects in a trapped section of piping, it can freeze and break the piping, possibly causing extensive water damage.

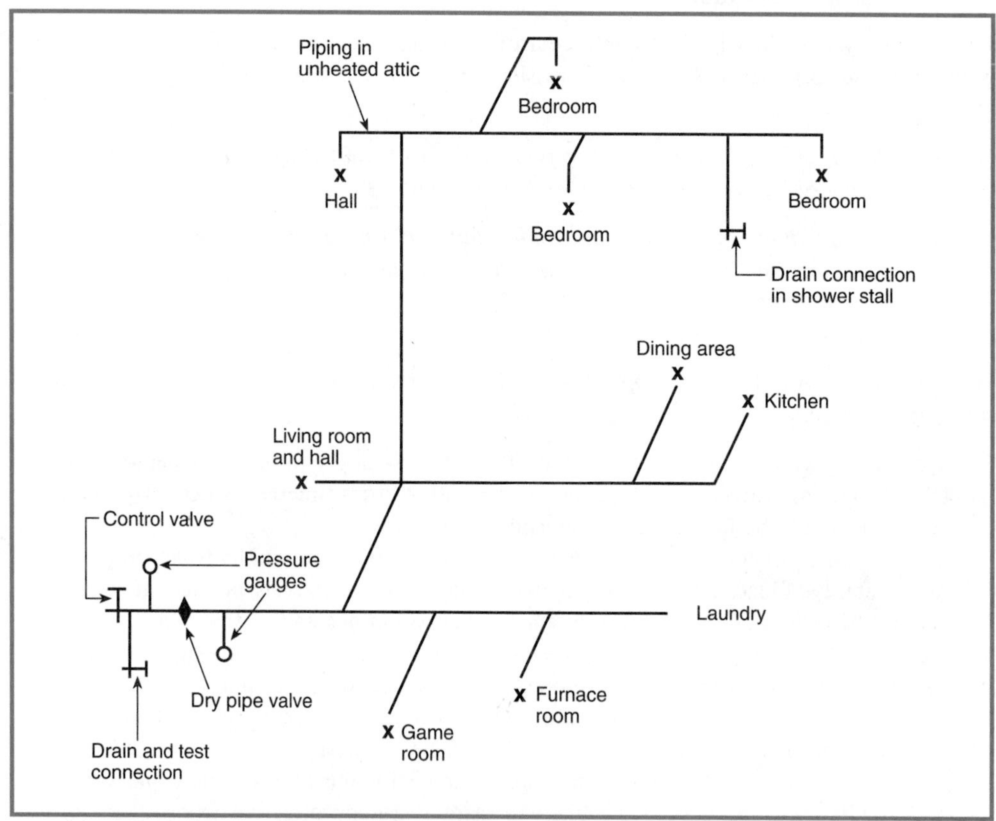

Exhibit 3.1 Drain connection for a trapped section of piping.

3-1.4* Where waterflow alarms are provided, inspector's test connections shall be installed at locations that allow flow testing of water supplies, connections, and alarm mechanisms. Where sprinklers used in the system have a nominal K-factor smaller than 5.6, the inspector's test shall be of the same size orifice as the smallest sprinkler.

A-3-1.4 These connections should be installed so that the valve can be opened fully and for a sufficient time period to ensure a proper test without causing water damage. The test connection should be designed and sized to verify the sufficiency of the water supply and alarm mechanisms.

This connection's installation provides a means of verifying the waterflow alarm's operation if installed and of verifying the flow through the system. Although a full-flow test is not required, a flowmeter, such as that used on lawn sprinkler systems, can be attached to this connection to determine the available flow rate. The 1999 edition was revised to specifically require the orifice of the inspector's test to be the same size as that of the sprinkler with the smallest orifice, and this ensures that the inspector's test simulates minimum fire flow conditions and that the waterflow alarm sounds at this flow.

3-2 Pressure Gauges

A pressure gauge shall be installed to indicate air pressure on dry systems and on water supply pressure tanks.

Residential sprinklers for use on a dry pipe system do not exist at the current time. Therefore, the requirement to monitor air pressure will not be applicable until dry pipe systems are used. The pressure tank's gauge indicates whether adequate operating pressure is available to the system.

3-3 Piping

Piping components manufactured to the material standards of Table 3-3.1 and Table 3-3.5 are not required to be listed even though testing laboratories have evaluated and listed many of these materials. The most commonly used piping materials in residential sprinkler systems are listed and are of nonmetallic materials, such as chlorinated polyvinyl chloride (CPVC) pipe and polybutylene (PB) pipe. Although Table 3-3.2 only identifies plastic pipe, it does not preclude the listing of metallic piping materials that are not identified in Table 3-3.1.

Although listed PB pipe is acceptable for use in sprinkler systems, PB pipe is no longer manufactured due to problems associated with its use in domestic plumbing systems. PB pipe's overall use in fire protection systems has been favorable. The development of other flexible pipe is being pursued as an alternative to PB pipe.

The 1999 edition provides reduced system working pressure criteria and temperature criteria for piping used in multipurpose piping systems. The original intent in the development of NFPA 13D was to provide standards for residential sprinkler systems that would be economically feasible to encourage installation and save lives. The installation of a multipurpose piping system could provide significant cost savings and potentially increase system reliability.

The pressure criterion was developed from the maximum pressure requirement of 80 psi (5.5 bar) for plumbing fixtures as provided by Section 604.8 of the *International Plumbing Code* and Section 608.2 of the *Uniform Plumbing Code*. The Technical Committee on Residential Sprinkler Systems believed that a safety factor should be included over and beyond 80 psi (5.5 bar) because of the associated life safety objectives. The technical committee chose a pressure rating of 130 psi (9.0 bar) at 120°F (49°C) for piping used in multipurpose systems where specific conditions exist. (See the exception to 3-3.1 and 3-3.5.) The temperature criterion of 120°F (49°C) was selected because it is included in Section 8.1 of UL 1821, *Standard for Safety-Thermoplastic Sprinkler Pipe and Fittings for Fire Protection Service*, based on attic temperature conditions from a 1977 UL study. A similar study conducted by the National Bureau of Standards, which is now the National Institute of Standards and Technology, in 1977 measured the attic temperatures of a house in Houston, TX. The maximum temperature at the top of the peak was 115°F (46°C) and 84°F (29°C) at the attic floor on the insulation's underside. This temperature criterion also prompted new requirements concerning the installation of plastic pipe in attics. (See 3-3.2.1.)

3-3.1* Pipe or tube used in sprinkler systems shall be of the materials specified in Table 3-3.1 or in accordance with 3-3.2 through 3-3.5. The chemical properties, physical properties, and dimensions of the materials in Table 3-3.1 shall be at least equivalent to the standards cited in the table and designed to withstand a working pressure of not less than 175 psi (12.1 bar).

Exception: When multipurpose piping system pressures do not exceed 130 psi (8.9 bar) and when such systems are not equipped with a fire department connection, piping shall be permitted to be designed to withstand a working pressure of not less than 130 psi (8.9 bar). When nonmetallic piping is used for this application, it shall be permitted to be rated at not less than 130 psi (8.9 bar) working pressure at not less than 120°F (49°C).

A-3-3.1 This standard anticipates the water supply for the system to be in compliance with the governing plumbing code for the jurisdiction. It is intended that any pipe material or diameter permitted by a plumbing code for one- or two-family dwellings and satisfying the hydraulic criteria of NFPA 13D is considered to be in compliance.

Because a number of systems conforming to NFPA 13D are installed retroactively, existing underground pipe and materials are to be used. A careful analysis of the existing pipe's size, condition, and type must be completed to ensure that it will provide the necessary flow and pressure to the system.

When new systems are installed, the underground pipe network only needs to comply with the requirements of the local plumbing code. The calculation procedures discussed in Chapter 4 and the associated commentary must be completed to ensure that the piping materials will allow for the proper system performance. With regard to the exception, see commentary to Section 3-3.

Table 3-3.1 Pipe or Tube Materials and Dimensions

Materials and Dimensions	Standard
Specification for Black and Hot-Dipped Zinc-Coated (Galvanized) Welded and Seamless Steel Pipe for Fire Protection Use	ASTM A 795
Specification for Welded and Seamless Pipe	ASTM A 53
Wrought Steel Pipe	ANSI B36.10M
Specification for Electric-Resistance-Welded Steel Pipe	ASTM A 135
Specification for Seamless Copper Tube [Copper Tube (Drawn, Seamless)]	ASTM B 75
Specification for Seamless Copper Water Tube	ASTM B 88
Specification for General Requirements for Wrought Seamless Copper and Copper-Alloy Tube	ASTM B 251
Fluxes for Soldering Applications of Copper and Copper-Alloy Tube	ASTM B 813
Specification for Filler Metals for Brazing and Braze Welding (BCuP, copper-phosphorus, or copper-phosphorus-silver brazing filler metal)	AWS A5.8
Specification for Solder Metal [alloy grades containing less than 0.2 percent lead as identified in ASTM B 32, Table 5, Section 1, and having a solidus temperature that exceeds 400°F (204°C)]	ASTM B 32

3-3.2* Other types of pipe or tube shall be permitted to be used, where investigated and listed for sprinkler systems by a testing and inspection agency laboratory. Listed piping materials including, but not limited to, chlorinated polyvinyl chloride (CPVC), polybutylene (PB), and steel differing from those provided in Table 3-3.1 shall be installed in accordance with their listings and the manufacturers' installation instructions. CPVC and PB pipe and tube shall comply with the portions of the American Society for Testing and Materials (ASTM) standards specified in Table 3-3.2 that apply to fire protection service in addition to the provisions of this paragraph.

Table 3-3.2 Specially Listed Pipe or Tube Materials and Dimensions

Materials and Dimensions	Standard
Nonmetallic Piping:	
Specification for Chlorinated Polyvinyl Chloride (CPVC) Pipe	ASTM F 442
Specification for Polybutylene (PB) Pipe	ASTM D 3309

Note: In addition to satisfying these minimum standards, specially listed pipe shall be required to comply with the provisions of 3-3.2.

A-3-3.2 Not all pipe or tube made to ASTM D 3309, *Standard Specification for Polybutylene (PB) Plastic Hot- and Cold-Water Distribution Systems,* and ASTM F 442, *Standard Specification for Chlorinated Poly (Vinyl Chloride) (CPVC) Plastic Pipe (SDR-PR),* as described in 3-3.1

and 3-3.2 is listed for fire sprinkler service. Listed pipe is identified by the logo of the listing agency.

This subsection and Table 3-3.2 specifically address the use of specially listed pipe. These pipe materials are now widely accepted and deserve specific recognition.

No material is prohibited from use as long as it is listed for use in sprinkler systems and is installed in accordance with its listing. This subsection is intended to encourage development of either more efficient or more cost-effective materials.

Not all pipe made to a particular standard is listed. Listed piping is identified by the logo of the listing agency. Similar piping that is required to be listed and that is manufactured with less exacting quality control must not be used in sprinkler systems. This provision precludes the use of nonmetallic pipe materials available at a building supply store, because these materials are not suitable for use in sprinkler systems.

At this time, two synthetic piping materials are listed for sprinkler system application—CPVC and PB pipe. The listings of these materials often include restrictions such as the pipe cannot be left exposed. (CPVC pipe can be installed exposed under special conditions.) The listings are specific and indicate for which standard (e.g., NFPA 13D, *Standard for the Installation of Sprinkler Systems in One- and Two-Family Dwellings and Manufactured Homes*, or NFPA 13R, *Standard for the Installation of Sprinkler Systems in Residential Occupancies up to and Including Four Stories in Height*) or hazard classification (e.g., Light Hazard in accordance with NFPA 13, *Standard for the Installation of Sprinkler Systems*) the pipe is listed. The user must refer to the extensive listing information to correctly install the system components. See the additional discussion on these materials in Chapter 2 of NFPA 13.

Although permitted by NFPA 13D, PB pipe is no longer manufactured due to problems associated with its use in domestic plumbing systems. PB pipe used in fire protection systems is subject to certain restrictions and installation practices that do not necessarily apply to its use in domestic plumbing systems. Overall, the use of PB pipe in fire protection systems has been favorable. The development of other flexible pipe is being pursued as an alternative to PB pipe.

3-3.2.1 When nonmetallic piping is installed in attics, adequate insulation shall be provided on the attic side of the piping to avoid exposure of the piping to temperatures in excess of the pipe's rated temperature.

At the request of the technical committee, Hughes Associates, Inc. performed a heat transfer analysis based on 145°F (63°C) attic temperatures and 100°F (38°C) temperatures in the room below. Six in. (152 mm) of R-30 insulation above the pipe resulted in a pipe temperature of 104°F (40°C). Two in. (51 mm) of the same insulation yielded a pipe temperature of 109°F (43°C). A lack of insulation resulted in a pipe temperature of 129°F (54°C). This analysis demonstrated the importance of adequate insulation for pipe protection in climates where the insulation might not be required because of cold weather. This lack of insulation is especially important in nonmetallic pipe in compliance with the conditions of the exception to 3-3.1 and 3-3.5.

3-3.3 Wherever the word *pipe* is used in this standard, it shall be understood also to mean tube.

For the purposes of NFPA 13D, the term *pipe* also means *tube*.

3-3.4 Schedule 10 steel pipe shall be permitted to be joined with mechanical groove couplings approved for service, with grooves rolled on the pipe by an approved groove-rolling machine.

3-3.5 Fittings used in sprinkler systems shall be of the materials listed in Table 3-3.5 or in accordance with 3-3.7. The chemical properties, physical properties, and dimensions of the materials specified in Table 3-3.5 shall be at least equivalent to the standards cited in the table. Fittings used in sprinkler systems shall be designed to withstand a working pressure of not less than 175 psi (12.1 bar).

Exception: When multipurpose piping system pressures do not exceed 130 psi (8.9 bar) and when such systems are not equipped with a fire department connection, fittings shall be permitted to be designed to withstand a working pressure of not less than 130 psi (8.9 bar). When nonmetallic fittings are used for this application, they shall be permitted to be rated at not less than 130 psi (8.9 bar) working pressure at not less than 120°F (49°C).

Table 3-3.5 Fitting Materials and Dimensions

Materials and Dimensions	Standard
Cast Iron:	
Gray Iron Threaded Fittings	ASME B16.4
Cast Iron Pipe Flanges and Flanged Fittings	ASME B16.1
Malleable Iron:	
Malleable Iron Threaded Fittings	ASME B16.3
Steel:	
Factory-Made Wrought Steel Buttweld Fittings	ASME B16.9
Buttwelding Ends	ASME B16.25
Specification for Piping Fittings of Wrought Carbon Steel and Alloy Steel for Moderate and Elevated Temperatures	ASTM A 234
Pipe Flanges and Flanged Fittings	ASME B16.5
Forged Fittings, Socket-Welding and Threaded	ASME B16.11
Copper:	
Wrought Copper and Copper Alloy Solder Joint Pressure Fittings	ASME B16.22
Cast Copper Alloy Solder Joint Pressure Fittings	ASME B16.18

Fittings of the types and materials indicated in Table 3-3.5 are not required to be listed provided they are manufactured to the indicated standards or to a standard meeting or exceeding those indicated.

With regard to the exception, see commentary to Section 3-3.

3-3.6 Joints for the connection of copper tube shall be brazed.

Exception: Soldered joints shall be permitted to be used for wet pipe copper tube systems.

Previous editions allowed 50-50 solder because it was the type normally used in domestic plumbing systems. The 1986 amendments to the *Federal Safe Drinking Water Act* prohibited the use of 50-50 solder in plumbing systems. While this act does not specifically apply to residential sprinkler systems, the technical committee considered it poor practice to use this leaded solder in close proximity to potable water systems. The solder alloys that conform to Table 5, Section 1 of ASTM B 32, *Standard Specification for Solder Metal*, have a higher melting point, contain no lead, and are permitted under the *Federal Safe Drinking Water Act*.

3-3.7* Other types of fittings shall be permitted to be used, but only where investigated and listed for sprinkler systems by a testing and inspection agency laboratory. Listed fittings including, but not limited to, CPVC, PB, and steel differing from those provided in Table 3-3.5 shall be installed in accordance with their listings and the manufacturers' installation instructions. CPVC and PB pipe and tube fittings shall comply with the portions of the ASTM standards specified in Table 3-3.7 that apply to fire protection service in addition to the provisions of this paragraph.

Table 3-3.7 Specially Listed Fittings and Dimensions

Materials and Dimensions	Standard
Specification for Schedule 80 CPVC Threaded Fittings	ASTM F 437
Specification for Schedule 40 CPVC Socket-Type Fittings	ASTM F 438
Specification for Schedule 80 CPVC Socket-Type Fittings	ASTM F 439

Note: In addition to satisfying these minimum standards, specially listed pipe fittings shall be required to comply with the provisions of 3-3.7.

A-3-3.7 Not all fittings made to ASTM F 437, *Standard Specification for Threaded Chlorinated Poly (Vinyl Chloride) (CPVC) Plastic Pipe Fittings, Schedule 80,* ASTM F 438, *Standard Specification for Socket-Type Chlorinated Poly (Vinyl Chloride) (CPVC) Plastic Pipe Fittings, Schedule 40,* and ASTM F 439, *Standard Specification for Socket-Type Chlorinated Poly (Vinyl Chloride) (CPVC) Plastic Pipe Fittings, Schedule 80,* as described in 3-3.5 and 3-3.7 are listed for fire sprinkler service. Listed fittings are identified by the logo of the listing agency.

The types of fittings acceptable are those that have been found satisfactory and listed by a testing and inspection agency laboratory. Although Table 3-3.7 identifies only plastic materials, specially listed metallic fittings are also available. These fittings are compatible with the pipe materials listed in Table 3-3.2. (See Sections 3-3 and 3-5 of NFPA 13 for additional information on pipe and fittings.)

3-4 Piping Support

3-4.1 Piping shall be supported from structural members using support methods comparable to those required by local plumbing codes.

Exception: Listed piping shall be supported in accordance with any listing limitations.

Unlike the hanging requirements for sprinkler systems installed according to NFPA 13, NFPA 13D simply requires adequate support by methods recognized in local plumbing codes for the piping materials. Generally, residential systems deal with pipe sizes of ¾ in. to 1¼ in. (19 mm to 33 mm), and pipe sizes are not expected to exceed 2 in. (51 mm). The dynamic loads and forces that can be generated by waterflow in a residential sprinkler system are considerably lower than those experienced in a sprinkler system in compliance with NFPA 13 or NFPA 13R, *Standard for the Installation of Sprinkler Systems in Residential Occupancies up to and Including Four Stories in Height*. Most plumbing code hangers provide adequate support for these piping materials. The exception recognizes that listed piping can have specific hanging requirements as a condition of its listing.

3-4.2 Piping laid on open joists or rafters shall be secured to prevent lateral movement.

3-4.3* Sprinkler piping shall be adequately secured to restrict the movement of piping upon sprinkler operation.

A-3-4.3 The reaction forces caused by the flow of water through the sprinkler could result in displacement of the sprinkler, thereby adversely affecting sprinkler discharge.

Many residential sprinkler designs use flush- or recessed-mounted configurations. The reaction force created by the discharge of water can cause the piping to lift, thereby resulting in a severe obstruction to the sprinkler spray pattern. This phenomenon is not considered in a plumbing code and requires special treatment.

3-5 Sprinklers

Since the listing of the first residential sprinkler, manufacturers have listed others, including both pendent and sidewall types. The newer residential sprinklers are more aesthetically pleasing in design, which should alleviate the homeowner's concern over the device's appearance.

3-5.1 Listed residential sprinklers shall be used. Listing shall be based on tests to establish the ability of the sprinklers to control residential fires under standardized fire test conditions. The standardized room fires shall be based on a residential array of furnishings and finishes.

Exception No. 1: Residential sprinklers shall not be used in dry pipe systems unless specifically listed for that purpose.

Exception No. 2: Listed dry-type sprinklers shall be permitted to be used in accordance with 4-3.2.

Both Underwriters Laboratories Inc. and Factory Mutual Research Corporation publish standards for testing and evaluating residential sprinklers.[1, 2] The protocol of these standards includes a room fire test, verification of the sprinkler's thermal sensitivity, and an evaluation of the water distribution characteristics. Standard spray sprinklers of the type used in NFPA 13 sprinkler systems are evaluated under a different set of criteria. The residential sprinkler test requirements were verified in testing at Factory Mutual Research Corporation [3] and in the Los Angeles and Charlotte, NC, residential sprinkler fire tests.[4, 5, 6, 7]

Exception Nos. 1 and 2 were added because of the concern for the expected delay in water delivery inherent in dry pipe systems. No residential sprinklers are available that are listed for use in a dry pipe sprinkler system. Residential sprinklers utilized in NFPA 13, NFPA 13D, and NFPA 13R, *Standard for the Installation of Sprinkler Systems in Residential Occupancies up to and Including Four Stories in Height*, are bound by this limitation.

3-5.2 Temperature Ratings.

The requirements of 3-5.2.1 through 3-5.2.3 shall be used for the selection of sprinkler temperature ratings.

The requirements of 3-5.2.1 through 3-5.2.3 apply to the environmentally controlled portions of the occupancy. The intent of the requirements is to use ordinary-temperature sprinklers in areas not subject to unusually high ambient temperatures.

3-5.2.1 Ordinary temperature–rated residential sprinklers [135°F to 170°F (57°C to 77°C)] shall be installed where maximum ambient ceiling temperatures do not exceed 100°F (38°C).

3-5.2.2 Intermediate temperature–rated residential sprinklers [175°F to 225°F (79°C to 107°C)] shall be installed where maximum ambient ceiling temperatures are between 101°F and 150°F (39°C and 66°C).

Conditions in a residence that can result in high ambient ceiling temperatures are attributable to track lighting units, fireplaces, wood-burning stoves, cooking equipment, and attics with insufficient ventilation.

Sprinkler spacing and positioning in the immediate vicinity of these areas must be considered. Steps must be taken to ensure that sprinklers are correctly selected for such areas. Recommendations and advice from the suppliers of heat-producing equipment and sprinklers should be solicited to allow for the placement of sprinklers in such areas. The necessary rules for applications of this concept are given in 3-5.2.3. These rules assist but do not completely cover all situations involving residential sprinkler placement. Certain limitations with regard to all types of fast-response sprinklers and their respective temperature ratings exist. Specifically, increases in the temperature ratings result in a decrease in sensitivity, and the ability to maintain the fast-response characteristic is diminished.

3-5.2.3 The following practices shall be observed where installing residential sprinklers.

Exception: Where higher expected ambient temperatures are otherwise determined.

(1) Sprinklers under glass or plastic skylights exposed to direct rays of the sun shall be of intermediate-temperature classification.
(2) Sprinklers in an unventilated concealed space under an uninsulated roof or in an unventilated attic shall be of intermediate-temperature classification.
(3) Sprinklers installed near specific heat sources that are identified in Table 3-5.2.3 shall be of ordinary- or intermediate-temperature rating, as indicated.

Exception: Where sprinklers are listed for positioning closer to a heat source than the minimum distance shown in Table 3-5.2.3, the closer minimum distances shall be permitted to be used.

Table 3-5.2.3 Minimum Distances for Ordinary- and Intermediate-Temperature Residential Sprinklers

	Minimum Distance from Edge of Source to Ordinary-Temperature Sprinkler		Minimum Distance from Edge of Source to Intermediate-Temperature Sprinkler	
Heat Source	**in.**	**mm**	**in.**	**mm**
Side of open or recessed fireplace	36	914	12	305
Front of recessed fire place	60	1524	36	914
Coal- or wood-burning stove	42	1067	12	305
Kitchen range	18	457	9	229
Wall oven	18	457	9	229
Hot air flues	18	457	9	229
Uninsulated heat ducts	18	457	9	229
Uninsulated hot water pipes	12	305	6	152
Side of ceiling- or wall-mounted hot air diffusers	24	607	12	305
Front of wall-mounted hot air diffusers	36	914	18	457
Hot water heater or furnace	6	152	3	76
Light fixture 0W–250W	6	152	3	76
Light fixture 250W–499W	12	305	6	152

Air temperature surrounding a sprinkler can exceed 100°F (38°C) when a skylight permits direct sun exposure to that area. Unventilated attics tend to develop rather high temperatures in the summer months. Thus, a similar precaution to prevent false activation should also be undertaken. Selecting the appropriate temperature rating for a given environment is always important. Proper selection helps to ensure that sprinklers will not operate prematurely and that an excessive number of sprinklers will not operate during a fire.

With regard to 3-5.2.3(3), positioning sprinklers closer than that allowed by Table

3-5.2.3 is likely to result in their false activation. Similar to other NFPA 13D provisions, if a sprinkler is listed so that it can be positioned closer to a heat source than that indicated by NFPA 13D, this smaller dimension can be used.

3-5.3 Operated or damaged sprinklers shall be replaced with sprinklers having the same performance characteristics as the original equipment.

Underwriters Laboratories Inc. has conducted a program in which field samples of automatic sprinklers were tested to determine their operating characteristics. According to the test program, a sprinkler that is damaged or painted after leaving the factory is unlikely to operate properly and should be replaced. Installing replacement sprinklers with performance characteristics differing from those of the original sprinklers will likely negatively impact overall system performance.

3-5.4 Painting and Ornamental Finishes.

It is extremely important that only manufacturers that have had their procedures listed apply paint or ornamental finishes to residential sprinklers. The installing contractor or homeowner does not have the facilities available to properly duplicate the manufacturer's procedures. Do-it-yourself applications will seriously impair the sprinkler's operation, render it inoperable, or impact the discharge pattern's characteristics.

3-5.4.1* Sprinkler frames shall be permitted to be factory painted or enameled as ornamental finish in accordance with 3-5.4.2; otherwise, sprinklers shall not be painted, and any sprinklers that have been painted shall be replaced with new, listed sprinklers.

Exception: Sprinklers painted with factory-applied coatings shall not be required to be replaced.

A-3-5.4.1 Decorative painting of a residential sprinkler is not to be confused with the temperature identification colors as specified in 2-2.3 of NFPA 13, *Standard for the Installation of Sprinkler Systems.*

3-5.4.2 Ornamental finishes shall not be applied to sprinklers by an individual other than the sprinkler manufacturer, and only sprinklers listed with such finishes shall be used.

3-5.5 Escutcheon Plates.

Where nonmetallic sprinkler ceiling plates (escutcheons) or recessed escutcheons (metallic or nonmetallic) are used, they shall be listed based on tests of the assembly as a residential sprinkler.

These devices are required to be tested and listed by the listing laboratory as part of the residential sprinkler assembly. The tests are intended to determine that the escutcheon will not interfere with the operation and discharge characteristics of the sprinkler.

3-6* Alarms

Local waterflow alarms with facilities for flow testing, such as alarm devices, shall be provided on all sprinkler systems.

Exception: Dwellings or manufactured homes having smoke detectors in accordance with NFPA 72, National Fire Alarm Code, shall not be required to be provided with a waterflow alarm.

A-3-6 Alarms should be of sufficient intensity to be clearly audible in all bedrooms over background noise levels while all intervening doors are closed. The tests of audibility level should be conducted with all household equipment that operates at night in full operation. Examples of such equipment are window air conditioners and room humidifiers. Where off-premises alarms are provided, the waterflow and the control valve position, as a minimum, should be monitored.

An exterior alarm can be of benefit in areas where a neighbor could alert the fire department or to enhance the ability for an assisted rescue by a passerby.

A waterflow test is normally conducted using the system drain. Figures A-2-2(a), A-2-2(b), and A-2-2(c) show examples of this arrangement.

NFPA 13D expects that smoke detectors are installed throughout the dwelling unit as indicated in the preface. When the detectors are present, the waterflow alarm is not required. However, the use of a waterflow alarm allows for the prompt notification of a sprinkler activation or other waterflow through the system if no occupants are present to witness such flow. If arranged properly, notification allows for immediate response to the waterflow situation and minimizes any water damage. As a reminder, when using the design option discussed in Chapter 5, the waterflow alarm must be installed.

NFPA 13D's basic philosophy is to provide a system that controls a fire for either 7 or 10 minutes, depending on the design employed, so occupants can safely evacuate. As a result, an alarm is needed to warn occupants and initiate evacuation. NFPA 13D suggests two methods by which notification can be done. If smoke detectors are installed in accordance with NFPA 72, *National Fire Alarm Code®*, NFPA 13D anticipates that they will give an adequate warning for evacuation. However, if smoke detectors are not installed, then the sprinkler system must be equipped with a waterflow device that sounds an audible alarm throughout the residence to initiate evacuation.

When determining whether the alarm device is of sufficient intensity, the following procedure can be used. To conduct audibility tests, the household occupants should position themselves on each level of the house inside a room, such as a bedroom, with the door closed. The alarm should be operated and the occupants should report whether or not it can be heard. The alarm must be loud enough to wake a person who is asleep.

References Cited in Commentary

National Fire Protection Association, 1 Batterymarch Park, P.O. Box 9101, Quincy, MA 02269-9101.

NFPA 13, *Standard for the Installation of Sprinkler Systems*, 1999 edition.

NFPA 13R, *Standard for the Installation of Sprinkler Systems in Residential Occupancies up to and Including Four Stories in Height*, 1999 edition.

NFPA 72, *National Fire Alarm Code®*, 1999 edition.

American Society for Testing and Materials, 100 Barr Harbor Drive, West Conshohocken, PA 19428-2959.

ASTM B 32, *Standard Specification for Solder Metal*, 1996 edition.

Underwriters Laboratories Inc., 333 Pfingsten Road, Northbrook, IL 60062.

UL 1821, *Standard for Safety-Thermoplastic Sprinkler Pipe and Fittings for Fire Protection Service*, 1st edition, 1994.

U.S. Government, U.S. Government Printing Office, Washington, DC 20402.

Federal Safe Drinking Water Act, 1974.

Building Officials and Code Administrators International, 4051 W. Flossmoor Road, Country Club Hills, IL 60478-5795.

International Plumbing Code, Section 604.8, 1997.

International Association of Plumbing and Mechanical Officials, 20001 E. Walnut Drive South, Walnut, CA 91789-2825.

Uniform Plumbing Code, Section 608.2, 1997.

Bibliography

1. Factory Mutual Research Corporation/Underwriters Laboratories Inc., *Standard for Residential Automatic Sprinklers for Fire Protection Service*, Northbrook, IL, April 1980. (This study developed a performance standard for testing residential sprinklers under room fire conditions that conform with the requirements of NFPA 13D.)
2. Underwriters Laboratories Inc., "Proposed Standard for Residential Automatic Sprinklers for Fire Protection Service," UL 1626, Northbrook, IL, October 1980. (This standard describes the requirements for the testing of residential sprinklers by a testing laboratory. Factory Mutual Research Corporation has a similar standard.)
3. Kung, H.C., Spaulding, R.D., and Hill, E.E., Jr., "Sprinkler Performance in Residential Fire Tests," Factory Mutual Research Corporation, Norwood, MA, December 1980. (This study includes fire tests on representative living room and bedroom residential configurations with combinations of furnishings, with open and closed windows, and with variations in sprinkler application rates and response sensitivity of sprinklers. This work has been closely evaluated by the NFPA 13D subcommittee during its conduct, and the subcommittee was responsible for directing many of the test conditions that were evaluated.)
4. Cote, A., and Moore, D., "Field Test and Evaluation of Residential Sprinkler Systems, Los Angeles Test Series," National Fire Protection Association, Boston, MA, April 1980. (This report was done for the NFPA 13D subcommittee and included a series of fire tests in actual dwellings with sprinkler systems installed in accordance with the data resulting from all prior test work and the criteria included in the proposed 1980 revisions to NFPA 13D so as to evaluate the effectiveness of the system under conditions approaching actual use.)

5. Moore, D., "Data Summary of the North Carolina Test Series of USFA Grant 79027 Field Test and Evaluation of Residential Sprinkler Systems," National Fire Protection Association, Boston, MA, September 1980. (This report was done for the NFPA 13D subcommittee and included a series of fire tests in actual mobile homes with sprinkler systems installed in accordance with the data resulting from prior test work and following criteria described in the 1980 edition of NFPA 13D in order to evaluate the effectiveness of the system under conditions approaching actual use.)
6. Kung, H.C., Spaulding, R.D., Hill, E.E., Jr., and Symonds, A.P., "Technical Report Field Evaluation of Residential Prototype Sprinkler, Los Angeles Fire Test Program," Factory Mutual Research Corporation, Norwood, MA, February 1982.
7. Cote, A.E., "Final Report on Field Test and Evaluation of Residential Sprinkler Systems," National Fire Protection Association, Quincy, MA, July 1982.

CHAPTER 4

System Design

4-1 Design Criteria

The design criteria for residential sprinkler systems are a radical departure from the conventional area/density methods of NFPA 13, *Standard for the Installation of Sprinkler Systems*. The goal for an NFPA 13D design is to obtain fire control with no more than two sprinklers. Chapter 4 illustrates how to accomplish this goal. The design criteria discussed in Section 4-1 were developed through laboratory and full-scale fire tests conducted at Factory Mutual Research Corporation and at test sites in Los Angeles and Charlotte, NC. The design criteria identified in Chapter 4 of NFPA 13D can be considered as baseline criteria. Although these criteria are the basis for several listed residential sprinklers, additional research and development efforts over the years have resulted in the creation of sprinklers that use differing design criteria.

4-1.1 Design Discharge.

The system shall provide a discharge of not less than 18 gpm (68 L/min) to any single operating sprinkler and not less than 13 gpm (49 L/min) per sprinkler to the number of design sprinklers, but the discharge shall not be less than the listing of the sprinkler. The minimum operating pressure of any residential sprinkler shall be 7 psi (0.5 bar).

This subsection requires a discharge of 18 gpm (68 L/min) for a single operating sprinkler and 13 gpm (49 L/min) per sprinkler for two sprinklers operating in a compartment. These flow rates can be extrapolated to a density of 0.125 gpm/ft^2 [5.1 (L/min)/m^2] for one sprinkler and 0.09 gpm/ft^2 [3.7 (L/min)/m^2] for two sprinklers at a maximum spacing of 144 ft^2 (13.4 m^2) per sprinkler. Test data indicate that the primary criterion for a residential system should be a flow rate rather than discharge density over an operating area.[1, 2, 3, 4]

The flow rates are based primarily on the Los Angeles and Charlotte, NC, fire tests.[1, 2, 3, 4] A high percentage (95 percent) of the full-scale tests that were conducted were successful. Actual fire experience has also demonstrated the ability of one sprinkler and under certain conditions two sprinklers to control the fire. Operation Life Safety has been tracking the number of residential sprinkler operations for over 10 years. Their records indicate that a high percentage of fires involving residential sprinklers usually only result in the operation of a single device.

4-1.2* Number of Design Sprinklers.

A-4-1.2 Residential sprinklers are currently listed for use under flat, smooth, horizontal ceilings only. Sloped, beamed, and pitched ceilings could require special design features such as larger flow, a design for three or more sprinklers to operate in the compartment, or both. Figures A-4-1.2(a) and A-4-1.2(b) show examples of a design configuration.

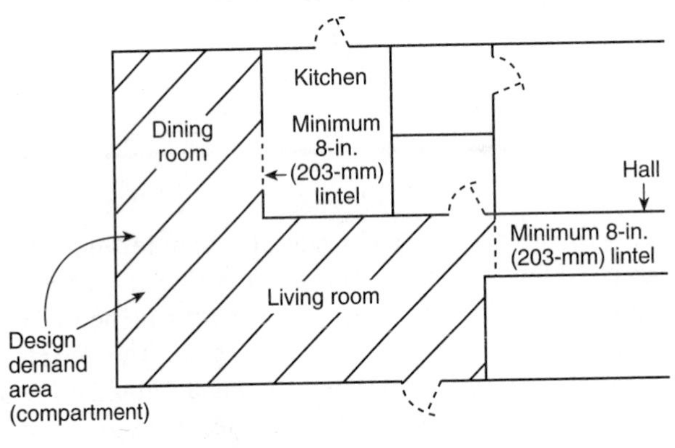

***Figure A-4-1.2(a)** Sprinkler design areas for typical residential occupancy—without lintel.*

4-1.2.1 The number of design sprinklers shall include all sprinklers within a compartment up to a maximum of two sprinklers under a flat, smooth, horizontal ceiling. For compartments containing two or more sprinklers, calculations shall be provided to verify the single operating sprinkler criteria and the multiple (two) operating sprinkler criteria.

Since 1980 when the first residential sprinkler was listed, the installation criteria have limited its use to flat, smooth, horizontal ceilings. While this type of ceiling is not representative of all ceilings in which residential sprinklers are used, it is a limitation of the device. Some manufacturers will provide guidance on installations where the ceilings are not horizontal, smooth, and flat. Some guidance for sloped and beamed ceilings is also given in A-4-2.3.

Even if more than two sprinklers are required to cover the floor area in a compartment, the maximum number of sprinklers that need to be included in the design calculations for water supply is two. This requirement is frequently misinterpreted to mean that the room size must be limited to make sure that only two sprinklers cover the floor area.

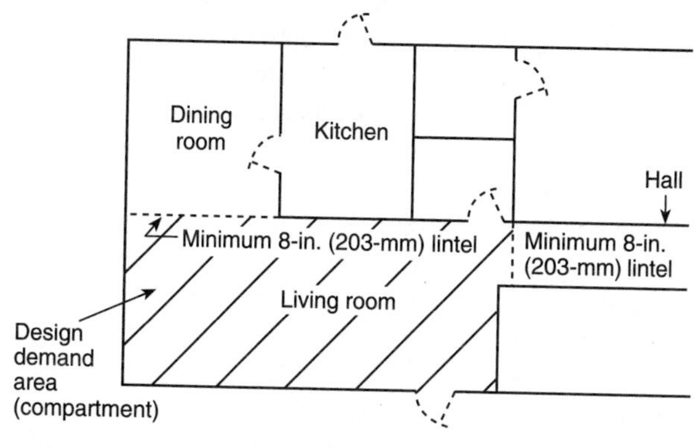

Figure A-4-1.2(b) *Sprinkler design areas for typical residential occupancy—with lintel.*

The calculation procedure also requires that both the single- and two-sprinkler discharge criteria be satisfied. For example, if the largest compartment contains three sprinklers, one calculation must be submitted to verify that the single-sprinkler criteria can be satisfied. A separate calculation must also be submitted to show that the two-sprinkler criteria can be satisfied.

Available data from fire tests of residential sprinklers do not support the four-sprinkler design criteria for dry systems that was included in the 1980 edition of NFPA 13D. Currently, residential sprinklers are not listed for use in dry pipe systems, and NFPA 13D provides no design criteria for dry pipe systems.

In the Los Angeles and Charlotte, NC, fire tests, [1, 2, 3, 4] all of the test fires used typical residential room configurations. These fires were controlled or extinguished with one or two sprinklers under specific sprinkler waterflow rates, spacing, and discharge (spray pattern) characteristics established by the test. When one of these factors, such as spacing, was not met, the test failed because an acceptance criterion, such as temperature (see commentary to 1-2), was exceeded or because the water supply was overtaxed when more than two sprinklers operated. Accordingly, residential system design should be based on two sprinklers operating unless each compartment is protected with only one sprinkler. With the two-sprinkler design, each sprinkler must produce a minimum flow of 13 gpm (49 L/min) when calculations are done in accordance with 4-4.3. With only one sprinkler in any room, the design is based on an 18-gpm (68-L/min) flow rate.

4-1.2.2 The definition of compartment as used in 4-1.2.1 for determining the number of design sprinklers shall be a space that is completely enclosed by walls and a ceiling. The compartment enclosure shall be permitted to have openings to an adjoining space, provided the openings have a minimum lintel depth of 8 in. (203 mm) from the ceiling.

A room compartment is determined by the enclosing walls and soffits or lintels along the ceiling over a compartment opening that have a minimum depth of 8 in. (203 mm). [See Figures A-4-1.2(a) and A-4-1.2(b).] The 8-in. (203-mm) lintel depth allows for the collection of heat in the compartment of fire origin and was confirmed by the Los Angeles test series. Large, beamed ceilings are not intended to be given credit as a compartment enclosure even if they have an 8-in. (203-mm) depth. The openings from the compartment are generally based on a 36-in. (914-mm) wide door opening. Larger openings or beamed ceilings allow for large furnishings to be located beneath the openings or the beams. A fire originating on the furniture could activate more than two sprinklers on both sides of the opening and overtax the water supply. As indicated in 4-1.2.1, the designer should select no more than two sprinklers to determine the area of sprinkler application. When a compartment is of a size that only one sprinkler is needed to provide proper coverage, the compartment boundaries define the area of sprinkler operation.

4-1.3 Water Demand.

The water demand for the system shall be determined by multiplying the design discharge specified in 4-1.1 by the number of design sprinklers specified in 4-1.2.1.

This calculation produces a water demand of 18 gpm (68 L/min) for a single sprinkler and 26 gpm (98 L/min) for two sprinklers when a device uses the generic design criteria specified in NFPA 13D. Residential sprinklers that are listed in accordance with 4-1.6 can discharge more or less water than that required in 4-1.3. Where sprinklers listed in accordance with 4-1.6 are used, the waterflow rates established by the listing evaluation should be used to determine pipe sizes and adequacy of the water supply.

4-1.4 Sprinkler Coverage.

4-1.4.1 Residential sprinklers shall be spaced so that the maximum area protected by a single sprinkler does not exceed 144 ft^2 (13.4 m^2).

NFPA 13D indicates that the design area for a single sprinkler should not exceed 144 ft^2 (13.4 m^2). The actual area covered by the sprinklers will normally be as included in the design and specified under 4-1.2. However, 4-1.6 allows for a greater spacing to be used if the sprinkler is listed for such coverage. This larger spacing usually requires a greater waterflow rate to achieve proper sprinkler performance.

4-1.4.2 The maximum distance between sprinklers shall not exceed 12 ft (3.7 m) on or between pipelines, and the maximum distance to a wall or partition shall not exceed 6 ft (1.8 m).

The 6-ft (1.8-m) spacing limit from walls is to ensure that the sprinklers' spray will strike the wall at a distance of not more than 18 in. (457 mm) below the ceiling for the single-sprinkler operating criteria and not more than 24 in. (610 mm) below the ceiling for the multiple-sprinkler operating criteria. The high wall-wetting criteria of a residential sprinkler are critical to its ability to maintain tenable conditions in the room of fire origin. Specially listed sprinklers in accordance with 4-1.6 could have spacing requirements in excess of those specified in 4-1.4.2.

4-1.4.3 The minimum distance between sprinklers within a compartment shall be 8 ft (2.4 m).

4-1.5 Operating Pressure.

The minimum operating pressure of any sprinkler shall be in accordance with the listing information of the sprinkler and shall provide the minimum flow rates specified in 4-1.1.

The 1996 edition of the standard required that residential sprinklers have a minimum operating pressure of 7 psi (0.5 bar). This provision is the same provision found in NFPA 13. This minimum was established because of the likelihood that sprinklers operating at pressures below 7 psi (0.5 bar) will not perform satisfactorily.

4-1.6 Application rates, design areas, areas of coverage, and minimum design pressures other than those specified in 4-1.1, 4-1.2.1, 4-1.4, and 4-1.5 shall be permitted to be used with special sprinklers that have been listed for such specific residential installation conditions. The minimum operating pressure of any residential sprinkler shall be 7 psi (0.5 bar).

The design criteria for spacing and sprinkler discharge in Section 4-1 were determined from the Los Angeles and Charlotte, NC, fire tests.[1, 2, 3, 4] Because these tests were conducted with prototype residential sprinklers, the Technical Committee on Residential Sprinkler Systems has recognized that future models of residential sprinklers might be able to achieve the intended life safety objectives with waterflow rates, design areas, areas of coverage, and minimum design pressures other than those determined in the Los Angeles and Charlotte, NC, test series. Such sprinklers are permitted to be used provided their performance has been verified through an evaluation process and listed for such use. (See definition of *Listed* in Section 1-3.)

Some residential sprinklers have been listed to cover areas up to 400 ft^2 (36 m^2), which is approximately 277 percent more than the prototype sprinkler. Although this larger area of coverage is advantageous because fewer sprinklers would be required for a large compartment, the use of such sprinklers requires a discharge of 30 gpm (114 L/min) for one sprinkler and 21 gpm (79 L/min) for two sprinklers. Greater operating pressures would also be required—for example, 30 psi (2.06 bar) for one sprinkler and 15 psi (1.03 bar) for two sprinklers. Also available are sprinklers that can discharge less than 18 gpm (68 L/min) for a single sprinkler and less than 13 gpm (49 L/min) for each of two operating sprinklers.

Examples of sprinklers that have been listed in accordance with 4-1.6 are provided in Table 4.1. This table is somewhat representative of the combinations of available devices as of August 1999.

In each case, these devices would have been evaluated in the test enclosure built to suit the proposed listed spacing criteria. Exhibit 4.1 illustrates the arrangement of the test enclosure in accordance with UL 1626, *Residential Sprinklers for Fire-Protection Service.*

Table 4.1 Sample of Listed Residential Sprinklers with Spacing and Flow Requirements Differing from those Specified by NFPA 13D

Sprinkler	Coverage Area (ft)	Supply for One Operating Sprinkler (gpm)	Supply for Two Operating Sprinklers (gpm)
AFAC, Inc. HR-1 pendent	14 × 14	14	11.5
Central Sprinkler Corp. LF pendent	20 × 20	16	13.5
Firematic Sprinkler Devices, Inc. URES 3 pendent	20 × 20	14	13.5
Globe Fire Sprinkler Corp. J (3-mm bulb) pendent	18 × 18	18	14
Grinnell Corp. F990 + flush pendent	20 × 20	22	18
Reliable Automatic Sprinkler Corp. ZX/RES/2 pendent	12 × 12	12	11
Star Sprinkler Corp. LD-2 HSW recessed	16 × 16	25	18
Viking Corp. H flush pendent	20 × 20	30	21

For SI units: 1 ft = 0.30 m; 1 gpm = 3.785 L/min.

4-2 Position of Sprinklers

4-2.1 Pendent and upright sprinklers shall be positioned so that the deflectors are within 1 in. to 4 in. (25.4 mm to 102 mm) from the ceiling.

Exception: Special residential sprinklers shall be installed in accordance with the listing limitations.

This positioning limitation is based on the results of the Los Angeles and Charlotte, NC, fire tests.[1, 2, 3, 4] When a residential sprinkler is tested and specifically listed for installation with a greater distance from the ceiling, the exception to 4-2.1 allows the larger distance to be used.

4-2.2 Sidewall sprinklers shall be positioned so that the deflectors are within 4 in. to 6 in. (102 mm to 152 mm) from the ceiling.

Exception: Special residential sprinklers shall be installed in accordance with the listing limitations.

Dead-air spaces in corners can affect a sprinkler's operation time. The 4-in. to 6-in. (102-mm to 152-mm) limitation ensures that the sprinkler will operate properly.

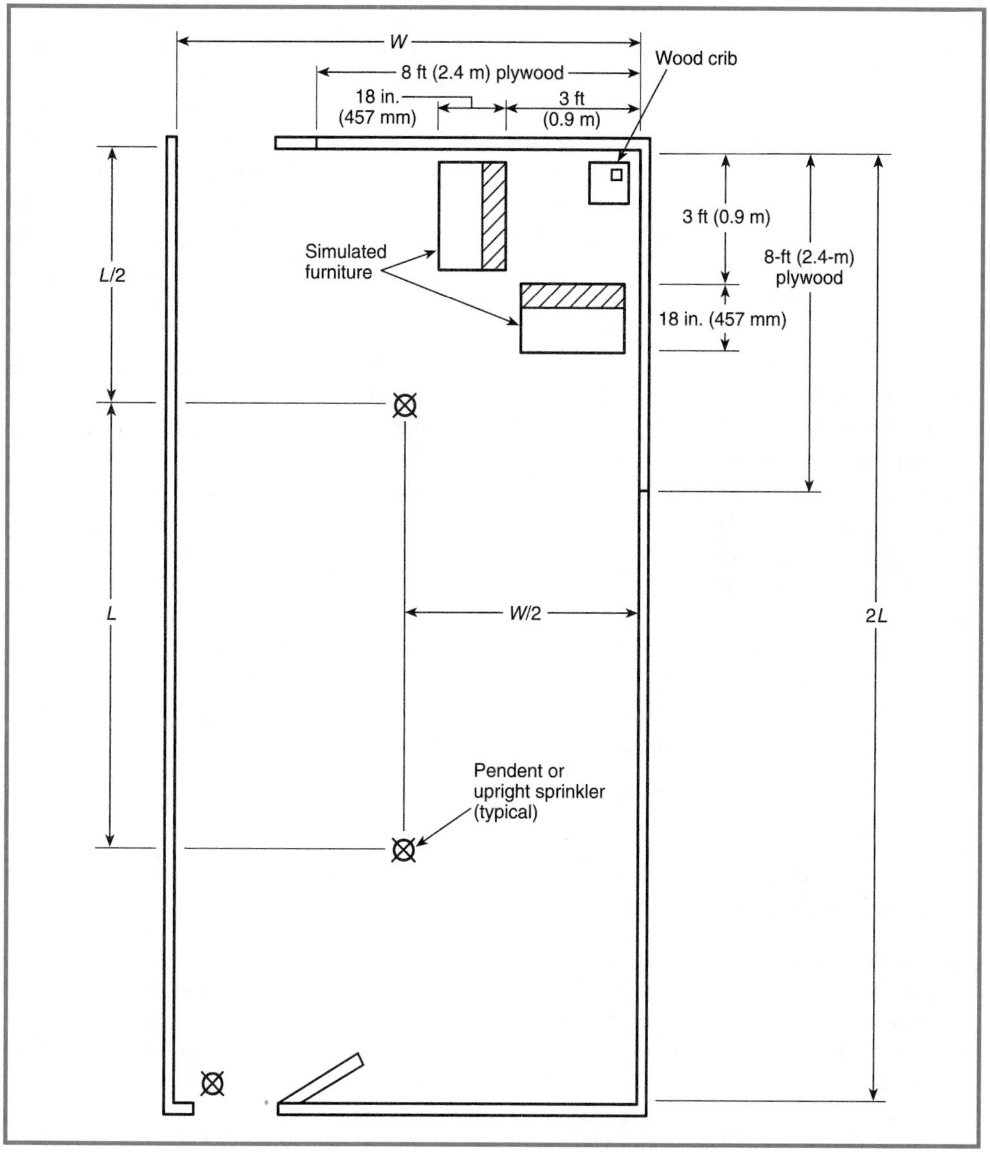

Exhibit 4.1 *UL 1626 fire test arrangement for pendent or upright sprinklers.*

4-2.3* Sprinklers shall be positioned so that the response time and discharge are not unduly affected by obstructions such as ceiling slope, beams, or light fixtures.

A-4-2.3 Fire testing has indicated the need to wet walls in the area protected by residential sprinklers at a level closer to the ceiling than that accomplished by standard sprinkler distribution. Where beams, light fixtures, sloped ceilings, and other obstructions occur, additional residential sprinklers are necessary to achieve proper response and distribution. In addition, for sloped ceilings, higher flow rates could be needed. Guidance should be obtained from the manufacturer.

A series of 33 full-scale tests were conducted in a test room with a floor area of 12 ft × 24 ft (3.6 m × 7.2 m) to determine the effect of cathedral (sloped) and beamed ceiling construction, and combinations of both, on fast-response residential sprinkler performance. The testing was performed using one pendent-type residential sprinkler model, two ceiling slopes (0 degrees and 14 degrees), and two beam configurations on a single enclosure size. In order to judge the effectiveness of sprinklers in controlling fires, two baseline tests, in which the ceiling was smooth and horizontal, were conducted with the pendent sprinklers installed and with a total water supply of 26 gpm (98 L/min) as required by this standard. The results of the baseline tests were compared with tests in which the ceiling was beamed or sloped, or both, and two pendent sprinklers were installed with the same water supply. Under the limited conditions used for testing, the comparison indicates that sloped or beamed ceilings, or a combination of both, represent a serious challenge to the fire protection afforded by fast-response residential sprinklers. However, further tests with beamed ceilings indicated that fire control equivalent to that obtained in the baseline tests can be obtained where one sprinkler is centered in each bay formed by the beams and a total water supply of 36 gpm (136 L/min) is available. Fire control equivalent to that obtained in the baseline tests was obtained for the smooth, sloped ceiling tests where three sprinklers were installed with a total water supply of 54 gpm (200 L/min). In a single smoldering-started fire test, the fire was suppressed.

Table A-4-2.3 and Figure A-4-2.3 provide guidance for the location of sprinklers near ceiling obstructions.

Small areas created by architectural features such as planter box windows, bay windows, and similar features can be evaluated as follows.

(a) Where no additional floor area is created by the architectural feature, no additional sprinkler protection is required.

(b) Where additional floor area is created by an architectural feature, no additional sprinkler protection is required, provided all of the following conditions are met.

(1) The floor area does not exceed 18 ft^2 (1.7 m^2).
(2) The floor area is not greater than 2 ft (0.65 m) in depth at the deepest point of the architectural feature to the plane of the primary wall where measured along the finished floor.
(3) The floor area is not greater than 9 ft (2.9 m) in length where measured along the plane of the primary wall.

Measurement from the deepest point of the architectural feature to the sprinkler should not exceed the maximum listed spacing of the sprinkler. The hydraulic design is not required to consider the area created by the architectural feature.

Location of sprinklers not in accordance with Table A-4-2.3 and Figure A-4-2.3 is likely to result in obstructed sprinkler discharge and could prevent the sprinkler from controlling the fire. Fire control characteristics of residential sprinklers depend on a sprinkler discharge pattern that has a higher wall-wetting characteristic than does the discharge of standard spray sprinklers. The obstruction spacing rules are necessary to allow the spray pattern to reach areas behind furnishings that can be located at the

Table A-4-2.3 Positioning of Sprinklers Near Ceiling Obstructions

Distance from Sprinkler to Side of Ceiling Obstruction		Maximum Distance from Sprinkler Deflector to Bottom of Ceiling Obstruction	
		in.	mm
<6 in.	<152 mm	Not permitted	
6 in. to <1 ft	152 mm to <305 mm	0	0
1 ft to <2 ft	0.32 m to <0.64 m	1	25.4
2 ft to <2 ft 6 in.	0.64 m to <0.80 m	2	51
2 ft 6 in. to <3 ft	0.80 m to < 0.97 m	3	76
3 ft to <3 ft 6 in.	0.97 m to <1.13 m	4	102
3 ft 6 in. to <4 ft	1.13 m to <1.29 m	6	152
4 ft to <4 ft 6 in.	1.29 m to <1.45 m	7	178
4 ft 6 in. to <5 ft	1.45 m to <1.61 m	9	229
5 ft to <5 ft 6 in.	1.61 m to <1.77 m	11	279
5 ft 6 in. to <6 ft	1.77 m to <1.93 m	14	356

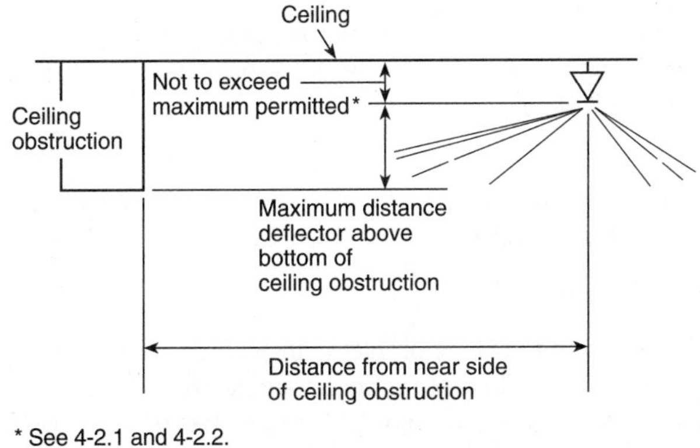

Figure A-4-2.3 *Position of deflector, upright or pendent, where located above bottom of ceiling obstruction.*

perimeter of a room. Chairs, couches, desks, and shelves are typically placed along a wall and can serve as the initial fuel package during a fire. This reason is one of the reasons why a higher horizontal discharge pattern is required for residential sprinklers.

Sprinklers with spray patterns other than those specifically addressed by NFPA 13D are allowed to be positioned closer to a beam or other similar obstruction than the distance specified by Table A-4-2.3. However, test data must be supplied to support the closer position.

The 1996 edition provided additional information with regard to the sprinkler positioning in relation to planter box windows, bay windows, and similar architectural features. Overall, these features are usually smaller in area than a closet and in some cases do not add floor space. If this is the case, these small architectural features are not anticipated to significantly increase the residence's fire hazard, nor will they adversely impact sprinkler performance.

Furthermore, adherence to the spacing rules of Table A-4-2.3 often cannot be achieved without placing a sprinkler within the bay window or planter box window. An alternative evaluation procedure was added because sprinkler location within these spaces raises concerns about system performance.

Small bay windows protected with sprinklers usually require piping run along an outside wall to feed the sprinkler in the bay window. Because the bay window is added onto the side of the house, the space is subject to greater temperature fluctuations and cannot always be protected by insulation. This fluctuation increases the potential of freezing during cold temperatures. Additionally, in areas where warmer climates prevail, solar heating of the window glass creates the potential for false actuation of the sprinkler.

An increased number of sprinklers operating within the compartment is also a concern. Activating more sprinklers than anticipated by the design in a compartment can overtax the water supply and reduce the effectiveness of the sprinkler system.

4-2.4 In basements where ceilings are not required for the protection of piping or where metallic pipe is installed, residential sprinklers shall be permitted to be positioned in a manner that anticipates future installation of a finished ceiling.

This provision was added to the 1994 edition and is intended to provide guidance when installing sprinklers in unfinished basements. NFPA 13D requires sprinklers in areas where the basement is not finished. If other than nonmetallic pipe is used, the deflectors can be installed as if the ceiling were installed.

4-2.5 In closets and storage areas that are required to be protected in accordance with Section 4-6 and are less than 5 ft (1.5 m) in height at the lowest ceiling, a single sprinkler located at the highest ceiling shall be permitted to protect a volume not larger than 300 ft^3 (8.93 m^3).

Closets less than 5 ft (1.5 m) in height at the lowest ceiling and less than 300 ft^3 (8.93 m^3) in volume only need a single sprinkler located at the highest ceiling to achieve fire control. These closet types are typically found under stairs and roofs. In some cases, multiple sprinklers are installed within a single closet or storage compartment in order to ensure proper floor coverage. Experience suggests that fire development within a small confined space, such as a closet, is not comparable with that of a larger space. Fire control in these smaller compartments will result primarily from the buildup of steam and the associated cooling of the space. Also, the presence of a sprinkler on lower ceiling surfaces within the closet can create the potential for unsatisfactory sprinkler activation and operation. This addition to the 1996 edition provided guidance in determining the closets that can be satisfactorily protected by a single sprinkler.

4-3 System Types

4-3.1* Wet Pipe Systems.

A wet pipe system shall be used where all piping is installed in areas not subject to freezing.

A-4-3.1 In areas subject to freezing, care should be taken to cover sprinkler piping completely in unheated attic spaces with insulation. Installation should follow the guidelines of the insulation manufacturer. Figures A-4-3.1(a) through A-4-3.1(e) show several methods that can be considered.

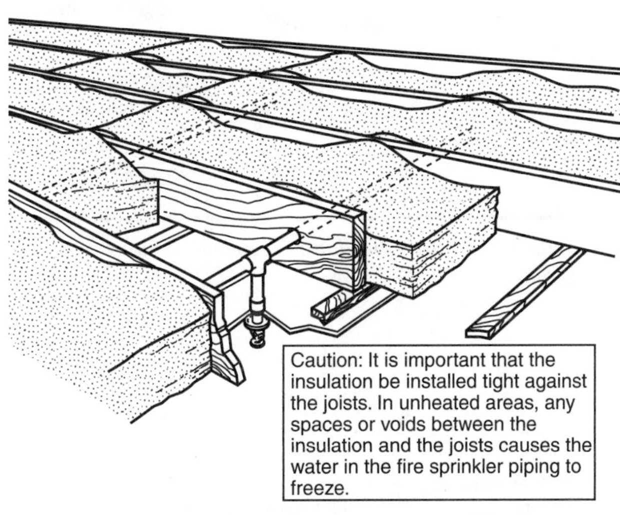

Caution: It is important that the insulation be installed tight against the joists. In unheated areas, any spaces or voids between the insulation and the joists causes the water in the fire sprinkler piping to freeze.

Figure A-4-3.1(a) *Insulation recommendations—Arrangement 1.*

The installation of sprinkler systems in one- and two-family dwellings typically requires that the system piping be installed in the attic space. During cold temperatures, system piping needs to be adequately insulated to protect it from freezing. The type of insulation used, the construction of the attic space, and anticipated low temperatures affect the means of protection. The associated concerns and recommendations on how the insulation could be installed are provided in A-4-3.1. Overall, the insulation is recommended to be installed according to the manufacturer's instructions.

4-3.2 Dry Pipe Systems.

Where system piping is located in unheated areas subject to freezing, a dry pipe or antifreeze system shall be used.

Exception: Listed standard dry pendent or dry sidewall sprinklers shall be permitted to be extended into unheated areas not intended for living purposes.

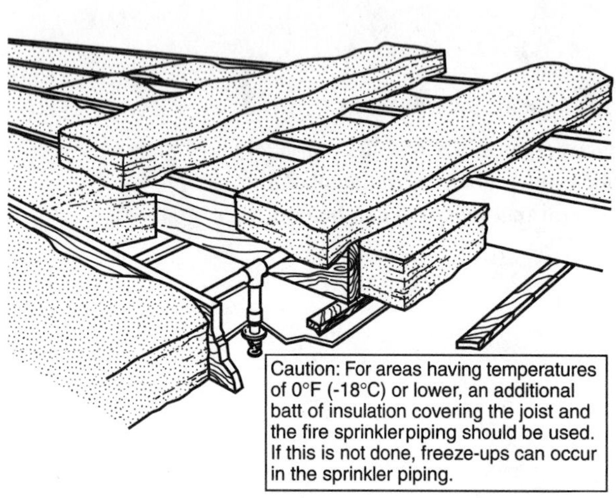
Caution: For areas having temperatures of 0°F (-18°C) or lower, an additional batt of insulation covering the joist and the fire sprinklerpiping should be used. If this is not done, freeze-ups can occur in the sprinkler piping.

Figure A-4-3.1(b) Insulation recommendations—Arrangement 2.

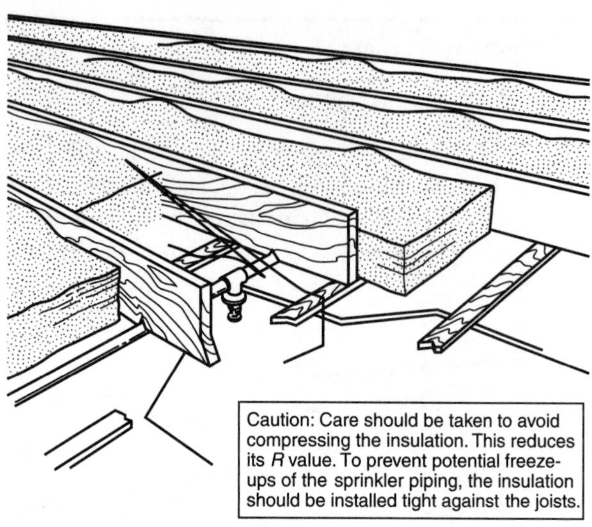
Caution: Care should be taken to avoid compressing the insulation. This reduces its R value. To prevent potential freeze-ups of the sprinkler piping, the insulation should be installed tight against the joists.

Figure A-4-3.1(d) Insulation recommendations—Arrangement 4.

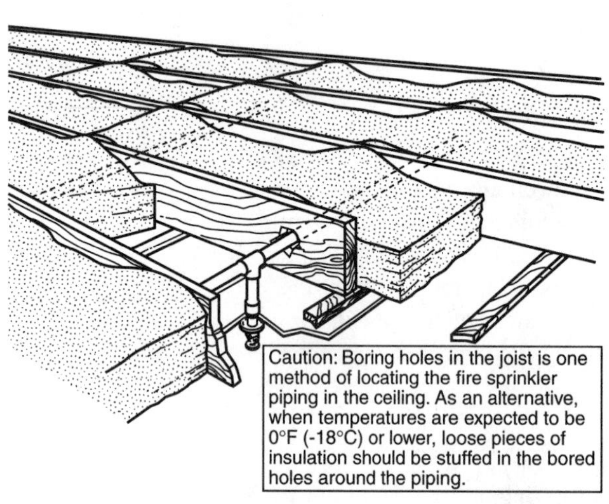
Caution: Boring holes in the joist is one method of locating the fire sprinkler piping in the ceiling. As an alternative, when temperatures are expected to be 0°F (-18°C) or lower, loose pieces of insulation should be stuffed in the bored holes around the piping.

Figure A-4-3.1(c) Insulation recommendations—Arrangement 3.

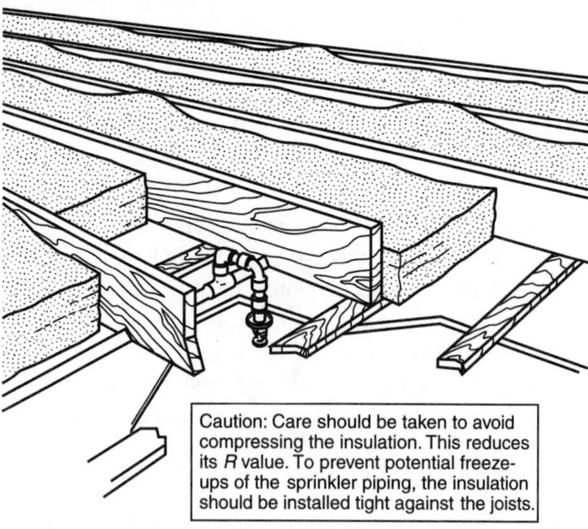
Caution: Care should be taken to avoid compressing the insulation. This reduces its R value. To prevent potential freeze-ups of the sprinkler piping, the insulation should be installed tight against the joists.

Figure A-4-3.1(e) Insulation recommendations—Arrangement 5.

No residential sprinkler is listed for dry pipe use, and, therefore, dry pipe systems cannot be installed in accordance with NFPA 13D.

4-3.3 Antifreeze Systems.

This subsection was added to the 1989 edition in order to allow the use of antifreeze systems. Although long recognized in NFPA 13, antifreeze systems were not specifically discussed nor specifically prohibited in NFPA 13D. This type of system has additional costs associated with it and requires more maintenance than a wet pipe system. Antifreeze systems also usually require the installation of backflow prevention equipment that further increases the system's costs and depletes the system's available pressure. The pressure losses caused by the backflow device need to be addressed by the system design.

Concerns regarding the use of antifreeze systems are that homeowners might not check antifreeze levels on a routine basis. Homeowners might not be able to acquire and use the types of antifreeze specified by 4-3.3.

If the precautions associated with an antifreeze system are understood and an active maintenance program is employed, then an antifreeze system can be safely and effectively used in accordance with NFPA 13D. See commentary on Section 4-5 of NFPA 13 for more information about antifreeze systems.

4-3.3.1 Definition. An antifreeze system is one employing automatic sprinklers attached to a piping system containing an antifreeze solution and connected to a water supply. The antifreeze solution, followed by water, discharges immediately from sprinklers opened by a fire.

4-3.3.2* Conformity with Health Regulations. The use of antifreeze solutions shall be in conformity with any state or local health regulations.

A-4-3.3.2 Antifreeze solutions can be used for maintaining automatic sprinkler protection in small, unheated areas. Antifreeze solutions are recommended only for systems not exceeding 40 gal (151 L).

Because of the cost of refilling the system or replenishing small leaks, small, dry valves should be used where more than 40 gal (151 L) are to be supplied.

Propylene glycol or other suitable material can be used as a substitute for priming water to prevent evaporation of the priming fluid and thus reduce ice formation within the system.

4-3.3.3 Antifreeze Solutions.

4-3.3.3.1 Where sprinkler systems are supplied by public water connections, the use of antifreeze solutions other than water solutions of pure glycerine (chemically pure or United States Pharmacopoeia 96.5 percent grade) or propylene glycol shall not be permitted. Suitable glycerine-water and propylene glycol-water mixtures are shown in Table 4-3.3.3.1.

4-3.3.3.2 Where public water is not connected to sprinklers, the commercially available materials in Table 4-3.3.3.2 shall be permitted to be used in antifreeze solutions.

Calcium chloride is no longer used as an antifreeze solution for sprinkler systems. Therefore, information related to calcium chloride was removed from the 1999 edition.

Table 4-3.3.3.1 Antifreeze Solutions to Be Used Where Public Water Is Connected to Sprinklers

Material	Solution (by Volume)	Specific Gravity at 60°F (15.6°C)	Freezing Point °F	Freezing Point °C
Glycerine	50% water	1.133	−15	−26.1
C.P. or U.S.P. Grade*	40% water	1.151	−22	−30.0
	30% water	1.165	−40	−40.0
Hydrometer scale 1.000 to 1.200				
Propylene glycol	70% water	1.027	+9	−12.8
	60% water	1.034	−6	−21.1
	50% water	1.041	−26	−32.2
	40% water	1.045	−60	−51.1
Hydrometer scale 1.000 to 1.200 (subdivisions, 0.002)				

*C.P.—Chemically pure; U.S.P.—United States Pharmacopoeia 96.5 percent.

Table 4-3.3.3.2 Antifreeze Solutions to Be Used Where Public Water Is Not Connected to Sprinklers

Material	Solution (by Volume)	Specific Gravity at 60°F (15.6°C)	Freezing Point °F	Freezing Point °C
Glycerine	(If glycerine is used, see Table 4-3.3.3.1.)			
Diethylene glycol	50% water	1.078	−13	−25.0
	45% water	1.081	−27	−32.8
	40% water	1.086	−42	−41.1
Hydrometer scale 1.000 to 1.120 (subdivisions, 0.002)				
Ethylene glycol	61% water	1.056	−10	−23.3
	56% water	1.063	−20	−28.9
	51% water	1.069	−30	−34.4
	47% water	1.073	−40	−40.0
Hydrometer scale 1.000 to 1.120 (subdivisions, 0.002)				
Propylene glycol	(If propylene glycol is used, see Table 4-3.3.3.1.)			

*Free from magnesium chloride and other impurities.

4-3.3.3.3* An antifreeze solution with a freezing point below the expected minimum temperature for the locality shall be prepared. The specific gravity of the prepared solution shall be checked by a hydrometer with a suitable scale.

A-4-3.3.3.3 Beyond certain limits, an increased proportion of antifreeze does not lower the freezing point of the solution *(see Figure A-4-3.3.3.3)*. Glycerine, diethylene glycol, ethylene glycol, and propylene glycol never should be used without mixing with water in the proper proportions, because these materials tend to thicken near 32°F (0°C).

Listed CPVC sprinkler pipe and fittings should be protected from freezing with glycerine only. The use of diethylene glycol, ethylene glycol, or propylene glycol are specifically prohibited. Laboratory testing shows that glycol-based antifreeze solutions present a chemical environment detrimental to CPVC. Listed PB sprinkler pipe and fittings can be protected with glycerine, diethylene glycol, ethylene glycol, or propylene glycol.

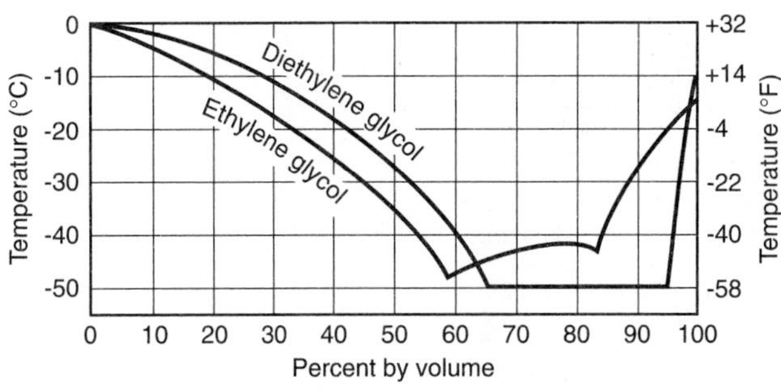

Figure A-4-3.3.3.3 *Freezing points of water solutions of ethylene glycol and diethylene glycol.*

Chlorinated polyvinyl chloride (CPVC) sprinkler piping is not compatible with some of the recognized antifreeze solutions. The glycol-based solutions react with CPVC. Only glycerin solutions are acceptable for CPVC pipe systems.

4-3.3.4* Arrangement of Supply Piping and Valves. All permitted antifreeze solutions are heavier than water. At the point of contact (interface), the heavier liquid is below the lighter liquid, which prevents diffusion of water into the unheated areas. In most cases, this necessitates the use of a 5-ft (1.5-m) drop pipe or U-loop as illustrated in Figure 4-3.3.4. The preferred arrangement is to have the sprinklers located below the interface between the water and the antifreeze solution.

If sprinklers are above the interface, a check valve with a $\frac{1}{32}$-in. (0.8-mm) hole in the clapper shall be provided in the U-loop. A water control valve and two small solution test valves shall be provided as illustrated in Figure 4-3.3.4. An acceptable arrangement of a filling cup is also shown.

Exception: Where the connection between the antifreeze system and the wet pipe system incorporates a backflow prevention device, an expansion chamber shall be provided to compensate for the expansion of the antifreeze solution.

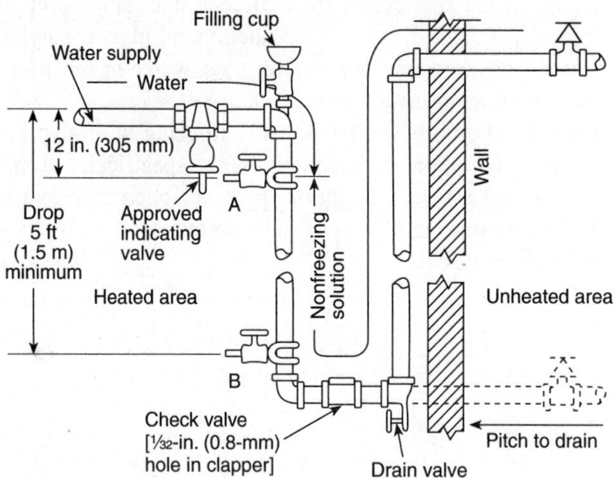
Filling cup
Water supply
Water
12 in. (305 mm)
Drop
5 ft
(1.5 m)
minimum
Approved
indicating
valve
A
Nonfreezing solution
Wall
Heated area
Unheated area
B
Check valve
[1⁄32-in. (0.8-mm)
hole in clapper]
Drain valve
Pitch to drain
Notes:
1. Check valve shall be permitted to be omitted where sprinklers are below the level of valve A.
2. The 1⁄32-in. (0.8-mm) hole in the check valve clapper is needed to allow for expansion of the solution during a temperature rise, thus preventing damage to sprinklers.

Figure 4-3.3.4 *Arrangement of supply piping and valves.*

A-4-3.3.4 To avoid leakage, the quality of materials and workmanship should be superior, the threads should be clean and sharp, and the joints should be tight. Only metal-faced valves should be used.

The expansion chamber is needed to prevent damage to the antifreeze system components when a backflow prevention device is installed in the supply to the antifreeze system.

4-3.3.5* Testing. Before freezing weather each year, the solution in the entire system shall be emptied into convenient containers and brought to the proper specific gravity by adding concentrated liquid as needed. The resulting solution shall be permitted to be used to refill the system.

A-4-3.3.5 Tests should be made by drawing a sample of the solution from valve B, as shown in Figure 4-3.3.4, two or three times during the freezing season, especially if it has been necessary to drain the building sprinkler system for reasons such as repairs or changes. A small hydrometer should be used so that a small sample is sufficient. Where water appears at valve B or where the test sample indicates that the solution has become weakened, the entire system should be emptied and then recharged as previously described.

4-4 Pipe Sizing

4-4.1 Piping shall be sized in accordance with 4-4.3 and 4-4.4. If more than one design discharge is required *(see 4-1.1)*, the pipe sizing procedure shall be repeated for each design discharge.

Exception No. 1: Where piping is sized hydraulically, calculations shall be made in accordance with the methods described in NFPA 13, Standard for the Installation of Sprinkler Systems.

Exception No. 2: For specially listed piping products, friction loss for pipe and fittings shall be permitted to be calculated based on the manufacturer's data, where available.

As indicated in 4-1.2, the single- and multiple-sprinkler design criteria must be verified. The flow–pressure combinations for each of these design points establish the size of branch lines and feed mains, respectively.

The dual discharge requirements of 4-1.1 and 4-1.2 result in two separate sets of calculations that need to be performed where two or more sprinklers are contained within a compartment.

Hydraulic calculations are required for grid-type or looped systems as indicated by Exception No. 1. This exception emphasizes the requirement in 4-4.5 that these types of systems should be hydraulically calculated in accordance with NFPA 13.

Tables 4.2 and 4.3 (see commentary following A-4-4.3) provide friction loss information for nonmetallic piping materials with regard to Exception No. 2.

4-4.2 Minimum Pipe Size.

Minimum pipe size, including that for copper, listed CPVC, and PB piping, shall be ¾ in. (19 mm).

Exception No. 1: The minimum size of steel pipe shall be 1 in. (25.4 mm).

Exception No. 2: ½-in. (12.7-mm) nonmetallic pipe and ½-in. (12.7-mm) copper pipe along with listed special fittings shall be permitted to be used only in network systems under the following conditions.

(a) Each sprinkler shall be supplied through a minimum of three separate paths from the supply shutoff valve assembly within the dwelling unit.

(b) Sprinkler supply lines shall not terminate in a dead end.

(c) Hydraulic calculations shall be prepared for each sprinkler flowing individually within the system and for each pair of sprinklers within the same compartment. The location of the most demanding single sprinkler and pair of sprinklers, including their pressure and flow requirements, shall be indicated on the plan review documents.

(d) The system shall be hydraulically calculated in accordance with the provisions of NFPA 13, Standard for the Installation of Sprinkler Systems. The friction loss straight through the fitting shall be included.

(e) The system shall be supplied from a potable water source, or it shall be equipped with a strainer at the connection to the supply line.

(f) The method of joining the pipe to fittings or to other pipe shall be covered by the listing.

(g) Insert fittings shall not be installed between sprinkler fitting inlet ports.

(h) The piping to the plumbing fixtures shall be of copper or listed pipe.

The requirements for the minimum pipe size of metallic pipe are the same as those in NFPA 13—that is, 1 in. (25.4 mm) for steel. The minimum size requirement decreases the possibility of pipe becoming obstructed due to scale and rust.

With regard to Exception No. 1, the minimum size for types of pipe or tube other than for copper or steel, including nonmetallic piping materials, would be included as part of the product's listing information. Because of the concerns with obstructing waterflow through small diameter pipe, sizes less than ¾ in. (19 mm) are unlikely to be listed for ferrous pipe.

Exception No. 2 was first added to NFPA 13D for the 1996 edition and allows the use of ½-in. (12.7-mm) nonmetallic pipe and copper tube. The previous limitation was ¾ in. (19 mm). A growing effort exists to make residential sprinkler systems a more affordable and attractive fire safety feature for the home. Allowing for smaller diameter pipe provides more options in this regard. However, in order to utilize the ½-in. (12.7-mm) pipe, a number of conditions must be met. The system must be hydraulically calculated in accordance with the provisions of NFPA 13 and friction loss through fittings must be accounted for in the design. These conditions ensure that a thorough hydraulic analysis of the system will be conducted. Additionally, due to small pipe sizes, pressure losses through fittings will have a greater overall impact and also must be accounted for in the design. Hydraulic calculations must be prepared for each sprinkler flowing individually within the system and for each pair of sprinklers flowing within the same compartment. This requirement applies to all NFPA 13D systems. The location of the most demanding single sprinkler and pair of sprinklers along with their pressure and flow requirements need to be indicated on the plan review documents. This information allows for easier plan review and system verification.

In addition to verifying the system's hydraulic performance, supply piping in a system utilizing ½-in. (12.7-mm) pipe must be arranged so that each sprinkler is supplied by at least three separate flow paths with no supply lines terminating in a dead end. This results in a more hydraulically advantageous arrangement. Methods for joining pipe must be addressed by product listings, and insert fittings are not to be installed between sprinkler fitting inlet ports. Due to the small internal diameters of insert fittings, the technical committee is of the opinion that small fittings would create too great a pressure drop and affect overall system reliability. Following listing requirements for joining of pipe serves to increase the system's overall reliability. Piping to any plumbing fixtures is required to be copper or listed. Finally, water sources are required to be potable or equipped with a strainer at the connection to the supply line. The small diameter piping is subject to a greater incidence of obstruction that must be guarded against. Overall, this type of system does not allow for a large margin of error in either its design or installation. However, when designed and installed correctly, the system will achieve the desired fire protection objectives.

4-4.3* To size piping for systems connected to a city water supply, the following general method shall be considered to be acceptable.

(a) The system flow rate shall be established in accordance with Section 4-1, and it shall be determined that the flow allowed by the water meter is adequate to supply the system demand and that the total demand flow does not exceed the maximum flow allowed by the piping system components.

(b) The water pressure in the street shall be determined.

(c) Pipe sizes shall be selected.

(d) Meter pressure losses, if any, shall be deducted. *[See Table 4-4.3(g).]* Higher pressure losses specified by the manufacturer shall be used in place of those specified in Table 4-4.3(e). Lower pressure losses shall be permitted to be used where supporting data is provided by the meter manufacturer.

(e) Pressure loss for elevation shall be deducted. (Building height above street in feet × 0.434 = psi. Building height above street in meters × 0.098 = bar.)

(f) Pressure losses from the city main to the inside control valve shall be deducted by multiplying the factor from Tables 4-4.3(a) or 4-4.3(b) by the total length(s) of pipe in feet (meters). [The total length includes equivalent length of fittings as determined by applying Tables 4-4.3(c), 4-4.3(d), 4-4.3(e), or 4-4.3(f).]

(g) Pressure losses for piping within the building shall be deducted by multiplying the factor from Tables 4-4.3(a) or 4-4.3(b) by the total length in feet (meters) of each size of pipe between the control valve and the farthest sprinkler.

(h) Valve and fitting pressure losses shall be deducted. The valves and fittings from the control valve to the farthest sprinkler shall be counted. The equivalent length for each valve and fitting as shown in Tables 4-4.3(c), 4-4.3(d), 4-4.3(e), or 4-4.3(f) shall be determined and the values added to obtain the total equivalent length for each pipe size. The equivalent length for each size shall be multiplied by the factor from Tables 4-4.3(a) or 4-4.3(b) and the values totaled.

(i) In multilevel buildings, steps 4-4.3(a) through 4-4.3(h) shall be repeated to size piping for each floor.

(j) If the remaining pressure is less than the operating pressure established by the testing laboratory for the sprinkler being used, the sprinkler system shall be redesigned. If the remaining pressure is higher than required, smaller piping shall be permitted to be used where justified by calculations.

(k) The remaining piping shall be sized the same as the piping up to and including the farthest sprinkler.

Exception No. 1: Where smaller pipe sizes are justified by calculations.

Exception No. 2: For sprinklered dwellings manufactured off-site, the minimum pressure needed to satisfy the system design criteria on the system side of the meter shall be specified on a data plate by the manufacturer. (See Section 2-3.)

A-4-4.3 The determination of public water supply pressure should take into account the probable minimum pressure conditions prevailing during such periods as during the night or during the

Table 4-4.3(a) Pressure Losses in psi/ft for Schedule 40 Steel Pipe (C = 120)

Pipe Size (in.)	Flow Rate (gpm)											
	10	12	14	16	18	20	25	30	35	40	45	50
1	0.04	0.05	0.07	0.09	0.11	0.13	0.20	0.28	0.37	0.47	0.58	0.71
1¼	0.01	0.01	0.02	0.02	0.03	0.03	0.05	0.07	0.10	0.12	0.15	0.19
1½	0.01	0.01	0.01	0.01	0.01	0.02	0.02	0.03	0.05	0.06	0.07	0.09
2	—	—	—	—	—	0.01	0.01	0.01	0.01	0.02	0.02	0.03

For SI units, 1 gal = 3.785 L; 1 psi = 0.0689 bar; 1 in. = 25.4 mm; 1 ft = 0.3048 m.

Table 4-4.3(b) Pressure Losses in psi/ft for Copper Tubing—Types K, L, and M (C = 150)

Tubing Size (in.)	Type	Flow Rate (gpm)											
		10	12	14	16	18	20	25	30	35	40	45	50
¾	M	0.08	0.12	0.16	0.20	0.25	0.30	0.46	0.64	0.85	—	—	—
	L	0.10	0.14	0.18	0.23	0.29	0.35	0.53	0.75	1.00	—	—	—
	K	0.13	0.18	0.24	0.30	0.38	0.46	0.69	0.97	1.28	—	—	—
1	M	0.02	0.03	0.04	0.06	0.07	0.08	0.13	0.18	0.24	0.30	0.38	0.46
	L	0.03	0.04	0.05	0.06	0.08	0.10	0.15	0.20	0.27	0.35	0.43	0.53
	K	0.03	0.04	0.06	0.07	0.09	0.11	0.17	0.24	0.31	0.40	0.50	0.61
1¼	M	0.01	0.01	0.02	0.02	0.03	0.03	0.05	0.07	0.09	0.11	0.15	0.17
	L	0.01	0.01	0.02	0.02	0.03	0.03	0.05	0.07	0.10	0.12	0.16	0.19
	K	0.01	0.01	0.02	0.02	0.03	0.04	0.06	0.08	0.11	0.13	0.17	0.20
1½	M	—	0.01	0.01	0.01	0.01	0.01	0.02	0.03	0.04	0.05	0.06	0.08
	L	—	0.01	0.01	0.01	0.01	0.01	0.02	0.03	0.04	0.05	0.07	0.08
	K	—	0.01	0.01	0.01	0.01	0.02	0.02	0.03	0.05	0.06	0.07	0.09
2	M	—	—	—	—	—	—	0.01	0.01	0.01	0.01	0.02	0.02
	L	—	—	—	—	—	—	0.01	0.01	0.01	0.01	0.02	0.02
	K	—	—	—	—	—	—	0.01	0.01	0.01	0.01	0.02	0.02

For SI units, 1 gal = 3.785 L; 1 psi = 0.0689 bar; 1 in. = 25.4 mm; 1 ft = 0.3048 m.

summer months when heavy usage can occur; the possibility of interruption by floods or ice conditions in winter also should be considered. *[See Figures A-4-4.3(a) and A-4-4.3(b).]*

Section 4-4 describes a simplified method of sizing pipe in a residence to ensure that the flow rate from the sprinklers in the design area will meet the design discharge criteria of 4-1.1. This approximation method is only permitted for tree-type systems connected to an automatic water supply. When the supply source is a city main, the main size must be at least 4 in. (102 mm) in diameter in order to utilize this procedure.

In performing the calculation for a specific system that has at least two sprinklers in any compartment, the calculation procedure must be repeated twice. When using a

Table 4-4.3(c) Equivalent Length of Fittings and Valves for Schedule 40 Steel Pipe in Feet

Diameter (in.)	45-Degree Elbows	90-Degree Elbows	Long-Radius Elbows	Tee or Cross (flow turned 90 degrees)	Tee Run	Gate Valve	Angle Valve	Globe Valve	Globe "Y" Pattern Valve	Cock Valve	Check Valve
1	1	2	2	5	2	0	12	28	15	4	5
1¼	1	3	2	6	2	0	15	35	18	5	7
1½	2	4	2	8	3	0	18	43	22	6	9
2	2	5	3	10	3	1	24	57	28	7	11

Table 4-4.3(d) Equivalent Length of Fittings and Valves for Type K Copper Tube in Feet

Diameter (in.)	45-Degree Elbows	90-Degree Elbows	Long-Radius Elbows	Tee or Cross (flow turned 90 degrees)	Tee Run	Gate Valve	Angle Valve	Globe Valve	Globe "Y" Pattern Valve	Cock Valve	Check Valve
¾	0	1	0	3	1	0	7	14	7	2	0
1	1	2	2	6	2	0	14	33	18	5	6
1¼	1	3	2	5	2	0	14	32	16	5	6
1½	2	4	2	8	3	0	18	43	22	6	9
2	2	6	3	12	4	1	28	66	33	8	13

Table 4-4.3(e) Equivalent Length of Fittings and Valves for Type L Copper Tube in Feet

Diameter (in.)	45-Degree Elbows	90-Degree Elbows	Long-Radius Elbows	Tee or Cross (flow turned 90 degrees)	Tee Run	Gate Valve	Angle Valve	Globe Valve	Globe "Y" Pattern Valve	Cock Valve	Check Valve
¾	0	2	0	4	1	0	8	18	10	3	0
1	1	3	3	7	2	0	16	38	20	5	7
1¼	1	3	2	6	2	0	15	35	18	5	7
1½	2	4	2	9	3	0	20	47	24	7	10
2	2	6	4	12	4	1	30	71	35	9	14

Table 4-4.3(f) Equivalent Length of Fittings and Valves for Type M Copper Tube in Feet

Diameter (in.)	45-Degree Elbows	90-Degree Elbows	Long-Radius Elbows	Tee or Cross (flow turned 90 degrees)	Tee Run	Gate Valve	Angle Valve	Globe Valve	Globe "Y" Pattern Valve	Cock Valve	Check Valve
¾	0	2	0	4	1	0	10	21	11	3	0
1	2	3	3	8	3	0	19	43	23	6	8
1¼	1	3	2	7	2	0	16	38	20	5	8
1½	2	5	2	9	3	0	21	50	26	7	11
2	3	7	4	13	5	1	32	75	37	9	14

Table 4-4.3(g) Pressure Losses in psi in Water Meters

Meter (in.)	Flow (gpm)					
	18	23	26	31	39	52
⅝	9	14	18	26	†	†
¾	4	8	9	13	†	†
1	2	3	3	4	6	10
1½	††	1	2	2	4	7
2	††	††	††	1	2	3

For SI units, 1 gpm = 3.785 L/min; 1 in. = 25.4 mm; 1 psi = 0.0689 bar.
†Above maximum rated flow of commonly available meters.
††Less than 1 psi (0.689 bar).

residential sprinkler that uses the generic waterflow requirements discussed in 4-1.1, the first calculation is made for a single sprinkler flowing at 18 gpm (68 L/min), and the second calculation is made for two sprinklers flowing at 26 gpm (98 L/min) [13 gpm (49 L/min) each]. The pressure required to produce this flow at the sprinkler will depend on the coefficient of discharge (K-factor), which must be obtained from the sprinkler manufacturer. In multilevel dwellings, calculations will need to be done for all levels of the building to ensure that pipe is sized for the variations in flow that can occur due to elevation differences between floors and differences in piping geometry between floors.

Additionally, in the 1996 edition, the first step of the calculation was expanded to take into account any maximum flow limits of system components. Some components such as meters, valves, backflow devices, and relief valves have associated flow limits. Exceeding these values could result in improper system operation. Therefore, the

	Individual Loss	Net Total
(1) Water pressure in street ________	________	________
(2) Arbitrarily select pipe size ________		
(3) Deduct meter loss (size) ________	________	________
(4) Deduct head loss for elevation (________ ft × 0.434) ________		
(5) Deduct pressure loss from city main to sprinkler system control valve* ________		
____ Pipe ________ ft	________	________
____ Valves ________ ft	________	________
____ Elbows ________ ft	________	________
____ Tee ________ ft	________	________
____ Total ________ ft × ________	________	________
(6) Deduct pressure loss for piping-control valve to furthest sprinkler*		

Size	Quantity	Description	Total Equivalent (ft)
_____	______	90-degree elbow	______
_____	______	45-degree elbow	______
_____	______	Tee	______
_____	______	Check valve	______
_____	______	Valve (______)	______
_____	______	Total	___ ft × ___ = ___

Size	Quantity	Description	Total Equivalent (ft)
_____	______	90-degree elbow	______
_____	______	45-degree elbow	______
_____	______	Tee	______
_____	______	Check valve	______
_____	______	Valve (______)	______
_____	______	Total	___ ft × ___ = ___

Remaining pressure for sprinkler operation ________________

For SI units, 1 ft = 0.3048 m; 1 psi = 0.0689 bar.

* Factors from Table 4-4.3(a), (b), (c), (d), and (e).

Figure A-4-4.3(a) *Calculation sheet.*

	Individual Loss	Net Total
Water pressure at supply outlet		
(1) Deduct head loss for elevation	______	______
(______ ft × 0.434)	______	______
(2) Deduct pressure loss from piping within building*	______	______
Remaining pressure for sprinkler operation	______	______

For SI units, 1 ft = 0.3048 m; 1 psi = 0.0689 bar.

* Factors from Table 4-4.3(a), (b), (c), (d), (e), (f), and (g).

Figure A-4-4.3(b) *Calculation sheet—elevated tank, booster pump, pump tank supply.*

designer must investigate whether the use of any system components is restricted by its flow limitation.

The second step is to determine the water pressure in the street in front of the property. This is done by obtaining information on the public water supply pressure from the local water company or by placing a pressure gauge on the water supply inlet connection in the dwelling.

After flow demands are determined, pipe sizes are arbitrarily selected and valves and fittings laid out to meet the piping arrangements described in Figures A-2-2(a), A-2-2(b), and A-2-2(c), and A-2-3(a), A-2-3(b), and A-2-3(c). Losses in piping, fittings, meters, and valves and for elevation are then determined in accordance with 4-4.3(d) through 4-4.3(h). These losses are deducted from the city pressure to determine the pressure at the sprinkler design area on each floor of the residence. If this pressure is less than the operating pressure determined in the calculation to meet flow demands (see commentary for 4-1.1), a redesign is necessary and pipe must be increased or the water supply must be improved. If the remaining pressure is higher than required to supply the demand rate at the sprinkler, the designer can either stop and submit the calculation for approval or reduce the pipe sizes to achieve a more cost-effective system. (See Tables 4.2 and 4.3 for PB and CPVC pipe pressure losses.)

A sample calculation for a typical residence is as follows. This example is based on the dwelling illustrated by Exhibit 4.2, which is also a plot plan showing an existing city main.

The addition of the requirement in 4-4.5 would require the city main to be sized at a minimum diameter of 4 in. (100 mm). Exhibits 4.3 and 4.4 show the sprinkler and piping layouts for the basement and first floor of this dwelling, respectively.

The following is assumed for the calculation:

(1) The sprinkler used is listed for application rate, design area, and area of coverage as indicated in 4-1.1, 4-1.2, and 4-1.4.

Table 4.2 Pressure Losses in psi/ft for PB Pipe

Tubing Size (in.)	Flow Rate (gpm)											
	10	12	14	16	18	20	25	30	35	40	45	50
¾	0.1655	0.2319	0.3084	0.3948	0.4909	0.5966	0.9014	1.2630	1.6799	2.1506		
1	0.0475	0.0666	0.0886	0.1134	0.1410	0.1714	0.2590	0.3629	0.4826	0.6179	0.7683	0.9336
1¼	0.0180	0.0253	0.0336	0.0430	0.0535	0.0650	0.0982	0.1375	0.1829	0.2342	0.2912	0.3539
1½	0.0080	0.0112	0.0149	0.0191	0.0237	0.0288	0.0436	0.0611	0.0812	0.1040	0.1293	0.1571
2	0.0022	0.0030	0.0040	0.0051	0.0064	0.0078	0.0118	0.0165	0.0219	0.0281	0.0349	0.0424

For SI units, 1 gal = 3.785 L; 1 psi = 0.0689 bar; 1 in. = 25.4 mm; 1 ft = 0.305 m.

Table 4.3 Pressure Losses in psi/ft for Listed CPVC Pipe (C = 150)

Pipe Size (in.)	Flow Rate (gpm)											
	10	12	14	16	18	20	25	30	35	40	45	50
1	0.0182	0.0255	0.0340	0.0435	0.0541	0.0657	0.0993	0.1391	0.1850	0.2368	0.2944	0.3578
1¼	0.0059	0.0082	0.0109	0.0140	0.0174	0.0211	0.0319	0.0447	0.0595	0.0761	0.0947	0.1150
1½	0.0030	0.0043	0.0057	0.0072	0.0090	0.0110	0.0166	0.0232	0.0308	0.0395	0.0491	0.0597
2	0.0010	0.0014	0.0019	0.0024	0.0030	0.0037	0.0056	0.0078	0.0104	0.0133	0.0165	0.0201
2½	0.0004	0.0006	0.0008	0.0010	0.0012	0.0015	0.0022	0.0031	0.0041	0.0053	0.0065	0.0080

For SI units; 1 gal = 3.785 L; 1 psi = 0.0689 bar; 1 in. = 25.4 mm; 1 ft = 0.305 m.

(2) The flow from one sprinkler is 18 gpm (68 L/min) and from two sprinklers is 13 gpm (49 L/ min) each [26 gpm (98 L/min) total] per 4-1.1.

(3) The sprinkler used has a coefficient (K- factor) of 3.9. Thus, the pressure needed for the system at the flowing sprinkler is calculated as follows:

$$Q = K\sqrt{p} \quad \text{or} \quad p = \left(\frac{Q}{K}\right)^2$$

For 18 gpm: $p = (18/3.9)^2 = 4.61^2 = 21.3$ psi
For 13 gpm: $p = (13/3.9)^2 = 3.33^2 = 11.1$ psi

(4) Type K copper tube is used.

(5) The water pressure in the street was determined to be 68 psi (4.7 bar).

The calculation procedure is outlined in Exhibits 4.5 and 4.6.

With regard to 4-4.3(h), the tables providing equivalent pipe length data for fittings and valves have been updated. Additional tables were created to address the different internal diameters of the three types of copper tubing.

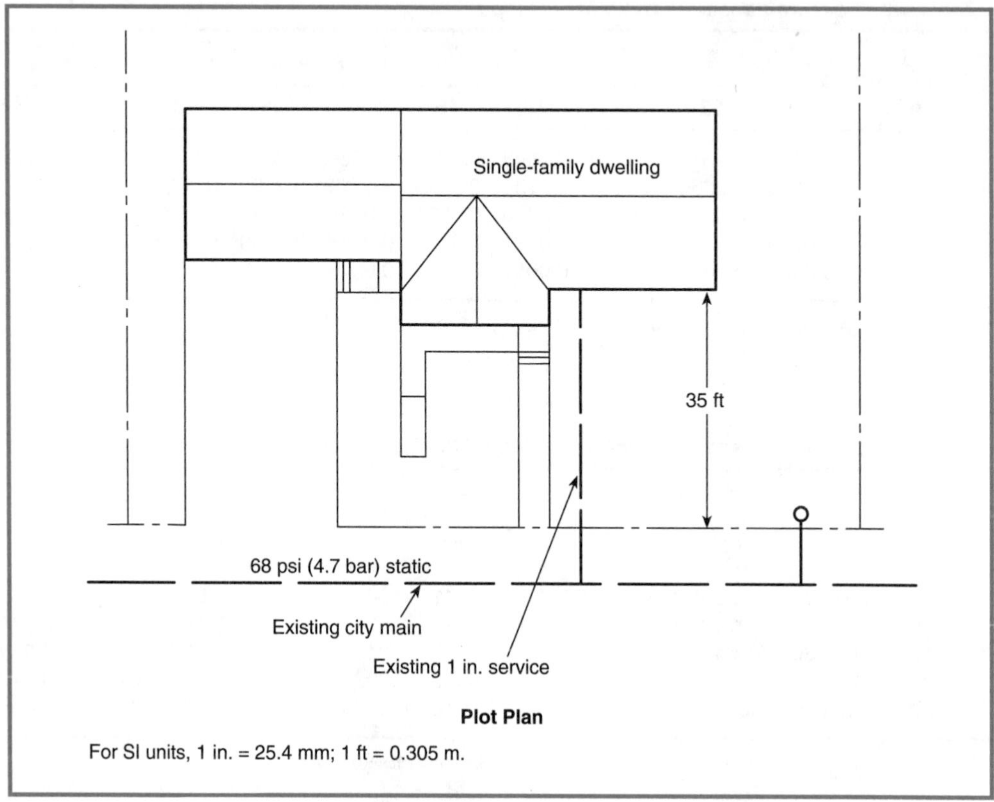

Exhibit 4.2 *Typical residence for sample sprinkler system hydraulic calculation.*

4-4.4 To size piping for systems with an elevated tank, pump, or pump-tank combination, the pressure at the water supply outlet shall be determined and steps (c), (e), (g), (h), (i), (j), and (k) of 4-4.3 shall be followed.

4-4.5 Hydraulic calculation procedures in accordance with NFPA 13, *Standard for the Installation of Sprinkler Systems,* shall be used for grid-type systems, looped-type systems, and systems connected to city water mains of less than 4 in. (100 mm) in diameter.

Because the approximation method of 4-4.3 relies on the static water supply pressure, it is restricted to city water mains with a 4-in. (100-mm) diameter or larger or to other water supply sources that utilize a captured supply.

4-5 Piping Configurations

Piping configurations shall be permitted to be looped, gridded, or straight run, or to be combinations thereof.

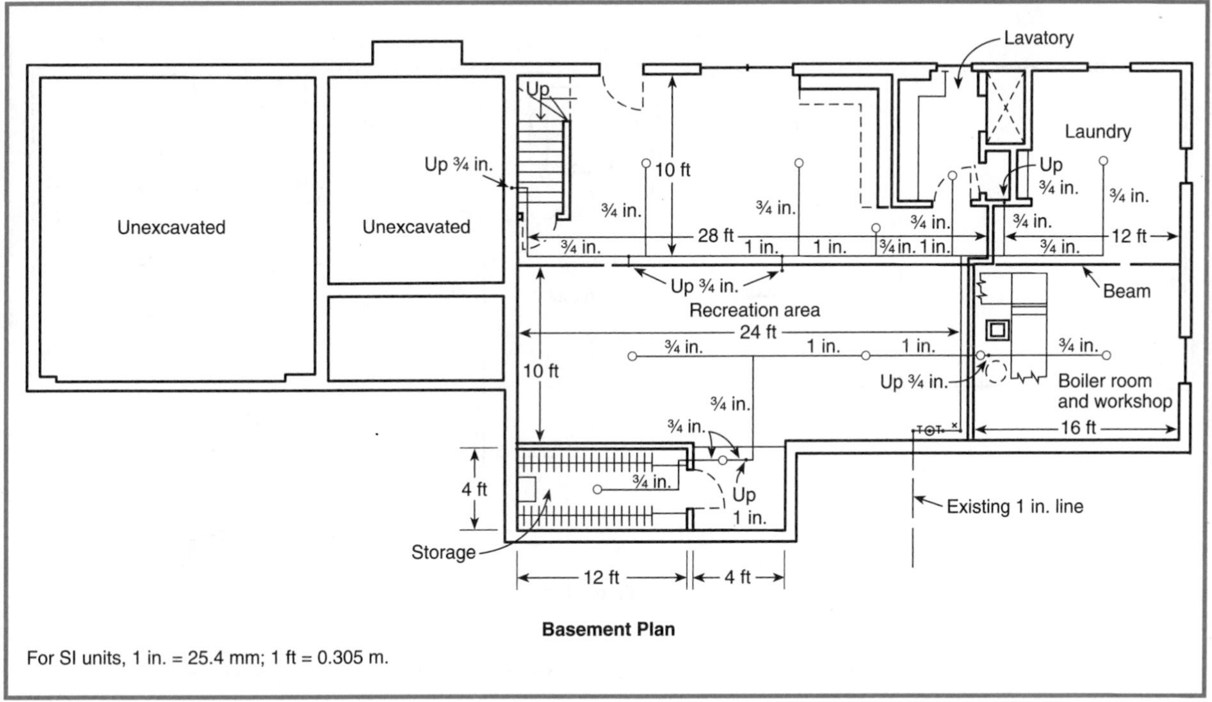

Exhibit 4.3 *Basement floor plan for sample sprinkler calculation.*

Where piping configurations are looped or gridded, calculations must be conducted in accordance with 4-4.5 and Exception No. 1 of 4-4.1. Therefore, hydraulic calculations to size the piping must be done in accordance with the method described in NFPA 13.

4-6 Location of Sprinklers

Sprinklers shall be installed in all areas.

Exception No. 1: Sprinklers shall not be required in bathrooms of 55 ft² (5.1 m²) and less.

Exception No. 2: Sprinklers shall not be required in clothes closets, linen closets, and pantries where the area of the space does not exceed 24 ft² (2.2 m²) and the least dimension does not exceed 3 ft (0.9 m) and the walls and ceilings are surfaced with noncombustible or limited-combustible materials as defined in NFPA 220, Standard on Types of Building Construction.

Exception No. 3: Sprinklers shall not be required in garages, open attached porches, carports, and similar structures.

Exception No. 4: Sprinklers shall not be required in attics, crawl spaces, and other concealed spaces that are not used or intended for living purposes.

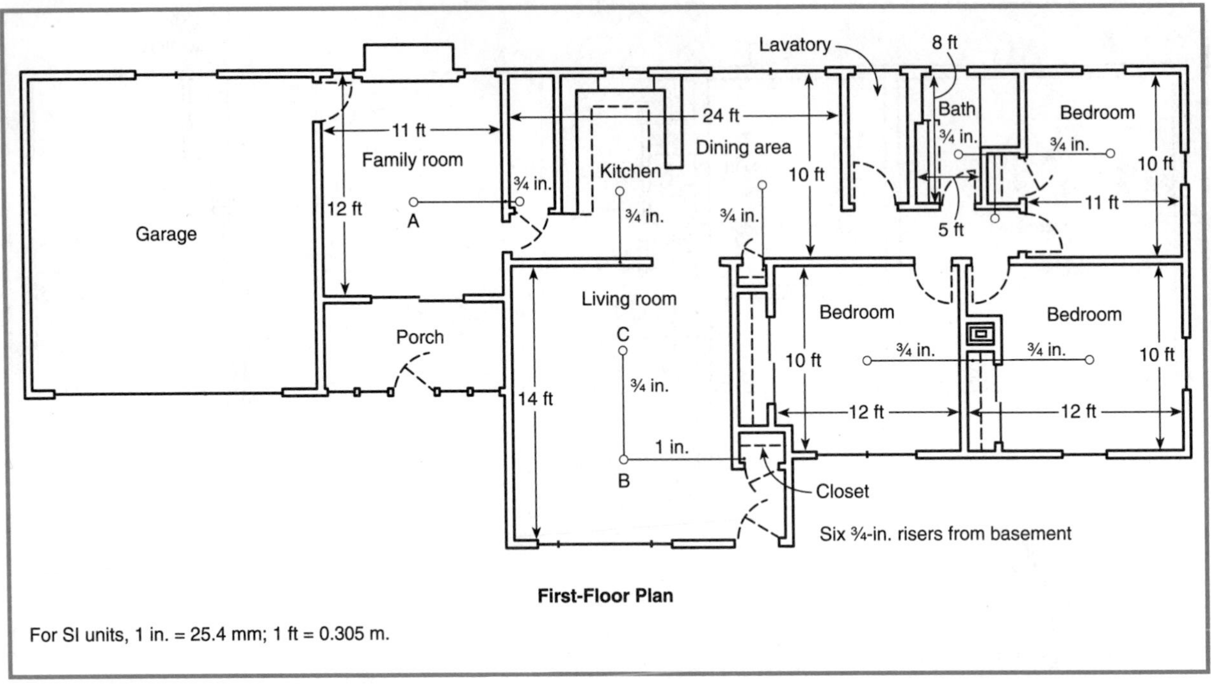

Exhibit 4.4 *First-floor plan for sample sprinkler calculation.*

Exception No. 5: Sprinklers shall not be required in entrance foyers that are not the only means of egress.

Selective sprinkler omission from certain areas indicates that a level of protection less than that anticipated by NFPA 13D is being considered. NFPA 13D recognizes the presence and availability of the levels of protection concept that is included in other fire protection codes and standards. Areas mentioned in Exception Nos. 1 through 5 are not selected at random but represent those areas in which fires do not result in a high percentage of fatalities. Table A-1-2(a) shows statistics for various fire deaths and their relation to the area of fire origin. In addition, the following is noted:

(1) Combustible fuel loading in most bathrooms is typically low (see Exception No. 1).
(2) Small closets are usually impractical places to install sprinklers because of their relatively small size (see Exception No. 2). (The use of the closet is then limited. When heat-producing equipment is contained in the closet, the exception is no longer valid.)
(3) Mandatory sprinklering of the areas in Exception Nos. 3, 4, and 5 would necessitate the use of dry pipe systems where freezing weather is encountered. The use of dry systems would detract from the system's rapid response within the occupied areas of the dwelling and thus detract from, rather than enhance, life safety. The added cost to cover these areas must also be considered. A dry pipe system would

Calculation 1 at 18 gpm (21.3 psi) from Sprinkler A

1st Floor: Type K Copper Tube		Individual loss	Net total
a. Water pressure in street			68
b. Arbitrarily selected pipe size 1 in. + ¾ in.			
c. Deduct meter loss (¾ in. size)		4.0	64
d. Deduct loss for elevation System control valve* (15 ft × 0.434)		6.5	57.5
e. Deduct loss from city main to control valve			
1 in. tube 48 ft			
3 valves 3 ft			
3 elbows 9 ft			
1 tee 5 ft			
Total 65 ft ×	0.09	5.9	51.6
f. Deduct loss for tubing: control valve to farthest sprinkler*			
¾-in. tube: 19 ft ×	0.38	7.2	
1-in. tube: 35 ft ×	0.09	3.1	41.3

Size	Quantity	Description	Total equivalent ft			
¾ in.	4	90-degree elbow	8			
		45-degree elbow				
		tee				
		check valve				
		valve (____)				
		Total	8	ft × 0.38	= 3.0	38.3

Size	Quantity	Description	Total equivalent ft			
1 in.	1	90-degree elbow	3			
		45-degree elbow				
	1	tee	5			
	1	check valve	4			
		valve (____)				
		Total	12	ft × 0.09	= 1.1	37.2

This pressure exceeds 21.3 psi and design is acceptable.
Note: Repeat calculation for basement to verify adequacy of basement piping size.

Remaining pressure for sprinkler operations.
* Factors from Tables 4-4.3(a), 4-4.3(c), and 4-4.3(d).

For SI units, 1 in. = 25.4 mm; 1 ft = 0.305 m.

Exhibit 4.5 *Calculation example no. 1.*

Calculation 2 at 26 gpm (11.1 psi) from Sprinklers B and C

1st Floor: Type K Copper Tube		Individual loss	Net total
a. Water pressure in street		____	68
b. Arbitrarily selected pipe size 1 in. + ¾ in.			
c. Deduct meter loss (¾ in. size)		9.0	59.0
d. Deduct loss for elevation System control valve* (15 ft × 0.434)		6.5	52.5
e. Deduct loss from city main to control valve			
1 in. tube 48 ft			
3 valves 3 ft			
3 elbows 9 ft			
1 tee 5 ft			
Total 65 ft ×	0.20	13.0	39.5
f. Deduct loss for tubing: control valve to farthest sprinkler*			
¾-in. tube: 12 ft ×	0.19	2.3	
1-in. tube: 14 ft ×	0.20	2.8	34.4

Size	Quantity	Description	Total equivalent ft		
	1	90-degree elbow	2		
	____	45-degree elbow	____		
¾ in.	____	tee	____		
	____	check valve	____		
	____	valve (____)	____		
		Total	2	ft × .27 = .54	33.8

Size	Quantity	Description	Total equivalent ft		
	1	90-degree elbow	3		
	____	45-degree elbow	____		
1 in.	1	tee	5		
	1	check valve	4		
	____	valve (____)	____		
		Total	12	ft × 0.20 = 2.4	31.4

This pressure exceeds 11.1 psi and design is acceptable.
Note: Repeat calculation for basement to verify adequacy of basement piping size.

Remaining pressure for sprinkler operations.
* Factors from Tables 4-4.3(a), 4-4.3(c), and 4-4.3(d).

For SI units, 1 in. = 25.4 mm; 1 ft = 0.305 m.

Exhibit 4.6 *Calculation example no. 2. The farthest sprinkler in this case is the farthest sprinkler in the only room having more than one sprinkler.*

be more costly. Furthermore, most building codes require a 1-hour fire-rated separation between garages and other portions of the dwelling.

References Cited in Commentary

National Fire Protection Association, 1 Batterymarch Park, P.O. Box 9101, Quincy, MA 02269-9101.

NFPA 13, *Standard for the Installation of Sprinkler Systems*, 1999 edition.

Underwriters Laboratories Inc., 333 Pfingsten Road, Northbrook, IL 60062.

UL 1626, *Residential Sprinklers for Fire-Protection Service*, 2nd edition, 1994.
Underwriters Laboratories Fire Protection Equipment Directory, 1999 edition.

Bibliography

1. Cote, A., and Moore, D., "Field Test and Evaluation of Residential Sprinkler Systems, Los Angeles Test Series," National Fire Protection Association, Boston, MA, April 1980. (This report was done for the NFPA 13D subcommittee and included a series of fire tests in actual dwellings with sprinkler systems installed in accordance with the data resulting from all prior test work and the criteria included in the proposed 1980 revisions to NFPA 13D so as to evaluate the effectiveness of the system under conditions approaching actual use.)
2. Moore, D., "Data Summary of the North Carolina Test Series of USFA Grant 79027 Field Test and Evaluation of Residential Sprinkler Systems," National Fire Protection Association, Boston, MA, September 1980. (This report was done for the NFPA 13D subcommittee and included a series of fire tests in actual mobile homes with sprinkler systems installed in accordance with the data resulting from prior test work and following criteria described in the 1980 edition of NFPA 13D in order to evaluate the effectiveness of the system under conditions approaching actual use.)
3. Kung, H.C., Spaulding, R.D., Hill, E.E., Jr., and Symonds, A.P., "Technical Report Field Evaluation of Residential Prototype Sprinkler, Los Angeles Fire Test Program," Factory Mutual Research Corporation, Norwood, MA, February 1982.
4. Cote, A.E., "Final Report on Field Test and Evaluation of Residential Sprinkler Systems," National Fire Protection Association, Quincy, MA, July 1982.

CHAPTER 5

Limited Area Dwellings

This chapter was added to the 1994 edition to address the use of a special design approach for the protection of specific kinds of small dwellings. This approach is an option to the traditional requirements described in other sections of NFPA 13D. Work on this concept began in April 1992. Prior to the 1994 edition, this concept was previously termed a *limited water supply sprinkler system* and a *manufactured home sprinkler system*, before being named the *limited area dwelling sprinkler system*.

Developmental testing for this system was sponsored by the U.S. Fire Administration. The testing involved a series of laboratory tests at Factory Mutual Research Corporation and Underwriters Laboratories Inc. and a demonstration test in Georgetown, SC. A number of goals were established for this system. The primary goal is the ability to provide sprinkler protection with a captured water supply of approximately 100 gal (378 L). Tenability criteria as discussed earlier in A-1-2 are the same for this system design approach, except that the criteria only need to be maintained for 7 minutes in some cases.

5-1* General

This chapter shall apply only as an alternative for one- and two-family dwellings, including manufactured homes, that do not exceed 2000 ft^2 (186 m^2) in area and that are one story in height with smooth ceilings with a slope not exceeding 10 degrees and a height not exceeding 8 ft (2.4 m) for horizontal ceilings and 9 ft (2.7 m) for sloped ceilings. All other requirements of this standard shall apply.

Exception: Where modified by this chapter.

A-5-1 The concept of this sprinkler system is the result of testing conducted by the U. S. Fire Administration. One of the goals of these tests was to determine if a smaller quantity of water

could be used to protect smaller dwellings. The purpose of the system is to enhance life safety and make installation economically feasible for smaller dwellings.

Design items in this chapter assume that egress times from smaller dwellings are minimal, since the units are single-story at grade level and occupants would be made aware of a fire in the unit almost immediately. This approach is considered in permitting the reduction of the water supply, in some instances, to 7 minutes.

A number of limitations on the dwelling are associated with this design concept. Namely, these limitations are:

- A maximum area of 2000 ft^2 (186 m^2)
- One story in height with no basement and no accessible attic
- A ceiling height of 8 ft (2.4 m) if ceiling is smooth, flat, and horizontal
- A ceiling height of 9 ft (2.7 m) if ceiling is smooth with a slope not in excess of 10 degrees

The 2000-ft^2 (186-m^2) area limit and single-story height limit are necessary for the 7-minute water supply. The amount of time necessary to egress from these smaller one-story dwellings is expected to be less than that of larger multiple-story structures.

Requirements that are not specifically addressed by Chapter 5 are to be in accordance with the other chapters.

5-2 Water Supply

Section 5-2 permits a reduction in the water supply duration. Subsection 5-4.1 requires a 10-gpm (37.8-L/min) flow for the single sprinkler criteria. Thus, a 100-gal (378-L) tank will satisfy this part of the requirement. Subsection 5-4.2 requires a total flow of 13 gpm (49.2 L/ min) resulting in a required water supply of 91 gal (344 L). These flow amounts are approximations and do not account for any additional flow that results from the hydraulic gain due to friction loss.

5-2.1 The water supply for sprinkler systems in limited area dwellings shall be capable of supporting the system demand for 10 minutes for one sprinkler operating as determined by 5-4.1 and 7 minutes for two sprinklers operating as determined by 5-4.2.

5-2.2 A listed strainer shall be provided in risers or feed mains that supply sprinklers with orifices less than ⅜ in. (9.5 mm) nominal diameter.

The presence of the strainer in the water supply line is an additional item that the homeowner must be responsible for maintaining. Without proper maintenance and cleaning, this device could impair the available water supply.

5-3 Components

5-3.1* Sprinklers.

Sprinklers for use in this type of system shall be specifically listed for use in the conditions described in Section 5-1. All sprinklers installed in a compartment shall have the same thermal response characteristics.

A-5-3.1 The criteria used to list a sprinkler for this use should include, but should not be limited to, the following:

(1) Temperature rating—155°F to 165°F (68°C to 74°C)
(2) RTI—nominal 60 (ft-s)$^{1/2}$
(3) Droplet size
(4) Distribution patterns
(5) Areas of coverage
(6) Flow rates
(7) Operating pressures
(8) K-factor = nominal 2.0
(9) Ceiling slope

All sprinklers within a compartment, as defined in 5-4.5, should be within 10°F (5.5°C) of one another in temperature rating.

The sprinkler used for this design approach is anticipated to be evaluated in a manner similar to other residential sprinklers—that is, the sprinkler characteristics need to be comparable to a listed residential sprinkler. Droplet size is not usually evaluated during the listing process for a residential sprinkler.

The sprinkler to be used with the design method of Chapter 5 differs from a residential sprinkler in the following two ways—the K-factor and the slope under which the sprinkler can be used. The limited area dwelling sprinkler system has a K-factor of 2.0, which is indicative of a 5⁄16-in. (7.9-mm) orifice. Currently, the smallest orifice residential sprinkler has a K-factor of 3.9, which corresponds to a 7⁄16-in. (11.1-mm) orifice. Subsection 5-3.1 refers to Section 5-1, which considers ceiling slope as an evaluation factor. The full-scale testing of the prototype device involved a sprinkler evaluation under a 10-degree slope.

See Exhibits 5.1 and 5.2 for placement of limited area dwelling sprinklers in test configurations.

5-3.2* Alarms.

Local waterflow alarms and facilities for flow testing them shall be provided on such sprinkler systems.

A-5-3.2 These alarms should be located within the dwelling and should be of sufficient intensity to be clearly audible in all bedrooms over background noise levels with all intervening doors closed.

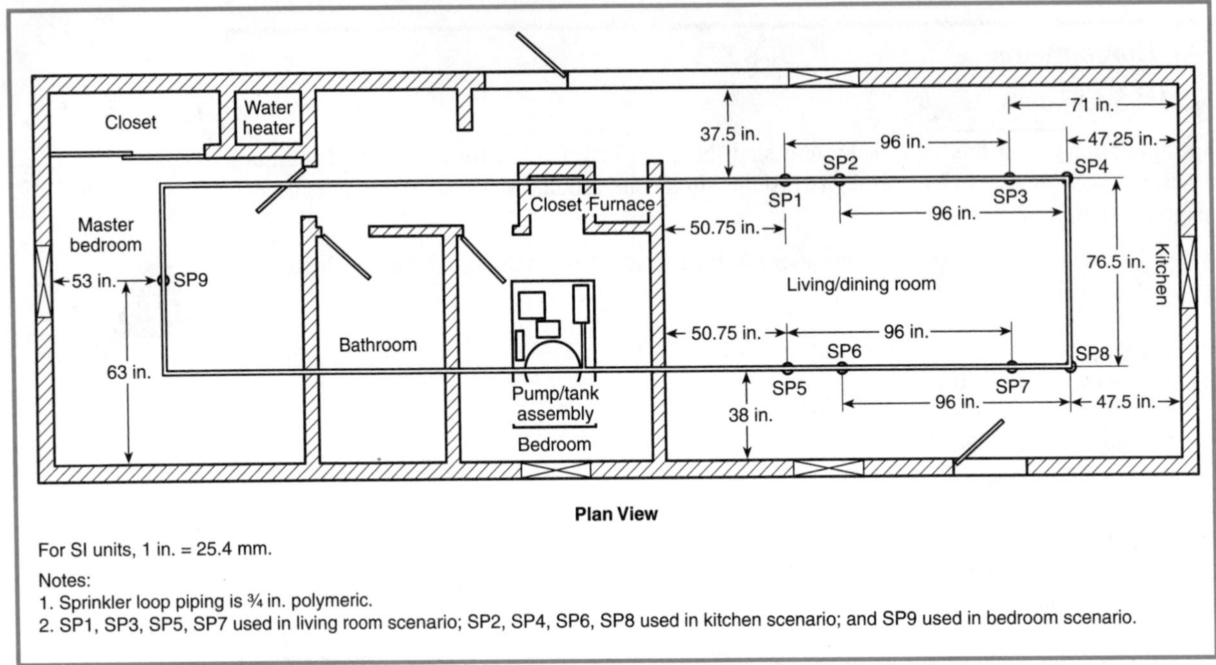

Exhibit 5.1 *Manufactured home sprinkler system layout.*

5-4 System Design

5-4.1 The water demand for the system shall be not less than 10 gpm (37.8 L/ min) at a flowing pressure of 25 psi (1.72 bar) at the sprinkler where only one sprinkler is installed in a compartment.

5-4.2 The water demand for the system shall be not less than 6.5 gpm (24.6 L/min) to each of two sprinklers at a flowing pressure of 11 psi (0.76 bar) at the sprinkler where two or more sprinklers are installed in a compartment. The single sprinkler demand point, as described in 5-4.1, also shall be verified.

The system design criteria for limited area dwellings allow a lower discharge rate for the sprinkler than that required in Chapter 4. However, when compared against the allowable areas of coverage, a higher density is actually available to the floor area. (See 5-4.3.) The reduced spacing of the sprinklers and the lower flow rate still provide sufficient fire control to permit occupant escape.

5-4.3* The area of coverage for sprinklers used in this system shall not exceed 64 ft^2 (5.9 m^2) per sprinkler.

Exception: For compartments not exceeding 100 ft^2 (9.3 m^2) and having no dimension exceeding 10 ft (3.1 m), a single sprinkler shall be permitted to cover this area.

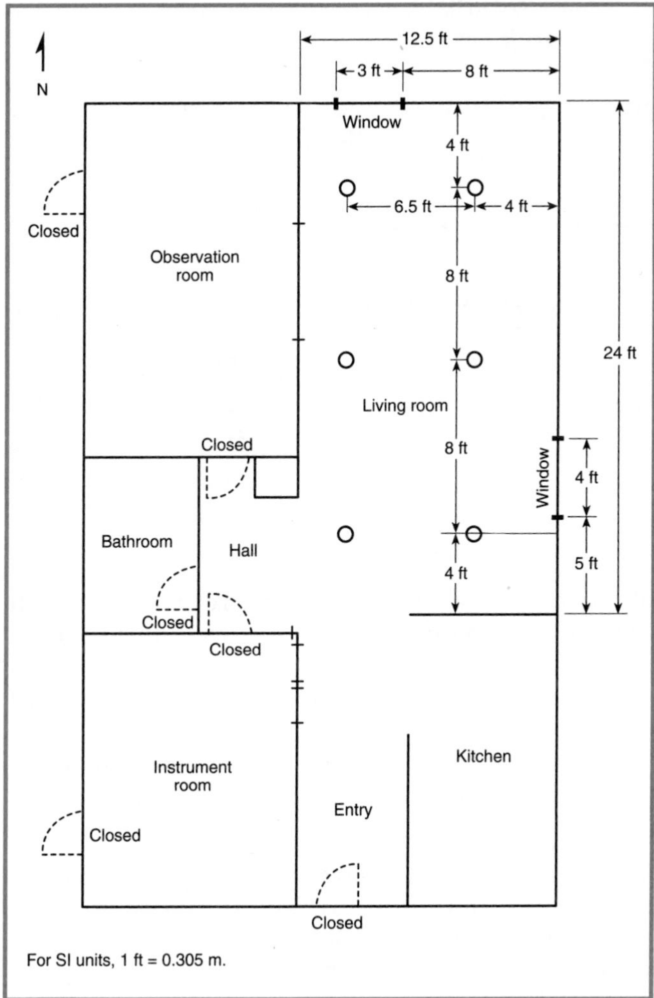

Exhibit 5.2 *Reduced-spacing sprinkler installation in living room.*

The reduced areas of coverage for a limited area dwelling sprinkler system include both advantages and disadvantages that should be considered. The advantage of reduced spacing translates to faster operation of the sprinkler, because sprinklers are physically closer to any point where a fire can originate in a room. The fire will also be more controllable under these circumstances, because the sprinkler will operate while the fire is smaller than if the sprinklers were located farther apart.

The disadvantage associated with reduced spacing is the need to provide a greater number of sprinklers in the dwelling unit. In an extreme case, the limited area dwelling unit system for a 16 ft × 20 ft (4.8 m × 6 m) compartment, would require up to six sprinklers. As discussed in the commentary to 4-1.6, listed sprinklers are available to protect this compartment size with a single device.

A-5-4.3 Sample areas of coverage for a sprinkler where listed for this design approach are provided in Table A-5-4.3.

Table A-5-4.3 Sample Areas of Coverage

Compartment Size		
ft	**m**	**Number of Sprinklers**
10 × 10	3 × 3	1
10 × 12	3 × 3.6	2
12 × 12	3.6 × 3.6	4
16 × 16	4.8 × 4.8	4
16 × 20	4.8 × 6	6

5-4.4 The maximum perpendicular distance to a wall or partition shall not exceed 5 ft (1.5 m), and the minimum distance between sprinklers shall not be less than 6 ft (1.8 m).

5-4.5 The definition of *compartment,* as used in 5-4.1, 5-4.2, and 5-4.3, shall be a space that is enclosed by walls and a ceiling.

The compartment enclosure shall be permitted to have openings to an adjoining space, provided the openings have a minimum lintel depth of 2 in. (51 mm) from the ceiling and do not exceed 20 ft^2 (1.86 m^2) from each compartment.

The testing used to establish the design criteria of Chapter 5 utilized a 2-in. (51-mm) lintel to determine the device's performance. This lintel size is typical of a manufactured home, which is one of the target dwellings for this design procedure. The 20-ft^2 (1.86-m^2) area is also typical of that found in a manufactured home.

The smaller lintel depth permits the control of the test fire without resulting in sprinkler operation in adjacent compartments. Reduced spacing of the sprinklers also permits the smaller lintel, because sprinklers are likely to operate earlier in the fire event, thus resulting in a lower amount of heat being generated.

5-5* Location of Sprinklers

Sprinklers shall be installed in all areas.

Exception No. 1: Sprinklers are not required in clothes closets, linen closets, and pantries where the area of space does not exceed 24 ft^2 (2.2 m^2) and the least dimension does not exceed 3 ft (0.9 m) and the walls and ceilings are surfaced with noncombustible or limited-combustible materials as defined by NFPA 220, Standard on Types of Building Construction.

Exception No. 2: Sprinklers are not required in garages, open attached porches, carports, and similar structures.

Exception No. 3: Sprinklers are not required in attics, crawl spaces, and other concealed spaces that are not used or intended for living purposes or storage.

A-5-5 Unlike Section 4-6, this chapter permits sprinklers to be omitted from only three areas. Foyers are required to be sprinklered, since these areas are usually located at the only entrance/exit for smaller homes. Fires at this point could block the only available exit and could operate sprinklers in the adjoining spaces, thereby taxing the available water supply.

Bathrooms are required to be sprinklered, since they are typically in the path of exit travel from bedrooms as well as from other living spaces. Fires originating in the bathroom and that would not be able to be controlled without sprinklers would likely operate sprinklers in the adjoining compartments, resulting in the operation of more than the number of design sprinklers and the depletion of the water supply.

These three areas are the only areas exempted from the sprinkler requirement for a limited area dwelling design. In larger dwellings, the expectation is that multiple means of escape are available and that bathrooms are not passed during dwelling evacuation.

CHAPTER 6

Referenced Publications

6-1

The following documents or portions thereof are referenced within this standard as mandatory requirements and shall be considered part of the requirements of this standard. The edition indicated for each referenced mandatory document is the current edition as of the date of the NFPA issuance of this standard. Some of these mandatory documents might also be referenced in this standard for specific informational purposes and, therefore, are also listed in Appendix B.

6-1.1 NFPA Publications.

National Fire Protection Association, 1 Batterymarch Park, P.O. Box 9101, Quincy, MA 02269-9101.
NFPA 13, *Standard for the Installation of Sprinkler Systems,* 1999 edition.
NFPA 72, *National Fire Alarm Code®,* 1999 edition.
NFPA 220, *Standard on Types of Building Construction,* 1999 edition.

6-1.2 Other Publications.

6-1.2.1 ANSI Publication. American National Standards Institute, Inc., 11 West 42nd Street, 13th floor, New York, NY 10036.
ANSI B36.10M, *Welded and Seamless Wrought Steel Pipe,* 1996.

6-1.2.2 ASME Publications. American Society of Mechanical Engineers, 345 East 47th Street, New York, NY 10017
ASME B16.1, *Cast Iron Pipe Flanges and Flanged Fittings,* 1989.
ASME B16.3, *Malleable Iron Threaded Fittings,* 1992.

ASME B16.4, *Gray Iron Threaded Fittings,* 1992.

ASME B16.5, *Pipe Flanges and Flanged Fittings,* 1996.

ASME B16.9, *Factory-Made Wrought Steel Buttwelding Fittings,* 1993.

ASME B16.11, *Forged Fittings, Socket-Welding and Threaded,* 1996.

ASME B16.18, *Cast Copper Alloy Solder Joint Pressure Fittings,* 1984.

ASME B16.22, *Wrought Copper and Copper Alloy Solder Joint Pressure Fittings,* 1995.

ASME B16.25, *Buttwelding Ends,* 1997.

6-1.2.3 ASTM Publications. American Society for Testing and Materials, 100 Barr Harbor Drive, West Conshohocken, PA 19428-2959.

ASTM A 53, *Standard Specification for Pipe, Steel, Black and Hot-Dipped, Zinc-Coated, Welded and Seamless,* 1998.

ASTM A 135, *Standard Specification for Electric-Resistance-Welded Steel Pipe,* 1997.

ASTM A 234, *Standard Specification for Piping Fittings of Wrought Carbon Steel and Alloy Steel for Moderate and Elevated Temperatures,* 1997.

ASTM A 795, *Standard Specification for Black and Hot-Dipped Zinc-Coated (Galvanized) Welded and Seamless Steel Pipe for Fire Protection Use,* 1997.

ASTM B 32, *Standard Specification for Solder Metal,* 1996.

ASTM B 75, *Standard Specification for Seamless Copper Tube,* 1995.

ASTM B 88, *Standard Specification for Seamless Copper Water Tube,* 1996.

ASTM B 251, *Standard Specification for General Requirements for Wrought Seamless Copper and Copper-Alloy Tube,* 1997.

ASTM B 813, *Standard Specification for Liquid and Paste Fluxes for Soldering Applications of Copper and Copper-Alloy Tube,* 1993.

ASTM D 3309, *Standard Specification for Polybutylene (PB) Plastic Hot- and Cold-Water Distribution Systems,* 1996.

ASTM F 437, *Standard Specification for Threaded Chlorinated Poly (Vinyl Chloride) (CPVC) Plastic Pipe Fittings, Schedule 80,* 1996.

ASTM F 438, *Standard Specification for Socket-Type Chlorinated Poly (Vinyl Chloride) (CPVC) Plastic Pipe Fittings, Schedule 40,* 1997.

ASTM F 439, *Standard Specification for Socket-Type Chlorinated Poly (Vinyl Chloride) (CPVC) Plastic Pipe Fittings, Schedule 80,* 1997.

ASTM F 442, *Standard Specification for Chlorinated Poly (Vinyl Chloride) (CPVC) Plastic Pipe (SDR-PR),* 1997.

6-1.2.4 AWS Publication. American Welding Society, 550 N.W. LeJeune Road, Miami, FL 33126.

AWS A5.8, *Specification for Filler Metals for Brazing and Braze Welding,* 1992.

APPENDIX A

Explanatory Material

Appendix A is not a part of the requirements of this NFPA document but is included for informational purposes only. This appendix contains explanatory material, numbered to correspond with the applicable text paragraphs.

The material contained in Appendix A of NFPA 13D is included within the text of this handbook and, therefore, is not repeated here.

APPENDIX B

Referenced Publications

B-1

The following documents or portions thereof are referenced within this standard for informational purposes only and are thus not considered part of the requirements of this standard unless also listed in Chapter 6. The edition indicated here for each reference is the current edition as of the date of the NFPA issuance of this standard.

B-1.1 NFPA Publications.

National Fire Protection Association, 1 Batterymarch Park, P.O. Box 9101, Quincy, MA 02269-9101.

NFPA 13, *Standard for the Installation of Sprinkler Systems,* 1999 edition.

NFPA 13R, *Standard for the Installation of Sprinkler Systems in Residential Occupancies up to and Including Four Stories in Height,* 1999 edition.

NFPA 20, *Standard for the Installation of Stationary Pumps for Fire Protection,* 1999 edition.

NFPA 25, *Standard for the Inspection, Testing, and Maintenance of Water-Based Fire Protection Systems,* 1998 edition.

NFPA 72, *National Fire Alarm Code®*, 1999 edition.

B-1.2 Other Publications.

B-1.2.1 ASTM Publications. American Society for Testing and Materials, 100 Barr Harbor Drive, West Conshohocken, PA 19428-2959.

ASTM D 3309, *Standard Specification for Polybutylene (PB) Plastic Hot- and Cold-Water Distribution Systems,* 1996.

ASTM F 437, *Standard Specification for Threaded Chlorinated Poly (Vinyl Chloride) (CPVC) Plastic Pipe Fittings, Schedule 80,* 1996.

ASTM F 438, *Standard Specification for Socket-Type Chlorinated Poly (Vinyl Chloride) (CPVC) Plastic Pipe Fittings, Schedule 40,* 1997.

ASTM F 439, *Standard Specification for Socket-Type Chlorinated Poly (Vinyl Chloride) (CPVC) Plastic Pipe Fittings, Schedule 80,* 1997.

ASTM F 442, *Standard Specification for Chlorinated Poly (Vinyl Chloride) (CPVC) Plastic Pipe (SDR-PR),* 1997.

ASTM SI 10, *Standard for Use of the International System of Units (SI): the Modernized Metric System,* 1997.

B-1.2.2 USFA Publication. U.S. Fire Administration, 16825 S. Seton Avenue., Emmitsburg, MD 21727.

Bill Jr., R. G., Kung, N. C., Brown, W. R., and Hill, E., "Effects of Cathedral and Beamed Ceiling Construction on Residential Sprinkler Performance," FMRC J.I. M3N5.RA(3). Prepared for U.S. Fire Administration, February 1988.

PART THREE

NFPA 13R, *Standard for the Installation of Sprinkler Systems in Residential Occupancies up to an Including Four Stories in Height,* with Commentary

Part Three of this handbook includes the complete text and illustrations of the 1999 edition of NFPA 13R. The text and illustrations from the standard are printed in black and are the official requirements of NFPA 13R. Line drawings and photographs from the standard are labeled "Figures."

Paragraphs that begin with the letter "A" are extracted from Appendix A of the standard. Although printed in black ink, this nonmandatory material is purely explanatory in nature. For ease of use, this handbook places Appendix A material immediately after the standard paragraph to which it refers.

Part Two also includes Formal Interpretations on NFPA 13R; these apply to previous and subsequent editions for which the requirements remain substantially unchanged. Formal Interpretations are not part of the standard and are printed in shaded gray boxes.

In addition to standard text, appendixes, and Formal Interpretations, Part Three includes commentary that provides the history and other background information for specific paragraphs in the standard. This insightful commentary takes the reader behind the scenes, into the reasons underlying the requirements.

To make commentary material readily identifiable, commentary text, captions, and tables are all printed in green. So that the reader can easily distinguish between line drawings and photographs from the standard and from the commentary, line drawings and photographs in the commentary are labeled "Exhibits." The distinction between figures in the standard and exhibits in the commentary is new for this edition of the *Automatic Sprinkler Systems Handbook*.

Preface to 1999 Edition of NFPA 13R

It is intended that this standard provide a method for those individuals wishing to install a sprinkler system for additional life safety and property protection. It is not the purpose of this standard to require the installation of an automatic sprinkler system. This standard assumes that one or more smoke detectors will be installed in accordance with NFPA 72, *National Fire Alarm Code®*.

CHAPTER 1

General Information

In 1989, the first edition of NFPA 13R, *Standard for the Installation of Sprinkler Systems in Residential Occupancies up to and Including Four Stories in Height*, was issued. Similar to NFPA 13D, *Standard for the Installation of Sprinkler Systems in One- and Two-Family Dwellings and Manufactured Homes*, NFPA 13R addressed cost-effective sprinkler protection with life safety as its primary goal and property protection as a secondary goal.

The development of NFPA 13D, along with continuing developments in residential sprinkler technology, sparked interest in expanding similar low-cost fire safety options to other types of low-rise residential occupancies such as apartments and condominiums. Many of the underlying fire safety objectives used to develop NFPA 13D were applied to these other larger residential occupancies. Fuel loading, basic room geometries, and subsequent fires in larger residential occupancies were observed to be similar to those anticipated in one- and two-family dwellings. Expanding public interest and the fact that some communities started developing their own installation standards for larger residential buildings made it clear that a national consensus standard such as NFPA 13R was needed.

The Technical Committee on Residential Sprinkler Systems intends that NFPA 13R provide an acceptable level of fire protection with respect to life safety and property protection. NFPA 13R provides a high, but not an absolute, level of life safety and a somewhat lesser degree of property protection. As with NFPA 13D systems, NFPA 13R permits the omission of sprinklers in certain areas of the building. A higher degree of life safety and property protection could be achieved by installing sprinklers throughout the premises in accordance with NFPA 13, *Standard for the Installation of Sprinkler Systems*, with residential sprinklers installed in the dwelling units. With sprinkler systems, life safety is generally more readily achievable than is property protection. In other words, a given system usually provides a higher degree of life safety than property protection. Where property protection is the primary objective, a

stronger water supply and more complete sprinkler coverage among other factors as identified by NFPA 13 need to be considered for the type of hazard in question.

Although NFPA 13R differs from NFPA 13 in several areas, such as extent of sprinkler coverage and the mandatory use of residential sprinklers with certain exceptions, the documents have many similarities. Overall, NFPA 13R offers a cost advantage with respect to providing life safety in certain types of low-rise residential occupancies because the document was developed with this objective.

Various building codes have adopted NFPA 13R. However, certain restrictions could be imposed on buildings protected by NFPA 13R sprinkler systems with respect to height and area limitations, protection of attic spaces, and acceptable construction materials. Sprinkler coverage elimination in certain areas results in a reduced level of property protection. Property protection is a major function of building codes, so provisions in addition to those of NFPA 13R could be required.

NFPA 13R assumes that one or more smoke detectors is installed in accordance with NFPA 72, *National Fire Alarm Code®*. Detectors are a key component of a balanced fire protection system for residential occupancies. Detectors are intended to provide improved protection against smoldering fires and from fires that originate in areas not required to be protected by automatic sprinklers. Section 2-6 identifies those areas not requiring sprinkler protection.

1-1* Scope

This standard covers the design and installation of automatic sprinkler systems for protection against fire hazards in residential occupancies up to and including four stories in height.

A-1-1 NFPA 13R is appropriate for use as an option to NFPA 13, *Standard for the Installation of Sprinkler Systems,* only in those residential occupancies, as defined in this standard, up to and including four stories in height. Residential portions of any other building should be protected with residential or quick-response sprinklers in accordance with 5-4.5 of NFPA 13. Other portions of such buildings should be protected in accordance with NFPA 13.

The criteria in this standard are based on full-scale fire tests of rooms containing typical furnishings found in residential living rooms, kitchens, and bedrooms. The furnishings were arranged as typically found in dwelling units in a manner similar to that shown in Figures A-1-1(a), A-1-1(b), and A-1-1(c). Sixty full-scale fire tests were conducted in a two-story dwelling in Los Angeles, California, and 16 tests were conducted in a 14-ft (4.3-m) wide mobile home in Charlotte, North Carolina. Sprinkler systems designed and installed according to this standard are expected to prevent flashover within the compartment of origin where sprinklers are installed in the compartment. A sprinkler system designed and installed according to this standard cannot, however, be expected to completely control a fire involving fuel loads that are significantly higher than average for dwelling units [10 lb/ft^2 (49 kg/m^2)], configurations of fuels other than those with typical residential occupancies, or conditions where the interior finish has an unusually high flame spread rating (greater than 225).

To be effective, sprinkler systems installed in accordance with this standard need to open the sprinklers closest to the fire before the fire exceeds the ability of the sprinkler discharge to extinguish or control the fire. Conditions that allow the fire to grow beyond that point before sprinkler activation or that interfere with the quality of water distribution can produce conditions beyond the capabilities of the sprinkler system described in this standard. Unusually high ceilings or ceiling configurations that tend to divert the rising hot gases from sprinkler locations or change the sprinkler discharge pattern from its standard pattern can produce fire conditions that cannot be extinguished or controlled by the systems described in this standard.

NFPA 13R is based in part on tests referenced in A-1-1. Those tests have demonstrated the ability of such residential systems to control fires that have growth rates similar to those involving residential furnishings. However, fire control depends on a sprinkler responding during the earliest stages of fire development. Curves indicating the energy release rates for different types of fires, including those using residential furnishings as the fuel package, are shown in Exhibit 1.1.[1] The solid-line curves indicate a common means of categorizing the available heat release data. (See NFPA 72 for additional information.)

NFPA 13R does not define the height of a story for use in determining overall building height. The building height is anticipated to be determined in accordance with the applicable building code or NFPA *101*®, *Life Safety Code*®. NFPA *101* and each of the three model building codes—that is, *Uniform Building Code, Standard Building Code,* and *National Building Code*—have their own method for determining building and story heights.

As noted in Section 1-3, the use of the term *residential occupancies* for purposes of applying NFPA 13R differs from *residential occupancies* as defined in NFPA *101*. The NFPA 13R definition of *residential occupancies* is consistent with NFPA *101* except that three categories of board and care occupancies and one- and two-family dwellings are excluded. The categories of board and care facilities are excluded due to a concern for overall building evacuation time and the fact that evacuation time in accordance with Chapters 22 and 23 of the 1997 edition of NFPA *101* is based on occupants reaching a point of safety that is not necessarily outside of the building. Because sprinklers are not required throughout the premises, NFPA 13R may not provide adequate protection for occupants who require an extended period of time to evacuate or who need to be protected in place. Based on the definitions in NFPA *101*, occupants who fall into the "slow" and the "impractical" evacuation capability are included. In general, this population group is not capable of self-preservation and must either be defended in place or must be assisted from the building. Because NFPA 13R allows the omission of sprinklers in certain areas, it is not well suited for those types of residential occupancies that contain population groups who are not readily capable of self-preservation.

One- and two-family dwellings are exempted from NFPA 13R because they are specifically addressed by NFPA 13D.

1-2* Purpose

The purpose of this standard is to provide design and installation requirements for a sprinkler system to aid in the detection and control of fires in residential occupancies and thus provide

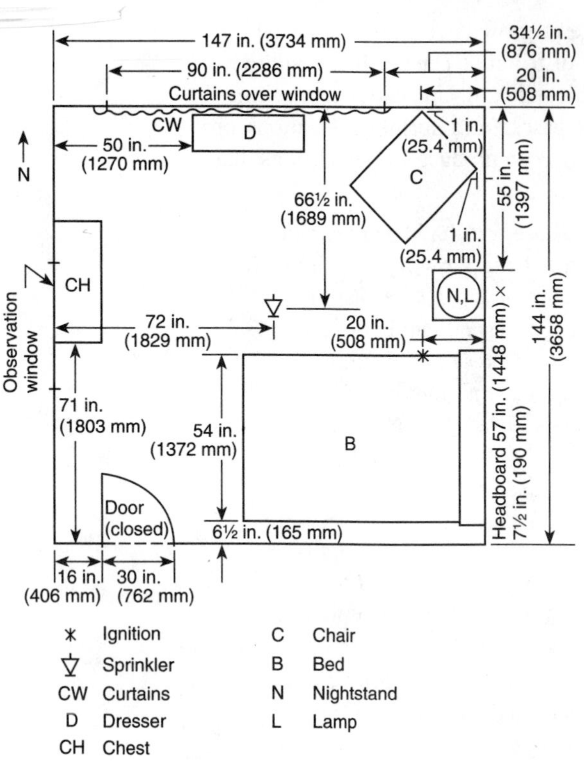

Figure A-1-1(a) Bedroom.

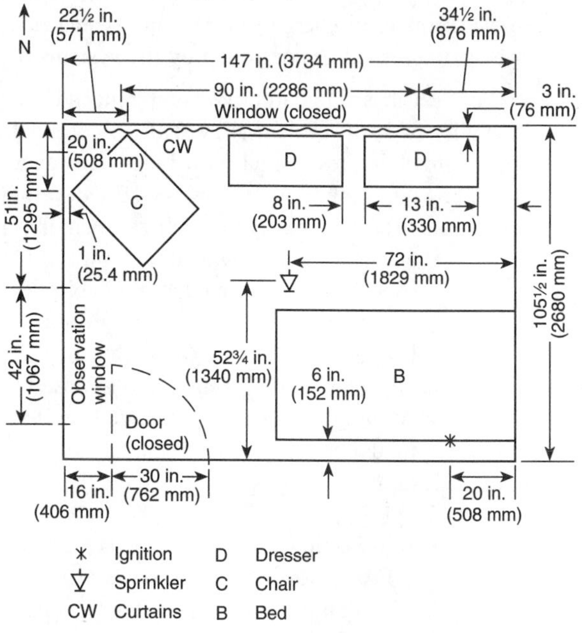

Figure A-1-1(b) Mobile home bedroom.

improved protection against injury, life loss, and property damage. A sprinkler system designed and installed in accordance with this standard is expected to prevent flashover (total involvement) in the room of fire origin, where sprinklered, and to improve the chance for occupants to escape or be evacuated.

Nothing in this standard is intended to restrict new technologies or alternative arrangements, provided that the level of safety prescribed by the standard is not reduced.

A-1-2 Various levels of sprinkler protection are available to provide life safety and property protection. This standard is designed to provide a high, but not absolute, level of life safety and a lesser level of property protection. Greater protection to both life and property could be achieved by sprinklering all areas in accordance with NFPA 13, *Standard for the Installation of Sprinkler Systems,* which permits the use of residential sprinklers in residential areas.

This standard recommends, but does not require, sprinklering of all areas in the building; it permits sprinklers to be omitted in certain areas. These areas have been proved by NFPA statistics to be those where the incidence of life loss from fires in residential occupancies is low. Such an approach provides a reasonable degree of fire safety. *(See Table A-1-2 for deaths and injuries in multifamily residential buildings.)*

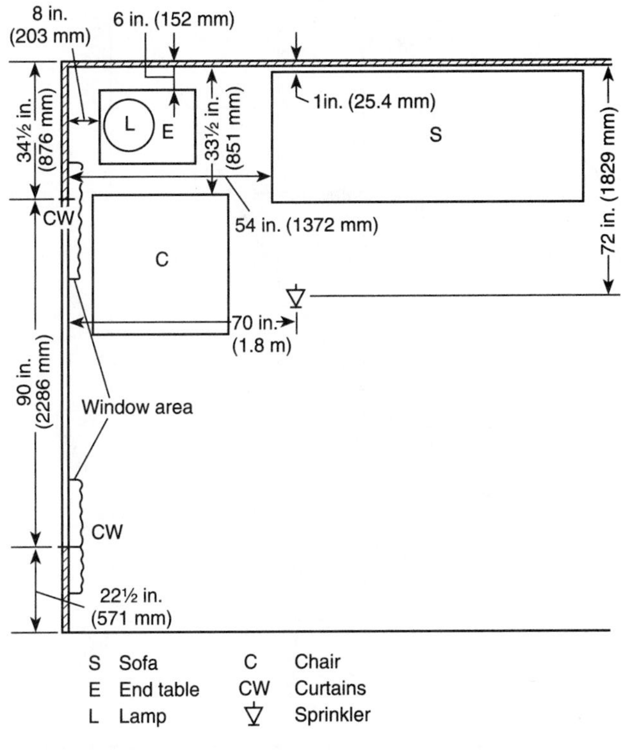

Figure A-1-1(c) Living room.

It should be recognized that the omission of sprinklers from certain areas could result in the development of untenable conditions in adjacent spaces. Where evacuation times could be delayed, additional sprinkler protection and other fire protection features, such as detection and compartmentation, could be necessary.

Both Underwriters Laboratories Inc. and Factory Mutual Research Corporation have developed test standards for the evaluation of residential sprinklers.[2, 3] These standards have significantly different test parameters than those that apply to nonresidential sprinklers such as the standard spray sprinkler. During a fire test, residential sprinklers are to perform as follows:

(1) Limit the maximum air temperature 3 in. (76 mm) below the ceiling to no greater than 600°F (315°C)
(2) Limit the maximum temperature at 5 ft 3 in. (1.6 m) above the floor to no greater than 200°F (93°C)
(3) Limit the temperature at the height of 5 ft 3 in. (1.6 m) above the floor to no greater than 130°F (54°C) for more than any continuous 2-minute period

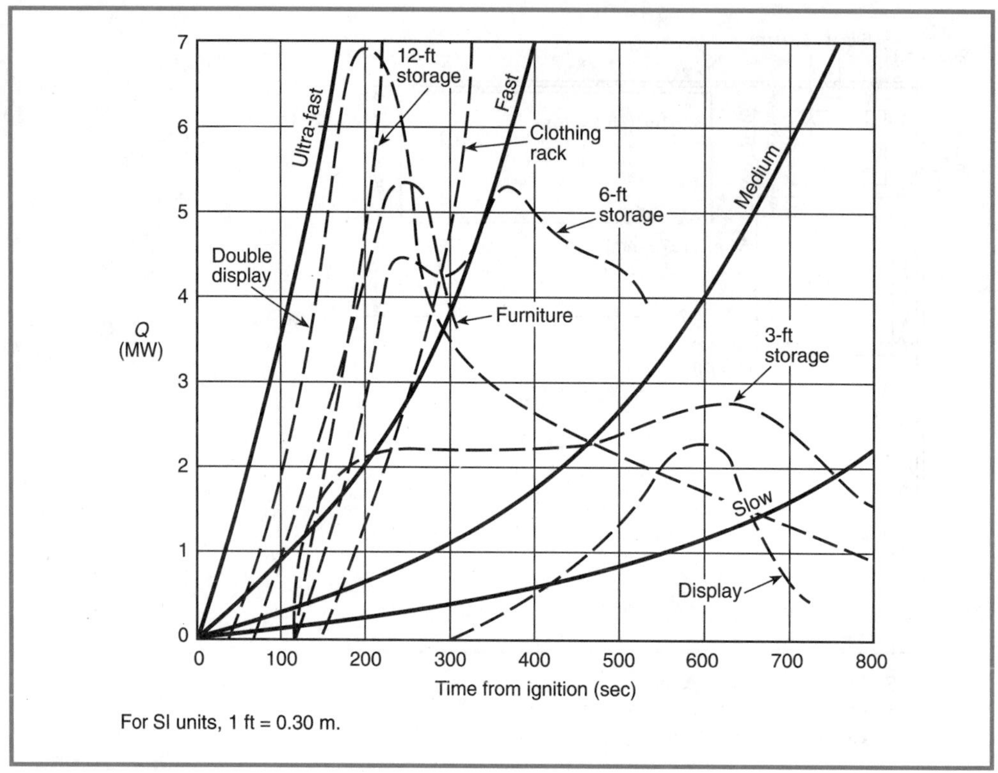

Exhibit 1.1 *Energy release time curves.*

(4) Limit the maximum temperature ¼ in. (6.4 mm) above the finished ceiling to no more than 500°F (260°C)
(5) Attain the above criteria with no more than two sprinklers operating
(6) Have sprinklers' water-spray discharge wet the walls of the test chamber within their area of coverage to at least 18 in. (457 mm) from the ceiling

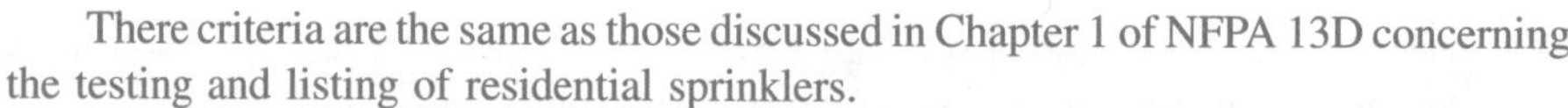

There criteria are the same as those discussed in Chapter 1 of NFPA 13D concerning the testing and listing of residential sprinklers.

The intent of the preceding criteria are to show the following:

(1) Flashover does not occur.
(2) The environment through which occupants must evacuate is tenable.
(3) Items (1) and (2) of this list can be accomplished with a limited water supply.
(4) Proper wall-wetting characteristics occur.

The basic laboratory test configuration is a 12 ft × 24 ft (3.7 m × 7.3 m) room with a combustible array simulating residential furnishings. The test facility is adjustable so that it can be configured to evaluate various areas of sprinkler coverage.

The concept of levels of protection as identified in A-1-2 was developed so that

Table A-1-2 Fires and Associated Deaths and Injuries in Apartments by Area of Origin; Annual Average of 1986–1990 Structure Fires Reported to U.S. Fire Departments

Area of Origin	Civilian Deaths	Civilian Percent	Fires	Percent	Injuries	Percent
Bedroom	309	33.9	17,960	15.8	1,714	27.2
Living room, family room, or den	308	33.8	10,500	9.3	1,272	20.2
Kitchen	114	12.5	46,900	41.4	1,973	31.2
Interior stairway	29	3.2	1,040	0.9	91	1.4
Hallway or corridor	23	2.6	3,130	2.8	165	2.6
Exterior balcony or open porch	17	1.8	1,880	1.7	69	1.1
Dining room	10	1.1	800	0.7	69	1.1
Closet	9	1.0	2,120	1.9	116	1.8
Multiple areas	9	1.0	780	0.7	38	0.6
Tool room or other supply storage room or area	8	0.9	1,250	1.1	53	0.8
Unclassified area	8	0.9	480	0.4	29	0.5
Exterior stairway	8	0.8	870	0.8	22	0.4
Bathroom	7	0.7	2,510	2.2	101	1.6
Heating equipment room or area	6	0.6	2,510	2.2	75	1.2
Exterior wall surface	5	0.5	2,150	1.9	26	0.4
Laundry room or area	4	0.4	3,380	3.0	89	1.4
Crawl space or substructure space	4	0.4	1,490	1.3	62	1.0
Wall assembly or concealed space	3	0.4	1,020	0.9	21	0.3
Attic or ceiling/roof assembly or concealed space	3	0.3	1,100	1.0	18	0.3
Ceiling/floor assembly or concealed space	3	0.3	560	0.5	18	0.3
Garage or carport*	3	0.3	1,290	1.1	36	0.6
Lobby or entrance way	3	0.3	670	0.6	31	0.5
Unclassified structural area	3	0.3	520	0.5	32	0.5
Unclassified storage area	3	0.3	430	0.4	22	0.3
Unclassified function area	3	0.3	250	0.2	13	0.2
Laboratory	2	0.3	80	0.1	3	0.0
Elevator or dumbwaiter	1	0.2	220	0.2	4	0.1
Sales or showroom area	1	0.2	110	0.1	3	0.1
Exterior roof surface	1	0.1	1,040	0.9	15	0.2
Unclassified means of egress	1	0.1	180	0.2	6	0.1
Office	1	0.1	120	0.1	4	0.1
Chimney	1	0.1	980	0.9	2	0.0
Personal service area	1	0.1	40	0.0	4	0.1
Library	1	0.1	10	0.0	0	0.0
Other known area	2	0.2	5,000	4.4	115	1.8
Total	912	100.0	113,390	100.0	6,313	100.0

Note: Fires are estimated to the nearest 10; civilian deaths and injuries are estimated to the nearest 1.

*Does not include dwelling garages coded as a separate property.

Source: 1986–1990 NFIRS and NFPA survey.

low-cost sprinkler systems would become a viable and attractive option for certain types of residential facilities and so that installing dry pipe sprinkler systems in cold climates would not be a necessity. The areas where sprinkler omission is permitted in accordance with Section 2-6 of NFPA 13R are those where statistical analysis supports such omission. (See Table A-1-2.)

The Technical Committee on Residential Sprinkler Systems believes that dry pipe systems are not as effective as wet systems for life safety because of the associated delay in the sprinkler discharge application. (This delay was noted in early Factory Mutual tests and substantiated in the fire tests done in Los Angeles and Charlotte, NC.[4, 5, 6, 7]) The key to achieving control of the residential fire is the rapid application of water. Residential systems also must be relatively simple systems and easy to maintain. These features are not characteristic of dry pipe systems.

Table A-1-2 contains data on life safety and not necessarily property protection. Reductions in building code criteria related to property protection need to be evaluated when the sprinkler system is installed in accordance with NFPA 13R. Areas of sprinkler omission would be of particular interest. Although fires that occur in nonsprinklered areas might not necessarily result in fatalities, such fires could result in serious property damage. (See Exception Nos. 1 through 5 of Section 2-6 for areas not requiring sprinkler protection.)

1-3 Definitions

Approved.* Acceptable to the authority having jurisdiction.

A-1-3 Approved. The National Fire Protection Association does not approve, inspect, or certify any installations, procedures, equipment, or materials; nor does it approve or evaluate testing laboratories. In determining the acceptability of installations, procedures, equipment, or materials, the authority having jurisdiction may base acceptance on compliance with NFPA or other appropriate standards. In the absence of such standards, said authority may require evidence of proper installation, procedure, or use. The authority having jurisdiction may also refer to the listings or labeling practices of an organization that is concerned with product evaluations and is thus in a position to determine compliance with appropriate standards for the current production of listed items.

A common error is the assumption that approved has the same meaning as the terms *listed* or *labeled*. For NFPA 13R's purposes, approved means that which is acceptable to the authority having jurisdiction. Although the authority having jurisdiction can use a listing or label to assist in approving an item, it should not be assumed that all approvals are based on a listing procedure. Some testing agencies use the terms *approved* or *approval* as their designation that a product is listed.

Subsection 2-2.2 requires that only listed or approved devices and materials as indicated by NFPA 13R be used in sprinkler systems. Other sections require that specific components—for example, residential sprinklers and nonmetallic sprinkler pipe—be listed.

Authority Having Jurisdiction.* The organization, office, or individual responsible for approving equipment, materials, an installation, or a procedure.

A-1-3 Authority Having Jurisdiction. The phrase "authority having jurisdiction" is used in NFPA documents in a broad manner, since jurisdictions and approval agencies vary, as do their responsibilities. Where public safety is primary, the authority having jurisdiction may be a federal, state, local, or other regional department or individual such as a fire chief; fire marshal; chief of a fire prevention bureau, labor department, or health department; building official; electrical inspector; or others having statutory authority. For insurance purposes, an insurance inspection department, rating bureau, or other insurance company representative may be the authority having jurisdiction. In many circumstances, the property owner or his or her designated agent assumes the role of the authority having jurisdiction; at government installations, the commanding officer or departmental official may be the authority having jurisdiction.

In the simplest terms, the authority having jurisdiction is the person or office enforcing the standard. The term authority having jurisdiction is commonly associated with the fire marshal or building official. However, the authority having jurisdiction can also be a contract officer, design engineer, insurance company representative, facility owner, or other agency representative responsible for verifying that the standard's requirements are achieved.

Check Valve. A valve that allows flow in one direction only.

Control Valve. An indicating valve employed to control (shut) a supply of water to a sprinkler system.

The control valve used should be an indicating type—that is, one that has some external means to indicate that the valve is in the open position.

Design Discharge. The rate of water discharged by an automatic sprinkler, expressed in gpm (L/min).

The term *design discharge* is used in the design approach for residential sprinklers. Although some sprinkler designs are based on their ability to achieve a certain density, residential sprinklers are listed to achieve a certain minimum flow rate. See 2-5.1.1 for more information on design discharge.

Dry System. A system employing automatic sprinklers attached to a piping system containing air under atmospheric or higher pressures. Loss of pressure from the opening of a sprinkler or detection of a fire condition causes the release of water into the piping system and out the opened sprinkler.

Because of the concern for delayed water delivery and the fact that no residential sprinklers are listed for dry system application, dry systems are not allowed for protection inside the dwelling unit in accordance with Exception No. 1 to 2-4.5.1. Exception No. 2 to 2-4.5.1 permits the use of a dry pipe system if other than residential sprinklers were used and if the conditions of the exception were satisfied. However, dry systems are permitted to be installed in areas outside the dwelling unit. See 2-5.2 with regard to the design criteria for areas outside the dwelling unit.

Dwelling Unit. One or more rooms, arranged for the use of one or more individuals living together, as in a single housekeeping unit, that normally have cooking, living, sanitary, and sleeping facilities.

For the purposes of NFPA 13R, a dwelling unit includes hotel and motel rooms and suites, dormitory rooms, apartments, condominiums, sleeping rooms and most common areas in board and care facilities, and similar living units. In general, a dwelling unit consists of any part of a structure that is normally occupied and has fuel loading that is indicative of a residential environment. A dwelling unit also includes the corridor that leads to the guest rooms in a hotel- or motel-type facility.

Labeled. Equipment or materials to which has been attached a label, symbol, or other identifying mark of an organization that is acceptable to the authority having jurisdiction and concerned with product evaluation, that maintains periodic inspection of production of labeled equipment or materials, and by whose labeling the manufacturer indicates compliance with appropriate standards or performance in a specified manner.

Listed.* Equipment, materials, or services included in a list published by an organization that is acceptable to the authority having jurisdiction and concerned with evaluation of products or services, that maintains periodic inspection of production of listed equipment or materials or periodic evaluation of services, and whose listing states that either the equipment, material, or service meets appropriate designated standards or has been tested and found suitable for a specified purpose.

A-1-3 Listed. The means for identifying listed equipment may vary for each organization concerned with product evaluation; some organizations do not recognize equipment as listed unless it is also labeled. The authority having jurisdiction should utilize the system employed by the listing organization to identify a listed product.

One prominent testing inspection agency laboratory uses the designation *classified* for pipe. Material with this designation meets the intent of the definition of *listed.* (Also see the commentary following the definition of *approved.*) Equipment critical to the system's proper performance and operation is typically listed. One such component is the sprinkler. Unless the component is evaluated for its performance under fire conditions and the quality control aspects of the manufacturing process can be assured, the device's level of reliability and expected performance cannot be affirmed.

Multipurpose Piping System. A piping system within a residential occupancy intended to serve both domestic and fire protection needs.

Preaction System. A sprinkler system employing automatic sprinklers that are attached to a piping system containing air that might or might not be under pressure, with a supplemental detection system installed in the same areas as the sprinklers. Actuation of the detection system opens a valve that allows water to flow into the sprinkler piping system and to be discharged from any sprinklers that are open.

Residential Occupancies. Occupancies, as specified in the scope of this standard, that include the following, as defined in NFPA *101*®, *Life Safety Code*®: (1) Apartment buildings, (2) Lodging and rooming houses, (3) Board and care facilities (slow evacuation type with 16 or fewer occupants and prompt evacuation type), and (4) Hotels, motels, and dormitories.

For purposes of applying NFPA 13R, the definition of *residential occupancies* differs from the definition used in NFPA *101*. Specifically, the definition in NFPA 13R does not include the following occupancies, which are considered residential occupancies by NFPA *101*:

(1) One- and two-family dwellings, including manufactured homes
(2) Small board and care facilities, which have sleeping accommodations for not more than 16 residents, with impractical evacuation capability
(3) Large board and care facilities with slow evacuation capability
(4) Large board and care facilities with impractical evacuation capability

The commentary to Section 1-1 provides more information on the differences between the occupancies.

Residential Sprinkler. A type of sprinkler that meets the definition of fast response as defined by NFPA 13, *Standard for the Installation of Sprinkler Systems,* that has been specifically investigated for its ability to enhance survivability in the room of fire origin, and that is listed for use in the protection of dwelling units.

Residential sprinklers are specifically listed by Underwriters Laboratories and Factory Mutual Research Corporation for residential service and have been tested for compliance with NFPA 13D.[2, 3] Their primary purpose is to achieve life safety in residential occupancies. Residential sprinklers contain fast-response operating elements and can be an upright, pendent, or sidewall configuration, but no upright devices are available at this time. The discussion on sprinkler sensitivity in the commentary to 1-4.5.1 of NFPA 13 provides additional information on sprinklers with fast-response operating elements.

Shall. Indicates a mandatory requirement.

Should. Indicates a recommendation or that which is advised but not required.

Sprinkler, Automatic. A fire suppression or control device that operates automatically when its heat-actuated element is heated to its thermal rating or above, allowing water to discharge over a specific area.

Depending on the type used, automatic sprinklers are intended to achieve fire control or fire suppression. Additionally, automatic sprinklers can be used for property protection or life safety, or both. Residential sprinklers are specifically listed as such and are intended to achieve life safety as their primary objective. NFPA 13 provides more information on the different types and styles of automatic sprinklers.

Sprinkler System. can integrated system of piping, connected to a water supply, with listed sprinklers that automatically initiate water discharge over a fire area. Where required, the sprinkler system also includes a control valve and a device for actuating an alarm when the system operates.

The definition of *sprinkler system* addresses water supplies and underground piping. When tanks, pumps, and underground piping installation and design requirements covered by standards other than NFPA 13 are used in connection with overhead sprinkler piping, they become an integral part of the sprinkler system. Because these components are considered critical to system performance, they are treated as part of the system.

Standard. A document, the main text of which contains only mandatory provisions using the word "shall" to indicate requirements and which is in a form generally suitable for mandatory reference by another standard or code or for adoption into law. Nonmandatory provisions shall be located in an appendix, footnote, or fine-print note and are not to be considered a part of the requirements of a standard.

System Working Pressure. The maximum anticipated static (non flowing) or flowing pressure applied to sprinkler system components exclusive of surge pressures.

The definition of *system working pressure* was added to the 1999 edition to clearly indicate that this pressure pertains to the installed system. Maximum static pressure could be the result of surge pressures held in a system by a check valve or back flow preventer.

Thermal Barrier. A material that limits the average temperature rise of the unexposed surface to not more than 250°F (121°C) after 15 minutes of fire exposure and that complies with the standard time-temperature curve of NFPA 251, *Standard Methods of Tests of Fire Endurance of Building Construction and Materials.*

Construction materials that provide a 15-minute thermal barrier must be used in bathrooms if sprinkler protection is not provided. (See Exception No. 1 of Section 2-6.) The primary method of determining whether a material qualifies is to check the assembly's finish rating. This information can usually be found in fire resistance directories that are published by listing organizations.

Waterflow Alarm. A sounding device activated by a waterflow detector or alarm check valve.

Waterflow Detector. An electric signaling indicator or alarm check valve actuated by waterflow in one direction only.

Wet System. A system employing automatic sprinklers that are attached to a piping system containing water and connected to a water supply so that water discharges immediately from sprinklers opened by a fire.

1-4* Units

Metric units of measurement in this standard are in accordance with the modernized metric system known as the International System of Units (SI). The liter and bar units are outside of but recognized by SI and are commonly used in international fire protection. These units are provided in Table 1-4 with their conversion factors.

Table 1-4 Metric Conversions

Name of Unit	Unit Symbol	Conversion Factor
liter	L	1 gal = 3.785 L
pascal	Pa	1 psi = 6894.757 Pa
bar	bar	1 psi = 0.0689 bar
bar	bar	1 bar = 105 Pa

A-1-4 For additional conversions and information, see ASTM SI 10, *Standard for Use of the International System of Units (SI): the Modern Metric System.*

1-4.1 Where a value for measurement as specified in this standard is followed by an equivalent value in other units, the first stated value shall be regarded as the requirement. A given equivalent value is considered to be approximate.

The International System of Units (SI) is typically rounded off to the nearest whole number.

1-4.2 SI units have been converted by multiplying the quantity by the conversion factor and then rounding the result to the appropriate number of significant digits.

1-5 Piping

1-5.1* Pipe or tube used in sprinkler systems shall be of the materials specified in Table 1-5.1 or in accordance with 1-5.2 through 1-5.5. The chemical properties, physical properties, and dimensions of the materials in Table 1-5.1 shall be at least equivalent to the standards cited in the table and designed to withstand a working pressure of not less than 175 psi (12.1 bar). When nonmetallic piping is used, it shall be rated at not less than 175 psi (12.1 bar) working pressure at not less than 120°F (49°C).

NFPA 13R recognizes the use of nonmetallic piping products. As with ferrous materials and copper products, nonmetallic pipe standards exist and are used as the basis for the manufacture and production of polybutylene (PB) and chlorinated polyvinyl chloride (CPVC) pipe. Subsection 1-5.1 sets performance criteria for the piping products that are currently allowed in NFPA 13R systems. Pipe materials are required to conform to specific industry standards, and the materials are required to be capable of withstanding a working pressure of 175 psi (12.1 bar). These criteria are applicable to most ferrous, copper, PB, and CPVC pipe.

A-1-5.1 This standard anticipates that the water supply for the system is in compliance with the governing plumbing code for the jurisdiction. It is intended that any pipe material or diameter

Table 1-5.1 Pipe or Tube Materials and Dimensions

Materials and Dimensions	Standard
Specification for Black and Hot-Dipped Zinc-Coated (Galvanized) Welded and Seamless Steel Pipe for Fire Protection Use	ASTM A 795
Specification for Welded and Seamless Pipe	ASTM A 53
Wrought Steel Pipe	ANSI B36.10M
Specification for Electric-Resistance-Welded Steel Pipe	ASTM A 135
Specification for Seamless Copper Water Tube [Copper Tube (Drawn, Seamless)]	ASTM B 88
Specification for General Requirements for Wrought Seamless Copper and Copper-Alloy Tube	ASTM B 251
Fluxes for Soldering Applications of Copper and Copper-Alloy Tube	ASTM B 813
Specification for Filler Metals for Brazing and Braze Welding (Classification BCuP-3 or BCuP-4)	AWS A5.8
Specification for Solder Metal [95-5 (Tin-Antimony-Grade 95TA)]	ASTM B 32

permitted for multiple family dwellings in the plumbing code and satisfying the hydraulic criteria of NFPA 13R is considered to be in compliance.

NFPA 13R permits the underground supply pipe to conform to local plumbing code requirements. Provided the pipe is sized to accommodate system flow rates and conforms to all local requirements, it is acceptable by NFPA 13R.

1-5.2* Other types of pipe or tube shall be permitted to be used where listed for sprinkler systems. Listed piping materials including but not limited to chlorinated polyvinyl chloride (CPVC), polybutylene (PB), and steel differing from those provided in Table 1-5.1 shall be installed in accordance with their listings and the manufacturers' installation instructions. CPVC and PB pipe shall comply with the portions of the American Society for Testing and Materials (ASTM) standards specified in Table 1-5.2 that apply to fire protection service in addition to the provisions of this paragraph.

Table 1-5.2 Specially Listed Pipe or Tube Materials and Dimensions

Materials and Dimensions	Standard
Nonmetallic Piping:	
Specification for Chlorinated Polyvinyl Chloride (CPVC) Pipe	ASTM F 442
Specification for Polybutylene (PB) Pipe	ASTM D 3309

Note: In addition to satisfying these minimum standards, specially listed pipe shall be required to comply with the provisions of 1-5.2.

A-1-5.2 Not all pipe or tube made to ASTM D 3309, *Standard Specification for Polybutylene (PB) Plastic Hot- and Cold-Water Distribution Systems,* and ASTM F 442, *Standard Specification* *for Chlorinated Poly (Vinyl Chloride) (CPVC) Plastic Pipe (SDR-PR),* as described in 1-5.1 and 1-5.2 is listed for fire sprinkler service. Listed pipe is identified by the logo of the listing agency.

NFPA 13R does not prohibit the use of any material, as long as it is listed for use in sprinkler systems. Subsection 1-5.2 is intended to encourage development of more effective materials in terms of either their function or cost.

NFPA 13R permits the use of alternative pipe or tube materials. Because alternative materials are available and used extensively, reference to such pipe or tube is included to aid the design professional, the installing contractor, and the authority having jurisdiction. Prior to installation, it must be verified that specially listed piping material is permitted to be used. Additionally, restrictions on its use and any special installation procedures must be identified.

At present the only two synthetic piping materials that are listed for use in sprinkler systems are CPVC and PB. Additionally, specially listed metallic pipe such as thin-wall steel pipe also exists. These materials often include restrictions with regard to their use and application. For example, PB pipe cannot be left exposed. The listing criteria are specific and indicate for which standard (e.g., NFPA 13D or NFPA 13R) or hazard classification (e.g., light hazard in accordance with NFPA 13) the pipe is listed. This listing information as well as the manufacturer's installation instructions need to be consulted for proper installation and application.

As noted at the end of 1-5.2 and stated in A-1-5.2, not all pipe is required to be listed. Listed piping is identified by the listing agency's logo. Similar piping manufactured with less exacting quality control must not be used in sprinkler systems. This qualification essentially means that "garden variety" PB or polyvinyl chloride (PVC) pipe materials found at a building supply store are not eligible for use in sprinkler systems. A 1993 fire investigation conducted by NFPA in a board and care facility uncovered a system installed with PVC plumbing pipe. Although it was not apparent that the use of PVC pipe contributed to the two fatalities in this fire, its use was indicative of the installer's general lack of knowledge about sprinkler system design.

PB pipe is no longer manufactured due to problems associated with its use in domestic plumbing systems, not because of its use in fire protection systems. The development of other flexible pipe is being pursued as an alternative to PB pipe.

1-5.3 Wherever the word pipe is used in this standard, it shall be understood also to mean *tube.*

For the purposes of NFPA 13R, the term *pipe* refers to any conduit for transporting water.

1-5.4 Pipe joined with mechanical grooved fittings shall be joined by a listed combination of fittings, gaskets, and grooves. Where grooves are cut or rolled on the pipe, they shall be dimensionally compatible with the fittings.

Exception: Steel pipe with wall thicknesses less than those of Schedule 30 [sizes 8 in. (203 mm) and larger] or Schedule 40 [sizes less than 8 in. (203 mm)] shall not be joined by fittings used with pipe having cut grooves.

When grooves are cut, material is lost, and the use of thin-wall steel pipe could result in too little material remaining between the inside diameter and the root diameter of the groove. Based on a listing from a testing laboratory, the exception permits an innovative threaded assembly for use with thin-walled steel pipe.

Subsection 2-4.4 requires the piping support to be in accordance with NFPA 13. (See Appendix G-2 of NFPA 13.)

1-5.5 Fittings used in sprinkler systems shall be of the materials listed in Table 1-5.5 or in accordance with 1-5.7. The chemical properties, physical properties, and dimensions of the materials specified in Table 1-5.5 shall be at least equivalent to the standards cited in the table. Fittings used in sprinkler systems shall be designed to withstand a working pressure of not less than 175 psi (12.1 bar). When nonmetallic fittings are used, they shall be rated at not less than 175 psi (12.1 bar) working pressure at not less than 120°F (49°C).

Table 1-5.5 Fitting Materials and Dimensions

Materials and Dimensions	Standard
Cast Iron:	
Gray Iron and Threaded Fittings (Class 125 and 250)	ASME B16.4
Cast Iron Pipe Flanges and Flanged Fittings	ASME B16.1
Malleable Iron:	
Malleable Iron Threaded Fittings	ASME B16.3
Steel:	
Factory-Made Wrought Steel Buttwelding Fittings	ASME B16.9
Buttwelding Ends	ASME B16.25
Specification for Piping Fittings of Wrought Carbon Steel and Alloy Steel for Moderate and Elevated Temperatures	ASTM A 234
Pipe Flanges and Flanged Fittings (Nickel Alloy and Other Special Alloys)	ASME B16.5
Forged Fittings, Socket-Welding and Threaded	ASME B16.11
Copper:	
Wrought Copper and Copper Alloy Solder Joint Pressure Fittings	ASME B16.22
Cast Copper Alloy Solder Joint Pressure Fittings	ASME B16.18

Fittings of the types and materials indicated in Table 1-5.5 must be manufactured to the standards indicated or to other standards that meet or exceed the intent of the standards referenced in NFPA 13R. Any fitting for use in a sprinkler system must be designed for a working pressure of at least 175 psi (12.1 bar). The same principles that apply to the quality control of piping materials also apply to pipe fittings.

1-5.6 Joints for the connection of copper shall be brazed.

Exception: Soldered joints (95-5 solder metal) shall be permitted for wet pipe copper tube systems.

NFPA 13R does not recognize 50-50 solder, which was prohibited for use in plumbing systems by the 1986 amendments to the Federal Safe Drinking Water Act. Although the prohibition does not specifically extend to residential sprinkler systems, the Technical Committee on Residential Sprinkler Systems considered it a good practice to discontinue using this leaded solder in such close proximity to a potable water system.

1-5.7* Other types of fittings shall be permitted to be used, but only where listed for sprinkler systems. Listed fittings including but not limited to CPVC, PB, and steel differing from those provided in Table 1-5.5 shall be installed in accordance with their listings and the manufacturers' installation instructions. CPVC and PB pipe fittings shall comply with the portions of the ASTM standards specified in Table 1-5.7 that apply to fire protection service in addition to the provisions of this paragraph.

Table 1-5.7 Specially Listed Fittings and Dimensions

Materials and Dimensions	Standard
Specification for Schedule 80 CPVC Threaded Fittings	ASTM F 437
Specification for Schedule 40 CPVC Socket-Type Fittings	ASTM F 438
Specification for Schedule 80 CPVC Socket-Type Fittings	ASTM F 439

Note: In addition to satisfying these minimum standards, specially listed pipe fittings are required to comply with the provisions of 1-5.7.

A-1-5.7 Not all fittings made to ASTM F 437, *Standard Specification for Threaded Chlorinated Poly (Vinyl Chloride) (CPVC) Plastic Pipe Fittings, Schedule 80,* ASTM F 438, *Standard Specification for Socket-Type Chlorinated Poly (Vinyl Chloride) (CPVC) Plastic Pipe Fittings, Schedule 40,* and ASTM F 439, *Standard Specification for Socket-Type Chlorinated Poly (Vinyl Chloride) (CPVC) Plastic Pipe Fittings, Schedule 80,* as described in 1-5.5 and 1-5.7 are listed for fire sprinkler service. Listed fittings are identified by the logo of the listing agency.

The types of pipe and fittings that are described in 1-5.7 and listed in Table 1-5.7 are those that have been found satisfactory either due to their service record or based on their evaluation by a listing agency. Sections 3-3 and 3-5 of NFPA 13 contain additional information on pipe and fittings, respectively. Subsection 1-5.7 of NFPA 13R allows for the development of new types of fittings. However, their performance must be evaluated.

The listing procedure for the materials in Tables 1-5.5 and 1-5.7 is performed as an assembly. The manufacturers of these materials have established a procedure for connecting fittings to the pipe to ensure that the system is reliable. Listing procedures are unique to each type of material. Special tools, bonding agents, and fittings are part of the tested assembly. The listing procedure's comprehensive methodology helps to ensure a reliable installation.

1-6 System Types

1-6.1 Wet Pipe Systems.

A wet pipe system shall be used where all piping is installed in areas not subject to freezing.

Wet pipe systems tend to be the most reliable and simplest of all sprinkler systems, because no equipment other than the sprinklers need operate to activate the system. Only those sprinklers operated by the fire's heat discharge water. Wet pipe systems are recommended wherever possible.

1-6.2* Protection of Piping.

A-1-6.2 Listed CPVC sprinkler pipe and fittings should be protected from freezing only with glycerine. The use of diethylene, ethylene, or propylene glycols is specifically prohibited. Laboratory testing shows that glycol-based antifreeze solutions present a chemical environment detrimental to CPVC. Listed polybutylene sprinkler pipe and fittings can be protected with glycerine, diethylene glycol, ethylene glycol, or propylene glycol.

The information in A-1-6.2 was added to ensure that the design professional and installing contractor are aware of the limitations associated with CPVC piping when used for antifreeze systems.

1-6.2.1* Provision shall be made to protect piping from freezing in unheated areas by use of one of the following methods:

(1) Antifreeze system
(2) Dry pipe system
(3) Preaction system
(4) Listed standard dry pendent, dry upright, or dry sidewall sprinklers extended from heated areas

A-1-6.2.1 Piping covered by insulation, as shown in Figures A-1-6.2.1(a) through A-1-6.2.1(e), is considered part of the area below the ceiling and not part of the unheated attic area.

With regard to antifreeze systems, state or local plumbing and health regulations may not allow the introduction of foreign materials into piping systems connected to public water. Where these regulations are in effect, the use of antifreeze and the type of solution permitted should be verified by local authorities. Antifreeze systems might be permitted where piping arrangements make contamination of public water highly unlikely. Where antifreeze systems are installed many health authorities require the use of a reduced-pressure zone (RPZ) backflow prevention device. In many instances, antifreeze systems offer the most attractive solution for a small unheated or cold storage area of a building.

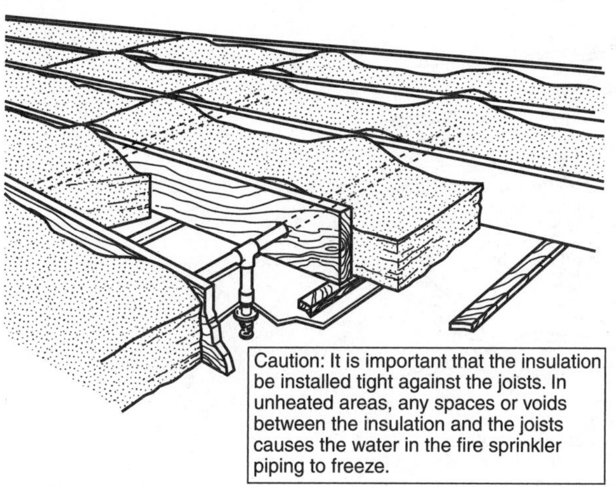

Caution: It is important that the insulation be installed tight against the joists. In unheated areas, any spaces or voids between the insulation and the joists causes the water in the fire sprinkler piping to freeze.

Figure A-1-6.2.1(a) Insulation recommendations—Arrangement 1.

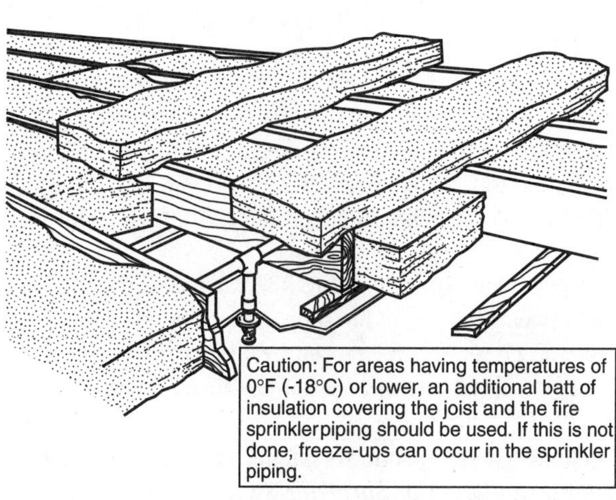

Caution: For areas having temperatures of 0°F (-18°C) or lower, an additional batt of insulation covering the joist and the fire sprinklerpiping should be used. If this is not done, freeze-ups can occur in the sprinkler piping.

Figure A-1-6.2.1(b) Insulation recommendations—Arrangement 2.

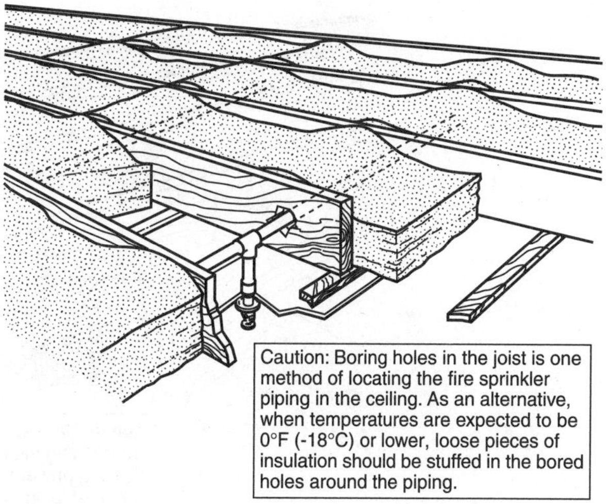

Caution: Boring holes in the joist is one method of locating the fire sprinkler piping in the ceiling. As an alternative, when temperatures are expected to be 0°F (-18°C) or lower, loose pieces of insulation should be stuffed in the bored holes around the piping.

Figure A-1-6.2.1(c) Insulation recommendations—Arrangement 3.

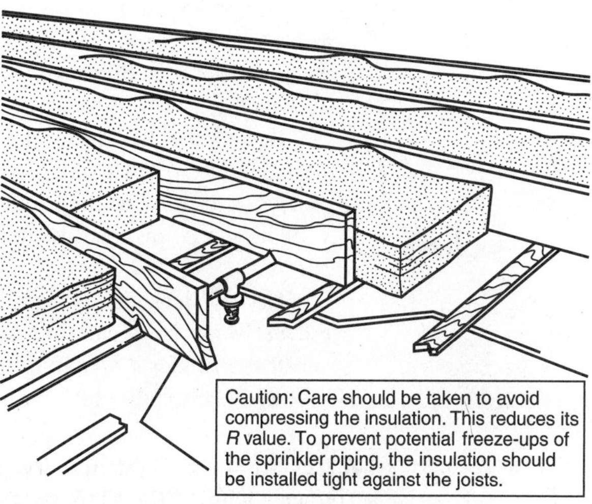

Figure A-1-6.2.1(d) Insulation recommendations—Arrangement 4.

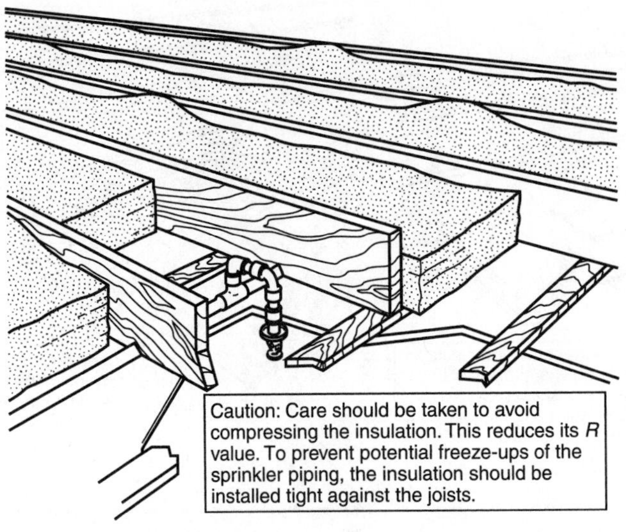
Caution: Care should be taken to avoid compressing the insulation. This reduces its R value. To prevent potential freeze-ups of the sprinkler piping, the insulation should be installed tight against the joists.

Figure A-1-6.2.1(e) *Insulation recommendations— Arrangement 5.*

Dry pipe systems are installed in lieu of wet pipe systems where piping is subject to freezing. Dry pipe systems should not be used for the purpose of reducing water damage from pipe breakage or leakage, because the systems operate too quickly to be of value for this purpose.

Because of the concern for delayed water delivery and the fact that no residential sprinklers are listed for dry system application, dry pipe systems are not allowed for protection inside dwelling units. See Exception No. 1 to 2-4.5.1. However, dry pipe systems can be installed in areas outside dwelling units in accordance with 2-5.2 and NFPA 13.

Dry pendent, dry upright, and dry sidewall sprinklers are specially designed to prevent water from entering the pipe between the sprinkler supply pipe (branch line) and the sprinkler's operating mechanism. These sprinklers can be used on wet pipe systems where individual sprinklers are extended into spaces subject to freezing.

1-6.2.2* Antifreeze systems, dry pipe systems, and preaction systems shall be installed in accordance with NFPA 13, *Standard for the Installation of Sprinkler Systems.*

A-1-6.2.2 Antifreeze solutions can be used for maintaining automatic sprinkler protection in small, unheated areas. Because of the cost of refilling the system or replenishing small leaks, antifreeze solutions are recommended only for systems not exceeding 40 gal (151 L).

Tables 4-5.2.1 and 4-5.2.2 in NFPA 13 list antifreeze concentrates that are permitted, depending on whether or not the sprinkler system is supplied by potable water. The

concentrates for those systems attached to potable water systems are chemically pure, food-grade additives. Even so, a backflow device is still required by most water utilities.

References Cited in Commentary

National Fire Protection Association, 1 Batterymarch Park, P.O. Box 9101, Quincy, MA 02269-9101.

NFPA 13, *Standard for the Installation of Sprinkler Systems*, 1999 edition.
NFPA 13D, *Standard for the Installation of Sprinkler Systems in One- and Two-Family Dwellings and Manufactured Homes*, 1999 edition.
NFPA 72, *National Fire Alarm Code®*, 1999 edition.
NFPA *101®*, *Life Safety Code®*, 1997 edition.

U.S. Government, U.S. Government Printing Office, Washington, DC 20402.

Federal Safe Drinking Water Act, 1974.

Bibliography

1. Nelson, H.E., "An Engineering Analysis of the Early Stages of Fire Development – The Fire at the Dupont Plaza Hotel and Casino Dec. 31, 1986," National Bureau of Standards, Gaithersburg, MD, 1987.
2. Factory Mutual Research Corporation, "Approval Standard – Residential Automatic Sprinklers, Class No. 2030," Norwood, MA, September 1983.
3. Underwriters Laboratories Inc., *Proposed First Edition of the Standard for Residential Sprinklers for Fire Protection Service*, UL 1626, Northbrook, IL, April 1986.
4. Cote, A.E., and Moore, D., "Field Test and Evaluation of Residential Sprinkler Systems, Los Angeles Test Series," National Fire Protection Association, Boston, MA, April 1980.
5. Moore, D., "Data Summary of the North Carolina Test Series of USFA Grant 79027 Field Test and Evaluation of Residential Sprinkler Systems," National Fire Protection Association, Boston, MA, September 1980.
6. Kung, H.C., Spaulding, R.D., Hill, E.E., Jr., and Symonds, A.P., "Technical Report, Field Evaluation of Residential Prototype Sprinkler, Los Angeles Fire Test Program," Factory Mutual Research Corporation, Norwood, MA, February 1982.
7. Cote, A.E., "Final Report on Field Test and Evaluation of Residential Sprinkler Systems," National Fire Protection Association, Quincy, MA, July 1982.

CHAPTER 2

Working Plans, Design, Installation, Acceptance Tests, and Maintenance

2-1 Working Plans and Acceptance Tests

2-1.1 Working Plans.

Working plans are prepared primarily for system installers. The plans also serve to protect the owner's interest, who usually is not knowledgeable in sprinkler system installations and consequently relies on others to check the plans for conformance to NFPA 13R. Various authorities having jurisdiction will likely be used to review the plans to ensure that the interests of society, the owner, and any other concerned parties, such as the insurance company, have been properly addressed.

When the installation is completed, it is checked against the plans to determine compliance with the approved plans and NFPA 13R. The owner should retain working plans and specifications to refer to in the future and to assist in the periodic maintenance program. In the event that future alterations are undertaken, a good deal of expense and time can be saved if accurate working plans and documentation are available since these systems are to be hydraulically calculated.

The symbols commonly used in working plans can be found in NFPA 170, *Standard for Fire Safety Symbols*. NFPA 13, *Standard for the Installation of Sprinkler Systems,* provides further information and examples of symbols and abbreviations commonly used on sprinkler system plans.

2-1.1.1 Working plans shall be submitted for approval to the authority having jurisdiction before any equipment is installed or remodeled. Deviations from approved plans shall require permission of the authority having jurisdiction.

2-1.1.2 Working plans shall be drawn to a specified scale on sheets of uniform size, shall provide a plan of each floor, shall be capable of being easily duplicated, and shall indicate the following:

(1) The name of owner and occupant
(2) The location, including street address
(3) The point of compass
(4) The ceiling construction
(5) The full height cross section
(6) The location of fire walls
(7) The location of partitions
(8) The occupancy of each area or room
(9) The location and size of concealed spaces, attics, closets, and bathrooms
(10) Any small enclosures in which no sprinklers are to be installed
(11) The size of the city main in the street, pressure, whether dead-end or circulating and, if dead-end, the direction and distance to the nearest circulating main, and the city main test results including elevation of the test hydrant
(12) The make, manufacturer, type, heat-response element, temperature rating, and nominal orifice size of the sprinkler
(13) The temperature rating and location of high-temperature sprinklers
(14) The number of sprinklers on each riser, per floor
(15) The kind and location of alarm bells
(16) The type of pipe and fittings
(17) The type of protection for nonmetallic pipe
(18)* The nominal pipe size with lengths shown to scale
(19) The location and size of riser nipples
(20) The types of fittings and joints and the locations of all welds and bends
(21) The types and locations of hangers, sleeves, and braces, and methods of securing sprinklers, where applicable
(22) All control valves, check valves, drain pipes, and test connections
(23) The underground pipe size, length, location, weight, material, and point of connection to the city main; type of valves, meters, and valve pits; and depth at which the top of the pipe is laid below grade
(24) In the case of hydraulically designed systems, the material to be included on the hydraulic data nameplate
(25) The name and address of the contractor

A-2-1.1.2(18) Where typical branch lines prevail, it will be necessary to size only one line.

The information required in 2.1.1.2(1) through 2-1.1.2(25) helps expedite the review process and saves time during the installation. Most of these items are easy to determine and result in a system design and installation that provides the protection level intended by NFPA 13R.

With regard to 2-1.1.2(13), consideration must be given to the sprinklers' location with respect to heat sources. The information in 2-4.5.2.3 and the manufacturer's literature contain guidance on positioning sprinklers near heat sources.

Where typical branch lines prevail, it will be necessary to size only one line with regard to 2-1.1.2(18).

With regard to 2-1.1.2(21), nonmetallic pipe has necessitated the securing of sprinklers to the ceiling. Refer to the manufacturer's listing information.

2-1.2 Approval of Sprinkler Systems.

2-1.2.1 The installer shall perform all required acceptance tests (see 2-1.3), complete the contractor's material and test certificate(s) (see Figure 2-1.2.1), and forward the certificate(s) to the authority having jurisdiction prior to asking for approval of the installation.

A contractor's material and test certificate acknowledges that materials used and tests made are in accordance with the requirements of the approved plans. The certificate also provides a record of the test results, which can be used for comparison with tests conducted as part of a system maintenance program. The certificate also verifies that the system design and installation are completed in accordance with NFPA 13R.

2-1.2.2 Where the authority having jurisdiction requires to be present during the conducting of acceptance tests, the installer shall provide advance notification of the time and date the testing will be performed.

Frequently the authority having jurisdiction wants to be present when acceptance tests are conducted. The installer should be aware of the policies or procedures of the authority having jurisdiction with respect to witnessing acceptance tests. When the authority having jurisdiction desires to witness the tests, the installer should provide sufficient advance notice. Failure to do so can require that the tests be repeated.

2-1.3 Acceptance Tests.

2-1.3.1 Flushing of Underground Connections.

2-1.3.1.1 Underground mains and lead-in connections to system risers shall be flushed before a connection is made to sprinkler piping in order to remove any foreign materials that have entered the underground piping during the course of the installation. For all systems, the flushing operation shall be continued until the water is clear.

Stones, gravel, blocks of wood, bottles, work tools, work clothes, and other objects have been found in piping when flushing was performed. Also, objects in underground piping quite remote from the sprinkler installation, which otherwise would remain stationary, will sometimes be forced into sprinkler system piping when sprinkler systems operate. Sprinkler systems can draw greater flows than normal domestic systems. Fire department pumpers, when taking suction from hydrants for pumping into sprinkler systems, tend to augment this effect by increasing the velocity of water flow through underground piping.

Because the size of sprinkler system piping typically gets smaller starting at the water supply's point of connection, objects that are forced from underground piping into the sprinkler system risers are likely to become lodged at a point in the system where they totally obstruct the water's passage.

2-1.3.1.2 Underground mains and lead-in connections shall be flushed at the hydraulically calculated water demand rate of the system.

Because only the design flow rate is anticipated, the calculated water demand rate is to be used for flushing. Overall field experience with NFPA 13 systems indicates that

Contractor's Material and Test Certificate for Aboveground Piping

PROCEDURE

Upon completion of work, inspection and tests shall be made by the contractor's representative and witnessed by an owner's representative. All defects shall be corrected and system left in service before contractor's personnel finally leave the job.

A certificate shall be filled out and signed by both representatives. Copies shall be prepared for approving authorities, owners, and contractor. It is understood the owner's representative's signature in no way prejudices any claim against contractor for faulty material, poor workmanship, or failure to comply with approving authority's requirements or local ordinances.

Property name	Date
Property address	

Plans	Accepted by approving authorities (names)		
	Address		
	Installation conforms to accepted plans	☐ Yes	☐ No
	Equipment used is approved If no, explain deviations	☐ Yes	☐ No
Instructions	Has person in charge of fire equipment been instructed as to location of control valves and care and maintenance of this new equipment? If no, explain	☐ Yes	☐ No
	Have copies of the following been left on the premises?	☐ Yes	☐ No
	1. System components instructions	☐ Yes	☐ No
	2. Care and maintenance instructions	☐ Yes	☐ No
	3. NFPA 25	☐ Yes	☐ No
Location of system	Supplies buildings		

Sprinklers	Make	Model	Year of manufacture	Orifice size	Quantity	Temperature rating

Pipe and fittings	Type of pipe ______ Type of fittings ______

Alarm valve or flow indicator	Alarm device			Maximum time to operate through test connection	
	Type	Make	Model	Minutes	Seconds

Dry pipe operating test	Dry valve			Q. O. D.		
	Make	Model	Serial no.	Make	Model	Serial no.

	Time to trip through test connection[1]		Water pressure	Air pressure	Trip point air pressure	Time water reached test outlet[1]		Alarm operated properly	
	Minutes	Seconds	psi	psi	psi	Minutes	Seconds	Yes	No
Without Q.O.D.									
With Q.O.D.									

If no, explain

[1] Measured from time inspector's test connection is opened

Figure 2-1.2.1 Contractor's material and test certificate for aboveground piping.

Deluge and preaction valves

Operation	☐ Pneumatic	☐ Electric	☐ Hydraulic
Piping supervised ☐ Yes ☐ No	Detecting media supervised ☐ Yes ☐ No		
Does valve operate from the manual trip, remote, or both control stations? ☐ Yes ☐ No			
Is there an accessible facility in each circuit for testing? ☐ Yes ☐ No	If no, explain		

Make	Model	Does each circuit operate supervision loss alarm?		Does each circuit operate valve release?		Maximum time to operate release	
		Yes	No	Yes	No	Minutes	Seconds

Pressure reducing valve test

Location and floor	Make and model	Setting	Static pressure		Residual pressure (flowing)		Flow rate
			Inlet (psi)	Outlet (psi)	Inlet (psi)	Outlet (psi)	Flow (gpm)

Test description

Hydrostatic: Hydrostatic tests shall be made at not less than 200 psi (13.6 bar) for 2 hours or 50 psi (3.4 bar) above static pressure in excess of 150 psi (10.2 bar) for 2 hours. Differential dry-pipe valve clappers shall be left open during the test to prevent damage. All aboveground piping leakage shall be stopped.

Pneumatic: Establish 40 psi (2.7 bar) air pressure and measure drop, which shall not exceed 1½ psi (0.1 bar) in 24 hours. Test pressure tanks at normal water level and air pressure and measure air pressure drop, which shall not exceed 1½ psi (0.1 bar) in 24 hours.

Tests

All piping hydrostatically tested at _____ psi (____ bar) for ____ hours Dry piping pneumatically tested ☐ Yes ☐ No Equipment operates properly ☐ Yes ☐ No	If no, state reason
Do you certify as the sprinkler contractor that additives and corrosive chemicals, sodium silicate or derivatives of sodium silicate, brine, or other corrosive chemicals were not used for testing systems or stopping leaks? ☐ Yes ☐ No	

Drain test	Reading of gauge located near water supply test connection: _____ psi (____ bar)	Residual pressure with valve in test connection open wide: _____ psi (____ bar)

Underground mains and lead in connections to system risers flushed before connection made to sprinkler piping Verified by copy of the U Form No. 85B ☐ Yes ☐ No Flushed by installer of underground sprinkler piping ☐ Yes ☐ No	Other Explain
If powder-driven fasteners are used in concrete, has representative sample testing be satisfactorily completed? ☐ Yes ☐ No	If no, explain

Blank testing gaskets

Number used	Locations	Number removed

Welding

Welding piping ☐ Yes ☐ No	
If yes. . .	
Do you certify as the sprinkler contractor that welding procedures comply with the requirements of at least AWS B2.1?	☐ Yes ☐ No
Do you certify that the welding was performed by welders qualified in compliance with the requirements of at least AWS B2.1?	☐ Yes ☐ No
Do you certify that the welding was carried out in compliance with a documented quality control procedure to ensure that all discs are retrieved, that openings in piping are smooth, that slag and other welding residue are removed, and that the internal diameters of piping are not penetrated?	☐ Yes ☐ No

Cutouts (discs)

Do you certify that you have a control feature to ensure that all cutouts (discs) are retrieved?	☐ Yes ☐ No

***Figure 2-1.2.1** Continued.*

Hydraulic data nameplate	Nameplate provided ☐ Yes ☐ No	If no, explain
Remarks	Date left in service with all control valves open	
Signatures	Name of sprinkler contractor	
	Tests witnessed by	
	For property owner (signed) Title Date	
	For sprinkler contractor (signed) Title Date	
Additional explanations and notes		

Figure 2-1.2.1 Continued.

flushing is normally accomplished at the maximum flow rate available from the water supply.

2-1.3.1.3 To avoid property damage, provision shall be made for the disposal of water issuing from test outlets.

2-1.3.2* Hydrostatic Pressure Tests. Hydrostatic pressure tests shall be provided in accordance with NFPA 13, *Standard for the Installation of Sprinkler Systems.*

Exception: Testing for leakage at a water pressure of 50 psi (3.4 bar) above the maximum system pressure shall be permitted for systems having fewer than 20 sprinklers and no fire department connection.

All NFPA 13R sprinkler systems are required to be hydrostatically tested to a pressure of at least 200 psi (13.8 bar) in accordance with NFPA 13, unless the system has 20 or fewer sprinklers with no fire department connection. This requirement ensures that pipe joints can withstand this working pressure without coming apart or leaking. Although the purpose of this test is to check the quality of the installation, it can also uncover defective components such as cracked fittings and leaky sprinklers.

The hydrostatic test pressure measurement is taken at the lowest elevation within the system or portion of the system being tested. Testing at the system's high point is not considered necessary because the system usually experiences the highest pressure at the lower elevation.

The measure of success for interior sprinkler piping under a hydrostatic test is no visible leakage. Often a very small bead of water forms on a fitting during the test. Unless the bead continues to grow and drip, it is not considered visible leakage.

The exception allows for lower hydrostatic pressure tests on small NFPA 13R systems. When a fire department connection is not installed and the system has no more than 20 sprinklers, the hydrostatic test pressure can be reduced to 50 psi (3.4 bar) above the maximum system design pressure. Where a fire department connection is not installed, the likelihood of the system experiencing very high pressures is rather remote.

In accordance with 2-4.2, a fire department connection is not required for systems installed in buildings in certain remote areas or in single-story buildings 2000 ft^2 (186 m^2) or less in area.

The special precautions required for hydrostatic testing of systems using rigid thermoplastic piping such as chlorinated polyvinyl chloride (CPVC) or flexible piping such as polybutylene (PB) are described in A-2-1.3.2.

A-2-1.3.2 Testing of a system can be accomplished by pressurizing the system with water and checking visually for leakage at each joint or coupling.

Where pressure testing systems have rigid thermoplastic piping, such as listed CPVC, or flexible piping, such as listed polybutylene, the sprinkler system should be filled with water. The air should be bled from the highest and farthest sprinklers before the test pressure is applied. Compressed air or compressed gas never should be used for pressure testing CPVC piping. Testing with air pressure is permitted for polybutylene piping where conducted in accordance with the testing procedures of 10-2.3 of NFPA 13, *Standard for the Installation of Sprinkler Systems.*

Fire department connections are not required for all systems covered by this standard but can be installed at the discretion of the owner. In these cases, hydrostatic tests in accordance with NFPA 13 are required.

Dry systems also should be tested by placing the system under air pressure. Any leak that results in a drop in system pressure greater than 2 psi (0.14 bar) in 24 hours should be corrected. Leaks should be identified using soapy water brushed on each joint or coupling. The presence of bubbles indicates a leak. This test should be made prior to concealing the piping.

2-2 Design and Installation Devices and Materials

2-2.1 Only new sprinklers shall be employed in the installation of sprinkler systems. At least three spare sprinklers of each type, temperature rating, and orifice size used in the system shall be kept on the premises. Replacement sprinklers shall have the same operating characteristics as the sprinklers being replaced.

The sprinkler must operate properly and effectively when a fire occurs. It should not leak, rupture, or operate for any reason other than in response to a fire. For these performance reasons and because of sprinklers' relatively low cost, new sprinklers are required. Additionally, the performance of used sprinklers is not easily verified, and their reliability can be questioned. The installation of new, listed residential sprinklers provides a higher degree of reliability.

Three spare sprinklers of each type used in the design are required to be kept on site. This provision helps ensure that spare sprinklers are immediately available and that the system can be put back into service with minimal delay. NFPA 13R also clarifies that replacement sprinklers must be the same type—that is, have the same thermal operating characteristics.

When a sprinkler needs to be replaced because it has operated or has been damaged, the orifice size, temperature rating, deflector configuration, and sprinkler type of the replacement sprinkler should be the same as the original sprinkler. Many sprinklers have unique operating characteristics, and some have special area of coverage limitations. The differences between the various residential sprinklers are not obvious, so care must be used when replacing them.

2-2.1.1 Where solvent cement is used as the pipe and fittings bonding agent, sprinklers shall not be installed in the fittings prior to the fittings being cemented in place.

This installation requirement was added to the 1999 edition to ensure that solvent cement does not drip into the piping drop and obstruct or clog the sprinkler.

2-2.2 Only listed or approved devices and materials as specified in this standard shall be used in sprinkler systems.

Any device critical to system operation or performance must be listed. Any device that does not directly affect system performance only needs to be approved. For example, the sprinkler must be listed because it is a critical device that initiates system operation. A drain valve only needs to be approved because it is not critical to system operation during a fire.

2-2.3 Sprinkler systems shall be designed for a maximum working pressure of 175 psi (12.1 bar).

Exception: Higher design pressures shall be permitted to be used where all system components are rated for pressures higher than 175 psi (12.1 bar).

Many sprinklers are currently rated for a working pressure of 175 psi (12.1 bar). The exception removes the 175-psi (12.1-bar) restriction in the event that higher pressure–rated sprinklers need to be used due to high system working pressures. Where the system working pressures are expected to be greater than 175 psi (12.1 bar), high-pressure components need to be used.

2-2.4* Water flow test connections shall be provided at locations that allow flow testing of water supplies, connections, and alarm mechanisms.

A-2-2.4 These connections should be installed so that the valve can be opened fully and for a sufficient time period to ensure a proper test without causing water damage. The test connection should be designed and sized to verify the sufficiency of the water supply and alarm mechanisms.

The test connection is used to verify operation of the waterflow device and gives an indication of the available water supply characteristics. The test connection can indicate the presence of large obstructions in the water supply or a closed control valve. However, the test is not sensitive enough to determine if a valve is partially closed or if a portion of pipe is only partially obstructed.

2-3 Water Supply

2-3.1 General Provisions.

Every automatic sprinkler system shall have at least one automatic water supply. Where stored water is used as the sole source of supply, the minimum quantity shall equal the water demand rate times 30 minutes. *(See 2-5.1.3.)*

The sprinkler system' effectiveness greatly depends on its water supply, which must be automatic and reliable. The adequacy of a public water supply is determined in accordance with 2-5.1.3, 2-5.2, and 2-5.3. Where a stored water supply such as a tank is used, the amount of water should provide at least 30-minute duration at the required flow rate. The 30-minute requirement parallels the NFPA 13 criteria for light hazard occupancies.

2-3.2* Water Supply Sources.

The following water supply sources shall be acceptable.

(1) A connection to a reliable waterworks system with or without a pump, as required. Fire pumps shall be installed in accordance with NFPA 20, *Standard for the Installation of Stationary Pumps for Fire Protection.*
(2) An elevated tank.
(3) A pressure tank installed in accordance with NFPA 13, *Standard for the Installation of Sprinkler Systems*, and NFPA 22, *Standard for Water Tanks for Private Fire Protection.*

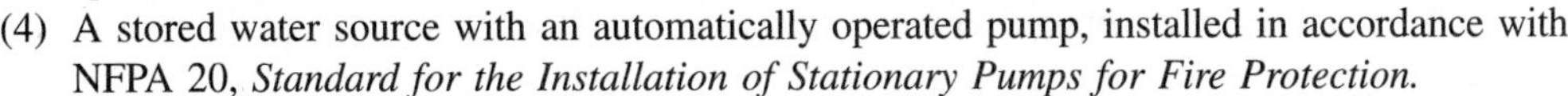

(4) A stored water source with an automatically operated pump, installed in accordance with NFPA 20, *Standard for the Installation of Stationary Pumps for Fire Protection.*

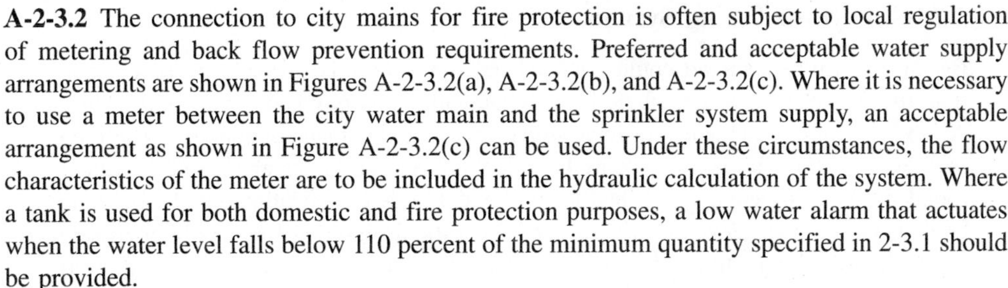

A-2-3.2 The connection to city mains for fire protection is often subject to local regulation of metering and back flow prevention requirements. Preferred and acceptable water supply arrangements are shown in Figures A-2-3.2(a), A-2-3.2(b), and A-2-3.2(c). Where it is necessary to use a meter between the city water main and the sprinkler system supply, an acceptable arrangement as shown in Figure A-2-3.2(c) can be used. Under these circumstances, the flow characteristics of the meter are to be included in the hydraulic calculation of the system. Where a tank is used for both domestic and fire protection purposes, a low water alarm that actuates when the water level falls below 110 percent of the minimum quantity specified in 2-3.1 should be provided.

A

Although editions of NFPA 13R prior to 1996 did not specifically state that fire pumps be installed in accordance with NFPA 20, *Standard for the Installation of Stationary*

Pumps for Fire Protection, this was always the intent. The 1996 edition made this requirement explicit.

Figures A-2-3.2(a), A-2-3.2(b), and A-2-3.2(c) show the preferred and acceptable arrangements of connections to city mains and include control valves, meters, domestic takeoffs, waterflow detectors, pressure gauges, and check valves appropriately located. When the local water company mandates back flow prevention devices, the devices will normally be located in the position where the rubber-faced check valve is shown on the diagrams. The piping arrangement, including valves and fittings, must be taken into account when performing the calculations for system adequacy.

From a fire protection standpoint, meters are not desirable because of their high friction loss characteristics. However, many water authorities require metering of the fire protection water supply connection. The friction loss must be accounted for through the system's hydraulic calculations. Table 4-4.3(g) of NFPA 13D, *Standard for the Installation of Sprinkler Systems in One- and Two-Family Dwellings and Manufactured Homes,* provides some friction loss values for water meters.

An elevated tank can serve as the water supply. With this arrangement the tank must have sufficient elevation to provide the required pressure for the system. Subsection 9-2.4 of NFPA 13 provides more information in this regard. When a pressure tank is used for water supply, the amount of water and the air pressure in the tank are determined using 9-2.3.3 and A-9-2.3.3 of NFPA 13.

A stored water source with an automatically operated pump must have a capacity sufficient to meet the system's demand as determined by 2-5.1.3 and 2-5.2. In addition, the stored water source must have adequate pressure determined in accordance with 2-5.3. A connection to the waterworks system might require a booster pump to provide the pressures required in accordance with 2-5.3. NFPA 13R is more restrictive than NFPA 13D with respect to a stored water source with a pump because NFPA 13R requires the pump to comply with NFPA 20. The Technical Committee on Residential Sprinkler Systems believes that the reliability features associated with NFPA 20 fire pump installations are necessary to achieve the objectives of NFPA 13R. A number of pump manufacturers and pump controller manufacturers have responded to this provision. Fire pumps with capacities as low as 25 gpm (95 L/ min) are now commercially available. Listed limited service fire pump controllers are also available for use with these pumps.

2-3.3 Multipurpose Piping System.

Because domestic demand could occur at the same time as sprinkler system operation, the system design must include such demand if this provision is utilized. Tables A-2-3.3.1(a) and A-2-3.3.1(b) provide guidance on water demand of domestic plumbing facilities and their impact on hydraulic calculations. The tables are based on values found in plumbing codes published by the model code organizations. If the building being protected is of new construction, the domestic design demand should be readily available from the plumbing design documents and should not require additional effort from the sprinkler system installer.

As indicated in Figure A-2-3.2(a), the preferred arrangement is to combine underground supply piping for domestic and sprinkler systems. The Technical Committee

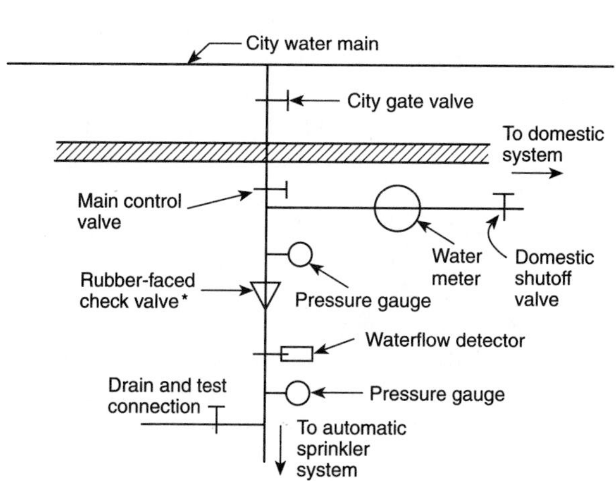

Figure A-2-3.2(a) *Preferable arrangement.*

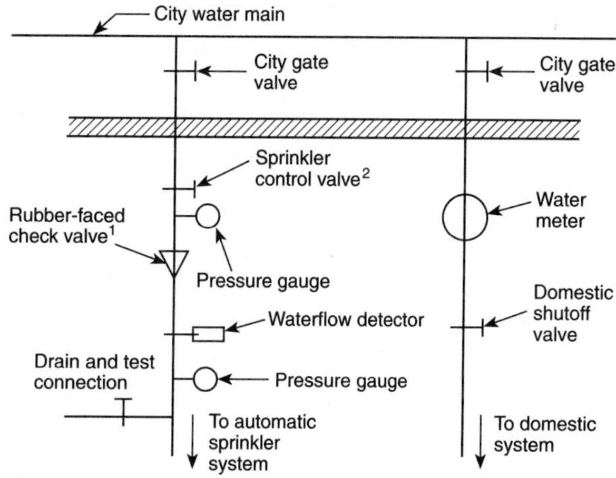

Figure A-2-3.2(b) *Acceptable arrangement with valve supervision (see 2-4.1.1, Exception).*

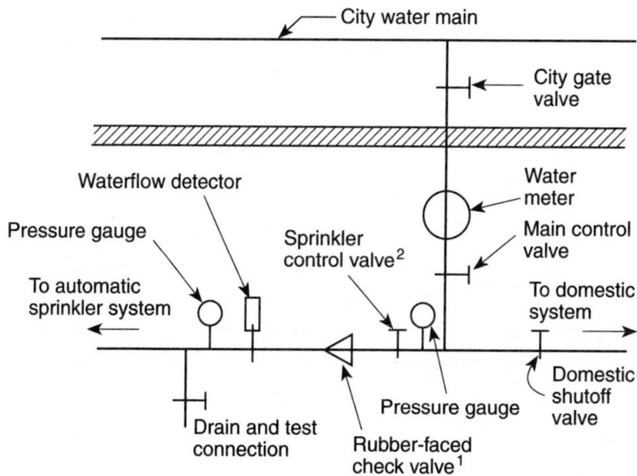

Figure A-2-3.2(c) *Acceptable arrangement with valve supervision (see 2-4.1.1, Exception).*

on Residential Sprinkler Systems feels that such an arrangement increases the sprinkler system's overall reliability. In a residential occupancy, if water is not available for normal usage such as for toilets, the impairment to the water supply system is likely to be investigated and repaired in a timely fashion. However, the technical committee also recognizes that local plumbing or health regulations can place restrictions on such combined systems because trapped piping sections at the end of a sprinkler system are frequent. This occurrence is not normally permitted by such codes. Furthermore, some plumbing regulations permit piping materials and methods of lesser quality than prescribed by NFPA 13R.

2-3.3.1* A common supply main to the building, serving both the sprinklers and domestic use, shall be permitted where the domestic design demand is added to the sprinkler system demand and the total demand flow does not exceed the maximum allowable flow of the piping system components.

Exception: Domestic design demand shall not be required to be added where provision is made to prevent flow on the domestic water system upon operation of sprinklers.

A-2-3.3.1 Tables A-2-3.3.1(a) and A-2-3.3.1(b) can be used to determine a domestic design demand. Using Table A-2-3.3.1(a), the total number of water supply fixture units downstream of any point in the piping serving both sprinkler and domestic needs is determined. Using Table A-2-3.3.1(b), the appropriate total flow allowance is determined and added to the sprinkler demand at the total pressure required for the sprinkler system at that point.

The 1996 edition expanded this requirement to take into account any maximum flow limits of system components. Some components such as meters, valves, backflow devices, and relief valves have associated flow limits. Exceeding these values could result in improper system operation. Therefore, the designer must investigate whether the use of any system components is restricted by its flow limitation.

2-3.3.2 Sprinkler systems with non fire protection connections shall comply with Section 3-6 of NFPA 13, *Standard for the Installation of Sprinkler Systems.*

For many years, making use of the sprinkler piping, which normally stands idle, for a more active purpose has been considered feasible. One such use is a circulating closed loop used for a heat pump system, in which water circulates through heating and air-conditioning equipment using sprinkler pipe as the primary conductors. Subsection 4-6.1 of NFPA 13 provides the basic criteria so that the auxiliary functions do not reduce sprinkler system's effectiveness.

2-4 System Components

2-4.1 Valve and Drains.

2-4.1.1 Where a common supply main is used to supply both domestic and sprinkler systems, a single, listed control valve shall be provided to shut off both the domestic and sprinkler

Table A-2-3.3.1(a) Fixture Load Values

Private Facilities (within individual dwelling units)	**Unit**
Bathroom group with flush tank (including lavatory, water closet, and bathtub with shower)	6
Bathroom group with flush valve	8
Bathtub	2
Dishwasher	1
Kitchen sink	2
Laundry trays	3
Lavatory	1
Shower stall	2
Washing machine	2
Water closet with flush valve	6
Water closet with flush tank	3
Public Facilities	**Unit**
Bathtub	4
Drinking fountain	0
Kitchen sink	4
Lavatory	2
Service sink	3
Shower head	4
Urinal with 1-in. (25.4-mm) flush valve	10
Urinal with ¾ -in. (19-mm) flush valve	5
Urinal with flush tank	3
Washing machine [8 lb (17.6 kg)]	3
Washing machine [16 lb (35.2 kg)]	4
Water closet with flush valve	10
Water closet with flush tank	5

systems, and a separate shutoff valve shall be provided for the domestic system only. *[See Figure A-2-3.2(a).]*

Exception: The sprinkler system piping shall be permitted to have a separate control valve where supervised by one of the following methods:

(a) Central station, proprietary, or remote station alarm service
(b) Local alarm service that causes the sounding of an audible signal at a constantly attended location
(c) Valves that are locked open

A control valve permits the system to be shut off for sprinkler replacement or other maintenance functions. The control valve must be in the open position at all other times. With one control valve serving both the domestic and sprinkler systems, the

Table A-2-3.3.1(b) Total Estimated Domestic Demand

Total Fixture Load Units [from Table A-2-3.3.1(a)]	Total Demand			
	For Systems with Predominately Flush Tanks		For Systems with Predominately Flush Valves	
	gpm	L/min	gpm	L/min
1	3	11.25	—	—
2	5	18.75	—	—
5	10	37.5	15	56
10	15	56	25	94
20	20	75	35	13
35	25	94	45	169
50	30	113	50	187
70	35	131	60	225
100	45	169	70	262
150	55	200	80	300
200	65	244	90	337
250	75	261	100	375
350	100	375	125	469
500	125	469	150	562
750	175	656	175	656
1000	200	750	200	750
1500	275	1031	275	1031
2000	325	1219	325	1219
3500	500	1875	500	1875

possibility of inadvertently shutting off the sprinkler system is greatly reduced, because inoperative plumbing will prompt immediate action to restore the system back into service. A separate valve for the sprinkler system is acceptable only when the valve is adequately supervised against inadvertent shutoff.

2-4.1.2 Each sprinkler system shall have a 1-in. (25.4-mm) or larger drain and test connection with a valve on the system side of the control valve.

These connections are used to drain the system when repairs are necessary and to indicate that the water supply is available at the system riser and on the system side of all check valves, the control valve, and underground piping. The maintenance program should include the proper documentation and test records to permit detection of possible water supply deterioration or valves that may have been closed.

2-4.1.3 Additional ½-in. (13-mm) drains shall be installed for each trapped portion of a dry system that is subject to freezing temperatures.

Because dry pipe systems are normally installed in areas subject to freezing, trapped areas must have drain valves to allow water to be promptly removed. Water can enter the system by either activation of the dry pipe valve tripping or condensation of moisture from the pressurized air in the system. Additional guidance on auxiliary drains for dry pipe systems can be found in Chapters 4 and 5 of NFPA 13. Condensate nipples are not required because the capacity of trapped sections is assumed to generally be small.

Auxiliary drains are not required for wet pipe systems because trapped sections of piping in a wet system rarely need to be drained. However, arrangements should be made to drain all portions of trapped piping if the following situations occur. For additional guidance on auxiliary drains, see 5-14.2.5 of NFPA 13.

(1) The building is used for seasonal occupancy.
(2) The water system is shut off during periods of nonoccupancy (a practice that is discouraged by other NFPA codes and standards).
(3) The authority having jurisdiction has permitted a wet pipe system to be installed in accordance with NFPA 13R.

2-4.2 Fire Department Connection.

At least one 1½-in. (38-mm) or 2½-in. (64-mm) fire department connection shall be provided.

Exception No. 1: Buildings located in remote areas that are inaccessible for fire department support.

Exception No. 2: Single-story buildings not exceeding 2000 ft^2 (186 m^2) in area.

Fire department connections are an important supplement to normal water supplies, because they allow the fire department to bypass a closed control valve in most instances. Even where gravity tanks, pressure tanks, or other stored water sources are the sole source of supply, fire department connections should be included. The connection allows the fire department to pump from engines, tankers, or some other water source into the system, thereby either increasing the supply or adding to the tank's supply prior to its depletion, or both.

Section 3-9 of NFPA 13 provides additional requirements on the arrangement of the fire department connection on wet pipe systems with a single riser, which is the most common system installed in accordance with NFPA 13R. The connection should be made to the riser on the system side of indicating, check, and alarm valves if provided. This placement reduces the likelihood that a valve can be closed and render the fire department connection ineffective. For this reason, the fire department connection should not have a shutoff valve.

Approved check valves are to be provided in the fire department connection as near as practicable to the point where it joins the system. The check valve should be installed to permit flow into the system and restrict waterflow from the system into the fire department connection. An approved automatic drip should be provided in the fire department connection where the piping is subject to freezing. The drip permits draining of water between the hose connection and the check valve. The drain should be located at the lowest point of the fire department connection piping to allow complete drainage.

All hose coupling threads on the fire department connection should match the threads used by the local fire department. The hose connections should be equipped with plugs or caps to prevent tampering with the fire department connection pipe.

The first two editions of NFPA 13R did not require the fire department connection unless the system had 20 or more sprinklers. Two exceptions now allow the connection to be omitted. Exception No. 1 is a practical consideration that allows the device to be omitted if there is no access for fire department vehicle support. Although this situation is not found in urban areas, it is likely to be common in rural areas, such as a ski resort or campground.

The limit of 2000 ft^2 (186 m^2) in Exception No. 2 is considered a reasonable size limitation. The limit represents the maximum size of a small structure where adequate accessibility is provided to enter any part of the building during fireground operations.

2-4.3 Pressure Gauges.

Pressure gauges shall be provided to indicate pressures on the supply and system sides of main check valves and dry pipe valves and to indicate pressure on water supply pressure tanks.

Pressure gauges allow for routine observation and verification of proper water pressure and, in dry systems, proper air pressure. This type of visual inspection is recommended in A-2-7 and required by NFPA 25, *Standard for the Inspection, Testing, and Maintenance of Water-Based Fire Protection Systems.*

2-4.4* Piping Support.

Piping hanging and bracing methods shall comply with NFPA 13, *Standard for the Installation of Sprinkler Systems.*

A-2-4.4 Sprinkler piping should be adequately secured to restrict the movement of piping upon sprinkler operation. The reaction forces caused by the flow of water through the sprinkler could result in displacement of the sprinkler, thereby adversely affecting sprinkler discharge. Listed CPVC pipe and listed PB pipe have specific requirements for piping support to include additional pipe bracing at sprinklers.

Chapter 6 of NFPA 13 provides the minimum requirements for supporting sprinkler system piping. Section 6-4 of NFPA 13 contains provisions for protection against seismic events. The use of nonmetallic pipe can require that sprinklers be secured to the ceiling. The pipe manufacturer's listing information should be consulted in this regard.

The Technical Committee on Residential Sprinkler Systems believes that the pipe sizes and pipe lengths used in NFPA 13R systems and the occurrence of short-term pressure surges and water hammer necessitate the need for the pipe hanging and bracing requirements of NFPA 13. The provisions of NFPA 13D are not considered to be adequate. In addition, flow rates through systems designed in accordance with NFPA 13R are similar to the flow rates anticipated by NFPA 13 for light hazard occupancies.

2-4.5 Sprinklers.

2-4.5.1 Listed residential sprinklers shall be used. Listing shall be based on tests to establish the ability of the sprinklers to control residential fires under standardized fire test conditions. The standardized room fires shall be based on a residential array of furnishings and finishes.

Exception No. 1: Residential sprinklers shall not be used in dry pipe systems unless specifically listed for that purpose.

Exception No. 2: Listed quick-response sprinklers shall be permitted to be installed in dwelling units meeting the definition of a compartment as defined in 2-5.1.2.2, provided no more than four sprinklers are located in the dwelling unit.

Underwriters Laboratories Inc. and Factory Mutual Research Corporation have standards for the testing and evaluation of residential sprinklers.[1, 2] The standards require that the sprinklers undergo room fire tests and have fast-response characteristics. Standard sprinklers of the type used in NFPA 13 sprinkler systems do not meet the criteria for residential sprinklers and should not be used except as specifically permitted by NFPA 13R. (See commentary to A-1-2 for a further discussion of the differences in the test criteria.) Residential sprinklers are permitted to be used in the corridor of a residential structure that serves the individual dwelling units. Exhibits 2.1 through 2.4 show different types of residential sprinklers.

Exception No. 1 is based on the concern for the expected delay in water delivery inherent with dry pipe systems and is included because no residential sprinklers are listed for this application.

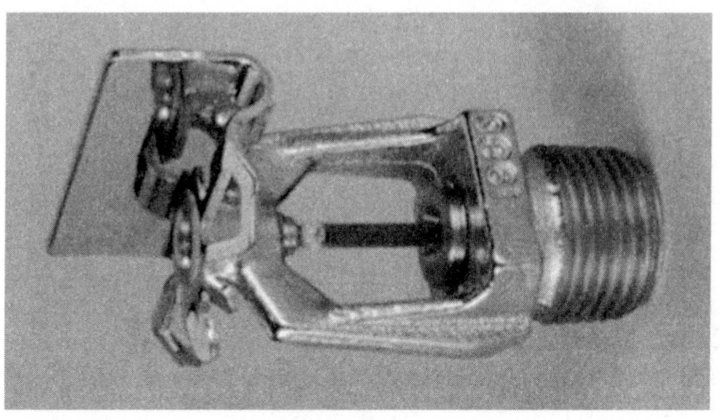

Exhibit 2.1 *Viking Model M-1 horizontal sidewall residential sprinkler.*

Exception No. 2 permits the use of quick-response sprinklers in applications where the desired performance can be achieved without the wall-wetting characteristics associated with residential sprinklers. The exception also permits systems to be installed

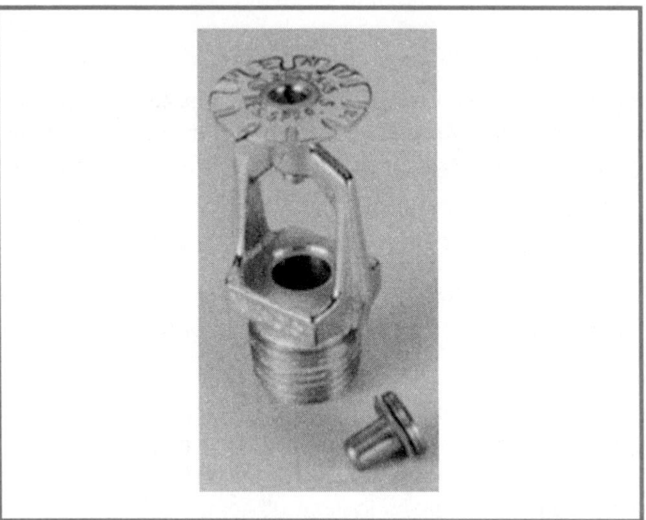

Exhibit 2.2 *Viking Model M-1 pendent-mounted residential sprinkler shown in an inverted position with glass bulb and orifice cover removed.*

Exhibit 2.3 *Viking Model M vertical sidewall quick-response sprinkler.*

in instances where room layout or ceiling configuration can restrict the use of residential sprinklers.[3] For example, a hotel room could be protected with quick-response extended coverage sidewall sprinklers, which are not listed as residential sprinklers. An arrangement using quick-response sprinklers is considered to provide an acceptable

Exhibit 2.4 Viking Model M pendent-mounted quick-response sprinkler.

level of protection for NFPA 13R systems. Exhibit 2.5 illustrates an example of the application of Exception No. 2.

Previous editions of NFPA 13R allowed four standard response sprinklers to be used. However, quick-response sprinklers provide a more effective means of achieving life safety than do standard response sprinklers. This philosophy also holds true in NFPA 13, which now requires that listed quick-response or residential sprinklers be installed in light hazard occupancies. Additionally, when the exception for the standard response sprinklers was first accepted in 1988, listed quick-response sprinklers were not as readily available as they are today.

2-4.5.2 Temperature Ratings. The requirements of 2-4.5.2.1 through 2-4.5.2.3 shall be used for the selection of sprinkler temperature ratings.

2-4.5.2.1 Ordinary temperature-rated residential sprinklers [135°F to 170°F (57°C to 77°C)] shall be installed where maximum ambient ceiling temperatures do not exceed 100°F (38°C).

The maximum ambient temperature in a residential environment should be no more than 35°F (19°C) less than the temperature rating of the installed residential sprinkler. If higher temperatures are expected, then higher temperature-rated sprinklers should be used. The temperature ratings of residential sprinklers currently range from 135°F to 170°F (57°C to 77°C).

2-4.5.2.2 Intermediate temperature–rated residential sprinklers [175°F to 225°F (79°C to 107°C)] shall be installed where maximum ambient ceiling temperatures are between 101°F and 150°F (39°C and 66°C).

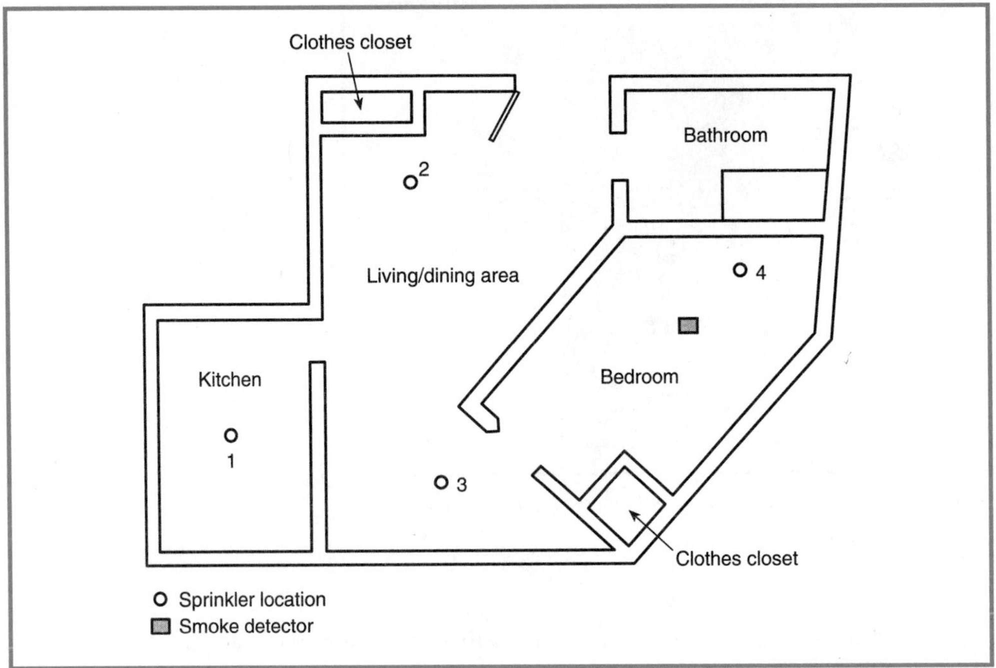

Exhibit 2.5 *Dwelling unit protected with quick-response sprinklers in accordance with Exception No. 2 to 2-4.5.1.*

Areas containing hot water heaters, heating equipment, and other similar heat-producing devices require some special treatment. In addition, sprinklers that are installed in the vicinity of a fireplace or wood-burning stove can also be exposed to relatively high temperatures. The requirements of 2-4.5.2.2 and 2-4.5.2.3, as well as the manufacturer's instructions, need to be followed with regard to sprinkler temperature classification and placement where high ambient temperatures are expected.

2-4.5.2.3 The following practices shall be observed when installing residential sprinklers.

Exception: Where higher expected ambient temperatures are otherwise determined.

(1) Sprinklers under glass or plastic skylights exposed to direct rays of the sun shall be of intermediate-temperature classification.
(2) Sprinklers in an unventilated concealed space under uninsulated roof or in an unventilated attic shall be of intermediate-temperature classification.
(3) Sprinklers installed near specific heat sources that are identified in Table 2-4.5.2.3 shall be of ordinary- or intermediate-temperature rating, as indicated.

Exception: Where sprinklers are listed for positioning closer to a heat source than the minimum distance shown in Table 2-4.5.2.3, the closer minimum distances shall be permitted to be used.

Table 2-4.5.2.3 Minimum Distances for Ordinary- and Intermediate-Temperature Residential Sprinklers

Heat Source	Minimum Distance from Edge of Source to Ordinary-Temperature Sprinkler		Minimum Distance from Edge of Source to Intermediate-Temperature Sprinkler	
	in.	mm	in.	mm
Side of open or recessed fireplace	36	914	12	305
Front of recessed fireplace	60	1524	36	914
Coal- or wood-burning stove	42	1067	12	305
Kitchen range	18	457	9	229
Wall oven	18	457	9	229
Hot air flues	18	457	9	229
Uninsulated heat ducts	18	457	9	229
Uninsulated hot water pipes	12	305	6	152
Side of ceiling- or wall-mounted hot air diffusers	24	607	12	305
Front of wall-mounted hot air diffusers	36	914	18	457
Hot water heater or furnace	6	152	3	76
Light fixture:				
0W–250W	6	152	3	76
250W–499W	12	305	6	152

With regard to 2-4.5.2.3(1), the temperature of the air surrounding a sprinkler can exceed 100°F (38°C) when a skylight permits direct sun exposure to that area. Also, unventilated attics tend to develop rather high temperatures in the summer months. Thus, a similar precaution to prevent false activation in attics should be undertaken. Selecting the appropriate temperature rating for a given environment is always important. The correct temperature rating helps to ensure that sprinklers will not operate prematurely and that an excessive number of sprinklers will not operate during the fire.

The 1996 edition of NFPA 13R was revised to include more comprehensive information regarding the placement of sprinklers near heat sources. Positioning sprinklers closer than that allowed by Table 2-4.5.2.3 is likely to result in false activation of the sprinkler. The exception allows for closer distances when sprinklers are specifically listed for this distance.

2-4.5.3 Operated or damaged sprinklers shall be replaced with sprinklers having the same performance characteristics as the original equipment.

2-4.5.4 Where residential sprinklers are installed within a compartment, as defined in 2-5.1.2.2, all sprinklers shall have the same temperature classification.

Exception: Different temperature classifications shall be permitted where required by 2-4.5.2.3.

A number of factors influence the operating time of a sprinkler, such as the sprinkler's location with respect to the fire, the air flow's temperature and velocity, the sprinkler's operating mechanism, and the sprinkler's position below the ceiling surface. A variation of any of the factors could result in a measurable difference in the operating time of the sprinkler. In order to establish some control over these related items, residential sprinklers with the same temperature classification are required in the compartment. This requirement helps to ensure that sprinklers located closest to the fire will operate first.

This requirement also allows the designer more flexibility when placing sprinklers in odd-shaped compartments. For example, a room can be protected with two residential sprinklers—one with an extended area of coverage and one with a 144-ft^2 (13-m^2) area of coverage. Finding sprinklers with these different areas of coverage and an identical temperature rating may not be possible. As long as devices with the same temperature classification are used, the concern with skipping or reverse order of operation is alleviated.

Sprinklers can be of different temperature classifications as permitted in 2-4.5.2.3.

2-4.5.5 Standard or quick-response spray sprinklers shall be used in areas outside the dwelling unit.

Exception: Residential sprinklers shall be permitted to be used in adjoining corridors or lobbies with flat, smooth ceilings and a height not exceeding 10 ft (3.0 m).

Standard sprinklers are listed, but not in accordance with the criteria for residential sprinklers. See commentary to A-1-2.

With regard to the exception, residential sprinklers are appropriate for use where the fire threat is similar to that expected in a dwelling unit. The fuel loading in a corridor adjoining the dwelling units is substantially lower than that found in the dwelling unit. Corridors and lobbies with ceiling heights of 10 ft (3.0 m) or less can be similar to hallways and living rooms within dwelling units. The criteria for a flat, smooth ceiling are consistent with the listing limitations of residential sprinklers.[3]

2-4.5.6 Operated or damaged sprinklers shall be replaced with sprinklers having the same performance characteristics as the original equipment.

Underwriters Laboratories Inc. has conducted a program in which to test field samples of automatic sprinklers to determine their operating characteristics. Results from this program have shown that, when a sprinkler has been damaged or painted after leaving the factory, it is unlikely to operate properly and should be replaced. Obviously, a sprinkler that has operated must also be replaced.

2-4.5.7 Where nonmetallic ceiling plates (escutcheons) are used, they shall be listed. Escutcheon plates used to create a recessed or flush-type sprinkler shall be part of a listed sprinkler assembly.

Although 2-5.1.7 indicates the minimum distances a sprinkler deflector should be placed below ceilings, these provisions are not always followed. Sprinklers with small frames that are used with incompatible escutcheons can have their performance adversely affected. Unless properly evaluated, the escutcheon can severely obstruct the spray pattern and delay the sprinkler's activation. Therefore, escutcheons need to be listed.

2-4.5.8 Painting and Ornamental Finishes.

2-4.5.8.1 Sprinkler frames shall be permitted to be factory painted or enameled as ornamental finish in accordance with 2-4.5.8.2; otherwise, sprinklers shall not be painted, and any sprinklers that have been painted shall be replaced with new, listed sprinklers.

Exception: Sprinklers painted with factory-applied coatings shall not be required to be replaced.

2-4.5.8.2* Ornamental finishes shall not be applied to sprinklers by an individual other than the sprinkler manufacturer, and only sprinklers listed with such finishes shall be used.

A-2-4.5.8.2 Decorative painting of a residential sprinkler is not to be confused with the temperature identification colors as referenced in 3-2.3 of NFPA 13, *Standard for the Installation of Sprinkler Systems.*

It is extremely important that only sprinkler manufacturers apply paint or ornamental finishes to their sprinklers. Painting done by others could seriously impair the sprinkler's operation or could render the sprinkler inoperable.

2-4.6 Alarms.

Local water flow alarms shall be provided on all sprinkler systems and shall be connected to the building fire alarm system, where provided.

NFPA 13R requires a water flow alarm for all systems to sound on the premises. The alarm's purpose is to indicate water flow through the sprinkler system. The water flow alarm is not required by NFPA 13R to be part of the building's evacuation alarm. Other supplemental alarm systems are not required by NFPA 13 but are likely to be required by other regulations such as NFPA *101®, Life Safety Code®*. Where other regulations require a fire alarm system in the building, then the water flow device usually needs to be connected to the building's fire alarm system.

2-5 System Design

Systems installed in accordance with NFPA 13R must be hydraulically designed. The calculation procedure must be as outlined in Chapter 8 of NFPA 13. The design criteria of NFPA 13R sprinkler systems usually consist of two parts—those areas within the dwelling unit and those areas outside the dwelling unit.

In many instances because of the size of the dwelling units, the systems can be designed with a maximum of four sprinklers operating. However, smaller dwelling units could require less.

2-5.1 Design Criteria—Inside Dwelling Unit.

Section 2-5 provides the design criteria for sprinklers protecting areas within the dwelling unit. Public corridors that serve areas other than dwelling units, lounges, and other areas outside the dwelling unit need to be protected in accordance with 2-5.2.

The information provided in 2-5.1.1, 2-5.1.4.1, 2-5.1.4.2, and 2-5.1.4.3 is based on the "prototype" residential sprinkler that was used in the Los Angeles and Charlotte, NC, development tests.[4,5,6] Listed residential sprinklers are still manufactured to these exact criteria, but other devices exist that cover larger areas, have different spacing criteria, and discharge greater or lesser amounts of water than that specified in 2-5.1. Residential sprinklers of this type are permitted to be used by 2-5.1.6.

2-5.1.1 Design Discharge. The system shall provide a discharge of not less than 18 gpm (68 L/min) to any single operating sprinkler and not less than 13 gpm (49 L/ min) per sprinkler to the number of design sprinklers, but the discharge shall not be less than the listing of the sprinkler.

Exception: Design discharge for sprinklers installed in accordance with Exception No. 2 to 2-4.5.1 shall be in accordance with sprinkler listing criteria.

This paragraph prescribes a minimum discharge of 18 gpm (68 L/min) for a single operating sprinkler and 13 gpm (49 L/min) per sprinkler for multiple sprinklers within a compartment. The definition of a *compartment* is provided in 2-5.1.2.2. The water flow criteria translate to a minimum density of 0.125 gpm/ft^2 [5.1 (L/min)/m^2] for one sprinkler and 0.09 gpm/ft^2 [(3.7(L/min)/m^2)] for multiple sprinklers, assuming the maximum area coverage of 144 ft^2 (13.4 m^2) per sprinkler. The test data indicate that the primary criterion should be flow rate rather than density for a residential sprinkler system.[4,5,6,7]

Because Exception No. 2 to 2-4.5.1 permits sprinklers other than residential sprinklers to be provided within certain dwelling units, such sprinklers must only be used in accordance with their listing. Depending on the type of sprinkler used, the basis of design can be a minimum operating pressure or a minimum flow rate.

2-5.1.2* Number of Design Sprinklers.

A-2-5.1.2 It is intended that the design area is to include up to four adjacent sprinklers that produce the greatest water demand within the compartment. *[See Figures A-2-5.1.2(a) and A-2-5.1.2(b).]*

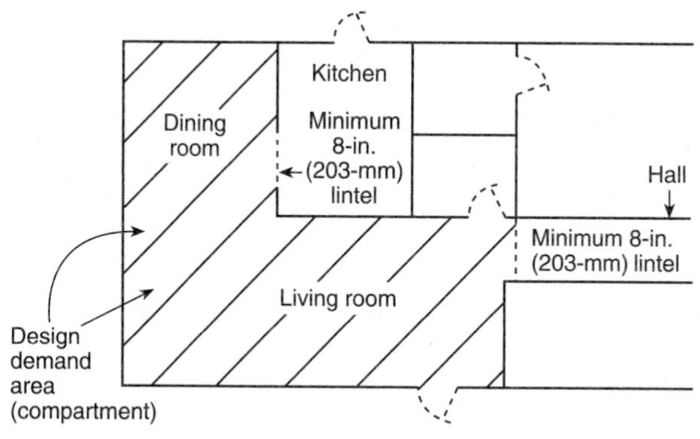

Figure A-2-5.1.2(a) Sprinkler design areas for typical residential occupancy—without lintel.

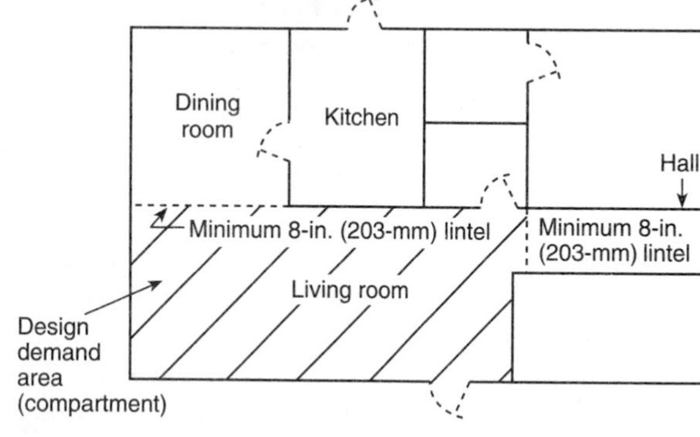

Figure A-2-5.1.2(b) Sprinkler design areas for typical residential occupancy—with lintel.

2-5.1.2.1* The number of design sprinklers shall include all sprinklers within a compartment up to a maximum of four sprinklers under a flat, smooth, horizontal ceiling. For compartments containing two or more sprinklers, calculations shall be provided to verify the single operating sprinkler criteria and the multiple (two, three, or four) operating sprinkler criteria.

A-2-5.1.2.1 Residential sprinklers are currently listed for use under flat, smooth, horizontal ceilings only. Sloped, beamed, and pitched ceilings could require special design features such as larger flows or a design for five or more sprinklers to operate in the compartment.

The NFPA 13D design criteria, which are based on a maximum of two sprinklers operating, are not considered adequate for NFPA 13R systems. The maximum four-sprinkler design criterion, required by NFPA 13R, incorporates a factor of safety that takes into consideration characteristics relative to larger multifamily residential occupancies, such as the wide range of occupant evacuation capabilities and a much larger number of occupants.

When all compartments are protected with fewer than four sprinklers, the maximum number of sprinklers within a single compartment must define the design area. This requirement means that some NFPA 13R systems might only require two or three sprinklers as part of the operating area. In a highly compartmented structure, only a single design sprinkler might be needed. A compartment is defined in 2-5.1.2.2 for purposes of applying this design rule.

When a compartment contains two or more sprinklers, the designer must perform multiple sets of calculations. The single sprinkler design point and the multiple—that is, up to a maximum of four—sprinkler criteria must both be verified. Commentary for 4-1.6 in NFPA 13D gives a number of examples of the listed flow rates for residential sprinklers.

2-5.1.2.2 The definition of *compartment* as used in 2-5.1.2.1 for determining the number of design sprinklers shall be a space that is completely enclosed by walls and a ceiling. The compartment enclosure shall be permitted to have openings to an adjoining space, provided the openings have a minimum lintel depth of 8 in. (203 mm) from the ceiling.

A room compartment is defined by the enclosing walls or a soffit or lintel that forms a definite barrier at the ceiling of a room opening. [See Figures A-2-5.1.2(a) and A-2-5.1.2(b).] In open-plan configurations, compartment boundaries can be determined by soffits or lintels over doorways as well as by walls or partitions. The 8-in. (203-mm) minimum depth ensures that heat will collect in the room of fire origin and activate the sprinklers over the fire rather than allowing sprinklers outside the room to activate.

It is not the intent of NFPA 13R that the number of sprinklers limit the size of a compartment—that is, a compartment can have more than four sprinklers installed in it. The size of the compartment or the number of sprinklers allowed in that compartment is not limited.

2-5.1.3 Water Demand. The water demand for the system shall be determined by multiplying the design discharge specified in 2-5.1.1 by the number of design sprinklers specified in 2-5.1.2.

This calculation produces a sprinkler system demand of 18 gpm (68 L/min) for a single sprinkler and between 26 gpm and 52 gpm (99 L/ min and 198 L/min) for multiple sprinklers. These values are used in performing the calculations to determine pipe sizing and water supply needs. Different sprinkler discharge values might need to be used depending on the type of sprinkler.

2-5.1.4 Sprinkler Coverage.

2-5.1.4.1 Residential sprinklers shall be spaced so that the maximum area protected by a single sprinkler does not exceed 144 ft^2 (13.4 m^2).

The design area for a single residential sprinkler should not exceed 144 ft^2 (13.4 m^2). Larger spacings are permitted when evaluated and listed for such spacing in accordance with 2-5.1.6.

2-5.1.4.2 The maximum distance between sprinklers shall not exceed 12 ft (3.7 m), and the maximum distance to a wall or partition shall not exceed 6 ft (1.8 m).

These limits act to provide adequate coverage to the floor and to ensure that the high wall-wetting characteristics of residential sprinklers are maintained.

2-5.1.4.3 The minimum distance between sprinklers within a compartment shall be 8 ft (2.4 m).

A minimum distance is necessary to prevent cold soldering of adjacent residential sprinklers. If sprinklers are located too close to one another, then the discharge from one sprinkler could cool the area surrounding an adjacent sprinkler and prevent its proper activation. Some residential sprinklers with large areas of coverage might require an even greater minimum separation between devices for the same reason.

2-5.1.5 Operating Pressure. The minimum operating pressure of any sprinkler shall be in accordance with the listing information of the sprinkler and shall provide the minimum flow rates specified in 2-5.1.1.

2-5.1.6 Application rates, design areas, areas of coverage, and minimum design pressures other than those specified in 2-5.1.1, 2-5.1.2, 2-5.1.4, and 2-5.1.5 shall be permitted to be used with special sprinklers that have been listed for such specific residential installation conditions. The minimum operating pressure of any residential sprinkler shall be 7 psi (0.5 bar).

The limitations on spacing were determined by the Los Angeles and Charlotte, NC, fire tests.[4, 5, 6, 7] Although these tests were conducted with prototype fast-response residential sprinklers, the Technical Committee on Residential Sprinkler Systems recognized that other residential sprinklers can be developed that will operate properly with different spacing limitations. In fact, some currently listed residential sprinklers have a coverage area of 20 ft × 20 ft (6.1 m × 6.1 m). Such sprinklers can be used provided that spacing differing from that prescribed in NFPA 13R is justified based on testing conducted by a testing and inspection agency/laboratory.

When coverage areas are extended, the minimum flow rates required by the sprinkler's listing must be used. As areas of coverage increase, the minimum flow rates also usually increase. However, improvements and refinements to sprinkler deflectors and their operating elements have produced listed residential sprinklers that can cover areas greater than 12 ft × 12 ft (3.7 m × 3.7 m) yet discharge quantities of water that are less than the 18-gpm (68-L/min) and 13-gpm (49-L/min) flows established in 2-5.1.1. The minimum operating pressure requirement of 7 psi (0.5 bar) has been included in NFPA 13R. In this regard, NFPA 13R is now in agreement with NFPA 13 and NFPA 13D.

2-5.1.7 Position of Residential Sprinklers.

2-5.1.7.1 Pendent and upright sprinklers shall be positioned so that the deflectors are within 1 in. to 4 in. (25.4 mm to 102 mm) from the ceiling.

Exception: Special residential sprinklers shall be installed in accordance with the listing limitations.

To allow the sprinkler to operate in a timely manner, the deflector's position below the ceiling is limited to ensure the sprinkler is placed where the thermal layer is expected to accumulate. This limitation is based on the results of the Los Angeles and Charlotte, NC, fire tests.[4, 5, 6, 7]

2-5.1.7.2 Sidewall sprinklers shall be positioned so that the deflectors are within 4 in. to 6 in. (102 mm to 152 mm) from the ceiling.

Exception: Special residential sprinklers shall be installed in accordance with the listing limitations.

Dead-air spaces in corners can affect a sprinkler's operation time. The 4-in (102-mm) limitation ensures that the sprinkler will operate properly.

2-5.1.7.3* Sprinklers shall be positioned so that the response time and discharge are not unduly affected by obstructions such as ceiling slope, beams, or light fixtures.

A-2-5.1.7.3 Fire testing has indicated the need to wet walls in the area protected by residential sprinklers at a level closer to the ceiling than that accomplished by standard sprinkler distribution. Where beams, light fixtures, sloped ceilings, and other obstructions occur, additional residential sprinklers could be necessary to achieve proper response and distribution, and a greater water supply could be necessary. Table A-2-5.1.7.3 and Figure A-2-5.1.7.3 provide guidance for the location of sprinklers near ceiling obstructions.

Table A-2-5.1.7.3 Positioning of Sprinklers to Avoid Obstructions to Discharge

Distance from Sprinkler to Side of Ceiling Obstruction		Maximum Distance from Sprinkler Deflector to Bottom of Ceiling Obstruction	
		in.	**mm**
<6 in.	<152 mm	Not permitted	
6 in. to <1 ft	152 mm to <305 mm	0	0
1 ft to <2 ft	0.32 m to <0.64 m	1	25.4
2 ft to <2 ft 6 in.	0.64 m to <0.80 m	2	51
2 ft 6 in. to <3 ft	0.80 m to <0.97 m	3	76
3 ft to <3 ft 6 in.	0.97 m to <1.13 m	4	102
3 ft 6 in. to <4 ft	1.13 m to <1.29 m	6	152
4 ft to <4 ft 6 in.	1.29 m to <1.45 m	7	178
4 ft 6 in. to <5 ft	1.45 m to <1.61 m	9	229
5 ft to <5 ft 6 in.	1.61 m to <1.77 m	11	279
5 ft 6 in. to <6 ft	1.77 m to <1.93 m	14	356

Small areas created by architectural features such as planter box windows, bay windows, and similar features can be evaluated as follows.

(a) Where no additional floor area is created by the architectural feature, no additional sprinkler protection is required.

(b) Where additional floor area is created by an architectural feature, no additional sprinkler protection is required, provided all of the following conditions are met.

(1) The floor area does not exceed 18 ft^2 (1.7 m^2).
(2) The floor area is not greater than 2 ft (0.65 m) in depth at the deepest point of the architectural feature to the plane of the primary wall where measured along the finished floor.
(3) The floor is not greater than 9 ft (2.9 m) in length where measured along the plane of the primary wall.

Measurement from the deepest point of the architectural feature to the sprinkler should not exceed the maximum listed spacing of the sprinkler. The hydraulic design is not required to consider the area created by the architectural feature.

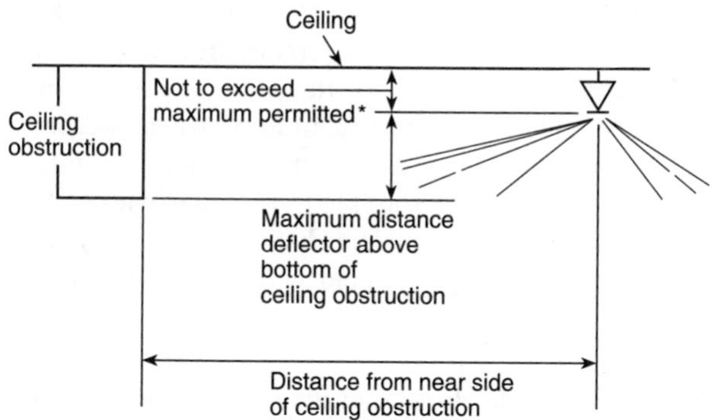

Figure A-2-5.1.7.3 *Position of deflector, upright or pendent, where located above bottom of ceiling obstruction.*

The installation of residential sprinklers under beamed, sloped, or cathedral ceilings is essentially prohibited. This prohibition is because of the lack of information about response times and distribution patterns under such conditions.

Sprinkler locations in violation of Table A-2-5.1.7.3 and Figure A-2-5.1.7.3 can cause obstruction of the discharge and can prevent the sprinkler from controlling a fire. Sprinklers designed to have different discharge characteristics might be permitted to be positioned closer to a beam than is permitted by Table A-2-5.1.7.3. However, test data must be supplied to support such different positioning.

Ceiling fans can create another obstruction problem. Although not specifically evaluated as a part of the listing for a residential sprinkler, a limited number of demonstration tests were conducted in Cobb County, GA, in 1985 to determine the effect of fans on sprinklers. Essentially, this demonstration indicated that no adverse effects occur as a result of the ceiling fan's rotation in either direction. The motor housing is the primary element that can impede the spray pattern and must be considered. The vertical obstruction rules in NFPA 13 can be used to minimize the motor's impact as an obstruction.

The 1996 edition of NFPA 13R included additional information on sprinkler positioning in relation to planter box windows, bay windows, and similar architectural features. Overall, these features are usually smaller in area than a closet and, in some cases, do not add floor space. Therefore, these small architectural features are not anticipated to significantly increase the fire hazard of a residence nor to adversely impact sprinkler performance.

Furthermore, adherence to the spacing rules of Table A-2-5.1.7.3 often cannot be achieved without placing a sprinkler within the bay window or planter box window. An alternative evaluation procedure has been added because sprinkler locations within these spaces raises concerns about system performance.

Small bay windows protected with sprinklers usually require piping run along an outside wall to feed the sprinkler in the bay window. Because the bay window is added onto the side of the house, the space is subject to greater temperature fluctuations that insulation cannot always protect against. This temperature fluctuation increases the potential of freezing during cold temperatures. Additionally, in areas where warmer climates prevail, solar heating of the window glass creates the potential for false actuation of the sprinkler. An increased number of sprinklers operating within the compartment is also a concern. Activating more sprinklers than anticipated by the design in a compartment can overtax the water supply and reduce the effectiveness of the sprinkler system.

2-5.1.8 Sprinklers in Closets and Storage Areas. In closets and storage areas that are required to be protected in accordance with Section 2-6 and are less than 5 ft (1.5 m) in height at the lowest ceiling, a single sprinkler located at the highest ceiling shall be permitted to protect a volume not larger than 300 ft^3 (8.93 m^3).

Closets less than 5 ft (1.5 m) in height at the lowest ceiling and less than 300 ft^3 (8.93 m^3) in volume only need a single sprinkler located at the highest ceiling to achieve fire control. These closet types are typically found under stairs and roofs. In some cases, multiple sprinklers are installed within a single closet or storage compartment in order to ensure proper floor coverage. Experience suggests that fire development within a small confined space, such as a closet, is not comparable with that of a larger space. Fire control in these smaller compartments will result primarily from the buildup of steam and the associated cooling of the space. Also, the presence of a sprinkler on lower ceiling surfaces within the closet can create the potential for unsatisfactory sprinkler activation and operation. This requirement identifies the closets that can be satisfactorily protected by a single sprinkler.

2-5.2 Design Criteria—Outside Dwelling Unit.

The design discharge, number of design sprinklers, water demand of the system, sprinkler coverage, and position of sprinklers for areas to be sprinklered outside the dwelling unit shall comply with specifications in NFPA 13, *Standard for the Installation of Sprinkler Systems.*

Exception No. 1: Where compartmented into areas of 500 ft^2 (46 m^2) or less by 30-minute fire-rated construction and the area is protected by standard or quick-response sprinklers not exceeding 130 ft^2 (12 m^2) per sprinkler, the system demand shall be permitted to be limited to the number of sprinklers in the compartment area but shall not be less than the demand for a total of four sprinklers. Openings from the compartments shall not be required to be protected, provided such openings have a lintel at least 8 in. (203 mm) in depth and the total area of such openings does not exceed 50 ft^2 (4.6 m^2) for each compartment. Discharge density shall be appropriate for the hazard classification as determined by NFPA 13.

Exception No. 2: Lobbies, other than in hotels and motels, foyers, corridors, and halls outside the dwelling unit, with flat, smooth ceilings not exceeding 10 ft (3.0 m) in height, shall be permitted to be protected with residential sprinklers, with a maximum system demand of four sprinklers.

Exception No. 3: Garage doors shall not be considered obstructions and shall be permitted to be ignored for placement and calculation of sprinklers.

Subsection 2-5.2 contains the design criteria for sprinklers protecting areas outside the dwelling unit, such as lobbies, corridors, halls, and foyers that are not contained within dwelling units; basements and storage areas; inside stairwells; and equipment, furnace, trash, laundry, and linen rooms. This subsection also refers the user of NFPA 13R to NFPA 13. However, it should be noted that 2-5.3 requires the system to be hydraulically calculated.

In many cases, a design criterion based on four sprinklers will be possible to maintain in accordance with the three exceptions.

Exception No. 1 provides an alternative to the design area criteria of NFPA 13. This alternative is an adaptation of the room design method of 7-2.3.3 of NFPA 13. The 130-ft^2 (12-m^2) area of coverage is based in part on the assumption that areas that will be protected in accordance with the exception will be ordinary hazard areas.

Exception No. 2 underscores the fact that residential sprinklers can only be used under certain conditions. These conditions include certain construction features such as maximum ceiling heights of 10 ft (3.0 m) and smooth ceiling configurations and anticipated fire hazards such as those typically presented in residential occupancies. Hotel and motel lobbies do not fall within the scope of this exception because the fire hazard in these spaces is no longer residential in nature. Such lobbies often have adjacent or connecting spaces used for gift shops, restaurants, and small displays.

Exception No. 3 addresses the practicality and cost of installing additional sprinklers in garages to minimize obstructions caused by overhead garage doors. While in the up position, the garage door will likely cause an obstruction to a sprinkler located above the door. However, garage doors are usually left in the closed position for the types of residential occupancies addressed by NFPA 13R. The likelihood of a deadly fire occurring in the garage while the door is in the open position falls within the level of protection concept employed by NFPA 13R.

2-5.3 Pipe Sizing.

Piping shall be sized using hydraulic calculation procedures in accordance with NFPA 13, *Standard for the Installation of Sprinkler Systems.*

Hydraulic calculations must be prepared in accordance with Chapter 8 of NFPA 13, and all appropriate provisions of NFPA 13 must apply. The different criteria for inside and outside dwelling units (see 2-5.1 and 2-5.2) as well as the dual discharge criteria of 2-5.1.1 require at least two and often three separate sets of calculations.

2-6 Location of Sprinklers

Sprinklers shall be installed in all areas.

Exception No. 1: Sprinklers shall not be required in bathrooms where the area does not exceed 55 ft² (5.1 m²) and the walls and ceilings, including walls and ceilings behind fixtures, are of noncombustible or limited-combustible materials providing a 15-minute thermal barrier. The area occupied by a noncombustible, full height, shower/bathtub enclosure shall not be required to be added to the floor area when determining the area of the bathroom.

Exception No. 2: Sprinklers shall not be required in clothes closets, linen closets, and pantries within the dwelling units where the area of the space does not exceed 24 ft² (2.2 m²), the least dimension does not exceed 3 ft (0.91 m), and the walls and ceilings are surfaced with noncombustible or limited-combustible materials as defined by NFPA 220, Standard on Types of Building Construction.

Exception No. 3: Sprinklers shall not be required in any porches, balconies, corridors, and stairs that are open and attached.

Exception No. 4: Sprinklers shall not be required in attics, penthouse equipment rooms, crawl spaces, floor/ceiling spaces, elevator shafts, and other concealed spaces that are not used or intended for living purposes or storage.

Exception No. 5: Sprinklers shall not be required in closets on exterior balconies regardless of size as long as there are no doors or unprotected penetrations from the closet directly into the dwelling unit.

Section 2-6 is a departure from NFPA 13. The location of sprinklers is the primary area where the life safety goal is paramount to the property protection goal when compared to NFPA 13. The areas listed in Exception Nos. 1 through 5 reflect spaces where fatalities are less likely to occur when a fire originates in one of them.

The basis for sprinkler omission in Exception Nos. 1 through 4 has been previously described under Section 1-2 and is based on NFPA statistics provided in Table A-1-2. The spaces in the exceptions are the ones shown to result in a low incidence of life loss from fires in dwellings.

With respect to Exception No. 1, the combustible load in bathrooms is normally extremely low, especially when the ceiling and wall materials are not likely to contribute to the fire's growth. This exception reflects the concern that a fire in an unsprinklered bathroom could enter unsprinklered concealed spaces. The additional requirement for floors, walls, and ceilings to provide a 15-minute thermal barrier was added to meet this concern. The 1996 edition revised Exception No. 1 to allow the area occupied by a noncombustible full height shower/bathtub enclosure to be excluded from the floor area of the bathroom. Including this area would require many small bathrooms to be sprinklered even though these fixtures do not significantly add to the fuel load of the bathroom.

With regard to Exception Nos. 3 and 4, mandatory sprinklering of these areas would necessitate the use of dry pipe systems in areas where freezing weather is encountered.

The word *open* in Exception No. 3 is intended to apply to areas that are open to the outside atmosphere. An example of an open corridor is one that empties to an outside stairway and has no doors attached to it. Another example is an exterior wraparound corridor that is typical in a two- or three-story motel.

One additional exception that users of NFPA 13R expect to see here but does not appear is for garages. The Technical Committee on Residential Sprinkler Systems is of the opinion that a large number of building occupants can reside in the residential occupancies addressed by NFPA 13R and that a fire originating in an unsprinklered garage could quickly get out of control prior to evacuation of the building's occupants.

With regard to Exception No. 5, closets on exterior balconies used either for storage or as a mechanical space do not require sprinklers provided that they meet the conditions of Exception Nos. 2 or 5. Where mechanical equipment in the closet penetrates into the living area, fire dampers or other fire penetration protection is required if the closet space is not sprinklered.

2-7* Maintenance

The owner shall be responsible for the condition of a sprinkler system and shall keep the system in normal operating condition. Sprinkler systems shall be inspected, tested, and maintained in accordance with NFPA 25, *Standard for the Inspection, Testing, and Maintenance of Water-Based Fire Protection Systems.*

A-2-7 The responsibility for properly maintaining a sprinkler system is that of the owner or manager, who should understand the sprinkler system operation. A minimum monthly maintenance program should include the following.

(1) Visual inspection of all sprinklers to ensure against obstruction of spray.
(2) Inspection of all valves to ensure that they are open.
(3) Testing of all waterflow devices.
(4) Testing of the alarm system, where installed.

NOTE: Where it is likely that the test will result in a fire department response, notification to the fire department should be made prior to the test.

(5) Operation of pumps, where employed. *(See NFPA 20, Standard for the Installation of Stationary Pumps for Fire Protection.)*
(6) Checking of the pressure of air used with dry systems.
(7) Checking of the water level in tanks.
(8) Special attention to ensure that sprinklers are not painted either at the time of installation or during subsequent redecoration. When sprinkler piping or areas next to sprinklers are being painted, the sprinklers should be protected by covering them with a bag, which should be removed immediately after painting is finished.

(For further information, see NFPA 25, Standard for the Inspection, Testing, and Maintenance of Water-Based Fire Protection Systems.)

Due to the passive nature of systems, their proper maintenance is important for effective sprinkler system performance in a fire. NFPA 13R mandates, therefore, that the owner be responsible for a proper preventive maintenance program involving inspection, testing, and maintenance. A system not maintained in normal operating condition is not in compliance with NFPA 13R.

NFPA 25 contains numerous requirements on the inspection, testing, and maintenance of automatic sprinklers and sprinkler system components. The monthly maintenance program outlined in A-2-7 represents what the Technical Committee on Residential Sprinkler Systems considers to be a minimum acceptable program.

References Cited in Commentary

National Fire Protection Association, 1 Batterymarch Park, P.O. Box 9101, Quincy, MA 02269-9101.

NFPA 13, *Standard for the Installation of Sprinkler Systems,* 1999 edition.

NFPA 13D, *Standard for the Installation of Sprinkler Systems in One- and Two- Family Dwellings and Manufactured Homes,* 1999 edition.

NFPA 20, *Standard for the Installation of Stationary Pumps for Fire Protection,* 1999 edition.

NFPA 25, *Standard for the Inspection, Testing, and Maintenance of Water-Based Fire Protection Systems,* 1998 edition.

NFPA *101*®, Life Safety Code®, 1997 edition.

NFPA 170, *Standard for Fire Safety Symbols,* 1999 edition.

Bibliography

1. Factory Mutual Research Corporation, "Approval Standard – Residential Automatic Sprinklers, Class No. 2030," Norwood, MA, September 1983.
2. Underwriters Laboratories Inc., *Proposed First Edition of the Standard for Residential Sprinklers for Fire Protection Service*, UL 1626, Northbrook, IL, April 1986.
3. Bill, R.G., Jr., Kung, H.C., Brown, W.R., and Hill, E.E., Jr., "Effects of Cathedral and Beamed Ceiling Construction on Residential Sprinkler Performance," Factory Mutual Research Corporation, Norwood, MA, February 1988.
4. Cote, A.E., and Moore, D., "Field Test and Evaluation of Residential Sprinkler Systems, Los Angeles Test Series," National Fire Protection Association, Boston, MA, April 1980.
5. Moore, D., "Data Summary of the North Carolina Test Series of USFA Grant 79027 Field Test and Evaluation of Residential Sprinkler Systems," National Fire Protection Association, Boston, MA, September 1980.
6. Kung, H.C., Spaulding, R.D., Hill, E.E., Jr., and Symonds, A.P., "Technical Report, Field Evaluation of Residential Prototype Sprinkler, Los Angeles Fire Test Program," Factory Mutual Research Corporation, Norwood, MA, February 1982.
7. Cote, A.E., "Final Report on Field Test and Evaluation of Residential Sprinkler Systems," National Fire Protection Association, Quincy, MA, July 1982.

CHAPTER 3

Referenced Publications

3-1

The following documents or portions thereof are referenced within this standard as mandatory requirements and shall be considered part of the requirements of this standard. The edition indicated for each referenced mandatory document is the current edition as of the date of the NFPA issuance of this standard. Some of these mandatory documents might also be referenced in this standard for specific informational purposes and, therefore, are also listed in Appendix B.

3-1.1 NFPA Publications.

National Fire Protection Association, 1 Batterymarch Park, P.O. Box 9101, Quincy, MA 02269-9101.

NFPA 13, *Standard for the Installation of Sprinkler Systems,* 1999 edition.

NFPA 20, *Standard for the Installation of Stationary Pumps for Fire Protection,* 1999 edition.

NFPA 22, *Standard for Water Tanks for Private Fire Protection,* 1998 edition.

NFPA 25, *Standard for the Inspection, Testing, and Maintenance of Water-Based Fire Protection Systems,* 1998 edition.

NFPA *101®*, *Life Safety Code®*, 1997 edition.

NFPA 220, *Standard on Types of Building Construction,* 1999 edition.

NFPA 251, *Standard Methods of Tests of Fire Endurance of Building Construction and Materials,* 1999 edition.

3-1.2 Other Publications.

3-1.2.1 ANSI Publication. American National Standards Institute, Inc., 11 West 42nd Street, 13th floor, New York, NY 10036.

ANSI B36.10M, *Welded and Seamless Wrought Steel Pipe,* 1996.

3-1.2.2 ASME Publications. American Society of Mechanical Engineers, 345 East 47th Street, New York, NY 10017.

ASME B16.1, *Cast Iron Pipe Flanges and Flanged Fittings,* 1989.

ASME B16.3, *Malleable Iron Threaded Fittings,* 1992.

ASME B16.4, *Gray Iron Threaded Fittings,* 1992.

ASME B16.5, *Pipe Flanges and Flanged Fittings,* 1996.

ASME B16.9, *Factory-Made Wrought Steel Buttwelding Fittings,* 1993.

ASME B16.11, *Forged Fittings, Socket-Welding and Threaded,* 1996.

ASME B16.18, *Cast Copper Alloy Solder Joint Pressure Fittings,* 1984.

ASME B16.22, *Wrought Copper and Copper Alloy Solder Joint Pressure Fittings,* 1995.

ASME B16.25, *Buttwelding Ends,* 1997.

3-1.2.3 ASTM Publications. American Society for Testing and Materials, 100 Barr Harbor Drive, West Conshohocken, PA 19428-2959.

ASTM A 53, *Standard Specification for Pipe, Steel, Black and Hot-Dipped, Zinc-Coated, Welded and Seamless,* 1998.

ASTM A 135, *Standard Specification for Electric-Resistance-Welded Steel Pipe,* 1997.

ASTM A 234, *Standard Specification for Piping Fittings of Wrought Carbon Steel and Alloy Steel for Moderate and Elevated Temperatures,* 1997.

ASTM A 795, *Standard Specification for Black and Hot-Dipped Zinc-Coated (Galvanized) Welded and Seamless Steel Pipe for Fire Protection Use,* 1997.

ASTM B 32, *Standard Specification for Solder Metal,* 1996.

ASTM B 88, *Standard Specification for Seamless Copper Water Tube,* 1996.

ASTM B 251, *Standard Specification for General Requirements for Wrought Seamless Copper and Copper-Alloy Tube,* 1997.

ASTM B 813, *Standard Specification for Liquid and Paste Fluxes for Soldering Applications of Copper and Copper-Alloy Tube,* 1993.

ASTM D 3309, *Standard Specification for Polybutylene (PB) Plastic Hot- and Cold-Water Distribution Systems,* 1996.

ASTM F 437, *Standard Specification for Threaded Chlorinated Poly (Vinyl Chloride) (CPVC) Plastic Pipe Fittings, Schedule 80,* 1996.

ASTM F 438, *Standard Specification for Socket-Type Chlorinated Poly (Vinyl Chloride) (CPVC) Plastic Pipe Fittings, Schedule 40,* 1997.

ASTM F 439, *Standard Specification for Socket-Type Chlorinated Poly (Vinyl Chloride) (CPVC) Plastic Pipe Fittings, Schedule 80,* 1997.

ASTM F 442, *Standard Specification for Chlorinated Poly (Vinyl Chloride) (CPVC) Plastic Pipe (SDR-PR),* 1997.

3-1.2.4 AWS Publication. American Welding Society, 550 N.W. LeJeune Road, Miami, FL 33126.

AWS A5.8, *Specification for Filler Metals for Brazing and Braze Welding,* 1992.

APPENDIX A

Explanatory Material

Appendix A is not a part of the requirements of this NFPA document but is included for informational purposes only. This appendix contains explanatory material, numbered to correspond with the applicable text paragraphs.

The material contained in Appendix A of NFPA 13R is included within the text of this handbook and, therefore, is not repeated here.

APPENDIX B

Referenced Publications

B-1

The following documents or portions thereof are referenced within this standard for informational purposes only and are thus not considered part of the requirements of this standard unless also listed in Chapter 3. The edition indicated here for each reference is the current edition as of the date of the NFPA issuance of this standard.

B-1.1 NFPA Publications.

National Fire Protection Association, 1 Batterymarch Park, P.O. Box 9101, Quincy, MA 02269-9101.

NFPA 13, *Standard for the Installation of Sprinkler Systems,* 1999 edition.

NFPA 20, *Standard for the Installation of Stationary Pumps for Fire Protection,* 1999 edition.

NFPA 25, *Standard for the Inspection, Testing, and Maintenance of Water-Based Fire Protection Systems,* 1998 edition.

B-1.2 Other Publications.

B-1.2.1 ASTM Publications. American Society for Testing and Materials, 100 Barr Harbor Drive, West Conshohocken, PA 19428-2959.

ASTM D 3309, *Standard Specification for Polybutylene (PB) Plastic Hot- and Cold-Water Distribution Systems,* 1995.

ASTM F 437, *Standard Specification for Threaded Chlorinated Poly (Vinyl Chloride) (CPVC) Plastic Pipe Fittings, Schedule 80,* 1996.

ASTM F 438, *Standard Specification for Socket-Type Chlorinated Poly (Vinyl Chloride) (CPVC) Plastic Pipe Fittings, Schedule 40,* 1997.

ASTM F 439, *Standard Specification for Socket-Type Chlorinated Poly (Vinyl Chloride) (CPVC) Plastic Pipe Fittings, Schedule 80,* 1997.

ASTM F 442, *Standard Specification for Chlorinated Poly (Vinyl Chloride) (CPVC) Plastic Pipe (SDR-PR),* 1997.

ASTM SI 10, *Standard for Use of the International System of Units (SI): the Modern Metric System,* 1997.

PART FOUR

A Brief History of Sprinklers, Sprinkler Systems, and the NFPA Sprinkler Standards

The following supplement is included in the handbook to provide additional information on sprinklers, sprinkler systems, and the NFPA sprinkler standards for the users of NFPA 13, 13D, and 13R. The supplement is not part of the standards.

Although the subject matter is the same as Supplement 1 in the 1996 edition of the *Automatic Sprinkler Systems Handbook*, the supplement in this edition has been significantly revised and expanded to include more perspective on technical advances.

Supplement author Rolf Jensen, P.E., has been a voting member of the NFPA Technical Committee on Automatic Sprinkler Systems for 39 years and currently continues as an emeritus member. He has chaired the residential and new technology task groups and has served on the hydraulic calculation and 1991 rewrite subcommittees. During his career in the sprinkler industry, he worked in testing of sprinklers and sprinkler system equipment and in fire research on sprinkler systems at UL. He has designed a sprinkler, alarm valve, and dry valve for a sprinkler manufacturer, designed sprinkler systems for numerous properties, and conducted research studies on residential sprinkler systems and for other types of sprinkler system applications. He has taught design of sprinkler systems to engineering students. He is currently retired and actively writing histories of the fire protection industry.

SUPPLEMENT

A Brief History of Sprinklers, Sprinkler Systems, and the NFPA Sprinkler Standards

Rolf Jensen, P.E.

Perhaps no system has been more important in the effort to protect lives and property from fire than the automatic sprinkler system. For more than 100 years, National Fire Protection Association (NFPA) sprinkler standards have guided sprinkler system technology, design, and installation. This supplement traces the history of the development of the early sprinklers and sprinkler systems, the origin of the first sprinkler standard, and the influences of changes in technology on sprinkler systems and NFPA sprinkler standards.

EARLY HISTORY: 1870–1900

In the beginning, four interdependent paths led to the development of the early sprinkler systems, and eventually initiated the creation of the first edition of NFPA 13 in 1896. NFPA 13's original title was *Rules and Regulations of the National Board of Fire Underwriters for Sprinkler Equipments, Automatic and Open Systems.*

The first path was the invention of the automatic sprinkler by Henry S. Parmelee in 1874. The first practical, usable sprinkler was manufactured in 1875.

The second path was the development of the original configurations and pipe size schedules. These were based on pipe arrangements by Parmelee and the Providence Steam and Gas Pipe Company (PSGP), which later became the Grinnell Corporation. The pipe schedules were refined later in tests on automatic sprinklers by C. J. H.

Woodbury and in experiments on the hydraulics of nozzles and fire streams by John R. Freeman. Both Woodbury and Freeman were with the Factory Mutuals.[a] [1, 2, 3]

The third path was the development of alarm valves and dry valves needed to control and detect waterflow and sound an alarm.

The fourth path was the recognition of the need for such systems initially by users, which were mostly textile manufacturers. Insurance companies subsequently recognized the benefit in the reduction of losses and catalyze the use of this important fire safety system by offering more insurance coverage or insurance rate reductions.

The Sprinkler

The automatic sprinkler of today is a far different device than the first sprinkler developed in 1874. (See Exhibit S1.1.) Several versions of the early Parmelee sprinklers were developed. The first version, Parmelee A, may not actually have been used, because it depended on a jute string to function. The second version, Parmelee 1, was actually installed in Parmelee's Mathushek Piano Factory in Providence, RI, possibly to achieve lower insurance rates but was replaced a year later with the Parmelee 3.

The Parmelee 1 sprinkler consisted of a perforated distributor and a valve, which was held in place by a spindle that rested against a lever. One end of the lever was pivoted and the other was attached to the casting with a heavy spring and a fusible link. Often referred to as a *salt shaker* sprinkler, the Parmelee 3 had a brass cap soldered in place over a distributor. The Parmelee 3's only similarity to modern sprinklers was that it was operated by a fusible element.[1, 3, 4]

Conceived in 1873 by Charles E. Buell although not installed until 1881, the Buell was the first sprinkler with an external deflector and the first to use a cap and seal to close the waterway. (See Exhibit S1.2.)

Regardless of the importance of the inventions of Parmelee, Buell, and others, it was Frederick Grinnell's sprinkler, patented in 1881, that was the first sensitive sprinkler. Grinnell followed with several improvements leading to the Grinnell Glass Disc Sprinkler in July 1890. This sprinkler was the earliest that had substantial similarity to those in modern use.[4]

Piping Systems

Early piping configurations evolved from perforated pipe systems that were supplied by a vertical riser for each floor because they were manually operated. Parmelee and Grinnell later originated a so-called "tree" system. In the system's original configuration, main feed pipes were placed about 20 ft (6.1 m) apart and 5-ft (1.5-m) long branch lines of ¾-in. pipe were spaced about 10 ft (3 m) apart on the feed pipe. The concept was that by placing each sprinkler on a dead-end pipe the sprinkler would not be cooled by water flowing past it in the event of a fire.

[a]Many organizations have been referred to as "the Factory Mutuals." Some of these organizations are insurance companies and others are research organizations or product testing and approval organizations. It is beyond the scope of this supplement to attempt to trace the history or organizational structure of any of the Factory Mutual organizations. The names used in various sections are those which were commonly used to identify each company or organization at a point in time.

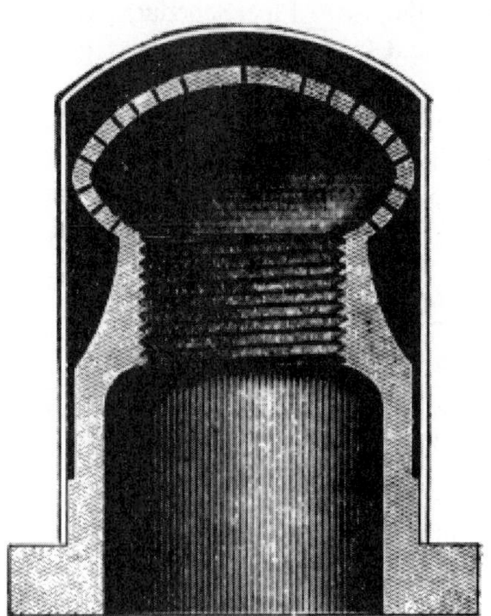

Exhibit S1.1
The first sprinkler, which was invented by Henry S. Parmelee.

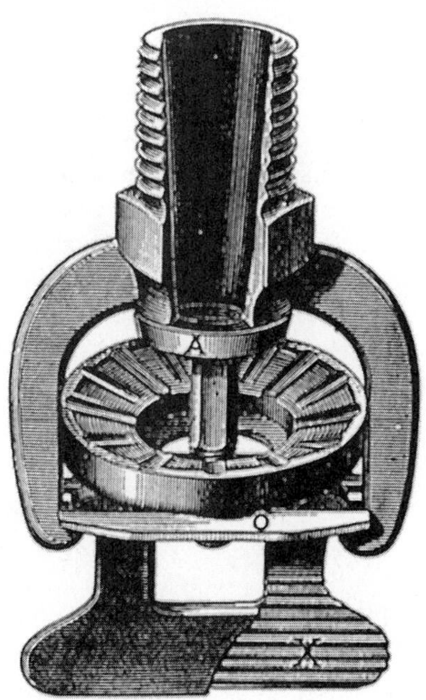

Exhibit S1.2 *Buell sprinkler.*

The first pipe schedule, the 1-3-6 Pipe Schedule, was developed by PSGP. Until 1896, a variety of pipe size and spacing rules were used mostly in New England. As an outgrowth of experiments on hydraulics (circa 1888-1895), John R. Freeman proposed a 1-2-3 Pipe Schedule and suggested a staggered arrangement of sprinklers under joisted construction for better distribution. The demand for uniformity, however, led to the adoption of a 1-2-4 Pipe Schedule in the first NFPA sprinkler standard in 1896. Table S1.1 compares these very early schedules. Also see Table S1.2.[1, 2, 3]

Table S1.1 Early Pipe Schedules

	Maximum Number of Sprinklers Allowed				
Nominal Pipe Size (in.)	**Pre-1892**	**Freeman 1892**	**Freeman 1895**	**NFPA 13 1896**	**NFPA 13 1896***
¾	1	1	1	1	1
1	3	2	2	2	2
1¼	6	3	3	4	4
1½	10	5	5	8	6
2	18	10	10	16	8
2½	28	20	20	28	16
3	48	40	36	48	28
3½	78	60	55	78	
4	115	120	80	110	
5	150	150	140	150	
6	200	300	200	200	

For SI units, 1 in. = 25.4 mm.

*More than six sprinklers on a branch line.

Valves

The early alarm valves used a mechanical attachment to the hinge pin to detect clapper movement and thus detect waterflow. The first valve was patented by J. C. Meloon of PSGP in Providence, RI, in 1881, followed by the Buell valve in 1884. The first grooved seat ring alarm valve was developed in England and patented in the United States in 1888 by R. Dowson and J. Taylor. The valve was manufactured by PSGP and later by PSGP's successor, the General Fire Extinguisher Co. None of the valves had retarding devices and most were not too reliable. [3]

Although several earlier dry valves were developed, the first to be used in practice was the "bellows type" invented by Grinnell in 1885. The first differential-type valve, the Grinnell #12 (1890), was the first to be widely used. [3]

The Need

Parmelee developed his first sprinkler for his piano factory. Many of the early sprinklers were developed by textile manufacturers who recognized they needed to cope with their fire problems. The owners of textile mills also started many of the early mutual insurance companies and most of the members of their boards of directors came from the textile industry. For example, Zachariah Allen, the owner of a woolen mill, founded the first

Factory Mutual Insurance Company in 1835. The origin of the Factory Mutual System can best be related to Allen's formation of a second mutual company in 1848 and his promotion of cooperation with other mutual companies as they were formed over the next thirty years mostly in New England. The most important encouragement to the development of automatic sprinkler systems undoubtedly came in 1880 when the companies of the Factory Mutual System began to request installation of sprinkler systems in textile mills. [1, 5]

The First Sprinkler Standard: 1896 [6]

In 1895, a group of men gathered in the Boston office of Everett U. Crosby, manager of the Underwriters' Bureau of New England, which later became the Fire Insurance Rating Bureau of New England. E. U. Crosby later also became the first secretary of the NFPA. It was obvious that these men representing fire insurance and sprinkler manufacturing interests had a common problem to solve. The group felt that automatic sprinklers, while proving their worth as extinguishing devices, were being installed in too many different ways. [7]

In his book *Men Against Fire,* Percy Bugbee, general manager of the NFPA from 1939 to 1969, lists the attendees at this meeting. "Attending that meeting in the Underwriters Bureau of New England office of the host, E. U. Crosby, were Uberto C. Crosby, Chairman of the Factory Improvement Committee of the New England Insurance Exchange; W. H. Stratton of the Factory Insurance Association; John R. Freeman of the Factory Mutual Insurance Companies; Frederick Grinnell of the Providence Steam and Gas Pipe Company; and F. Eliot Cabot of the Boston Board of Fire Underwriters." [7]

Grinnell told of the successful performance of sprinkler systems, and Freeman reported on Factory Mutual's success in insuring sprinklered properties. Thus the representatives of the stock insurance companies could see that action was required. In December 1895, a small group of inspection bureau men met in New York to prepare some sprinkler rules. The group met again in March 1896 to complete the work. The first "Report of Committee on Automatic Sprinkler Protection" was adopted at a conference held in New York on March 18 and 19, 1896. [6, 7, 8]

This first NFPA sprinkler standard came into being concurrent with the formation of the NFPA on November 6, 1896.[b] Several of the original officers of the NFPA were also the first sprinkler committee members.[c] In effect the NFPA may have been founded by stock insurance bureaus to meet Factory Mutual competition. [7]

[b]Concurrently, the Underwriters Electrical Bureau, which later became Underwriters' Laboratories, Inc. (UL), was formed in Chicago by William H. Merrill. A few years later, the National Board of Fire Underwriters (NBFU) provided financial support, and the name was changed to the Electrical Bureau of Fire Underwriters. In about 1900, the NFPA formed a Committee on Devices and Materials to cover the field of fire protection appliances with W. C. Robinson as chairman and Merrill as secretary. Robinson and Fitzhugh Taylor soon joined Merrill to be in charge of fire protection engineering and initiated the testing of automatic sprinkler equipment by UL. [9]

[c]The NFPA Technical Correlating Committee on Automatic Sprinkler Systems as it is currently named has had several names over time. Though the name has varied, the term *automatic sprinkler* has always been part of the name. In a similar way the structure of the committee has changed as technology has required the committee to consider varied approaches to the types of sprinklers in use, the variety of hardware needed or used to build a system, and the applications for which sprinkler systems were used.

The 1896 sprinkler standard was concerned with sprinklers, valves, hangers, piping, and pumps. Occupancy groups, calculations, and other requirements came later. Revisions were made in 58 successive editions. The most important of these revisions are as follows:

(1) New pipe schedules in 1905
(2) Separate standard on Class B systems in 1930
(3) Class B standard merged into the standard in 1931
(4) Three occupancy classes in 1940
(5) Spray sprinklers in 1953 to 1958
(6) Hydraulic calculations in 1966
(7) Design curves in 1973 and 1974
(8) Early suppression fast-response sprinklers (ESFR) in 1989
(9) Complete reorganization in 1991
(10) The standards 100-year anniversary in 1996
(11) Consolidation of all sprinkler system design requirements in 1999

This first standard contained requirements for sprinkler positioning (upright was preferred), distance of deflectors from bottom of joists or ceilings [3 in. to 10 in. (76 mm to 250 mm)], and full sprinkler coverage throughout a building. The standard also required a sprinkler orifice to produce 12 gpm (45 L/min) at 5 psi (0.34 bar) (K = 5.4) and a minimum static pressure of 25 psi (1.7 bar) at the highest sprinklers. Only "tree" systems were permitted with center/center or side/center feed recommended. Other requirements controlled spacing for mill construction, joists, bays, and pitched roofs.

Two water supplies, which could be steam pumps, public supply, elevated tanks, or pressure tanks, were deemed essential. The rules for rotary and steam fire pumps were in the sprinkler standard as were those for elevated and pressure tanks. The use of meters in a water supply connection required special consent. Pipe had to be securely supported; however, no detailed requirements were given.

Though dry systems were discouraged, they were preferred to system shutdown in cold weather. If there were more than 500 sprinklers, "water should be supplied through two or more dry valves—system divided horizontally," according to the standard. [6]

Some of the sprinklers illustrated in the 1896 sprinkler standard included the Buell of 1892 (see Exhibit S1.2), the Mackey of 1887, the Neracher of 1888, the Harkness of 1890, the New York and New Haven of 1889, the early Grinnells, and the Kane of 1888. See Exhibits S1.3 through S1.8.

In this era many highly inventive approaches were made towards automatic sprinkler design. Most early designers believed that the key to successful design was distribution of water and an automatic, thermally sensitive releasing element.

Most early sprinklers were a form of link and lever. In a link and lever design, the sprinkler cap is held closed against the water pressure by the mechanical advantage (generally about 20:1) of the sprinkler levers which, in turn, are held closed by the soldered link. In these early sprinklers, the first true eutectic alloys (the first solder-type sprinklers) and the first center strut sprinklers appeared. Even then, the potential for the thermal characteristics or low thermal inertia of quick-response sprinklers was present

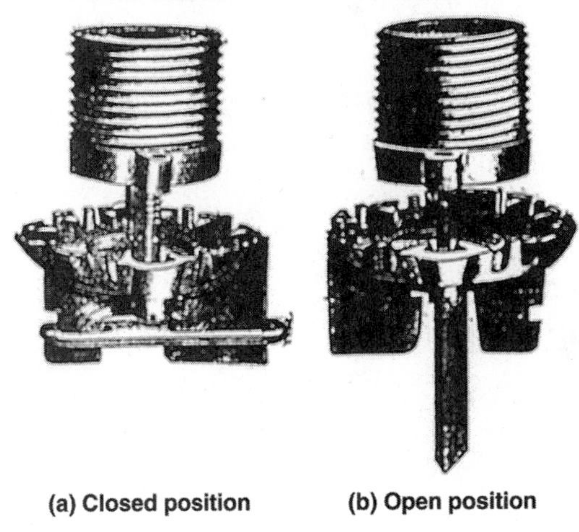

Exhibit S1.3 Mackey sprinkler.

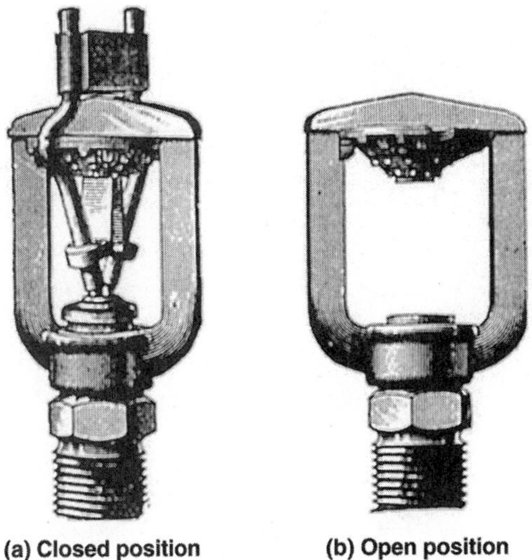

Exhibit S1.4 Neracher sprinkler.

in some designs that isolated the link load from the frame load and the link or solder from the water.

These early designs were conceived long before the mechanics of a sprinkler frame were fully understood, which was revealed in a 1933 patent application that was issued in 1939 to Albert J. Lipsinger of Grinnell. Lipsinger was the inventor of the Grinnell Duraspeed sprinkler. See Exhibit S1.9.

Hydraulic calculations had not yet been developed in 1896, so design was based on preferred piping arrangements with no more than six sprinklers on a branch line. All

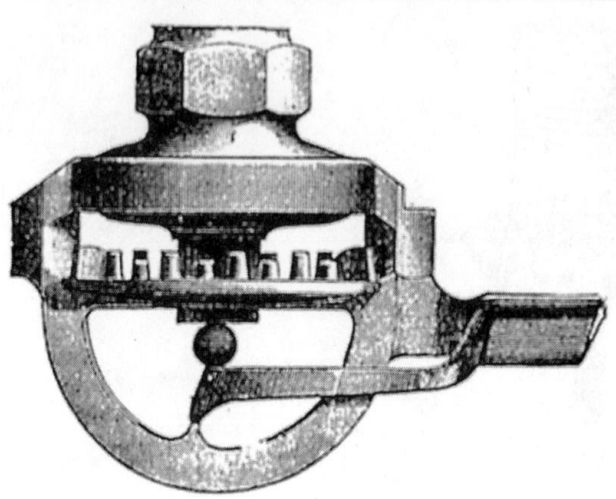

Exhibit S1.5 *Harkness sprinkler.*

Exhibit S1.6 *Elbow head of New York and New Haven sprinkler.*

of the piping arrangements were tree systems. Because of hydraulic calculations, a contemporary designer can arrange piping in loops, compound loops, and grids and then size the piping to take full advantage of an existing water supply.

THE NEXT 50 YEARS: 1900 THROUGH WORLD WAR II

In 1905, a substantial change was made in the pipe schedules of NFPA 13 (see Table S1.2). The restrictions for the number of sprinklers on a branch line were removed and new rules were added to limit the number of sprinklers fed by a given pipe size, whether on branch lines or collectively. [10]

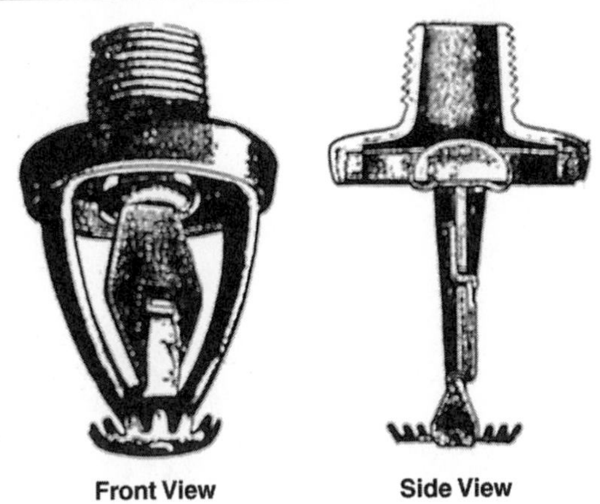

Exhibit S1.7 Early Grinnell sprinklers.

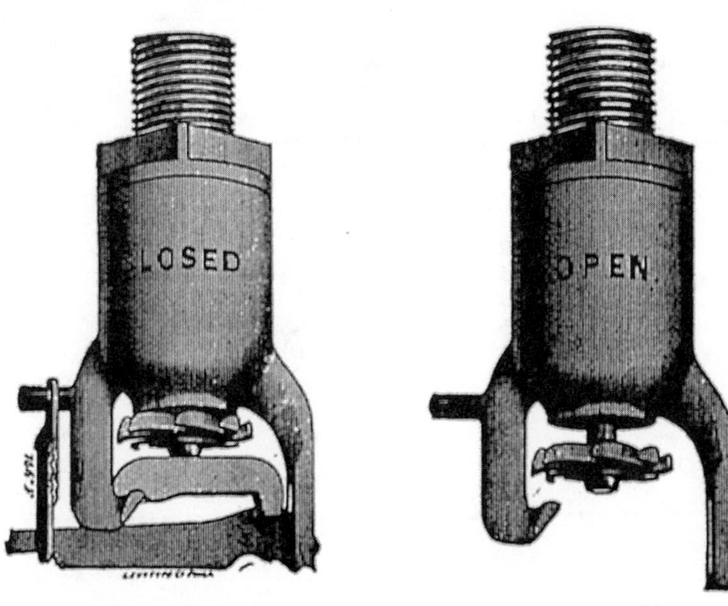

Exhibit S1.8 Kane sprinkler.

In the early standard editions, the need to securely support pipe was recognized. Hanger location rules appeared by 1920. Permitted types of hangers were U-type hangers made of wrought iron or malleable cast-iron, ring clips, or approved adjustable hangers. Flat U-type hangers were permitted if the metal was at least $^3/_{16}$ in. (0.48 mm) thick.

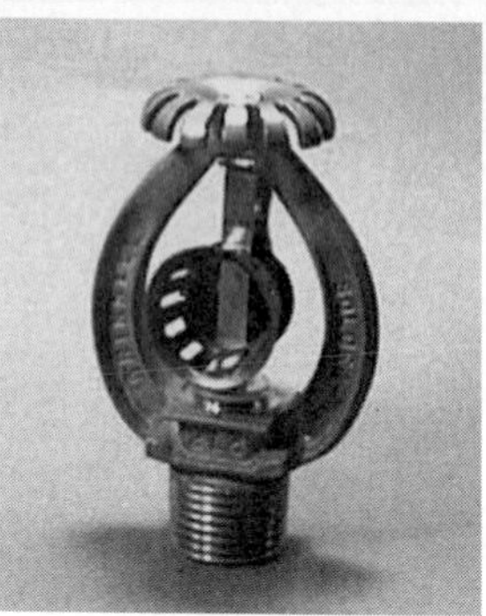

Exhibit S1.9 Duraspeed sprinkler. (Courtesy of Grinnell Corporation.)

Table S1.2 Sprinkler Pipe Schedules

Nominal Pipe Size (in.)	Maximum Number of Sprinklers Allowed						
	1940 Standard			1905 to 1940 1-2-3 System		Original NFPA Standard 1896 to 1905 1-2-4 System	Providence Steam and Gas Pipe Co. 1878 to 1896 1-3-6 System
	Light Hazard	Ordinary Hazard	Extra Hazard	Class A	Class B*		
¾	Eliminated			1	1	1	1
1	2	2	1	2	2	2	3
1¼	3	3	2	3	3	4	6
1½	5	5	5	5	5	8	10
2	10	10	8	10	10	16	18
2½	40	20	15	20		28	28
3		40	27	36		48	48
3½		65	40	55		78	78
4		100	55	80		110	115
5		160	90	140		150	150
6		250	150	200		200	200
8				400			

For SI units, 1 in. = 25.4 mm.

*Added in 1931.

Drive screws were only allowed in a horizontal position such as in the side of a beam. [3]

In 1926, UL completed a series of tests on the distribution of water by automatic sprinkler systems. The results were published in the report "The Distribution of Water by Automatic Sprinkler Systems." These tests were under the direction of a subcommittee of the NFPA Committee on Automatic Sprinklers and representatives of the NBFU, the National Automatic Sprinkler Association, and Underwriters' Laboratories. The purpose was to evaluate the adequacy of the existing pipe schedules and spacing requirements of the NFPA sprinkler standard. Although the tests clearly showed that the existing pipe schedules created a wide variation in the discharge from the sprinklers on a branch line, no action was taken to modify the standard until the 1940 edition. [11, 12]

In 1930, a separate standard was created for Class B systems, which later became light hazard. Even though this standard was combined with NFPA 13 in 1931, it was the first time that multiple hazard categories were recognized.

The following major new concepts appeared in the 1940 edition:

(1) Three hazard categories (light, ordinary, and extra)
(2) Looped piping systems
(3) Changes in the staggered spacing rules
(4) Elimination of the use of ¾-in. (1.9-mm) pipe
(5) Expanded rules for deluge systems

The changes in hazard categories came from studies by Henry Fiske on the fire performance of sprinkler systems. Fiske was one of the three editors of the *Crosby-Fiske-Forster Handbook.*[d] The spacing and pipe size rules in 1940 were also influenced by the previously mentioned 1926 UL water distribution tests. [8, 12]

During the 1920–1947 era, constant changes in sprinklers and sprinkler system equipment were made and test methods at the UL and FM laboratories were improved as the sprinkler industry developed. Developments such as centrifugal fire pumps, steel water tanks, and rate-of-rise supplemental operation of deluge systems took place. Major sprinkler improvements were made such as the first glass bulb sprinkler in 1921 and the Duraspeed sprinkler in 1933. Also during this era, UL and FM began to publish formal standards to describe the test methods by which sprinklers and sprinkler equipment are tested.

Hydraulic Calculations of Sprinkler and Water Spray Systems by Clyde Wood was published by the "Automatic" Sprinkler Corporation of America in the 1940s. General use of hydraulic calculations to design systems, however, did not begin until the middle 1950s. Much of the mathematics of hydraulic calculations came from knowledge developed by John R. Freeman in the very late 1800s or early 1900s. Wood reduced the mathematics to practice in the 1940s and concurrently published a complete set of tables to identify friction factors used in the flow calculations. In a talk before NFPA in 1953, Joe Johnson proposed that hydraulic calculations be included in the sprinkler standard. A hydraulic design method[e] was first included in NFPA 13 in 1966. The method is now mandatory for most system design.

[d]The original *Fire Protection Handbook* was published as the *Handbook of the Underwriters' Bureau of New England.* This became the *Crosby-Fiske Handbook (second to seventh editions)* and then the *Crosby-Fiske-Forster Handbook (eighth and later editions).* This handbook is now the *Fire Protection Handbook* published by NFPA. [8, 13].

[e]The industry and NFPA 13 still base hydraulic calculations of sprinkler systems on the work of John R. Freeman and Wood. Their work applied the Hazen-Williams formula. Later the Hardy-Cross method was adopted for gridded systems. The method was used mostly because the calculations are much easier to compute though less accurate than they would be using classic fluid mechanics. [14]

POST–WORLD WAR II ERA: 1947–1972

In 1947, shortly after the end of World War II, FM opened a new fire testing facility in Norwood, MA. The two fire testing buildings had an area of 2400 ft^2 (40 ft × 60 ft, 34 ft high) (223 m^2) and 3600 ft^2 (60 ft × 60 ft, 18 ft high) (334 m^2). UL opened their Northbrook, IL, testing station with two fire test buildings in 1954. One building had an area of 3600 ft^2 (60 ft × 60 ft, 16 ft 9 in. high) (334 m^2). The other building had an area of 2442 ft^2 (37 ft × 66 ft, 23 ft high) (227 m^2). The era of testing sprinkler performance on actual fires began with these facilities. Prior to this time, sprinklers were only tested for the manner in which water was distributed—no testing was done to evaluate the fire-extinguishing potential.

The Rockwood Sprinkler Company developed the T-1 and T-2 sprinklers—the first of the spray sprinklers—in approximately 1948.[f] Subsequent testing by FM established the value of this new approach. Another result of the spray sprinkler introduction was that both FM and UL introduced new methods for evaluating water distribution and included a fire testing requirement in their sprinkler testing requirements. All sprinkler manufacturers thereafter developed sprinklers with spray deflectors in both upright and pendent configurations. (The "old style" sprinkler, previously in general use, was used universally for both upright and pendent positions.) [15]

NFPA 13 required the use of spray sprinklers by 1955. The sprinkler types in use stayed constant until the introduction of a variety of new technical approaches started to appear in about 1981. At this point the definitions and terminology of sprinklers and sprinkler systems began to change rapidly, which is discussed in later sections.

In 1958, the ordinary hazard occupancy group was split into three subgroups, apparently for insurance underwriting, to define their different water supplies. The subgrouping of ordinary hazard occupancies appears only in the water supply rules. Extra hazard occupancies were all in one class.

A chapter was added to NFPA 13 in 1966 to describe methods for the hydraulic calculation of system pipe sizes as an acceptable alternative to the rules for pipe schedule systems. This addition was made largely because then new storage standards required calculated systems. (See "Fire Testing for the Storage Standards.")

In the late 1960s, Gordon Price, chairman of the NFPA sprinkler committee, and Chester Schirmer, later to become chairman, were instructed by the NFPA Board of Directors to have the sprinkler committee work diligently to "keep the costs of automatic sprinkler protection economically justifiable to encourage the installation of these systems." Cost effectiveness thus became an important criterion for the committee in subsequent changes to the standard and remains so to this date. [17]

Permissive language was included in the 1961 edition of NFPA 13 that provided for the use of any type of piping system or joining method that was evaluated and listed

[f]Prior to 1953 all sprinklers, whether upright or pendent, were designed to discharge 40 percent of the water upward and 60 percent downward. The FM tests in the 1948-1953 era established that better fire control was achieved when all water was discharged downward in a finely divided spray pattern. In the 1953 edition, a new chapter permitted the use of the new spray sprinkler. In 1955, the new spray sprinkler was named the standard sprinkler and the old conventional or regular sprinkler with the 40/60 discharge pattern was renamed the "old style" sprinkler. [15, 16]

by the approval laboratories. These guidelines permitted the use of piping systems other than wrought iron or steel for the first time. By 1968, NFPA 13 specifically provided for the use of listed copper piping systems. Shop-welding was permitted in the mid-60s, flexible couplings were permitted in the mid-70s, and lightweight steel piping systems were introduced in 1973. Also in 1973, valves controlling water supplies to sprinkler systems were required to be supervised in the open position. The following six methods were permitted:

(1) Central station alarm system
(2) Proprietary station alarm system
(3) Remote station alarm system
(4) A constantly attended local alarm system or other alarm service
(5) Valves locked open
(6) Valves sealed in the open position

Guidance for the protection of piping against damage due to earthquakes appeared in the 1947 edition and was improved in the 1951 edition. By then, NFPA 13 contained criteria for placement of longitudinal and lateral braces, maximum slenderness ratio of braces, and use of flexible couplings on risers. The stated intent was to brace the piping to withstand the horizontal force of half the weight of the water-filled piping and valves. Diagrams were included to show locations of braces. Also, tables were included for maximum lengths of angles, rods, flats, and pipe used as braces. The key was to restrain and support the piping with flexibility.

Several major changes occurred in the 1970s as the industry moved from large, heavy flanged or screwed piping systems based on a pipe schedule approach, to hydraulically calculated systems using lightweight piping with screwed fittings or welded piping and fittings and grooved couplings. During this time period there were major improvements in hanging systems and material-handling methods. The changes in installation and fabrication methods had a significant impact on reducing or at least maintaining the installed cost of sprinkler systems in an era of substantial worldwide inflation.

In the 1972 edition, the committee introduced a chapter on system design for life safety and fire protection in high-rise buildings. This chapter was in response to a renewed recognition that such buildings constituted a unique fire problem and that fire fighting and evacuation in these buildings was difficult at best. These requirements were also in response to many building code changes that required such buildings to be sprinklered. The design methods permitted the combined use of risers for both sprinkler and standpipe systems and created a conflict about whether the NFPA sprinkler or standpipe standards should decide water supply requirements. This conflict was not fully resolved until 1983.

Fire Testing for the Storage Standards[g]

Beginning with the establishment of the FM (1947) and UL (1954) fire testing facilities, several series of fire tests were conducted to establish sprinkler system design characteris-

[g]Although the terms full scale and large scale are used somewhat interchangeably, it should be recognized that the original fire test facilities of UL and FM were about 2500 ft^2 to 3500 ft^2 (230 m^2 to 325 m^2). FM

tics for certain types of storage hazards. Included among these tests were rack and palletized storage of whiskey in barrels, palletized storage of rubber tires, and palletized storage of a variety of commodities normally found in general storage warehouses. These tests were extensively reported to the sponsoring agencies, which included the insurance companies and testing organizations of the Factory Mutual System, the NBFU, the Factory Insurance Association (FIA), and a number of cooperating industries and companies. Based on the test data and on insurance considerations, NBFU, FIA, and FM published several standards for general fire protection and sprinkler system design including rack whiskey warehouses, palletized whiskey warehouses, palletized rubber tire storage, and general storage warehouses.

The FM and UL tests also led to the publication of NFPA 231, *Standard for General Storage,* in 1965. NFPA 231 included an appendix with sprinkler system design requirements for palletized storage. The design requirements for storage heights less than 12 ft (3.6 m) were also included in the 1965 edition of NFPA 13. NFPA 13 thus became the first NFPA standard containing sprinkler system design requirements partly based on full-scale fire testing. From 1970 onward, revisions in NFPA 13 and NFPA 231 continued to be based on a combination of test data and insurance considerations.

In 1967, FM constructed a 50,000-ft^2 (250 ft × 200 ft) [4645-m^2 (76.2 m × 61 m)] fire testing facility at West Gloucester, RI. The building has two testing areas. One section is 100 ft × 200 ft × 30 ft (30 m × 61 m × 9 m) high and the other section is 200 ft × 150 ft × 60 ft (61 m × 46 m × 18.3 m) high. Both have flat ceilings. The primary purpose of this facility was full-scale fire testing of a variety of storage commodities and configurations. The need for such a test facility was clear—larger warehouses were being developed for the storage of a variety of products on pallets, in racks, in bulk, and bundled in a variety of new and imaginative ways. The insurance industry was faced with an increasing number of situations where conventional sprinkler system design could not cope with the magnitude of the fires. Storage areas reaching well over 500,000 ft^2 (46,450 m^2), racks up to 60 ft (18.3 m) in height, automated racking systems, special racks for carpets, hanging garments, aerosol storage, and many more specialized storage methods came into use. Industry was experiencing numerous large losses in warehouses protected by sprinkler systems previously thought to be adequate.

In August 1967, Roger M. L. Russell of the FIA persuaded the rack manufacturers, sprinkler industry, industrial and commercial users, and insurance interests to join forces to financially sponsor a large-scale fire testing program at the new FM research facility. Subsequently, the Rack Storage Fire Protection Committee was formed, consisting mostly of the research sponsors. The Rack Storage Test Planning Committee and its steering committee planned an extensive series of fire tests. Tests were initially conducted at an 8-ft (2.4-m) height to establish representative commodity classes. Following these tests, full-scale fire tests were conducted at storage heights from 15 ft to 50 ft (4.6 m to 15.2 m) with a variety of commodities.

The NFPA Technical Committee on Rack Storage of Materials was organized under

began testing in a 50,000-ft^2 (4645-m^2) facility in 1967, and UL began testing in a 14,400-ft^2 (1338-m^2) facility in 1996.

the chairmanship of Russell in 1965. The committee developed the area/density curves for the 1971 edition of NFPA 231C, *Standard for Rack Storage of Materials,* based on analyses of the test results. The resulting standard became the first NFPA sprinkler standard entirely supported by full-scale fire test data. This committee continued a policy that all changes to the standard must be supported by full-scale fire test data. This committee was merged into the NFPA 13 committee in 1998.

Additional full-scale fire testing continued throughout the late 1970s and into the 1990s. NFPA 231C was changed as additional test data became available. Substantive changes were made to NFPA 231C in regard to storage of empty pallets, recognition of sprinkler system design methods for storage heights greater than 25 ft (7.6 m) (1973), modified commodity classes, means to protect plastic storage (1986), and others. In 1974 a companion standard was developed for rubber tire storage (NFPA 231D, *Standard for Storage of Rubber Tires*), and in 1984 new standards were added for rolled paper (NFPA 231F, *Standard for the Storage of Rolled Paper*) and cotton storage (NFPA 231E, *Recommended Practice for the Storage of Baled Cotton*).

The storage standards were also substantially influenced by the development of quick-response sprinkler technology. System design based on the use of early suppression fast-response (ESFR) sprinklers, extra large orifice (ELO) sprinklers, and large drop sprinklers were included in the storage standards starting in 1986. See "New Technology." [18]

The Impact of Hydraulic Calculations

After the introduction of hydraulic calculations in the 1966 edition of NFPA 13, economics made it apparent that hydraulic design criteria were needed for NFPA 13 systems in all hazard categories. The original area/density design curves were developed by a water supply subcommittee of NFPA 13, consisting of Jack Wood and Lin McCool. The curves were based on analyses of various piping configurations including piping variations such as center/center, side/center, and end, all with long and short branch lines. The concept was that the existing pipe schedule systems had performed well in most hazards, so they were evaluated to determine the comparable hydraulic design. Curves were then prepared having a slope similar to those planned in the storage standards. The curves became part of NFPA 13 in 1972 for light hazard and ordinary hazard occupancies. In 1974 the curves in NFPA 13 and NFPA 231 were further modified, which provided a smooth transition between the standards based on hazard severity.

In 1978, area/density design curves for hydraulically designed extra hazard systems were included in the appendix of NFPA 13. These design curves were included in the body of the standard and thus became mandatory requirements in the 1983 edition.

The concept of sizing system piping and water supplies based on density[h] and area of expected sprinkler operation was introduced in the 1972 edition of NFPA 13. This concept was further modified in the 1974 edition to permit design based on the largest room being considered, provided it was enclosed with fire-rated construction equal to the duration of the water supply.

[h]NFPA 13 uses the term *density* to define the water application rate of a sprinkler system in gpm/ft^2 [(L/min)/m^2] as determined by hydraulic calculations.

A small but important change was made in NFPA 13 in 1973. A section was added to permit the use of special sprinklers for "larger protection areas or distances between sprinklers . . . when such installations are made in accordance with approvals or listings of a nationally recognized testing laboratory." Although this requirement was modified substantially in subsequent editions, this change permitted the sprinkler industry to begin to innovate in sprinkler design. What followed was the development of fast-response, extended coverage, residential, large orifice, extra large orifice, and flow control sprinklers.

Residential and Quick-Response Sprinklers

At a meeting in Cleveland, OH, in 1967, Gordon Price suggested that faster responding sprinklers could greatly improve sprinkler system performance. Subsequently, a quick-response subcommittee was formed within the sprinkler committee. The subcommittee held several meetings through 1971 and concluded that a quick-response sprinkler was feasible and valuable for light hazard life safety situations.

In 1972, Rolf Jensen, a member of the sprinkler committee, was involved in the design of the fire protection systems for Water Tower Place, a 75-story, multiple-use, high-rise building in Chicago in which sprinklers were proposed for use in the hotel and residential portions of the building. Jensen wrote a specification for a quick-response apartment sprinkler that proposed that the sprinkler be fire tested in a room configuration using a simulated urethane-foamed plastic chair. The measurement criteria were formulated to assure that the sprinkler was evaluated for its ability to limit the accumulation of combustion products so that a person could survive in the test room environment for a reasonable period of time. The measurement criteria at the 5 ft 6 in.- (1.7-m) and 3-ft (0.9-m) levels were as follows:

(1) Temperature was to be controlled to a maximum of 150°F (65.6°C).
(2) Carbon-monoxide was to be controlled to 900 ppm before sprinkler operation. and 400 ppm after sprinkler operation.
(3) Smoke obscuration was to be controlled to a maximum Ringleman number of 4.
(4) Oxygen depletion was to be controlled to not less than 15 percent.

Though later refined and improved, these measurement criteria became the means by which residential sprinkler performance would later be evaluated in tests by UL, in research by the United States Fire Administration (USFA) and its various subcontractors, and in research on the development of residential sprinkler systems by Factory Mutual Research Corporation (FMRC). With modifications, these criteria also became the basis for the room fire test standards for residential sprinklers used by UL and FM. All residential sprinklers are thus tested to evaluate their ability to enhance survivability in the room of origin. See Exhibit S1.10.

In July 1973, the NFPA 13 Subcommittee on Residential Sprinkler Systems was formed. The subcommittee's work led to the publication of NFPA 13D, *Standard for the Installation of Sprinkler Systems in One- and Two-Family Dwellings and Mobile Homes,* in 1975. Much of this first edition was based on the collective experience of the committee members; little was based on real-world fire testing.

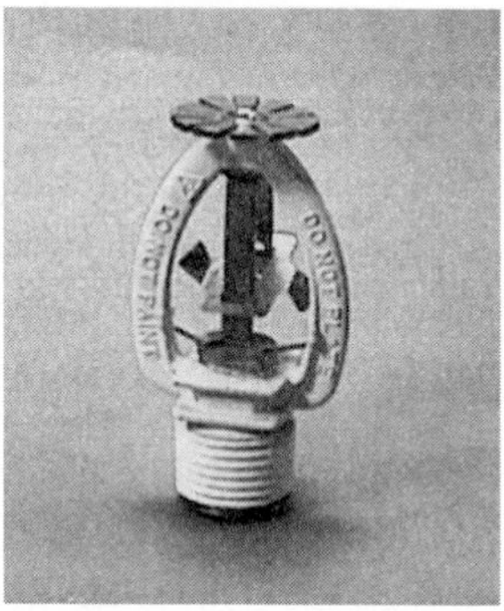

Exhibit S1.10 *An early residential sprinkler. (Courtesy of Grinnell Corporation.)*

Once published, NFPA 13D encouraged the USFA to sponsor extensive test programs to evaluate what impact residential sprinkler systems could have on the loss of life then occurring in dwellings. The USFA sponsored programs to develop a prototype residential sprinkler, to refine the methods by which residential sprinklers are tested, and to develop methods for the design of installed systems.

One study was to establish factors that influenced homeowners' acceptance of these systems, a second was to evaluate design features such as waterflow rates and number of sprinklers to operate, a third was to check design for flaming fires, and a fourth was to evaluate design on smoldering fires. These studies suggested that good design could be based on waterflow rates as low as 0.05 gpm/ft^2 [(2.04 L/min)/m^2]. The NFPA 13 subcommittee then asked for tests involving flaming ventilated fire configurations in a simulated living room with combustible walls and ceilings. The tests using standard thermal response sprinklers were not successful. Additional tests using simulated fast-response sprinklers (none were yet commercially available) were completed. Due to the efforts of Dr. Hsiang-Cheng Kung of FMRC, the tests proved that control of the environment in a residential fire scenario with combustible walls and ceilings could only be achieved with fast-response sprinklers having water distribution characteristics that assured wall wetting to a specified distance below the ceiling.

Under sponsorship of the USFA, FM, and the NFPA, the Los Angeles Fire Department and the residential sprinkler subcommittee joined in an effort to test prototype residential sprinkler systems in houses in Los Angeles destined for demolition. The accumulated data from these tests and the previous USFA programs became the basis for the 1980 edition of NFPA 13D. These development tests of a prototype sprinkler and sprinkler system established that a system designed by the methods described in NFPA 13D would provide a survivable environment both in the room of fire origin and in other areas of a one- or two-story dwelling permitting sufficient time to allow occupants to escape. At the time the 1980 edition of NFPA 13D was published by the NFPA, no residential sprinkler was commercially available even though the standard required their use. The first UL-listed residential sprinkler became available in 1981. [18, 19]

USFA sponsorship of residential sprinkler testing programs continued at both FM and UL. Substantial refinements continue to be made in subsequent editions of NFPA 13D.

The sprinkler committee recognized a need for improved fire safety for dwelling units in hotels, apartments, condominiums, and nursing homes and developed requirements for the 1985 edition of NFPA 13. System design was based on the use of residential sprinklers and the operation of four sprinklers. This concept became the basis for NFPA 13R, *Standard for the Installation of Sprinkler Systems in Residential Occupancies up to and Including Four Stories in Height,* which was developed in 1989 for residential facilities up to four stories in height. NFPA 13R provides a greater level of reliability and fire protection than NFPA 13D but not as much as NFPA 13. An important philosophy in the evolution of residential sprinkler systems was that both NFPA 13D and NFPA 13R permit water supplies smaller than those required by NFPA 13, omission of sprinklers in certain areas, and design based on fewer sprinklers than is required for NFPA 13 systems. These and other cost savings achieved in NFPA 13D and 13R systems are intended to encourage users to install sprinklers in residential facilities with *life safety* as the primary goal.

New Technology

In 1974, NFPA adopted a new standards-making system to better achieve due process and public exposure in consensus standards. This new method of processing changes and improvements to NFPA standards allowed the general public to submit proposed changes to NFPA standards and to comment on the manner by which technical committees disposed of these changes. Although the new procedures opened the standards-making process to full public disclosure and participation, it also placed a substantial burden on the technical committees in the processing of the public proposals and comments. Less time was thus available to the committees to develop their own changes in response to changes in technology.

During this same era, changes in building codes and the extensive development of high-rise, multiple-use, and large-scale buildings of all types demanded that sprinkler systems be installed for facilities not previously commonly sprinklered. As a result, many new technologies in sprinkler design, sprinkler systems, and related equipment were developed largely through fire research. Fabrication methods and hydraulic design approaches were also changing rapidly.

The NFPA sprinkler committee had a difficult time keeping up with both public proposals and the rapid changes in technology. Faced with this challenge, the committee chairman created a system of balanced task groups organized by technical disciplines and the needs of the committee. Each of these subgroups concentrated on a specific portion of NFPA 13, such as hydraulic calculations or earthquake requirements for pipe hangers.

The subgroup on new technology was charged with reviewing new technological challenges and proposing changes in an orderly fashion. (This group did not normally process routine proposals or comments.) The new technology group addressed such changes as quick-response, ESFR, extended coverage, extended coverage sidewall, large drop, large orifice, and extra large orifice sprinklers. They also addressed new piping methods such as lightweight steel and plastic.

Starting in the mid-1970s, sprinkler manufacturers and fire research organizations focused their research on sprinkler technology. Included in the research efforts were programs to develop quick-response sprinklers, programs to measure sprinkler responsiveness, and continuing research on sprinkler protection methods for storage hazards. One result was the previously mentioned development of fast-response residential sprinklers. Between 1975 and 1980, Heskestad and Smith of FMRC had developed a plunge test apparatus for measuring sprinkler sensitivity and proposed a formula for the measurement of the response time index (RTI) of a sprinkler. [20]

In the early 1980s FMRC also developed a new concept for fire control of storage hazards using ESFR sprinklers. In their research, they developed a large-capacity calorimeter and a special water applicator in order to measure the actual delivered density (ADD) of a sprinkler and the required delivered density (RDD) of a sprinkler. RTI, ADD, and RDD were discovered to be the three most important factors to achieve early fire suppression (the ESFR approach). [21]

FMRC continued extensive testing of this concept during the 1980s and 1990s, which included numerous tests of rack and palletized storage arrangements using a variety of commodities at variable storage heights. This research led to a new concept in sprinkler design for high-challenge fire hazards using ESFR sprinklers. In this approach, design is based on 12 sprinklers operating at a minimum pressure of 50 psi (3.4 bar). The concepts were included in NFPA 13 in 1989 and NFPA 231C in 1991.[i]

Continuing research at FM and UL refined these new design concepts for sprinkler system design for high-challenge fires and storage occupancies. Much of this resulted from sponsorship and funding of fire research by FMRC, other insurance interests, industrial and commercial users, the National Institute for Standards and Technology (NIST), National Fire Protection Research Foundation, and the sprinkler industry. Other research was directed to the development of fire modeling methods whereby fire growth could be better predicted, which is a necessity for more efficient system design.

In response, sprinkler manufacturers developed many new and innovative types of sprinklers that function more efficiently, effectively, and quickly to control fires. Many new sprinkler designs resulted with a wide variety of orifice sizes, thermal elements, special distribution patterns, and operating pressure restrictions. A typical sprinkler manufacturer could supply approximately 250 varieties of sprinklers in 1980. By 1996 the number had exceeded 2000 for some manufacturers. This large increase in the types of sprinklers in use has resulted in an increased level of complexity when sprinklers must be replaced. In the 1999 edition of NFPA 13, the committee included a new marking system that requires each sprinkler to be marked by a one- or two-character symbol followed by up to four numbers to identify each variation in orifice size, shape, deflector characteristic, and thermal sensitivity. This marking system becomes mandatory in January 2001.

Also during the 1980s, substantial changes were made to the methods of testing

[i]This brief history does not intend to describe the specifics of these research programs, which have been widely published by FMRC and others. [20, 21]

sprinklers at the UL and FM approval laboratories. For example, the UL sprinkler standard *Automatic Sprinklers for Fire-Protection Service,* UL 199, was first published in 1919 and next revised in 1966. From 1967 to 1977, eight more revisions were made in order to stay current with technological changes. The task of monitoring these changes so they were harmonized with the NFPA sprinkler standards was assigned to the new technology subcommittee. [16]

These developments in new technology led to a continuing series of changes in the sprinkler requirements of the NFPA sprinkler and storage standards during the 1980s. For example, in 1985 a chapter was added to NFPA 13 to provide requirements for large drop sprinklers. Similar requirements were added to NFPA 231C in 1986. With the appearance of many new types of special sprinklers, the NFPA 13 committee enlarged the section of NFPA 13 that permitted use of such sprinklers to provide more explicit direction for sprinkler testing and evaluation by UL and FM. Many new sprinklers, such as standard and sidewall extended coverage sprinklers, had to be included in the standard. Thus a rule was adopted in the 1987 edition to limit the maximum area covered by a single sprinkler to 400 ft^2 (37.2 m^2). The committee's reasoning was based on a concern for the size of fire that could develop if a single sprinkler failed to function.

Also during the 1980s the National Fire Protection Research Foundation established a technical advisory committee to plan and review research efforts by FMRC, UL, and other contractors to introduce quick-response sprinkler technology into NFPA 13. These included studies by FMRC, NIST, UL, Worcester Polytechnic Institute (WPI), and others. Although these studies showed that quick-response technology had application across the full range of hazards for which sprinkler systems were designed, they have not yet been fully implemented in NFPA 13. As with the FMRC research on ESFR sprinklers, a full treatment of this research is beyond the scope of this brief history.

Fire research efforts continue even today. In 1996, UL commissioned a new large-scale fire test facility of 14,400 ft^2 (120 ft × 120 ft × 54 ft high) [1338 m^2 (36.6 m × 36.6 m × 16.5 m high)], having a ceiling that can be adjusted from 5 ft to 48 ft (1.5 m to 14.6 m). Both UL and FMRC are continuing to explore better methods of measuring sprinkler distribution patterns with and without a fire plume present. Both testing laboratories also continue to conduct full-scale tests to better refine methods of sprinkler protection for many types of storage and other hazards. [22, 23]

THE LAST 10 YEARS

Prior to the 1974 edition of NFPA 13, little guidance for protection of sprinkler piping due to earthquakes was available. Earthquake guidance rules first appeared in 1974. Thereafter, few changes were made until 1987 when a seismic map was included and guidance for hanging and bracing of piping was added.

The 1989 Loma Prieta and 1994 Northridge earthquakes in California permitted studies to be made of sprinkler system performance in actual seismic events. These analyses contributed to a series of enhancements in the earthquake protection rules in the 1989, 1994, and 1996 editions. Requirements for sway bracing, swing joints, and flexible couplings were modified and a new velocity-based seismic map was added in

1994. Methods for installation of piping, clearances to the building structure (generally and at seismic joints), and hanger strap requirements were included.

Even with the revised requirements, NFPA 13 still addresses how to provide earthquake protection rather then when. The fundamental approach of NFPA 13 is to permit evaluation of how anticipated horizontal loads affect braces, their connections, and their fasteners. While NFPA 13 retains the baseline criterion that bracing is designed to withstand a horizontal load equal to one-half the weight of the water-filled pipe and valves, it allows the use of higher or lower loads based on site-specific conditions.

As a result of the substantial changes in the design and installation of sprinkler systems over the preceding generation, NFPA 13 was completely reorganized in 1991. Numerous changes and improvements were accomplished to achieve a more user friendly standard. Overall NFPA 13 was arranged to be in the order in which information is used in the design of a system. NFPA 13 was organized in the following order:

(1) General information
(2) System components and hardware
(3) System requirements
(4) Installation requirements
(5) Design approaches
(6) Plans and calculations
(7) Water supplies
(8) System acceptance
(9) Maintenance

Definitions for the wide variety of sprinklers in use were improved and consolidated in the first chapter. The rules on the location and spacing of sprinklers were clarified and definitions of *obstructed construction* and *unobstructed construction* were adopted. The obstruction rules were organized to define obstructions based on the degree to which they materially affect or impede the sprinkler's ability to control or suppress a fire. Obstructions were categorized depending on whether they occur at the ceiling or below the sprinkler and whether they impede the sprinkler discharge horizontally or vertically.

Other important changes included combining ordinary hazard (Group 2 and Group 3) occupancies and their design curves because the difference in required water application rates was insignificant for hydraulically designed systems and nonexistent for pipe schedule systems. Hydraulic calculations for pipe sizing were effectively made mandatory for all new systems covering an area less than 5000 ft^2 (464.5 m^2), with some minor exceptions.

In the 1996 100th anniversary edition, a few significant changes were made. Quick-response types of sprinklers were defined as sprinklers that have an RTI of 50 (meters-seconds) as measured under standard test conditions. RTI is a measure of the sensitivity of the sprinkler's element as installed in a specific sprinkler. RTI is usually determined by plunging a sprinkler into a heated laminar-airflow within a test oven. The factors that determine the response of a sprinkler's thermal element to a fire include RTI, temperature rating, sprinkler position, rate of fire growth, and radiation. [20]

Certain design rules were changed in 1996 to encourage the use of quick-response technology by proving a substantial reduction in the required design area when fast-response sprinklers were used. A 30-percent increase in the design area was also required when standard upright and pendent sidewall, and large drop sprinklers are installed below sloped ceilings. These changes all lead in a direction whereby sprinkler systems of the future can be designed to extinguish, not just control, fires.

Of equal importance, the installation rules and the rules for obstructions to sprinkler discharge were rewritten to be specific for all types of sprinklers, including special sprinklers. A new chapter was added for design of marine systems, the sprinkler systems installed onboard ships.

CONSOLIDATION: THE 1999 EDITION

The increased use of sprinkler systems over the past 25 years and significant improvements in sprinkler technology led to a standard based on fire control and encouraged the use of fast-response technology. Recognition of these improvements also motivated many writers of fire safety codes and standards to include mandates for sprinkler system installation. In some cases sprinkler design criteria were included in other NFPA standards that either supplemented or in some cases overrode NFPA 13.

The NFPA Standards Council observed that "sprinkler design criteria has been spread out among numerous installation and occupancy documents, creating the potential for redundancies, conflicts and inconsistencies." A Standards Council task force chaired by Jennifer Nelson was charged to study the problem and recommend a solution. After initial public exposure of the proposed concept at the NFPA meeting in November 1996, the task force met and a report was issued in March 1997. Additional public hearings were held in May 1997. After these meetings, the Standards Council decided to move forward with a complete reorganization of the Sprinkler Project. A mandate was given to the sprinkler committee and other NFPA technical committees to consolidate all sprinkler design and installation standards into NFPA 13, NFPA 13D, or NFPA 13R.

The newly formed NFPA 13 Technical Correlating Committee on Automatic Sprinkler Systems (see Exhibit S1.11 for new committee organization) and its associated technical committees began their work in November 1997 and held numerous meetings in 1998 and 1999. The work was correlated with ongoing changes in the storage standards and in other NFPA standards on water-based extinguishing systems. Although the initial report of the Standards Council task force suggested that about 45 documents contained sprinkler system design criteria in 1997, subsequent analysis indicated that the number was much higher.

The 1999 edition of NFPA 13 has the sprinkler system design requirements from 39 other NFPA standards integrated into the standard so that a sprinkler system designer can find most necessary design criteria in one document. The included material was extracted from the following NFPA standards:

NFPA 30, *Flammable and Combustible Liquids Code*

NFPA 30B, *Code for the Manufacture and Storage of Aerosol Products*

NFPA 33, *Standard for Spray Application Using Flammable or Combustible Materials*

NFPA 36, *Standard for Solvent Extraction Plants*

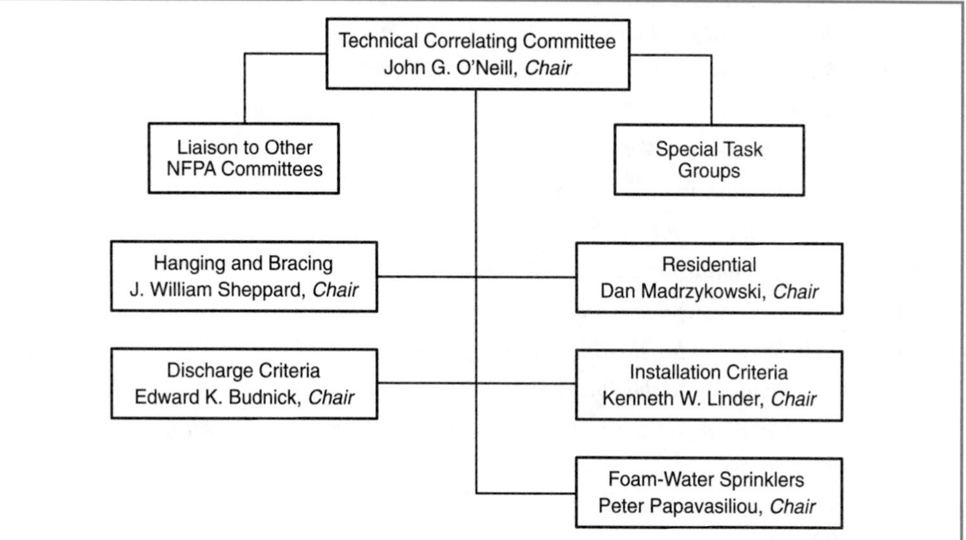

Exhibit S1.11 *Organization of NFPA 13's Technical Correlating Committee on Automatic Sprinkler Systems.*

NFPA 40, *Standard for the Storage and Handling of Cellulose Nitrate Motion Picture Film*

NFPA 42, *Code for the Storage of Pyroxylin Plastic*

NFPA 45, *Standard on Fire Protection for Laboratories Using Chemicals*

NFPA 51, *Standard for the Design and Installation of Oxygen-Fuel Gas Systems for Welding, Cutting, and Allied Processes*

NFPA 51A, *Standard for Acetylene Cylinder Charging Plants*

NFPA 55, *Standard for the Storage, Use, and Handling of Compressed and Liquefied Gases in Portable Cylinders*

NFPA 59, *Standard for the Storage and Handling of Liquefied Petroleum Gases at Utility Gas Plants*

NFPA 59A, *Standard for the Production, Storage, and Handling of Liquefied Natural Gas (LNG)*

NFPA 75, *Standard for the Protection of Electronic Computer/Data Processing Equipment*

NFPA 82, *Standard on Incinerators and Waste and Linen Handling Systems and Equipment*

NFPA 86C, *Standard for Industrial Furnaces Using a Special Processing Atmosphere*

NFPA 96, *Standard for Ventilation Control and Fire Protection of Commercial Cooking Operations*

NFPA *101®*, *Life Safety Code®*

NFPA 130, *Standard for Fixed Guideway Transit Systems*

NFPA 150, *Standard on Fire Safety in Racetrack Stables*

NFPA 203, *Guide on Roof Coverings and Roof Deck Constructions*

NFPA 214, *Standard on Water-Cooling Towers*

NFPA 231, *Standard for General Storage*

NFPA 231C, *Standard for Rack Storage of Materials*

NFPA 231D, *Standard for Storage of Rubber Tires*

NFPA 231E, *Recommended Practice for the Storage of Baled Cotton*

NFPA 231F, *Standard for the Storage of Roll Paper*

NFPA 232, *Standard for the Protection of Records*

NFPA 307, *Standard for the Construction and Fire Protection of Marine Terminals, Piers, and Wharves*

NFPA 318, *Standard for the Protection of Cleanrooms*

NFPA 409, *Standard on Aircraft Hangars*

NFPA 415, *Standard on Airport Terminal Buildings, Fueling Ramp Drainage, and Loading Walkways*

NFPA 423, *Standard for Construction and Protection of Aircraft Engine Test Facilities*

NFPA 430, *Code for the Storage of Liquid and Solid Oxidizers*

NFPA 432, *Code for the Storage of Organic Peroxide Formulations*

NFPA 803, *Standard for Fire Protection for Light Water Nuclear Power Plants*

NFPA 804, *Standard for Fire Protection for Advanced Light Water Reactor Electric Generating Plants*

NFPA 850, *Recommended Practice for Fire Protection for Electric Generating Plants and High Voltage Direct Current Converter Stations*

NFPA 851, *Recommended Practice for Fire Protection for Hydroelectric Generating Plants*

NFPA 909, *Standard for the Protection of Cultural Resources, Including Museums, Libraries, Places of Worship, and Historic Properties*

In the process of integrating the new material, the correlating committee also reorganized the standard into the following chapters:

(1) General information
(2) Classification of occupancies and commodities
(3) System components and hardware
(4) System requirements
(5) Installation requirements
(6) Hanging, bracing, and restraint of systems
(7) Design approaches
(8) Plans and calculations
(9) Water supplies

(10) System acceptance
(11) Marine systems
(12) System inspection, testing, and maintenance

These changes better organized the material and made the new edition more user friendly.

Since the sprinkler committee's inception, 14 men have served as the chair for terms ranging from 3 to 24 years. (See Table S1.3.) The original driving force came from the insurance industry, but the interests that participate now are from a broad spectrum of interest groups including insurance, users, sprinkler industry, government, and special experts. (See Table S1.4.)

Table S1.3 NFPA Sprinkler Committee Chairmen

Term	Chairman
1896–1901	U. C. Crosby
1902–1904	W. A. Anderson
1905–1909	W. C. Robinson
1910–1913	R. P. Boone
1914–1925	C. L. Scofield
1924–1927	F. B. Quackenboss
1928–1931	E. P. Boone
1932–1934	H. P. Smith
1935–1941	C. W. Johnson
1942–1956	K. W. Adkins
1957–1969	G. F. Price
1969–1972	I. L. Lamar
1973–1996	C. W. Schirmer
1997–	J. G. O'Neill

Table S1.4 NFPA 13's Voting Membership Interests

Interest	1960 Technical Committees	1969 Technical Committees	1974 Technical Committees	1989 Technical Committee	1999 Technical Committees	1999 Correlating Committee
Insurance	16	13	9	4	14	3
Sprinkler industry	5	5	4	8	30	5
Consulting and miscellaneous	0	2	4	6	14	4
Industry	3	3	4	3	10	1
Fire service	2	1	1	3	4	3
Test lab	1	2	2	2	8	11
Government	0	0	0	3	5	1
Total	27	26	24	29	85	28

Moreover NFPA 13 would not be where it is today without the contributions of many other people. Some have served as committee members for extensive periods, some have worked in research and testing, and many have participated by making proposals for changes or improvements in the NFPA standards or by commenting on proposed changes. All have contributed to the evolution of this important standard, which is the best overall way of achieving cost-effective fire safety whether the goal is life safety, property protection, or assurance of business operation. NFPA 13 is a true story of success through concensus standards.

References Cited in Supplement

National Fire Protection Association, 1 Batterymarch Park, P.O. Box 9101, Quincy, MA 02269-9101.

NFPA 13, *Standard for the Installation of Sprinkler Systems,* 1999 edition.
NFPA 13D, *Standard for the Installation of Sprinkler Systems in One- and Two-Family Dwellings and Manufactured Homes,* 1999 edition.
NFPA 13R, *Standard for the Installation of Sprinkler Systems in Residential Occupancies up to and Including Four Stories in Height,* 1999 edition.
NFPA 30, *Flammable and Combustible Liquids Code*, 1996 edition.
NFPA 30B, *Code for the Manufacture and Storage of Aerosol Products*, 1998 edition.
NFPA 33, *Standard for Spray Application Using Flammable or Combustible Materials*, 1995 edition.
NFPA 36, *Standard for Solvent Extraction Plants*, 1997 edition.
NFPA 40, *Standard for the Storage and Handling of Cellulose Nitrate Motion Picture Film*, 1997 edition.
NFPA 42, *Code for the Storage of Pyroxylin Plastic*, 1997 edition.
NFPA 45, *Standard on Fire Protection for Laboratories Using Chemicals*, 1996 edition.
NFPA 51, *Standard for the Design and Installation of Oxygen-Fuel Gas Systems for Welding, Cutting, and Allied Processes*, 1997 edition.
NFPA 51A, *Standard for Acetylene Cylinder Charging Plants*, 1996 edition.
NFPA 55, *Standard for the Storage, Use, and Handling of Compressed and Liquefied Gases in Portable Cylinders*, 1998 edition.
NFPA 59, *Standard for the Storage and Handling of Liquefied Petroleum Gases at Utility Gas Plants*, 1998 edition.
NFPA 59A, *Standard for the Production, Storage, and Handling of Liquefied Natural Gas (LNG)*, 1996 edition.
NFPA 75, *Standard for the Protection of Electronic Computer/Data Processing Equipment*, 1999 edition.
NFPA 82, *Standard on Incinerators and Waste and Linen Handling Systems and Equipment*, 1999 edition.
NFPA 86C, *Standard for Industrial Furnaces Using a Special Processing Atmosphere*, 1999 edition.
NFPA 96, *Standard for Ventilation Control and Fire Protection of Commercial Cooking Operations*, 1998 edition.
NFPA *101*®, *Life Safety Code*®, 1997 edition.

NFPA 130, *Standard for Fixed Guideway Transit Systems*, 1997 edition.
NFPA 150, *Standard on Fire Safety in Racetrack Stables*, 1995 edition.
NFPA 203, *Guide on Roof Coverings and Roof Deck Constructions*, 1995 edition.
NFPA 214, *Standard on Water-Cooling Towers*, 1996 edition.
NFPA 230, *Standard for Fire Protection of Storage*, 1999 edition.
NFPA 231D, *Standard for Storage of Rubber Tires*, 1998 edition.
NFPA 232, *Standard for the Protection of Records*, 1995 edition.
NFPA 307, *Standard for the Construction and Fire Protection of Marine Terminals, Piers, and Wharves*, 1995 edition.
NFPA 318, *Standard for the Protection of Cleanrooms*, 1998 edition.
NFPA 409, *Standard on Aircraft Hangars*, 1995 edition.
NFPA 415, *Standard on Airport Terminal Buildings, Fueling Ramp Drainage, and Loading Walkways*, 1997 edition.
NFPA 423, *Standard for Construction and Protection of Aircraft Engine Test Facilities*, 1999 edition.
NFPA 430, *Code for the Storage of Liquid and Solid Oxidizers*, 1995 edition.
NFPA 432, *Code for the Storage of Organic Peroxide Formulations*, 1997 edition.
NFPA 803, *Standard for Fire Protection for Light Water Nuclear Power Plants*, 1998 edition.
NFPA 804, *Standard for Fire Protection for Advanced Light Water Reactor Electric Generating Plants*, 1995 edition.
NFPA 850, *Recommended Practice for Fire Protection for Electric Generating Plants and High Voltage Direct Current Converter Stations*, 1996 edition.
NFPA 851, *Recommended Practice for Fire Protection for Hydroelectric Generating Plants*, 1996 edition.
NFPA 909, *Standard for the Protection of Cultural Resources, Including Museums, Libraries, Places of Worship, and Historic Properties*, 1997 edition.

Bibliography

1. *The Factory Mutuals 1835-1935*, Manufacturer's Mutual Insurance Co., Providence, RI, 1935.
2. John R. Freeman, "Experiments Relating to Hydraulics of Fire Streams," *ASME Transactions,* 1889.
3. Gorham Dana, S.B., *Automatic Sprinkler Protection,* third edition, 1923.
4. Jerome S. Pepi, *The Early History of Grinnell Corporation and the Fire Sprinkler Industry,* Grinnell Corporation, Providence, RI, January 1996.
5. Edward B. French, *Factory Mutual Insurance,* Arkwright Mutual Fire Insurance Company, Boston, MA, 1912.
6. *NFPA sprinkler standards,* National Fire Protection Association, Quincy, MA, 1896–1999.
7. Percy Bugbee, *Men Against Fire,* National Fire Protection Association, Boston, MA, 1971.
8. E. U. Crosby, *Handbook of The Underwriters' Bureau of New England*, Boston, MA, 1896.

9. Merwin Brandon, *Reminiscences of Underwriters' Laboratories, Inc.,* Chicago, IL, 1964.
10. *NFPA Handbook of Fire Protection,* National Fire Protection Association, Boston, MA, 1914.
11. R. W. Hendricks, *The Distribution of Water by Automatic Sprinkler Systems,* Underwriters' Laboratories, Inc., Chicago, IL, 1926.
12. Henry A. Fiske, "Sprinkler Control—Pipe Sizes and Spacing," *NFPA Quarterly,* Boston, MA, 1927.
13. NFPA handbooks, National Fire Protection Association, Quincy, MA, 1940–present.
14. *Technical Bulletin #410,* Crane Co., Chicago, IL, 1957.
15. Norman J. Thompson, "Proving Spray Sprinkler Efficiency," *NFPA Quarterly,* Boston, MA, 1954.
16. UL199, *Standard for Automatic Sprinklers,* Underwriters' Laboratories, Inc., Chicago, IL, 1919.
17. Private phone conversation with Chester W. Schirmer, June 1999.
18. Russell B. Fleming, P.E., *Quick Response Sprinklers—A Technical Analysis,* National Fire Protection Research Foundation, Quincy, MA, 1985.
19. Arthur E. Cote, P.E., "Field Test and Evaluation of Residential Sprinkler Systems," National Fire Protection Association, Quincy, MA, July 1982.
20. Heskestad and Smith, "Plunge Test for Determination of Sprinkler Sensitivity," Factory Mutual Research Corporation, Norwood, MA, 1980.
21. James L. Lee, "Early Suppression Fast Response (ESFR) Program," Factory Mutual Research Corporation, Norwood, MA, 1984.
22. William M. Carey, "Report of the Quick Response Sprinkler Project, Group 2 Performance Tests, Phase I-RDD Tests," Underwriters Laboratories, Inc., Northbrook, IL, May 1988.
23. William M. Carey, "Report of the Quick Response Sprinkler Project, Group 2 Performance Tests, Phase II-ADD Tests," Underwriters Laboratories, Inc., Northbrook, IL, January 1989.

Index

D

E

I

J

L

P

Q

R

Y

Tentative Interim Amendment

NFPA 13

Standard for the Installation of Sprinkler Systems

1999 Edition

Reference7-9.2.2
TIA 99-1 (NFPA 13)

Pursuant to Section 5 of the NFPA Regulations Governing Committee Projects, the National Fire Protection Association has issued the following Tentative Interim Amendment to NFPA 13, *Standard for the Installation of Sprinkler Systems,* 1999 edition. The TIA was processed by the Automatic Sprinkler Committee, and was issued by the Standards Council on April 27, 2000 with an effective date of May 22, 2000.

Tentative Interim Amendment is tentative because it has not been processed through the entire standards-making procedures. It is interim because it is effective only between editions of the standard. A TIA automatically becomes a proposal of the proponent for the next edition of the standard; as such, it then is subject to all of the procedures of the standards-making process.

1. *Revise the last sentence of 7-9.2.2 to read as follows:*

   Calculations shall be provided to verify the single (one) operating sprinkler criteria and the multiple (four) operating sprinkler criteria, but not less than to provide a minimum of 0.1 gpm/sq ft (4.1 mm/min) over the design area.

2. *Add an exception to read as follows:*

   Exception: For modifications or additions to existing systems equipped with residential sprinklers, the listed discharge criteria shall be permitted to be used.

For Your Ready Reference

Sprinkler Sizes

Nominal K-factor (gpm/psi$^{1/2}$)	Nominal K-factor (l/min/bar$^{1/2}$)	Nominal Orifice Size (Inches)	Nominal Orifice Size (mm)	Orifice Size Description
1.4	20	1/4	–	Small
1.9	27	5/16	–	Small
2.8	40	3/8	–	Small
4.2	60	7/16	10	Small
5.6	80	1/2	15	Standard
8.0	115	17/32	20	Large
11.2	160	5/8	–	Extra Large
14.0	200	3/4	–	Very Extra Large
16.8	240	–	–	–
19.6	280	–	–	–
22.4	320	–	–	–
25.2	360	–	–	–
28.0	400	–	–	–

Sprinkler Type Application based upon Occupancy Hazard

Light Hazard	Ordinary Hazard	Extra Hazard	Storage	Dwelling Units and Adjoining Corridors for NFPA 13 and 13R Applications	One- and Two-Family Dwellings and Manufactured Homes
Standard Spray[1] Sidewall Spray[1,8] Extended Coverage[1,7,9] Extended Coverage, Sidewall[1,7,9]	Standard Spray Sidewall Spray[5,8] Extended Coverage[6] Extended Coverage, Sidewall[6]	Standard Spray (standard response only)[2]	Standard Spray (standard response only)[3,4] ESFR[5,10] Large Drop	Standard Spray[1] Sidewall Spray[1,8] Extended Coverage[1,7,9] Extended Coverage, Sidewall[1,7,9] Residential[5]	Residential[5]

1. Quick response sprinklers are required but standard response sprinklers are permitted on existing systems that are being modified are being modified. See 5-3.1.5.
2. Quick response sprinklers not permitted with area/density design method.
3. QR sprinklers permitted for storage applications only where specifically listed for such use.
4. See 5-4.1.2 for limitations on orifice size.
5. Limited to wet pipe systems only.
6. Where specifically listed for use in Ordinary Hazard occupancies.
7. Limited to smooth flat ceilings with slopes not exceeding 16.7% unless otherwise listed.
8. Limited to smooth flat ceilings.
9. See 5-4.3 for limitations on ceiling construction.
10. See 5-4.6.3 for limitations on ceiling construction.

Table 7-2.3.1.1† Hose Stream Demand and Water Supply Duration Requirements for Hydraulically Calculated Systems

Occupancy or Commodity Classification	Inside Hose (gpm)	Total Combined Inside and Outside Hose (gpm)	Duration (minutes)
Light hazard	0, 50, or 100	100	30
Ordinary hazard	0, 50, or 100	250	60-90
Extra hazard	0, 50, or 100	500	90-120
Rack storage, Class I, II, and III commodities up to 12 ft (3.7 m) in height	0, 50, or 100	250	90
Rack storage, Class IV commodities up to 10 ft (3.1 m) in height	0, 50, or 100	250	90
Rack storage, Class IV commodities up to 12 ft (3.7 m) in height	0, 50, or 100	500	90
Rack storage, Class I, II, and III commodities over 12 ft (3.7 m) in height	0, 50, or 100	500	90
Rack storage, Class IV commodities over 12 ft (3.7 m) in height and plastic commodities	0, 50, or 100	500	120
General storage, Class I, II, and III commodities over 12 ft (3.7 m) up to 20 ft (6.1 m)	0, 50, or 100	500	90
General storage, Class IV commodities over 12 ft (3.7 m) up to 20 ft (6.1 m)	0, 50, or 100	500	120
General storage, Class I, II, and III commodities over 20 ft (6.1 m) up to 30 ft (9.1 m)	0, 50, or 100	500	120
General storage, Class IV commodities over 20 ft (6.1 m) up to 30 ft (9.1 m)	0, 50, or 100	500	150
General storage, group A plastics ≤ 5 ft (1.5 m)	0, 50, or 100	250	90
General storage, group A plastics over 5 ft (1.5 m) up to 20 ft (6.1 m)	0, 50, or 100	500	120
General storage, group A plastics over 20 ft (6.1 m) up to 25 ft (7.6 m)	0, 50, or 100	500	150
For SI units, 1 gpm = 3.785 L/min.			

Figure 7-2.3.1.2 Area/Density Curves

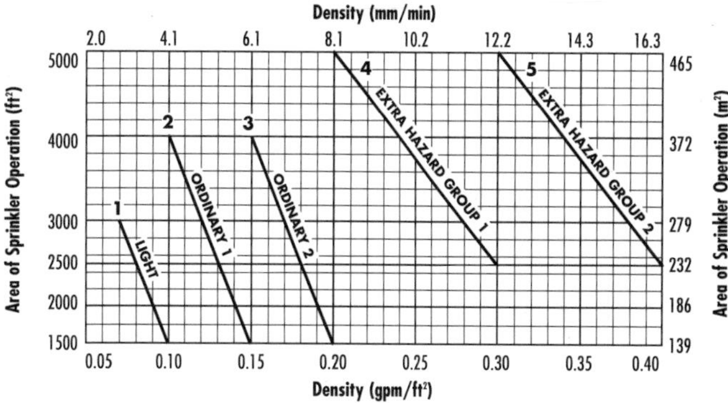